D1341942

Physical Chemistry

Keith J. Laidler
University of Ottawa
Ottawa, Ontario

John H. Meiser
Ball State University
Muncie, Indiana

The Benjamin/Cummings Publishing Company, Inc.
Menlo Park, California • Reading, Massachusetts
London • Amsterdam • Don Mills, Ontario • Sydney

Sponsoring Editor: Philip Hagopian
Production Editor: Greg Hubit
Technical Illustrator: Reproduction Drawings Limited
Copy Editor: Betty Adam
Text Designer: Christy Butterfield

Library of Congress Cataloging in Publication Data

Laidler, Keith James, 1916–
Physical chemistry.

Bibliography
Includes index.
1. Chemistry, Physical and theoretical.
I. Meiser, John H. II. Title.
QD453.2.L338 541.3 81-21737
ISBN 0-8053-5682-7 AACR2

hij-HA-8987

The Benjamin/Cummings Publishing Company, Inc.
2727 Sand Hill Road
Menlo Park, California 94025

Preface

After having taught physical chemistry for many years, we have found that the textbooks available were either too complex without sufficient verbal explanation or lacked the depth and range of topics required for the training of chemists. We therefore set out to write a text that can be easily understood by students who are meeting physical chemistry for the first time and which would still have the mathematical rigor and range necessary for a solid foundation. The students are assumed to have a basic knowledge of chemistry, physics, and mathematics such as is provided by the courses usually given to science students in their first year at a university. Since this book is rather comprehensive, it covers more than can probably be included in a one-year course, and it may be useful in more advanced courses and as a general reference book for those working in fields that require a basic knowledge of the subject.

Several special aids are provided for the student in this book: The *Preview* of each chapter describes the material to be presented in a brief narrative that gives a sense of unity to the material of the chapter. As an aid for the student, all new terms are in *italics* or in **boldface** type for special emphasis. Particular attention should be paid to these terms as well as to the equations that appear in boldface type. There are many worked-out examples in the text in which we have emphasized the dimensionality of the units of the quantities. Key equations that appear in the chapter occur in a concise listing at the end of each chapter. More is said about end-of-chapter material later. A unique feature of this text is Appendix A, in which are listed all the SI quantities and units generally useful to chemists. In addition, the mathematical relationships provided in Appendix C should prove useful as a handy reference.

Organization and Flexibility

We have treated the various branches of physical chemistry in what seems to us to be the most logical order, but this is a matter of personal preference and other teachers may prefer a different order. The book has been written with some flexibility in mind, and it may help if we suggest some alternatives. The subject matter may be divided into the following topics:

A. Chapters 1–6: General properties of gases, liquids, and solutions; thermodynamics; physical and chemical equilibrium

B. Chapters 7–8: Electrochemistry of solutions

C. Chapters 9–10: Chemical kinetics

D. Chapters 11–14: Quantum chemistry; spectroscopy; statistical mechanics

E. Chapters 15–19: Some special topics: solids, liquids, surfaces, transport properties, and macromolecules

Our sequence has the advantage that the more difficult topics of Chapters 11–14 can come at the beginning of the second half of the course. The book also lends itself without difficulty to various alternative sequences, such as the following:

A	B	C
Chapters 1–6	Chapters 1–6	Chapters 1–6
Chapters 9–10	Chapters 11–14	Chapters 11–14
Chapters 7–8	Chapters 7–8	Chapters 9–10
Chapters 11–14	Chapters 9–10	Chapters 7–8
Chapters 15–19	Chapters 15–19	Chapters 15–19

Aside from this, the order of topics in some of the sections, such as those in Chapters 15–19, can be easily varied.

End-of-Chapter Material

The *Key Equation* section lists equations with which the student should be thoroughly familiar. This listing should not be construed as including the only equations that are important but rather as foundation expressions that are widely applicable to chemical problems. The *Problems* have been organized according to subject matter and have been graded, the more difficult problems being indicated with an asterisk. The *Supplementary Problems* are generally more difficult and are listed separately. Answers to all problems are included at the back of the book, with detailed solutions provided in a separate *Solutions Manual for Physical Chemistry*.

Units and Symbols

We have tried to adhere strictly to the Système International d' Unités (SI) and to the recommendations of the International Union of Pure and Applied Chemistry (IUPAC); the reader is referred to Appendix A for an outline of these units and recommendations. The essential feature of these recommendations is that the methods of quantity calculus are used; a symbol represents a physical quantity, which is the product of a pure number (the *value* of the quantity) and a unit. Sometimes, as in taking a logarithm or making a plot, one needs the *value* of a quantity, and IUPAC has made no recommendation for a symbol to denote such a value. We have made the innovation of using a superscript u (for *unitless*) to denote such a value. For example, the symbol K stands for an equilibrium constant, which in general has units, and we write K^u for the *value* of the equilibrium constant. However, for some of the later topics (e.g., kinetics) we decided that the superscripts complicated the notation unduly, and that by the time these topics were treated the student would understand the point sufficiently well and would not need the help of the superscripts. Thus the logarithm of a rate constant is written simply as $\ln k$, and the reader is to understand that one takes the logarithm of the *value* of k (i.e., of k^u).

Acknowledgments

We are particularly grateful to a number of colleagues for help and advice, in particular: Drs. R. Norman Jones and D. A. Ramsay of the National Research Council of Canada (spectroscopy); Drs. Robert A. Smith of the University of Ottawa and David E. Koltenbah (quantum mechanics). Our correspondence with Dr. J. Lee, University of Manchester, on units and IUPAC conventions has been of great help to us in clarifying some difficult problems and in avoiding pitfalls.

In addition, we have benefited greatly from the comments and constructive criticisms of the following teachers of physical chemistry, who reviewed the manuscript during various stages of its development: Steven A. Adelman (Purdue University), Seymour M. Blinder (University of Michigan), Robert J. Doedens (University of California, Irvine), Dane R. Jones (California State Polytechnic University, San Luis Obispo), Arthur G. Keenan (University of Miami), John R. Luoma (Cleveland State University), Claude F. Meares (University of California, Davis), Lee G. Pedersen (University of North Carolina), who also provided a review of the problem sets; Steven M. Schildcrout (Youngstown State University), Leonard D. Spicer (University of Utah), Thomas E. Taylor (Texas A&M University).

Our first editor at Benjamin/Cummings, Mary Forkner, made a number of very valuable suggestions as to the general format of the book and she was indefatigable in securing comments from reviewers. Her successor, Phil Hagopian, has also been of considerable help, and we are grateful to Betty Adam for her painstaking copy editing. Mention must be made of our typists, especially Monique Auger, Joanne Beckham, Cass Guthridge, and Renata Kamal, who bore with us during this production. Finally, our thanks go to Laura L. Nelson who has helped to make the final text more error-free than would otherwise have been possible.

K.J.L.
J.H.M.

Contents

The Nature of Physical Chemistry and the Behavior of Gases

Preview

Physical chemistry is the application of the methods of physics to chemical problems. It can be organized into *thermodynamics, quantum mechanics, kinetics,* and *statistical thermodynamics.* Basic concepts particularly from the branch of physics called *classical mechanics* are important to these areas and include the relation between *work* and *kinetic energy* changes.

As we begin to learn the language of physical chemistry, particular attention should be paid to definitions or special terms, which in this book are printed in boldface type. Our main item of interest is the *system* that we focus on and its *surroundings.*

Gases are easier to treat than liquids and solids. Two important equations relating to a fixed amount of gas follow:

Boyle's Law: $PV = \text{constant}_1$, (at constant T)

Gay-Lussac's law: $\dfrac{V}{T} = \text{constant}_2$, (at constant P)

These expressions combine, using *Avogadro's hypothesis* that the *amount of substance n* (SI unit: mole) is proportional to the volume at a fixed T and P, to give the *ideal gas law*:

$$PV = nRT$$

where R is the *gas constant.* A gas that obeys this equation is called an *ideal gas.*

Experimental observations as embodied in these laws are important but so too is the development of a theoretical explanation for the observations. An important development in this regard is the calculation of the *pressure of a one-dimensional gas* from the kinetic-molecular theory. The relation of kinetic energy to temperature, namely

$$\bar{\epsilon}_k = \tfrac{3}{2}\mathbf{k}T \qquad (\mathbf{k} = \text{Boltzmann constant})$$

allows a theoretical derivation of the ideal gas law and of laws found experimentally.

Molecular collisions between gas molecules play an important role in many concepts. *Collision numbers* tell us how often collisions occur in unit volume between like or unlike molecules in unit time. Related to collisions is the idea of *mean free path,* which is the average distance gas molecules travel between collisions.

Real gases differ in their behavior from ideal gases and this difference can be expressed using the *compression factor* $Z = PV/nRT$ where $Z = 1$ if the real gas behavior is identical to that of an ideal gas. Values of Z above or below unity indicate deviations from ideal behavior. Real gases also show *critical phenomena* and *liquefaction,* phenomena that are impossible for an ideal gas. The concept that there is complete *continuity of states* in the transformation from the gas to the liquid state is important in the treatment of the condensation of gas. The *van der Waals equation*, in which the pressure of the ideal gas is modified to account for attractive forces between gas particles and in which the ideal volume is reduced to allow for the actual size of the gas particles, has proved to be an important expression for describing real gases.

1 The Nature of Physical Chemistry and the Behavior of Gases

Humans are exceedingly complex creatures, and they live in a very complicated universe. In searching for a place in their environment, they have developed a number of intellectual disciplines through which they have gained some insight into themselves and their surroundings. They are not content merely to acquire the means of putting their environment to practical use, but they also have an insatiable desire to discover the basic principles that govern the behavior of all matter. These endeavors have led to the development of bodies of knowledge that were formerly known as *natural philosophy* but that are now generally known as *science*.

1.1 The Nature of Physical Chemistry

In this book, we are concerned with the branch of science known as *physical chemistry*. Physical chemistry is the application of the methods of physics to chemical problems. It includes the qualitative and quantitative study, both experimental and theoretical, of the general principles determining the behavior of matter, particularly the transformation of one substance into another. Although the physical chemist uses many of the methods of the physicist, he applies them to chemical structures and chemical processes. Physical chemistry is not so much concerned with the description of chemical substances and their reactions—this is the concern of organic and inorganic chemistry—as with theoretical principles and with quantitative problems.

Two approaches are possible in a physicochemical study. In what might be called a *systemic* approach, the investigation begins with the very basic constituents of matter—the fundamental particles—and proceeds conceptually to construct larger systems from them. The adjective *microscopic* (Greek *micros*, small) is used to refer to these tiny constituents. In this way, increasingly complex phenomena can be interpreted on the basis of the elementary particles and their interactions.

In the second approach, the study starts with investigations of *macroscopic* material (Greek *macros*, large), such as a sample of liquid or solid that is easily observable with the eye. Measurements are made of macroscopic properties such as pressure, temperature, and volume. In this *phenomenological* approach, more detailed studies of microscopic behavior are made only insofar as they are needed to understand the macroscopic behavior in terms of the microscopic.

Physical chemistry encompasses the structure of matter at equilibrium and also the processes of chemical change. Its three principal subject areas are thermodynamics, quantum chemistry, and chemical kinetics; other topics, such as electrochemistry, have aspects that lie in all of these three categories. *Thermodynamics,* as applied to chemical problems, is primarily concerned with the position of chemical equilibrium, with the direction of chemical change, and with the associated changes in energy. *Quantum chemistry* theoretically describes bonding at a molecular level. In its exact treatments, it deals only with the simplest of atomic and molecular systems, but it can be extended in an approximate way to deal with bonding in much more complex molecular structures. *Chemical kinetics* is concerned with the rates and mechanisms with which processes occur as equilibrium is approached.

An intermediate area, known as *statistical thermodynamics,* links the three main areas of thermodynamics, quantum chemistry, and kinetics and also provides a basic relationship between the microscopic and macroscopic worlds. Related to this area is nonequilibrium statistical mechanics, which is becoming an increasingly important part of modern physical chemistry. This area includes problems in such areas as the theory of dynamics in liquids and light scattering. This latter area is beyond the scope of this book.

1.2 Some Concepts from Classical Mechanics

We will often calculate the work done or a change in energy when a chemical process takes place. It is important to know how these are related; so, our study of physical chemistry begins with some fundamental macroscopic principles in mechanics and energy.

Work

Work can take many forms, but any type of work can be resolved through dimensional analysis as the application of a force through a distance. If a force $\boldsymbol{F}$ (a vector indicated by boldface type) acts through an infinitesimal distance $d\boldsymbol{l}$ ($\boldsymbol{l}$ is the position vector), the work is

$$dw = \boldsymbol{F} \cdot d\boldsymbol{l} \tag{1.1}$$

If the applied force is not in the direction of motion but makes an angle θ with this direction, as shown in Figure 1.1, the work is the component $F \cos \theta$ in the direction of the motion multiplied by the distance traveled, dl:

$$dw = F \cos \theta \, dl \tag{1.2}$$

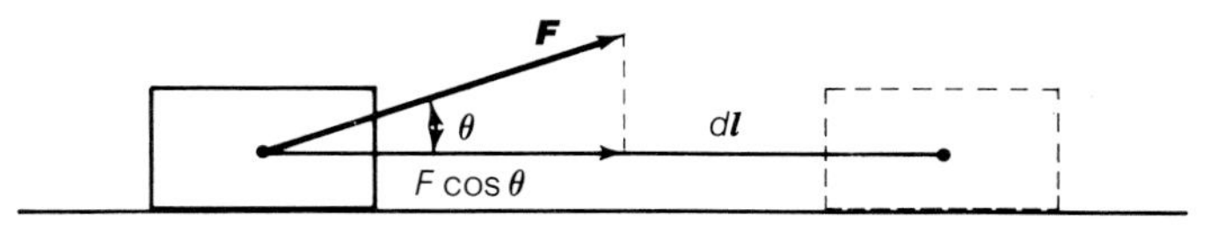

FIGURE 1.1 Work is the applied force in the direction of motion multiplied by dl.

Equation 1.2 can then be integrated to determine the work in a single direction. The force $\boldsymbol{F}$ can also be resolved into three components, F_x, F_y, F_z, one along each of the three-dimensional axes. For instance, for a constant force F_x in the X-direction,

$$w = \int_{x_0}^{x} F_x\, dx = F_x(x - x_0) \qquad (x_0 = \text{initial value of } x) \tag{1.3}$$

Several very important cases exist where the force does not remain constant. As an example, **Hooke's law** states that for an idealized spring

$$F = -k_h x \tag{1.4}$$

where x is the displacement from a position ($x_0 = 0$) at which F is initially zero and k_h (known as a *force constant*) relates the displacement to the force. The work done on the spring to extend it is found from Eq. 1.3:

$$w = \int_0^x -k_h x\, dx = -\frac{k_h}{2}x^2 \tag{1.5}$$

A particle vibrating under the influence of a restoring force that obeys Hooke's law is called a **harmonic oscillator**. These relationships apply to a good approximation to the stretching of a chemical bond.

Kinetic and Potential Energy

The energy possessed by a moving body by virtue of its motion is called its **kinetic energy** and can be expressed as

$$E_k = \tfrac{1}{2}m\boldsymbol{u}^2 \tag{1.6}$$

where $\boldsymbol{u}$ ($= d\boldsymbol{l}/dt$) is the velocity (i.e., the instantaneous rate of change of the position vector $\boldsymbol{l}$ with respect to time), and m is the mass. An important relation between work and kinetic energy for a point mass can be demonstrated by casting Eq. 1.1 into an integral over time:

$$w = \int_{l_0}^{l} \boldsymbol{F}(\boldsymbol{l}) \cdot d\boldsymbol{l} = \int_{t_0}^{t} \boldsymbol{F}(\boldsymbol{l}) \cdot \frac{d\boldsymbol{l}}{dt}\, dt = \int_{t_0}^{t} \boldsymbol{F}(\boldsymbol{l}) \cdot \boldsymbol{u}\, dt \tag{1.7}$$

Substitution of Newton's second law,

$$\boldsymbol{F} = m\boldsymbol{a} = m\frac{d\boldsymbol{u}}{dt} \tag{1.8}$$

where **a** is the acceleration, yields

$$w = \int_{t_0}^{t} m\frac{d\boldsymbol{u}}{dt} \cdot \boldsymbol{u}\, dt = m\int_{u_0}^{u} \boldsymbol{u} \cdot d\boldsymbol{u} \tag{1.9}$$

After integration and substitution of the limits if $\boldsymbol{u}$ is in the same direction as $d\boldsymbol{u}$, then $\cos\theta = 1$, and the expression becomes, with the definition of kinetic energy (Eq. 1.6),

$$w = \int_{l_0}^{l} \boldsymbol{F}(\boldsymbol{l}) \cdot d\boldsymbol{l} = \tfrac{1}{2}mu_1^2 - \tfrac{1}{2}mu_0^2 = E_{k_1} - E_{k_0} \tag{1.10}$$

Thus we find that the difference in kinetic energy of the initial and final states of a point body is the work performed in the process.

Another useful expression is found if we assume that the force is conservative. Since the integral in Eq. 1.10 is a function of l alone, it can therefore be used to define a new function of l that can be written as

$$\boldsymbol{F}(\boldsymbol{l}) \cdot d\boldsymbol{l} = -dE_p(l) \tag{1.11}$$

This new function $E_p(l)$ is the **potential energy**, which is the energy a body possesses by virtue of its position.

For the case of a system that obeys Hooke's law, the potential energy for a mass in position x is usually defined as the work done against a force in moving the mass to the position from one at which the potential energy is arbitrarily taken as zero:

$$E_p = \int_0^x -F\,dx = \int_0^x k_h x\,dx = \tfrac{1}{2}k_h x^2 \tag{1.12}$$

Thus, the potential energy rises parabolically on either side of the equilibrium position. There is no naturally defined zero of potential energy. This means that absolute potential energy values cannot be given but only values that relate to an arbitrarily defined zero energy.

An expression similar to Eq. 1.10 but now involving potential energy can be obtained by substituting Eq. 1.11 into Eq. 1.10:

$$w = \int_{l_0}^{l} \boldsymbol{F}(l) \cdot d\boldsymbol{l} = -E_{p_1} + E_{p_0} = E_{k_1} - E_{k_0} \tag{1.13}$$

Rearrangement gives

$$E_{p_0} + E_{k_0} = E_{p_1} + E_{k_1} \tag{1.14}$$

which states that the sum of the potential and kinetic energies, $E_p + E_k$, remains constant in a transformation. Although Eq. 1.14 was derived for a body moving between two locations, it is easy to extend the idea to two interacting particles. We then find that the sum of the kinetic energy of two or more bodies in an elastic collision (no energy lost to internal motion of the bodies) is equal to the sum after impact. Expressions such as Eq. 1.14 are known as *conservation laws* and will be used later in the development of kinetic theory.

1.3 Systems, States, and Equilibrium

Physical chemists attempt to define very precisely the object of their study, which is called the *system*. It may be solid, liquid, gaseous, or any combination of these. The study may be concerned with a large number of individual components that comprise a macroscopic system. Alternatively, if the study focuses on individual atoms and molecules, a microscopic system is involved. We may summarize by saying that the system is a particular segment of the world (with definite boundaries) on which we focus our attention. Outside the system are the *surroundings*, and the system plus the surroundings compose a *universe*. In an *open system* there can be transfer of heat and also material. If no material can pass between the system and the surroundings, but there can be transfer of heat, the system is said to be a *closed*

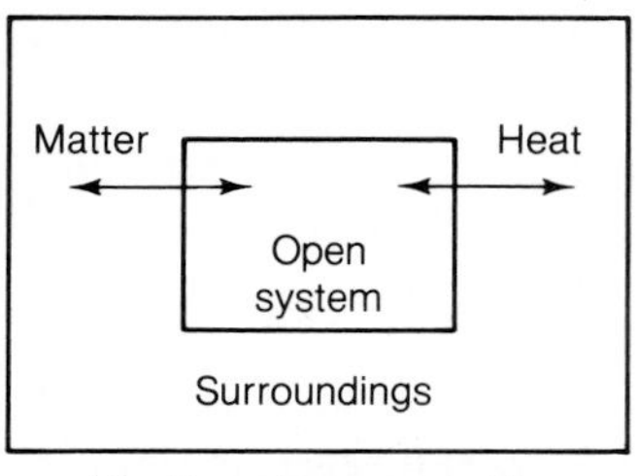

(a) Boundary permeable to matter and heat

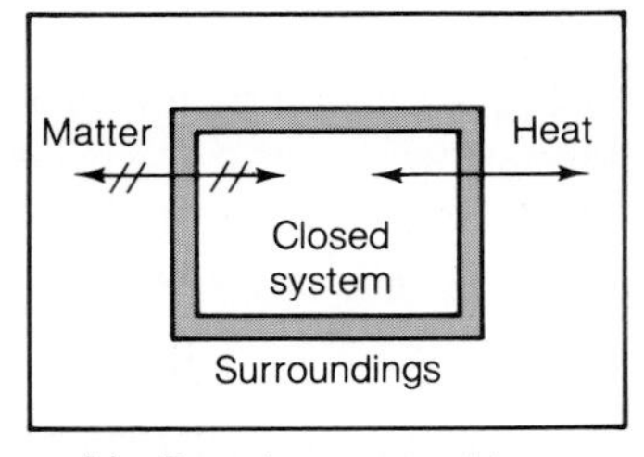

(b) Boundary permeable to heat but impermeable to matter

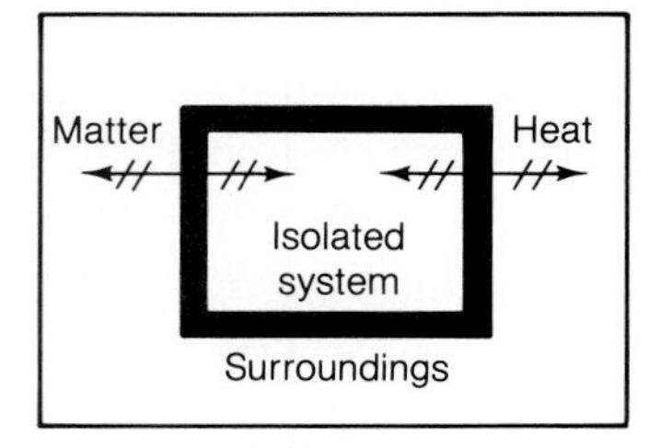

(c) Boundary impermeable to matter and heat

FIGURE 1.2 Relationship of heat and matter flow in open, closed, and isolated systems.

system. Finally, a system is said to be *isolated* if neither matter nor heat is permitted to exchange across the boundary. This could be accomplished by surrounding the system by an insulating container. These three possibilities are illustrated in Figure 1.2.

Physical chemists generally concern themselves with the measurement of the properties of a system, properties such as pressure, temperature, and volume. These properties may be of two types. If the value of the property *does not change with the quantity of matter present* (i.e., if it does not change when the system is subdivided), we say that the property is an *intensive* property: Examples are pressure, temperature, and refractive index. If the property *does change with the quantity of matter present*, that property is called an *extensive* property. Volume and mass are extensive. The ratio of two extensive properties is an intensive property. There is a familiar example of this; the density of a sample is an intensive quantity obtained by the division of mass by volume, two extensive variables.

A certain minimum number of properties have to be measured in order to determine the condition or state of a macroscopic system completely. For a given amount of material it is then generally possible to write an equation that describes the state in terms of intensive variables. Empirical data are summarized in terms of variables that are experimentally defined. For example, if our system consists of gas, we normally could describe its state by specifying three independent properties such as amount of substance, temperature, and pressure. The volume of gas is another property that will change as temperature and pressure are altered, but this fourth variable is fixed by an equation (known as an *equation of state*) that connects these four properties. In some cases it is important to specify the shape or extent of the surface. Therefore, we cannot state unequivocally that a predetermined number of independent variables will always be sufficient to specify the state of an arbitrary system. However, if the variables that specify the state of the system do not change with time, then we say the system is in **equilibrium**. Thus, a state of equilibrium exists when there is no change with time in any of the system's macroscopic properties.

1.4 Thermal Equilibrium

It is common experience that when two objects at different temperatures are placed in contact with each other for long enough, their temperatures will become equal; they are then in equilibrium with respect to temperature. The concept of *heat* enters here. We observe that the flow of heat from a warmer body serves to increase the temperature of a colder body. However, heat is not temperature.

The Concept of Temperature and Its Measurement

The physiological sensations that we accept as indications of whether an object is hot or cold cannot serve us quantitatively; they are relative and qualitative. The first thermometer using the freezing point and boiling point of water as references was introduced by the Danish astronomer Olaus Rømer (1644–1710). On the old centigrade scale [Latin *centum*, hundred; *gradus*, step; also called the Celsius scale named in honor of the Swedish astronomer Anders Celsius (1701–1744)] the freezing point of water at 1 atmosphere (atm) pressure was fixed at exactly 0°C, and the boiling point at exactly 100°C. We shall see later that the Celsius scale is now defined somewhat differently.

The construction of most thermometers is based on the fact that a column of mercury changes its length when its temperature is changed. The length of a solid metal rod or the volume of a gas at constant pressure could equally well be used in principle. Indeed, for any thermometric property, whether a length change is involved or not, the old centigrade temperature θ was related to two defined temperatures. In the case of the mercury column, we assign its length the value l_{100} when it is at thermal equilibrium with boiling water vapor at 1 atm pressure. The achievement of equilibrium with melting ice exposed to 1 atm pressure serves to establish the value of l_0 for this length. Assuming a linear relationship between the temperature θ and the thermometric property (length in this case), and assuming 100 divisions between the fixed marks, allows us to write

$$\theta = \frac{(l - l_0)}{(l_{100} - l_0)} (100°) \qquad (1.15)$$

where l is the length at temperature θ, and l_0 and l_{100} are the lengths at the ice and boiling water temperatures, respectively. Some thermometric properties do not depend on a length, such as in a quartz thermometer where the resonance frequency response of quartz crystal is used as the thermometric property. An equation of the form of Eq. 1.15 still, however, applies.

1.5 Boyle's Law

On page 6 we implied that the temperature of a gaseous system could be determined through its relationship to pressure and volume. To determine how this can be done, we consider the pressure-volume relation at constant temperature. A gas contained within a closed vessel exerts a force on the walls of the vessel. This force is related to the pressure P of gas (the force F divided by the wall area A) and is a scalar quantity; that is, it is independent of direction. For convenience the standard atmosphere has been defined as the pressure exerted by a column of mercury 760 mm high at 0°C. A practical unit of pressure has been the torr, which is the pressure exerted by a 1-mm column of mercury. In SI units,* the standard atmospheric

*See Appendix A for a discussion of SI units and the recommendations of the International Union of Pure and Applied Chemistry.

pressure is 101 325 Pa, where the abbreviation Pa stands for the SI unit of pressure, the pascal (kg m^{-1} s^{-2} = N m^{-2}). In this system 133.32 Pa is equal to the pressure produced by a column of mercury 1 millimetre (mm) in height.

The British chemist Robert Boyle (1627–1691) set out to study quantitatively the relation between pressure and volume. His definitive work* came in 1662 and may be summarized in the following statement, known as **Boyle's law**:

> **The pressure of a fixed amount of gas varies inversely with the volume if the temperature is maintained constant.**

Mathematically, Boyle's law can be stated as

$$P \propto \frac{1}{V}, \quad \frac{1}{P} \propto V, \quad P = \text{constant}/V, \quad \text{or}$$
$$PV = \text{constant} \quad (\text{valid at constant } T) \tag{1.16}$$

A plot of $1/P$ against V for some of Boyle's original data is shown in Figure 1.3. The advantage of this plot over a P against V plot is that the linear relationship makes it easier to see deviations from the law. Boyle's law is surprisingly accurate for many gases at moderate pressures.

1.6 Gay-Lussac's (Charles's) Law

The French physicist Guillaume Amontons (1663–1705) measured the influence of temperature on the pressure of a fixed volume of a number of different gases and predicted that as the air cooled, the pressure should become zero at some low temperature, which he estimated to be −240°C. He thus anticipated the work of Jacques Alexandre Charles (1746–1823), who a century later independently derived the direct proportionality between the volume of a gas and the temperature. Since Charles never published his work, it was left to the French chemist Joseph Louis Gay-Lussac (1778–1850), proceeding independently, to make a more careful study using mercury to confine the gas and to report that all gases showed the same dependence of V on θ. He expanded the idea of an absolute zero of temperature and calculated its value to be −273°C. Thus for a particular value of the temperature θ and a fixed volume of gas V_0 at 0°C, we have the linear relation

$$V = V_0(1 + \alpha\theta) \tag{1.17}$$

where α is the *cubic expansion coefficient*. The modern value of α equals 1/273.15. Plots of the volume against temperature for several different gases are shown in Figure 1.4. It can be seen that the curves of the experimentally determined region can be extrapolated to zero volume where θ is −273.15°C. This fact immediately suggests that the addition of 273.15° to the Celsius temperature would serve to define a new temperature scale T that would not have negative numbers:

$$T = \theta + 273.15° \tag{1.18}$$

Thus it would be more convenient to measure temperatures on this new scale in which the temperature interval of the degree remains the same as on the Celsius

* Robert Boyle, *Collected Works* (Vol. 1), London: Thomas Birch, 1722.

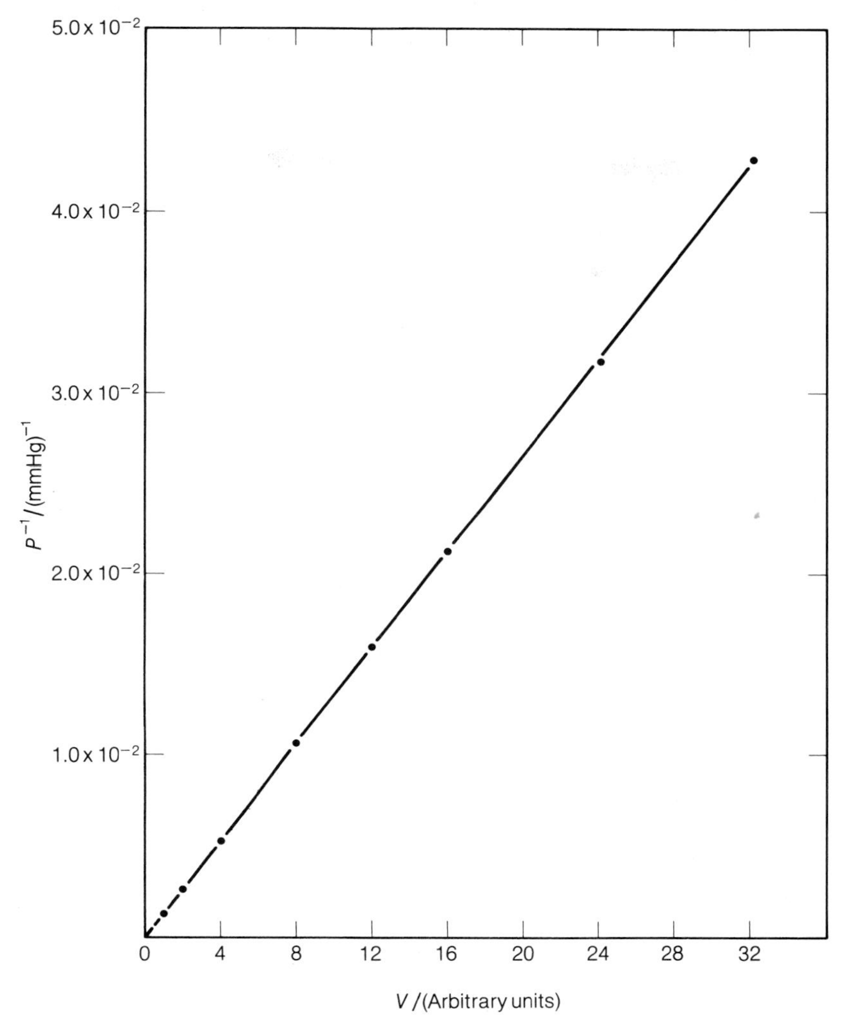

FIGURE 1.3 A plot of $1/P$ against V for Boyle's original data. This plot, passing through the origin, shows that PV = constant.

scale. The scale so defined is named the *absolute temperature* scale or the **Kelvin temperature** scale after William Thomson, Lord Kelvin of Largs (1824–1907). A numerical value followed by the symbol K defines values on this scale.

Variously called **Charles's law** or **Gay-Lussac's law**, the law may be written in the form of Boyle's law by utilizing the absolute temperature:

$$V \propto T, \qquad V = \text{constant} \cdot T, \quad \text{or}$$
$$\frac{V}{T} = \text{constant} \qquad \text{(valid at constant } P\text{)} \tag{1.19}$$

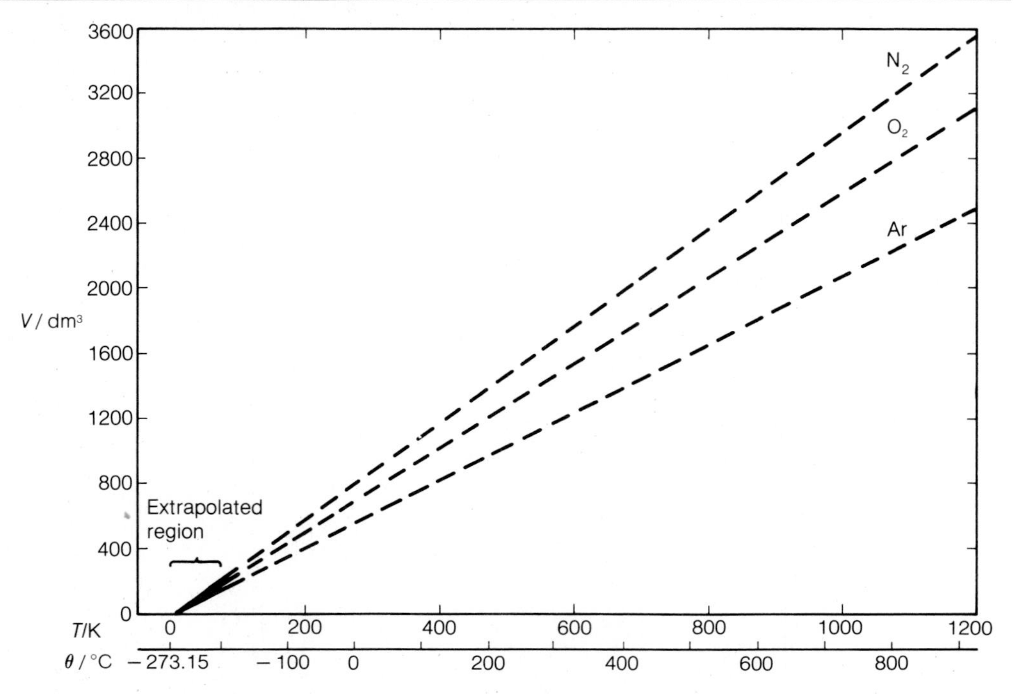

FIGURE 1.4 A plot of volume against temperature for argon, nitrogen, and oxygen. The individual curves show the effect of a change in molecular mass for the three gases. In each case one kilogram of gas is used.

The behavior of many gases near atmospheric pressure is approximated quite well by this law at moderate to high temperatures.

1.7 The Ideal Gas Thermometer

The work of Gay-Lussac provided an important advance in the development of science, but temperature was still somewhat dependent on the working substance used in its determination. We now look for a means to measure temperature that does not depend on the properties of any one substance.

Experimentally one finds that Eq. 1.19 holds true for real gases at moderate to high temperatures only as the pressure is reduced to zero. Thus we could rewrite Gay-Lussac's law, Eq. 1.19, as

$$\lim_{P \to 0} V = CT \tag{1.20}$$

where C is a constant. This expression may serve as the basis of a new temperature scale. Thus if the temperature and volume of a fixed amount of gas held at some low pressure are T_1 and V_1, respectively, before the addition of heat, the ratio of the temperature T_2 to T_1 after the addition of heat is given by the ratio of the initial volume and the final volume, V_2, of the gas. Thus,

$$\frac{T_2}{T_1} = \frac{\lim_{P \to 0} V_2}{\lim_{P \to 0} V_1} \tag{1.21}$$

However, the low-pressure limit of a gas volume is infinite, and this relationship is therefore impractical. Instead, we can cast the expression into the form

$$\frac{T_2}{T_1} = \frac{\lim_{P\to 0} (PV)_2}{\lim_{P\to 0} (PV)_1} \tag{1.22}$$

We thus have a gas thermometer that will work equally well for any gas. If a gas can be imagined to obey Eq. 1.19 or Eq. 1.22 exactly for all values of P, then we have defined an **ideal gas** and the thermometer using such a gas is known as the *ideal gas thermometer*.

Although considerable work went into defining a two-reference-point temperature scale as described earlier, low-temperature work has demonstrated the need for a temperature scale based only on one experimental point along with the absolute zero. In 1954, the decision was made to redefine the absolute scale and the Celsius scale. The absolute zero is zero kelvin, written as 0 K. Since the most careful measurements of the ice point (water-ice equilibrium at 1 atm pressure) varied over several hundredths of a kelvin, the **triple point** *of water* (water-ice-water vapor equilibrium) *is defined as* 273.16 *K exactly*. The ice point is then almost exactly 273.15 K and the boiling point is simply another temperature to be measured experimentally; it is almost exactly 373.15 K or 100°C. *The temperature in kelvins is by definition obtained by adding exactly* 273.15° *to the temperature in degrees Celsius*. As far as temperature *intervals* are concerned, the degree Celsius is the same as the kelvin.

Using the defined value of the triple point for T_1, a working definition for the new Celsius scale becomes

$$T_2 = 273.16° \frac{\lim_{P\to 0} (PV)_{T_2}}{\lim_{P\to 0} (PV)_{\text{triple point}}} \tag{1.23}$$

where $\lim_{P\to 0} (PV)_T = 0$ when $T_2 = 0$.

1.8 The Equation of State for an Ideal Gas

The Gas Constant and the Mole Concept

In Sections 1.5 and 1.6, we found that pressure, volume, and temperature are related. Experimentally, these three properties of a gas cannot be arbitrarily chosen to describe the state of a fixed amount of gas. (The effects of gravitational, electric, and magnetic fields are neglected in this treatment.) For a fixed amount of gas these basic properties are related by an equation of state. However, one finds experimentally that the linear relationship of Boyle's law (Eq. 1.16) for P-V data is attained only at very low pressures. Hence, in the limit of zero pressure, all gases should obey Boyle's law to the same degree of accuracy. Thus Boyle's law could be better written as

$$\lim_{P\to 0} (PV) = C' \tag{1.24}$$

In the same manner, Gay-Lussac's law can be written in the form seen in Eq. 1.20. On combining these two expressions, we obtain

$$\lim_{P\to 0} (PV) = C''T \tag{1.25}$$

In order to determine the value of the constant C'', we can utilize an important hypothesis proposed in 1811 by the Italian physicist Amedeo Avogadro (1776–1856) that a given volume of any gas (at a fixed temperature and pressure) must contain the same number of independent units. Furthermore, he specified that the particles of gas could be atoms or combinations of atoms, coining the word *molecule* for the latter case. One may state **Avogadro's hypothesis** as

$$V \propto n \quad \text{or} \quad V = C'''n \qquad \text{(valid at constant } P \text{ and } T\text{)} \tag{1.26}$$

where n is the amount of substance, the SI unit for which is the *mole.*

One **mole** is the amount of any substance containing the same number of elementary entities (atoms, molecules, ions, and so forth) as there are in exactly 0.012 kg of carbon-12 (see also Appendix A). The number of elementary entities is related to the amount of substance by the Avogadro constant; this is given the symbol L [after Joseph Loschmidt (1821–1895) who first measured its magnitude] and has the value $6.022\,045 \times 10^{23}$ mol^{-1}. The *numerical value* of the Avogadro constant, $6.022\,045 \times 10^{23}$, is the *number* of elementary entities in a mole.

Since Eqs. 1.20, 1.24, and 1.26 show that the volume of a gas depends on T, $1/P$, and n, respectively, these three expressions may be combined as

$$\lim_{P\to 0} (PV) = nRT \tag{1.27}$$

If we keep in mind the limitations imposed by the $\lim_{P\to 0}$ requirement, the equation may be approximated by

$$PV = nRT \tag{1.28}$$

where R is the *universal gas constant*. The value of R will depend on the units of measurements of P and V. In terms of the variables involved, R must be of the form

$$R = \frac{PV}{nT} = \frac{\text{(force/area)} \cdot \text{volume}}{\text{mol} \cdot \text{K}} = \text{force} \cdot \text{length} \cdot \text{K}^{-1} \cdot \text{mol}^{-1}$$
$$= \text{energy} \cdot \text{K}^{-1} \cdot \text{mol}^{-1}$$

Various units of R are required for different purposes, and some of the most often used values of R are given in Table 1.1. Equation 1.28, the **equation of state of an ideal gas,** is one of the most important expressions in physical chemistry. What it states is that in any sample of gas behaving ideally, if one of the four variables (amount, pressure, volume, or temperature) is allowed to change, the values of the other three variables will adjust to maintain a constant value of R.

TABLE 1.1 Numerical Values of R as Commonly Used

8.314 41 J K^{-1} mol^{-1} (SI unit)
0.082 057 atm dm^3 K^{-1} mol^{-1}*
$8.314\,4 \times 10^7$ erg K^{-1} mol^{-1}
1.987 19 cal K^{-1} mol^{-1}

* In SI the volume is expressed in cubic decimeters (dm^3). The more familiar unit, the liter, is now defined as 1 dm^3.

A useful relation can be obtained by rearranging Eq. 1.28 into the form

$$P = \frac{n}{V}RT = \frac{m/M}{V}RT = \frac{\rho}{M}RT \tag{1.29}$$

or

$$M = \rho\frac{RT}{P}$$

where n, the amount of substance, is the mass m divided by the *molar mass** M, and m/V is the density ρ of the gas.

EXAMPLE

Calculate the average molar mass of air at sea level and 0°C if the density of air is 1.29 kg m^{-3}.

SOLUTION

At sea level the pressure may be taken equal to 1 atm or 101 325 Pa. Using Eq. 1.29,

$$M = \frac{\rho RT}{P} = \frac{1.29(\text{kg m}^{-3}) \times 8.314(\text{J K}^{-1}\ \text{mol}^{-1}) \times 273.15(\text{K})}{101\ 325\ \text{N m}^{-2}\ (\text{or Pa})}$$

$$= \frac{1.29(\text{kg m}^{-3}) \times 8.314(\text{kg m}^2\text{s}^{-2}\ \text{K}^{-1}\ \text{mol}^{-1}) \times 273.15(\text{K})}{101\ 325(\text{kg m s}^{-2}\ \text{m}^{-2})}$$

$$= 0.0289\ \text{kg mol}^{-1} = 28.9\ \text{g mol}^{-1}$$

■

1.9 The Kinetic-Molecular Theory of Ideal Gases

An experimental study of the behavior of a gas, such as that carried out by Boyle, cannot determine the nature of the gas or why the gas obeys particular laws. In order to understand gases, one could first propose some hypothesis about the nature of gases. Such hypotheses are often referred to as constituting a "model" for a gas. The properties of the gas that are deduced from this model are then compared to the experimental properties of the gas. The validity of the model lies in its ability to predict the behavior of gases.

We will simply state three postulates of the kinetic-molecular model, and show how this model leads to the ideal gas laws. This model of an idealized gas will be seen to fit the behavior of many real gases.

1. The gas is assumed to be composed of individual molecules whose actual dimensions are small in comparison to the distances between them.
2. These molecules are in constant motion and therefore have kinetic energy.
3. Neither attractive nor repulsive forces exist between the molecules.

---

* The molar mass is mass divided by the amount of substance. The relative molecular mass of a substance M_r was formerly called the molecular weight and is dimensionless.

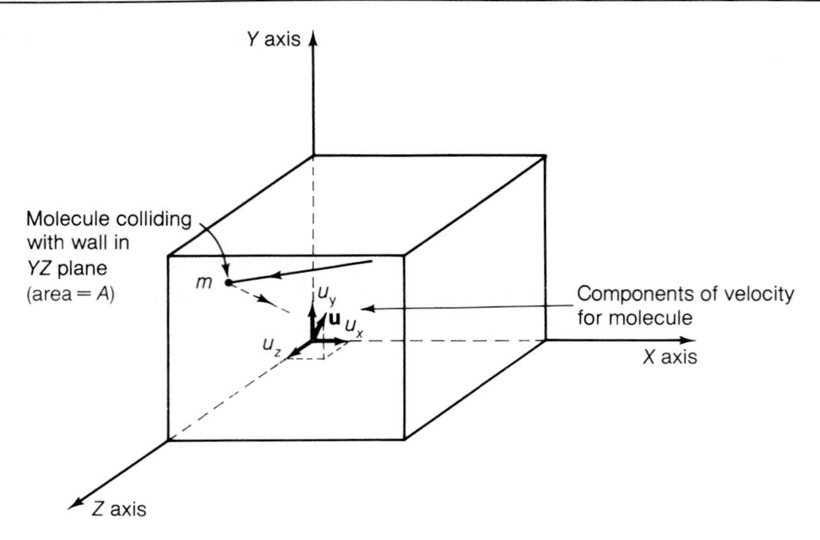

FIGURE 1.5 Container showing coordinates and velocity components for gas particle of mass m.

In order to see how this model predicts the observed behavior quantitatively, we must arrive at an equation relating the pressure and volume of a gas to the important characteristics of the gas, namely the number of molecules present, their mass, and the velocity with which they move. We will first focus our attention on a single molecule of mass m confined in an otherwise empty container of volume V. The particle traverses the container with velocity $\boldsymbol{u}$ (a vector quantity indicating both speed and direction). Because of the particle's X-component of velocity (i.e., the component u_x as shown in Figure 1.5) the molecule will traverse the container of length x in the X-direction, collide with the wall YZ, and then rebound. In the impact the molecule exerts a force F_w on the wall. This force is exactly counterbalanced by the force F exerted on the molecule by the wall. The force F is equal to the change of momentum $\boldsymbol{p}$ of the molecule in the given direction per unit time, in agreement with Newton's second law of motion,

$$F = \frac{d\boldsymbol{p}}{dt} = m\boldsymbol{a} = \frac{d(m\boldsymbol{u})}{dt} = m\frac{d\boldsymbol{u}}{dt} \tag{1.30}$$

The molecule's momentum in the X-direction when it strikes the wall is mu_x. Since we assume that the collision is perfectly elastic, the molecule bounces off the wall with a velocity $-u_x$ in the opposite direction.

The change in velocity of the molecule on each collision is

$$\begin{aligned}\Delta u_x &= [-u_x(\text{after collision})] - [u_x(\text{before collision})]\\ &= -2u_x\end{aligned} \tag{1.31}$$

The corresponding change of momentum is $-2mu_x$. Since each collision with this wall occurs only after the molecule travels a distance of $2x$ (i.e., one round trip), the number of collisions a molecule makes in unit time may be calculated by dividing the distance u_x the molecule travels in unit time by $2x$. The result is

$$\text{number of collisions in unit time} = \frac{u_x}{2x} \tag{1.32}$$

The change of momentum per unit time is thus

$$F = m\frac{d\mathbf{u}}{dt} = (-2u_x m)\left(\frac{u_x}{2x}\right) = -\frac{mu_x^2}{x} \tag{1.33}$$

The force F_w exerted on the wall by the particle is exactly equal in magnitude to this but with opposite sign:

$$F_w = -F = \frac{mu_x^2}{x} \tag{1.34}$$

Since the pressure is force per unit area and the area A is yz, we may write the pressure in the X-direction, P_x, as

$$\begin{aligned} P_x &= \frac{F_w}{A} = \frac{F_w}{yz} \\ &= \frac{mu_x^2}{xyz} = \frac{mu_x^2}{V} \end{aligned} \tag{1.35}$$

since $xyz = V$ is the volume of the container.

The Pressure of a Gas Derived from Kinetic Theory

Attention has so far been confined to one molecule that has been assumed to travel at constant velocity. For an assembly of N molecules there will be a distribution of molecular velocities since even if the molecules all began with the same velocity, collisions would occur altering the original velocity. If we define u_i^2 as the square of the velocity component in the X-direction of molecule i and take the average of this over molecules rather than the sum over u_i^2, we have

$$\overline{u_x^2} = \frac{u_1^2 + u_2^2 + u_3^2 + \cdots + u_N^2}{N} = \frac{\sum_{i=1}^{N} u_i^2}{N} \tag{1.36}$$

where $\overline{u_x^2}$ is the mean of the squares of the normal component of velocity in the X-direction. The pressure expressed in Eq. 1.35 should, therefore, be written as

$$\boldsymbol{P_x = \frac{Nm\overline{u_x^2}}{V}} \tag{1.37}$$

This is the equation for the pressure of a one-dimensional gas. For the components of velocity in the Y-direction and in the Z-direction, we would obtain expressions similar to Eq. 1.37 but now involving $\overline{u_y^2}$ and $\overline{u_z^2}$, respectively. It is more convenient to write these expressions in terms of the magnitude of the velocity u rather than in terms of the squares of the velocity components. The word *speed* is used for the magnitude of the velocity; speed is defined as the positive square root of u^2 and is related to the velocity components by the Pythagorean theorem:

$$|u| = \sqrt{u^2} = \sqrt{u_x^2 + u_y^2 + u_z^2} \tag{1.38}$$

If we average over all molecules, we obtain

$$\overline{u^2} = \overline{u_x^2} + \overline{u_y^2} + \overline{u_z^2} \tag{1.39}$$

Since there is no reason for one direction to be favored over the others, the mean of the u_x^2 values will be the same as the mean of the u_y^2 values and the mean of the u_z^2 values. Hence, the sum of the means is equal to $\overline{u^2}$ and each mean is equal to $\frac{1}{3}\overline{u^2}$; that is,

$$\overline{u_x^2} = \overline{u_y^2} = \overline{u_z^2} = \frac{1}{3}\overline{u^2} \quad (1.40)$$

Substitution of Eq. 1.40 into Eq. 1.37 gives the final expression for the pressure on any wall:

$$P = \frac{Nm\overline{u^2}}{3V} \quad \text{or} \quad PV = \frac{1}{3}Nm\overline{u^2} \quad (1.41)$$

This is the fundamental equation as derived from the simple kinetic theory of gases. We see that Eq. 1.41 is in the form of Boyle's law and is consistent with Charles's law if $m\overline{u^2}$ is directly proportional to the absolute temperature. In order to make this relation exact, we use Eq. 1.28 and substitute nRT for PV in Eq. 1.41:

$$nRT = \frac{1}{3}Nm\overline{u^2} \quad (1.42)$$

or

$$\frac{1}{3}Lm\overline{u^2} = RT \quad (1.43)$$

since N/n is equal to the Avogadro constant L.

Kinetic Energy and Temperature

We have seen how the kinetic molecular theory of gases may be used to explain the experimental form of two gas laws. It can also shed light on the nature of kinetic energy. In order to determine the exact relation between $\overline{u^2}$ and T, the mechanical variable u of Eq. 1.41 must be related to the temperature, which is not a mechanical variable. For our purpose of determining the relation between kinetic energy and temperature, Eq. 1.41 may be converted into another useful form by recognizing that the average kinetic energy $\overline{\epsilon}_k$ per molecule is

$$\overline{\epsilon}_k = \frac{1}{2}m\overline{u^2} \quad (1.44)$$

Substitution of this expression into Eq. 1.41 gives

$$PV = \frac{1}{3}N \cdot 2\overline{\epsilon}_k = \frac{2}{3}N\overline{\epsilon}_k \quad (1.45)$$

Thus, at constant pressure, the volume of a gas is proportional to the number of gas molecules and the average kinetic energy of the molecules. Since

$$N = nL \quad (1.46)$$

substitution into Eq. 1.45 yields

$$PV = \frac{2}{3}nL\overline{\epsilon}_k \quad (1.47)$$

Since $L\overline{\epsilon}_k$ is the total kinetic energy E_k per mole of gas,

$$PV = \frac{2}{3}nE_k \quad (1.48)$$

The connection between E_k and the temperature is provided by the empirical ideal gas law, $PV = nRT$. By equating the right-hand sides of the last two equations we obtain

$$\tfrac{2}{3}nE_k = nRT \quad \text{or} \quad E_k = \tfrac{3}{2}RT \tag{1.49}$$

The average kinetic energy per molecule $\bar{\epsilon}_k$ is obtained by dividing both sides by the Avogadro constant L:

$$\bar{\epsilon}_k = \tfrac{3}{2}\mathbf{k}T \tag{1.50}$$

where $\mathbf{k} = R/L$. Named after the Austrian physicist Ludwig Edward Boltzmann (1844–1906), the **Boltzmann constant*** $\mathbf{k}$ is the gas constant per molecule. Thus the average kinetic energy of the molecules is proportional to the absolute temperature. Since $\bar{\epsilon}_k$ in Eq. 1.50 is independent of the kind of substance, the average molecular kinetic energy of all substances is the same at a fixed temperature.

An interesting aspect of this fact occurs when we consider a number of different gases all at the same temperature and pressure. Then Eq. 1.41 may be written as

$$\frac{N_1 m_1 \overline{u_1^2}}{3V_1} = \frac{N_2 m_2 \overline{u_2^2}}{3V_2} = \cdots = \frac{N_i m_i \overline{u_i^2}}{3V_i} \tag{1.51}$$

or

$$\frac{N_1}{3V_1} = \frac{N_2}{3V_2} = \cdots = \frac{N_i}{3V_i} \tag{1.52}$$

Thus $N_1 = N_2 = \cdots = N_i$ when the volumes are equal. In other words, equal volumes of gases at the same pressure and temperature contain equal numbers of molecules. This is just a statement of Avogadro's hypothesis already seen in Eq. 1.26.

Dalton's Law of Partial Pressures

The studies of the English chemist John Dalton (1766–1844) showed in 1801 that the *total pressure* observed for a mixture of gases is equal to the *sum of the pressures that each individual component gas would exert* had it alone occupied the container at the same temperature. This is known as **Dalton's law of partial pressures**. Of course, in order for it to be obeyed, no chemical reactions between component gases may occur and the component gases must behave ideally.

The term **partial pressure** is used to express the *pressure exerted by one component of the gas mixture*. Thus

$$P_t = P_1 + P_2 + \cdots + P_i \tag{1.53}$$

where P_t is the total pressure. Then, using a form of Eq. 1.28, we may write

$$P_t = \frac{n_1RT}{V} + \frac{n_2RT}{V} + \cdots + \frac{n_iRT}{V}$$
$$= (n_1 + n_2 + \cdots + n_i)\frac{RT}{V} \tag{1.54}$$

where the P_i's are the partial pressures and the n_i's are the amounts of the individual gases. Dalton's law can be shown to be predicted by the simple kinetic molec-

*$\mathbf{k} = 1.380\,622 \times 10^{-23}$ J K^{-1}

ular theory (Eq. 1.41) by writing expressions of the form $P_i = N_i m_i \overline{u_i^2}/3V$ for each gas. Thus

$$P_t = \sum_1^i P_i = \frac{N_1 m_1 \overline{u_1^2}}{3V} + \frac{N_2 m_2 \overline{u_2^2}}{3V} + \cdots + \frac{N_i m_i \overline{u_i^2}}{3V} \tag{1.55}$$

Dalton's law immediately follows from the kinetic theory of gases since the average kinetic energy from Eq. 1.49 is $\frac{3}{2}RT$ and is the same for all gases at a fixed temperature.

Application of Dalton's law is particularly useful when a gas is generated and subsequently collected over water. The total gas pressure consists of the pressure of the water vapor present in addition to the pressure of the gas that is generated.

Graham's Law of Effusion

The Scottish physical chemist Thomas Graham (1805–1869) measured the movement of gases through plaster of Paris plugs, fine tubes, and small orifices in plates where the passages for the gas are small as compared with the average distance that the gas molecules travel between collisions (see next subsection). Such movement is known as **effusion**. In 1831 Graham showed that the *rate of effusion* of a gas was *inversely proportional to the square root of its density* ρ. Later, in 1848, he showed that the *rate of effusion was inversely proportional to the square root of the molar mass M*. This is known as *Graham's law of effusion*. Thus, in the case of the gases oxygen and hydrogen at equal pressures, oxygen molecules are $\frac{32}{2} = 16$ times more dense than those of hydrogen. Therefore, hydrogen diffuses 4 times as fast as oxygen:

$$\frac{\text{rate } (H_2)}{\text{rate } (O_2)} = \sqrt{\frac{\rho(O_2)}{\rho(H_2)}} = \sqrt{\frac{M(O_2)}{M(H_2)}} = \sqrt{\frac{32}{2}} = \frac{4}{1}$$

Graham's law also can be explained on the basis of the simple kinetic molecular theory. Our starting point is Eq. 1.43, which shows that the **root-mean-square velocity** $\sqrt{\overline{u^2}}$ is proportional to the absolute temperature. However, the speed with which gases effuse is proportional to the mean velocity $\bar{u}$ and not $\sqrt{\overline{u^2}}$. The mean velocity differs from the root-mean-square velocity by a constant factor of 0.92 (see Section 14.2); consequently, proportionality relationships deduced for $\sqrt{\overline{u^2}}$ are equally valid for $\bar{u}$. Hence, since the mean-square velocity is proportional to the absolute temperature (Eq. 1.43), the root-mean-square velocity and the mean velocity are each proportional to $T^{1/2}$. Again we can see from Eq. 1.43 that, at constant temperature, the quantity $m\overline{u^2}$ is constant and the mean square velocity, therefore, is inversely proportional to the molecular mass,

$$\overline{u^2} = \frac{\text{constant}_1}{m} \tag{1.56}$$

The mean square velocity is also inversely proportional to the molar mass M, which is simply mL:

$$\overline{u^2} = \frac{\text{constant}_2}{M} \tag{1.57}$$

TABLE 1.2 Average Speeds of Gas Molecules at 293.15 K

Gas	Mean Speed, $\overline{u}$/m s^{-1}
Ammonia, NH_3	582.7
Argon, Ar	380.8
Carbon dioxide, CO_2	454.5
Chlorine, Cl_2	285.6
Helium, He	1204.0
Hydrogen, H_2	1692.0
Oxygen, O_2	425.1
Water, H_2O	566.5

From this it follows that the root-mean-square velocity $\sqrt{\overline{u^2}}$, and consequently the mean speed $\overline{u}$, is inversely proportional to the square root of the molar mass:

$$\overline{u} = \frac{\text{constant}}{M^{1/2}} \tag{1.58}$$

This, therefore, provides a theoretical explanation for Graham's law of effusion. Table 1.2 gives some average speeds of gases at 293.15 K.

These results from simple kinetic molecular theory, being confirmed through empirical observation, give additional support to the original postulates of the theory.

Molecular Collisions

The ability to relate pressure and temperature to molecular quantities in Eqs. 1.41 and 1.50 is a major accomplishment of the kinetic theory of gases. We now apply the kinetic theory to the investigation of the collisions of molecules in order to have a clearer understanding of the interactions in a gas. We will be interested in three aspects of molecular interaction: the number of collisions experienced by a molecule per unit time, the total number of collisions in a unit volume per unit time, and how far the molecules travel between collisions.

In this development we consider that the molecules behave as rigid spheres and that there are two kinds of molecules, A and B, with diameters d_A and d_B. Suppose that a molecule of A travels with an average speed of $\overline{u}_A$ in a container that contains both A and B molecules. We will first assume that the molecules of B are stationary and later remove this restriction. As shown in Figure 1.6, a collision will occur each time the distance between the center of a molecule A and that of a molecule B becomes equal to $d_{AB} = (d_A + d_B)/2$. A convenient way of visualizing this is to construct around the center of A an imaginary sphere of radius d_{AB}, which is the sum of the two radii. In unit time this imaginary sphere, represented by the dashed circle in Figure 1.6, will sweep out a volume of $\pi d_{AB}^2 \overline{u}_A$, and if the center of a B molecule

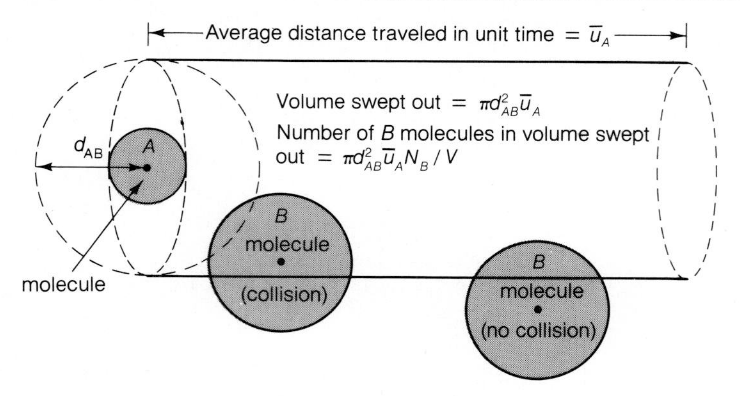

FIGURE 1.6 Collisions between gas molecules A and B. We construct around molecule A a sphere having a radius d_{AB} equal to the sum of the radii of A and B.

is in this volume, there will be a collision. If N_B is the total number of B molecules in the system, the number per unit volume is N_B/V, and the number of centers of B molecules in the volume swept out is $\pi d_{AB}^2 \bar{u}_A N_B/V$. This number is the number Z_A of collisions experienced by the one molecule of A in unit time:

$$Z_A = \frac{\pi d_{AB}^2 \bar{u}_A N_B}{V} \qquad \textbf{(SI unit: s}^{-1}\textbf{)} \tag{1.59}$$

This number is known as the **collision frequency**. If there is a total of N_A/V molecules of A per unit volume in addition to the B molecules, the total number Z_{AB} of A-B collisions per unit volume per unit time is Z_A multiplied by N_A/V:

$$Z_{AB} = \frac{\pi d_{AB}^2 \bar{u}_A N_A N_B}{V^2} \qquad \textbf{(SI unit: m}^{-3}\textbf{ s}^{-1}\textbf{)} \tag{1.60}$$

This quantity, the number of collisions divided by volume and by time, is known as the **collision number**.

If only A molecules are present, we are concerned with the collision number Z_{AA} for the collisions of A molecules with one another. The expression for this resembles that for Z_{AB} (Eq. 1.60) but a factor $\frac{1}{2}$ has to be introduced:

$$Z_{AA} = \frac{\frac{1}{2}\pi d_A^2 \bar{u}_A N_A^2}{V^2} \qquad \text{(SI unit: m}^{-3}\text{ s}^{-1}\text{)} \tag{1.61}$$

The reason for the factor $\frac{1}{2}$ is that otherwise we would count each collision twice. For example, we would count a collision of A_1 with A_2 and A_2 with A_1 as two separate collisions instead of one.

An error in this treatment of collisions is that we have only considered the average speed $\bar{u}_A$ of the A molecules. More correctly we should consider the relative speeds $\bar{u}_{AB}$ or $\bar{u}_{AA}$ of the molecules.

When we have a mixture containing two kinds of molecules A and B of different masses, the $\bar{u}_A$ and $\bar{u}_B$ values are different. The average relative speed $\bar{u}_{AB}$ is equal to $(\bar{u}_A^2 + \bar{u}_B^2)^{1/2}$, and we modify Eqs. 1.59 and 1.60 by replacing $\bar{u}_A$ by $\bar{u}_{AB}$:

$$Z_A = \frac{\pi d_{AB}^2 (\bar{u}_A^2 + \bar{u}_B^2)^{1/2} N_B}{V} \qquad \textbf{(SI unit: s}^{-1}\textbf{)} \tag{1.62}$$

$$Z_{AB} = \frac{\pi d_{AB}^2(\overline{u}_A^2 + \overline{u}_B^2)^{1/2} N_A N_B}{V^2} \quad \text{(SI unit: m}^{-3}\text{s}^{-1}) \tag{1.63}$$

If we have only A molecules, the average relative speed of two A molecules is

$$\overline{u}_{AA} = \sqrt{2}\,\overline{u}_A \tag{1.64}$$

We must therefore replace $\overline{u}_A$ in Eq. 1.61 by $\overline{u}_{AA}$ or $\sqrt{2}\,\overline{u}_A$, and we obtain for the collision number the more correct expression

$$Z_{AA} = \frac{\frac{\sqrt{2}}{2}\pi d_A^2 \overline{u}_A N_A^2}{V^2} \quad \text{(SI unit: m}^{-3}\text{ s}^{-1}) \tag{1.65}$$

The corresponding expression for Z_A, the number of collisions of 1 molecule of A in unit time, when only A molecules are present, is

$$Z_A = \frac{\sqrt{2}\pi d_A^2 \overline{u}_A N_A}{V} \quad \text{(SI unit: s}^{-1}) \tag{1.66}$$

EXAMPLE

Nitrogen and oxygen are held in a 1.00 m^3 container maintained at 300 K at partial pressures of P_{N_2} = 80 kPa and P_{O_2} = 21 kPa. If the collision diameters are

$$d_{N_2} = 3.74 \times 10^{-10}\text{ m} \quad \text{and} \quad d_{O_2} = 3.57 \times 10^{-10}\text{ m}$$

calculate Z_A, the average number of collisions experienced in unit time by one molecule of nitrogen and by one molecule of oxygen. Also calculate Z_{AB}, the average number of collisions per unit volume per unit time. Do this last calculation both at 300 K and 3000 K on the assumption that the values for d and N do not change. At 300 K, $(\overline{u}_{N_2}^2 + \overline{u}_{O_2}^2)^{1/2}$ is 625 m s^{-1}; at 3000 K, it is 2062 m s^{-1}.

SOLUTION

The values of N_{N_2} and N_{O_2} are calculated from the ideal gas law:

$$PV = nRT = \frac{NRT}{L}$$

or for N_2

$$N_{N_2} = \frac{LPV}{RT}$$

$$N_{N_2} = \frac{6.022 \times 10^{23}(\text{mol}^{-1}) \times 80\,000(\text{Pa}) \times 1(\text{m}^3)}{8.314(\text{J K}^{-1}\text{ mol}^{-1}) \times 300(\text{K})}$$
$$= 1.93 \times 10^{25}$$

For O_2,

$$N_{O_2} = 5.07 \times 10^{24} \tag{1.63}$$

The total number of collisions with unlike molecules is given by Eq. 1.62:

$$Z_{N_2} = \pi\left[\left(\frac{3.74 + 3.57}{2}\right) \times 10^{-10}(\text{m})\right]^2 \times 625(\text{m s}^{-1}) \times 5.07 \times 10^{24} \times 1(\text{m}^{-3})$$
$$= 1.39 \times 10^9\text{ s}^{-1}$$

$$Z_{O_2} = \pi[3.66 \times 10^{-10}(\text{m})]^2 \times 652(\text{m s}^{-1}) \times 1.93 \times 10^{25} \times 1(\text{m}^{-3})$$
$$= 5.29 \times 10^9 \text{ s}^{-1}$$

Using Eq. 1.63 we have for the total number of collisions per cubic metre per second

$$Z_{N_2,O_2} = \pi(3.66 \times 10^{-10})^2(652)(1.93 \times 10^{25})(5.07 \times 10^{24})$$
$$= 2.68 \times 10^{34} \text{ m}^{-3} \text{ s}^{-1}$$

At 3000 K, $Z_{N_2,O_2} = 8.49 \times 10^{34}$ m^{-3} s^{-1}. From this example it can be seen that the effect of T on Z is not large, since, from Eq. 1.43 and the later discussion, T enters as $\sqrt{T}$. The effect of d is much more pronounced since it enters as d^2. ■

A property of particular significance in the treatment of certain properties is the **mean free path** λ. This is the average distance that a molecule travels between two successive collisions. We have seen that for a single gas the number of collisions that a molecule makes in unit time, Z_A, is $\sqrt{2}\pi d_A^2 \overline{u}_A N_A/V$ (Eq. 1.66). In unit time the molecule has traveled, on the average, a distance of $\overline{u}_A$. The mean free path is therefore

$$\lambda = \frac{\text{distance traveled in unit time}}{\text{number of collisions in unit time}} = \frac{\overline{u}_A}{\sqrt{2}\pi d_A^2 \overline{u}_A N_A/V} \tag{1.67}$$

$$= \frac{V}{\sqrt{2}\pi d_A^2 N_A} \tag{1.68}$$

The magnitude of the d_A's obviously is of great importance in connection with the kinetic theory of gases, since it is the only molecular property that we need to know in order to calculate collision numbers and the mean free path.

EXAMPLE
Molecular oxygen has a collision diameter of 3.57×10^{-10} m. Calculate λ for oxygen at 300 K and 101.325 kPa.

SOLUTION
Since $PV = nRT = (N/L)RT$, λ may be written as

$$\lambda = \frac{RT}{\sqrt{2}\pi d^2 LP}$$

$$= \frac{8.314(\text{J K}^{-1}\text{ mol}^{-1}) \times 300(\text{K})}{\sqrt{2}\pi[3.57 \times 10^{-10}(\text{m})]^2 \times 6.022 \times 10^{23}(\text{mol}^{-1}) \times 101\,325(\text{Pa})}$$

$$= 7.22 \times 10^{-8} \text{ m}$$

since J/Pa = m^3. ■

1.10 The Barometric Distribution Law

We have mentioned that a change in the properties of a system can be brought about by an applied potential field. In this section we will consider the effect of a gravitational field. In a laboratory experiment the effect of the gravitational field can be ignored. However, for a large-scale system such as our atmosphere or an ocean, gravity can cause appreciable variation in properties. An example is the large increase in hydrostatic pressure at greater ocean depths.

The effect of gravity on the pressure can be determined by considering a column of fluid (either liquid or gas) at constant temperature as shown in Figure 1.7. The column has cross-sectional area A and is subjected to a gravitational field that gives the individual particles an acceleration g in the downward direction. Because of the field, the particles will experience different forces, and as a result the pressure will vary with height.

The force at level z due to the weight of fluid above z is F_z, and at level $z + dz$ it is F_{z+dz}, as shown in Figure 1.7. The force due to the weight within the volume element $A\,dz$ is dF. Thus we may write

$$F_z = dF + F_{z+dz} \tag{1.69}$$

Equation 1.69 may be written in terms of the pressure P, since $F_z = PA$ and $F_{z+dz} = (P + dP)A$. After eliminating terms we obtain

$$A\,dP = -dF \tag{1.70}$$

The force dF is the mass in the volume element $A\,dz$ multiplied by the standard gravitational acceleration g (9.806 65 m s^{-2}). If the density (i.e., the mass per unit volume) is ρ, the mass of the element is $\rho A\,dz$ and its weight is $\rho g A\,dz$. Thus, on substitution for dF,

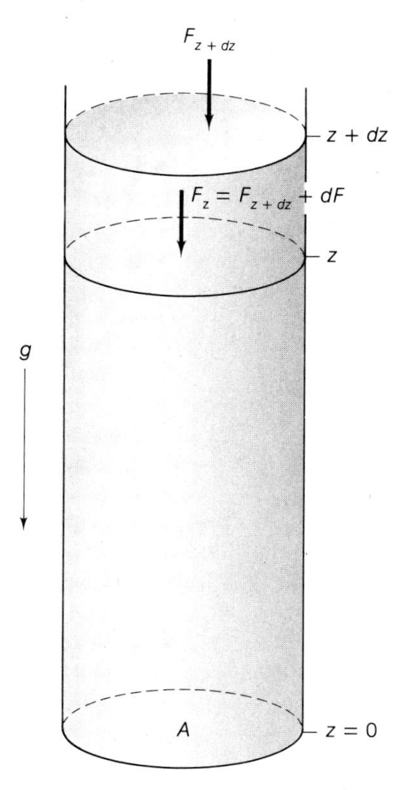

FIGURE 1.7 Distribution of a gas in a gravity field.

$$A\,dP = -\rho g A\,dz \tag{1.71}$$

or

$$dP = -\rho g\,dz \tag{1.72}$$

The change in pressure is thus proportional to the length of the column, and since dz is positive, the pressure decreases with an increase in height.

In general the density depends on the pressure. For liquids, however, the density is practically independent of pressure; for liquids, Eq. 1.72 can be integrated at once:

$$\int_{P_0}^{P} dP = P - P_0 = -\rho g \int_0^z dz \tag{1.73}$$

where P_0 is the reference pressure at the base of the column and P is the pressure at height z. This quantity $P - P_0$ is the familiar *hydrostatic pressure* in liquids.

In a vessel of usual laboratory size, the effect of gravity on the pressure of a gas is negligibly small. On a larger scale, however, such as in our atmosphere, there is a marked variation in pressure, and we must now consider the effect of pressure on the density of the gas. For an ideal gas, from Eq. 1.29, ρ is equal to PM/RT, and substitution into Eq. 1.72 gives

$$\frac{dP}{P} = -\left(\frac{Mg}{RT}\right) dz \tag{1.74}$$

Integration of this expression, with the boundary condition that $P = P_0$ when $z = 0$, gives

$$\ln\frac{P}{P_0} = -\frac{Mgz}{RT} \tag{1.75}$$

or

$$\boldsymbol{P = P_0 e^{-Mgz/RT}} \tag{1.76}$$

This expression describes the distribution of gas molecules in the atmosphere as a function of their molar mass, height, temperature, and the acceleration due to gravity. It is known as the **barometric distribution law.**

The distribution function in Eq. 1.74, which is in differential form, is more informative when the sign is transposed to the dP term:

$$-\frac{dP}{P} = \frac{Mg\,dz}{RT} \tag{1.77}$$

Here $-dP/P$ represents a relative decrease in pressure; it is a constant, Mg/RT, multiplied by the increase in height. This means that it does not matter where the origin is chosen; the function will decrease the same amount over each increment of height.

The fact that the relative decrease in pressure is proportional to Mg/RT shows that, for a given gas, a smaller relative pressure change is expected at high temperatures than at low temperatures. In a similar manner, at a given temperature a gas having a higher molar mass is expected to have a larger relative decrease in pressure than gases with lower molar masses.

Equation 1.76 applies equally to the partial pressures of the individual components in a gas. It follows from the previous treatment that in the upper reaches of the earth's atmosphere the partial pressure will be relatively higher for a very light gas. The distribution function accounts satisfactorily for the gross details of the atmosphere, although winds and temperature variations lead to nonequilibrium conditions.

Since Mgz is the gravitational potential energy, E_p, Eq. 1.76 can be written as

$$P = P_0 e^{-E_p/RT} \tag{1.78}$$

Since the density ρ is directly proportional to the pressure, we also may write

$$\rho = \rho_0 e^{-E_p/RT} \tag{1.79}$$

where ρ_0 represents the density at the reference state height of $z = 0$.

These equations, in which the property varies exponentially with $-E_p/RT$, are special cases of the **Boltzmann distribution law**. We deal with this law in more detail in Chapter 14, where we will meet a number of its important applications in physical chemistry.

1.11 Real Gases

The Compression Factor

The gas laws that we have treated in the preceding sections hold fairly well for most gases over a limited range of pressures and temperatures. However, when the range and the accuracy of experimental measurements were extended and improved, real gases were found to deviate from the expected behavior of an ideal gas. For instance, the PV_m product does not have the same value for all gases nor is the pressure dependence the same for different gases. (V_m represents the **molar volume**, the volume occupied by 1 mol of gas.) Figure 1.8 shows the deviations of N_2 and Ar from the expected behavior of an ideal gas under particular *isothermal* (constant temperature) conditions. It is difficult, however, to determine easily the relative deviation from ideal conditions from a graph of this sort or even from a P against $1/V$ plot. A more convenient technique often used to show the deviation from ideal behavior involves the use of graphs or tables of the **compression** (or compressibility) **factor** Z, defined by

$$Z = \frac{PV}{nRT} = \frac{PV_m}{RT} \tag{1.80}$$

and normally presented as a function of pressure or volume.

For the ideal gas, $Z = 1$. Therefore, departures from the value of unity indicate nonideal behavior. Since each gas will have different interactions between its molecules, the behavior of Z can be expected to be quite varied. In Figure 1.9, a plot of Z versus P for several gases shows the variation of deviations typically found. The shapes of the initial negative slopes of these curves can be related to attractive forces prevalent among the molecules in the gas, whereas initial positive slopes indicate that repulsive forces between molecules predominate.

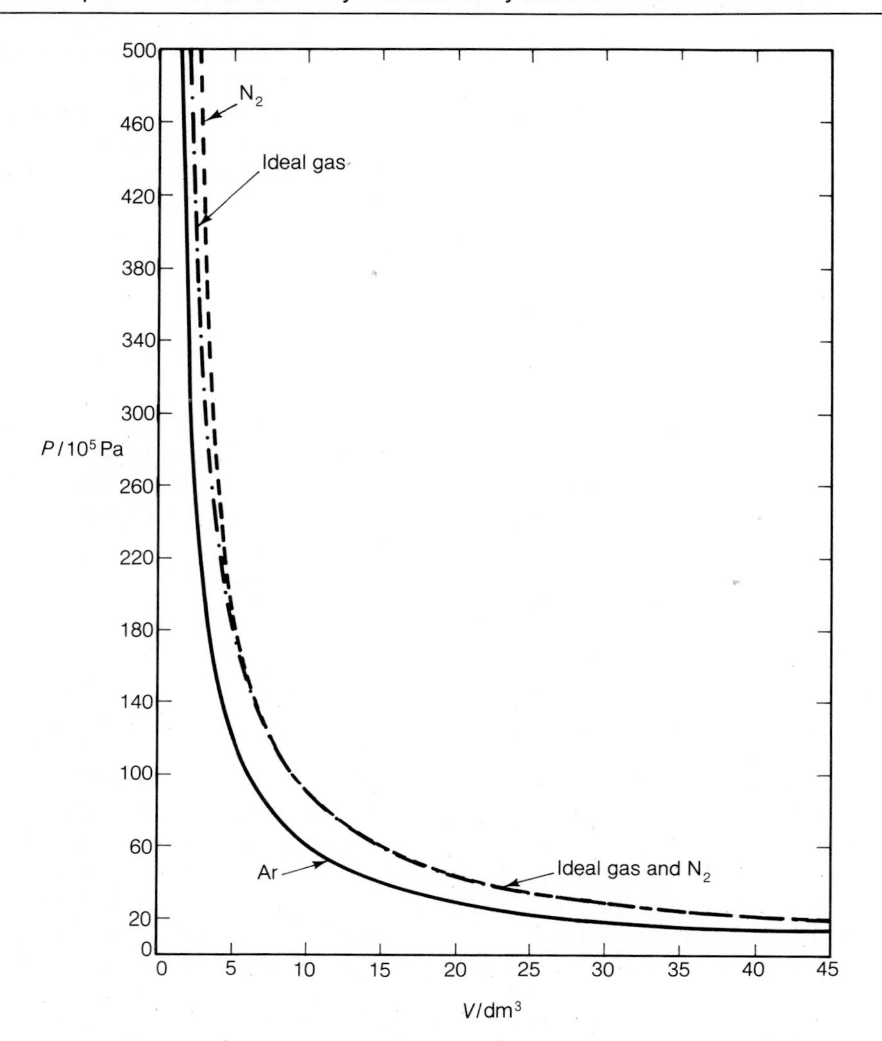

FIGURE 1.8 Plots of pressure against volume for nitrogen and argon at 300 K. Nitrogen follows the ideal gas law very closely, but argon shows significant deviations.

Condensation of Gases: The Critical Point

The Irish physical chemist Thomas Andrews (1813–1885) studied the behavior of carbon dioxide under pressure at varying temperatures. Using a sample of liquid carbon dioxide (CO_2 can be liquefied at room temperature using sufficiently high pressure), he gradually raised its temperature while maintaining the pressure. To his surprise, the boundary between the gas and liquid regions disappeared at 31.1°C. Further increase in pressure could not bring about a return to the liquid state. This experiment led Andrews to suggest* that a **critical temperature** T_c existed for each gas. Above this temperature pressure alone could not liquefy the gas. In other words, T_c is the highest temperature at which a liquid can exist as a distinct

* T. Andrews, *Phil, Mag.*, *41*, 39, 150–153 (1870).

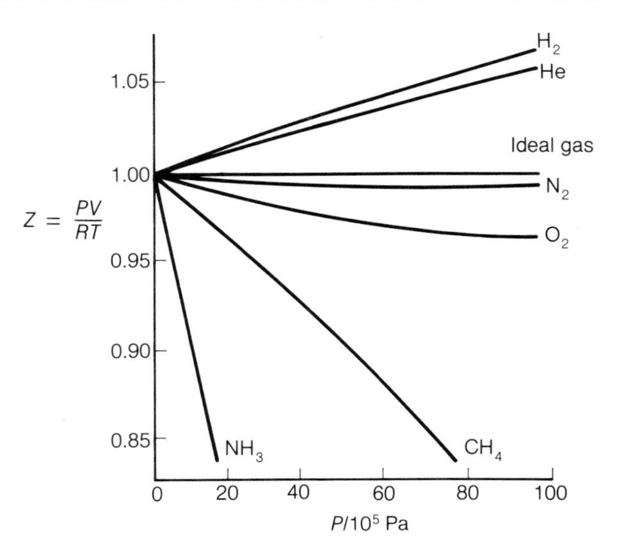

FIGURE 1.9 Plot of the compression factor, Z, against pressure for several gases at 273 K.

phase or region. He introduced the word *vapor* for the gas that is in equilibrium with its liquid below the critical temperature.

Investigation of other real gases showed similar behavior. In Figure 1.10 the isotherms (lines of constant temperature) are shown for a typical real gas. For the higher temperatures T_5 and T_6, the appearance of the isotherms is much like the hyperbolic curves expected of an ideal gas. However, below T_c the appearance of the curves is quite different. The horizontal portions along T_1, T_2, and T_3 are called *tie lines* and contain a new feature not explained by the ideal gas law. The significance of the tie lines can be seen by considering a point such as y on the T_2 isotherm. Since y lies in the stippled region representing the coexistence of liquid and vapor, any variation in volume of the system between the values x and z serves merely to change the liquid-vapor ratio. The pressure of the system corresponding to the tie line pressure is the saturated vapor pressure of the liquefied gas. The right endpoint z of the tie line represents the molar volume in the gas phase; the left endpoint x represents the molar volume in the liquid phase. The distance between the endpoints and y is related to the percentage of each phase present.

If a sample of gas is compressed along isotherm T_3, starting at point A, the PV curve is approximately the Boyle's law isotherm until point B is reached. As soon as we move into the stippled region, liquid and vapor coexist as pointed out previously. Liquefaction begins at point B and ends at point C when all the gas is converted to liquid. As we pass out of the two-phase region, only liquid is present along CD. The steepness of the isotherm indicates the rather low ability of the liquid to be compressed compared to that of the gas.

As the isotherms approach that of T_c, the tie lines become successively shorter until at the **critical point** they cease to exist and only one phase is present. The pressure and volume of the substance corresponding to this critical point are called the **critical pressure** P_c and **critical volume** V_c, respectively. The critical point is a point of inflection; therefore, *the equations*

$$\left(\frac{\partial P}{\partial V}\right)_{T_c} = 0 \quad \text{and} \quad \left(\frac{\partial^2 P}{\partial V^2}\right)_{T_c} = 0 \tag{1.81}$$

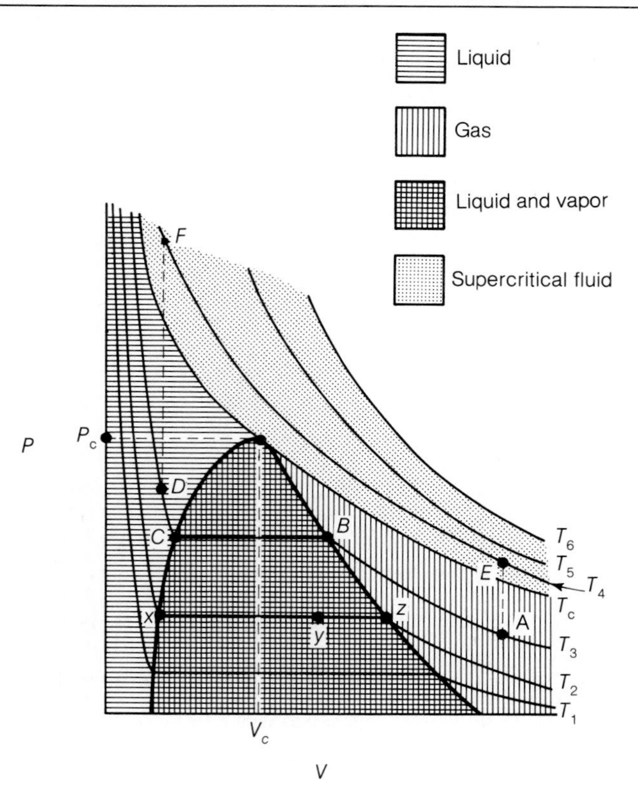

FIGURE 1.10 Isotherms for a typical real gas.

serve to define the critical point. The notation ∂ indicates partial differentiation and merely means that another variable that influences the relation is being held constant as indicated by the subscript on the parentheses. This subject will be discussed more fully later.

Since there is no distinction between liquid and gas phases above the critical point and no second phase is formed regardless of the pressure of the system, the term *fluid* is used instead of liquid or vapor. Indeed, at the critical point, the gas and liquid densities, molar volumes, and indexes of refraction are identical. Normally, one cannot precisely determine the critical point visually. An application of the **law of the rectilinear diameter** may be utilized to determine the value of the critical point accurately. The law states that the average density $\overline{\rho}$ of a pure substance, $\overline{\rho} = (\rho_l + \rho_v)/2$, is a linear function of the temperature. The application of this law to data for CO_2 is shown in Figure 1.11. We see that the intersection of the essentially straight line with the curved line of densities makes it easy to determine the critical temperature.

Furthermore, since gas and liquid can coexist only in the isolated stippled region, it must be possible to pass from a single-phase gas region to a single-phase liquid region without noticing a phase change. In order to illustrate this point, consider 1 mol of liquid contained in a sealed vessel at the condition represented by point D in Figure 1.10; the temperature is T_3. We now heat the liquid above the critical temperature to point F. With the vessel thermostatted at T_4, the volume is

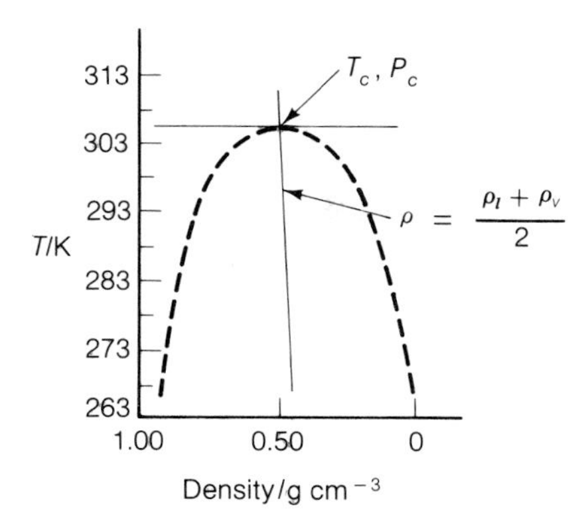

FIGURE 1.11 Density of the liquid and vapor states of CO_2 near the critical point showing the application of the law of rectilinear diameter.

allowed to increase to E. There is no change of phase during these processes. Then the temperature is dropped to the isotherm T_3 at point A. Again no phase change is involved. The system now is in the gaseous state without ever having undergone a discontinuous change of phase. There is thus a complete *continuity of states* in which the transformation from the gas to the liquid state occurs continuously.

In the light of the previous discussion, the distinction between the liquid and gaseous states can be made only when two phases coexist.

1.12 Equations of State

The van der Waals Equation of State

The Dutch physicist Johannes Diderik van der Waals (1837–1923) found that two simple modifications in the ideal gas equation could account for the two-phase equilibrium when a gas is liquefied. The first of these modifications involves the volume of the molecules used in the ideal gas expression. From simple kinetic theory, the molecules are point particles; that is, they occupy no space. Actually, however, individual molecules do have a finite size and do occupy space. If V is the volume available for the ideal gas to move within its container, then the measured or observed volume of the container must be reduced by a volume b, called the *covolume*, which is approximately four times the volume occupied by the individual molecules. This volume will be proportional to the amount of substance n and may be written as nb. A term of the form $V - nb$ should therefore replace V_{ideal} in the ideal gas law.

In the same manner that the ideal volume is substituted by the actual volume, the pressure term can be modified by considering that in real gases there are intermolecular attractive forces not accounted for in the simple theory. (Repulsive forces are neglected in this treatment. See Chapter 16 for a fuller discussion of intermolecular forces.) Although the attractive forces are relatively small, they account for the ultimate liquefaction of gases when they are cooled sufficiently. With a force of attraction present, part of the pressure expected from the ideal gas law calculation is reduced in overcoming the force of intermolecular attraction. Thus the observed pressure is less than the ideal pressure, and a correction term must be added to the observed pressure. This attractive force is proportional to the square of the density of the gaseous molecules. The density is proportional to the amount of substance n and inversely proportional to the volume V. The square of the density is thus proportional to n^2/V^2, and van der Waals therefore suggested that a term of the form an^2/V^2 be added to the observed pressure. His total equation then becomes

$$\left(P + \frac{an^2}{V^2}\right)(V - nb) = nRT \quad \text{or} \quad \left(P + \frac{a}{V_m^2}\right)(V_m - b) = RT \tag{1.82}$$

where a and b are the *van der Waals constants*. They are *empirical* constants; that is, their values are chosen to give the best agreement between the points experimentally observed and the points calculated from the van der Waals equation. Once chosen, the constants in the van der Waals equation allow the determination of the liquid-gas interface (PVT surface).

The van der Waals equation may be solved for P giving

$$P = \frac{RT}{V_m - b} - \frac{a}{V_m^2} \tag{1.83}$$

Substitution of constants a and b (values of a and b for several gases are given in Table 1.3) allows the determination of the volume for a particular isotherm. Figure 1.12 was obtained using $a = 0.679\,9$ Pa m^6 mol^{-2} and $b = 0.056\,4 \times 10^{-3}$ m^3 mol^{-1}, the van der Waals constants for SO_2. In this figure, the gas-liquid equilibrium region for SO_2 has been superimposed on the van der Waals isotherms as a dashed line.

TABLE 1.3 Van der Waals Constants for Some Gases

Substance	$\frac{a}{\text{Pa m}^6\text{ mol}^{-2}}$	$\frac{b}{\text{m}^3\text{ mol}^{-1} \times 10^{-3}}$
H_2	0.0247	0.0266
He	0.0034	0.0237
N_2	0.1408	0.0391
O_2	0.1378	0.0318
Cl_2	0.6579	0.0562
Ar	0.2351	0.0398
CO	0.1505	0.0398
NO	0.1358	0.0279
CO_2	0.3637	0.0427
N_2O	0.3830	0.0441
SO_2	0.6799	0.0564
H_2O	0.5532	0.0305
NH_3	0.4225	0.0371
CH_4	0.2280	0.0428

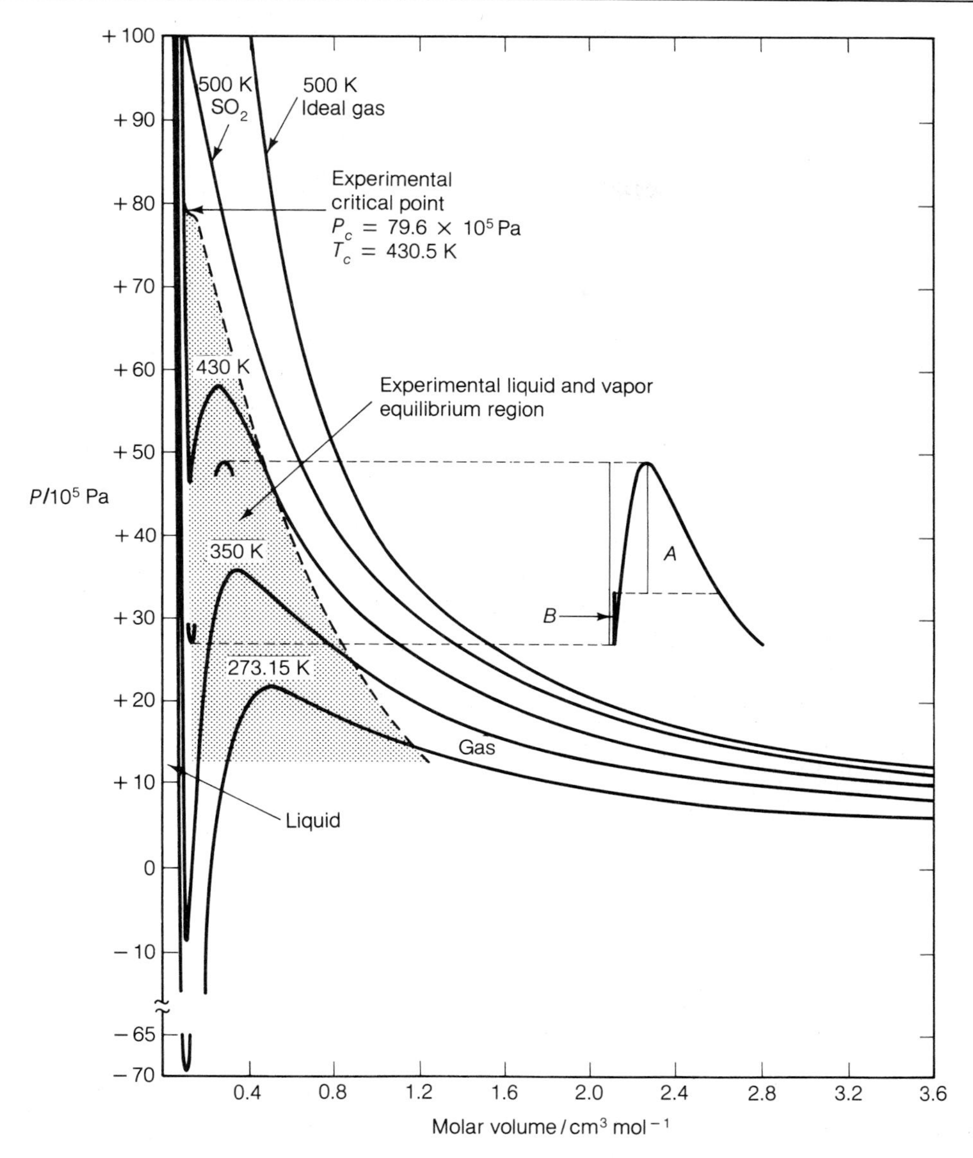

FIGURE 1.12 Isotherms for SO_2 showing the behavior predicted by the van der Waals equation. The insert shows the 400 K isotherm marked to indicate regions of supersaturation (A) and of less than expected pressure (B).

Below T_c there are three values of V for each value of P. To explain this, Eq. 1.82 may be multiplied by V^2 and expanded as a cubic equation in V:

$$PV^3 - (nbP + nRT)V^2 + n^2aV - n^3ab = 0 \tag{1.84}$$

Mathematically, a cubic equation may have three real roots, or one real and two complex roots. From the dashed line in the insert in Figure 1.12 we see that there are three real roots for pressure below T_c and only one above T_c. The three real roots give an oscillatory behavior to the curve; this is at variance with the normal experimental fact that the pressure remains constant along a tie line. However, some physical significance can be attached to two of the regions in the S-shaped part of the curve. The regions of the kind marked A correspond to situations in which a higher vapor pressure occurs than the liquefaction pressure. This is known as *supersaturation* and may be achieved experimentally if the vapor is entirely dust free. The region marked B corresponds to having the liquid under less pressure than its vapor pressure. Here bubbles of vapor should form spontaneously to offset the difference in pressures. Both metastable effects are due to the surface tension of the liquid.

Note that negative pressures may be achieved as seen from the lowest isotherms in Figure 1.12. This is equivalent to saying that the liquid is under tension. These last three phenomena have all been experimentally demonstrated.

In spite of its simplicity and the easily understood significance of its constants, the van der Waals equation can be used to treat critical phenomena as well as a wide range of properties. It has therefore become one of the most widely used approximate expressions for work with gases. Van der Waals was awarded the Nobel Prize in 1910 for his development of the simple model that predicts the physics of gas imperfections and condensations.

The Law of Corresponding States

In the last subsection, it was shown that both gases and liquids may be characterized by the critical constants T_c, P_c, and V_c. A useful rule is that the normal boiling point of a liquid is usually about two-thirds of its critical temperature. The relationship between critical constants of two *different* substances is found from the equations linking the van der Waals constants a and b to the critical constants. Equation 1.84 may be rewritten for 1 mol of gas ($n = 1$) as

$$V_m^3 - \left(b + \frac{RT}{P}\right)V_m^2 + \frac{a}{P}V_m - \frac{ab}{P} = 0 \tag{1.85}$$

At T_c the volume has three real roots that are all identical. This may be expressed as

$$(V_m - V_c)^3 = 0 \tag{1.86}$$

or

$$V_m^3 - 3V_cV_m^2 + 3V_c^2V_m - V_c^3 = 0 \tag{1.87}$$

Since Eqs. 1.85 and 1.87 describe the same condition when P and T are replaced by P_c and T_c in Eq. 1.85, we may equate coefficients of like powers of V_m. From the coefficients of V_m^2 we have

$$3V_c = b + \frac{RT_c}{P_c} \tag{1.88}$$

From terms in V_m,

$$3V_c^2 = \frac{a}{P_c} \tag{1.89}$$

and finally, from the constant terms,

$$V_c^3 = \frac{ab}{P_c} \tag{1.90}$$

From these last three equations, we obtain

$$a = 3P_cV_c^2, \qquad b = \frac{V_c}{3}, \qquad R = \frac{8P_cV_c}{3T_c} \tag{1.91}$$

Although the van der Waals constants may be evaluated from these equations, the method of choice is to determine a and b empirically from experimental PVT data.

Alternatively the same results may be determined using the expressions in Eq. 1.81. Application of the mathematical conditions in these equations to the van der Waals equation will eventually yield Eq. 1.91. See Problem 26 (p. 40) for a similar application.

If the expressions obtained in Eq. 1.91 are inserted into the van der Waals equation for 1 mol of gas, we obtain

$$\left(\frac{P}{P_c} + \frac{3V_c^2}{V_m^2}\right)\left(\frac{V_m}{V_c} - \frac{1}{3}\right) = \frac{8T}{3T_c} \tag{1.92}$$

It is convenient at this point to replace each of the ratios P/P_c, V/V_c, and T/T_c by P_r, V_r, and T_r, respectively; these represent the *reduced pressure* P_r, *reduced volume* V_r, and *reduced temperature* T_r and are dimensionless variables. Equation 1.92 then takes the form

$$\left(P_r + \frac{3}{V_r^2}\right)\left(V_r - \frac{1}{3}\right) = \frac{8}{3}T_r \tag{1.93}$$

It is thus seen that all gases obey the same equation of state within the accuracy of the van der Waals relation when there are no arbitrary constants specific to the individual gas. This is a statement of the **law of corresponding states.**

As an illustration, two gases having the same reduced temperature and reduced pressure are in corresponding states and should occupy the same reduced volume. Thus, if 1 mol of He is held at 3.43×10^5 kPa and 15.75 K, and 1 mol of CO_2 is held at 110.95×10^5 kPa and 912 K, they are in corresponding states (in both cases, $P/P_c = 1.5$ and $T/T_c = 3$) and hence should occupy the same reduced volume. The law's usefulness lies particularly in engineering where its range of validity is sufficient for many applications. The ability of the law to predict experimental behavior is nicely shown in Figure 1.13, where the reduced pressure is plotted against the compression factor for 10 different gases at various reduced temperatures.

Other Equations of State

There are two other major equations of state in common use. P. A. Daniel Berthelot (1865–1927) developed the equation

$$\left(P + \frac{a}{V_m^2T}\right)(V_m - B) = RT \tag{1.94}$$

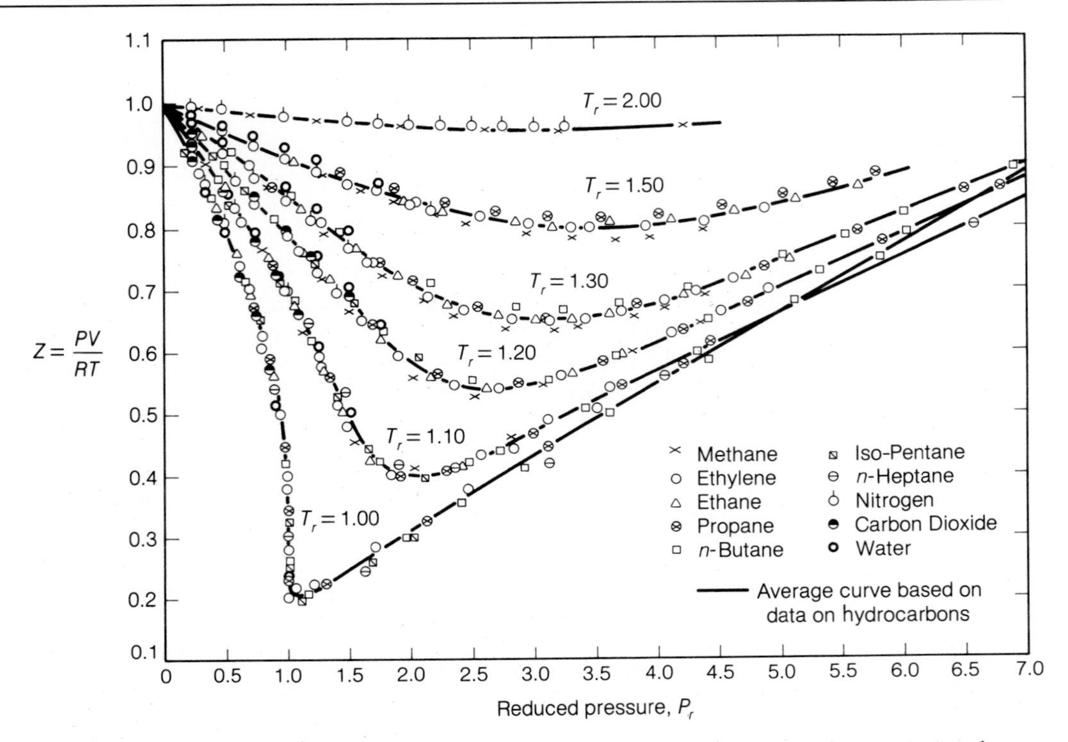

FIGURE 1.13 Compression factor versus reduced pressure for ten gases. Reprinted with permission from Goug-Jen Su, *Industrial and Engineering Chemistry, 38,* 803 (1946). Copyright 1946, American Chemical Society.

which is the van der Waals equation modified for the temperature dependence of the attractive term. It may also be written slightly modified in terms of the reduced variables as

$$P = \left(\frac{RT}{V_m}\right)\left(1 + \frac{9}{128T_r} - \frac{27}{64T_r^3}P_r\right) \tag{1.95}$$

where it provides high accuracy at low pressures and temperatures.

Another major equation of state is the one introduced by C. Dieterici. The Dieterici equation* involves the transcendental number e (the base of the natural logarithms) and is therefore less convenient to apply than the previous equation; however it gives a better representation than the other expressions near the critical point. It may be written as

$$(Pe^{a/V_mRT})(V_m - b) = RT \tag{1.96}$$

where a and b are constants not necessarily equal to the van der Waals constants. In reduced form, Eq. 1.96 becomes

$$P_r = \frac{T_r}{2V_r - 1}\exp\left(2 - \frac{2}{T_rV_r}\right) \tag{1.97}$$

* C. Dieterici, *Ann. Physik, 65,* 829 (1898); *69,* 685 (1899).

Several other equations have been proposed. Otto Redlich (b. 1896) and Joseph N. S. Kwong (b. 1916)* introduced the equation

$$\left[P + \frac{n^2 a}{T^{1/2} V(V + nb)}\right](V - bn) = nRT \tag{1.98}$$

which provides a simple and accurate two-parameter expression applicable over a wide range of temperature and pressures.

Another expression that has gained some popularity is the Benedict-Webb-Rubin† equation of state that relates pressure, molar density, and temperature. It has been used to predict fairly accurately the thermodynamic properties of complex hydrocarbon systems.

1.13 The Virial Equation

The advantage of the equations discussed in the last sections is that the constants are kept to a minimum and relate to theoretically defined parameters. Another technique is to use a large number of constants to fit the behavior of a gas almost exactly, but the resulting equation is then less practical for general use and particularly for thermodynamic applications. Furthermore, as the number of constants is increased, it becomes more difficult to correlate the constants with physical parameters. However, two such expressions are of such general usefulness that they are discussed here.

The Dutch physicist Heike Kamerlingh-Onnes (1853–1926) suggested in 1901 that a power series, called a **virial equation**, be used to account for the deviations from linearity shown by real gases. The general form of the power series for Z as a function of P is

$$Z(P, T) = \frac{PV_m}{RT} = 1 + B'(T)P + C'(T)P^2 + D'(T)P^3 + \cdots \tag{1.99}$$

However, this does not represent the data as well as a series in $1/V_m$ where the odd powers greater than unity are omitted. Thus the form of the equation of state of real gases presented by Kamerlingh-Onnes is

$$\frac{PV}{nRT} = 1 + \frac{B(T)n}{V} + \frac{C(T)n^2}{V^2} + \frac{D(T)n^4}{V^4} + \cdots \tag{1.100}$$

where the coefficients $B'(T)$, $C'(T)$, $D'(T)$ and $B(T)$, $C(T)$, $D(T)$ are called the second, third, and fourth *virial coefficients*, respectively, and the notation indicates that they are functions of temperature. When Eq. 1.100 is multiplied through by R, the first term on the right is seen to be R. Sometimes, therefore, R is called the *first virial coefficient*. For mixtures the coefficients are functions of both temperature and composition, and they are found experimentally from low-pressure PVT data by a graphical procedure. To illustrate how this is done, Eq. 1.100 is rewritten in terms of the molar density $\rho_m \equiv n/V$.

* O. Redlich and J. N. S. Kwong, *Chem. Rev.*, *44*, 233 (1949).

† M. Benedict, G. B. Webb, and L. C. Rubin, *J. Chem. Phys.*, *10*, 747 (1942). Also see T. G. Kaufmann, *I & E C Fundamentals*, *7*, 115 (1968).

$$\frac{1}{\rho_m}\left(\frac{P}{\rho_m RT} - 1\right) = B(T) + C(T)\rho_m + \cdots \tag{1.101}$$

The left-hand side of Eq. 1.101 is plotted against ρ for fixed T, yielding a value of B at the intercept where $\rho = 0$. At the *Boyle temperature*, $B(T) = 0$. Another way of stating this is that the derivative $d(PV)/dP$ becomes zero as $P \rightarrow 0$. The slope of the curve at $\rho = 0$ gives the value of C for that particular temperature. In Figure 1.14 a plot of the second virial coefficient $B(T)$ is made showing the dependence of B on temperature for several gases.

The importance of the virial coefficients lies in the fact that through the methods of statistical mechanics an equation of state of a real gas may be developed in the virial form. The empirically derived coefficients can thus be related to their theoretical counterparts, which (it turns out) are the intermolecular potential energies. In this interpretation, the second virial coefficients, for instance, are due to molecular pair interactions; the other coefficients are due to higher-order interactions.

The virial equation is not particularly useful at high pressures or near the critical point. Furthermore, if one is to proceed on a theoretical basis rather than an empirical one, the calculation of the constants from statistical mechanics is difficult because the potential functions are not well known and the evaluation of the multiple integrals involved is very difficult.

The final expression to be considered here is the equation proposed by the American chemists James Alexander Beattie (b. 1895) and Oscar C. Bridgeman (b. 1897).*

$$\boldsymbol{P = \frac{RT[1 - (c/V_m T^3)]}{V_m^2}(V_m + B) - \frac{A}{V_m^2}} \tag{1.102}$$

where

TABLE 1.4 Constants for Use in the Beattie-Bridgeman Equation with $R = 8.3143 \text{ J K}^{-1} \text{ mol}^{-1}$

Gas	A_0 (Pa m^6 mol^{-2})	a (10^{-6} m^3 mol^{-1})	B_0 (10^{-6} m^3 mol^{-1})	b (10^{-6} m^3 mol^{-1})	c (10 m^3 K^3 mol^{-1})
He	0.00219	59.84	14.00	0.0	0.0040
Ne	0.02153	21.96	20.60	0.0	0.101
A	0.13078	23.28	39.31	0.0	5.99
H_2	0.02001	−5.06	20.96	−43.59	0.0504
N_2	0.1362	26.17	50.46	−6.91	4.20
O_2	0.1511	25.62	46.24	4.208	4.80
Air	0.13184	19.31	46.11	−11.01	4.34
CO_2	0.50728	71.32	104.76	72.35	66.00
CH_4	0.23071	18.55	55.87	−15.87	12.83
$(C_2H_5)_2O$	3.1692	124.26	454.46	119.54	33.33

*J. A. Beattie and O. C. Bridgeman, *J. Am. Chem. Soc.*, *49*, 1665 (1927); *50*, 3133, 3151 (1928).

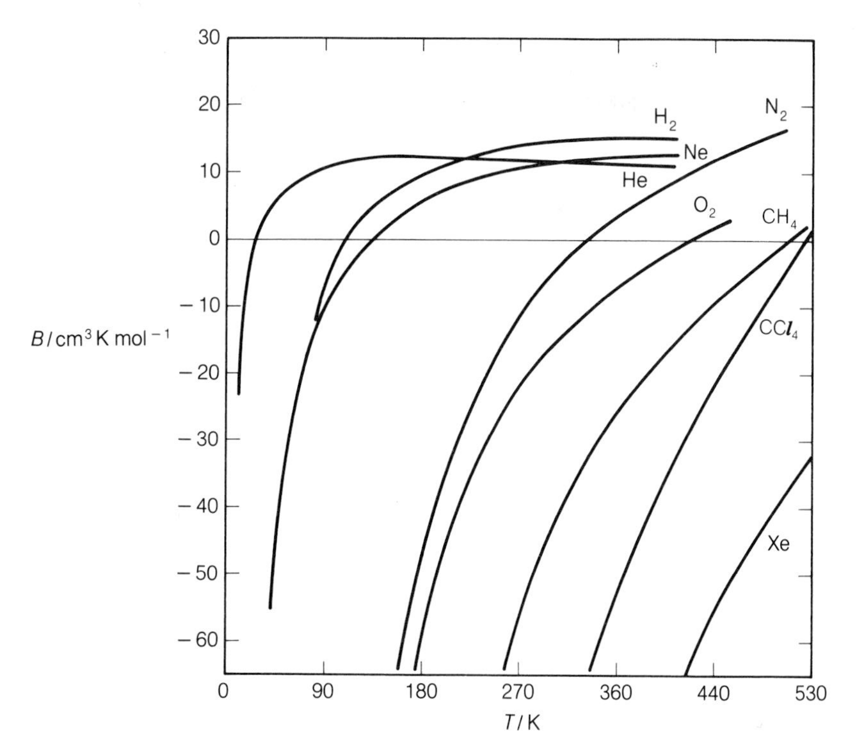

FIGURE 1.14 Plot of the second virial coefficient, $B(T)$, against T for several gases.

$$A = A_0\left(1 - \frac{a}{V_m}\right) \qquad B = B_0\left(1 - \frac{b}{V_m}\right)$$

and a, b, A_0, B_0, and c are empirically determined constants. The *Beattie-Bridgeman equation* utilized five constants in addition to R and is well suited for precise work, especially in the high-pressure range. Table 1.4 gives the Beattie-Bridgeman* constants for 10 gases.

Key Equations

Definition of *kinetic energy*:

$$E_k = \tfrac{1}{2}mu^2$$

Potential energy for a body obeying Hooke's law:

$$E_p = \tfrac{1}{2}k_h x^2$$

*J. A. Beattie and O. C. Bridgeman, *Proc. Am. Acad. Arts Sci.*, *63*, 229 (1928).

Boyle's law:

$$P \propto \frac{1}{V} \quad \text{or} \quad PV = \text{constant} \qquad (\text{at constant } T)$$

Gay-Lussac's (Charles's) law:

$$V \propto T \quad \text{or} \quad \frac{V}{T} = \text{constant} \qquad (\text{at constant } P)$$

Equation of state of an ideal gas:

$$PV = nRT$$

Pressure of a gas derived from the *kinetic-molecular theory*:

$$P = \frac{Nm\overline{u^2}}{3V}$$

where $\overline{u^2}$ is the mean square speed.

Relation of *kinetic energy* to *temperature*:

$$\bar{\epsilon}_k = \tfrac{3}{2}\mathbf{k}T$$

where **k** is the *Boltzmann constant.*

Dalton's law of partial pressures:

$$P_t = (n_1 + n_2 + \cdots + n_i)\frac{RT}{V}$$

Graham's law of effusion:

$$\frac{\text{rate (gas}_1)}{\text{rate (gas}_2)} = \sqrt{\frac{\rho\ (\text{gas}_2)}{\rho\ (\text{gas}_1)}} = \sqrt{\frac{M\ (\text{gas}_2)}{M\ (\text{gas}_1)}}$$

where ρ is the density and M is the molar mass.

Collision frequency (SI unit: s^{-1}):

$$Z_A = \frac{\sqrt{2}\pi d_A^2 \bar{u}_A N_A}{V}$$

Collision number (SI unit: $m^{-3}\ s^{-1}$):

$$Z_{AA} = \frac{\frac{\sqrt{2}}{2}\pi d_A^2 \bar{u}_A N_A^2}{V^2} \quad \text{and} \quad Z_{AB} = \frac{\pi d_{AB}^2(\bar{u}_A^2 + \bar{u}_B^2)^{1/2} N_A N_B}{V^2}$$

Mean free path:

$$\lambda = \frac{V}{\sqrt{2}\pi d_A^2 N_A}$$

Barometric distribution law:

$$P = P_0 e^{-E_p/RT}$$

Compression factor:

$$Z = \frac{PV}{nRT} = \frac{PV_m}{RT}$$

Van der Waals equation:

$$\left(P + \frac{an^2}{V^2}\right)(V - nb) = nRT$$

where a and b are the *van der Waals constants.*

Problems

A Classical Mechanics and Thermal Equilibrium

1. Calculate the amount of work required to accelerate a 1000-kg car (typical of a VW Rabbit) to 88 km hr^{-1} (55 miles hr^{-1}). Compare this value to the amount of work required for a 1600-kg car (typical of a Mercury Monarch) under the same conditions.

2. Assume that a rod of copper is used to determine the temperature of some system. The rod's length at 0°C is 27.5 cm, and at the temperature of the system it is 28.1 cm. What is the temperature of the system? The linear expansion of copper is given by an equation of the form $l_t = l_0(1 + \alpha t + \beta t^2)$ where $\alpha = 0.160 \times 10^{-4}$ K^{-1}, $\beta = 0.10 \times 10^{-7}$ K^{-2}, l_0 is the length at 0°C, and l_t is the length at t°C.

3. Atoms can transfer kinetic energy in a collision. If an atom has a mass of 1×10^{-24} g and travels with a velocity of 500 m s^{-1}, what is the maximum kinetic energy that can be transferred from the moving atom in a head-on elastic collision to a stationary atom of mass 1×10^{-23} g?

4. Power is defined as the rate at which work is done. The unit of power is the *watt* (W) $\equiv$ 1 J s^{-1}. What is the power that a man can expend if all his food consumption of 8000 kJ a day ($\approx$ 2000 kcal) is his only source of energy and it is used entirely for work?

B Gas Laws

5. The standard atmosphere of pressure is the force per unit area exerted by a 760-mm column of mercury, the density of which is 13.595 1 g cm^{-3}. If the gravitational acceleration is 9.806 65 m s^{-2}, calculate the pressure of 1 atm in kPa.

6. An ideal gas occupies a volume of 0.30 dm^3 at a pressure of 1.80×10^5 Pa. What is the new volume of the gas maintained at the same temperature if the pressure is reduced to 1.15×10^5 Pa?

7. If the gas in Problem 6 were initially at 330 K, what will be the final volume if the temperature is raised to 550 K at constant pressure?

8. Calculate the concentration in mol dm^{-3} of an ideal gas at 298.15 K and at (a) 101.325 kPa (1 atm), (b) 1×10^{-4} Pa ($\approx 10^{-9}$ atm). In each case, determine the number of molecules in 1 dm^3.

9. A Dumas experiment to determine molar mass is conducted in which a gas sample's P, θ, and V are determined. If a 1.08-g sample is held in 0.250 dm^3 at 303 K and 101.3 kPa,

a. What would the sample's volume be at 273.15 K, at constant pressure?

b. What is the molar mass of the sample?

10. A gas that behaves ideally has a density of 1.92 g dm^{-3} at 150 kPa and 298 K. What is the molar mass of the sample?

11. The density of air at 101.325 kPa and 298.15 K is 1.159 g dm^{-3}. Assuming that air behaves as an ideal gas, calculate its molar mass.

12. A 0.20-dm^3 sample of H_2 is collected over water at a temperature of 298.15 K and at a pressure of 99.99 kPa. What is the pressure of hydrogen in the dry state at 298.15 K? The vapor pressure of water at 298.15 K is 3.17 kPa.

13. Balloons now are used to move huge trees from their cutting place on mountain slopes to conventional transportation. Calculate the volume of a balloon needed if it is desired to have a lifting force of 1000 kg when the temperature is 290 K at 0.940 atm. The balloon is to be filled with helium. Assume that air is 80 mol % N_2 and 20 mol % O_2.

***14.** A gas mixture containing 5 mol % butane and 95 mol % argon (such as is used in Geiger-Müller counter tubes) is to be prepared by allowing gaseous butane to fill an evacuated cylinder at 1 atm pressure. The 40.0-dm^3 cylinder is then weighed. Calculate the mass of argon that gives the desired composition if the temperature is maintained at 25°C. Calculate the total pressure of the final mixture. The atomic mass of argon is 39.9 g mol^{-1}.

C Graham's Law, Molecular Collisions, and Kinetic Theory

15. It takes gas A 2.3 times as long to effuse through an orifice as the same amount of nitrogen. What is the molar mass of gas A?

16. What is the total kinetic energy of 0.5 mol of an ideal monatomic gas confined to 8 dm^3 at 200 kPa?

17. Nitrogen gas is maintained at 152 kPa in a 2-dm^3 vessel at 298.15 K. If its molar mass is 28.0134 g mol^{-1}, calculate

- **a.** The amount of N_2 present.
- **b.** The number of molecules present.
- **c.** The root-mean-square speed of the molecules.
- **d.** The average translational kinetic energy of each molecule.
- **e.** The total translational kinetic energy in the system.

18. By what factor are the root-mean-square speeds changed if a gas is heated from 300 K to 400 K?

***19.** The collision diameter of N_2 is 3.74×10^{-10} m at 298.15 K and 101.325 kPa. Its average speed is 474.6 m s^{-1}. Calculate the mean free path, the average number of collisions Z_A experienced by one molecule in unit time, and the average number of collisions Z_{AA} per unit volume per unit time for N_2.

***20.** Express the mean free path of a gas in terms of the variables pressure and temperature, which are more easily measured than the volume.

21. Hydrogen gas has a molecular collision diameter of 0.258 nm. Calculate the mean free path of hydrogen at 298.15 K and (a) 133.32 Pa, (b) 101.325 kPa, and (c) 1.0×10^8 Pa.

22. In interstellar space it is estimated that atomic hydrogen exists at a concentration of one particle per cubic metre. If the collision diameter is 2.5×10^{-10} m, calculate λ.

D Real Gases

23. Draw the van der Waals PV isotherm over the same range of P and V as in Figure 1.10 at 350 K and 450 K for Cl_2 using the values in Table 1.3.

24. Compare the pressures predicted for 0.8 dm^3 of Cl_2 weighing 17.5 g at 273.15 K using (a) the ideal gas equation and (b) the van der Waals equation.

25. The critical temperature T_c of nitrous oxide (N_2O) is 36.5°C, and its critical pressure P_c is 71.7 atm. Suppose that 1 mol of N_2O is compressed to 54 atm at 356 K. Calculate the reduced temperature and pressure, and use Figure 1.13, interpolating as necessary, to estimate the volume occupied by 1 mol of the gas at 54 atm and 356 K.

***26.** For the Dieterici equation, derive the relationship of a and b to the critical volume and temperature. [*Hint:* Remember that at the critical point $(\partial P/\partial V)_T = 0$ and $(\partial^2 P/\partial V^2)_T = 0$.]

***27.** A general requirement of all equations of state is that they reduce to the ideal gas equation (Eq. 1.26) in the limit of low pressures. Show that this is true for the van der Waals equation.

28. The van der Waals constants for C_2H_6 in the older literature are found to be

$$a = 5.49 \text{ atm } \ell^2 \text{ mol}^{-2} \quad \text{and } b = 0.0638 \ \ell \text{ mol}^{-1}$$

Express these constants in SI units (ℓ = liter = dm^3).

***29.** Compare the values obtained for the pressure of 3.00 mol CO_2 at 298.15 K held in a 8.25-dm^3 bulb using the ideal gas, van der Waals, Dieterici, and Beattie-Bridgeman equations. For CO_2 the Dieterici equation constants are

$$a = 0.462 \text{ Pa m}^6 \text{ mol}^{-2},$$

$b = 4.63 \times 10^{-5}\ m^3\ mol^{-1}$

30. A gas obeys the van der Waals equation with $P_c = 3.040 \times 10^6$ Pa ($\approx$ 30 atm) and $T_c = 473$ K. Calculate the value of the van der Waals constant b for this gas.

E Supplementary Problems

31. Expand the Dieterici equation in powers of V_m^{-1} in order to cast it into the virial form. Find the second and third virial coefficients. Then show that at low densities the Dieterici and van der Waals equations give essentially the same result for P.

32. An ideal gas thermometer and a mercury thermometer are calibrated at 0°C and at 100°C. The thermal expansion coefficient for mercury is

$$\alpha = \frac{1}{V_0}\left(\frac{\partial V}{\partial T}\right)_P = 1.817 \times 10^{-4} + 5.90 \times 10^{-9}\theta + 3.45 \times 10^{-10}\theta^2$$

where θ is the Celsius temperature and $V_0 = V$ at $\theta = 0$. What temperature would appear on the mercury scale when the ideal gas scale reads 50°C?

33. The gravitational constant g decreases by 0.010 m s^{-2} km^{-1} of altitude.

a. Modify the barometric equation to take this variation into account. Assume that the temperature remains constant.

b. Calculate the pressure of nitrogen at an altitude of 100 km assuming that sea-level pressure is 1 atm and that the temperature of 298.15 K is constant.

34. Calculate the value of Avogadro's constant from a study made by Perrin [*Ann. Chim. Phys., 18,* 1 (1909)] in which he measured as a function of height the distribution of bright yellow colloidal gamboge (a gum resin) particles suspended in water. The necessary data at 15°C are

height, $z/10^{-6}$ m	5	35
N, relative number of gamboge particles at height z	100	47

$\rho_{gamboge} = 1.206$ g cm^{-3}

$\rho_{water} = 0.999$ g cm^{-3}

radius of gamboge particles, $r = 0.212 \times 10^{-6}$ m

35. Vacuum technology has become increasingly more important in many scientific and industrial applications. The unit torr, defined as $\frac{1}{760}$ atm, is commonly used in the measurement of low pressures.

a. Find the relation between the older unit 1 mmHg and the torr.

b. Calculate at 298.15 K the number of molecules present at 10^{-6} torr and at 10^{-15} torr (approximately the best vacuum obtainable). The density of mercury is 13.5951 g cm^{-3}. The acceleration of gravity is defined as 9.806 65 m s^{-2}.

F Essay Questions

1. In light of the van der Waals equation, explain the liquefaction of gases.

2. State the postulates of the kinetic molecular theory of gases.

Suggested Reading

The references at the end of each chapter are to specialized books where more information can be obtained on the particular subjects of the chapter.

H. A. Boorse and L. Motz (Eds.), *The World of the Atom,* New York: Basic Books, 1966.

J. H. Dymond and E. B. Smith, *The Virial Coefficients of Gases, A Critical Compilation,* Oxford: Clarendon Press, 1969.

O. A. Hougen, K. M. Watson, and R. A. Ragatz, *Chemical Process Principles: Part II, Thermodynamics* (2nd ed.), New York: Wiley, 1959, Chapter 14.

A. A. Vasserman, Ya. Z. Kazavchinskii, and V. A. Rabinovich, *Thermophysical Properties of Air and Air Components,* Israel Program for Scientific Translation, 1971.

Chapter 2

The First Law of Thermodynamics

Preview

In Chapter 1 we have been mainly concerned with the relationships between the pressure, volume, and temperature of a gas, and we have used the kinetic-molecular theory to explain these effects. Chapters 2 and 3 deal with *thermodynamics,* which deals with general relationships that are valid quite independently of the existence of molecules. Thermodynamics is concerned with transformations of various forms of energy, and it leads to very far-reaching conclusions.

The *First Law of Thermodynamics,* dealt with in this chapter, states that energy cannot be created or destroyed but can only be converted into other forms. It says that the change in the internal energy of a system ΔU is the sum of the heat q supplied to the system and the work w done on it. We will see that if a *constant* pressure P causes a system to sweep out a certain volume, the work done by that pressure is equal to the pressure multiplied by that volume.

The *internal energy* U is an important property for processes that occur at constant volume. For processes at constant pressure the fundamental property is the *enthalpy* H, which is the internal energy plus the product of the pressure and the volume.

The field of *thermochemistry,* which is based on the First Law, is concerned primarily with heat changes when a chemical reaction occurs under precisely defined conditions. The concept of standard states is introduced to define these conditions. An important consequence of the First Law is that one can write balanced equations for chemical reactions, together with their enthalpy changes, and manipulate them algebraically so as to obtain enthalpy changes for other reactions. When we do this, we are applying *Hess's law,* which is valuable for obtaining and compiling thermochemical information. A vast amount of thermochemical information is summarized in the form of *standard enthalpies of formation,* which are the enthalpy changes when compounds are formed from the elements in their standard states.

For an ideal gas the internal energy and enthalpy depend only on the temperature and not on the pressure or volume. This is not true for a real gas; as a consequence, if a real gas expands freely, doing no work, its temperature changes. This is the *Joule-Thompson effect.* The quantity $(\partial U/\partial V)_T$ is known as the *internal pressure,* and it arises from the attractions between the molecules of a real gas.

2 The First Law of Thermodynamics

In Chapter 1 we have been mainly concerned with *macroscopic* properties such as pressure, volume, and temperature. We have seen how some of the relationships between these properties of ideal and real gases can be interpreted in terms of the behavior of the molecules—that is, of the *microscopic* properties. This kinetic-molecular theory is of great value, but it is possible to interpret many of the relationships between macroscopic properties without any reference to the behavior of molecules, or even to the very existence of molecules. This is what is done in the most formal treatments of the science of *thermodynamics*, which is concerned with the general relationships between the various forms of energy, including heat. In this book we shall present thermodynamics in a less formal way, and from time to time we shall clarify some of the basic ideas by showing how they arise from the molecular behavior. In presenting the subject we will follow the principles and symbolism recommended in a recent IUPAC report*.

At first it might appear that the rigorous study of thermodynamics, with no regard to microscopic behavior, would not lead us very far. The contrary, however, is true; it has proved possible to develop some very far-reaching conclusions on the basis of purely thermodynamic arguments, and these conclusions are all the more convincing because they do not depend on the truth or falsity of any theories of atomic and molecular structure. Pure thermodynamics starts with a small number of assumptions that are based on very well-established experimental results and makes logical deductions from them, finally giving a set of relationships that are bound to be true provided that the original premises are true.

The student should realize that all of us find thermodynamics a somewhat difficult subject the first time we meet it. There is an amusing story told of the distinguished German physicist Arnold Sommerfeld (1868–1951), who had written lucid books on every subject in physics except thermodynamics. When asked why he did not write on that field, he replied somewhat as follows:

> Thermodynamics is a funny subject. The first time you go through the subject, you don't understand it at all. The second time you go through it, you think you understand it, except for one or two small points. The third time you go through it, you *know* you don't understand it, but by that time you are so used to the subject that it doesn't bother you any more.

Students who initially find trouble with the subject should be heartened by the knowledge that this happens to all of us, even to brilliant people like Sommerfeld. If the student will persist with his study of thermodynamics until he is "so used to the subject that it doesn't bother him any more," he will find that he is amply rewarded by gaining a much deeper understanding of the workings of nature.

---

* *Pure and Applied Chemistry, 51,* 393 (1979).

2.1 Origins of the First Law

There are three laws of thermodynamics, of which the *First Law* is essentially the principle of conservation of energy. This law was formulated after it was found that expenditure of work causes the production of heat. The first quantitative experiments along these lines were carried out by Benjamin Thompson, Count Rumford (1753–1814). While supervising the boring of cannon at the Munich Arsenal, he became interested in the generation of heat during the operation. He suggested in 1798* that the heat arose from the work expended and obtained a numerical value for the amount of heat generated by a given amount of work. More precise experiments were carried out in the 1840s by the English scientist James Prescott Joule (1818–1887), and in his honor the modern unit of energy, work, and heat is called the joule (J). The thermodynamic calorie is now *defined* as 4.184 J.

These experiments led to the conclusion that

The energy of the universe remains constant,

which is a compact statement of the **First Law of Thermodynamics.** Both work and heat are quantities that describe the transfer of energy from one system to another. If two systems are at different temperatures, heat can pass from one to the other in the form of heat or by transfer of matter. Energy can also be transferred from one place to another in the form of work, which means that there has been motion in part of the system. No matter how these transfers occur, the total energy of the universe remains the same.

2.2 Thermodynamic Systems

In Section 1.3 we emphasized the important distinction between a *system* and its *surroundings*. We also explained the differences among *open* systems, *closed* systems, and *isolated* systems. The distinction between a system and its surroundings is particularly important in thermodynamics, since we are constantly concerned with transfer of heat between the system and the surroundings. We are also concerned with the work done by the system on its surroundings or by the surroundings on the system. In all cases the system must be carefully defined.

2.3 States and State Functions

Certain of the macroscopic properties have fixed values for a particular *state* of the system, whereas others do not. For example, suppose that we maintain 1 g of water in a vessel at 25°C and 1 atm pressure (101.325 kPa); it will have a volume of close to 1 cm^3. These quantities, 1 g of H_2O, 25°C, 101.325 kPa, and 1 cm^3, all specify the state of the system; whenever we satisfy these four conditions, we have the water in the same state, and as long as it is in that state, it will have these particular

*B. Thompson (Count Rumford), *Philosophical Transactions, 88,* 80 (1798).

specifications. These macroscopic properties that we have mentioned (mass, pressure, temperature, and volume) are known as *state functions* or *state variables*.

One very important characteristic of a state function is that once we have specified the state of a system by giving the values of *some* of the state functions, the values of all other state functions are fixed. Thus, in the example just given, once we have specified the mass, temperature, and pressure of the water, the volume is fixed. So, too, is the total energy of the system, and energy is therefore another state function.

Another important characteristic of a state function is that when the state of a system is changed, the change in any state function depends only on the initial and final states of the system, and not on the path followed in making the change. For example, if we heat the water from 25°C to 26°C, the change in temperature is equal to the difference between the initial and final temperatures:

$$\Delta T = T_{\text{final}} - T_{\text{initial}} = 1°\text{C} \tag{2.1}$$

The way in which the temperature change is brought about has no effect on this result.

This example may seem trivial, but it is to be emphasized that not all functions have this characteristic. For example, raising the temperature of water from 25°C to 26°C can be done in various ways, the simplest being to add heat. Alternatively, we could stir the water vigorously with a paddle until the desired temperature rise had been achieved; this means that we are doing work on the system. We could also add some heat and do some work in addition. This shows that heat and work are *not* state functions.

In the meantime, it is useful to consider an analogy. Suppose that there is a point A on the earth's surface that is 1000 m above sea level and another point B that is 4000 m above sea level. The difference, 3000 m, is the height of B with respect to A. In other words, the difference in height Δh can be expressed as

$$\Delta h = h_B - h_A \tag{2.2}$$

where h_A and h_B are the heights of A and B above sea level. Height above sea level is thus a state function, the difference Δh being in no way dependent on the path chosen. However, the distance we have to travel in order to go from A to B *is* dependent on the path; we can go by the shortest route or take a longer route. Distance traveled is therefore not a state function.

2.4 Equilibrium States and Reversibility

Thermodynamics is directly concerned only with *equilibrium states,* in which the state functions have constant values throughout the system. It provides us with information about the circumstances under which nonequilibrium states will move toward equilibrium, but by itself it tells us nothing about the nonequilibrium states.

Suppose, for example, that we have a gas confined in a cylinder having a frictionless movable piston (Figure 2.1). If the piston is motionless, the state of the gas can be specified by giving the values of pressure, volume, and temperature. However, if the gas is compressed very rapidly, it passes through states for which pressure and temperature cannot be specified, there being a variation of these

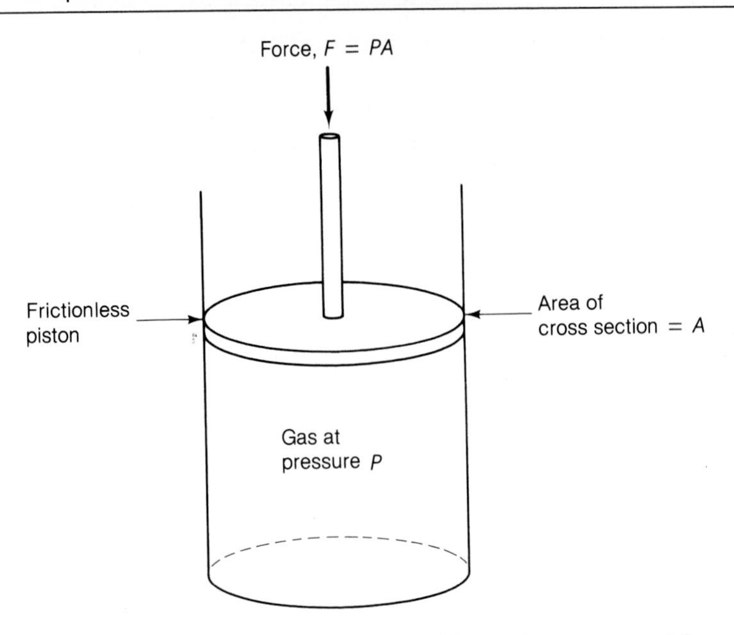

FIGURE 2.1 A gas at pressure P maintained at equilibrium by an external force, F, equal to PA, where A is the area of cross section of the piston.

properties throughout the gas; the gas near to the piston is at first more compressed and heated than the gas at the far end of the cylinder. The gas then would be said to be in a nonequilibrium state; pure thermodynamics could not deal with such a state, although it could tell us what kind of a change would spontaneously occur in order for equilibrium to be attained.

The criteria for equilibrium are very important and may be summarized as follows:

1. The mechanical properties must be uniform throughout the system and constant in time.

2. The chemical composition of the system must be uniform, with no net chemical change taking place.

3. The temperature of the system must be uniform and must be the same as the temperature of its surroundings.

The first of these criteria means that the force acting on the system must be exactly balanced by the force exerted by the system; otherwise the volume would be changing. If we consider the system illustrated in Figure 2.1, we see that for the system to be at equilibrium the force F exerted on the piston must exactly balance the pressure P of the gas; if A is the area of the piston,

$$PA = F \tag{2.3}$$

If we increase the force, the gas will be compressed; if we decrease it, the gas will expand.

Suppose that we increase the force F by an *infinitesimal* amount dF. The pressure that we are exerting on the gas will now be infinitesimally greater than the pressure of the gas (i.e., it will be $P + dP$). The gas will therefore be compressed.

We can make dP as small as we like, and at all stages during the infinitely slow compression we are therefore maintaining the gas in a state of equilibrium. We refer to a process of this kind as a **reversible** process. If we reduce the pressure to $P - dP$, the gas will expand infinitely slowly—that is, reversibly. Reversible processes play very imporant roles in thermodynamic arguments. All processes that occur naturally are, however, irreversible; since they do not occur infinitely slowly, there is necessarily some departure from true equilibrium.

2.5 Energy, Heat, and Work

We come now to the statement of the First Law of Thermodynamics, according to which the total amount of energy in the universe is conserved. If we add heat q to a system, and no other change occurs, the *internal energy* U increases by an amount that is exactly equal to the heat supplied:

$$\Delta U = q \tag{2.4}$$

If an amount of work w is performed *on* the system,* and no heat is transferred, the system gains internal energy by an amount equal to the work done:

$$\Delta U = w \tag{2.5}$$

In general, if heat q is supplied to the system, and an amount of work w is also performed on the system, the increase in internal energy is given by

$$\Delta U = q + w \tag{2.6}$$

This is a statement of the First Law of Thermodynamics. In applying this equation it is, of course, necessary to employ the same units for ΔU, q, and w. The SI unit is the joule ($\text{J} = \text{kg m}^2 \text{ s}^{-2}$); it is the energy corresponding to a force of one newton ($\text{N} = \text{kg m s}^{-2}$) operating over a distance of one metre. In this book we shall use joules entirely, although up to now most thermodynamic values have been given in calories in the literature.

We should note that Eq. 2.6 leaves the internal energy U indefinite, in that we are dealing with only the energy change ΔU. In practice this is usually adequate; we do not need to know absolute values.

The internal energy U is a state function of the system; that is, it depends only on the state of the system and not on how the system achieved its particular state. Earlier we saw that a change from one state to another, namely from 25°C to 26°C, can be achieved by adding heat, by doing work, or by a combination of the two. It is found experimentally that however we bring about the temperature rise, the sum $q + w$ is always the same. In other words, for a particular change in state the quantity ΔU, equal to $q + w$, is independent of the way in which the change is brought about. This behavior is characteristic of a state function. This example demonstrates that heat q and work w *are not state functions* since the change can be brought about by various divisions of the energy between heat and work; only the sum $q + w$ is fixed.

---

*The IUPAC recommendation is to use the symbol w for the work done *on* the system. The reader is warned that many treatments use the symbol w for the work done *by* the system.

The distinction between state functions such as U and quantities such as q and w that are not state functions may be considered from another point of view. Whether or not a function is a state function is related to the mathematical concept of *exact* and *inexact differentials*. The definite integral of a state function such as U, that is,

$$\int_{U_1}^{U_1} dU$$

is a quantity $U_2 - U_1 = \Delta U$, which is independent of the path by which the process occurs. On the other hand, the integral of an inexact differential such as heat and work; that is,

$$\int_1^2 đq \quad \text{or} \quad \int_1^2 đw$$

is a quantity that is not fixed but depends on the process by which the change from state 1 to state 2 occurs; we have used the symbol $đ$ to indicate an inexact differential. It would therefore be wrong to write these integrals as $q_2 - q_1 = \Delta q$ or $w_2 - w_1 = \Delta w$; the quantities Δq and Δw have no meaning. Heat and work make themselves evident only during a change from one state to another and have no significance when a system remains in a particular state; they are properties of the path and not of the state. A state function such as the internal energy U, on the other hand, has a significance in relation to a particular state.

If U were not a state function, we could have violations of the principle of conservation of energy, something that has never been observed. To see how a violation could occur, consider two states A and B, and suppose that there are two alternative paths from A and B. Suppose that for one of these paths ΔU is 10 J; and for the other, 30 J:

$$\Delta U_1 = 10 \text{ J} \qquad \Delta U_2 = 30 \text{ J}$$

We could then go from A to B and expend 10 J of heat by going by the first path. If we returned from B to A by the second path, we would gain 30 J. Thus, we would have the system in its original state, and would have a net gain of 20 J. Energy would therefore have been created from nothing. The process could be continued indefinitely, with a gain of energy at each completion of the cycle. Many attempts have been made to create energy in this way, by the construction of "perpetual motion machines of the first kind," but all have ended in failure. This inability to make perpetual motion machines provides convincing evidence that energy cannot be created or destroyed.

In purely thermodynamic studies it is not necessary to consider what internal energy really consists of; however, most of us like to have some answer to this question, in terms of molecular energies. There are contributions to the internal energy of a substance from

1. The kinetic energy of motion of the individual molecules.
2. The potential energy that arises from interactions between molecules.
3. The kinetic and potential energy of the nuclei and electrons within the individual molecules.

The precise treatment of these factors is very difficult, and it is a great strength of thermodynamics that we can make use of the concept of internal energy without having to deal with it on a detailed molecular basis.

The Nature of Work

There are various ways in which a system may do work or by which work may be done on a system. For example, if we pass a current through a solution and electrolyze it, we are performing one form of work—*electrical work;* conversely, an electrochemical cell *performs* work. Other forms of work are chemical work, osmotic work, and mechanical work. *Chemical work* is usually involved when larger molecules are synthesized from smaller ones, as in living organisms. *Osmotic work* is the work required to transport and concentrate chemical substances; it is involved, for example, when seawater is purified by reverse osmosis (p. 205) and in the formation of the gastric juice, where the acid concentration is much higher than that of the surroundings. *Mechanical work* is performed whenever a system undergoes a volume change and when a weight is lifted.

The simplest way in which work is done is when an external force brings about a compression of a system. Suppose, for example, that we have an arrangement in which a gas or liquid is maintained at *constant* pressure P, which it exerts against a movable piston (Figure 2.2). In order for the system to be at equilibrium we must apply a force F to the piston, the force being related to the pressure by the relationship

$$F = PA \tag{2.7}$$

where A is the area of the piston. Suppose that the force is increased by an infinitesimal amount dF, so that the piston moves infinitely slowly, the process

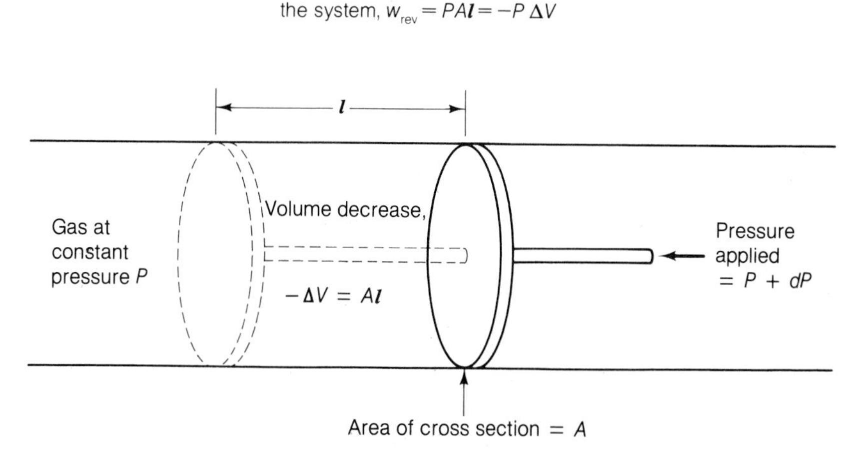

FIGURE 2.2 The reversible work done by a constant pressure P moving a piston. A simple way for a gas to be at constant pressure is to have a vapor in equilibrium with its liquid.

being reversible. If the piston moves to the left a distance l, the reversible work w_{rev} done on the system is

$$w_{rev} = Fl = PAl \tag{2.8}$$

However, Al is the volume swept out by the movement of the piston, that is, the decrease in volume of the gas, which is $-\Delta V$. The work done on the system is thus

$$w_{rev} = -P\,\Delta V \tag{2.9}$$

In our example this is a positive quantity since we have compressed the gas and ΔV is negative. If the gas had expanded ΔV would have been positive and the work done *on* the system would have been negative; that is, the gas would have done a positive amount of work on the surroundings.

If the pressure P varies during a volume change, we must obtain the work done by a process of integration. The work done on the system while an external pressure P moves the piston so that the volume of the gas changes by an infinitesimal volume dV is

$$dw_{rev} = -P\,dV \tag{2.10}$$

If, as illustrated in Figure 2.3, the volume changes from a value V_1 to a value V_2, the reversible work done on the system is

$$w_{rev} = -\int_{V_1}^{V_2} P\,dV \tag{2.11}$$

In the example shown in Figure 2.3, $V_1 > V_2$ (i.e., we have compressed the gas) and this work is positive. Only if P is constant is it permissible to integrate this directly to give

$$w_{rev} = -P\int_{V_1}^{V_2} dV = -P(V_2 - V_1) = -P\,\Delta V \tag{2.12}$$

(compare Eq. 2.9). If P is not constant, we must express it as a function of V before performing the integration.

We have already noted that work done is not a state function, and this may be further stressed with reference to the mechanical work of expansion. The previous derivation has shown that the work is related to the *process* carried out rather than to the initial and final states. We can consider the *reversible expansion* of a gas from volume V_1 to volume V_2, and can also consider an *irreversible* process, in which case *less* work will be done *by* the system. This is illustrated in Figure 2.4. The diagram to the left shows the *expansion* of a gas, in which the pressure is falling as the volume increases. The reversible work done *by* the system is given by the integral

$$-w_{rev} = \int_{V_1}^{V_2} P\,dV \tag{2.13}$$

which is represented by the shaded area in the figure. Suppose instead that we performed the process irreversibly, by instantaneously dropping the external pressure to the final pressure P_2. The work done by the system is now against this pressure P_2 throughout the whole expansion, and is given by

$$-w_{irr} = P_2(V_2 - V_1) \tag{2.14}$$

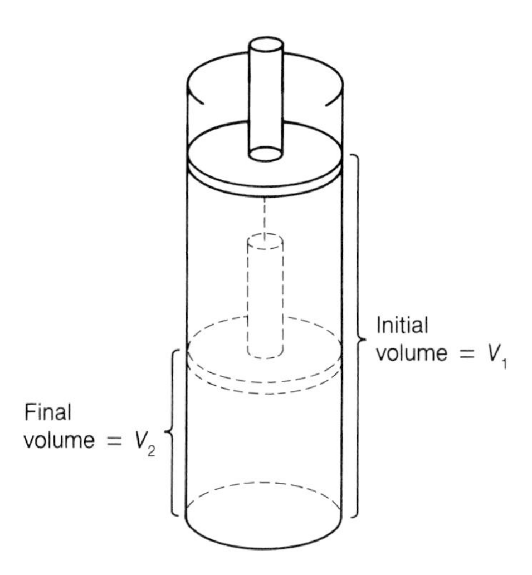

FIGURE 2.3 The reversible work performed when there is a volume decrease from V_1 to V_2.

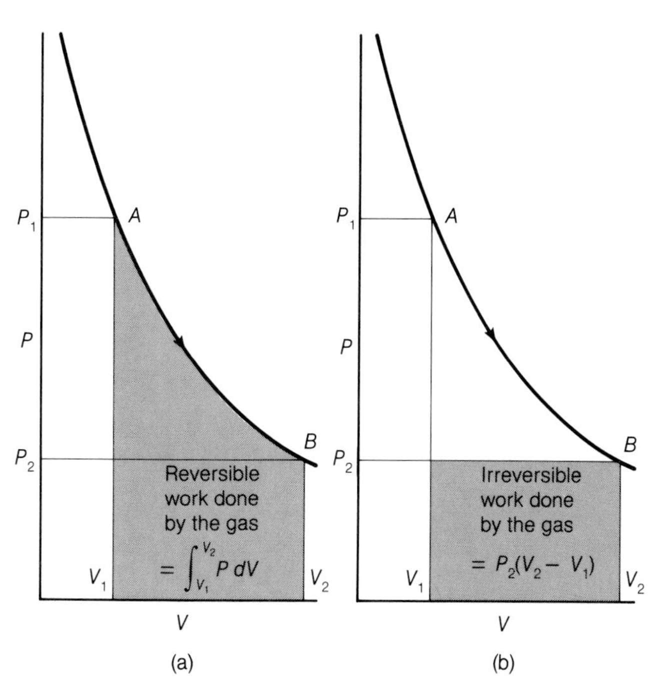

FIGURE 2.4 The left-hand diagram (a) illustrates the reversible work of expansion from V_1 to V_2. The right-hand diagram (b) shows the irreversible work that would be performed by the system if the external pressure were suddenly dropped to the final value P_2.

This work done by the system is represented by the shaded area in Figure 2.4b and is *less* than the reversible work. Thus, although in both processes the state of the system has changed from A to B, the work done is different.

This argument leads us to another important point. The work done by the system in a *reversible* expansion from A to B represents the *maximum work that the system can perform* in changing from A to B.

EXAMPLE

Suppose that water at its boiling point is maintained in a cylinder that has a frictionless piston. For equilibrium to be established, the pressure that must be applied to the piston is 1 atm (101.325 kPa). Suppose that we now reduce the external pressure by an infinitesimal amount in order to have a reversible expansion. If the piston sweeps out a volume of 2.00 dm^3, what is the work done by the system?

SOLUTION

The external pressure remains constant at 101.325 kPa, and therefore the reversible work done by the system is

$$-w_{rev} = P\,\Delta V = 101\ 325\ \text{Pa} \times 2.00\ \text{dm}^3 = 202.65\ \text{Pa m}^3$$

Since Pa $\equiv$ kg m^{-1} s^{-2} (see Appendix A), the units are kg m^2 s^{-2} $\equiv$ J; thus the work done by the system is 202.65 J. ■

For many purposes it is convenient to express the First Law of Thermodynamics with respect to an infinitesimal change. In place of Eq. 2.6, we have

$$dU = đq + đw \tag{2.15}$$

However, if only PV work is involved, $đw$ may be written as $-P\,dV$, where dV is the infinitesimal increase in volume; thus,

$$dU = đq - P\,dV \tag{2.16}$$

Processes at Constant Volume

It follows from this equation that if an infinitesimal process occurs at constant volume, and only PV work is involved,

$$dU = dq_V \tag{2.17}$$

where the subscript V indicates that the heat is supplied at constant volume. This equation integrates to

$$\Delta U = U_2 - U_1 = q_V \tag{2.18}$$

The increase of internal energy of a system in a rigid container (i.e., at constant volume) is thus equal to the heat q_V that is supplied to it.

Processes at Constant Pressure: Enthalpy

In most chemical systems we are concerned with processes occurring in open vessels, which means that they occur at constant pressure rather than at constant volume. The relationships valid for constant-pressure processes may readily be deduced from Eq. 2.16. For an infinitesimal process at constant pressure the heat absorbed dq_P is given by

$$dq_P = dU + P\,dV \tag{2.19}$$

provided that no work other than PV work is performed. If the process involves a change from state 1 to state 2, this equation integrates as follows:

$$q_P = \int_{U_1}^{U_2} dU + \int_{V_1}^{V_2} P\,dV \tag{2.20}$$

Since P is constant,

$$q_P = \int_{U_1}^{U_2} dU + P\int_{V_1}^{V_2} dV \tag{2.21}$$

$$= (U_2 - U_1) + P(V_2 - V_1) = (U_2 + PV_2) - (U_1 + PV_1) \tag{2.22}$$

This relationship suggests that it would be convenient to give a name to the quantity $U + PV$, and it is known as the **enthalpy,*** with the symbol H:

$$\boldsymbol{H \equiv U + PV} \tag{2.23}$$

We thus have

$$q_P = H_2 - H_1 = \Delta H \tag{2.24}$$

This equation is valid only if the work is all PV work. Under these circumstances the increase in enthalpy ΔH of a system is equal to the heat q_P that is supplied to it at constant pressure. Since U, P, and V are all state functions, it follows from Eq. 2.23 that *enthalpy is also a state function.*

A chemical process occurring at constant pressure for which q_P and ΔH are *positive* is one in which a positive amount of heat is *absorbed* by the system. Such processes are known as **endothermic processes** (Greek *endo,* inside; *therme,* heat). Conversely, processes in which heat is *evolved* (q_P and ΔH are *negative*) are known as **exothermic processes** (Greek *exo,* outside).

Heat Capacity

The amount of heat required to raise the temperature of any substance by 1° Celsius is known as its **heat capacity,** and is given the symbol C; its SI unit is J K^{-1}. The word *specific* before the name of any extensive physical quantity refers to the quantity *per unit mass.* The term *specific heat capacity* is thus the amount of heat required to raise the temperature of unit mass of a material by 1° Celsius; if the unit mass is 1 kg, the unit is J K^{-1} kg^{-1}, which is the SI unit for specific heat capacity. The word *molar* before the name of a quantity refers to the quantity divided by the amount of substance. The SI unit for the *molar heat capacity* is J K^{-1} mol^{-1}.

Since heat is not a state function, neither is the heat capacity. It is therefore always necessary, when stating a heat capacity, to specify the process by which the temperature is raised by 1°C. Two heat capacities are of special importance:

*Sometimes it is known as the *heat content,* but this term can be misleading.

1. The heat capacity related to a process occurring at *constant volume;* this is denoted by C_V and is defined by

$$C_V \equiv \frac{dq_V}{dT} \tag{2.25}$$

where q_V is the heat supplied at constant volume. Since q_V is equal to ΔU, it follows that

$$C_V = \left(\frac{\partial U}{\partial T}\right)_V \tag{2.26}$$

If we are working with 1 mol of the substance, this heat capacity is the *molar heat capacity at constant volume,* and is represented by the symbol $C_{V,m}$.*

2. The heat capacity at *constant pressure* is C_P and is defined by

$$C_P \equiv \frac{dq_P}{dT} = \left(\frac{\partial H}{\partial T}\right)_P \tag{2.27}$$

The molar quantity is represented by the symbol $C_{P,m}$.

The heat required to raise the temperature of 1 mol of material from T_1 to T_2 at *constant volume* is

$$q_{V,m} = \int_{T_1}^{T_2} C_{V,m}\, dT \tag{2.28}$$

If $C_{V,m}$ is independent of temperature, this integrates to

$$q_{V,m} = C_{V,m}(T_2 - T_1) \tag{2.29}$$

Similarly, for a process at *constant pressure*

$$q_{P,m} = \int_{T_1}^{T_2} C_{P,m}\, dT \tag{2.30}$$

This integrates to

$$q_{P,m} = C_{P,m}(T_2 - T_1) \tag{2.31}$$

if C_P is independent of temperature. The expressions in Eqs. 2.29 and 2.31 represent ΔU_m and ΔH_m, respectively, per mole of material.

For liquids and solids, ΔU_m and ΔH_m are very close to one another. Consequently, $C_{V,m}$ and $C_{P,m}$ are essentially the same for solids and liquids. For gases, however, the $\Delta(PV)$ term is appreciable, and there is a significant difference between $C_{V,m}$ and $C_{P,m}$. For an ideal gas, which obeys the equation

$$PV = nRT \tag{2.32}$$

the relationship between C_V and C_P can be derived as follows. For 1 mol of gas

$$H_m = U_m + PV_m = U_m + RT \tag{2.33}$$

and therefore

*The subscript m may be omitted when there is no danger of ambiguity.

$$\frac{dH_m}{dT} = \frac{dU_m}{dT} + \frac{d(RT)}{dT} \tag{2.34}$$

Thus,

$$C_{P,m} = C_{V,m} + \frac{d(RT)}{dT} \tag{2.35}$$

or

$$C_{P,m} = C_{V,m} + R \tag{2.36}$$

The general relationship between $C_{P,m}$ and $C_{V,m}$ for a gas that is not necessarily ideal is obtained in Section 2.8; see Eq. 2.117.

2.6 Thermochemistry

We have seen that the heat supplied to a system at constant pressure is equal to the enthalpy increase. The study of enthalpy changes in chemical processes is known as *thermochemistry*.

Extent of Reaction

In dealing with enthalpy changes in chemical processes it is very convenient to make use of a quantity known as the **extent of reaction;** it is given the symbol ξ. This quantity was introduced in 1922 by the Belgian thermodynamicist T. de Donder, and IUPAC recommends that enthalpy changes be considered with reference to it. The extent of reaction must be related to a specified stoichiometric equation for a reaction. A chemical reaction can be written in general as

$$a\text{A} + b\text{B} + \cdots \longrightarrow y\text{Y} + z\text{Z} + \cdots$$

It can also be written as

$$(-\nu_\text{A})\text{A} + (-\nu_\text{B})\text{B} + \cdots \longrightarrow \nu_\text{Y}\text{Y} + \nu_\text{Z}\text{Z} + \cdots$$

where ν_A, ν_B, ν_Y, and ν_Z are known as *stoichiometric coefficients.* By definition the stoichiometric coefficient is *positive for a product and negative for a reactant.*

The extent of reaction is then defined by

$$\xi \equiv \frac{n_i - n_{i,0}}{\nu_i} \tag{2.37}$$

where $n_{i,0}$ is the initial amount of the substance i and n_i is the amount at any time. What makes the extent of reaction so useful is that it is *the same for every reactant and product.* Thus the extent of reaction is the amount of any product formed divided by its stoichiometric coefficient:

$$\xi = \frac{\Delta n_\text{Y}}{\nu_\text{Y}} = \frac{\Delta n_\text{Z}}{\nu_\text{Z}} \tag{2.38}$$

It is also the change in the amount of any reactant (a negative quantity), divided by its stoichiometric coefficient (also a negative quantity):

$$\xi = \frac{\Delta n_A}{\nu_A} = \frac{\Delta n_B}{\nu_B} \qquad (2.39)$$

These quantities are all equal.

EXAMPLE

When 10 mol of nitrogen and 20 mol of hydrogen are passed through a catalytic converter, after a certain time 5 mol of ammonia are produced. Calculate the amounts of nitrogen and hydrogen that remain unreacted. Calculate also the extent of reaction

a. on the basis of the stoichiometric equation

$$N_2 + 3H_2 \longrightarrow 2NH_3$$

b. on the basis of the stoichiometric equation

$$\tfrac{1}{2}N_2 + \tfrac{3}{2}H_2 \longrightarrow NH_3$$

SOLUTION

The amounts are

	N_2	H_2	NH_3	
Initially	10	20	0	mol
Finally	7.5	12.5	5	mol

a. The extent of reaction is the amount of ammonia formed, 5 mol, divided by the stoichiometric coefficient for NH_3:

$$\xi = \tfrac{5}{2} = 2.5 \text{ mol}$$

The same answer is obtained if we divide the amounts of N_2 and H_2 consumed, 2.5 mol and 7.5 mol, by the respective stoichiometric coefficients:

$$\xi = \tfrac{2.5}{1} = \tfrac{7.5}{3} = 2.5 \text{ mol}$$

b. The extent of reaction is now doubled, since the stoichiometric coefficients are halved:

$$\xi = \underset{N_2}{\tfrac{2.5}{0.5}} = \underset{H_2}{\tfrac{7.5}{1.5}} = \underset{NH_3}{\tfrac{5}{1}} = 5.0 \text{ mol}$$

■

The SI unit for the extent of reaction is the mole, and the mole referred to relates to the stoichiometric equation. For example, if the equation is specified to be

$$N_2 + 3H_2 \longrightarrow 2NH_3$$

the mole relates to N_2, $3H_2$, or $2NH_3$. If, therefore, the ΔH for this reaction is stated to be -46.0 kJ mol^{-1}, it is to be understood that this value refers to the removal of 1 mol of N_2 and 1 mol of $3H_2$, which is the same as 3 mol of H_2. It also refers to the formation of 1 mol of $2NH_3$, which is the same as 2 mol of NH_3. In other words, the ΔH value relates to the reaction *as written in the stoichiometric equation*.

This recommended IUPAC procedure, which we shall use throughout this book, avoids the necessity of saying, for example, -46.0 kJ per mol of nitrogen, or 23.0 kJ per mol of ammonia. It cannot be emphasized too strongly that when this IUPAC procedure is used, the *stoichiometric equation must be specified*.

Standard States

The enthalpy change that occurs in a chemical process depends on the states of the reactants and products. Consider, for example, the complete combustion of ethanol, in which 1 mol is oxidized to carbon dioxide and water:

$$C_2H_5OH + 3O_2 \longrightarrow 2CO_2 + 3H_2O$$

The enthalpy change in this reaction depends on whether we start with liquid ethanol or with ethanol in the vapor phase. It also depends on whether liquid or gaseous water is produced in the reaction. Another factor is the pressure of the reactants and products. Also, the enthalpy change in a reaction varies with the temperature at which the process occurs. In giving a value for an enthalpy change it is therefore necessary to specify (1) the state of matter of the reactants and products (gaseous, liquid, or solid; if the last, the allotropic form), (2) the pressure, and (3) the temperature. If the reaction occurs in solution, the concentrations must also be specified.

It has proved convenient in thermodynamic work to define certain **standard states** and to quote data for reactions involving these standard states. By general agreement the standard state of a substance is the form in which it is most stable at 25.00°C (298.15 K) and 1 atm (760 mmHg or 101.325 kPa) pressure. For example, the standard state of oxygen is the gas, and we specify this by writing $O_2(g)$. Since mercury, water, and ethanol are liquids at 25°C, their standard states are $Hg(l)$, $H_2O(l)$, and $C_2H_5OH(l)$. The standard state of carbon is graphite. These standard states should be specified if there is any ambiguity; for example,

$$C_2H_5OH(l) + 3O_2(g, 1\ \text{atm}) \longrightarrow 2CO_2(g, 1\ \text{atm}) + 3H_2O(l).$$

It is quite legitimate, of course, to consider an enthalpy change for a process not involving standard states; for example,

$$C_2H_5OH(g, 1\ \text{atm}) + 3O_2(g, 1\ \text{atm}) \longrightarrow 2CO_2(g, 1\ \text{atm}) + 3H_2O(g, 1\ \text{atm})$$

If a reaction involves species in solution, their standard state is 1 mol kg^{-1} (1 m)*; for example,

$$H^+(1\ m) + OH^-(1\ m) \longrightarrow H_2O(l)$$

Enthalpy changes depend somewhat on the temperature at which the process occurs. Standard thermodynamic data are commonly quoted for a temperature of 25.00°C (298.15 K), and this can be given as a subscript or in parentheses; thus

$$C_2H_5OH(l) + 3O_2(g) \longrightarrow 2CO_2(g) + 3H_2O(l)$$
$$\Delta_c H^\circ(298\ \text{K}) = 1357.7\ \text{kJ mol}^{-1}$$

The superscript ° on the ΔH° specifies that we are dealing with standard states, so that a pressure of 1 atm is assumed and need not be stated. The subscript c is commonly used to indicate complete combustion, and the modern practice is to

* A solution having 1 mol of solute in 1 kg of solvent is known as a 1-molal (1 m) solution: The **molality** of a solution is the *amount of substance per kilogram of solvent.*

attach such subscripts to the Δ and not to the H. As emphasized in our discussion of extent of reaction, the value 1357.7 kJ mol^{-1} relates to the combustion of 1 mol of ethanol, since that is what appears in the equation.

Standard thermodynamic values can be given for a temperature other than 25°C; for example we could give a value for $\Delta H°$ (100°C), and it would be understood that the pressure was again 1 atm and that reactants and products were in their standard states but at 100°C.

Measurement of Enthalpy Changes

The enthalpy changes occurring in chemical processes may be measured by three main methods:

1. *Direct Calorimetry.* Some reactions occur to completion and without side reactions, and it is therefore possible to measure their ΔH^0 values by causing the reactions to occur in a calorimeter. The neutralization of an aqueous solution of a strong acid by a solution of a strong base is an example of such a process, the reaction that occurs being

$$H^+(aq) + OH^-(aq) \longrightarrow H_2O(l)$$

Combustion processes also frequently occur to completion with simple stoichiometry. When an organic compound is burnt in excess of oxygen, the carbon is practically all converted into CO_2 and the hydrogen into H_2O, while the nitrogen is usually present as N_2 in the final products. Often such combustions of organic compounds occur cleanly, and much thermochemical information has been obtained by burning organic compounds in calorimeters.

2. *Indirect Calorimetry: Use of Hess's Law.* The majority of reactions are accompanied by side reactions, and their enthalpy changes therefore cannot be measured directly. For many of these the enthalpy changes can be calculated from the values for other reactions, by making use of **Hess's law,** named after Germain Henri Hess (1802–1850). According to this law, it is permissible to write stoichiometric equations, together with the enthalpy changes, and to treat them as mathematical equations, thereby obtaining a thermochemically valid result. For example, suppose that a substance A reacts with B according to the equation

$$1. \quad A + B \longrightarrow X \qquad \Delta H_1 = -10 \text{ kJ mol}^{-1}$$

Suppose that X reacts with an additional molecule of A to give another product Y:

$$2. \quad A + X \longrightarrow Y \qquad \Delta H_2 = -20 \text{ kJ mol}^{-1}$$

According to Hess's Law, it is permissible to add these two equations and obtain

$$3. \quad 2A + B \longrightarrow Y \qquad \Delta H_3 = \Delta H_1 + \Delta H_2 = -30 \text{ kJ mol}^{-1}$$

The law follows at once from the principle of conservation of energy and from the fact that enthalpy is a state function. Thus, if reactions 1 and 2 occur, there is a net *evolution* of 30 kJ when 1 mol of Y is produced. In principle we could reconvert Y into 2A + B by the reverse of reaction 3. If the heat required to do this were different from 30 kJ, we should have obtained the starting materials with a net gain or loss of heat, and this would violate the principle of conservation of energy.

EXAMPLE

The enthalpy changes in the complex combustion of crystalline α-D-glucose and maltose at 298 K, with the formation of gaseous CO_2 and liquid H_2O, are:

	$\Delta_c H°/\text{kJ mol}^{-1}$
α-D-Glucose, $C_6H_{12}O_6(c)$	−2809.1
Maltose, $C_{12}H_{22}O_{11}(c)$	−5645.5

Calculate the enthalpy change accompanying the conversion of 1 mol of crystalline glucose into crystalline maltose.

SOLUTION

The enthalpy changes given relate to the processes

1. $C_6H_{12}O_6(c) + 6O_2(g) \longrightarrow 6CO_2(g) + 6H_2O(l)$

$$\Delta_c H° = -2809.1 \text{ kJ mol}^{-1}$$

2. $C_{12}H_{22}O_{11}(c) + 12O_2(g) \longrightarrow 12CO_2(g) + 11H_2O(l)$

$$\Delta_c H° = -5645.5 \text{ kJ mol}^{-1}$$

We are asked to convert 1 mol of glucose into maltose; the reaction is

$$C_6H_{12}O_6(c) \longrightarrow \tfrac{1}{2}C_{12}H_{22}O_{11}(c) + \tfrac{1}{2}H_2O(l)$$

Reaction 2 can be written in reverse and divided by 2:

2′. $6CO_2(g) + \tfrac{11}{2}H_2O(l) \longrightarrow \tfrac{1}{2}C_{12}H_{22}O_{11}(c) + 6O_2(g)$

$$\Delta H° = \frac{5645.5}{2} = 2822.8 \text{ kJ mol}^{-1}$$

If we add reactions 1 and 2′, we obtain the required equation, with

$$\Delta H° = -2809.1 + 2822.8 = 13.7 \text{ kJ mol}^{-1}$$ ■

3. *Variation of Equilibrium Constant with Temperature.* A third general method of measuring $\Delta H°$ will only be mentioned here very briefly, since it is based on the Second Law of Thermodynamics and is considered in Section 4.8. This method is based on the equation for the variation of the equilibrium constant K with the temperature:

$$\frac{d \ln K}{d(1/T)} = -\frac{\Delta H°}{R} = -\frac{\Delta H°/\text{J mol}^{-1}}{8.314} \tag{2.40}$$

If, therefore, we measure K at a series of temperatures and plot ln K against $1/T$, the slope of the line at any temperature will be $-\Delta H°/8.314 \text{ J mol}^{-1}$, and hence $\Delta H°$ can be calculated. Whenever an equilibrium constant for a reaction can be measured satisfactorily at various temperatures, this method thus provides a very useful way of obtaining $\Delta H°$. The method cannot be used for reactions that go essentially to completion, in which case a reliable K cannot be obtained, or for reactions that are complicated by side reactions.

Calorimetry

The heats evolved in combustion processes are determined in a *bomb calorimeter*, a schematic diagram of which is shown in Figure 2.5. A weighed sample of the material to be burnt is placed in the cup supported in the reaction vessel, or bomb, which is designed to withstand high pressures. The bomb is filled with oxygen at a pressure of perhaps 25 atm, this being more than enough to cause complete combustion. The reaction is initiated by passing an electric current through the ignition wire. Heat is evolved rapidly in the combustion process and is determined by the temperature rise of the water surrounding the bomb. A correction must be made for the heat produced in the ignition wire, and it is customary to calibrate the apparatus by burning a sample having a known heat of combustion. In this type of calorimeter the bomb and water bath are carefully insulated from the surroundings so that no heat can flow in or out. Such a calorimeter is known as an *adiabatic* calorimeter. Heats of combustion can be measured with an accuracy of better than 0.01%. Such high precision is necessary, since heats evolved in combustion are large, and sometimes we are more interested in the differences between the values for two compounds than in the absolute values.

Many other experimental techniques have been developed for the measurement of heats of reactions. Sometimes the heat changes occurring in chemical reactions are exceedingly small, and it is then necessary to employ very sensitive calorimeters. Such instruments are known as *microcalorimeters*. The prefix *micro* refers to the amount of heat and not to the physical dimensions of the instrument, some microcalorimeters being extremely large. In a special technique known as *heat-burst calorimetry* a heat pulse is measured during the course of a reaction.

Another type of microcalorimeter is the *continuous flow calorimeter*, which permits two reactant solutions to be thermally equilibrated during passage through separate platinum tubes and then brought together in a mixing chamber; a thermopile measures the heat change in the reaction.

Relationship between ΔU and ΔH

Bomb calorimeters and other calorimeters in which the volume is constant give the internal energy change ΔU. Other calorimeters operate at constant pressure and therefore give ΔH values. Whether ΔU or ΔH is determined, the other quantity is easily calculated from the stoichiometric equation for the reaction. From Eq. 2.23 we see that ΔH and ΔU are related by

$$\Delta H = \Delta U + \Delta(PV) \tag{2.41}$$

If all reactants and products are solids or liquids, the change in volume if a reaction occurs at constant pressure is quite small. Usually 1 mol of a solid or liquid has a volume of less than 1 dm^3, and the volume change in a reaction will always be less than 1% (i.e., less than 0.01 dm^3). At 1 atm pressure, with $\Delta V = 0.01\ dm^3$,

$$\Delta(PV) = 101\ 325\ \text{Pa} \times 10^{-5}\ \text{m}^3\ \text{mol}^{-1} = 1.013\ \text{J mol}^{-1}$$

This is quite negligible compared with most heats of reaction, which are of the order of kilojoules, and is much less than the experimental error of most determinations.

If gases are involved in the reaction, however, either as reactants or products, ΔU and ΔH may differ significantly, as illustrated in the following example.

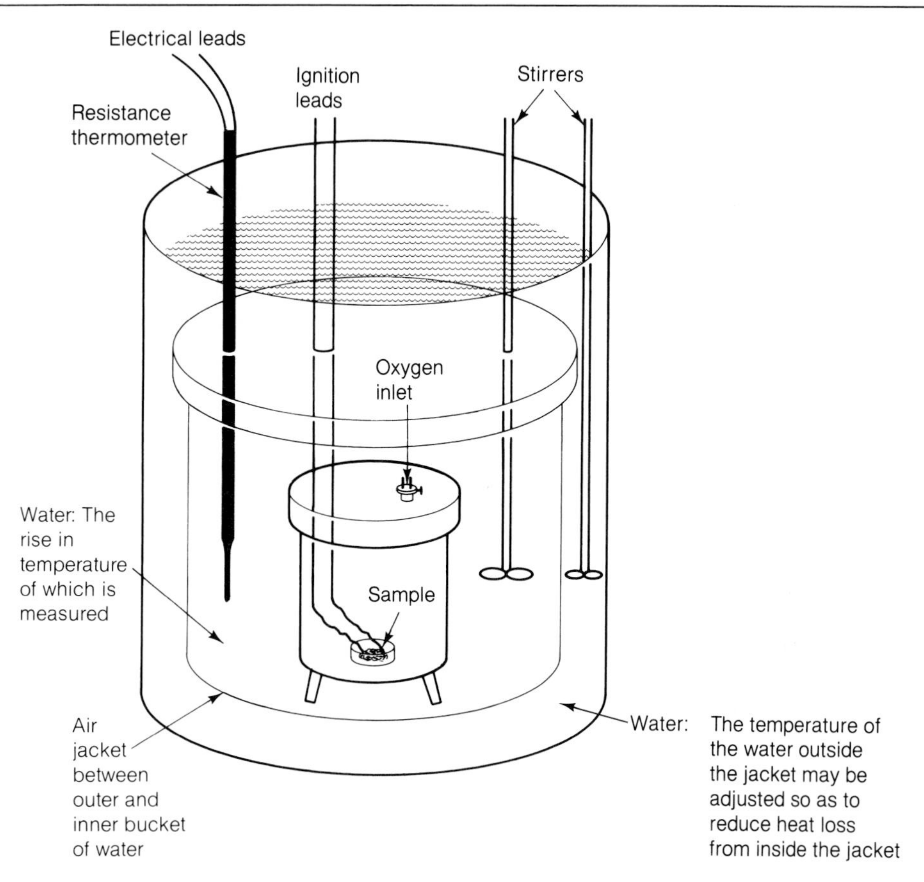

FIGURE 2.5 Schematic diagram of a bomb calorimeter.

EXAMPLE

For the complete combustion of ethanol,

$$C_2H_5OH(l) + 3O_2(g) \longrightarrow 2CO_2(g) + 3H_2O(l)$$

the amount of heat produced, as measured in a bomb calorimeter, is 1364.47 kJ mol^{-1} at 25°C. Calculate $\Delta_c H$ for the reaction.

SOLUTION

Since the bomb calorimeter operates at constant volume, $\Delta_c U = -1364.47$ kJ mol^{-1}. The product of reaction contains 2 mol of gas and the reactants, 3 mol; the change Δn is therefore -1 mol. Assuming the ideal gas law to apply, $\Delta(PV)$ is equal to ΔnRT, and therefore to

$$(-1)RT = -8.314 \times 298.15 \text{ J mol}^{-1} = -2.48 \text{ kJ mol}^{-1}$$

Therefore,

$$\Delta_c H = -1364.47 + (-2.48) = -1366.95 \text{ kJ mol}^{-1}$$

The difference between ΔU and ΔH is now large enough to be experimentally significant. ■

Temperature Dependence of Enthalpies of Reaction

Enthalpy changes are commonly tabulated at 25°C, and it is frequently necessary to have the values at other temperatures. These can be calculated from the heat capacities of the reactants and products. The enthalpy change in a reaction can be written as

$$\Delta H = H(\text{products}) - H(\text{reactants}) \tag{2.42}$$

Partial differentiation with respect to temperature at constant pressure gives*

$$\left(\frac{\partial\, \Delta H}{\partial T}\right)_P = \left[\left(\frac{\partial H(\text{products})}{\partial T}\right)_P - \left(\frac{\partial H(\text{reactants})}{\partial T}\right)_P\right] \tag{2.43}$$

$$= C_P(\text{products}) - C_P(\text{reactants}) = \Delta C_P \tag{2.44}$$

Similarly,

$$\left(\frac{\partial\, \Delta U}{\partial T}\right)_V = \Delta C_V \tag{2.45}$$

For small changes in temperature the heat capacities, and hence ΔC_P and ΔC_V, may be taken as constant. In that case Eq. 2.44 can be integrated between two temperatures T_1 and T_2 to give

$$\Delta\,(\Delta H) = \Delta H_2 - \Delta H_1 = \Delta C_P(T_2 - T_1) \tag{2.46}$$

If there is a large difference between the temperatures T_1 and T_2, this procedure is not satisfactory, and it is necessary to take into account the variation of C_P with temperature. This is often done by expressing the molar value $C_{P,m}$ as a power series:

$$C_{P,m} = a + bT + cT^2 + \cdots \tag{2.47}$$

To a good approximation the values can be calculated over a wide temperature range by using only the first three terms of this expansion. For hydrogen, for example, the $C_{P,m}$ values are fitted to within 0.5% over the temperature range from 273 K to 1500 K if the following constants are used:

$$a = 29.07 \text{ J K}^{-1} \text{ mol}^{-1} \qquad b = -0.836 \times 10^{-3} \text{ J K}^{-2} \text{ mol}^{-1}$$

$$c = 20.1 \times 10^{-7} \text{ J K}^{-3} \text{ mol}^{-1}$$

These values lead to

$$C_{P,m} = 28.99 \text{ J K}^{-1} \text{ mol}^{-1} \text{ at } 273 \text{ K} \quad \text{and} \quad C_{P,m} = 32.34 \text{ J K}^{-1} \text{ mol}^{-1} \text{ at } 1500 \text{ K}$$

Alternatively, and somewhat more satisfactorily, we can use an equation of the form

$$C_{P,m} = d + eT + fT^{-2} \tag{2.48}$$

Some values of d, e, and f are given in Table 2.1.

---

* These relationships (Eqs. 2.43, 2.44) were first deduced by the German physicist Gustav Robert Kirchhoff (1824–1887), *Ann. Physik*, *103*, 454 (1858).

TABLE 2.1 Parameters for the Equation $C_{P,m} = d + eT + fT^{-2}$

Substance	State	$\frac{d}{\text{J K}^{-1}\text{ mol}^{-1}}$	$\frac{e}{\text{J K}^{-2}\text{ mol}^{-1}}$	$\frac{f}{\text{J K mol}^{-1}}$
He, Ne, Ar, Kr, Xe	Gas	20.79	0	0
H_2	Gas	27.28	3.26×10^{-3}	5.0×10^{4}
O_2	Gas	29.96	4.18×10^{-3}	-1.67×10^{5}
N_2	Gas	28.58	3.76×10^{-3}	-5.0×10^{4}
CO	Gas	28.41	4.10×10^{-3}	-4.6×10^{4}
CO_2	Gas	44.22	8.79×10^{-3}	-8.62×10^{5}
H_2O	Vapor	30.54	10.29×10^{-3}	0
H_2O	Liquid	75.48	0	0
C (graphite)	Solid	16.86	4.77×10^{-3}	-8.54×10^{5}
NaCl	Solid	45.94	16.32×10^{-3}	0

If each of the $C_{P,m}$ values for products and reactants is written in the form of Eq. 2.48, the $\Delta C_{P,m}$ for the reaction will have the same form:

$$\Delta C_{P,m} = \Delta d + \Delta eT + \Delta fT^{-2} \tag{2.49}$$

Integration of Eq. 2.44 between the limits T_1 and T_2 leads to

$$\Delta H_m(T_2) - \Delta H_m(T_1) = \int_{T_1}^{T_2} \Delta C_P \, dT \tag{2.50}$$

If $\Delta H(T_1)$ is known for $T_1 = 25°\text{C}$, the $\Delta H_m(T_2)$ at any temperature T_2 is thus given by this equation, and substitution of Eq. 2.49 leads to

$$\Delta H_m(T_2) = \Delta H_m(T_1) + \int_{T_1}^{T_2} (\Delta d + \Delta eT + \Delta fT^{-2}) \, dT \tag{2.51}$$

$$= \Delta H_m(T_1) + \Delta d(T_2 - T_1) + \frac{1}{2}\Delta e(T_2^2 - T_1^2) - \Delta f\left(\frac{1}{T_2} - \frac{1}{T_1}\right) \tag{2.52}$$

EXAMPLE

Consider the gas-phase reaction

$$2CO(g) + O_2(g) \longrightarrow 2CO_2(g)$$

A bomb-calorimetric study of this reaction at 25°C leads to $\Delta H° = -565.98 \text{ kJ mol}^{-1}$. Calculate $\Delta H°$ for this reaction at 2000 K.

SOLUTION

From the values in Table 2.1 we obtain

$$\Delta d = d(\text{products}) - d(\text{reactants})$$

$$= (2 \times 44.22) - (2 \times 28.41) - 29.96 = 1.66 \text{ J K}^{-1}\text{ mol}^{-1}$$

$$\Delta e = e(\text{products}) - e(\text{reactants})$$
$$= (2 \times 8.79 \times 10^{-3}) - (2 \times 4.10 \times 10^{-3}) - 4.18 \times 10^{-3}$$
$$= 5.29 \times 10^{-3}\ \text{J K}^{-2}\ \text{mol}^{-1}$$
$$\Delta f = f(\text{products}) - f(\text{reactants})$$
$$= [2 \times (-8.62 \times 10^5)] + (2 \times 0.46 \times 10^5) + 1.67 \times 10^5$$
$$= -14.65 \times 10^5\ \text{J K mol}^{-1}$$

Then, from Eq. 2.52,

$$\Delta H°(2000\ \text{K})/\text{J mol}^{-1} = -565\,980 + 1.66(2000 - 298)$$
$$+ (\tfrac{1}{2})\ 5.20 \times 10^{-3}(2000^2 - 298^2)$$
$$+ 14.65 \times 10^5(\tfrac{1}{2000} - \tfrac{1}{298})$$
$$= -\ 565\,980 + 2825 + 10\,169 - 4183$$
$$\Delta H°(2000\ \text{K}) = -557\,169\ \text{J mol}^{-1} = -557.17\ \text{kJ mol}^{-1}$$

Note that when numerical values are given, it is permissible to drop the subscript m from $\Delta H°$, since the unit kJ mol^{-1} avoids ambiguity. Remember that the mole referred to always relates to the reaction as written (i.e., to extent of reaction). ■

Enthalpies of Formation

The total number of known chemical reactions is enormous, and it would be very inconvenient if one had to tabulate enthalpies of reaction for all of them. We can avoid having to do this by tabulating *molar enthalpies of formation* of chemical compounds, which are the enthalpy changes associated with the formation of 1 mol of the substance from the elements in their standard states. From these enthalpies of formation it is possible to calculate enthalpy changes in chemical reactions.

We have seen that the standard state of each element and compound is taken to be the most stable form in which it occurs at 1 atm pressure and at 25°C. Suppose that we form methane, at 1 atm and 25°C, from C(graphite) and $H_2(g)$, which are the standard states; the stoichiometric equation is

$$\text{C(graphite)} + 2\text{H}_2(g) \longrightarrow \text{CH}_4(g)$$

It does not matter that we cannot make this reaction occur cleanly and, therefore, that we cannot directly measure its enthalpy change; as seen previously, indirect methods can be used. In such ways it is found that $\Delta H°$ for this reaction is −74.81 kJ mol^{-1}, and this quantity is known as the *standard molar enthalpy of formation* $\Delta_f H°$ of methane. The term **standard enthalpy of formation** refers to the enthalpy change when the compound in its standard state is formed from the elements in their standard states; it must not be used in any other sense. Obviously, the standard enthalpy of formation of any *element* in its standard state is zero.

Enthalpies of formation of organic compounds are commonly obtained from their enthalpies of combustion, by application of Hess's law. When, for example, 1 mol of methane is burned in excess of oxygen, 802.37 kJ of heat is evolved, and we can therefore write

1. $CH_4(g) + 2O_2(g) \rightarrow CO_2(g) + 2H_2O(g)$ $\quad \Delta_c H^\circ = -802.37 \text{ kJ mol}^{-1}$

In addition, we have the following data:

2. $C(\text{graphite}) + O_2(g) \rightarrow CO_2(g)$ $\quad \Delta H^\circ = -393.50 \text{ kJ mol}^{-1}$

3. $2H_2(g) + O_2(g) \rightarrow 2H_2O(g)$ $\quad \Delta H^\circ = 2(-241.83) \text{ kJ mol}^{-1}$

If we add reactions 2 and 3 and subtract reaction 1, the result is

$$C(\text{graphite}) + 2H_2(g) \rightarrow CH_4(g)$$

$$\Delta_f H^\circ(CH_4) = 2(-241.84) - 393.50 - (-802.37) = -74.80 \text{ kJ mol}^{-1}$$

Enthalpies of formation of many other compounds can be deduced in a similar way.

Table 2.2 gives some enthalpies of formation.* The values, of course, depend on the state in which the substance occurs, and this is indicated in the table; the value for liquid ethanol, for example, is a little different from that for ethanol in aqueous solution.

Included in Table 2.2 are enthalpies of formation of individual ions. There is an arbitrariness about these values, because thermodynamic quantities can never be determined experimentally for individual ions; it is always necessary to work with assemblies of positive and negative ions having a net charge of zero. For example, $\Delta_f H^\circ$ for HCl in aqueous solution is -167.15 kJ mol^{-1}, but there is no way that one can make experimental determinations on the individual H^+ and Cl^- ions. The procedure followed is to take $\Delta_f H^\circ$ for the proton in its standard state (1 mol kg^{-1}) to be zero; it then follows that, on this basis, the value of $\Delta_f H^\circ$ for the Cl^- ion is -167.15 kJ mol^{-1}. Then, since the $\Delta_f H^\circ$ value for NaCl in aqueous solution is -407.27 kJ mol^{-1}, we have

$$\Delta_f H^\circ(Na^+) = \Delta H^\circ(NaCl, aq) - \Delta_f H^\circ(Cl^-) = -407.27 + 167.15$$
$$= -240.12 \text{ kJ mol}^{-1}$$

In this way a whole set of values can be built up. Such values are often known as *conventional standard enthalpies of formation;* the word *conventional* refers to the value of zero for the proton. In spite of the arbitrariness of the procedure, correct values are always obtained when one uses these conventional values in making calculations for reactions; this follows from the fact that there is always a balancing of charges in a chemical reaction.

Enthalpies of formation allow us to calculate enthalpies of any reaction, provided that we know the $\Delta_f H^\circ$ values for all the reactants and products. The ΔH° for any reaction is the difference between the sum of the $\Delta_f H^\circ$ values for all the products and the sum of the $\Delta_f H^\circ$ values for all the reactants:

$$\Delta_r H^\circ = \Sigma\, \Delta_f H^\circ(\text{products}) - \Sigma\, \Delta_f H^\circ(\text{reactants}) \tag{2.53}$$

EXAMPLE

Calculate, from the data in Table 2.2, ΔH° for the hydrolysis of urea to give carbon dioxide and ammonia in aqueous solution:

$$H_2NCONH_2(aq) + H_2O(l) \rightarrow CO_2(aq) + 2NH_3(aq)$$

*The table also includes, for convenience, values of Gibbs energies of formation; these are considered in Chapter 3.

TABLE 2.2 Standard Enthalpies and Gibbs Energies of Formation at 25.0°C*

Compound	Formula	State	$\Delta_f H°$ / kJ mol^{-1}	$\Delta_f G°$ / kJ mol^{-1}
Inorganic Compounds				
Water	H_2O	*l*	−285.85	−237.19
Water	H_2O	*g*	−241.84	−228.57
Hydrogen chloride	HCl	*aq*	−167.15	−131.25
Sodium chloride	$NaCl$	*c*	−410.9	−384.1
Carbon monoxide	CO	*g*	−110.54	−137.15
Carbon dioxide	CO_2	*g*	−393.51	−394.34
Carbon dioxide	CO_2	*aq*	−413.80	−386.02
Ammonia	NH_3	*aq*	−80.71	−26.57
Phosphoric acid	H_3PO_4	*aq*	−1288.34	−1142.65
Aqueous ions (conventional values)				
Hydrogen ion	H^+	*aq*	0	0
Sodium ion	Na^+	*aq*	−240.12	−261.88
Potassium ion	K^+	*aq*	−252.38	−283.26
Calcium ion	Ca^{2+}	*aq*	−543.08	−553.1
Magnesium ion	Mg^{2+}	*aq*	−461.91	−456.1
Hydroxide ion	OH^-	*aq*	−229.99	−157.28
Chloride ion	Cl^-	*aq*	−167.15	−131.25
Phosphate ion	PO_4^{3-}	*aq*	−1277.38	−1018.8
Hydrogen phosphate ion	HPO_4^{2-}	*aq*	−1292.14	−1089.26
Dihydrogen phosphate ion	$H_2PO_4^-$	*aq*	−1296.29	−1088.55
Nitrite ion	NO_2^-	*aq*	−104.6	−37.2
Nitrate ion	NO_3^-	*aq*	−207.36	−111.34

*Standard states; 1 atm pressure; 1 *m* for substances in aqueous solution, *c* = crystal; *l* = liquid; *g* = gas; *aq* = aqueous solution.

The data in this table were collected from various sources, in particular: D. D. Wagman et al., "Selected Values of Chemical Thermodynamic Properties," National Bureau of Standard Technical Note 270–3 (1968); and R. C. Wilhoit, "Selected Values of Thermodynamic Properties," being the Appendix (pp. 305–317) of H. D. Brown (Ed.), *Biochemical Microcalorimetry*, New York: Academic Press, 1969. These publications contain many additional values. Wilhoit's compilation also includes some values for systems buffered at pH 7.

SOLUTION

From the data in Table 2.2 we have

1. $C(\text{graphite}) + 2H_2(g) + \frac{1}{2}O_2(g) + N_2(g) \rightarrow H_2NCONH_2(aq) \qquad \Delta_f H° = -317.77 \text{ kJ mol}^{-1}$

2. $H_2(g) + \frac{1}{2}O_2(g) \rightarrow H_2O(l) \qquad \Delta_f H° = -285.85 \text{ kJ mol}^{-1}$

TABLE 2.2 (Continued)

Compound	Formula	State	$\Delta_f H^\circ$ / kJ mol^{-1}	$\Delta_f G^\circ$ / kJ mol^{-1}
Organic Molecules and Ions				
Methane	CH_4	*g*	− 74.81	− 50.75
Ethane	C_2H_6	*g*	− 84.68	− 32.89
Propane	C_3H_8	*g*	− 103.85	− 23.47
Ethylene	C_2H_4	*g*	52.26	68.12
Acetylene	C_2H_2	*g*	226.73	200.92
Methanol	CH_3OH	*l*	− 238.66	− 166.36
Ethanol	C_2H_5OH	*l*	− 277.69	− 174.89
Ethanol	C_2H_5OH	*aq*	− 287.02	− 181.75
Formic acid	HCOOH	*l*	− 424.72	− 361.41
Formate ion	$HCOO^-$	*aq*	− 425.55	− 351.0
Acetic acid	CH_3COOH	*l*	− 484.5	− 389.9
Acetate ion	CH_3COO^-	*aq*	− 486.01	− 369.41
Formaldehyde	HCHO	*g*	− 117	− 113
Acetaldehyde	CH_3CHO	*l*	− 192.30	− 128.20
Acetone	CH_3COCH_3	*l*	− 246.81	− 153.55
Dimethyl ether	$(CH_3)_2O$	*g*	− 184.05	− 112.68
Diethyl ether	$(C_2H_5)_2O$	*l*	− 279.57	—
Methylamine	CH_3NH_2	*l*	− 47.28	35.6
Ethylamine	$C_2H_5NH_2$	*l*	− 74.06	—
Glucose	$C_6H_{12}O_6$	*aq*	−1263.07	− 914.54
Sucrose	$C_{12}H_{22}O_{11}$	*aq*	−2215.8	−1551.4
Glycine	H_2NCH_2COOH	*aq*	− 472.16	− 370.79
Urea	H_2NCONH_2	*aq*	− 317.77	− 202.80

3. $C(\text{graphite}) + O_2(g) \rightarrow CO_2(aq)$ $\qquad \Delta_f H^\circ = -413.80 \text{ kJ mol}^{-1}$

4. $\frac{1}{2}N_2(g) + \frac{3}{2}H_2(g) \rightarrow NH_3(aq)$ $\qquad \Delta_f H^\circ = -80.71 \text{ kJ mol}^{-1}$

4′. $N_2(g) + 3H_2(g) \rightarrow 2NH_3(aq)$ $\qquad \Delta_f H^\circ = 2(-80.71) \text{ kJ mol}^{-1}$

Subtraction of reactions 1 + 2 from reactions 3 + 4′ then leads to the desired equation, and the enthalpy change in the reaction is thus

$$\Delta H^\circ = -413{,}80 + 2 \times (-80.71) + 285.85 + 317.77 = 28.32 \text{ kJ mol}^{-1}$$

It is not necessary, of course, to write out the reactions; Eq. 2.53 may be used directly. ∎

Bond Strengths

One important aspect of thermochemistry relates to the energies of different chemical bonds. As a very simple example, consider the case of methane, CH_4. The standard enthalpy of formation of methane is -74.81 kJ mol^{-1}:

1. $C(\text{graphite}) + 2H_2(g) \rightarrow CH_4(g) \qquad \Delta_f H° = -74.81 \text{ kJ mol}^{-1}$

We also know the following thermochemical values:

2. $C(\text{graphite}) \rightarrow C(\text{gaseous atoms}) \qquad \Delta H° = 716.7 \text{ kJ mol}^{-1}$

3. $\frac{1}{2}H_2(g) \rightarrow H(\text{gaseous atoms}) \qquad \Delta H° = 218.0 \text{ kJ mol}^{-1}$

The former is the enthalpy of sublimation of graphite, and the latter is one-half of the heat of dissociation of hydrogen. We may now apply Hess's law in the following manner:

$$
\begin{array}{lll}
1. & CH_4(g) \rightarrow C(\text{graphite}) + 2H_2(g) & \Delta H° = 74.8 \text{ kJ mol}^{-1} \\
2. & C(\text{graphite}) \rightarrow C(\text{gaseous atoms}) & \Delta H° = 716.7 \text{ kJ mol}^{-1} \\
4 \times 3. & 2H_2(g) \rightarrow 4H\ (\text{gaseous atoms}) & \Delta H° = 872.0 \text{ kJ mol}^{-1} \\
\text{Adding:} & CH_4(g) \rightarrow C + 4H(\text{gaseous atoms}) & \Delta H° = 1663.5 \text{ kJ mol}^{-1}
\end{array}
$$

This quantity, 1663.5 kJ mol^{-1}, is known as the *heat of atomization* of methane; it is the heat that has to be supplied to 1 mol of methane in order to dissociate all the molecules into gaseous atoms. Since each CH_4 molecule has four C—H bonds, we can divide 1663.5 by 4, obtaining 415.9 kJ mol^{-1}, and call this the C—H **bond strength** or *bond enthalpy;* it is an average quantity.

A similar procedure with ethane, C_2H_6, leads to a heat of atomization of 2829.2 kJ mol^{-1}. This molecule contains one C—C bond, and six C—H bonds. If we subtract $6 \times 415.9 = 2495.4$ kJ mol^{-1} as the contribution of the C—H bonds, we are left with 333.8 kJ mol^{-1} as the C—C bond strength.

However, if we calculate the heats of atomization of the higher paraffin hydrocarbons using these values, we find that the agreement with experiment is by no means perfect. In other words, there is not a strict *additivity* of bond strengths. The reason for this is that chemical bonds in a given molecule are not isolated from each other, but they interact. On the whole, heats of atomization are more satisfactorily predicted if we use the following bond energies rather than the ones deduced from the data for CH_4 and C_2H_6:

C—H 413 kJ mol^{-1}

C—C 348 kJ mol^{-1}

By the use of similar procedures for molecules containing different kinds of bonds, it is possible to arrive at a set of bond strengths that will allow us to make approximate estimates of heats of atomization and heats of formation. Such a set is shown in Table 2.3. Values of this kind have proved very useful in deducing thermochemical information when the experimental heats of formation are not available. These simple additive procedures can be improved in various ways.

TABLE 2.3 Bond Strengths

Bond	Bond Strength $kJ\ mol^{-1}$
H—H	436
C—C	348
C—H	413
C═C	682
C≡C	962
N—H	391
O—H	463
C—O	351
C═O	732

These values are for gaseous molecules; adjustments must be made for substances in condensed phases.

It is important to distinguish clearly between *bond strengths,* the additive quantities we have just considered, and **bond dissociation energies.** The distinction can be illustrated by the case of methane. We can consider the successive removal of hydrogen atoms from methane in the gas phase:

1. $CH_4 \rightarrow CH_3 + H \qquad \Delta H_1^\circ = 431.8\ kJ\ mol^{-1}$
2. $CH_3 \rightarrow CH_2 + H \qquad \Delta H_2^\circ = 471.1\ kJ\ mol^{-1}$
3. $CH_2 \rightarrow CH + H \qquad \Delta H_3^\circ = 421.7\ kJ\ mol^{-1}$
4. $CH \rightarrow C + H \qquad \Delta H_4^\circ = 338.8\ kJ\ mol^{-1}$

Only the first and last of these bond dissociation energies are known with any degree of reliability, and the values for the second and third reactions are estimates. What is certain is that the sum of the four values must be 1663.5 $kJ\ mol^{-1}$, the heat of atomization of CH_4. Although all four C—H bonds in methane are the same, the four dissociation energies are not all the same, because there are adjustments in the electron distributions as each successive hydrogen atom is removed. The additive bond strength is the average of these four dissociation energies.

For the gaseous water molecule we have similarly

1. $HOH \rightarrow H + OH \qquad \Delta H^\circ = 498.7\ kJ\ mol^{-1}$
2. $OH \rightarrow H + O \qquad \Delta H^\circ = 428.2\ kJ\ mol^{-1}$

1. + 2. $H_2O \rightarrow 2H + O \qquad \Delta H^\circ = 926.9\ kJ\ mol^{-1}$

Note that it is easier to remove the second hydrogen atom than the first. The bond strength is 926.9/2 = 463.5 $kJ\ mol^{-1}$, which is the mean of the two dissociation energies.

Only in the case of diatomic molecules can the bond strength be identified with the bond dissociation energy; for example,

$$H_2 \rightarrow 2H \qquad \Delta H^\circ = 435.9 \ \text{kJ mol}^{-1}$$

which is both the dissociation energy and the bond strength.

2.7 Ideal Gas Relationships

The various transformations that can be brought about on ideal gases have played a very important part in the development of thermodynamics. There are good reasons for devoting careful study to ideal gases. In the first place, ideal gases are the simplest systems to deal with, and they therefore provide us with valuable and not too difficult exercises for testing our understanding of the subject. In addition, some of the simple conclusions that we can draw for ideal gases can readily be adapted to more complicated systems, such as solutions; a direct application of thermodynamics to solutions would be difficult if we did not have the ideal gas equations to guide our way.

Reversible Compression at Constant Pressure

As a first example, we will consider the reversible compression of an ideal gas at constant pressure. We are going to find how much work is done on the system during this process, how much heat is lost, and what the changes are in internal energy and enthalpy. Suppose that we have 1 mol of an ideal gas confined in a cylinder with a piston, at a pressure P_1, a molar volume $V_{m,1}$, and an absolute temperature T_1. The **isotherm** (i.e., the PV relationship) for this temperature is shown in the upper curve in Figure 2.6a and the initial state is represented by point A. We now remove heat from the system reversibly, at the constant pressure P_1, until the volume has fallen to $V_{m,2}$ (point B). This could be done by lowering the temperature of the surroundings by infinitesimal amounts, until the temperature of the system is T_2; the isotherm for T_2 is the lower curve in Figure 2.6a.

As far as work and heat changes are concerned, we could not obtain values unless we had specified the path taken, since work and heat are not state functions. In this example, we have specified that the compression is reversible and is occurring at constant pressure. The work done on the system is

$$w_{\text{rev}} = -\int_{V_{m,1}}^{V_{m,2}} P_1 \, dV = P_1(V_{m,1} - V_{m,2}) \tag{2.54}$$

Use of the ideal gas law $PV_m = RT$ (for 1 mol of gas) leads to an alternative expression:

$$w_{\text{rev}} = P_1\left(\frac{RT_1}{P_1} - \frac{RT_2}{P_1}\right) \tag{2.55}$$

$$= R(T_1 - T_2) \tag{2.56}$$

Since $V_1 > V_2$, and $T_1 > T_2$, a positive amount of work has been done on the system. This work is represented by the shaded area in Figure 2.6a. If the system had

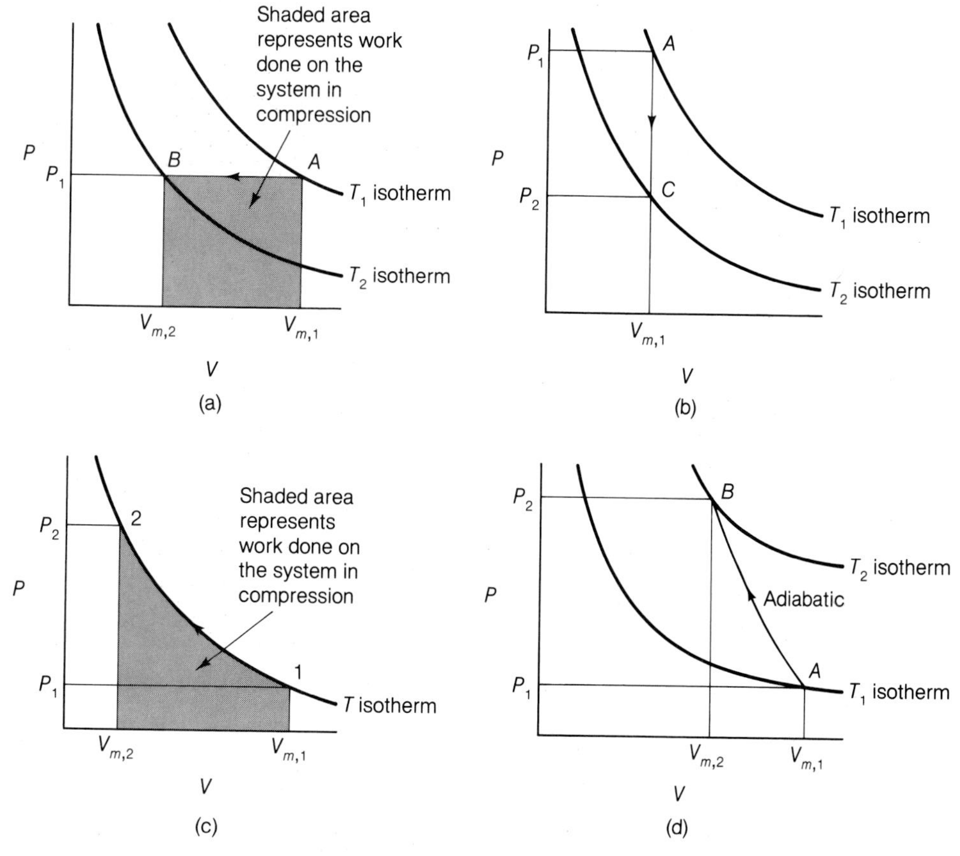

FIGURE 2.6 Pressure-volume relationships for an ideal gas. (a) Isotherms at two temperatures, T_1 and T_2. The gas is compressed reversibly at constant pressure from state A to state B. (b) Isotherms at two temperatures, T_1 and T_2. The gas at state A is cooled at constant volume to bring it to state C. (c) An isotherm, showing a reversible isothermal compression from state 1 to state 2. (d) Isotherms and an adiabatic. The gas at state A is compressed reversibly and adiabatically to State B.

expanded isothermally and at constant pressure, from state B to state A, the shaded area would represent the work done by the system.

The heat absorbed by the system during the process $A \rightarrow B$ is given by

$$q_{P,m} = \int_{T_1}^{T_2} C_{P,m}\, dT \tag{2.57}$$

since the process occurs at constant pressure. For an ideal gas, $C_{P,m}$ is independent of temperature and this expression therefore integrates to

$$q_{P,m} = C_{P,m}(T_2 - T_1) \tag{2.58}$$

Since $T_1 > T_2$, a negative amount of heat is absorbed (i.e., heat is released by the system). This amount of heat $q_{P,m}$, being absorbed at constant pressure, is the molar enthalpy change, which is also negative:

$$\Delta H_m = C_{P,m}(T_2 - T_1) \tag{2.59}$$

The molar internal energy change ΔU_m (also negative for this process) is obtained by use of the First Law:

$$\Delta U_m = q + w = C_{P,m}(T_2 - T_1) + R(T_1 - T_2) \tag{2.60}$$

$$= (C_{P,m} - R)(T_2 - T_1) = C_{V,m}(T_2 - T_1) \tag{2.61}$$

using Eq. 2.36. It can easily be verified that these expressions for ΔH_m and ΔU_m are consistent with the relationship

$$\Delta H_m = \Delta U_m + \Delta(PV) \tag{2.62}$$

Reversible Pressure Change at Constant Volume

Suppose, instead, that we take 1 mol of ideal gas from the initial state P_1, $V_{m,1}$, T_1 to the final state P_2, $V_{m,1}$, T_2 as shown in Figure 2.6b. The pressure P_1 is taken to be higher than P_2, and to accomplish this at constant volume we must remove heat until the temperature is T_2. Again, we bring about the change reversibly.

The work done on the system is the area below the line AC in Figure 2.6b and is zero. This is confirmed by considering the integral

$$w_{\text{rev}} = -\int_{V_{m,1}}^{V_{m,1}} P\,dV = 0 \tag{2.63}$$

Since the process occurs at constant volume, the heat absorbed is given by

$$q_{V,m} = \int_{T_1}^{T_2} C_{V,m}\,dT = C_{V,m}(T_2 - T_1) \tag{2.64}$$

which is negative since $T_1 > T_2$. This expression is also ΔU_m:

$$\boldsymbol{\Delta U_m = C_{V,m}(T_2 - T_1)} \tag{2.65}$$

It can be verified that Eqs. 2.63, 2.64, and 2.65 are consistent with the First Law. The value of ΔH_m is obtained as follows:

$$\Delta H_m = \Delta U_m + \Delta(PV_m) = \Delta U_m + \Delta(RT) \tag{2.66}$$

$$= C_{V,m}(T_2 - T_1) + R(T_2 - T_1) = (C_{V,m} + R)(T_2 - T_1) \tag{2.67}$$

$$\boldsymbol{= C_{P,m}(T_2 - T_1)} \tag{2.68}$$

It is interesting to compare the changes that occur in going from A to B (volume decrease at constant pressure, Figure 2.6a) with those when we go from A to C (pressure decrease at constant volume, Figure 2.6b). The work and heat values are different in the two cases. However, the ΔU_m and ΔH_m values are the same. This means that the internal energy is the same at point B on the T_2 isotherm as it is at point C; the same is true of the enthalpy. This result can be proved for any two points on an isotherm. We thus reach the very important conclusion that the *internal energy and enthalpy of an ideal gas depend only on the temperature and remain constant as we move along any isotherm.*

Reversible Isothermal Compression

Another process of very great importance is that of compression along an isotherm (i.e., at constant temperature). Such a process is illustrated in Figure 2.6c, the temperature being written simply as T. The initial conditions are P_1, $V_{m,1}$ and the final are P_2, $V_{m,2}$ with $V_{m,1} > V_{m,2}$. We have just seen that for an isothermal process in an ideal gas

$$\Delta U_m = 0 \quad \text{and} \quad \Delta H_m = 0 \tag{2.69}$$

The work done on the system in a reversible compression is

$$w_{\text{rev}} = -\int_{V_{m,1}}^{V_{m,2}} P\,dV \tag{2.70}$$

Since P is varying, we must express it in terms of V_m by use of the ideal gas equation, which for 1 mol is $PV_m = RT$; thus

$$w_{\text{rev}} = -\int \frac{RT}{V_m}\,dV = -RT \ln V \Big|_{V_{m,1}}^{V_{m,2}} \tag{2.71}$$

$$= \boldsymbol{RT \ln \frac{V_{m,1}}{V_{m,2}}} \tag{2.72}$$

Since $V_{m,1} > V_{m,2}$, this is a positive quantity. The heat absorbed is found by use of the equation for the First Law:

$$\Delta U_m = q_{\text{rev}} + w_{\text{rev}} \tag{2.73}$$

Therefore,

$$\boldsymbol{q_{\text{rev}} = \Delta U_m - w_{\text{rev}} = 0 - w_{\text{rev}} = RT \ln \frac{V_{m,2}}{V_{m,1}}} \tag{2.74}$$

This is negative; that is, heat is evolved during the compression. It is easy to understand why this should be the case. When we compress a gas, we do work on it and supply energy to it; if the temperature is to remain constant, heat must be evolved.

If 1 mol of gas has a volume $V_{m,1}$, the concentration is

$$c_1 = \frac{1}{V_{m,1}} \tag{2.75}$$

Similarly,

$$c_2 = \frac{1}{V_{m,2}} \tag{2.76}$$

The ratio of volumes is therefore the inverse ratio of the concentrations:

$$\frac{V_{m,2}}{V_{m,1}} = \frac{c_1}{c_2} \tag{2.77}$$

Equation 2.72 for the work done in the isothermal reversible expansion of 1 mol of an ideal gas can therefore be written alternatively as

$$w_{\text{rev},m} = RT \ln \frac{c_2}{c_1} \tag{2.78}$$

For n mol,

$$w_{\text{rev}} = nRT \ln \frac{c_2}{c_1} \tag{2.79}$$

This is a useful form of the equation, because it will be seen that certain types of solutions, known as *ideal* solutions, obey exactly the same relationship.

EXAMPLE

Calculate the work done by the system when 6 mol of an ideal gas at 25.0°C are allowed to expand isothermally and reversibly from an initial volume of 5 dm³ to a final volume of 15 dm³.

SOLUTION

From Eq. 2.72 for 6 mol of gas, work done by the system is

$$-w_{rev} = -6RT \ln \tfrac{5}{15} = 6RT \ln \tfrac{15}{5}$$

Since R is 8.314 J K^{-1} mol^{-1},

$$-w_{rev} = 6 \times 8.314 \times 298 \ln 3 = 16\ 400 \text{ J} = 16.4 \text{ kJ}$$

■

EXAMPLE

Gastric juice in humans has an acid concentration of about 10^{-1} M* (pH ≈ 1) and it is formed from other body fluids, such as blood, which have an acid concentration of about 4.0×10^{-8} M (pH ≈ 7.4). On the average, about 3 dm³ of gastric juice are produced per day. Calculate the minimum work required to produce this quantity at 37°C, assuming the behavior to be ideal.

SOLUTION

Equation 2.79 gives us the reversible work required to produce n mol of acid, and 3 dm³ of 10^{-1} M acid contains 0.3 mol, so that the reversible work is

$$w_{rev} = 0.3 \times 8.314 \times 310 \ln \frac{10^{-1}}{4.0 \times 10^{-8}} = 11\ 391 \text{ J} = 11.4 \text{ kJ}$$

The actual work required will, of course, be greater than this because the process, being a natural one, does not occur reversibly. ■

Reversible Adiabatic Compression

The final process that we will consider is the compression of an ideal gas contained in a vessel whose walls are perfectly insulating, so that no heat can pass through them. Such processes are said to be **adiabatic** (Greek *adiabatos,* impassable, which in turn is derived from the Greek prefix *a*-, not, and the words *dia,* through, and *bainein,* to go).

The pressure-volume diagram for the process is shown in Figure 2.6d. Since work is performed on the gas in order to compress it, and no heat can leave the system, the final temperature T_2 must be higher than the initial temperature T_1. The figure shows the T_1 and T_2 isotherms, as well as the adiabatic curve AB. We will now consider n mol of the ideal gas.

We first need the equation for the adiabatic AB. According to the First Law

$$dU = dq - P\,dV \tag{2.80}$$

*The symbol M (molar) represents mol dm^{-3}.

Since the process is adiabatic, $dq = 0$, and therefore

$$dU + P\,dV = 0 \tag{2.81}$$

Also, for n mol, $dU = nC_{V,m}\,dT$, and thus

$$nC_{V,m}\,dT + P\,dV = 0 \tag{2.82}$$

For n mol of an ideal gas

$$PV = nRT \tag{2.83}$$

Elimination of P between Eqs. 2.82 and 2.83 leads to

$$C_{V,m}\frac{dT}{T} + R\frac{dV}{V} = 0 \tag{2.84}$$

For the adiabatic AB we integrate this equation between the temperatures T_1 and T_2 and the volumes V_1 and V_2:

$$C_{V,m}\int_{T_1}^{T_2}\frac{dT}{T} + R\int_{V_1}^{V_2}\frac{dV}{V} = 0 \tag{2.85}$$

and, therefore,

$$C_{V,m}\ln\frac{T_2}{T_1} + R\ln\frac{V_2}{V_1} = 0 \tag{2.86}$$

Since $R = C_{P,m} - C_{V,m}$, this equation may be written as

$$\ln\frac{T_2}{T_1} + \frac{(C_{P,m} - C_{V,m})}{C_{V,m}}\ln\frac{V_2}{V_1} = 0 \tag{2.87}$$

The ratio of $C_{P,m}$ to $C_{V,m}$ is often written as γ:

$$\gamma \equiv \frac{C_{P,m}}{C_{V,m}} \tag{2.88}$$

and Eq. 2.87 can therefore be written as

$$\ln\frac{T_2}{T_1} + (\gamma - 1)\ln\frac{V_2}{V_1} = 0 \tag{2.89}$$

or

$$\frac{T_2}{T_1} = \left(\frac{V_1}{V_2}\right)^{\gamma-1} \tag{2.90}$$

We can eliminate the temperature by making use of the ideal gas relationship

$$\frac{T_2}{T_1} = \frac{P_2V_2}{P_1V_1} \tag{2.91}$$

Equating the right-hand sides of Eqs. 2.90 and 2.91 gives

$$\frac{P_2}{P_1} = \left(\frac{V_1}{V_2}\right)^{\gamma} \tag{2.92}$$

or

$$P_1V_1^\gamma = P_2V_2^\gamma \tag{2.93}$$

This is to be contrasted with the Boyle's law equation $P_1V_1 = P_2V_2$ for the isothermal process. Because γ is necessarily greater than unity, the adiabatic is steeper than the isotherm, as shown in Figure 2.6d.

We now consider the various changes in the thermodynamic quantities when the process $A \rightarrow B$ in Figure 2.6d is undergone by an ideal gas. Since the process is adiabatic,

$$q = 0 \tag{2.94}$$

Both U and H remain unchanged as we move along the T_1 isothermal, and the same is true of the T_2 isothermal. The changes are (Eqs. 2.29 and 2.31)

$$\Delta U = C_V(T_2 - T_1) \tag{2.95}$$

and

$$\Delta H = C_P(T_2 - T_1) \tag{2.96}$$

Since $\Delta U = q + w$, and $q = 0$, the work done on the system during the adiabatic compression is

$$w = C_V(T_2 - T_1) = nC_{V,m}(T_2 - T_1) \tag{2.97}$$

2.8 Real Gases*

We have seen that the internal energy U of an ideal gas is a function of temperature but remains constant as we move along any isotherm. For an ideal gas, therefore, the following conditions hold:

1. $$PV = nRT \tag{2.98}$$

2. $$\left(\frac{\partial U}{\partial V}\right)_T = 0 \tag{2.99}$$

The second of these conditions follows from the first. For a nonideal gas, neither of these conditions is satisfied. We have seen in Chapter 1 that for many real gases the van der Waals equation (Eq. 1.82) is obeyed.

The Joule-Thomson Experiment

In 1845 Joule published the results of an experiment that was designed to determine whether $(\partial U/\partial V)_T = 0$ for a real gas.† His apparatus, which is shown in Figure 2.7a, consisted of two containers, connected through a stopcock, the whole apparatus being placed in a water bath, the temperature of which could be measured very accurately. He filled one container with air at 22 atm pressure and evacuated the

---

*This section could be omitted on first reading; for a mathematical review see Appendix C.

†J. P. Joule, *Philosophical Magazine*, 26, 369 (1845).

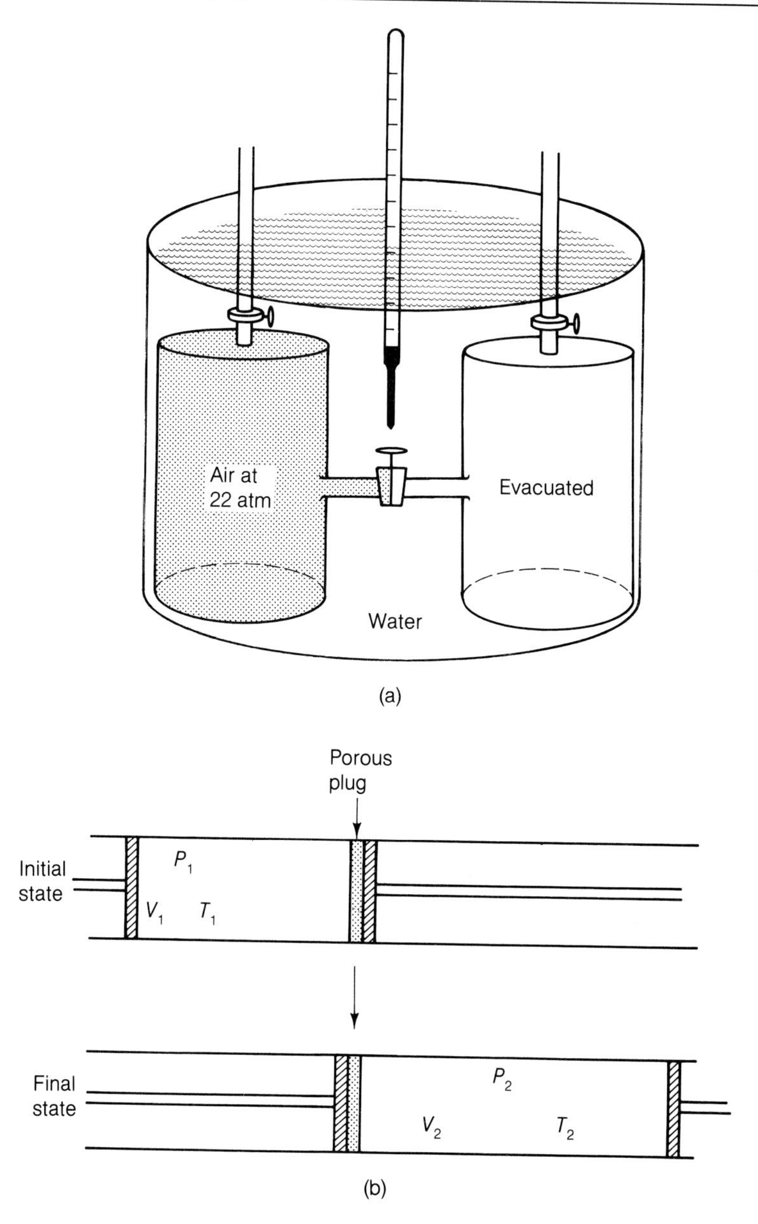

FIGURE 2.7 (a) Joule's apparatus for attempting to measure a temperature change on the free expansion of a gas. (b) The Joule-Thomson experiment: The gas is forced through a membrane and the temperature change measured.

other. On opening the stopcock connecting the two containers, he was unable to detect any change in temperature.

The expansion in Joule's experiment was an irreversible process. The gas underwent no detectable change in internal energy, since no work was done and there was no measurable exchange of heat with the surroundings. The change dU can be expressed as (see Appendix C)

$$dU = \left(\frac{\partial U}{\partial V}\right)_T dV + \left(\frac{\partial U}{\partial T}\right)_V dT \tag{2.100}$$

As far as Joule could determine, $dU = 0$ and, therefore,

$$\left(\frac{\partial U}{\partial V}\right)_T = -\left(\frac{\partial U}{\partial T}\right)_V \left(\frac{\partial T}{\partial V}\right)_U \tag{2.101}$$

$$= -C_V \left(\frac{\partial T}{\partial V}\right)_U \tag{2.102}$$

by Eq. 2.26. Joule's experiment indicated no temperature change, [i.e., $(\partial T/\partial V)_U = 0$] and, therefore,

$$\left(\frac{\partial U}{\partial V}\right)_T = 0 \tag{2.103}$$

As far as Joule could detect, therefore, the gas was behaving ideally.

His experiment, however, was not very satisfactory, because the heat capacity of the water was extremely large compared with the heat capacity of the gas, so that any temperature rise would be very small and hard to detect. William Thomson (Lord Kelvin) suggested a better procedure and, with Joule, carried out a series of experiments between 1852 and 1862.* Their apparatus, shown schematically in Figure 2.7b, consists essentially of a cylinder containing a porous plug. By means of a piston, the gas on one side of the plug was forced completely through the plug at constant pressure P_1, the second piston moving back and maintaining the gas at constant pressure P_2, which is lower than P_1. Since the whole system is thermally isolated, the process is adiabatic (i.e., $q = 0$). The temperatures T_1 and T_2 were measured on the two sides of the plug.

Suppose that the initial volume is V_1. The work done in moving the piston to the porous plug, thereby forcing the gas through the plug, is P_1V_1. If the piston in the right-hand chamber starts at the plug, and if the final volume is V_2, the work done by the gas in expanding is P_2V_2. The net work done on the gas is thus

$$w = P_1V_1 - P_2V_2 \tag{2.104}$$

Since $q = 0$, the change in internal energy is

$$\Delta U = U_2 - U_1 = q + w = w = P_1V_1 - P_2V_2 \tag{2.105}$$

Thus

$$U_2 + P_2V_2 = U_1 + P_1V_1 \tag{2.106}$$

or, from the definition of enthalpy (Eq. 2.23),

$$H_2 = H_1 \tag{2.107}$$

The Joule-Thomson expansion thus occurs at constant enthalpy: $\Delta H = 0$.

Joule and Thomson were able to detect a temperature change ΔT as the gas undergoes a pressure change P through the porous plug. The *Joule-Thomson coefficient* μ is defined as

---

*J. P. Joule and W. Thomson, *Proc. Roy. Soc.*, 143, 357 (1853) and later publications.

$$\mu \equiv \left(\frac{\partial T}{\partial P}\right)_H \approx \frac{\Delta T}{\Delta P} \tag{2.108}$$

When there is an expansion, ΔP is negative; if there is a cooling on expansion (i.e., ΔT is negative when ΔP is negative), the Joule-Thomson coefficient μ is positive. Conversely, a negative μ corresponds to a rise in temperature on expansion. Most gases at ordinary temperatures cool on expansion. Hydrogen is exceptional, in that its Joule-Thomson coefficient is negative above 193 K (i.e., it warms on expansion); below 193 K, the coefficient is positive. The temperature 193 K, at which $\mu = 0$, is known as the *Joule-Thomson inversion temperature* for hydrogen.

Since the Joule-Thomson expansion occurs at constant enthalpy, the total differential is

$$dH = \left(\frac{\partial H}{\partial P}\right)_T dP + \left(\frac{\partial H}{\partial T}\right)_P dT = 0 \tag{2.109}$$

from which it follows that

$$\left(\frac{\partial H}{\partial P}\right)_T = -\left(\frac{\partial H}{\partial T}\right)_P \left(\frac{\partial T}{\partial P}\right)_H = -C_P\mu \tag{2.110}$$

For an ideal gas, $\mu = 0$ and the enthalpy is therefore independent of pressure, as we have seen previously. For a real gas, μ is in general different from zero, and H shows some variation with P.

An equation for the difference between the heat capacities C_P and C_V, and applicable to gases, liquids, and solids, is obtained as follows:

$$C_P - C_V = \left(\frac{\partial H}{\partial T}\right)_P - \left(\frac{\partial U}{\partial T}\right)_V \tag{2.111}$$

$$= \left(\frac{\partial U}{\partial T}\right)_P + P\left(\frac{\partial V}{\partial T}\right)_P - \left(\frac{\partial U}{\partial T}\right)_V \tag{2.112}$$

We also have the following relationships from the total differentials of U and V:

$$dU = \left(\frac{\partial U}{\partial V}\right)_T dV + \left(\frac{\partial U}{\partial T}\right)_V dT \tag{2.113}$$

and

$$dV = \left(\frac{\partial V}{\partial T}\right)_P dT + \left(\frac{\partial V}{\partial P}\right)_T dP \tag{2.114}$$

Substitution of this expression for dV into Eq. 2.113 gives

$$dU = \left(\frac{\partial U}{\partial V}\right)_T \left(\frac{\partial V}{\partial T}\right)_P dT + \left(\frac{\partial U}{\partial V}\right)_T \left(\frac{\partial V}{\partial P}\right)_T dP + \left(\frac{\partial U}{\partial T}\right)_V dT \tag{2.115}$$

At constant pressure, we may eliminate the second term on the right-hand side:

$$\left(\frac{\partial U}{\partial T}\right)_P = \left(\frac{\partial U}{\partial V}\right)_T \left(\frac{\partial V}{\partial T}\right)_P + \left(\frac{\partial U}{\partial T}\right)_V \tag{2.116}$$

Substitution of this expression into Eq. 2.112 gives

$$C_P - C_V = \left[P + \left(\frac{\partial U}{\partial V}\right)_T\right]\left(\frac{\partial V}{\partial T}\right)_P \qquad (2.117)$$

For an ideal gas, $(\partial U/\partial V)_T = 0$ and thus

$$C_P - C_V = P\left(\frac{\partial V}{\partial T}\right)_P \qquad (2.118)$$

For 1 mol of an ideal gas, $PV_m = RT$ and, therefore, $(\partial V_m/\partial T)_P = R/P$; hence,

$$C_{P,m} - C_{V,m} = R \qquad (2.119)$$

as we proved earlier in Eq. 2.36.

In general, the term $P(\partial V/\partial T)_P$ in Eq. 2.117 represents the contribution to C_P caused by the change in volume of the system acting against the *external* pressure P. The other term, $(\partial U/\partial V)_T\,(\partial V/\partial T)_P$, is an additional contribution due to $(\partial U/\partial V)_T$, which is an effective pressure arising from the attractive or repulsive forces between the molecules; the quantity $(\partial U/\partial V)_T$ is known as the **internal pressure.** In an ideal gas the internal pressure is zero, and in real gases the term is usually small compared to the external pressure P. Liquids and solids, on the other hand, have strong attractive or *cohesive* forces, and the internal pressure $(\partial U/\partial V)_T$ may therefore be large compared to the external pressure.

Key Equations

The First Law of Thermodynamics:

$$\Delta U = q + w$$

where ΔU = change in internal energy; q = heat given to system; w = work done on system.

Reversible PV work done on a gas:

$$w_{\text{rev}} = -\int_{V_1}^{V_2} P\,dV$$

where V_1 = initial volume; V_2 = final volume.

Definition of enthalpy H:

$$H \equiv U + PV$$

Heat capacities:

At constant volume, $C_V = \dfrac{dq_V}{dT} = \left(\dfrac{\partial U}{\partial T}\right)_V$

At constant pressure, $C_P = \frac{dq_P}{dT} = \left(\frac{\partial H}{\partial T}\right)_P$

For an ideal gas:

$$PV = nRT \quad \text{and} \quad \left(\frac{\partial U}{\partial V}\right)_T = 0$$

Isothermal reversible compression of 1 mol of an ideal gas:

$$w_{rev} = -q_{rev} = RT \ln \frac{V_1}{V_2}$$

Problems

A Energy, Heat, and Work

1. A bird weighing 1.5 kg leaves the ground and flies to a height of 75 metres, where it attains a velocity of 20 m s^{-1}. What change in energy is involved in the process? (Acceleration of gravity = 9.81 m s^{-2}).

2. A chemical reaction occurs at 300 K in a gas mixture that behaves ideally, and the total amount of gas increases by 0.27 mol. If ΔU = 9.4 kJ, what is ΔH?

3. The densities of ice and water at 0°C are 0.9168 and 0.9998 g cm^{-3}, respectively. If ΔH for the fusion process at atmospheric pressure is 6.025 kJ mol^{-1}, what is ΔU? How much work is done on the system?

4. The density of liquid water at 100°C is 0.9584 g cm^{-3}, whereas that of steam is 0.000 596 g cm^{-3}. If the enthalpy of evaporation of water at atmospheric pressure is 40.63 kJ mol^{-1}, what is ΔU? How much work is done by the system?

5. The latent heat of fusion of water at 0°C is 6.025 kJ mol^{-1} and the molar heat capacities ($C_{P,m}$) of water and ice are 75.3 and 37.7 J K^{-1} mol^{-1}, respectively. The C_P values can be taken to be independent of temperature. Calculate ΔH for the freezing of 1 mol of supercooled water at −10.0°C.

6. A sample of liquid acetone weighing 0.700 g was burned in a bomb calorimeter for which the heat capacity (including the sample) is 6937 J K^{-1}. The observed temperature rise was from 25.00°C to 26.69°C.

a. Calculate ΔU for the combustion of 1 mol of acetone.

b. Calculate ΔH for the combustion of 1 mol of acetone.

7. A sample of liquid benzene weighing 0.633 g is burned in a bomb calorimeter at 25°C, and 26.54 kJ of heat are evolved.

a. Calculate ΔU per mole of benzene.

b. Calculate ΔH per mole of benzene.

8. An average man weighs about 70 kg and produces about 10 460 kJ of heat per day.

a. Suppose that a man were an isolated system and that his heat capacity is 4.18 J K^{-1} g^{-1}; if his temperature were 37°C at a given time, what would be his temperature 24 h later?

b. Man is in fact an open system, and the main mechanism for maintaining his temperature constant is evaporation of water. If the enthalpy of vaporization of water at 37°C is 43.4 kJ mol^{-1}, how much water needs to be evaporated per day to keep the temperature constant?

9. In an open beaker at 25°C and 1 atm pressure, 100 g of zinc are caused to react with dilute sulfuric acid. Calculate the work done by the liberated hydrogen gas, assuming it to behave ideally. What would be the work done if the reaction took place in a sealed vessel?

10. A chemical reaction is caused to occur in a bulb to which is attached a capillary tube of cross-sectional area 2.5 mm^2. The tube is open to the atmosphere (pressure = 101.3 kPa), and during the reaction the rise in the capillary is 2.4 cm. Calculate the work done by the reaction.

11. When 1 cal of heat is given to 1 g of water at 14.5°C, the temperature rises to 15.5°C. Calculate the molar heat capacity of water at 15°C.

12. A vessel containing 1 kg of water at 25°C is heated until it boils. How much heat is supplied? How long would it take a 1-kW heater to supply this amount of heat? Assume the heat capacity calculated in Problem 11 to apply over the temperature range.

B Thermochemistry

13. Deduce the standard enthalpy change for the process

$$2CH_4(g) \rightarrow C_2H_6(g) + H_2(g)$$

from the data in Table 2.2.

14. A sample of liquid methanol weighing 5.27 g was burned in a (constant volume) bomb calorimeter at 25°C, and 119.50 kJ of heat was evolved (after correction for standard conditions).

a. Calculate $\Delta_c H°$ for the combustion of 1 mol of methanol.

b. Use this value and the data in Table 2.2 for $H_2O(l)$ and $CO_2(g)$ to obtain a value for $\Delta_f H°$ (CH_3OH, l), and compare with the value given in the table.

c. If the enthalpy of vaporization of methanol is 35.27 kJ mol^{-1}, calculate $\Delta_f H°$ for $CH_3OH(g)$.

15. Calculate the heat of combustion ($\Delta_c H°$) of ethane from the data given in Table 2.2.

16. Suggest a practicable method for determining the heat of formation $\Delta_f H°$ of gaseous carbon monoxide at 25°C. (*Note:* Burning graphite in a limited supply of oxygen is *not* satisfactory, since the product will be a mixture of unburned graphite, CO, and CO_2.)

17. If the enthalpy of combustion $\Delta_c H°$ of gaseous cyclopropane, C_3H_6, is −2091.2 kJ mol^{-1} at 25°C, calculate the standard enthalpy of formation $\Delta_f H°$.

18. From the data in Table 2.2, calculate $\Delta H°$ for the reaction

$$C_2H_4(g) + H_2O(l) \longrightarrow C_2H_5OH(l)$$

at 25°C.

19. The bacterium *Acetobacter suboxydans* obtains energy for growth by oxidizing ethanol in two stages, as follows:

(a) $C_2H_5OH(l) + \frac{1}{2}O_2(g) \longrightarrow CH_3CHO(l) + H_2O(l)$

(b) $CH_3CHO(l) + \frac{1}{2}O_2(g) \longrightarrow CH_3COOH(l)$

The enthalpy increases in the complete combustion (to CO_2 and liquid H_2O) of the three compounds are

	$\Delta_c H°$/kJ mol^{-1}
Ethanol (l)	−1370.7
Acetaldehyde (l)	−1167.3
Acetic acid (l)	− 876.1

Calculate the $\Delta H°$ values for reactions **(a)** and **(b)**.

20. Calculate ΔH for the reaction

$$C_2H_5OH(l) + O_2(g) \longrightarrow CH_3CO_2H(l) + H_2O(l)$$

making use of the enthalpies of formation given in Table 2.2. Is the result consistent with the results obtained for Problems 19?

21. The disaccharide α-maltose can be hydrolyzed to glucose according to the equation

$$C_{12}H_{22}O_{11}(aq) + H_2O(l) \longrightarrow 2C_6H_{12}O_6(aq)$$

Using data in Table 2.2 and the following values, calculate the standard enthalpy change in this reaction:

$$\Delta_f H°(C_6H_{12}O_6,\ aq) = -1263.1 \text{ kJ mol}^{-1}$$

$$\Delta_f H°(C_{12}H_{22}O_{11},\ \text{aq}) = -2238.3 \text{ kJ mol}^{-1}$$

22. The standard enthalpy of formation of the fumarate ion is −777.4 kJ mol^{-1}. If the standard enthalpy change of the reaction fumarate$^{2-}(aq)$ + $H_2(g) \longrightarrow$ succinate$^{2-}(aq)$ is −131.4 kJ mol $^{-1}$, calculate the enthalpy of formation of the succinate ion.

23. The $\Delta H°$ for the mutarotation of glucose in aqueous solution,

$$\alpha\text{-D-glucose}(aq) \longrightarrow \beta\text{-D-glucose}(aq)$$

has been measured in a microcalorimeter and found to be −1.16 kJ mol^{-1}. The enthalpies of solution of the two forms of glucose have been determined to be

$$\alpha\text{-D-glucose}(s) \longrightarrow \alpha\text{-D-glucose}(aq) \qquad \Delta H° = 10.72 \text{ kJ mol}^{-1}$$

$$\beta\text{-D-glucose}(s) \longrightarrow \beta\text{-D-glucose}(aq) \qquad \Delta H° = 4.68 \text{ kJ mol}^{-1}$$

Calculate ΔH° for the mutrotation of solid α-D-glucose to solid β-D-glucose.

***24.** From the data in Tables 2.1 and 2.2, calculate the enthalpy change in the reaction

$$C(graphite) + O_2(g) \longrightarrow CO_2(g)$$

at 1000 K.

25. From the bond strengths in Table 2.3, estimate the enthalpy of formation of gaseous propane, C_3H_8, using the following additional data:

$$C(graphite) \longrightarrow C(g) \quad \Delta H^\circ = 716.7 \text{ kJ mol}^{-1}$$

$$H_2(g) \longrightarrow 2H(g) \quad \Delta H^\circ = 436.0 \text{ kJ mol}^{-1}$$

C Ideal Gases

26. In a volume of 11.2 dm^3 at 273 K, 2 mol of oxygen gas, which can be regarded as ideal with $C_p = 29.4$ J K^{-1} mol^{-1} (independent of temperature), are maintained.

a. What is the pressure of the gas?
b. What is PV?
c. What is C_v?

27. Suppose that the gas in Problem 26 is heated reversibly to 373 K at constant volume:

a. How much work is done on the system?
b. What is the increase in internal energy U?
c. How much heat was added to the system?
d. What is the final pressure?
e. What is the final value of PV?
f. What is the increase in enthalpy H?

28. Suppose that the gas in Problem 26 is heated reversibly to 373 K at constant pressure.

a. What is the final volume?
b. How much work is done on the system?
c. How much heat is supplied to the system?
d. What is the increase in enthalpy?
e. What is the increase in internal energy?

29. Suppose that the gas in Problem 26 is reversibly compressed to half its volume at constant T (273 K).

a. What is the change in U?
b. What is the final pressure?
c. How much work is done on the system?
d. How much heat flows out of the system?
e. What is the change in H?

30. With temperature maintained at 0°C, 2 mol of an ideal gas are allowed to expand against a piston that supports 2 atm pressure. The initial pressure of the gas is 10 atm and the final pressure 2 atm.

a. How much energy is transferred to the surroundings during the expansion?
b. What is the change in the internal energy and the enthalpy of the gas?
(c) How much heat has been absorbed by the gas?

31. Suppose that the gas in Problem 30 is allowed to expand *reversibly* and *isothermally* from the initial pressure of 10 atm to the final pressure of 2 atm.

a. How much work is done by the gas?
b. What are ΔU and ΔH?
c. How much heat is absorbed by the gas?

***32.** A sample of hydrogen gas, which may be assumed to be ideal, is initially at 3.0 atm pressure and a temperature of 25.0°C and has a volume of 1.5 dm^3. It is expanded reversibly and adiabatically until the volume is 5.0 dm^3. The heat capacity at constant pressure of H_2 is 28.80 J K^{-1} mol^{-1} and may be assumed to be independent of temperature.

a. Calculate the final pressure and temperature after the expansion.
b. Calculate ΔU and ΔH for the process.

***33.** Initially 0.1 mol of methane is at 1 atm pressure and 80°C. The gas behaves ideally and the value of γ $(= C_p/C_v) = 1.31$. The gas is allowed to expand reversibly and adiabatically to a pressure of 0.1 atm.

a. What are the initial and final volumes of the gas?
b. What is the final temperature?
c. Calculate ΔU and ΔH for the process.

34. A gas behaves ideally and its C_v is given by

$$C_v/\text{J K}^{-1}\text{ mol}^{-1} = 21.52 + 8.2 \times 10^{-3}T/\text{K}$$

a. What is $C_{P,m}$ as a function of T?
b. A sample of this gas is initially at T_1 = 300 K, P_1 = 10 atm, and V_1 = 1 dm^3. It is allowed to expand until P_2 = 1 atm and V_2 = 10 dm^3. What are ΔU and ΔH for this process? Could the process be carried out adiabatically?

35. Prove that for an ideal gas two reversible adiabatic curves on a PV diagram cannot intersect.

36. An ideal gas is defined as one that obeys the relationship $PV = nRT$. We showed in Section 2.7 that for such gases

$$\left(\frac{\partial U}{\partial V}\right)_T = 0 \quad \text{and} \quad \left(\frac{\partial H}{\partial P}\right)_T = 0$$

Prove that for an ideal gas C_v and C_P are independent of volume and pressure.

37. One mole of an ideal gas underwent a *reversible isothermal expansion* until its volume was doubled. If the gas performed 1 kJ of work, what was its temperature?

38. A gas that behaves ideally was allowed to expand *reversibly* and *adiabatically* to twice its volume. Its initial temperature was 25.00°C, $C_{V,m} = \frac{5}{2}R$. Calculate ΔU_m and ΔH_m for the expansion process.

***39.** With $C_{V,m} = \frac{3}{2}R$, 1 mol of an ideal monatomic gas undergoes a reversible process in which the volume is doubled and in which 1 kJ of heat is absorbed by the gas. The initial pressure is 1 atm and the initial temperature is 300 K. The enthalpy change ΔH is 1.5 kJ.

a. Calculate the final pressure and temperature.

b. Calculate ΔU and w for the process.

D Real Gases

***40.** One mole of a gas at 300 K is compressed isothermally and reversibly from an initial volume of 10 dm^3 to a final volume of 0.2 dm^3. Calculate the work done on the system if

a. The gas is ideal.

b. The equation of state of the gas is $P(V - b) = RT$, with $b = 0.03$ dm^3.

Explain the difference between the two values.

***41.** One mole of a gas at 100 K is compressed isothermally from an initial volume of 20 dm^3 to a final volume of 5 dm^3. Calculate the work done on the system if

a. The gas is ideal.

b. The equation of state is

$$\left(P + \frac{3.8\ \text{dm}^6\ \text{atm}}{V^2}\right) V = RT$$

[This equation is obeyed approximately at low temperatures, whereas $P(V - b) = RT$ (see Problem 40) is obeyed more closely at higher temperatures.] Account for the difference between the values in (a) and (b).

E Supplementary Problems

42. For an ideal gas, $PV_m = RT$ and therefore $(\partial T/\partial P)_V = V_m/R$. Derive the corresponding relationship for a van der Waals gas.

43. Prove that

$$C_v = -\left(\frac{\partial U}{\partial V}\right)_T \left(\frac{\partial V}{\partial T}\right)_U$$

44. Prove that for an ideal gas the rate of change of the pressure dP/dt is related to the rates of change of the volume and temperature by

$$\frac{1}{P}\frac{dP}{dt} = -\frac{1}{V}\frac{dV}{dt} + \frac{1}{T}\frac{dT}{dt}$$

45. Derive the expression

$$dP = \frac{P\,dV_m}{V_m - b} - \frac{ab}{V_m^3(V_m - b)}\,dV_m + \frac{P\,dT}{T} + \frac{a\,dT}{V^2T}$$

for 1 mol of a van der Waals gas.

46. If a substance is burned at constant volume with no heat loss, so that the heat evolved is all used to heat the product gases, the temperature attained is known as the *adiabatic flame temperature.* Calculate this quantity for methane burned at 25°C in the amount of oxygen required to give complete combustion to CO_2 and H_2O. Use the data in Table 2.2 and the following approximate expressions for the heat capacities:

$$C_{P,m}(CO_2)/\text{J K}^{-1}\ \text{mol}^{-1} = 44.22 + 8.79 \times 10^{-3}T/\text{K}$$

$$C_{P,m}(H_2O)/\text{J K}^{-1}\ \text{mol}^{-1} = 30.54 + 1.03 \times 10^{-2}T/\text{K}$$

47. Initially 5 mol of nitrogen are at a temperature of 25°C and a pressure of 10 atm. The gas may be assumed to be ideal; $C_{V,m} = 20.8$ J K^{-1} mol^{-1} and is independent of temperature. Suppose that the pressure is *suddenly* dropped to 1 atm; calculate the final temperature, ΔU, and ΔH.

E Essay Questions

1. Explain clearly what is meant by a thermodynamically reversible process. Why is the reversible work done by a system the maximum work?

2. Explain the thermodynamic meaning of a *system*, distinguishing among open, closed, and isolated systems. Which one of these systems is (a) a fish swimming in the sea or (b) an egg?

Suggested Reading

J. A. V. Butler, *The Fundamentals of Chemical Thermodynamics,* London: Macmillan (4th ed.), 1951.

E. F. Caldin, *An Introduction to Chemical Thermodynamics,* Oxford: Clarendon Press, 1961.

I. M. Klotz and R. M. Rosenberg, *Chemical Thermodynamics,* New York: Benjamin, 1972.

B. H. Mahan, *Elementary Chemical Thermodynamics,* New York: W. A. Benjamin, 1963.

C. T. Mortimer, *Reaction Heats and Bond Strengths,* Reading, Mass: Addison-Wesley, 1962.

K. S. Pitzer and L. Brewer, *Thermodynamics,* New York McGraw-Hill: 1961.

E. B. Smith, *Basic Chemical Thermodynamics* (2nd ed.), Oxford: Clarendon Press, 1977.

Chapter 3

The Second and Third Laws of Thermodynamics

Preview

The *Second Law of Thermodynamics* deals with whether a chemical or physical process occurs spontaneously (i.e., naturally). Many processes—such as the flow of heat from a lower to a higher temperature—do not violate the First Law, but nevertheless they do not occur. The Second Law defines the conditions under which processes take place.

The property that is of fundamental importance in this connection is the *entropy*. If a process occurs reversibly from state A to state B, with infinitesimal amounts of heat dq_{rev} absorbed at each stage, the entropy change is defined as

$$\Delta S_{A\to B} \equiv \int_A^B \frac{dq_{\text{rev}}}{T}$$

This relationship provides a way of determining entropy changes from thermal data.

The great importance of the entropy is that if a process $A \to B$ is to occur spontaneously, the total entropy of the system plus the surroundings must *increase*. Entropy provides a measure of *probability;* a process that occurs spontaneously must involve a change from a state of lower probability to one of higher probability.

The *Third Law of Thermodynamics* is concerned with the absolute values of entropies, which are based on the values at the absolute zero. This law is important for chemical and physical studies near the absolute zero, but it is less relevant to most chemical studies, in which entropy *changes* are of prime importance.

It is usually inconvenient to consider the entropy of the surroundings as well as the entropy of the system itself. It has proved possible to define thermodynamic functions that are related only to the *system* but that nevertheless provide information about spontaneity of reaction. For a process occurring in an open vessel (i.e., at *constant pressure*), such a function is the *Gibbs energy* G, defined as equal to $H - TS$. For a system at constant temperature and pressure this function has the property of having a minimum value when the system is at equilibrium. For a reaction to proceed spontaneously, there must be a *decrease* in G. The decrease in G during a process at constant T and P is the amount of work, other than PV work, that can be performed.

For processes occurring in closed vessels (i.e., at constant volume), there is a corresponding function, the *Helmholtz energy* A, equal to $U - TS$. A process at constant volume tends toward a state of lower A, and at equilibrium A has a minimum value.

An important thermodynamic relationship is the *Gibbs-Helmholtz equation,* which is concerned with the effect of *temperature* on the Gibbs energy. In Chapter 4 we will see that this relationship leads to a useful procedure for obtaining enthalpies of reaction from the temperature dependence of the equilibrium constant.

3 The Second and Third Laws of Thermodynamics

The First Law of Thermodynamics is concerned with the conservation of energy and with the interrelationship of work and heat. A second very important problem with which thermodynamics deals is *whether a chemical or physical change can take place spontaneously*. This particular aspect is the concern of the **Second Law of Thermodynamics.**

There are several well-known examples of processes that do not violate the First Law but that do not occur naturally. Suppose, for example, that a cylinder is separated into two compartments by means of a diaphragm (Figure 3.1), with a gas at high pressure on one side and the gas at low pressure on the other. If the diaphragm is ruptured, there will be an *equalization of pressure.* However, the reverse of this process does not occur; if we start with gas at uniform pressure, it is highly unlikely that we will obtain gas at high pressure on one side and gas at low pressure on the other side. The first process, in which the gas pressures equalize, is known as a *natural* or **spontaneous process.** The reverse process, which does not occur but can only be imagined, is known as an *unnatural process.* Note that the natural, spontaneous, process is *irreversible* in the sense in which we used the term in Section 2.4. We could instead equalize gas pressures by reversibly expanding the gas at the higher pressure, in which case it would perform work. The reverse process could then be carried out, but only at the expense of work that we would have to perform on the system. Such a process would not be a spontaneous one.

Another example of a process that occurs naturally and spontaneously in one direction but not in the other is the *equalization of temperature.* Suppose that we bring together a hot solid and a cold one. Heat will pass from the hot to the cold solid until the temperature is equalized; this is a spontaneous process. We know from common experience that heat will not flow in the opposite direction, from a cold to a hot body. There would be no violation of the First Law if this occurred, but there would be a violation of the Second Law.

Our third example relates to a *chemical reaction.* There are many reactions that will go spontaneously in one direction but not in the other. One of them is the reaction between hydrogen and oxygen to form water:

$$2H_2 + O_2 = 2H_2O$$

If we bring together two parts of hydrogen and one of oxygen, we can readily cause their reaction to occur essentially to completion; the amounts of hydrogen and oxy-

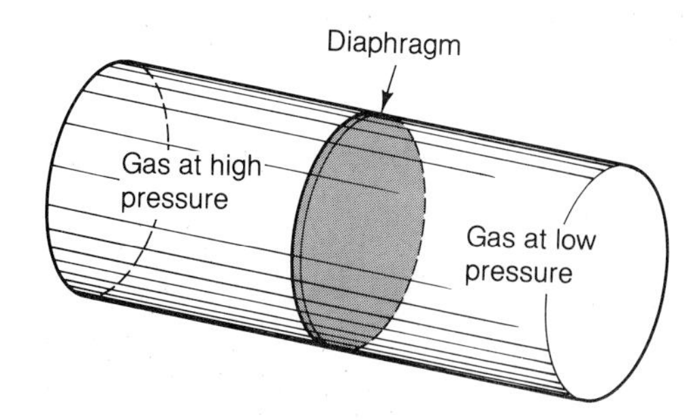

FIGURE 3.1 A cylinder separated into two compartments, with gases at different pressures on the two sides of the diaphragm.

gen that remain are quite undetectable.* The reaction is accompanied by the evolution of considerable heat (ΔH° is negative). There would be no violation of the First Law if we were to return this heat to water and reconvert it into hydrogen and oxygen, but in practice this cannot be done. The reaction from left to right is spontaneous; that from right to left does not occur naturally.

It is obviously a matter of great importance to know what the factors are that determine *the direction in which a process will occur spontaneously.* This amounts to asking what factors determine the *position of equilibrium,* because a system will move spontaneously toward the state of equilibrium. These matters are the concern of the Second Law of Thermodynamics.

3.1 The Carnot Cycle

The arguments relating to the Second Law of Thermodynamics are based upon the **Carnot cycle** of operations; this was first suggested in 1824 by the French physicist Sadi Carnot (1796–1832).† Suppose that we have 1 mol of an ideal gas contained in a cylinder with a piston and having an initial pressure of P_1, an initial volume of V_1, and an initial temperature of T_h (h standing for *hotter*). We refer to this gas as being in state A. Figure 3.2 is a pressure-volume diagram in which is indicated the initial state A.

Now we will bring about four reversible changes in the system, which will eventually bring it back to the initial state A. We first bring about an *isothermal* expansion $A \rightarrow B$, the pressure and volume changing to P_2 and V_2 and the temperature remaining at T_h; we can imagine the cylinder to be immersed in a bath of liquid at temperature T_h. Second, we bring about an adiabatic expansion (i.e., one in which no heat is allowed to leave or enter the system); we could accomplish this by surrounding the cylinder with insulating material. Since the gas does work during expansion, and no heat is supplied, the temperature must fall; we call the final temperature T_c (c for *cooler*) and the pressure and volume P_3 and V_3, respectively. Third, we compress the gas isothermally (at temperature T_c) until the pressure and volume are P_4 and V_4. Finally, the gas is compressed adiabatically until it returns to its original state A (P_1, V_1, T_h). The performance of work on the system, with no heat transfer permitted, raises the temperature from T_c to T_h.

Let us consider these four steps in further detail. In particular we want to know the ΔU values for each step, the amounts of heat absorbed (q), and the work done (w). The expressions for these quantities are summarized in Table 3.1.

1. *Step $A \rightarrow B$* is the reversible isothermal expansion at T_h. In Section 2.7 we proved that for the isothermal expansion of an ideal gas there is no change of internal energy:

$$\Delta U_{A\rightarrow B} = 0 \qquad (3.1)$$

---

*The reaction is exceedingly slow at room temperature, but it can be brought about by various catalysts, such as platinum; by passing a spark through the mixture; or by a small flame.

†S. Carnot, *Reflexions sur la Puissance Motrice du Feu,* Paris, 1824. The modern form and interpretation of the Carnot cycle are quite different from those of Carnot, who was unaware of the equivalence of heat and work. For an historical account, see M. Kerker, *Sci. Monthly, 85,* 143 (1957), and S. S. Wilson, *Scientific American, 245,* 134 (1981).

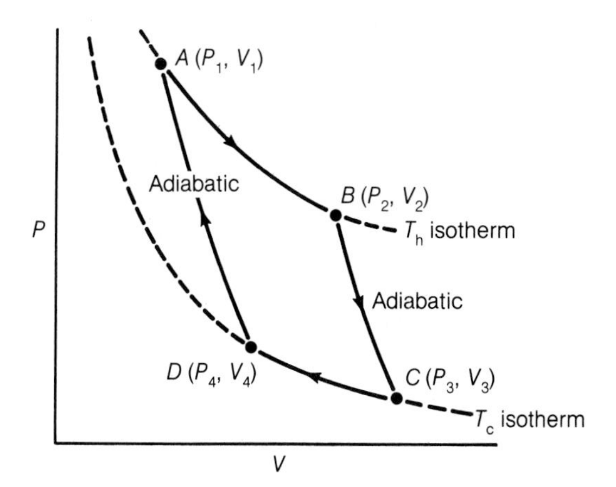

FIGURE 3.2 Pressure-volume diagram for the Carnot cycle; AB and CD are isotherms, and BC and DA are adiabatics (no heat transfer).

We also showed (Eq. 2.72) that the work done on the system in an isothermal reversible process is $RT \ln(V_{\text{initial}}/V_{\text{final}})$:

$$w_{A\to B} = RT_h \ln \frac{V_1}{V_2} \quad \text{(for 1 mol)} \tag{3.2}$$

Since by the First Law

$$\Delta U_{A\to B} = q_{A\to B} + w_{A\to B} \tag{3.3}$$

it follows that

$$q_{A\to B} = RT_h \ln \frac{V_2}{V_1} \tag{3.4}$$

TABLE 3.1 Values of ΔU, q, and w for the Four Reversible Steps in the Carnot Cycle

Step	ΔU	q_{rev}	w_{rev}
$A \to B$	0	$RT_h \ln \frac{V_2}{V_1}$	$RT_h \ln \frac{V_1}{V_2}$
$B \to C$	$C_v(T_c - T_h)$	0	$C_v(T_c - T_h)$
$C \to D$	0	$RT_c \ln \frac{V_4}{V_3}$	$RT_c \ln \frac{V_3}{V_4}$
$D \to A$	$C_v(T_h - T_c)$	0	$C_v(T_h - T_c)$
Net	0	$R(T_h - T_c) \ln \frac{V_2}{V_1}$ $\left(\text{since } \frac{V_1}{V_2} = \frac{V_4}{V_3}\right)$	$R(T_h - T_c) \ln \frac{V_1}{V_2}$

2. *Step* $B \rightarrow C$ involves surrounding the cylinder by an insulating jacket and allowing the system to expand reversibly and adiabatically to a volume of V_3. Since the process is adiabatic,

$$q_{B \rightarrow C} = 0 \tag{3.5}$$

We saw in Eq. 2.29 that ΔU for an adiabatic process involving 1 mol of gas is

$$\Delta U = C_v(T_{\text{final}} - T_{\text{initial}}) \tag{3.6}$$

so that for the process $B \rightarrow C$

$$\Delta U_{B \rightarrow C} = C_v(T_c - T_h) \tag{3.7}$$

Application of the First Law then gives

$$w_{B \rightarrow C} = C_v(T_c - T_h) \tag{3.8}$$

3. *Step* $C \rightarrow D$ involves placing the cylinder in a heat bath at temperature T_c and compressing the gas reversibly until the volume and pressure are V_4 and P_4, respectively. The state D must lie on the adiabatic that passes through A (Figure 3.2). Since process $C \rightarrow D$ is isothermal,

$$\Delta U_{C \rightarrow D} = 0 \tag{3.9}$$

The work done on the system is

$$w_{C \rightarrow D} = RT_c \ln \frac{V_3}{V_4} \tag{3.10}$$

and this is a positive quantity since $V_3 > V_4$ (i.e., we must do work to compress the gas). By the First Law,

$$q_{C \rightarrow D} = RT_c \ln \frac{V_4}{V_3} \tag{3.11}$$

which means that a negative amount of heat is absorbed (i.e., heat is actually rejected).

4. The gas is finally compressed reversibly and adiabatically from D to A. The heat absorbed is zero:

$$q_{D \rightarrow A} = 0 \tag{3.12}$$

The $\Delta U_{D \rightarrow A}$ value is

$$\Delta U_{D \rightarrow A} = C_v(T_h - T_c) \tag{3.13}$$

By the First Law,

$$w_{D \rightarrow A} = C_v(T_h - T_c) \tag{3.14}$$

Table 3.1 gives, as well as the individual contributions, the net contributions for the entire cycle. We see that ΔU for the cycle is zero; the contributions for the isotherms are zero, whereas those for the adiabatics are equal and opposite to each other. This result that ΔU is zero for the entire cycle is necessary in view of the fact that the internal energy is a state function; in completing the cycle the system is returned to its original state, and the internal energy is therefore unchanged.

Equation 2.90 applies to an adiabatic process, and if we apply this equation to the two processes $B \rightarrow C$ and $D \rightarrow A$ in Figure 3.2, we have

$$\frac{T_h}{T_c} = \left(\frac{V_4}{V_1}\right)^{\gamma-1} \quad \text{and} \quad \frac{T_h}{T_c} = \left(\frac{V_3}{V_2}\right)^{\gamma-1} \tag{3.15}$$

and therefore

$$\frac{V_4}{V_1} = \frac{V_3}{V_2} \tag{3.16}$$

The net q_{rev} value (see Table 3.1) is

$$q_{\text{rev}} = RT_h \ln \frac{V_2}{V_1} + RT_c \ln \frac{V_4}{V_3} \tag{3.17}$$

and since $V_4/V_3 = V_1/V_2$, this becomes

$$q_{\text{rev}} = R(T_h - T_c) \ln \frac{V_2}{V_1} \tag{3.18}$$

which is a positive quantity. Since $q = -w$, by the First Law, it follows that

$$w_{\text{rev}} = R(T_h - T_c) \ln \frac{V_1}{V_2} \tag{3.19}$$

This is a negative quantity (i.e., a positive amount of work has been done by the system).

Note that the net work done *by* the system, and hence the net heat absorbed, is represented by the area within the Carnot diagram. This is illustrated in Figure 3.3. Diagram (a) shows the processes $A \rightarrow B$ and $B \rightarrow C$; both are expansions, and the work done by the system is represented by the area below the lines, which is shaded. Diagram (b) shows the processes $C \rightarrow D$ and $D \rightarrow A$, in which work is done on the system in the amount shown by the shaded area. The *net* work done

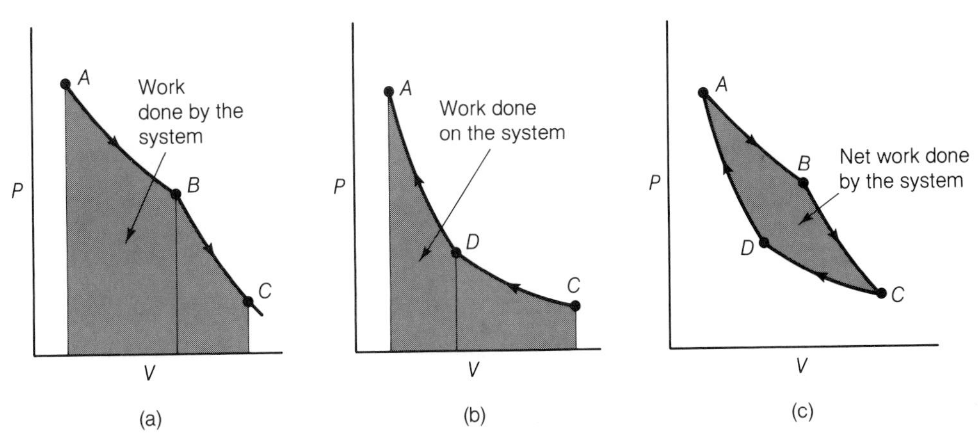

FIGURE 3.3 Diagram (a) shows the work done *by* the system in going from $A \rightarrow B$ and then from $B \rightarrow C$; (b) shows the work done *on* the system in the return process, via D. The *net* work, obtained by subtracting the shaded area in (b) from that in (a), is thus the area enclosed by the cycle (c).

by the system is thus represented by the area in (a) minus the area in (b) and is thus the area within the cycle shown in (c).

The important thing to note about the Carnot cycle is that we have returned the system to its original state by processes in the course of which a net amount of work has been done by the system. This work has been performed at the expense of heat absorbed, as required by the First Law. Since work and heat are not state functions, net work can be done even though the system returns to its original state.

Efficiency of a Reversible Carnot Engine

The *efficiency* of the reversible Carnot engine can be defined as *the work done by the system* during the cycle, *divided by the work that would have been done if all the heat absorbed at the higher temperature had been converted into work.* Thus

$$\text{efficiency} = \frac{w}{q_h} = \frac{R(T_h - T_c)\ln(V_2/V_1)}{RT_h \ln(V_2/V_1)} \tag{3.20}$$

$$= \frac{T_h - T_c}{T_h} \tag{3.21}$$

This efficiency is unity (i.e., 100%) only if the lower temperature T_c is zero (i.e., if the heat is rejected at the absolute zero). This result provides a definition of the absolute zero. The efficiency is also the net heat absorbed, $q_h + q_c$, divided by the heat absorbed at the higher temperature, q_h:

$$\text{efficiency} = \frac{q_h + q_c}{q_h} \tag{3.22}$$

From Eqs. 3.21 and 3.22 it follows that, for the reversible engine,

$$\frac{T_h - T_c}{T_h} = \frac{q_h + q_c}{q_h} \quad \text{or} \quad -\frac{T_h}{T_c} = \frac{q_h}{q_c} \tag{3.23}$$

Carnot's Theorem

Carnot's cycle was discussed previously for an ideal gas. Similar reversible cycles can be performed with other materials, including solids and liquids, and the efficiencies of these cycles determined. The importance of the reversible cycle for the ideal gas is that it gives us an extremely simple expression for the efficiency, namely $(T_h - T_c)/T_h$. We will now show, by a theorem due to Carnot, that *the efficiency of all reversible cycles operating between the temperatures T_h and T_c is the same,* namely $(T_h - T_c)/T_h$. This result leads to important quantitative formulations of the Second Law of Thermodynamics.

Carnot's theorem employs the method of *reductio ad absurdum*. It supposes that there are two reversible engines that operate between T_h and T_c but have different efficiencies, and then it shows that this would lead to the possibility of heat flowing from a lower to a higher temperature, which is contrary to experience. Suppose that there are two engines A and B operating reversibly between T_h and T_c and that A has a higher efficiency than B. By making B use a larger amount of material, it is possible to arrange for it to do exactly the same amount of work in a cycle. We can then imagine the engines to be coupled together so that A forces B to work in reverse, so that it rejects heat at T_h and absorbs it at T_c. During each complete cycle the more efficient engine A draws a quantity of heat q_h from the heat reservoir at T_h and rejects $-q_c$ at T_h. The less efficient engine B, which is being driven backward, rejects q_h' at T_h and absorbs $-q_c'$ at T_c. (If it were operating in its normal manner, it would absorb q_h' at T_h and reject $-q_c'$ at T_c.)

Since the engines were adjusted to perform equal amounts of work in a cycle, engine A only just operates B in reverse, with no extra energy available. The work performed by A is $q_h + q_c$, while that performed *on* B is $q_h' + q_c'$, and these quantities are equal:

$$q_h + q_c = q_h' + q_c' \tag{3.24}$$

Since engine A is more efficient than B,

$$\frac{q_h + q_c}{q_h} > \frac{q_h' + q_c'}{q_h'} \tag{3.25}$$

From Eqs. 3.24 and 3.25 it follows that

$$q_h' > q_h \tag{3.26}$$

and

$$q_c > q_c' \tag{3.27}$$

During the operation of the cycle, in which A has forced B to work backward, A has absorbed q_h at T_h and B has rejected q_h' at T_h; the combined system $A + B$ has therefore absorbed $q_h - q_h'$ at T_h, and since q_h' is greater than q_h (Eq. 3.26), this is a negative quantity (i.e., the system has rejected a positive amount of heat at the higher temperature). At the lower temperature T_c, A has absorbed q_c while B has absorbed $-q_c'$; the combined system has therefore absorbed $q_c - q_c'$ at this temperature, and this according to Eq. 3.27 is a positive quantity (it is in fact equal to $q_h' - q_h$ by Eq. 3.24). The $A + B$ system has thus, in performing the cycle, absorbed heat at a lower temperature and rejected it at a higher temperature. It is contrary to experience that heat can flow uphill in this way, during a complete cycle of operations in which the system ends up in its initial state. It must therefore be concluded that the original postulate is invalid; there cannot be two reversible engines, A and B, operating reversibly between two fixed temperatures and having different efficiencies. Thus, the efficiencies of all reversible engines must be the same as that for the ideal gas reversible engine, namely

$$\frac{T_h - T_c}{T_h}$$

This conclusion is not a necessary consequence of the First Law of Thermodynamics. If engine A were more efficient than engine B, energy would not have been created nor destroyed if engine A drove engine B backward, because the net work done would be equivalent to the heat extracted from the reservoir. However, the removal of heat from a reservoir and its conversion into work without any other changes in the system has never been observed. If this could occur, it would be possible, for example, for a ship to propel itself by removing heat from the surrounding water and converting it into work; no fuel would be needed. Such a continuous extraction of work from the environment has been called *perpetual motion of the second kind*. We have seen that the First Law forbids perpetual motion of the first kind, which involves the creation of energy from nothing. The Second Law forbids the operation of engines in which energy is continuously extracted from the environment. We shall see later, when we consider *heat pumps* (p. 132), that it is possible to extract some heat from a cooler environment provided that some work is done by the system at the same time.

The Thermodynamic Scale of Temperature

In our consideration of the Carnot cycle we have made use of the ideal gas law $PV = nRT$ in order to write expressions such as $RT_h \ln(V_1/V_2)$ (see Table 3.1) for the heat and work contributions in the isothermal steps. This led us to the useful Eq. 3.23, which we obtained by equating the alternative expressions (Eqs. 3.21 and 3.22) for the efficiency.

Equation 3.22 can be arrived at without use of the expressions for q_h and q_c; Eq. 3.22 merely states that the efficiency is the net heat absorbed (i.e., the work done) divided by the work that would have been done if all the heat absorbed at the higher temperature had been converted into work, with no heat returned to the cooler reservoir. If we proceed in this way, we can *define* an absolute temperature scale by saying that the ratio of the two absolute temperatures is equal to the ratio of the heat absorbed to the heat rejected in the operation of a reversible Carnot cycle. In other words, if we have two reservoirs at different temperatures and want to know the ratio of the temperatures, we operate a reversible Carnot engine employing these two reservoirs and measure its efficiency, which we *postulate* to be $(T_h - T_c)/T_h$; we can thus calculate the ratio T_h/T_c. Since, by Carnot's theorem, all reversible engines operating between two temperatures have the same efficiency, it does not matter what substances we use in our reversible Carnot engine. This absolute temperature scale therefore does not rely on the properties of any substance, such as the hypothetical ideal gas.

This definition of the absolute scale of temperature based on the Carnot cycle was first suggested by Lord Kelvin, who showed that his thermodynamic scale is identical to the ideal gas scale.

The Generalized Cycle: The Concept of Entropy

From Eq. 3.23 for the efficiency of a reversible Carnot cycle, it follows that

$$\frac{q_h}{T_h} + \frac{q_c}{T_c} = 0 \tag{3.28}$$

This equation applies to any reversible cycle that has distinct isothermal and adiabatic parts, and it can be put into a more general form to apply to any reversible cycle. Consider the cycle represented by ABA in Figure 3.4. During the operation of the engine from A to B and back to A, there is heat exchange between the system and its environment at various temperatures. The whole cycle may be split up into elements, such as ab shown in the figure. The distance between a and b should be infinitesimally small, but it is enlarged in the diagram. During the change from a to b, the pressure, volume, and temperature have all increased, and a quantity of heat has been absorbed. Let the temperature corresponding to a be T and that corresponding to b be $T + dT$. The isothermal corresponding to $T + dT$ is shown as bc

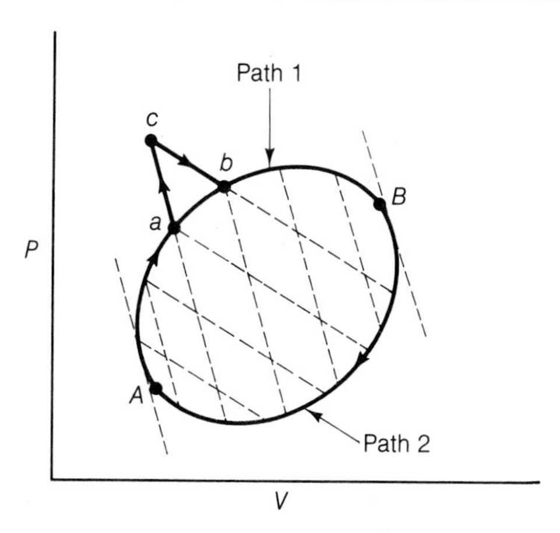

FIGURE 3.4 A generalized cycle ABA. The cycle can be traversed via infinitesimal adiabatics and isothermals; thus we can go from a to b by the adiabatic ac followed by the isothermal cb.

and the adiabatic at a as ac; the two intersect at c. The change from a to b may therefore be carried out by means of an adiabatic change ac, followed by an isothermal change cb. During the isothermal process an amount of heat dq has been absorbed by the system. If the whole cycle is completed in this manner, quantities of heat dq_1, dq_2, etc., will have been absorbed by the system during isothermal changes carried out at the temperatures T_1, T_2, etc. Equation 3.28 is therefore replaced by

$$\sum_i \frac{dq_i}{T_i} = \frac{dq_1}{T_1} + \frac{dq_2}{T_2} + \frac{dq_3}{T_3} + \cdots = 0 \tag{3.29}$$

the summation being made round the entire cycle. Since the ab elements are all infinitesimal, the cycle consisting of reversible isothermal and adiabatic steps is equivalent to the original cycle $A \rightarrow B \rightarrow A$. It follows that for the original cycle

$$\oint \frac{dq_{\text{rev}}}{T} = 0 \tag{3.30}$$

where the symbol $\oint$ denotes integration over a complete cycle.

This result that the integral of dq_{rev}/T over the entire cycle is equal to zero is a very important one. We have seen that certain functions are *state functions,* which means that their value is a true property of the system; the change in a state function when we pass from state A to state B is independent of the path, and therefore a state function will not change when we traverse a complete cycle. For example, pressure P, volume V, temperature T, and internal energy U, are state functions; thus we can write

$$\oint dP = 0; \qquad \oint dV = 0; \qquad \oint dT = 0; \qquad \oint dU = 0 \tag{3.31}$$

The relationship expressed in Eq. 3.30 is therefore an important one, and it is convenient to write dq_{rev}/T as dS, so that

$$\oint \frac{dq_{\text{rev}}}{T} \equiv \oint dS = 0 \tag{3.32}$$

The property S is known as the *entropy* of the system, and it is a state function. The word **entropy** was coined in 1854 by the German physicist Rudolf Julius Emmanuel Clausius* (1822–1888); the word literally means *to give a direction* (Greek *en*, in; *trope*, change, transformation).

Since entropy is a state function, its value is independent of the path by which the state is reached. Thus if we consider a reversible change from state A to state B and back again (Figure 3.4), it follows from Eq. 3.32 that

$$\int_A^B dS + \int_B^A dS = 0 \tag{3.33}$$

or

$$\Delta S^{(1)}_{A\to B} + \Delta S^{(2)}_{B\to A} = 0 \tag{3.34}$$

where $\Delta S^{(1)}_{A\to B}$ denotes the change of entropy in going from A to B by path (1) and $\Delta S^{(2)}_{B\to A}$ is the change in going from B to A by path (2). The change of entropy in going from B to A by path (2) is the negative of the change in going from A to B by this path,

$$\Delta S^{(2)}_{B\to A} = -\Delta S^{(2)}_{A\to B} \tag{3.35}$$

It therefore follows from Eqs. 3.34 and 3.35 that

$$\Delta S^{(1)}_{A\to B} = \Delta S^{(2)}_{A\to B} \tag{3.36}$$

That is, the change of entropy is the same whatever path is followed.

3.2 Irreversible Processes

The treatment of thermodynamically reversible processes is of great importance in connection with the Second Law. However, in practice we are concerned with thermodynamically irreversible processes, because these are the processes that occur naturally. It is therefore important to consider the relationships that apply to irreversible processes.

A simple example of an irreversible process is the transfer of heat from a warmer to a colder body. Suppose that we have two reservoirs, a warm one at a temperature T_h and a cooler one at a temperature T_c. We might imagine connecting these together by a metal rod, as shown in Figure 3.5a, and waiting until an amount of heat q has flowed from the hotter to the cooler reservoir. To simplify the argument, let us suppose that the reservoirs are so large that the transfer of heat does not change their temperatures appreciably.

In order to calculate the entropy changes in the two reservoirs after this irreversible process has occurred, we must devise a way of transferring the heat reversibly. We can make use of an ideal gas to carry out the heat transfer process, as shown in Figure 3.5b. The gas is contained in a cylinder with a piston, and we first place it in the warm reservoir, at temperature T_h, and expand it reversibly and isothermally until it has taken up heat equal to q. The gas is then removed from the

*R. J. E. Clausius, *Annalen der Physik*, 93, 481 (1854).

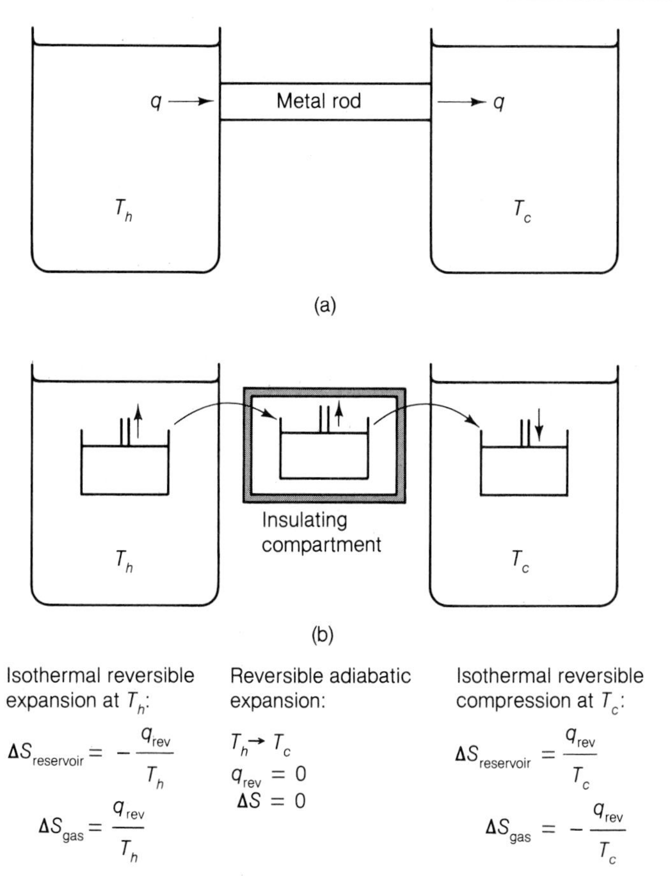

FIGURE 3.5 The transfer of heat q from a hot reservoir at temperature T_h to a cooler one at temperature T_c: (a) irreversible transfer through a metal rod; (b) reversible transfer by use of an ideal gas.

hot reservoir, placed in an insulating container, and allowed to expand reversibly and adiabatically until its temperature has fallen to T_c. Finally, the gas is placed in contact with the colder reservoir at T_c and compressed isothermally until it has given up heat equal to q.

The entropy changes that have occurred in the two reservoirs and in the gas are shown in Figure 3.5. We see that the two reservoirs have experienced a net entropy change of

$$\Delta S_{\text{reservoirs}} = -\frac{q}{T_h} + \frac{q}{T_c} \tag{3.37}$$

and this is a positive quantity since $T_h > T_c$. The gas has experienced an exactly equal and opposite entropy change:

$$\Delta S_{\text{gas}} = \frac{q}{T_h} - \frac{q}{T_c} \tag{3.38}$$

There is thus no overall entropy change, as is necessarily the case for reversible changes in an isolated system.

On the other hand, for the irreversible change in which the reservoirs are in thermal contact (Figure 3.5a) there is no compensating entropy decrease in the expansion of an ideal gas. The entropy increase in the two reservoirs is the same as for the reversible process (Eq. 3.37), and this is the overall entropy increase.

This result, that a spontaneous (and therefore irreversible) process occurs with an overall increase of entropy in the system plus its surroundings, is universally true. The proof of this is based on the fact that the efficiency of a Carnot cycle in which some of the steps are irreversible must be less than that of a purely reversible cycle, since the maximum work is performed by systems that are undergoing reversible processes. Thus in place of Eq. 3.23 we have, for an irreversible cycle,

$$\frac{q_h^{\text{irr}} + q_c^{\text{irr}}}{q_h^{\text{irr}}} < \frac{T_h - T_c}{T_h} \tag{3.39}$$

This relationship reduces to

$$\frac{q_h^{\text{irr}}}{T_h} + \frac{q_c^{\text{irr}}}{T_c} < 0 \tag{3.40}$$

so that in general, for any cycle that is not completely reversible,

$$\oint \frac{dq_{\text{irr}}}{T} < 0 \tag{3.41}$$

This is known as the **inequality of Clausius,** after R. J. E. Clausius,* who suggested this relationship in 1854.

Consider an irreversible change from state A to state B in an *isolated* system, as represented by the dashed line in Figure 3.6. Suppose that the conditions are then changed; the system is no longer isolated. It is finally returned to its initial state A by a reversible path represented by the solid line in Figure 3.6. During this reversible process the system is not isolated and can exchange heat and work with the environment. Since the entire cycle $A \rightarrow B \rightarrow A$ is in part reversible, Eq. 3.41 applies, which means that

$$\int_A^B \frac{dq_{\text{irr}}}{T} + \int_B^A \frac{dq_{\text{rev}}}{T} < 0 \tag{3.42}$$

The first integral is equal to zero since the system was isolated during the irreversible process, so that any heat change in one part of the system is exactly compensated by an equal and opposite change in another part. The second integral is the entropy change when the process $B \rightarrow A$ occurs, so that

$$\Delta S_{B \rightarrow A} < 0 \tag{3.43}$$

It thus follows that

$$\Delta S_{A \rightarrow B} > 0 \tag{3.44}$$

The entropy of the final state B is thus always greater than that of the initial state A if the process $A \rightarrow B$ occurs irreversibly in an isolated system.

*R. J. E. Clausius, *Annalen der Physik, 93,* 481 (1854).

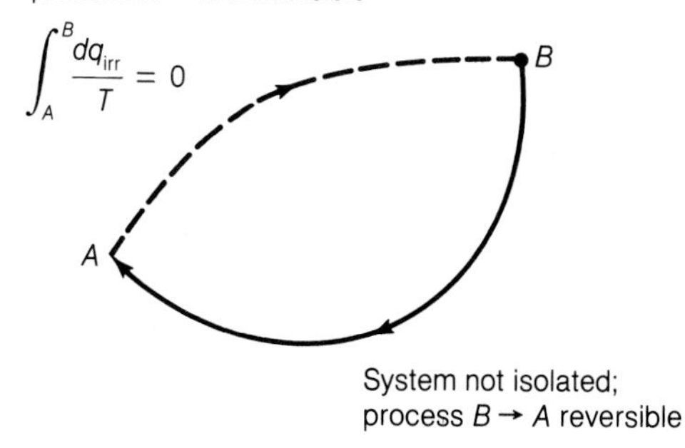

FIGURE 3.6 A cycle process in two stages: (1) The system is isolated and changes its state from A to B by an irreversible process (dashed line); (2) The system is not isolated and changes its state by a reversible process (solid line).

Any change that occurs in nature is spontaneous and is therefore accompanied by a net increase in entropy. This conclusion led Clausius to his famous concise statement of the laws of thermodynamics:

The energy of the universe is a constant; the entropy of the universe tends always towards a maximum.

3.3 The Calculation of Entropy Changes

We have seen that entropy is a state function, which means that an entropy change $\Delta S_{A \rightarrow B}$ when a system changes from state A to state B is independent of the path. This entropy change is given by

$$\Delta S_{A \rightarrow B} = \int_A^B \frac{dq_{\text{rev}}}{T} \tag{3.45}$$

for the transition $A \rightarrow B$ by a reversible path. This integral is the negative of the second integral in Eq. 3.42, so that we have

$$\int_A^B \frac{dq_{\text{irr}}}{T} < \Delta S_{A \rightarrow B} \tag{3.46}$$

That is, whereas $\Delta S_{A \rightarrow B}$ is independent of the path, the integral

$$\int_A^B \frac{dq}{T}$$

may be *equal to or less than* the entropy change: It is equal to the entropy change if the process is reversible and less than the entropy change if the process is irreversible.

It follows that if a system changes from state A to state B by an irreversible process, we cannot calculate the entropy change from the heat transfers that occur. Instead we must contrive a way of bringing about the change by purely reversible processes. Some examples of how this is done will now be considered.

Changes of State of Aggregation

The melting of a solid and the vaporization of a liquid are examples of changes of state of aggregation. If we keep the pressure constant, a solid will melt at a fixed temperature, the melting point T_m at which solid and liquid are at equilibrium. As long as both solid and liquid are present, heat can be added to the system without changing the temperature; the heat absorbed is known as the *latent heat of melting* (or *fusion*) $\Delta_{fus}H$ of the solid. Since the change occurs at constant pressure, this heat is an enthalpy change and is the difference in enthalpy between liquid and solid. Thus,

$$\Delta_{fus}H = H_{liquid} - H_{solid} \tag{3.47}$$

It is easy to heat a solid sufficiently slowly at its melting point that the equilibrium between liquid and solid is hardly disturbed. The process is therefore reversible, since it follows a path of successive equilibrium states, and the latent heat of melting is thus a reversible heat. Since the temperature remains constant, the integral becomes simply the heat divided by the temperature:

$$\int_A^B \frac{dq_{rev}}{T} = \frac{q_{rev}^{(A \to B)}}{T} \tag{3.48}$$

The entropy of melting is thus

$$\Delta_{fus}S = \frac{\Delta_{fus}H}{T_{fus}} \tag{3.49}$$

For example, $\Delta_{fus}H$ for ice is 6.02 kJ mol^{-1} and the melting point is 273.15 K, so that

$$\Delta_{fus}S = \frac{6020 \text{ J mol}^{-1}}{273.15 \text{ K}} = 22.0 \text{ J K}^{-1} \text{ mol}^{-1}$$

The entropy of vaporization can be dealt with in the same way, since when a liquid is vaporized without any rise in temperature, the equilibrium between liquid and vapor remains undisturbed. Thus for water at 100°C,

$$\Delta_{vap}S = \frac{\Delta_{vap}H}{T_{vap}} = \frac{40\,600 \text{ J mol}^{-1}}{373.15 \text{ K}} = 108.8 \text{ J K}^{-1} \text{ mol}^{-1}$$

The same procedure can be used for a transition from one allotropic form to another, provided that the process occurs at a temperature and pressure at which the two forms are in equilibrium. Gray tin and white tin, for example, are in equilibrium at 1 atm pressure and 286.0 K, and $\Delta_{trs}H = 2.09$ kJ mol^{-1}. The entropy change is thus

$$\Delta_{trs}S = \frac{\Delta_{trs}H}{T_{trs}} = \frac{2090 \text{ J mol}^{-1}}{286.0 \text{ K}} = 7.31 \text{ J K}^{-1} \text{ mol}^{-1}$$

Ideal Gases

A particularly simple process is the isothermal expansion of an ideal gas. Suppose that an ideal gas changes its volume from V_1 to V_2 at constant temperature. In order to calculate the entropy change we must consider the reversible expansion; since entropy is a state function, ΔS is the same however the isothermal expansion from V_1 to V_2 occurs.

We have seen (Eq. 2.74) that if n mol of an ideal gas undergoes a reversible isothermal expansion, at temperature T, from volume V_1 to volume V_2, the heat absorbed is

$$q_{rev} = nRT \ln \frac{V_2}{V_1} \tag{3.50}$$

Since the temperature is constant, ΔS is simply the reversible heat absorbed divided by the temperature:

$$\Delta S = nR \ln \frac{V_2}{V_1} \tag{3.51}$$

If a volume change occurs in an ideal gas with a change in temperature, we proceed as follows. Suppose that the volume changes from V_1 to V_2 and that the temperature changes from T_1 to T_2. Again, we imagine a reversible change, knowing that ΔS will be the same whether the change is reversible or not:

$$dq^{rev} = dU + P\,dV \tag{3.52}$$

$$= C_V\,dT + \frac{nRT\,dV}{V} \tag{3.53}$$

$$= nC_{V,m}\,dT + \frac{nRT\,dV}{V} \tag{3.54}$$

Then

$$dS = \frac{dq^{rev}}{T} = nC_{V,m}\frac{dT}{T} + nR\frac{dV}{V} \tag{3.55}$$

Integration leads to

$$\Delta S = S_2 - S_1 = n\int_{T_1}^{T_2} C_{V,m}\,dT/T + nR\int_{V_1}^{V_2} dV/V \tag{3.56}$$

If $C_{V,m}$ is independent of temperature,

$$\Delta S = nC_{V,m} \ln \frac{T_2}{T_1} + nR \ln \frac{V_2}{V_1} \tag{3.57}$$

If the temperature of a gas is changed by a substantial amount, it will be necessary to take into account the variation of the heat capacity with the temperature. The procedure is best illustrated by an example.

EXAMPLE
A mole of hydrogen gas is heated from 300 K to 1000 K at constant volume. The gas may be treated as ideal with

$$C_{P,m}/\text{J K}^{-1}\text{ mol}^{-1} = 27.28 + 3.26 \times 10^{-3}T/\text{K} + 5.0 \times 10^4(T/K)^{-2}$$

Calculate the entropy change.

SOLUTION
From Eq. 3.55, with V constant and $n = 1$,

$$\Delta S_m = \int_{300}^{1000} \frac{C_{V,m}}{T}\, dT$$

But

$$C_{V,m} = C_{P,m} - R$$
$$= (18.97 + 3.26 \times 10^{-3}T/\text{K} + 5.0 \times 10^4(T/\text{K})^{-2}\text{ J K}^{-1}\text{ mol}^{-1}$$

Therefore

$$\Delta S/\text{J K}^{-1}\text{ mol}^{-1} = \int_{300}^{1000} \left(\frac{18.97}{T/\text{K}} + 3.26 \times 10^{-3} + 5.0 \times 10^4 T^{-3}/\text{K}^{-3}\right) dT$$
$$= 18.97 \ln \frac{1000}{300} + (3.26 \times 10^{-3} \times 700) - 2.5 \times 10^4$$
$$\times \left(\frac{1}{1000^2} - \frac{1}{300^2}\right)$$

$$\therefore \quad \Delta S = 22.84 + 2.28 + 0.25 = 25.37\text{ J K}^{-1}\text{ mol}^{-1}$$ ■

Entropy of Mixing

Suppose that we have two ideal gases at equal pressures, separated by a partition as shown in Figure 3.7. If the partition is removed, the gases will mix with no change of temperature. If we want to calculate the entropy change, we must imagine the mixing to occur reversibly. This can be done by allowing the first gas to expand reversibly from its initial volume V_1 to the final volume $V_1 + V_2$; the entropy change is

$$\Delta S_1 = n_1 R \ln \frac{V_1 + V_2}{V_1} \tag{3.58}$$

Similarly, for the second gas

$$\Delta S_2 = n_2 R \ln \frac{V_1 + V_2}{V_2} \tag{3.59}$$

Since the pressures and temperatures are the same,

$$\frac{n_1}{V_1} = \frac{n_2}{V_2} \tag{3.60}$$

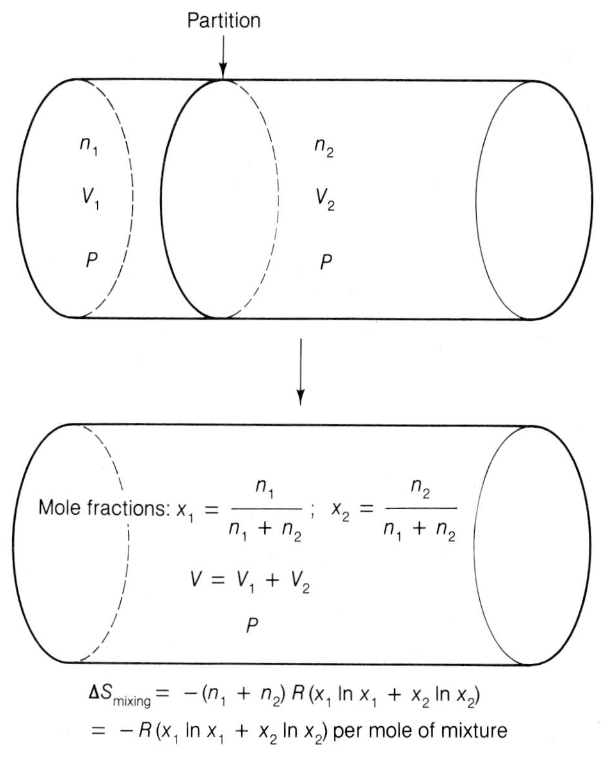

FIGURE 3.7 The mixing of ideal gases at equal pressures and temperatures.

and the mole fractions x_1 and x_2 of the two gases in the final mixture are

$$x_1 = \frac{n_1}{n_1 + n_2} = \frac{V_1}{V_1 + V_2} \tag{3.61}$$

and

$$x_2 = \frac{n_2}{n_1 + n_2} = \frac{V_2}{V_1 + V_2} \tag{3.62}$$

The total entropy change $\Delta S_1 + \Delta S_2$ is thus

$$\Delta S = n_1 R \ln \frac{1}{x_1} + n_2 R \ln \frac{1}{x_2} \tag{3.63}$$

$$= -(n_1 + n_2) R (x_1 \ln x_1 + x_2 \ln x_2) \tag{3.64}$$

The entropy change per mole of mixture is thus

$$\Delta S = -R(x_1 \ln x_1 + x_2 \ln x_2) \tag{3.65}$$

For any number of gases, initially at the same pressure, the entropy change per mole of mixture is

$$\Delta S = -R(x_1 \ln x_1 + x_2 \ln x_2 + x_3 \ln x_3 + \cdots) \tag{3.66}$$

If the initial pressures of two ideal gases are not the same, Eqs. 3.58 and 3.59 are still applicable, and the entropy increase is

$$\Delta S = n_1 R \ln\left(\frac{V_1 + V_2}{V_1}\right) + n_2 R \ln\left(\frac{V_1 + V_2}{V_2}\right) \tag{3.67}$$

Solids and Liquids

Entropy changes when solids or liquids are heated or cooled can readily be calculated provided that we know the relevant heat capacities and also that we know the latent heats for any phase transitions that occur. The entropy contributions arising from volume changes, which are small in these cases, can usually be neglected.*

EXAMPLE
Calculate the entropy change when 1 mol of ice is heated from 250 K to 300 K. Take the heat capacities ($C_{P,m}$) of water and ice to be constant at 75.3 and 37.7 J K^{-1} mol^{-1}, respectively, and the latent heat of fusion of ice as 6.02 kJ mol^{-1}.

SOLUTION
The entropy change when 1 mol of ice is heated from 250 K to 273.15 K is

$$\Delta S_1/\text{J K}^{-1}\ \text{mol}^{-1} = \int_{250}^{273} \frac{37.7}{T}\, dT = 37.7 \ln \frac{273.15}{250}$$

$$\Delta S_1 = 3.34\ \text{J K}^{-1}\ \text{mol}^{-1}$$

For the melting at 273.15 K,

$$\Delta S_2 = \frac{6020\ \text{J mol}^{-1}}{273.15\ \text{K}} = 22.04\ \text{J K}^{-1}\ \text{mol}^{-1}$$

For the heating from 273.15 K to 300 K;

$$\Delta S_3 = \int_{273.15}^{300} \frac{75.3}{T}\, dT = 75.3 \ln \frac{300}{273.15}$$

$$= 7.06\ \text{J K}^{-1}\ \text{mol}^{-1}$$

The total entropy change is

$$\Delta S = 3.34 + 22.04 + 7.06 = 32.44\ \text{J K}^{-1}\ \text{mol}^{-1}$$ ■

EXAMPLE
One mole of supercooled water at −10°C and 1 atm pressure turns into ice. Calculate the entropy change in the system and in the surroundings and the net entropy change, using the data given in the previous example.

SOLUTION
Supercooled water at −10°C and the surroundings at −10°C are not at equilibrium, and the freezing is therefore not reversible. To calculate the entropy change we must devise a series of reversible processes by which supercooled water at −10°C

*For further details on empirical methods of determining entropy values for a wide range of compounds, see A. A. Bondi, *Physical Properties of Molecular Crystals, Liquids and Glass,* New York: Wiley, 1968.

is converted into ice at −10°C. We can (a) heat the water reversibly to 0°C, (b) bring about the reversible freezing, and then (c) cool the ice reversibly to −10°C.

a. The entropy change in heating the supercooled water from 263.15 K to 273.15 K is

$$\Delta S_1/\text{J K}^{-1}\text{ mol}^{-1} = \int_{263.15}^{273.15} \frac{75.3}{T}\, dT = 75.3 \ln \frac{273.15}{263.15}$$

$$\Delta S_1 = 2.81 \text{ J K}^{-1}\text{ mol}^{-1}$$

b. The entropy change in the reversible freezing of water at 0°C is

$$\Delta S_2 = -\frac{q_{\text{fusion}}}{T} = -\frac{6020 \text{ J mol}^{-1}}{273.15 \text{ K}} = -22.04 \text{ J K}^{-1}\text{ mol}^{-1}$$

c. The entropy change in cooling the ice from 273.15 K to 263.15 K is

$$\Delta S_3 = 37.7 \int_{273.15}^{263.15} \frac{dT}{T} = 37.7 \ln \frac{263.15}{273.15}$$

$$= -1.41 \text{ J K}^{-1}\text{ mol}^{-1}$$

The entropy change in the water when it froze at −10°C is therefore

$$\Delta S_{\text{syst}} = \Delta S_1 + \Delta S_2 + \Delta S_3$$

$$= 2.81 - 22.04 - 1.41$$

$$= -20.64 \text{ J K}^{-1}\text{ mol}^{-1}$$

The entropy increase in the surroundings is obtained by first calculating the net amount of heat that has been transferred to the surroundings. This is the sum of three terms corresponding to the three steps:

a. Heat lost by the surroundings in heating the water from −10°C is 10 K × 75.3 J K^{-1} mol^{-1} = 753 J mol^{-1}; the surroundings thus gain −753 J mol^{-1}.
b. Heat gained by the surroundings when the water freezes at 0°C is 6020 J mol^{-1}.
c. Heat gained by the surroundings when the ice is cooled to −10°C is 10 K × 37.7 J K^{-1} mol^{-1} = 377 J mol^{-1}.

The net heat transferred to the surroundings when the water freezes at −10°C is therefore

$$-753 + 6020 + 377 = 5644 \text{ J mol}^{-1}$$

This heat is taken up by the surroundings at the constant temperature of −10°C, and the entropy increase is

$$\Delta S_{\text{surr}} = \frac{5644}{263.15} = 21.45 \text{ J K}^{-1}\text{ mol}^{-1}$$

The overall entropy change in the system and the surroundings is therefore

$$\Delta S_{\text{overall}} = \Delta S_{\text{syst}} + \Delta S_{\text{surr}} = -20.66 + 21.45 = 0.81 \text{ J K}^{-1}\text{ mol}^{-1}$$

A net entropy increase in the system and surroundings is, of course, what we expect for an irreversible process. ■

When liquids are mixed, the entropy change is sometimes given by Eq. 3.66, which we derived for ideal gases. The condition for this equation to apply to liquids is that the intermolecular forces between the different components must all be equal. We have seen (Section 2.8) that for a gas to behave ideally there must be a complete absence of forces between the molecules; the internal pressure $(\partial U/\partial V)_T$ must be zero. The cohesive forces between molecules in a liquid can never be zero, but if the various intermolecular forces are all equal, Eq. 3.66 will be obeyed. For a two-component system, for example, the intermolecular forces between A and A, B and B, and A and B, must be the same.

Further details about the thermodynamics of solutions will be considered in later chapters.

3.4 Molecular Interpretation of Entropy

It has been emphasized earlier that thermodynamics is a branch of science that can be developed without any regard to the molecular nature of matter. The logical arguments employed do not require any knowledge of molecules, but many of us find it helpful, in the understanding of thermodynamics, to interpret its principles in the light of molecular structure.

In specifying a thermodynamic state we ignore the question of the positions and velocities of individual atoms and molecules. However, any macroscopic property is in reality a consequence of the position and motion of these particles. At any instant we could *in principle* define the *microscopic* state of a system, which means that we would specify the position and momentum of each atom. An instant later, even though the system might remain in the same *macroscopic* state, the *microscopic* state would be completely different, since at ordinary temperatures molecules are changing their positions at speeds of the order of 10^3 m s^{-1}. A system at equilibrium thus remains in the same macroscopic state, even though its microscopic state is changing rapidly.

There is an enormous number of microscopic states that are consistent with any given macroscopic state. This concept leads at once to a molecular interpretation of entropy: It is *a measure of how many different microscopic states are consistent with a given macroscopic state.* When a system moves spontaneously from one state to another, it goes to a state in which there are more microscopic states. We can express this differently by saying that when a spontaneous change takes place, there is an increase in *disorder.* In other words, *entropy is a measure of disorder;* an *increase in entropy means an increase in disorder.*

A deck of playing cards provides us with a useful analogy. A deck of 52 cards may be arranged in a particular specified order (suits separate, cards arranged from ace to king) or in a completely shuffled and disordered state. There are many sequences (analogous to microscopic states) that correspond to the shuffled and disordered (macroscopic) state, whereas there is only one microscopic state of the specified order, in which there is less disorder. Thus, the shuffled state has higher entropy than the unshuffled. If we start with an ordered deck and shuffle it, the deck moves toward a state of greater randomness or disorder; the entropy increases. The reason the random state is approached when the ordered deck is shuffled is simply that there are many microscopic states consistent with the shuffled condition, and only one consistent with the ordered condition. The chance of producing an ordered deck by shuffling a disordered one is obviously very small. This is true for 52 cards; when we are dealing with a much larger number of molecules (e.g., 6.022×10^{23} in a mole), the likelihood of a net decrease in entropy is obviously much more remote.

In the light of these ideas, it is easy to predict what kinds of entropy changes will occur when various processes take place. If, for example, we raise the temperature of a gas, the range of molecular speeds becomes more extended; a larger proportion of the molecules have speeds that differ from the most probable value. There is thus *more disorder at a high temperature,* and the *entropy is greater.*

Entropy also increases if a solid melts. The entropy change on melting is the enthalpy of fusion $\Delta_{fus}H$ divided by the melting point T_m, and since $\Delta_{fus}H$ must be positive, there is always an entropy increase on melting. This is understandable on a molecular basis, since in the solid the molecules occupy fixed sites, while in a liquid there is much less restriction as to position. Similarly, for the evaporation of a liquid at the boiling point T_b, the entropy change $\Delta_{vap}H/T_b$ must be positive since the latent heat of vaporization $\Delta_{vap}H$ must be positive.* In the liquid the attractive forces between molecules are much greater than in the vapor, and there is a large increase in disorder in going from the liquid to the vapor state; there are many more microscopic states for the gas compared with the liquid. The conversion of a solid into a gas is also accompanied by an entropy increase, for the same reason.

Entropy changes in chemical reactions can also be understood on a molecular basis. Consider, for example, the process

$$H_2 \rightarrow 2H$$

If we convert 1 mol of hydrogen molecules into 2 mol of hydrogen atoms, there is a considerable increase in entropy. The reason for this is that there are more microscopic states (more disorder) associated with the separated hydrogen atoms than with the molecules, in which the atoms are paired together. Again, an analogy is provided by a deck of cards. The hydrogen atoms are like a completely shuffled deck, while the molecular system is like a deck in which aces, twos, etc., are paired. The latter restriction means fewer permissible states and, therefore, a lower entropy.

In general, for a gaseous chemical reaction there is an increase of entropy in the system if there is an increase in the number of molecules. The dissociation of ammonia, for example,

$$2NH_3 \rightarrow N_2 + 3H_2$$

is accompanied by an entropy increase, because we are imposing a smaller restriction on the system by pairing the atoms as N_2 and H_2, as compared with organizing them as NH_3 molecules.

The situation with reactions in solution is, however, a good deal more complicated. It might be thought, for example, that a process of the type

$$MX \rightarrow M^+ + X^-$$

*For many liquids, $\Delta_{vap}S$ is around 88 J K^{-1} mol^{-1}. This is the basis of Trouton's rule (Eq. 5.18), according to which $\Delta_{vap}S$/(calories per mole) is about 88 J K^{-1} mol^{-1}/4.184 J cal^{-1} = 21 times the boiling point in kelvins.

occurring in aqueous solution, would be accompanied by an entropy increase, by analogy with the dissociation of H_2 into 2H. However, there is now an additional factor, arising from the fact that ions interact with surrounding water molecules, which tend to orient themselves in such a way that there is electrostatic attraction between the ion and the dipolar water molecules. This effect is known as *electrostriction* or more simply as the *binding* of water molecules.* This electrostriction leads to a considerable reduction in entropy, since the bound water molecules have a restricted freedom of motion. As a result, ionization processes always involve an entropy *decrease*.

An interesting example is provided by the attachment of adenosine triphosphate (ATP) to myosin, a protein that is an important constituent of muscle and that plays an important role in muscular contraction. Myosin is an extended protein that bears a number of positive charges, whereas ATP under normal physiological conditions bears four negative charges. These charges bring about a lowering of the entropy of the water molecules, but when the ATP and myosin molecules come together, there is some charge neutralization and a consequent increase in entropy because of the release of water molecules. This entropy increase associated with the binding of ATP to myosin plays a significant role in connection with the mechanism of muscular contraction.

The entropy change that occurs on the contraction of a muscle is also of interest. A stretched strip of muscle, or a stretched piece of rubber, contracts spontaneously. When stretched, muscle or rubber is in a state of lower entropy than in the contracted state. Both muscle and rubber consist of very long molecules. If a long molecule is stretched as far as possible without breaking bonds, there are few conformations available to it. However, if the ends of the molecule are brought closer together, the molecules can then assume a large number of conformations, and the entropy will therefore be higher. In 1913 the British physiologist Archibald Vivian Hill made accurate measurements of the heat produced when a muscle contracts and found it to be extremely small. The same is true of a piece of rubber. When muscle or rubber contracts, there is therefore very little entropy change in the surroundings, so that the overall entropy change is essentially that of the material itself, which is positive. We can therefore understand why muscular contraction, or the contraction of a stretched piece of rubber, occurs spontaneously. Processes of this kind that are largely controlled by the entropy change in the system are referred to as *entropic processes*.

3.5 The Third Law of Thermodynamics

The definition of entropy,

$$\Delta S = \int_A^B \frac{dq_{\text{rev}}}{T} \tag{3.68}$$

defines only changes in entropy, not entropy itself. The question of absolute entropy values is dealt with by the **Third Law of Thermodynamics,** which was first

* This matter is discussed in more detail in Section 7.9; see especially Figure 7.17.

formulated in 1906 by the German physical chemist Walther Hermann Nernst (1864–1941). This law is related to the experimental study of the behavior of matter at very low temperatures, a subject known as **cryogenics** (Greek *kryos*, frost; *genes*, become).*

Cryogenics: The Approach to Absolute Zero

Various techniques are used for producing low temperatures. The most familiar one, used in commercial refrigerators, is based on the fact that under certain circumstances gases become cooler when they expand, as a result of the work done by the gas in overcoming the mutual attraction of the molecules. This is an application of the Joule-Thomson effect that we considered in Section 2.8. Liquid nitrogen, which boils at 77 K, is manufactured commercially by the application of this principle, a cascade process being employed; further details are given in Section 3.10 (Figure 3.15). By performing successive expansions first with nitrogen, then with hydrogen, and finally with helium, the Dutch physicist Heike Kamerlingh-Onnes (1853–1926) liquefied helium, the last gas to be liquefied, in 1908.† He thus opened a temperature region to somewhat below 1 K for study and eventual exploitation.

The attainment of still lower temperatures requires the use of another principle. It was suggested independently in 1926 by the American chemist William Francis Giauque and the Dutch chemist Peter Joseph Wilhelm Debye (1884–1966) that one can make use of the temperature changes occurring during magnetization and demagnetization procedures. Certain salts, such as those of the rare earths, have high paramagnetic susceptibilities. The cations act as little magnets, which line up when a magnetic field is applied, and the substance is then in a state of lower entropy. When the magnetic field is decreased, the magnets adopt a more random arrangement, and the entropy increases. That this can occur was demonstrated in 1933 by Giauque.‡

Figure 3.8 illustrates a procedure that can be employed to achieve a low temperature. A paramagnetic salt, such as gadolinium sulfate octahydrate, is placed between the poles of an electromagnet and is cooled to about 1 K, which can be done by the expansion techniques mentioned previously. The magnetic field is then applied, and the heat produced is allowed to flow into the surrounding liquid helium (Step 1). In Step 2 the system is then isolated and the magnetic field removed; this adiabatic process leads to a cooling. Temperatures of about 0.005 K are produced in this way.

The attainment of still lower temperatures, down to about 10^{-6} K, is achieved by making use of nuclear magnetic properties. The nuclear magnets are about 2000 times smaller than the electron magnet in a paramagnetic substance such as gadolinium sulfate; yet there is a significant difference between the nuclear entropies even at temperatures as low as 10^{-6} K.

The facts that the absolute zero of temperature cannot be attained and that it becomes more and more difficult to approach that temperature suggest that the

---

* For a very readable account of low-temperature work, see K. Mendelssohn, *The Quest for Absolute Zero*. New York: McGraw-Hill, 1966.

†H. Kamerlingh-Onnes, *Leiden Commun.*, 108 (1911).

‡W. F. Giauque and D. P. MacDougall, *Phys. Rev. 43*, 7689 (1933).

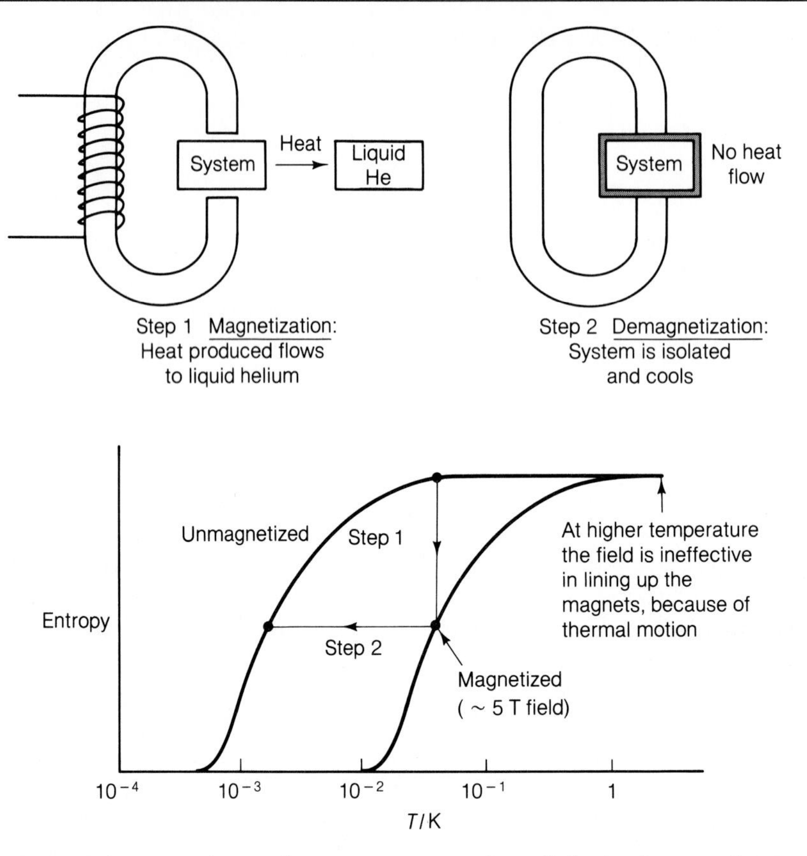

FIGURE 3.8 The production of very low temperatures by adiabatic demagnetization: (a) the magnetization and demagnetization steps; (b) the variation of entropy with temperature for the magnetized and demagnetized material.

entropies of all materials at the absolute zero are the same. However, this statement must be qualified slightly, by adding that the substance must be in its thermodynamically most stable state. For example, many substances are frozen into a metastable glassy state as the temperature is reduced. This state will persist indefinitely because of the slowness with which processes occur at low temperatures. We must therefore exclude noncrystalline systems and conclude that the *entropy of all perfect crystalline substances must be the same at the absolute zero.* This is the **heat theorem** proposed by Nernst in 1906, and it has become known as the **Third Law of Thermodynamics.**

Absolute Entropies

As a result of the Nernst heat theorem it is convenient to adopt the convention of assigning a value of zero to every crystalline substance at the absolute zero. Entropies can then be determined at other temperatures, by considering a series of reversible processes by which the temperature is raised from the absolute zero to the temperature in question. Table 3.2 lists some absolute entropies obtained in this way.

TABLE 3.2 Absolute Entropies at 25°C

Substance	State	Absolute Entropy, $S°$/ J K^{-1} mol^{-1}
Carbon	Graphite*	5.69
Hydrogen	Gas*	130.5
Oxygen	Gas*	205.0
Nitrogen	Gas*	191.6
Carbon dioxide	Gas	213.7
Water	Liquid	69.9
Ammonia	Gas	192.5
Ethane	Gas	229.7
Ethylene	Gas	219.7
Methanol	Liquid	126.8
Ethanol	Liquid	160.7
Acetic acid	Liquid	159.8
Acetaldehyde	Gas	265.7
Urea	Solid	104.6

*These are the standard states of the elements.
These data are from F. D. Rossini et al., *Selected Values of Chemical Thermodynamic Properties*, National Bureau of Standards, Circular 500 (1952); and F. D. Rossini et al., *Selected Values of Physical and Thermodynamic Properties of Hydrocarbons and Related Compounds*, American Petroleum Institute Research Project 44 (1953).

If the absolute entropies of all the substances in a chemical reaction are known, it is then a simple matter to calculate the entropy change in the reaction; the relationship is

$$\Delta S = \Sigma S(\text{products}) - \Sigma S(\text{reactants}) \tag{3.69}$$

3.6 Conditions for Equilibrium

The Second Law, in stating that any spontaneous process must be accompanied by an increase in the total entropy, gives us at once a condition for equilibrium, since a system at equilibrium cannot undergo any spontaneous change. Suppose, with reference to Figure 3.9a, that a system is at equilibrium in state A and that an infinitesimal change takes it to another state B where it is still at equilibrium. The change $A \rightarrow B$ cannot involve a total entropy increase, since otherwise the change would be spontaneous and equilibrium could not exist at A; by the same argument the change $B \rightarrow A$ cannot involve a total entropy increase. It follows that the states A and B must have the same total entropies. The *condition for equilibrium* is therefore

$$dS^{\text{total}} = dS^{\text{syst}} + dS^{\text{surr}} = 0 \tag{3.70}$$

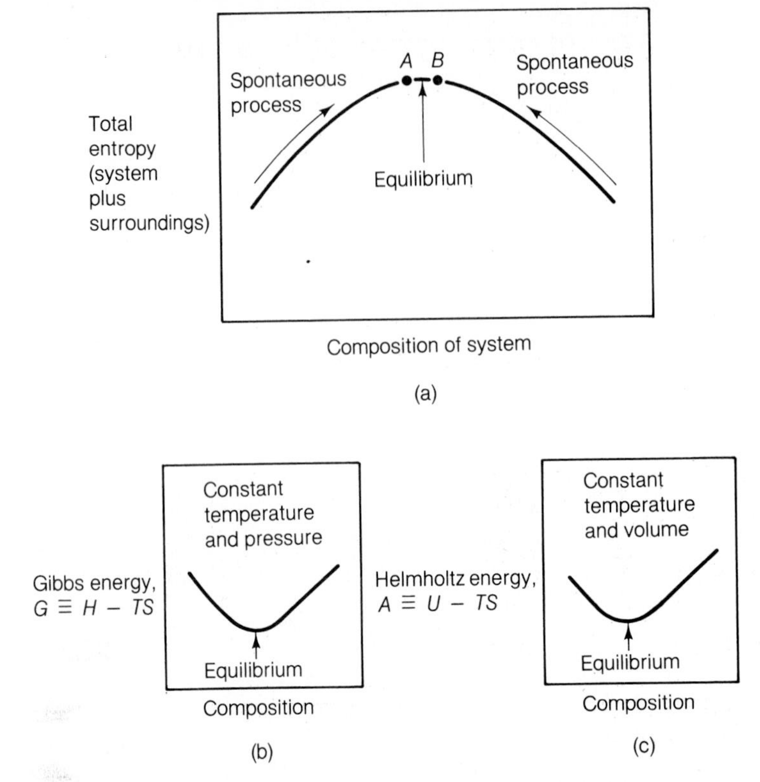

FIGURE 3.9 Conditions for chemical equilibrium: (a) the system moves toward a state of maximum total entropy; (b) at constant T and P, the system moves toward a state of minimum Gibbs energy; (c) at constant T and V, the system moves toward a state of minimum Helmholtz energy.

where S^{syst} is the entropy of the system and S^{surr} is the entropy of the surroundings. This is shown in Figure 3.9a, and we see that the *position of equilibrium must correspond to a state of maximum total entropy,* since the total entropy increases in any spontaneous process.

It is more convenient to define equilibrium with reference to *changes in the system only,* without explicitly considering the environment. Suppose that the system and the surroundings are at the same temperature,

$$T^{syst} = T^{surr} \tag{3.71}$$

Suppose that a process occurs spontaneously in the system and that an amount of heat dq leaves the system and enters the surroundings. This amount entering the surroundings may be written as dq^{surr}, and it is equal to the heat change $-dq^{syst}$ in the system:

$$dq^{surr} = -dq^{syst} \tag{3.72}$$

We now come to a very important point about dq^{surr}, which arises on account of the vastness of the surroundings. As a result of this, the surroundings experience no volume change when heat is transferred to them, and dq^{surr} is therefore equal to the change dU^{surr} in the internal energy, which is a state function. It therefore does not matter whether the heat dq^{surr} enters the surroundings reversibly or irreversibly; the heat dq^{surr}_{irrev} is still equal to the increase in internal energy of the surroundings:

$$dq^{surr}_{irrev} = dq^{surr}_{rev} = dU^{surr} \tag{3.73}$$

The entropy change in the surroundings is

$$dS^{surr} = \frac{dq^{surr}}{T^{surr}} \tag{3.74}$$

But $T^{surr} = T^{syst}$ and $dq^{surr} = -dq^{syst}$; therefore

$$dS^{surr} = -\frac{dq^{syst}}{T^{syst}} \tag{3.75}$$

Instead of Eq. 3.70 we can therefore write the equilibrium condition as

$$dS^{total} = dS^{syst} - \frac{dq^{syst}}{T^{syst}} = 0 \tag{3.76}$$

We now have expressed everything in terms of the system, and to simplify the notation from now on we will drop the superscript "syst." The condition for equilibrium will thus be written simply as

$$dS - \frac{dq}{T} = 0 \tag{3.77}$$

with the understanding that it is *the system we are referring to.* Alternatively, we can write

$$dq - T\,dS = 0 \tag{3.78}$$

as the condition for equilibrium.

Constant Temperature and Pressure: The Gibbs Energy

Chemical processes commonly occur in open vessels at constant pressure, in which case dq can be equated to dH, the enthalpy change. Equation 3.78 therefore becomes

$$dH - T\,dS = 0 \tag{3.79}$$

In view of this relationship the American physicist Josiah Willard Gibbs (1839–1903) defined a new thermodynamic function that is now known as the *Gibbs function* or **Gibbs energy*** and is given the symbol G:

$$G \equiv H - TS \tag{3.80}$$

At constant temperature

$$dG = dH - T\,dS \tag{3.81}$$

---

*It has long been known as the Gibbs *free* energy. However, IUPAC has recommended that the *free* be dropped and that we call it the *Gibbs function* or the *Gibbs energy*. The same recommendation applies to the Helmholtz energy.

and it follows from Eq. 3.79 that the *condition for equilibrium at constant T and P is*

$$dG = 0 \tag{3.82}$$

Since G is composed of H, T, and S, which are state functions, it is also a state function.

This condition for equilibrium is represented in Figure 3.9b. We see that systems tend to move toward a state of *minimum* Gibbs energy, and this is easily understood if we follow through the preceding arguments beginning with the inequality

$$dS^{\text{total}} = dS^{\text{syst}} + dS^{\text{surr}} > 0 \tag{3.83}$$

which applies to the irreversible case. Instead of Eq. 3.82 we then find that

$$dG < 0 \tag{3.84}$$

In other words, *in spontaneous processes at constant T and P, systems move toward a state of minimum Gibbs energy.*

Constant Temperature and Volume: The Helmholtz Energy

The argument for constant volume conditions is very similar; the quantity dq in Eq. 3.78 now is equated to dU:

$$dU - T\,dS = 0 \tag{3.85}$$

The quantity $U - TS$, also a state function, is called the *Helmholtz function* or **Helmholtz energy**, after the German physiologist and physicist Ludwig Ferdinand von Helmholtz (1821–1894), and is given the symbol A:

$$A \equiv U - TS \tag{3.86}$$

Equation 3.85 therefore can be written as

$$dA = 0 \tag{3.87}$$

and this is the condition for equilibrium at *constant T and V*. Under these conditions, systems tend to move toward a state of *minimum Helmholtz energy*, as shown in Figure 3.9c.

3.7 The Gibbs Energy

Molecular Interpretation

Although, as previously emphasized, thermodynamic arguments can be developed without any reference to the existence and behavior of atoms and molecules, it is nevertheless instructive to interpret the arguments in terms of molecular structure.

Consider first the dissociation of hydrogen molecules into hydrogen atoms,

$$H_2 \rightleftharpoons 2H$$

This process from left to right occurs only to a very slight extent at ordinary temperatures, but if we start with hydrogen atoms, the combination will occur spontaneously. We have seen that at constant temperature and pressure a natural or spontaneous process is one in which there is a decrease in Gibbs energy; the system approaches an equilibrium state in which the Gibbs energy is at a minimum. Therefore, if the process

$$2H \rightarrow H_2$$

occurs at ordinary temperatures, ΔG is negative. Let us now consider how this can be interpreted in terms of enthalpy and entropy changes, in the light of the molecular structures.

We know that when hydrogen atoms are brought together and combine, there is evolution of heat, which means that the enthalpy goes to a lower level; that is

$$\Delta H(2H \rightarrow H_2) < 0$$

The entropy change is also negative when hydrogen atoms combine, because the atoms have a less ordered arrangement than the molecules:

$$\Delta S(2H \rightarrow H_2) < 0$$

The Gibbs energy change for the combination process at constant temperature is made up as follows:

$$\Delta G = \underset{<0}{\Delta H} - T \underset{<0}{\Delta S} \tag{3.88}$$

If T is small enough, ΔG will be negative. This is the situation at room temperature; indeed up to quite high temperatures the negative ΔH term dominates the situation and ΔG is negative, which means that the process occurs spontaneously.

If, however, we go to very high temperatures,* the $T\Delta S$ term will become dominant; since ΔS is negative and $T\,\Delta S$ is *subtracted* from ΔH, the net value of ΔG becomes positive when T is large enough. Therefore we predict that at very high temperatures hydrogen atoms will not spontaneously combine; instead hydrogen molecules will spontaneously dissociate into atoms. This is indeed the case.

In this example the ΔH and $T\,\Delta S$ terms (except at very high temperatures) work in opposite directions; both are negative. In almost all reactions, ΔH and $T\,\Delta S$ work against each other. In the reaction

$$2H_2 + O_2 \rightarrow 2H_2O$$

ΔH is negative (the reaction is exothermic) and $T\,\Delta S$ is negative; there is a decrease in the number of molecules and an increase of order. Thus, at a fixed temperature,

$$\Delta G = \underset{<0}{\Delta H} - T \underset{<0}{\Delta S} \tag{3.89}$$

At ordinary temperatures, $T\,\Delta S$ is negligible compared with ΔH; ΔG is therefore negative and reaction occurs spontaneously from left to right. As the temperature is raised, $T\,\Delta S$ becomes more negative, and at sufficiently high temperatures $\Delta H - T\,\Delta S$ becomes positive. The spontaneous reaction is then from left to right.

It follows from the relationship

$$\Delta G = \Delta H - T\,\Delta S \tag{3.90}$$

*It must be emphasized that ΔH and ΔS do vary somewhat with temperature, so that this discussion is oversimplified.

that temperature is a weighting factor that determines the relative importance of enthalpy and entropy. At the absolute zero, $\Delta G = \Delta H$, and the direction of spontaneous change is determined solely by the enthalpy change. At very high temperatures, on the other hand, the entropy is the driving force that determines the direction of spontaneous change.

EXAMPLE

a. Liquid water at 100°C is in equilibrium with water vapor at 1 atm pressure. If the enthalpy change associated with the vaporization of liquid water at 100°C is 40.60 kJ mol^{-1}, what are ΔG and ΔS?

b. Suppose that water at 100°C is in contact with water vapor at 0.9 atm. Calculate ΔG and ΔS for the vaporization process.

SOLUTION

(See Figure 3.10)

a. Since liquid water at 100°C is in equilibrium with water vapor at 1 atm pressure,

$$\Delta G = 0$$

Since $\Delta H = 40.60 \text{ kJ mol}^{-1}$, and

$$\Delta G = \Delta H - T\,\Delta S$$

it follows that

$$\Delta S = \frac{40\,600 \text{ J mol}^{-1}}{373.15 \text{ K}} = 108.9 \text{ J K}^{-1} \text{ mol}^{-1}$$

b. The entropy increase for the expansion of 1 mol of gas from 1 atm pressure to 0.9 atm is

$$\Delta S = R \ln \frac{V_2}{V_1} = R \ln \frac{P_1}{P_2}$$

$$\Delta S/\text{J K}^{-1} \text{ mol}^{-1} = 8.314 \ln \frac{1.0}{0.9}$$

$$\Delta S = 0.876 \text{ J K}^{-1} \text{ mol}^{-1}$$

The entropy increase when 1 mol of liquid water evaporates to give vapor at 0.9 atm pressure is thus

$$\Delta S = 108.9 + 0.876 = 109.7 \text{ J K}^{-1} \text{ mol}^{-1}$$

The value of $T\,\Delta S$ is

$$109.7 \times 373.15 = 40.96 \text{ kJ mol}^{-1}$$

The value of ΔH has not changed in this process, and the value of the Gibbs energy change is thus

$$\Delta G = \Delta H - T\,\Delta S = 40.60 - 40.96 = -0.36 \text{ kJ mol}^{-1}$$

Since this is a negative quantity, the vaporization process is spontaneous. ■

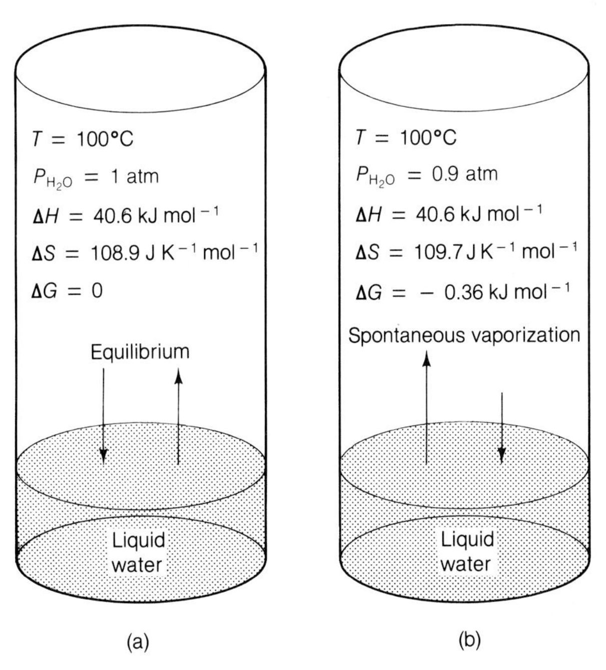

FIGURE 3.10 The vaporization of water at 100°C. In (a), liquid water at 100°C is in equilibrium with water vapor at 1 atm pressure. In (b), liquid water at 100°C is in contact with water vapor at 0.9 atm pressure, and there is spontaneous vaporization.

Gibbs Energies of Formation

In Section 2.6 we dealt with enthalpy changes in chemical reactions and found that it was very convenient to tabulate enthalpies of formation of compounds.

The same procedure is followed with Gibbs energies. The *standard Gibbs energy of formation* of any compound is then simply the Gibbs energy change $\Delta_f G^\circ$ that accompanies the formation of the compound in its standard state from its elements in their standard states. We can then calculate the standard Gibbs energy change for any reaction, ΔG°, by adding the Gibbs energies of formation of all the products and subtracting the sum of the Gibbs energies of formation of all the reactants:

$$\Delta G^\circ = \Sigma\, \Delta_f G^\circ(\text{products}) - \Sigma\, \Delta_f G^\circ(\text{reactants}) \tag{3.91}$$

Table 2.2 (p. 66) lists Gibbs energies of formation of a number of compounds, and of ions in aqueous solution. It is impossible to measure Gibbs energies of formation of individual ions since experiments are always done with systems involving ions of opposite signs. To overcome this difficulty the same procedure is adopted as with enthalpies; the arbitrary assumption is made that the Gibbs energy of formation of the proton in water is zero, and the Gibbs energies of formation of all the other ions are calculated on that basis. The ionic values obtained in this way are known as *conventional* Gibbs energies of formation.

A negative ΔG° for a reaction means that the process is spontaneous; a compound having a negative $\Delta_f G^\circ$ is therefore *thermodynamically* stable with respect to its elements. A compound whose standard Gibbs energy of formation is negative is known as an *exergonic** compound (compare *exothermic,* for a compound formed with a negative $\Delta_f H°$). Conversely, a compound having a positive $\Delta_f G°$ value is known as an *endergonic* compound (compare *endothermic*). Most compounds are exergonic.

The terms *exergonic* and *endergonic* are also employed with respect to other processes. Thus any reaction having a negative $\Delta G°$ value (i.e., accompanied by a liberation of Gibbs energy) is said to be *exergonic.* A reaction having a positive $\Delta G°$ is said to be *endergonic.*

Gibbs Energy and Reversible Work

When an ideal gas is reversibly compressed at constant temperature, the work done is the increase in Gibbs energy. We have seen in Section 2.7 that the reversible work done in compressing n mol of an ideal gas, at temperature T, from a volume V_1 to a volume V_2 is

$$w_{\text{rev}} = nRT \ln \frac{V_1}{V_2} \tag{3.92}$$

During this isothermal process there is no change in internal energy; the internal energy of an ideal gas is a function of temperature only and not of pressure or volume. It follows from the First Law that the heat absorbed by the system is the negative of the work done on the system:

$$q_{\text{rev}} = nRT \ln \frac{V_2}{V_1} \tag{3.93}$$

The entropy change (numerically a decrease since $V_1 > V_2$) is therefore

$$\Delta S = \frac{q_{\text{rev}}}{T} = nR \ln \frac{V_2}{V_1} \tag{3.94}$$

There is no change in enthalpy; the enthalpy for an ideal gas is a function only of temperature. The change in Gibbs energy is thus

$$\Delta G = \Delta H - T\,\Delta S \tag{3.95}$$

$$= nRT \ln \frac{V_1}{V_2} \tag{3.96}$$

The reversible work done on the system (Eq. 3.92) is thus the change in its Gibbs energy.

An even more important relationship between Gibbs energy and work arises for processes occurring at constant temperature and pressure. Work can be classified into two types: work that arises from a volume change that occurs when a process takes place, and any other type of work. For example, when an electrochemical cell operates, there may be a small volume change; work may be performed on the surroundings, or the surroundings may do work on the system. Of much more interest and practical importance, however, is the electrical work that results from the operation of the cell. We will call the work arising from the volume change the

*From the Greek *ergon,* work.

PV work and give it the symbol w_{PV}. Any other kind of work we will call the non-*PV* work, $w_{\text{non-}PV}$. The total work is thus

$$w = w_{PV} + w_{\text{non-}PV} \tag{3.97}$$

Another kind of non-*PV* work is osmotic work. Non-*PV* work is sometimes called the *net work.*

We will now derive the important result that the *non-PV work is equal to the change in Gibbs energy for a reversible process occurring at constant temperature and pressure.* We start with the definition of Gibbs energy (Eq. 3.80):

$$G \equiv H - TS \equiv U + PV - TS \tag{3.98}$$

For any change,

$$dG = dU + P\,dV + V\,dP - T\,dS - S\,dT \tag{3.99}$$

At constant T and P,

$$dG = dU + P\,dV - T\,dS \tag{3.100}$$

From the First Law, dU is equal to $dq_p + dw$:

$$dG = dq_p + dw + P\,dV - T\,dS \tag{3.101}$$

For a process in which the system undergoes a volume change dV, the *PV* work is $-P\,dV$ and the total work is thus

$$dw = dw_{PV} + dw_{\text{non-}PV} = -P\,dV + dw_{\text{non-}PV} \tag{3.102}$$

Together with Eq. 3.101 this gives

$$dG = dq_p + dw_{\text{non-}PV} - T\,dS \tag{3.103}$$

However, since the process is reversible, $dq_p = T\,dS$ and therefore

$$dG = dw_{\text{non-}PV} \quad \text{or} \quad \Delta G = w_{\text{non-}PV} \tag{3.104}$$

This result has many important applications in physical chemistry. In Section 8.3 we shall see it used to derive the emf of a reversible electrochemical cell.

Another important result, which we will leave for the reader to derive (see Problem 41), is that at a given temperature the *change in Helmholtz energy, for all processes irrespective of changes in P and V, is equal to the total work (PV + non-PV).* Because of this relationship the Helmholtz function has often been called the *work function.*

3.8 Some Thermodynamic Relationships*

On the basis of the principles so far developed, it is possible to derive a number of relationships between different thermodynamic quantities. Some of the most important of these will now be obtained.

---

* This section can be omitted on first reading; see Appendix C for some mathematical relationships used in this section.

Maxwell Relations

For an infinitesimal process involving only PV work, we can combine the First and Second Laws in the equation

$$dU = dw + dq = -P\,dV + T\,dS \tag{3.105}$$

For dH we have similarly

$$dH = d(U + PV) = dU + d(PV) = dU + P\,dV + V\,dP \tag{3.106}$$

$$= -P\,dV + T\,dS + P\,dV + V\,dP = V\,dP + T\,dS \tag{3.107}$$

In making use of the relationship $d(PV) = P\,dV + V\,dP$, we have performed a *Legendre transformation*, named after the French mathematician Adrien Marie Legendre (1752–1853).

Similarly, for the Helmholtz and Gibbs energies, we have

$$dA = d(U - TS) = dU - d(TS) = dU - T\,ds - S\,dT \tag{3.108}$$

$$= -P\,dV + T\,dS - T\,dS - S\,dT = -P\,dV - S\,dT \tag{3.109}$$

and

$$dG = d(H - TS) = dH - T\,dS - S\,dT \tag{3.110}$$

$$= V\,dP + T\,dS - T\,dS - S\,dT = V\,dP - S\,dT \tag{3.111}$$

We can now combine the expressions we have obtained for dU, dH, dA, and dG with general relationships from differential calculus:

$$dU = -P\,dV + T\,dS = \left(\frac{\partial U}{\partial V}\right)_S dV + \left(\frac{\partial U}{\partial S}\right)_V dS \tag{3.112}$$

$$dH = V\,dP + T\,dS = \left(\frac{\partial H}{\partial P}\right)_S dP + \left(\frac{\partial H}{\partial S}\right)_P dS \tag{3.113}$$

$$dA = -P\,dV - S\,dT = \left(\frac{\partial A}{\partial V}\right)_T dV + \left(\frac{\partial A}{\partial T}\right)_V dT \tag{3.114}$$

$$dG = V\,dP - S\,dT = \left(\frac{\partial G}{\partial P}\right)_T dP + \left(\frac{\partial G}{\partial T}\right)_P dT \tag{3.115}$$

Important relationships are now obtained by equating coefficients; thus from Eq. 3.112 we have

$$\left(\frac{\partial U}{\partial V}\right)_S = -P \qquad \left(\frac{\partial U}{\partial S}\right)_V = T \tag{3.116}$$

Similarly, from Eqs. 3.113, 3.114, and 3.115,

$$\left(\frac{\partial H}{\partial P}\right)_S = V \qquad \left(\frac{\partial H}{\partial S}\right)_P = T \tag{3.117}$$

$$\left(\frac{\partial A}{\partial V}\right)_T = -P \qquad \left(\frac{\partial A}{\partial T}\right)_V = -S \tag{3.118}$$

$$\left(\frac{\partial G}{\partial P}\right)_T = V \qquad \left(\frac{\partial G}{\partial T}\right)_P = -S \tag{3.119}$$

A mnemonic device for obtaining these eight relationships is illustrated in Figure 3.11.

According to a theorem due to the Swiss mathematician Leonhard Euler (1707–1783), the order of differentiation does not matter. Thus, if f is a function of the variables x and y,

$$\frac{\partial}{\partial x}\left(\frac{\partial f}{\partial y}\right)_x = \frac{\partial}{\partial y}\left(\frac{\partial f}{\partial x}\right)_y \tag{3.120}$$

Application of this *Euler reciprocity theorem* to Eqs. 3.108 to 3.111 and use of Eqs. 3.112 to 3.115 lead to a number of useful relationships. For example, in Eq. 3.112, U is a function of V and S, so that by Euler's theorem

$$\frac{\partial}{\partial V}\left(\frac{\partial U}{\partial S}\right)_V = \frac{\partial}{\partial S}\left(\frac{\partial U}{\partial V}\right)_S \tag{3.121}$$

and introduction of Eq. 3.116 gives

$$\left(\frac{\partial T}{\partial V}\right)_S = -\left(\frac{\partial P}{\partial S}\right)_V \tag{3.122}$$

Similarly, from Eqs. 3.113 to 3.115 we obtain

$$\left(\frac{\partial T}{\partial P}\right)_S = \left(\frac{\partial V}{\partial S}\right)_P \tag{3.123}$$

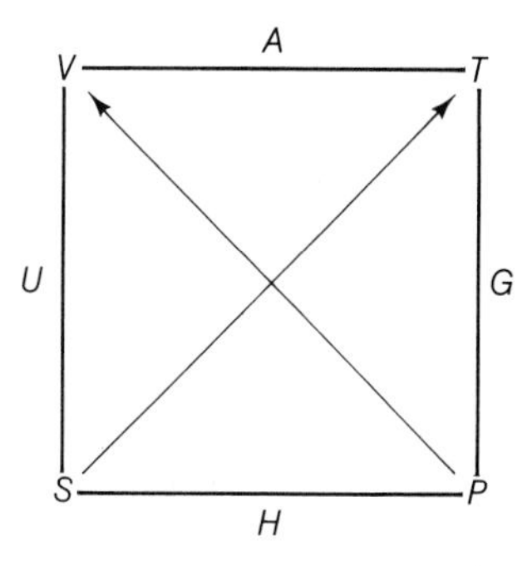

FIGURE 3.11 A mnemonic device for obtaining Eqs. 3.116–3.119 and the Maxwell relations 3.122–3.125. Each of the four thermodynamic potentials U, A, G, and H is flanked by two properties (e.g., U by V and S) to which it has a special relationship. The direction of the arrows indicates the sign of the equation. (a) For equations 3.116–3.119, any thermodynamic potential can be differentiated with respect to one of its neighboring properties, the other being held constant. The result is obtained by following the arrow. For example,

$$\left(\frac{\partial U}{\partial S}\right)_V = T \quad \text{and} \quad \left(\frac{\partial U}{\partial V}\right)_S = -P$$

(b) For Eqs. 3.122 and 3.125, any partial derivative of a property with respect to a neighboring property (e.g., $(\partial V/\partial T)_p$) is related to the corresponding derivative at the other side of the square [e.g., $(\partial S/\partial P)_T$], the arrows indicating the signs (in this example, negative).

$$\left(\frac{\partial P}{\partial T}\right)_V = \left(\frac{\partial S}{\partial V}\right)_T \tag{3.124}$$

$$\left(\frac{\partial V}{\partial T}\right)_P = -\left(\frac{\partial S}{\partial P}\right)_T \tag{3.125}$$

These are known as **Maxwell relations** and are particularly useful for obtaining quantities that are not easily measurable directly. For example, it would be difficult to measure $(\partial S/\partial P)_T$, but $(\partial V/\partial T)_P$ is easily obtained.

A useful device for obtaining these four Maxwell relations is included in Figure 3.11.

Thermodynamic Equations of State

We can also derive equations that give U and H in terms of P, V, and T and that are therefore called *thermodynamic equations of state.* From the definition of A, which equals $U - TS$, we have

$$\left(\frac{\partial U}{\partial V}\right)_T = \left[\frac{\partial(A + TS)}{\partial V}\right]_T \tag{3.126}$$

$$= \left(\frac{\partial A}{\partial V}\right)_T + T\left(\frac{\partial S}{\partial V}\right)_T \tag{3.127}$$

Then, from Eqs. 3.118 and 3.124 the thermodynamic equation of state for U is

$$\left(\frac{\partial U}{\partial V}\right)_T = -P + T\left(\frac{\partial P}{\partial T}\right)_V \tag{3.128}$$

The corresponding thermodynamic equation of state for H is obtained as follows:

$$\left(\frac{\partial H}{\partial P}\right)_T = \left[\frac{\partial(G + TS)}{\partial P}\right]_T = \left(\frac{\partial G}{\partial P}\right)_T + T\left(\frac{\partial S}{\partial P}\right)_T \tag{3.129}$$

and by use of Eqs. 3.119 and 3.125 we have the thermodynamic equation of state for H

$$\left(\frac{\partial H}{\partial P}\right)_T = V - T\left(\frac{\partial V}{\partial T}\right)_P \tag{3.130}$$

Some Applications of the Thermodynamic Relationships

A great many further relationships between the thermodynamic quantities can be obtained on the basis of the equations we have derived in the last few pages. Only a few examples can be included here.

One useful application is related to the theory of the Joule-Thomson effect, which we considered in Section 2.8. The Joule-Thomson coefficient μ is defined by Eq. 2.108:

$$\mu \equiv \left(\frac{\partial T}{\partial P}\right)_H \tag{3.131}$$

and C_P is defined by Eq. 2.27:

$$C_P \equiv \left(\frac{\partial H}{\partial T}\right)_P \tag{3.132}$$

But, from the theory of partial derivatives (Appendix C),

$$\left(\frac{\partial T}{\partial P}\right)_H = -\left(\frac{\partial H}{\partial P}\right)_T \left(\frac{\partial T}{\partial H}\right)_P \tag{3.133}$$

so that

$$\mu = -\frac{1}{C_P}\left(\frac{\partial H}{\partial P}\right)_T \tag{3.134}$$

Use of the thermodynamic equation of state, Eq. 3.130, then gives

$$\mu = \frac{T\,(\partial V/\partial T)_P - V}{C_P} \tag{3.135}$$

For an ideal gas

$$V = \frac{nRT}{P} \tag{3.136}$$

and

$$\left(\frac{\partial V}{\partial T}\right)_P = \frac{nR}{P} = \frac{V}{T} \tag{3.137}$$

so that the numerator in Eq. 3.135 is equal to zero. In the Joule-Thomson experiment there is therefore no temperature change for an ideal gas. For real gases, the numerator in Eq. 3.135 is in general other than zero. At the inversion temperature for a gas $\mu = 0$, and the condition for the inversion temperature is, therefore,

$$T\left(\frac{\partial V}{\partial T}\right)_P = V \tag{3.138}$$

The **cubic expansion coefficient** (formerly called the *thermal expansivity*) of a substance is defined as

$$\alpha \equiv \frac{1}{V}\left(\frac{\partial V}{\partial T}\right)_P \tag{3.139}$$

so that Eq. 3.135 can alternatively be written as

$$\mu = \frac{\alpha VT - V}{C_P} = \frac{V(\alpha T - 1)}{C_P} \tag{3.140}$$

The condition for the inversion temperature is therefore that

$$\alpha = T^{-1} \tag{3.141}$$

Another example of the application of the thermodynamic relationships is concerned with the van der Waals equation of state for 1 mol of gas:

$$\left(P + \frac{a}{V_m^2}\right)(V_m - b) = RT \tag{3.142}$$

We saw in Section 2.8 that the *internal pressure* of a gas is $(\partial U/\partial V)_T$, and we will now prove that if a gas obeys the van der Waals equation, the internal pressure is a/V_m^2; in other words, we will prove that

$$\left(\frac{\partial U}{\partial V}\right)_T = \frac{a}{V_m^2} \tag{3.143}$$

We do this by starting with the thermodynamic equation of state, Eq. 3.128, for the internal energy.

$$\left(\frac{\partial U}{\partial V}\right)_T = -P + T\left(\frac{\partial P}{\partial T}\right)_V \tag{3.144}$$

Equation 3.142 can be written as

$$P = \frac{RT}{V_m - b} - \frac{a}{V_m^2} \tag{3.145}$$

and therefore

$$\left(\frac{\partial P}{\partial T}\right)_V = \frac{R}{V_m - b} = \frac{1}{T}\left(P + \frac{a}{V_m^2}\right) \tag{3.146}$$

Substitution of this expression into Eq. 3.144 gives

$$\left(\frac{\partial U}{\partial V}\right)_T = \frac{a}{V_m^2} \tag{3.147}$$

and this is the internal pressure.

Fugacity

We have seen (Eq. 3.96) that for an isothermal process involving 1 mol of an ideal gas,

$$\Delta G_m = RT \ln \frac{V_1}{V_2} = RT \ln \frac{P_2}{P_1} \tag{3.148}$$

If the initial pressure P_1 is 1 atm, the gas is in a standard state and we write its molar Gibbs energy as G_m°; the Gibbs energy at any pressure P is then

$$G_m = G_m^\circ + RT \ln(P/\text{atm}) \tag{3.149}$$

If the gas is not ideal, this expression no longer applies. In order to have a parallel treatment of real gases, the American chemist Gilbert Newton Lewis (1875–1946) introduced a new function known as the **fugacity** (Latin *fugare,* to fly) and given the symbol f. The fugacity is *the pressure adjusted for lack of ideality*; if a gas is behaving ideally, the fugacity is equal to the pressure.

The fugacity is such that, to parallel Eq. 3.148,

$$\Delta G_m = RT \ln \frac{f_2}{f_1} \tag{3.150}$$

For the ideal gas we took the standard state to correspond to 1 atm pressure and so obtained Eq. 3.149. Similarly, for the nonideal gas we define the standard state

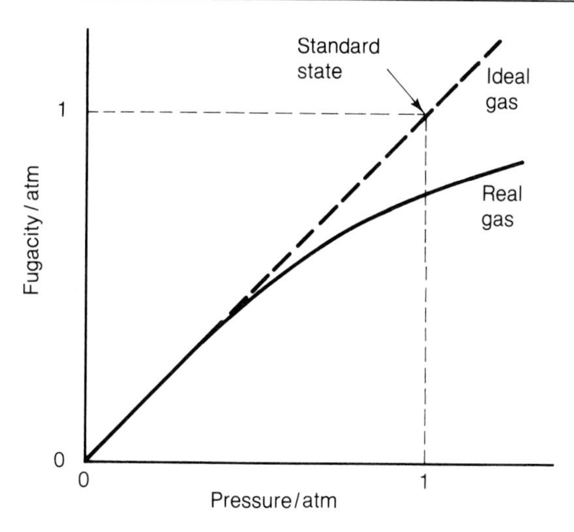

FIGURE 3.12 Fugacity of a real gas as a function of pressure. The standard state is the state at which the fugacity would be equal to 1 atm if the gas remained ideal from low pressures to 1 atm pressure.

to correspond to *unit fugacity* (i.e., $f_1 = 1$ atm) so that we obtain

$$G_m = G_m^\circ + RT \ln(f/\text{atm}) \tag{3.151}$$

The ratio of the fugacity of a substance in any state to the fugacity in the standard state is known as the **activity** and given the symbol a. Since for a gas the fugacity in the standard state is by definition 1 atm, the *fugacity of a gas is numerically equal to its activity.* Equation 3.151 can thus be written as

$$G_m = G_m^\circ + RT \ln a \tag{3.152}$$

Figure 3.12 shows schematically how the fugacity of a pure gas may vary with its pressure. At sufficiently low pressure, every gas behaves ideally, because the intermolecular forces are negligible and because the effective volume of the molecules is insignificant compared with the total volume. Therefore, as shown in Figure 3.12, we can draw a tangent to the curve at low pressures, and this line will represent the behavior of the gas if it were ideal. This line has a slope of unity; the fugacity of an ideal gas is 1 atm when the pressure is 1 atm.

It might be thought that the best procedure would have been to choose the standard state of a nonideal gas to correspond to some low pressure, such as 10^{-6} atm, at which the behavior is ideal. It has proved more convenient, however, to make use of the hypothetical ideal gas line in Figure 3.12 and to choose the point corresponding to 1 atm as the standard state. The true fugacity at 1 atm pressure is different from 1 atm (it is less in Figure 3.12), but we choose the *standard state* as the *state at which the fugacity would be equal to* 1 *atm if the gas remained ideal from low pressures to* 1 *atm pressure.*

The fugacity of a pure gas or a gas in a mixture can be evaluated if adequate P-V-T data are available. For 1 mol of gas at constant temperature,

$$dG = V_m \, dP \tag{3.153}$$

so that, for two states 1 and 2,

$$\Delta G_m = G_{m,2} - G_{m,1} = \int_{P_1}^{P_2} V_m \, dP \tag{3.154}$$

The quantity RT/P can be added to and subtracted from the integrand to give

$$\Delta G_m = \int_{P_1}^{P_2} \left[\frac{RT}{P} + \left(V_m - \frac{RT}{P} \right) \right] dP \tag{3.155}$$

$$= RT \ln \frac{P_2}{P_1} + \int_{P_1}^{P_2} \left(V_m - \frac{RT}{P} \right) dP \tag{3.156}$$

With Eq. 3.150 we have

$$RT \ln \frac{f_2}{f_1} = RT \ln \frac{P_2}{P_1} + \int_{P_1}^{P_2} \left(V_m - \frac{RT}{P} \right) dP \tag{3.157}$$

and therefore

$$RT \ln \frac{f_2 / P_2}{f_1 / P_1} = \int_{P_1}^{P_2} \left(V_m - \frac{RT}{P} \right) dP \tag{3.158}$$

If P_1 is a sufficiently low pressure, f_1/P_1 is equal to unity, so that

$$RT \ln \frac{f_2}{P_2} = \int_{P_1}^{P_2} \left(V_m - \frac{RT}{P} \right) dP \tag{3.159}$$

Suppose, for example, that we have reliable pressure-volume data, covering a wide range of pressures and going down to low pressures, at a particular temperature T. We can then calculate $(V_m - RT)/P$ and plot this quantity against P, as shown schematically in Figure 3.13. We can then obtain the value of the shaded area in Figure 3.13, from $P_1 = 0$ to any value P_2. This area is the integral on the right-hand side of Eq. 3.158, and the fugacity f_2 can therefore be calculated. Various analytical procedures have been used for evaluating these integrals. Use is sometimes made of the law of corresponding states (Section 1.12); to the extent that this law is valid, all gases fit a single set of curves, and for a given gas all that is necessary is to know its critical constants.

EXAMPLE

Oxygen at pressures that are not too high obeys the equation

$$P(V_m - b) = RT$$

where $b = 0.0211$ dm^3 mol^{-1}.

a. Calculate the fugacity of oxygen gas at 25°C and 1 atm pressure.

b. At what pressure is the fugacity equal to 1 atm?

SOLUTION

For this equation of state

$$V_m - \frac{RT}{P} = b$$

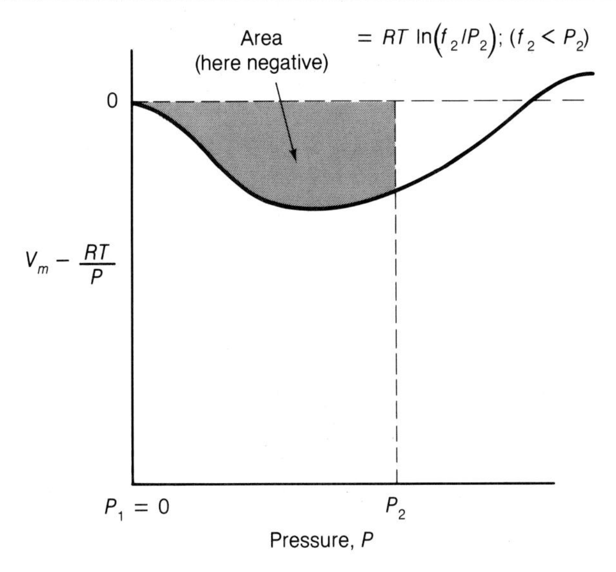

FIGURE 3.13 Schematic plot of $V_m - RT/P$ for 1 mol of a real gas.

so that

$$\int_{P_1}^{P_2}\left(V_m - \frac{RT}{P}\right) dP = \int_{P_1}^{P_2} b\, dP = b(P_2 - P_1)$$

a. Equation 3.159 then gives, with $P_1 = 0$, $P_2 = 1$ atm, and $b = 0.0211$ dm^3 mol^{-1},

$$RT \ln f_2^u = 0.0211 \text{ atm dm}^3 \text{ mol}^{-1}$$

$$= 0.0211 \times 1.013\,25 \times 10^5 \times 10^{-3} \text{ Pa m}^3 \text{ mol}^{-1}$$

$$\ln f_2^u = \frac{0.0211 \times 1.013\,25 \times 10^2(\text{Pa m}^3 \text{ mol}^{-1})}{8.314(\text{J K}^{-1} \text{ mol}^{-1})298.15 \text{ K}}$$

$$= 8.63 \times 10^{-4}$$

$$f_2 = 1.0009 \text{ atm}$$

b. $\ln f^u = \ln P^u + \dfrac{bP}{RT}$

The pressure at which $f = 1$ atm ($\ln f^u = 0$) is thus given by

$$\ln P^u = -\frac{bP}{RT} \quad \text{or} \quad P = e^{-bP/RT}$$

Since bP/RT is very small, we can expand the exponential and keep only the first term:

$$P = 1 - \frac{bP}{RT}$$

or

$$P = \frac{RT}{RT + b}$$

$$= \frac{0.082\,05 \times 298.15}{0.082\,05 \times 298.15 + 0.0211} \text{*}$$

$$= 0.9991 \text{ atm}$$

■

3.9 The Gibbs-Helmholtz Equation

The variation with temperature of the Gibbs energy change in a chemical process is a matter of great importance, and we shall see in Chapter 4 that it leads to an extremely useful way of determining the enthalpy change in a reaction.

Equation 3.119 gives the variation of G with T:

$$\left(\frac{\partial G}{\partial T}\right)_P = -S \tag{3.160}$$

For a change from one state to another, therefore,

$$\left(\frac{\partial \Delta G}{\partial T}\right)_P = -\Delta S \tag{3.161}$$

However, since

$$\Delta G = \Delta H - T\,\Delta S \tag{3.162}$$

$$\Delta S = \frac{\Delta H - \Delta G}{T} \tag{3.163}$$

so that, with Eq. 3.161,

$$\left(\frac{\partial \Delta G}{\partial T}\right)_P - \frac{\Delta G}{T} = -\frac{\Delta H}{T} \tag{3.164}$$

From differential calculus (Appendix C)

$$\frac{\partial}{\partial T}\left(\frac{\Delta G}{T}\right) = \frac{1}{T}\frac{\partial \Delta G}{\partial T} - \frac{\Delta G}{T^2} \tag{3.165}$$

so that the left-hand side of Eq. 3.164 can be written as

$$T\left[\frac{\partial}{\partial T}\left(\frac{\Delta G}{T}\right)\right]_P$$

Thus

$$T\left[\frac{\partial}{\partial T}\left(\frac{\Delta G}{T}\right)\right]_P = -\frac{\Delta H}{T} \tag{3.166}$$

---

* $R = 8.314 \text{ J K}^{-1} \text{ mol}^{-1} = 0.082\,05 \text{ atm dm}^3 \text{ K}^{-1} \text{ mol}^{-1}$; the latter value is often convenient when pressures are given in atmospheres.

or

$$\left[\frac{\partial}{\partial T}\left(\frac{\Delta G}{T}\right)\right]_P = -\frac{\Delta H}{T^2} \tag{3.167}$$

This important thermodynamic relationship is known as the **Gibbs-Helmholtz equation.** If the reactants and products are in their standard states, the equation takes the form

$$\left[\frac{\partial}{\partial T}\left(\frac{\Delta G^\circ}{T}\right)\right]_P = -\frac{\Delta H^\circ}{T^2} \tag{3.168}$$

3.10 Thermodynamic Limitations to Energy Conversion

The laws of thermodynamics have many practical applications to the interconversion of the various forms of energy, problems that are becoming of increasing technical and economic importance. Both the First Law and the Second Law place limits on how much useful energy or work can be obtained from a given source.

First Law Efficiencies

The First Law is merely a statement of the principle of conservation of energy, and its application is very straightforward; any energy that does not serve the purpose intended must be subtracted from the total in order to obtain the amount of useful energy. The efficiency with which the energy contained in any fuel is converted into useful energy varies very widely.* When wood or coal is burned in an open fireplace, about 80% of the heat escapes up the chimney; only about 20% enters the room. A good home furnace, on the other hand, can convert about 75% of the energy in the fuel into useful heat. However, many home furnaces operate at lower efficiencies, of perhaps 50 or 55%. These low efficiencies are simply due to the fact that much of the heat produced in the combustion of the fuel passes to the outside of the building.

Second Law Efficiencies

An entirely different type of problem is encountered with energy-conversion devices for which there is a Carnot, or Second-Law, limitation. We have seen that for a *reversible* engine operating between two temperatures T_h and T_c the efficiency is

$$\frac{T_h - T_c}{T_h}$$

In practice, since the behavior cannot be reversible, a lower efficiency will be obtained. The higher temperature T_h in a modern steam turbine may be 811 K (1000°F) and the lower temperature T_c may be 311 K (100°F). The Carnot efficiency is thus

*Typical efficiency values are given in an article by C. M. Summers, *Scientific American, 225,* 149 (September 1971).

$$\frac{811 - 311}{811} = 0.62 = 62\%$$

However, because the two temperatures cannot be held constant and because the behavior is not reversible, the efficiency actually obtained is more like 47%. This is considerably greater than the efficiency of the old steam engines, which operated at a much lower T_h; their efficiencies were often less than 10%.

If we want to calculate the overall efficiency for the conversion of the energy of a fuel into electricity, we must consider the efficiencies of the three processes involved:

1. Conversion of the energy of a fuel into heat; in a modern boiler this efficiency is typically about 88%.
2. Conversion of the heat into mechanical energy; efficiency is 47% (as noted previously).
3. Conversion of the mechanical energy into electricity; modern generators have a very high efficiency of about 99%.

The overall efficiency is thus

$$0.88 \times 0.47 \times 0.99 = 0.41 = 41\%$$

Nuclear power plants operate at lower efficiencies, largely because of the lower Carnot efficiency. Nuclear reactors are usually run at lower temperatures than boilers burning fossil fuel. A typical value of T_h is 623 K, and if T_c is 311 K, the Carnot efficiency is

$$\frac{623 - 311}{623} = 0.50 = 50\%$$

In practice this is further reduced to about 37% because the process is not reversible; the overall efficiency is 29%.

Refrigeration and Liquefaction

The Carnot limitation to efficiency also becomes important when we consider a refrigerator or a device for liquefying gases. The principle of operation of a refrigerator is shown schematically in Figure 3.14a. A refrigerator consists essentially of a compressor, which can pump vapor out of an evaporator and pump it into a condenser, where it liquefies. The fluid employed is one that has a high latent heat of evaporation. For many years, ammonia, carbon dioxide, and sulfur dioxide were used, but fluorinated hydrocarbons are more commonly used at the present time. Evaporation of the liquid produces cooling, and condensation leads to release of heat. The work done by the compressor thus transfers heat from the evaporator to the condenser.

The thermodynamics of refrigeration were first worked out in 1895 by the German chemist Karl von Linde (1842–1934). The essential principles are illustrated in Figure 3.14b. Suppose that the refrigerator is operating reversibly and that the condenser is at a temperature of 20°C and the evaporator at a temperature of 0°C. We thus have a Carnot engine operating in reverse, and the ratio of the absolute tem-

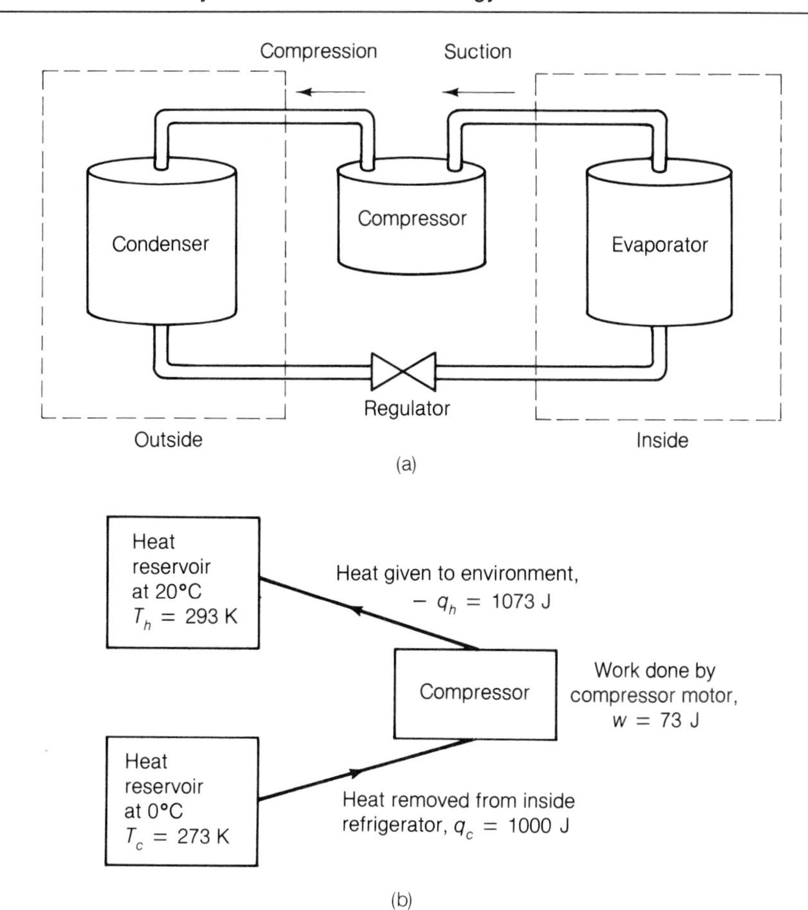

FIGURE 3.14 (a) Schematic diagram of a domestic refrigerator. (b) Analysis of the operation of a domestic refrigerator, on the assumption of reversible behavior.

peratures, 293/273, is the ratio of the heat liberated at the higher temperature to the heat absorbed at the lower temperature:

$$\frac{293}{273} = -\frac{q_h}{q_c}$$

Thus if the evaporator removes 1000 J from inside the refrigerator, at 273, the amount of heat $-q_h$ discharged by the condenser into its environment at 293 is

$$-q_h = \frac{293}{273} \times 1000 = 1073 \text{ J}$$

The difference, 73 J, is the work that has to be done by the compressor. We can define the *performance factor* of the refrigerator as the heat removed from the environment at the lower temperature divided by the work done by the compressor; the performance factor in this example, for reversible behavior, is thus

$$\frac{1000}{73} = 13.7$$

In general, the maximum performance factor for a refrigerator is given by

$$\text{max performance factor} = \frac{T_c}{T_h - T_c} \tag{3.169}$$

In practice the performance is considerably less than this since the cycle does not operate reversibly; in order to maintain the inside of the refrigerator at 0°C the temperature of the evaporator will have to be significantly less than 0°C, and the condenser will be significantly warmer than its surroundings.

When much lower temperatures are required, it is necessary to employ *cascade* processes. The first practical way of liquefying air, based on work done earlier by Linde, was devised in 1877 by Raoul Pierre Pictet (1846–1929), a Swiss chemist and refrigeration engineer. The critical temperature of air (approximately 20% O_2, 79% N_2, and 1% Ar) is about −141°C, and this temperature must be reached before pressure will bring about liquefaction. Ammonia has a critical temperature of 132.9°C, and it can therefore be liquefied by the same type of expansion and compression cycle as in Figure 3.14; this process provides a vessel with liquid ammonia at −34°C. A stream of ethylene compressed to 19 atm is passed through this bath and cooled to about −31°C, which is below its critical point (9.6°C); on being throttled through a valve, two-thirds of it is liquefied and collects in a bath. When it boils away at atmospheric pressure, it cools to −104°C. The next stage employs methane at 25 atm, which is cooled by the liquid ethylene and produces, on liquefaction and subsequent evaporation at 1 atm, liquid methane at −161°C in bath. This suffices to liquefy the nitrogen after the latter has been compressed by 18.6 atm.

Heat Pumps

The heat pump works on exactly the same principle as the refrigerator, but the purpose is to bring about heating instead of cooling. Figure 3.15a shows schematically the type of arrangement; the building is being maintained at 30°C with an external temperature of −15°C. The temperature ratio is 303/258 = 1.17; therefore, in order for 1000 J of heat to leave the condenser (Figure 3.15b),

$$\frac{1000}{1.17} = 855 \text{ J}$$

must be extracted from the external environment. The difference, 145 J, is the work that must be done by the compressor, assuming all processes to be reversible. The performance factor can be defined as the heat provided to the building divided by the work done by the compressor, and in this example it is thus

$$\frac{1000}{145} = 6.9$$

In general, the performance factor of a heat pump, operating reversibly, is given by

$$\text{max performance factor} = \frac{T_h}{T_h - T_c} \tag{3.170}$$

In practice, the performance factor will be much less, since the behavior will be far from reversible.

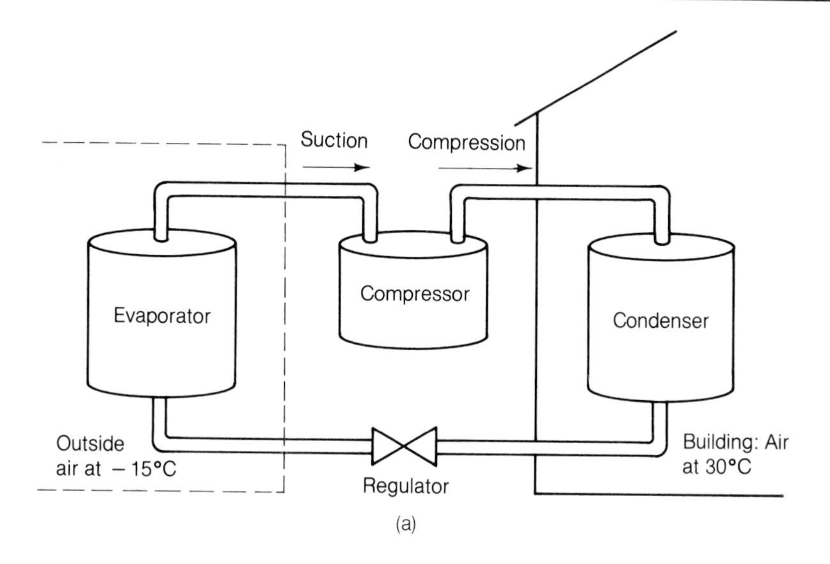

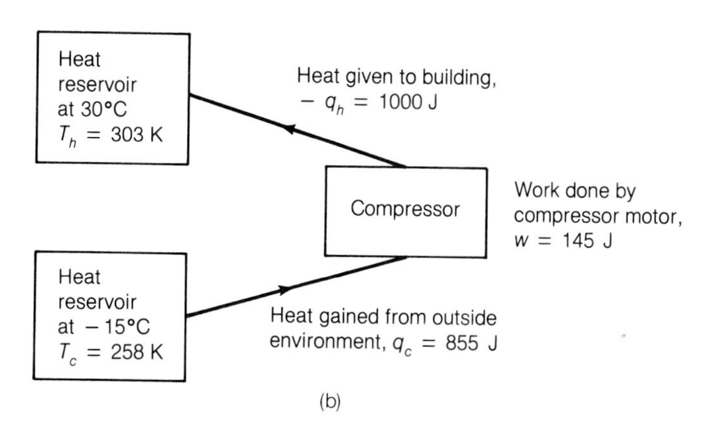

FIGURE 3.15 (a) Schematic diagram of a heat pump used for heating a building to 30°C, with an external temperature of −15°C. (b) Analysis of the operation of the heat pump.

The principle of the heat pump has been known for over a century, but it is only in recent years that the idea has been tried out to any extent. It is obvious that in theory the heat pump could lead to a considerable saving of energy, but unfortunately there are a number of practical problems. One of these is that if the evaporator is maintained in the atmosphere outside the building, the air near to the evaporator will undergo considerable cooling, and the performance factor will fall. A better procedure is to maintain the evaporator in a rapidly flowing river, in which case the external temperature remains fairly constant. Heat pumps have therefore been employed in particular for the heating of buildings that are close to rivers, such as the River Thames at London.

Chemical Conversion

The thermodynamics of the conversion of the chemical energy of a fuel into heat and work is best considered with reference to the energy and enthalpy changes and the Gibbs and Helmholtz energy changes involved. Consider, for example, the combustion of 1 mol of isooctane (2,2,4-trimethylpentane), an important constituent of gasoline:

$$C_8H_{18}(g) + 12\tfrac{1}{2}O_2(g) \rightarrow 8CO_2(g) + 9H_2O(g)$$

Measurement of the heat of this reaction, in a calorimeter at constant volume, leads to

$$\Delta U = -5109 \text{ kJ mol}^{-1}$$

The stoichiometric sum* $\Sigma\nu$ for the process is $8 + 9 - (12\frac{1}{2} + 1) = 3.5$, with the result that Eq. 2.33 becomes

$$\begin{aligned}\Delta H &= \Delta U + \Sigma\nu\, RT \\ &= -5\,109\,000 \text{ J mol}^{-1} + 3.5 \times 8.314 \text{ J K}^{-1} \text{ mol}^{-1} \times 209.15 \text{ K} \\ &= -5\,100\,000 \text{ J mol}^{-1} = -5100 \text{ kJ mol}^{-1}\end{aligned}$$

From the absolute entropies of the reactants and products it is found that, at 298 K,

$$\Delta S = 422 \text{ J K}^{-1} \text{ mol}^{-1}$$

We thus obtain

$$\begin{aligned}\Delta A = \Delta U - T\,\Delta S &= -5\,109\,000 \text{ J mol}^{-1} - 422 \text{ J K}^{-1} \text{ mol}^{-1} \times 298.15 \text{ K} \\ &= -5\,235\,000 \text{ J mol}^{-1} = -5235 \text{ kJ mol}^{-1}\end{aligned}$$

and

$$\begin{aligned}\Delta G = \Delta H - T\,\Delta S &= -5\,100\,000 \text{ J mol}^{-1} - 422 \text{ J K}^{-1} \text{ mol}^{-1} \times 298.15 \text{ K} \\ &= -5\,226\,000 \text{ J mol}^{-1} = -5226 \text{ kJ mol}^{-1}\end{aligned}$$

These values, with the negative signs removed, are the maximum amounts of work that could be obtained by the combustion of 1 mol of isooctane, at constant volume and constant pressure, respectively.

Note that the maximum amount of work that could be obtained from the oxidation of 1 mol of isooctane at constant volume, 5235 kJ, is actually greater than the heat liberated at constant volume, 5109 kJ. There is, of course, no violation of the First Law; the system is not isolated and heat will enter from the environment; there is an entropy increase in the system and an entropy decrease in the environment. In practice, however, it will never be possible to obtain anything like 5235 kJ of work from 1 mol of isooctane. If the fuel is simply burned in a calorimeter, all the energy is released as heat, and no work is done. If it is burned in an internal combustion engine, a good deal of heat would again be released; a typical efficiency for an automobile engine is 25%, so that it might be possible to obtain about 1300 kJ of work from 1 mol. To obtain more work we must devise a process in which less heat is evolved. A possible arrangement is a fuel cell (Section 8.7), in which the isooctane is catalytically oxidized at the surface of an electrode, with the production of an electric potential. Much research is going on at the present time with the object of increasing the efficiencies of such devices.

---

*The *stoichiometric sum* $\Sigma\nu$ is the sum of the stoichiometric coefficients (Section 2.6) in a reaction. Since stoichiometric coefficients are negative for reactants and positive for products the stoichiometric sum is the change in the number of molecules for the reaction as written; e.g. for $A + B \longrightarrow Z$, $\Sigma\nu = -1$.

Another example is provided by the reaction between hydrogen and oxygen:

$$2H_2(g) + O_2(g) \rightarrow 2H_2O(l)$$

The thermodynamic data are

$$\Delta U = -562.86 \text{ kJ mol}^{-1} \qquad \Delta H = -570.30 \text{ kJ mol}^{-1}$$

$$\Delta A = -466.94 \text{ kJ mol}^{-1} \qquad \Delta G = -474.38 \text{ kJ mol}^{-1}$$

Note that now the maximum work per mole obtainable at constant volume, 466.94 kJ, is *less* than the energy released; this is because there is an entropy *loss* on the formation of liquid water from gaseous hydrogen and oxygen. In a fuel cell at constant pressure it would be theoretically possible to obtain 474.38 kJ mol^{-1} of electrical work, somewhat less than the heat evolved (570.30 kJ mol^{-1}) in the combustion at constant pressure. In practice the efficiency of a modern fuel cell is about 60%, so that about 280 kJ of work might be produced from the reaction of 2 mol of hydrogen with 1 mol of oxygen.

Key Equations

Definition of entropy change ΔS:

$$\Delta S_{A \to B} \equiv \int_A^B \frac{dq_{\text{rev}}}{T}$$

For any completely reversible cycle,

$$\oint \frac{dq_{\text{rev}}}{T} \equiv \oint dS = 0$$

If any part of the cycle is irreversible,

$$\oint \frac{dq_{\text{irr}}}{T} < 0 \qquad \text{(the inequality of Clausius)}$$

Entropy of mixing of ideal gases, per mole of mixture:

$$\Delta S_{\text{mix}} = -R(x_1 \ln x_1 + x_2 \ln x_2)$$

where x_1 and x_2 are the mole fractions.

Definition of Helmholtz energy A: $A \equiv U - TS$

Definition of Gibbs energy G: $G \equiv H - TS$

Conditions for equilibrium:

Constant V and T: $dA = 0$

Constant P and T: $dG = 0$

$$dG = w_{\text{non-}PV} \qquad (w_{\text{non-}PV} = \text{non-}PV \text{ work})$$

Important relationships:

$$\left(\frac{\partial U}{\partial V}\right)_S = -P \qquad \left(\frac{\partial U}{\partial S}\right)_V = T$$

$$\left(\frac{\partial H}{\partial V}\right)_S = V \qquad \left(\frac{\partial H}{\partial S}\right)_P = T$$

$$\left(\frac{\partial A}{\partial V}\right)_T = -P \qquad \left(\frac{\partial A}{\partial T}\right)_V = -S$$

$$\left(\frac{\partial G}{\partial P}\right)_T = V \qquad \left(\frac{\partial G}{\partial T}\right)_P = -S$$

Maxwell's relations:

$$\left(\frac{\partial T}{\partial V}\right)_S = -\left(\frac{\partial P}{\partial S}\right)_V \qquad \left(\frac{\partial T}{\partial P}\right)_S = \left(\frac{\partial V}{\partial S}\right)_P$$

$$\left(\frac{\partial P}{\partial T}\right)_V = \left(\frac{\partial S}{\partial V}\right)_T \qquad \left(\frac{\partial V}{\partial T}\right)_P = -\left(\frac{\partial S}{\partial P}\right)_T$$

The *Gibbs-Helmholtz equation:*

$$\left[\frac{\partial}{\partial T}\left(\frac{\Delta G}{T}\right)\right]_P = -\frac{\Delta H}{T^2}$$

Problems

A The Carnot Cycle (see also Section 3d)

1. The accompanying diagram represents a reversible Carnot cycle for an ideal gas:

a. What is the thermodynamic efficiency of the engine?

b. How much heat is rejected at the lower temperature, 200 K, during the isothermal compression?

c. What is the entropy increase during the isothermal expansion at 1000 K?

d. What is the entropy decrease during the isothermal compression at 200 K?

e. What is the entropy change during the adiabatic expansion $B \rightarrow C$?

f. What is the overall entropy change for the entire cycle?

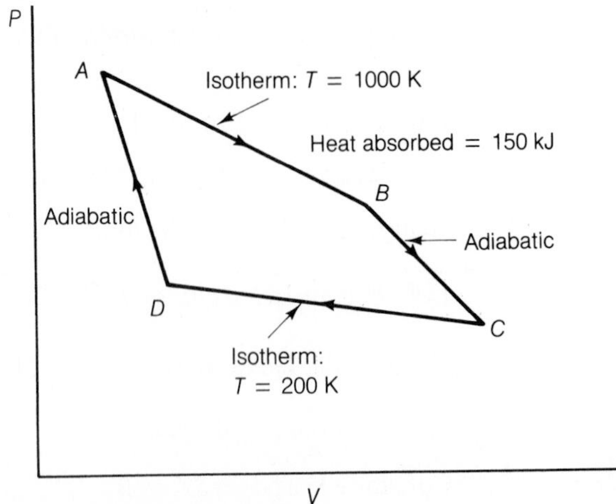

g. What is the increase in Gibbs energy during the process $A \rightarrow B$?
h. If $C_p = 200$ J K^{-1}, what is the increase in Gibbs energy in the process $D \rightarrow A$?

2. An engine operates between 125°C and 40°C. What is the minimum amount of heat that must be withdrawn from the reservoir to obtain 1500 J of work?

***3. a.** Figure 3.2 shows a Carnot cycle in the form of a pressure-volume diagram. Sketch the corresponding entropy-temperature diagram, labeling the individual steps $A \rightarrow B$ (isotherm at T_h), $B \rightarrow C$ (adiabatic), $C \rightarrow D$ (isotherm at T_c), and $D \rightarrow A$ (adiabatic).
b. Suppose that a reversible Carnot engine operates between 300 K and a higher temperature T_h. If the engine produces 10 kJ of work per cycle and the entropy change in the isothermal expansion at T_h is 100 J K^{-1}, what are q_h, q_c, and T_h?

B Entropy Changes

4. Calculate the entropies of vaporization in J K^{-1} mol^{-1} of the following substances, from their boiling points and enthalpies of vaporization:

	Boiling Point/K	$\Delta_{vap}H$/kJ mol^{-1}
C_6H_6	353	30.8
$CHCl_3$	334	29.4
H_2O	373	40.6
C_2H_5OH	351	38.5

In terms of the structures of the liquids, suggest reasons for the higher values observed for H_2O and C_2H_5OH.

5. Calculate the standard entropies of formation of (a) liquid methanol and (b) solid urea, making use of the absolute entropies listed in Table 3.2.

6. One mole of an ideal gas, with $C_{V,m} = \frac{3}{2}R$, is heated (a) at constant pressure and (b) at constant volume, from 298 K to 353 K. Calculate ΔS for the system in each case.

7. One mole each of N_2 and O_2 and $\frac{1}{2}$ mol of H_2, at 25°C and 1 atm, are mixed isothermally; the final total pressure is 1 atm. Calculate ΔS, on the assumption of ideal behavior.

8. Initially 1 mol of O_2 is contained in a 1-dm^3 vessel, and 5 mol of N_2 are in a 2-dm^3 vessel; the two vessels are connected by a tube with a stopcock. If the stopcock is opened and the gases mix, what is the entropy change?

9. Calculate the entropy of mixing per mole of air, taking the composition by volume to be 79% N_2, 20% O_2, and 1% Ar.

10. From the data given in Table 3.2, calculate the standard entropy of formation $\Delta_f S°$ of liquid ethanol at 25°C.

11. a. One mole of an ideal gas at 25°C is allowed to expand reversibly and isothermally from 1 dm^3 to 10 dm^3. What is ΔS for the gas, and what is ΔS for its surroundings?
b. The same gas is expanded adiabatically and irreversibly from 1 dm^3 to 10 dm^3 with no work done. What is the final temperature of the gas? What is ΔS for the gas, and what is ΔS for the surroundings? What is the net ΔS?

12. Predict the signs of the entropy changes in the following reactions when they occur in aqueous solution;

a. Hydrolysis of urea:

$$H_2NCONH_2 + H_2O \rightarrow CO_2 + 2NH_3$$

b. $H^+ + OH^- \rightarrow H_2O$
c. $CH_3COOH \rightarrow CH_3COO^- + H^+$
d. $CH_2BrCOOCH_3 + S_2O_3^{2-} \rightarrow CH_2(S_2O_3^-)COOCH_3 + Br^-$

13. Obtain a general expression, in terms of the molar heat capacity $C_{P,m}$ and temperatures T_1 and T_2, for the entropy increase of n mol of a gas (not necessarily ideal) that is heated at constant pressure so that its temperature changes from T_1 to T_2. To what does your expression reduce if the gas is ideal?

14. Initially 5 mol of an ideal gas, with $C_{V,m} = 12.5$ J K^{-1} mol^{-1}, are at a volume of 5 dm^3 and a temperature of 300 K. If the gas is heated to 373 K and the volume changed to 10 dm^3, what is the entropy change?

***15.** At 100°C 200 g of mercury is added to 80 g of water at 20°C in a vessel that has a water equivalent of 20 g. The specific heat capacities of water and mercury may be taken as constant at 4.18 and 0.140 J K^{-1} g^{-1}, respectively. Calculate the entropy change of (a) the mercury; (b) the water and vessel; (c) the mercury, water, and vessel together.

***16.** At 0°C 20 g of ice are added to 50 g of water at 30°C in a vessel that has a water equivalent of 20 g. Calculate the entropy changes in the system and in the surroundings. The heat of fusion of ice at 0°C is 6.02 kJ mol^{-1}, and the specific heat capacities of water and ice may be taken as constant at 4.184 and 2.094 J K^{-1} g^{-1}, respectively, and independent of temperature.

***17.** Calculate the increase in entropy of 1 mol of nitrogen if it is heated from 300 K to 1000 K at a constant pressure of 1 atm; use the C_p data in Table 2.1.

***18.** The entropy change for the isothermal expansion of an ideal gas at 300 K from a particular state A to a state B is 50 J K^{-1}. When an expansion was performed, the work done by the system was 6 kJ. Was the process reversible or irreversible? If the latter, calculate the *degree of irreversibility* (i.e., the ratio of the work done to the reversible work).

C Gibbs and Helmholtz Energies

19. Calculate $\Delta G°$ at 25°C for the following fermentation reaction:

$$\underset{\text{glucose}}{C_6H_{12}O_6(aq)} \rightarrow \underset{\text{ethanol}}{2C_2H_5OH(aq)} + 2CO_2(g)$$

The standard Gibbs energies of formation of glucose, ethanol, and carbon dioxide are given in Table 2.2.

20. The latent heat of vaporization of water at 100°C is 40.6 kJ mol^{-1} and when 1 mol of water is vaporized at 100°C and 1 atm pressure, the volume increase is 30.19 dm^3. Calculate the work done by the system, the change in internal energy ΔU, the change in Gibbs energy ΔG, and the entropy change ΔS.

21. On page 104 we worked out the ΔS values for the freezing of ice at 0°C and at −10°C. What are the corresponding ΔG values?

22. At 25°C 1 mol of an ideal gas is expanded isothermally from 2 to 20 dm^3. Calculate ΔU, ΔH, ΔS, ΔA, and ΔG. Do the values depend on whether the process is reversible or irreversible?

23. The values of ΔH and ΔS for a chemical reaction are −85.2 kJ mol^{-1} and −170.2 J K^{-1} mol^{-1}, respectively, and the values can be taken to be independent of temperature.

a. Calculate ΔG for the reaction at 300 K, 600 K, and 1000 K.
b. At what temperature would ΔG be zero?

24. The heat of vaporization of water at 25°C is 44.01 kJ mol^{-1}, and the equilibrium vapor pressure at that temperature is 0.0313 atm. Calculate ΔS, ΔH, and ΔG when 1 mol of liquid water at 25°C is converted into vapor at 25°C and a pressure of 10^{-5} atm, assuming the vapor to behave ideally.

25. For each of the following processes, state which of the quantities ΔU, ΔH, ΔS, ΔA, and ΔG are equal to zero:

a. Isothermal reversible expansion of an ideal gas.
b. Adiabatic reversible expansion of a nonideal gas.
c. Adiabatic expansion of an ideal gas through a throttling valve.
d. Adiabatic expansion of a nonideal gas through a throttling valve.
e. Vaporization of liquid water at 80°C and 1 atm pressure.
f. Vaporization of liquid water at 100°C and 1 atm pressure.
g. Reaction between H_2 and O_2 in a thermally insulated bomb.
h. Reaction between H_2SO_4 and NaOH in dilute aqueous solution at constant temperature and pressure.

26. Calculate the change ΔG_m in the Gibbs energy of 1 mol of liquid mercury initially at 1 atm pressure if a pressure of 1000 atm is applied to it. The process occurs at the constant temperature of 25°C, and the mercury may be assumed to be incompressible and to have a density of 13.5 g cm^{-3}.

27. The entropy of argon is given to a good approximation by the expression

$$S_m/\text{J K}^{-1}\text{ mol}^{-1} = 36.36 + 20.79 \ln (T/\text{K})$$

Calculate the change in Gibbs energy of 1 mol of argon if it is heated at constant pressure from 25°C to 50°C.

28. Initially at 300 K and 1 atm pressure, 1 mol of an ideal gas undergoes an irreversible isothermal expansion in which its volume is doubled, and the work it performs is 500 J mol^{-1}. What are the values of q, ΔU, ΔH, ΔG, and ΔS? What would q and w be if the expansion occurred reversibly?

***29.** At 100°C 1 mol of liquid water is allowed to expand isothermally into an evacuated vessel of such a volume that the final pressure is 0.5 atm. The amount of heat absorbed in the process was found to be 30 kJ mol^{-1}. What are w, ΔU, ΔH, ΔS, and ΔG?

***30.** Water vapor can be maintained at 100°C and 2 atm pressure for a time, but it is in a state of metastable equilibrium and is said to be *supersaturated*. Such a system will undergo spontaneous condensation; the process is

$$H_2O(g, 100°\text{C}, 2\text{ atm}) \rightarrow H_2O(l, 100°\text{C}, 2\text{ atm})$$

Calculate ΔH_m, ΔS_m, and ΔG_m. The molar enthalpy of vaporization $\Delta_{vap}H_m$ is 40.60 kJ mol^{-1}; assume the vapor to behave ideally and liquid water to be incompressible.

*31. Initially at 300 K and 10 atm pressure, 1 mol of a gas is allowed to expand adiabatically against a constant pressure of 4 atm until equilibrium is reached. Assume the gas to be ideal with

$$C_{P,m}/\text{J K}^{-1}\text{ mol}^{-1} = 28.58 + 1.76 \times 10^{-2}T/\text{K}$$

and calculate ΔU, ΔH, and ΔS.

D Energy Conversion

32. At 100 atm pressure water boils at 312°C, whereas at 5 atm it boils at 152°C. Compare the Carnot efficiencies of 100- and 5-atm steam engines, if T_c is 30°C.

33. A cooling system is designed to maintain a refrigerator at −4°C in a room at 20°C. If 10^4 J of heat leaks into the refrigerator each minute, and the system works at 40% of its maximum thermodynamic efficiency, what is the power requirement in watts? [1 watt (W) = 1 J s^{-1}.]

34. A heat pump is employed to maintain the temperature of a house at 25°C. Calculate the maximum performance factor of the pump when the external temperature is (a) 20°C, (b) 0°C, and (c) −20°C.

35. A typical automobile engine works with a cylinder temperature of 2000°C and an exit temperature of 800°C. A typical octane fuel (molar mass = 114.2 g mol^{-1}) has an enthalpy of combustion of −5500 kJ mol^{-1}, and 1 dm^3 (0.264 U.S. gal) has a mass of 0.80 kg. Calculate the maximum amount of work that could be performed by the combustion of 10 dm^3 of the fuel.

36. The temperature of a building is maintained at 20°C by means of a heat pump, and on a particular day the external temperature is 10°C. The work is supplied to the heat pump by a heat engine that burns fuel at 1000°C and operates at 20°C. Calculate the performance factor for the system (i.e., the ratio of the heat delivered to the building to the heat produced by the fuel in the heat engine). Assume perfect efficiencies of the pump and the engine.

37. Suppose that a refrigerator cools to 0°C, discharges heat at 25°C, and operates with 40% efficiency.

a. How much work would be required to freeze 1 kg of water ($\Delta_f H = -6.02$ kJ mol^{-1})?

b. How much heat would be discharged during the process?

E Supplementary Problems

38. Prove that, for a gas obeying the van der Waals equation

$$\left(P + \frac{a}{V_m^2}\right)(V_m - b) = RT, \qquad \left(\frac{\partial U}{\partial V_m}\right)_T = \frac{a}{V_m^2}$$

39. Obtain an expression for the Joule-Thomson coefficient μ for a gas obeying the equation of state

$$P(V_m - b) = RT$$

in terms of R, T, P_m, V_m, and $C_{P,m}$.

40. Derive the following equations:

a. $C_p = -T\left(\frac{\partial^2 G}{\partial T^2}\right)_P$

b. $\left(\frac{\partial C_p}{\partial P}\right)_T = -T\left(\frac{\partial^2 V}{\partial T^2}\right)_P$

41. Starting with the definition of the Helmholtz energy, $A \equiv U - TS$, prove that the change in Helmholtz energy for a process at constant temperature is the total work (PV and non-PV). (This relationship holds without any restriction as to volume or pressure changes.)

42. Prove that if a gas obeys Boyle's law and if in addition $(\partial U/\partial V)_T = 0$, it must obey the equation of state PV = constant × T.

43. Derive the relationship

$$\left(\frac{\partial S}{\partial V}\right)_U = \frac{P}{T}$$

and confirm that it applies to an ideal gas.

F Essay Questions

1. The frying of a hen's egg is a spontaneous reaction and has a negative Gibbs energy change. The process can apparently be reversed by feeding the fried egg to a hen and waiting for it to lay another egg. Does this constitute a violation of the Second Law? Discuss.*

2. Consider the following statements:

a. In a reversible process there is no change in the entropy.

*In answering this question, a student commented that a hen would never eat a fried egg. We suspect she would if she were hungry and had no alternative. In any case, let us postulate a hen sufficiently eccentric to eat a fried egg.

b. In a reversible process the entropy change is

$$\frac{dq_{\text{rev}}}{T}$$

How must these statements be qualified so that they are correct and not contradictory?

3. Consider the following statements:

a. The solution of certain salts in water involves a decrease in entropy.

b. For any process to occur spontaneously there must be an increase in entropy.

Qualify these statements so that they are correct and not contradictory, and suggest a molecular explanation for the behavior.

4. A phase transition, such as the melting of a solid, can occur reversibly and therefore $\Delta S = 0$. But it is often stated that melting involves an entropy increase. Reconcile these two statements.

Suggested Reading

See Suggested Reading for Chapter 2 (p. 85).

K. Denbigh, *The Principles of Chemical Equilibrium*, Cambridge: University Press, 1961.

D. K. C. MacDonald, *Near Zero: An Introduction to Low Temperature Physics*, Garden City, N.Y.: Doubleday, 1961.

K. Mendelssohn, *The Quest for Absolute Zero*, New York: McGraw-Hill, 1966.

H. J. Morowitz, *Entropy for Biologists*, New York: Academic Press, 1970.

P. M. Morse, *Thermal Physics*, New York: W. A. Benjamin, 1962.

C. M. Summers, "The Conversion of Energy," *Scientific American*, 225, 149 (September 1971).

M. Tribus and E. C. McIrvine, "Energy and Information," *Scientific American*, 225, 179 (September 1971).

An excellent account of the application of thermodynamic principles to practical problems of energy conservation is to be found in *Efficient Use of Energy* (A.I.P. Conference Proceedings No. 25), published in 1975 by the American Institute of Physics, 335 East 45th Street, New York, N.Y. 10017.

Chemical Equilibrium

Preview

In Chapter 3 we saw that certain thermodynamic functions give conditions for *equilibrium.* In particular, if a system is maintained at constant temperature and pressure, it settles down at an equilibrium state in which its Gibbs energy is a minimum. This is true of both physical and chemical processes, and in the present chapter we treat chemical processes in greater detail.

It has been found empirically that the equilibrium condition for a chemical process depends only on the *stoichiometry* and not at all on the mechanism by which the process occurs. For a general reaction of stoichiometry

$$aA + bB + \cdots \rightleftharpoons \cdots + yY + zZ,$$

involving gases that behave ideally, equilibrium is established at a given temperature when the ratio

$$\left(\frac{P_Y^y P_Z^z \cdots}{P_A^a P_B^b \cdots}\right)_{eq}$$

has a particular value. The P's are the partial pressures, and this ratio is the *equilibrium constant* with respect to pressure and given the symbol K_P. The equilibrium constant can also be expressed in terms of concentrations and then has the symbol K_c. In this chapter the equilibrium constant expressions are deduced on the basis of thermodynamic arguments.

The thermodynamic equilibrium constant K^u is related to the standard Gibbs energy change for the process by

$$\Delta G^\circ = -RT \ln K^u$$

This thermodynamic equilibrium constant K^u is a dimensionless quantity. If the standard state for ΔG° is 1 atm pressure, K^u is K_P^u, which is the dimensionless form of K_P in which pressures are in atmospheres; if the standard state is 1 mol dm^{-3}, the K^u is K_c^u, which is the dimensionless form of K_c in which concentrations are expressed in mol dm^{-3}.

Deviations from ideal behavior are dealt with by using *activities* instead of concentrations. Two or more chemical reactions can be *coupled* together, by either a common reactant or a catalyst.

The variation of an equilibrium constant with the temperature provides a useful way of obtaining enthalpy changes for chemical reactions.

4 Chemical Equilibrium

The concept of equilibrium had its origin very early in the history of chemistry. The first quantitative work was carried out in 1862 by the French chemists P. E. M. Berthollet (1827–1907) and Péan de St. Gilles, who studied the equilibrium

$$CH_3COOH + C_2H_5OH \rightleftharpoons CH_3COOC_2H_5 + H_2O$$

They showed that at a fixed temperature the concentration ratio

$$\frac{[CH_3COOC_2H_5][H_2O]}{[CH_3COOH][C_2H_2OH]}$$

is always the same after equilibrium is reached. Much later work has established that for any reaction

$$a\text{A} + b\text{B} + \cdots \rightleftharpoons y\text{Y} + z\text{Z} + \cdots$$

the ratio

$$\left(\frac{\cdots [\text{Y}]^y[\text{Z}]^z}{[\text{A}]^a[\text{B}]^b \cdots}\right)_{\text{eq}}$$

is constant at a given temperature. This ratio is known as the **equilibrium constant** and given the symbol K_c.

In 1864 the Norwegians Cato Maximillian Guldberg (1836–1902) and Peter Waage (1833–1900) arrived at the correct equilibrium expression on the basis of an argument that is now known, rather paradoxically, to have no general validity. Their procedure for the reaction written previously was to write the rates in forward and reverse directions as

$$v_1 = k_1[\text{A}]^a[\text{B}]^b \cdots \quad \text{and} \quad v_{-1} = k_{-1}[\text{Y}]^y[\text{Z}]^z \cdots \tag{4.1}$$

At equilibrium, when there is no net concentration change, these rates are equal and therefore

$$\left(\frac{\cdots [\text{Y}]^y[\text{Z}]^z}{[\text{A}]^a[\text{B}]^b \cdots}\right)_{\text{eq}} = \frac{k_1}{k_{-1}} = K_c \tag{4.2}$$

Although this procedure always gives the correct result, it is unsatisfactory, because the rate equations are not necessarily those given in Eq. 4.1. The rate equations in fact depend on the mechanism, as we will see in Chapters 9 and 10. The equilibrium expression, however, is independent of the mechanism and is correctly given by Eq. 4.2.

The only satisfactory theoretical basis for the equilibrium equation is provided by thermodynamics. The first thermodynamic treatment was given in 1874–1878 by Willard Gibbs, and shortly afterward the Dutch chemist Jacobus Henricus van't Hoff (1852–1911) independently published a similar treatment.

4.1 Chemical Equilibrium Involving Ideal Gases

Consider a gas-phase reaction

$$a\text{A} + b\text{B} = y\text{Y} + z\text{Z}$$

in which all the reactants and products are ideal gases. We have seen in Eq. 3.96 that when the volume of n mol of an ideal gas changes from V_1 to V_2 at constant temperature T, the Gibbs energy change is

$$\Delta G = nRT \ln \frac{V_1}{V_2} \tag{4.3}$$

Since the pressures P_1 and P_2 are inversely proportional to the volumes V_1 and V_2, we also have

$$\Delta G = nRT \ln \frac{P_2}{P_1} \tag{4.4}$$

The standard state for a gas is 1 atm, and if P_1 is taken to be 1 atm, Eq. 4.4 becomes

$$\Delta G = nRT \ln (P_2/\text{atm}) = nRT \ln P_2^u \tag{4.5}$$

where for convenience we have written the *value* of the pressure, a dimensionless quantity, as P_2^u. The Gibbs energy at pressure P (we will now drop the subscript 2) is thus greater than that at 1 atm by $nRT \ln P^u$. If the Gibbs energy at 1 atm is denoted as $G°$, that at pressure P is thus given by

$$G = G° + nRT \ln P^u \tag{4.6}$$

For 1 mol of gas this equation becomes

$$G_m = G_m^\circ + RT \ln P^u \tag{4.7}$$

The molar Gibbs energy is known as the **chemical potential** and is given the symbol μ (mu); thus

$$\mu = \mu° + RT \ln P^u \tag{4.8}$$

If n_A mol of A are present in a mixture of A, B, Y, etc., the chemical potential μ_A of A is defined as

$$\boldsymbol{\mu}_\text{A} \equiv \left(\frac{\partial \boldsymbol{G}}{\partial \boldsymbol{n}_\text{A}}\right)_{T,P,n_\text{B},n_\text{Y}\ldots} \tag{4.9}$$

The subscripts indicate that the temperature, pressure, and amounts of all components other than A itself are held constant.

In the reaction under consideration we have a mol of A, the Gibbs energy of which is given by

$$G_\text{A} = a\mu_\text{A} = a\mu_\text{A}^\circ + aRT \ln P_\text{A}^u \tag{4.10}$$

where μ_A° is the standard chemical potential of A (i.e., the Gibbs energy of 1 mol of A at a pressure of 1 atm). Similarly, for B, Y, and Z,

$$G_\text{B} = b\mu_\text{B} = b\mu_\text{B}^\circ + bRT \ln P_\text{B}^u \tag{4.11}$$

$$G_Y = y\mu_Y = y\mu_Y^\circ + yRT \ln P_Y^u \tag{4.12}$$

$$G_Z = z\mu_Z = z\mu_Z^\circ + zRT \ln P_Z^u \tag{4.13}$$

The increase in Gibbs energy when a mol of A (at pressure P_A) react with b mol of B (at pressure P_B) to give y mol of Y (at pressure P_Y) and z mol of Z (at pressure P_Z) is thus

$$\Delta G = y\mu_Y + z\mu_Z - a\mu_A - b\mu_B \tag{4.14}$$

$$= y\mu_X^\circ + z\mu_Z^\circ - a\mu_A^\circ - \mathrm{b}\mu_B^\circ + RT \ln\left(\frac{P_Y^y P_Z^z}{P_A^a P_B^b}\right)^u \tag{4.15}$$

$$= \Delta G^\circ + RT \ln\left(\frac{P_Y^y P_Z^z}{P_A^a P_B^b}\right)^u \tag{4.16}$$

where ΔG°, equal to $y\mu_Y^\circ + z\mu_Z^\circ - a\mu_A^\circ - b\mu_B^\circ$ is the *standard Gibbs energy change*. The *standard state* for this ΔG° is 1 atm pressure.

If the pressures, P_A, P_B, P_Y, P_Z correspond to equilibrium pressures, the Gibbs energy change ΔG is equal to zero, and therefore

$$\Delta G^\circ = -RT \ln\left(\frac{P_Y^y P_Z^z}{P_A^a P_B^b}\right)^u_{\mathrm{eq}} \tag{4.17}$$

Since ΔG° is independent of pressure, the ratio of pressures in Eq. 4.17 is also a constant, which is the thermodynamic equilibrium constant K_P^u:

$$\left(\frac{P_Y^y P_Z^z}{P_A^a P_B^b}\right)^u_{\mathrm{eq}} = K_P^u \tag{4.18}$$

Note that this equilibrium constant K_P^u is a *dimensionless* quantity; it has the same *value* as the pressure equilibrium constant K_P, which in general has units and which is defined by

$$K_P = \left(\frac{P_Y^y P_Z^z}{P_A^a P_B^b}\right)_{\mathrm{eq}} \tag{4.19}$$

From Eqs. 4.17 and 4.18 we have

$$\boldsymbol{\Delta G^\circ = -RT \ln K_P^u} \tag{4.20}$$

If the initial and final pressures do not correspond to equilibrium, Eq. 4.16 applies and can be written as

$$\Delta G = -RT \ln K_P^u + RT \ln\left(\frac{P_Y^y P_Z^z}{P_A^a P_B^b}\right)^u \tag{4.21}$$

The relationship between ΔG° and K_P^u, as expressed by Eq. 4.20, is illustrated in Figure 4.1.

Equilibrium Constant in Concentration Units

It is often convenient to use concentrations instead of pressures in dealing with equilibrium problems. For an ideal gas

$$P = \frac{nRT}{V} = cRT \tag{4.22}$$

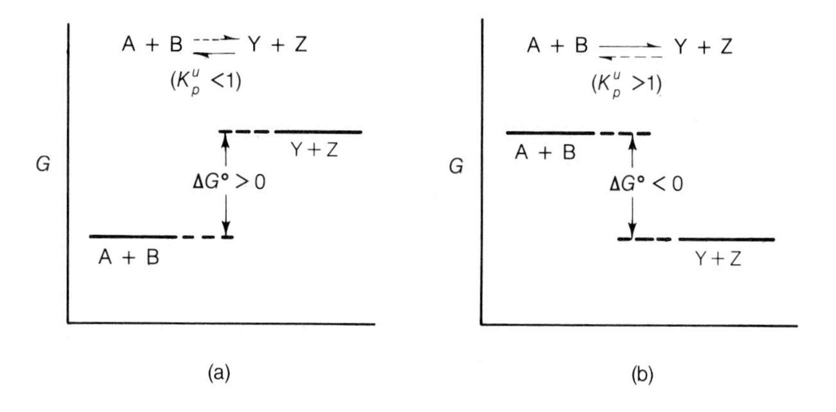

FIGURE 4.1 A standard Gibbs energy diagram, illustrating the relationship between ΔG° for a reaction and the equilibrium constant K_P. (a) ΔG° is positive and $K_P^\circ < 1$; (b) ΔG° is negative and $K_P^\circ > 1$; (c) ΔG° is zero and $K_P^\circ = 1$.

Substitution of the expressions for P_Y, P_Z, P_A, and P_B into Eq. 4.19 leads to

$$K_P = \left(\frac{c_Y^y c_Z^z}{c_A^a c_B^b}\right)_{eq} (RT)^{y+z-a-b} \tag{4.23}$$

This is more conveniently written as

$$K_P = \left(\frac{[Y]^y[Z]^z}{[A]^a[B]^b}\right)_{eq} (RT)^{\Sigma\nu} \tag{4.24}$$

where $\Sigma\nu$, the stoichiometric sum, is the difference between the coefficients in the stoichiometric equation. If we write the equilibrium constant in terms of concentrations as

$$K_c = \left(\frac{[Y]^y[Z]^z}{[A]^a[B]^b}\right)_{eq} \tag{4.25}$$

we see that K_c and K_P are related by

$$K_P = K_c(RT)^{\Sigma\nu} \tag{4.26}$$

From K_c we can calculate a ΔG°, using the relationship

$$\mathbf{\Delta G^\circ = -RT \ln K_c^u} \tag{4.27}$$

where K_c^u is the numerical value of K_c and is thus the *standard* equilibrium constant in terms of concentration.

If the concentrations are in mol dm^{-3}, the standard state for $\Delta G°$ is 1 mol dm^{-3}. In general, the change of standard state will change the value of $\Delta G°$, and it is important always to indicate the standard state. By analogy with Eq. 4.16 we also have

$$\Delta G = \Delta G° + RT \ln\left(\frac{[Y]^y[Z]^z}{[A]^a[B]^b}\right)^u \tag{4.28}$$

The superscript u to the ratio again indicates that we have made it dimensionless. This is the Gibbs energy change when a mol of A, at concentration [A], reacts with b mol of B, at concentration [B], to produce y mol of Y, at concentration [Y], and z mol of Z, at concentration [Z].

Alternatively, we can work with mole fractions. The pressure P_A of substance A is related to the total pressure P by

$$P_A = x_A P \tag{4.29}$$

where x_A is the mole fraction of A; similar equations apply to B, Y, and Z. Substitution of these expressions into Eq. 4.18 gives

$$K_P = \left(\frac{x_Y^y x_Z^z}{x_A^a x_B^b}\right)_{eq} P^{y+z-a-b} \tag{4.30}$$

Thus, if we write the equilibrium constant in terms of mole fractions as K_x, namely

$$K_x = \left(\frac{x_Y^y x_Z^z}{x_A^a x_B^b}\right)_{eq} \tag{4.31}$$

we see that K_P and K_x are related by

$$K_P = K_x P^{\Sigma\nu} \tag{4.32}$$

Since, for ideal gases, K_P is a true constant at constant temperature, it follows that K_x will be pressure dependent unless $\Sigma\nu = 0$. The ratio K_x is thus only constant at constant temperature and constant total pressure. The functions K_P and K_c, on the other hand, are (for an ideal gas) functions of temperature only.

From K_x, which is dimensionless, we can calculate a Gibbs energy change,

$$\Delta G° = -RT \ln K_x \tag{4.33}$$

the standard state for which is unit mole fraction. Also,

$$\Delta G = \Delta G° + RT \ln\left(\frac{x_Y^y x_Z^z}{x_A^a x_B^b}\right) \tag{4.34}$$

This is the Gibbs energy change when a mol of A at mole fraction x_A reacts with b mol of B at mole fraction x_B to yield y mol of Y at mole fraction x_Y plus z mol of Z at mole fraction x_Z.

Units of the Equilibrium Constant

The *practical* equilibrium constant in general has units, which depend on the type of reaction. Consider, for example, a reaction of the type

$$A + B \rightleftharpoons Z$$

in which there is a decrease, by one, in the number of molecules. The equilibrium constant for the standard state of 1 mol dm^{-3} is then

$$K_c = \frac{[Z]}{[A][B]} \tag{4.35}$$

and since the units of [A], [B], and [Z] are mol dm^{-3}, the units of K_c are dm^3 mol^{-1}. If there is no change in the number of molecules, as in a reaction of the type

$$A + B \rightleftharpoons Y + Z$$

the units cancel out, so that K_c is dimensionless. In general, if the stoichiometric sum is $\Sigma\nu$, the units are

$$\text{mol}^{\Sigma\nu}\ \text{dm}^{-3\Sigma\nu}$$

Another important point about equilibrium constants is that their value depends on how the stoichiometric equation is written. Consider, for example, the dissociation

1. $A \rightleftharpoons 2Z$

the equilibrium constant K_c for which is

$$K_c = \frac{[Z]^2}{[A]} \tag{4.36}$$

If the concentrations are expressed as mol dm^{-3}, the units of K_c are mol dm^{-3}. Alternatively, we could write the reaction as

2. $\frac{1}{2}A \rightleftharpoons Z$

and express the equilibrium constant as

$$K_c' = \frac{[Z]}{[A]^{1/2}} \tag{4.37}$$

This latter constant K_c' is obviously the square root of K_c:

$$K_c' = K_c^{1/2} \tag{4.38}$$

If the concentrations are in mol dm^{-3}, the units of K_c' are mol$^{1/2}$ dm$^{-3/2}$. The standard Gibbs energy change for reaction 1 is

$$\Delta G_1^\circ = -RT \ln K_c^\circ \tag{4.39}$$

while that for reaction 2 is

$$\Delta G_2^\circ = -RT \ln K_c^{\circ\prime} \tag{4.40}$$

Since

$$\ln K_c^{\circ\prime} = \tfrac{1}{2} \ln K_c^\circ \tag{4.41}$$

we see that

$$\Delta G_2^\circ = \tfrac{1}{2}\, \Delta G_1^\circ \tag{4.42}$$

This is as it should be, since reaction 2 to which ΔG_2° refers is for the conversion of

0.5 mol of A into X; for reaction 1 we are referring to the conversion of 1 mol of A into 2X.

EXAMPLE

The Gibbs energies of formation of $NO_2(g)$ and $N_2O_4(g)$ are 51.30 and 102.00 kJ mol^{-1}, respectively (standard state: 1 atm and 25°C).

a. Assume ideal behavior and calculate, for the reaction $N_2O_4 \rightleftharpoons 2NO_2$, K_P (standard state: 1 atm) and K_c (standard state: 1 mol dm^{-3}).
b. Calculate K_x at 1 atm pressure.
c. At what pressure is N_2O_4 50% dissociated?
d. What is $\Delta G°$ if the standard state is 1 mol dm^{-3}?

SOLUTION

For the reaction

$$N_2O_4(g) \rightleftharpoons 2NO_2(g)$$

$$\Delta G°(\text{standard state: 1 atm}) = 2 \times 51.30 - 102.0 = 0.60 \text{ kJ mol}^{-1}$$

a. Therefore, from Eq. 4.20,

$$\ln(K_P/\text{atm}) = -\frac{600}{8.314 \times 298.15} = -0.242$$

$$K_P = 0.785 \text{ atm}$$

From Eq. 4.26, since $\Sigma\nu = 1$

$$K_c = \frac{K_P}{RT} = \frac{0.785(\text{atm}) \times 1.013\ 25 \times 10^5(\text{Pa atm}^{-1})}{8.314(\text{J K}^{-1}\text{ mol}^{-1}) \times 298.15(\text{K})}$$

$$= 32.1 \text{ mol m}^{-3}$$

$$= 3.21 \times 10^{-2} \text{ mol dm}^{-3}$$

b. From Eq. 4.32, with $\Sigma\nu = 1$,

$$K_x = K_P P^{-1} = 0.785 \qquad \text{at } P = 1 \text{ atm}$$

c. Suppose that we start with 1 mol of N_2O_4 and that α mol have become converted into NO_2; the amounts at equilibrium are

$$\begin{array}{ccc} N_2O_4 & \rightleftharpoons & 2NO_2 \\ 1-\alpha & & 2\alpha \end{array}$$

and the total amount is $(1 + \alpha)$ mol. If P is the total pressure, the partial pressures are

$$N_2O_4: \quad \frac{1-\alpha}{1+\alpha}P$$

$$NO_2: \quad \frac{2\alpha}{1+\alpha}P$$

The equilibrium equation is thus

$$K_P^u = K_P/\text{atm} = 0.785 = \frac{[2\alpha/(1+\alpha)]^2(P/\text{atm})^2}{[(1-\alpha)/(1+\alpha)](P/\text{atm})} = \frac{4\alpha^2 P/\text{atm}}{1-\alpha^2}$$

If $\alpha = 0.5$,

$$0.785 = \frac{4 \times 0.25 \times P/\text{atm}}{1 - 0.25} = \frac{P/\text{atm}}{0.75}$$

and therefore

$$P = 0.589 \text{ atm}$$

d. K_c, with the concentrations in mol dm^{-3}, is 3.21×10^{-2} mol dm^{-3}; then

$$\Delta G°/\text{J mol}^{-1} = -8.314 \times 298.15 \ln(3.21 \times 10^{-2})$$

$$\Delta G° = 8524 \text{ J mol}^{-1} = 8.52 \text{ kJ mol}^{-1}$$

■

EXAMPLE

At 3000 K the equilibrium partial pressures of CO_2, CO, and O_2 are 0.6, 0.4, and 0.2 atm, respectively. For the reaction $2CO_2 \rightleftharpoons 2CO + O_2$,

a. Calculate $\Delta G°$(standard state: 1 atm).
b. Calculate K_x at 1 atm.

SOLUTION

$$K_P = \frac{(0.4)^2(0.2)}{(0.6)^2} = 0.0889 \text{ atm}$$

a. $\Delta G°/\text{J mol}^{-1} = -8.314 \times 3000 \ln 0.0889$

$$\Delta G° = 60\,370 \text{ J mol}^{-1} = 60.37 \text{ kJ mol}^{-1}$$

b. $\Sigma\nu = 1$

so that

$$K_x = \frac{K_P}{1 \text{ atm}} = 0.0889$$

■

4.2 Equilibrium in Nonideal Gaseous Systems

We saw in Section 3.8 that the thermodynamics of real gases are dealt with by making use of *fugacities* or *activities*, which play the same role for real gases that pressures do for ideal gases.

The procedure for treating chemical equilibria involving nonideal gases is to follow through exactly the same treatment as for ideal gases, replacing pressures by activities that are dimensionless. Then, instead of Eq. 4.20 we have

$$\Delta G° = -RT \ln K_a^u \tag{4.43}$$

where K_a^u is the dimensionless equilibrium constant in terms of activities:

$$K_a^u = \left(\frac{a_X^x a_Y^y}{a_A^a a_B^b}\right)_{eq}^u \tag{4.44}$$

Similarly, in place of Eq. 4.16, we have

$$\Delta G = \Delta G^\circ + RT \ln\left(\frac{a_X^x a_Y^y}{a_A^a a_B^b}\right)^u = -RT \ln K_a^u + RT \ln\left(\frac{a_X^x a_Y^y}{a_A^a a_B^b}\right)^u \tag{4.45}$$

4.3 Chemical Equilibrium in Solution

We have seen that chemical equilibria in ideal gases can be formulated in terms of the molar concentrations of the reactants and products. The same relationships frequently apply to equilibria involving substances present in solution. Equations involving mole fractions also often apply to a good approximation to reactions in solution. Alternatively, molalities (mol kg^{-1}) may be used instead of concentrations. When any of these relationships apply, the behavior is said to be *ideal*.

When molalities are used, deviations from ideality are taken care of by replacing the molality m by the activity a, which is the molality multiplied by the activity coefficient γ. The Gibbs energy is then given in terms of the dimensionless activity a^u by

$$G = G^\circ + RT \ln a^u = G^\circ + RT \ln \gamma m^u \tag{4.46}$$

where G° is the value at unit activity. The standard equilibrium constant for a reaction

$$a\text{A} + b\text{B} \rightleftharpoons y\text{Y} + z\text{Z}$$

is then given by

$$K_a^u = \left(\frac{a_Y^y a_Z^z}{a_A^a a_B^b}\right)^u = \left(\frac{m_Y^x m_Z^y}{m_A^a m_B^b}\right)^u \cdot \left(\frac{\gamma_Y^y \gamma_Z^z}{\gamma_A^a \gamma_B^b}\right) \tag{4.47}$$

where the superscripts ^{u} indicate that we have made the ratios dimensionless. A similar procedure is followed if we work with concentrations or with mole fractions. The symbol y is used for an activity coefficient used with concentration, while f is used for an activity coefficient used with mole fraction.

The theory of activities of species in solution and the way activities are determined experimentally are considered further in Section 5.9. For uncharged species in solution the behavior is often close to ideal, and equilibrium constants are then sometimes satisfactorily expressed in terms of molalities, concentrations, or mole fractions, without the use of activity coefficients. For ions, however, there may be serious deviations from ideality, and activities must be used. The activity coefficients of ions are considered in Sections 7.10 and 8.5.

The usual procedure in dealing with *solvent* species is to use the mole fraction. The Gibbs energy is expressed as

$$G = G^\circ + RT \ln x_1 f_1 \tag{4.48}$$

where x_1 is the mole fraction of the solvent and f_1 its activity coefficient.

EXAMPLE

The equilibrium constant K_c for the reaction

$$\text{fructose-1,6-diphosphate} \rightleftharpoons \text{glyceraldehyde-3-phosphate} + \text{dihydroxyacetone phosphate}$$

is 8.9×10^{-5} M at 25°C, and we can assume the behavior to be ideal.

a. Calculate $\Delta G°$ for the process (standard state: 1 M).

b. Suppose that we have a mixture that is initially 0.01 M in fructose-1, 6-diphosphate and 10^{-5} M in both glyceraldehyde-3-phosphate and dihydroxyacetone phosphate. What is ΔG? In which direction will reaction occur?

SOLUTION

a.
$$\Delta G° = -RT \ln K^u$$
$$\Delta G°/\text{J mol}^{-1} = -8.314 \times 298.15 \ln (8.9 \times 10^{-5})$$
$$\therefore \quad \Delta G° = 23\,120 \text{ J mol}^{-1} = 23.12 \text{ kJ mol}^{-1}$$

b.
$$\Delta G = \Delta G° + RT \ln \frac{[\text{G-3-P}][\text{DHAP}]/\text{mol dm}^{-3}}{[\text{FDF}]}$$
$$\Delta G/\text{J mol}^{-1} = 23\,120 + 8.314 \times 298.15 \ln \frac{10^{-5}10^{-5}}{10^{-2}}$$
$$\therefore \quad \Delta G = (23\,120 - 45\,660) \text{ J mol}^{-1} = -22\,540 \text{ J mol}^{-1} = -22.54 \text{ kJ mol}^{-1}$$

Because ΔG is negative, reaction proceeds spontaneously from left to right. If the initial concentrations had been unity, the positive $\Delta G°$ value would have forced the reaction to the left; however the low concentrations of products have reversed the sign and the reaction proceeds to the right.

Another way of seeing that the reaction will shift to the right is to note that the concentration ratio

$$\frac{[\text{G-3-P}][\text{DHAP}]}{[\text{FDF}]} = \frac{10^{-5}10^{-5}}{10^{-2}} = 10^{-8} \text{ mol dm}^{-3}$$

is *less* than the equilibrium constant 8.9×10^{-5} mol dm^{-3}. ■

4.4 Heterogeneous Equilibrium

When equilibrium involves substances in different phases, such as different states of matter, we speak of **heterogeneous equilibrium**. A simple example is provided by the dissociation of solid calcium carbonate into solid calcium oxide and gaseous carbon dioxide:

$$CaCO_3(s) \rightleftharpoons CaO(s) + CO_2(g)$$

As before, we may write

$$\frac{[CaO][CO_2]}{[CaCO_3]} = K_c' \tag{4.49}$$

The concentration of a solid, however, has a special feature. As always it is the amount of substance divided by the volume. However, in the case of a given solid, the amount contained in a given volume is a fixed quantity. The concentrations $[CaO]$ and $[CaCO_3]$ are therefore constants, independent of the total quantity of solid present.

Since in Eq. 4.49 the concentrations [CaO] and [$CaCO_3$] are constants, we can incorporate them into the constant K_c' and simply write

$$[CO_2] = K_c \tag{4.50}$$

This is the procedure adopted for pure solids and pure liquids; their concentrations are always incorporated into the equilibrium constant. Equation 4.50 can be expressed more exactly in terms of the activity of CO_2:

$$a_{CO_2} = K_c \tag{4.51}$$

It is important to note that if solids are present in solid solution or liquids are present mixed with other liquids, their concentrations are obviously not fixed, and they must not be incorporated into the equilibrium constant in this way.

Another example of heterogeneous equilibrium is that between a salt and its saturated solution. Suppose, for example, that the sparingly soluble salt silver chloride, AgCl, is placed in contact with water. The equilibrium between the solid and its saturated solution, in which there is complete dissociation into Ag^+ and Cl^- ions, is

$$AgCl(s) \rightleftharpoons Ag^+(aq) + Cl^-(aq)$$

The equilibrium equation could be written as

$$\frac{[Ag^+][Cl^-]}{[AgCl]} = K_c' \tag{4.52}$$

Again, however, the concentration [AgCl] is a constant and the convention is to combine it with K_c' to give

$$[Ag^+][Cl^-] = K_s \tag{4.53}$$

This constant K_s is known as the **solubility product**. For greater precision, activities are used:

$$a_{Ag^+}\, a_{Cl^-} = K_s \tag{4.54}$$

4.5 Tests for Chemical Equilibrium

Any reaction

$$aA + bB + \cdots \rightleftharpoons \cdots yY + zZ$$

will eventually reach a state of equilibrium at which

$$\left(\frac{\cdots [Y]^y[Z]^z}{[A]^a[B]^b \cdots}\right)_{eq} = K_c \tag{4.55}$$

It is obviously of great practical importance to determine whether a chemical system is at equilibrium.

It is not enough to establish that the composition of the system is not changing as time goes on. Reaction may be proceeding so slowly that no detectable change will occur over a long period of time. A good example is the reaction

$$2H_2 + O_2 \longrightarrow 2H_2O$$

The equilibrium for this reaction lies almost completely to the right. However, the reaction is so slow that if we bring hydrogen and oxygen together at room temperature, there will be no detectable change even over many hundreds of years. In fact, even after 5×10^9 years (the estimated age of the solar system) only an insignificant amount of this reaction will have taken place!

Obviously, more practical tests for equilibrium are required. One of these consists of adding a substance that speeds up reaction. For example, if to the hydrogen-oxygen mixture we introduce a lighted match, or if we add certain substances known as **catalysts** (such as powdered platinum), the reaction will occur with explosive violence. This shows that the system was not at equilibrium, but it appeared to be at equilibrium because of the slowness of the reaction. The function of catalysts in speeding up reactions will be considered later (Section 10.9); here we may simply note that catalysts do not affect the position of equilibrium but merely decrease the time required for equilibrium to be attained.

A second and more fundamental test for equilibrium is as follows: If a system

$$A + B \rightleftharpoons Y + Z$$

is truly at equilibrium, the addition of a small amount of A (or B) will cause the equilibrium to shift to the right, with the formation of more Y and Z. Similarly, the addition of more Y or Z will cause a shift to the left. If the system is not in a state of true equilibrium, either these shifts will not occur at all or they will not occur in the manner predicted by the equilibrium equations.

4.6 Shifts of Equilibrium at Constant Temperature

One of the important consequences of the theory of equilibrium relates to the way in which equilibria shift as the volume of the system changes. The equilibrium constants K_p and K_c are functions of temperature only and do not change when we vary the volume. However, as a result of this constancy of the equilibrium constant, the *position* of equilibrium does shift as the volume changes. Consider the equilibrium

$$AB \rightleftharpoons A^+ + B^-$$

in which AB is dissociating into ions. The equilibrium constant K_c is

$$K_c = \frac{[A^+][B^-]}{[AB]} \tag{4.56}$$

If we add A^+ to the system, the ratio will temporarily be increased; the equilibrium will have to shift from right to left to maintain K_c constant. Similarly, addition of B^- will cause a shift to the left. Addition of AB will cause a shift to the right.

The effect of changing the *volume* can be seen by expressing each concentration in Eq. 4.56 as the ratio of the amount of each substance to the volume;

$$[AB] = \frac{n_{AB}}{V} \qquad [A^+] = \frac{n_{A^+}}{V} \qquad [B^-] = \frac{n_{B^-}}{V} \tag{4.57}$$

so that Eq. 4.56 becomes

$$\frac{(n_{A^+}/V)(n_{B^-}/V)}{n_{AB}/V} = K_c \tag{4.58}$$

or

$$\frac{n_{A^+}n_{B^-}}{n_{AB}} \cdot \frac{1}{V} = K_c \tag{4.59}$$

Suppose that we dilute the system, by adding solvent and making V larger. Since K_c must remain constant, the ratio

$$\frac{n_{A^+}n_{B^-}}{n_{AB}}$$

must become larger, in proportion to V; that is, there will be more dissociation into ions at the larger volume. This effect was predicted in 1884 by the **principle of Le Châtelier**, formulated by the French chemist Henri Louis Le Châtelier (1850–1936).

4.7 Coupling of Reactions

For a chemical reaction

1. $A + B \rightleftharpoons X + Y$

removal of either of the products X or Y will lead to a shift of equilibrium to the right. Frequently a product is removed by the occurrence of another reaction; for example, X might isomerize into Z:

2. $X \rightleftharpoons Z$

We can write an equilibrium constant for reaction 1,

$$K_1 = \left(\frac{[X][Y]}{[A][B]}\right)_{eq} \tag{4.60}$$

and a corresponding standard Gibbs energy change:

$$\Delta G_1^\circ = -RT \ln K_1 \tag{4.61}$$

Similarly for the second reaction

$$K_2 = \left(\frac{[Z]}{[X]}\right)_{eq} \tag{4.62}$$

and

$$\Delta G_2^\circ = -RT \ln K_2 \tag{4.63}$$

We can add reactions 1 and 2 to obtain

3. $A + B \rightleftharpoons Y + Z$

The equilibrium constant K_3 for the combined reaction 3 is

$$K_3 = \left(\frac{[Y][Z]}{[A][B]}\right)_{eq} \tag{4.64}$$

The same result is obtained by multiplying K_1 and K_2:

$$K_1K_2 = \left(\frac{[X][Y]}{[A][B]}\right)_{eq} \times \left(\frac{[Z]}{[X]}\right)_{eq} = \left(\frac{[Y][Z]}{[A][B]}\right)_{eq} = K_3 \tag{4.65}$$

We can take natural logarithms of both sides of this equation, and then multiply by $-RT$:

$$-RT \ln K_3 = -RT \ln K_1 - RT \ln K_2 \tag{4.66}$$

so that

$$\Delta G_3^\circ = \Delta G_1^\circ + \Delta G_2^\circ \tag{4.67}$$

The result is easily shown to be quite general; if we add reactions, the ΔG° of the resulting reaction is the sum of the ΔG° values of the component reactions. This is an extension of Hess's law (Section 2.6) to standard Gibbs energy changes.

As a result of this **coupling** of chemical reactions, it is quite possible for a reaction to occur to a considerable extent even though it has a positive value of ΔG°. Thus, for the scheme just given, suppose that reaction 1 has a positive value:

$$\Delta G_1^\circ > 0 \tag{4.68}$$

Reaction 1 by itself will therefore not occur to a considerable extent ($K_1 < 1$). Suppose however, that reaction 2 has a negative value. It is then possible for ΔG_3° to be negative, so that A + B may react to a considerable extent to give X + Y. In terms of the Le Châtelier principle, this simply means that X is constantly removed by the isomerization reaction 2, so that the equilibrium of reaction 1 is shifted to the right.

There is another way in which chemical reactions can be coupled. In the cases we have considered, the product of one reaction is removed in a second reaction. Alternatively, two reactions might be coupled by a *catalyst*. Suppose, for example, that we have two independent reactions, having no common reactants or products:

1. $A + B \rightleftharpoons X + Y$
2. $C \rightleftharpoons Z$

There might exist some catalyst that brings about the reaction

3. $A + B + C \rightleftharpoons X + Y + Z$

and this catalyst would couple the reactions 1 and 2. This type of coupling of reactions is quite common in biological systems.

4.8 Temperature Dependence of Equilibrium Constants

In Section 3.9 we derived the Gibbs-Helmholtz equation, Eq. 3.168:

$$\left[\frac{\partial}{\partial T}\left(\frac{\Delta G^\circ}{T}\right)\right]_P = -\frac{\Delta H^\circ}{T^2} \tag{4.69}$$

Since $\Delta G^\circ = -RT \ln K_P^u$ (Eq. 4.20), this at once leads to an equation that gives the temperature dependence of K_P^u; thus

$$\left[\frac{\partial}{\partial T}(-R \ln K_P^u)\right]_P = -\frac{\Delta H^\circ}{T^2} \tag{4.70}$$

or

$$\left(\frac{\partial \ln K_P^u}{\partial T}\right)_P = \frac{\Delta H^\circ}{RT^2} \tag{4.71}$$

The equilibrium constant K_P is affected very little by the pressure (see Section 4.9), and the equation is usually written as

$$\frac{d \ln K_P^u}{dT} = \frac{\Delta H^\circ}{RT^2} \tag{4.72}$$

Since $d(1/T) = -dT/T^2$, the equation can also be written as

$$\frac{d \ln K_P^u}{d(1/T)} = -\frac{\Delta H^\circ}{R} \tag{4.73}$$

This equation tells us that a plot of ln K_P^u against $1/T$ will have a slope equal to $\Delta H^\circ/R$. In Figure 4.2a the slope is varying with temperature so that ΔH° is a function of temperature. Frequently, however, plots of ln K^u against $1/T$ are linear, as in Figure 4.2b, which means that ΔH° is independent of temperature over the range investigated. When this is so, we can integrate Eq. 4.73 as follows:

$$\int d \ln K_P^u = -\int \frac{\Delta H^\circ}{R}\, d\left(\frac{1}{T}\right) \tag{4.74}$$

and thus

$$\ln K_P^u = -\frac{\Delta H^\circ}{RT} + I \tag{4.75}$$

where I is the constant of integration. However, from Eq. 4.20

$$\ln K_P^u = -\frac{\Delta G^\circ}{RT} = -\frac{\Delta H^\circ}{RT} + \frac{\Delta S^\circ}{R} \tag{4.76}$$

The intercept on the ln K_P^u axis, at $1/T = 0$, is therefore $\Delta S^\circ/R$.

Equation 4.72 gives the temperature dependence of K_P, and we can obtain analogous equations for the temperature dependence of K_x, the equilibrium constant in terms of mole fractions, and of K_c, the equilibrium constant in terms of concentrations. Equation 4.32 relates K_P and K_x,

$$K_P = K_x P^{\Sigma\nu} \tag{4.77}$$

and from Eq. 4.71 it follows that

$$\left(\frac{\partial \ln K_x^u}{\partial T}\right)_P = \frac{\Delta H^\circ}{RT^2} \tag{4.78}$$

Since in general K_x is pressure-dependent, it is now necessary to express the condition that the pressure must be held constant.

For ideal gases the concentration equilibrium constant K_c is related to K_P by Eq. 4.26:

$$K_P = K_c(RT)^{\Sigma\nu} \tag{4.79}$$

and therefore

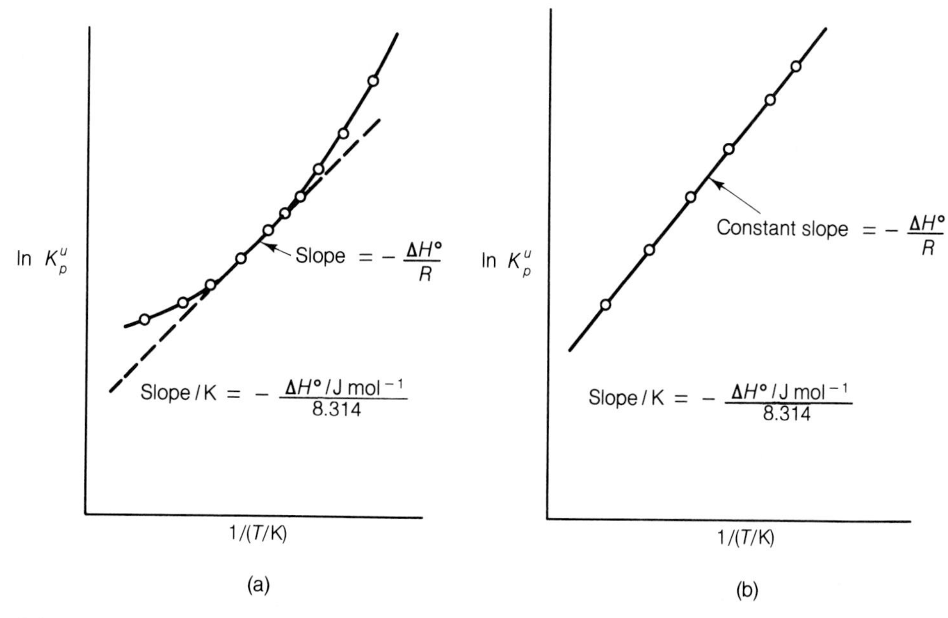

FIGURE 4.2 Schematic plots of ln K_P^u against $1/T$. (a) $\Delta H°$ is temperature dependent; (b) $\Delta H°$ is independent of temperature, in which case Eq. 4.76 applies.

$$\ln K_P^u = \ln K_c^u + \Sigma\nu \ln R^u + \Sigma\nu \ln T^u \tag{4.80}$$

where R^u and T^u are the numerical values of R and T, respectively. Partial differentiation at constant pressure gives

$$\left(\frac{\partial \ln K_P^u}{\partial T}\right)_P = \left(\frac{\partial \ln K_c^u}{\partial T}\right)_P + \frac{\Sigma\nu}{T} \tag{4.81}$$

By Eq. 4.71 the left-hand side of this equation is $\Delta H°/RT^2$, and therefore

$$\left(\frac{\partial \ln K_c^u}{\partial T}\right)_P = \frac{\Delta H°}{RT^2} - \frac{\Sigma\nu}{T} = \frac{\Delta H° - \Sigma\nu\, RT}{RT^2} \tag{4.82}$$

However, $\Delta H° - \Sigma\nu\, RT$ for ideal gases is equal to $\Delta U°$ and therefore

$$\boldsymbol{\frac{d \ln K_c^u}{dT} = \frac{\Delta U°}{RT^2}} \tag{4.83}$$

Again, since K_c^u changes only slightly with pressure, it is usually not necessary to specify constant pressure conditions. For reactions in solution, Eq. 4.83 is the one usually employed, but $\Delta U°$ and $\Delta H°$ are then very close to each other.

EXAMPLE

The equilibrium constant for an association reaction

$$A + B \rightleftharpoons AB$$

is 1.80×10^3 dm^3 mol^{-1} at 25°C and 3.45×10^3 dm^3 mol^{-1} at 40°C. Assuming $\Delta H°$ to be independent of temperature, calculate $\Delta H°$ and $\Delta S°$.

SOLUTION

If the lower temperature, 298.15 K, is written as T_1 and the corresponding dimensionless equilibrium constant as K_1^u,

$$\text{(a)}\quad \ln K_1^u = -\frac{\Delta H^\circ}{RT_1} + \frac{\Delta S^\circ}{R}$$

Similarly, for the higher temperature 313.15 K,

$$\text{(b)}\quad \ln K_2^u = -\frac{\Delta H^\circ}{RT_2} + \frac{\Delta S^\circ}{R}$$

Thus

$$\ln \frac{K_2^u}{K_1^u} = \frac{\Delta H^\circ}{R} \cdot \frac{T_2 - T_1}{T_1 T_2}$$

In our particular example,

$$\ln \frac{3.45}{1.80} = \frac{(\Delta H^\circ/\text{J mol}^{-1})}{8.314} \cdot \frac{15.00}{298.15 \times 313.15}$$

from which it follows that

$$\Delta H^\circ = 33.67 \text{ kJ mol}^{-1}$$

The entropy change ΔS° can be calculated by inserting the values of K^u, ΔH°, and T into Eq. (a). For example,

$$\ln(3.45 \times 10^3) = 8.146 = -\frac{33\,670}{8.314 \times 313.15} + \frac{\Delta S^\circ/\text{J K}^{-1}\text{ mol}^{-1}}{8.314}$$

so that

$$\Delta S^\circ = 175.2 \text{ J K}^{-1} \text{ mol}^{-1}$$ ■

The procedure is somewhat more complicated when ΔH° is a function of temperature. We have seen in Section 2.6 that for a gas the heat capacity at constant pressure can often be expressed as

$$C_P = d + eT + fT^{-2} \tag{4.84}$$

and that as a result the enthalpy change at a temperature T_2 is related to that at T_1 by an equation of the form (see Eq. 2.52)

$$\Delta H_m(T_2) = \Delta H_m(T_1) + \Delta d(T_2 - T_1) + \frac{1}{2}\Delta e(T_2^2 - T_1^2) - \Delta f\left(\frac{1}{T_2} - \frac{1}{T_1}\right) \tag{4.85}$$

If T_1 is 25°C and $\Delta H_m^\circ(25°\text{C})$ refers to specified standard states, ΔH_m° at any temperature T is

$$\Delta H_m^\circ(T) = \Delta H_m^\circ(25°\text{C}) + \Delta d(T - 298.15\text{ K}) + \frac{1}{2}\Delta e[T^2 - (298.15\text{ K})^2] - \Delta f\left(\frac{1}{T} - \frac{1}{298.15\text{ K}}\right) \tag{4.86}$$

Substitution of this into Eq. 4.71, and integration, gives

$$\ln K_P^u = \frac{1}{R} \int \left[\frac{\Delta H_m^\circ}{T^2} + \frac{\Delta d}{T} - \frac{(298.15 \text{ K}) \Delta d}{T^2} + \frac{\Delta e}{2} - \frac{(298.15 \text{ K})^2 \Delta e}{2T^2} - \frac{\Delta f}{T^3} + \frac{\Delta f}{(298.15 \text{ K})T^2} \right] dT + I \quad (4.87)$$

where I is the constant of integration. This equation is readily integrated, so that if ΔH_m°, Δd, Δe, and Δf are known for a particular system, and K is known at one temperature (thus allowing I to be determined), K can be calculated at any other temperature.

4.9 Pressure Dependence of Equilibrium Constants

We have seen that for equilibria involving ideal gases the equilibrium constants K_P and K_c are independent of the total pressure. A shift in equilibrium is brought about by a change in pressure when $\Sigma\nu$ is other than zero, but the equilibrium constant itself does not change. For nonideal systems, however—and this includes nonideal gases and all solids and liquids—there is in general a change in the equilibrium constant itself as the pressure is varied.

The theory of this can be deduced from Eq. 3.119:

$$\left(\frac{\partial G}{\partial P}\right)_T = V \quad (4.88)$$

The change ΔG° for a chemical reaction, in which the reactants in their standard states are transformed into the products in their standard states, is thus

$$\left(\frac{\partial\, \Delta G^\circ}{\partial P}\right)_T = \Delta V^\circ \quad (4.89)$$

where ΔV° is the corresponding volume change. Substitution of $-RT \ln K^u$ for ΔG° then leads to

$$\left(\frac{\partial \ln K^u}{\partial P}\right)_T = -\frac{\Delta V^\circ}{RT} \quad (4.90)$$

Schematic plots of $\ln K^u$ against P are shown in Figure 4.3. In Figure 4.3a the volume change ΔV° is itself a function of pressure, and the slope of the plot varies as the pressure changes. In Figure 4.3b the plot is linear, which means that ΔV° is independent of pressure. In that case, Eq. 4.90 can be integrated as follows:

$$\int d \ln K^u = -\int \frac{\Delta V^\circ}{RT}\, dP \quad (4.91)$$

so that

$$\ln K^u = -\frac{\Delta V^\circ}{RT} P + \text{constant} \quad (4.92)$$

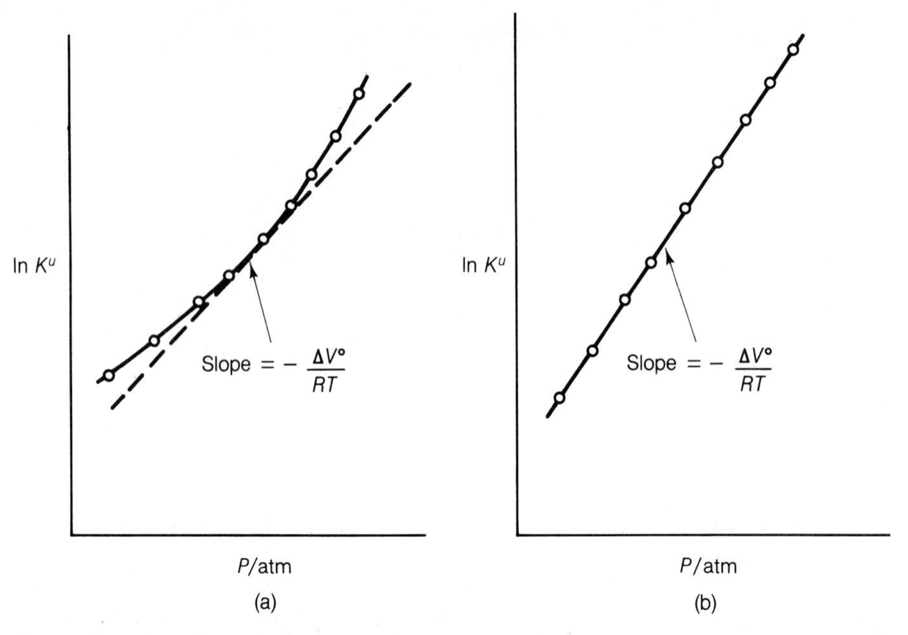

FIGURE 4.3 Plots of ln K^u against pressure, P. (a) $\Delta V°$ dependent on pressure; (b) $\Delta V°$ independent of pressure, in which case Eq. 4.75 applies.

In practice, the change of an equilibrium constant with pressure is fairly small, and quite high pressures have to be used in order to produce an observable effect. Suppose, for example, that $\Delta V°$ for a reaction were 10 $cm^3\ mol^{-1}$ and independent of pressure. The value of $\Delta V°/RT$ at 25°C is then

$$\frac{10^{-5}(m^3\ mol^{-1}) \times 1.013\ 25 \times 10^5(Pa\ atm^{-1})}{8.314(J\ K^{-1}\ mol^{-1}) \times 298.15(K)} = 4.09 \times 10^{-4}\ atm^{-1}$$

A pressure of $1/4.09 \times 10^{-4} = 2445$ atm will therefore change ln K^u by one unit (i.e., will change K by a factor of 2.718).

EXAMPLE

The equilibrium constant for the reaction

$$2NO_2 \rightleftharpoons N_2O_4$$

in carbon tetrachloride solution and 22°C is increased by a factor of 3.77 when the pressure is increased from 1 atm to 1500 atm. Calculate $\Delta V°$, on the assumption that it is independent of pressure.

SOLUTION

An increase by a factor of 3.77 means that ln K^u has increased by 1.327 as P increases by 1500 − 1 = 1499 atm. The slope of a plot of ln K^u versus P is thus

$$\frac{1.327}{1499}\ atm^{-1}$$

and this is equal to $-\Delta V°/RT$. Thus

$$\Delta V^\circ = -\frac{(1.327/1499)(\text{atm}^{-1}) \times 8.314 \times 293.15(\text{J mol}^{-1})}{1.013\ 25 \times 10^5(\text{Pa atm}^{-1})}$$

$$= -2.13 \times 10^{-5}\ \text{m}^3\ \text{mol}^{-1}$$

■

Key Equations

For a reaction $aA + bB + \cdots \rightleftharpoons \cdots yY + zZ$:

$$\left(\frac{\ldots [Y]^y[Z]^z}{[A]^a[B]^b \ldots}\right)_{eq} = K_c$$

where K_c is the equilibrium constant.

Definition of *chemical potential* for species A:

$$\mu_A \equiv \left(\frac{\partial G}{\partial n_A}\right)_{T,P,n_B,n_Y,\ldots}$$

Relationship between *standard Gibbs energy change and equilibrium constant:*

$$\Delta G^\circ = -RT \ln K^u$$

$$\Delta G = \Delta G^\circ + RT \ln\left(\frac{\ldots [Y]^y[Z]^z}{[A]^a[B]^b \ldots}\right)^u$$

Temperature dependence of equilibrium constants:

$$\frac{d \ln K_P^u}{d(1/T)} = -\frac{\Delta H^\circ}{R}$$

$$\frac{d \ln K_c^u}{d(1/T)} = -\frac{\Delta U^\circ}{R}$$

Problems

A Equilibrium Constants

1. A reaction occurs according to the equation

$$2A \rightleftharpoons Y + 2Z$$

If in a volume of 5 dm^3 we start with 4 mol of pure A and find that 1 mol of A remains at equilibrium, what is the equilibrium constant K_c?

2. The equilibrium constant for a reaction

$$A + B \rightleftharpoons Y + Z$$

is 0.1. What amount of A must be mixed with 3 mol of B to yield, at equilibrium, 2 mol of Y?

3. The equilibrium constant for the reaction

$$A + 2B \rightleftharpoons Z$$

is 0.25 $dm^6\ mol^{-2}$. In a volume of 5 dm^3, what amount of A must be mixed with 4 mol of B to yield 1 mol of Z at equilibrium?

4. The equilibrium constant K_c for the reaction

$$2SO_3(g) \rightleftharpoons 2SO_2(g) + O_2(g)$$

is 0.0271 mol dm^{-3} at 1100 K. Calculate K_P at that temperature.

5. When gaseous iodine is heated, dissociation occurs:

$$I_2 \rightleftharpoons 2I$$

It was found that when 0.0061 mol of iodine was placed in a volume of 0.5 dm^3 at 900 K, the degree of dissociation was 0.0274. Calculate K_c and K_P at that temperature.

6. It has been observed with the ammonia equilibrium

$$N_2 + 3H_2 \rightleftharpoons 2NH_3$$

that under certain conditions the addition of nitrogen to an equilibrium mixture, with the *temperature and pressure held constant*, causes further dissociation of ammonia. Explain how this is possible. Under what particular conditions would you expect this to occur? Would it be possible for added hydrogen to produce the same effect?

7. Nitrogen dioxide, NO_2, exists in equilibrium with dinitrogen tetroxide, N_2O_4:

$$N_2O_4(g) \rightleftharpoons 2NO_2(g)$$

At 25.0°C and a pressure of 0.589 atm the density of the gas is 1.477 g dm^{-3}. Calculate the degree of dissociation under those conditions and the equilibrium constants K_c, K_P, and K_x. What shift in equilibrium would occur if the pressure were increased by the addition of helium gas?

8. At 25.0°C the equilibrium

$$2NOBr(g) \rightleftharpoons 2NO(g) + Br_2(g)$$

is rapidly established. When 1.10 g of NOBr is present in a 1.0-dm^3 vessel at 25.0°C the pressure is 0.35 atm. Calculate the equilibrium constants K_c, K_P, and K_x.

9. At 100°C and 2 atm pressure the degree of dissociation of phosgene is 6.30×10^{-5}. Calculate K_P, K_c, and K_x for the dissociation

$$COCl_2(g) \rightleftharpoons CO(g) + Cl_2(g)$$

10. In a study of the equilibrium

$$H_2 + I_2 \rightleftharpoons 2HI$$

1 mol of H_2 and 3 mol of I_2 gave rise at equilibrium to x mol of HI. Addition of a further 2 mol of H_2 gave an additional x mol of HI. What is x? What is K at the temperature of the experiment?

***11.** The equilibrium constant for the reaction

$$H_2(g) + I_2(g) \rightleftharpoons 2HI(g)$$

is 20.0 at 40.0°C, and the vapor pressure of solid iodine is 0.10 atm at that temperature. If 12.7 g of solid iodine are placed in a 10-dm^3 vessel at 40.0°C, what is the minimum amount of hydrogen gas that must be introduced in order to remove all the solid iodine?

B Equilibrium Constants and Gibbs Energy Changes

12. The equilibrium constant for the reaction

$$(C_6H_5COOH)_2 \rightleftharpoons 2C_6H_5COOH$$

in benzene solution at 10°C is 2.19×10^{-3} mol dm^{-3}.

a. Calculate $\Delta G°$ for the dissociation of the dimer.
b. If 0.1 mol of benzoic acid is present in 1 dm^3 of benzene at 10°C, what are the concentrations of the monomer and of the dimer?

13. At 3000 K the equilibrium partial pressures of CO_2, CO, and O_2 are 0.6, 0.4, and 0.2 atm, respectively. Calculate $\Delta G°$ at 3000 K for the reaction

$$2CO_2(g) \rightleftharpoons 2CO(g) + O_2(g)$$

14. The conversion of malate into fumarate

1. malate(aq) $\rightleftharpoons$ fumarate(aq) + $H_2O(l)$

is endergonic at body temperature, 37°C; $\Delta G°$ is 2.93 kJ mol^{-1}. In metabolism the reaction is coupled with

2. fumarate (aq) $\rightleftharpoons$ aspartate(aq)

for which $\Delta G°$ is −15.5 kJ mol^{-1} at 37°C.

a. Calculate K_c for reaction 1.
b. Calculate K_c for reaction 2.
c. Calculate K_c and $\Delta G°$ for the coupled reaction 1 + 2.

15. From the data in Table 2.2, deduce the $\Delta G°$ and K_P values for the following reactions at 25.0°C:

a. $N_2(g) + 3H_2(g) \rightleftharpoons 2NH_3(g)$
b. $2H_2(g) + C_2H_2(g) \rightleftharpoons C_2H_6(g)$
c. $H_2(g) + C_2H_4(g) \rightleftharpoons C_2H_6(g)$
d. $2CH_4(g) \rightleftharpoons C_2H_6(g) + H_2(g)$

16. At 25.0°C the equilibrium constant for the reaction

$$CO(g) + H_2O(g) \rightleftharpoons CO_2(g) + H_2(g)$$

is 1.00×10^{-5}, and $\Delta S^\circ = -41.8 \text{ J K}^{-1} \text{ mol}^{-1}$.

a. Calculate ΔG° and ΔH° at 25.0°C.
b. Suppose that 2 mol of CO and 2 mol of H_2O are introduced into a 10-dm³ vessel at 25.0°C. What are the amounts of CO, H_2O, CO_2, and H_2 at equilibrium?

17. For the reaction

1. glutamate + $NH_4^+ \rightleftharpoons$ glutamine

the ΔG° value at 37.0°C is 15.7 kJ mol^{-1}. (Standard state = 1 mol dm^{-3}.) In biological systems the enzyme glutamine synthetase couples this reaction with

2. ATP $\rightleftharpoons$ ADP + phosphate

for which $\Delta G^\circ = -31.0$ kJ mol^{-1}. Calculate the equilibrium constant at 37.0°C for these two reactions and for the coupled reaction

3. glutamate + NH_4^+ + ATP $\rightleftharpoons$ glutamine + ADP + phosphate

C Temperature Dependence of Equilibrium Constants

18. A gas reaction

$$A \rightleftharpoons B + C$$

is endothermic and its equilibrium constant K_P is 1 atm at 25°C.

a. What is ΔG° at 25°C (standard state: 1 atm)?
b. Is ΔS°, with the same standard state, positive or negative?
c. For the standard state of 1 M, what are K_c and ΔG°?
d. Will K_P at 40°C be greater than or less than 1 atm?
e. Will ΔG° at 40°C (standard state: 1 atm) be positive or negative?

19. A solution reaction

$$A + B \rightleftharpoons X + Y$$

is endothermic, and K_c at 25° is 10.

a. Is the formation of X + Y exergonic at 25°C?
b. Will raising the temperature increase the equilibrium yield of X + Y?
c. Is ΔS° positive or negative?

20. From the data given in Table 2.2, for the reaction

$$C_2H_4(g) + H_2(g) \rightleftharpoons C_2H_6(g)$$

calculate the following:

a. ΔG°, ΔH°, and ΔS° at 25°C; what is the standard state?
b. K_P at 25°C.
c. K_c at 25°C (standard state: 1 M).
d. ΔG° at 25°C (standard state: 1 M).
e. ΔS° at 25°C (standard state: 1 M).
f. K_P at 100°C, on the assumption that ΔH° and ΔS° are temperature-independent.

21. From the data in Table 2.2, for the reaction

$$2H_2(g) + O_2(g) \rightleftharpoons 2H_2O(g)$$

calculate the following:

a. ΔG°, ΔH°, and ΔS° at 25°C (standard state: 1 atm).
b. K_P at 25°C.
c. ΔG° and K_P at 2000°C, on the assumption that ΔH° and ΔS° are temperature independent.

22. The hydrolysis of adenosine triphosphate to give adenosine diphosphate and phosphate can be represented by

$$ATP \rightleftharpoons ADP + P$$

The following values have been obtained for the reaction at 37°C (standard state: 1 M):

$$\Delta G^\circ = -31.0 \text{ kJ mol}^{-1}$$

$$\Delta H^\circ = -20.1 \text{ kJ mol}^{-1}$$

a. Calculate ΔS°.
b. Calculate K_c at 37°C.
c. On the assumption that ΔH° and ΔS° are temperature independent, calculate ΔG° and K_c at 25°C.

23. Thermodynamic data for n-pentane (g) and neopentane (g) (standard state: 1 atm and 25°C) are as follows

	Enthalpy of Formation, ΔH_f° kJ mol^{-1}	Entropy, S° J K^{-1} mol^{-1}
n-Pentane(g)	−146.44	349.0
Neopentane(g)	−165.98	306.4

a. Calculate ΔG° for n-pentane → neopentane.
b. Pure n-pentane is in a vessel at 1 atm and 25°C, and a catalyst is added to bring about the equilibrium between n-pentane and neopentane. Calculate the final partial pressures of the two isomers.

24. a. The equilibrium constant K_c is increased by a factor of 3 when the temperature is raised

from 25.0°C to 40.0°C. Calculate the standard enthalpy change.
b. What is the standard enthalpy change if instead K_c is *decreased* by a factor of 3 under the same conditions?

25. The ionization constant $[H^+][OH^-]$ for the dissociation of water,

$$H_2O \rightleftharpoons H^+ + OH^-$$

is 1.00×10^{-14} mol² dm^{-6} at 25.0°C and 1.45×10^{-14} mol² dm^{-6} at 30°C. Deduce $\Delta H°$ for the process.

26. The equilibrium constant K_P for the reaction

$$I_2(g) + \text{cyclopentane}(g) \rightleftharpoons 2HI(g) + \text{cyclopentadiene}(g)$$

varies with temperature according to the equation

$$\log_{10}(K_P/\text{atm}) = 7.55 - \frac{4844}{T/K}$$

a. Calculate K_P, $\Delta G°$, $\Delta H°$, $\Delta S°$ (standard state: 1 atm) at 400°C.
b. Calculate K_c and $\Delta G°$ (standard state: 1 M) at 400°C.
c. If I_2 and cyclopentene are initially at 400°C and at concentrations of 0.1 M, calculate the final equilibrium concentrations of I_2, cyclopentene, HI, and cyclopentadiene.

27. From the data in Table 2.2, for the synthesis of methanol,

$$CO(g) + 2H_2(g) \rightleftharpoons CH_3OH(l)$$

calculate $\Delta H°$, $\Delta G°$, and $\Delta S°$ and the equilibrium constant at 25°C.

28. The bacterium *nitrobacter* plays an important role in the "nitrogen cycle" by oxidizing nitrite to nitrate. It obtains the energy it requires for growth from the reaction

$$NO_2^-(aq) + \tfrac{1}{2}O_2(g) \longrightarrow NO_3^-(aq)$$

Calculate $\Delta H°$, $\Delta G°$, and $\Delta S°$ for this reaction from the following data, at 25°C:

	$\Delta H_f°$ / kJ mol^{-1}	$\Delta G_f°$ / kJ mol^{-1}
$NO_2^-(aq)$	−104.6	−37.2
$NO_3^-(aq)$	−207.4	−111.3

29. When the reaction

$$\text{glucose-1-phosphate}(aq) \rightleftharpoons \text{glucose-6-phosphate}(aq)$$

is at equilibrium at 25°C, the amount of glucose-6-phosphate present is 95% of the total.

a. Calculate $\Delta G°$ at 25°C.
b. Calculate ΔG for reaction in the presence of $10^{-2}M$ glucose-1-phosphate and $10^{-4}M$ glucose-6-phosphate. In which direction does reaction occur under these conditions?

30. From the data in Table 2.2, for the reaction

$$CO_2(g) + H_2(g) \rightleftharpoons CO(g) + H_2O(g)$$

calculate the following:
a. $\Delta H°$, $\Delta G°$, and ΔS (standard state: 1 atm and 25°C).
b. The equilibrium constant at 25°C.
c. From the heat capacity data in Table 2.1, obtain an expression for $\Delta H°$ as a function of temperature.
d. Obtain an expression for ln K_P^u as a function of temperature.
e. Calculate K_P at 1000 K.

31. Irving Langmuir [*J. Am. Chem. Soc., 28*, 1357 (1906)] studied the dissociation of CO_2 into CO and O_2 by bringing the gas at 1 atm pressure into contact with a heated platinum wire. He obtained the following results:

T/K	Percent Dissociation
1395	0.0140
1443	0.0250
1498	0.0471

Calculate K_P for $2CO_2(g) \rightleftharpoons 2CO(g) + O_2(g)$ at each temperature, and estimate $\Delta H°$, $\Delta G°$, and $\Delta S°$ at 1395 K.

32. G. Stark and M. Bodenstein [*Z. Electrochem., 16*, 961 (1910)] carried out experiments in which they sealed iodine in a glass bulb and measured the vapor pressure using a quartz-fiber manometer. The following are some of the results they obtained:

volume of bulb = 249.8 cm³
amount of iodine = 1.958 mmol

Temperature/°C	Pressure/mmHg
800	558.0
1000	748.0
1200	1019.2

a. Calculate the degree of dissociation at each temperature.
b. Calculate K_c at each temperature, for the process $I_2 \rightleftharpoons 2I$.
c. Calculate K_P at each temperature.
d. Obtain values for $\Delta H°$ and $\Delta U°$ at 1000°C.
e. Calculate $\Delta G°$ and $\Delta S°$ at 1000°C.

D Supplementary Problems

33. Suppose that a large molecule, such as a protein, contains n sites to which a molecule A (a *ligand*) can become attached. Assume that the sites are equivalent and independent, so that the reactions $M + A \rightleftharpoons MA$, $MA + A \rightleftharpoons MA_2$, etc., all have the same equilibrium constant K_s. Show that the average number of occupied sites per molecule is

$$\bar{\nu} = \frac{nK_s[A]}{1 + K_s[A]}$$

34. Modify the derivation in Problem 33 so as to deal with sites that are not all equivalent; the equilibrium constants for the attachments of successive ligands are each different:

$$M + A \rightleftharpoons MA \qquad K_1 = \frac{[MA]}{[M][A]}$$

$$MA + A \rightleftharpoons MA_2 \qquad K_2 = \frac{[MA_2]}{[MA][A]}$$

$$MA_{n-1} + A \rightleftharpoons MA_n \qquad K_n = \frac{[MA_n]}{[MA_{n-1}][A]}$$

Show that the average number of molecules of A bound per molecule M is

$$\bar{\nu} = \frac{K_1[A] + 2K_1K_2[A]^2 + \cdots + n(K_1K_2K_3 \cdots K_n)[A]^n}{1 + K_1[A] + K_1K_2[A]^2 + \cdots + (K_1K_2K_3 \cdots K_n)[A]^n}$$

This equation is important in biology and biochemistry and is often called the *Adair equation*, after the British biophysical chemist G. S. Adair.

35. Now show that the Adair equation, derived in Problem 34, reduces to the equation obtained in Problem 33 when the sites are equivalent and independent. [It is not correct simply to put $K_1 = K_2 = K_3 \cdots = K_n$; certain statistical factors must be introduced. Thus if K_s is the equilibrium constant for the binding at a *given* site, $K_1 = nK_s$ since there are n ways for A to become attached to a given molecule and one way for it to come off. Similarly $K_2 = (n - 1)K_s/2$; $n - 1$ ways on and 2 ways off. Continue this argument and develop an expression for $\bar{\nu}$ that will factorize into $nK_s[A]/(1 + K_s[A])$. Suggest a method of testing the equilibrium obtained and for arriving at a value of n from experimental data.]

36. Another special case of the equation derived in Problem 34 is if the binding on one site affects that on another. An extreme case is highly cooperative binding, in which the binding of A on one site influences the other sites so that they fill up immediately. This means that K_n is much greater than K_1, K_2, etc. Show that now

$$\bar{\nu} = \frac{nK[A]^n}{1 + K[A]^n}$$

where K is the product of $K_1, K_2, \cdots K_n$. The British physiologist A. V. Hill suggested that binding problems can be treated by plotting

$$\ln \frac{\theta}{1 - \theta} \quad \text{against} \quad \ln[A]$$

where θ is the fraction of sites that are occupied. Consider the significance of such Hill plots, especially their shapes and slopes, with reference to the equations obtained in Problems 33 to 35.

37. Protein denaturations are sometimes irreversible but may be reversible under a narrow range of conditions. At pH 2.0, at temperatures ranging from about 40°C to 50°C, there is an equilibrium between the active form P and the deactivated form D of the enzyme trypsin:

$$P \rightleftharpoons D$$

Thermodynamic values are $\Delta H° = 283$ kJ mol^{-1} and $\Delta S° = 891$ J K^{-1} mol^{-1}. Assume these values to be independent of temperature over this narrow range, and calculate $\Delta G°$ and K values at 40.0°C, 42.0°C, 44.0°C, 46.0°C, 48.0°C, and 50.0°C. At what temperature will there be equal concentrations of P and D?

Note that the high thermodynamic values lead to a considerable change in K over this 10°C range.

E Essay Questions

1. Give an account of the effect of temperature on equilibrium constants, and explain how such experimental studies lead to thermodynamic data.

2. Give an account of the effect of pressure on (a) the position of equilibrium and (b) the equilibrium constant.

3. Explain what experimental studies might be made to decide whether a chemical system is at equilibrium or not.

4. Give an account of the coupling of chemical reactions.

Suggested Reading

See listing of Chapter 3.

Chapter 5

Phases and Solutions

Preview

Most substances can exist in more than one *phase* or *state of aggregation*. This chapter examines the conditions for equilibrium at various temperatures and pressures. The criterion for equilibrium between two phases is that the Gibbs energy is the same in the two phases. This condition leads to insight into the thermodynamics of vapor pressure. In particular, the *Clapeyron equation* and the *Clausius-Clapeyron* equation, which are concerned with the variation of vapor pressure with temperature, are two important expressions for treating liquid-vapor systems.

Trouton's rule, according to which most non-hydrogen-bonded compounds have values of $\Delta_{vap}S_m$ in the neighborhood of 88 J K^{-1} mol^{-1}, has proved to be very useful for estimating heats of vaporization.

It is useful to deal with ideal solutions first. Such solutions obey *Raoult's law,* according to which the vapor pressure of each component of a solution is proportional to its mole fraction. For real solutions Raoult's law applies to the *solvent* when the concentration of solute is small, but deviations occur at higher concentrations. At low solute concentrations the *solute* obeys *Henry's law,* according to which the solute vapor pressure is proportional to the solute concentration.

The contribution of each component in a solution is described in terms of *partial molar quantities.* Any extensive thermodynamic property $X = X(T, P, n_i)$ has the partial molar value X_i defined as

$$X_i \equiv \left(\frac{\partial X}{\partial n_i}\right)_{T,P,n_j}$$

All the thermodynamic equations apply to these partial molar quantities. The partial molar Gibbs energy is known as the *chemical potential* μ_i, and for multicomponent systems the chemical potential plays a thermodynamic role equivalent to that of the Gibbs energy.

The concepts of *activity* a and *activity coefficient* f or γ are introduced to handle nonideal solutions. In the *rational system,* activities for the solvent are defined so that $a_i = 1$ and $f_i = 1$ as $x_i \rightarrow 1$. In the *practical system,* activities for the solute are defined so that $a_i = 1$ and $\gamma_i = 1$ as $n_i \rightarrow 0$.

Colligative properties, which depend only on the number of molecules of solute present, are a consequence of vapor pressure lowering as expressed by Raoult's law. Aside from *vapor pressure lowering* the colligative properties are *boiling point elevation, freezing point depression,* and *osmotic pressure*.

5 Phases and Solutions

In Chapter 4 we have been concerned primarily with the study of equilibrium in reacting systems. Up to this point little has been said concerning the nature of the equilibria that can exist within a pure material, or in a solution between a solute and its solvent. No formal chemical reaction need occur in such equilibria. For instance, the melting of ice, the boiling of water, and the dissolving of sugar in coffee are all examples of **phase** (Greek *phasis,* appearance) **changes** or changes in the state of aggregation. For each of these changes an equilibrium relation defines the behavior of the system, in much the same way as the equilibrium constant does for chemically reacting systems. In this chapter we will investigate what constitutes a phase, and the criteria of equilibrium as applied to phases, for both pure substances and solutions. Then the thermodynamics needed to treat the behavior of solutions will be explored. In Chapter 6, the ideas developed in this chapter will be applied to a systematic study in a less mathematical vein of systems consisting of many components and/or phases.

5.1 Phase Recognition

The vaporization of a liquid into its vapor state is an example of a phase change. For pure water we say that both the liquid and the gaseous states are single phases. A single phase is uniform throughout both in chemical composition and physical state, and it is said to be *homogeneous.* Note that subdivision does not produce new phases; a block of ice reduced to crushed ice still consists of only one phase.

In contrast to this, a *heterogeneous* system consists of more than one phase; the phases are distinguished from each other through separation by distinct boundaries. A familiar example is provided by ice cubes in water; there two phases coexist, one solid and one liquid.

Two solids of different substances may be mixed, and if each retains its characterisic boundary, the mixture will consist of two phases. Even if two solids were melted together to give the outward appearance of a single uniform melt, more than one phase would exist if the phase boundaries in the solid can be discerned microscopically.

Mixtures of liquids also can occur in one or more phases. For example, carbon tetrachloride forms a separate layer when mixed with water. In this system, two phases coexist; whereas, for example, only one phase exists in the ethyl alcohol-water system. Although systems with many liquid phases are known, generally only one or two liquid phases are present.

Although solid and liquid mixtures may consist of a number of phases, gases can exist in only one phase at normal pressures since gases mix in all proportions to give a uniform mixture.

Phase Distinctions in the Water System

We will first investigate the equilibria present in the ice, liquid water, water-vapor system. The variables associated with the equilibrium criterion of the Gibbs energy are pressure and temperature, and it is therefore natural to depict the phase equilibria on a pressure-temperature diagram. This is done in Figure 5.1.

The areas marked *solid, liquid,* and *vapor* are regions where only one phase may exist. Where these single-phase regions are indicated, arbitrary values of P and T may exist within the lines defining the phase limits.

The solid lines on the diagram give the conditions under which the two adjoining phases are in equilibrium. Thus, the curve TT_c gives the vapor pressure of water up to the critical point T_c. This line, therefore, defines the pressures and temperatures at which the gas and liquid phases can coexist at equilibrium. The extension of this line to B gives the equilibrium conditions for supercooled water, an example of a substance in a *metastable state* (i.e., not the thermodynamically most stable state). This state can be achieved because the rate of formation of ice has been reduced by using a very clean sample of water, thereby reducing the number of nucleation sites. In a similar manner, the line TC gives the conditions under which solid and liquid coexist at equilibrium; that is, it represents the melting point of ice at different pressures of the solid. The final line TA represents the equilibrium between the vapor and solid and shows the vapor pressure of the solid as a function of temperature. Above 2×10^5 kPa (approximately 2000 atm) different crystalline forms of ice may exist, giving rise to a phenomenon known as *polymorphism* (Greek *pollor*, many; *morphe,* shape, form).

It should be emphasized that water is not truly representative of most materials, in that the slope of the line TC is negative. Only a few materials behave as water does in this respect, among them antimony.

At the **triple point** T, all these phases coexist at the same vapor pressure, 0.61 kPa or 4.58/760 atm. This is an *invariant* point for the water system. Only under this set of conditions can the three phases coexist. The triple-point temperature is a defining point in the modern temperature scale and, by definition, is exactly 273.16 K.

Phase Equilibria in a One-Component System: Water

The pressure-temperature diagram for water is well known, but what dictates the dependence of the equilibrium pressure on temperature for different phases? Only a thermodynamic analysis can give a quantitative expression for this relationship. In a reacting system, equilibrium is established when the change in Gibbs energy is zero. In like manner, for two phases of the same pure substance to be in equilibrium under given conditions of temperature and pressure, the Gibbs energy of one phase must be equal to that of the other. In other words, the change in Gibbs energy between the two phases is zero. Specifically, when equilibrium occurs between ice and water

$$G(s) = G(l) \tag{5.1}$$

and when there is equilibrium between water and steam

$$G(l) = G(g) \tag{5.2}$$

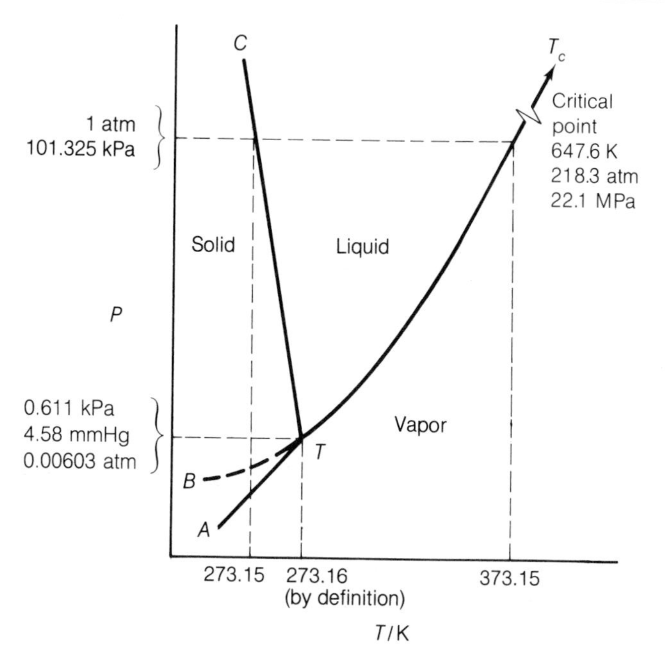

FIGURE 5.1 Phase diagram for water (not drawn to scale).

where *s* stands for solid, *l* for liquid, and *g* for vapor. At the triple point, both of these conditions apply. If only one phase is present under the *P-T* conditions, it will be the phase with the lowest value of *G*.

With Eqs. 5.1 and 5.2 as our criteria of equilibrium we now examine the equilibrium existing at 1 atm in the water system. Moving horizontally at 1 atm it is seen from Figure 5.1 that any one of three phases may be present, depending on the temperature of the system. This may be depicted in another way by plotting the variation of *G* with *T* for water in Figure 5.2. The curvatures for the ice, liquid, and vapor regions are all different, but the ice-liquid curves intersect at 0°C, the normal melting point, and the liquid-vapor curves intersect at 100°C, the normal boiling point. At these intersections Eqs. 5.1 and 5.2 apply and the respective phases are in equilibrium.

For an explanation of the slopes of the lines depicting individual phase equilibria in Figure 5.2, we utilize two of the Maxwell relations from Chapter 3, namely Eq. 3.119 written for 1 mol of substance:

$$\left(\frac{\partial G_m}{\partial P}\right)_T = V_m \quad \text{and} \quad \left(\frac{\partial G_m}{\partial T}\right)_P = -S_m \tag{5.3}$$

Since the second of these equations has a negative sign and the curves in Figure 5.2 have negative slopes, the entropy is positive. Moreover, the relative slopes show that $(S_m)_g > (S_m)_l > (S_m)_s$. Furthermore, the curves depicting the variation of *G* with *T* for each phase do not have the same curvature because $C_{P,m}$ for each phase is different. From the second derivative of *G* with respect to *T*,

$$\left(\frac{\partial^2 G_m}{\partial T^2}\right)_P = -\left(\frac{\partial S_m}{\partial T}\right)_P = -\frac{C_{P,m}}{T} \tag{5.4}$$

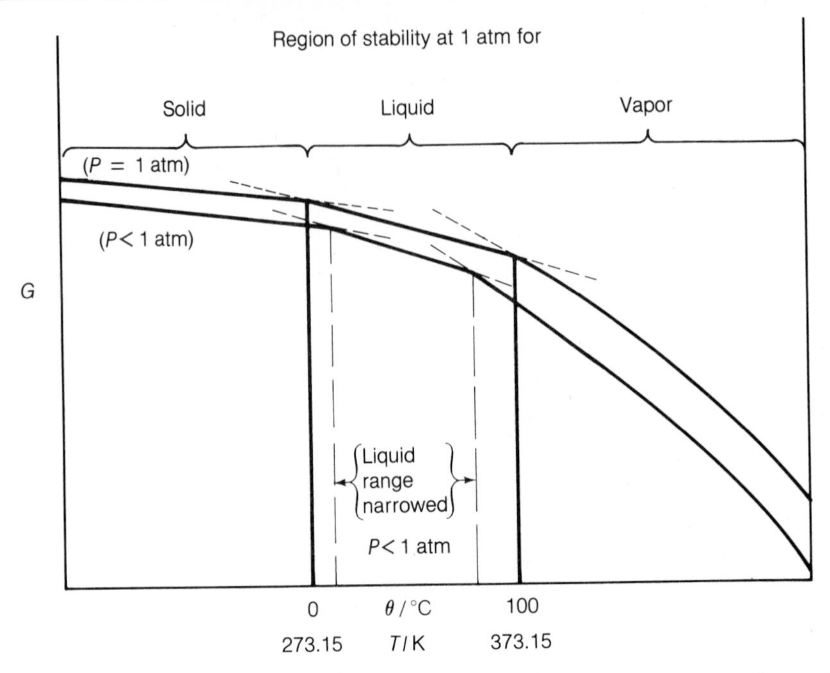

FIGURE 5.2 The variation of G with T for water. The upper solid line represents the variation of G at 1 atm pressure, and the lower line represents the variation of G at a reduced pressure. The figure shows the narrowing of the liquid range with a decrease in pressure.

we see that the curvature increases as T increases since $C_{P,m}$ does not vary greatly with T.

From an analysis of the left-hand expression in Eq. 5.3, G must decrease as the pressure is decreased, since V_m is always positive. The effect of reducing the pressure from 1 atm is indicated by a narrowing of the liquid range as shown by the vertical dashed lines in Figure 5.2. Notice that the decrease in Gibbs energy for the gas phase is much greater than for the other two phases. This is because the molar volume of the gas $V_m(g)$ is much larger compared to that of the liquid or solid. Note too that the reduction of pressure reduces the boiling point of all liquids. This is the basic reason for vacuum distillation of organic compounds: to cause the distillation in a temperature range over which the liquid is stable. Also in Figure 5.2 there is observed an increase in melting point for water as the pressure is reduced. The situation described here for water is different for most other substances, for which a reduction in pressure normally decreases the melting point of the substance.

Now consider an increase in pressure; we expect the Gibbs energy of a phase to increase. In a normal substance, $V_m(l) > V_m(s)$, and thus the increase in pressure will cause a larger increase in Gibbs energy for the liquid than for the solid. In order to have equilibrium between the two phases at a higher pressure, ΔG must equal 0 and, to accomplish this, the melting point must shift to a higher value as shown in Figure 5.3a. (The phase with the lowest Gibbs energy at a particular temperature is the stable phase.) Just the opposite result is obtained for water. Since the molar volume of ice is larger than that of liquid water, the increase in G will be greater for the solid form; thus the melting point is decreased as shown in Figure 5.3b. Indeed this decrease in melting point of water is utilized by the ice skater; the pressure on

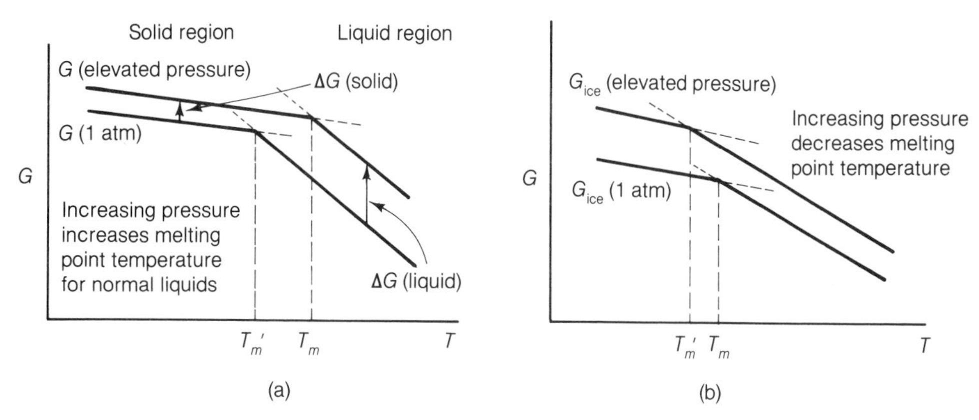

FIGURE 5.3 The effect of increased pressure on solid-liquid phase transitions. (a) Normal solid-liquid transitions; (b) ice-water transition. Dashed lines indicate hypothetical value of G of phases at temperatures where the phases are not stable. The phase with the lowest value of G is the most stable at a particular temperature.

the skate blade melts a thin layer of ice so that the blade glides on a film of water.

If the pressure is reduced far enough below that shown in Figure 5.2, the system will reach the triple point and all three phases will be at equilibrium. Below this pressure the curve for the gas will intersect that for the solid phase and the liquid phase is bypassed. The practical result is that the solid will pass directly into the gaseous state, a process called **sublimation.** Again the Gibbs energies of the two phases must be equal for equilibrium to exist.

This behavior may be depicted in a qualitative way by plotting G as a function of P and T, as in Figure 5.4. In this figure each plane surface represents the variation of a particular phase—gas, liquid, or solid—with pressure and temperature. The surface for the solid is generally fairly planar, which indicates only a minor variation in G. The surface for the liquid shows more slope and hence more change as P or T changes; the plane for the gas shows the most variation for G. Figure 5.2 may be viewed simply as a section of Figure 5.4 taken parallel to the P axis at 1 atm.

The intersection of two planes results in a line along which the two phases are in equilibrium. At some pressure, the gas, liquid, and solid curves will intersect at the triple point. This point is invariant for the system; that is, it can exist at only one pressure and temperature. Finally, a figure similar to Figure 5.1 may be obtained by projecting the intersection of the planes to the surface $G = 0$, thus giving the familiar equilibrium phase diagram for a typical material. To proceed further we need to make these observations quantitative.

5.2 Vaporization and Vapor Pressure

Thermodynamics of Vapor Pressure: The Clapeyron Equation

In Section 5.1 we found that the criterion for two phases of a pure substance to coexist is that their Gibbs energies must be equal at constant temperature and pressure. However, if either T or P is varied, with the other held constant, one of the

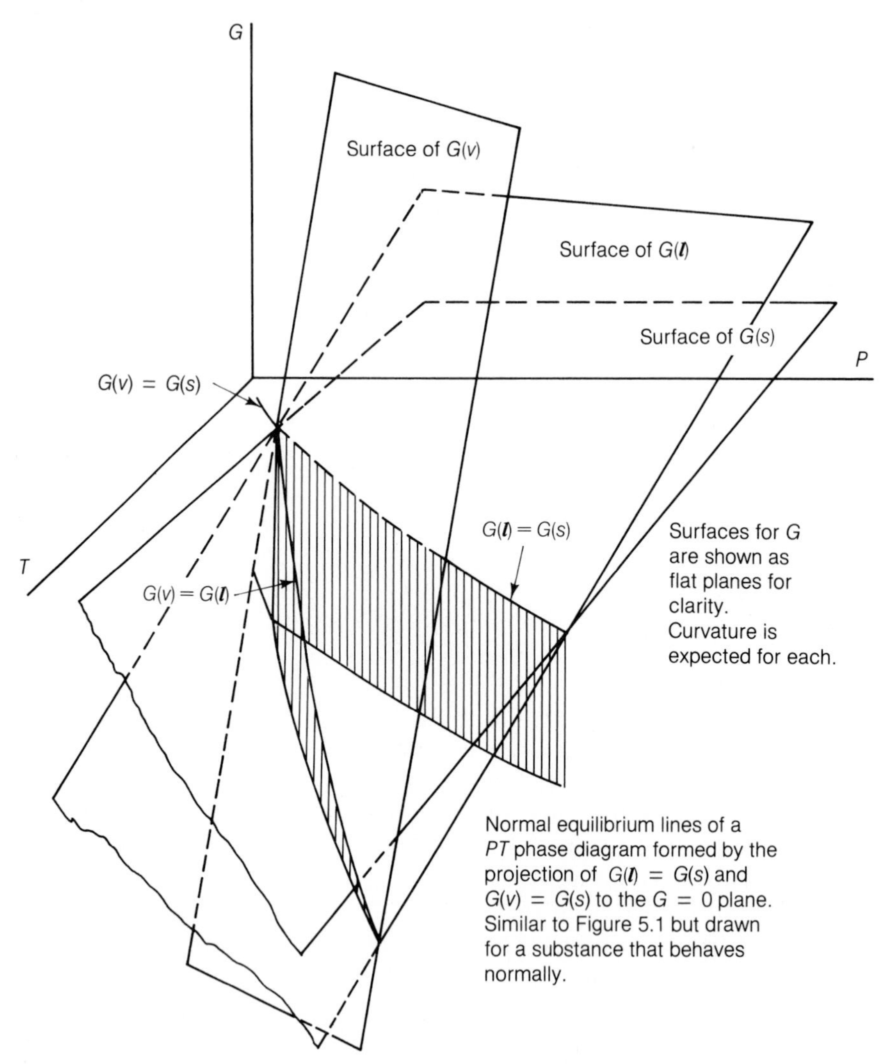

FIGURE 5.4 Variation of G as a function of pressure and temperature for the three states of matter, solid (s), liquid (l), and vapor (v).

phases will disappear. Thus, if we are at a particular point on one of the phase-equilibria lines on a P-T diagram (gas-liquid, liquid-solid, solid-gas, or even solid-solid), our problem becomes how to vary P and T while maintaining equilibrium. In 1834 Benoit Clapeyron (1799–1864) published a solution for the case of a liquid and its vapor. We investigate the vapor pressure because, since the vapor generally behaves as an ideal gas, it provides, through thermodynamics, one of the most important means of determining the properties of either a solid or a liquid.

We begin with the statement of phase equilibrium, Eq. 5.2, written for a single pure substance in the liquid and vapor states. If the pressure and temperature are changed infinitesimally in such a way that equilibrium is maintained,

$$dG_v = dG_l \tag{5.5}$$

Since this expression depends only on T and P, the total derivatives may be written as

$$\left(\frac{\partial G_v}{\partial P}\right)_T dP + \left(\frac{\partial G_v}{\partial T}\right)_P dT = \left(\frac{\partial G_l}{\partial P}\right)_T dP + \left(\frac{\partial G_l}{\partial T}\right)_P dT \tag{5.6}$$

From Eq. 5.3 we may substitute for these partials, obtaining

$$V_m(v)\, dP - S_m(v)\, dT = V_m(l)\, dP - S_m(l)\, dT \tag{5.7}$$

or

$$\frac{dP}{dT} = \frac{S_m(v) - S_m(l)}{V_m(v) - V_m(l)} = \frac{\Delta S_m}{\Delta V_m} \tag{5.8}$$

If the molar enthalpy change at constant pressure for the phase transformation is ΔH_m, the term ΔS_m may be written as $\Delta H_m/T$ and Eq. 5.8 becomes

$$\frac{dP}{dT} = \frac{\Delta H_m}{T\,\Delta V_m} \tag{5.9}$$

The general form of Eq. 5.9 is known as the **Clapeyron equation** and may be applied to vaporization, sublimation, fusion, or solid phase transitions of a pure substance. It is thus valid for the general process of equilibrium between any two phases of the same substance.

In order to use the Clapeyron equation the molar enthalpy of the process must be known along with the equilibrium vapor pressure. If the two phases are condensed phases (solid or liquid), then the pressure is the mechanical pressure established. (See p. 178.) To integrate this expression exactly, both ΔH_m and ΔV_m must be known as functions of temperature and pressure; however, they may be considered constant over short temperature ranges.

It should be pointed out that the enthalpies of sublimation, fusion, and vaporization are related at constant temperature by the expression

$$\Delta_{sub}H_m = \Delta_{fus}H_m + \Delta_{vap}H_m \tag{5.10}$$

This is as expected since enthalpy is a state function and the same amount of heat is required to vaporize a solid directly as to go through an intermediate melting stage to reach the final vapor state.

EXAMPLE

What is the expected boiling point of water at 98.7 kPa (approximately 740 mm Hg, a typical barometric pressure at 40°N latitude)? The heat of vaporization is 2258 J g^{-1}, the molar volume of liquid water is 18.78 $cm^3\ mol^{-1}$, and the molar volume of steam is 30.199 $dm^3\ mol^{-1}$, all values referring to 373.1 K and 101.325 kPa.

SOLUTION

Apply Eq. 5.9 to find the change in boiling point for 1 Pa and then multiply by the difference between the given pressure and the standard atmosphere:

$$\frac{dP}{dT} = \frac{\Delta_{vap}H_m}{T_b[V_m(v) - V_m(l)]}$$

where T_b refers to the temperature at boiling.

$$\frac{dP}{dT} = \frac{2258(\text{J g}^{-1}) \times 18.01(\text{g mol}^{-1})}{373.1(\text{K}) \times [30.199(\text{dm}^3\ \text{mol}^{-1}) - 0.019(\text{dm}^3\ \text{mol}^{-1})]}$$

$$= 3.614\ \text{J K}^{-1}\ \text{dm}^{-3} = 3.614 \times 10^3\ \text{J m}^{-3}\ \text{K}^{-1}$$

$$= 3.614 \times 10^3\ \text{Pa K}^{-1} = 3.614\ \text{kPa K}^{-1}$$

or

$$\frac{dT}{dP} = 2.767 \times 10^{-4}\ \text{K Pa}^{-1}$$

For a decrease of 101.325 kPa − 98.7 kPa = 2.625 kPa, there is a decrease in temperature of

$$2.767 \times 10^{-4}\ \text{K Pa}^{-1} \times 2625\ \text{Pa} = 0.73\ \text{K}$$

Therefore the new boiling point is

$$373.15\ \text{K} - 0.73\ \text{K} = 372.42\ \text{K}$$

■

EXAMPLE

Determine the change in the freezing point of ice with increasing pressure. The molar volume of water is 18.02 cm^3 mol^{-1} and the molar volume of ice is 19.63 cm^3 mol^{-1} at 273.15 K. The molar heat of fusion $\Delta_{\text{fus}} H_m = 6.008 \times 10^3$ J mol^{-1}.

SOLUTION

Equation 5.9 applies to all phase equilibria. For a fusion process we have

$$\frac{dP}{dT} = \frac{\Delta_{\text{fus}} H_m}{T_m \Delta_{\text{fus}} V_m}$$

where the subscript fus refers to the value of the variables at the melting point. Here, since the pressure is not the equilibrium value, P refers to the applied pressure maintained mechanically or through an inert gas. Also $\Delta_{\text{fus}} V_m$ is the molar volume difference $V_m(v) - V_m(l)$ and is assumed to be approximately constant over a moderate pressure range.

$$\frac{dP}{dT} = \frac{6.008 \times 10^3(\text{J mol}^{-1})}{273.15(\text{K}) \times [0.018\,02(\text{dm}^3\ \text{mol}^{-1}) - 0.019\,63(\text{dm}^3\ \text{mol}^{-1})]}$$

$$= -13\,660\ \text{J K}^{-1}\ \text{dm}^{-3} = -13\,660\ \text{kPa K}^{-1}$$

$$\frac{dT}{dP} = -7.32 \times 10^{-5}\ \text{K kPa}^{-1}$$

This is a 0.74 K decrease in temperature per 100 atm increase in pressure. ■

This example gives a quantitative explanation of the negative slope along the line in Figure 5.1, arising from the increase in volume of water as it freezes.

The Clausius-Clapeyron Equation

Some 30 years after Clapeyron introduced his equation, Clausius introduced a modification that improved the versatility of the expression. When one of the phases in equilibrium is a vapor phase, we assume that $V_m(v)$ is so much larger than $V_m(l)$ that we may neglect $V_m(l)$ in comparison to $V_m(v)$ when the pressure is near 1 atm. (For water, the volume of vapor is at least a thousand times that of a liquid.) The second assumption is to replace $V_m(v)$ by its equivalent from the ideal gas law RT/P. Equation 5.9 then becomes

$$\frac{dP}{dT} = \frac{\Delta_{\text{vap}} H_m P}{RT^2} \tag{5.11}$$

or if the standard pressure is 1 atm,

$$\boldsymbol{\frac{dP}{P\,dT} = \frac{d\,\ln(P/\text{atm})}{dT} = \frac{\Delta_{\text{vap}} H_m}{RT^2}} \tag{5.12}$$

This expression is known as the **Clausius-Clapeyron equation.** For pressure expressed in other units, P^u is used to indicate the numerical value of the pressure. In order to make use of Eq. 5.12, integration is performed assuming $\Delta_{\text{vap}} H_m$ to be independent of temperature and pressure. We thus obtain

$$\int d\,\ln P^u = \frac{\Delta_{\text{vap}} H_m}{R} \int T^{-2}\,dT \tag{5.13}$$

$$\ln P^u = -\frac{\Delta_{\text{vap}} H_m}{RT} + C \tag{5.14}$$

where C is a constant of integration. A plot of $\ln P^u$ against $1/T$ should be linear. The slope of the line is $-\Delta_{\text{vap}} H_m/R$. A plot of $R \ln P^u$ against $1/T$ could also be made in which case the intercept on the $R \ln P^u$ axis (i.e., where the temperature is infinitely high) will give the value of $\Delta_{\text{vap}} S_m$. This is shown in Figure 5.5 for five liquids, the dotted lines representing the extrapolation of data. As might be expected over a wide range of temperature, deviations from linearity occur as a result of the temperature variation of $\Delta_{\text{vap}} H_m$ and of nonideal gas behavior of the vapor.

Another useful form of Eq. 5.14 may be obtained by integrating between specific limits:

$$\int_{P_1}^{P_2} d\,\ln P^u = \frac{\Delta_{\text{vap}} H_m}{R} \int_{T_1}^{T_2} T^{-2}\,dT \tag{5.15}$$

$$\boldsymbol{\ln \frac{P_2}{P_1} = \frac{\Delta_{\text{vap}} H_m}{R} \left(\frac{1}{T_1} - \frac{1}{T_2} \right)} \tag{5.16}$$

It is immediately evident that one may calculate $\Delta_{\text{vap}} H_m$ (or $\Delta_{\text{sub}} H_m$ for sublimation from a similar expression) by measuring the vapor pressure of a substance at two different temperatures. It is generally convenient to let $P_1 = 1$ atm so that $T_1 = T_b$. Even if curvature exists in the plot, the enthalpy change may be calculated by drawing a tangent to the curve at the temperature of interest. Fairly good results are obtained if the equilibrium vapor density is not too high.

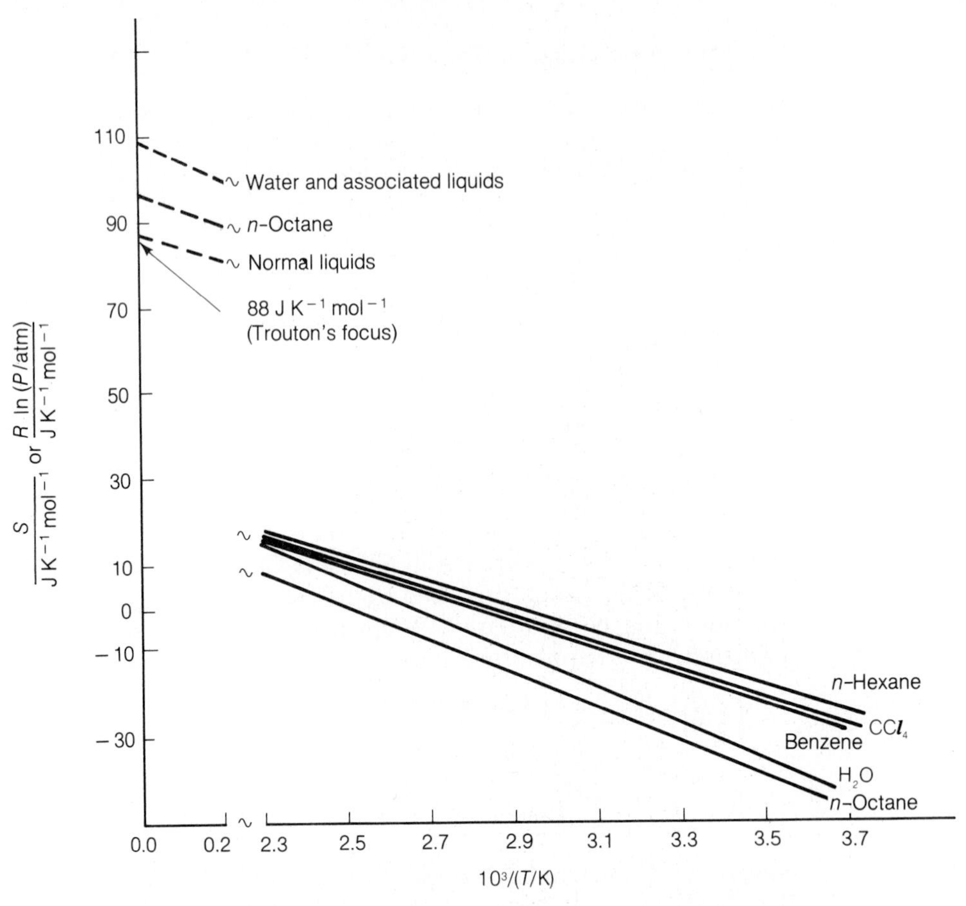

FIGURE 5.5 Plot of $R \ln(P/\text{atm})$ against $1/T$ for five liquids. The right-hand portion gives experimental points where the slope is $-\Delta_{\text{vap}}H_m$. The left-hand portion shows the extrapolation of the data to Trouton's focus discussed on page 177. For normal liquids, the intercept is 88 J K^{-1} mol^{-1}.

Enthalpy and Entropy of Vaporization: Trouton's Rule

In Section 3.3, we saw that entropy values may be obtained from the expression

$$\Delta_{\text{trs}} S = \frac{\Delta_{\text{trs}} H}{T_{\text{trs}}} \tag{5.17}$$

where trs specifies a particular transition. There are no easy generalizations to suggest even approximate values of the three quantities appearing in Eq. 5.17 when the transition process is fusion although the melting point is easily determined experimentally. However, the entropies of vaporization $\Delta_{\text{vap}} S_m$ of most non-hydrogen-bonded compounds have values of $\Delta_{\text{vap}} S_m$ in the neighborhood of 88 J K^{-1} mol^{-1}. This generalization is known as **Trouton's rule** and was pointed out in 1884:

$$\frac{\boldsymbol{\Delta_{\text{vap}} H_m}}{\boldsymbol{T_b}} = \boldsymbol{\Delta_{\text{vap}} S_m = 88 \text{ J K}^{-1} \text{ mol}^{-1}} \tag{5.18}$$

TABLE 5.1 Enthalpies and Entropies of Vaporization and Fusion

	Liquid ⇌ Vapor			Solid ⇌ Liquid		
Substance	$\frac{T_b}{K}$	$\frac{\Delta_{vap}H_m}{kJ\ mol^{-1}}$	$\frac{\Delta_{vap}S_m}{J\ K^{-1}\ mol^{-1}}$	$\frac{T_m}{K}$	$\frac{\Delta_{fus}H_m}{kJ\ mol^{-1}}$	$\frac{\Delta_{fus}S_m}{J\ K^{-1}\ mol^{-1}}$
He	4.20	0.084	19.66	3.45	0.021	6.28
H_2	20.38	0.904	44.35	13.95	0.117	8.37
N_2	77.33	1.777	72.13	63.14	0.720	11.38
O_2	90.18	6.820	75.60	54.39	0.444	8.16
H_2O	373.15	40.656	108.951	273.15	6.009	22.096
SO_2	263.13	24.916	94.68	197.48	7.401	37.45
CH_4	111.16	8.180	73.26	190.67	.941	10.38
C_2H_6	184.52	14.715	79.75	89.88	2.858	31.80
CH_3OH	337.85	35.27	104.39	175.25	3.167	18.07
C_2H_5OH	351.65	38.58	109.70	158.55	5.021	31.67
n-C_4H_{10}	272.65	22.40	82.13	134.80	4.661	34.572
C_6H_6	353.25	30.76	87.07	278.68	10.590	35.296
C_7H_8 (toluene)	383.77	33.48	87.19			
CH_3COOH	391.45	24.35	61.92	289.76	11.72	40.42

The intersection of lines on the $R \ln (P/\text{atm})$ axis in Figure 5.5 at approximately 88 J mol^{-1} K^{-1} is known as *Trouton's focus.* Table 5.1 lists for a number of substances the values of the enthalpy and entropy of both fusion and vaporization. The average of $\Delta_{vap}S_m$ for a large number of liquids that are not appreciably hydrogen-bonded bears out the value given in Eq. 5.18.

The effect of hydrogen bonding is seen in the case of water. The abnormally low value for acetic acid may be explained by its appreciable association in the vapor state. Allowing for its apparent molecular weight of about 100 in the vapor state, acetic acid will have a value of $\Delta_{vap} S_m$ of approximately 100 J K^{-1} mol^{-1}, in line with other associated liquids.

An alternative rule, known as the *Hildebrand rule** and named after the American physical chemist Joel Henry Hildebrand (b. 1881), states that the entropies of vaporization of unassociated liquids are equal, not at their boiling points but at temperatures at which the vapors occupy equal volumes.

The accuracy of the Hildebrand rule indicates that, under the conditions of comparison, all liquids with fairly symmetrical molecules possess equal amounts of configurational entropy. This means that, on the molecular level, maximum molecular disorder exists in a liquid that adheres to the rule.

*J. H. Hildebrand, *J. Am. Chem. Soc., 37,* 970 (1915); *40,* 45 (1918); O. L. I. Brown, *J. Am. Chem. Soc., 74,* 5096 (1952).

EXAMPLE

Estimate the enthalpy of vaporization of CS_2 if its boiling point is 319.40 K.

SOLUTION

From Eq. 5.18,

$$\Delta_{vap} H_m = \Delta_{vap} S_m \times T_b$$

$$\Delta_{vap} H_m = 88 \text{ J K}^{-1} \text{ mol}^{-1} \times 319.40 \text{ K} = 28.11 \text{ kJ mol}^{-1}$$

The experimental value is 28.40 kJ mol^{-1}. ■

Normally, boiling points are recorded at 101.325 kPa (1 atm), whereas they seldom are obtained experimentally under this exact pressure. A useful expression for correcting the boiling point to the standard pressure was derived in 1887 by James Mason Crafts (1839–1917) by combining Eq. 5.11 with Trouton's rule. Assuming that dP/dT is approximately $\Delta P/\Delta T$, we obtain

$$\frac{\Delta P}{\Delta T} = \frac{\Delta_{vap} H_m}{T_b} \times \frac{P}{RT_b} \tag{5.19}$$

where T_b is the normal boiling point. For ordinary liquids we may substitute $\Delta_{vap} S_m = 88 \text{ J K}^{-1} \text{ mol}^{-1}$ for $\Delta_{vap} H_m / T_b$.

$$\frac{\Delta P}{\Delta T} \approx \frac{88(\text{J K}^{-1} \text{ mol}^{-1}) \times 101.3(\text{kPa})}{8.314(\text{J K}^{-1} \text{ mol}^{-1})\ T_b\ (\text{K})} = \frac{1072}{T_b} \text{ kPa K}^{-1} \tag{5.20}$$

or

$$\Delta T \approx 9.3 \times 10^{-4}\, T_b\, \Delta P/\text{kPa}$$

For associated liquids, a numerical coefficient of 7.5×10^{-4} gives better results than 9.3×10^{-4}, which is used for normal liquids.

Variation of Vapor Pressure with External Pressure

In the preceding sections the saturated vapor pressure of a pure liquid was considered as a function of temperature alone. However, if an external pressure is applied in addition to the saturated vapor pressure, the vapor pressure becomes a function of pressure as well as of temperature. The increased pressure may be applied by adding an inert gas that is insoluble in the liquid or through the action of a piston that is permeable to the gas.

For a closed system, with the total pressure equal to the pressure exerted on the liquid, Eq. 5.7 may be written in a form known as the *Gibbs equation*:

$$V_m(v)\, dP = V_m(l)\, dP_t \quad \text{or} \quad \frac{dP}{dP_t} = \frac{V_m(l)}{V_m(v)} \tag{5.21}$$

where P is the pressure of the vapor under the total pressure P_t. If we assume that the vapor phase behaves ideally, we may substitute $V_m(v) = RT/P$, obtaining

$$\frac{d \ln P^u}{dP_t} = \frac{V_m(l)}{RT} \tag{5.22}$$

Since the molar volume of a liquid does not change significantly with pressure, we may assume $V_m(l)$ to be constant through the range of pressure and obtain, by integrating,

$$\ln \frac{P}{P_v} = \frac{V_m(l)}{RT}(P_t - P_v) \tag{5.23}$$

where P_v is the saturated vapor pressure of the pure liquid without an external pressure. Note that in this case the vapor pressure is now dependent on the total pressure as well as on the temperature. This is because the two phases exist under different pressures.

5.3 Classification of Transitions in Single-Component Systems

We have seen in some detail the nature of transitions involving vaporization, fusion, and sublimation. Other types of phase transitions also occur, such as polymorphic transitions in solids (i.e., changes from one crystalline structure to another).

In order to classify the transitions, a scheme was devised by Paul Ehrenfest (1880–1933). According to his classification, phase transitions are said to be *first-order phase transitions* if the molar Gibbs energy is continuous from one phase to another, but for which the derivatives of the Gibbs energy such as

$$S_m = -\left(\frac{\partial G_m}{\partial T}\right)_P \quad \text{and} \quad V_m = \left(\frac{\partial G_m}{\partial P}\right)_T \tag{5.24}$$

are discontinuous. For example, at the melting point of ice, even though the Gibbs energies of the two forms are equal, there is a change in the slope of the G against T curve as seen in Figure 5.2. There is consequently a break in the S against T curve, since $S = -(\partial G/\partial T)_P$. The observed finite value of ΔS at 273.15 K is related to the change in enthalpy for the transition. In a similar manner there is a sudden discontinuous change ΔV in the volume since the densities of the solid and liquid phases are different.

Although the Ehrenfest classification, involving finite jumps in entropy and volume, works well for first-order transitions such as vaporization, most higher-order phase transitions have discontinuities at which the second or higher derivatives are infinite. The more general theory developed by Laszlo Tisza (b. 1907) is applicable to such transitions. According to Tisza's theory, a *second-order phase transition* is one in which the discontinuity appears in the thermodynamic properties that are expressed as the second derivative of the Gibbs energy. Thus C_p would show an infinite value at the transition temperature since

$$C_p = T\left(\frac{\partial S}{\partial T}\right)_P = -T\left(\frac{\partial^2 G}{\partial T^2}\right)_P \tag{5.25}$$

This theory adequately explains, among other things, the onset of superfluidity in liquid helium.

5.4 Ideal Solutions: Raoult's and Henry's Laws

In the earlier sections of this chapter we have been primarily concerned with phase equilibria of one-component systems. When we consider a two-component system with variable composition, a way must be found to represent the composition variable. Since three-dimensional plots generally are difficult to work with, most variable composition equilibria are represented either at constant temperature, with pressure and composition as the variables, or at constant pressure, with temperature and composition as the variables.

We generally speak of solutions when discussing systems of variable composition. A *solution* is any homogeneous phase that contains more than one component. Although there is no fundamental difference between components in a solution, we call the component that constitutes the larger proportion of the solution the *solvent*. The component in lesser proportion is called the *solute*.

In Chapter 1 we studied the behavior of gases and characterized them as ideal or nonideal on the basis of whether or not they obeyed a particularly simple mathematical expression, the ideal gas law. When examining the behavior of liquid-vapor equilibria of solutions, we are naturally inclined to ask the question: Is there an ideal solution? We look for what we may call an ideal solution with the hope that the concept may lead us to a better understanding of solutions in general, just as the concept of the ideal gas has helped in understanding real gases.

Begin by imagining a solution of molecules of A and B. If the molecular sizes are the same, and the intermolecular attractions of A for A and of B for B are the same as the attraction of A for B, then we may expect the most simple behavior possible from a solution. Thus the solution is considered to be ideal when there is a complete uniformity of intermolecular forces, arising from similarity in molecular size and structure; a thermodynamic definition will be given later.

The partial vapor pressures of the individual components within the solution are a good measure of the behavior of the individual components. This is because the partial vapor pressure measures the escaping tendency of a molecule from solution, which is in turn a measure of the cohesive forces present in solution. Thus, measurements of the vapor pressure of each component as a function of pressure, temperature, and mole fraction lead to a good understanding of the system.

François Marie Raoult* (1830–1916) formed a generalization for an ideal solution known as **Raoult's law.** According to Raoult's law, the vapor pressure P_1 of solvent 1 is equal to its mole fraction in the solution multiplied by the vapor pressure P_1^* of the pure solvent. The superscript * indicates that the substance is at the temperature of the solution but not necessarily at the normal reference temperature. A similar expression holds for substance 2 in the mixture. Mathematically this may be stated as

$$P_1 = x_1 P_1^*; \qquad P_2 = x_2 P_2^* \tag{5.26}$$

If the solution has partial vapor pressures that follow Eq. 5.26, we say that the solution obeys Raoult's law and behaves ideally.

---

*F. M. Raoult, *Comptes rendus, 104,* 1430 (1887).

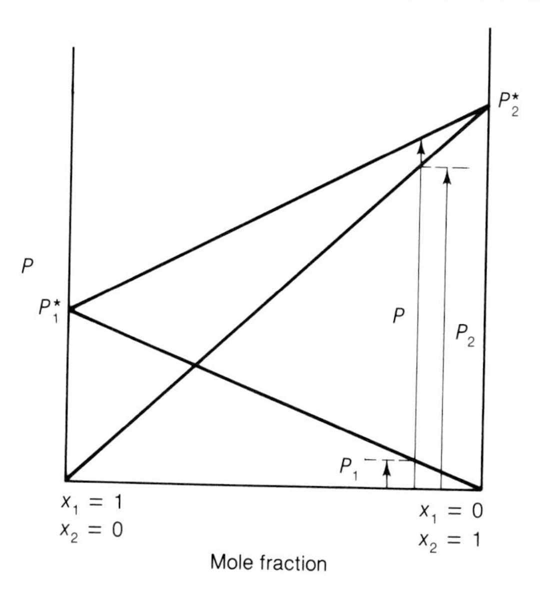

FIGURE 5.6 Vapor pressure of two liquids obeying Raoult's law. The total pressure at constant temperature is obtained from $P = x_A P_A^* + (1 - x_A)P_B^*$.

This behavior is shown in Figure 5.6. A number of pairs of liquids obey Raoult's law over a wide range of compositions. Among these solutions are benzene-toluene, ethylene bromide-ethylene chloride, carbon tetrachloride-trichloroethylene, and acetic acid-isobornyl acetate. It should be noted that air is generally present above the solution at a pressure that makes up the difference between the total vapor pressure P_t and atmospheric pressure. The fact that air is present can generally be ignored in the vapor phase.

Deviations from Raoult's law do occur and may be explained if we consider again the interaction between molecules 1 and 2. If the strength of the interaction between like molecules, 1–1 or 2–2, is stronger than that between 1 and 2, the tendency will be to force both components into the vapor phase. This increases the pressure above what is predicted by Raoult's law and is known as a *positive deviation*. As an illustration, consider a drop of water initially dispersed in a container of oil. The attraction of oil molecules and of water molecules for their own kind is great enough that the dispersed water is re-formed into a droplet and is excluded from solution. In the same manner, when 1–1 or 2–2 interactions are strong, both 1 and 2 are excluded from solutions and enter the vapor state. Figure 5.7 illustrates positive deviations from Raoult's law for a binary system of chloroform-ethanol at 318 K. Many other binary systems consisting of dissimilar liquids show positive deviations. In addition to the aforementioned, negative deviations occur when the attractions between components 1 and 2 are strong. This may be visualized as a holding back of molecules that would otherwise go into the vapor state. Figure 5.8 shows one example of *negative deviation* from Raoult's law.

In examining Figures 5.7 and 5.8 we notice that, in the regions where the solutions are dilute, the vapor pressure of the solvent in greatest concentration approaches ideal behavior, which is shown by the dashed lines. Therefore in the limit of infinite dilution the vapor pressure of the *solvent* obeys Raoult's law.

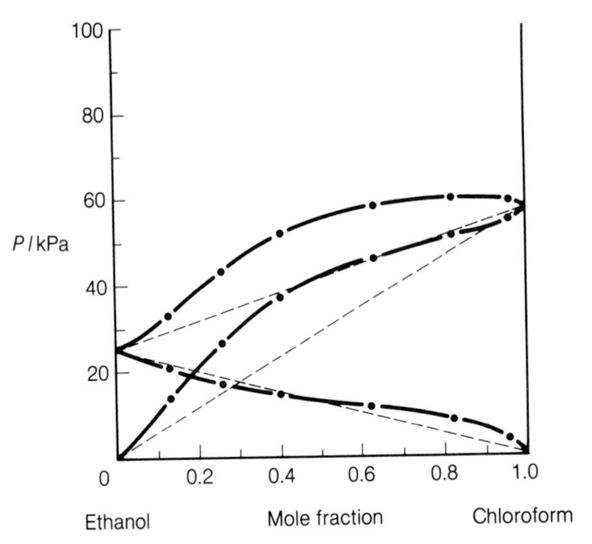

FIGURE 5.7 Vapor pressure of the system ethanol-chloroform at 318 K showing positive deviations from Raoult's law. Dashed lines show Raoult's law behavior.

Another property of binary systems was discovered by the English physical chemist William Henry (1774–1836). He found that the mass of gas m_2 dissolved by a given volume of solvent at constant temperature, is proportional to the pressure of the gas in equilibrium with the solution. Mathematically stated,

$$m_2 = k_2 P_2 \tag{5.27}$$

where generally the subscript 2 is used to refer to the solute (the subscript 1 is used to refer to the solvent) and k is the *Henry's law constant*. Most gases obey Henry's law when the temperatures are not too low and the pressures are moderate.

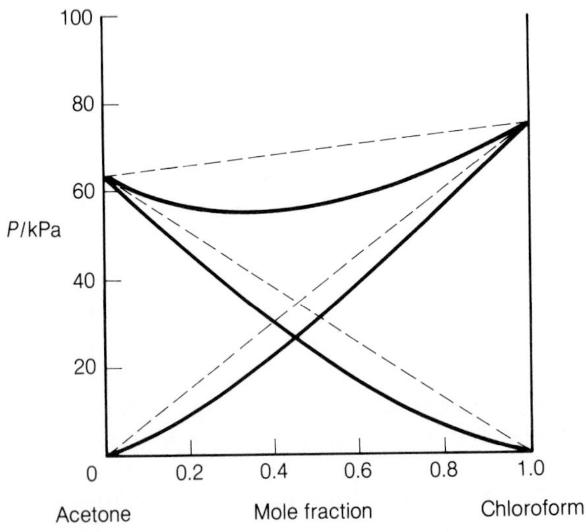

FIGURE 5.8 Vapor pressure of the system acetone-chloroform at 328 K showing negative deviations from Raoult's law. Dashed lines show Raoult's law behavior.

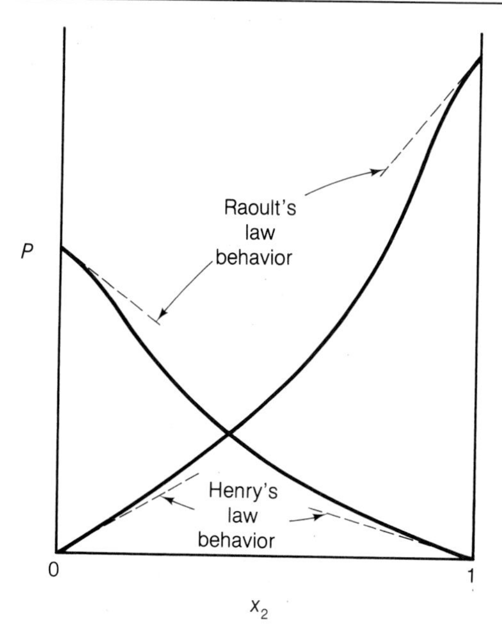

FIGURE 5.9 A hypothetical binary liquid mixture exhibiting negative deviation from Raoult's law, showing regions where Henry's and Raoult's laws are obeyed.

If several gases from a mixture of gases dissolve in a solution, Henry's law applies to each gas independently, regardless of the pressure of the other gases present in the mixture.

Although Eq. 5.27 is the historical form of Henry's law, since mass per unit volume is equal to a concentration term we may write Eq. 5.27 as

$$P_2 = k'x_2 \quad \text{or} \quad P_2 = k''c_2 \tag{5.28}$$

where for dilute solutions the concentration c_2 is proportional to the mole fraction of dissolved substance. Equation 5.28, in either form, is now often referred to as **Henry's law.**

Since Raoult's law may be applied to the volatile solute as well as to the volatile solvent, the pressure P_2 of the volatile solute in equilibrium with a solution in which the solute has a mole fraction of x_2 is given by

$$P_2 = x_2 P_2^* \tag{5.29}$$

In this case, P_2^* is the vapor pressure of the liquified gas at the temperature of the solution. From the last two equations it is easily seen that either a dissolved gas may be viewed in terms of its solubility under the pressure P_2 or P_2 may be taken as the vapor pressure of the volatile solute.

Thus Henry's law may also be applied to dilute solutions of a binary liquid system. It is found that in the limit of infinite dilution most liquid solvents obey Raoult's law but that under the same conditions the solute obeys Henry's law. Figure 5.9 shows a hypothetical binary liquid system exhibiting negative deviations from Raoult's law. The regions where the behavior of the system approaches that predicted by Henry's law and Raoult's law are shown by the dashed lines.

5.5 Partial Molar Quantities

Previously, in Chapter 4, we examined chemical system equilibrium, each component being considered to be fixed in composition. This corresponds to a closed system, as discussed in Section 1.3. We made the tacit assumption that the system under investigation is an ideal one in which each component has the same thermodynamic properties as it would have if it were the only reagent present. The thermodynamics developed thus far is adequate to deal with such systems. But, after examining Raoult's and Henry's laws and the deviations therefrom, we must develop a thermodynamic treatment capable of handling situations in which the thermodynamic quantities not only vary with the composition of the system but also allow for nonideal behavior. In order to consider the changing composition of a solution, it generally is found convenient to consider the changes that occur when different amounts of any component are added to or subtracted from the solution.

The type of problem that we are confronted with in dealing with solutions may best be considered by utilizing what might be the most easily visualized property of a solution, its volume. For example, it is well known that upon mixing 50 cm^3 of water with 50 cm^3 of methanol the total volume of solution is approximately 95 cm^3 instead of the expected 100 cm^3. Although the molar volume of both water and methanol may be uniquely and unambiguously determined by experimentation, there exists no unique and unambiguous method to determine the molar volume of either in solution. This is in spite of the fact that the total volume is easily determined. If we were to assign a fraction to the total volume to each component, there is no way to determine what part of the contraction that occurs upon mixing is due to the water and what part to the alcohol. Similar difficulties are present for all the other thermodynamic properties.

In our search for a method of handling variable compositions, we avoid the problem just stated by the invention of partial molar quantities. The following treatment is applicable to any extensive thermodynamic quantity such as the enthalpy, internal energy, or the Gibbs energy. Each of these extensive functions depends on the amounts of every component in the system, along with the two natural independent variables for the particular function. Explicitly, for the Gibbs energy, $G = G(T, P, n_1, n_2, \ldots, n_i)$. Initially we will proceed in our development using the volume as the extensive function and then later focus on the Gibbs energy.

Let us imagine that we start with a large volume of solution and to it add an infinitesimal amount of component 1, dn_1, without changing the amounts of the other components. If this is done at constant temperature and pressure, the corresponding increment in the volume V is dV. Since $V = V(n_1, n_2, n_3, \ldots)$, at constant temperature and pressure, the increase in V for the general case is given by

$$dV = \left(\frac{\partial V}{\partial n_1}\right)_{T,P,n_2,n_3,\ldots} dn_1 + \left(\frac{\partial V}{\partial n_2}\right)_{T,P,n_1,n_3,\ldots} dn_2 + \cdots \tag{5.30}$$

where the second and following terms on the right are zero when only dn_1 is changed.

The increase in volume per mole of component 1 is known as the **partial molar volume** of component 1. It is given the symbol V_1 and is written as

$$V_1 \equiv \left(\frac{\partial V}{\partial n_1}\right)_{T, P, n_2, n_3 \ldots} \tag{5.31}$$

Analogous definitions apply to the other components. Note that the same symbol V_1 is used both for the molar volume of pure component 1 and for the partial molar volume. This is logical since the latter becomes the molar volume when the component is pure. If there is any danger of confusion, the molar volume of pure component 1 will be designated as V_1^*.

We may also interpret the partial molar volume as the increase in V at the specified composition, temperature, and pressure when 1 mol of component 1 is added to such a large amount of solution that the addition does not appreciably change the concentration.

In either case, the definition for the partial molar volume, Eq. 5.31, may be used to rewrite Eq. 5.30 as

$$dV = V_1\, dn_1 + V_2\, dn_2 + \cdots \tag{5.32}$$

This expression may be integrated under the condition of constant composition so that the V_i's are constant. We can visualize the physical process of this integration as the addition of infinitesimals of each component in the same proportions in which they exist in the final solution. We obtain

$$V = n_1 V_1 + n_2 V_2 + \cdots \tag{5.33}$$

Although this expression was obtained under the restriction of constant composition, it is generally applicable, since each term in it is a function of state, which in turn is independent of the way in which the solution is formed.

If the total differential of Eq. 5.33 is now taken at constant temperature and pressure, we obtain

$$dV = n_1\, dV_1 + n_2\, dV_2 + \cdots + V_1\, dn_1 + V_2\, dn_2 + \cdots \tag{5.34}$$

Upon subtracting Eq. 5.32, we have

$$n_1\, dV_1 + n_2\, dV_2 + \cdots = 0 \tag{5.35}$$

If we now divide by the total amount of the components, $n_1 + n_2 + \cdots + n_i$, Eq. 5.35 may be expressed in terms of the mole fractions, x_i:

$$x_1\, dV_1 + x_2\, dV_2 + \cdots = 0 \tag{5.36}$$

This expression is one form of the **Gibbs-Duhem equation.** For a two-component system, Eqs. 5.35 and 5.36 may be put into the form

$$dV_1 = -\frac{n_1}{n_2}\, dV_2 \quad \text{or} \quad dV_1 = -\frac{x_2}{x_1}\, dV_2 = \frac{x_2}{x_2 - 1}\, dV_2 \tag{5.37}$$

If the variation of either V_1 or V_2 with concentration is known, the other may be calculated from the last equations. For example, by integration we have

$$\int dV_1 = \int \frac{x_2}{x_2 - 1}\, dV_2 \tag{5.38}$$

A plot of $x_2/(x_2 - 1)$ against V_2 gives the change in V_1 between the limits of integration. The molar volume V_1^* of pure component 1 may be used as the starting point of the integration to some final concentration.

Although these expressions are written in terms of volume, they apply equally well to any partial molar quantity.

Relation of Partial Molar Quantities to Normal Thermodynamic Properties*

The introduction of partial molar quantities is designed to handle open systems. We now inquire whether or not the thermodynamic equations developed in the earlier chapters are applicable to partial molar quantities.

We may begin by differentiating the definition of the Gibbs energy, $G = H - TS$, with respect to the amounts n_i of the various components:

$$\left(\frac{\partial G}{\partial n_1}\right)_{T,P,n_j} = \left(\frac{\partial H}{\partial n_1}\right)_{T,P,n_j} - T\left(\frac{\partial S}{\partial n_1}\right)_{T,P,n_j} \tag{5.39}$$

Since each of these partials is taken with respect to n_1, with T, P, and n_j held constant, they are in the generalized form of the expression for partial molar quantities. We may write therefore

$$G_1 = H_1 - TS_1 \tag{5.40}$$

For the partial molar enthalpy we obtain in a similar manner

$$H_1 = U_1 + PV_1 \tag{5.41}$$

Expressions could be written for each of the other extensive functions, resulting in similar equations. Since Eqs. 5.40 and 5.41 are of the same form as those for constant composition systems, our task is to show that this is a general result.

In order to show this, we combine the First and Second Laws of Thermodynamics in terms of the Gibbs energy. From Eq. 3.111 we have

$$dG = V\,dP - S\,dT \tag{5.42}$$

which is valid for all substances of constant composition, including solutions. It immediately follows that

$$\left(\frac{\partial G}{\partial P}\right)_{T,n_i} = V \quad \text{and} \quad \left(\frac{\partial G}{\partial T}\right)_{P,n_i} = -S \tag{5.43}$$

We now differentiate the expression on the right with respect to n_1:

$$\left[\frac{\partial}{\partial n_1}\left(\frac{\partial G}{\partial T}\right)_{P,n_i}\right]_{T,P,n_j} = -\left(\frac{\partial S}{\partial n_1}\right)_{T,P,n_j} \equiv -S_1 \tag{5.44}$$

Since the order of differentiation is immaterial, the left-hand side of Eq. 5.44 may be written as

$$\left[\frac{\partial}{\partial T}\left(\frac{\partial G}{\partial n_1}\right)_{T,P,n_j}\right]_{P,n_i} = \left(\frac{\partial G_1}{\partial T}\right)_{P,n_i} \equiv -S_1 \tag{5.45}$$

*This section may be omitted in the first reading.

Similarly, the left-hand expression of Eq. 5.43 leads to

$$\left(\frac{\partial G_1}{\partial P}\right)_{T,n_i} = V_1 \tag{5.46}$$

The partial molar Gibbs energy is still a state function and may be represented as a function of temperature and pressure at constant composition. Therefore upon differentiation

$$dG_1 = \left(\frac{\partial G_1}{\partial P}\right)_{T,n_i} dP + \left(\frac{\partial G_1}{\partial T}\right)_{P,n_i} dT \tag{5.47}$$

or upon substitution of Eqs. 5.45 and 5.46 we have

$$dG_1 = V_1\,dP - S_1\,dT \tag{5.48}$$

which is identical in form with the combined statement of the First and Second Laws, Eq. (5.42).

If we now add Eqs. 5.40 and 5.41

$$G_1 = U_1 + PV_1 - TS_1 \tag{5.49}$$

and take the total differential, we have

$$dG_1 = dU_1 + P\,dV_1 + V_1\,dP - T\,dS_1 - S_1\,dT \tag{5.50}$$

Subtracting Eq. 5.48 leaves

$$dU_1 = T\,dS_1 - P\,dV_1 \tag{5.51}$$

which is the partial molar internal energy analog of the original combined expression of the First and Second Laws for systems of fixed composition. All the earlier thermodynamic reactions for systems of fixed composition were derived from it, along with the definitions for H, A, and G. Therefore, since Eq. 5.51 and Eqs. 5.40 and 5.41 are identical in form to their counterparts previously derived, every relation developed for a system of fixed composition expressed in molar quantities is applicable to each component of the system expressed in terms of the partial molar quantities.

5.6 The Chemical Potential

The thermodynamic functions such as V, U, H, and G are extensive properties, since their values depend on the amount of substance present. Thus a total differential of the Gibbs function $G = G(T, P, n_i)$ is written as

$$dG = \left(\frac{\partial G}{\partial T}\right)_{P,n_i} dT + \left(\frac{\partial G}{\partial P}\right)_{T,n_i} dP + \sum_i \left(\frac{\partial G}{\partial n_i}\right)_{T,P,n_j} dn_i \tag{5.52}$$

Substitution of Eq. 5.43 into Eq. 5.52 gives

$$dG = -S\,dT + V\,dP + \sum_i \left(\frac{\partial G}{\partial n_i}\right)_{T,P,n_j} dn_i \tag{5.53}$$

We now introduce a symbolism, used by Gibbs, in which the coefficient of the form $(\partial G/\partial n_i)_{T,P,n_j}$ (the partial molar Gibbs energy) is called the **chemical potential** μ_i, for the ith component. Therefore, Eq. 5.53 may be written in the form

$$\boldsymbol{dG = -S\ dT + V\ dP + \sum_i \mu_i\ dn_i} \tag{5.54}$$

In a similar manner, the total differential of the internal energy may be written as

$$dU = \left(\frac{\partial U}{\partial S}\right)_{V,n_j} dS + \left(\frac{\partial U}{\partial V}\right)_{S,n_j} dV + \sum_i \left(\frac{\partial U}{\partial n_i}\right)_{S,V,n_j} dn_i \tag{5.55}$$

Again, substitution for the first two partials (Eq. 3.116) gives

$$dU = T\ dS - P\ dV + \sum_i \left(\frac{\partial U}{\partial n_i}\right)_{S,V,n_j} dn_i \tag{5.56}$$

However, since $G \equiv U + PV - TS$ for any closed system, we may differentiate this expression,

$$dG = dU + P\ dV + V\ dP - T\ dS - S\ dT \tag{5.57}$$

and substitute Eq. 5.56 into it:

$$dG = -S\ dT + V\ dP + \sum_i \left(\frac{\partial U}{\partial n_i}\right)_{S,V,n_j} dn_i \tag{5.58}$$

Upon comparison of Eqs. 5.53 and 5.58 we see that

$$\mu_i = \left(\frac{\partial G}{\partial n_i}\right)_{T,P,n_j} = \left(\frac{\partial U}{\partial n_i}\right)_{S,V,n_j} \tag{5.59}$$

A similar treatment of the other thermodynamic functions shows that

$$\begin{aligned} \mu_i &= \left(\frac{\partial G}{\partial n_i}\right)_{T,P,n_j} = \left(\frac{\partial U}{\partial n_i}\right)_{S,V,n_j} = \left(\frac{\partial H}{\partial n_i}\right)_{S,P,n_j} \\ &= \left(\frac{\partial A}{\partial n_i}\right)_{T,V,n_j} = -T\left(\frac{\partial S}{\partial n_i}\right)_{U,V,n_j} \end{aligned} \tag{5.60}$$

Note that the constant quantities are the natural variables for each function.

We now see that a special name is warranted for μ_i rather than identifying it with a particular thermodynamic function. It should be pointed out however that the chemical potential is most commonly associated with the Gibbs energy because we most often work with systems at constant temperature and pressure.

One may wonder why the word *potential* is used here. If we were to integrate Eq. 5.54 under the condition of constant temperature, pressure, and composition, the μ_i would remain constant. Hence, the total Gibbs energy increase would be proportional to the size of the system. This is analogous to the electrical concept of a fixed voltage being a *potential* or *capacity factor*.

One of the most common uses of the chemical potential is as the criterion of equilibrium for a component distributed between two or more phases. Let us investigate this use. From Eq. 5.54 we have, under conditions of constant temperature and pressure,

$$dG = \sum_i \mu_i \, dn_i \tag{5.61}$$

This expression allows the calculation of the Gibbs energy for the change in both the amount of substance present in a phase and also in the number of the phase's components. Therefore, this equation applies to phases that are open to the transport of matter. If a single phase is closed, and no matter is transferred across its boundary ($dG = 0$ for a closed system),

$$\sum_i \mu_i \, dn_i = 0 \tag{5.62}$$

However, if a system consisting of several phases in contact is closed but matter is transferred between phases, the condition for equilibrium at constant T and P becomes

$$dG = dG^{\alpha} + dG^{\beta} + dG^{\gamma} + \cdots = 0 \tag{5.63}$$

where α, β, and γ refer to the different phases that are in contact. Since an expression of the form of Eq. 5.62 exists for each individual term, we may write

$$dG = \sum_i \mu_i^{\alpha} \, dn_i^{\alpha} + \sum_i \mu_i^{\beta} \, dn_i^{\beta} + \sum_i \mu_i^{\gamma} \, dn_i^{\gamma} + \cdots = 0 \tag{5.64}$$

Now suppose that dn_i mol of component i are transferred from phase α to phase β in the closed system, without any mass crossing the system boundaries. The equilibrium condition would require that

$$dG = -\mu_i^{\alpha} \, dn_i + \mu_i^{\beta} \, dn_i = 0 \tag{5.65}$$

$$\boldsymbol{\mu_i^{\alpha} = \mu_i^{\beta}} \tag{5.66}$$

This can be generalized by stating that the equilibrium condition for transport of matter between phases, as well as chemical equilibrium between phases, requires that the value of the chemical potential μ_i for each component i be the same in every phase at constant T and P. Thus, for a one-component system, the requirement for equilibrium is that the chemical potential of component i is the same in the two phases.

To further emphasize the importance of Eq. 5.65 for a nonequilibrium situation, we write it in the form

$$dG = (\mu_i^{\beta} - \mu_i^{\alpha}) \, dn_i \tag{5.67}$$

If μ_i^{β} is less than μ_i^{α}, dG is negative and a transfer of matter occurs with a decrease of the Gibbs energy of the system. The transfer occurs spontaneously by the flow of substance i from a region of high μ_i to a region of low μ_i. The flow of matter continues until μ_i is constant throughout the system, that is, until $dG = 0$. This is the driving force for certain processes. Diffusion is an example; see Chapter 18.

It might be emphasized that the conditions of constant T and P are required to maintain thermal and mechanical equilibrium. Thus if all phases were not at the same temperature, heat could flow from one phase to another, and equilibrium would not exist. In the same manner, if one phase were under a pressure different from the rest, it could change its volume, thus destroying the condition of equilibrium.

5.7 Thermodynamics of Solutions

Raoult's Law Revisited

We may approach Raoult's law in a thermodynamic vein through Eq. 5.66. For any component i of a solution in equilibrium with its vapor, we may write

$$\mu_{i,\mathrm{sol}} = \mu_{i,\mathrm{vap}} \tag{5.68}$$

If the vapor behaves ideally, the Gibbs energy for each component is given by Eq. 3.149:

$$G_i = G_i^\circ + n_i RT \ln P_i^u \tag{5.69}$$

Since $\mu_i = G_i/n_i$, we may write

$$\mu_{i,\mathrm{vap}} = \mu_{i,\mathrm{vap}}^\circ + RT \ln P_i^u \tag{5.70}$$

where $\mu_{i,\mathrm{vap}}^\circ$ is the chemical potential of the vapor when $P_i = 1$ atm at the temperature T of the system. But at equilibrium, since Eq. 5.68 holds, substitution gives

$$\mu_{i,\mathrm{sol}} = \mu_{i,\mathrm{vap}}^\circ + RT \ln P_i^u \tag{5.71}$$

However, for pure liquid in equilibrium with its vapor,

$$\mu_i^* = \mu_{i,\mathrm{vap}}^\circ + RT \ln (P_i^*)^u \tag{5.72}$$

where the superscript * represents the value for the pure material. By subtraction of the last two equations, we obtain an expression for the difference between the chemical potentials of the solution and the pure material:

$$\mu_{i,\mathrm{sol}} - \mu_i^* = RT \ln \frac{P_i}{P_i^*} \tag{5.73}$$

The fugacity of a nonideal gas was defined (Eq. 3.152) as numerically equal to its activity, since the standard state of fugacity is by definition 1 atm. Thus the chemical potential may be written in terms of a dimensionless activity:

$$\mu_i = \mu_i^* + RT \ln f_i^u = \mu_i^* + RT \ln a_i \tag{5.74}$$

If the gas behaves ideally, as is required for exact adherence to Raoult's law, the partial pressure is equal to its fugacity. Then, by comparison with Eq. 5.73, we may substitute the pressure for the fugacity when the gas behaves ideally,

$$a_i = \frac{f_i}{f_i^\circ} = \frac{P_i}{P_i^*} \tag{5.75}$$

where f_i° is the fugacity in the standard state and is 1 atm and P_i^* is the reference state for pure component i. The relative activity a_i of a component in solution is just the ratio of the partial pressure of component i above its solution compared to the vapor pressure of pure component i at the temperature of the system. By comparison with Raoult's law written in the form

$$x_i = \frac{P_i}{P_i^*} \tag{5.76}$$

it is seen that

$$a_i = x_i \tag{5.77}$$

Equation 5.77 may be used as a definition of an ideal solution.

EXAMPLE

Calculate the activities and activity coefficients for an acetone-chloroform solution in which $x_2 = 0.6$. The vapor pressure of pure chloroform at 323 K is $P_2^* = 98.6$ kPa and the vapor pressure above the solution is $P_2 = 53.3$ kPa. For acetone, the corresponding values are $P_1^* = 84.0$ kPa and $P_1 = 26.6$ kPa.

SOLUTION

The activities, from Eq. 5.75, are

$$a_1 = \frac{26.6}{84.0} = 0.317, \qquad a_2 = \frac{53.3}{98.6} = 0.540$$

The activity coefficients are determined from the definition $f_i = a_i/x_i$ where f_i is the activity coefficient. (See Section 4.3.) Therefore,

$$f_1 = \frac{0.317}{0.400} = 0.792, \qquad f_2 = \frac{0.540}{0.600} = 0.900$$

Note that these values are for a solution exhibiting negative behavior from Raoult's law. ■

An interesting extension of Eq. 5.76 results in the original historical form of Raoult's law. If the addition of component 2 lowers the vapor pressure of component 1, then the difference $P^* - P$ is the lowering of the vapor pressure. Dividing by P^* gives a relative vapor pressure lowering, which is equal to the mole fraction of the solute. The mathematical form of this statement may be shown to be derived directly from Eq. 5.76, by subtracting unity from both sides. Assuming a two-component system, we have

$$\left(\frac{P_1}{P_1^*}\right) - 1 = x_1 - 1, \qquad \frac{P_1^* - P_1}{P_1^*} = x_2 \tag{5.78}$$

This form of Raoult's law is especially useful for solutions of relatively involatile solutes in a volatile solution.

Equation 5.78 may be written so that the molar mass of the solute 2 may be determined. Since

$$x_2 = \frac{n_2}{n_1 + n_2} \quad \text{and} \quad n_i = \frac{W_i}{M_i}$$

where W_i is the mass and M_i is the molar mass, we write

$$\frac{P_1^* - P_1}{P_1^*} = \frac{n_2}{n_1 + n_2} = \frac{W_2/M_2}{(W_1/M_1) + (W_2/M_2)} \tag{5.79}$$

For a dilute solution, n_2 may be neglected in the denominator and we obtain

$$\frac{P_1^* - P_1}{P_1^*} = \frac{n_2}{n_1} = \frac{W_2}{M_2} \cdot \frac{M_1}{W_1} \tag{5.80}$$

Ideal Solutions

In view of our foregoing discussion in which Eq. 5.71 leads to a statement of Raoult's law, we may inquire as to the values of other thermodynamic functions for mixtures.

The value of the change in volume upon mixing is our first consideration. We have already seen from Eq. 5.46 and the fact that $\mu = G_1$ that

$$\left(\frac{\partial \mu}{\partial P}\right)_{T,n_i} = V_1 \tag{5.81}$$

If we look at the difference in chemical potentials upon mixing, namely $\mu_i - \mu_i^*$, and substitute into the last expression, we have

$$\left(\frac{\partial(\mu_i - \mu_i^*)}{\partial P}\right)_{T,n_i} = V_i - V_i^* \tag{5.82}$$

Then using the definition of the activity, Eq. 5.74, we have

$$RT\left(\frac{\partial \ln a_i}{\partial P}\right)_{T,n_i} = V_i - V_i^* \tag{5.83}$$

Inserting the definition of an ideal solution from Eq. 5.77, namely $a_i = x_i$, the partial derivative in Eq. 5.83 is found to be zero. Consequently,

$$V_i = V_i^* \tag{5.84}$$

Thus, the partial molal volume of a component in an ideal solution is equal to the molar volume of the pure component. There is therefore no change in volume on mixing ($\Delta_{mix}V^{id} = 0$); that is, the volume of the solution is equal to the sum of the molar volumes of the pure components.

The enthalpy of formation of an ideal solution may be investigated in a similar manner starting from the Gibbs-Helmholtz equation (Eq. 3.167). Again inserting our definition for activity, $\mu_i - \mu_i^* = RT \ln a_i$, we have

$$R\left(\frac{\partial \ln a_i}{\partial(1/T)}\right)_{P,n_i} = H_i - H_i^* \tag{5.85}$$

But if $a_i = x_i$ as required for an ideal solution, the partial derivative is zero and

$$\Delta_{mix}H^{id} = \sum_i (H_i - H_i^*) = 0 \tag{5.86}$$

or

$$H_i = H_i^* \tag{5.87}$$

As with the volume, the change in enthalpy of mixing is the same as though the components in an ideal solution were pure. Thus there is no heat of solution ($\Delta_{\text{mix}} H^{\text{id}} = 0$) when an ideal solution is formed from its components.

The value of the Gibbs energy change for mixing cannot be zero as with the two previous functions since the formation of a mixture from a solution is a spontaneous process and must show a decrease.

The Gibbs energy of mixing $\Delta_{\text{mix}} G$ for any solution is

$$\Delta_{\text{mix}} G = G - G^* = \sum_i n_i \mu_i - \sum_i n_i \mu_i^* \tag{5.88}$$

From the definition of activity in Eq. 5.74 and after multiplication by n_i, we obtain

$$n_i \mu_i - n_i \mu_i^* = n_i \, RT \ln a_i \tag{5.89}$$

Combining this with Eq. 5.88 gives

$$\Delta_{\text{mix}} G = RT(n_1 \ln a_1 + n_2 \ln a_2 + \cdots) \tag{5.90}$$

If the solution is ideal, then $a_i = x_i$. We have, therefore,

$$\Delta_{\text{mix}} G^{\text{id}} = RT(n_1 \ln x_1 + n_2 \ln x_2 + \cdots) \tag{5.91}$$

This can be put into a more convenient form by using the substitution

$$n_i = x_i \sum_i n_i \qquad \text{where} \quad \sum_i n_i = n_1 + n_2 + \cdots$$

is the total amount of material in the mixture and x_i is the mole fraction of component i. Then,

$$\Delta_{\text{mix}} G^{\text{id}} = \sum_i n_i RT(x_1 \ln x_1 + x_2 \ln x_2 + \cdots) = \sum_i n_i RT \sum_i x_i \ln x_i \tag{5.92}$$

which is the Gibbs energy of mixing in terms of the mole fractions of the components. Note that each term on the right-hand side is negative since the logarithm of a fraction is negative. Thus the sum is always negative.

For a two-component mixture, Eq. 5.92 may be written as

$$\boldsymbol{\Delta_{\text{mix}} G^{\text{id}} = \sum_i n_i RT[x_1 \ln x_1 + (1 - x_1) \ln(1 - x_1)]} \tag{5.93}$$

A plot of this function is shown in Figure 5.10, where it is seen that the curve is symmetrical about $x_1 = 0.5$. For more complex systems, the lowest value will occur when the mole fraction is equal to $1/n$ where n is the number of components.

The final thermodynamic function to be considered here is the entropy. Using Eq. 5.43, differentiate $\Delta_{\text{mix}} G$ with respect to temperature and obtain $\Delta_{\text{mix}} S$ directly:

$$\begin{aligned}\left(\frac{\partial \, \Delta_{\text{mix}} G}{\partial T}\right)_{P,n_i} &= \left(\frac{\partial G}{\partial T}\right)_{P,n_i} - \frac{\partial}{\partial T}(n_1 \mu_1^* + n_2 \mu_2^* + \cdots)_{P,n_i} \\ &= -S + n_1 S_1^* + n_2 S_2^* + \cdots = -\Delta_{\text{mix}} S \end{aligned} \tag{5.94}$$

Differentiating the right-hand side of Eq. 5.92 with respect to T and substituting it into the last expression gives for an ideal mixture

$$\Delta_{\text{mix}} S^{\text{id}} = \sum_i n_i R \sum_i x_i \ln x_i \tag{5.95}$$

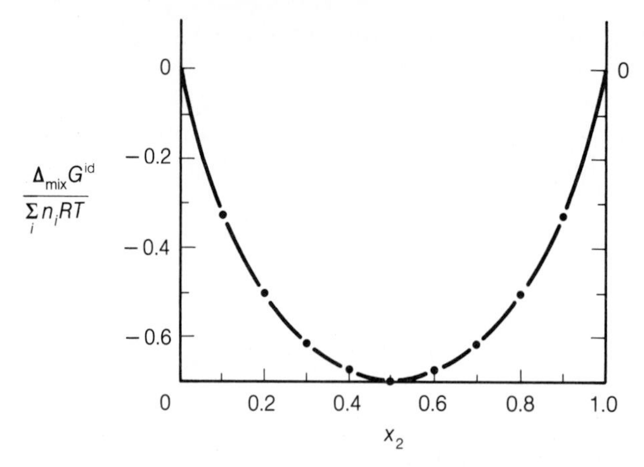

FIGURE 5.10 $\Delta_{mix}G/\sum_i n_i RT$ against x for a binary ideal mixture.

Thus the entropy of mixing for an ideal solution is independent of temperature and pressure. Figure 5.11 is a plot of this function for a binary system.

Another interesting aspect of mixing for an ideal solution can be seen by writing

$$\Delta_{mix}G = \Delta_{mix}H - T\,\Delta_{mix}S \tag{5.96}$$

Since we have seen that $\Delta_{mix}H^{id} = 0$,

$$-\Delta_{mix}G^{id} = T\,\Delta_{mix}S^{id} \tag{5.97}$$

Thus the driving force for mixing is purely an entropy effect. Since the mixed state is a more random state, it is therefore a more probable state. However, this need not mean that the driving force is large. If $\Delta_{mix}G$ were to be positive, the material would stay as distinct phases; this explains the lack of miscibility found in some liquid systems.

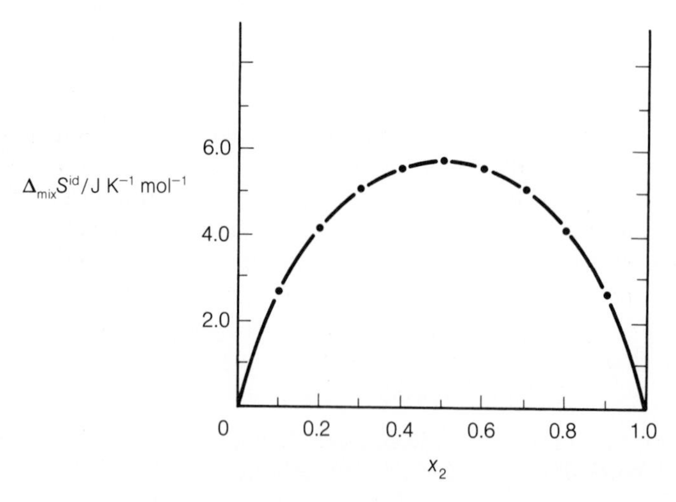

FIGURE 5.11 $\Delta_{mix}S$ of 1 mol of an ideal solution from the pure components.

Nonideal Solutions; Activity and Activity Coefficients

We now approach the problem of how solutions are treated when they are not ideal. We have already seen from Eq. 3.152 and the discussion of the last section that

$$\mu_i = \mu_i^\circ + RT \ln a_i \tag{5.98}$$

There are two ways of defining μ_i° that lead to different systems of activities. In both systems the activity remains a measure of the chemical potential.

In the *rational system*, μ_i° is identified with the chemical potential of the pure liquid μ_i^*:

$$\mu_i = \mu_i^* + RT \ln a_i \tag{5.99}$$

As $x_i \to 1$ and the system approaches pure i, μ_i must approach μ_i^*. Hence, $\ln a_i = 0$ as $x_i \to 1$. In other words,

$$a_i = 1 \quad \text{as} \quad x_i \to 1 \tag{5.100}$$

and the activity of the pure liquid, a dimensionless quantity, is equal to unity. For an ideal liquid solution a_i may be replaced by x_i, as seen in Eq. 5.77:

$$\mu_{i,\text{id}} = \mu_i^* + RT \ln x_i \tag{5.101}$$

Subtraction of Eq. 5.101 from 5.99 gives the difference between a nonideal and ideal solution:

$$\mu_i - \mu_{i,\text{id}} = RT \ln \frac{a_i}{x_i} \tag{5.102}$$

We may now define the *rational activity coefficient* of i, f_i, as

$$\boldsymbol{f_i = \frac{a_i}{x_i} \quad \text{and} \quad f_i = 1 \quad \text{as} \quad x_i \to 1} \tag{5.103}$$

from which it is seen that f_i is a measure of the extent of deviation from ideal behavior. This is basically the system employed earlier.

For a solution, it is most useful to use the rational system for the solvent when the solvent has a mole fraction near unity. Although this same system may be used for solutes, the *practical system* may be used for the solute (if the solute is present in small amounts). In this system

$$\mu_{i,\text{id}} = \mu_i^\ominus + RT \ln m_i^u \tag{5.104}$$

where m_i^u is the value of the molality (i.e., $m_i/\text{mol kg}^{-1}$) and μ_i° in Eq. 5.98 is identified with $\mu_i^\ominus$ (the value μ_i° would have in a hypothetical state of unit molality if the solution behaved as an ideal dilute solution, i.e., obeyed Henry's law, over the range of $m_i = 0$ to 1). This identification defines the practical system of activities. In the same manner in which we obtained Eq. 5.101, setting $\mu_i^\ominus = \mu_i^\circ$ in Eq. 5.98 gives us

$$\mu_i = \mu_i^\ominus + RT \ln a_i \tag{5.105}$$

and then, subtracting Eq. 5.104, we have

$$\mu_i - \mu_{i,\text{id}} = RT \ln \frac{a_i}{m_i} \tag{5.106}$$

From this, the *practical activity coefficient* is defined by

$$\gamma_i = \frac{a_i}{m_i} \quad \text{and} \quad \gamma_i = 1 \quad \text{as} \quad m_i \to 0 \tag{5.107}$$

Thus γ_i like f_i is a measure of the departure of substance i from the behavior expected in an ideal solution.

The practical system of activities defined by Eq. 5.105 measures the chemical potential of the solute relative to the chemical potential of the hypothetical ideal solution of unit molality. Equation 5.106 applies to both volatile and nonvolatile solutes.

EXAMPLE
The solubility of oxygen in water at 1 atm pressure and 298.15 K is 0.001 15 mol kg^{-1} of water. Under these conditions, calculate μ_i° for a saturated solution of oxygen in water.

SOLUTION
Since $O_2(g, 1 \text{ atm}) \rightleftharpoons O_2(aq, 0.001\ 15\ m)$, the chemical potentials of the two sides are equal. Since $\mu_{O_2}(g)$ for formation equals zero, $\mu_{O_2}(aq) = 0$. Substitution of this into Eq. (5.104) gives

$$\mu_{O_2}(g) = 0 = \mu^\circ_{O_2}(aq) + RT \ln m_i^u$$

$$\mu^\circ_{O_2}(aq) = -8.314(298.15) \ln 0.001\ 15 = 16.78 \text{ kJ mol}^{-1}$$

This is the hypothetical chemical potential of O_2 if the solution behaved according to Henry's law. The hypothetical chemical potential is useful in biological reactions for establishing the position of equilibrium when oxygen, dissolved in the solution, is also involved in the reaction. ■

5.8 The Colligative Properties

The properties of dilute solutions that depend on only the number of solute molecules and not on the type of species present are called **colligative properties** (Latin *colligatus*, bound together). All colligative properties (such as *boiling point elevation*, *freezing point depression*, and *osmotic pressure*) ultimately can be related to a lowering of the vapor pressure $P^* - P$ for dilute solutions of nonvolatile solutes.

Freezing Point Depression

The boiling point elevation and freezing point depression are related to the vapor pressure through the chemical potential. The situation is simplified by recognizing that the nonvolatile solute does not appear in the gas phase, and consequently the curve of μ for the gas phase is unchanged. An additional simplifying assumption is that the solid phase contains only the solid solvent, since when freezing occurs, only the solvent usually becomes solid. This leaves the curve for μ for the solid

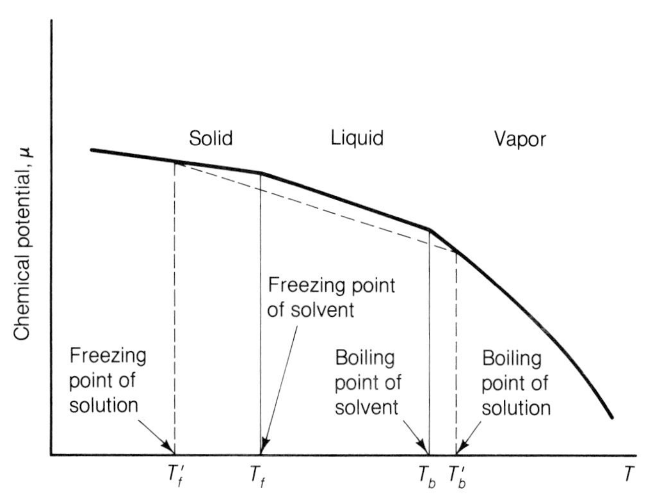

FIGURE 5.12 Variation of the chemical potential. Note the enhanced decrease in freezing point from the normal value compared to the increase in boiling point.

unchanged. From the last section we combine Eqs. 5.73 and 5.76 and use the convention that the subscript 1 refers to the solvent. Thus we have

$$\mu_1^l - \mu_1^{*,l} = RT \ln x_1 \tag{5.108}$$

where μ_1^l now refers to the chemical potential of the liquid solvent in solution, and x_1 is its mole fraction. As a result, the freezing point of the solution is different from that of the pure solvent, as shown in Figure 5.12. Drawn for constant pressure, the figure is similar to Figure 5.2. The solid lines show normal behavior of the pure solvent. In view of Eq. 5.108, the superimposed dashed curve represents the effects of lowering the vapor pressure, as a result of the term $RT \ln x_1$. The size of the phenomenon is larger for the freezing point depression than for the boiling point elevation in a solution of the same concentration.

Another way to illustrate the change is to use the ordinary phase diagram plot of P against T for water shown in Figure 5.13. This plot takes the form of Figure 5.1 but again has dashed lines superimposed to indicate the effects of vapor pressure lowering. The intersections of the horizontal line at the reference point of 1 atm with the solid and dashed vertical lines give the freezing points and boiling points.

Now consider a solution in equilibrium with pure solid solvent. Our aim is to discover how T depends on x_1. The chemical potential is a function of both T and P and will be written to emphasize this point. Since by our previous hypothesis the solid is pure, the chemical potential of the pure solid $\mu_1^s(T, P)$ is independent of the mole fraction of the solute. On the other hand, the chemical potential of the liquid solvent $\mu_1^l(T, P, x_1)$ is a function of the mole fraction of solvent, x_1. The equilibrium condition is

$$\mu_1^l(T, P, x_1) = \mu_1^s(T, P) \tag{5.109}$$

For an ideal solution, we may substitute from Eq. 5.108 for μ_1^l and obtain

$$\mu_1^{*,l}(T, P) + RT \ln x_1 = \mu_1^s(T, P) \tag{5.110}$$

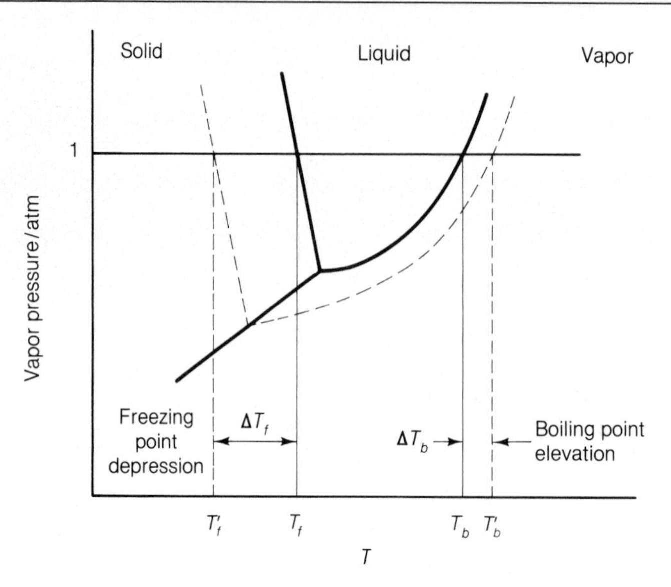

FIGURE 5.13 Phase diagram (P against T) of a solvent (water)-nonvolatile solute system showing depression of freezing point and elevation of boiling point (not drawn to scale).

or

$$\ln x_1 = \frac{-(\mu_1^{*,l}(T, P) - \mu_1^{s}(T, P))}{RT} \tag{5.111}$$

The difference in chemical potentials for pure liquid and pure solid expressed in the numerator is simply $\Delta_{\text{fus}}G_m$ (i.e., the molar Gibbs energy of fusion of the solvent at T and 1 atm of pressure). Thus we may write

$$\ln x_1 = \frac{-\Delta_{\text{fus}}G_m}{RT} \tag{5.112}$$

Differentiation with respect to T gives

$$\frac{d \ln x_1}{dT} = -\frac{1}{R}\left[\frac{\partial(\Delta_{\text{fus}}G_m/T)}{\partial T}\right]_P \tag{5.113}$$

Substitution of the Gibbs-Helmholtz equation, Eq. 3.167, yields

$$\frac{d \ln x_1}{dT} = \frac{\Delta_{\text{fus}}H_m}{RT^2} \tag{5.114}$$

If $\Delta_{\text{fus}}H_m$, the heat of fusion for pure solvent, is independent of T over a moderate range of temperature, we may integrate Eq. 5.114 from T_f^*, the freezing point of pure solvent at $x_1 = 1$, to T, the temperature at which solid solvent is in equilibrium with liquid solvent of mole fraction x_1. The result is

$$\ln x_1 = \frac{\Delta_{\text{fus}}H_m}{R}\left(\frac{1}{T_f^*} - \frac{1}{T}\right) \tag{5.115}$$

This expression gives the mole fraction of the solvent in relation to the freezing point of an ideal solution and to the freezing point of the pure solvent.

If the solution is dilute, this relation may be simplified by expressing x_1 in terms of x_2, the mole fraction of the solute:

$$\ln(1 - x_2) = \frac{\Delta_{\text{fus}}H_m}{R}\left(\frac{T - T_f^*}{TT_f^*}\right) \tag{5.116}$$

The term $\ln(1 - x_2)$ may be expanded in a power series, and if x_2 is small, only the first term need be kept:

$$\ln(1 - x_2) = -x_2 - \tfrac{1}{2}x_2^2 - \tfrac{1}{3}x_2^3 - \cdots = -x_2 \tag{5.117}$$

in which only the first term is kept. The freezing point depression is $T_f^* - T = \Delta_{\text{fus}}T$. Since $\Delta_{\text{fus}}T$ is small in comparison to T_f^*, we may set the product $TT_f^* \approx T_f^{*2}$. These changes convert Eq. 5.116 to

$$x_2 = \frac{\Delta_{\text{fus}}H_m}{R} \cdot \frac{\Delta_{\text{fus}}T}{T_f^{*2}} \tag{5.118}$$

Since the most important application of this expression is for determining molar masses of dissolved solutes, an alternate form is useful. The mole fraction of solute is $x_2 \equiv n_2/(n_1 + n_2)$, and in dilute solution is approximately n_2/n_1. The molality m_2 is the amount n_2 of solute divided by the mass W_1 of solvent: $m_2 = n_2/W_1$. For the solvent, the amount present is its mass W_1 divided by its molar mass M_1: $m_1 = W_1/M_1$. Then $x_2 \approx m_2W_1/W_1/M_1 = m_2M_1$. Rearrangement of Eq. 5.118 and substitution for x_2 yields

$$\Delta_{fus}T \approx \frac{M_1RT_f^{*2}}{\Delta_{fus}H_m} \cdot m_2 \tag{5.119}$$

This can be further simplified by introducing the *freezing point depression* or *cryoscopic constant* K_f defined as

$$\boldsymbol{K_f = \frac{M_1RT_f^{*2}}{\Delta_{\text{fus}}H_m}} \tag{5.120}$$

EXAMPLE

Find the value of K_f for the solvent p-dichlorobenzene from the following data:

$$M = 147.01 \text{ g mol}^{-1}; \qquad T_f^* = 326.28 \text{ K}; \qquad \Delta_{\text{fus}}H_m = 17.88 \text{ kJ mol}^{-1}$$

SOLUTION

The value of K_f depends only on the properties of the pure solvent. Therefore,

$$K_f = \frac{0.147\,01(\text{kg mol}^{-1}) \times 8.314(\text{J K}^{-1}\text{ mol}^{-1}) \times 326.28^2(\text{K}^2)}{17\,880(\text{J mol}^{-1})}$$

$$= 7.28 \text{ K kg mol}^{-1}$$

■

With the definition of K_f, Eq. 5.119 may be expressed as

$$\boldsymbol{\Delta_{\text{fus}}T = K_fm_2} \tag{5.121}$$

This is a particularly simple relation between the freezing point depression and the solute's molality in a dilute ideal solution. Table 5.2 lists K_f for several substances.

TABLE 5.2 Freezing Point Depression Constants K_f

Solvent	$\dfrac{K_f}{\text{K kg mol}^{-1}}$	Melting Point, T_m/K
Water	1.86	273.15
Acetic acid	3.90	289.75
Dioxane	4.71	284.85
Benzene	4.90	278.60
Phenol	7.40	316.15
Camphor	37.7	451.55

If the depression constant is known, the molar mass of the dissolved substance may be found. Since $m_2 = W_2/M_2W_1$, we have after rearrangement

$$M_2 = \frac{K_f W_2}{\Delta_{\text{fus}} T W_1} \tag{5.122}$$

EXAMPLE

A solution contains 1.5 g of solute in 30 g of benzene and its freezing point is 3.74°C. The freezing point of pure benzene is 5.48°C. Calculate the molar mass of the solute.

SOLUTION

From Table 5.2, K_f for benzene is 5.12 K kg mol^{-1}. The value of $\Delta_{\text{fus}}T = 5.48 - 3.74 = 1.74$ K. Substitution into Eq. 5.122 gives

$$M_2 = \frac{5.12(\text{K kg mol}^{-1}) \times 1.5(\text{g})}{1.74(\text{K}) \times 30(\text{g})}$$

$$= 0.147 \text{ kg mol}^{-1} = 147 \text{ g mol}^{-1}$$

■

It should be pointed out that the foregoing treatment does not apply to a solid solution in which a homogeneous solid containing both solute and solvent solidifies.

Ideal Solubility and the Freezing Point Depression

We may digress from our main development briefly to investigate some implications of Eq. 5.115. This expression gives the temperature variation of the solubility x_1 of a pure solid in an ideal solution. This expression applies to the equilibrium

$$\text{pure solid 1} \rightleftharpoons \text{ideal solution of 1 and 2} \tag{5.123}$$

Thus, in one sense the system could be viewed as a saturated solution of solid 1 in solvent 2 with some excess solid 1 present. In this perspective, solid component 1 is now thought of as the solute in the solvent compound 2. Furthermore T_f^* is now

TABLE 5.3 Mole-Fraction Solubilities of Several Solutes in Different Solvents

Solvent \ Solute	Naphthalene	Phenanthrene	*p*-Dibromobenzene
Temperature	298.1 K	298.1 K	298.1 K
Benzene	0.241	0.207	0.217
Hexane	0.090	0.042	0.086
Carbon tetrachloride	0.205	0.086	0.193
Toluene	0.224	0.255	0.224
Ideal solution	0.268	0.234	0.248

simply the temperature at which the solubility is measured. The basic idea is that the mole fraction solubility of a substance in an ideal solution only depends on the properties of that substance. Stated differently, the mole fraction solubility of a substance in an ideal solution is the same in all solvents.

In Table 5.3 are listed the mole fraction solubilities of several solutes. In the row marked *ideal solution* are the solubilities as calculated from Eq. 5.115.

EXAMPLE

Calculate the ideal solubility of phenanthrene at 298.1 K in solvent in which it forms an ideal solution.

SOLUTION

The heat of fusion of phenanthrene is 18.64 kJ mol^{-1} at the melting point of 369.4 K. Using Eq. 5.115 we find

$$\ln x_1 = \frac{18\,640}{8.314}\left(\frac{1}{369.4} - \frac{1}{298.1}\right) = -1.45$$

$$x_1 = 0.234$$

Comparisons are available in Table 5.3. ■

Two observations may be made with respect to the solubility of all solutes from Eq. 5.115. First, the *solubility of all solutes should increase as the temperature is increased.* Second, the solubility in a series of similar solutes will decrease as the melting point increases. The first prediction is universally true, whereas the second applies if the heats of fusion lie fairly close to each other. Some deviations do occur and generally a low melting point and a low heat of fusion will favor increased solubility. Hydrogen bonding also plays an important role in causing deviations. In spite of the deviations from ideal solubility values in some cases, techniques have been developed that allow accurate prediction of values. However, further discussion is beyond the scope of this book.*

*The interested reader should consult the Hildebrand books in the suggested readings.

Boiling Point Elevation

The physical changes occurring on the addition of a nonvolatile solute to a boiling solution have already been shown in Figures 5.12 and 5.13. The condition for equilibrium of a solution with the vapor of the pure solvent is

$$\mu_1^l(T, P, x_1) = \mu_1^{v}(T, P) \tag{5.124}$$

The development is the same as in the section on freezing point depression. Following the same development through Eq. 5.115 we have

$$\ln x_1 = \frac{\Delta_{\text{vap}} H_m}{R}\left(\frac{1}{T} - \frac{1}{T_b^*}\right) \tag{5.125}$$

where T is the boiling point of the solution, T_b^* is now the normal boiling point of pure solvent, and $\Delta_{\text{vap}}H_m$ is the heat of vaporization. The development may be continued in the same manner as before and we obtain

$$\boldsymbol{\Delta_{\text{vap}} T = K_b m_2} \tag{5.126}$$

where

$$K_b = M_1 R T_b^{*2}/\Delta_{\text{vap}} H_m \quad \text{and} \quad \Delta_{\text{vap}} T \text{ is } T - T_b^*.$$

In Table 5.4 are listed several *boiling point elevation* or *ebullioscopic constants*, K_b. Generally, the higher the molar mass of the solute, the larger the value of K_b. These data may be used to determine molar mass in the same manner as for freezing point depression constants. However, since the boiling point is a function of the total pressure, K_b is also a function of temperature. Its variation may be calculated using the Clausius-Clapeyron equation, and it is generally small. If the solution does not behave ideally, the mole fraction can be written in terms of the activity.

Osmotic Pressure

In 1748 the French physicist Jean Antoine Nollet (1700–1770) placed wine in an animal bladder and immersed the sealed bladder in pure water. Water entered the bladder and the bladder expanded and eventually burst. The bladder acted as a *semipermeable membrane.* Such a membrane permits the solvent molecules, but not the solute molecules, to pass through it. The concentration of the solvent molecules is greater on the pure solvent side of the membrane than on the solution side, since,

TABLE 5.4 Boiling Point Elevation Constants K_b

Solvent	$\frac{K_b}{\text{K kg mol}^{-1}}$	Boiling Point, T_b/K
Water	0.51	373.15
Acetic acid	3.07	391.45
Acetone	1.71	329.25
Benzene	2.53	353.35
Ethanol	1.22	351.65

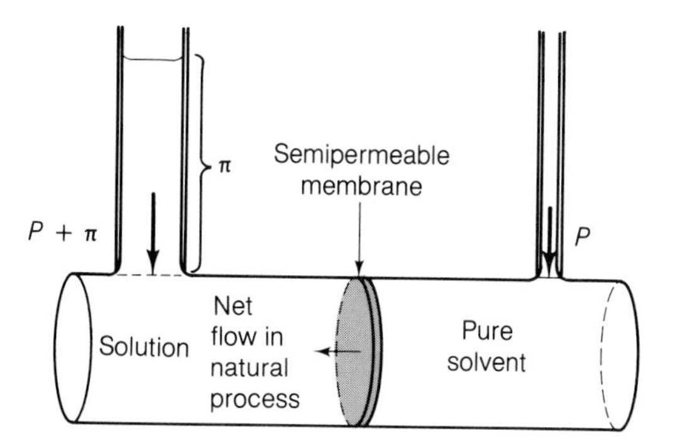

FIGURE 5.14 Schematic representation of apparatus for measuring osmotic pressure. At equilibrium the natural process of flow of pure solvent into solution is just stopped by the osmotic pressure π due to the mass of the column of fluid. In actual osmotic pressure determations an external pressure π is applied to stop the flow.

on the solution side, some of the volume is occupied by solute molecules. There is therefore a tendency for molecules of the solvent to pass through the membrane. This tendency increases with increasing concentration of the solute. Although there is a flow of solvent molecules in both directions through the membrane, there is a more rapid flow from the pure solvent into the solution than from the solution into the pure solvent. The overall effect is a net flow of solvent into the solution, thus diluting its concentration and expanding its original volume.

Much later in 1887 Wilhelm Friedrick Philipp Pfeffer (1845–1920) utilized colloidal cupric ferrocyanide as the semipermeable membrane to study this effect quantitatively.

This phenomenon may be investigated through use of the apparatus shown in Figure 5.14. The solution and the pure solvent are separated from each other by a semipermeable membrane. The natural flow of solvent molecules can be stopped by applying a hydrostatic pressure to the solution side. The effect of the pressure is to increase the tendency of the solvent molecules to flow from solution into pure solvent. The particular pressure that causes the net flow to be reduced to zero is known as the **osmotic pressure** π of the solution.

The equilibrium requirement for the system in Figure 5.14 is that the chemical potentials on the two sides of the membrane be equal. This can only be achieved through a pressure difference on the two sides of the membrane. If the pure solvent is under atmospheric pressure P, the solution is under the pressure $P + \pi$. The chemical potentials of the two sides are

$$\mu(T, P + \pi, x) = \mu^*(T, P) \tag{5.127}$$

where x_1 is the mole fraction of the solvent. Replacing the left-hand side with the explicit dependency on the mole fraction (Eq. 5.101), we have

$$\mu^*(T, P + \pi) + RT \ln x_1 = \mu^*(T, P) \tag{5.128}$$

We must now express the variation of μ as a function of pressure. One of the Maxwell relations is $(\partial G_1/\partial P)_{T,n_i} = V_1$ and since $G_1 = \mu$, it follows that $d\mu^* = V_1^* \, dP$. Integrating between the limits of P and $P + \pi$,

$$\mu^*(T, P + \pi) - \mu^*(T, P) = \int_P^{P+\pi} V_1^* \, dP \tag{5.129}$$

where V_1^* is the molar volume of the pure solvent. Substitution into Eq. 5.128 yields

$$\int_P^{P+\pi} V_1^* \, dP + RT \ln x_1 = 0 \tag{5.130}$$

On the assumption that the solvent is incompressible, V_1^* is independent of pressure and our expression, after integration, reduces to

$$V_1^* \pi + RT \ln x_1 = 0 \tag{5.131}$$

We need to express this relationship in terms of the solute concentration. Since $\ln x_1 = \ln(1 - x_2)$, and if the solution is dilute, $x_2 << 1$, the logarithm may be expanded in a series. Only the first term of the series $-x_2$ makes an important contribution. From its definition, x_2 may be expressed in terms of the amount of solute n_2 and solvent n_1 present. We have, therefore,

$$\ln(1 - x_2) = -x_2 = -\frac{n_2}{n_1 + n_2} \approx -\frac{n_2}{n_1} \tag{5.132}$$

since $n_2 << n_1$ in dilute solution. Equation 5.131 then becomes

$$\pi = \frac{n_2 RT}{n_1 V_1^*} \tag{5.133}$$

If a solution is dilute, as required in Eq. 5.133, the contribution of the molar volume of n_2 is small so that $n_1 V_1^* \approx V$, the volume of the solution. Using this, our expression for π becomes

$$\boldsymbol{\pi = \frac{n_2 RT}{V} \quad \text{or} \quad \pi = cRT} \tag{5.134}$$

where $c = n_2/V$ is the concentration of the solute. This is **van't Hoff's equation for osmotic pressure** and may be used for molar mass determinations.

EXAMPLE

In a study of the osmotic pressure of hemoglobin at 276.15 K, the pressure was found to be equal to that of a column of water 3.51 cm in height. The concentration was 1 g per 0.100 dm^3. Calculate the molar mass.

SOLUTION

The pressure must first be converted to Pa. This can be done using 13.59 as the relative density of mercury with reference to water. Also, 1 mmHg = 133.32 Pa. Therefore,

$$P \text{ of } 3.51 \text{ cm } H_2O = \frac{3.51(\text{cm } H_2O) \times 1333.2(\text{Pa/cmHg})}{13.59(\text{cm}H_2O/\text{cmHg})} = 344.3 \text{ Pa}$$

From Eq. 5.134, and since J = kg m^2 s^{-2} and Pa = kg m^{-1} s^{-2},

$$c = \frac{\pi}{RT} = \frac{344.3(\text{Pa})}{8.314(\text{J K}^{-1}\text{ mol}^{-1}) \times 276.1(\text{K})} = 0.150 \text{ mol m}^{-3}$$

$$= 1.50 \times 10^{-4} \text{ mol dm}^{-3}$$

Since the concentration of the solution is 10 g dm^{-3}, 10 g is 1.50×10^{-4} mol. Consequently,

$$\text{molar mass} = \frac{10 \text{ g}}{1.50 \times 10^{-4} \text{ mol}} = 66\ 700 \text{ g mol}^{-1}$$

■

Besides being a valuable technique for high-molecular-mass materials, osmotic pressure effects are important in physiological systems. Another area for large-scale industrial application is in water desalination. Seawater contains approximately 35 000 parts per million (ppm) of dissolved salts consisting mostly of sodium chloride. The osmotic pressure of seawater due to these salts is approximately 30 MPa (300 atm). A hydrostatic pressure in excess of 30 MPa applied to seawater will separate pure water through a suitable semipermeable membrane. This process is known as *reverse osmosis*.

Key Equations

Clapeyron equation:

$$\frac{dP}{dT} = \frac{\Delta H_m}{T\,\Delta V_m}$$

Clausius-Clapeyron equation:

$$\frac{d \ln P}{dT} = \frac{\Delta_{\text{vap}} H_m}{RT^2}$$

Trouton's rule:

$$\frac{\Delta_{\text{vap}} H_m}{T_b} = \Delta_{\text{vap}} S_m \approx 88 \text{ J K}^{-1} \text{ mol}^{-1}$$

Raoult's law:

$$P_1 = x_1 P_1^* \qquad P_2 = x_2 P_2^*$$

Henry's law:

$$P_2 = k' x_2 \quad \text{or} \quad P_2 = k'' c_2$$

Definition of *partial molar quantity:*

$$X_i \equiv \left(\frac{\partial X}{\partial n_i}\right)_{T,P,n_j}$$

Definition of *chemical potential:*

$$\mu_i \equiv \left(\frac{\partial G}{\partial n_i}\right)_{T,P,n_j}$$

Depression of the freezing point:

$$\Delta_{\text{fus}}T \approx \frac{M_1 R T_f^{*2}}{\Delta_{\text{fus}}H_m} \cdot m_2 = K_f m_2$$

where K_f is the *freezing point depression constant.*

Elevation of the boiling point:

$$\Delta_{\text{vap}}T \approx \frac{M_1 R T_b^{*2}}{\Delta_{\text{vap}}H_m} \cdot m_2 = K_b m_2$$

Osmotic pressure:

$$\pi = \frac{n_2 RT}{n_1 V_1^*} \approx cRT$$

Problems

A Thermodynamics of Vapor Pressure

1. Calculate the heat of vaporization of water at 373.15 K and 101.325 kPa using the Clausius-Clapeyron equation. The vapor pressure of water is 3.17 kPa at 298.15. Compare your answer to the *CRC Handbook** value.

2. Estimate the vapor pressure of iodine under an external pressure of 101.3×10^6 Pa at 313.15 K. The density of iodine is 4.93 g cm^{-3}. The vapor pressure at 101.3 kPa is 133 Pa.

3. The cubic expansion coefficient is given by $\alpha = 1/V(\partial V/\partial T)_P$. According to Ehrenfest's or Tisza's theory, find the order of the transition. Suggest what a plot of α against T would look like near the transition point.

4. The vapor pressure of n-propanol is 1.94 kPa at 293 K and 31.86 kPa at 343 K. What is the enthalpy of vaporization?

**Handbook of Chemistry and Physics*, 60th ed., R. C. Weast, Ed., CRC Press, Boca Raton, Florida, 1980.

5. The compound 2-hydroxybiphenyl (o-phenylphenol) boils at 286°C under 101.325 kPa and at 145°C under a reduced pressure of (14/760)(101.325) kPa. Calculate the value of the molar enthalpy of vaporization. Compare this value to that given in the *CRC Handbook.*

6. Using Trouton's rule, estimate the molar enthalpy of vaporization of n-hexane, the normal boiling point of which is 342.10 K. Compare the value obtained to the value 31.912 kJ mol^{-1} obtained in vapor pressure studies.

7. 2-Propanone (acetone) boils at 329.35 K at 1 atm of pressure. Estimate its boiling point at 98.5 kPa using Crafts' rule.

8. The boiling point of water at 102.7 kPa is 373.52 K. Calculate the value at 101.325 kPa using Crafts' rule.

***9.** Derive an equation for the temperature dependence of the vapor pressure of a liquid (analogous to the integrated form of the Clausius-Clapeyron equation) assuming that the vapor has the equation of state $PV = RT + M$ where M is a constant.

B Raoult's Law, Equivalence of Units, and Partial Molar Quantities

10. Benzene and toluene form nearly ideal solutions. If, at 300 K, P^*(toluene) = 3.572 kPa and P^*(benzene) = 9.657 kPa, compute the vapor pressure of a solution containing 0.60 mol fraction of toluene. What is the mole fraction of toluene in the vapor over this liquid?

11. Often it is important to express one unit of concentration in terms of another. Derive a general expression to find the mole fraction x_2 in a two-component system where the molality is given as m_2.

***12.** The familiar term *molarity* is now discouraged by IUPAC because of the danger of confusion with *molality*. In its place *concentration* is defined as the amount of substance 2, n_2, dissolved in unit volume of solution. Derive a general relation to find x_2 from the concentration c_2. Let the solution density be ρ.

***13.** Derive a general expression to relate the molality m to concentration c_2.

***14.** The volume of a solution of NaCl in water is given by the expression

$$V/\text{cm}^3 = 1002.874 + 17.8213m + 0.873\ 91m^2 - 0.047\ 225m^3$$

where m is the molality. Assume that $m \propto n_{NaCl}$ and that $n_{H_2O} = 55.508$ mol, where $V^*_{H_2O} = 18.068$ cm^3. Derive an analytical expression for the partial molar volume of H_2O in the solution.

15. The partial molal volume of component 2 in a solution may be written as

$$V_2 = \left(\frac{\partial V}{\partial n_2}\right)_{n_1} = \frac{M_2}{\rho} - (M_1 n_1 + M_2 n_2)\frac{1}{\rho^2}\left(\frac{\partial \rho}{\partial n_2}\right)_{n_1}$$

where n_1 and M_1 are amount and molar mass of component 1 and n_2 and M_2 represent the same quantities for component 2. The density is ρ. Rewrite the previous expression in terms of the mole fractions x_1 and x_2.

***16.** Mikhail and Kimel, *J. Chem. Eng. Data, 6*, 533 (1961), give the density of a water-methanol solution in g cm^{-3} at 298 K related to the mole fraction x_2 of the methanol through the equation

$$\rho/\text{g cm}^{-3} = 0.9971 - 0.289\ 30x_2 + 0.299\ 07x_2^2 - 0.608\ 76x_2^3 + 0.594\ 38x_2^4 - 0.205\ 81x_2^5$$

Using the equation developed in Problem 15, calculate V_2 at 298 K when $x_2 = 0.100$.

17. The vapor pressure of pure ethylene dibromide is 172 mmHg and that of pure propylene dibromide is 128 mmHg at 358 K and 1 atm. If these two components follow Raoult's law, estimate the total vapor pressure in kPa and vapor composition in equilibrium with a solution that is 0.600 mol fraction propylene dibromide.

***18.** Henry's law constants k' for N_2 and O_2 in water at 20°C and 1 atm pressure are 7.58×10^4 atm and 3.88×10^4 atm, respectively. If the density of water at 20.0°C is 0.9982 g cm^{-3}, calculate (a) the equilibrium mole fraction and (b) the concentration of N_2 and O_2 in water exposed to air at 20°C and 1 atm total pressure. Assume air is 80 mol % N_2 and 20 mol % O_2.

C Thermodynamics of Solutions

19. In a molar mass determination, 18.04 g of the sugar mannitol was dissolved in 100 g of water. The vapor pressure of the solution at 298 K was 2.291 kPa, having been lowered by 0.041 kPa from the value for pure water. Calculate the molar mass of mannitol.

20. A liquid has a vapor pressure of 40.00 kPa at 298.15 K. When 0.080 kg of an involatile solute is dissolved in 1 mol of the liquid, the new vapor pressure is 26.66 kPa. What is the molar mass of the solute? Assume that the solution is ideal.

***21.** Components 1 and 2 form an ideal solution. The pressure of pure component 1 is 13.3 kPa at 298 K, the corresponding vapor pressure of component 2 is approximately zero. If the addition of 1.00 g of component 2 to 10.00 g of component 1 reduces the total vapor pressure to 12.6 kPa, find the ratio of the molar mass of component 2 to that of component 1.

22. Pure naphthalene has a melting point of 353.35 K. Estimate the purity of a sample of naphthalene in mol %, if its freezing point is 351.85 K (K_f=7.0 K kg mol^{-1}).

23. Calculate the activity and activity coefficients for 0.33 mol fraction toluene in benzene. The vapor pressure of pure benzene is 9.657 kPa at 298 K. $P_2^* = 3.572$ kPa for toluene. The vapor pressure for benzene above the solution is $P_1 = 6.677$ kPa and for toluene $P_2 = 1.214$ kPa.

24. Calculate the mole fraction, activity, and activity coefficients for water when 11.5 g NaCl are dissolved in 100 g water at 298 K. The vapor pressure is 95.325 kPa.

25. Determine the range for the Gibbs energy of mixing for an ideal 50/50 mixture at 300 K. How does this value limit $\Delta_{mix}H$?

26. The mole fraction of a nonvolatile solute dissolved in water is 0.010. If the vapor pressure of pure water at 293 K is 2.339 kPa and that of the solution is 2.269 kPa, calculate the activity and activity coefficient of water.

D Colligative Properties

27. Using Henry's law, determine the difference between the freezing point of pure water and water saturated with air. For N_2 at 298.15 K,

$$(k'')^{-1} = 2.17 \times 10^{-8} \text{ mol dm}^{-3} \text{ Pa}^{-1}$$

For O_2 at 298.15 K,

$$(k'')^{-1} = 1.02 \times 10^{-8} \text{ mol dm}^{-3} \text{ Pa}^{-1}$$

28. Using van't Hoff's equation, calculate the osmotic pressure developed if 6 g of urea, $(NH_2)_2CO$, is dissolved in 1.00 dm^3 of solution at 27°C.

29. The apparent value of K_f in 1.50-molal aqueous sucrose ($C_{12}H_{22}O_{11}$) solution is 2.17 K kg mol^{-1}. The solution does not behave ideally; calculate its activity and activity coefficient ($\Delta_{fus}H° = 6009.5$ J mol^{-1}).

30. A 0.85-g sample is dissolved in 0.150 kg of bromobenzene. Determine the molar mass of the solute if the solution boils at 429.0 K at 1 atm pressure. The normal boiling point of bromobenzene is 428.1 K and the boiling point elevation constant is 6.26 K kg mol^{-1}.

E Supplementary Problems

31. If in a colligative properties experiment a solute dissociates, a term known as *van't Hoff's i factor*, which is the total concentration divided by the initial concentration, must be included as a factor. Thus, for the lowering of the freezing point, $\Delta_{fus}T = imK_f$. Derive an expression that relates i to the degree of dissociation α and to ν, the number of particles that would be produced if the solute were completely dissociated. Then calculate van't Hoff's i factor and α for a 0.010-m solution of HCl that freezes at 273.114 K.

32. In an osmotic pressure experiment to determine the molar mass of a sugar, the following data were taken at 20°C:

π/atm	2.59	5.06	7.61	12.75	18.13	23.72
$\frac{m_2}{V}$/g dm^{-3}	33.5	65.7	96.5	155	209	259

Estimate the molar mass of the sugar. If the sugar is sucrose, what is the percentage error and why?

33. A nonideal solution contains n_A of substance A and n_B of substance B and the mole fractions of A and B are x_A and x_B. The Gibbs energy of the solution is given by the equation

$$G = n_A\mu_A^\circ + n_B\mu_B^\circ + RT(n_A \ln x_A + n_B \ln x_B) + \frac{Cn_An_B}{n_A + n_B}$$

where C is a constant and describes the pair interaction.

a. Derive an equation for μ_A in the solution in terms of the quantities on the right-hand side. {*Hint:* $(\partial \ln x_A/\partial n_A)_{n_B} = (1/n_A) - [1/(n_A + n_B)]$.}

b. Derive a similar expression for the activity coefficient of A. Specify the conditions when the activity coefficient equals unity.

F Essay Questions

1. Describe the form of a typical $P\theta$ diagram and how the diagram may be generated for a one-component system using the Gibbs energy. What is the requirement of stability for each region in the $P\theta$ diagram?

2. Detail the steps in going from the Clapeyron equation to the Clausius-Clapeyron equation. What specific assumptions are made?

3. Explain why Trouton's rule, according to which the entropy of vaporization is 88 J K^{-1} mol^{-1}, holds fairly closely for normal liquids.

4. Describe three colligative properties and comment on their relative merits for the determination of molar masses of proteins.

5. Show mathematically how the chemical potential is the driving force of diffusion for component A between two phases α and β.

6. Why do positive and negative deviations from Raoult's law occur?

Suggested Reading

A. A. Bondi, *Physical Properties of Molecular Crystals, Liquids and Gases,* New York: Wiley, 1968.

L. S. Darken and R. W. Gurry, *Physical Chemistry of Metals.* New York: McGraw-Hill, 1953.

D. Eisenberg and W. Kauzmann, *The Structure and Properties of Water,* New York: Oxford University Press, 1969.

A. Findlay, *Phase Rule* (9th ed. revised by A. N. Campbell and N. O. Smith). New York: Dover Pub., 1951.

J. C. W. Frazer, "The Laws of Dilute Solutions," in *A Treatise on Physical Chemistry* (2nd ed.) H. S. Taylor (Ed.), New York: D. Van Nostrand Co., 1931.

J. H. Hildebrand and R. L. Scott, *Regular Solutions.* Englewood Cliffs, N.J.: Prentice-Hall, 1962.

J. H. Hildebrand and R. L. Scott, *The Solubility of Nonelectrolytes* (3rd ed.). New York: Reinhold Pub. Corp., 1950.

J. H. Hildebrand, J. M. Prausnitz, and R. L. Scott, *Regular and Related Solutions, The Solubility of Gases, Liquids, and Solids.* New York: Van Nostrand Reinhold Co., 1970.

W. J. Moore, *Physical Chemistry* (4th ed.). Englewood Cliffs, N.J.: Prentice-Hall, 1972.

R. A. Swalin, *Thermodynamics of Solids* (2nd ed.). New York: Wiley, 1972.

R. H. Wagner and L. D. Moore, "Determination of Osmotic Pressure," in A. Weissberger (Ed.), *Physical Methods of Organic Chemistry,* Part 1 (3rd ed.), New York: Interscience Pub., 1959.

Chapter 6

Phase Equilibria

Preview

In this chapter we apply the principles developed in Chapter 5 to a number of examples. The *number of components* c in a system is the smallest number of independent chemical constituents needed to fix the composition of *every* phase in the system. At fixed temperature and pressure the number of components and the *number of phases* p are related through the *phase rule*

$$f = c - p + 2$$

where f is the *number of degrees of freedom.* The value of f is important because it is the number of intensive variables, such as temperature, pressure, and concentrations, that can be independently varied without changing the number of phases. Systems may thus be *invariant* (unchangeable with respect to the number of components, phases, T and P), *univariant* (only one of the variables may change), *bivariant* (two variables may change), etc.

In binary-liquid systems there is a finite temperature and pressure range over which liquid and vapor coexist. An expression can be obtained for the total pressure P of the vapor in terms of the individual vapor pressures of the pure liquids and the mole fractions in the vapor of one of the components. When P is plotted against this mole fraction, the curve is concave upward and is called the *vapor curve.* The curve representing the liquid and liquid-vapor boundary is called the *liquid curve.*

The existence of the region between the two curves allows *distillation* to be carried out. A tie line between the liquid and vapor curves determines the ratio of liquid to vapor through the use of the *lever rule.* Both positive and negative deviations from Raoult's law occur, and there is the possibility of *azeotropes,* which are mixtures corresponding to a maximum or minimum in the boiling point curve. When there are azeotropes, it is impossible by normal distillation to separate both pure components of a liquid mixture.

It is often convenient to ignore the vapor in a mixture and consider only the liquid-solid equilibrium. Often there is a single composition that has the lowest melting point in the phase diagram. This is called the *eutectic composition,* and the temperature of melting is the *eutectic temperature.* The method of *thermal analysis* is useful for constructing such phase diagrams. Sometimes a chemical reaction occurs when there is a phase change; such reactions are known as *phase reactions* or as *peritectic reactions.* The point in the phase diagram at which they occur is known as a *peritectic point.*

This chapter also deals with equilibria involving three components (i.e., ternary systems).

6 Phase Equilibria

Chapter 5 was mainly concerned with general principles related to equilibria involving vapors, liquids, solids, and solutions. In the present chapter we apply these principles to a number of additional examples, and we will see that they lead to useful ways of presenting and classifying experimental data. The approach we take has been found useful in a wide range of scientific and technical problems.

6.1 Equilibrium Between Phases

Number of Components

In Section 5.1 we considered what constituted a phase (p), which is easily distinguished since it is homogeneous and distinct part of a system separated by definite boundaries from the rest of the system.

We have not concerned ourselves with discussing what constitutes the **number of components** c in a phase diagram, since we have only dealt at most with two nonreacting constituents. However this idea needs careful development. In the sense used here the number of components is the smallest number of independent chemical constituents needed to fix the composition of *every* phase in the system. If we consider water, it consists of only one component even though it may exist in three different phases: solid, liquid, and vapor. However, if sodium chloride is now dissolved in the water, the system becomes a two-component system and remains so even though under some conditions the salt may separate and form a pure solid phase within the system. One also might argue that since the salt dissociates into ions, additional components are present. However this is not the case. Even though there are two more constituents, namely Na^+ and Cl^- ions, there is a material balance since the reduction in amount of NaCl dissociated must equal the amounts of cations or anions formed. There is in addition a requirement of electroneutrality since the number of Na^+ ions must equal the number of Cl^- ions. Consequently, the material balance and electroneutrality conditions reduce the number of constituents, H_2O, NaCl, Na^+, and Cl^-, to two; that is, $c = 4 - 1 - 1 = 2$. Thus, although there is no unique set of components from the possible constituents, the *number* of components is unique.

If a chemical reaction can take place between constituents of a solution, the number of constituents is reduced by the number of equilibrium conditions. Considering the system

$$PCl_5 \rightleftharpoons PCl_3 + Cl_2 \tag{6.1}$$

we recognize three distinct chemical species. This number of species is reduced by the independent equilibrium condition and, therefore $c = 3 - 1 = 2$. Another reduction in the number of components is possible if we start with pure PCl_5, in which case $[PCl_3] = [Cl_2]$. In this case the number of components is unity because of the additional mathematical relation.

EXAMPLE

How many components are present when ethanol and acetic acid are mixed?

SOLUTION
At first sight we might predict two. However, these components react and ethyl acetate and water are also present at equilibrium. This raises the number of constituents from two to four.

$$\text{HOAc} + \text{EtOH} \rightleftharpoons \text{EtOAc} + \text{HOH}$$

But now the equilibrium condition is applied and reduces by one the number of components (i.e., the number of constituents whose value is required to specify the system); thus, $c = 4 - 1 = 3$. Since EtOAc and HOH must be formed in equal amounts, another mathematical condition exists reducing the number of components to two. ■

There are many systems in which it might be difficult to decide how to apply the equilibrium condition to arrive at the correct number of components. For example, we know that an equilibrium condition can be written for water, hydrogen, and oxygen. If these three constituents are present in the absence of a catalyst, we have a three-component system because the equilibrium that could reduce the number of constituents, $H_2 + \frac{1}{2}O_2 \rightleftharpoons H_2O$, for all practical purposes is not achieved. There are other instances where our criterion of establishing equilibrium is not so clear-cut as this. For example, a mixture of water, hydrogen and oxygen is a three-component system at room temperature in the absence of a catalyst, but at higher temperatures, because the equilibrium becomes established, it consists of two components. Where does the shift from a three-component system to a two-component system occur? To resolve this problem we must consider the time required to make the measurement of the specific property of the system. If the equilibrium shifts with a change in the measured variable in a time very short in comparison to the time required for the measurement, the equilibrium condition is applicable. In our example the number of components at high temperature is two. However, if the time required to make the measurement of the property is very short in comparison to the time required for equilibrium to be reestablished (that will be the case at low temperatures), then effectively there is no equilibrium in the H_2O–H_2–O_2 system and it remains a three-component system.

Degrees of Freedom

Another important idea for understanding phase diagrams is the concept of the *number of* **degrees of freedom** f or the *variance* of a system. The number of degrees of freedom is the number of intensive variables, such as temperature, pressure, and concentration, that can be independently varied without changing the number of phases. Since some variables are fixed by the values chosen for the independent variables and by the requirement of equilibrium, it is only the number of those that remain unencumbered that is referred to as the number of degrees of freedom. Stated differently, the number of degrees of freedom is the number of variables that must be fixed in order for the condition of a system at equilibrium to be completely specified.

If the system has one degree of freedom, we say it is *univariant*. Thus pure water is univariant since, at any temperature, the pressure of vapor in equilibrium with liquid is fixed and only one variable may be varied independently. If a system has two degrees of freedom, it is a *bivariant* system, etc.

The Phase Rule

The application to a system of the thermodynamic techniques developed in Chapter 5 can be laborious if each system must be treated as an individual case. However, the general conditions for equilibrium between phases, the **phase rule,** was deduced theoretically by J. Willard Gibbs* in 1875–1876. We assume in this development that the equilibrium between phases is not influenced by gravity, electrical forces, magnetic forces, or surface forces but only by pressure, temperature, and composition. Under these conditions the phase rule is given by the expression

$$f = c - p + 2 \tag{6.2}$$

where the term 2 is for the two variables, temperature and pressure.

This result may be derived by expanding upon the development in Section 5.6. We showed there that, for equilibrium to exist throughout a system consisting of more than one phase, the chemical potential of any given component must have the same value in every phase. From Eq. 5.66 we have that $\mu_i^\alpha = \mu_i^\beta$ at constant T and P. Extending this to a system consisting of p phases, we have

$$\mu_1^\alpha = \mu_1^\beta = \mu_1^\gamma = \cdots = \mu_1^p \tag{6.3}$$

There will be an equation similar to Eq. 6.3 for each component present in the system. This leads to an array of information needed to specify the system, along with the temperature and pressure:

$$\begin{aligned}
\mu_1^\alpha &= \mu_1^\beta = \mu_1^\gamma = \cdots = \mu_1^p \\
\mu_2^\alpha &= \mu_2^\beta = \mu_2^\gamma = \cdots = \mu_2^p \\
\mu_3^\alpha &= \mu_3^\beta = \mu_3^\gamma = \cdots = \mu_3^p \\
\mu_c^\alpha &= \mu_c^\beta = \mu_c^\gamma = \cdots = \mu_c^p
\end{aligned} \tag{6.4}$$

Our task now is to reduce the information of this array to the bare minimum required to describe the system at constant T and P. In any phase containing c components, the composition is completely specified by $c - 1$ concentration terms if they are expressed in *mole fractions* or *weights percent*. This is the case since the last term is automatically determined from the remaining mole fraction or remaining weight percent. Thus, for a system of p phases, $p(c - 1)$ concentration terms are required to define the composition completely, and an additional two terms are required for the temperature and pressure of the system. Thus, $p(c - 1) + 2$ terms are required. But from Eq. 6.4 we see that each line contains $p - 1$ independent equations specifying the state of the system. Each equality sign is a condition imposed on the system that reduces by one the pieces of information required. Since there are $c(p - 1)$ of these expressions defining $c(p - 1)$ independent variables, the total number of variables left to be defined (i.e., the number of degrees of freedom) is

$$f = [p(c - 1) + 2] - [c(p - 1)] = c - p + 2 \tag{6.5}$$

This is of course the phase rule given in Eq. 6.2.

---

*J. W. Gibbs, *Trans. Conn. Acad. Arts. Sci.*, 1876–1878, in *The Collected Works of J. Willard Gibbs* (Vol. 1), New Haven: Yale University Press, reprinted 1948.

6.2 One-Component Systems

In Section 5.1 we saw an example of phase equilibrium in a one-component system, water. We will now apply the phase rule to another one-component system, sulfur. The general features of the phase diagram of sulfur are shown in Figure 6.1, with P plotted against T. There are several features in this figure not present in Figure 5.1 for water. The stable form of sulfur at 1 atm pressure and room temperature is a crystalline form called (*ortho*)*rhombic* sulfur. (See Section 15.2 for the details on the arrangement of atoms in this form of sulfur.) As rhombic sulfur is heated slowly at 1 atm, it transforms to a different crystalline form called *monoclinic* sulfur at 368.55 K. This is an example of polymorphism (see Section 5.1). The name **allotrope** (Greek *allos*, other; *tropia*, turning) refers to each of the crystalline forms when this type of transformation occurs in elements.

At any particular pressure along line BC there is a definite temperature, called the *transition point*, at which one form will change reversibly to the other. In sulfur the transition point at pressures less than 1 atm (101.325 kPa) lies below the melting point of the solid. Each crystalline form possesses a definite range of stable existence. *Enantiotropy* (Greek *enantios*, opposite; *tropia*, turning) is the name given to this phenomenon and the crystalline forms are said to be *enantiotropic*.

Monoclinic sulfur melts to the liquid along line CE. However if the rhombic form is rapidly heated at 1 atm pressure, the transformation temperature of 368.55 K is bypassed and the rhombic form melts directly to liquid sulfur at 387 K. When the rhombic sulfur is in equilibrium with liquid (dashed lines in Figure 6.1), we have an example of a metastable equilibrium since this equilibrium position lies in the region of the more thermodynamically stable monoclinic form of sulfur. This is not a true equilibrium, but it appears to be one because the process of change into the more favored form is slow. In this region both the rhombic and liquid forms of sulfur can decrease their Gibbs energy by converting to the favored monoclinic form.

The intersection of the two lines BC and CE is brought about by the difference in density of the two crystalline forms. The density of rhombic sulfur is greater than that of the monoclinic form, which has a density greater than the liquid. Therefore the lines BC and EC have positive slopes. The variation of vapor pressure with temperature may be handled using the Clapeyron equation. There are three triple points of stable equilibrium in this system:

Point B at 368.55 K: $S_{\text{rhombic}} \rightleftharpoons S_{\text{monoclinic}} \rightleftharpoons S_{\text{vapor}}$

Point E at 392.15 K: $S_{\text{monoclinic}} \rightleftharpoons S_{\text{liquid}} \rightleftharpoons S_{\text{vapor}}$

Point C at 424 K: $S_{\text{rhombic}} \rightleftharpoons S_{\text{monoclinic}} \rightleftharpoons S_{\text{liquid}}$

At these three points (B, C, E), there is one component, $c = 1$, and three phases, $p = 3$. By the phase rule then, $f = c - p + 2$, and we have

$$f = 1 - 3 + 2 = 0 \tag{6.6}$$

From this we see that there are no degrees of freedom; the system is *invariant*. This means that the equilibrium of the three phases automatically fixes both the temperature and pressure of the system, and no variable may be changed without reducing the number of phases present.

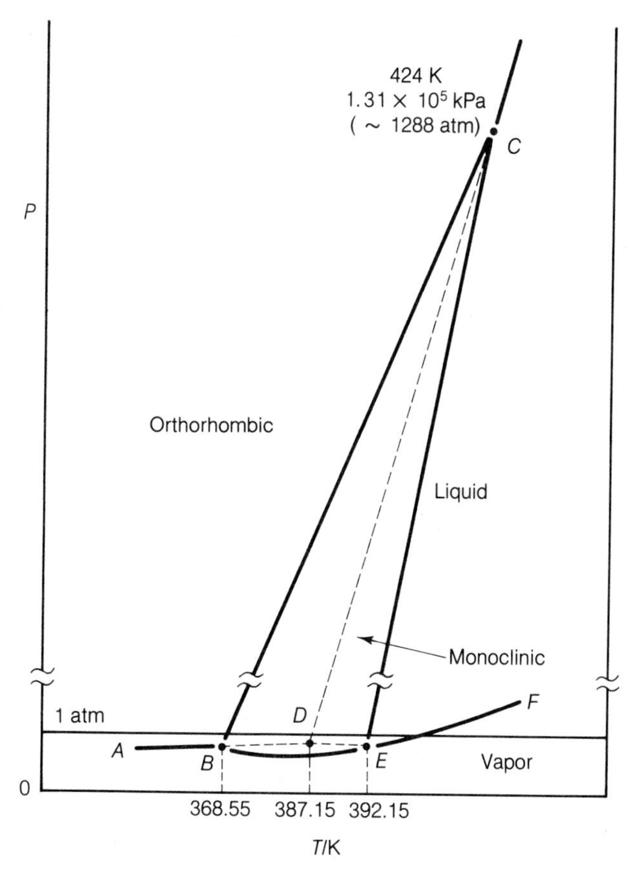

FIGURE 6.1 Phase diagram for sulfur. Temperature along abscissa refers to transitions at 1 atm.

The curves AB, BC, BE, CE, and EF describe the equilibria between two phases. The phase rule requires that along these curves

$$f = c - p + 2 = 1 - 2 + 2 = 1 \tag{6.7}$$

The system is thus *univariant.* Thus under conditions of two phases present either temperature or pressure may be varied, but once one of them is fixed along the respective equilibrium line, the final state of the system is completely defined.

In the four regions where single phases exist (namely rhombic, monoclinic, liquid, or vapor) we have $f = 1 - 1 + 2 = 2$ and the system is *bivariant.* To define the state of the system completely the two variables temperature and pressure must both be specified.

Point D is a metastable triple point with two-phase equilibrium occurring along lines BD, CD, and DE. The phase rule predicts the number of degrees of freedom regardless of the fact that the system is metastable, since the metastable system, within the time scale of the measurements, behaves as if it were at equilibrium. Thus we find invariant and univariant metastable systems as we did earlier for the corresponding stable equilibria.

This is the first example of metastable equilibrium that we have seen. In the past we always assumed that a phase change would take place when the external conditions were suitable. However, it long has been known that water can be cooled 9.4 K below its normal freezing point without solidification occurring, a process called **supercooling** discovered in 1724 by Fahrenheit (1686–1736). Supercooled water is thus another example of a metastable system, and it becomes *unstable* in the presence of solid ice.

6.3 Binary Systems Involving Vapor

Liquid-Vapor Equilibria of Two-Component Systems

Raoult's law (Section 5.4) may be applied to binary solutions of volatile components to predict their total pressure. A curve such as Figure 5.6 in which the vapor pressure is plotted against x (the mole fraction of one of the liquid components) is incapable of describing states of the system that are completely gaseous at lower pressures. In order to describe the boundary between the liquid-vapor and pure vapor regions, we introduce the mole fraction of component 1 in the vapor, y_1. This is just the ratio of the partial pressure of component 1 to the total pressure,

$$y_1 = \frac{P_1}{P} \tag{6.8}$$

Our task then is to describe the total pressure in terms of the mole fraction of component 1 in the vapor state and the individual pressures of the pure components.

Raoult's law, $P_1 = x_1 P_1^*$, and the expression for P

$$P = P_1 + P_2 = x_1 P_1^* + (1 - x_1)P_2^* = P_2^* + (P_1^* - P_2^*)x_1 \tag{6.9}$$

may be substituted into Eq. 6.8 and the resulting equation solved for x_1:

$$x_1 = \frac{y_1 P_2^*}{P_1^* + (P_2^* - P_1^*)y_1} \tag{6.10}$$

This expression may now be substituted back into Eq. 6.9 with the elimination of x_1. The result is

$$P = \frac{P_1^* P_2^*}{P_1^* + (P_2^* - P_1^*)y_1} \tag{6.11}$$

When P is plotted against y_1, or y_2, as in the vapor curve shown in Figure 6.2, the curve is concave upward; P becomes P_1^* when $y_1 = 1$ and P_2^* when $y_1 = 0$.

In Figure 6.2 the pressure-composition curve for the mixture isobutyl alcohol and isoamyl alcohol is plotted. This mixture behaves almost ideally (i.e., almost as predicted from Raoult's law). The upper curve is a straight line giving the boundary between pure liquid and liquid-vapor regions. This line is called the *liquid curve* and is merely a plot of the total vapor pressure for the liquid mixture against x_2. Above this line, vapor will condense to a liquid.

The lower curve is a plot of P against y_2 from Eq. 6.11 and is called the *vapor curve*. It gives the phase boundary between the liquid-vapor and pure vapor regions. Below this line, liquid cannot exist in equilibrium.

In the region marked *Liquid*, only one phase exists. There are two components so that from the phase rule, $f = 2 - 1 + 2 = 3$. Since Figure 6.2 is drawn at a specific

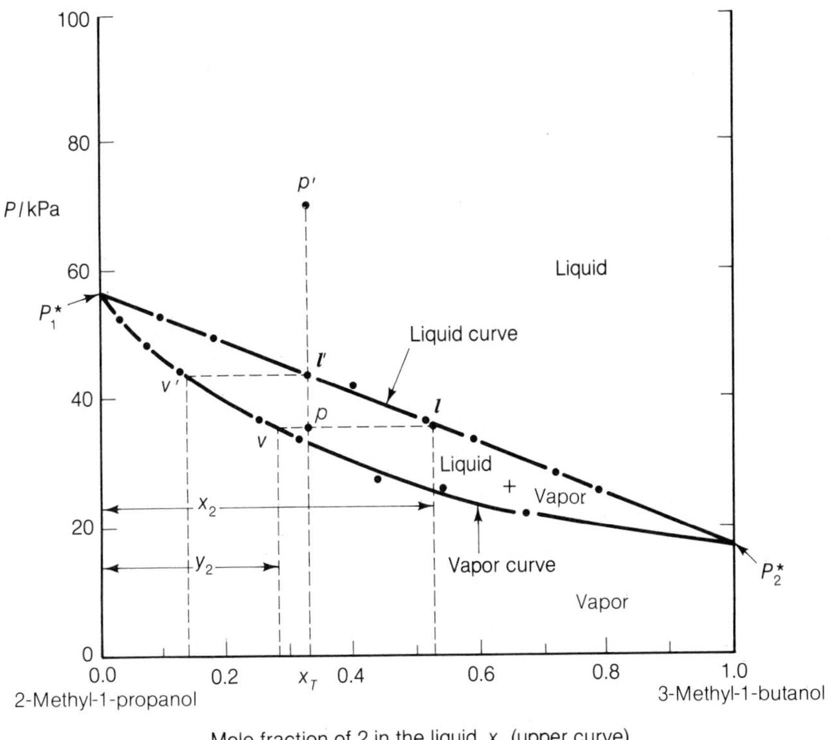

FIGURE 6.2 Pressure-composition diagram for two volatile components showing liquid-vapor equilibrium as a function of vapor pressure and mole fraction for the system isobutyl alcohol (component 1)–isoamyl alcohol (component 2) at 323.1 K.

temperature, one degree of freedom is already determined. Only two more variables are needed to specify the state of the system completely. A choice of P and x_1 then completes the description. In the region marked *Vapor*, specific values of P and y_1 will define the system.

In the region marked *Liquid + Vapor*, there are two phases present and the phase rule requires specification of only one variable since T is fixed. This variable may be P, x_1, or y_1. To demonstrate this, consider point p in Figure 6.2. The value x_T at point p corresponds to the mole fraction of component 2 in the entire system. To find the compositions of both liquid and vapor in equilibrium at this pressure, we draw a horizontal line through the point p corresponding to constant pressure. This line, called a **tie line**, intersects the liquid curve at l and the vapor curve at v. The compositions x_2 and y_2 corresponding to l and v give the mole fractions of component 2 in the liquid phase and vapor phase, respectively. The intersection of the vertical composition line representing x_2 or y_2 with the respective liquid or vapor curves gives the pressure. The tie line at that pressure gives the other composition variable. Thus the system is fully defined by specifying only P or x_1, or y_1 in the two-phase region, since $x_1 = 1 - x_2$.

It is often necessary to determine the relative amounts of the components in the two phases at equilibrium. When point p in Figure 6.2 lies close to point v, the amount of substance in the vapor state is large in comparison to the amount in the liquid state. Conversely, if p were close to l, the liquid phase would predominate. To make this observation quantitative, we will obtain an expression for the ratio of the amount of substance in the liquid state, n_l, to the amount in the vapor state, n_v.

The total amount present is given by the mass balance

$$n = n_l + n_v \tag{6.12}$$

A similar expression can be written for each component; thus

$$n_1 = n_{l,1} + n_{v,1} \tag{6.13}$$

In terms of mole fractions (see Figure 6.2 for x_T)

$$x_T = \frac{n_1}{n}, \qquad x_1 = \frac{n_{l,1}}{n_l}, \qquad y_1 = \frac{n_{v,1}}{n_v} \tag{6.14}$$

Substituting these expressions into Eq. 6.13 gives

$$nx_T = n_l x_1 + n_v y_1 \tag{6.15}$$

Substitution for n from Eq. 6.12 gives, after rearrangement,

$$\frac{n_l}{n_v} = \frac{y_1 - x_T}{x_T - x_1} = \frac{\overline{pv}}{\overline{lp}} \tag{6.16}$$

where $\overline{pv}$ is the length of the line segment between p and v, and $\overline{lp}$ is the length of the line segment between l and p. This is a statement of the **lever rule.**

An interesting feature of curves similar to those in Figure 6.2 is that the vapor always contains relatively more of the more volatile component than does the liquid. This applies to all liquid mixtures and is useful in interpreting even those systems not obeying Raoult's law. Assuming that the vapors behave ideally, we have from Eq. 6.8 and Dalton's law

$$y_1 = \frac{P_1}{P_1 + P_2} \tag{6.17}$$

If the liquid mixture in contact with the vapor follows Raoult's law, we may express the mole fraction in the vapor phase in terms of the mole fractions in the liquid state. Thus

$$y_1 = \frac{x_1 P_1^*}{x_1 P_1^* + x_2 P_2^*} \tag{6.18}$$

Taking the ratio of y_1 to x_1 we have, from Eq. 6.18,

$$\frac{\text{mole fraction of component 1 in vapor}}{\text{mole fraction of component 1 in liquid}} = \frac{y_1}{x_1} = \frac{1}{x_1 + x_2(P_2^*/P_1^*)} \tag{6.19}$$

In a binary solution, $x_1 + x_2 = 1$. Therefore the ratio of the mole fraction in the vapor state to that in the liquid state can be unity only if the pressure $P_1^* = P_2^*$. If $P_2^* > P_1^*$, the denominator is greater than 1 and therefore $y_1 < x_1$. Thus if the liquid of pure component 2 has a higher vapor pressure than that of pure component 1

($P_2^* > P_1^*$), the vapor will contain relatively more of component 2 than does the liquid that is in equilibrium with it.

Application of the aforementioned principle is found in *isothermal distillation*. In this process the temperature of a liquid mixture is held constant and the vapor is progressively removed. As the process proceeds, the composition of the liquid is altered. As an illustration, consider a mixture whose composition is represented by point p' in Figure 6.2. As the pressure on the liquid is reduced by removing vapor, the solution does not reach equilibrium with the vapor until the pressure at point l' is reached. Then the vapor has composition v'. This vapor is much richer in the more volatile component than is the liquid. Hence, as this vapor is removed, the remaining solution will become concentrated in the less volatile component, changing composition along line $\overline{l'l}$. This method is not very commonly used for separation, but the method is particularly useful if the mixture were to decompose upon heating when distilled at atmospheric pressure. It also is useful when one of the components is much more volatile than the other. For instance, in the preparation of freeze-dried coffee, water is removed from the frozen coffee solution by the sublimation of ice leaving behind a rigid, porous structure that is readily soluble.

Liquid-Vapor Equilibrium in Systems Not Obeying Raoult's Law

When Raoult's law holds reasonably well for completely miscible liquid pairs, the total vapor pressure in a binary solution will be *intermediate* between the vapor pressures of the two pure components. The pressure-composition diagram will be similar to that in Figure 6.2.

When there are large, positive deviations from Raoult's law (Section 5.5), there will be a *maximum* in the total vapor pressure curve. The liquid-vapor equilibrium will then be established, as shown in Figure 6.3 for the ethylene chloride-ethanol system. This curve shows a maximum at 38 mol % ethanol at a constant temperature of 313.1 K.

In a similar manner, there may be a *minimum* in the total vapor-pressure curve for *negative deviations* from Raoult's law. The liquid-vapor equilibrium is established as shown in Figure 6.4 for the acetone-chloroform system. Here a minimum occurs at approximately 58 mol % chloroform at 308 K.

Temperature-Composition Diagrams: Boiling Point Curves

The plots of vapor pressure against composition in the last two subsections have been made at constant temperature. It is more common to work at constant pressure, and temperature-composition diagrams are then required to detail the behavior of the liquid-vapor equilibrium. We first consider a full representation of a binary system with respect to T, P, and composition. Such a description requires a three-dimensional figure.

For a typical completely miscible binary system the temperature and composition are represented along the two horizontal axes, and pressure along the vertical axis. A solid figure will be formed from the volume representing the region of liquid-vapor equilibrium. It lies between a concave and convex surface meeting in edges lying in the two faces of the figure representing the T-P planes at $x = 0$ and $x = 1$. This is shown in Figure 6.5 where the right and left faces of the figure are

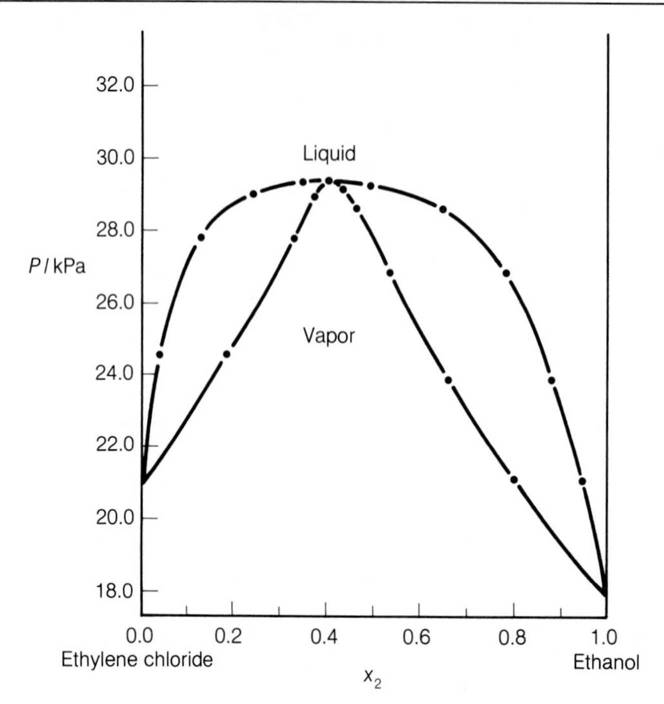

FIGURE 6.3 Liquid-vapor equilibrium as a function of vapor pressure and mole fraction for the system ethylene chloride–ethanol at 313.1 K.

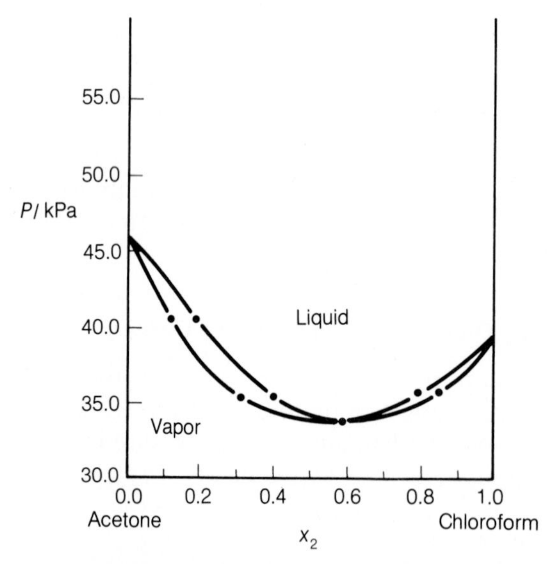

FIGURE 6.4 Liquid-vapor equilibrium as a function of vapor pressure and mole fraction for the system acetone–chloroform at 308 K.

the respective vapor-pressure curves of the pure liquids. Note that, in the region above the envelope of liquid-vapor equilibrium, only liquid exists; below the envelope, only vapor exists.

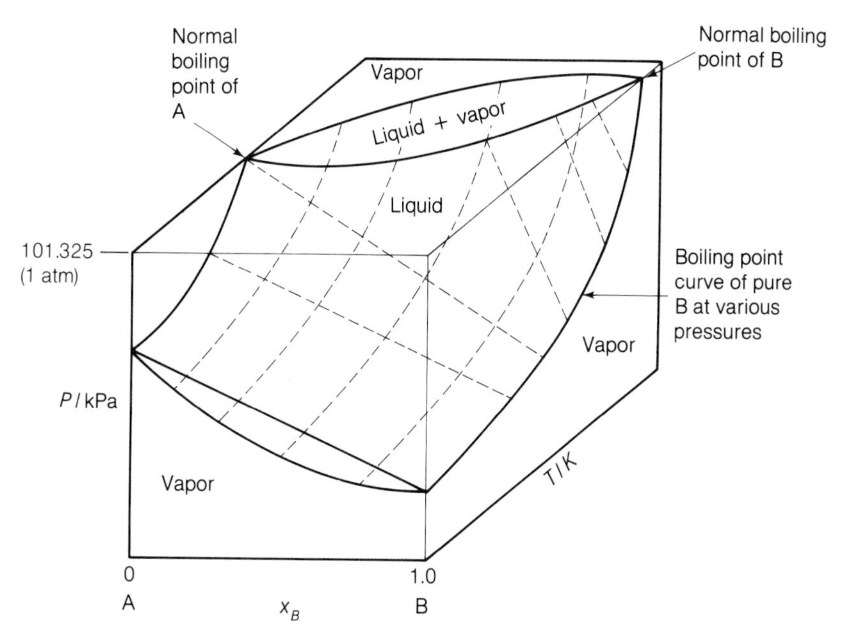

FIGURE 6.5 Typical three-dimensional phase diagram for a binary system obeying Raoult's law.

A typical temperature-composition plane (similar to the isobaric section at 1 atm pressure in Figure 6.5) is shown in Figure 6.6 for a binary system consisting of 2-butanone and *trans*-dichloroethylene. In comparison to the simple double curve or lens in Figure 6.2, the temperature-composition curve will have the component with the higher vapor pressure at the low end of the lens in the T-x plot. The vapor-composition curve is now the upper curve because the vapor is more stable at high temperatures.

Although the binary system may behave ideally, the straight line from a P against x plot for the total pressure of the liquids is not a straight line in the temperature-composition plot. (The vapor pressure does not increase proportionally with T.) This curve is normally determined experimentally by measuring the boiling points and vapor compositions corresponding to a range of liquid mixtures.

Distillation

Changes in state brought about by an increase in temperature may be utilized to separate the components in a binary mixture. To see how this can be accomplished, examine Figure 6.7. The boiling points of all the mixtures are intermediate between the boiling points of the pure components. Point p in Figure 6.7 describes the mixture in the liquid state, and we will consider what happens as the temperature is raised. The liquid system does not reach equilibrium with its vapor until point l is reached on the liquid curve. At this point the vapor has the composition v, which is much richer in the more volatile component than the original liquid. As vapor v is removed, the composition of the remaining liquid has a lower mole fraction of component 2. Boiling will not continue (i.e., equilibrium will not be

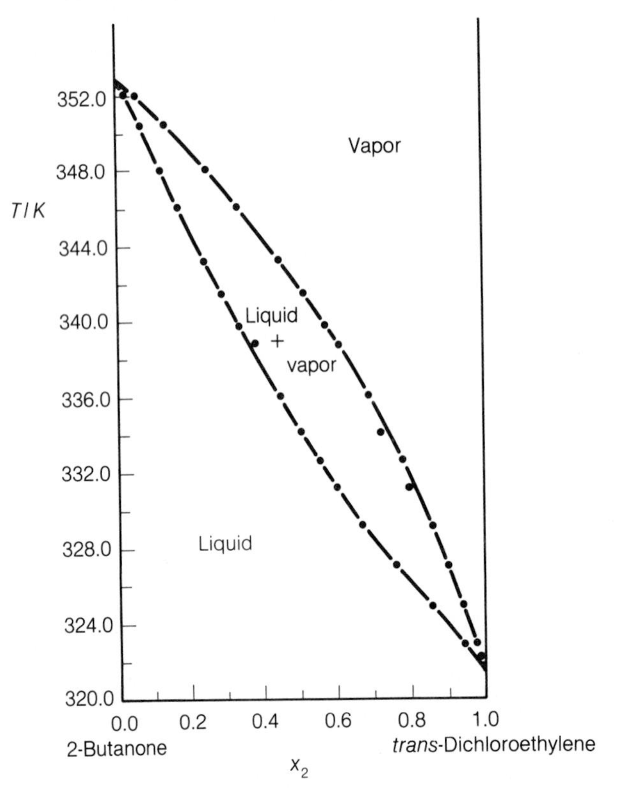

FIGURE 6.6 Temperature-vs-composition diagram for 2-butanone and *trans*-dichloroethylene at 101.325 kPa.

maintained) unless the temperature is raised. As the temperature is raised to θ_1 (point p'), the composition of the liquid will have moved from l to R along line $\overline{lR}$. In a similar way the composition of the vapor will have moved from v to D along line $\overline{vD}$. The lever rule may be used to determine the amounts of the two phases present. Finally, at any position higher than the vapor curve, say at p'', the entire state consists of vapor.

With composition and pressure already fixed, the phase rule allows only one degree of freedom, namely temperature, in the single-phase regions for point p. As we have seen on page 217, the system is invariant in the two-phase region when one degree of freedom is taken by the constant pressure and the other by either temperature, or by either of the composition variables.

Within a closed system, the movement of the system variables from p along the line to p' leaves the composition of the overall system unchanged. The amount of vapor formed at p' is given by the lever rule and the vapor has composition D. The vapor D may be removed and condensed to point l'. This liquid l' will have a much higher concentration of the more volatile component than the original mixture of composition p. In a similar manner the residual liquid R, boiling at θ_1, will be enriched in the higher boiling component. The condensate from D, now at point l', can be evaporated again at the lower temperature θ_3. The resulting composition of the vapor at v' will be much closer to that of the pure component than before.

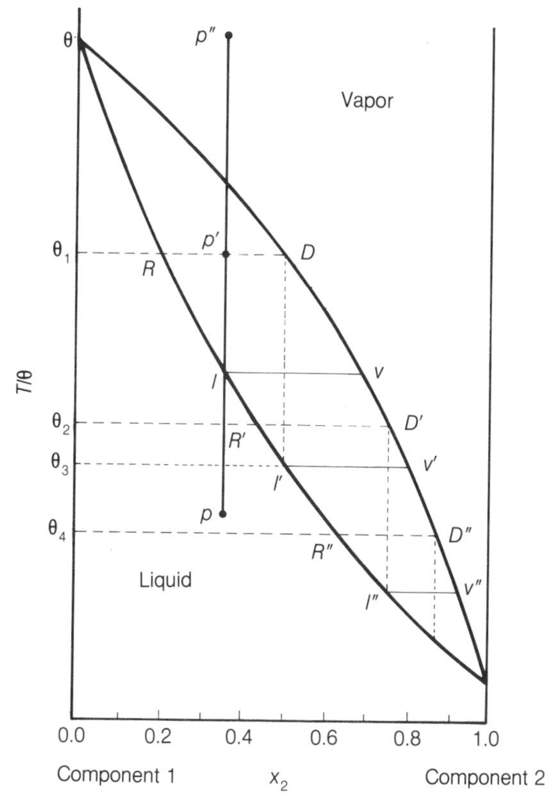

FIGURE 6.7 Temperature-composition diagram showing the technique for separation of a mixture into its pure components.

As more vapor is collected, assume that the composition of the vapor reaches point D' at θ_2. The D' distillate may be collected and condensed to l'' and reevaporated, and the process repeated as shown in Figure 6.7 until the pure component of higher vapor pressure is achieved. Using the same technique, the component boiling at the higher temperature also may be separated as the residual liquid.

The *batch-type* operation just described would be obviously a very time-consuming process. In practice, the process is automated using a variety of techniques. Perhaps the easiest of these to picture is the use of a *fractionating column* of the *bubble-cap* variety shown in Figure 6.8.

In this device preheated liquid mixture enters the column about halfway up. The less volatile components drop to the bottom boiler or still, A, where they are reheated. As the vapor ascends the column, the higher-boiling components begin to condense while the lower-boiling materials proceed to higher stages. Thus a temperature gradient is established, with the highest temperature at the bottom of the column and the lowest at the top from which the lowest-boiling solution may be removed. At each level in the column, such as B, vapor from the level or *plate* below bubbles through a thin film of liquid C, at a temperature slightly lower than that of the vapor coming through the bubble cap, D. Partial condensation of the vapor occurs. The lower-boiling mixture remains as vapor and moves to the next

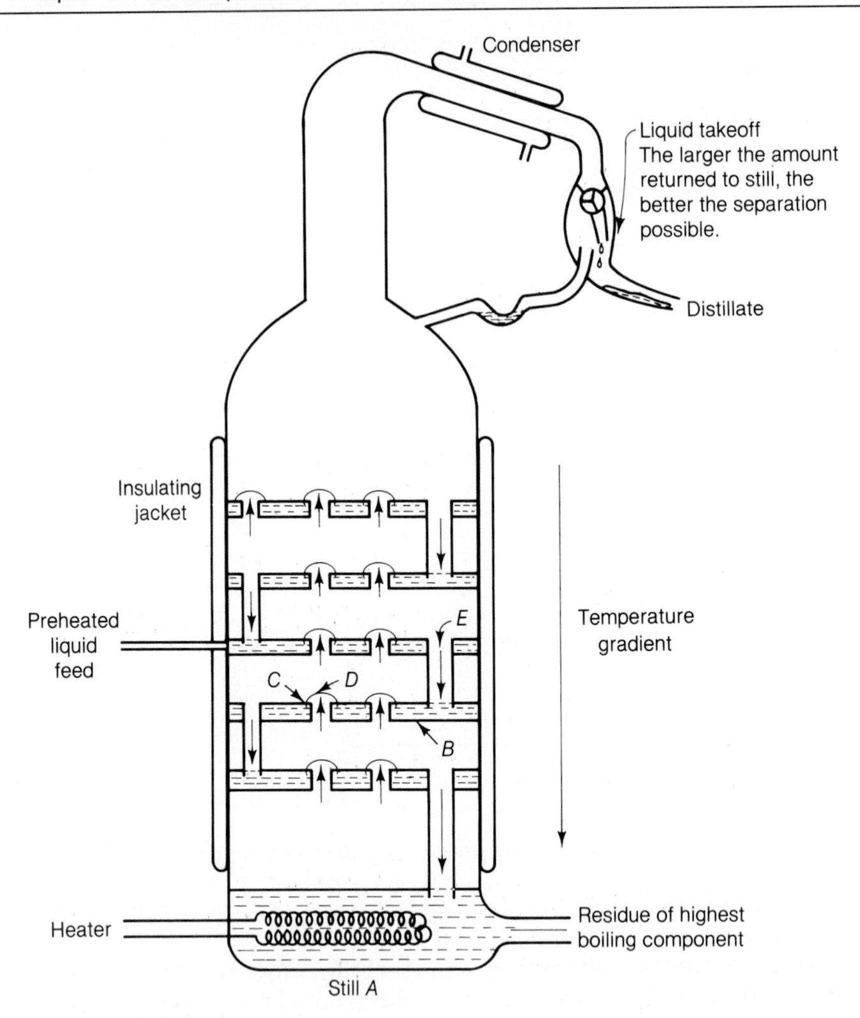

FIGURE 6.8 A bubble-cap fractionating column.

plate. Thus the vapor leaving each plate is enriched in the more volatile component compared to the entering vapor from the plate below. The action of the vapor condensing and then reevaporating is the same as described for the behavior shown in Figure 6.7. Excess liquid at each plate is returned to the plates below via overflow tubes, E.

The intimate contact required between liquid and vapor to achieve the equilibrium at each plate may be achieved in other ways. In the *Hempel column,* glass beads or other packing is used to increase the surface area of the liquid in the column. The vapor must then pass through and over more liquid on its path to the top of the column. Direct flow of the liquid back to the boiler must be avoided because this type of pathway does not provide the intimate contact needed between liquid and vapor to approximate equilibrium.

The efficiency of the column is generally measured in terms of its number of *theoretical plates.* We describe the height equivalent of a theoretical plate (HETP), h, as the distance between the vapor in the column and the liquid with which it would

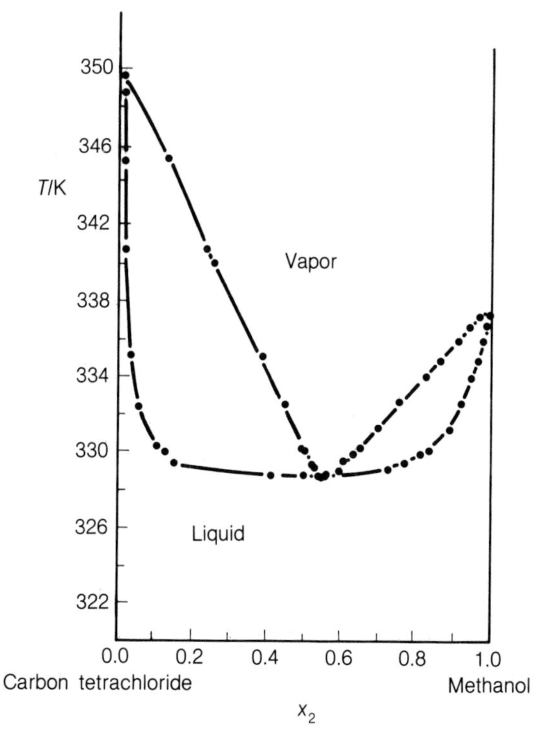

FIGURE 6.9 Temperature-composition diagram for carbon tetrachloride-methanol at 101.325 kPa.

be exactly in equilibrium. This separation comes about in practice because equilibrium is not established at every position.

If two components obey Raoult's law, the closeness of their boiling points dictates how many theoretical plates are needed for separation. For widely separated boiling points a few plates will suffice, whereas if the boiling points are close, many theoretical plates will be needed. Although the word *equilibrium* is sometimes used to refer to an operating column, it is more correct to say that the column is in a **steady state.** In a thermodynamic sense, true equilibrium is not established throughout since a uniformity of temperature does not exist and because the counterflow of liquid and vapor in itself is another nonequilibrium condition.

Azeotropes

The separation of two liquids just described relates to liquids conforming to Raoult's law. If a plot of vapor pressure against composition shows a maximum, the boiling point curve will show a minimum. An example of this is shown in Figure 6.9 for carbon tetrachloride-methanol at 1 atm. This should be compared with Figure 5.7, which is the vapor pressure curve for two compounds showing a maximum in the vapor pressure. In the same way, a minimum in the vapor pressure curve will result in a maximum in the boiling point curve. This behavior is shown in Figure 6.10 for the system tetrahydrofuran–*cis*-dichloroethylene at 101.325 kPa. Mixtures corresponding to either a minimum or maximum in the boiling point curves are called **azeotropes** (Greek *a*, without; *zein,* to boil; *trope,* change).

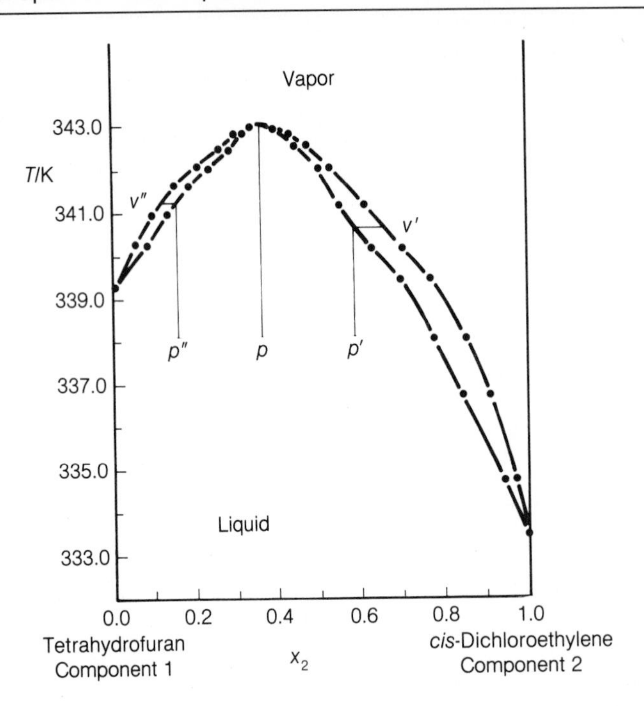

FIGURE 6.10 Temperature against composition diagram for the system tetrahydrofuran-*cis*-dichloroethylene at 101.325 kPa.

Separation of azeotropic mixtures into two pure components by direct distillation is impossible. However, one pure component may be separated as well as the azeotropic composition. As an example, consider Figure 6.10. If we heat a mixture having the azeotropic composition p, the mixture will boil unchanged; no separation is accomplished. If we now boil a mixture having composition p', the first vapor to appear will have composition v' and is richer in *cis*-dichloroethylene, component 2. Continued fractionation, as described before, will eventually produce pure component 2 in the distillate with azeotrope as the residual liquid in the pot. Boiling a mixture at point p'' results in vapor having composition v'', which is much richer in tetrahydrofuran, component 1, than before. Fractionation can eventually yield pure component 1 at the head of the column and leave the azeotropic mixture in the pot.

Similar results are obtained with a minimum boiling azeotrope except that the azeotropic composition is obtained at the head of the column and pure component remains behind in the still. Examples of minimum boiling azeotropes are far more numerous than those exhibiting a maximum in their curves.

In Tables 6.1 and 6.2 are listed a number of minimum and maximum boiling azeotropes at 101.325 kPa (1 atm). Azeotropes have sometimes been mistaken for pure compounds because they boil at a constant temperature. However, for an azeotrope a variation in pressure changes not only the boiling temperature but also the composition of the mixture, and this easily distinguishes it from a true compound. This is demonstrated for HCl in Table 6.3.

One of the more useful applications of azeotropes in binary systems is the preparation of a constant composition mixture. In the case of H_2O and HCl, the

TABLE 6.1 Azeotropes with Minimum Boiling Points at 1 Atm Pressure

Component 1	T_b/K	Component 2	T_b/K	Azeotrope Wt % 1	Azeotrope T_b/K
H_2O	373.15	Ethyl alcohol	351.45	4.0	351.32
H_2O	373.15	Isopropyl alcohol	355.65	12.0	353.25
H_2O	373.15	1-Chlorohexane	407.65	29.7	364.95
H_2O	373.15	Acetophenone	474.75	81.5	372.25
Carbon disulfide	319.35	Iodomethane	315.7	18.6	314.35
Methanol	337.85	Pentane	309.3	7	304.00
Acetic acid	391.25	Heptane	371.4	33	364.87
Ethyl alcohol	351.45	Benzene	353.25	31.7	341.05

TABLE 6.2 Azeotropes with Maximum Boiling Points at 1 Atm Pressure

Component 1	T_b/K	Component 2	T_b/K	Azeotrope Wt % 1	Azeotrope T_b/K
H_2O	373.15	HNO_3	359.15	85.6	393.85
Chloroform	334.35	Methyl acetate	330.25	64.35	337.89
Acetic acid	391.25	Butyl alcohol	390.25	43.	393.45
Ethyl alcohol	351.45	Butylamine	350.95	82.2	355.35

TABLE 6.3 Influence of Pressure on Azeotropic Temperature and Composition

Pressure/kPa	Wt % HCl in H_2O	T_b/K
66.660	20.916	370.728
93.324	20.360	379.574
101.325	20.222	381.734
106.656	20.155	383.157

constancy of the composition of this azeotrope allows its use as a standard solution of known composition.

Distillation of Immiscible Liquids: Steam Distillation

We have just considered the distillation of miscible liquids. Now consider the behavior of two liquids whose mutual solubility is so small that they may be considered immiscible. In this case each liquid exerts the same pressure as though it were the only liquid present. Thus the total pressure above the mixture at a particular temperature is simply the sum of the vapor pressures of the two components and remains so until one of the components disappears. This fact makes possible a distillation quite different from the type discussed previously.

Since the two vapor pressures are added together, any total pressure is reached at a much lower temperature than the boiling point of either component. As the distillation proceeds, both components are distilled in a definite ratio by weight. Thus if P_T is the total pressure, and P_A^* and P_B^* are the vapor pressures of pure liquids A and B, respectively, then

$$P_T = P_A^* + P_B^* \tag{6.20}$$

If n_A and n_B are the amounts of each component present in the vapor, the composition of the vapor is

$$\frac{n_A}{n_B} = \frac{P_A^*}{P_B^*} \tag{6.21}$$

Since the ratio of partial pressures is a constant at a particular temperature, the ratio n_A/n_B must also be a constant. Thus the distillate is of constant composition as long as both liquids are present, and it boils at a constant temperature.

Water is often one component when this type of distillation is used for purifying organic compounds. This process, called **steam distillation,** is frequently used for substances that would decompose when boiled at atmospheric pressure. What makes this process attractive is the high yield of organic materials brought about by the low molar mass of water and its convenient boiling point, in contrast to the relatively high molar masses of most organic substances.

If a liquid is partially miscible with water, it too may be distilled as long as the solubility is not too great. However, one important difference arises when the composition of distillate is calculated. The actual partial pressures must replace P_A^* and P_B^* in Eq. 6.21. The relative molecular mass of an unknown substance may also be estimated if the masses and vapor pressures of the two components are known.

Distillation of Partially Miscible Liquids

Many pairs of liquids are miscible only to a limited extent; an example is the water-butanol system. Generally partial miscibility at low temperatures is caused by large positive deviations from Raoult's law, in which case we expect to find a minimum in the boiling point-composition curve. As the pressure on the system is reduced, the boiling point curve generally intersects the liquid-liquid equilibrium curve, resulting in the curve for a typical system shown in Figure 6.11.

Any composition in the range from 0 to x_a and from x_c to 1 will show the same behavior upon boiling as already demonstrated for minimum boiling azeotropes, with one exception. Two layers are formed if liquid at point p is evaporated and its vapor v condensed. The two liquids at p' are L_1 and L_2, in amounts given by the lever rule. L_1 has the composition given by f and L_2 that given by g. If solution of overall composition p' is boiled at T_e, three phases will be in equilibrium: liquid phase L_1 having composition x_a, liquid phase L_2 having composition x_c, and vapor having composition y_b. From the phase rule the system is invariant at the specified pressure. Thus, as vapor of composition y_b is removed, the composition of the two liquid phases does not change, only the relative amounts of the two layers. In this particular case, continued distillation will cause all the L_2 layer to be consumed.

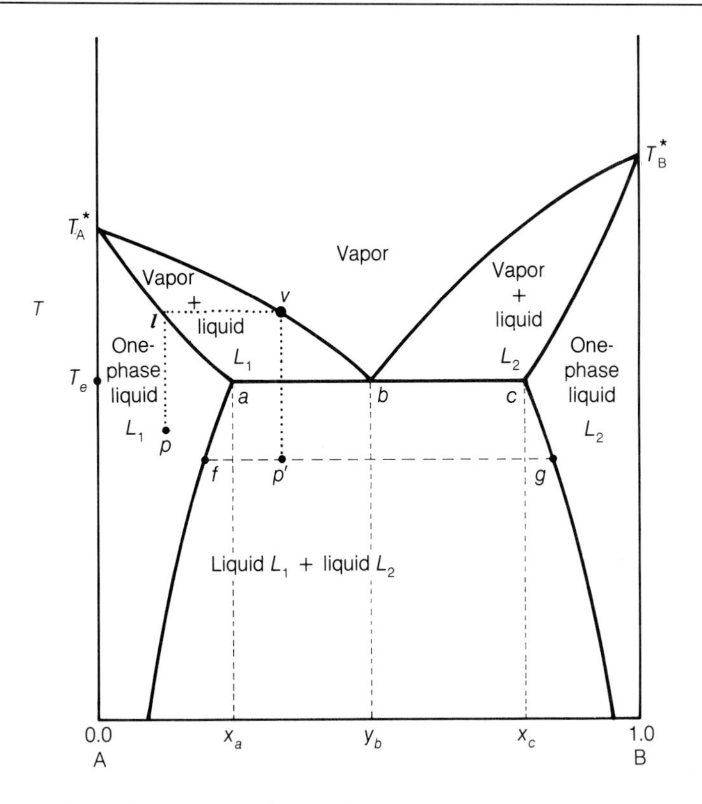

FIGURE 6.11 Liquid-vapor equilibrium for a system with partial miscibility in the liquid phase.

When it is exhausted, liquid L_1 at a and vapor at b are left. The temperature may be increased at this same pressure; then the liquid composition changes along curve $\overline{al}$ and the vapor along $\overline{bv}$. The last drop of liquid disappears when l and v are reached, and only vapor remains since this is the original composition of liquid from which the liquid p' was produced.

If the composition of the liquid layers lies in the range b to c, then as distillation occurs, vapor at b will be formed at the expense of the L_1 layer. The rest of the distillation will be similar to that already described.

6.4 Condensed Binary Systems

Two-Liquid Components

The vapor phase in an equilibrium system is often of little or no interest compared to the interaction in the liquid and solid phases. It is customary, therefore, in these cases to disregard the vapor phase and to fix a total pressure, normally 1 atm. The system may then be studied in open vessels. When only solid and liquid phases are considered, we speak of a *condensed system*. There is a minor drawback in this procedure in that the pressure is generally not the equilibrium value, and the system is consequently not in true thermodynamic equilibrium. However, for liquid-solid systems this does not have much effect on the behavior.

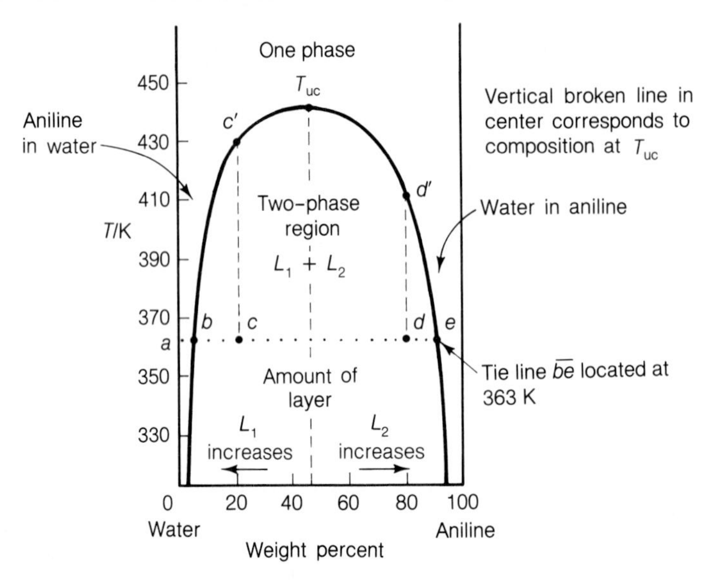

FIGURE 6.12 Solubility of water and aniline as a function of temperature. The **upper consolute** temperature, T_{uc}, is 441 K. The lines $\overline{cc}'$ and $\overline{dd}'$ are isopleths at 20 and 80 wt% aniline, respectively.

Since we have just discussed the distillation of two immiscible liquids, it seems appropriate to investigate the types of behavior exhibited by a condensed system consisting of two liquids that are only partially miscible.

The water-aniline system provides a simple example of partial miscibility (see Figure 6.12). If a small amount of aniline is added to pure water at any temperature below 441 K, the aniline dissolves in the water. If we work at a constant temperature of 363 K, pure water is present at point *a* and only one phase is present as aniline is added. However, as more aniline is added, point *b* on the solubility curve is reached and then, in addition to phase L_1 of composition *b*, a slight amount of a second liquid phase L_2 appears having composition *e*. The composition of the L_1 layer is a solution of aniline in water and that of the L_2 layer is a solution of water in aniline. As more aniline is added, the second liquid layer L_2 becomes more evident and continually increases with the addition of aniline until the composition is given by point *e*. Beyond point *e*, only one phase is present. The same type of behavior is observed as water is added to pure aniline. The composition of any point in the two-phase region along the tie line between points *b* and *e* is composed of varying proportions of solution L_1 and of solution L_2. These solutions are called **conjugate solutions**. The composition of these layers depends on the temperature. The amount of the individual layers present may be determined using the lever rule (Section 6.3).

EXAMPLE
Calculate the ratio of the mass of the water-rich layer to that of the aniline-rich layer, for a 20-wt % water-aniline mixture at 363 K.

SOLUTION
The compositions along the tie line $\overline{be}$ are maintained at 363 K. The composition at *c* is 20%; for L_1 at *b*, it is 8%; and for L_2 at *e*, it is 90% (see Figure 6.12).

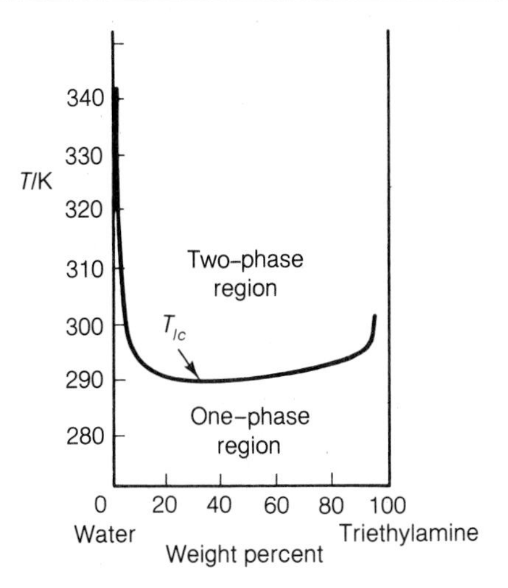

FIGURE 6.13 Solubility of water and triethylamine as a function of temperature (T_{lc} = 291.6 K).

Using the lever rule, we have

$$\frac{\text{mass of water-rich layer, } L_1}{\text{mass of aniline-rich layer, } L_2} = \frac{\overline{ce}}{\overline{bc}} = \frac{90 - 20}{20 - 8} = 5.83$$

■

Constant composition lines, vertical on the diagram in Figure 6.12, are known as **isopleths** (Greek *iso*, the same; *plethora*, fullness). As the temperature is increased from point c along the isopleth cc' or indeed from any point left of the vertical dashed line joining T_{uc}, the solubility of the aniline in water layer, L_1, grows as does the solubility of the water in aniline layer, L_2. As a result of the change in solubility, the predominant layer L_1 increases at the expense of the L_2 layer. Similar behavior is observed for the L_2 layer when the temperature is increased from point d. A different behavior is observed with the *critical composition*, which is the composition corresponding to the highest temperature T_{uc} at which two layers may coexist. (This temperature is known as the *upper consolute temperature* or *critical solution temperature.*) If the curve is symmetrical, the relative size of the layers remains constant as the temperature is raised along the dotted line. Above T_{uc}, only one phase exists.

The increased solubility with temperature can be explained by the fact that the forces holding different types of molecules apart are counteracted by the thermal kinetic energy of the molecules. It is curious, therefore, that for some systems a *lower consolute temperature* exists; Figure 6.13 shows an example of this behavior in the water-triethylamine system. The lower consolute temperature is 291.65 K, and above this temperature two immiscible layers exist. In this case the large positive deviations from Raoult's law responsible for the immiscibility may be just balanced at the lower temperature by large negative deviations from Raoult's law, which are normally associated with compound formation.

The final type of liquid-liquid equilibrium is exhibited by the water-nicotine system. In this case, shown in Figure 6.14, the two-phase region is enclosed and has both an upper and lower consolute temperature at atmospheric pressure. It has been shown for this system that an increase of pressure can cause total solubility.

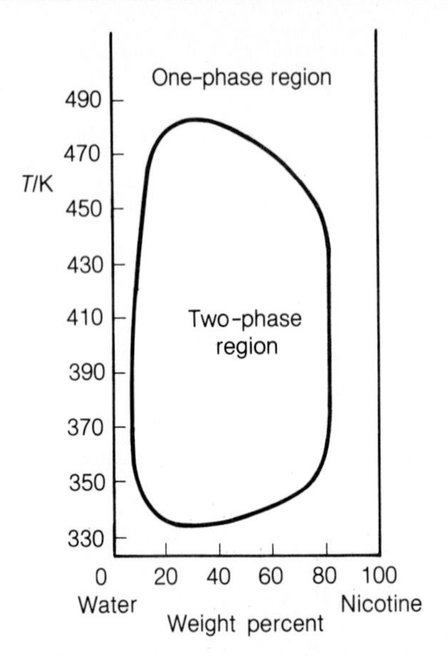

FIGURE 6.14 Solubility of water and nicotine as a function of temperature.

An interesting example of liquid-liquid solubility in two materials normally solid, and its application to a practical problem, is afforded by the Pb–Zn system. Figure 6.15 gives the phase diagram for this system to 1178 K, above which boiling occurs. Ignoring the details in the right- and left-hand extremes of the diagram, we see that miscibility occurs above the upper consolute temperature at 1071.1 K. The rather limited solubility of zinc in lead may be used in the metallurgical separation of dissolved silver in lead. The zinc is added to the melted lead, which has an economically recoverable amount of silver dissolved in it. The melt is agitated to effect thorough mixing. The zinc is then allowed to rise to the surface of the lead and is skimmed off. Because of the much higher solubility of silver in Zn than in Pb, most of the silver will now be in the zinc. The zinc may be boiled off from this liquid to give the desired silver.

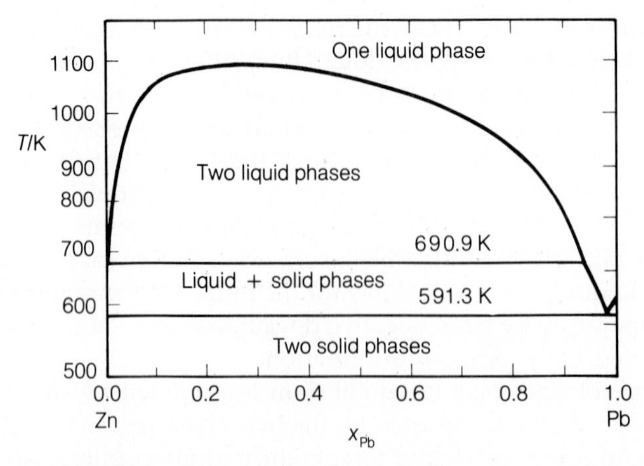

FIGURE 6.15 The zinc-lead system. This is an example of a two-phase liquid system.

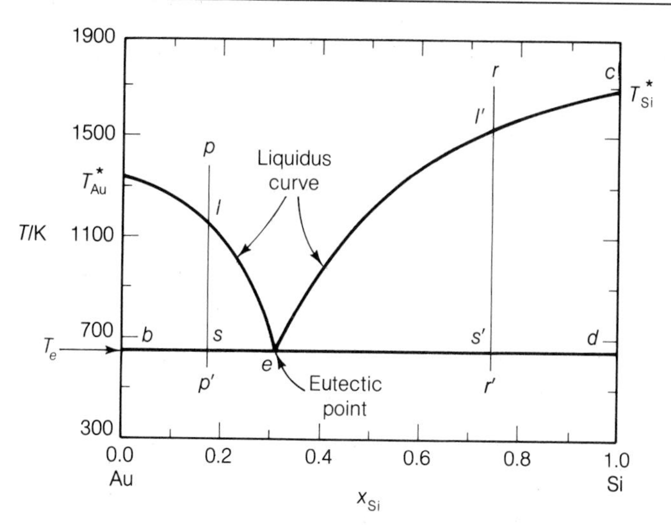

FIGURE 6.16 The gold-silicon system. This is an example of a simple eutectic system without miscibility. T_e is the eutectic temperature.

Solid-Liquid Equilibrium: Simple Eutectic Phase Diagrams

When a single-liquid melt formed from two immiscible solids is cooled sufficiently, a solid is formed. The temperature at which solid is first formed is the freezing point of the solution, and this is dependent on the composition. Such a case is provided by the gold-silicon system shown in Figure 6.16. A curve that represents the boundary between liquid only and the liquid plus solid phase is known as a **liquidus** curve. If we begin with pure gold, the liquidus curve will start at $x_{Au} = 1$ and drop toward the center of the figure. The curve from the silicon side behaves in a similar manner. The temperature of intersection of the two curves, T_e, is called the **eutectic** (Greek *eu*, easily; *tecktos*, molten) temperature. The eutectic composition x_e has the lowest melting point in the phase diagram.

At the eutectic point, three phases are in equilibrium: solid Au, solid Si, and liquid. At fixed pressure the eutectic point is invariant. This means that the temperature is fixed until one of the phases disappears. The relationship between phases is easy to follow if we isobarically cool the liquid represented by point p in a single-liquid-phase region. As the temperature is lowered, the first solid appears at l at the same temperature as the liquid. Since this is an almost completely immiscible system, the solid is practically pure gold. As the temperature is dropped further, more crystals of pure gold form. The *composition of the liquid* follows the line le, and the overall composition between liquid and solid is given by the lever rule. In this region of liquid plus solid, either the temperature or the composition may be varied. When point s is reached, three phases are in equilibrium: solid Au, solid Si, and liquid of composition x_e. Since the system is invariant, the liquid phase is entirely converted into the two solids before the temperature may drop lower. Finally, at point p', only two solids, Au and Si, exist. The right-hand side of the diagram may be treated in the same way.

A mixture of eutectic composition will melt sharply at the eutectic temperature to form a liquid of the same composition. Materials of eutectic composition are generally fine grained and uniformly dispersed, as revealed by high magnification microscopy. Microscopic observation of a noneutectic melt cooled to solidification reveals individual crystals of either solid dissolved in a uniform matrix of eutectic composition.

The mole fraction of silicon in the region near pure silicon varies with temperature according to Eq. 5.115:

$$\ln x_{Si} = -\frac{\Delta_{fus}H_{Si}}{R}\left(\frac{1}{T} - \frac{1}{T_0}\right) \tag{6.22}$$

6.5 Thermal Analysis

The careful determination of phase boundaries, particularly in complicated metallic systems, is quite difficult and requires considerable effort. One method that has proved useful for phase determination is the technique of **thermal analysis.** In this technique a series of mixtures of known composition is prepared. Each sample is heated above its melting point and, where possible, is made homogeneous. Then the rate of cooling of each sample is followed very closely. Figure 6.17 shows a series of cooling curves in a plot of T against time and shows how individual points are used to form a phase diagram similar to Figure 6.16. For the two pure materials the rate of cooling of the liquid melt is fairly rapid. When the melting temperature is reached, there is generally a little supercooling that is evidenced by a slight jog in the curve. This is shown in curve 1. The curve returns to the melting point and remains there until all the liquid is converted to solid. The temperature then drops more rapidly for the solid than for the liquid since the heat capacity of a solid is generally lower than that of a liquid. Thus it requires less heat removal to cool the sample a fixed number of degrees.

Curve 2 represents a mixture of some B in A. The mixture cools rapidly until point l is reached. This point appears on the liquidus curve. Liquid and solid are in equilibrium as the mixture cools more slowly along line $\overline{ls}$. This is because of the heat that is released on solidification. At point s, a horizontal region appears called the *eutectic halt.* The liquid still present in the system must completely solidify before the temperature can drop farther. Once all the liquid is converted to solid, the temperature will drop. As with the pure material, the cooling of two solids is much more rapid than if liquid were present. The descriptions of curves 3 and 5 are the same as for curve 2 except for the lengths of the line $\overline{l's'}$ and $\overline{l''s''}$ and for the time that the system stays at the eutectic halt. This period at the eutectic halt provides a means to establish the eutectic temperature. In general the eutectic halt will lengthen, and lines like $\overline{l's'}$ will shorten, as the composition approaches the eutectic composition. The reason for this is that for the eutectic composition the cooling is rapid until the eutectic temperature is reached. After all the liquid is converted to solid, the mixture can then cool further. The temperature for each composition at which a change occurs in the cooling curve is then used to establish a point on the phase diagram, as is shown in the right-hand portion of Figure 6.17.

Nonmetallic Simple Eutectic Systems

Several simple nonmetallic eutectic systems also deserve special attention. The system H_2O–NaCl is one of these and is shown in Figure 6.18. The curve AE is the freezing point curve for water, and EB is the solubility curve for sodium chloride.

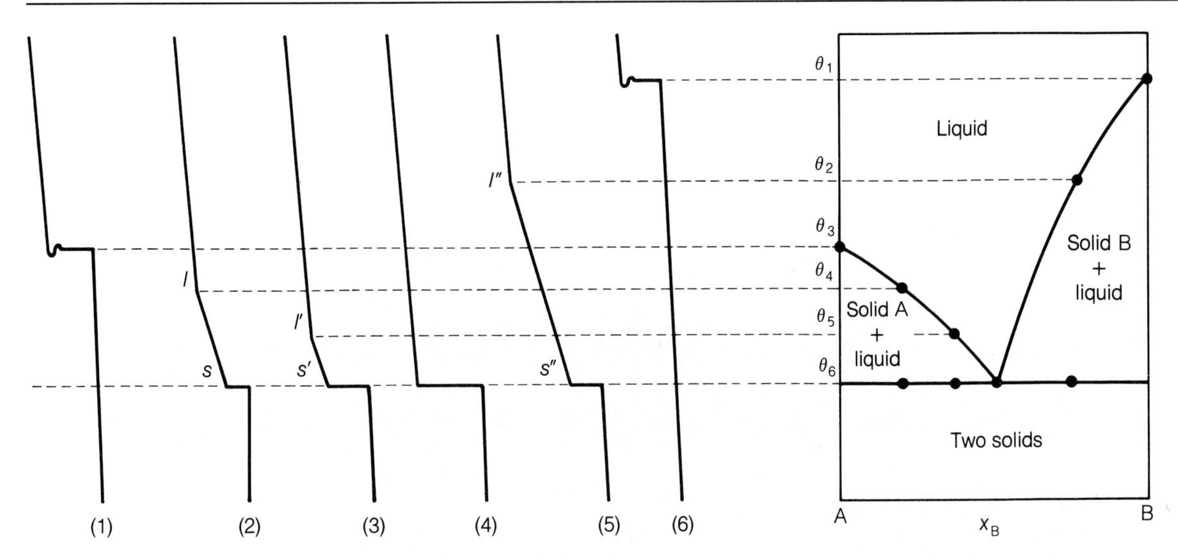

FIGURE 6.17 Use of thermal analysis. Demonstration of how thermal analysis can be used to determine a phase diagram. Cooling curves (1) and (6) represent behavior for pure A and B, respectively.

Suppose that we wish to prepare a constant temperature bath at some temperature below that of melting ice. If NaCl is added to ice, the ice melts. Indeed, if this is done in an insulated container, ice continues to melt with the addition of

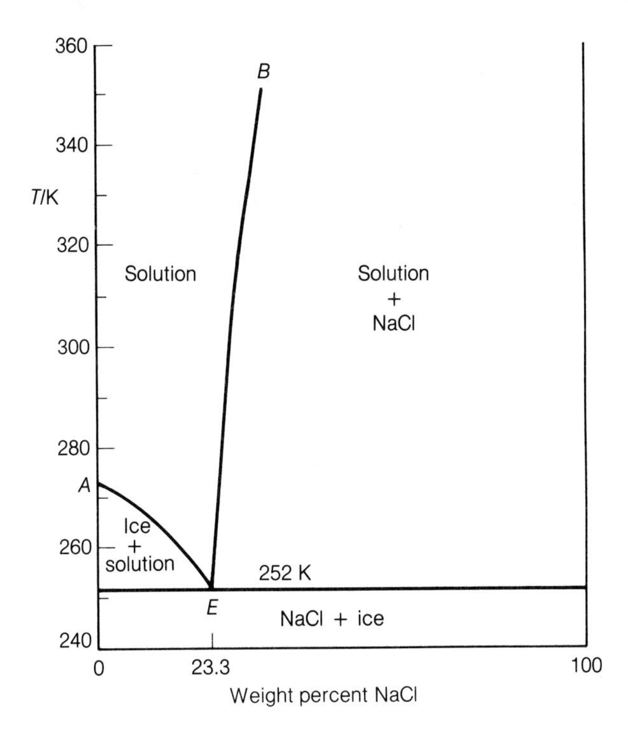

FIGURE 6.18 Lowering of the freezing point in the NaCl-water system (not drawn exactly to scale). This type of diagram is typical of the salt-water interactions given in Table 6.4.

NaCl until 252.0 K is reached. Then the temperature of the system and any vessel immersed in it will remain invariant until all the ice has been melted by heat from an outside source.

The eutectic point is again seen to be associated with two solid phases (ice and salt), similar to the two solid phases at the eutectic point in the metallic systems that we have considered. In Table 6.4 several eutectic compositions involving different salts and water are presented.

Organic compounds can also form simple eutectic mixtures. The *method of mixed melting points* is seen as an application of eutectics in organic chemistry. To confirm the identity of a substance, the sample is mixed with a pure sample of the compound thought to be identical to it. If the two substances are the same, the melting point will not change. If they are not the same, the temperature will drop, since the sample lowers the freezing point, as shown in Figure 6.17.

Solid Solutions

Only one type of situation is known in which the mixture of two different substances results in an increase of melting point. This is the case in which the two substances are **isomorphous** (Greek *iso,* the same; *morphos,* form). In terms of metallic alloys this behavior is a result of the complete mutual solubility of the binary components. This can occur when the sizes of the two atoms of the two components are about the same. Then atoms of one type may replace the atoms of the other type and form a *substitutional alloy.* An example of this behavior is found in the Mo–V system, the phase diagram for which is shown in Figure 6.19. Addition of molybdenum to vanadium will raise the melting point. The Cu–Ni system also forms a solid solution. Copper melts at 1356 K, and addition of nickel raises the temperature until for $x_{Ni} = 1$ the temperature reaches 1726 K. An alloy known as *constantan,* consisting of 60 wt % Cu and 40 wt % Ni, has special interest since it is useful as one component of a thermocouple for the determination of temperature.

TABLE 6.4 Eutectics Involving Salts and Ice

Salt	Eutectic Temperature/K	Eutectic Composition/ Wt % Salt
Ammonium chloride	257.7	19.7
Potassium chloride	262.4	19.7
Sodium bromide	245.1	40.3
Sodium chloride	252.0	23.3
Sodium iodide	241.6	39.0
Sodium nitrate	257.7	44.8
Sodium sulfate	272.0	3.84

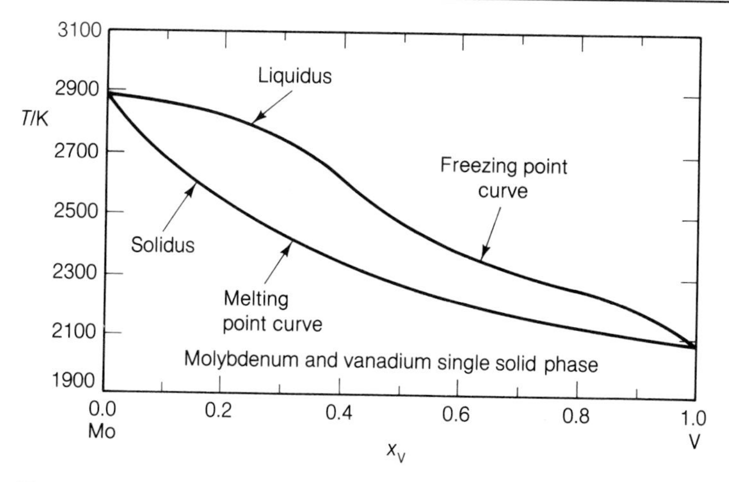

FIGURE 6.19 The molybdenum-vanadium system. This is an example of a solid solution. The **solidus** curve establishes the equilibrium between solid and solid + liquid regions.

Partial Miscibility

The two previous sections covered two extremes, namely complete immiscibility of components and complete solubility of components. Most actual cases show limited solubility of one component in the other. Figure 6.20 is an example of this limited solubility for both components. Tin dissolves lead to a maximum of approximately 2.5 mol % or 1.45 wt %. Tin is more soluble in lead, dissolving to a maximum of 29 mol % at 466 K. The two-phase solid region is composed of these two solid alloys in proportions dictated by the lever rule. The situation is analogous

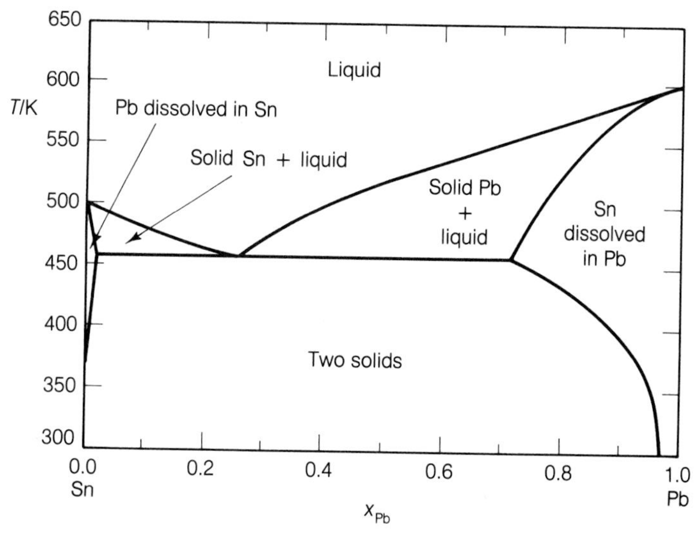

FIGURE 6.20 The tin-lead system. An example of a simple eutectic system with limited miscibility of the components.

to that represented in Figure 6.11 concerning liquid-vapor equilibrium where partial miscibility occurs in the liquid phase.

This is an important system from a practical standpoint, because in the electrical industry and in plumbing the low-temperature eutectic is used to make commercial solders. Actually the tin content should be about 20 mol % rather than the eutectic composition, to improve "wiping" characteristics.

A second type of system in which partial miscibility occurs involves a transition point. An example is provided by the Mn_2O_3–Al_2O_3 system, the phase diagram for which is shown in Figure 6.21. This system gives rise to one series of *spinels* (Latin *spina,* thorn), which are of variable composition with the general formula AB_2O_4 in which A may be magnesium, iron, zinc, manganese, or nickel and B may be aluminum, chromium, or iron. These form octahedral crystals. (The gem ruby is a magnesium aluminum spinel.) Also formed in this system are the *corundums,* which are abrasive materials of high aluminum oxide content that form hexagonal crystals. The gem sapphire belongs to this group of substances. A transition temperature exists along *abc* at which spinel, corundum, and liquid of composition *a* coexist and the system is invariant. At any temperature above the transition temperature the spinel phase disappears. Cooling of the liquid and corundum phases in the *a* to *b* composition range results in the formation of the spinel phase and coexistence of two solid phases. Cooling in the *b* to *c* range initially results in the disappearance of corundum and formation of spinel along with the liquid. Further cooling results in solid spinel only. However, as the temperature falls further, corundum makes an appearance again.

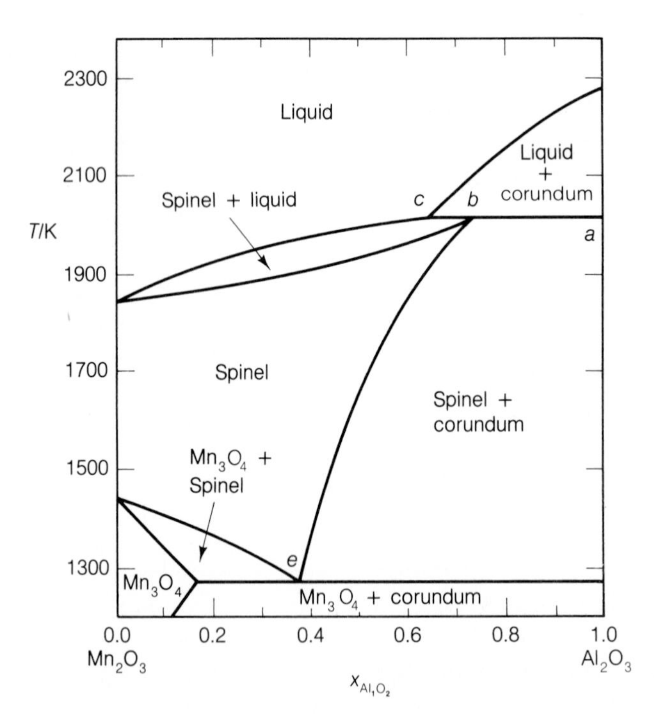

FIGURE 6.21 The system Mn_2O_3–Al_2O_3 exposed to air.

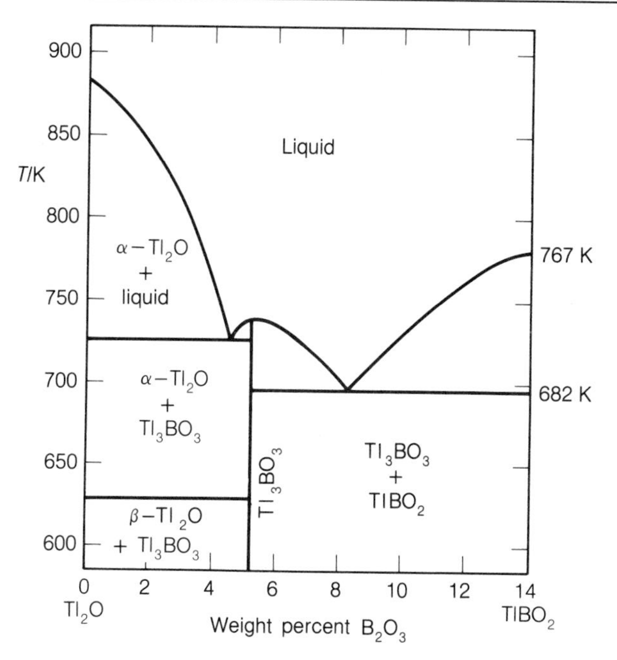

FIGURE 6.22 The Tl_2O–$TlBO_2$ system.

Another feature of this system deserves mention. The region near 1300 K and 0.4 mol fraction Al_2O_3 appears to be similar to what has been described as a eutectic point. However, where liquid would be expected in a normal eutectic system, this region is entirely solid. An invariant point such as *e* surrounded solely by crystalline phases is called a *eutectoid*. At the eutectoid, phase reactions occur on change of heat resulting in a change in proportions of the solid phases exactly analogous to that at a eutectic point.

Compound Formation

Sometimes there are such strong interactions between components that an actual compound is formed. Two types of behavior can then be found. In the first type, the compound formed melts into liquid having the same composition as the compound. This process is called **congruent melting.** In the second type, when the compound melts, the liquid does not contain melt of the same composition as the compound. This process is called **incongruent melting.**

In Figure 6.22 for the system Tl_2O–$TlBO_2$ the compound Tl_3BO_3 is formed and melts congruently at 725 K. A eutectic occurs at 8.2 wt % B_2O_3. Note that the X axis is plotted as weight percent B_2O_3 over a very limited range. The easiest way to interpret this figure is to cut it in half mentally along the line representing pure Tl_3BO_3. Then each half is treated as in Figure 6.16. The left-hand portion introduces a new feature that is often found not only in ceramic systems such as this but also in metallic systems. At 627 K a reversible transformation occurs in the crystalline structure of Tl_2O. A form called β is stable below 627 K and the second form α is stable above the transition temperature all the way to the melting point.

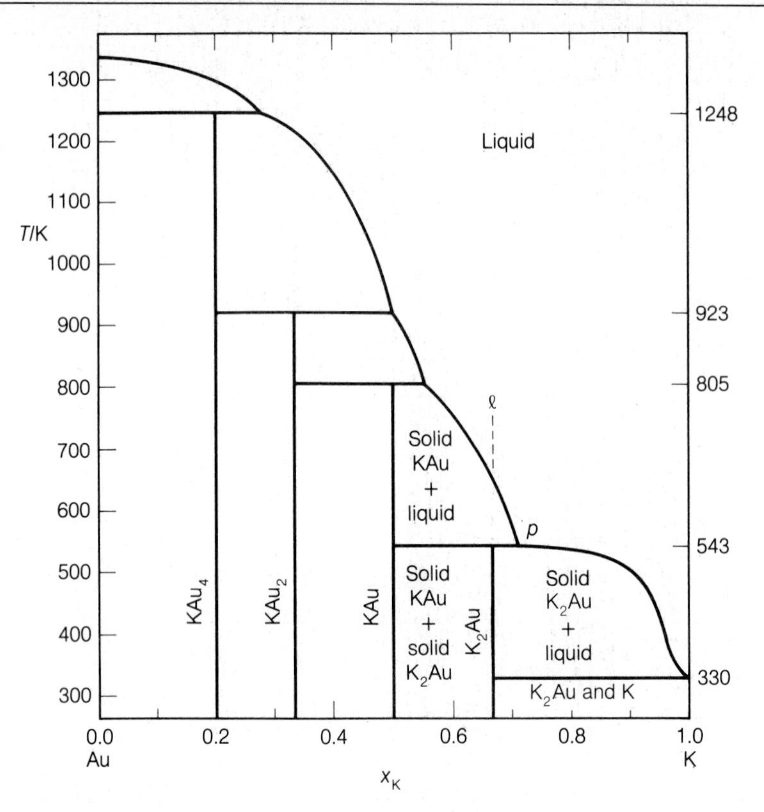

FIGURE 6.23 The Au–K system.

It appears that there are now two points in this system that were previously described as eutectic points. However, only the lowest one is referred to as the eutectic point; others are called **monotectic points.**

In contrast to congruent melting, incongruent melting occurs for each of the compounds in the Au–K system shown in Figure 6.23. The composition of each compound formed is given by the formula alongside the line representing that compound. The composition in the region around K_2Au is given in detail. From this example, and the following discussion, the details in the other regions may be supplied.

If we examine the compound K_2Au as it is heated, we find that liquid of composition p is formed at 543 K:

$$K_2Au(s) \rightleftharpoons KAu(s) + \text{liquid(composition } p)$$

Since the liquid is richer in potassium than is solid KAu, some solid KAu will remain as solid. Thus the reaction is known as a **phase reaction** or, more commonly, a **peritectic reaction.** (Greek *peri*, around; *tektos*, melting). The point p is known as the **peritectic point.** This reaction is reversible if liquid of the same total composition as K_2Au is cooled. Solid KAu begins to separate at l. More solid KAu forms until the temperature of 543 K is reached. As heat is removed, the reverse of the peritectic reaction just shown occurs. From the lever rule, approximately 25% of the material initially exists as particles of solid KAu surrounded by liquid of composition p. Thus the KAu is consumed as the reaction proceeds. As the last trace of liquid and KAu disappears, the temperature is free to drop and only K_2Au is present.

If the starting melt has a composition that is slightly rich in Au compared to K_2Au (i.e., lies to the left of K_2Au), cooling into the two-solid region will produce crystals of KAu surrounded by the compound K_2Au. It is difficult to establish equilibrium in these systems unless a long annealing period is used to homogenize the sample.

6.6 Ternary Systems

When we consider a three-component, one-phase system, the phase rule allows for $f = c - p + 2 = 3 - 1 + 2 = 4$ degrees of freedom. These four independent variables are generally taken as pressure, temperature, and two composition variables, since only two mole fractions are necessary to define the composition. Thus the composition of a three-component system can be represented in two dimensions with T and P constant.

The most convenient technique for plotting such a phase diagram is due to Hendrik William Bakhuis Roozeboom (1854–1907).* The composition is determined using the fact that, from any point within an equilateral triangle, *the sum of the distances perpendicular to each side is equal to the height of the triangle.* The height is set equal to 100% and is divided into 10 equal parts. A network of small equilateral triangles is formed by drawing lines parallel to the three sides through the 10 equal divisions. Each apex of the equilateral triangle in Figure 6.24 represents one of the three pure components, namely 100% A, 100% B, or 100% C. The three sides of the triangle represent the three possible binary systems and 0% of the third component. Thus any point on the line $\overline{BC}$ represents 0% A. A line parallel to $\overline{BC}$ through P represents all possible compositions of B and of C in combination with 30% A. Here the percentage of A is read from the value of the length of the line $\overline{A'P}$. Compositions along the other two sides are read in a similar manner. Since the distance perpendicular to a given side of the triangle represents the percentage of the component in the opposite apex, the compositions at P are 30% A, 50% B, and 20% C. Any composition of a ternary system can thus be represented within the equilateral triangle.

Liquid-Liquid Ternary Equilibrium

A simple example for demonstrating the behavior of a three-component liquid system is the system toluene-water-acetic acid. In this system, toluene and acetic acid are completely miscible in all proportions. The same is true for water and acetic acid. However, toluene and water are only slightly soluble in each other. Their limited solubility causes two liquids to form as shown along the base of the triangle at points p and q in Figure 6.25. Added acetic acid will dissolve, distributing itself between the two liquid layers. Therefore, two conjugate ternary solutions are formed in equilibrium. With temperature and pressure fixed in the two-phase

*H. W. Bakhius Roozeboom, *Die heterogenen Gleichgewichte vom Standpunkte der Phasenlehre,* Braunschweigr, 1901–1904.

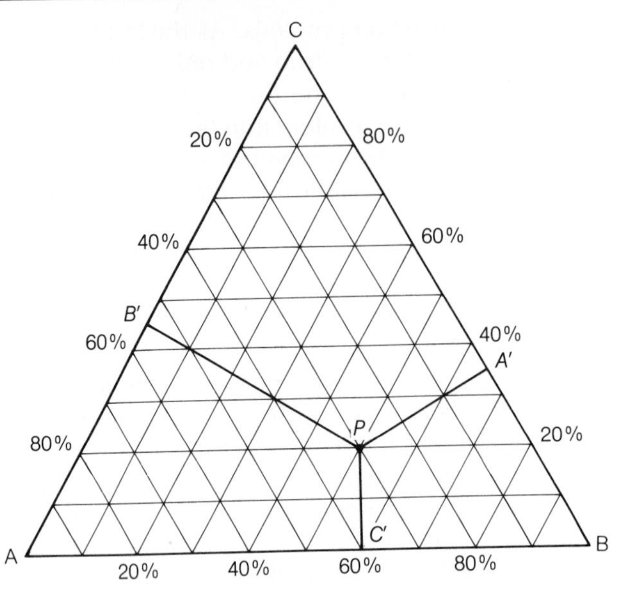

FIGURE 6.24 The triangular diagram for representing a three-component system at constant temperature and pressure.

region, only one degree of freedom remains and that is given by the composition of one of the conjugate solutions.

Because of the difference in solubility of acetic acid in the two layers, however, the tie lines connecting conjugate solutions are not parallel to the toluene-water base. This is shown for the tie lines $\overline{p'q'}$, $\overline{p''q''}$, etc. This type of curve is called *binodal*. The relative amounts of the two liquids are given by the lever rule. As the two liquid solutions become more nearly the same, the tie lines become shorter, finally reducing in length to a point. This point generally does not occur at the top of the solubility curve and is called an *isothermal critical* point or *plait* point, p^* in the diagram. At p^* both layers are present in approximately the same proportion, whereas at p'' only a trace of water remains in the toluene layer. This curve becomes more complicated if the other sets of components are only partially miscible.

Solid-Liquid Equilibrium in Three-Component Systems

The *common-ion effect* may be explained by use of phase diagrams. Water and two salts with an ion in common form a three-component system. A typical phase diagram for such a system is shown in Figure 6.26. Such systems as $NaCl$–KCl–H_2O and NH_4Cl–$(NH_4)_2SO_4$–H_2O give this type of equilibrium diagram. We now will see how each salt influences the solubility of the other and how one salt may actually be separated.

In the figure, A, B, and C represent nonreactive pure components with C being the liquid. Point *a* gives the maximum solubility of A in C when B is not present. Point *c* gives the maximum solubility of B in C in the absence of A. Points along the line $\overline{Aa}$ represent various amount of solid A in equilibrium with saturated solution

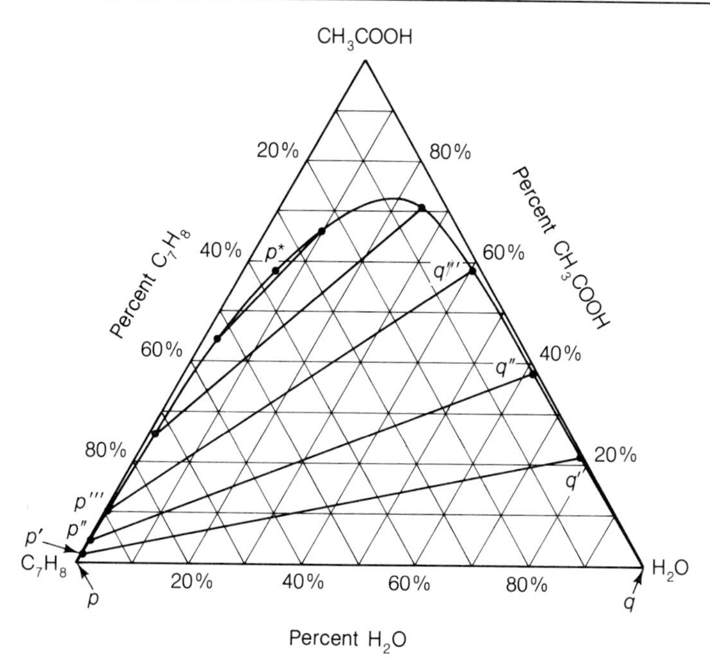

FIGURE 6.25 The ternary system acetic acid-toluene-water at a fixed temperature and pressure. The formation and representation of two conjugate solutions is shown.

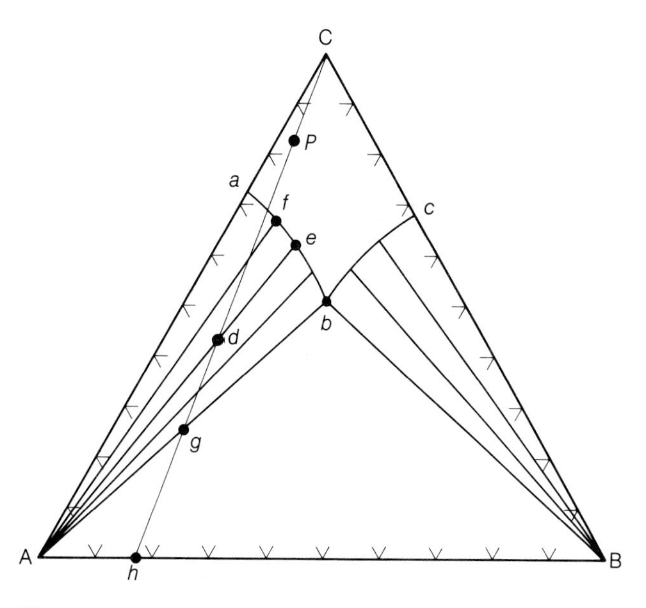

FIGURE 6.26 Phase equilibrium for two solids and a liquid showing the common-ion effect.

a. Solutions having composition between *a* and C are unsaturated solutions of A and C. When B is added to a mixture of A and C, the solubility of A usually decreases as shown by the line $\overline{ab}$. In like manner, addition of A to a solution B in

C usually decreases the solubility along the line cb. This is the effect normally called the *common-ion effect*. The meeting of these curves at b represents a solution that is saturated with respect to both salts. In the region AbB, three phases coexist, the two pure solids A and B and a saturated solution of composition b. In the tie line regions, the pure solid and saturated solution are in equilibrium. Thus, if point d gives the composition of the mixture, the amount of solid phase present is given by the length of the line $\overline{de}$ and the amount of saturated solution is given by the length of the line $\overline{d\mathrm{A}}$. Considering the regions of the three-component systems with one solid phase present in contact with liquid, we find a bivariant system (no vapor phase present and the system at constant pressure). Thus at any given temperature the concentration of the solution may be changed. However, at the point b, two solid phases are present and the system is isothermally invariant and must have a definite composition.

Now consider what happens to an unsaturated solution P as it is isothermally evaporated. The overall or state composition moves along the line $\overline{\mathrm{C}h}$. Pure A begins to crystallize at point f, with the composition of the solution moving along $\overline{fb}$. At point g the solution composition is b and B begins to crystallize. As evaporation and hence removal of C continue, both solid A and solid B are deposited until at h all the solution is gone.

The process of recrystallization can also be interpreted from Figure 6.26. Let A be the solid to be purified from the only soluble impurity B. If the original composition of the solid mixture of A and B is h, water is added to achieve the overall composition d. The mixture is heated, thus changing the state to an unsaturated liquid and effecting complete solubility. (Complete solubility can usually be brought about by increasing the temperature sufficiently.) When the liquid is cooled, the impurity B stays in the liquid phase as pure A crystallizes out along the line equivalent to $\overline{ab}$ at the higher temperature. Thus when the solution returns to room temperature, the crystals of pure A may be filtered off.

Key Equation

The phase rule:

$$f = c - p + 2$$

where

f is the number of degrees of freedom,
c is the number of components,
p is the number of phases.

Problems

A Number of Components and Degrees of Freedom

1. What is the composition of the two-phase region in Figure 6.14? How many degrees of freedom exist in this region?

2. Determine the number of degrees of freedom for the following systems:

a. A solution of potassium chloride in water at the equilibrium pressure.

b. A solution of potassium chloride and sodium chloride at 298 K at 1 atm pressure.

c. Ice in a solution of water and alcohol.

3. How many components are present in a water solution of Na acetate?

4. How many components are present in the system $CaCO_3$–CaO–CO_2?

5. A certain substance exists in two solid phases *A* and *B* and also in the liquid and gaseous states. Construct a *P-T* phase diagram indicating the regions of stable existence for each phase from the following triple-point data:

T/K	P/kPa	Phases in Equilibrium
200	100	*A*, *B*, gas
300	300	*A*, *B*, liquid
400	400	*B*, liquid, gas

B Use of the Lever Rule; Distillation

6. Answer the following questions, using the accompanying figure.

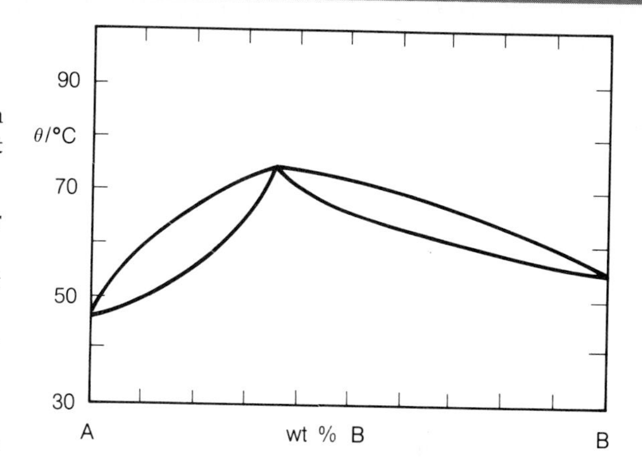

a. A liquid mixture consists of 33 g of component A and 99 g of component B. At what temperature would the mixture begin to boil?

b. Under the conditions in (a), what is the composition of the vapor when boiling first occurs?

c. If the distillation is continued until the boiling point is raised by 5°C, what would be the composition of the liquid left in the still?

d. Under the conditions in (c), what is the composition and mass of the two components collected over the initial 5°C interval?

7. From the data of Figure 6.14, calculate the ratio of the mass of the water-rich layer to that of the nicotine-rich layer, for a 40 wt % water-nicotine mixture at 350 K.

8. The ratio of the mass of chlorobenzene to that of water collected in a steam distillation is 1.93 when the mixture was boiled at 343.85 K and 56.434 kPa. If the vapor pressure of water at this temperature is 43.102 kPa, calculate the molar mass of chlorobenzene.

9. Obtain an expression for the ratio of masses of the materials distilled in a steam distillation in terms of the molar masses and the partial pressures of the two components.

10. Under atmospheric pressure 1 kg of pure naphthalene is to be prepared by steam distillation at 372.4 K. What mass of steam is required to perform this purification? The vapor pressure of pure water at 372.4 K is 98.805 kPa.

11. The vapor pressure of water at 343.85 K is 43.102 kPa. A certain mixture of chlorobenzene and water boils at 343.85 K under a reduced pressure of 56.434 kPa. What is the composition of the distillate?

12. Calculate the composition of the vapor in equilibrium at 323 K with a liquid solution of 0.6 mol fraction isobutyl alcohol and 0.4 mol fraction isoamyl alcohol. The vapor pressure of pure isobutyl alcohol is 7.46 kPa and that of pure isoamyl alcohol is 2.33 kPa both at 323 K.

13. The thermal expansion coefficient $\alpha(=(1/V)\cdot(\partial V/\partial T)_P)$ is often used when predicting changes in vapor pressure induced by temperature changes. From the relation $\rho = m/V$, show that $\alpha = -(\partial \ln \rho/\partial T)_P$.

14. At 293.15 K the density of water is 0.998 234 g cm^{-3} and at 294.15 K it is 0.998 022 g cm^{-3} under 1 atm of pressure. Estimate the value of α for water at 1 atm.

***15.** A sealed reaction vessel is completely filled with liquid water at 293.15 K and 1 atm. If the temperature is raised 6 K and the walls of the vessel remain rigid, what is the pressure in the container if the average value of $\alpha = 2.85 \times 10^{-4}$ K^{-1} and the compressibility coefficient $\kappa[=-(1/V)(\partial V/\partial P)_T]$ is 4.49×10^{-5} atm^{-1}?

C Data from Phase Diagrams of Condensed Systems

16. In Figure 6.16, a solution having composition p is cooled to just above the eutectic temperature (point s is about $0.18x_{Si}$ and x_e is $0.31x_{Si}$); calculate the composition of the solid that separates and that of the liquid that remains.

***17.** The melting points and heats of fusion of gold and silicon are

	Au	Si
T/K	1337	1683
$\Delta_{fus}H$/J mol^{-1}	12 677.5	39 622.5

For the data, calculate the solid-liquid equilibrium lines and estimate the eutectic composition graphically. Compare the result with the values given by Figure 6.16.

18. The following questions refer to Figure 6.26:

a. If liquid C were added to the system, what changes would occur if the system originally contained 80% salt A and 20% salt B?

b. What changes would occur if the system originally contained 50% salt A and 50% salt B upon the addition of liquid?

c. If liquid is added to an unsaturated solution of salt A and salt B in solution of composition lying at e, what changes would occur?

19. In diagram at the top of p. 247, due to B. S. R. Sastry and F. A. Hammel, *J. Am. Ceram. Soc.*, 42 [5] 218 (1959), identify the composition of all the areas. Identify the phenomenon associated with each lettered position.

20. The following information is obtained from cooling curve data on the partial system Fe_2O_3–Y_2O_3 [J. W. Nielsen and E. F. Dearborn, *Phys. Chem. Solids*, 5, 203 (1958)]:

Composition of Melt /mol % Y_2O_3	Temperature of Break/°C	Temperature of Halt/°C
0		1550
5	1540	1440
10	1515	1440
15	1450	1440
20	1520	1440
25	1560	1440
30	1620	1575/1440
40	1705	1575
50		1720

Sketch the simplest melting point diagram consistent with these data. Label the phase regions and give the composition of any compounds formed.

21. A preliminary thermal analysis of the Fe–Au system showed two solid phases of composition 8.1 mol % Au and 25.5 mol % Au in equilibrium at 1168°C with liquid of composition 43 mol % Au. Construct the simplest melting point diagram consistent with this information and label all the phase regions. Sketch the cooling curves for the composition 10 mol % Au, 30 mol % A, and 60 mol % Au, and make them consistent with the fact that there is an $\alpha \rightarrow \gamma$ phase transition in iron at 903°C and that the γ phase field extends to 45 mol % Au at this temperature. Iron melts at 1536°C and gold at 1063°C.

22. The aluminum-selenium system was determined from thermal analysis. Al_2Se_3 melts congruently at approximately 950°C and forms an eutectic both with aluminum and with selenium at a very low concentration of the alloying element and at a temperature close to the melting point of the base element. Draw a diagram from this information and give the composition of the phases. Aluminum melts at 659.7°C and selenium melts at approximately 217°C.

***23.** The metals Al and Ca form the compounds Al_4Ca and Al_2Ca. The solids Al, Ca, Al_4Ca, and Al_2Ca essentially are immiscible in each other but are completely miscible as liquids. Maximum Ca solubility in Al is about 2% and occurs at 616°C. Al melts at 659.7°C and Ca melts at 848°C. Compound Al_2Ca melts congruently at 1079°C and gives a sim-

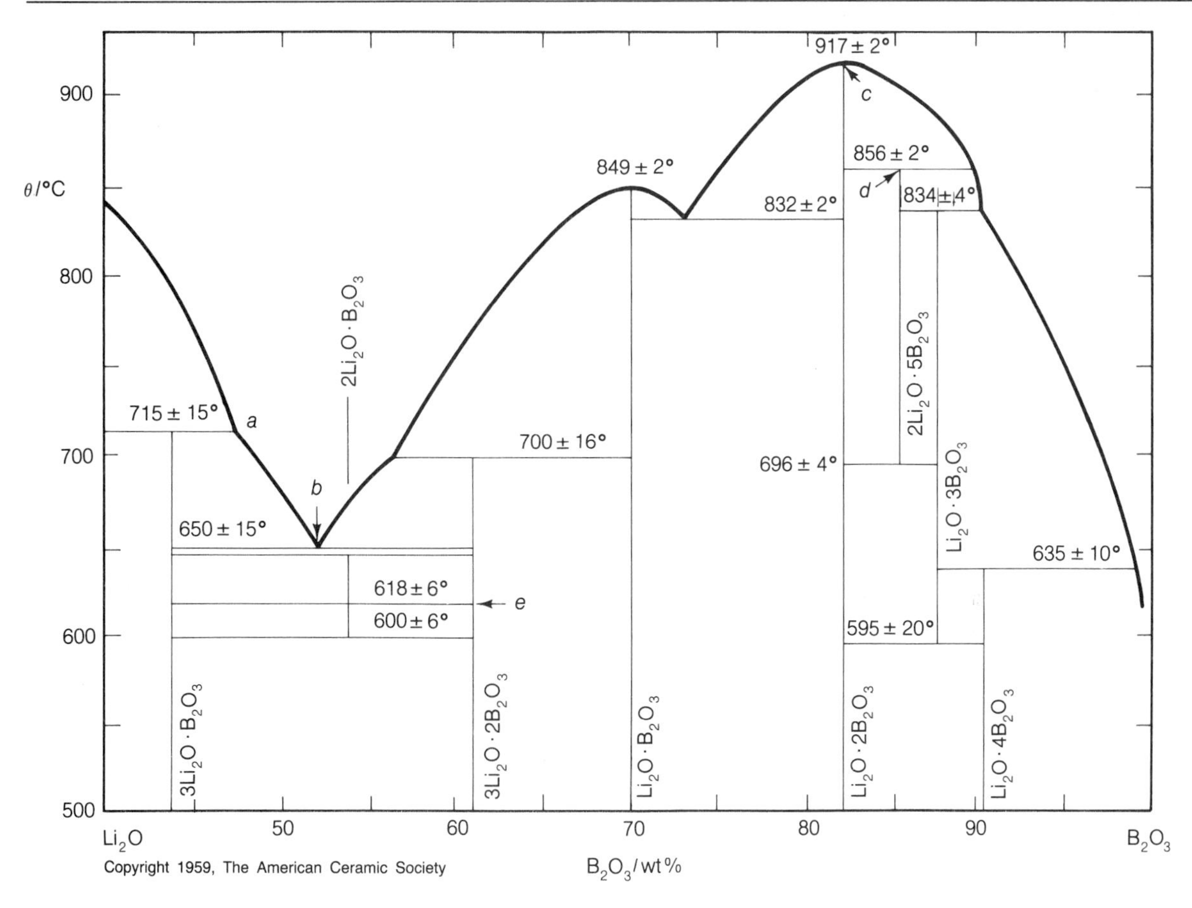

ple eutectic with Ca at 545°C. Compound Al_4Ca decomposes at 700°C to give Al_2Ca and a melt, the peritectic lying at 10 mol %. A monotectic exists at 616°C. At approximately 450°C a transition occurs between α-Ca and β-Ca.

a. Draw the simplest phase diagram consistent with this information and label all phase regions.

b. Sketch cooling curves for melts of composition 15 mol % Ca and 80 mol % Ca.

***24.** The extent of dehydration of a salt such as $CuSO_4$ can often be followed by measuring the vapor pressure over the hydrated salt. The system H_2O–$CuSO_4$ is shown in the accompanying figure as an example of such a system. Label the areas as to the phase(s) present. Then describe the sequence of phase changes if a dilute solution of copper sulfate is dehydrated at 275 K, ending with anhydrous copper sulfate. What would a vacuum gauge read starting with pure water during the dehydration process at 298.15 K? Sketch a plot of P/mmHg against $CuSO_4$/wt%. Relevant data are:

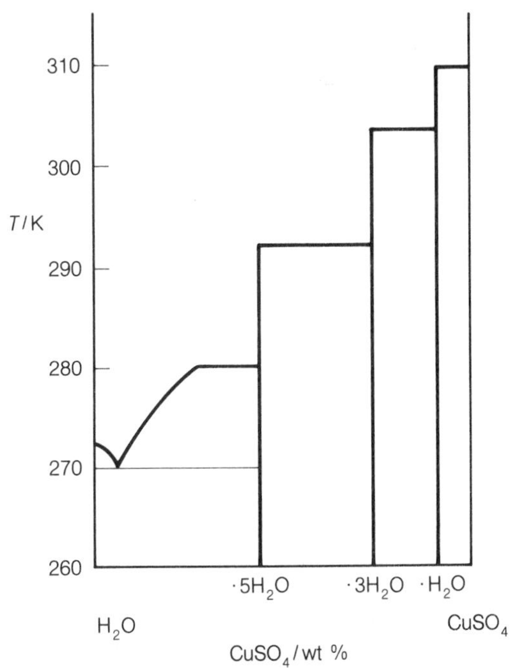

The vapor pressure of $CuSO_4$–H_2O at 298.15 K:	P/mmHg
Vapor + saturated solution + $CuSO_4 \cdot 5H_2O$	16
Vapor + $CuSO_4 \cdot 5H_2O$ + $CuSO_4 \cdot 3H_2O$	7.85
Vapor + $CuSO_4 \cdot 3H_2O$ + $CuSO_4 \cdot H_2O$	4.32
Vapor + $CuSO_4 \cdot H_2O$ + $CuSO_4$	0.017
Vapor pressure of water	23.8

25. The isobaric solubility diagram for the system acetic acid-toluene-water is shown in Figure 6.25. What phase(s) and their composition(s) will be present if 0.2 mol of toluene is added to a system consisting of 0.5 mol of water and 0.3 mol of acetic acid? Give the relative amounts of each phase.

***26.** The data in the accompanying table are approximate for the isobaric-isothermal system SnO_2–CaO–MgO at 298.15 K and 1 atm. Sketch a reasonable phase diagram in mol % with SnO_2 at the apex of the triangle. Label all phase regions; the results are known as composition triangles.

Material	In Equilibrium with Solid Phases
SnO_2	$(MgO)_2SnO_2$, $CaOSnO_2$
$(MgO)_2SnO_2$	SnO_2, $(CaO)SnO_2$, MgO
MgO	CaO, $(CaO)_2SnO_2$
CaO	MgO, $(CaO)_2SnO_2$
$(CaO)_2SnO_2$	CaO, MgO, $CaOSnO_2$
$CaOSnO_2$	$(CaO)_2SnO_2$, MgO, $(MgO)_2SnO_2$, SnO_2

D Supplementary Problems

27. Sketch the P against T diagram for phosphorus from the following information. White phosphorus melts at 311 K and 0.2 mmHg; red phosphorus melts at 763 K and 43 atm. The white form is more dense than the liquid and the red form is less dense than the liquid. The white form has a vapor pressure that is everywhere greater than that of the red. Label the areas on the plot, and explain which triple point(s) is (are) stable or metastable.

28. Giguère and Turrell, *J. Am. Chem. Soc.*, *102*, 5476 (1980), describe three ionic hydrates formed between HF and H_2O. Sketch the H_2O–HF phase diagram in mol % HF from the following information. $HF \cdot H_2O$ melts at −35.2°C, $2HF \cdot H_2O$ decomposes by a peritectic reaction at −75°C, and $4HF \cdot H_2O$ melts at −98.2°C. HF melts at −83.1°C. Label the composition of all regions. The eutectic occurs at −111°C with monotectics at −71°C, −77°C, and −102°C.

29. In the system A–B a line of three-phase equilibrium occurs at 900 K as determined by thermal analysis. A second three-phase equilibrium occurs at 500 K. Only one halt is observed for any one cooling curve. The compound AB_2 is known and melts at 600 K. If A melts at 1200 K and B at 700 K, sketch the simplest phase diagram consistent with the given data. Label each region.

30. In organic chemistry it is a common procedure to separate a mixture of an organic liquid in water by adding a salt to it. This is known as "salting out."

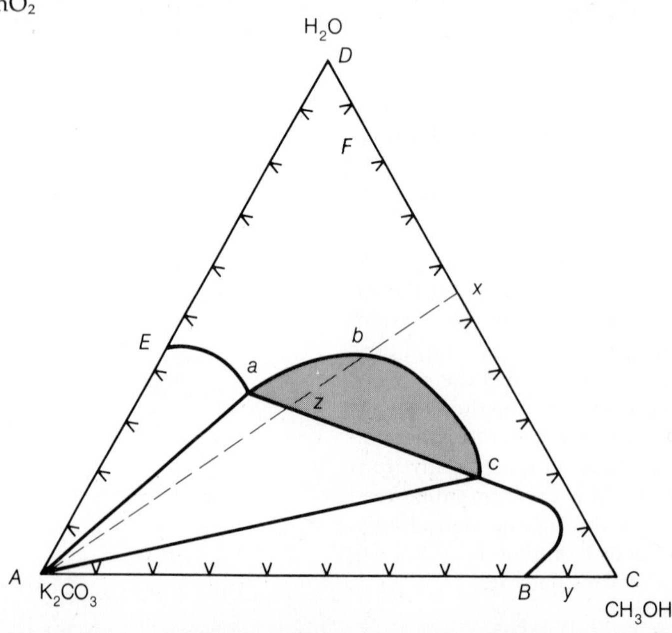

The ternary system K_2CO_3–H_2O–CH_3OH is typical. The system is distinguished by the appearance of the two-liquid region *abc*.

a. Describe the phase(s) present in each region of the diagram.
b. What would occur as solid K_2CO_3 is added to a solution of H_2O and CH_3OH of composition x?
c. How can the organic-rich phase in (b) be separated?
d. How can K_2CO_3 be precipitated from a solution having composition y?
e. Describe in detail the sequence of events when a solution of composition F is evaporated.

E Essay Questions

1. How is thermal analysis used to determine the liquid-solid equilibria and the eutectic temperature?

2. Explain what is meant by a *metastable system*.

3. Outline how isothermal distillation may be used to prepare a pure sample.

4. Detail the process by which a pure sample is obtained using a fractionating column.

5. What is the difference on a molecular level between a maximum and minimum boiling azeotrope? How do the plots of P against x and T against x differ?

Suggested Reading

G. W. Castellan, *Physical Chemistry* (2nd ed.), Reading, Mass.: Addison-Wesley, 1971.

A. Findlay, *Phase Rule* (revised and enlarged 9th ed., by A. N. Campbell and N. O. Smith), New York: Dover Pub., 1951.

R. J. Forbes, *A Short History of the Art of Distillation*, Leiden: E. J. Brill, 1970.

J. H. Hildebrand, J. M. Prausnitz, and R. L. Scott, *Regular and Related Solutions*, New York: Van Nostrand Reinhold Co., 1970.

W. Hume-Rothery, R. E. Smallman, and C. W. Haworth, *The Structure of Metals and Alloys*, The Metals and Metallurgy Trust of the Institute of Metals and the Institution of Metallurgists, London, 1969.

C. S. Robinson and E. R. Gilliland, *Fractional Distillation*, New York: McGraw-Hill, 1950.

Much of the literature has been reviewed and compiled in several areas. The following are convenient sources of much of the work.

Azeotropic Data, Advances in Chemistry Series No. 6, American Chemical Society, Washington, D.C., 1952.

Azeotropic Data, Advances in Chemistry Series No. 35, American Chemical Society, Washington, D.C., 1962.

R. P. Elliott, *Constitution of Binary Alloys* (1st Suppl.), New York: McGraw-Hill, 1965.

W. Guertler, M. Guertler, and E. Anastasiadias, *A Compendium of Constitutional Ternary Diagrams of Metallic Systems*, U.S. Department of Commerce, and National Science Foundation, Washington, D.C. Translated from German by Israel Program for Scientific Translations, 1969.

M. Hansen, *Constitution of Binary Alloys* (2nd ed.), New York: McGraw-Hill, 1958.

E. M. Levin, R. Robbins, and H. F. McMurdie, *Phase Diagrams for Ceramists*, The American Ceramic Society, Inc., 1964; 1969 Supplement (Figures 2067–4149); E. M. Levin and H. F. McMurdie, 1975 Supplement (Figures 4150–4999).

E. Rudy, *Compendium of Phase Diagram Data*, Air Force Materials Laboratory, Metals and Ceramics Division, Wright-Patterson AFT, Dayton, Ohio, 1969.

F. A. Shunk, *Constitution of Binary Alloys* (2nd Suppl.), New York: McGraw-Hill, 1969.

Chapter 7

Solutions of Electrolytes

Preview

Electrochemistry is concerned with the properties of solutions of electrolytes and with processes that occur at electrodes. A useful starting point is provided by *Faraday's laws of electrolysis*, which relate the mass of a substance deposited at an electrode to the quantity of electricity (current × time) passed through the solution and to the relative atomic or molecular mass of the substance.

Important properties of solutions of electrolytes are the *resistance*, the *conductance*, and the *electrolytic conductivity*. To compare solutions of different concentrations c, it is convenient to introduce the *molar conductivity* Λ (formerly called the *equivalent conductivity*), which is the conductivity when one mole of the electrolyte is dissolved in a solvent.

Different theories are required for *weak electrolytes*, which are only slightly dissociated, and for *strong electrolytes*, which are completely dissociated. The former are satisfactorily treated by a theory due to Arrhenius, which considers how the dissociation equilibrium shifts as the concentration of the electrolyte is varied. The theory finds its quantitative expression in *Ostwald's dilution law*, which relates the molar conductivity Λ at any concentration to that at infinite dilution Λ_0. The ratio Λ/Λ_0 is the *degree of dissociation* α.

This theory does not apply to strong electrolytes, which are dealt with satisfactorily by an entirely different treatment due mainly to *Debye and Hückel*. This theory focuses attention on the distribution of positive and negative ions in solution as a result of the electrostatic forces. In the near neighborhood of each ion there are more ions of opposite sign than of the same sign, and it is convenient to speak of an *ionic atmosphere* as existing round each ion. The *thickness of the ionic atmosphere* decreases as the concentration is raised. When an ion moves through the solution as a result of an applied potential field, the atmosphere lags behind and causes a retardation of the motion; this is known as the *relaxation* or the *asymmetry* effect. The moving ionic atmosphere drags solvent molecules with it, and as a result there is an additional retardation; this is known as the *electrophoretic effect*.

The *Law of Independent Migration of Ions* states that the molar conductivity of an electrolyte is the sum of the individual ionic conductivities. The *mobility* of an ion is the speed with which it moves in a unit potential gradient, and it is proportional to the ionic conductivity. The *transport number* of an ion is the fraction of the current carried by an ion. Transport numbers can be measured experimentally, and they allow the molar conductivity to be split into the individual ionic conductivities. A simple theory due to Max Born, which regards an ion as a conducting sphere, leads to satisfactory estimates of enthalpies and entropies of hydration of ions. The activity coefficients of ions are treated by the *Debye-Hückel limiting law*, according to which the activity coefficient y_i of an ion of the ith type is proportional to the square root of the ionic strength of the solution.

An important matter, particularly in biological systems, is the distribution of electrolytes across membranes that are permeable to some ions but not to others. This problem is treated by the theory of the *Donnan equilibrium*.

7 Solutions of Electrolytes

At the atomic level, electrical factors determine the structures and reactions of chemical substances. The important branch of physical chemistry known as *electrochemistry* is particularly concerned with the properties of solutions of electrolytes and with processes occurring when electrodes are immersed in these solutions. The present chapter deals with the electrochemistry of solutions, a topic that is of special interest since the subject of physical chemistry was largely initiated by studies in this area.

A very great deal can be learned about the behavior of ions in water and other solvents by investigations of electrical effects. Thus, measurements of the conductivities of aqueous solutions at various concentrations have led to understanding of the extent to which substances are ionized in water, of the association of ions with surrounding water molecules, and of the way in which ions move in water.

It is convenient to begin our study of electrochemistry with a brief survey of the development of concepts related to electricity. It has been known since ancient times that amber, when rubbed, acquires the property of attracting light objects such as small pieces of paper or pitch. In 1600 Sir William Gilbert (1544–1603) coined the word *electrics,* from the Greek word for amber (*elektron*), to describe substances that acquired this power to attract. It was subsequently found that materials such as glass, when rubbed with silk, exerted forces opposed to those from amber. Two types of electricity were thus distinguished: *resinous* (from substances like amber) and *vitreous* (from substances like glass).

In 1747 Benjamin Franklin (1706–1790) proposed a "one-fluid" theory of electricity. According to him, if bodies such as amber and fur are rubbed together, one of them acquires a surplus of "electric fluid" and the other, a deficit of "electric fluid." Quite arbitrarily, Franklin established the convention, still used today, that the type of electricity produced on glass, when rubbed with silk, is positive; conversely, the resinous type of electricity is negative. The positive electricity was supposed to involve an excess of electric fluid; the negative electricity, a deficit. In reality, negative electricity corresponds to a *surplus of electrons,* whereas positive electricity involves a deficit. Franklin's convention, however, is now so well entrenched that it would be futile to try to change it.

Until 1791 the study of electricity was entirely concerned with *static* electricity, in the form of charges developed by friction. In that year Luigi Galvani (1737–1798) brought the nerve of a frog's leg into contact with an electrostatic machine and observed a sharp convulsion of the leg muscles. Later he found that the twitching could be produced simply by bringing the nerve ending and the end of the leg into contact by means of a metal strip. Galvani believed that this effect could only be produced by living tissues. Then in 1794 the Italian physicist Alessandro Volta (1745–1827) discovered that the same type of electricity could be produced from inanimate objects. In 1800 he constructed his famous "Voltaic pile," which consisted of a number of consecutive plates of silver and zinc, separated by cloth soaked in salt solution. From the terminals of the pile he produced the shocks and sparks that had previously only been observed with frictional machines.

Very shortly after Volta's experiment, the British chemists William Nicholson (1753–1815) and Sir Anthony Carlisle (1768–1842) decomposed water by means of an electric current and observed that oxygen appeared at one pole and hydrogen at the other. The word **electrolysis** (Greek *lysis,* setting free) was coined by the English scientist Michael Faraday (1791–1867) to describe such splitting by an electric current. He gave the name **cathode** (Greek *cathodos,* way down) to the electrode at which the hydrogen collects and **anode** (Greek *ana,* up; *hodos,* way) to the other electrode. The important studies of Faraday, which put the subject of electrolysis on a quantitative basis, are considered later.

Electrical Units

The *electrostatic force* F between two charges Q_1 and Q_2 separated by a distance r in a vacuum is

$$F = \frac{Q_1 Q_2}{4\pi\epsilon_0 r^2} \tag{7.1}$$

The constant ϵ_0 is the *permittivity of a vacuum* and has the value 8.854×10^{-12} C^2 J^{-1} m^{-1}. The factor 4π is introduced into this expression in order that the Gauss and Poisson equations (see p. 264) are free of this factor.* If the charges are in a medium having a *relative permittivity,* or *dielectric constant*, of ϵ, the equation for the force is

$$F = \frac{Q_1 Q_2}{4\pi\epsilon_0 \epsilon r^2} \tag{7.2}$$

For example, water at 25°C has a dielectric constant of about 78, with the result that the electrostatic forces between ions are reduced by this factor. The SI unit of charge Q is the coulomb, C, and that of distance r is the metre, m. The unit of force F is the newton, N, which is joule per metre, J m^{-1}.

The *electric field* E at any point is the force exerted on a unit charge (1 C) at that point. The field strength at a distance r from a charge Q, in a medium of dielectric constant ϵ, is thus

$$E = \frac{Q}{4\pi\epsilon_0 \epsilon r^2} \tag{7.3}$$

*See Appendix A for a further discussion of the term $4\pi\epsilon_0$.

The SI unit of field strength is therefore newton per coulomb (N C^{-1}). However since joule = volt coulomb = newton metre (see Appendix A), the SI unit of field strength is usually expressed as volt per metre (V m^{-1}).

Just as a mechanical force is the negative gradient of a potential, the *electric field strength* is the negative gradient of an *electric potential* ϕ. Thus the field strength $\boldsymbol{E}$ at a distance r from a charge Q is given by

$$E = -\frac{d\phi}{dr} \tag{7.4}$$

and the electric potential is thus

$$\phi = -\int E\,dr = \frac{Q}{4\pi\epsilon_0 \epsilon r} \tag{7.5}$$

The SI unit of the electric potential is the volt (V).*

Further details about electrical units, including a comparison of SI units with the electrostatic units, are to be found in Appendix A.

7.1 Faraday's Laws of Electrolysis

During the years 1833 and 1834 Michael Faraday published the results of an extended series of investigations on the relationships between the quantity of electricity passing through a solution and the amount of material liberated at the electrodes.† He formulated **Faraday's laws of electrolysis,** which may be summarized as follows:

1. **The mass of an element produced at an electrode is proportional to the *quantity of electricity* Q passed through the liquid; the SI unit of Q is the coulomb (C). The quantity of electricity is defined as equal to the current I (SI unit = ampere, A) multiplied by the time t (SI unit = second, s):**

$$Q = It \tag{7.6}$$

2. **The mass of an element liberated at an electrode is proportional to the *equivalent weight* of the element.**

The SI recommendation, supported by the International Union of Pure and Applied Chemistry and other international scientific bodies, is to abandon the use of the word *equivalent* and to refer to *moles* instead. However, the concept of the equivalent is still employed, even though the name has been dropped. Suppose, for example, that we pass 1 A of electricity for 1 h‡ through a dilute sulfuric acid solution and through solutions of silver nitrate, $AgNO_3$, and cupric sulfate, $CuSO_4$. The masses liberated at the respective cathodes are

*In terms of basic SI units, volt = kg m^2 s^{-3} A^{-1}.

† M. Faraday, *Phil. Trans. Roy. Soc., A, 124,* 77 (1834).

‡The symbol for hour is h.

0.038 g of H_2 (relative atomic mass, $A_r = 1.008$; relative molecular mass, $M_r = 2.016$)

4.025 g of Ag ($A_r = 107.9$)

1.186 g of Cu ($A_r = 63.6$)

These masses 0.038, 4.025, and 1.186 are approximately in the ratio 1.008/107.9/31.8. Therefore, the amount of electricity that liberates 1 mol of Ag liberates 0.5 mol of H_2 and 0.5 mol of Cu. These quantities were formerly referred to as 1 equivalent, but the modern practice is to speak instead of 1 mol of $\frac{1}{2}H_2$ and 1 mol of $\frac{1}{2}Cu$. These are the quantities liberated by 1 mol of electrons:

$$e^- + H^+ \longrightarrow \tfrac{1}{2}H_2 \quad \text{and} \quad e^- + \tfrac{1}{2}Cu^{2+} \longrightarrow \tfrac{1}{2}Cu$$

In what follows, when we refer to 1 mol we may mean 1 mol of a fraction of an entity (e.g., to $\frac{1}{2}Cu$ or $\frac{1}{2}SO_4^{2-}$).

The proportionality factor that relates the amount of substance deposited to the quantity of electricity passed through the solution is known as the **Faraday constant** and given the symbol F. In modern terms, the charge carried by 1 mol of ions bearing z unit charges is zF. According to the latest measurements the Faraday constant F is equal to 96 485 coulombs per mole (C mol^{-1}); in this discussion the figure will be rounded to 96 500 C mol^{-1}. In other words, 96 500 C will liberate 1 mol of Ag, 1 mol of $\frac{1}{2}H_2$ (i.e., 0.5 mol of H_2), etc. If a constant current I is passed for a period of time t, the amount of substance deposited is $It/(96\ 500\ \text{C mol}^{-1})$.

It follows that the (negative) charge on 1 mol of electrons is 96 500 C. The charge on one electron is this quantity divided by the Avogadro constant L and is thus

$$\frac{96\ 500\ \text{C mol}^{-1}}{6.022 \times 10^{23}\ \text{mol}^{-1}} = 1.602 \times 10^{-19}\ \text{C}$$

EXAMPLE

An aqueous solution of gold (III) nitrate, $Au(NO_3)_3$, was electrolyzed with a current of 0.0250 A until 1.200 g of Au (atomic mass 197.0 g mol^{-1}) had been deposited at the cathode. Calculate (a) the quantity of electricity passed, (b) the duration of the experiment, and (c) the volume of O_2 (at STP) liberated at the anode. The reaction at the cathode is

$$3e^- + Au^{3+} \longrightarrow Au \quad \text{or} \quad e^- + \tfrac{1}{3}Au^{3+} \longrightarrow \tfrac{1}{3}Au$$

The equation for the liberation of O_2 is

$$2H_2O \longrightarrow O_2 + 4H^+ + 4e^- \quad \text{or} \quad \tfrac{1}{2}H_2O \longrightarrow \tfrac{1}{4}O_2 + H^+ + e^-$$

SOLUTION

a. From the cathode reaction, 96 500 C of electricity liberates 1 mol of $\frac{1}{3}Au$ (i.e., 197.0/3 g). The quantity of electricity required to produce 1.200 g of Au is thus

$$\frac{96\ 500 \times 1.200}{197.0/3} = 1.76 \times 10^3\ \text{C}$$

b. Since the current was 0.250 A, the time required was

$$\frac{1.76 \times 10^3(\text{A s})}{0.025(\text{A})} = 7.04 \times 10^4 \text{ s}$$

c. From the anode reaction, 96 500 C liberates $\frac{1}{4}$ mol O_2, so that 1.76×10^3 C produces

$$\frac{1.76 \times 10^3}{96\ 500 \times 4} = 4.56 \times 10^{-3} \text{ mol } O_2$$

The volume of this at STP is

$$4.56 \times 10^{-3}(\text{mol}) \times 22.4(\text{dm}^3 \text{ mol}^{-1}) = 0.102 \text{ dm}^3$$

■

Faraday's work on electrolysis was of great importance in that it was the first to suggest a relationship between matter and electricity. His work indicated that *atoms*, recently postulated by Dalton, might contain electrically charged particles. Later work led to the conclusion that an electric current is a stream of electrons and that electrons are universal components of atoms.

7.2 Molar Conductivity

Important information about the nature of solutions has been provided by measurements of their conductivities. A solution of sucrose in water does not dissociate into ions and has the same electrical conductivity as water itself. Such substances are known as *nonelectrolytes*. A solution of sodium chloride or acetic acid, on the other hand, forms ions in solution and has a much higher conductivity. These substances are known as *electrolytes*.

Some electrolytes in aqueous solutions of the same molality conduct better than others. This suggests that some electrolytes exist as ions in solution to a greater extent than do others. More detailed investigations have indicated that certain substances such as the salts sodium chloride and cupric sulfate, as well as some acids such as hydrochloric acid, occur almost entirely as ions when they are in aqueous solution. Such substances are known as **strong electrolytes.** Other substances, including acetic acid and ammonia, are present only partially as ions (e.g., acetic acid exists in solution partly as CH_3COOH and partly as $CH_3COO^- + H^+$). These substances are known as **weak electrolytes.**

Further insight into these matters is provided by studies of the way in which the electrical conductivities of solutions vary with the concentration of solute.

According to **Ohm's law,** the resistance R of a slab of material is equal to the electric potential difference V divided by the electric current I:

$$R = \frac{V}{I} \tag{7.7}$$

The SI unit of potential is the volt (V) and that for current is the ampere (A). The unit of electrical resistance is the ohm,* given the symbol Ω (omega). The reciprocal

*In terms of base SI units, ohm $\equiv$ kg m^2 s^{-3} A^{-2}.

of the resistance is the *electrical conductance* G, the SI unit of which is the *siemens* ($S \equiv \Omega^{-1}$). The electrical conductance of material of length l and cross-sectional area A is proportional to A and inversely proportional to l:

$$G \text{ (conductance)} = \kappa \frac{A}{l} \tag{7.8}$$

The proportionality constant κ, which is the conductance of a unit cube (see Figure 7.1a), is known as the *conductivity* (formerly as the specific conductivity); for the particular case of a solution of an electrolyte, it is known as the *electrolytic conductivity*. Its SI unit is Ω^{-1} m^{-1} ($\equiv$ S m^{-1}), but the unit more commonly used is Ω^{-1} cm^{-1} ($\equiv$ S cm^{-1}).

The electrolytic conductivity is not a suitable quantity for comparing the conductivities of different solutions. If a solution of one electrolyte is much more concentrated than another, it may have a higher conductivity simply because it contains more ions. What is needed instead is a property in which there has been some compensation for the difference in concentrations. The German physicist Friedrich Wilhelm Georg Kohlrausch* (1840–1910) clarified the early theory by introducing the concept of the *equivalent conductivity*. Such a property is now called the **molar conductivity** and given the symbol Λ (lambda). It is defined as the *electrolytic conductivity* κ *divided by concentration* c,

$$\Lambda = \frac{\kappa}{c} \tag{7.9}$$

The meaning of Λ can be visualized as follows. Suppose that we construct a cell having parallel plates a unit distance apart, the plates being of such an area that for a particular solution 1 mol of electrolyte (e.g., HCl, $\frac{1}{2}CuSO_4$) is present in it (see Figure 7.1b). The molar conductivity Λ is the conductance (i.e., the reciprocal of the resistance) across the plates. We need not actually construct cells for each solution for which we need to know the molar conductivity, since we can calculate it from the electrolytic conductivity κ. If the concentration of the solution is c, the volume of the hypothetical molar conductivity cell must be (1 mol)/c in order for 1 mol to be between the plates (Figure 7.1b). The molar conductivity Λ is therefore greater than the electrolytic conductivity κ by a factor of $1/c$. In using Eq. 7.9, care must be taken with units, as shown by the following example.

EXAMPLE

The electrolytic conductivity of a 0.1 *M* solution of acetic acid (corrected for the conductivity of the water) was found to be 5.3×10^{-4} Ω^{-1} cm^{-1}. Calculate the molar conductivity.

SOLUTION

From Eq. 7.9,

$$\Lambda = \frac{5.3 \times 10^{-4}(\Omega^{-1}\ \text{cm}^{-1})}{0.1(\text{mol dm}^{-3})} = \frac{5.3 \times 10^{-4}(\Omega^{-1}\ \text{cm}^{-1})}{10^{-4}(\text{mol cm}^{-3})}$$

$$= 5.3\ \Omega^{-1}\ \text{cm}^2\ \text{mol}^{-1}$$ ■

*F. W. G. Kohlrausch, *Ann. Phys.*, 26, 161 (1885) and many other publications.

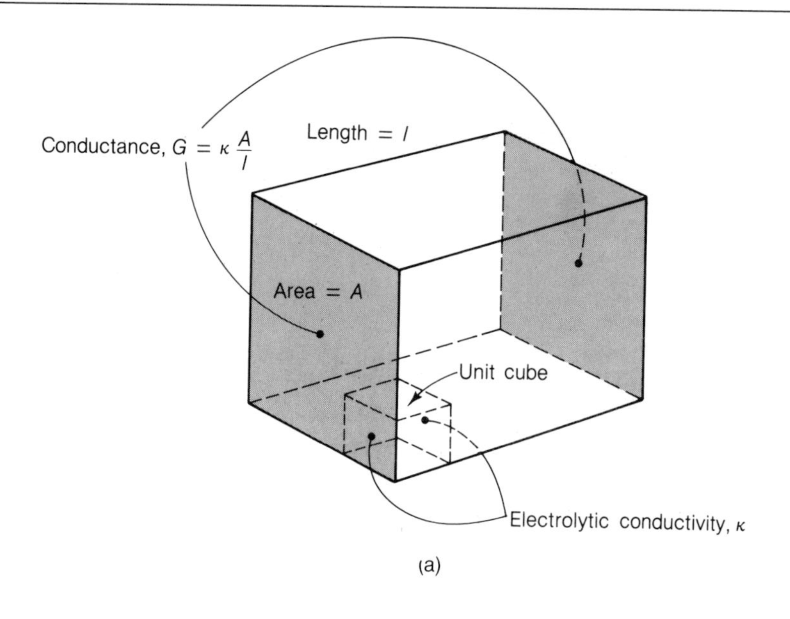

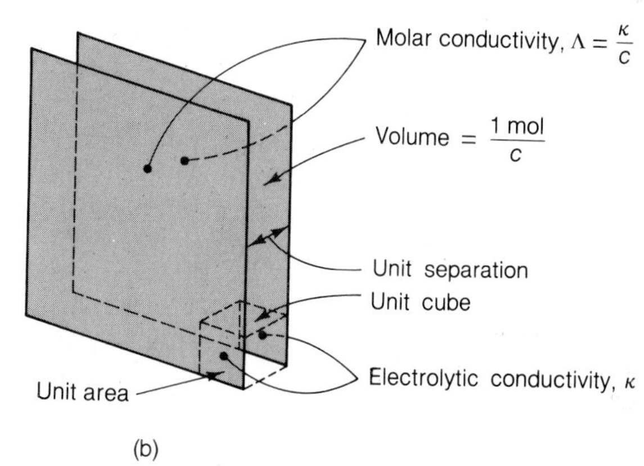

FIGURE 7.1 (a) The relationship between conductance and electrolytic conductivity. (b) The relationship between molar conductivity and electrolytic conductivity.

The importance of the molar conductivity is that it gives information about the conductivity of the ions produced in solution by 1 mol of a substance. In all cases the molar conductivity diminishes as the concentration is raised, and two patterns of behavior can be distinguished. From Figure 7.2 we see that for strong electrolytes the molar conductivity falls only slightly as the concentration is raised. On the other hand, the weak electrolytes produce fewer ions and exhibit a much more pronounced fall of Λ with increasing concentration, as shown by the lower curve in Figure 7.2.

We can extrapolate the curves back to zero concentration and obtain a quantity known as Λ_0, the *molar conductivity at infinite dilution,* or zero concentration. With weak electrolytes this extrapolation may be unreliable, and an indirect method, explained on page 281, is usually employed.

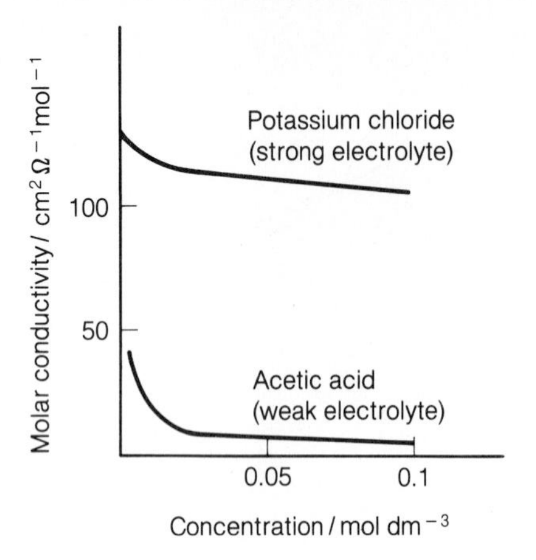

FIGURE 7.2 The variations of molar conductivity Λ with concentration for strong and weak electrolytes.

7.3 Weak Electrolytes: The Arrhenius Theory

The conductivity data of Kohlrausch and the fact that the heat of neutralization of a strong acid by a strong base in dilute aqueous solution was practically the same for all strong acids and strong bases (about 54.7 kJ mol^{-1} at 25°C) led the Swedish scientist Svante August Arrhenius* (1859–1927) to propose in 1884 a new theory of electrolyte solutions. According to this theory, there exists an equilibrium in solution between undissociated molecules AB and the ions A^+ and B^-:

$$AB \rightleftharpoons A^+ + B^- \tag{7.10}$$

The **degree of dissociation,** that is, the fraction of AB in the form $A^+ + B^-$, is Λ/Λ_0. It is convenient to denote the ratio of Λ at any concentration to Λ_0 by the symbol α:

$$\alpha = \frac{\Lambda}{\Lambda_0} \tag{7.11}$$

At high concentrations there is less dissociation, while at infinite dilution there is complete dissociation and the molar conductivity then is Λ_0.

According to this theory, *all* strong acids and bases are almost completely dissociated at *all* concentrations. The theory explains the constant heat of neutralization of strong acids and bases since the neutralization involves simply the reaction $H^+ + OH^- \longrightarrow H_2O$. However, as we shall see, Arrhenius's theory to explain molar conductivity is valid only for weak electrolytes; another explanation is required for the lowering of Λ with increasing concentration of strong electrolytes.

Soon after Arrhenius's theory was proposed, van't Hoff made osmotic pressure measurements that gave it considerable support. Van't Hoff† found that the osmotic pressures of solutions of electrolytes were always considerably higher than

*S. Arrhenius, *Recherches sur la conductibilité galvanique des electrolytes.* Doctoral Dissertation, Univ. of Uppsala, 1884 (presented June 6, 1883).

†J. H. van't Hoff, *Z. physikal. Chem.*, *5*, 174 (1890).

predicted by the osmotic pressure equation for nonelectrolytes (Eq. 5.134). He proposed the modified equation

$$\pi = icRT \tag{7.12}$$

where i is known as the *van't Hoff factor*. For strong electrolytes, the van't Hoff factor is approximately equal to the number of ions formed from one molecule; thus for NaCl, HCl, etc., $i = 2$; for Na_2SO_4, $BaCl_2$, etc., $i = 3$; and so forth. For weak electrolytes the van't Hoff factor i involves the degree of dissociation. Suppose that one molecule of a weak electrolyte would produce, if there were complete dissociation, ν ions. The number of ions actually produced is thus $\nu\alpha$, and the number of undissociated molecules is $1-\alpha$, so that the total number of particles produced from 1 molecule is*

$$i = 1 - \alpha + \nu\alpha \tag{7.13}$$

Therefore

$$\alpha = \frac{i - 1}{\nu - 1} \tag{7.14}$$

It is thus possible to calculate α from osmotic pressure data. The fact that the values so obtained are in good agreement with Λ/Λ_0 is excellent support for the Arrhenius theory.

Ostwald's Dilution Law

In 1888 the ideas of Arrhenius were expressed quantitatively by the Russian-German physical chemist Friedrich Wilhelm Ostwald (1853–1932) in terms of a *dilution law.*† Consider an electrolyte AB that exists in solution partly as the undissociated species AB and partly as the ions A^+ and B^-:

$$AB \rightleftharpoons A^+ + B^-$$

The equilibrium constant, on the assumption of ideal behavior, is

$$K_c = \frac{[A^+][B^-]}{[AB]} \tag{7.15}$$

Suppose that an amount n of the electrolyte is present in a volume V and that the fraction dissociated is α; the fraction not dissociated is $1 - \alpha$. The amounts of the three species present at equilibrium, and the corresponding concentrations, are therefore

	AB $\rightleftharpoons$	A^+ +	B^-
Amounts present at equilibrium:	$n(1-\alpha)$	$n\alpha$	$n\alpha$
Concentrations at equilibrium:	$\frac{n(1-\alpha)}{V}$	$\frac{n\alpha}{V}$	$\frac{n\alpha}{V}$

* S. Arrhenius, *Z. physikal. Chem.*, *1*, 631 (1887).
† F. W. Ostwald, *Z. physikal. Chem.*, *2*, 270 (1888).

The equilibrium constant is

$$K_c = \frac{(n\alpha/V)^2}{[n(1-\alpha)]/V} = \frac{n\alpha^2}{V(1-\alpha)} \tag{7.16}$$

Therefore, for a given amount of substance the degree of dissociation α must vary with the volume V as follows:

$$\frac{\alpha^2}{1-\alpha} = \text{constant} \times V \tag{7.17}$$

Alternatively, since $n/V = c$, we can write

$$\frac{c\alpha^2}{1-\alpha} = K \tag{7.18}$$

The larger the V, the lower the concentration c and the larger the degree of dissociation. Thus, if we start with a solution containing 1 mol of electrolyte and dilute it, the degree of dissociation increases and the amounts of the ionized species increase. As V becomes very large (the concentration approaching zero), the degree of dissociation α approaches unity; that is, dissociation approaches 100% as infinite dilution is approached. The experimental value of Λ_0 corresponds to complete dissociation; at finite concentrations the molar conductivity Λ is therefore lower by the factor α $(=\Lambda/\Lambda_0)$. The dilution law can thus be expressed as

$$KV = \frac{n(\Lambda/\Lambda_0)^2}{1-(\Lambda/\Lambda_0)} \tag{7.19}$$

or

$$\mathbf{K = \frac{c(\Lambda/\Lambda_0)^2}{1-(\Lambda/\Lambda_0)}} \tag{7.20}$$

Equation 7.20 has been found to give a satisfactory interpretation of the variation of Λ with c for a number of weak electrolytes.

7.4 Strong Electrolytes

For strong electrolytes there are some features that are inconsistent with the Arrhenius theory. For example, the Ostwald dilution law was not satisfactorily obeyed by acids stronger than acetic acid. Furthermore, values for the conductivity ratio α $(=\Lambda/\Lambda_0)$, assumed to be the degree of dissociation, were sometimes significantly different from those obtained from the van't Hoff factor (Eq. 7.12). Also, the values of K calculated from the Ostwald equation sometimes, for strong electrolytes, varied markedly with the concentration.

We saw earlier that the constancy of heats of neutralization of strong acids and bases was originally good evidence for the Arrhenius theory. But the heats were *too* constant to be consistent with the theory, because at a given concentration there should have been small but significant differences in the degrees of ionization of acids such as HCl, HNO_3, and H_2SO_4, and these differences should have given measurable differences between the heats of neutralization. These differences could not be observed, a result that suggests that these acids are completely dissociated at all concentrations. Supporting this contention is the fact that the absorption spectra of solutions of strong electrolytes showed no evidence for the existence of such undissociated molecules. The fall in Λ with increasing concentration must therefore, for strong electrolytes, be attributed to some cause other than a decrease in the degree of dissociation.

Debye-Hückel Theory

A possible explanation for the decrease in molar conductivity with the increase in concentration of a strong electrolyte was suggested as early as 1894 by the Dutchman Johannes Jacobus van Laar (1860–1938). He pointed out that the strong electrostatic forces present in an ionic solution must have an important effect on the behavior of the ions.*

In 1923, P. J. W. Debye and E. Hückel published a very important mathematical treatment of the problem. According to these theories, the decrease in the molar conductivity of a strong electrolyte is due to the mutual interference of the ions, which becomes more pronounced as the concentration increases. Because of the strong attractive forces between ions of opposite signs, the arrangement of ions in solution is not completely random. In the immediate neighborhood of any positive ion, there tend to be more negative than positive ions, whereas for a negative ion there are more positive than negative ions. This is shown schematically in Figure 7.3 for a sodium chloride solution. In *solid* sodium chloride there is a regular array of sodium and chloride ions. As shown in Figure 7.3a, each sodium ion has six chloride ions as its nearest neighbors, and each chloride ion has six sodium ions. When the sodium chloride is dissolved in water, this ordering is still preserved to a very slight extent (Figure 7.3b). The ions are much further apart than in the solid; the electrical attractions are therefore much smaller and the thermal motions cause irregularity. The small amount of ordering that does exist, however, is sufficient to exert an important effect on the conductivity of the solution.

The way in which this ionic distribution affects the conductivity is as follows. If an electric potential is applied, a positive ion will move toward the negative electrode and must drag along with it an entourage of negative ions. The more concentrated the solution, the closer these negative ions are to the positive ion under consideration, and the greater is the drag. The ionic "atmosphere" around a moving ion is therefore not symmetrical; the charge density behind is greater than that in front, and this will result in a retardation in the motion of the ion. This influence on the speed of an ion is called the **relaxation,** or **asymmetry,** effect.

A second factor that retards the motion of an ion in solution is the tendency of the applied potential to move the ionic atmosphere itself. This in turn will tend to drag solvent molecules, because of the attractive forces between ions and solvent molecules. As a result, the ion at the center of the ionic atmosphere is required to move upstream, and this is an additional retarding influence. This effect is known as the **electrophoretic** effect.

---

*J. J. van Laar, *Z. Physikal. Chem.*, *15*, 457 (1894); *18*, 245 (1895).

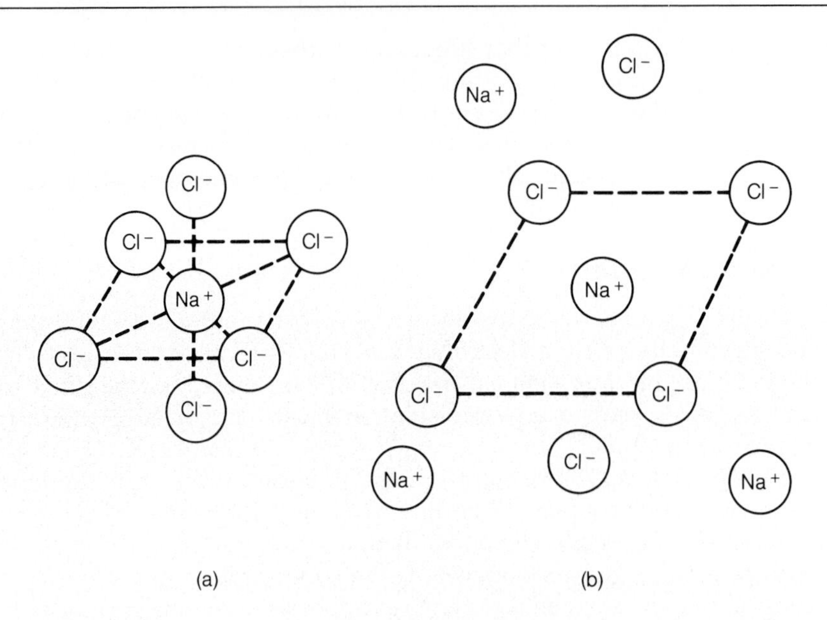

FIGURE 7.3 The distribution of chloride ions around a sodium ion (a) in the crystal lattice and (b) in a solution of sodium chloride. In the solution the interionic distances are greater, and the distribution is not so regular; but near to the sodium ion there are more chloride ions than sodium ions.

The Ionic Atmosphere

The Debye-Hückel theory is very complicated mathematically and a detailed treatment is outside the scope of this book. We will give a brief account of the theoretical treatment of the ionic atmosphere, followed by a qualitative discussion of the way in which the theory of the ionic atmosphere explains conductivity behavior.

Figure 7.4 shows a positive ion, of charge $z_c e$, situated at a point A; e is the unit positive charge (the negative of the electronic charge) and z_c is the valence of the ion. As a result of thermal motion there will sometimes be an excess of positive ions in the volume element and sometimes an excess of negative ions. On the average, the charge density will be negative because of the electrostatic forces. In other words, the probability that there is a negative ion in the volume element is greater than the probability that there is a positive ion. The negative charge density will be greater if r is small than if it is large, and we will see later that it is possible to define an effective thickness of the ionic atmosphere.

Suppose that the average electric potential in the volume element dV is ϕ, and let z_+ and z_- be the positive numerical values of the ionic valencies; for example, for Na^+, $z_+ = 1$; for Cl^-, $z_- = 1$. The work required to bring a positive ion of charge z_+e from infinity up to this volume element is $z_+e\phi$, a positive quantity. Because the central positive ion attracts the negative ion, the work required to bring a negative ion of charge $-z_-e$ is $-z_-e\phi$, a negative quantity.

The time-average numbers of the positive and negative ions present in the

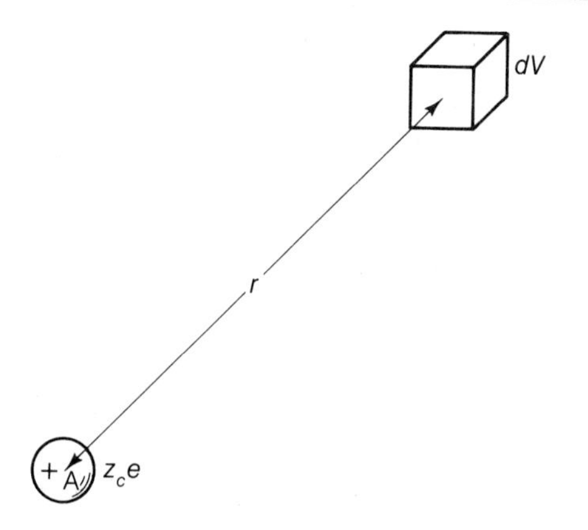

FIGURE 7.4 An ion in solution of charge $z_c e$, with an element of volume dV situated at a distance r from it.

volume element are given by the Boltzmann principle (see Section 14.2, Eq. 14.34):*

$$dN_+ = N_+ \mathrm{e}^{-z_+ e\phi/\mathbf{k}T} \tag{7.21}$$

and

$$dN_- = N_- \mathrm{e}^{-(-z_- e\phi)/\mathbf{k}T} \tag{7.22}$$

where N_+ and N_- are the total numbers of positive and negative ions, respectively, per unit volume of solution; **k** is the Boltzmann constant; and T is the absolute temperature. The charge density in the volume element (i.e., the net charge per unit volume) is given by

$$\rho = \frac{e(z_+ dN_+ - z_- dN_-)}{dV} = e(N_+ z_+ \mathrm{e}^{-z_+ e\phi/\mathbf{k}T} - N_- z_- \mathrm{e}^{z_- e\phi/\mathbf{k}T}) \tag{7.23}$$

For a univalent electrolyte (one in which both ions are univalent) z_+ and z_- are unity, and N_+ and N_- must be equal because the entire solution is neutral; Eq. 7.23 then becomes

$$\rho = Ne(\mathrm{e}^{-e\phi/\mathbf{k}T} - \mathrm{e}^{e\phi/\mathbf{k}T}) \tag{7.24}$$

where $N = N_+ = N_-$. If $e\phi/\mathbf{k}T$ is sufficiently small (which requires that the volume element is not too close to the central ion, so that ϕ is small), the exponential terms are given by†

$$\mathrm{e}^{-e\phi/\mathbf{k}T} \approx 1 - \frac{e\phi}{\mathbf{k}T} \quad \text{and} \quad \mathrm{e}^{e\phi/\mathbf{k}T} \approx 1 + \frac{e\phi}{\mathbf{k}T} \tag{7.25}$$

*We use roman e here for the exponential function to distinguish it from the charge on the electron e.

†The expansion of e^x is

$$\mathrm{e}^x = 1 + x + \frac{x^2}{2!} + \frac{x^3}{3!} + \frac{x^4}{4!} + \cdots$$

and when x is sufficiently small, we can neglect terms beyond x: thus $\mathrm{e}^x = 1 + x$ and $\mathrm{e}^{-x} = 1 - x$.

Equation 7.24 thus reduces to

$$\rho = -\frac{2Ne^2\phi}{\mathbf{k}T} \tag{7.26}$$

In the more general case, in which there are a number of different types of ions, the expression is

$$\rho = -\frac{e^2\phi}{\mathbf{k}T}\sum_i N_i z_i \tag{7.27}$$

where N_i and z_i represent the number per unit volume and the positive value of the valence of the ions of the ith type. The summation is taken over all types of ions present.

Equation 7.27 relates the charge density to the average potential ϕ. In order to obtain these quantities separately it is necessary to have another relationship between ρ and ϕ.

This is provided in the following way. From Eq. 7.3 we have the magnitude E of the electric field E at a point P, a distance r from the origin. The direction of the vector field due to a positive charge at the origin is the same as the direction of the vector from the origin to the point P (see Figure 7.5a). To make the connection we will be interested in determining the flux of the electric field (i.e., the number of lines of force) through an element of area on a surface perpendicular to the electric field vector. Consider a closed surface S surrounding a charge Q (Figure 7.5b). At any point on the surface the field strength is $\mathbf{E}$, and for an element of area $d\mathbf{S}$ the scalar product $\mathbf{E} \cdot d\mathbf{S}$ is defined as

$$\mathbf{E} \cdot d\mathbf{S} = E \cos\theta \, dS \tag{7.28}$$

where θ is the angle between $\mathbf{E}$ and the normal to $d\mathbf{S}$. This scalar product is the flux of the electric field through $d\mathbf{S}$. The element of solid angle $d\omega$ subtended by dS at the charge Q is $dS(\cos\theta)/r^2$, and the total solid angle $d\omega$ subtended by the entire closed surface S at any point within the surface is 4π. These considerations lead to an expression known as **Gauss's theorem:**

$$\int \mathbf{E} \cdot d\mathbf{S} = \frac{Q}{\epsilon_0} \tag{7.29}$$

In our case the distribution of charge can vary, so Q is replaced by the charge density ρ integrated over the volume enclosed by the surface; that is,

$$\int \mathbf{E} \cdot d\mathbf{S} = \frac{1}{\epsilon_0}\int \rho \, dV \tag{7.30}$$

It is known from vector analysis that the divergence (div) of a vector $\mathbf{E}$

$$\mathbf{div}\ E = \nabla E = \frac{\partial E_x}{\partial x} + \frac{\partial E_y}{\partial y} + \frac{\partial E_z}{\partial z} \tag{7.31}$$

is equal to the flux of E from dV per unit volume. Here ∇ is known as the **del operator** and when applied to a vector it represents the operation $\partial/\partial x + \partial/\partial y + \partial/\partial z$ on its components in Cartesian coordinates. If Eq. 7.31 is integrated over the system of charges enclosed by the surface and equated to Eq. 7.30, we have

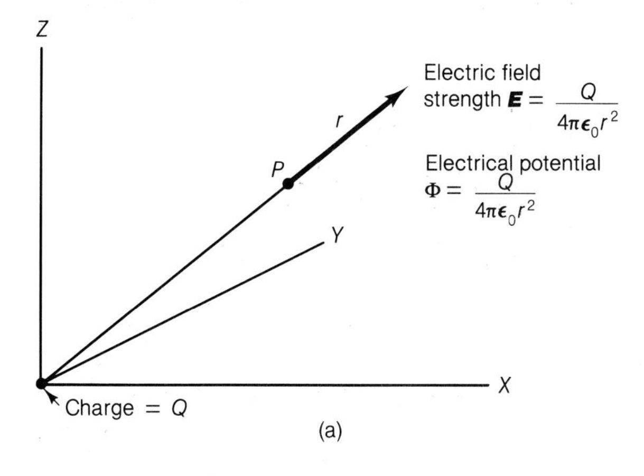

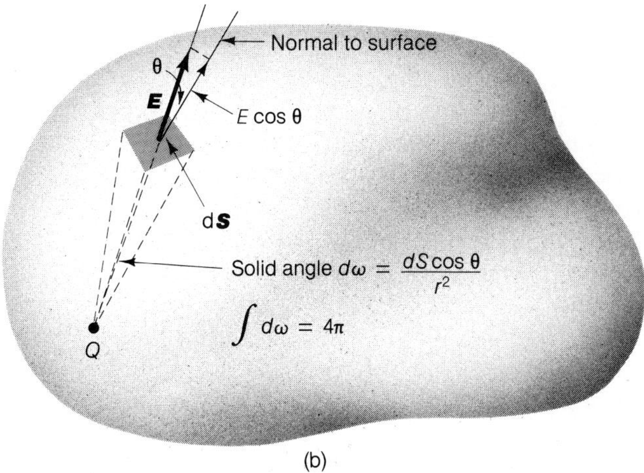

FIGURE 7.5 (a) The electric field strength **E** and the electric potential Φ, at a distance r from a charge Q. (b) A diagram to illustrate Gauss's theorem. A charge Q is surrounded by a surface S.

$$\int \operatorname{div} \boldsymbol{E}\, dV = \int \boldsymbol{E} \cdot d\boldsymbol{S} = \frac{1}{\epsilon_0} \int \rho\, dV \tag{7.32}$$

The equality on the left is known as the *divergence theorem.* Since the integrals are equal for any volume,

$$\operatorname{div} \boldsymbol{E} = \frac{\rho}{\epsilon_0} \tag{7.33}$$

We now relate E to the potential ϕ. From Eq. 7.4 the electric field strength can be represented as the gradient of an electric potential and in vector notation

$$\boldsymbol{E} = -\text{grad}\ \phi = -\nabla\phi \tag{7.34}$$

Substitution into Eq. 7.33 gives

$$\operatorname{div} \boldsymbol{E} = -\operatorname{div}\ \text{grad}\ \phi = \frac{\rho}{\epsilon_0} \tag{7.35}$$

The operator div grad is often written as ∇^2 and read as "del squared"; it is known as the *Laplacian operator*. Equation 7.35 can thus be written as

$$\nabla^2 \phi = -\frac{\rho}{\epsilon_0} \tag{7.36}$$

This is the **Poisson equation** where

$$\nabla^2 = \frac{\partial^2}{\partial x^2} + \frac{\partial^2}{\partial y^2} + \frac{\partial^2}{\partial z^2} \tag{7.37}$$

For a spherically symmetrical field, such as produced by an isolated charge, the Laplacian operator may be written in terms of spherical polar coordinates $[f(r, \theta, \phi)]$. The Laplacian operator applied to Eq. 7.36 then gives

$$\frac{1}{r^2}\frac{\partial}{\partial r}\left(r^2 \frac{\partial \phi}{\partial r}\right) = -\frac{\rho}{\epsilon_0} \tag{7.38}$$

because no dependency on the angles θ and ϕ is found.

The preceding equations are for a vacuum, which has a permittivity ϵ_0 equal to 8.854×10^{-12} C^2 J^{-1} m^{-1}. If the potential ϕ is in a medium having a relative permittivity, or dielectric constant, of ϵ, the absolute permittivity is $\epsilon_0\epsilon$ and must replace ϵ_0 in the preceding equations.

Equation 7.38 is the needed relation to determine ρ and ϕ in conjunction with Eq. 7.27. Insertion of the expression for ρ gives

$$\frac{1}{r^2}\frac{\partial}{\partial r}\left(r^2 \frac{\partial \phi}{\partial r}\right) = \kappa^2 \phi \tag{7.39}$$

where the quantity κ (not to be confused with electrolytic conductivity) is given by

$$\kappa^2 = \frac{e^2}{\epsilon_0 \epsilon \mathbf{k} T} \sum_i N_i z_i^2 \tag{7.40}$$

The general solution of Eq. 7.39 is

$$\phi = \frac{A e^{-\kappa r}}{r} + \frac{B e^{\kappa r}}{r} \tag{7.41}$$

where A and B are constants. Since ϕ must approach zero as r becomes very large, and $e^{\kappa r}/r$ becomes very large as r becomes large, the multiplying factor B must be zero, and Eq. 7.41 thus becomes

$$\phi = \frac{A e^{-\kappa r}}{r} \tag{7.42}$$

The constant A can be evaluated by considering the situation when the solution is infinitely dilute, when the central ion can be considered to be isolated. In such a solution $\sum_i N_i z_i^2$ approaches zero, so that κ also approaches zero and the potential is therefore

$$\phi = \frac{A}{r} \tag{7.43}$$

In these circumstances, however, the potential at a distance r will simply be the potential due to the ion itself, since there is no interference by the atmosphere. The potential at a distance r due to an ion of charge z_c is

$$\phi = \frac{z_c e}{4\pi\epsilon_0 \epsilon r} \tag{7.44}$$

Equating these two expressions for ϕ leads to

$$A = \frac{z_c e}{4\pi\epsilon_0 \epsilon} \tag{7.45}$$

and insertion of this in Eq. 7.42 gives

$$\phi = \frac{z_c e \, e^{-\kappa r}}{4\pi\epsilon_0 \epsilon r} \tag{7.46}$$

If κ is sufficiently small (i.e., if the solution is sufficiently dilute), the exponential $e^{-\kappa r}$ is approximately $1 - \kappa r$, and Eq. 7.46 therefore becomes

$$\phi = \frac{z_c e}{4\pi\epsilon_0 \epsilon r} - \frac{z_c e \kappa}{4\pi\epsilon_0 \epsilon} \tag{7.47}$$

The first term on the right-hand side of this equation is the potential at a distance r, due to the central ion itself. The second term is therefore the potential produced by the ionic atmosphere:

$$\boldsymbol{\phi_a = -\frac{z_c e \kappa}{4\pi\epsilon_0 \epsilon}} \tag{7.48}$$

There is now no dependence on r; this potential is therefore uniform and exists at the central ion. If the ionic atmosphere were replaced by a charge $-z_c e$ situated at a distance $1/\kappa$ from the central ion, the effect due to it at the central ion would be exactly the same as that produced by the ionic atmosphere. The distance $1/\kappa$ is therefore referred to as the **thickness of the ionic atmosphere.**

The quantity κ is given by Eq. 7.40; the thickness of the ionic atmosphere is thus

$$\frac{1}{\kappa} = \left(\frac{\epsilon_0 \epsilon \mathbf{k} T}{e^2 \sum_i N_i z_i^2} \right)^{1/2} \tag{7.49}$$

Instead of the number of ions N_i per unit volume, it is usually more convenient to use the concentration c_i; $N_i = c_i L$, where L is the Avogadro constant, and thus

$$\boldsymbol{\frac{1}{\kappa} = \left(\frac{\epsilon_0 \epsilon \mathbf{k} T}{e^2 \sum_i c_i z_i^2 L} \right)^{1/2}} \tag{7.50}$$

This equation allows values of the thickness of the ionic atmosphere to be estimated.

EXAMPLE

Estimate the thickness of the ionic atmosphere for a solution of (a) 0.01 M NaCl and (b) 0.001 M $ZnCl_2$, both in water at 25°C, with $\epsilon = 78$.

SOLUTION

a. The summation in Eq. 7.50 is

$$\sum_i c_i z_i^2 = 0.01 \times 1^2 + 0.01 \times 1^2 = 0.02 \text{ mol dm}^{-3}$$

Insertion of the appropriate values into Eq. 7.50 then gives

$$\frac{1}{\kappa} = \left[\frac{8.854 \times 10^{-12}(\text{C}^2\text{ N}^{-1}\text{m}^{-2}) \times 78 \times 1.381 \times 10^{-23}(\text{J K}^{-1}) \times 300(\text{K})}{(1.602 \times 10^{-19})^2(\text{C}^2) \times 0.02(\text{mol dm}^{-3}) \times 10^3(\text{dm}^3\text{ m}^{-3}) \times 6.022 \times 10^{23}(\text{mol}^{-1})}\right]^{1/2}$$

$$= 3.04 \times 10^{-9} \text{ (J N}^{-1}\text{ m)}^{1/2}$$

Since $\text{J} = \text{kg m}^2\text{ s}^{-2}$ and $\text{N} = \text{kg m s}^{-2}$, the units are metres; the thickness is thus 3.04 nm (nanometres).

b. The summation is now

$$\sum_i c_i z_i^2 = 0.001 \times 2^2 + 0.002 \times 1^2 = 0.006 \text{ mol dm}^{-3}$$

and the result is

$$\frac{1}{\kappa} = 5.56 \times 10^{-9} \text{ m} = 5.56 \text{ nm}$$

■

Table 7.1 shows values of $1/\kappa$ for various types of electrolytes at different molar concentrations in aqueous solution. We see from Eq. 7.50 that the thickness of the ionic atmosphere is inversely proportional to the square root of the concentration; the atmosphere moves further from the central ion as the solution is diluted.

Mechanism of Conductivity

The treatment of conductivity is very difficult mathematically and only a very general account of the main ideas can be given here.

The effect of the ionic atmosphere is to exert a drag on the movement of a given ion. If the ion is stationary, the atmosphere is arranged symmetrically about it and does not tend to move it in either direction (see Figure 7.6a). However, if a potential that tends to move the ion to the right is applied, the atmosphere will decay to some

TABLE 7.1 Thickness of Ionic Atmospheres in Water at 25°C

	Molar Concentration		
Type of Electrolyte	0.10 *M*	0.01 *M*	0.001 *M*
Uni-univalent	0.962 nm	3.04 nm	9.62 nm
Uni-bivalent and bi-univalent	0.556 nm	1.76 nm	5.56 nm
Bi-bivalent	0.481 nm	1.52 nm	4.81 nm
Uni-tervalent and ter-univalent	0.392 nm	1.24 nm	3.92 nm

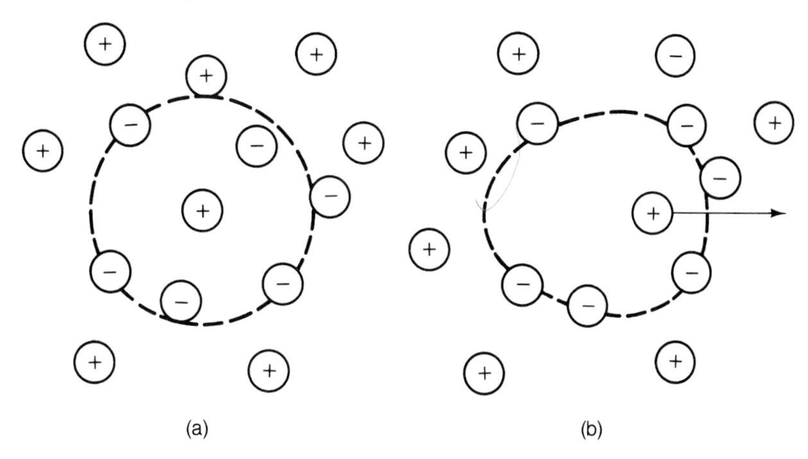

FIGURE 7.6 (a) A stationary central ion with a spherically symmetrical ion atmosphere. (b) A positive ion moving to the right. The ion atmosphere behind it is relaxing, while that in front is building up. The distribution of negative ions round the positive ion is now asymmetric.

extent on the left of the ion and build up more on the right (Figure 7.6b). Since it takes time for these *relaxation* processes to occur, there will be an excess of ionic atmosphere to the left of the ion (i.e., behind it) and a deficit to the right (in front of it). This asymmetry of the atmosphere will have the effect of dragging the central ion back. This is the **relaxation** or **asymmetry** effect.

There is a second reason why the existence of the ionic atmosphere impedes the motion of an ion. Ions are attracted to solvent molecules mainly by ion-dipole forces; therefore when they move, they drag solvent along with them. The ionic atmosphere, having a charge opposite to that of the central ion, moves in the opposite direction to it and therefore drags solvent in the opposite direction. This means that the central ion has to travel upstream, and it therefore travels more slowly than if there were no effect of this kind. This is the **electrophoretic** effect.

Debye and Hückel carried through a theoretical treatment that led them to expressions for the forces exerted upon the central ion by the relaxation effect. They supposed the ions to travel through the solution in straight lines, neglecting the zigzag Brownian motion brought about by the collisions of surrounding solvent molecules. This theory was improved in 1926 by the Norwegian-American physical chemist Lars Onsager* (1903–1976), who took Brownian motion into account and whose expression for the *relaxation force* f_r was

$$f_r = \frac{e^2 z_i \kappa}{24\pi\epsilon_0 \epsilon \mathbf{k} T} w V' \tag{7.51}$$

where V' is the applied potential gradient and w is a number whose magnitude depends on the type of electrolyte; for a uni-univalent electrolyte, w is $2 - \sqrt{2} = 0.586$.

The *electrophoretic force* f_e was given by Debye and Hückel as

$$f_e = \frac{e z_i \kappa}{6\pi\eta} K_c V' \tag{7.52}$$

*L. Onsager, *Physik. Z.*, *27*, 388 (1926).

where K_c is the coefficient of frictional resistance of the central ion with reference to the solvent and η is the viscosity of the medium. The viscosity enters into this expression because we are concerned with the motion of the ion past the solvent molecules, which depends on the viscosity of the solvent.

The final expression obtained on this basis for the molar conductivity Λ of an electrolyte solution of concentration c is

$$\Lambda = \Lambda_0 - (P + Q\Lambda_0)\sqrt{c} \tag{7.53}$$

where P and Q can be expressed in terms of various constants and properties of the system. For the particular case of a *symmetrical* electrolyte [i.e., one for which the two ions have equal and opposite signs ($z_+ = -z_- = z$)], P and Q are given by

$$P = \frac{zeF}{3\pi\eta}\left(\frac{2z^2e^2L}{\epsilon_0\epsilon \mathbf{k}T}\right)^{1/2} \tag{7.54}$$

and

$$Q = \frac{z^2e^2w}{24\pi\epsilon_0\epsilon \mathbf{k}T}\left(\frac{2z^2e^2L}{\pi\epsilon_0\epsilon \mathbf{k}T}\right)^{1/2} \tag{7.55}$$

Equation 7.53 is known as the **Debye-Hückel-Onsager equation**. It is based on the assumption of complete dissociation. With weak electrolytes such as acetic acid the ions are sufficiently far apart that the relaxation and electrophoretic effects are negligible; the Arrhenius-Ostwald treatment (Section 7.3) then applies. For electrolytes of intermediate strength a combination of the Debye-Hückel-Onsager and Arrhenius-Ostwald theories must be used.

A number of experimental tests have been made of the Debye-Hückel-Onsager equation. Equation 7.53 can be tested by seeing whether a plot of Λ against $\sqrt{c}$ is linear and whether the slopes and intercepts of the lines are consistent with Eqs. 7.54 and 7.55. For aqueous solutions of uni-univalent electrolytes, the theoretical equation is found to be obeyed very satisfactorily up to a concentration of about 2×10^{-3} mol dm^{-3}; at higher concentrations, deviations are found. The corresponding equations for other types of electrolytes in water are also obeyed satisfactorily at very low concentrations, but deviations are found at lower concentrations than with uni-univalent electrolytes.

Various attempts have been made to improve the Debye-Hückel treatment. One of these attempts involved taking ion association into account, a matter that is dealt with in the next subsection. Another attempt, made by Gronwall, La Mer, and co-workers,* was based on including higher terms in the expansions of the exponentials in Eq. 7.24. This led to improvement for ions of higher valence, but unfortunately the resulting equations are complicated and difficult to apply to the experimental results.

Another weakness of the Debye-Hückel theory is that the ions are treated as point charges; no allowance is made for the fact that they occupy a finite volume and cannot come close to each other. This difficulty was dealt with by E. Wicke and Manfred Eigen† in a manner similar to that of introducing the excluded volume b into the van der Waals equation for a real gas (Section 1.12).

*T. H. Gronwall, V. K. La Mer, and K. Sandved, *Physik, Z.*, *29*, 358 (1928); V. K. La Mer, T. H. Gronwall, and L. J. Greiff, *J. Phys. Chem.*, *35*, 2245 (1931).

†E. Wicke and M. Eigen, *Z. Elektrochem.*, *56*, 551 (1952); *57*, 319 (1953); *Z. Naturforsch.*, *8a*, 161 (1953).

Ion Association*

A common type of deviation from the Debye-Hückel-Onsager conductivity equation is for the negative slopes of the Λ versus $\sqrt{c}$ plots to be greater than predicted by the theory; that is, the experimental conductivities are lower than predicted by the theory. This result, and also certain anomalies found with ion activities (Section 7.10), led to the suggestion that the assumption of complete dissociation is not strictly valid for strong electrolytes. In the case of an electrolyte such as sodium chloride it is quite correct to say that there are no covalent NaCl molecules present in aqueous solution. However, because of the strong electrostatic attractions between Na^+ and Cl^- ions, it is possible for pairs of ions to become *associated* in solution, just as Na^+Cl^- ion pairs are present in the vapor of sodium chloride. In the case of an electrolyte such as NaCl, in which the ions have equal charges, the ion pairs have no net charge and therefore make no contribution to the conductivity, thus accounting for the anomalously low conductivity. It is important to distinguish clearly between ion association and covalent bond formation; the latter is more permanent, whereas in ion association there is a constant interchange between the various ions in the solution.

The term *ion association* was first employed by Neils Bjerrum (1879–1958), who in 1926 developed a theory to explain the anomalies found with strong electrolytes.† The basis of the theory is that ions of opposite signs separated in solution by a distance r^* form an associated ion pair held together by coulombic forces. The calculation of Bjerrum's distance r^* is based on the Boltzmann equations (Eqs. 7.21 and 7.22) and is somewhat involved. The final result is that, with $z_c = 1$,

$$r^* = \frac{e^2}{8\pi\epsilon_0\epsilon \mathbf{k}T} \tag{7.56}$$

The value of this at 25°C, with $\epsilon = 78.3$, is equal to 3.58×10^{-10} m = 0.358 nm. Substitution of this distance r^* into the electrostatic potential energy of two ions (based on Eq. 7.44 with $z_c = z_i = 1$) gives (see Problem 29)

$$E^* = 2\mathbf{k}T \tag{7.57}$$

Therefore, the electrostatic potential energy at this distance r^* is four times the mean kinetic energy per degree of freedom. The energy at this distance is therefore sufficient for there to be significant ion association, which will be dynamic in character in that there will be a rapid exchange with the surrounding ions.

At distances smaller than 0.358 nm the probability of ionic association increases rapidly. Therefore, to a good approximation one can say that if the ions are closer than 0.358 nm, they can be considered to be "undissociated" and to make no contribution to the conductivity. On the basis of these ideas, Bjerrum made some estimates of the degree of ionic association under various conditions.

The theory is easily extended to ions of higher valences. If the charges on the ions are z_c and z_i, Eq. 7.56 is replaced by

$$r^* = \frac{z_c z_i e^2}{8\pi\epsilon_0\epsilon \mathbf{k}T} \tag{7.58}$$

---

*For a review, see G. Kortüm, *Treatise on Electrochemistry*, Amsterdam: Elsevier Pub. Co., 1965, Chapter 6.

†N. Bjerrum, *Kgl. Danske Videnskab. Selskab; Mat.–Fys. Medd.*, *7*, 9 (1926).

and in aqueous systems association can then be regarded as occurring when the ions approach to within a distance of $|z_c z_i|$0.358 nm. Univalent ions can rarely approach one another as closely as 0.358 nm, and ion association is therefore of little importance for such ions in aqueous solution. With ions of higher valence, ion association is much more important.

It follows from Eq. 7.58 that r^* values will be greater and ionic association therefore more important in solvents of low dielectric constant. For example, dioxane has a dielectric constant of 2.2 at 25°C, and the r^* value for a uni-univalent electrolyte is 12.7 nm; ion association is therefore expected to be important.

With salts having ions of different valences, such as Na_2SO_4, ion association will lead to the formation of species such as $Na^+SO_4^{2-}$. Again there will be a reduction in conductivity, since these species will carry less current than the free ions. Bjerrum's theory has been extended to deal with such unsymmetrical electrolytes and also with the formation of triple ions.

Conductivity at High Frequencies and Potentials

An important consequence of the existence of the ionic atmosphere is that the conductivity should depend on the frequency if an alternating potential is applied to the solution. Suppose that the alternating potential is of sufficiently high frequency that the time of oscillation is small compared with the time it takes for the ionic atmosphere to relax. There will then not be time for the atmosphere to relax behind the ion and to form in front of it; the ion will be virtually stationary and its ionic atmosphere will remain symmetrical. Therefore, as the frequency of the potential increases, the relaxation and electrophoretic effects will become less and less important, and there will be an increase in the molar conductivity.

A related effect is observed if conductivities are measured at very high potential gradients. For example, if the applied potential is 20 000 V cm^{-1}, an ion will move at a speed of about 1 m s^{-1} and will travel several times the thickness of the effective ionic atmosphere in the time of relaxation of the atmosphere. Consequently, the moving ion is essentially free from the effect of the ionic atmosphere, which does not have time to build up around it to any extent. Therefore, at sufficiently high voltages, the relaxation and electrophoretic effects will diminish and eventually disappear and the molar conductivity will increase. This effect was first observed experimentally by the German physicist Max Carl Wien* (1866–1938) in 1927 and is known as the **Wien effect.** Unfortunately, molar conductivities of weak electrolytes at high potentials are anomalously large, and it appears that very high potentials bring about a dissociation of the molecules into ions. This phenomenon is known as the **dissociation field effect.**

7.5 Independent Migration of Ions

In principle the plots of Λ against concentration (Figure 7.2) can be extrapolated back to zero concentration to give the Λ_0 value. In practice this extrapolation can only satisfactorily be made with strong electrolytes. With weak electrolytes there is

*M. C. Wien, *Ann. Physik* (4) *83*, 327 (1927); *85*, 795 (1928); (5) *1*, 408 (1929).

TABLE 7.2 Molar Conductivities at Infinite Dilution for Various Sodium and Potassium Salts in Aqueous Solution at 25°C

Electrolyte	$\frac{\Lambda_0}{\Omega^{-1}\ \text{cm}^2\ \text{mol}^{-1}}$	Electrolyte	$\frac{\Lambda_0}{\Omega^{-1}\ \text{cm}^2\ \text{mol}^{-1}}$	Difference
KCl	149.9	NaCl	126.5	23.4
KI	150.3	NaI	126.9	23.4
$\frac{1}{2}K_2SO_4$	153.5	$\frac{1}{2}Na_2SO_4$	130.1	23.4

a strong dependence of Λ on c at low concentrations and therefore the extrapolations do not lead to reliable Λ_0 values. An indirect method for obtaining Λ_0 for weak electrolytes is considered later.

In 1875 Kohlrausch made a number of determinations of the Λ_0 values and observed that they exhibited certain regularities.* Some values are given in Table 7.2 for corresponding sodium and potassium salts. We see that the difference between the molar conductivities of a potassium and a sodium salt of the same anion is independent of the nature of the anion. Similar results were obtained for a variety of pairs of salts with common cations or anions, in both aqueous and nonaqueous solvents.

This behavior was explained in terms of **Kohlrausch's law of independent migration of ions.** Each ion is assumed to make its own contribution to the molar conductivity, irrespective of the nature of the other ion with which it is associated. In other words,

$$\boldsymbol{\Lambda_0 = \lambda_+^\circ + \lambda_-^\circ} \tag{7.59}$$

where λ_+° and λ_-° are the ion conductivities of cation and anion, respectively, at infinite dilution. Thus for potassium chloride (see Table 7.2)

$$\Lambda_0(\text{KCl}) = \lambda_{\text{K}^+}^\circ + \lambda_{\text{Cl}^-}^\circ = 149.9\ \Omega^{-1}\ \text{cm}^2\ \text{mol}^{-1} \tag{7.60}$$

and for sodium chloride

$$\Lambda_0(\text{NaCl}) = \lambda_{\text{Na}^+}^\circ + \lambda_{\text{Cl}^-}^\circ = 126.5\ \Omega^{-1}\ \text{cm}^2\ \text{mol}^{-1} \tag{7.61}$$

The difference, $23.4\ \Omega^{-1}\ \text{cm}^2\ \text{mol}^{-1}$, is thus the difference between the λ_+° values for K^+ and Na^+:

$$\lambda_{\text{K}^+}^\circ - \lambda_{\text{Na}^+}^\circ = 23.4\ \Omega^{-1}\ \text{cm}^2\ \text{mol}^{-1} \tag{7.62}$$

and this will be the same whatever the nature of the anion.

*F. Kohlrausch and O. N. A. Grotrian, *Ann. Phys., 154,* 1 (1875); F. Kohlrausch, *Gott. Nachr.,* 213 (1876); *Ann. Phys., 6,* 145 (1879).

Ionic Mobilities

It is not possible from the Λ_0 values alone to determine the individual λ_+° and λ_-° values; however, once one λ_+° or λ_-° value (such as $\lambda_{Na^+}^\circ$) has been determined, the rest can be calculated. The λ_+° and λ_-° values are proportional to the speeds with which the ions move under standard conditions, and this is the basis of the methods by which the individual conductivities are determined. The **mobility** of an ion, u, is defined as the speed with which the ion moves under a unit potential gradient. Suppose that the potential drop across the opposite faces of a unit cube is V and that the concentration of univalent positive ions in the cube is c_+ (see Figure 7.7). If the mobility of the positive ions is u_+, the speed of the ions is u_+V. All positive ions within a distance of u_+V of the negative plate will therefore reach that plate in unit time, and the number of such ions is u_+Vc_+. The charge they carry is Fu_+Vc_+, and since this is carried in unit time, the current is Fu_+Vc_+. The electrolytic conductivity due to the positive ions, κ_+, is thus

$$\kappa_+ = \frac{\text{current}}{\text{potential drop}} = \frac{Fu_+Vc_+}{V} = Fu_+c_+ \tag{7.63}$$

The molar conductivity due to these positive ions is therefore

$$\boldsymbol{\lambda_+^\circ = \frac{\kappa_+}{c_+} = Fu_+} \tag{7.64}$$

The SI unit of mobility is m s^{-1}/V m^{-1} = m^2 V^{-1} s^{-1}, but it is more common to use the unit cm^2 V^{-1} s^{-1}.

EXAMPLE

The mobility of a sodium ion in water at 25°C is 5.19×10^{-4} cm^2 V^{-1} s^{-1}. Calculate the molar conductivity of the sodium ion.

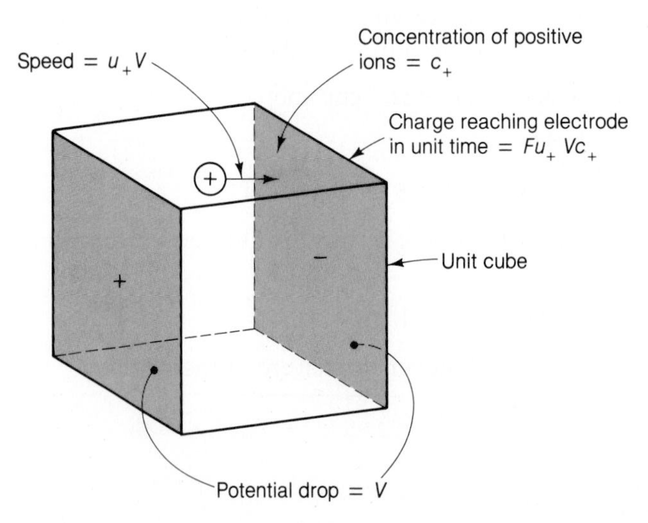

FIGURE 7.7 A unit cube containing a solution in which the concentration of positive ions is c_+, and where there is a potential drop of V between the opposite faces.

SOLUTION

From Eq. 7.64,

$$\lambda^\circ_{Na^+} = 96\,500 \text{ A s mol}^{-1} \times 5.19 \times 10^{-4} \text{ cm}^2 \text{ V}^{-1} \text{ s}^{-1}$$
$$= 50.1 \text{ A V}^{-1} \text{ cm}^2 \text{ mol}^{-1}$$

A V^{-1} is Ω^{-1}; therefore

$$\lambda^\circ_{Na^+} = 50.1 \ \Omega^{-1} \text{ cm}^2 \text{ mol}^{-1}$$

7.6 Transport Numbers

In order to split the Λ° values into the λ°_+ and λ°_- values for the individual ions, we make use of a property known variously as the **transport number,** the *transference number,* or the *migration number.* It is the *fraction of the current carried by each ion* present in solution.

Consider an electrolyte of formula A_aB_b that ionizes as follows:

$$A_aB_b \rightleftharpoons a\,A^{z_+ +} + b\,B^{z_- -}$$

The quantity of electricity carried by a mol of the ions $A^{z_+ +}$, whose charge number is z_+, is aFz_+, and the quantity of electricity crossing a given cross-sectional area in unit time is aFz_+u_+, where u_+ is the mobility. This quantity is the current carried by the positive ions when 1 mol of the electrolyte is present in the solution. Similarly, the current carried by b mol of the negative ions is bFz_-u_-. The fraction of the current carried by the positive ions is therefore

$$t_+ = \frac{aFz_+u_+}{aFz_+u_+ + bFz_-u_-} = \frac{az_+u_+}{az_+u_+ + bz_-u_-} \tag{7.65}$$

This is the transport number of the positive ions. Similarly, the transport number of the negative ions is

$$t_- = \frac{bz_-u_-}{az_+u_+ + bz_-u_-} \tag{7.66}$$

In order for the solution to be electrically neutral it is necessary that

$$az_+ = bz_- \tag{7.67}$$

and Eq. 7.65 and 7.66 then reduce to

$$t_+ = \frac{u_+}{u_+ + u_-} \quad \textbf{and} \quad t_- = \frac{u_-}{u_+ + u_-} \tag{7.68}$$

The individual molar ion conductivities are given by

$$\lambda(A^{z_+ +}) = Fz_+u_+ \quad \text{and} \quad \lambda(B^{z_- -}) = Fz_-u_- \tag{7.69}$$

The molar conductivity of the electrolyte is

$$\Lambda(A_aB_b) = F(az_+u_+ + bz_-u_-) \tag{7.70}$$

It then follows that the transport number t_+ is related to the individual ion conductivities by

$$t_+ = \frac{\lambda(A^{z_+ +})/z_+}{\lambda(A^{z_+ +})/z_+ + \lambda(B^{z_- -})/z_-} \tag{7.71}$$

Together with Eq. 7.67 this gives

$$t_+ = \frac{a\lambda(A^{z_+ +})}{a\lambda(A^{z_+ +}) + b\lambda(B^{z_- -})} = \frac{a\lambda(A^{z_+ +})}{\Lambda(A_aB_b)} \tag{7.72}$$

Similarly, for t_-,

$$t_- = \frac{b\lambda(B^{z_- -})}{a\lambda(A^{z_+ +}) + b\lambda(B^{z_- -})} = \frac{b\lambda(B^{z_- -})}{\Lambda(A_aB_b)} \tag{7.73}$$

If t_+ and t_- can be measured over a range of concentrations, the values at zero concentration, t_+° and t_-°, can be obtained by extrapolation. These values allow $\Lambda^\circ(A_aB_b)$ to be split into the individual ion conductivities $\lambda^\circ(A^{z_+ +})$ and $\lambda^\circ(B^{z_- -})$ by use of Eqs. 7.72 and 7.73.

At first sight it may appear surprising that Faraday's laws, according to which equivalent quantities of different ions are liberated at the two electrodes, can be reconciled with the fact that the ions are moving at different speeds toward the electrodes. Figure 7.8, however, shows how these two facts can be reconciled. The diagram shows in a very schematic way an electrolysis cell in which there are equal numbers of positive and negative ions of unit charge. The situation before electrolysis occurs is shown in Figure 7.8a. Suppose that the cations only were able to move, the anions having zero mobility; after some motion has occurred, the situation will be as represented in Figure 7.8b. At each electrode two ions remain unpaired and are discharged, two electrons at the same time traveling in the outer circuit from the anode to the cathode. Thus, although only the cations have moved through the bulk of the solution, equivalent amounts have been discharged at the two electrodes. It is easy to extend this argument to the case in which both ions are moving but at different speeds. Thus Figure 7.8c shows the situation when the speed of the cation is three times that of the anion, four ions being discharged at each electrode.

This problem was considered by Hittorf, and Figure 7.8d is reproduced from one of his publications.

Hittorf Method

Two experimental methods have been employed for the determination of transport numbers. One of them, developed by the German physicist Johann Wilhelm Hittorf (1824–1914) in 1853, involves measuring the changes of concentration in the vicinity of the electrodes. In the second, the **moving boundary method,** a study is made of the rate of movement, under the influence of a current, of the boundary between two solutions. This method is described on page 279. A third method, involving the measurement of the electromotive force of certain cells, is considered in Section 8.5.

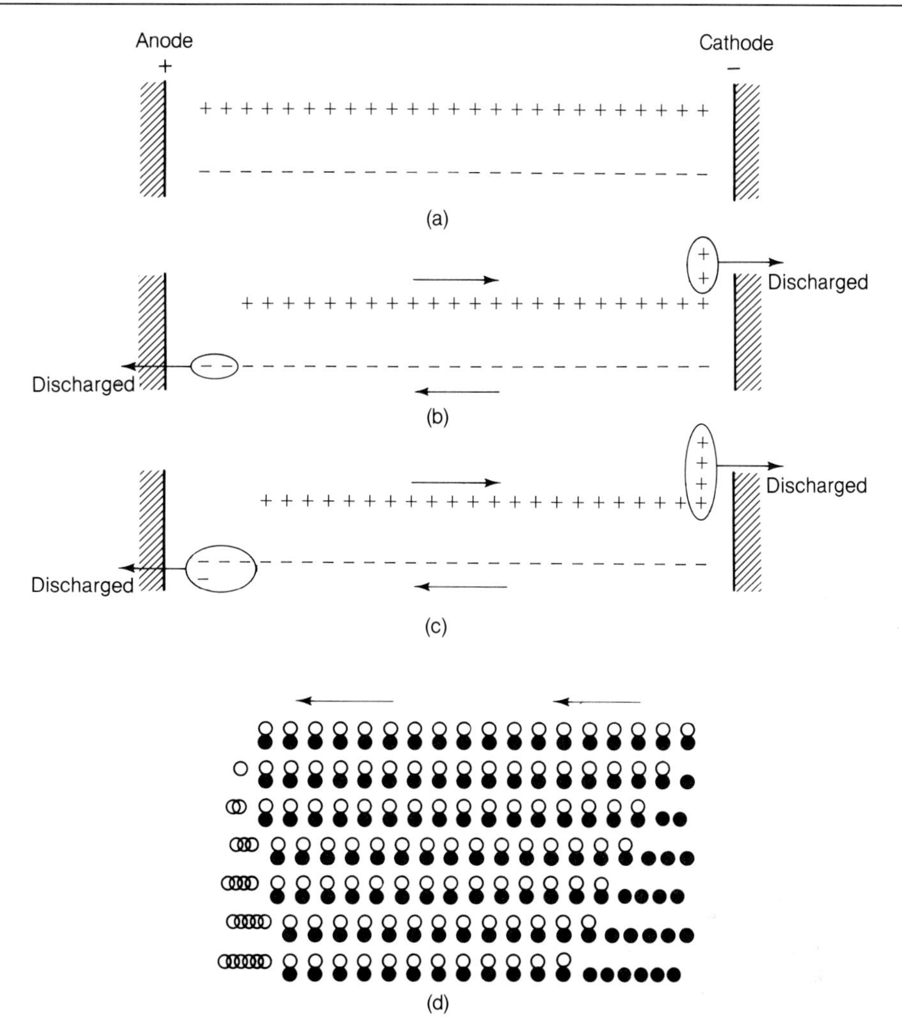

FIGURE 7.8 Schematic representation of the movement of ions during an electrolysis experiment, showing that equivalent amounts are neutralized at the electrodes in spite of differences in ionic velocities. Diagram (d) is reproduced from an original publication by Hittorf.

A very simple type of apparatus used in the **Hittorf method** is shown in Figure 7.9. The solution to be electrolyzed is placed in the cell, and a small current is passed between the electrodes for a short period of time. Solution then is run out through the stopcocks, and the samples are analyzed for concentration changes.

The theory of the method is as follows. Suppose that the solution contains the ions M^+ and A^-, which are not necessarily univalent but are denoted as such for simplicity. The fraction of the total current carried by the cations is t_+, and that carried by the anions is t_-. Thus, when 96 500 C of electricity passes through the solution, 96 500 t_+ C are carried in one direction by t_+ mol of M^+ ions, and 96 500 t_- C are carried in the other direction by t_- mol of A^- ions. At the same time 1 mol of each ion is discharged at an electrode.

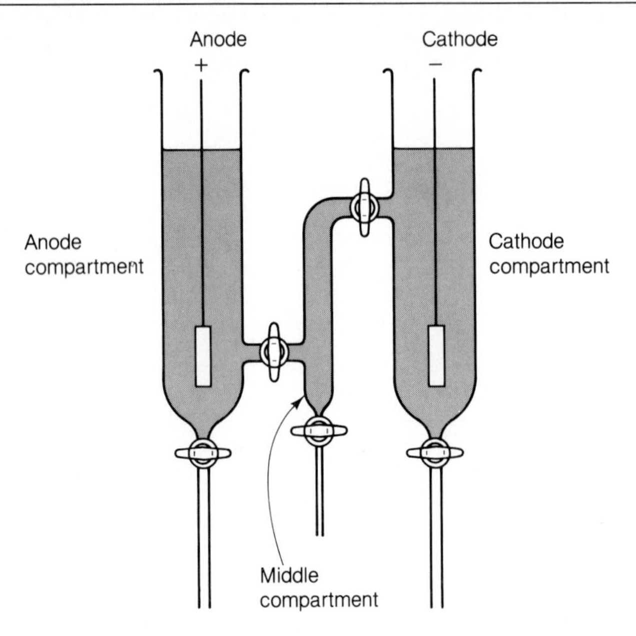

FIGURE 7.9 Simple type of apparatus used for measuring transference numbers by the Hittorf method.

Suppose that, as represented in Figure 7.10, the cell containing the electrolyte is divided into three compartments by hypothetical partitions. One is a compartment near the cathode, one is near the anode, and the middle compartment is one in which no concentration change occurs. It is supposed that the two ions are discharged at the electrodes, which are unchanged in the process. The changes that occur when 96 500 C is passed through the solution are shown in Figure 7.10. The results are

At the cathode: A loss of $1 - t_+ = t_-$ mol of M^+, and a loss of t_- mol of A^- (i.e., a loss of t_- mol of the electrolyte MA).
In the middle compartment: No concentration change.
At the anode: Loss of $1 - t_- = t_+$ mol of A^-; loss of t_+ mol of M^+; the net loss is therefore t_+ mol of MA.

It follows that

$$\frac{\text{amount lost from anode compartment}}{\text{amount lost from cathode compartment}} = \frac{t_+}{t_-} \tag{7.74}$$

Then, since $t_+ + t_- = 1$, the values of t_+ and t_- can be calculated from the experimental results. Alternatively,

$$\frac{\text{amount lost from anode compartment}}{\text{amount deposited}} = t_+ \tag{7.75}$$

and

$$\frac{\text{amount lost from cathode compartment}}{\text{amount deposited}} = t_- \tag{7.76}$$

Any of these three equations allows t_+ and t_- to be determined.

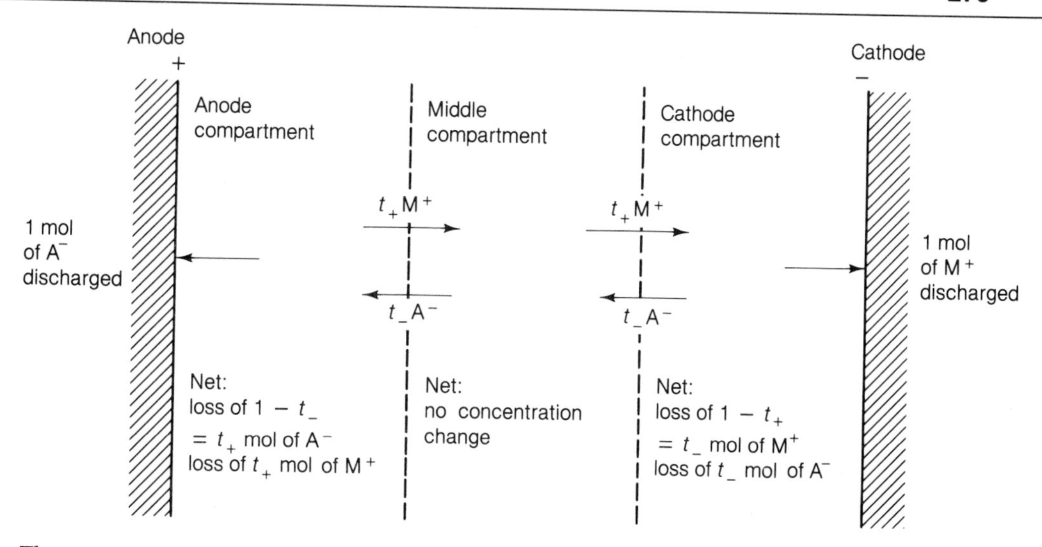

FIGURE 7.10 The concentration changes in three compartments of a conductivity cell (Hittorf method).

In the preceding discussion we have assumed that the electrodes are inert (i.e., are not attacked as electrolysis proceeds) and that the ions M^+ and A^- are deposited. If instead, for example, the anode were to pass into solution during electrolysis, the concentration changes would be correspondingly different; the treatment can readily be modified for this and other situations.

Moving Boundary Method

The moving boundary method was developed in 1886 by the British physicist Sir Oliver Joseph Lodge (1851–1940)* and in 1893 by the British physicist Sir William Cecil Dampier (formerly Whetham) (1867–1952).†

The method is illustrated in Figure 7.11. Suppose that it is necessary to measure the transport numbers of the ions in the electrolyte MA. Two other electrolytes M′A and MA′ are selected as "indicators"; each has an ion in common with MA, and the electrolytes are such that M'^+ moves more slowly than M^+ and A'^- moves more slowly than A^-. The solution of MA is placed in the electrolysis tube with the solution of M′A on one side of it and that of MA′ on the other; the electrode in M′A is the anode and that in MA′ is the cathode. As the current flows, the boundaries remain distinct, since the M'^+ ions cannot overtake the M^+ ions at the boundary a, and the A'^- ions cannot overtake the A^- ions at boundary b. The slower ions M'^+ and A'^- are known as the *following* ions. The boundaries a and b move, as shown in Figure 7.11a, to positions a' and b'. The distances aa' and bb' are proportional to the ionic velocities, and therefore

$$\frac{aa'}{aa' + bb'} = t_+ \quad \text{and} \quad \frac{bb'}{aa' + bb'} = t_- \tag{7.77}$$

*O. J. Lodge, *British Association Reports*, 389 (1886).

†W. C. D. Whetham, *Phil. Trans. Roy. Soc.*, *184*, 337 (1893); *186*, 507 (1895); *Phil. Mag.*, *38*, 392 (1894); *A Treatise on the Theory of Solution*, Cambridge University Press, 1902.

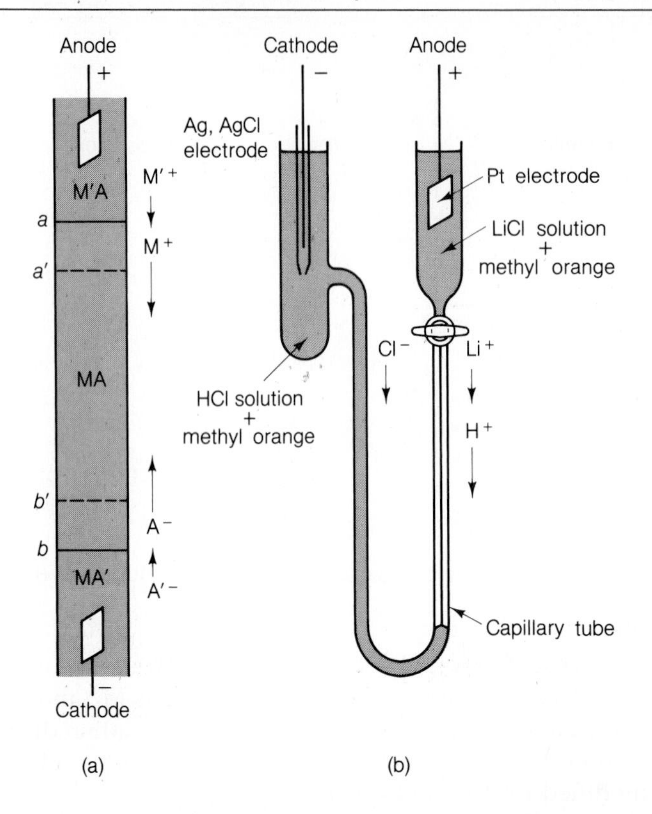

FIGURE 7.11 The determination of transport numbers by the moving boundary method. (a) Schematic diagram showing the movement of the ions and the boundaries. (b) Simple apparatus for measuring the transport number of H^+, using Li^+ as the following cation. A clear boundary is formed between the receding acid solution and the LiCl solution and is easily observed if methyl orange is present.

Alternatively, the movement of only one boundary may be followed, as shown by the example in Figure 7.11b. If Q is the quantity of electricity passed through the system, t_+Q is the quantity of electricity carried by the positive ions, and the amount of positive ions to which this corresponds is t_+Q/F. If the concentration of positive ions is c, the amount t_+Q/F occupies a volume of t_+Q/Fc. This volume is equal to the area of cross section of the tube A multiplied by the distance aa' through which the boundary moves. Thus it follows that

$$aa' = \frac{t_+Q}{FcA} \tag{7.78}$$

Since aa', Q, F, c, and A are known, the transport number t_+ can be calculated.

7.7 Ion Conductivities

With the use of the transport numbers the Λ_0 values can be split into λ_+° and λ_-°, the contributions for the individual ions; the transport numbers extrapolated to infinite dilution are

$$t_+ = \frac{\lambda_+^\circ}{\Lambda_0} \quad \text{and} \quad t_- = \frac{\lambda_-^\circ}{\Lambda_0} \tag{7.79}$$

Once a single λ_+° or λ_-° value has been determined, a complete set can be calculated from the available Λ_0 values. For example, suppose that a transport number study on NaCl led to $\lambda_{Na^+}^\circ$ and $\lambda_{Cl^-}^\circ$. If Λ_0 for KCl is known, the value of $\lambda_{K^+}^\circ$ [$=\Lambda_0(KCl) - \lambda_{Cl^-}^\circ$] can be obtained; from $\lambda_{K^+}^\circ$ and $\Lambda_0(KBr)$ the value of $\lambda_{Br^-}^\circ$ can be calculated, and so on. A set of values obtained in this way is given in Table 7.3. Symbols such as $\frac{1}{2}Ca^{2+}$, $\frac{1}{2}SO_4^{2-}$ are used in the case of the polyvalent ions.

An important use for individual ion conductivities is in determining the Λ_0 values for weak electrolytes. We have seen (see Figure 7.2) that for a weak electrolyte the extrapolation to zero concentration is rather a long one, with the result that a reliable figure cannot be obtained. For acetic acid, however, Λ_0 is $\lambda_{H^+}^\circ + \lambda_{CH_3COO^-}^\circ$; from the values in Table 7.3 the value of Λ_0 is thus $349.8 + 40.9 = 390.7\ \Omega^{-1}\ cm^2\ mol^{-1}$.

An equivalent but alternative procedure avoids the necessity of splitting the Λ_0 values into the individual ionic contributions. The molar conductivity Λ_0 of any electrolyte MA can be expressed, for example, as

$$\Lambda_0(MA) = \Lambda_0(MCl) + \Lambda_0(NaA) - \Lambda_0(NaCl) \tag{7.80}$$

If MCl, NaA and NaCl are all strong electrolytes, their Λ_0 values are readily obtained by extrapolation and therefore permit the calculation of Λ_0 for MA, which

TABLE 7.3 Individual Molar Ionic Conductivities in Water at 25°C

Cation	λ_+° / Ω^{-1} cm^2 mol^{-1}	Anion	λ_-° / Ω^{-1} cm^2 mol^{-1}
H^+	349.8	OH^-	198.6
Li^+	38.6	F^-	55.4
Na^+	50.1	Cl^-	76.4
K^+	73.5	Br^-	78.1
Rb^+	77.8	I^-	76.8
Cs^+	77.2	CH_3COO^-	40.9
Ag^+	61.9	$\frac{1}{2}SO_4^{2-}$	80.0
Tl^+	74.7	$\frac{1}{2}CO_3^{2-}$	69.3
$\frac{1}{2}Mg^{2+}$	53.1		
$\frac{1}{2}Ca^{2+}$	59.5		
$\frac{1}{2}Sr^{2+}$	59.5		
$\frac{1}{2}Ba^{2+}$	63.6		
$\frac{1}{2}Cu^{2+}$	56.6		
$\frac{1}{2}Zn^{2+}$	52.8		
$\frac{1}{3}La^{3+}$	69.7		

To calculate the mobility u in cm^2 V^{-1} s^{-1}, divide the λ° by 96 500 (see Eq. 7.64 and the example on p. 274).

may be a weak electrolyte. The following example illustrates the use of this method for acetic acid:

$$\Lambda_0(CH_3COOH) = \Lambda_0(HCl) + \Lambda_0(CH_3COONa) - \Lambda_0(NaCl) \tag{7.81}$$

$$= 426.2 + 91.0 - 126.5 = 390.7\ \Omega^{-1}\ cm^2\ mol^{-1} \quad \text{at } 25°C$$

This procedure is also useful for a highly insoluble salt, for which direct determinations of Λ_0 values might be impractical.

Ionic Solvation

The magnitudes of the individual ion conductivities (Table 7.3) are of considerable interest. There is no simple dependence of the conductivity on the size of the ion. It might be thought that the smallest ions, such as Li^+, would be the fastest moving, but the values for Li^+, K^+, Rb^+, and Cs^+ show that this is by no means true; K^+ moves faster than either Li^+ or Na^+. The explanation is that Li^+, because of its small size, becomes strongly attached to about four surrounding water molecules by ion-dipole and other forces, so that when the current passes, it is $Li(H_2O)_4^+$ and not Li^+ that moves. With Na^+ the binding of water molecules is less strong, and with K^+ it is still weaker. This matter of the solvation of ions is discussed further in Section 7.9.

Mobilities of Hydrogen and Hydroxide Ions

Of particular interest is the very high conductivity of the hydrogen ion. This ion has an abnormally high conductivity in a number of hydroxylic solvents such as water, methanol, and ethanol, but it behaves more normally in nonhydroxylic solvents such as nitrobenzene and liquid ammonia. At first sight the high values might appear to be due to the small size of the proton. However, there is a powerful electrostatic attraction between a water molecule and the proton, which because of its small size can come very close to the water molecule. As a consequence the equilibrium

$$H^+ + H_2O \rightleftharpoons H_3O^+$$

lies very much to the right. In other words, there are very few free protons in water; the ions exist as H_3O^+ ions, which are hydrated by other water molecules (see Figure 7.19). Thus there remains a difficulty in explaining the very high conductivity and mobility of the hydrogen ion. By virtue of its size it would be expected to move about as fast as the Na^+ ion, which in fact is the case in the nonhydroxylic solvents.

In order to explain the high mobilities of hydrogen ions in hydroxylic solvents such as water, a special mechanism must be invoked. As well as moving through the solution in the way that other ions do, the H_3O^+ ion can also transfer its proton to a neighboring water molecule:

$$H_3O^+ + H_2O \longrightarrow H_2O + H_3O^+$$

The resulting H_3O^+ ion can now transfer a proton to another H_2O molecule. In other words, protons, although not free in solution, can be passed from one water molecule to another. Calculations from the known structure of water show that the proton must jump a distance of 86 pm from an H_3O^+ ion to a water molecule but that as a result the proton moves effectively through a distance of 310 pm. The conductivity by this mechanism therefore will be much greater than by the normal mechanism. This proton transfer must be accompanied by some rotation of H_3O^+ and H_2O molecules in order for them to be positioned correctly for the next proton transfer. Similar types of mechanisms have been proposed for the hydroxylic solvents. In methyl alcohol, for example, the process is

$$\underset{\text{H}\quad\text{H}}{\overset{\text{CH}_3}{\text{O}^+}} + \underset{\quad\text{H}}{\overset{\text{CH}_3}{\text{O}}} \longrightarrow \underset{\text{H}\quad}{\overset{\text{CH}_3}{\text{O}}} + \underset{\text{H}\quad\text{H}}{\overset{\text{CH}_3}{\text{O}^+}}$$

Mechanisms of this type bear some resemblance to a mechanism that was suggested in 1805 by the German physicist Christian Johann Dietrich von Grotthuss (1785–1822)* as a general explanation of conductance and are frequently referred to as **Grotthuss mechanisms.**

Ionic Mobilities and Diffusion Coefficients

Ions also migrate in the absence of an electric field, if there is a concentration gradient. The migration of a substance when there is a concentration gradient is known as **diffusion,** a topic that will be considered in further detail in Chapter 18. It will there be seen that the tendency of a substance to move in a concentration gradient is measured in terms of a *diffusion coefficient D.* In 1888 the German physicist Walter Herman Nernst (1864–1941) showed that the relationship between the diffusion coefficient and the mobility u of an ion is

$$D = \frac{\mathbf{k}T}{Q}u \tag{7.82}$$

where Q is the charge on the ion. This equation is derived in Section 18.2. Since the molar ionic conductivity λ is equal to Fu and the charge Q is equal to $F|z|/L$, this equation can be written as

$$D = \frac{L\mathbf{k}T\lambda}{F^2|z|} = \frac{RT\lambda}{F^2|z|} \tag{7.83}$$

These relationships show that diffusion experiments, as well as conductivity measurements, can provide information about mobilities and hence about molar conductivities. Equations 7.82 and 7.83 relate to a single ionic species but, in practice, experiments must involve at least two types of ions. Diffusion experiments with a uni-univalent electrolyte A^+B^-, such as NaCl, will be concerned with the diffusion of the two ions A^+ and B^-. The overall diffusion constant D must then relate to the individual diffusion constants D_{A^+} and D_{B^-}, and Nernst showed that for uni-univalent electrolytes the proper average is

$$D = \frac{2D_{A^+}D_{B^-}}{D_{A^+} + D_{B^-}} \tag{7.84}$$

*C. J. D. von Grotthuss, *Mémoire sur la décomposition de l'eau et des corps, qu'elle tient en dissolution, à l'aide de l'électricité galvanique,* Rome, 1805; Milan, 1806; *Ann. Chim., 58,* 54 (1806); *Phil. Mag., 25,* 330 (1806).

Walden's Rule

An important relationship between molar conductivity and viscosity was discovered in 1906 by Paul Walden (1863–1957). In the course of a study of the conductivity of tetramethylammonium iodide in various solvents, Walden noticed that the product of the molar conductivity at infinite dilution and the viscosity η of the solvent was approximately constant:

$$\Lambda_0 \eta = \text{constant} \tag{7.85}$$

Similarly, it has been shown that the product of individual ion conductivities and viscosity is also a constant. Such relationships are most satisfactory for ions that are approximately spherical.

7.8 Thermodynamics of Ions

In thermodynamic work it is customary to obtain values of standard enthalpies and Gibbs energies of species. These quantities relate to the formation of 1 mol of the substance from the elements in their standard states and usually refer to 25.0°C. In the case of entropies it is common to obtain absolute values based on the Third Law of Thermodynamics.

It is a comparatively straightforward matter to determine standard enthalpies and Gibbs energies of formation and absolute entropies, for pairs of ions in solution, for example, for a solution of sodium chloride. We cannot, however, carry out experiments on ions of one kind. The conventional procedure is to set the value for H^+ as zero. This allows complete sets of values to be built up, and one then speaks of *conventional standard enthalpies of formation*, of *conventional Gibbs energies of formation*, and of *conventional absolute entropies*.

A few such values for ions have been included in Table 2.1 (p. 63); more complete compilations are given in Table 7.4. This table includes values of enthalpies, Gibbs energies, and entropies of hydration; the subscript hyd is used to identify these quantities. These values relate the thermodynamic properties of the ions in water to their properties in the gas phase and are therefore important in leading to an understanding of the effect of the surrounding water molecules. The Gibbs energy of hydration is the change in Gibbs energy when an ion is transferred from the gas phase into aqueous solution. Again, the convention is that the values for the hydrogen ion H^+ are all zero.

Whereas absolute thermodynamic values for individual ions cannot be measured directly, they can be estimated on the basis of theory. Unfortunately, owing to the large number of interactions involved, the theoretical treatment of an ion in aqueous solution is very difficult. The following absolute values are generally agreed to be not far from the truth, for the proton:

$$\Delta_{\text{hyd}}H^\circ(\text{absolute}) = -1090.8 \text{ kJ mol}^{-1}$$

$$\Delta_{\text{hyd}}S^\circ(\text{absolute}) = -131.8 \text{ J K}^{-1} \text{ mol}^{-1}$$

$$\Delta_{\text{hyd}}G^\circ(\text{absolute}) = -1051.4 \text{ kJ mol}^{-1}$$

Once these values are accepted, absolute values for the other ions can be calculated from their conventional values.

TABLE 7.4 Conventional Standard Enthalpies, Gibbs Energies, and Entropies of Formation and Hydration of Individual Ions at 25° C

Ion	$\Delta_f H°$ / kJ mol^{-1}	$\Delta_{hyd} H°$ / kJ mol^{-1}	$\Delta_f G°$ / kJ mol^{-1}	$\Delta_{hyd} G°$ / kJ mol^{-1}	$S°$ / J K^{-1} mol^{-1}	$\Delta_{hyd} S°$ / J K^{-1} mol^{-1}
H^+	0	0	0	0	0	0
Li^+	−278.7	576.1	−293.7	579.1	14.2	10.0
Na^+	−239.7	685.3	−261.9	679.1	60.2	21.3
K^+	−251.0	769.9	−282.4	751.4	102.5	56.9
Rb^+	—	—	−282.4	774.0	124.3	69.0
Cs^+	—	—	−282.0	806.3	133.1	72.0
Ag^+	105.9	615.5	77.1	610.9	77.0	15.5
Mg^{2+}	—	—	−456.1	274.1	−118.0	−49.0
Ca^{2+}	−543.1	589.1	−553.1	586.6	−55.2	7.5
Sr^{2+}	—	—	−557.3	732.6	−39.2	14.2
Ba^{2+}	−538.5	879.1	−560.7	861.5	—	—
Al^{3+}	—	—	−481.2	−1346.4	−313.4	−136.8
F^-	—	—	−276.6	−1524.2	−9.6	−264.0
Cl^-	−167.4	−1469.0	−131.4	−1407.1	55.2	−207.1
Br^-	−120.9	−1454.8	−102.9	−1393.3	80.8	−191.6
I^-	−56.1	−1397.0	−51.9	−1346.8	109.2	−168.6
OH^-	−230.1	−1552.3	—	—	—	—

Standard enthalpies of formation and standard Gibbs energies of formation are from a National Bureau of Standards compilation "Selected Values of Chemical Thermodynamical Properties," N.B.S. Circular 500 (1952); heats of hydration, Gibbs energies of hydration, and entropies of hydration are from D. R. Rossinsky, *Chem. Rev.*, 65, 467 (1965).

EXAMPLE

Calculate the absolute enthalpies of hydration of Li^+, I^-, and Cu^{2+} on the basis of a value of −1090.8 kJ mol^{-1} for the proton, using the values listed in Table 7.4.

SOLUTION:

The enthalpy of hydration of a uni-univalent electrolyte M^+X^- is independent of whether conventional or absolute ionic values are used. Thus, for H^+I^-, if we lower the value of H^+ by 1090.8, we must raise that for I^- by the same amount. Thus the absolute value for I^- will be

$$-1397.0 + 1090.8 = -306.2 \text{ kJ mol}^{-1}$$

Similarly, the value for Li^+ must be lowered by 1090.8; the absolute value is thus

$$576.1 - 1090.8 = -514.7 \text{ kJ mol}^{-1}$$

To obtain the value for Cu^{2+}, we may consider a salt like CuI_2 ($Cu^{2+} + 2I^-$). Since we have raised the value for *each* I^- ion by 1090.8, we must lower the Cu^{2+} value by 2×1090.8; thus, $83.3 - (2 \times 1090.8) = -2098.3$ kJ mol^{-1}. ■

7.9 Theories of Ions in Solution

Born's Model

A simple interpretation of the thermodynamic quantities for ions in solution was suggested by the German-British physicist Max Born (1882–1970).* In Born's model the solvent is assumed to be a continuous dielectric and the ion a conducting sphere. Born obtained an expression for the work of charging such a sphere, which is the Gibbs energy change during the charging process. The total reversible work in transporting increments until the sphere has a charge of ze is, according to Born's model,

$$w_{\text{rev}} = \frac{z^2e^2}{8\pi\epsilon_0\epsilon r} \tag{7.86}$$

This work, being non-PV work, is the electrostatic contribution to the Gibbs energy of the ion:

$$G^\circ_{\text{es}} = \frac{z^2e^2}{8\pi\epsilon_0\epsilon r} \tag{7.87}$$

If the same charging process is carried out in vacuo ($\epsilon = 1$), the electrostatic Gibbs energy is

$$G^\circ_{\text{es}}(\text{vacuum}) = \frac{z^2e^2}{8\pi\epsilon_0 r} \tag{7.88}$$

The electrostatic Gibbs energy of hydration is therefore

$$\Delta_{\text{hyd}}G^\circ = G^\circ_{\text{es}} - G^\circ_{\text{es}}(\text{vacuum}) = \frac{z^2e^2}{8\pi\epsilon_0 r}\left(\frac{1}{\epsilon} - 1\right) \tag{7.89}$$

For water at 25°C, $\epsilon = 78$, and Eq. 7.89 leads to

$$\Delta_{\text{hyd}}G^\circ = -68.6\frac{z^2}{r/\text{nm}}\ \text{J mol}^{-1} \tag{7.90}$$

Figure 7.12 shows a plot of absolute $\Delta_{\text{hyd}}G^\circ$ values against z^2/r. In view of the simplistic nature of the model, the agreement with the predictions of the treatments is not unsatisfactory; the theory accounts quite well for the main effects. Note, however, that the agreement with the theory is worse the higher the charge and the smaller the radius. On the whole the percentage difference is greatest with trivalent ions and least with univalent ions. Figure 7.12 also shows the prediction for $\epsilon = 2$, which is approximately the dielectric constant of water when it is subjected to an intense field. The line for $\epsilon = 2$ is closer to the points for ions of high charge and small radius. The significance of this will be discussed later.

The Born equation also leads to a simple interpretation of entropies of hydration and of absolute entropies of ions. The thermodynamic relationship between entropy and Gibbs energy (see Eq. 3.119) is

*M. Born, *Z. Physik.*, *1*, 45 (1920).

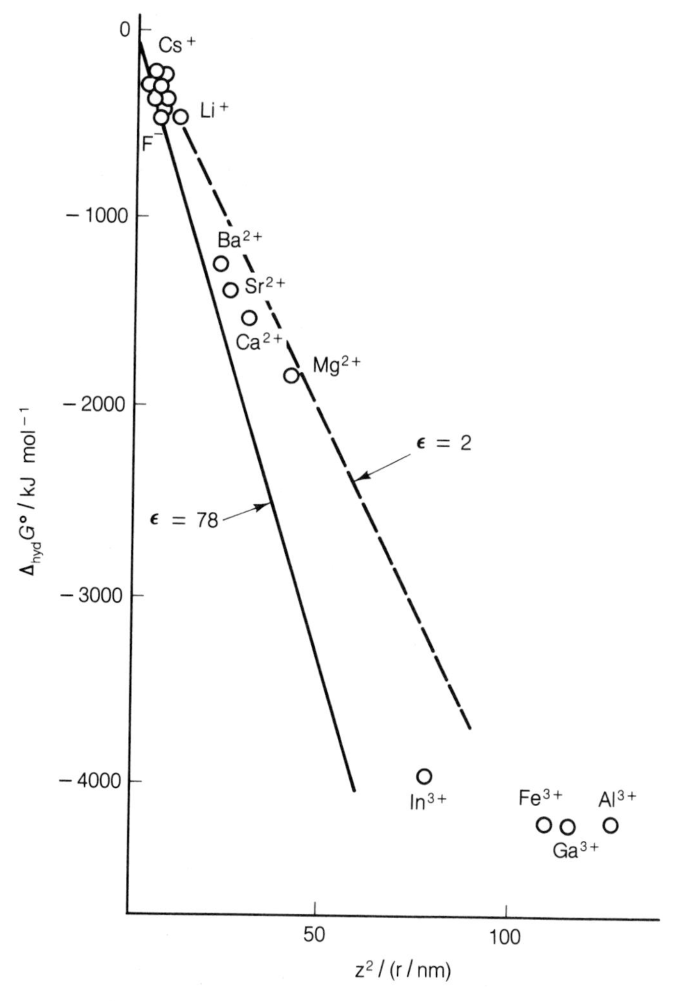

FIGURE 7.12 A plot of the absolute Gibbs energy of hydration of ions against z^2/r. The firm line is the theoretical line for a dielectric constant of 78; the dashed line is for a dielectric constant of 2.

$$S = -\left(\frac{\partial G}{\partial T}\right)_P \tag{7.91}$$

The only quantity in Eq. 7.89 that is temperature dependent is the dielectric constant ϵ, so that

$$S^\circ_{es} = \Delta_{hyd} S^\circ = \frac{z^2 e^2}{8\pi\epsilon_0 \epsilon r} \frac{\partial \ln \epsilon}{\partial T} \tag{7.92}$$

The reason that the theory leads to the same expression for the absolute entropy of the ion and for its entropy of hydration is that it gives zero entropy for the ion in the gas phase, since the Gibbs energy expression in Eq. 7.88 contains no temperature-dependent terms. For water at 25°C, ϵ is approximately 78 and $(\partial \ln \epsilon)/\partial T$ is about -0.0046 K^{-1} over a considerable temperature range. Insertion of these values into Eq. 7.92 leads to

$$S^\circ = \Delta_{\text{hyd}} S^\circ = -4.10 \frac{z^2}{r/\text{nm}} \text{ J K}^{-1} \text{ mol}^{-1} \tag{7.93}$$

Figure 7.13 shows a plot of the experimental $\Delta_{\text{hyd}} S^\circ$ values, together with a line of slope −4.10. The line has been drawn through a value of −42 J K^{-1} mol^{-1} at $z^2/r = 0$, since it is estimated that there will be a $\Delta_{\text{hyd}} S^\circ$ value of about −42 J K^{-1} mol^{-1} due to nonelectrostatic effects. In particular, there is an entropy loss resulting from the fact that the ion is confined in a small solvent "cage," an effect that is ignored in the electrostatic treatment.

Again, the simple model of Born accounts satisfactorily for the main effects. The deviations are primarily for small ions.

More Advanced Theories

Various attempts have been made to improve Born's simple treatment while still regarding the solvent as continuous. One way of doing this is to consider how the effective dielectric constant of a solvent varies in the neighborhood of an ion. The dielectric behavior of a liquid is related to the tendency of ions to orient themselves in an electric field. At very high field strengths the molecules become fully aligned in the field and there can be no further orientation. The dielectric constant then falls to a very low value of about 2, an effect that is known as *dielectric saturation*.

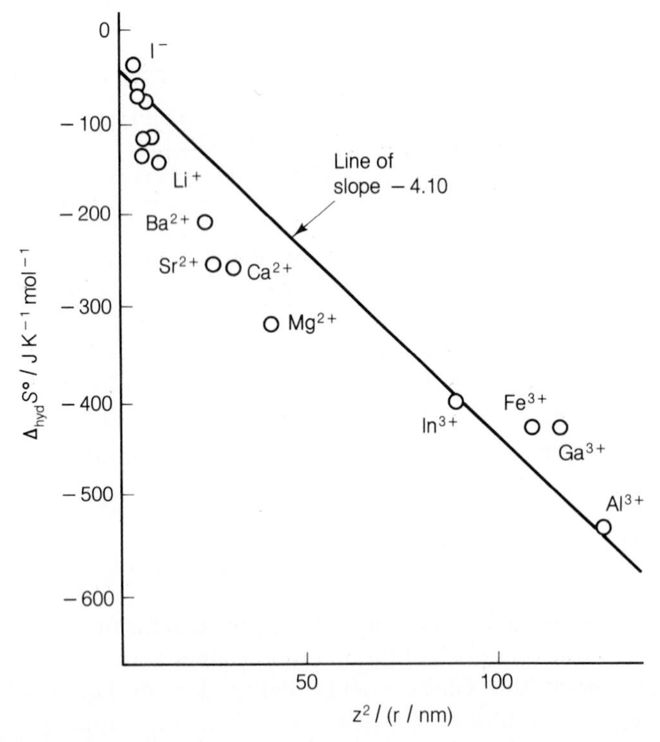

FIGURE 7.13 Plot of entropy of hydration against $z^2/(r/\text{nm})$.

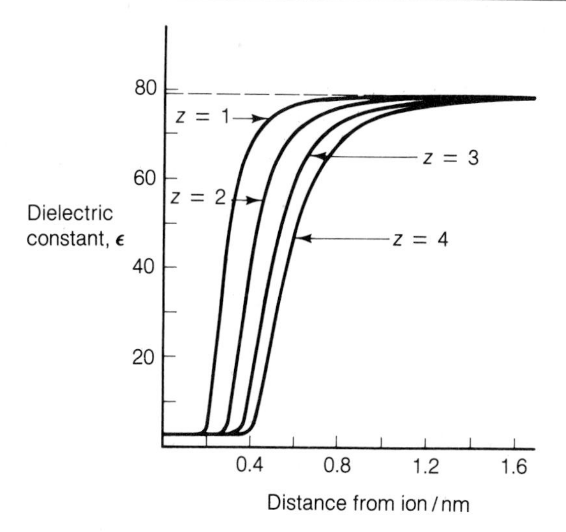

FIGURE 7.14 The variation of dielectric constant of water in the neighborhood of ions.

In this way it is possible to estimate that the effective dielectric constant in the neighborhood of ions of various types will be as shown in Figure 7.14. For example, for a ferric (Fe^{3+}) ion the dielectric constant has a value of about 1.78 up to a distance of about 0.3 nm from the center of the ion, owing to the high field produced by the ion. At greater distances, where the field is less, the effective dielectric constant rises toward its limiting value of 78.

When the Born treatments of Gibbs energies and entropies of hydration of ions are modified by taking these dielectric saturation effects into account, the agreement with experiment is considerably better. The experimental values in Figure 7.12 are consistent with a dielectric constant of somewhat less than the true value of 78, and this can be understood in terms of the reduction in dielectric constant arising from dielectric saturation.

Qualitative Treatments

On the basis of these quantitative theories it is possible to arrive at useful qualitative pictures of ions in solution. The main attractive force between an ion and a surrounding water molecule is the ion-dipole attraction. The way in which a water molecule is expected to orient itself in the neighborhood of a positive or negative ion is shown in Figure 7.15. There are two possible orientations for a negative ion, and it appears that both may play a role. Quite strong binding can result from ion-dipole forces; the binding energy for an Na^+ ion and a neighboring water molecule, for example, is about 80 kJ mol^{-1}.

The smaller the ion, the greater the binding energy between the ion and a water molecule, because the water molecule can approach the ion more closely. Thus, the binding energy is particularly strong for the Li^+ ion, and calculations suggest that a Li^+ ion in water is surrounded tetrahedrally by four water molecules oriented as in Figure 7.15a; this is shown in Figure 7.16. Thus we can say that Li^+ has a *hydration number* of 4. The four water molecules are sufficiently strongly attached to the Li^+ ion that during electrolysis they are dragged along with it. For this reason (Table

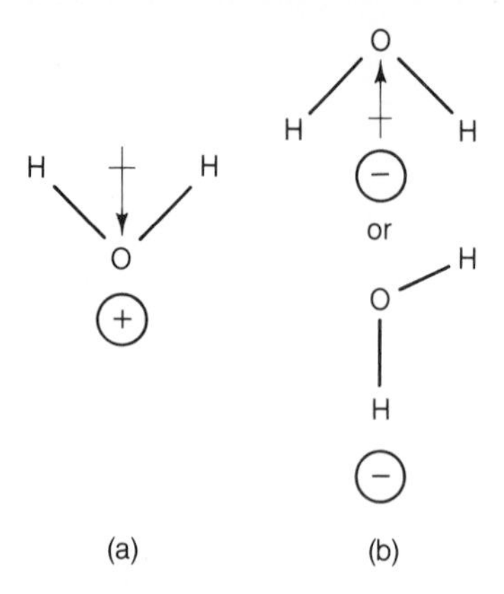

FIGURE 7.15 Orientation of a water molecule. (a) Close to a positive ion. (b) Close to a negative ion.

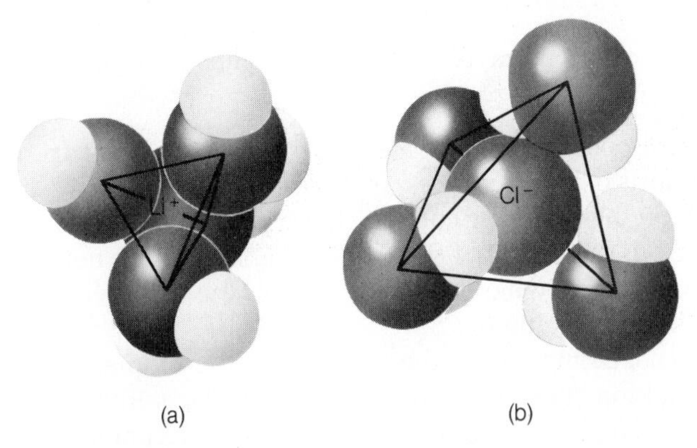

FIGURE 7.16 The tetrahedral arrangement of four water molecules. (a) Around a Li^+ ion. (b) Around a Cl^- ion that can probably also have five or six water molecules around it.

7.3) Li^+ has a lower mobility than Na^+ and K^+; the latter ions, being larger, have a smaller tendency to drag water molecules. However, the situation is not so clear-cut as just implied. A moving Li^+ ion drags not only its four neighboring water molecules but also some of the water molecules that are further away from it.

A very useful treatment of ions in solution has been developed by the American physical chemist Henry S. Frank and his co-workers.* They have concluded that for ions in aqueous solution it is possible to distinguish three different zones in the neighborhood of the solute molecule, as shown in Figure 7.17. In the immediate vicinity of the ion there is a shell of water molecules that are more or less immobilized by the very high field due to the ion. These water molecules can be described as constituting the *inner hydration shell* and are sometimes described as an *iceberg*, since their structure has some icelike characteristics. However, this analogy should not be pressed too far: ice is less dense than liquid water, whereas the iceberg around the ion is more compressed than normal liquid water.

---

*H. S. Frank and M. W. Evans, *J. Chem. Phys.*, *13*, 507 (1945); H. S. Frank and W. Y. Wen, *Disc. Faraday Soc.*, *24*, 133 (1957).

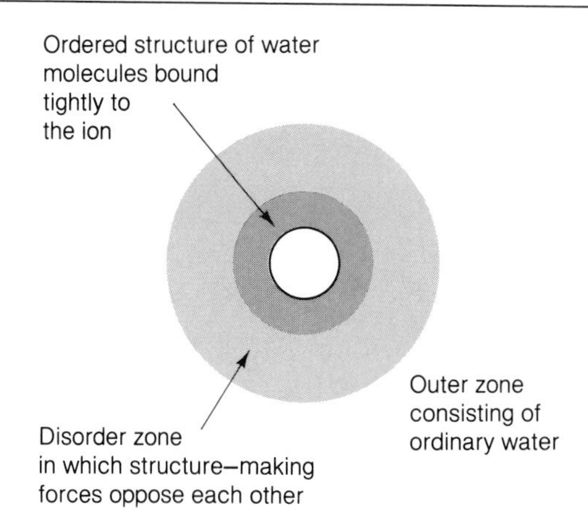

FIGURE 7.17 H. S. Frank's concept of the structure of water in the neighborhood of an ion.

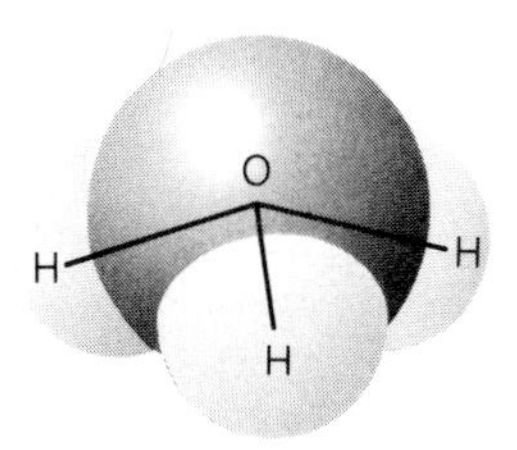

FIGURE 7.18 The structure of the oxonium (H_3O^+) ion.

Surrounding this iceberg there is a second region, which Frank and his co-workers refer to as a region of *structure breaking*. Here the water molecules are oriented more randomly than in ordinary water, where there is considerable ordering due to hydrogen bonding. The occurrence of this structure-breaking region results from two competing orienting influences on each water molecule. One of these is the normal structural effect of the neighboring water molecules; the other is the orienting influence upon the dipole of the spherically symmetrical ionic field.

The third region around the ion comprises all the water sufficiently far from the ion that its effect is not felt.

The behavior of the proton H^+ in water deserves some special discussion. Because of its small size, the proton attaches itself very strongly to a water molecule,

$$H^+ + H_2O \rightleftharpoons H_3O^+$$

The equilibrium for this process, which is strongly exothermic, lies very far to the right. Thus, the hydrogen ion is frequently regarded as existing as the H_3O^+ ion, which is known as the **oxonium ion.** Experimental studies and theoretical calculations have indicated that the angle between the O-H bonds in H_3O^+ is about 115°. The ion is thus almost, but not quite, flat, as shown in Figure 7.18.

However, the state of the hydrogen ion in water is more complicated than implied by this description. Because of ion-dipole attractions, other molecules are held quite closely to the H_3O^+ species. The most recent calculations tend to support the view that three water molecules are held particularly strongly, in the manner

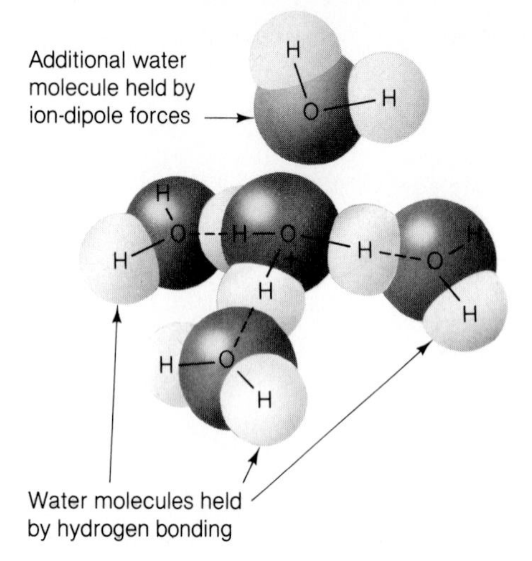

FIGURE 7.19 A possible structure for the hydrated proton. The hydrated H_3O^+ ion (e.g., $H_9O_4^+$) is called the *hydronium ion*.

shown in Figure 7.19. These three molecules are held by hydrogen bonds involving three hydrogen atoms of the H_3O^+ species. The hydrated proton may thus be written as $H_3O^+ \cdot 3H_2O$, or as $H_9O_4^+$. Alternatively, some think that the species $H_9O_4^+$ has the same kind of structure as the hydrated Li^+ ion (Figure 7.15a) in which the proton is surrounded tetrahedrally by four water molecules. Other water molecules will also be held to $H_9O_4^+$ but not so strongly. Thus in Figure 7.19 there is shown an additional H_2O molecule held by ion-dipole forces but not by hydrogen bonding, and it therefore is not held so strongly as the three other water molecules. IUPAC has recommended that the term **hydronium ion** be employed for these hydrated H_3O^+ ions, in contrast to *oxonium ion* for H_3O^+ itself.

The situation is hardly clear-cut; we can write the hydrated proton as H^+, H_3O^+, $H_9O_4^+$, or $H_{11}O_5^+$. It is satisfactory to write it simply as H^+, as long as we remember that the proton is strongly hydrated.

7.10 Activity Coefficients

The electrostatic interactions between ions, besides having an important effect on the conductivities of solutions of strong electrolytes, have an effect on the thermodynamic properties of ions. This matter is most conveniently dealt with in terms of **activity coefficients.** Several experimental methods are available for determining the activity coefficients of ions—see Sections 7.11 and 8.5 (emf measurements). First we will apply the Debye-Hückel treatment to ionic activity coefficients.

Debye-Hückel Limiting Law

If there were no electrostatic interactions, the behavior would be ideal, and the Gibbs energy of an ion of type i would be given by the relationship

$$G_i = G_i^\circ + \mathbf{k}T \ln c_i \tag{7.94}$$

where c_i is the concentration of the ion. In reality the Gibbs energy is

$$G_i = G_i^\circ + \mathbf{k}T \ln c_i y_i \tag{7.95}$$

$$= G_i^\circ + \mathbf{k}T \ln c_i + \mathbf{k}T \ln y_i \tag{7.96}$$

The additional term $\mathbf{k}T \ln y_i$ is due to the presence of the ionic atmosphere; y is the activity coefficient used with concentrations.

We have seen that the work of charging an isolated ion, on the basis of the Born model, Eq. 7.86, is

$$w = \frac{z_i^2 e^2}{8\pi\epsilon_0 \epsilon r} \tag{7.97}$$

We also need the work of charging the ionic atmosphere, because this is the required correction to the Gibbs energy; this work is equal to $\mathbf{k}T \ln y_i$. If ϕ is the potential due to the ionic atmosphere, the work of transporting a charge dQ to the ion is $\phi\, dQ$; the net work of charging the ionic atmosphere is thus

$$w = \int_0^{z_i e} \phi\, dQ \tag{7.98}$$

The potential ϕ corresponding to the charge Q is given by Eq. 7.48 as $-Q\kappa/4\pi\epsilon_0\epsilon$, where $1/\kappa$ is the radius of the ionic atmosphere. The work is thus

$$w = -\int_0^{z_i e} \frac{Q\kappa}{4\pi\epsilon_0\epsilon}\, dQ = -\frac{Q^2\kappa}{8\pi\epsilon_0\epsilon}\bigg|_0^{z_i e} = -\frac{z_i^2 e^2 \kappa}{8\pi\epsilon_0\epsilon} \tag{7.99}$$

Note that Eq. 7.99 can be obtained from Eq. 7.97 by replacing r by $1/\kappa$ and changing the sign. The reciprocal $1/\kappa$ plays the same role for the atmosphere as does r for the ion, and the change of sign is required because the net charge on the atmosphere is opposite to that on the ion.

The expression for the activity coefficient y_i is therefore

$$\mathbf{k}T \ln y_i = -\frac{z_i^2 e^2 \kappa}{8\pi\epsilon_0\epsilon} \tag{7.100}$$

whence

$$\ln y_i = -\frac{z_i^2 e^2 \kappa}{8\pi\epsilon_0\epsilon \mathbf{k}T} \tag{7.101}$$

or

$$\log_{10} y_i = -\frac{z_i^2 e^2 \kappa}{8 \times 2.303\pi\epsilon_0\epsilon \mathbf{k}T} \tag{7.102}$$

Equation 7.40 shows that κ is proportional to the square root of the quantity $\sum_i N_i z_i^2$, where N_i is the number of the ions of the ith type per unit volume and z_i is its valency.

The **ionic strength** I of a solution is defined as

$$I = \frac{1}{2} \sum_i c_i z_i^2 \tag{7.103}$$

where c_i is the *molar* concentration of the ions of type i.* The ionic strength is proportional to $\sum_i N_i z_i^2$, and the reciprocal of the radius of the ionic atmosphere, κ, is thus proportional to $\sqrt{I}$ (see Eq. 7.49). Equation 7.102 may thus be written as

$$\log_{10} y_i = -z_i^2 B\sqrt{I} \tag{7.104}$$

where B is a quantity that depends on properties such as ϵ and T. When water is the solvent at 25°C, the value of B is 0.51 $mol^{-1/2}$ $dm^{3/2}$.

Experimentally we cannot measure the activity coefficient or indeed any thermodynamic property of a simple ion, since at least two types of ions must be present in any solution. To circumvent this difficulty we define a **mean activity coefficient** $y_\pm$ in terms of the individual values for y_+ and y_- by the relationship

$$y_\pm^{\nu_+ + \nu_-} = y_+^{\nu_+} y_-^{\nu_-} \tag{7.105}$$

where ν_+ and ν_- are the numbers of ions of the two kinds produced by the electrolyte. For example, for $ZnCl_2$, $\nu_+ = 1$ and $\nu_- = 2$. For a uni-univalent electrolyte ($\nu_+ = \nu_- = 1$) the mean activity coefficient is the geometric mean $(y_+ y_-)^{1/2}$ of the individual values.

In order to express $y_\pm$ in terms of the ionic strength we proceed as follows. From Eq. 7.105

$$(\nu_+ + \nu_-)\log_{10} y_\pm = \nu_+ \log_{10} y_+ + \nu_- \log_{10} y_- \tag{7.106}$$

Insertion of the expression in Eq. 7.104 for $\log y_+$ and $\log y_-$ gives

$$(\nu_+ + \nu_-)\log_{10} y_\pm = -(\nu_+ z_+^2 + \nu_- z_-^2)B\sqrt{I} \tag{7.107}$$

For electrical neutrality†

$$\nu_+ z_+ = \nu_- |z_-| \quad \text{or} \quad \nu_+^2 z_+^2 = \nu_-^2 z_-^2 \tag{7.108}$$

and therefore

$$(\nu_+ + \nu_-)\log_{10} y_\pm = -\nu_+^2 z_+^2\left(\frac{1}{\nu_+} + \frac{1}{\nu_-}\right)B\sqrt{I} \tag{7.109}$$

Thus

$$\log_{10} y_\pm = -\frac{\nu_+ z_+^2}{\nu_-}B\sqrt{I} = -z_+|z_-|B\sqrt{I} \tag{7.110}$$

For aqueous solutions at 25°C,

$$\mathbf{\log_{10} y_\pm = -0.51 z_+|z_-|\sqrt{I/\text{mol dm}^{-3}}} \tag{7.111}$$

Equation 7.111 is known as the **Debye-Hückel limiting law** (DHLL).

---

*For a uni-univalent electrolyte such as NaCl the ionic strength is equal to the molar concentration. Thus for a 1 *M* solution $c_+ = 1$ and $c_- = 1$, $z_+ = 1$ and $z_- = -1$; hence

$$\frac{1}{2}\sum_i c_i z_i^2 = \frac{1}{2}(1 + 1) = 1\ M$$

For a 1 *M* solution of an uni-bivalent electrolyte such as K_2SO_4, $c_+ = 2$, $c_- = 1$, $z_+ = 1$, and $z_- = -2$; hence the ionic strength is $\frac{1}{2}(2 \times 1 + 1 \times 4) = 3\ M$. Similarly, for a 1 *M* solution of a uni-trivalent electrolyte such as Na_3PO_4, the ionic strength is 6 *M*.

†Note that in this treatment we use the positive sign for the valency of negative ions (e.g., $|z_{Cl^-}| = +1$).

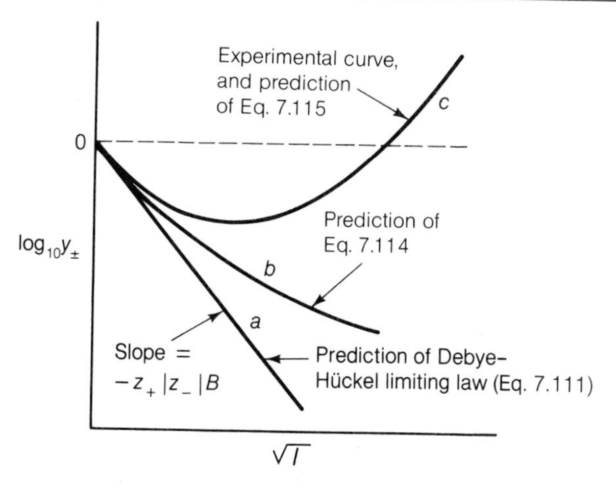

FIGURE 7.20 Variation of $\log_{10} y_\pm$ with the square root of the ionic strength.

Deviations from the Debye-Hückel Limiting Law

Experimentally the DHLL is found to apply satisfactorily only at extremely low concentrations*; at higher concentrations there are very significant deviations, as shown schematically in Figure 7.20. Equation 7.111 predicts that a plot of $\log_{10} y_\pm$ against $\sqrt{I}$ will have a negative slope, the magnitude of the slope being $-z_+|z_-|B$. The results with a number of electrolytes have shown that at very low I values $\log_{10} y_\pm$ does indeed fall linearly with I, with the correct slope; for this reason Eq. 7.111 is very satisfactory as a *limiting* law. At higher $\sqrt{I}$ values, however, the value of $\log_{10} y_\pm$ becomes significantly less negative than predicted by the law and, at sufficiently high ionic strengths, may actually attain positive values. The significance of this in connection with the solubilities of salts is considered in Section 7.11.

Various theories have been put forward to explain these deviations from the DHLL, but here they can be considered only briefly. One factor that has been neglected in the development of the equations is the fact that the ions occupy space; thus far they have been treated as point charges with no restrictions on how closely they can come together. If the theory is modified in such a way that the centers of the ions cannot approach one another closer than the distance a, the expression for the activity coefficient of an individual ion (compare Eq. 7.101) becomes

$$\ln y_i = -\frac{z_c^2 e^2 \kappa}{8\pi\epsilon_0\epsilon \mathbf{k}T} \cdot \frac{1}{1 + \kappa a} \tag{7.112}$$

Since κ is proportional to $\sqrt{I}$, this equation can be written as

$$\log_{10} y_i = -\frac{z_c^2 B\sqrt{I}}{1 + aB'\sqrt{I}} \tag{7.113}$$

*For example, for a uni-univalent electrolyte such as KCl, the law is obeyed satisfactorily up to a concentration of about 0.01 M; for other types of electrolytes, deviations appear at even lower concentrations.

(compare Eq. 7.104), where B is the same constant as used previously and B' is a new constant. The corresponding equation for the mean activity coefficient $y_{\pm}$ is

$$\log_{10} y_{\pm} = -\frac{z_{+}|z_{-}|B\sqrt{I}}{1 + aB'\sqrt{I}} \tag{7.114}$$

(compare Eq. 7.110). Equation 7.114 leads to the curve shown as b in Figure 7.20; however, it cannot explain the positive values of $\log_{10} y_{\pm}$ that are obtained at high ionic strengths.

In order to explain these positive values it is necessary to take account of the orientation of solvent molecules by the ionic atmosphere. This was considered by Hückel*, who showed that it gives a term linear in I in the expression for $\log_{10} y_{\pm}$:

$$\log_{10} y_{\pm} = -\frac{Bz_{+}|z_{-}|\sqrt{I}}{1 + aB'\sqrt{I}} + CI \tag{7.115}$$

where C is a constant. At sufficiently high ionic strengths the last term (CI) predominates, and $\log_{10} y_{\pm}$ is approximately linear in I, as found experimentally. The CI term is often known as the "salting-out" term since, as seen in Section 7.11, it accounts for the lowered solubilities of salts at high ionic strengths.

7.11 Ionic Equilibria

Equilibrium is usually established very rapidly between ionic species in solution. In this section an account will be given of some of the more important ways in which activity coefficients can be determined when equilibria are established in solution. Only some special topics are presented here; for a detailed account of how calculations can be made for fairly complicated systems, the reader is referred to the book by J. N. Butler in *Suggested Reading* (p. 306).

Activity Coefficients from Equilibrium Constant Measurements

Equilibrium constant determinations can provide values of activity coefficients. The procedure may be illustrated with reference to the dissociation of acetic acid,

$$CH_3COOH \rightleftharpoons H^+ + CH_3COO^-$$

The true equilibrium constant is

$$K = \frac{[H^+][CH_3COO^-]}{[CH_3COOH]} \cdot \frac{y_{+}y_{-}}{y_u} \tag{7.116}$$

where y_{+} and y_{-} are the activity coefficients of the ions and y_u is that of the undissociated acid. In reasonably dilute solution the undissociated acid will behave ideally ($y_u = 1$), but y_{+} and y_{-} may be significantly different from unity because of the electrostatic interactions. Replacement of $y_{+}y_{-}$ by $y_{\pm}^2$ and taking logarithms of Eq. 7.116 leads to†

---

*E. Hückel, *Physik. Z.*, *26*, 93 (1925).

†In Eqs. 7.117 and 7.118 we again use the superscript " to indicate the numerical value of the quantity.

$$\log_{10}\left(\frac{[H^+][CH_3COO^-]}{[CH_3COOH]}\right)^u = \log_{10} K^u - 2\log_{10} y_\pm \tag{7.117}$$

The left-hand side can be written as

$$\log_{10}\left(\frac{c\alpha^2}{1-\alpha}\right)^u = \log_{10} K^u - 2\log_{10} y_\pm \tag{7.118}$$

where c is the concentration and α is the degree of dissociation, which can be determined from conductivity measurements. (See Eq. 7.11). Values of the left-hand side of this equation can therefore be calculated for a variety of concentrations, and these values are equal to $\log_{10} K^u - 2\log_{10} y_\pm$.

If the Debye-Hückel limiting law applies, $\log_{10} y_\pm$ is given by Eq. 7.111. If the solution contains only acetic acid, the ionic strength I is given by

$$I = \tfrac{1}{2}[(c\alpha)(1)^2 + (c\alpha)(-1)^2] = c\alpha \tag{7.119}$$

If other ions are present, their contributions must be added. It is convenient to plot the left-hand side of Eq. 7.118 against the ionic strength, as shown schematically in Figure 7.21. If there were exact agreement with the DHLL, the points would lie on a line of slope $2B$ (i.e., 1.018 for water at 25°C). Extrapolation to zero ionic strength gives the value of K. Once this value has been obtained, subtraction of $\log_{10} K^u$ from the experimental curve at any ionic strength gives $-\log_{10} y_\pm$, and $y_\pm$ can therefore

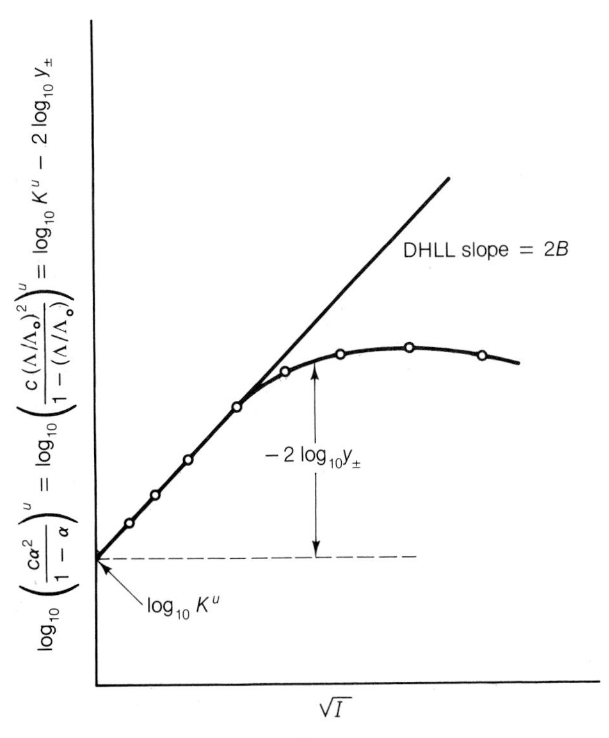

FIGURE 7.21 A schematic plot against $\sqrt{I}$ of $\log_{10}[c\alpha^2/(1-\alpha)]^u$, where α, the degree of dissociation, may be obtained from conductivity measurements.

be calculated at that ionic strength. The individual y_+ and y_- values cannot, of course, be determined.

Activity Coefficients from Solubility Product Measurements

When we write solubility products, we generally ignore activity coefficients; the solubility products are expressed as products of concentrations instead of activities. The solubility product for silver chloride should more accurately be written as

$$K_s = [Ag^+][Cl^-]y_+ y_- \tag{7.120}$$

where y_+ and y_- are the activity coefficients of Ag^+ and Cl^-, respectively. The product $y_+ y_-$ is equal to $y_\pm^2$, where $y_\pm$ is the mean activity coefficient, and therefore

$$K_s = [Ag^+][Cl^-]y_\pm^2 \tag{7.121}$$

One matter of interest that can be understood in terms of this equation is the effect of inert electrolytes on solubilities. An inert electrolyte is one that does not contain a common ion (Ag^+ or Cl^- in this instance) and also does not contain any ion that will complicate the situation by forming a precipitate with either the Ag^+ or the Cl^- ions. In other words, the added inert electrolyte does not bring about a chemical effect; its influence arises only because of its ionic strength.

The influence of ionic strength I on the activity coefficient of an ion is given according to the DHLL by the equation

$$\log_{10} y_\pm = -z_+|z_-|B\sqrt{I} \tag{7.122}$$

Figure 7.20 shows that this equation satisfactorily accounts for the drop in $y_\pm$ that occurs at very low ionic strengths but that considerable deviations occur at higher ones; the value of $\log y_\pm$ becomes positive (i.e., $y_\pm$ is greater than unity) at sufficiently high values of I.

It follows as a result of this behavior that there are two qualitatively different ionic-strength effects on solubilities, one arising at low I values when the $y_\pm$ falls with increasing I and the other being found when $y_\pm$ increases with increasing I. Thus, at low ionic strengths the product $[Ag^+][Cl^-]$ will increase with increasing I, because the product $[Ag^+][Cl^-]y_\pm^2$ remains constant and $y_\pm$ decreases. Under these conditions, added salt increases solubility, and we speak of *salting in*.

At higher ionic strengths, however, $y_\pm$ rises as I increases, and $[Ag^+][Cl^-]y_\pm^2$ therefore diminishes. Thus there is a decrease in solubility, and we speak of *salting out*. Of particular interest are the salting-in and salting-out effects found with protein molecules, a matter of considerable practical importance since proteins are conveniently classified in terms of their solubility behavior.

We saw earlier that measurements of equilibrium constants over a range of ionic strengths allow activity coefficients to be obtained. The same can be done with measurements of solubility. We will outline the method for a sparingly soluble uni-univalent salt AB, for which the solubility equilibrium is

$$AB(s) \rightleftharpoons A^+ + B^-$$

The solubility product is

$$K_s = a_{A^+}a_{B^-} = [A^+][B^-]y_\pm^2 \tag{7.123}$$

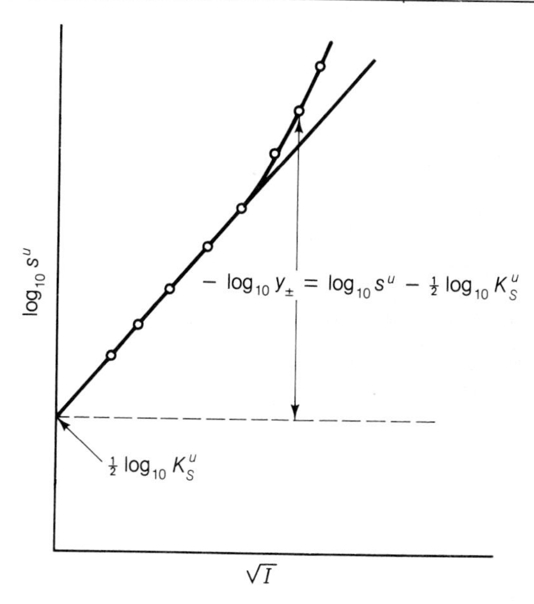

FIGURE 7.22 A schematic plot of $\log_{10} s^u$ against the square root of the ionic strength, showing how K_s^u and $y_\pm$ are obtained.

Thus

$$\log_{10}([A^+][B^-])^u = \log_{10} K_s^u - 2\log_{10} y_\pm \tag{7.124}$$

For a solution in which no common ions are present, the solubility is

$$s = [A^+] = [B^-] \tag{7.125}$$

and therefore

$$\log_{10}(s^2)^u = \log K_s^u - 2\log_{10} y_\pm \tag{7.126}$$

$$\log_{10} s^u = \tfrac{1}{2}\log_{10} K_s^u - \log_{10} y_\pm \tag{7.127}$$

Insofar as the DHLL is obeyed, a plot of $\log_{10} s^u$ against $\sqrt{I/\text{mol dm}^{-3}}$ will therefore be a straight line of slope $B = 0.51$ in water at 25°C.

Figure 7.22 shows the type of plot that is obtained in this way. At sufficiently low ionic strengths the points lie on a line of slope B. Extrapolation to zero ionic strength, where $\log_{10} y_\pm = 0$, therefore gives $\frac{1}{2}\log_{10} K_s^u$, from which the true thermodynamic solubility product K_s is obtained. The value of $\log y_\pm$ is then given by the difference between $\frac{1}{2}\log_{10} K_s^u$ and $\log_{10} s^u$ at that ionic strength, as shown in the figure.

7.12 The Donnan Equilibrium

A matter of considerable importance is the ionic equilibrium that exists between two solutions separated by a membrane. Complications arise with ions that are too large to diffuse through the membrane, and the diffusible ions then reach a special type of equilibrium, known as the **Donnan equilibrium.** Its theory was first worked out in 1911 by the British physical chemist Frederick George Donnan (1870–1956).*

Consider first the equilibrium established when all ions can diffuse through the membrane. A very simple case is when solutions of sodium chloride, of volume 1 dm^3, are separated by the membrane, as shown in Figure 7.23a. Suppose that at equilibrium the concentrations are $[Na^+]_1$ and $[Cl^-]_1$ on the left-hand side and $[Na^+]_2$ and $[Cl^-]_2$ on the right-hand side. Intuitively, we know that in this case these concentrations must be all the same at equilibrium. In thermodynamic terms we can arrive at this conclusion by noting that at equilibrium

$$\Delta G = \Delta G_{Na^+} + \Delta G_{Cl^-} = 0 \tag{7.128}$$

where the terms are the Gibbs energy differences across the membrane (e.g., ΔG is the change in Gibbs energy in going from left to right). The expressions for the individual molar ionic Gibbs energy terms are

$$\Delta G_{Na^+} = RT \ln \frac{[Na^+]_2}{[Na^+]_1} \quad \text{and} \quad \Delta G_{Cl^-} = RT \ln \frac{[Cl^-]_2}{[Cl^-]_1} \tag{7.129}$$

Thus

$$RT \ln \frac{[Na^+]_2}{[Na^+]_1} + RT \ln \frac{[Cl^-]_2}{[Cl^-]_1} = 0 \tag{7.130}$$

from which it follows that

$$\frac{[Na^+]_2[Cl^-]_2}{[Na^+]_1[Cl^-]_1} = 1 \tag{7.131}$$

Since, for electrical neutrality, $[Na^+]_1 = [Cl^-]_1$ and $[Na^+]_2 = [Cl^-]_2$, the result is that

$$[Na^+]_1 = [Na^+]_2 = [Cl^-]_1 = [Cl^-]_2 \tag{7.132}$$

In other words, at equilibrium we have equal concentrations of the electrolyte on each side of the membrane.

This rather trivial case is useful as an introduction to the Donnan equilibrium, since Eq. 7.131 is still obeyed even if the system contains a nondiffusible ion in addition to sodium and chloride ions. For example, suppose that we initially have the situation represented in Figure 7.23b. On the left-hand side, there are sodium ions and nondiffusible anions P^-; on the right-hand side there are sodium and chloride ions. Since there are no chloride ions on the left-hand side, spontaneous diffusion of chloride ions from right to left will occur. Since there must always be electrical neutrality on each side of the membrane, an equal number of sodium ions must also diffuse from right to left. Figure 7.23c shows the situation at equilibrium; x mol dm^{-3} of $[Na^+]$ and $[Cl^-]$ have diffused from right to left; the initial concentrations on the two sides are c_1 and c_2.

Application of Eq. 7.131 then leads to

$$(c_2 - x)^2 = (c_1 + x)x \tag{7.133}$$

*F. G. Donnan, Z. *Elektrochem.*, *17*, 572 (1911).

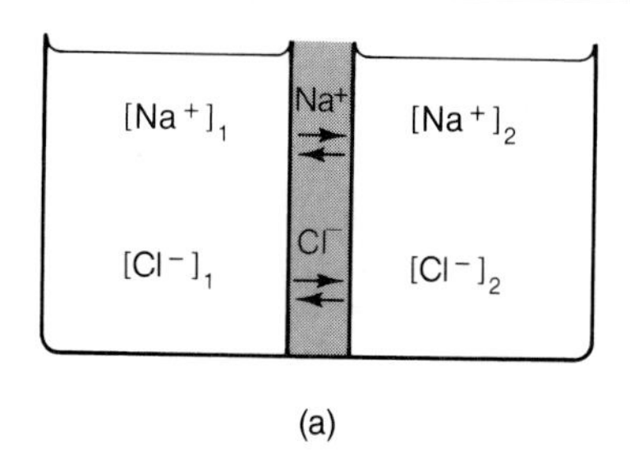

(a)

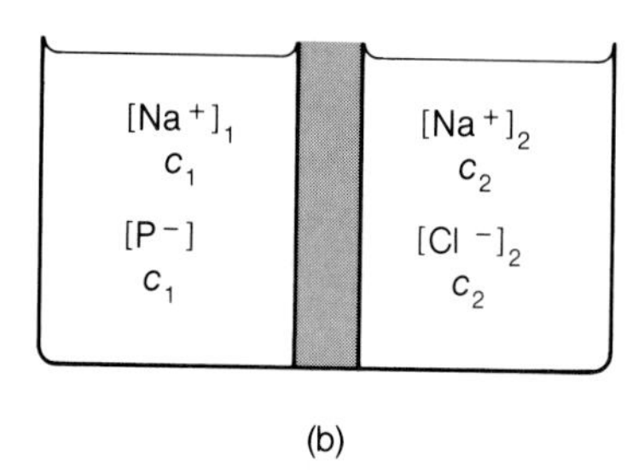

(b)

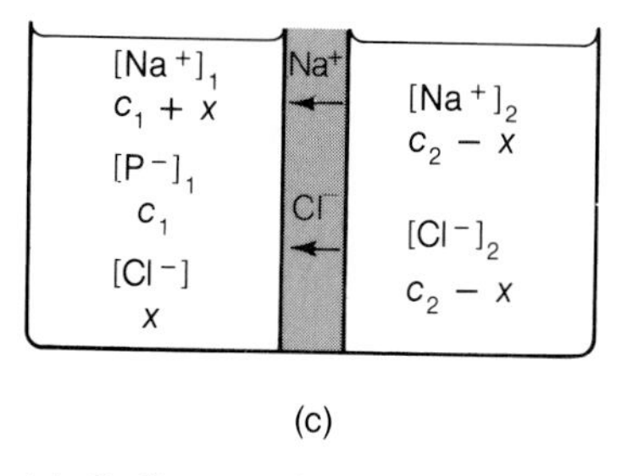

(c)

FIGURE 7.23 (a) Sodium and chloride ions separated by a membrane. (b) Na^+P^- and Na^+Cl^- separated by a membrane: initial conditions. (c) The final Donnan equilibrium conditions arising from (b).

whence

$$x = \frac{c_2^2}{c_1 + 2c_2} \tag{7.134}$$

As an example, suppose that $c_1 = 0.01\ M$ and $c_2 = 0.05\ M$. Use of Eq. 7.134 leads to the result that $x = 0.023\ M$. The final concentrations of the Na^+ and Cl^- ions on the right-hand side are thus $0.027\ M$; the Na^+ concentration on the left-hand side is $0.033\ M$, while the Cl^- concentration on the left is $0.023\ M$.

Equations for more complicated Donnan equilibria (e.g., those involving divalent ions) can easily be worked out using the same principles. Equilibria of the Donnan type are relevant to many types of biological systems; the theory is particularly important with reference to the passage of ions across the membranes of nerve fibers. However, under physiological conditions the significance of the Donnan effect is not easily assessed, because of the complication of *active transport,* a phenomenon in which ions are transported against concentration gradients by processes requiring the expenditure of energy. One straightforward example of the Donnan effect in which there is little or no active transport is found with the erythrocytes (red blood cells). Here the concentration of chloride ions within the cells is significantly smaller than the concentration in the plasma surrounding the cells. This effect can be attributed to the much higher concentration of protein anions retained within the erythrocyte; hemoglobin accounts for a third of the dry weight of the cell.

Under certain circumstances, the establishment of the Donnan equilibrium can lead to other effects, such as changes in pH. Suppose, for example, that an electrolyte NaP (where P is a large anion) is on one side of a membrane, with pure water on the other. The Na^+ ions will tend to cross the membrane and, to restore the electrostatic balance, H^+ ions will cross in the other direction, leaving an excess of OH^- ions. Dissociation of water molecules will occur as required. There will therefore be a lowering of pH on the NaP side of the membrane and a raising on the other side.

Key Equations

Definition of molar conductivity Λ:

$$\Lambda \equiv \frac{\kappa}{c}$$

where

κ = electrolytic conductivity;

c = concentration.

Ostwald's dilution law:

$$\frac{c(\Lambda/\Lambda_0)^2}{1 - (\Lambda/\Lambda_0)} = K$$

where

Λ_0 = molar conductivity at infinite dilution;

Λ/Λ_0 = degree of dissociation;

K = equilibrium constant.

Law of independent migration of ions:

$$\Lambda = \lambda_+ + \lambda_-$$

where λ_+ and λ_- are the individual ion conductivities.

Definitions of transport numbers:

$$t_+ = \frac{u_+}{u_+ + u_-} \qquad t_- = \frac{u_-}{u_+ + u_-}$$

where u_+ and u_- are the ion mobilities.

Definition of ionic strength I:

$$I \equiv \frac{1}{2} \sum_i c_i z_i^2$$

where

c_i = concentration of ion of ith type;

z_i = its charge number.

Debye-Hückel limiting law for activity coefficient y_i:

$$\log_{10} y_i = -z_i B \sqrt{I}$$

For the mean activity coefficient $y_\pm$,

$$\log_{10} y_\pm = -z_+|z_-|B\sqrt{I}$$
$$= -0.51 z_+|z_-|\sqrt{I/\text{mol dm}^{-3}}$$

for water at 25°C.

Problems

A Faraday's Laws, Molar Conductivity, and Weak Electrolytes

1. A constant current was passed through a solution of cupric sulfate, $CuSO_4$, for 1 hr, and 0.040 g of copper was deposited. Calculate the current (A_r of Cu = 63.5; Faraday constant = 96 500 C mol^{-1}).

2. After passage of a constant current for 45 min, 7.19 mg of silver (A_r = 107.9) was deposited from a solution of silver nitrate. Calculate the current.

3. The following are the molar conductivities of chloroacetic acid in aqueous solution at 25°C:

Concentration/10^{-4} M	625	312.5	156.3	78.1	39.1	19.6	9.0
Λ/Ω^{-1} cm^2 mol^{-1}	53.1	72.4	96.8	127.7	164.0	205.8	249.0

Plot Λ against c. If $\Lambda_0 = 362\ \Omega^{-1}$ cm^2 mol^{-1}, are these values in accord with the Ostwald dilution law? What is the value of the dissociation constant? (See also Problem 28.)

4. The electrolytic conductivity of a saturated solution of silver chloride, AgCl, in pure water at 25°C is $1.26 \times 10^{-6}\ \Omega^{-1}$ cm^{-1} higher than that for the water used. Calculate the solubility of AgCl in water if the molar ionic conductivities are Ag^+, 61.9 Ω^{-1} cm^2 mol^{-1}; Cl^-, 76.4 Ω^{-1} cm^2 mol^{-1}.

***5.** The electrolytic conductivity of a 0.001 M solution of Na_2SO_4 is 2.6×10^{-4} Ω^{-1} cm^{-1}; if the solution is saturated with $CaSO_4$, the conductivity becomes 7.0×10^{-4} Ω^{-1} cm^{-1}. Calculate the solubility product for $CaSO_4$ using the following molar conductivities at these concentrations: $\lambda(Na^+) = 50.1$ Ω^{-1} cm^2 mol^{-1}; $\lambda(\frac{1}{2}Ca^{2+}) = 59.5$ Ω^{-1} cm^2 mol^{-1}.

***6.** A conductivity cell when standardized with 0.01 M KCl was found to have a resistance of 189 Ω. With 0.01 M ammonia solution the resistance was 2460 Ω. Calculate the base dissociation constant of ammonia, given the following molar conductivities at these concentrations: $\lambda(K^+) = 73.5$ Ω^{-1} cm^2 mol^{-1}; $\lambda(Cl^-) = 76.4$ Ω^{-1} cm^2 mol^{-1}; $\lambda(NH_4^+) = 73.4$ Ω^{-1} cm^2 mol^{-1}; $\lambda(OH^-) = 198.6$ Ω^{-1} cm^2 mol^{-1}.

B Debye-Hückel Theory and Transport of Electrolytes

7. The thickness of the ionic atmosphere ($1/\kappa$) for a univalent electrolyte is 0.964 nm at a concentration of 0.10 M in water at 25°C ($\epsilon = 78$). Estimate the thickness (a) in water at a concentration of 0.0001 M and (b) in a solvent of $\epsilon = 38$ at a concentration of 0.10 M.

8. The molar conductivities of 0.001 M solutions of potassium chloride, sodium chloride, and potassium sulfate ($\frac{1}{2}K_2SO_4$) are 149.9, 126.5, and 153.3 Ω^{-1} cm^2 mol^{-1}, respectively. Calculate an approximate value for the molar conductivity of a solution of sodium sulfate of the same concentration.

9. The molar conductivity at 18°C of a 0.0100 M aqueous solution of ammonia is 9.6 Ω^{-1} cm^2 mol^{-1}. For NH_4Cl, $\Lambda_0 = 129.8$ Ω^{-1} cm^2 mol^{-1} and the molar ionic conductivities of OH^- and Cl^- are 174 and 65.6 Ω^{-1} cm^2 mol^{-1}, respectively. Calculate Λ_0 for NH_3 and the degree of ionization in 0.01 M solution.

10. A solution of LiCl was electrolyzed in a Hittorf cell. After a current of 0.79 A had been passed for 2 h, the mass ot LiCl in the anode compartment had decreased by 0.793 g.

a. Calculate the transport numbers of the Li^+ and Cl^- ions.

b. If $\Lambda°$(LiCl) is 115.0 Ω^{-1} cm^2 mol^{-1}, what are the molar ionic conductivities and the ionic mobilities?

11. A solution of cadmium iodide, CdI_2, having a molality of 7.545×10^{-3} mol kg^{-1}, was electrolyzed in a Hittorf cell. The mass of cadmium deposited at the cathode was 0.03462 g. Solution weighing 152.64 g was withdrawn from the anode compartment and was found to contain 0.3718 g of cadmium iodide. Calculate the transport numbers of Cd^{2+} and I^-.

12. The transport numbers for HCl at infinite dilution are estimated to be $t_+ = 0.821$ and $t_- = 0.179$, and the molar conductivity is 426.16 Ω^{-1} cm^2 mol^{-1}. Calculate the mobilities of the hydrogen and chloride ions.

13. If a potential gradient of 100 V cm^{-1} is applied to a 0.01 M solution of NaCl, what are the speeds of the Na^+ and Cl^- ions? Take the ionic conductivities to be those listed in Table 7.3.

***14.** A solution of LiCl at a concentration of 0.01 M is contained in a tube having a cross-sectional area of 5 cm^2. Calculate the speeds of the Li^+ and Cl^- ions if a current of 1 A is passed. Use the ion conductivities listed in Table 7.3.

15. What is the work required to separate in vacuum two particles, one with the charge of the proton, from another particle with the same charge of opposite sign? Do the calculation for an initial distance of (a) 1.0 nm to an infinite distance apart and (b) from 1.0 mm to an infinite distance apart. (c) In (a) how much work would be required if the charge is moved to a distance of 0.1 m? The charge on a proton is 1.6×10^{-19} C.

C Thermodynamics and Activities of Ions

16. The following are some conventional standard enthalpies of ions in aqueous solution at 25°C:

Ion	$\Delta_f H°$/kJ mol^{-1}
H^+	0
Na^+	−239.7
Ca^{2+}	−543.1
Zn^{2+}	−152.3
Cl^-	−167.4
Br^-	−120.9

Calculate the enthalpy of formation in aqueous solution of 1 mol of NaCl, $CaCl_2$, and $ZnBr_2$, assuming complete dissociation.

17. One estimate for the absolute Gibbs energy of hydration of the H^+ ion in aqueous solution is

-1051.4 kJ mol^{-1}. On this basis, calculate the absolute Gibbs energies of hydration of the following ions, whose conventional standard Gibbs energies of hydration are as follows:

Ion	$\Delta_{hyd}G°(\text{conv})/\text{kJ mol}^{-1}$
H^+	0
Na^+	679.1
Mg^{2+}	274.1
Al^{3+}	-1346.4
Cl^-	-1407.1
Br^-	-1393.3

18. Calculate the ionic strengths of 0.1 *M* solutions of KNO_3, K_2SO_4, $ZnSO_4$, $ZnCl_2$, and $K_4Fe(CN)_6$; assume complete dissociation and neglect hydrolysis.

19. Calculate the mean activity coefficient $y_\pm$ for the Ba^{2+} and SO_4^{2-} ions in a saturated solution of $BaSO_4$ ($K_{sp} = 9.2 \times 10^{-11}$ mol^2 dm^{-6}) in 0.2 *M* K_2SO_4, assuming the Debye-Hückel limiting law to apply.

20. The solubility of AgCl in water at 25°C is 1.274×10^{-5} mol dm^{-3}. On the assumption that the Debye-Hückel limiting law applies,

a. Calculate $\Delta G°$ for the process
$AgCl(s) \rightarrow Ag^+(aq) + Cl^-(aq)$.
b. Calculate the solubility of AgCl in an 0.005 *M* solution of K_2SO_4.

21. Employ Eq. 7.114 to make plots of $\log y_\pm$ against $\sqrt{I}$ for a uni-univalent electrolyte in water at 25°C, with $B = 0.51$ mol^{-1} dm$^{3/2}$ and $B' = 0.33 \times 10^{10}$ mol^{-1} dm$^{3/2}$ m^{-1}, and for the following values of the interionic distance a:

$$a = 0,\ 0.1,\ 0.2,\ 0.4,\ \text{and } 0.8 \text{ nm}$$

22. Estimate the change in Gibbs energy ΔG when 1 mol of K^+ ions (radius 0.133 nm) is transported from aqueous solution ($\epsilon = 78$) to the lipid environment of a cell membrane ($\epsilon = 4$) at 25.0°C.

23. At 18°C the electrolytic conductivity of a saturated solution of CaF_2 is 3.86×10^{-5} Ω^{-1} cm^{-1}, and that of pure water is 1.5×10^{-6} Ω^{-1} cm^{-1}. The molar ionic conductivities of $\frac{1}{2}Ca^{2+}$ and F^- are 51.1 Ω^{-1} cm^2 mol^{-1} and 47.0 Ω^{-1} cm^2 mol^{-1}, respectively. Calculate the solubility of CaF_2 in pure water at 18°C and the solubility product.

24. A 0.1 *M* solution of sodium palmitate, $C_{15}H_{31}COONa$, is separated from a 0.2 *M* solution of sodium chloride by a membrane that is permeable to Na^+ and Cl^- ions but not to palmitate ions. Calculate the concentrations of Na^+ and Cl^- ions on the two sides of the membrane after equilibrium has become established.

25. What concentrations of the following have the same ionic strength as 0.1 *M* NaCl?

$CuSO_4$, $Ni(NO_3)_2$, $Al_2(SO_4)_3$, Na_3PO_4

Assume complete dissociation and neglect hydrolysis.

***26.** The solubility product of PbF_2 at 25.0°C is 4.0×10^{-9} mol^3 dm^{-9}. Assuming the Debye-Hückel limiting law to apply, calculate the solubility of PbF_2 in (a) pure water and (b) 0.01 *M* NaF.

***27.** Calculate the solubility of silver acetate in water at 25.0°C, assuming the DHLL to apply; the solubility product is 4.0×10^{-3}mol^2 dm^{-6}.

D Supplementary Problems

28. Equation 7.20 is one form of Ostwald's dilution law. Show how it can be *linearized* (i.e., convert it into a form that will allow experimental values of Λ at various concentrations to be tested by means of a straight-line plot). Explain how Λ_0 and K can be obtained from the plot.

Kraus and Callis, *J. Am. Chem. Soc.*, *45*, 2624 (1923), obtained the following results for the dissociation of tetramethyl tin chloride, $(CH_3)_3SnCl$, in ethyl alcohol solution at 25.0°C:

Concentration, $c/10^{-4}$ mol dm^{-3}	1.566	2.600	6.219	10.441
Electrolytic conductivity, $\kappa/10^{-6}$ Ω^{-1} cm^{-1}	1.788	2.418	4.009	5.336

By the use of the linear plot you have devised, determine Λ_0 and K.

29. According to Bjerrum's theory of ion association, the number of ions of type i present in a spherical shell of thickness dr and distance r from a central ion is

$$dN_i = N_i e^{-z_i z_c e^2/4\pi\epsilon_0\epsilon r \mathbf{k}T} 4\pi r^2 dr$$

where z_i and z_c are the charge numbers of the ion of type i and of the central ion and e, ϵ_0, ϵ, and $\mathbf{k}$ have their usual significance. Plot the exponential in this expression and also $4\pi r^2$ against r for a uni-univalent electrolyte in water at 25.0°C ($\epsilon = 78.3$). Allow r to have values from 0 to 1 nm. Plot also the product of these functions, which is $(dN_i/N_i)\,dr$ and is the probability of finding an ion of type i at a distance between r and $r + dr$ of the central ion.

By differentiation, obtain a value r^* for which the probability is a minimum, and calculate the value for water at 25.0°C. The electrostatic potential is given to a good approximation by the first term in Eq. 7.47. Obtain an expression, in terms of $\mathbf{k}T$, for the electrostatic energy between the two univalent ions at this minimum distance, and evaluate this energy at 25.0°C.

30. Problem 22 was concerned with the Gibbs energy change when 1 mol of K^+ ions are transported from water to a lipid. Estimate the electrostatic contribution to the entropy change when this occurs, assuming the dielectric constant of the lipid to be temperature independent, and the following values for water at 25.0°C: $\epsilon = 78$; $\partial \ln \epsilon/\partial T = -0.0046$ K^{-1}. Suggest a qualitative explanation for the sign of the value you obtain.

31. Assuming the Born equation (Eq. 7.86) to apply, make an estimate of the reversible work of charging 1 mol of Na^+Cl^- in aqueous solution at 25.0°C ($\epsilon = 78$), under the following conditions: (a) The electrolyte is present at infinite dilution. (b) The electrolyte is present at such a concentration that the mean activity coefficient is 0.70. The ionic radii are 95 pm for Na^+ and 181 pm for Cl^-.

E Essay Questions

1. State Faraday's two laws of electrolysis and discuss their significance in connection with the electrical nature of matter.

2. Discuss the main ideas that lie behind the Debye-Hückel theory, as applied to the conductivities of solutions of strong electrolytes.

3. Outline two important methods for determining transport numbers of ions.

4. Explain why Li^+ has a lower ionic conductivity than Na^+ and why the value for H^+ is so much higher than the values for both of these ions.

5. Describe briefly the type of hydration found with the following ions in aqueous solution: Li^+, Br^-, H^+, OH^-.

6. What modifications to the Debye-Hückel limiting law are required to explain the influence of ionic strength on solubilities?

Suggested Reading

J. N. Butler, *Ionic Equilibrium: A Mathematical Approach*, Reading, Mass.: Addison-Wesley, 1964.

B. E. Conway, *Ionic Hydration in Chemistry and Biophysics*, Amsterdam: Elsevier North Holland, 1980.

B. E. Conway and R. G. Barradas, (Eds.), *Chemical Physics of Ionic Solutions*, New York: John Wiley, 1966.

C. W. Davies, *Ion Association*, London: Butterworth, 1962.

D. Eisenberg and W. J. Kauzmann, *The Structure and Properties of Water*, Oxford: Clarendon Press, 1969.

R. M. Fuoss and F. Accascina, *Electrolytic Conductance*, New York: Interscience, 1959.

S. Glasstone, *An Introduction to Electrochemistry*, New York: D. Van Nostrand Co., 1942.

H. S. Harned and B. B. Owen, *The Physical Chemistry of Electrolytic Solutions*, New York: Reinhold Pub. Co., 1950.

J. L. Kavanau, *Water and Solute-Water Interactions*, San Francisco: Holden-Day, 1964.

G. Kortüm, *Treatise on Electrochemistry*, Amsterdam: Elsevier Pub. Co., 1965.

E. H. Lyons, *Introduction to Electrochemistry*, Boston: D. C. Heath, 1967.

W. J. Moore, *Physical Chemistry* (4th ed.), Englewood Cliffs, N.J.: Prentice-Hall, 1972.

Lars Onsager, "The Motion of Ions: Principles and Concepts," *Science*, *166*, 1359 (1969).

J. Robbins, *Ions in Solution*, Oxford: Clarendon Press, 1972.

R. A. Robinson and R. H. Stokes, *Electrolyte Solutions* (2nd ed.), London: Butterworth, 1959.

Electrochemical Cells

Preview

In this chapter we discuss *galvanic cells,* in which a chemical reaction produces an electric potential difference between two electrodes. In the Daniell cell, zinc and copper electrodes are immersed in solutions of Zn^{2+} and Cu^{2+} ions that are separated by a membrane. The chemical reactions that occur at the two electrodes cause a flow of electrons in the outer circuit.

It is impossible to measure the electromotive force (emf) of a single electrode, and the convention is to use a *standard hydrogen electrode* as the left-hand electrode in a cell. With another standard electrode on the right-hand side, the cell emf is then taken to be the standard electrode potential for the right-hand electrode. Such standard electrode potentials are very useful for predicting the direction of a chemical reaction.

There is an important thermodynamic relationship between the Gibbs energy change and the emf of a galvanic cell. This equation is useful in the determination of equilibrium constants. The *Nernst equation* shows how the cell emf varies with the concentrations of solutions in the cell. The *Nernst potential* is the potential difference established across a membrane when two different solutions are on opposite sides of it.

An important type of cell is a *redox cell,* in which oxidation-reduction processes occur at electrodes. It is often convenient to relate the emfs of such cells to a standard pH value, usually taken to be 7.

There are a number of important practical applications of emf measurements that can be used for the determination of pH, activity coefficients, equilibrium constants, solubility products, and transport numbers.

Certain electrochemical cells are also useful devices for the generation of electric power. *Fuel cells,* for example, are cells in which the reacting substances are continuously fed into the system. *Photogalvanic cells* are electrochemical cells in which radiation induces a chemical process that gives rise to an electric current.

8 Electrochemical Cells

An *electrochemical cell,* also called a *voltaic cell* or a *galvanic cell,* is a device in which a chemical reaction occurs with the production of an electric potential difference between two electrodes. There can thus be a flow of current, which can lead to the performance of mechanical work, and the electrochemical cell thus transforms chemical energy into work. Besides being of considerable practical importance, electrochemical cells are valuable laboratory instruments, since they provide some extremely useful scientific data. For example, they lead to thermodynamic quantities such as enthalpies and Gibbs energies, and they allow us to determine transport numbers and activity coefficients for ions in solution. The present chapter deals with the general principles of electrochemical cells and with some of their more important applications.

8.1 The Daniell Cell

As originally developed, the *Daniell cell* consisted of a zinc electrode immersed in dilute sulfuric acid and a copper electrode immersed in a cupric sulfate solution. It was later found that the cell gave a more stable voltage if the sulfuric acid was replaced by zinc sulfate solution. The expression 'Daniell cell' today usually refers to such an arrangement, which is shown schematically in Figure 8.1. The voltage produced by such a cell depends on the concentrations of the Zn^{2+} and Cu^{2+} ions in the two solutions. If the molalities of the two solutions are 1 mol kg^{-1} (1 m), the cell is called a *standard cell.*

When such a cell is set up, there is a flow of electrons from the zinc to the copper electrode in the outer circuit. This means that positive current is moving from left to right in the cell itself. By convention, a potential difference corresponding to an external flow of electrons from the left-hand electrode to the right-hand electrode is said to be a *positive* potential difference.

The processes that occur when this cell operates are shown in Figure 8.1. Since positive electricity in the form of positive ions moves from left to right within the cell, zinc metal dissolves to form Zn^{2+} ions,

$$Zn \longrightarrow Zn^{2+} + 2e^-$$

Some of these zinc ions pass through the membrane into the right-hand solution, and at the right-hand electrode cupric ions interact with electrons to form metallic copper:

$$Cu^{2+} + 2e^- \longrightarrow Cu$$

Every time a zinc atom dissolves, and a copper atom is deposited, two electrons travel round the outer circuit.

Cells of this kind can be made to behave in a reversible fashion, by balancing their potentials by an external potential so that no current flows. This can be done by means of a potentiometer, the principle of which is illustrated in Figure 8.2a. A current is passed through a uniform slide wire AB, along which there is a linear potential drop, the magnitude of which can be adjusted by a rheostat. The cell

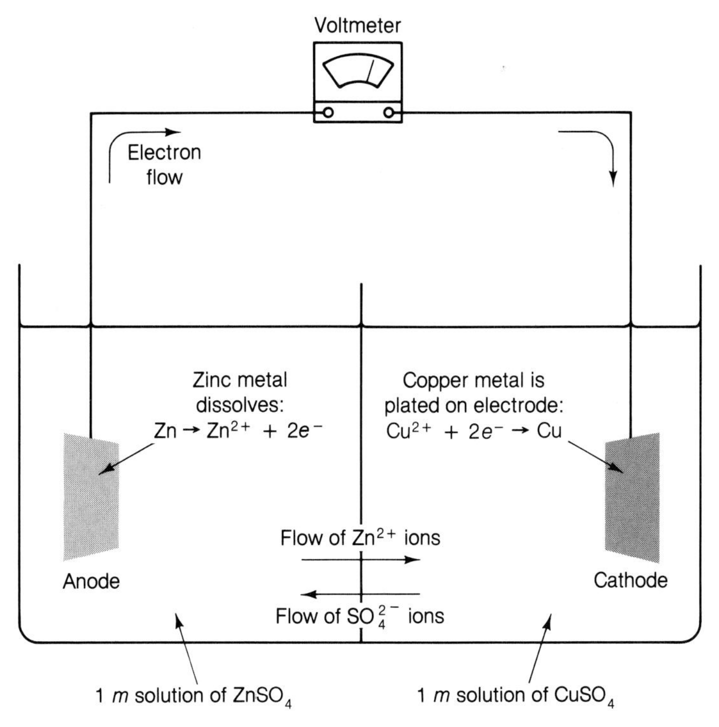

FIGURE 8.1 The standard Daniell cell.

under investigation is connected through a tap-key switch and a galvanometer to one end (A) of the slide wire and to a movable contact C. The contact is moved along the slide wire until, when the switch is depressed, no current passes through the galvanometer. The potential difference produced by the cell is then exactly balanced by the potential difference between points A and C. The potentiometer wire can be calibrated by use of a standard cell, such as the Weston cell (Figure 8.2b).

When the electric potential of a cell is exactly balanced in this way, the cell is operating *reversibly* and its potential is then referred to as the *electromotive force* (*emf*) of the cell. If the counter-potential in the slide wire is slightly less than the emf of the cell, there is a small flow of electrons from left to right in the outer circuit. If the counter-potential is adjusted to be slightly greater than the cell emf, the cell is forced to operate in reverse; zinc is deposited at the left-hand electrode,

$$Zn^{2+} + 2e^- \longrightarrow Zn$$

and copper dissolves at the right,

$$Cu \longrightarrow Cu^{2+} + 2e^-$$

The fact that the electrons normally flow from the zinc to the copper electrode indicates that the tendency for $Zn \rightarrow Zn^{2+} + 2e^-$ to occur is greater than for the reaction $Cu \rightarrow Cu^{2+} + 2e^-$, which is forced to occur in the reverse direction. The magnitude of the emf developed is a measure of the relative tendencies of the two processes. The emf varies with the concentrations of the Zn^{2+} and Cu^{2+} ions in the two solutions. Thus the tendency for $Zn \rightarrow Zn^{2+} + 2e^-$ to occur is smaller when the

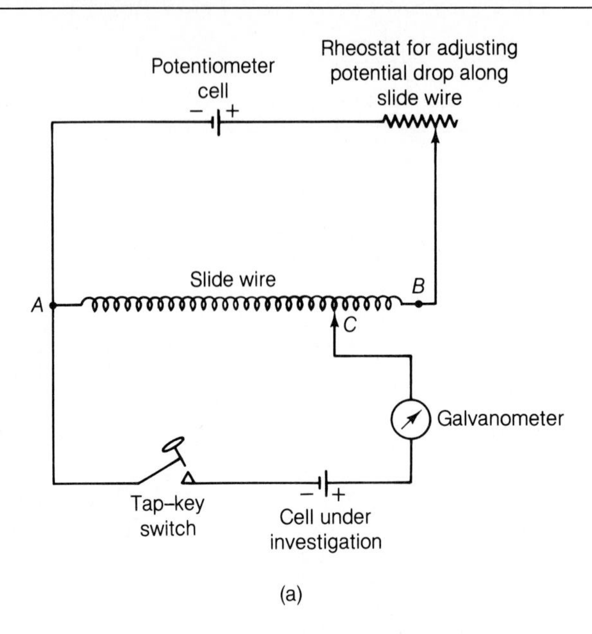

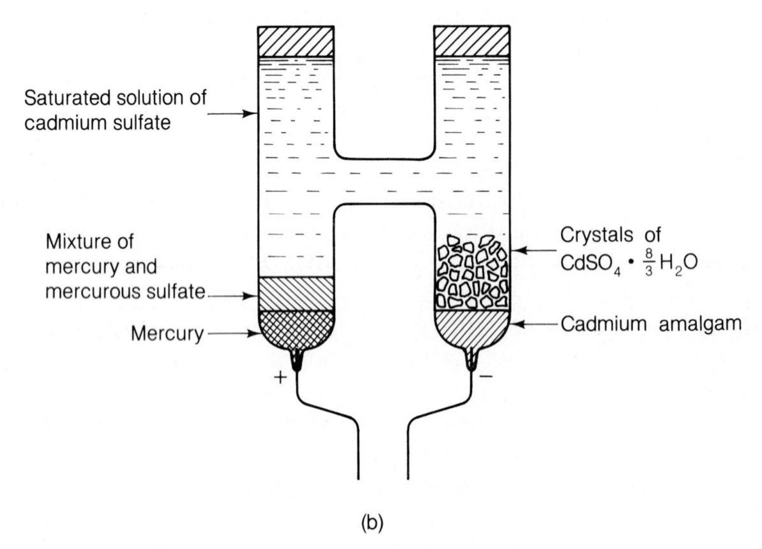

FIGURE 8.2 (a) A potentiometer circuit used for determining the reversible emf of a cell. (b) The Weston standard cell, which is the most widely used standard cell. At 25°C its emf is 1.018 32 V and it has a very small temperature coefficient (see Problem 20).

concentration of Zn^{2+} is large, while the tendency for $2e^- + Cu^{2+} \rightarrow Cu$ to occur increases when the concentration of Cu^{2+} is increased. The precise relationship between emf and the concentrations will be considered in Section 8.3.

8.2 Standard Electrode Potentials

It would be very convenient if we could measure the potential of a single electrode, such as the right-hand electrode in Figure 8.1, which we will write as

$Cu^{2+}|Cu$

The potential of such an electrode would be a measure of the tendency of the process

$$Cu^{2+} + 2e^- \longrightarrow Cu$$

to occur. However, there is no way to measure the emf of a single electrode, since in order to obtain an emf there must be two electrodes, with an emf associated with each one. The procedure used is to choose one electrode as a standard and to measure emf values of other electrodes with reference to that standard.

The Standard Hydrogen Electrode

The electrode chosen as the ultimate standard is the standard hydrogen electrode, which is illustrated in Figure 8.3. It consists of a platinum electrode immersed in a 1 *m* solution of hydrogen ions maintained at 25°C and 1 atm pressure. Hydrogen gas is bubbled over the electrode and passes into solution, forming hydrogen ions and electrons:

$$H_2 \longrightarrow 2H^+ + 2e^-$$

The emf corresponding to this electrode is arbitrarily assigned to have the value of zero, and this electrode is used as a standard for other electrodes.

There are two conventions in common use, and the student should be aware of both methods of procedure. The standard hydrogen electrode can be either the left-hand or the right-hand electrode. In the convention adopted by the International Union of Pure and Applied Chemistry (IUPAC) the hydrogen electrode is placed on the *left-hand side*, and the emf of the other electrode is taken to be that of the cell. Such emf values, under standard conditions, are known as **standard electrode potentials** or *standard reduction potentials* and are given the symbol $E°$. Alternatively, the standard hydrogen electrode may be placed on the right-hand side; the potential so obtained is known as the *standard oxidation potential*. The latter potentials are the standard electrode potentials with the signs reversed; the only difference is that the cells have been turned round.

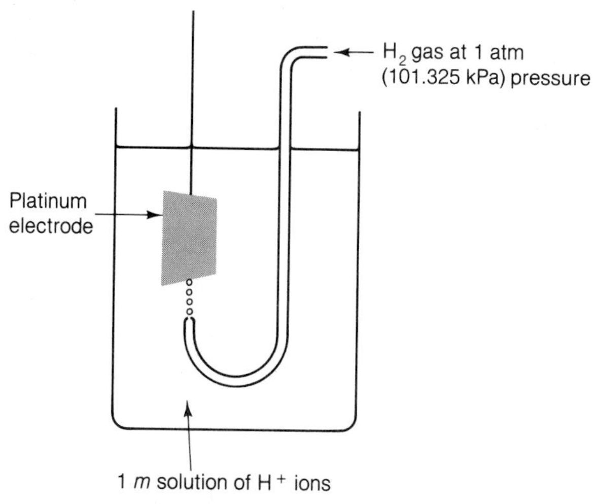

FIGURE 8.3 The standard hydrogen electrode.

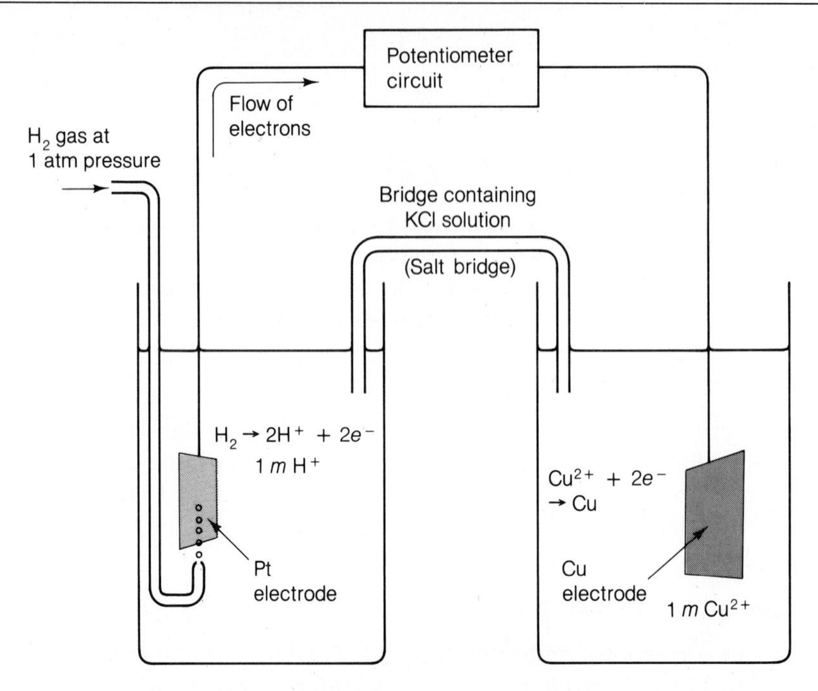

FIGURE 8.4 A voltaic cell in which a standard hydrogen electrode has been combined with a copper electrode immersed in a 1 *m* solution of cupric ions; the two solutions are connected by a potassium chloride bridge.

To illustrate the *standard electrode (reduction) potentials* (IUPAC convention), consider the cell shown in Figure 8.4. The left-hand electrode is the standard hydrogen electrode, with a hydrogen gas pressure of 1 atm, and the acid solution is 1 *m* in H^+ ions. The right-hand electrode is the $Cu^{2+}|Cu$ electrode, and the concentration of Cu^{2+} ions is 1 *m*. The two solutions are connected by a "salt bridge" such as a potassium chloride solution, which conducts electricity but does not allow bulk mixing of the two solutions. Alternatively, an agar gel containing KCl is commonly used. This procedure is somewhat more reliable than separating the two solutions by a porous partition, which itself sets up a small emf; the salt bridge minimizes this effect.

The voltaic cell shown in Figure 8.4 can be represented as follows:

$$\text{Pt}, H_2(1\ \text{atm})|H^+(1\ m)||Cu^{2+}(1\ m)|Cu$$

where the double vertical lines represent the salt bridge. The observed emf is +0.34 V; the sign is positive by convention since electrons flow from left to right in the outer circuit. There is therefore a greater tendency for the process

$$Cu^{2+} + 2e^- \longrightarrow Cu$$

to occur than for

$$2H^+ + 2e^- \longrightarrow H_2$$

to occur; the latter process is forced to go in the reverse direction. By the IUPAC convention, the $Cu^{2+}|Cu$ electrode is on the right, and the standard electrode potential $E°$ of this electrode is +0.34 V. Because this is a measure of the tendency for the cupric ions to be *reduced* by the process

$$Cu^{2+} + 2e^- \longrightarrow Cu$$

these electrode potentials are also known as *standard reduction potentials.*

The opposite convention is to set up the cell in the reverse manner to that depicted in Figure 8.4; such a cell could be represented as

$$Cu|Cu^{2+}||H^+|Pt,\ H_2(1\ atm)$$

where again the double vertical lines represent the salt bridge. The emf of this cell, under standard conditions ($[Cu^{2+}] = [H^+] = 1\ m$; hydrogen gas pressure = 1 atm), is obviously −0.34 V. We can write the electrode reactions in such a way as to correspond to the flow of electrons in the conventional direction:

$$Cu \longrightarrow Cu^{2+} + 2e^- \qquad 2H^+ + 2e^- \longrightarrow H_2$$

The value of −0.34 V is therefore a measure of the tendency of the reaction

$$Cu \longrightarrow Cu^{2+} + 2e^-$$

to occur. Because of the negative sign, this process is *not* spontaneous; and since it is an oxidation, −0.34 V is hence the *standard oxidation potential* of the $Cu|Cu^{2+}$ electrode.

In the remainder of this book, standard electrode (reduction) potentials will be employed, as recommended by the IUPAC. Table 8.1* gives a list of such potentials; the reactions are written as reduction processes (e.g., $Cu^{2+} + 2e^- \rightarrow Cu$). A table of standard oxidation potentials would have all the signs reversed and the corresponding reactions would be oxidations; for example,

$$Cu \longrightarrow Cu^{2+} + 2e^-$$

By combining the standard electrode potentials for two electrodes we can deduce the emf of a cell involving the two electrodes, neglecting the hydrogen electrode. Consider, for example, the following items in Table 8.1:

$$Cu^{2+} + 2e^- \longrightarrow Cu \qquad E° = 0.34\ V$$

$$Zn^{2+} + 2e^- \longrightarrow Zn \qquad E° = -0.76\ V$$

These values are the emf values for the following cells, in which the electrode processes are shown:

$$Pt,\ H_2|H^+(1\ m)||Cu^{2+}(1\ m)|Cu \qquad E° = 0.34\ V$$

$$H_2 \longrightarrow 2H^+ + 2e^- \qquad Cu^{2+} + 2e^- \longrightarrow Cu$$

$$Pt,\ H_2|H^+(1\ m)||Zn^{2+}(1\ m)|Zn \qquad E° = -0.76\ V$$

$$H_2 \longrightarrow 2H^+ + 2e^- \qquad Zn^{2+} + 2e^- \longrightarrow Zn$$

We could connect the two cells together as follows:

$$Zn|Zn^{2+}(1\ m)||H^+(1\ m)|H_2,\ Pt—Pt,\ H_2|H^+(1\ m)||Cu^{2+}(1\ m)|Cu$$

*Some of the values in this table were determined indirectly from other experimental results, since hypothetical electrodes (e.g., $Li\ |\ Li^+$) are impossible to set up.

TABLE 8.1 Standard Electrode (Reduction) Potentials

Half-Reaction	Standard Electrode Potential, $E°/V$
$F_2 + 2e^- \longrightarrow 2F^-$	2.87
$H_2O_2 + 2H^+ + 2e^- \longrightarrow 2H_2O$	1.77
$Au^+ + e^- \longrightarrow Au$	1.68
$MnO_4^- + 8H^+ + 5e^- \longrightarrow Mn^{2+} + 4H_2O$	1.52
$Cl_2 + 2e^- \longrightarrow 2Cl^-$	1.3595
$Cr_2O_7^{2-} + 14H^+ + 6e^- \longrightarrow 2Cr^{3+} + 7H_2O$	1.33
$MnO_2 + 4H^+ + 2e^- \longrightarrow Mn^{2+} + 2H_2O$	1.23
$O_2 + 4H^+ + 4e^- \longrightarrow 2H_2O$	1.23
$Pt^{2+} + 2e^- \longrightarrow Pt$	1.20
$Br_2 + 2e^- \longrightarrow 2Br^-$	1.0652
$Hg^{2+} + 2e^- \longrightarrow Hg$	0.854
$Ag^+ + e^- \longrightarrow Ag$	0.7996
$Hg_2^{2+} + 2e^- \longrightarrow 2Hg$	0.79
$Fe^{3+} + e^- \longrightarrow Fe^{2+}$	0.771
$O_2 + 2H^+ + 2e^- \longrightarrow H_2O_2$	0.682
$I_2 + 2e^- \longrightarrow 2I^-$	0.5355
$Cu^{2+} + 2e^- \longrightarrow Cu$	0.337
$Hg_2Cl_2 + 2e^- \longrightarrow 2Hg + 2Cl^-$	0.26808
$AgCl(s) + e^- \longrightarrow Ag + Cl^-$	0.2224
$Cu^{2+} + e^- \longrightarrow Cu^+$	0.153
$Sn^{4+} + 2e^- \longrightarrow Sn^{2+}$	0.15
$2H^+ + 2e^- \longrightarrow H_2$	0.00 (by definition)
$Pb^{2+} + 2e^- \longrightarrow Pb$	−0.126
$Sn^{2+} + 2e^- \longrightarrow Sn$	−0.136
$Ni^{2+} + 2e^- \longrightarrow Ni$	−0.25
$Co^{2+} + 2e^- \longrightarrow Co$	−0.28
$Fe^{2+} + 2e^- \longrightarrow Fe$	−0.4402
$Cr^{3+} + 3e^- \longrightarrow Cr$	−0.744
$Zn^{2+} + 2e^- \longrightarrow Zn$	−0.7628
$Al^{3+} + 3e^- \longrightarrow Al$	−1.662
$Mg^{2+} + 2e^- \longrightarrow Mg$	−2.363
$Na^+ + e^- \longrightarrow Na$	−2.7142
$Ca^{2+} + 2e^- \longrightarrow Ca$	−2.866
$K^+ + e^- \longrightarrow K$	−2.925
$Li^+ + e^- \longrightarrow Li$	−3.045

The standard oxidation potentials are the negatives of the values given here; the reactions are written in the opposite direction.

and the emf would then be $E° = 0.34 - (-0.76) = 1.10$ V (i.e., the right-hand electrode potential minus the left-hand potential). We could also eliminate the hydrogen electrodes altogether and set up the cell

$$Zn|Zn^{2+}(1\ m)||Cu^{2+}(1\ m)|Cu$$

The emf of this would be the same, 1.10 V, since we have merely eliminated two identical hydrogen electrodes working in opposition to each other. This cell is, of course, just the standard Daniell cell (see Figure 8.1).

Note that in writing the individual cell reactions it makes no difference whether they are written with one or with more electrons. Thus the hydrogen electrode reaction can be written as either

$$2H^+ + 2e^- \longrightarrow H_2 \quad \text{or} \quad H^+ + e^- \longrightarrow \tfrac{1}{2}H_2$$

However, in considering the overall process we must obviously balance out the electrons. Thus for the cell

$$Pt, H_2|H^+||Cu^{2+}|Cu$$

the individual reactions can be written as

$$H_2 \longrightarrow 2H^+ + 2e^- \quad \text{and} \quad 2e^- + Cu^{2+} \longrightarrow Cu$$

so that the overall process is

$$H_2 + Cu^{2+} \longrightarrow Cu + 2H^+$$

This process is accompanied by the passage of *two* electrons around the outer circuit. We could equally well write the reactions as

$$\tfrac{1}{2}H_2 \longrightarrow H^+ + e^-$$
$$\tfrac{1}{2}Cu^{2+} + e^- \longrightarrow \tfrac{1}{2}Cu$$
$$\overline{\tfrac{1}{2}H_2 + \tfrac{1}{2}Cu^{2+} \longrightarrow \tfrac{1}{2}Cu + H^+}$$

This tells us that every time 0.5 mol of Cu^{2+} disappears and 0.5 mol of Cu appears, 1 mol of electrons pass from the left-hand electrode to the right-hand electrode.

Other Standard Electrodes

The standard hydrogen electrode is not the most convenient electrode because of the necessity of bubbling hydrogen over the platinum electrode. Several other electrodes are commonly used as secondary standard electrodes. One of these is the standard silver-silver chloride electrode, in which a silver electrode is in contact with solid silver chloride, which is a highly insoluble salt. The whole is immersed in potassium chloride solution in which the chloride-ion concentration is 1 *m*. This electrode can be represented as

$$Ag|AgCl|Cl^-(1\ m)$$

We can set up a cell involving this electrode and the hydrogen electrode,

$$Pt, H_2|H^+(1\ m)||Cl^-(1\ m)|AgCl|Ag$$

with a salt bridge connecting the two solutions. The emf is found to be 0.2224 V. The individual reactions are

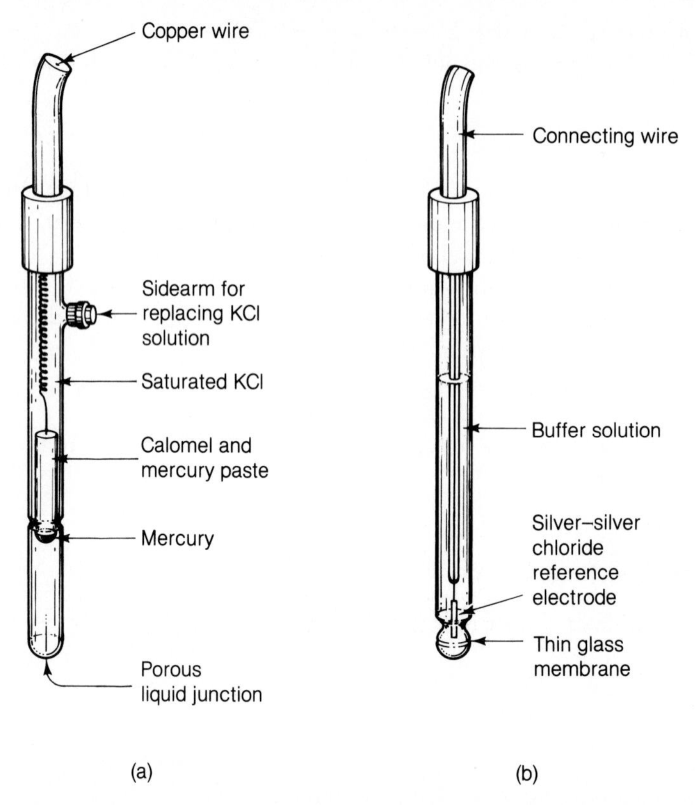

FIGURE 8.5 The calomel electrode (a), and the glass electrode (b). The pH meter, commonly used in chemical and biological laboratories, often employs a glass electrode that is immersed in the unknown solution and is used with a reference calomel electrode.

$$\tfrac{1}{2}H_2 \longrightarrow H^+ + e^-$$

$$e^- + AgCl \longrightarrow Ag + Cl^-$$

and the overall process is

$$\tfrac{1}{2}H_2 + AgCl \longrightarrow H^+ + Cl^- + Ag$$

The standard electrode potential for the silver-silver chloride electrode is thus 0.2224 V.

Another commonly used electrode is the *calomel electrode,* illustrated in Figure 8.5a. In this, mercury is in contact with mercurous chloride (calomel, Hg_2Cl_2) immersed either in a 1 *m* solution of potassium chloride or in a saturated solution of potassium chloride. If the cell

$$\text{Pt}, H_2|H^+(1\ m)\|Cl^-(1\ m)|Hg_2Cl_2|Hg$$

is set up, the individual reactions are

$$\tfrac{1}{2}H_2 \longrightarrow H^+ + e^-$$

$$e^- + \tfrac{1}{2}Hg_2Cl_2 \longrightarrow Hg + Cl^-$$

and the overall process is

$$\tfrac{1}{2}H_2 + \tfrac{1}{2}Hg_2Cl_2 \longrightarrow H^+ + Cl^- + Hg$$

The emf at 25°C is 0.3338 V, which is the standard electrode potential $E°$. If a saturated solution of KCl is used with the calomel electrode, the standard electrode potential is 0.2415 V.

Another electrode commonly used as a secondary standard is the *glass electrode*, illustrated in Figure 8.5b. In its simplest form this consists of a tube terminating in a thin-walled glass bulb; the glass is reasonably permeable to hydrogen ions. The glass bulb contains an 0.1 m hydrochloric acid solution and a tiny silver-silver chloride electrode. The theory of the glass electrode is somewhat complicated, but when the bulb is inserted into an acid solution, it behaves like a hydrogen electrode. This electrode is particularly convenient for making pH determinations.

Ion-Selective Electrodes*

The glass electrode was devised in 1906 by the German biologist M. Cremer† and was the prototype of a considerable number of membrane electrodes that have been developed subsequently. The importance of membrane electrodes is that some of them are highly selective to particular ions. For example, membrane electrodes have been constructed that are 10^4 times as responsive to Na^+ than to K^+ ions, and such electrodes are of great value for chemical analysis. Electrodes that are selective for more than 50 ions are now available, and most of them are membrane electrodes.

The principle of the membrane electrode is illustrated in Figure 8.6a. The sample solution is separated from an internal solution by an ion-selective membrane, and an internal reference electrode is placed within the internal solution. An external reference electrode, such as a silver-silver chloride electrode, is also immersed in the sample solution, and a measurement is made of the reversible emf of the assembly. The glass electrode is illustrated in Figure 8.5b, and some other assemblies are shown in Figures 8.6b–d.

The theory of membrane electrodes is quite complicated and is different in detail for each type of electrode. It is not necessary for the ion to which the electrode is sensitive actually to be transported through the membrane. What occurs at the membrane is a combination of an ion-exchange process at the solution-membrane interface and the movement of various cations at the interface. It is not necessary for the ion to be measured to be especially mobile or for the same ion to be present in the membrane. A complete theoretical treatment of a membrane electrode requires a consideration of the Donnan equilibrium that is established (Section 7.12),

---

*For reviews, see G. J. Hills, "Membrane Electrodes," Chapter 9 of D. J. G. Ives and G. J. Janz (Eds.), *Reference Electrodes: Theory and Practice,* New York: Academic Press, 1961; G. Eisenman (Ed.), *Glass Electrodes for Hydrogen and Other Cations, Principles and Practice,* New York: Dekker, 1967, G. A. Rechnitz, *Accounts Chem. Res., 3,* 69 (1970); R. A. Durst, *American Scientist, 59,* 353 (1971); R. P. Buck, "Potentiometry: pH Measurements and Ion-Selective Electrodes," Chapter 2 of A. Weissberger and B. W. Rossiter (Eds.), *Techniques of Chemistry* (Vol. 1), Part IIA. New York: Wiley-Interscience, 1971; N. Lakshminarayanaiah, *Membrane Electrodes.* New York: Academic Press, 1976; J. Koryta (Ed.), *Ion Selective Electrodes,* New York: Wiley, 1980.

†M. Cremer, *Z. Biol. 47,* 562 (1906).

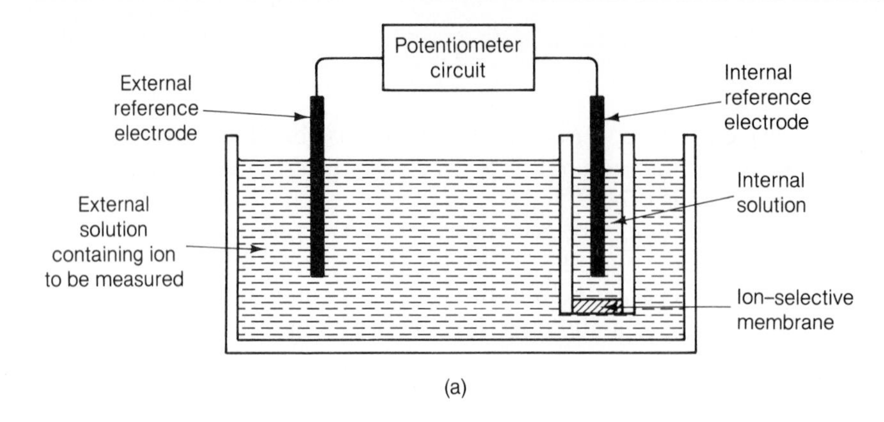

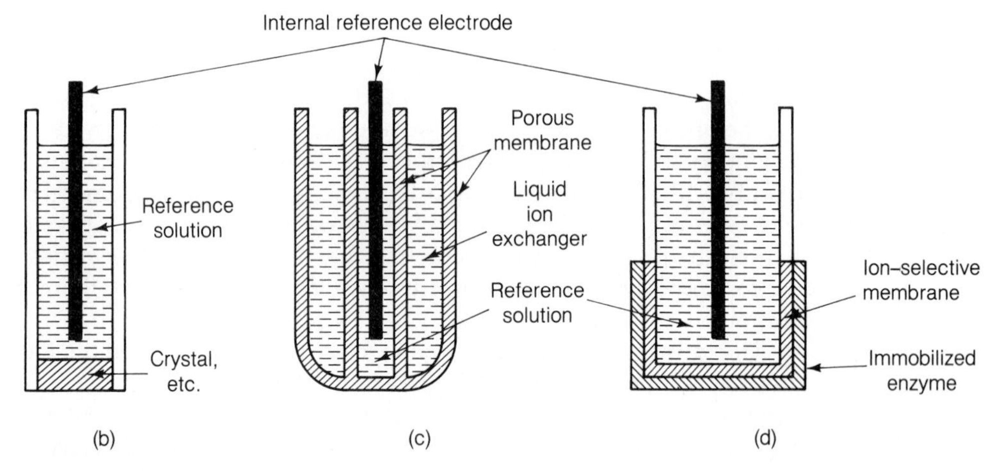

FIGURE 8.6 (a) The principle of the membrane electrode. (b) A membrane electrode in which the membrane is a single crystal, or a mixed crystal, or a matrix impregnated with a precipitate. (c) A membrane electrode in which the membrane is a liquid ion exchanger. (d) An enzyme-substrate electrode, which makes use of the ability of an enzyme to react selectively with an organic ion.

of the Nernst potential (Section 8.3), and of complications arising from deviations from equilibrium. Since so many factors are involved, the development of ion-selective electrodes is necessarily done on the basis of a good deal of empiricism.

8.3 Thermodynamics of Electrochemical Cells

During the last century studies were made of the relationship between the emf of a cell and the thermodynamics of the chemical reactions occurring in the cell. These studies made important contributions to the understanding of the basic principles of thermodynamics. An early contribution was made by Joule* who, with very

*J. P. Joule, *Phil. Mag.*, *18*, 308 (1841); *Proc. Roy. Soc.*, *4*, 280 (1843); an account of the work was communicated to the Royal Society by Faraday in 1840.

simple apparatus but with accurate temperature and current measurements, found in 1840 that

> The calorific effects of equal quantities of transmitted electricity are proportional to the resistance opposed to its passage, whatever may be the length, thickness, shape, or kind of metal which closes the circuit; and also that, *caeteris paribus*, these effects are in the duplicate ratio of the quantities of transmitted electricity, and, consequently, also in the duplicate ratio of the velocity of transmission.

By "quantity of transmitted electricity," Joule meant the current; by "duplicate ratio," the square. His conclusion was therefore that the heat produced was proportional to the square of the current I^2 and to the resistance R. Since it is also proportional to the time t, Joule had shown that the heat is proportional to

$$I^2Rt$$

Since the resistance R is the potential drop V divided by the current I (Ohm's law), it follows that the heat is proportional to

$$IVt$$

These conclusions have been confirmed by many later investigations.

The SI unit of heat is the joule, that of current is the ampere, and that of electric potential the volt; in these units the proportionality factor relating heat to IVt is unity:

$$q = IVt \tag{8.1}$$

This is readily seen by expressing the joule and the volt in terms of the base SI units (see Appendix A, Table A-1); thus

$$\text{J} \equiv \text{kg m}^2\ \text{s}^{-2}$$

$$\text{V} \equiv \text{kg m}^2\ \text{s}^{-3}\ \text{A}^{-1}$$

The product IVt is

$$\text{A} \times \text{kg m}^2\ \text{s}^{-3}\ \text{A}^{-1} \times \text{s} \equiv \text{kg m}^2\ \text{s}^{-2} \equiv \text{J}$$

Joule's conclusion that the heat generated in a wire is IVt is quite correct, but later he and others went wrong. In 1852 he concluded* that there is a correspondence between the heat of reaction of a cell and the electrical work. This error was also made by Helmholtz† and by William Thomson‡ (later Lord Kelvin). Thomson's conclusion appeared to be supported by his calculation of the emf of the Daniell cell from the heat of the reaction; his value, 1.074 V, is practically the measured value, but this agreement is accidental.§

It remained for Willard Gibbs‖ to draw the correct conclusion, in 1878, that the work done in an electrochemical cell is equal to the decrease in what is now known

---

*J. P. Joule, *Phil. Mag.*, 3, 481 (1952).

†H. Helmholtz, *Uber die Erhaltung der Kraft*, Berlin, 1847, p. 45.

‡W. Thomson, *Phil. Mag.*, 2, 429 (1851).

§It results from the fact that there is only a small entropy change in the Daniell cell ($\partial E/\partial T = 0$, see Eq. 8.23), so that $\Delta H \approx \Delta G$.

‖J. W. Gibbs, *Trans. Connecticut Acad.*, 3, 439 (1878). Helmholtz later gave the correct relationship in *Sitzb. Berlin Acad.*, 1, 21 (1882).

as the Gibbs energy. This is an example of the deduction we have already made, in Section 3.7, that non-PV work (i.e., available work) is equal to the decrease in Gibbs energy.

This may be illustrated for the standard cell

$$\text{Pt},\text{H}_2|\text{H}^+(1\ m)||\text{Cu}^{2+}(1\ m)|\text{Cu}$$

for which the emf ($E°$) is +0.337 V. The overall reaction is

$$\text{H}_2 + \text{Cu}^{2+} \rightarrow 2\text{H}^+ + \text{Cu}$$

Every time 1 mol of H_2 reacts with 1 mol of Cu^{2+}, 2 mol of electrons pass through the outer circuit. According to Faraday's laws, this means the transfer of 2 × 96 500 C of electricity. The emf developed is +0.337 V, and the passage of 2 × 96 500 C across this potential drop means that

$$2 \times 96\ 500 \times 0.337 \text{ C V} = 65\ 030 \text{ J}$$

of work has been done by the system. Thus, for this cell process,

$$\Delta G° = -65\ 030 \text{ J}$$

In general, for any standard-cell reaction associated with the passage of z electrons and an emf of $E°$, the change in Gibbs energy is

$$\mathbf{\Delta G° = -zFE°} \tag{8.2}$$

Since this Gibbs energy change is calculated from the $E°$ value, which relates to a cell in which the molalities are unity, it is a *standard* Gibbs energy change, as indicated by the superscript °.

The same argument applies to any cell; if the emf is E, the Gibbs energy change is

$$\Delta G = -zFE \tag{8.3}$$

The ΔG is the change in Gibbs energy when the reaction occurs with the concentrations having the values employed in the cell. Note that if E is positive, ΔG is negative; a positive E means that the cell is operating spontaneously with the reactions occurring in the forward direction (e.g., $H_2 + Cu^{2+} \rightarrow 2H^+ + Cu$), and this requires ΔG to be negative.

For any reaction

$$a\text{A} + b\text{B} + \cdots \longrightarrow \cdots + y\text{Y} + z\text{Z}$$

the Gibbs energy change that occurs when a mol of A at a concentration [A] reacts with b mol of B at [B], etc., is given by*

$$\mathbf{\Delta G = -RT\left[\ln K^u - \ln\left(\frac{\ldots [Y]^y[Z]^z}{[A]^a[B]^b \ldots}\right)^u\right]} \tag{8.4}$$

If the initial and final concentrations are unity, ΔG is $\Delta G°$ and is given by

$$\mathbf{\Delta G° = -RT \ln K^u} \tag{8.5}$$

*This is the approximate relationship in which concentrations rather than activities are used. In Eq. 8.4 and subsequent equations we are again using the superscript ^{u} to indicate the *numerical value* of the quantities such as K and the ratio $[Y]^y[Z]^z/[A]^a[B]^b$.

For any cell involving standard electrodes, such as the standard Daniell cell

$$Zn|Zn^{2+}(1\ m)||Cu^{2+}(1\ m)|Cu$$

Eq. 8.2 applies and therefore

$$E^\circ = \frac{RT}{zF} \ln K^u \tag{8.6}$$

At 25°C this becomes

$$E^\circ/\mathrm{V} = \frac{0.059\ 16}{z} \log_{10} K^u \tag{8.7}$$

These equations provide a very important method for calculating Gibbs energy changes and equilibrium constants. The extension of the method to the calculation of ΔH° and ΔS° values is considered on page 330.

EXAMPLE

Calculate the equilibrium constant at 25°C for the reaction occurring in the Daniell cell, if the standard emf is 1.100 V.

SOLUTION

The reaction is

$$Zn + Cu^{2+} \rightleftharpoons Zn^{2+} + Cu$$

and $z = 2$. From Eq. 8.7,

$$\log_{10} K^u = \frac{2 \times 1.100}{0.059\ 16} = 37.19$$

and thus

$$K = 1.5 \times 10^{37}$$ ■

EXAMPLE

Using the data in Table 8.1, calculate the equilibrium constant for the reaction

$$H_2 + 2Fe^{3+} \rightleftharpoons 2H^+ + 2Fe^{2+}$$

SOLUTION

From Table 8.1, the standard electrode potentials are

$$2H^+ + 2e^- \longrightarrow H_2 \qquad E^\circ = 0$$

$$Fe^{3+} + e^- \longrightarrow Fe^{2+} \qquad E^\circ = 0.771\ \mathrm{V}$$

The E° value for the process

$$H_2 + 2Fe^{3+} \longrightarrow 2H^+ + 2Fe^{2+}$$

for which $z = 2$ is thus 0.771 V. Therefore

$$0.771 = \frac{0.059\ 16}{2} \log_{10} K^u$$

and

$$K = 1.16 \times 10^{26} \text{ mol}^2 \text{ dm}^{-6}$$

■

It should be emphasized that in making this calculation *we must not* multiply the value of 0.771 V by 2; the emf of 0.771 V applies equally well to the process

$$2Fe^{3+} + 2e^- \longrightarrow 2Fe^{2+}$$

If the problem had been to calculate the equilibrium constant K' for the process

$$\tfrac{1}{2}H_2 + Fe^{3+} \rightleftharpoons H^+ + Fe^{2+}$$

the reactions would have been written as

$$H^+ + e^- \longrightarrow \tfrac{1}{2}H_2 \qquad E° = 0$$

$$Fe^{3+} + e^- \longrightarrow Fe^{2+} \qquad E° = 0.771 \text{ V}$$

and again $E° = 0.771$ V. In this case $z = 1$ and

$$0.771 = 0.059\,16 \log_{10} (K')^u \quad \text{or} \quad \log_{10} (K')^u = \frac{0.771}{0.059\,16} = 13.05$$

and therefore

$$K' = 1.08 \times 10^{13}$$

K' is, of course, the square root of K.

EXAMPLE

Calculate $E°$ for the process

$$Cu^+ + e^- \longrightarrow Cu$$

making use of the following $E°$ values:

1. $Cu^{2+} + e^- \longrightarrow Cu^+ \qquad E_1° = 0.153$ V
2. $Cu^{2+} + 2e^- \longrightarrow Cu \qquad E_2° = 0.337$ V

SOLUTION

The $\Delta G°$ values for these two reactions are

$$Cu^{2+} + e^- \longrightarrow Cu^+ \qquad \Delta G_1° = -zE_1°F = -1 \times 0.153 \times 96\,500 \text{ J mol}^{-1}$$

$$Cu^{2+} + 2e^- \longrightarrow Cu \qquad \Delta G_2° = -zE_2°F = -2 \times 0.337 \times 96\,500 \text{ J mol}^{-1}$$

The reaction $Cu^+ + e^- \rightarrow Cu$ is obtained by subtracting reaction 1 from reaction 2 and the $\Delta G°$ value for $Cu^+ + e^- \rightarrow Cu$ is therefore obtained by subtracting $\Delta G_1°$ from $\Delta G_2°$:

$$\Delta G° = [-2 \times 0.337 \times 96\,500 - (-1 \times 0.153 \times 96\,500)] \text{ J mol}^{-1}$$

$$= (0.153 - 0.674)96\,500 \text{ J mol}^{-1}$$

$$= -0.521 \times 96\,500 \text{ J mol}^{-1}$$

Since, for $Cu^+ + e^- \rightarrow Cu$, $z = 1$, it follows that

$$E^\circ = 0.521 \text{ V}$$

■

Note that it is incorrect, in working the previous example, simply to combine the E° values directly.

In view of this the student may wonder why it is legitimate to calculate E° values for overall cell reactions by simply combining the E° values for individual electrodes. Consider, for example, the following E° values:

$$Fe^{3+} + e^- \longrightarrow Fe^{2+} \qquad E^\circ = 0.771 \text{ V}$$

$$I_2 + 2e^- \longrightarrow 2I^- \qquad E^\circ = 0.536 \text{ V}$$

The procedure we have been adopting is to combine these two E° values:

$$2Fe^{3+} + 2I^- \longrightarrow 2Fe^{2+} + I_2 \qquad E^\circ = 0.771 - 0.536 = 0.235 \text{ V}$$

The fact that this is justified can be seen by writing the ΔG° values:

$$Fe^{3+} + e^- \longrightarrow Fe^{2+} \qquad \Delta G^\circ = -1 \times 0.771 \times 96\,500 \text{ J mol}^{-1}$$

$$I_2 + 2e^- \longrightarrow 2I^- \qquad \Delta G^\circ = -2 \times 0.536 \times 96\,500 \text{ J mol}^{-1}$$

We combine the two equations by multiplying the first by two and subtracting the second:

$$2Fe^{3+} + 2I^- \longrightarrow 2Fe^{2+} + I_2$$

$$\Delta G^\circ = [2(-1 \times 0.771 \times 96\,500) - (-2 \times 0.536 \times 96\,500)] \text{ J mol}^{-1}$$

$$= -2 \times 0.235 \times 96\,500 \text{ J mol}^{-1}$$

Thus $E^\circ = 0.235$ V, and this is simply $0.771 - 0.536$. We are justified in simply subtracting E° values to find E° for an overall reaction, in which there are no electrons left over. However, to obtain an E° for a half-reaction (as in last example), we can in general *not* simply combine E° values but must calculate the ΔG° values as just done.

The Nernst Equation

So far we have limited our discussion to standard electrode potentials E° and to E° values for cells in which the active species are present at 1 m concentrations. The corresponding standard Gibbs energies have been written as ΔG°.

Let us now remove this restriction and consider cells in which the concentrations are other than unity. Consider, for example, the cell

$$Pt, H_2|H^+||Cu^{2+}|Cu$$

in which a hydrogen electrode has been combined with a copper electrode immersed in a Cu^{2+} solution, the concentration of which is other than unity. The overall cell reaction is

$$H_2 + Cu^{2+} \longrightarrow 2H^+ + Cu$$

and the Gibbs energy change is (see Eq. 8.4)

$$\Delta G = -RT\left[\ln K^u - \ln\left(\frac{[H^+]^2}{[Cu^{2+}]}\right)^u\right] \tag{8.8}$$

However $\Delta G^\circ = -RT \ln K^u = -zE^\circ F$, and therefore

$$\Delta G = -zE^\circ F + RT \ln\left(\frac{[H^+]^2}{[Cu^{2+}]}\right)^u \tag{8.9}$$

Since $\Delta G = -zEF$, where E is the emf of this cell, we obtain

$$-zEF = -zE^\circ F + RT \ln\left(\frac{[H^+]^2}{[Cu^{2+}]}\right)^u \tag{8.10}$$

and therefore

$$E = E^\circ - \frac{RT}{zF} \ln\left(\frac{[H^+]^2}{[Cu^{2+}]}\right)^u \tag{8.11}$$

In general, we may consider any cell for which the overall reactions have the general form

$$a\text{A} + b\text{B} + \cdots \longrightarrow \cdots + y\text{Y} + z\text{Z}$$

ΔG is given by Eq. 8.3 and $\Delta G^\circ = -RT \ln K^u = -zE^\circ F$, and therefore

$$\Delta G = -zE^\circ F + RT \ln\left(\frac{\ldots [\text{Y}]^y[\text{Z}]^z}{[\text{A}]^a[B]^b \ldots}\right)^u \tag{8.12}$$

Note that, as in an equilibrium constant, products are in the numerator and reactants in the denominator. Since $\Delta G = -zEF$, Eq. 8.12 leads to the following general expression for the emf:

$$\boldsymbol{E = E^\circ - \frac{RT}{zF} \ln\left(\frac{\ldots [\text{Y}]^y[\text{Z}]^z}{[\text{A}]^a[\text{B}]^b \ldots}\right)^u} \tag{8.13}$$

This relationship was first given in 1889 by the German physical chemist Walter H. Nernst (1864–1941) and is known as the **Nernst equation.***

Suppose that we apply the Nernst equation to the cell

$$\text{Zn}|\text{Zn}^{2+}||\text{Ni}^{2+}|\text{Ni}$$

for which the overall reaction is

$$\text{Zn} + \text{Ni}^{2+} \longrightarrow \text{Zn}^{2+} + \text{Ni}$$

The standard electrode potentials (see Table 8.1) are

$$\text{Ni}^{2+} + 2e^- \longrightarrow \text{Ni} \qquad E^\circ = -0.25 \text{ V}$$

$$\text{Zn}^{2+} + 2e^- \longrightarrow \text{Zn} \qquad E^\circ = -0.76 \text{ V}$$

and E° for the overall process is $-0.25 - (-0.76) = 0.51$ V. The Nernst equation is thus

*W. Nernst, *Z. Physik. Chem.*, *4*, 129 (1889).

$$E = 0.51\ (\text{V}) - \frac{RT}{zF} \ln \frac{[\text{Zn}^{2+}]}{[\text{Ni}^{2+}]} \tag{8.14}$$

(As always, the concentrations of solid species such as Zn and Ni are incorporated into the equilibrium constant and are therefore not included explicitly in the equation.) At 25°C this equation becomes

$$E/V = 0.51 - \frac{0.059\ 16}{2} \log_{10} \frac{[\text{Zn}^{2+}]}{[\text{Ni}^{2+}]} \tag{8.15}$$

since $z = 2$. We see from this equation that increasing the ratio $[Zn^{2+}]/[Ni^{2+}]$ decreases the cell emf; this is understandable in view of the fact that a positive emf means that the cell is producing Zn^{2+} and that Ni^{2+} ions are being removed.

EXAMPLE

Calculate the emf of the cell

$$\text{Co}|\text{Co}^{2+}||\text{Ni}^{2+}|\text{Ni}$$

if the concentrations are

(a) $[Ni^{2+}] = 1\ m$ and $[Co^{2+}] = 0.1\ m$

(b) $[Ni^{2+}] = 0.01\ m$ and $[Co^{2+}] = 1.0\ m$

SOLUTION

The cell reaction is

$$\text{Co} + \text{Ni}^{2+} \longrightarrow \text{Co}^{2+} + \text{Ni}$$

and from Table 8.1 the standard electrode potentials are

$$\text{Ni}^{2+} + 2e^- \longrightarrow \text{Ni} \qquad E° = -0.25\ \text{V}$$

$$\text{Co}^{2+} + 2e^- \longrightarrow \text{Co} \qquad E° = -0.28\ \text{V}$$

The standard emf is thus $-0.25 - (-0.28) = 0.03$ V, and $z = 2$. The cell emf at the first concentrations specified (a) is

$$E = 0.03 - \frac{0.059\ 16}{2} \log_{10} \frac{[\text{Co}^{2+}]}{[\text{Ni}^{2+}]}$$

$$= 0.03 - \frac{0.059\ 16}{2} \log_{10} 0.1 = 0.03 + 0.03 = 0.06\ \text{V}$$

In (b),

$$E = 0.03 - \frac{0.059\ 16}{2} \log_{10} \frac{1}{0.01}$$

$$= 0.03 - 0.059 = -0.029\ \text{V}$$

We see that the cell operates in opposite directions in the two cases. ■

Nernst Potentials

If two electrolyte solutions of different concentrations are separated by a membrane, in general there will be an electric potential difference across the membrane. Various situations are possible. Suppose, for example, that solutions of potassium chloride were separated by a membrane, as shown in Figure 8.7. If both ions could cross the membrane, the concentrations would eventually become equal and there would be no potential difference (Figure 8.7a). If, however, only the potassium ions can cross, and the membrane is not permeable to the solvent,* there will be very little change in the concentrations on the two sides of the membrane. Some K^+ ions will cross from the more concentrated side (the left-hand side in Figure 8.7b), and as a result the left-hand side will have a negative potential with respect to the right-hand side. The effect of this will be to prevent more K^+ ions from crossing. An equilibrium will therefore be established at which the electric potential will exactly balance the tendency of the concentrations to become equal. This potential is known as the **Nernst potential.**

The Gibbs energy difference ΔG_e arising from the potential difference $\Delta\Phi$ is

$$\Delta G_e = zF\,\Delta\Phi \tag{8.16}$$

where z is the charge on the permeable ions ($z = 1$ in this case). In our particular example (Figure 8.7b), the potential is higher on the right-hand side ($\Delta\Phi > 0$), because a few K^+ ions have crossed, and there will thus be a higher Gibbs energy, as far as K^+ ions are concerned, on the right-hand side ($\Delta G_e > 0$). The Gibbs energy difference arising from the concentration difference is

$$\Delta G_c = RT \ln \frac{c_2}{c_1} \tag{8.17}$$

At equilibrium there is no net ΔG across the membrane and therefore

$$\Delta G_e + \Delta G_c = zF\,\Delta\Phi + RT \ln \frac{c_2}{c_1} = 0 \tag{8.18}$$

Thus

$$\boldsymbol{\Delta\Phi = \frac{RT}{zF} \ln \frac{c_1}{c_2}} \tag{8.19}$$

EXAMPLE

Mammalian muscle cells are freely permeable to K^+ ions but much less permeable to Na^+ and Cl^- ions. Typical concentrations of K^+ ions are

Inside the cell: $[K^+] = 155$ mM

Outside the cell: $[K^+] = 4$ mM

Calculate the Nernst potential at 310 K (37°C) on the assumption that the membrane is impermeable to Na^+ and Cl^-.

*This restriction is made so that there will be no complications due to osmotic effects.

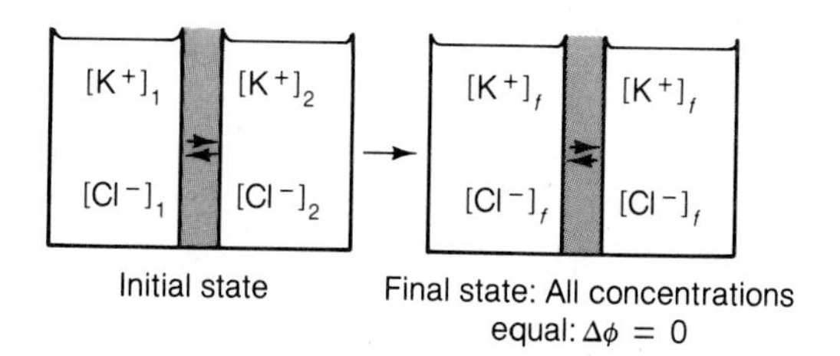

(a) Membrane permeable to both ions

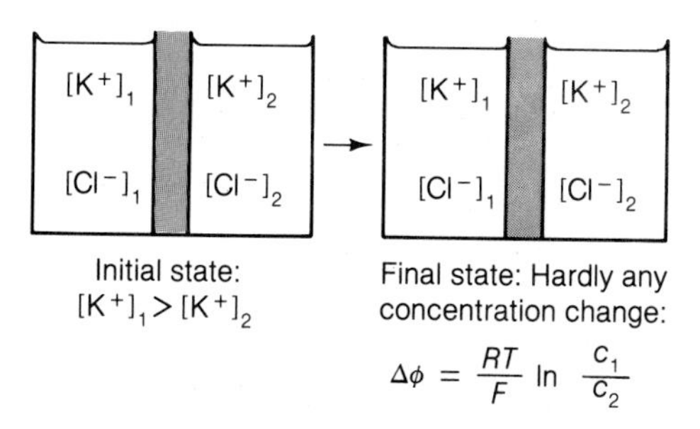

(b) Membrane permeable to K^+ only

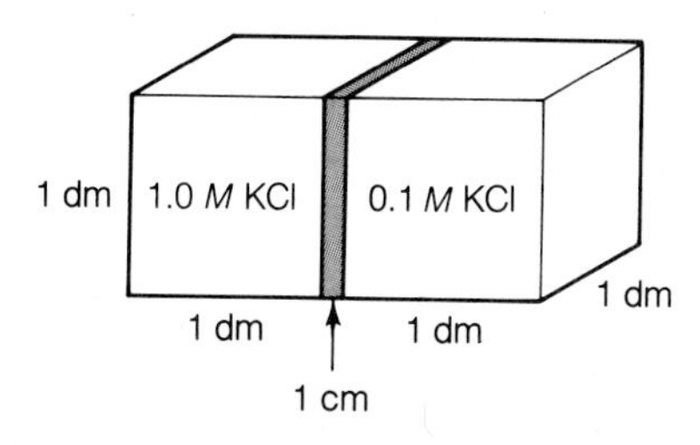

(c) Two cubic cells separated by a membrane

FIGURE 8.7 (a) Solutions of KCl separated by a membrane that is permeable to both ions. (b) Solutions of KCl separated by a membrane that is permeable only to K^+. (c) Two 1-dm^3 cubes separated by a membrane of area 1 dm^2 and thickness 1 cm.

SOLUTION

From Eq. 8.19 with $z = 1$,

$$\Delta\Phi = \frac{8.314(\text{J K}^{-1}\ \text{mol}^{-1}) \times 310(\text{K})}{96\,500(\text{C mol}^{-1})} \ln \frac{4}{155}$$

$$= 0.0267 \ln \frac{4}{155} = -0.0976 \text{ V} = -97.6 \text{ mV}$$

The potential is negative inside the cell and positive outside. In reality, such potentials are more like −85 mV, because there is a certain amount of diffusion of Na^+ and Cl^-, and there is also a biological pumping mechanism. ■

It is important to realize that in situations of this kind the Nernst potential is due to the transfer of only an exceedingly small fraction of the diffusible ions, so that there is no detectable concentration change. Suppose, for example, that two solutions 1 dm^3 in volume are separated by a membrane 1 dm^2 in area (A) and 1 cm in thickness (l), and suppose that 0.1 M and 1.0 M solutions of KCl are present on the two sides (Figure 8.7c), the membrane again being permeable only to the K^+ ions. The capacitance C of the membrane is

$$C = \frac{\epsilon_0 \epsilon A}{l} \tag{8.20}$$

where ϵ, the dielectric constant of the membrane, will be taken to have a value of 3 (this is typical of biological and other organic membranes). The capacitance in this example is therefore

$$C = \frac{8.854 \times 10^{-12}(\text{C}^2\ \text{N}^{-1}\ \text{m}^{-2}) \times 3 \times 10^{-2}(\text{m}^2)}{10^{-2}(\text{m})}$$

$$= 2.66 \times 10^{-11}\ \text{C}^2\ \text{N}^{-1}\ \text{m}^{-1} = 2.66 \times 10^{-11}\ \text{F}$$

where F is the *farad*, the unit of electric capacitance. In terms of base SI units:

$$\text{F} \equiv \text{C V}^{-1} \equiv \text{A s (kg m}^2\ \text{s}^{-3}\ \text{A}^{-1})^{-1} \equiv \text{A}^2\ \text{s}^4\ \text{kg}^{-1}\ \text{m}^{-2}$$

$$\text{C}^2\ \text{N}^{-1}\ \text{m}^{-1} \equiv (\text{A s})^2(\text{kg m s}^{-2})^{-1}\ \text{m}^{-1} \equiv \text{A}^2\ \text{s}^4\ \text{kg}^{-1}\ \text{m}^{-2} \equiv \text{F}$$

The Nernst potential at 25°C is

$$\Delta\Phi = \frac{8.314 \times 298.15}{96\ 500} \ln 10 = 0.059\ \text{V}$$

With a capacitance of 2.66×10^{-11} F, the net charge on each side of the wall, required to maintain this potential, is

$$Q = 2.66 \times 10^{-11}\ \text{F} \times 0.059\ \text{V}$$

$$= 1.57 \times 10^{-12}\ \text{C}$$

The number of K^+ ions required to produce this charge is

$$\frac{Q}{e} = \frac{1.57 \times 10^{-12}\ \text{C}}{1.602 \times 10^{-19}\ \text{C}} = 9.8 \times 10^6$$

However, 1 dm^3 of the 0.1 M solution contains $0.1 \times 6.022 \times 10^{23} = 6.022 \times 10^{22}$ ions. The fraction involved in establishing the potential difference is therefore exceedingly small ($<10^{-14}$) and no concentration change could be detected. The fraction is larger if the system is smaller, but even for a biological cell it is no more than 10^{-6} (see Problem 15).

The situation that we have described relates to a membrane permeable to only one ion. If more than one ion can pass through the membrane, the situation is more complex. The distribution of K^+, Na^+, and Cl^- ions across a biological membrane provides an interesting example. In the example on page 327 we saw that mainly as a result of the K^+ ions, which are freely permeable, the Nernst potential is about -85 mV. The Cl^- ions are also somewhat permeable, and typical concentrations are

Inside the cell: $[Cl^-] = 4.3$ mM

Outside the cell: $[Cl^-] = 104$ mM

The Nernst potential corresponding to this distribution is

$$\Delta\Phi = 0.0267 \ln \frac{4.3}{104} = -0.0851 \text{ V} = -85.1 \text{ mV}$$

which is close to the observed potential. The Cl^- ions are therefore more or less at equilibrium. Typical concentrations for Na^+ ions are

Inside the cell: $[Na^+] = 12$ mM

Outside the cell: $[Na^+] = 145$ mM

If these ions were permeable, their Nernst potential would be

$$\Delta\Phi = 0.0267 \ln \frac{145}{12} = 66.5 \text{ mV}$$

This is of the opposite sign to the true potential, and this distribution of Na^+ ions only arises because of their inability to cross the membrane. These ionic distributions are of great importance in connection with nerve impulses, which are activated by changes in permeability of the nerve membranes.

Potential differences across membranes are set up when a Donnan equilibrium (Section 7.12) is established. The procedure for calculating them is best illustrated by an example.

EXAMPLE

A 0.10 M solution of sodium palmitate is separated from an equal volume of a 0.20 M solution of sodium chloride by a membrane that is permeable to Na^+ and Cl^- but not to palmitate ions. Calculate the final concentrations and the Nernst potential at 25°C, assuming ideal behavior.

SOLUTION

Suppose that x mol dm^{-3} of Na^+ and Cl^- ions move to the palmitate side; the final concentrations are

Palmitate side: $[Na^+] = (0.1 + x)\ M$; $[Cl^-] = x\ M$

Other side: $[Na^+] = (0.2 - x)\ M$; $[Cl^-] = (0.2 - x)\ M$

At equilibrium

$$(0.2 - x)^2 = x(0.1 + x)$$

$$x = 0.08\ M$$

The final concentrations are therefore

Palmitate side: $[Na^+] = 0.18\ M$; $[Cl^-] = 0.08\ M$

Other side: $[Na^+] = [Cl^-] = 0.12\ M$

The Nernst potential arising from the distribution of Na^+ ions is, at 25°C,

$$\Delta\Phi = \frac{8.314 \times 298.15}{96\ 500} \ln \frac{0.18}{0.12} = 0.0104 \text{ V} = 10.4 \text{ mV}$$

The palmitate side, having the higher concentration of Na^+, is the negative side, since Na^+ ions will tend to cross from that side. The Nernst potential calculated from the distribution of Cl^- ions is exactly the same:

$$\Delta\Phi = \frac{8.314 \times 298.15}{96\ 500} \ln \frac{0.12}{0.08} = 10.4 \text{ mV}$$ ■

Strictly speaking, $\Delta\Phi$ values measured by placing platinum electrodes in two solutions separated by a membrane are not precisely the same as these calculated Nernst potentials. The measured values are for the potential drop between the Pt electrodes, but the potential drop between the electrode and the solution is not quite the same for the two solutions. However, the error is about the same as the error in the calculations, and in electrophysiological and other studies it is usually assumed that the measured values can be equated to those obtained in the calculations.

Temperature Coefficients of Cell emfs

Since a Gibbs energy change can be obtained from the standard emf of a reversible cell (Eq. 8.2), the $\Delta S°$ and $\Delta H°$ values can be calculated if emf measurements are made over a range of temperature.

The basic relationship is Eq. 3.119:

$$S = -\left(\frac{\partial G}{\partial T}\right)_P \tag{8.21}$$

and for an overall reaction

$$\Delta S = -\left(\frac{\partial\ \Delta G}{\partial T}\right)_P \tag{8.22}$$

Introduction of Eq. 8.2 gives

$$\mathbf{\Delta S = zF\left(\frac{\partial E}{\partial T}\right)_P} \tag{8.23}$$

The enthalpy change is thus

$$\Delta H = \Delta G + T\,\Delta S \tag{8.24}$$

$$\mathbf{\Delta H = -zF\left(E - T\frac{\partial E}{\partial T}\right)} \tag{8.25}$$

The measurement of emf values at various temperatures provides a very convenient method of obtaining thermodynamic values for chemical reactions and has frequently been employed. For the results to be reliable the temperature coefficients should be known to three significant figures, and this requires careful temperature and emf measurements.

EXAMPLE

The emf of the cell

$$\text{Pt, } H_2(1 \text{ atm})|HCl(0.01\ m)|AgCl(s)|Ag$$

is 0.2002 V at 25°C, and $\partial E/\partial T$ is -8.665×10^{-5} V K^{-1}. Write the cell reaction and calculate ΔG, ΔS, and ΔH at 25°C.

SOLUTION

The electrode reactions are

$$\tfrac{1}{2}H_2 \longrightarrow H^+ + e^- \quad \text{and} \quad e^- + AgCl(s) \longrightarrow Ag + Cl^-$$

The cell reaction is

$$\tfrac{1}{2}H_2 + AgCl(s) \longrightarrow Ag + H^+ + Cl^-$$

The Gibbs energy change is

$$\Delta G = -96\ 500 \times 0.2002 = -19\ 320 \text{ J mol}^{-1}$$

The entropy change is obtained by use of Eq. 8.23:

$$\Delta S = 96\ 500 \times -8.665 \times 10^{-5} = -8.36 \text{ J K}^{-1} \text{ mol}^{-1}$$

The enthalpy change can be calculated by use of Eq. 8.25 or more easily from the ΔG and ΔS values:

$$\begin{aligned}\Delta H &= \Delta G + T\,\Delta S \\ &= -19\ 320 + (-8.36 \times 298.15) \\ &= -21\ 800 \text{ J mol}^{-1}\end{aligned}$$

■

8.4 Types of Electrochemical Cells

In the cells we have considered so far there is a net chemical change. Such electrochemical cells are classified as **chemical cells.** There are also cells in which the driving force, instead of being a chemical reaction, is a dilution process. Such cells are known as **concentration cells.** The changes in concentration can occur either in the electrolyte or at the electrodes. Examples of concentration changes at electrodes are found with electrodes made of amalgams or consisting of alloys and with gas electrodes (e.g., the Pt, H_2 electrode) when there are different gas pressures at the two electrodes.

Figure 8.8 shows a classification of electrochemical cells. A subclassification of chemical and concentration cells relates to whether or not there is a boundary between two solutions. If there is not, as in the cell

$$\text{Pt, } H_2|HCl|AgCl(s)|Ag$$

we have a *cell without transference.* If there is, as in the Daniell cell (Figure 8.1), the cell is known as a *cell with transference.* In the latter case there is a potential difference between the solutions—which can be minimized by use of a salt bridge—and there are irreversible changes in the two solutions as the cell is operated.

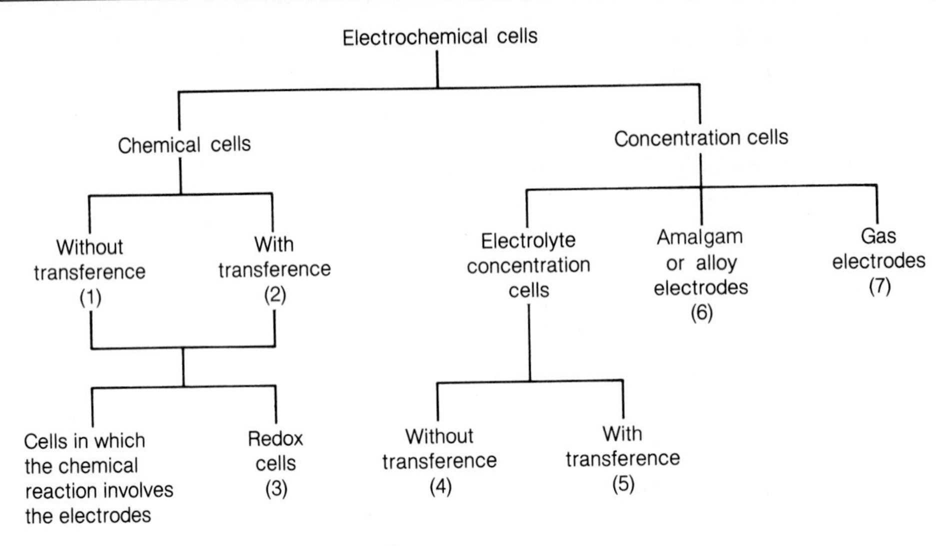

FIGURE 8.8 Classification of electrochemical cells.

1. Pt, $H_2|HCl|AgCl|Ag$
2. $Zn|Zn^{2+} \vdots Cu^{2+}|Cu$
3. Pt, $H_2|H^+Fe^{2+}$, $Fe^{3+}|Pt$
4. Pt, $H_2|HCl(m_1)|AgCl|Ag - Ag|AgCl|HCl(m_2)|Pt$, H_2
5. Pt, $H_2|H^+(m_1) \vdots H^+(m_2)|Pt$, H_2
6. Na in Hg at $c_1|Na^+|$Na in Hg at c_2
7. Pt, $H_2(P_1)|H^+|Pt$, $H_2(P_2)$

Concentration Cells

A simple example of a concentration cell is obtained by connecting two hydrogen electrodes by means of a salt bridge:

$$\text{Pt, } H_2|HCl(m_1)||HCl(m_2)|H_2\text{, Pt}$$

The salt bridge could be a tube containing saturated potassium chloride solution. The reaction at the left-hand electrode is

$$\tfrac{1}{2}H_2 \longrightarrow H^+(m_1) + e^-$$

while that at the right-hand electrode is

$$H^+(m_2) + e^- \longrightarrow \tfrac{1}{2}H_2$$

The net process is therefore

$$H^+(m_2) \longrightarrow H^+(m_1)$$

and is simply the transfer of hydrogen ions from a solution of molality m_2 to one of molality m_1. If m_2 is greater than m_1, the process will actually occur in this direction, and a positive emf is produced; if m_2 is less than m_1, the emf is negative and electrons flow from the right-hand to the left-hand electrode.

The Gibbs energy change associated with the transfer of H^+ ions from a molality m_2 to a molality m_1 is

$$\Delta G = RT \ln \frac{m_1}{m_2} \tag{8.26}$$

Since $z = 1$, the emf produced is

$$E = \frac{RT}{F} \ln \frac{m_2}{m_1} \tag{8.27}$$

and is positive when $m_2 > m_1$.

EXAMPLE

Calculate the emf at 25°C of a concentration cell of this type in which the molalities are 0.2 m and 3.0 m.

SOLUTION

The emf is given by

$$E = \frac{RT}{F} \ln \frac{m_2}{m_1} = 0.059\ 16 \log_{10} \frac{3.0}{0.2} = 0.0696 \text{ V}$$ ■

Redox Cells

Since, when cells operate, there is an electron transfer at the electrodes, oxidations and reductions are occurring. In all the cells considered so far these oxidations and reductions have involved the electrodes themselves—for example, the H_2 gas in the hydrogen electrode. There is also an important class of oxidation-reduction cells, known as **redox cells,** in which both the oxidized and reduced species are in solution; their interconversion is effected by an inert electrode such as one of platinum.

Consider, for example, the cell

$$\text{Pt, } H_2|H^+(1\ m)|Fe^{2+}, Fe^{3+}|\text{Pt}$$

The left-hand electrode is the standard hydrogen electrode. The right-hand electrode consists simply of a platinum electrode immersed in a solution containing both Fe^{2+} and Fe^{3+} ions. The platinum electrode is able to catalyze the interconversion of these ions, and the reaction at this electrode is

$$e^- + Fe^{3+} \longrightarrow Fe^{2+}$$

Since the reaction at the hydrogen electrode is

$$\tfrac{1}{2}H_2 \longrightarrow H^+ + e^-$$

the overall process is

$$Fe^{3+} + \tfrac{1}{2}H_2 \longrightarrow Fe^{2+} + H^+$$

The emf of this cell represents the ease with which the Fe^{3+} is reduced to Fe^{2+}. The emf of the cell is

$$E = E° - \frac{RT}{F} \ln \frac{[Fe^{2+}]}{[Fe^{3+}]} \tag{8.28}$$

(Note that the $E°$ relates to $P_{H_2} = 1$ atm and $[H^+] = [Fe^{2+}] = [Fe^{3+}] = 1\ m$.) The $E°$ for this system is +0.77 V (Table 8.1).

The interconversion of oxidized and reduced forms frequently involves also the participation of hydrogen ions. Thus, the half-reaction for the reduction of fumarate ions to succinate ions is

$$\begin{matrix} CHCOO^- \\ \| \\ CHCOO^- \end{matrix} + 2H^+ + 2e^- \longrightarrow \begin{matrix} CH_2COO^- \\ | \\ CH_2COO^- \end{matrix}$$

fumarate succinate

If we wished to study this system, we could set up the following cell:

$$Pt, H_2|H^+(1\ m)||F^{2-}, S^{2-}, H^+(m_{H^+})|Pt$$

where F^{2-} and S^{2-} represent fumarate and succinate, respectively. The hydrogen-ion molality in the right-hand solution is not necessarily 1 m; it will here be denoted as m_{H^+}. If we combine the standard hydrogen electrode,

$$H_2 \longrightarrow 2H^+(1\ m) + 2e^-$$

we obtain, for the overall cell reaction,

$$F^{2-} + 2H^+(m_{H^+}) + H_2 \longrightarrow S^{2-} + 2H^+(1\ m)$$

The equation for the emf is thus, since $z = 2$,

$$E = E° - \frac{RT}{2F} \ln \left(\frac{[S^{2-}]}{[F^{2-}]m_{H^+}^2} \right)^u \tag{8.29}$$

The $E°$ for this system is related to a standard Gibbs energy change $\Delta G°$ by the usual equation

$$\Delta G° = -zFE° \tag{8.30}$$

This standard Gibbs energy change is related to the equilibrium constant:

$$K = \frac{[S^{2-}]}{[F^{2-}]m_{H^+}^2} \tag{8.31}$$

However, it is frequently convenient to deal with the modified equilibrium constant

$$K' = \frac{[S^{2-}]}{[F^{2-}]} \tag{8.32}$$

at some specified hydrogen-ion concentration. Often this standard concentration is taken to be 10^{-7} M, corresponding to a pH of 7. In that case

$$K' = (10^{-7})^2K = 10^{-14}K \tag{8.33}$$

In many cases the K' value corresponds to a fairly well-balanced equilibrium at pH 7, whereas K will be larger by the factor 10^{14}; the K' value and the corresponding $\Delta G°'$ at pH 7 therefore give a clearer indication of the situation at that pH.

Equation 8.29 for the emf can be written as

$$E = E° - \frac{RT}{2F} \ln \frac{[S^{2-}]}{[F^{2-}]} + \frac{2.303RT}{F} \log m_{H^+} \tag{8.34}$$

$$= E° - \frac{RT}{2F} \ln \frac{[S^{2-}]}{[F^{2-}]} + 0.059\ 16 \log m_{H^+} \tag{8.35}$$

$$= E^\circ - \frac{RT}{2F} \ln \frac{[S^{2-}]}{[F^{2-}]} - 0.059\ 16\ \text{pH} \tag{8.36}$$

where pH is the pH of the solution in which the $S^{2-}:F^{2-}$ system is maintained. Then, if we define a modified standard potential $E^{\circ\prime}$ by

$$E = E^{\circ\prime} - \frac{RT}{2F} \ln \frac{[S^{2-}]}{[F^{2-}]} \tag{8.37}$$

it follows that

$$E^{\circ\prime} = E^\circ - 0.059\ 16\ \text{pH} \tag{8.38}$$

Different relationships apply to other reaction types.

8.5 Applications of emf Measurements*

A number of physical measurements are conveniently made by setting up appropriate electrochemical cells. Since emf values can be determined very accurately, such techniques are frequently employed. A few of them will be mentioned briefly.

pH Determinations

Since the emf of a cell such as that shown in Figure 8.4 depends on the hydrogen-ion concentration, pH values can be determined by dipping hydrogen electrodes into solutions and measuring the emf with reference to another electrode. In commercial pH meters the electrode immersed in the unknown solution is often a glass electrode, and the other electrode may be the silver-silver chloride or the calomel electrode. In accordance with the Nernst equation the emf varies logarithmically with the hydrogen-ion concentration and therefore varies linearly with the pH. Commercial instruments are calibrated so as to give a direct reading of the pH.

Activity Coefficients

Up to now we have expressed the Gibbs energy changes and emf values of cells in terms of molalities. This is an approximation and the errors become more serious as concentrations are increased. For a correct formulation, *activities* must be employed, and the emf measurements over a range of concentrations lead to values for the activity coefficients.

Consider the cell

$$\text{Pt, } H_2|HCl|AgCl|Ag$$

The overall process is

$$\tfrac{1}{2}H_2 + AgCl \longrightarrow Ag + H^+ + Cl^-$$

and the Gibbs energy change is

$$\Delta G = \Delta G^\circ + RT \ln [a_+ a_-]^u \tag{8.39}$$

*The remainder of this chapter could be omitted on first reading.

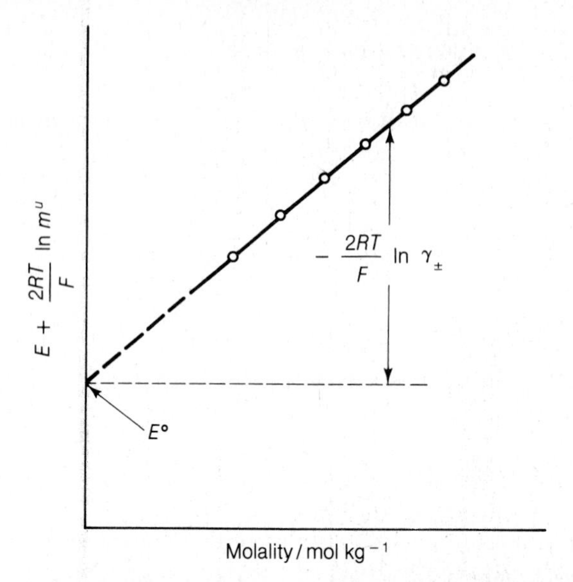

FIGURE 8.9 A plot that will provide the value of $E°$ for a cell and the mean activity coefficients at various concentrations.

where a_+ and a_- are the activities of the H^+ and Cl^- ions, and $\Delta G°$ is the standard Gibbs energy change, when the activities are unity. The emf is, since $z = 1$,

$$E = E° - \frac{RT}{F} \ln [a_+a_-]^u \tag{8.40}$$

Since each of the activities a_+ and a_- is the molality m multiplied by the activity coefficient γ_+ or γ_-, we can write

$$E = E° - \frac{RT}{F} \ln [m^u]^2 - \frac{RT}{F} \ln \gamma_+ \gamma_- \tag{8.41}$$

$$= E° - \frac{2RT}{F} \ln m^u - \frac{2RT}{F} \ln \gamma_\pm \tag{8.42}$$

where $\gamma_\pm$, equal to $\sqrt{\gamma_+ \gamma_-}$, is the **mean activity coefficient**. This equation can be written as

$$E + \frac{2RT}{F} \ln m^u = E° - \frac{2RT}{F} \ln \gamma_\pm \tag{8.43}$$

and if E is measured over a range of molalities of HCl, the quantity on the left-hand side can be calculated at various molalities. If this quantity is plotted against m, as shown schematically in Figure 8.9, the value extrapolated to zero m gives $E°$, since at zero m the activity coefficient $\gamma_\pm$ is unity so that the final term vanishes. At any molality the ordinate minus $E°$ then yields

$$-\frac{2RT}{F} \ln \gamma_\pm$$

from which the activity coefficient $\gamma_\pm$ can be calculated.

Equilibrium Constants

The emf determination of equilibrium constants may be illustrated with reference to the measurement of the dissociation constant of an electrolyte HA such as acetic acid. Suppose that the following cell is set up:

$$\text{Pt, } H_2|HA(m_1),\ NaA(m_2),\ NaCl(m_3)|AgCl|Ag$$

The essential feature of this cell is that the solution contains NaA, which being a salt is essentially completely dissociated and therefore provides a known concentration of A^- ions. The emf of the cell provides a measure of the hydrogen-ion concentration (since the hydrogen electrode is used); since A^- and H^+ are known, and the total amount of HA is known, the amount of undissociated HA is known, and hence the dissociation constant can be obtained.

In more detail, the result is obtained as follows. The reactions at the two electrodes and the overall reaction are

$$\begin{array}{l} \frac{1}{2}H_2 \longrightarrow H^+ + e^- \\ e^- + AgCl \longrightarrow Ag + Cl^- \\ \hline \frac{1}{2}H_2 + AgCl \longrightarrow H^+ + Cl^- + Ag \end{array}$$

The emf of the cell is

$$E = E^\circ - \frac{RT}{F} \ln [a_{H^+} a_{Cl^-}]^u \tag{8.44}$$

$$= E^\circ - \frac{RT}{F} \ln [m_{H^+} m_{Cl^-}]^u - \frac{RT}{F} \ln \gamma_{H^+} \gamma_{Cl^-} \tag{8.45}$$

E° is the standard electrode potential of the silver-silver chloride electrode (0.2224 V at 25°C).

The dissociation constant for the acid HA is

$$K_a = \frac{m_{H^+} m_{A^-}}{m_{HA}} \cdot \frac{\gamma_{H^+} \gamma_{A^-}}{\gamma_{HA}} \tag{8.46}$$

and Eq. 8.45 can be written as

$$E = E^\circ - \frac{RT}{F} \ln \left[\frac{m_{HA} m_{Cl^-}}{m_{A^-}}\right]^u - \frac{RT}{F} \ln \frac{\gamma_{HA} \gamma_{Cl^-}}{\gamma_{A^-}} - \frac{RT}{F} \ln K_a^u \tag{8.47}$$

or as

$$(E - E^\circ)\frac{F}{RT} + \ln \left[\frac{m_{HA} m_{Cl^-}}{m_{A^-}}\right]^u = -\ln \frac{\gamma_{HA} \gamma_{Cl^-}}{\gamma_{A^-}} - \ln K_a^u \tag{8.48}$$

The molality m_{Cl^-} is equal to m_3, the molality of the NaCl solution. The value of m_{HA} is equal to $m_1 - m_{H^+}$; and the value of m_{A^-} is $m_2 + m_{H^+}$. The molality m_{H^+} can be sufficiently well estimated from an approximate value of the dissociation constant; it is usually much less than m_1 and m_2 so that not much error arises from a rough estimate. The quantities on the left-hand side of Eq. 8.48 are therefore known at various values of the molalities m_1, m_2, and m_3, and the left-hand side can be plotted against the ionic strength I, as shown schematically in Figure 8.10. At zero ionic

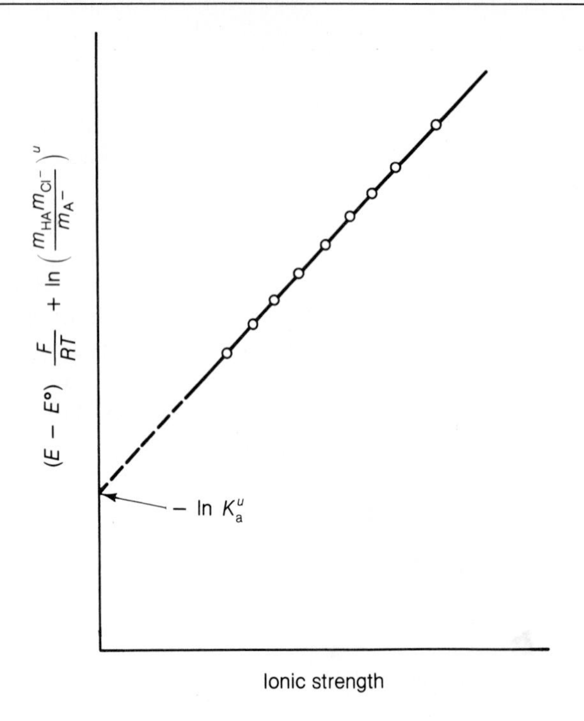

FIGURE 8.10 Plot of the function shown on the left-hand side of Eq. 8.48 against the ionic strength.

strength the activity coefficients become unity, and the first term on the right-hand side of Eq. 8.48 is therefore zero; extrapolation to $I = 0$ thus gives $-\ln K_a^u$.

Similar methods can be employed for the determination of the dissociation constants of bases and of polybasic acids.

Solubility Products

The determination of solubility products by emf measurements may be exemplified by the use of the following cell, which gives the solubility product of silver chloride:

$$Cl_2(1\ \text{atm})|HCl|AgCl|Ag$$

The electrode processes are

$$Cl^- \longrightarrow \tfrac{1}{2}Cl_2 + e^- \quad \text{and} \quad e^- + AgCl(s) \longrightarrow Ag + Cl^-$$

and the overall process is

$$AgCl(s) \longrightarrow Ag + \tfrac{1}{2}Cl_2$$

However, the $AgCl(s)$ is in equilibrium with Ag^+ and Cl^- ions present in solution, and we can write the overall process as

$$Ag^+ + Cl^- \longrightarrow Ag + \tfrac{1}{2}Cl_2$$

The emf corresponding to this process is*

$$E = E^\circ + \frac{RT}{F} \ln \left[a_{Ag^+} a_{Cl^-}\right]^u \tag{8.49}$$

$$= E^\circ + \frac{RT}{F} \ln K_s^u \tag{8.50}$$

where E° is given by

$$E^\circ = E^\circ(Ag^+ + e^- \longrightarrow Ag) - E^\circ(\tfrac{1}{2}Cl_2 + e^- \longrightarrow Cl^-) \tag{8.51}$$

(see Table 8.1). At 25°C the value of E° is $0.771 - 1.3595 = -0.589$ V, and the measured emf of the cell is -1.14 V; thus

$$-1.14 = -0.589 + \frac{RT}{F} \ln K_s^u$$

and

$$K_s = 4.85 \times 10^{-10} \text{ mol}^2 \text{ dm}^{-6}$$

Similar cells can be devised for other cells. The general principle, for a salt AB, is to use the following type of cell:

B|soluble salt of B^- ions|AB(*s*)|A

The emf method is a valuable one for measuring solubility products for salts of very low solubility, for which direct solubility measurements cannot be made with high accuracy. A practical difficulty with the emf method is that sometimes the electrodes do not operate reversibly.

8.6 Fuel Cells†

The first fuel cell was constructed in 1839 by the British physicist and lawyer Sir William Robert Grove (1811–1896); it had platinum electrodes with hydrogen bubbled over one electrode and oxygen over the other. For many years little was done to develop fuel cells for commercial purposes, but since the 1960s there has been a considerable revival of interest in this problem, particularly in view of present energy shortages. Fuel cells have been used as sources of auxiliary power in spacecraft, and consideration is being given to their use in automobiles in order to minimize air pollution and noise.

Fuel cells employ the same electrochemical principles as conventional cells. Their distinguishing feature is that the reacting substances are continuously fed into the system, so that fuel cells, unlike conventional cells, do not have to be discarded when the chemicals are consumed. The simplest type of fuel cell uses

---

*It is of interest that E does not depend on the concentration of HCl used in the cell, since if a_{Cl^-} is large, a_{Ag^+} is correspondingly small.

†For further details, see J. O'M. Bockris and S. Srinivasan, *Fuel Cells: Their Technology*, New York: McGraw-Hill, 1969; A. K. Vijh, *J. Chem. Education*, *47*, 680 (1970).

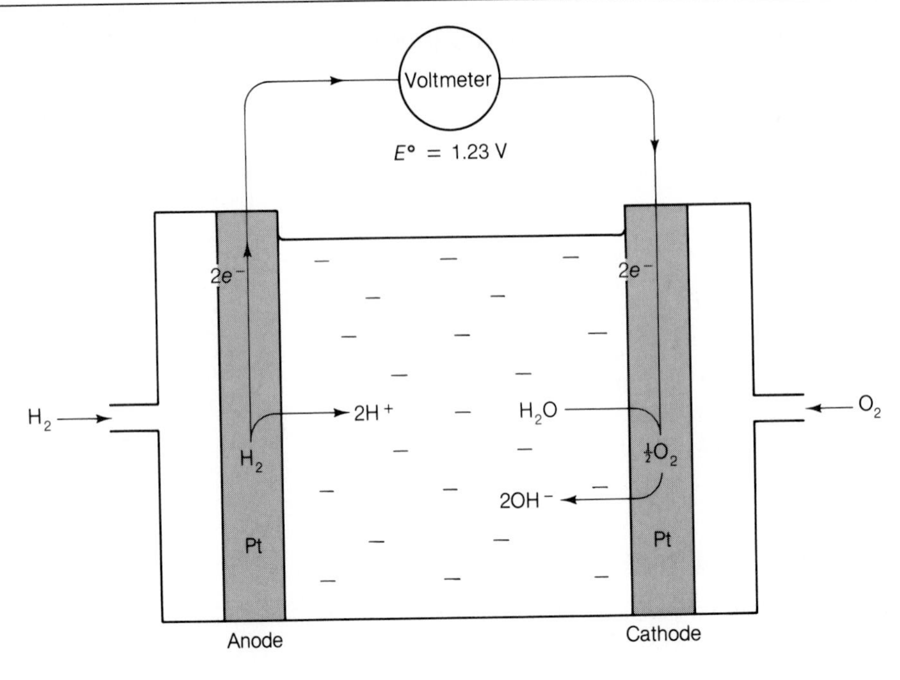

Anode reaction: $H_2 \rightarrow 2H^+ + 2e^-$

Cathode reaction: $\frac{1}{2}O_2 + H_2O + 2e^- \rightarrow 2OH^-$

Overall reaction: $H_2 + \frac{1}{2}O_2 \rightarrow H_2O \quad \Delta G° = -237.18 \text{ kJ mol}^{-1}$

FIGURE 8.11 Diagrammatic representation of a hydrogen-oxygen fuel cell, showing the reactions occurring at the two electrodes and the overall reaction.

hydrogen and oxygen as fuel. Figure 8.11 is a schematic representation of a hydrogen-oxygen fuel cell and shows the reactions occurring at the two electrodes. Various electrolyte solutions, such as sulfuric acid, phosphoric acid, and potassium hydroxide solutions, have been employed in such cells.

The overall reaction in this type of fuel cell is

$$H_2(g) + \tfrac{1}{2}O_2(g) \longrightarrow H_2O(l)$$

This process corresponds to the transfer of two electrons. The standard Gibbs energy change in this reaction [i.e., the standard Gibbs energy of formation of $H_2O(l)$], is -237.18 kJ mol^{-1} at 25°C, and the theoretical reversible emf, from Eq. 8.2, is

$$E° = \frac{237\ 180}{2 \times 96\ 500} = 1.23 \text{ V}$$

In practice voltages are less than this, because of deviations from reversible behavior. The extent of these deviations depends on the materials used as electrodes. Much present research is devoted to the development of improved electrodes. If these difficulties can be overcome, fuel cells will offer a considerable advantage over other methods of obtaining energy from fuels. As discussed in Section 3.10, furnaces in which fuels are burned suffer from second-law limitations to their efficiency. There are no such Carnot limitations to the efficiencies of fuel cells.

Because of the hazards of using hydrogen and the cost of preparing it from other fuels, attention is being given to the development of fuel cells using other materials, such as hydrocarbons and alcohols. There are again practical difficulties with thermodynamic reversibility, and these are being slowly overcome. A very effective fuel cell would be one that accepts conventional fuels such as coal, oil, or gasoline, as anodic fuels. However, such fuels bring about considerable pollution of the electrodes and lead to diminishing efficiency, but it is possible that future research will overcome these difficulties.

Attempts are also being made to develop biochemical fuel cells that would use as fuels the products of biological processes. For example, at the bottom of the Black Sea, bacteria obtain oxygen from sulfates and produce large amounts of hydrogen sulfide that could be used as anodic fuel. It will probably be many years before these speculative possibilities are turned into practical devices.

8.7 Photogalvanic Cells*

Certain materials, such as selenium, produce electricity directly when they are irradiated, and devices that employ this effect are known as *photovoltaic cells.* Such cells are to be distinguished from **photogalvanic cells,** in which irradiation induces a chemical process that in turn gives rise to an electric current. The generation of electricity by the chemical system follows the same principle as in an ordinary electrochemical cell; the special feature of a photogalvanic cell is that the reaction is brought about photochemically.

A number of devices of this kind have been set up on the laboratory scale, but little has as yet been achieved in the direction of producing photogalvanic cells that will be of practical use in the utilization of solar energy. One reaction that has been studied in the laboratory is the light-induced reaction between the purple dye thionine and Fe^{2+} ions in aqueous solution. If we represent the thionine ion as T^+, and its reduced colorless form as TH^+, the overall reaction is

$$\underset{\text{purple}}{T^+} + Fe^{2+} + H^+ \xrightarrow{h\nu} \underset{\text{colorless}}{TH^+} + Fe^{3+}$$

In the dark the equilibrium for this reaction lies over to the left. Irradiation with visible light causes this equilibrium to shift considerably to the right, and the solution becomes colorless. The most effective wavelength for bringing about this equilibrium shift is 478 nm, corresponding to the yellow-green region of the spectrum. In the dark the reaction has a positive standard Gibbs energy change, but absorption of photons by the thionine molecules provides energy for displacement of the equilibrium.

A cell that utilizes this equilibrium shift is shown schematically in Figure 8.12. There are two compartments separated by a membrane that is impermeable to

*For further details, see R. Gomer, *Electrochimica Acta, 20,* 13 (1975); W. D. K. Clark and J. A. Eckert, *Solar Energy, 17,* 147 (1975); N. N. Lichtin, "Photogalvanic Process," in J. R. Bolton (Ed.), *Solar Power and Fuels,* New York: Academic Press, 1977.

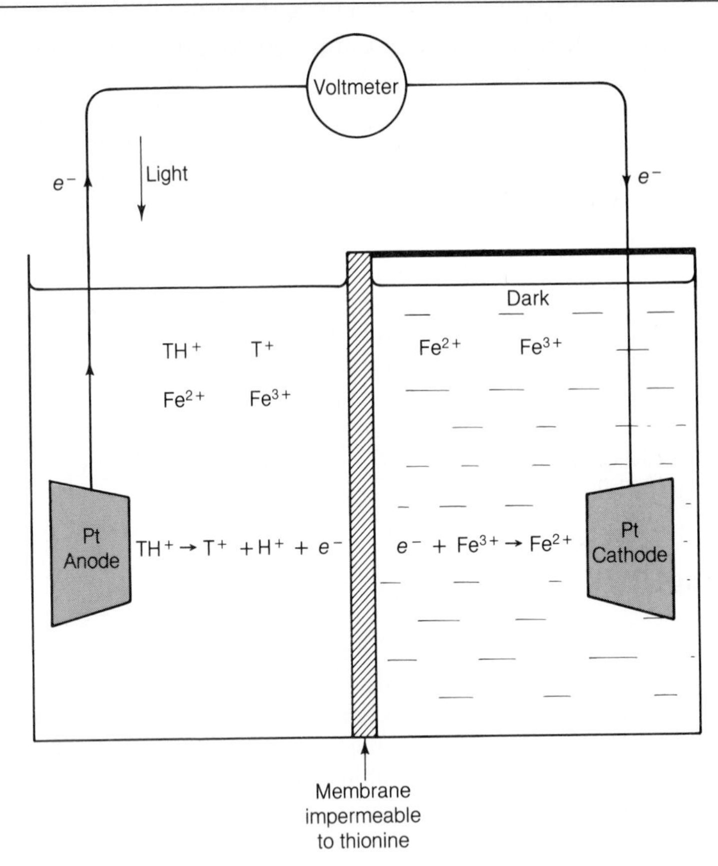

FIGURE 8.12 Schematic representation of a photogalvanic cell.

thionine but permeable to Fe^{2+} and Fe^{3+}. The left-hand compartment, which contains thionine as well as Fe^{2+} and Fe^{3+}, is irradiated. The right-hand compartment contains no thionine and is kept in darkness. The irradiation of the left-hand compartment causes the ratio $[TH^+]/[T^+]$ to be abnormally large, and the process

$$TH^+ \longrightarrow T^+ + H^+ + e^-$$

will therefore occur at the electrode. In the dark right-hand solution, the process

$$e^- + Fe^{3+} \longrightarrow Fe^{2+}$$

will occur at the electrode. The overall reaction giving rise to the emf is thus

$$TH^+ + Fe^{3+} \longrightarrow T^+ + Fe^{2+} + H^+$$

and is the reverse of the reaction that is occurring as a result of the irradiation. As long as the light shines on the thionine solution, the current flows; but when it is turned off, the reaction rapidly reverts to equilibrium and the current ceases. The theoretical standard emf for the system is 0.47 V. The reversible potential obtained depends on the concentrations, in accordance with the Nernst equation, and these concentrations depend on the initial concentrations of the materials and also on the intensity of the radiation. In practice the voltages obtained are only 2 to 3% of the theoretical values because of deviations from reversibility.

There are formidable difficulties that may make it impractical to make effective use of solar radiation with photogalvanic cells. The most important of these relates to the small proportion of light that is absorbed. With thionine the maximum absorption is at a wavelength of 478 nm, and even when monochromatic light of this wavelength is used only about 0.1% of the light is absorbed in a typical experiment. With sunlight, covering a wide range of wavelengths, the fraction absorbed is reduced by a further factor of 10 or more, since most of the visible and near-ultraviolet radiation lies outside the absorption region of the dye.

Photogalvanic cells thus represent an interesting laboratory device, but only if completely new systems are used will they become of practical importance.

Key Equations

Thermodynamics of an electrochemical cell:

$$\Delta G = -zEF$$

where z = charge number of cell reaction; E = emf; F = Faraday constant.

Nernst equation:

$$E = E^\circ - \frac{RT}{zF} \ln \left(\frac{[Y]^y [Z]^z}{[A]^a [B]^b} \right)^u$$

for a reaction $aA + bB \rightleftharpoons yY + zZ$.

Nernst potential:

$$\Delta \Phi = \frac{RT}{zF} \ln \frac{c_1}{c_2}$$

Emf of a concentration cell:

$$E = \frac{RT}{zF} \ln \frac{m_2}{m_1}$$

Problems

A Electrode Reactions and Electrode Potentials

1. Write the electrode reactions, the overall reaction, and the expression for the emf for each of the following reversible cells:

a. Pt, H_2(1 atm)|HCl|Pt, Cl_2(1 atm)

b. Hg|Hg_2Cl_2(s)|HCl|Pt,|H_2(1 atm)

c. Ag|AgCl(s)|KCl⋮Hg_2Cl_2(s)|Hg

d. Pt, H_2(1 atm)|HI(*aq*)|AuI(*s*)|Au

e. Ag|AgCl(*s*)|KCl(c_1)⦙KCl(c_2)|AgCl(*s*)|Ag

2. At 25°C and pH 7, a solution containing compound A and its reduced form AH_2 has a standard electrode potential of −0.6 V. A solution containing B and BH_2 has a standard potential of −0.16 V. If a cell were constructed with these systems as half-cells,

a. Is AH_2 oxidized by B or is BH_2 oxidized by A under standard conditions?

b. What is the reversible emf of the cell?

c. What would be the effect of pH on the equilibrium ratio $[B][AH_2]/[A][BH_2]$?

3. Calculate the standard electrode potential for the reaction $Cr^{2+} + 2e^- \rightarrow Cr$. The necessary $E°$ values are

1. $Cr^{3+} + 3e^- \longrightarrow Cr \qquad E° = -0.74$ V
2. $Cr^{3+} + e^- \longrightarrow Cr^{2+} \qquad E° = -0.41$ V

B Thermodynamics of Electrochemical Cells

4. Calculate the equilibrium constant at 25°C for the reaction

$$2Fe^{3+}(aq) + 2I^-(aq) \rightleftharpoons 2Fe^{2+}(aq) + I_2(s)$$

using the standard electrode potentials given in Table 8.1.

5. From data in Table 8.1, calculate the equilibrium constant at 25°C for the reaction

$$Sn + Fe^{2+} \rightleftharpoons Sn^{2+} + Fe$$

6. The standard electrode potential at 25°C for

$$\text{cytochrome c}(Fe^{3+}) + e^- \longrightarrow \text{cytochrome c}(Fe^{2+})$$

is 0.25 V. Calculate $\Delta G°$ for the process

$$\tfrac{1}{2}H_2(g) + \text{cytochrome c}(Fe^{3+}) \longrightarrow H^+ + \text{cytochrome c}(Fe^{2+})$$

7. Using the values given in Table 8.1, calculate the standard Gibbs energy change $\Delta G°$ for the reaction

$$H_2 + \tfrac{1}{2}O_2 \longrightarrow H_2O$$

***8.** From the data in Table 8.1, calculate the equilibrium constant at 25°C for the reaction

$$2Cu^+ \longrightarrow Cu^{2+} + Cu$$

What will be produced if Cu_2O is dissolved in dilute H_2SO_4?

C Nernst Equation and Nernst Potentials

9. Calculate the emf for the following cell at 25°C:

Pt, H_2(1 atm)|HCl(0.5 *m*)⦙HCl(1.0 *m*)|Pt, H_2(1 atm)

10. The pyruvate-lactate system has an $E°'$ value of −0.185 V at 25°C and pH 7.0. What will be the potential of this system if the oxidation has gone to 90% completion?

11. a. From the data in Table 8.1, calculate the standard electrode potential for the half-reaction

$$Fe^{3+} + 3e^- \rightleftharpoons Fe$$

b. Calculate the emf at 25°C of the cell

Pt|Sn^{2+}(0.1 *m*), Sn^{4+}(0.01 *m*)‖Fe^{3+}(0.5 *m*)|Fe

12. Calculate the concentration of I_3^- in a standard solution of iodine in 0.5 *M* KI, making use of the following standard electrode potentials:

$I_2 + 2e^- \longrightarrow 2I^- \qquad E° = 0.5355$ V

$I_3^- + 2e^- \longrightarrow 3I^- \qquad E° = 0.5365$ V

The concentration of I^- in the standard solution can be assumed to be 0.5 *M*.

13. Calculate the Nernst potential at 25°C arising from the equilibrium established in Problem 24 in Chapter 7.

14. Calculate the emf of the cell

Pt, H_2(1 atm)|HCl(0.1 *m*)⦙HCl(0.2 *m*)|Pt, H_2(10 atm)

***15.** A typical biological cell has a volume of 10^{-9} cm^3, a surface area of 10^{-6} cm^2, and a membrane thickness of 10^{-6} cm; the dielectric constant of the membrane may be taken as 3. Suppose that the concentration of K^+ ions inside the cell is 0.155 *M* and that the Nernst potential across the cell wall is 0.085 V. Calculate (a) the net charge on either side of the wall, and (b) the fraction of the K^+ ions in the cell that are required to produce this charge.

***16.** Calculate the emf at 25.0°C of the cell

Pt, H_2(1 atm)|0.001 *m* H_2SO_4|$CrSO_4$(*s*)|Cr

given the following standard electrode potential:

$CrSO_4(s) + 2e^- \longrightarrow Cr + SO_4^{2-}$

$E° = -0.40$ V

a. First make the calculation neglecting activity coefficient corrections.

b. Then make the calculation using activity coefficients estimated on the basis of the Debye-Hückel limiting law.

***17.** Write the individual electrode reactions and the overall reaction for

$$Cu|CuCl_2(aq)|AgCl(s)|Ag$$

If the emf of the cell is 0.191 V when the concentration of $CuCl_2$ is 1.0×10^{-4} M and is -0.074 V when the concentration is 0.20 M, make an estimate of the mean activity coefficient in the latter solution.

***18. a.** Write both electrode reactions and the overall reaction for the cell

$$Tl|TlCl(s)|CdCl_2(0.01\ m)|Cd$$

b. Calculate the E and $E°$ for this cell at 298.15 K from the following information:

$$Tl^+ + e^- \longrightarrow Tl \qquad E° = -0.34 \text{ V}$$

$$Cd^{2+} + 2e^- \longrightarrow Cd \qquad E° = -0.40 \text{ V}$$

The solubility product for $TlCl = 1.6 \times 10^{-3}$ $mol^2\ dm^{-6}$ at 298.15 K.

D Temperature Dependence of Cell emfs

19. a. Calculate the standard emf $E°$ for the reaction

$$\text{fumarate}^{2-} + \text{lactate}^- \longrightarrow \text{succinate}^{2-} + \text{pyruvate}^-$$

on the basis of the following information:

$$\text{fumarate}^{2-} + 2H^+ + 2e^- \longrightarrow \text{succinate}^{2-} \qquad E°' = 0.031 \text{ V}$$

$$\text{pyruvate}^- + 2H^+ + 2e^- \longrightarrow \text{lactate}^- \qquad E°' = -0.185 \text{ V}$$

The $E°'$ values relate to pH 7. The temperature coefficient $\partial E°/\partial T$ for this cell is 2.18×10^{-5} V K^{-1}.

b. Calculate $\Delta G°$, $\Delta H°$, and $\Delta S°$ at 25.0°C.

20. The Weston standard cell (see Figure 8.2b) is

$$\text{Cd amalgam}|CdSO_4 \cdot \tfrac{8}{3}H_2O(s)|Hg_2SO_4(s), \text{ Hg}$$
saturated solution

a. Write the cell reaction.

b. At 25°C, its emf is 1.018 32 V and $\partial E°/\partial T = -5.00 \times 10^{-5}$ V K^{-1}. Calculate $\Delta G°$, $\Delta H°$, and $\Delta S°$.

***21. a.** Estimate the Gibbs energy of the formation of the fumarate ion, using data in Problem 19 and the following values:

$$\Delta_f G°(\text{succinate}, aq) = -690.44 \text{ kJ mol}^{-1}$$

$$\Delta_f G°(\text{acetaldehyde}, aq) = -139.08 \text{ kJ mol}^{-1}$$

$$\Delta_f G°(\text{ethanol}, aq) = -181.75 \text{ kJ mol}^{-1}$$

$$\text{acetaldehyde} + 2H^+ + 2e^- \longrightarrow \text{ethanol} \qquad E°' = -0.197 \text{ V}$$

b. If the $\partial E/\partial T$ value for the process

$$\text{fumarate} + \text{ethanol} \longrightarrow \text{succinate} + \text{acetaldehyde}$$

is -1.45×10^{-4} V K^{-1}, estimate the enthalpy of formation of the fumarate ion from the following values:

$$\Delta_f H°(\text{succinate}, aq) = -908.68 \text{ kJ mol}^{-1}$$

$$\Delta_f H°(\text{acetaldehyde}, aq) = -210.66 \text{ kJ mol}^{-1}$$

$$\Delta_f H°(\text{ethanol}, aq) = -287.02 \text{ kJ mol}^{-1}$$

***22. a.** Calculate the emf at 298.15 K for the cell

$$Tl|TlBr|HBr(\text{unit activity})|H_2(1 \text{ atm})|Pt$$

b. Calculate ΔH for the cell reaction

$$Tl|Tl^+(\text{unit activity}), H^+(\text{unit activity})|H_2(1 \text{ atm})|Pt$$

For the half-cell

$$Tl^+ + e^- \longrightarrow Tl \qquad E° = -0.34 \text{ V}$$

$$\frac{dE°}{dT} = -0.003 \text{ V/K} \quad \text{and} \quad K_{sp}(TlBr) = 10^{-4}$$

E Applications of emf Measurements

23. Calculate the solubility product and the solubility of AgBr at 25°C on the basis of the following standard electrode potentials:

$$AgBr(s) + e^- \longrightarrow Ag + Br^- \qquad E° = 0.0713 \text{ V}$$

$$Ag^+ + e^- \longrightarrow Ag \qquad E° = 0.7991 \text{ V}$$

24. The emf of a cell

$$Pt, H_2(1 \text{ atm})|HCl|AgCl,Ag$$

was found to be 0.517 V at 25°C. Calculate the pH of the HCl solution.

25. The emf of the cell

$$Ag|AgI(aq)|AgI, Ag$$

is −0.9509 V at 25°C. Calculate the solubility and the solubility product of AgI at that temperature.

26. The following thermodynamic data apply to the complete oxidation of butane at 25°:

$$C_4H_{10}(g) + \tfrac{13}{2}O_2(g) \longrightarrow 4CO_2(g) + 5H_2O(l)$$

$$\Delta H^\circ = -2877 \text{ kJ mol}^{-1}$$

$$\Delta S^\circ = -432.7 \text{ J K}^{-1}\text{mol}^{-1}$$

Suppose that a completely efficient fuel cell could be set up utilizing this reaction. Calculate (a) the maximum electrical work and (b) the maximum total work that could be obtained at 25°C.

F Supplementary Questions

27. Suppose that the cell in Problem 14 is set up but that the two solutions are separated by a membrane that is permeable to H^+ ions but impermeable to Cl^- ions. What will be the emf of the cell at 25.0°C?

28. Suppose that the cell

$$Ag|AgCl(s)|HCl(0.10\ m)\,\vdots\,HCl(0.01\ m)|AgCl(s)|Ag$$

is set up and that the membrane separating the two solutions is permeable only to H^+ ions. What is the emf of the cell at 25.0°C?

29. a. Now consider the cell

$$Pt|H_2(1\text{ atm})|HCl(m_1)\,\vdots\,HCl(m_2)|Pt|H_2(1\text{ atm})$$

in which the solutions are separated by a partition that is permeable to both H^+ and Cl^-. The ratio of the speeds with which these ions pass through the membrane is the ratio of their transport numbers t_+ and t_-. Derive an expression for the emf of this cell.

b. If when $m_1 = 0.01\ m$ and $m_2 = 0.01\ m$ the emf is 0.0190 V, what are the transport numbers of the H^+ and Cl^- ions?

30. The metal M forms a soluble nitrate and a very slightly soluble chloride. The cell

$$M|M^+(0.1\ m),\ HNO_3(0.2\ m)|H_2(1\text{ atm})|Pt$$

has a measured $E = -0.40$ V at 298.15 K. When sufficient solid KCl is added to make the solution of the cell 0.20 m in K^+, the emf changes to −0.15 V at 298.15 K as MCl precipitates. Calculate the K_{sp} of MCl, taking all activity coefficients to be unity.

G Essay Questions

1. List three common reference electrodes and their cell reactions.

2. Show how, starting with a value of ΔG, ΔH can be obtained from emf measurements.

3. Work out an additional example for each of the electrochemical cells listed in Figure 8.8.

Suggested Reading

Also see *Suggested Reading* for Chapter 7.

J. O'M. Bockris and S. Srinivasan, *Fuel Cells: Their Technology*, New York: McGraw-Hill, 1969.

J. R. Bolton (Ed.), *Solar Power and Fuels*, New York: Academic Press, 1977.

B. E. Conway, *Electrochemical Data*, Amsterdam: Elsevier Pub. Co., 1952.

A. J. de Bethune and N. A. S. Loud, *Standard Aqueous Electrode Potentials and Temperature Coefficients at 25°C*, Stokie, Ill. C. A. Hampel, 1964.

D. J. G. Ives and G. J. Janz (Eds.), *Reference Electrodes; Theory and Practice*, New York: Academic Press, 1961.

Kinetics of Elementary Reactions

Preview

Chemical kinetics is concerned with the rates and mechanisms of chemical reactions. This chapter is particularly concerned with *elementary reactions,* which proceed in a single step, and we will learn something about the energetics and dynamics of such reactions.

If a reaction is of stoichiometry

$$a\mathrm{A} + b\mathrm{B} \longrightarrow y\mathrm{Y} + z\mathrm{Z}$$

the rate of consumption of A is defined as $-d[\mathrm{A}]/dt$, and the rate of formation of Y is defined as $d[\mathrm{Y}]/dt$. These quantities are not necessarily the same for different reactants and products, and it is convenient to have a quantity that applies to the reaction and is the same for all reactants and products. This is done by making use of the concept of *extent of reaction* ξ.

For some reactions the rate is related to reactant concentrations by an equation of the form

$$v = k[\mathrm{A}]^{\alpha}[\mathrm{B}]^{\beta} \ldots$$

where α and β are known as *partial orders:* α is the order with respect to A and β the order with respect to B. The sum of all the partial orders is known as the *overall order.* The coefficient k that appears in this equation is known as the *rate constant* or the rate coefficient.

The effect of temperature T on a rate constant is expressed by the *Arrhenius equation*

$$k = Ae^{-E/RT}$$

where R is the gas constant, E is the *activation energy,* and A is the *preexponential factor.* In principle, activation energies can be calculated by quantum mechanics, but in practice the results are not reliable. The course of a reaction is conveniently mapped by means of a *potential-energy surface,* in which energy is plotted against suitable parameters that describe the reacting system.

The preexponential factor A has been interpreted on the basis of the kinetic molecular theory of collisions in an ideal gas. It is more satisfactorily interpreted on the basis of *transition state* theory, which focuses attention on the *activated complex,* a reaction intermediate that lies at the highest point of the energy profile connecting the initial and the final states. In transition state theory the activation complexes are assumed to be in equilibrium with the reactants. This theory leads to the concepts of the *entropy of activation* $\Delta S^{\ddagger}$ and the *enthalpy of activation* $\Delta H^{\ddagger}$. These are the changes in entropy and enthalpy that occur when the activated complexes are formed from the reactants.

9 Kinetics of Elementary Reactions

Thermodynamics tells us the direction in which a process will occur, but it tells us nothing about its rate. *Chemical kinetics* is the branch of physical chemistry that deals with the *rates* of chemical reactions and of the factors on which the rates depend. One reason for the great importance of this subject is that it provides most of the information needed to arrive at the *mechanism* of a reaction.

One of the pioneers in this field was the British chemist A. G. Vernon Harcourt (1834–1919) who together with the mathematician William Esson (1838–1916) made, in the 1860s, the first detailed analysis of the time course of a chemical reaction. Later Arrhenius elucidated the effect of temperature on reaction rates. Subsequent to these investigations a vast amount of work has been done on reaction rates. It is now realized that some reactions go in a single step; they are known as *elementary* reactions and are dealt with in this chapter. Other reactions go in more than one step and are said to be *composite, stepwise,* or *complex;* such reactions are treated in Chapter 10.

For further details about definitions and symbols in chemical kinetics, the reader is referred to a recent IUPAC report.*

9.1 Rates of Consumption and Formation

Most kinetic investigations are concerned with rates of change of concentrations of reactants and products. Consider, for example, a reaction

$$A + 3B \longrightarrow 2Y$$

Figure 9.1 shows schematically the variations in concentrations of A, B, and Y if we start the kinetic experiment with a mixture of A and B but no Y. At any time t we can draw a tangent to the curve representing the consumption of A, and the *rate of consumption* of A at that time is

$$v_A = -\frac{d[A]}{dt} \tag{9.1}$$

As a special case, we may draw a tangent at $t = 0$, corresponding to the beginning of the reaction; the negative of the slope is the *initial* rate of consumption of A. The *rate of formation* of Y, v_Y, is given by

$$v_Y = \frac{d[Y]}{dt} \tag{9.2}$$

In this particular reaction the stoichiometric coefficients are different for the three species, A, B, and Y, and the rates of change of their concentrations are correspondingly different. Thus, the rate of consumption of B, v_B, is three times the

*K. J. Laidler, "Symbolism and Terminology in Chemical Kinetics," *Pure and Applied Chemistry 53*, 753 (1981); this is a report of the IUPAC Subcommittee on Chemical Kinetics.

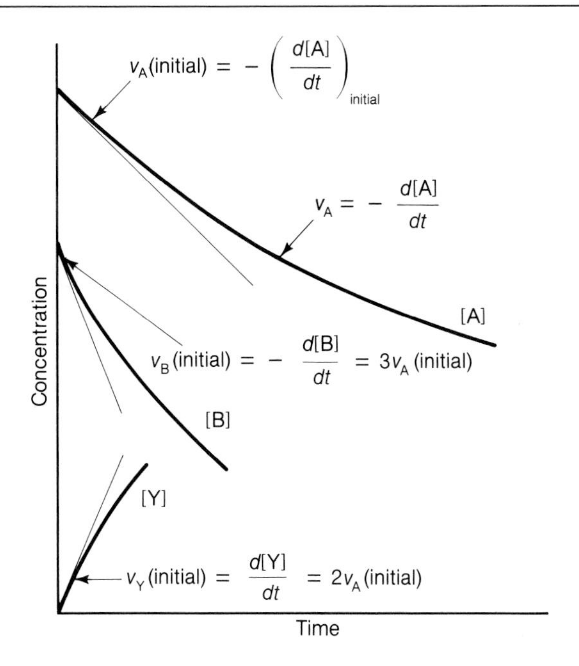

FIGURE 9.1 The variations with time of the concentrations of A, B, and Y for a reaction of the type $A + 3B \rightarrow 2Y$.

rate of consumption of A, v_A, and the rate of formation of Y, v_Y, is twice the rate of consumption of A:

$$v_A = \tfrac{1}{3}v_B = \tfrac{1}{2}v_Y \tag{9.3}$$

It would be ambiguous to refer to v_A, v_B, or v_Y as the rate of reaction.

9.2 Rate of Reaction

In Section 2.6 we introduced the concept of *extent of reaction* ξ, and this forms the basis of the definition of rate of reaction. The *rate of reaction* is now defined as the time derivative of the extent of reaction divided by the volume:

$$v \equiv \frac{1}{V}\frac{d\xi}{dt} = \frac{1}{V\nu_i}\frac{dn_i}{dt} \tag{9.4}$$

If the volume is constant, dn_i/V can be replaced by the concentration change dc_i, and therefore

$$v = \frac{1}{\nu_i}\frac{dc_i}{dt} \tag{9.5}$$

This quantity is independent of which reactant or product species is chosen. For a reaction

$$aA + bB \longrightarrow yY + zZ$$

occurring at constant volume, the rate of reaction is

$$v = -\frac{1}{a}\frac{d[\mathrm{A}]}{dt} = -\frac{1}{b}\frac{d[\mathrm{B}]}{dt} = \frac{1}{y}\frac{d[\mathrm{Y}]}{dt} = \frac{1}{z}\frac{d[\mathrm{Z}]}{dt} \tag{9.6}$$

The time derivatives of the concentrations (with a negative sign for reactants) are the rates of consumption or formation (see Figure 9.1) and thus

$$v = \frac{v_{\mathrm{A}}}{a} = \frac{v_{\mathrm{B}}}{b} = \frac{v_{\mathrm{Y}}}{y} = \frac{v_{\mathrm{Z}}}{z} \tag{9.7}$$

A distinction must be made between v without lettered subscript, meaning rate of reaction, and v with lettered subscript (e.g., v_{A}) meaning rate of consumption or formation. Since the stoichiometric coefficients and therefore extents of reaction depend on the way in which the reaction is written (e.g., $H_2 + Br_2 \longrightarrow 2HBr$ or $\frac{1}{2}H_2 + \frac{1}{2}Br_2 \longrightarrow HBr$), whenever rates of reaction are given the stoichiometric equation must be stated.

9.3 Empirical Rate Equations

For *some* reactions the rate of consumption or formation can be expressed empirically by an equation of the form

$$v_{\mathrm{A}} = k_{\mathrm{A}}[\mathrm{A}]^{\alpha}[\mathrm{B}]^{\beta} \tag{9.8}$$

where k_A, α, and β are independent of concentration and of time. Similarly, for a product Z, where k_{Z} is not necessarily the same as k_{A},

$$v_{\mathrm{Z}} = k_{\mathrm{Z}}[\mathrm{A}]^{\alpha}[\mathrm{B}]^{\beta} \tag{9.9}$$

When these equations apply, the rate of reaction must also be given by an equation of the same form:

$$v = k[\mathrm{A}]^{\alpha}[\mathrm{B}]^{\beta} \tag{9.10}$$

In these equations, k_{A}, k_{Z}, and k are not necessarily the same, being related by stoichiometric coefficients; thus if the stoichiometric equation is

$$\mathrm{A} + 2\mathrm{B} \longrightarrow 3\mathrm{Z}$$

$$k = k_{\mathrm{A}} = \tfrac{1}{2}k_{\mathrm{B}} = \tfrac{1}{3}k_{\mathrm{Z}} \tag{9.11}$$

Order of Reaction

The exponent α in the previous equations is known as the *order of reaction with respect to A* and can be referred to as a *partial order*. Similarly, the partial order β is the order with respect to B. These orders are *purely experimental* quantities and are not necessarily integral. The sum of all the partial orders, $\alpha + \beta + \ldots$, is referred to as the *overall order* and is usually given the symbol n.

A very simple case is when the rate is proportional to the first power of the concentration of a single reactant:

$$v = k[\mathrm{A}] \tag{9.12}$$

Such a reaction is said to be of the *first order*. An example is the conversion of cyclopropane into propylene:

$$\underset{H_2C\text{———}CH_2}{\overset{CH_2}{\triangle}} \longrightarrow CH_3\text{—}CH\text{=}CH_2$$

The rate of this reaction is proportional to the first power of the cyclopropane concentration.

There are many examples of *second-order* reactions. The reaction

$$H_2 + I_2 \rightleftharpoons 2HI$$

is second order in both directions. The term *chemical flux* or *chemiflux** (symbol ϕ) is convenient for expressing the rate of a *single chemical process* in *one direction only*. For this reaction the chemiflux from left to right is proportional to the product of the concentrations of H_2 and I_2:

$$\phi_1 = \phi_{-H_2,2HI} = k_1[H_2][I_2] \tag{9.13}$$

where k_1 is a constant at a given temperature. The reaction from left to right is said to be first order in H_2 and first order in I_2, and its overall order is two. The reverse reaction is also second order; the chemiflux from right to left is proportional to the square of the HI concentration:

$$\phi_{-1} = \phi_{-2HI,H_2} = k_{-1}[HI]^2 \tag{9.14}$$

The rate of a reaction must be proportional to the product of two concentrations [A] and [B] if the reaction simply involves collisions between A and B molecules. Similarly, the kinetics must be third order if a reaction proceeds in one stage and involves collisions between three molecules, A, B, and C. There are a few reactions of the third order, but reactions of higher order are unknown. Collisions in which three or more molecules all come together at the same time are very unlikely; reaction will instead proceed more rapidly by a composite mechanism involving two or more elementary processes, each of which is only first or second order.

There is no simple connection between the stoichiometric equation for a reaction and the order of the reaction. An example that illustrates this is the decomposition of gaseous acetaldehyde, for which the equation is

$$CH_3CHO \longrightarrow CH_4 + CO$$

We might think that because there is one molecule on the left-hand side of this equation, the reaction should be first order. In fact, the order is three-halves:

$$v = k[CH_3CHO]^{3/2} \tag{9.15}$$

We will see in Section 10.5 that this reaction occurs by a composite free-radical mechanism of a particular type that leads to three-halves-order behavior.

*V. Gold, *Nouveau Journal de Chemie*, 3, 69 (1979).

Reactions Having No Order

Not all reactions behave in the manner described by Eq. 9.8, and the term *order* should not be used for those that do not. For example, as we will discuss in Section 10.9, reactions catalyzed by enzymes frequently follow a law of the form

$$v = \frac{V[\mathrm{S}]}{K_m + [\mathrm{S}]} \tag{9.16}$$

where V and K_m are constants, and [S] is the concentration of the substance, known as the *substrate*, that is undergoing catalyzed reaction. This equation does not correspond to a simple order, but under two limiting conditions an order may be assigned. Thus, if the substrate concentration is sufficiently low that $[\mathrm{S}] << K_m$, Eq. 9.16 becomes

$$v = \frac{V}{K_m}[\mathrm{S}] \tag{9.17}$$

The reaction is then first order with respect to S. Also, when [S] is sufficiently large that $[\mathrm{S}] >> K_m$, the equation reduces to

$$v = V \tag{9.18}$$

The rate is then independent of [S] (i.e., is proportional to [S] to the zero power) and the reaction is said to be *zero order*.

Rate Constants and Rate Coefficients

The constant k that appears in rate equations that are special cases of Eq. 9.8 is known as the **rate constant** or the *rate coefficient*. It is convenient to use the former term when the reaction is believed to be elementary and the latter term when the reaction is known to occur in more than one stage.

The units of the rate constant or coefficient vary with the order of the reaction. Suppose, for example, that a reaction is of the first order; that is,

$$v = k[\mathrm{A}] \tag{9.19}$$

If the units of v are mol dm^{-3} s^{-1}, and those of [A] are mol dm^{-3}, the unit of the rate constant k is s^{-1}. For a second-order reaction, for which

$$v = k[\mathrm{A}]^2 \quad \text{or} \quad k[\mathrm{A}][\mathrm{B}] \tag{9.20}$$

the units of k will be

$$\frac{\text{mol dm}^{-3}\ \text{s}^{-1}}{(\text{mol dm}^{-3})^2} = \text{dm}^3\ \text{mol}^{-1}\ \text{s}^{-1}$$

The units corresponding to other orders can easily be worked out.

We have seen that the rate of change of a concentration in general depends on the reactant or product with which we are concerned. The rate constant also reflects this dependence. For example, for the dissociation of ethane into methyl radicals,

$$C_2H_6 \longrightarrow 2CH_3$$

the rate of formation of methyl radicals, v_{CH_3}, is twice the rate of consumption of ethane, $v_{C_2H_6}$:

$$v_{C_2H_6}\left(= -\frac{d[C_2H_6]}{dt}\right) = \frac{1}{2}v_{CH_3}\left(= \frac{1}{2}\frac{d[CH_3]}{dt}\right) \tag{9.21}$$

The reaction is first order in ethane under certain conditions, and the rate coefficient for the consumption of ethane, $k_{C_2H_6}$, is one-half of that for the appearance of methyl radicals:

$$k_{C_2H_6} = \tfrac{1}{2}k_{CH_3} \tag{9.22}$$

In cases such as this it is obviously important to specify the species to which a rate constant applies.

If a reaction shows time-independent stoichiometry, it is possible to evaluate a rate of reaction (Eq. 9.7). For the dissociation of ethane into methyl radicals, the rate of reaction v is equal to the rate of consumption of C_2H_6 and is one-half the rate of formation of methyl radicals. Whenever possible, the rate constant should be evaluated with respect to the rate of reaction.

9.4 Analysis of Kinetic Results

The first task in the kinetic investigation of a chemical reaction is to measure rates under a variety of experimental conditions and to determine how the rates are affected by the concentrations of reactants, products of reaction, and other substances (e.g., inhibitors) that may affect the rate.

There are two main methods for dealing with such problems; they are the method of integration and the differential method. In the **method of integration** we start with a rate equation that we think may be applicable. For example, if the reaction is believed to be a first-order reaction, we start with

$$-\frac{dc}{dt} = kc \tag{9.23}$$

where c is the concentration of reactant. By integration we convert this into an equation giving c as a function of t and then compare this with the experimental variation of c with t. If there is a good fit, we can then, by simple graphical procedures, determine the value of the rate constant. If the fit is not good, we must try another rate equation and go through the same procedure until the fit is satisfactory. The method is something of a hit-and-miss one, but it is nevertheless very useful, especially when no special complications arise.

The second method, the **differential method,** employs the rate equation in its differential, unintegrated, form. Values of dc/dt are obtained from a plot of c against t by taking slopes, and these are directly compared with the rate equation. The main difficulty with this method is that slopes cannot always be obtained very accurately. In spite of this drawback the method is on the whole the more reliable one, and unlike the integration method it does not lead to any particular difficulties when there are complexities in the kinetic behavior.

These two methods will now be considered in further detail.

Method of Integration

A *first-order reaction* may be of the type

$$\mathrm{A} \longrightarrow \mathrm{Z}$$

Suppose that at the beginning of the reaction ($t = 0$) the concentration of A is a_0, and that of Z is zero. If after time t the concentration of Z is x, that of A is $a_0 - x$. The rate of change of the concentration of Z is dx/dt, and thus for a first-order reaction

$$\frac{dx}{dt} = k(a_0 - x) \tag{9.24}$$

Separation of the variables gives

$$\frac{dx}{a_0 - x} = k\,dt \tag{9.25}$$

and integration gives

$$-\ln(a_0 - x) = kt + I \tag{9.26}$$

where I is the constant of integration. This constant may be evaluated using the boundary condition that $x = 0$ when $t = 0$; hence

$$-\ln a_0 = I \tag{9.27}$$

and insertion of this into Eq. 9.26 leads to

$$\ln \frac{a_0}{a_0 - x} = kt \tag{9.28}$$

This equation can also be written as

$$x = a_0(1 - e^{-kt}) \tag{9.29}$$

and as

$$a_0 - x = a_0 e^{-kt} \tag{9.30}$$

Equation 9.30 shows that the concentration of reactant, $a_0 - x$, decreases exponentially with time, from an initial value of a_0 to a final value of zero.

The first-order equations can be tested and the constant evaluated using a graphical procedure. It follows from Eq. 9.28 that a plot of

$$\ln \frac{a_0}{a_0 - x} \quad \text{against } t$$

will give a straight line if the order is the first; this is shown schematically in Figure 9.2a. The rate constant is the slope. We may simply plot $\ln(a_0 - x)$ against t, as shown in Figure 9.2b. We may also plot the common logarithms, in which case the slopes are $k/2.303$ or $-k/2.303$.

Second-order reactions can be treated in a similar fashion. There are now two possibilities: The rate may be proportional to the product of two equal concentrations or to the product of two different ones. The first case must occur when a

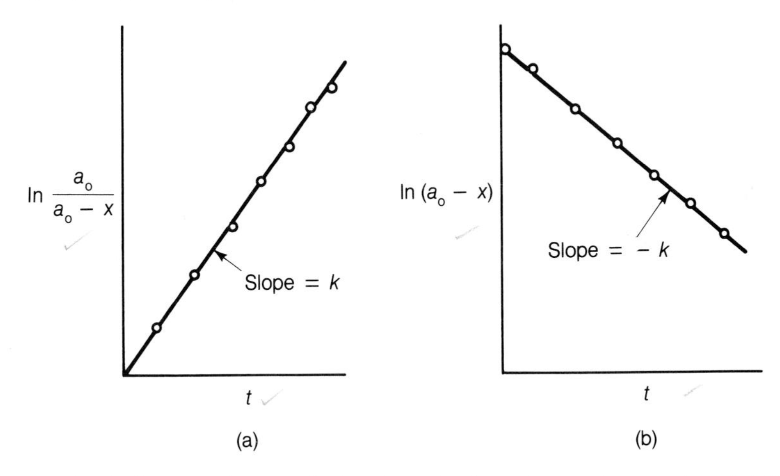

FIGURE 9.2 Method of integration; analysis of results for a first-order reaction. (a) Plot of $\ln[a_o/(a_o - x)]$ against t. (b) Plot of $\ln(a_o - x)$ against t.

single reactant is involved, as in the process

$$2A \longrightarrow Z$$

It may also be found in reactions between two different substances,

$$A + B \longrightarrow Z$$

provided that their initial concentrations are the same.

In such situations the rate may be expressed as

$$\frac{dx}{dt} = k(a_0 - x)^2 \tag{9.31}$$

where x is the amount of A that has reacted in unit volume at time t and a_0 is the initial amount of A. Separation of the variables leads to

$$\frac{dx}{(a_0 - x)^2} = k\,dt \tag{9.32}$$

which integrates to

$$\frac{1}{a_0 - x} = kt + I \tag{9.33}$$

where I is the constant of integration. The boundary condition is that $x = 0$ when $t = 0$; therefore

$$I = \frac{1}{a_0} \tag{9.34}$$

Hence

$$\frac{x}{a_0(a_0 - x)} = kt \tag{9.35}$$

The variation of x with t is no longer an exponential one.

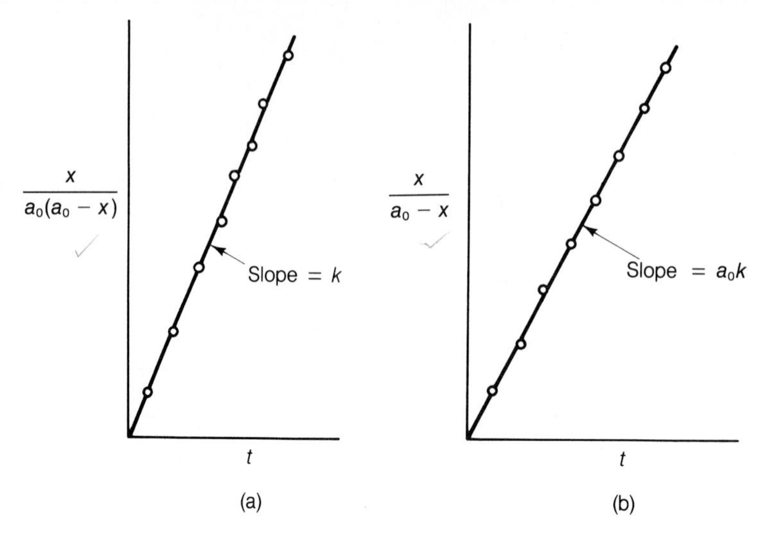

FIGURE 9.3 Method of integration; analysis of results for a second-order reaction. (a) Plot of $x/a_0(a_0 - x)$ against t. (b) Plot of $x/(a_0 - x)$ against t.

Graphical methods can again be employed to test this equation and to obtain the rate constant k. One procedure is to plot $x/a_0(a_0 - x)$ against t; if the equation is obeyed, the points will lie on a straight line passing through the origin (see Figure 9.3), and the slope will be k. Alternatively, we may plot $x/(a_0 - x)$ against t, in which case the slope is a_0k.

If the rate is proportional to the concentrations of two different reactants, and these concentrations are not initially the same, the integration proceeds differently. Suppose that the initial concentrations are a_0 and b_0; the rate after an amount x (per unit volume) has reacted is

$$\frac{dx}{dt} = k(a_0 - x)(b_0 - x) \tag{9.36}$$

The result of the integration, with the boundary condition $t = 0$, $x = 0$, is

$$\frac{1}{a_0 - b_0} \ln \frac{b_0(a_0 - x)}{a_0(b_0 - x)} = kt \tag{9.37}$$

This equation can be tested by plotting the left-hand side against t; if a straight line is obtained, its slope is k.

These are the most common orders; reactions of other orders can be treated in a similar manner.* Table 9.1 summarizes some of the results.

The main disadvantage of the method of integration is that the integrated expressions, giving the variation of x with t, are often quite similar for different types of reactions. For example, the time course of a simple second-order reaction is closely similar to that of a first-order reaction inhibited by products; unless the experiments are done very accurately, there is danger of confusion. We can test for

*Solutions for many systems are given by C. Capellos and B. H. J. Bielski, *Kinetic Systems*, New York: Wiley-Interscience, 1972.

TABLE 9.1 Rate Equations and Half-Lives

Order	Rate Equation: Differential Form	Rate Equation: Integrated Form	Units of Rate Constant	Half-Life, $t_{1/2}$
0	$\frac{dx}{dt} = k$	$k = \frac{x}{t}$	mol $dm^{-3}s^{-1}$	$\frac{a_0}{2k}$
1	$\frac{dx}{dt} = k(a_0 - x)$	$k = \frac{1}{t} \ln \frac{a_0}{a_0 - x}$	s^{-1}	$\frac{\ln 2}{k}$
2	$\frac{dx}{dt} = k(a_0 - x)^2$	$k = \frac{1}{t} \frac{x}{a_0(a_0 - x)}$	$dm^3\ mol^{-1}s^{-1}$	$\frac{1}{ka_0}$
2	$\frac{dx}{dt} = k(a_0 - x)(b_0 - x)$ (reactants at different concentrations)	$k = \frac{1}{t(a_0 - b_0)} \ln \frac{b_0(a_0 - x)}{a_0(b_0 - x)}$	$dm^3\ mol^{-1}s^{-1}$	—
n	$\frac{dx}{dt} = k(a_0 - x)^n$	$k = \frac{1}{t(n-1)} \left[\frac{1}{(a_0 - x)^{n-1}} - \frac{1}{a_0^{n-1}} \right]$	$mol^{1-n}\ dm^{3n-3}\ s^{-1}$	$\frac{2^{n-1} - 1}{k(n-1)a_0^{n-1}}$

inhibition by products by measuring the rate after the deliberate introduction of products of reaction.

Half-Life

For a given reaction the **half-life** $t_{1/2}$ of a particular reactant is the time required for its concentration to reach a value that is halfway between its initial and final values. The value of the half-life is always inversely proportional to the rate constant and in general depends on reactant concentrations. For a first-order reaction the rate equation is Eq. 9.28, and the half-life is obtained by putting x equal to $a_0/2$:

$$\ln \frac{a_0}{a_0 - a_0/2} = kt_{1/2} \tag{9.38}$$

and therefore

$$t_{1/2} = \frac{\ln 2}{k} \tag{9.39}$$

In this case, the half-life is independent of the initial concentration. Since, for a first-order reaction, there is only one reactant, the half-life of the reactant can also be called the half-life of the *reaction*.

For a second-order reaction involving a single reactant or two reactants of equal initial concentrations, the rate equation is Eq. 9.35, and setting $x = a_0/2$ leads to

$$t_{1/2} = \frac{1}{a_0 k} \tag{9.40}$$

The half-life is now inversely proportional to the concentration of the reactant.

In the general case of a reactant of the nth order involving equal initial reactant concentrations a_0, the half-life, as seen in Table 9.1, is inversely proportional to a_0^{n-1}. For reactions of order other than unity the half-life is the same for different reactants only if they are initially present in their stoichiometric ratios; only in that case is it permissible to speak of the half-life of the *reaction*.

The order of a reaction can be determined by determining half-lives at two different initial concentrations a_1 and a_2. The half-lives are related by

$$\frac{t_{1/2}(1)}{t_{1/2}(2)} = \left(\frac{a_2}{a_1}\right)^{n-1} \tag{9.41}$$

$$n = 1 + \frac{\log[t_{1/2}(1)/t_{1/2}(2)]}{\log(a_2/a_1)} \tag{9.42}$$

and n can readily be calculated. This method can give misleading results if the reaction is not of simple order or if there are complications such as inhibition by products.

Since the half-lives of all reactions have the same units, they provide a useful way of comparing the rates of reactions of different orders. Rate constants, as we have seen, have different units for different reaction orders. Thus if two reactions have different orders, we cannot draw conclusions about their relative rates from their rate constants.

Radioactive disintegrations follow first-order kinetics and therefore have half-lives that are independent of the radioactive substance present. An atom whose nucleus has a specified number of protons and neutrons is known as a *nuclide;* if it is radioactive, it is referred to as a *radionuclide.* A radionuclide disintegrates at a rate that is a function only of the constitution of the nucleus; unlike ordinary chemical processes, radioactive disintegration cannot be influenced by any chemical or physical means, such as changing the temperature. The first-order rate constant for a radionuclide is known as its *decay constant.*

EXAMPLE

The half-life of radium, $^{226}_{88}Ra$ is 1600 years. How many disintegrations per second would be undergone by 1 g of radium?

SOLUTION

The half-life in seconds is

$$1600 \times 365.25 \times 24 \times 60 \times 60 = 5.049 \times 10^{10}\ \text{s}$$

The decay constant is thus, from Eq. 9.39,

$$k = \frac{\ln 2}{t_{1/2}} = \frac{0.693}{5.049 \times 10^{10}} = 1.37 \times 10^{-11}\ \text{s}^{-1}$$

The number of nuclei present in 1 g of radium is

$$\frac{6.022 \times 10^{23}}{226} = 2.666 \times 10^{21}\ \text{g}^{-1}.$$

The number of disintegrations is therefore

$$1.37 \times 10^{-11}\ \text{s}^{-1} \times 2.67 \times 10^{21}\ \text{g}^{-1} = 3.66 \times 10^{10}\ \text{g}^{-1}\ \text{s}^{-1}$$ ■

The kinetic characteristics of radionuclides are used to specify their amounts. The old unit for the amount of a radionuclide was the curie (Ci), which is defined as the amount producing exactly 3.7×10^{10} disintegrations per second; we have seen in the preceding example that this is approximately the number of disintegrations produced per second by 1 g of radium. In a 1975 IUPAC recommendation the curie was replaced by the becquerel (Bq), named after the French physicist Antoine Henri Becquerel (1852–1908). The becquerel is defined as the amount of radioactive substance giving one disintegration per second; thus

$$1 \text{ Ci} \equiv 3.7 \times 10^{10} \text{ Bq}$$

Differential Method

In the differential method, which was first suggested in 1884 by the Dutch physical chemist J. H. van't Hoff (1852–1911), the procedure is to determine rates directly by measuring tangents to the experimental concentration-time curves and to introduce these into the equations in their differential forms.

The theory of the method is as follows. The instantaneous rate of a reaction of the nth order involving only one reacting substance is proportional to the nth power of its concentration:

$$v = -\frac{da}{dt} = ka^n \tag{9.43}$$

Therefore

$$\log_{10} v = \log_{10} k + n \log_{10} a \tag{9.44}$$

A plot of $\log_{10} v$ against $\log_{10} a$ therefore will give a straight line if the reaction is of simple order; the slope k is the order n.

Reactions Having No Simple Order

Many reactions do not admit to the assignment of an order. Although such reactions can be treated by the method of integration, this procedure is seldom entirely satisfactory, owing to the difficulty of distinguishing between various possibilities. In such cases the best method is usually the differential method; rates of change of concentrations are measured accurately in the initial stages of the reaction, and runs are carried out at a series of initial concentrations. In this way, a plot of rate against concentration can be prepared and the dependence of rate on concentration can then be determined by various methods; one that is frequently used for enzyme reactions will be considered in Section 10.9.

Opposing Reactions

One complication is that a reaction may proceed to a state of equilibrium that differs appreciably from completion. The simplest case is when both forward and reverse reactions are of the first order:

$$\text{A} \underset{k_{-1}}{\overset{k_1}{\rightleftharpoons}} \text{Z}$$

If the experiment is started using pure A, of concentration a_0, and if after time t the concentration of Z is x, that of A is $a_0 - x$. The rate of reaction 1 if it occurred in isolation (i.e., the chemiflux into Z), is then equal to $k_1(a_0 - x)$, while the chemiflux into A is $k_{-1}x$; the net rate of change of concentration of Z is thus

$$\frac{dx}{dt} = k_1(a_0 - x) - k_{-1}x \tag{9.45}$$

If x_e is the concentration of Z at equilibrium, when the net rate is zero,

$$k_1(a_0 - x_e) - k_{-1}x_e = 0 \tag{9.46}$$

Elimination of k_{-1} between Eq. 9.45 and 9.46 gives rise to

$$\frac{dx}{dt} = \frac{k_1 a_0}{x_e}(x_e - x) \tag{9.47}$$

Integration of this equation, subject to the boundary condition that $x = 0$ when $t = 0$, gives

$$k_1 t = \frac{x_e}{a_0} \ln \frac{x_e}{x_e - x} \tag{9.48}$$

The amount of x present at equilibrium, x_e, can be measured directly. A procedure for obtaining k_1 is therefore to obtain values of x at various values of t and to plot

$$\frac{x_e}{a_0} \ln \frac{x_e}{x_e - x}$$

against t. The slope of the line then gives the value of k_1. The constant k_{-1} for the reverse reaction can be obtained by use of the fact that the equilibrium constant is k_1/k_{-1}.

Equations for more complicated kinetic situations have also been worked out (see footnote on p. 356).

9.5 Techniques for Very Fast Reactions

Some reactions are so fast that special techniques have to be employed. Such techniques are of two main types. The first employs essentially the same principles as are used for slow reactions, the methods being modified to make them suitable for more rapid reactions; the second type is of a different character and involves special principles.

The reasons why conventional techniques lead to difficulties for very rapid reactions are as follows:

1. The time that it usually takes to mix reactants, or to bring them to a specified temperature, may be significant in comparison with the half-life of the reaction. An appreciable error therefore will be made, since the initial time cannot be determined accurately.

2. The time that it takes to make a measurement of concentration is significant compared with the half-life.

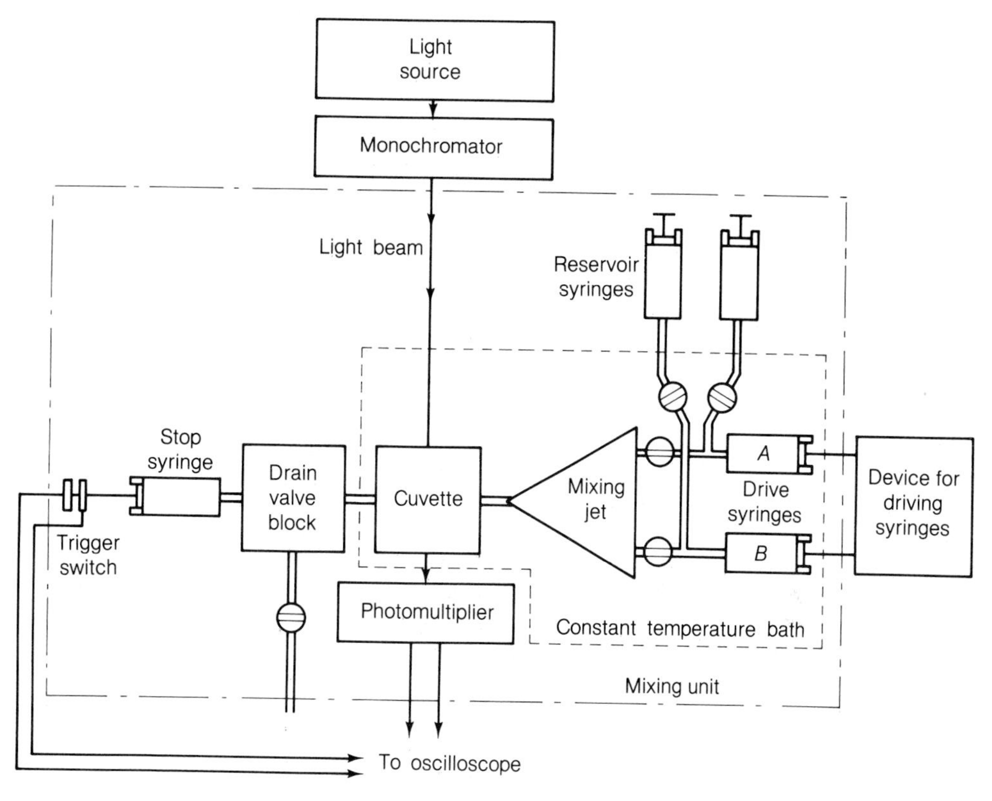

FIGURE 9.4 Schematic diagram of stopped-flow apparatus. Solutions in the two drive syringes A and B are rapidly forced through a mixing jet into a cuvette. When the flow is stopped, the oscilloscope is triggered and records light absorption as a function of time.

Flow Techniques

The first difficulty can sometimes be surmounted by using special techniques for bringing the reactants very rapidly into the reaction vessel and for mixing them very rapidly. With the use of conventional techniques, it takes from several seconds to a minute to bring solutions into a reaction vessel and to have them completely mixed and at the temperature of the surroundings. This time can be reduced greatly by using a rapid flow, and *flow techniques* are frequently employed for rapid reactions. One particular modification of these methods is the **stopped-flow** technique, shown schematically in Figure 9.4. This particular apparatus is designed for the study of a reaction between two substances in solution. A solution of one of the substances is maintained initially in the syringe A, and a solution of the other is in syringe B. The plungers of the syringes can be forced down rapidly and a rapid stream of two solutions passes into the mixing system. This is designed in such a way that jets of the two solutions impinge on one another and give very rapid mixing; with a suitable design of the mixing chamber it is possible for mixing to be essentially complete in 0.001 s. From this mixing chamber the solution passes at once into the reaction cuvette; alternatively, the two may be combined.

If a reaction is rapid, it is not possible to carry out chemical analyses at various stages. This difficulty must be resolved by employing techniques that allow properties to be determined instantaneously. For reactions in solution, spectrophotometric methods are commonly employed. If the products absorb differently from the reactants at a particular wavelength, we can pass monochromatic light of this wavelength through the reaction vessel; using a photoelectric device with suitable electronic circuits, we can display the output on a recorder or oscilloscope. If the reaction is not too fast, a pen-and-ink recorder may respond sufficiently rapidly; otherwise, an oscilloscope can give a record of absorption against time, and a photograph of the record can be taken. Fluorescence, electrical conductivity, and optical rotation are also convenient properties to measure in such high-speed studies.

Relaxation Methods

The flow techniques just described are limited by the speed with which it is possible to mix solutions. There is no difficulty, using optical or other techniques, in following the course of a very rapid reaction, but for hydrodynamic reasons it is impossible to mix two solutions in less than about 10^{-3} s. If the half-life is less than this, the reaction will be largely completed by the time that it takes for mixing to be achieved; any rate measurement made will be of the rate of mixing, not of the rate of reaction. The neutralization of an acid by a base, that is, the reaction

$$H^+ + OH^- \longrightarrow H_2O$$

under ordinary conditions has a half-life of 10^{-6} s or less (see Problem 6), and its rate therefore cannot be measured by any technique involving the mixing of solutions.

These technical problems were overcome by the development of a group of methods known as **relaxation methods;** the pioneer worker in this field was the German physical chemist Manfred Eigen. These methods differ fundamentally from conventional kinetic methods in that we start with the system at equilibrium under a given set of conditions. We then change these conditions very rapidly; the system is no longer at equilibrium, and it *relaxes* to a new state of equilibrium. The speed with which it relaxes can be measured, usually by spectrophotometry, and we can then calculate the rate constants.

There are various ways in which the conditions are disturbed. One is by changing the hydrostatic pressure. Another, the most common technique, is to increase the temperature suddenly, usually by the rapid discharge of a capacitor; this method is called the **temperature-jump** or **T-jump method.** It is possible to raise the temperature of a tiny cell containing a reaction mixture by a few degrees in less than 10^{-7} s, which is sufficiently rapid to allow us to study even the fastest chemical processes.

The principle of the method is illustrated in Figure 9.5. Suppose that the reaction is of the simple type

$$A \underset{k_{-1}}{\overset{k_1}{\rightleftharpoons}} Z$$

At the initial state of equilibrium, the product Z is at a certain concentration, and it stays at this concentration until the temperature jump occurs, when the concentration changes to another value that will be higher or lower than the initial value

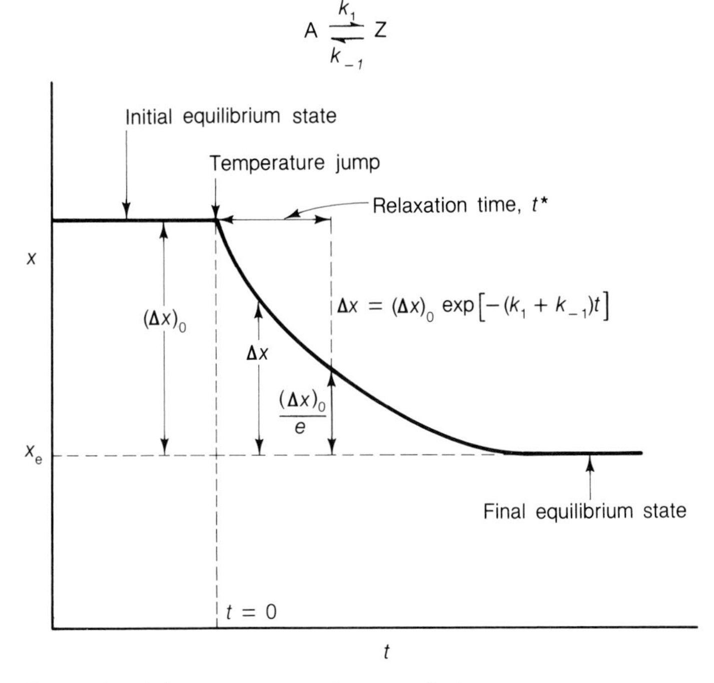

FIGURE 9.5 Principle of the temperature-jump technique.

according to the sign of ΔH° for the reaction. From the shape of the curve during the relaxation phase, we can obtain the sum of the rate constants, $k_1 + k_{-1}$, as shown by the following treatment.

Let a_0 be the sum of the concentrations of A and Z, and let x be the concentration of Z at any time; the concentration of A is $a_0 - x$. The kinetic equation is thus

$$\frac{dx}{dt} = k_1(a_0 - x) - k_{-1}x \tag{9.49}$$

If x_e is the concentration of Z at equilibrium,

$$k_1 (a_0 - x_e) - k_{-1}x_e = 0 \tag{9.50}$$

The deviation of x from equilibrium Δx is equal to $x - x_e$, and therefore

$$\frac{d\,\Delta x}{dt} = \frac{dx}{dt} = k_1(a_0 - x) - k_{-1}x \tag{9.51}$$

Subtraction of the expression in Eq. 9.50 gives

$$\frac{d\,\Delta x}{dt} = k_1(x_e - x) + k_{-1}(x_e - x) \tag{9.52}$$

$$= -(k_1 + k_{-1})\,\Delta x \tag{9.53}$$

The quantity Δx thus varies with time in the same manner as does the concentration of a reactant in a first-order reaction. Integration of Eq. 9.53, subject to the boundary condition that $\Delta x = (\Delta x)_0$ when $t = 0$, leads to

$$\ln \frac{(\Delta x)_0}{\Delta x} = (k_1 + k_{-1})t \tag{9.54}$$

We can define a relaxation time t^* as the time corresponding to

$$\frac{(\Delta x)_0}{\Delta x} = e \tag{9.55}$$

or to

$$\ln \frac{(\Delta x)_0}{\Delta x} = 1 \tag{9.56}$$

The relaxation time is thus the time at which the distance from equilibrium is $1/e$ of the initial distance (see Figure 9.5). From Eq. 9.54 we see that

$$t^* = \frac{1}{k_1 + k_{-1}} \tag{9.57}$$

If, therefore, we determine t^* experimentally for such a system, we can calculate $k_1 + k_{-1}$. However, the ratio k_1/k_{-1} is the equilibrium constant and it can be determined directly; hence the individual constants k_1 and k_{-1} can be calculated.

Relaxation curves for other types of reactions can similarly be analyzed to give rate constants. Note that the rate constants obtained refer to the second temperature, after the T-jump has occurred. Thus, if we want rate constants at 25°C, and the T-jump is 7°C, we should start at 18°C.

During recent years a considerable number of investigations have been made using this technique. An important reaction studied in this way is the dissociation of water and combination of hydrogen and hydroxide ions:

$$2H_2O \rightleftharpoons H_3O^+ + OH^-$$

The reaction was followed by measuring conductivity; after the T-jump the conductivity increased with time, with a relaxation half-life of 3.7×10^{-5} s at 23°C. From this value it can be calculated that the second-order rate constant for the combination of H_3O^+ and OH^- ions is 1.3×10^{11} dm^3 mol^{-1} s^{-1}, which is a remarkably high value. The rate constant for the reverse dissociation, which is very small, can be calculated from this value and the equilibrium constant. A number of other hydrogen-ion transfer processes have also been studied using the same technique.

9.6 Molecular Kinetics

An important aspect of chemical kinetics is concerned with how rates depend on *temperature*. The conclusions from such studies lead to considerable insight into the molecular nature of chemical reactions, and as a result the field is frequently referred to as **molecular kinetics.**

In this chapter we will be concerned only with temperature effects on **elementary reactions,** which are reactions that occur in a single stage, with no identifiable reaction intermediates. Elementary reactions are to be contrasted with **composite reactions,** which go in more than one stage.

Molecularity and Order

Once a process has been identified as an elementary one, an important question that arises is: How many molecules enter into reaction? This number is referred to as the **molecularity** of the reaction. We have seen that from the variation of rate with concentration we can frequently determine a *reaction order*. This number, a purely experimental one, should be sharply distinguished from *molecularity*, which represents a deduction as to the number of molecules. It is permissible to speak of the *order* of a composite reaction, provided that the rate is proportional simply to concentrations raised to certain powers. It is meaningless, on the other hand, to speak of the *molecularity* if the mechanism is a composite one.

With certain exceptions, discussed later, we can assume that the order of an *elementary reaction* indicates the number of molecules that enter into reaction (i.e., that *the order and the molecularity are the same*). For example, if an elementary reaction is first order with respect to a reactant A and first order with respect to another substance B, we often find that the reaction is bimolecular, a molecule of A and a molecule of B entering into the reaction.

Sometimes, however, this procedure leads to incorrect conclusions. Suppose, for example, that one reactant is present in large excess, so that its concentration does not change appreciably as the reaction proceeds; moreover (for example, if it is the solvent) its concentration may be the same in different kinetic runs. If this is the case, the kinetic investigation will not reveal any dependence of the rate on the concentration of this substance, which would therefore not be considered to be entering into reaction. This situation is frequently found in reactions in solution where the solvent may be a reactant. For example, in hydrolysis reactions in aqueous solution, a water molecule may undergo reaction with a solute molecule. Unless special procedures are employed, the kinetic results will not reveal the participation of the solvent. However, its participation is indicated if it appears in the stoichiometric equation.

Another case in which the kinetic study may not reveal that a substance enters into reaction is when a *catalyst* is involved. A **catalyst,** by definition, is a *substance that influences the rate of reaction without itself being used up;* it may be regarded as being *both a reactant and a product of reaction* (see Section 10.9). The concentration of a catalyst therefore remains constant during reaction, and the kinetic analysis of a single run will not reveal the participation of the catalyst in the reaction. However, the fact that it does enter into reaction may be shown by measuring the rate at a variety of catalyst concentrations; generally a linear dependence is found.

It follows that the decision about the molecularity of an elementary reaction must involve not only a careful kinetic study in which as many factors as possible are varied but also a consideration of other aspects of the reaction.

9.7 The Arrhenius Law

It was found empirically that the rate constant k is related to the absolute temperature T by the equation

$$k = Ae^{-B/T} \tag{9.58}$$

where A and B are constants. This relationship was expressed by van't Hoff and Arrhenius in the form

$$k = Ae^{-E/RT} \tag{9.59}$$

where R is the gas constant, equal to 8.314 J K^{-1} mol^{-1} and E is known as the *activation energy*. The equation was arrived at in 1884 by van't Hoff,* who argued on the basis of the variation of the equilibrium constant with the temperature and pointed out that a similar relationship should hold for the rate constant of a reaction. This idea was extended by Arrhenius, who successfully applied it to a large number of reactions and, as a result, Eq. 9.59 is generally referred to as the **Arrhenius law.**

The arguments of van't Hoff are briefly as follows. The temperature dependence of a standard equilibrium constant K_c^u is given by Eq. 4.83:

$$\frac{d \ln K_c^u}{dT} = \frac{\Delta U^\circ}{RT^2} \tag{9.60}$$

where ΔU° is the standard internal energy change in the reaction. For a reaction such as

$$A + B \underset{k_{-1}}{\overset{k_1}{\rightleftharpoons}} Y + Z$$

the equilibrium constant K_c is equal to the ratio of the rate constants k_1 and k_{-1},

$$K_c = \frac{k_1}{k_{-1}} \tag{9.61}$$

Equation 9.60 can therefore be written as†

$$\frac{d \ln k_1}{dT} - \frac{d \ln k_{-1}}{dT} = \frac{\Delta U^\circ}{RT^2} \tag{9.62}$$

and this may be split into the two equations

$$\frac{d \ln k_1}{dT} = \text{constant} + \frac{E_1}{RT^2} \tag{9.63}$$

$$\frac{d \ln k_{-1}}{dT} = \text{constant} + \frac{E_{-1}}{RT^2} \tag{9.64}$$

where $E_1 - E_{-1}$ is equal to ΔU°. Experimentally, it is found that the constants appearing in Eqs. 9.63 and 9.64 can be set equal to zero, and integration of these equations gives rise to

$$k_1 = A_1 e^{-E_1/RT} \tag{9.65}$$

$$k_{-1} = A_{-1} e^{-E_{-1}/RT} \tag{9.66}$$

---

*J. H. van't Hoff, *Etudes de dynamique chimique,* Amsterdam: F. Muller, 1884.

†In this chapter, in order to simplify notation we drop the device of using superscripts u to indicate the value of a quantity. It is to be understood by the reader that one takes the logarithm of the *value* of the quantity rather than the quantity itself.

The quantities A_1 and A_{-1} are best known as the **preexponential factors*** of the reactions, and E_1 and E_{-1} are known as the **activation energies**, or **energies of activation.**

Arrhenius's approach to the law was a little different from that of van't Hoff. He pointed out that for ordinary chemical reactions the majority of collisions between the reactant molecules are ineffective; the energy is insufficient. In a small fraction of the collisions, however, the energy is great enough to allow reaction to occur. According to the Boltzmann principle (see Section 14.2) the fraction of collisions in which the energy is in excess of a particular value E is

$$e^{-E/RT}$$

This fraction is larger the higher the temperature T and the lower the energy E. The rate constant should therefore be proportional to this fraction.

In order to test the Arrhenius equation (Eq. 9.59), we first take logarithms of both sides:

$$\ln k = \ln A - \frac{E}{RT} \tag{9.67}$$

If the Arrhenius law applies, a plot of $\ln k$ against $1/T$ will be a straight line, and the slope (which has the unit of K) will be $-E/R$. Alternatively, we can take common logarithms:

$$\log_{10} k = \log_{10} A - \frac{E}{2.303RT} \tag{9.68}$$

and the slope of a plot of $\log_{10} k$ against $1/T$ is $-E/2.303R$, or

$$\frac{\text{slope}}{\text{K}} = -\frac{E/\text{J mol}^{-1}}{19.14} = -\frac{E/\text{cal mol}^{-1}}{4.575} \tag{9.69}$$

An example of such an Arrhenius plot is shown in Figure 9.6. It follows from Eq. 9.59 that the preexponential factor has the same units as the rate constant itself (e.g., s^{-1} for a first-order reaction, $dm^3\ mol^{-1}\ s^{-1}$ for a second-order reaction). The SI unit of activation energy is $J\ mol^{-1}$, and $kJ\ mol^{-1}$ is very commonly used, but many values in the scientific literature are in $cal\ mol^{-1}$ or $kcal\ mol^{-1}$.

The Arrhenius law has a surprisingly wide applicability. It is obeyed not only by the rate constants of elementary reactions but frequently also by the rates of much more complicated processes. For example, the law is obeyed by the chirping of crickets (see Figure 9.7), the creeping of ants, the flashing of fireflies, the α brain-wave rhythm, the rate of aging, and even by psychological processes such as the rates of counting and forgetting.† The reason for the applicability of the law to such processes is that they are controlled by chemical reactions.

---

*The term *frequency factor* is very commonly used, but A does not in general have the dimensions of frequency.

†For details, with Arrhenius plots, of these processes see K. J. Laidler, *J. Chem. Education, 49,* 343 (1972).

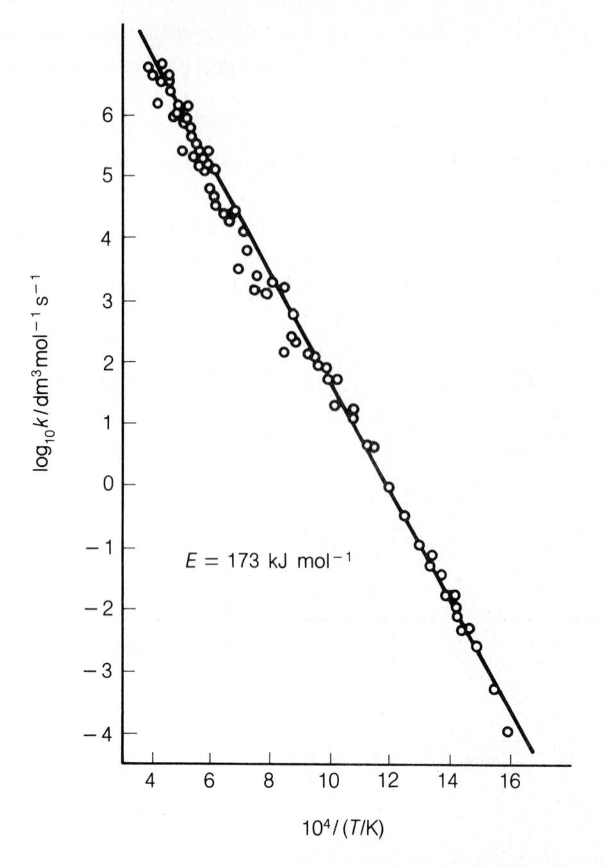

FIGURE 9.6 An Arrhenius plot for the thermal decomposition of acetylene, a second-order reaction. This plot covers a very wide temperature range, from 600 K to 2500 K, and the rate constants range over a factor of more than 10^{10}. The rates were measured by 11 different investigators, using a variety of experimental techniques [M. H. Back, *Can. J. Chem.*, *49*, 2199 (1971)].

EXAMPLE

A second-order reaction in solution has a rate constant of $5.7 \times 10^{-5}\ \mathrm{dm^3\ mol^{-1}\ s^{-1}}$ at 25.0°C and of $1.64 \times 10^{-4}\ \mathrm{dm^3\ mol^{-1}\ s^{-1}}$ at 40.0°C. Calculate the activation energy and the preexponential factor, assuming the Arrhenius law to apply.

SOLUTION

To solve this type of problem it is convenient (but not necessary) to sketch an Arrhenius plot; this is shown in Figure 9.8, in which common logarithms are plotted. It is to be emphasized that, in taking the logarithms of the rate constants and the reciprocals of the rates, it is necessary to use enough significant figures as we are dealing with relatively small differences between the values.

The slope of the line in Figure 9.8 is

$$\frac{-4.244 - (-3.785)}{(3.3540 - 3.1934) \times 10^{-3}} = -\frac{0.459}{0.1606} \times 10^3\ \mathrm{K}$$

$$= -2.858 \times 10^3\ \mathrm{K}$$

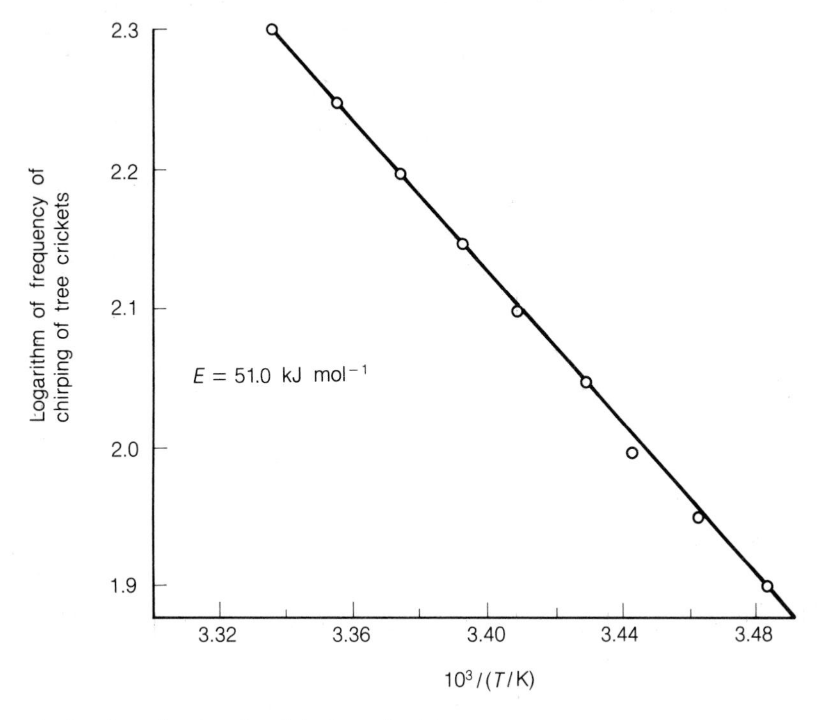

FIGURE 9.7 Arrhenius plot for the chirping of crickets.

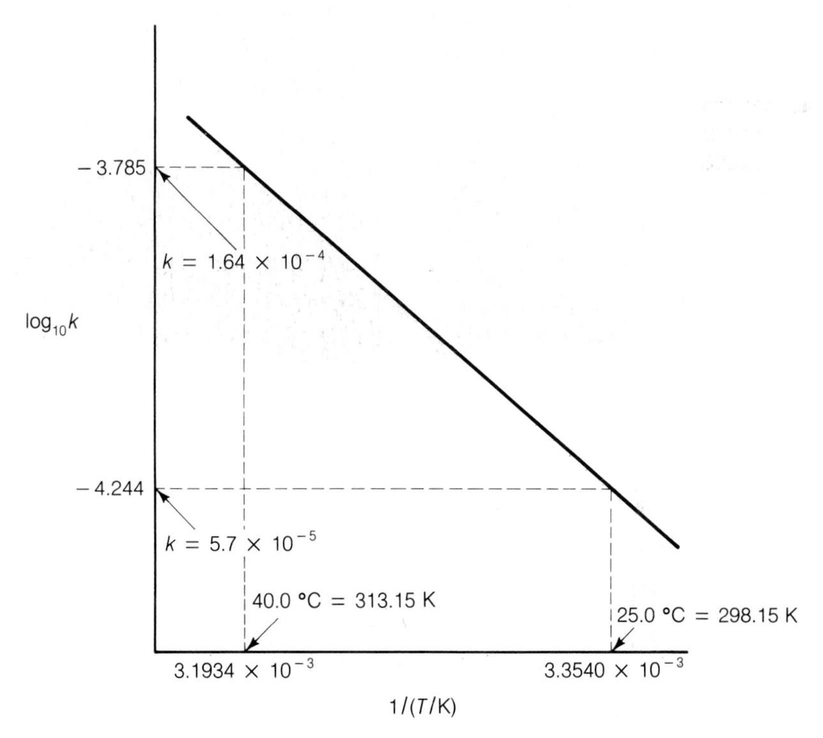

FIGURE 9.8 Schematic Arrhenius plot for the example given on page 368.

The energy of activation is the slope multiplied by -19.14 J K^{-1} mol^{-1}:

$$E = 19.14 \times 2.858 \times 10^3 = 54.70 \times 10^3 \text{ J mol}^{-1}$$
$$= 54.7 \text{ kJ mol}^{-1}$$

The preexponential factor A can be obtained from Eq. 9.68. The rate constant at either temperature can be used to evaluate A; if we use 25.0°C, we obtain (see Figure 9.8)

$$-4.244 = \log_{10} A - \frac{54\,700}{19.14 \times 298.15}$$

or

$$-4.244 = \log_{10} A - 9.586$$

and thus

$$\log_{10} A = 5.342$$
$$A = 2.20 \times 10^5 \text{ dm}^3 \text{ mol}^{-1} \text{ s}^{-1}$$

The units of A are the same as those of the rate constant. ■

Activation Energy

The activation energy is looked upon differently in the theories of van't Hoff and Arrhenius. Van't Hoff placed emphasis on the energy levels of the reactants and products and of species occurring during the course of reactions. Figure 9.9 shows a potential-energy diagram for a reaction

$$A + B \rightleftharpoons Y + Z$$

In this diagram the internal energy for the products Y + Z is larger than that for the reactants A + B, by an amount ΔU. In general there is not much difference between energy and enthalpy changes, so that ΔH will also be positive—that is, the reaction is endothermic.

During the course of the reaction between a molecule of A and a molecule of B the potential energy very often passes through a maximum, as shown in Figure 9.9. The height of this maximum plays a very important role in all theories of the rates of reactions. The molecular species having the maximum energy is called the **activated complex,** and its *state* is called the **transition state** and is usually denoted by the symbol ‡. The energy E_1 of this complex with respect to the energy of A + B is the activation energy for the reaction in the left-to-right direction. The energy E_{-1} of the activated complex with respect to Y + Z is the activation energy for the reverse reaction. We see from Figure 9.9 that

$$E_1 - E_{-1} = \Delta U \tag{9.70}$$

This equation is very useful. Sometimes we can measure both E_1 and E_{-1} and then have a value for ΔU, from which ΔH can easily be obtained by making a small correction (see Eq. 2.41). More often we know ΔH and hence ΔU and can measure one activation energy conveniently; the other is then obtained from Eq. 9.70.

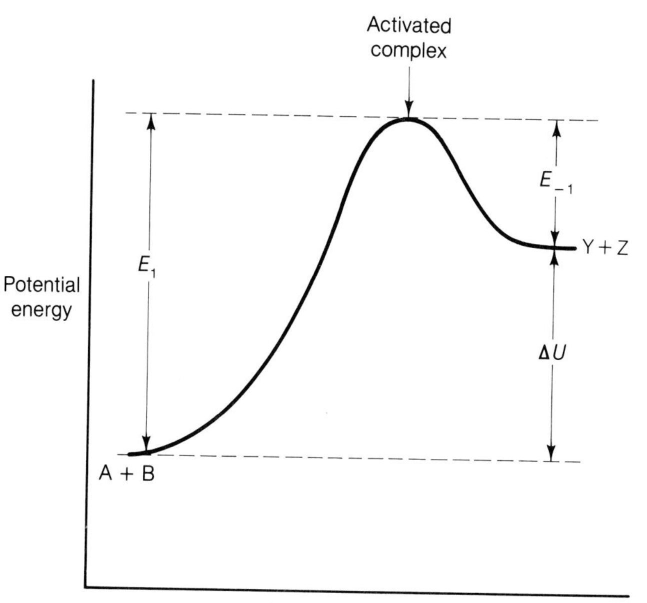

FIGURE 9.9 Potential-energy diagram for a chemical reaction $A + B \rightleftharpoons Y + Z$.

This concept of an energy barrier for a reaction is a very useful and important one. It follows that for an endothermic reaction the activation energy must be at least equal to the endothermicity; for our example, E_1 must be at least equal to ΔU, since E_{-1} cannot be negative. Arrhenius's discussion of the activation energy is closely related to this concept; for reaction between A and B to occur, the molecules must come together with at least the energy E_1, in order to surmount the barrier.

Why is there usually an energy barrier to reaction? Why does not the curve in Figure 9.9 rise smoothly from one level to the other, without passing through a maximum? An important clue is provided by the fact that certain special types of reactions do occur with zero activation energy. These are reactions in which there is simply a pairing of electrons, without the breaking of any chemical bond. A free methyl radical, for example, has an odd electron; its Lewis structure is

$$\begin{array}{c} \text{H} \\ \text{H}:\ddot{\text{C}}\cdot \\ \text{H} \end{array}$$

When methyl radicals combine to form ethane,

$$2CH_3 \longrightarrow C_2H_6$$

they do so with zero activation energy, and the energy diagram is as shown in Figure 9.10. The activation energy for the reverse reaction, the dissociation of C_2H_6, is obviously equal to the dissociation energy of the bond.

A zero activation energy, however, only occurs in reactions where there is no breaking of a chemical bond. In most reactions at least one bond is broken, and at least one bond is formed. For example, in the hydrolysis reaction

$$C_2H_5X + H_2O \longrightarrow C_2H_5OH + HX$$

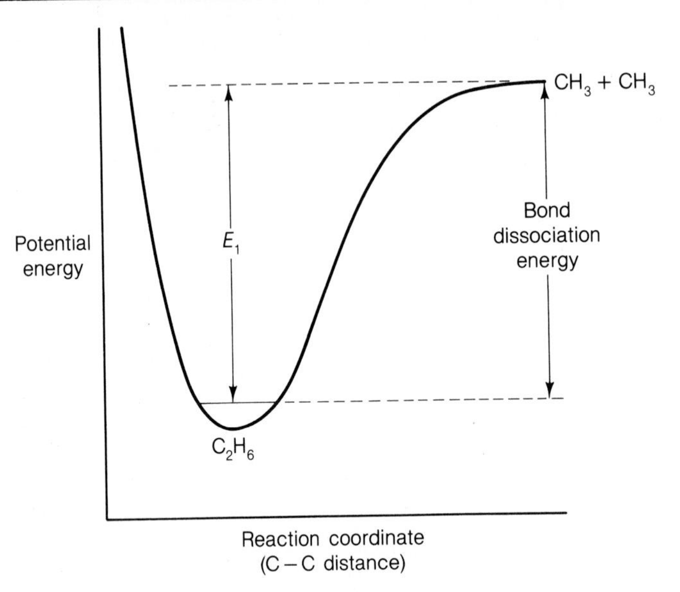

FIGURE 9.10 Potential-energy diagram for the $2CH_3 \rightarrow C_2H_6$ system.

the C—X bond is being broken during reaction at the same time that the O—C bond is being formed. Even though the overall process is exothermic, the energy released in the formation of the O—C bond is not transferred completely efficiently into the energy required to break the C—X bond, and there is an energy barrier.

9.8 Potential-Energy Surfaces

The theoretical treatment of the dynamics of a chemical reaction is a matter of considerable difficulty, even for the simplest of systems. A very important advance was made in 1931 by the American physical chemist Henry Eyring and the Hungarian-British physical chemist (later sociologist) Michael Polanyi (1891–1976), who developed the method of *potential-energy surfaces*,* which resemble maps of the reaction system.

It is convenient to describe the method with reference to one of the simplest of all chemical reactions, the reaction between a hydrogen atom and a hydrogen molecule. It is possible to measure the rate of a reaction of this type by labeling one or more of the atoms; a deuterium atom, for example, may be caused to react with a hydrogen molecule, in which case the products will be DH + H. Otherwise, it is possible to make atomic hydrogen react with pure *para*-hydrogen,† in which case

* H. Eyring and M. Polanyi, *Z. Physikal. Chem.*, *B*, *12*, 279 (1931); excerpts from this article, translated into English, are included in M. H. Back and K. J. Laidler (Eds.), *Selected Readings in Chemical Kinetics*, Oxford: Pergamon Press, 1967.

† *Para*-hydrogen is a form of H_2 in which the nuclear spins are opposed; in *ortho*-hydrogen the spins are parallel. *Para*-hydrogen can be prepared pure and has physical properties that differ slightly from the *ortho* form. Ordinary hydrogen is a mixture of the two forms.

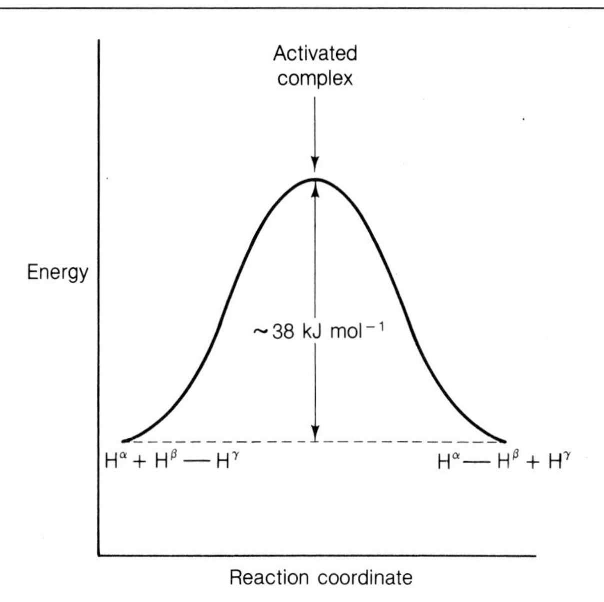

FIGURE 9.11 The energy barrier for the reaction $H + H_2 \rightarrow H_2 + H$.

there is conversion of the *para*-hydrogen into the equilibrium mixture of the *ortho* and *para* forms. In either case we may represent the reaction as follows:

$$H^\alpha + H^\beta—H^\gamma \longrightarrow H^\alpha—H^\beta + H^\gamma$$

This process is the simplest reaction in which there is breaking of one bond and making of another. Many reactions are of this type, and a study of this very simple process has provided great insight into the mechanisms of much more complicated processes.

It is found from experiment that the activation energy of this reaction is about 38 kJ mol^{-1} (9 kcal mol^{-1}). The energy levels of the reactants and the products are the same, but the activation energy of 38 kJ mol^{-1} implies that an energy barrier of this height must be crossed during the passage of the system from the initial to the final state. This is represented schematically in Figure 9.11, in which the energy of the system is plotted against a reaction coordinate that represents the extent to which the individual reaction process has occurred.

The first treatment of chemical reactions from the standpoint of molecular structure was carried out in 1928 by F. London,* who suggested that the properties of reaction intermediates can be calculated using the same quantum-mechanical methods that are used for calculating the energies of stable molecules (see Chapter 12). We can imagine a series of steps in which H^α is brought closer and closer to the $H^\beta—H^\gamma$ molecule. We suppose that the atom H^α and the molecule $H^\beta—H^\gamma$ possess sufficient energy between them to come close together and to give rise to reaction. As the atom approaches the molecule, there is an electronic interaction between

*F. London, *Probleme der modernen Physik* (Sommerfeld Festschrift), Leipzig: S. Hirtel, 1928, p. 104; Z. *Elektrochem.*, *35*, 552 (1929).

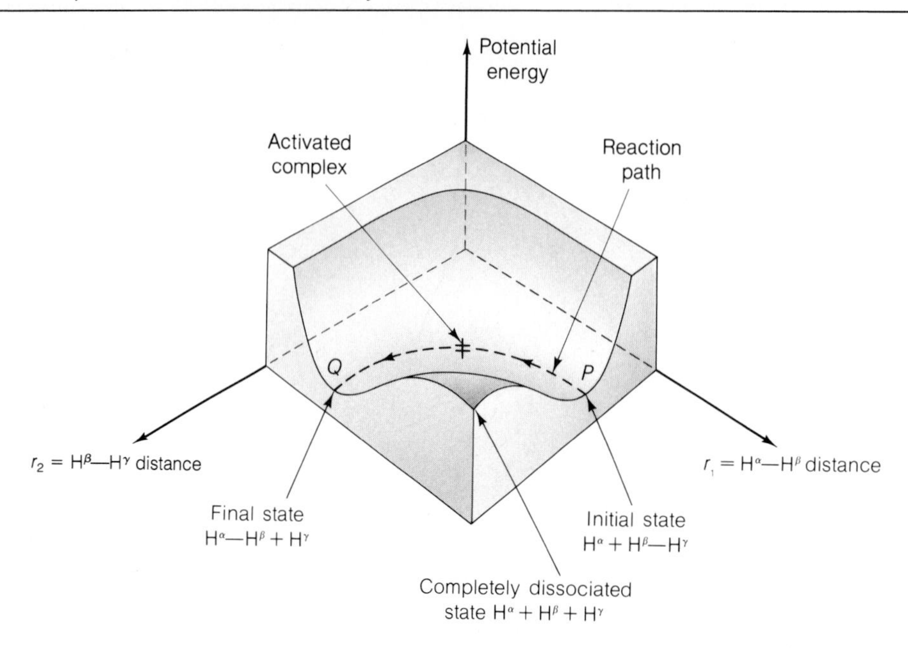

FIGURE 9.12 Potential-energy surface for the reaction $H^\alpha + H^\beta - H^\gamma \rightarrow H^\alpha - H^\beta + H^\gamma$.

them, and the potential energy of the system at first increases and later decreases. Since this particular system is symmetrical, the maximum energy corresponds to a symmetrical *activated complex* in which the distance between H^α and H^β is equal to the distance between H^β and H^γ. After this complex has formed, the energy gradually decreases as the system approaches the state corresponding to the molecule $H^\alpha—H^\beta$ and the separated atom H^γ.

The most stable complex formed by the interaction between a hydrogen atom and a hydrogen molecule is linear. We will now consider how the energy of such a linear complex varies with the distances between the atoms. The distance between H^α and H^β is designated as r_1, and the distance between H^β and H^γ as r_2; since the complex is linear, the distance between H^α and H^γ is $r_1 + r_2$. The energy of such a linear system of atoms can be represented in a three-dimensional diagram in which the energy is plotted against r_1 and r_2. Such a diagram is shown schematically in Figure 9.12. On the left-hand face of this diagram the distance r_2 may be considered to be sufficiently great that we are dealing simply with the diatomic molecule $H^\alpha—H^\beta$. Similarly, on the right-hand face of the diagram there is a curve for the diatomic molecule $H^\beta—H^\gamma$, the distance r_1 now being large. The course of reaction is shown by the arrows in the diagram.

To determine the course of such a reaction it is necessary to make quantum-mechanical calculations corresponding to a number of points in the interior of the diagram. Such calculations were made by Eyring and Polanyi. These calculations showed that, running in from the points P and Q on the diagram, there are two valleys that meet in the interior of the diagram at a *col*, or *saddle point*. This result may be represented in a different type of diagram, such as that in Figure 9.13, where the energy levels are shown by means of contour lines. To go from point P to the point Q using the minimum amount of energy, the system will travel along the valley, over the col, and down into the second valley. This reaction path is

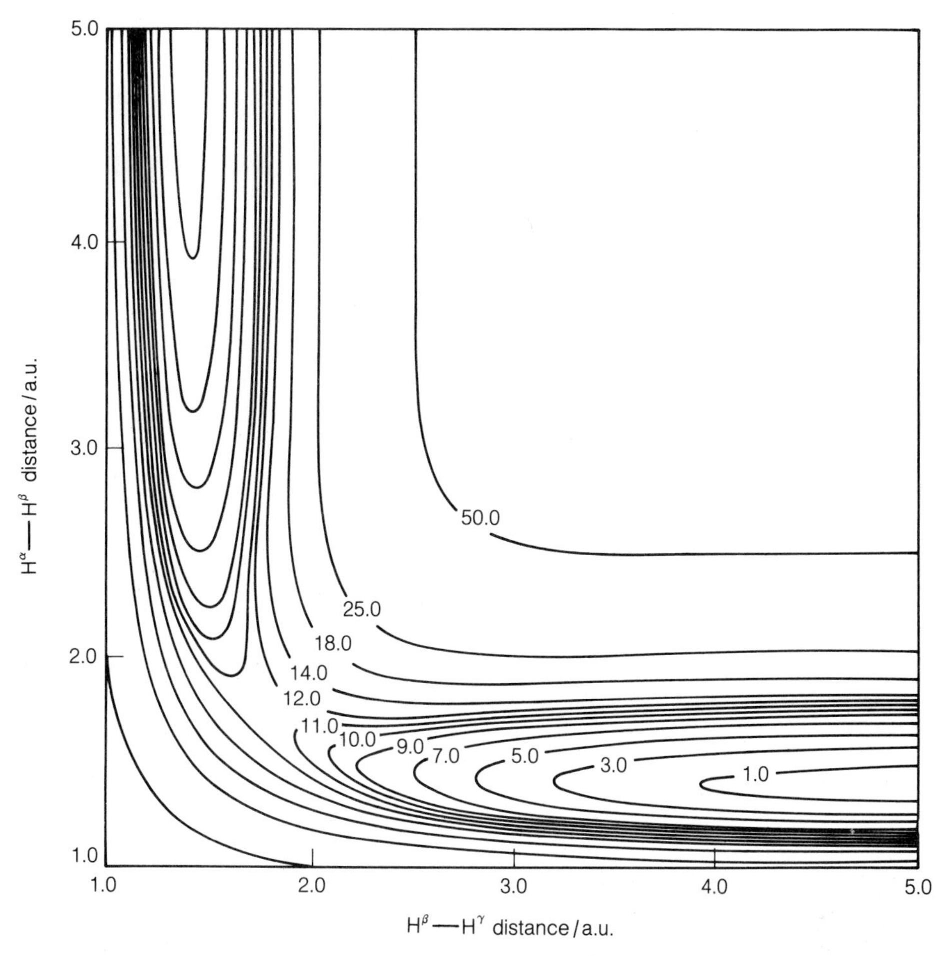

FIGURE 9.13 Potential-energy surface for the reaction $H^\alpha + H^\beta - H^\gamma \rightarrow H^\alpha - H^\beta + H^\gamma$, showing contour lines connecting positions of equal energy. This surface was calculated by I. Shavitt, R. M. Stevens, F. L. Minn, and M. Karplus, *J. Chem. Phys.*, *48*, 6 (1968), and it is one of the more reliable surfaces for this reaction. The energies are shown in kcal mol^{-1} (1 kcal = 4.184 J); 1 a.u. (atomic unit) = 0.05292 nm.

represented by the dashed line shown in the contour diagram, and for energetic reasons the majority of the reaction systems will follow this path. The energy corresponding to the col is from one point of view a maximum energy and from another point of view a minimum energy. It is a minimum energy in the sense that the system cannot use less energy in going from P to Q; it is a maximum energy in the sense that, as the system travels along its most economical path, the col represents the highest point in the path. The height of this col represents the activation energy of the system.

When the reaction is not a symmetrical one but involves three atoms that are not identical, the results are similar, but the potential-energy surface is no longer symmetrical with respect to the two axes.

Note that the mechanism of this reaction is very different from one corresponding to the complete dissociation of the molecule H^β—H^γ, followed in a separate stage by the combination of H^α with H^β. The activation energy that would be required for this process is the energy of dissociation of the hydrogen molecule, which is 431.8 kJ mol^{-1} (103.2 kcal mol^{-1}). It is evident that by moving along the valleys the system can achieve reaction at the expense of very much less energy than would be required if the mechanism involved complete dissociation. The physical explanation of this is that the energy released by the making of the second bond (the bond between H^α and H^β) continuously contributes toward the energy requirement for the breaking of the first bond, the bond between H^β and H^γ.

The calculation of potential-energy surfaces, by the use of pure quantum mechanics, is a matter of considerable difficulty even for the $H + H_2$ system. Subsequent to the work of Eyring and Polanyi in 1929 many attempts have been made, on the basis of the variational principle (Section 11.13); but even when a large amount of work is done, the results are not very reliable. For example, in 1968 Shavitt, Stevens, Minn, and Karplus* published the results of elaborate calculations in which the molecular eigenfunction involved 15 atomic orbitals; their surface is shown in Figure 9.13. The activation energy calculated in this way was over 15 kJ mol^{-1} (4 kcal mol^{-1}) higher than the observed value. Since rates depend exponentially on activation energies, errors of this magnitude lead to very large errors in calculated rates. The situation with more complicated reactions is even less satisfactory.

Because of this, considerable effort has gone into the calculation of potential-energy surfaces and hence of activation energies by methods in which empirical information is used, sometimes with quantum-mechanical theory used in addition.

The interested reader is referred to more advanced texts (see *Suggested Reading*, p. 402) for further details.

9.9 The Preexponential Factor

According to the Arrhenius equation (Eq. 9.59), the rate of a reaction at a given temperature is controlled entirely by the two quantities E and A. We saw in the last section that the calculation of the activation energy E presents very considerable difficulty. The situation regarding the preexponential factor A is somewhat more encouraging, in that it is possible to make reasonable estimates of the value of this quantity, taking due account of the type of reaction and of the various factors involved.

Hard-Sphere Collision Theory

The first attempt to calculate the preexponential factor was based on the kinetic theory of collisions, the assumption being that the molecules are hard spheres. For a single gas A, the **collision number** (i.e., the total number of collisions per unit time per unit volume) is (see Section 1.9)

$$Z_{AA} = \frac{1}{2}\sqrt{2}\pi d^2 \bar{u}\, N_A^2 \qquad (9.71)$$

*I. Shavitt, R. M. Stevens, F. L. Minn, and M. Karplus, *J. Chem. Phys.*, *48*, 6 (1968).

where d is the molecular diameter, N_A is the number of molecules per unit volume, and $\bar{u}$ is the mean molecular velocity, given in Table 14.2:

$$\bar{u} = \sqrt{\frac{8\mathbf{k}T}{\pi m}} \tag{9.72}$$

The collision number is therefore

$$Z_{AA} = 2d^2N_A^2\sqrt{\frac{\pi\mathbf{k}T}{m}} \tag{9.73}$$

Its SI unit is m^{-3} s^{-1}. The corresponding expression for the collision number Z_{AB} for collisions between two unlike molecules A and B, of masses m_A and m_B, is

$$Z_{AB} = N_AN_Bd_{AB}^2\left(8\pi\mathbf{k}T\frac{m_A + m_B}{m_Am_B}\right)^{1/2} \tag{9.74}$$

Here d_{AB} is the average of the diameters, or the sum of the radii. The quantity d_{AB}^2 is known as the **collision cross section** and given the symbol σ (Greek lowercase letter sigma).

According to the hard-sphere kinetic theory of reactions, these collision numbers multiplied by the Arrhenius factor $e^{-E/RT}$ give the rate of formation of the products of reaction, in molecules per unit volume per unit time; thus for reaction between A and B,

$$v_{AB} = N_AN_Bd_{AB}^2\left(8\pi\mathbf{k}T\frac{m_A + m_B}{m_Am_B}\right)^{1/2}e^{-E/RT} \tag{9.75}$$

Division by N_AN_B gives a rate constant in molecular units (SI unit: m^3 s^{-1}); it can be put into molar units by multiplication by the Avogadro constant L:

$$k = Ld_{AB}^2\left(8\pi\mathbf{k}T\frac{m_A + m_B}{m_Am_B}\right)^{1/2}e^{-E/RT} \tag{9.76}$$

The preexponential factor in this expression is called the **collision frequency factor** and given the symbol Z_{AB} (or Z_{AA} if there is only one type of molecule):

$$Z_{AB} = Ld_{AB}^2\left(8\pi\mathbf{k}T\frac{m_A + m_B}{m_Am_B}\right)^{1/2} \tag{9.77}$$

The development of this treatment of reactions was first made by the German chemist M. Trautz* in 1916 and independently in 1918 by the British chemist William Cudmore McCullagh Lewis (1885–1956).† Lewis applied this treatment to the reaction

$$2HI \longrightarrow H_2 + I_2$$

and calculated a preexponential factor of 3.5×10^{-7} dm^3 mol^{-1} s^{-1}, which was in excellent agreement with the experimental value of 3.52×10^{-7} dm^3 mol^{-1} s^{-1}.

*M. Trautz, *Z. anorgan. Chem.*, *96*, 1 (1916).

†W. C. McC. Lewis, *J. Chem. Soc.* (London), *113*, 471 (1918); an excerpt from this paper is included in M. H. Back and K. J. Laidler (Eds.), *Selected Readings in Chemical Kinetics*, Oxford: Pergamon Press, 1967.

Paradoxically, this good agreement was unfortunate, because it led to undue confidence in the theory and delayed the development of the subject for many years. It later became evident that there are many reactions for which there are large discrepancies between the observed preexponential factors and the collision frequency factors. For example, it is often found that when the reactant molecules are of some complexity, the observed preexponential factors are lower by several powers of 10 than the factors calculated by the simple collision theory. Deviations from theory are also encountered with solution reactions involving ions or dipolar substances as reactants.

The source of these discrepancies lies in the use of the hard-sphere kinetic theory of gases in order to evaluate the frequency of collisions. This theory is quite satisfactory for the treatment of energy transfer processes such as occur in viscous flow, but for a theory of chemical reactivity a more precise definition of a collision is needed. If two molecules are to undergo a chemical reaction, they not only must collide with sufficient mutual energy but must come together with such a mutual orientation that the required bonds can be broken and made. The kinetic theory, by treating the reacting molecules as hard spheres, counts every collision as an effective one; if the molecules are complex, on the other hand, in only a small fraction of the collisions will the molecules come together in the right way for reaction to occur.

Some workers remedied the situation by retaining the kinetic theory of gases but introducing into the preexponential factor a steric factor P that was supposed to represent the fraction of the total number of collisions that are effective from the point of view of the mutual orientation of the molecules. The rate constant would then be written as

$$k = PZ_{AB}e^{-E/RT} \tag{9.78}$$

This procedure certainly introduces some improvement, but the evaluation of P cannot be done in an entirely satisfactory way. Moreover, there are certain factors other than orientation that enter into the magnitude of the preexponential factor, and these cannot easily be estimated.

Transition State Theory

Since effects of this kind cannot be treated satisfactorily by a modification of hard-sphere collision theory, an alternative approach to the problem is necessary. A much more satisfactory treatment was presented in 1935 by Henry Eyring* and is known as *activated complex theory* or as **transition state theory.** A useful way of approaching this theory is to regard it as a logical extension of the argument of van't Hoff with regard to the energy of activation. We have seen in Section 9.7 that van't Hoff regarded the energy change of a reaction as being the difference between two terms, one of which exercises the sole control over the reaction in the forward

*H. Eyring, *J. Chem. Phys.*, *3*, 107 (1935); a more comprehensive treatment is in W. F. K. Wynne-Jones and H. Eyring, *J. Chem. Phys.*, *3*, 492 (1935), and this article is reproduced in full in M. H. Back and K. J. Laidler (Eds.), *Selected Readings in Chemical Kinetics*, Oxford: Pergamon Press, 1967; a very similar treatment was presented by M. G. Evans and M. Polanyi, *Trans. Faraday Soc.*, *31*, 875 (1935).

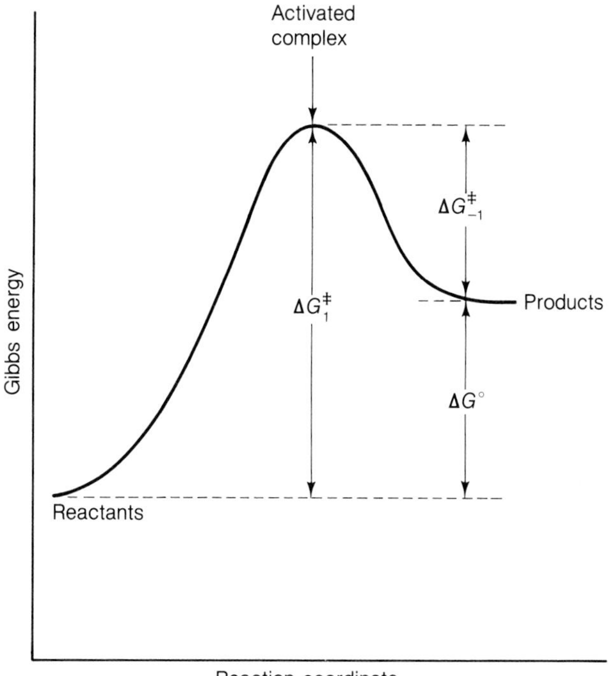

FIGURE 9.14 Gibbs energy diagram for a reaction.

direction and the other in the reverse direction. We can extend this argument to the standard Gibbs energy change ΔG° in a chemical reaction. This quantity can be broken up into two terms, as shown in Figure 9.14. In going to the final state a reaction system must pass over a Gibbs energy barrier, and once at the top of this barrier the reaction can proceed without the expenditure of any additional Gibbs energy. The reaction from left to right will, therefore, depend solely on $\Delta G_1^\ddagger$; that from right to left, solely on $\Delta G_{-1}^\ddagger$.

This type of theory may be formulated as follows, by analogy with the van't Hoff approach. The standard (i.e., dimensionless) equilibrium constant K_c of the reaction is related to the Gibbs energy change by

$$\ln K_c = -\frac{\Delta G^\circ}{RT} \tag{9.79}$$

This equilibrium constant K_c is the ratio of k_1/k_{-1} of the rate constants and therefore

$$\ln k_1 - \ln k_{-1} = -\frac{\Delta G^\circ}{RT} \tag{9.80}$$

$$= -\frac{\Delta G_1^\ddagger}{RT} + \frac{\Delta G^\ddagger_{-1}}{RT} \tag{9.81}$$

In the latter equation, ΔG° has been split into the two terms $\Delta G_1^\ddagger$ and $\Delta G_{-1}^\ddagger$. This equation may be split into the two equations

$$\ln k_1 = \ln \nu - \frac{\Delta G_1^\ddagger}{RT} \tag{9.82}$$

$$\ln k_{-1} = \ln \nu - \frac{\Delta G_{-1}^\ddagger}{RT} \tag{9.83}$$

Here ln ν is a quantity that must appear in both equations. Equation 9.82 may be written as

$$k_1 = \nu e^{-\Delta G_1^\ddagger/RT} \tag{9.84}$$

Since the Gibbs energy changes may be written in terms of entropy and enthalpy changes,

$$\Delta G = \Delta H - T\,\Delta S \tag{9.85}$$

Eq. 9.84 may be written as

$$k_1 = \nu e^{\Delta S_1^\ddagger/R} e^{-\Delta H_1^\ddagger/RT} \tag{9.86}$$

The quantity $\Delta G_1^\ddagger$ appearing in Eq. 9.84 is known as the **Gibbs energy of activation** (formerly as the free energy of activation), and $\Delta S_1^\ddagger$ and $\Delta H_1^\ddagger$ are known as the **entropy of activation** and as the **enthalpy of activation,** respectively.

If we compare Eq. 9.86 with Eq. 9.78, we see that there is a close similarity as far as the heat and energy terms are concerned; E in Eq. 9.78 is the energy required to reach the activated state and $\Delta H_1^\ddagger$ is the corresponding enthalpy change, and these are very similar in magnitude. The expressions differ, however, in that instead of the PZ of Eq. 9.78, the new equation involves an entropy term. It is theoretically superior to the kinetic theory in this respect, because the kinetic theory fails to give any interpretation of the fact that the ratio of the rate constants of two opposing reactions, being an equilibrium constant, must involve an entropy term. Moreover, transition state theory gives a satisfactory interpretation of orientation and solvent effects, since it interprets these in terms of entropy changes when the reactant molecules come together.

We will not give a derivation of Eyring's equation but will simply state that he found the factor ν in Eq. 9.84 and 9.86 to be $\mathbf{k}T/h$, where $\mathbf{k}$ is the Boltzmann constant and h is the Planck constant. The Eyring equation is therefore

$$k = \frac{\mathbf{k}T}{h} e^{\Delta S^\ddagger/R} e^{-\Delta H^\ddagger/RT} = \frac{\mathbf{k}T}{h} e^{-\Delta G^\ddagger/RT} \tag{9.87}$$

Another convenient form of the equation is

$$k = \frac{\mathbf{k}T}{h} K^\ddagger \tag{9.88}$$

where $K^\ddagger$ is the equilibrium constant for the process

$$A + B \rightarrow X^\ddagger$$

in which activated complexes are formed from the reactants. An important feature of transition state theory is that the activated complexes $X^\ddagger$ are assumed to be in equilibrium with the reactants A and B. This assumption can be shown to be satisfactory for all except extremely rapid reactions.

Equation 9.87 may be expressed in a form that involves the experimental energy of activation E instead of the heat of activation $\Delta H^{\ddagger}$. Since $K^{\ddagger}$ is a concentration equilibrium constant, its variation with temperature is given by Eq. 4.83:

$$\frac{d \ln K^{\ddagger}}{dT} = \frac{\Delta U^{\ddagger}}{RT^2} \tag{9.89}$$

where $\Delta U^{\ddagger}$ is the increase in internal energy in passing from the initial state to the activated state. Differentiation of the logarithmic form of Eq. 9.88 gives

$$\frac{d \ln k}{dT} = \frac{1}{T} + \frac{d \ln K^{\ddagger}}{dT} \tag{9.90}$$

and together with Eq. 9.89 this gives

$$\frac{d \ln k}{dT} = \frac{1}{T} + \frac{\Delta U^{\ddagger}}{RT^2} = \frac{RT + \Delta U^{\ddagger}}{RT^2} \tag{9.91}$$

The experimental energy of activation E is thus given by

$$E = RT + \Delta U^{\ddagger} \tag{9.92}$$

The relationship between $\Delta U^{\ddagger}$ and $\Delta H^{\ddagger}$ is

$$\Delta H^{\ddagger} = \Delta U^{\ddagger} + P\,\Delta V^{\ddagger} \tag{9.93}$$

where $\Delta V^{\ddagger}$ is the increase in volume in going from the initial state to the activated state. Substitution of this in Eq. 9.92 gives

$$E = \Delta H^{\ddagger} - P\,\Delta V^{\ddagger} + RT \tag{9.94}$$

For unimolecular gas reactions there is no change in the number of molecules as the activated complex is formed, and $\Delta V^{\ddagger}$ is therefore zero; $\Delta V^{\ddagger}$ is also small for reactions in solution. In these cases

$$E = \Delta H^{\ddagger} + RT \tag{9.95}$$

and the rate equation may therefore be written as

$$k = \frac{\mathbf{k}T}{h} e^{\Delta S^{\ddagger}/R}\, e^{-(E-RT)/RT} \tag{9.96}$$

or as

$$k = e\left(\frac{\mathbf{k}T}{h}\right) e^{\Delta S^{\ddagger}/R}\, e^{-E/RT} \tag{9.97}$$

Thus, for reactions in solution and unimolecular gas reactions the preexponential factor A is given by

$$A = e\left(\frac{\mathbf{k}T}{h}\right) e^{\Delta S^{\ddagger}/R} \tag{9.98}$$

For gas reactions the general relationship is

$$P\,\Delta V^{\ddagger} = \Sigma \nu RT \tag{9.99}$$

where $\Sigma \nu$ is the increase in the number of molecules when the activated complex is formed from the reactants (i.e., is the stoichiometric sum for the process $A + B + \cdots \rightarrow X^{\ddagger}$). In a bimolecular reaction two molecules become one, so that $\Sigma\nu$ is equal to -1; in this case the experimental energy of activation is related to the enthalpy of activation by the relationship

$$E_{exp} = \Delta H^{\ddagger} + 2RT \tag{9.100}$$

From this it follows that the rate constant may be written as

$$k = e^2\left(\frac{\mathbf{k}T}{h}\right) e^{\Delta S^{\ddagger}/R}\, e^{-E/RT} \tag{9.101}$$

For agreement with the simple hard-sphere theory of collisions the entropy of activation, relative to the units of $dm^3\ mol^{-1}\ s^{-1}$, must be about $-60\ J\ K^{-1}\ mol^{-1}$ (see Problem 19). If, on the other hand, the standard state is 1 mol cm^{-3} (i.e., if the rate constant units are $cm^3\ mol^{-1}\ s^{-1}$), the entropy of activation must be approximately zero for agreement with the collision theory.

EXAMPLE

From the data given in the example on page 368, calculate the Gibbs energy of activation at 25.0°C, the entropy of activation, and the enthalpy of activation.

SOLUTION

The Gibbs energy of activation is related to the rate constant by Eq. 9.87, which in logarithmic form is

$$\log_{10} k = \log_{10}\frac{\mathbf{k}T}{h} - \frac{\Delta G^{\ddagger}/\mathrm{J\ mol^{-1}}}{19.14T/\mathrm{K}}$$

At 25.0°C the value of $\mathbf{k}T/h$ is $6.214 \times 10^{12}\ s^{-1}$. Then, inserting values into the last equation,

$$-4.244 = \log_{10}(6.214 \times 10^{12}) - \frac{\Delta G^{\ddagger}/\mathrm{J\ mol^{-1}}}{19.14 \times 298.15}$$

or

$$-4.244 = 12.793 - \frac{\Delta G^{\ddagger}/\mathrm{J\ mol^{-1}}}{5707}$$

and thus

$$\Delta G^{\ddagger} = 97\,230\ \mathrm{J\ mol^{-1}} = 97.2\ \mathrm{kJ\ mol^{-1}}$$

The entropy of activation can be calculated from the preexponential factor, using Eq. 9.98, which in logarithmic form is

$$\log_{10} A = \log_{10} e + \log_{10}\frac{\mathbf{k}T}{h} + \frac{\Delta S^{\ddagger}/\mathrm{J\ K^{-1}\ mol^{-1}}}{19.14}$$

The value of $\log_{10} A$ was calculated to be 5.342; thus

$$5.342 = \log_{10} 2.718 + 12.793 + \frac{\Delta S^{\ddagger}}{19.14\ \mathrm{J\ K^{-1}\ mol^{-1}}}$$

$$= 0.4342 + 12.793 + \frac{\Delta S^{\ddagger}}{19.14\ \mathrm{J\ K^{-1}\ mol^{-1}}}$$

and thus

$$\Delta S^{\ddagger} = -150.9 \text{ J K}^{-1} \text{ mol}^{-1}$$

From Eq. 9.95

$$\Delta H^{\ddagger} = E - RT = 54\,700 - 8.314 \times 298.15$$

$$= 52\,220 \text{ J mol}^{-1} = 52.2 \text{ kJ mol}^{-1}$$

Alternatively, $\Delta H^{\ddagger}$ could have been calculated from $\Delta G^{\ddagger}$ and $\Delta S^{\ddagger}$:

$$\Delta H^{\ddagger} = \Delta G^{\ddagger} + T\,\Delta S^{\ddagger}$$

$$= 97\,230 - 150.9 \times 298.15$$

$$= 52\,240 \text{ J mol}^{-1} = 52.2 \text{ kJ mol}^{-1}$$

■

9.10 Reactions in Solution

We will now consider some of the factors that determine the magnitudes of the rates of reactions in solution. This topic can be divided into two parts, one concerned with the activation energy and the other with the entropy of activation. As mentioned earlier, very little can be said, from a fundamental standpoint, about the magnitudes of activation energies, and this is especially true of reactions in solution.

Rather more can be said about entropies of activation (or preexponential factors). No exact treatment is possible at the present owing to the inherent complexity of the problem of reactions in solution. A number of important generalizations are, however, well recognized, and it is now possible to make a fairly reliable estimate of the entropy of activation of a solution reaction, and hence of the preexponential factor, in terms of the various factors involved in the reaction. Conversely, it is possible, from the knowledge of values of entropies of activation, to draw some inferences about the nature of a reaction (such as its ionic character) in cases where this is not already known.

Reactions in solution are of a variety of types. There are certain reactions, involving nonpolar molecules, in which it seems that the solvent plays a relatively subsidiary role. Some reactions of this kind occur in the gas phase as well as in solution, and when they do so, they usually occur with much the same rates as in solution and with similar entropies and energies of activation. In such reactions the interactions between reactant molecules and solvent molecules are not of great importance, and the solvent can be regarded as merely filling up space between the reactant molecules. Theory and experiments with mechanical models suggest that in such cases the number of effective collisions between reactant molecules is hardly affected by the presence of solvent. If the reactant molecules are fairly simple, the orientation effect is small, and the hard-sphere collision theory is not far from the truth. The preexponential factor may then be described as a "normal" one; its magnitude for a bimolecular reaction is of the order 10^9 to 10^{11} $\text{dm}^3 \text{ mol}^{-1} \text{ s}^{-1}$. If the reacting molecules are more complex, there will be a loss of entropy in forming the activated complex and a lower preexponential factor.

More often, however, the solvent does not act as an inert space filler but is involved in a significant way in the reaction itself. This is particularly the case when the reactants are ions or neutral molecules and when the solvent also is polar. Changes in polarity during the course of reaction will then cause a reorientation of solvent molecules and will have important effects on the entropies of activation.

For reactions between ions the electrostatic effects are very important, and the preexponential factors of such reactions depend on the ionic charges. For reactions between ions of opposite signs, the preexponential factors are much higher than for reactions between neutral molecules. If the ions are of the same sign, the preexponential factors are abnormally low. In terms of simple collision theory these effects can be explained in terms of electrostatic forces. The frequency of collisions between reactants of opposite signs is greater than for neutral molecules, whereas the frequency is smaller for ions of the same sign. These effects can be explained more satisfactorily in terms of the entropy changes that occur when the activated complexes are formed. We will now consider this matter by a treatment that will lead to a useful prediction of the influence of the solvent dielectric constant on the rates of ionic reactions.

Influence of Solvent Dielectric Constant

We can make estimates of the Gibbs energy change when two reacting ions form an activated complex and then obtain the rate constant by the use of Eq. 9.87. In reactions between ions the electrostatic interactions make the most important contribution to the Gibbs energy of activation. The simplest treatment of electrostatic interactions makes the same assumptions as the Born model for ions in solution (Section 7.9). The charged ions are considered to be conducting spheres, and the solvent is regarded as a continuous dielectric, having a fixed dielectric constant ϵ. Such a treatment represents a gross oversimplification, but it has proved surprisingly useful in that it leads to conclusions that are semiquantitatively correct.

A very simple model for a reaction between two ions in solution is represented in Figure 9.15. The reacting molecules are regarded as conducting spheres; the radii are r_A and r_B and the charges are z_Ae and z_Be; e is the electronic charge and z_A and z_B are whole numbers (positive or negative) that indicate the numbers of charges on the ion. Initially the ions are at an infinite distance from each other, and in the transition state they are considered to be intact (i.e., there is no "smearing" of charge and they are at a distance d_{AB} apart). This model is frequently referred to as the *double-sphere model*. When the ions are at a distance x apart, the force acting between them is equal to

$$f = \frac{z_A z_B e^2}{4\pi\epsilon_0 \epsilon x^2} \tag{9.102}$$

The work that must be done in moving them together a distance dx is

$$dw = -\frac{z_A z_B e^2}{4\pi\epsilon_0 \epsilon x^2}\,dx \tag{9.103}$$

(The negative sign is used because x decreases by dx when the ions move together by a distance dx.) The work that is done in moving the ions from an initial distance of infinity to a final distance of d_{AB} is therefore

$$w = -\int_{\infty}^{d_{AB}} \frac{z_A z_B e^2}{4\pi\epsilon_0 \epsilon x^2}\, dx \tag{9.104}$$

$$= \frac{z_A z_B e^2}{4\pi\epsilon_0 \epsilon d_{AB}} \tag{9.105}$$

If the signs on the ions are the same, this work is positive; if they are different, it is negative. This work is equal to the electrostatic contribution to the Gibbs energy increase as the ions are moved up to each other.

There is also a nonelectrostatic term $\Delta G^{\ddagger}_{nes}$. The Gibbs energy of activation per mole may therefore be written as

$$\Delta G^{\ddagger} = \Delta G^{\ddagger}_{nes} + \frac{L z_A z_B e^2}{4\pi\epsilon_0 \epsilon d_{AB}} \tag{9.106}$$

The last term has been multiplied by the Avogadro constant L so as to give the value per mole. Introduction of this equation into Eq. 9.87 gives

$$k = \frac{\mathbf{k}T}{h} \exp\left(-\frac{\Delta G^{\ddagger}_{nes}}{RT}\right) \exp\left(-\frac{z_A z_B e^2}{4\pi\epsilon_0 \epsilon d_{AB} \mathbf{k}T}\right) \tag{9.107}$$

since $R/L = \mathbf{k}$. Taking natural logarithms

$$\ln k = \ln \frac{\mathbf{k}T}{h} - \frac{\Delta G^{\ddagger}_{nes}}{RT} - \frac{z_A z_B e^2}{4\pi\epsilon_0 \epsilon d_{AB} \mathbf{k}T} \tag{9.108}$$

which may be written as

$$\ln k = \ln k_0 - \frac{z_A z_B e^2}{4\pi\epsilon_0 \epsilon d_{AB} \mathbf{k}T} \tag{9.109}$$

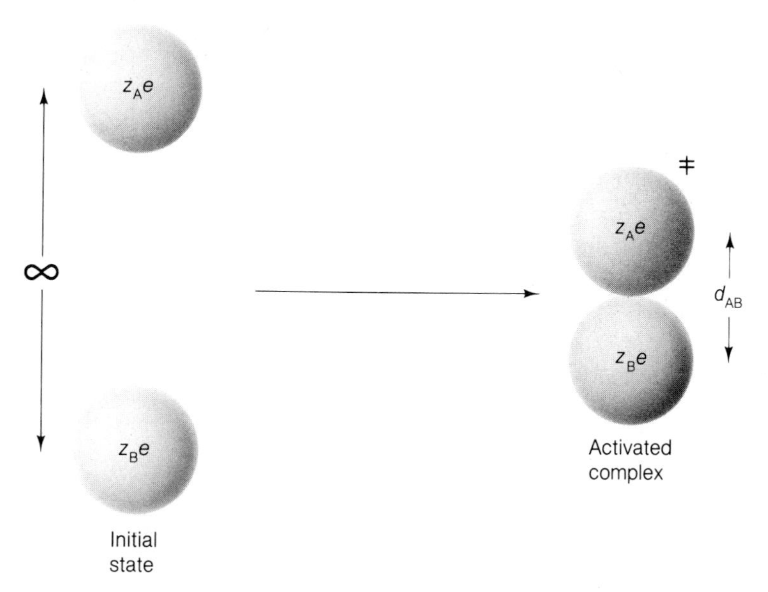

FIGURE 9.15 A simple model for a reaction between two ions, of charges, $z_A e$ and $z_B e$, in a medium of dielectric constant ϵ. This is known as the "double-sphere" model.

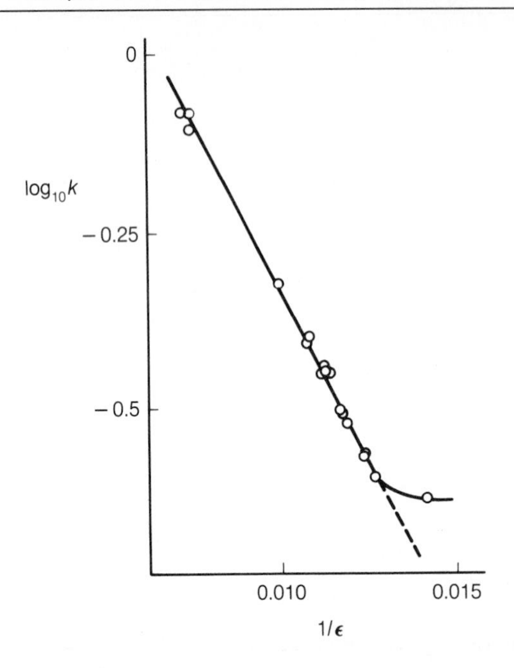

FIGURE 9.16 A plot of $\log_{10}k$ against the reciprocal of the dielectric constant, for the reaction between bromoacetate and thiosulfate ions in aqueous solution.

The rate constant k_0 is the value of k in a medium of infinite dielectric constant, when the final term in Eq. 9.109 becomes zero (i.e., when there are no electrostatic forces).

Equation 9.109 leads to the prediction that the logarithm of the rate constant of a reaction between ions should vary linearly with the reciprocal of the dielectric constant. An example of a test of Eq. 9.109 is shown in Figure 9.16. Deviations from linearity can be explained as due to failure of the simple approximations involved in deriving Eq. 9.109 and in some cases to a change in reaction mechanism as the solvent is varied.

The slope of the line obtained by plotting ln k against $1/\epsilon$ is given by Eq. 9.109 as $z_A z_B e^2/4\pi\epsilon_0 d_{AB}\mathbf{k}T$. Since everything in this expression is known except d_{AB}, it is possible to calculate d_{AB} from the experimental slope. This has been done in a number of cases, and the values obtained, of the order of a few tenths of a nanometre, have always been entirely reasonable; for the data shown in Figure 9.16 the value of d_{AB} is 0.51 nm.

We can extend this treatment to give an interpretation of the magnitudes of the preexponential factors of ionic reactions. We have seen that the electrostatic contribution to the molar Gibbs energy of activation is

$$\Delta G^{\ddagger}_{es} = \frac{L z_A z_B e^2}{4\pi\epsilon_0 \epsilon d_{AB}} \tag{9.110}$$

The thermodynamic relationship between entropy and Gibbs energy is given by Eq. 3.119:

$$S = -\left(\frac{\partial G}{\partial T}\right)_P \tag{9.111}$$

and the electrostatic contribution to the entropy of activation is

$$\Delta S^{\ddagger}_{es} = -\left(\frac{\partial\, \Delta G^{\ddagger}_{es}}{\partial T}\right)_P \tag{9.112}$$

The only quantity in Eq. 9.110 that is temperature-dependent is ϵ and it therefore follows that

$$\Delta S^{\ddagger}_{es} = \frac{L z_A z_B e^2}{4\pi\epsilon_0 \epsilon^2 d_{AB}} \left(\frac{\partial \epsilon}{\partial T}\right)_p \tag{9.113}$$

$$= \frac{L z_A z_B e^2}{4\pi\epsilon_0 \epsilon d_{AB}} \left(\frac{\partial \ln \epsilon}{\partial T}\right)_P \tag{9.114}$$

In aqueous solution ϵ is about 80 and $(\partial \ln \epsilon/\partial T)_P$ remains constant at $-0.0046\ \mathrm{K^{-1}}$ over a considerable temperature range. If d_{AB} is taken as equal to 0.2 nm, it follows from Eq. 9.114 that the entropy of activation is given by

$$\Delta S^{\ddagger}_{es} \approx -42\ z_A z_B\ \mathrm{J\ K^{-1}\ mol^{-1}} \tag{9.115}$$

The entropy of activation in aqueous solution should thus decrease by about 42 $\mathrm{J\ K^{-1}\ mol^{-1}}$ for each unit of $z_A z_B$. Moreover, since the preexponential factor is proportional to $e^{\Delta S^{\ddagger}/R}$, which equals to $10^{\Delta S^{\ddagger}/2.303R}$ or $10^{\Delta S^{\ddagger}/19.12}$, it follows that the factor should decrease by a factor of $10^{42/19.12}$ (i.e., by about one-hundredfold) for each unit of $z_A z_B$. Table 9.2 shows that these relationships are obeyed in a very approximate manner. The treatment is too crude to allow detailed predictions to be made, but it is evidently along the right lines.

The physical model that lies behind these relationships is represented schematically in Figure 9.17. In Figure 9.17a the ions are shown as having single positive charges, and the activated complex therefore bears a double positive charge. The frequency of collision between the two ions will be reduced on account of the

TABLE 9.2 Some Observed and Predicted Preexponential Factors and Entropies of Activation

	Experimental Values		Estimated Values	
Reactants	A / $\mathrm{dm^3\ mol^{-1}\ s^{-1}}$	$\Delta S^{\ddagger}$ / $\mathrm{J\ K^{-1}\ mol^{-1}}$	A / $\mathrm{dm^3\ mol^{-1}\ s^{-1}}$	$\Delta S^{\ddagger}$ / $\mathrm{J\ K^{-1}\ mol^{-1}}$
$[Cr(H_2O)_6]^{3+} + CNS^-$	$\sim 10^{19}$	~ 126	10^{19}	126
$Co(NH_3)_5Br^{2+} + OH^-$	5×10^{17}	92	10^{17}	84
$CH_2BrCOOCH_3 + S_2O_3^{2-}$	1×10^{14}	25	10^{13}	0
$CH_2ClCOO^- + OH^-$	6×10^{10}	−50	10^{11}	−42
$ClO^- + ClO_2^-$	9×10^{8}	−84	10^{11}	−42
$CH_2BrCOO^- + S_2O_3^{2-}$	1×10^{9}	−71	10^{9}	−84
$Co(NH_3)_5Br^{2+} + Hg^{2+}$	1×10^{8}	−100	10^{5}	−167
$S_2O_4^{2-} + S_2O_4^{2-}$	2×10^{4}	−167	10^{5}	−167
$S_2O_3^{2-} + SO_3^{2-}$	2×10^{6}	−126	10^{5}	−167

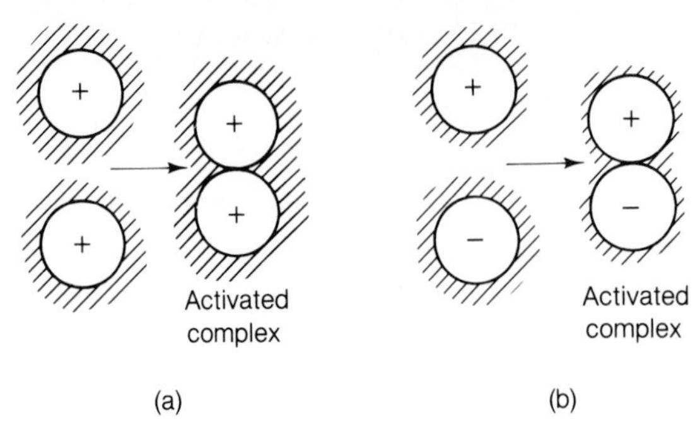

FIGURE 9.17 An interpretation of entropies of activation in terms of the electrostriction in the activated complex. In (a) there is more electrostriction in the activated complex, and there is therefore a decrease in entropy. In (b) there is less electrostriction in the activated complex.

electrostatic repulsions between the like charges, but this is by no means the entire effect. When the ions of like sign come together, they form a complex of higher charge, and the activated complex therefore exerts more electrostriction* on the surrounding water molecules than is exerted by the reactant ions. Since electrostriction leads to entropy loss, the result is that a reaction between like charges has a negative entropy of activation and a low preexponential factor.

If the reaction is such that oppositely charged ions come together (Figure 9.17b), there is less charge on the activated complex than on the reactants. There is then a decrease in electrostriction when the activated complex is formed, a positive entropy of activation, and a high preexponential factor.

Similar effects are found, but to a lesser extent, when reactions involve the approach of dipolar molecules.

Influence of Ionic Strength

The influence of ionic strength on the rates of reaction between ions may be considered with reference to a reaction of the general type.

$$A + B \rightarrow X^{\ddagger} \rightarrow \text{products}$$

The theory of ionic-strength effects was first worked out by J. N. Brønsted.† We will give a more modern version, in terms of the concept of the activated complex.

The basis of the treatment is that the rate of a reaction is proportional to the *concentration* of the activated complexes $X^{\ddagger}$ and not to their activity:

$$v = k'[X^{\ddagger}] \tag{9.116}$$

The equilibrium between the activated complexes and the reactants A and B may be expressed as

$$K^{\ddagger} = \frac{a^{\ddagger}}{a_A a_B} = \frac{[X^{\ddagger}]}{[A][B]} \frac{y^{\ddagger}}{y_A y_B} \tag{9.117}$$

---

*Electrostriction has been discussed in Section 3.4.

†J. N. Brønsted, *Z. Physik. Chem.*, *102*, 169 (1922); see also N. Bjerrum, *Z. Physik. Chem.*, *108*, 82 (1924); J. A. Christiansen, *Z. Physik. Chem.*, *113*, 35 (1924); G. Scatchard, *Chem. Rev.*, *10*, 229 (1932).

where the a's are the activities and the y's the activity coefficients. Introduction of Eq. 9.117 into Eq. 9.116 gives rise to

$$v = k[A][B] = k_0[A][B]\,\frac{y_A y_B}{y^\ddagger} \tag{9.118}$$

where k is the second-order rate constant and $k_0 = k'K^\ddagger$. Taking logarithms,

$$\log_{10} k = \log_{10} k_0 + \log_{10} \frac{y_A y_B}{y^\ddagger} \tag{9.119}$$

According to the Debye-Hückel limiting law (Section 7.10) the activity coefficient of an ion is related to its valency z and the ionic strength I by Eq. 7.104:

$$\log_{10} y = -Bz^2\sqrt{I} \tag{9.120}$$

Introduction of Eq. 9.120 into the rate equation (Eq. 9.119) gives

$$\log_{10} k = \log_{10} k_0 + \log_{10} y_A + \log_{10} y_B - \log_{10} y^\ddagger \tag{9.121}$$

$$= \log_{10} k_0 - B[z_A^2 + z_B^2 - (z_A + z_B)^2]\sqrt{I} \tag{9.122}$$

$$= \log_{10} k_0 + 2Bz_A z_A\sqrt{I} \tag{9.123}$$

The value of B is approximately 0.51 $dm^{-3/2}$ $mol^{-1/2}$ for aqueous solutions at 25°C; Eq. 9.123 may therefore be written as

$$\log_{10} k = \log_{10} k_0 + 1.02\, z_A z_B\sqrt{I/\text{mol dm}^{-3}} \tag{9.124}$$

This equation has been tested a considerable number of times. The procedure has usually been to measure the rates of ionic reactions in media of varying ionic strength; according to Eq. 9.124, a plot of $\log_{10} k$ against $\sqrt{I/\text{mol dm}^{-3}}$ will give a straight line of slope 1.02 $z_A z_B$. Figure 9.18 shows a plot of results for reactions of various types; the lines drawn are those with theoretical slopes, and the points lie close to them. If one of the reactants is a neutral molecule, $z_A z_B$ is zero, and the rate constant is expected to be independent of the ionic strength; this is true, for example, for the base-catalyzed hydrolysis of ethyl acetate, shown in Figure 9.18.

The investigations of ionic-strength effects have provided valuable support for the applicability of the Debye-Hückel limiting law and for the validity of the assumption that rate is proportional to the concentration of activated complexes, not to their activities (see Problem 27).

Influence of Hydrostatic Pressure*

Studies of the effect of hydrostatic pressure on reactions in solution provide important information about the detailed mechanisms of reactions. The theory of pressure effects on rates was first formulated in 1901 by van't Hoff,† who started with Eq. 4.90 for the effect of pressure on equilibrium constants:

$$\left(\frac{\partial \ln K^u}{\partial P}\right)_T = -\frac{\Delta V^\circ}{RT} \tag{9.125}$$

---

*S. D. Hamann, *Physico-Chemical Effects of Pressure,* London: Butterworth, 1957, Chapter 9.

†J. H. van't Hoff, *Vorlesungen übe Theoretische und Physikalische Chemie* (Vol. 1). Braunschweig: F. Vreweg and Sohn, 1901, p. 236.

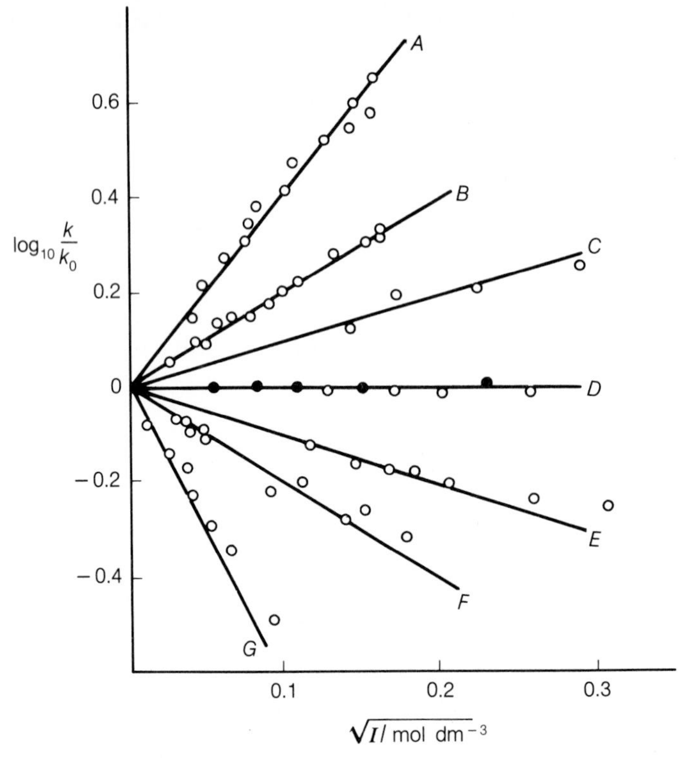

FIGURE 9.18 Plots of $\log_{10}(k/k_o)$ against the square root of the ionic strength, for ionic reactions of various types. The lines are drawn with slopes equal to $z_A z_B$. The reactions follow:

A: $Co(NH_3)_5Br^{2+} + Hg^{2+}$ ($z_A z_B = 4$)
B: $S_2O_8^{2-} + I^-$ ($z_A z_B = 2$)
C: $CO(OC_2H_5)N{:}NO_2^- + OH^-$ ($z_A z_B = 1$)
D: $[Cr(urea)_6]^{3+} + H_2O$ (open circles) ($z_A z_B = 0$)
$CH_3COOC_2H_5 + OH^-$ (closed circles) ($z_A z_B = 0$)
E: $H^+ + Br^- + H_2O_2$ ($z_A z_B = -1$)
F: $Co(NH_3)_5Br^{2+} + OH^-$ ($z_A z_B = -2$)
G: $Fe^{2+} + Co(C_2O_4)_3^{3-}$ ($z_A z_B = -6$)

He employed an argument very similar to the one he used to arrive at the Arrhenius law, pointing out that since K is the ratio of rate constants, the latter must vary with pressure in a similar way.

An alternative derivation is in terms of transition state theory. If we apply Eq. 9.125 to the equilibrium constant $K^\ddagger$ for the equilibrium between reactants and activated complexes, the result is

$$\left(\frac{\partial \ln K^\ddagger}{\partial P}\right)_T = -\frac{\Delta V^\ddagger}{RT} \tag{9.126}$$

Here $\Delta V^\ddagger$ is the volume change in going from the initial state A + B to the activated state $X^\ddagger$; this volume change is known as the **volume of activation.** According to Eq. 9.88 the rate constant k is proportional to $K^\ddagger$, and therefore

$$\left(\frac{\partial \ln k}{\partial P}\right)_T = -\frac{\Delta V^\ddagger}{RT} \tag{9.127}$$

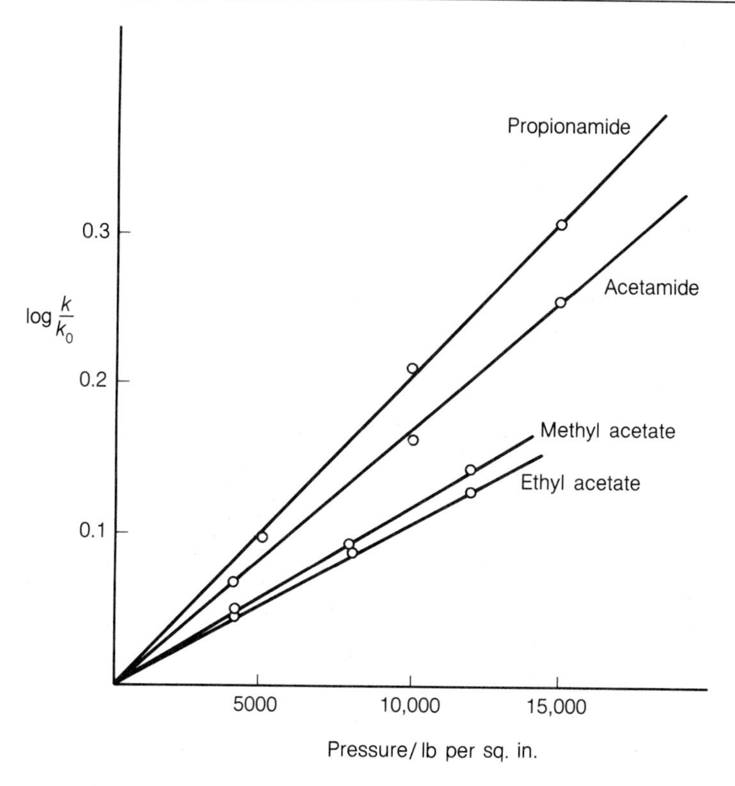

FIGURE 9.19 Plots of $\log_{10}(k/k_0)$ against the hydrostatic pressure, for the alkaline hydrolyses of some esters and amides [K. J. Laidler and D. Chen, *Trans. Faraday Soc.*, *54*, 1020 (1958)].

A rate constant will therefore increase with increasing pressure if $\Delta V^{\ddagger}$ is negative (i.e., if the activated complex has a smaller volume than the reactants). Conversely, pressure has an adverse effect on rates if there is a volume increase when the activated complex is formed. By use of Eq. 9.127, values of $\Delta V^{\ddagger}$ can be determined from experimental determinations of rates at different pressures. In practice, it is usually necessary to use fairly high pressures for this purpose (at least 100 atm, or 10^4 kPa) since otherwise the changes of rate are too small for accurate $\Delta V^{\ddagger}$ values to be obtained.

According to Eq. 9.127, if $\ln k$ is plotted against pressure, the slope at any pressure is $-\Delta V^{\ddagger}/RT$. In some cases the plots are linear, which means that $\Delta V^{\ddagger}$ is independent of pressure. If this is so, Eq. 9.127 can be integrated to give

$$\ln k = \ln k_0 - \frac{\Delta V^{\ddagger}}{RT} P \tag{9.128}$$

where k_0 is the rate constant at zero pressure (this is always very close to the value at atmospheric pressure). Examples of such linear plots are shown in Figure 9.19.

The interpretation of volumes of activation for reactions in solution involves the consideration of several factors, for example, the structural changes in the reactant molecules as they form the activated complex. Often, however, even more profound volume changes occur in the solvent itself. If, for example, a reaction occurs between like charges in aqueous solution (Figure 9.17a), there is an increase in electrostriction when the activated complex is formed. Since the bound water molecules occupy less volume than water molecules in ordinary water, there will be a negative $\Delta V^\ddagger$ for a reaction of this type, which also has a negative $\Delta S^\ddagger$. Conversely, for a reaction between oppositely charged ions, there is a decrease in electrostriction leading to a positive $\Delta V^\ddagger$ and a positive $\Delta S^\ddagger$. There is, in fact, quite a good correlation between $\Delta V^\ddagger$ and $\Delta S^\ddagger$ values for reactions in which electrostatic effects are important.

Diffusion-Controlled Reactions

If a rapid bimolecular reaction in solution is initiated by mixing solutions of the reactants, the observed rate may depend on the rate with which the solutions mix. This effect is known as *mixing control* or **macroscopic diffusion control.**

We have seen in Section 9.5 that the effect of mixing can be eliminated by the use of relaxation methods. However, even when this is done, the rate of reaction may be influenced by the rate with which the reactant molecules diffuse toward each other. This effect is known as **microscopic diffusion control** or as *encounter control.* If the rate we measure is almost exactly equal to the rate of diffusion, we speak of *full microscopic diffusion control* or of *full encounter control.* An example is provided by the combination of H^+ and OH^- ions in solution, which we saw in Section 9.5 to have a rate constant of 1.3×10^{11} dm^3 mol^{-1} s^{-1} at 23°C. For some reactions the rates of chemical reaction and of diffusion are similar to each other, and we then speak of *partial microscopic diffusion control* or of *partial encounter control.*

Linear Gibbs Energy Relationships

A number of relationships have been suggested in connection with the rate constants of reactions in solution. One of the most useful of these is an equation proposed by the American chemist Louis P. Hammett,* which relates equilibrium and rate constants for the reactions of *meta*- and *para*-substituted benzene derivatives; for example, one might compare the rate constants for two types of reactions involving a given set of substituents in the *meta* position. According to Hammett's relationship, a rate or equilibrium constant for the reaction of one compound is related to that for the unsubstituted ("parent") compound in terms of two parameters ρ and σ. For rate constants the relationship is

$$\log_{10} k = \log_{10} k_0 + \sigma\rho \tag{9.129}$$

where k_0 is the rate constant for the parent compound. For equilibrium constants

$$\log_{10} K = \log_{10} K_0 + \sigma\rho \tag{9.130}$$

where K_0 is for the parent compound. Of the two parameters, σ depends only on the *substituent,* whereas ρ is a *reaction constant,* varying with the nature of the reaction and the external conditions such as the solvent. A value of unity is arbitrarily chosen for ρ for the ionization equilibrium constant of benzoic acid in aqueous solution and for the substituted benzoic acids. It follows that σ is the logarithm of the ratio of the dissociation constant of a substituted benzoic acid to that of benzoic acid itself.

---

*L. P. Hammett, *Physical Organic Chemistry,* New York: McGraw-Hill, 1940, pp. 184–199.

The Hammett relationships imply linear relationships between the Gibbs energies, of reaction or of activation, for different series of reactions. We have seen (Eq. 9.87) that the rate constant is related to the Gibbs energy of activation by

$$k = \frac{\mathbf{k}T}{h} e^{-\Delta G^{\ddagger}/RT} \tag{9.131}$$

and therefore

$$\log_{10} k = \log_{10} \frac{\mathbf{k}T}{h} - \frac{\Delta G^{\ddagger}}{2.303RT} \tag{9.132}$$

Equation 9.129 may therefore be written as

$$\Delta G^{\ddagger} = \Delta G_0^{\ddagger} - 2.303RT\rho\sigma \tag{9.133}$$

This equation, with a particular value of ρ, applies to any reaction involving a reactant having a series of substituents. For a second series of homologous reactions, with the reaction constant ρ',

$$\Delta G'^{\ddagger} = \Delta G_0'^{\ddagger} - 2.303RT\rho'\sigma \tag{9.134}$$

Equations 9.133 and 9.134 may be written as

$$\frac{\Delta G^{\ddagger}}{\rho} = \frac{\Delta G_0^{\ddagger}}{\rho} - 2.303RT\sigma \tag{9.135}$$

and

$$\frac{\Delta G'^{\ddagger}}{\rho'} = \frac{\Delta G_0'^{\ddagger}}{\rho'} - 2.303RT\sigma \tag{9.136}$$

Subtraction gives

$$\frac{\Delta G^{\ddagger}}{\rho} - \frac{\Delta G'^{\ddagger}}{\rho'} = \frac{\Delta G_0^{\ddagger}}{\rho} - \frac{\Delta G_0'^{\ddagger}}{\rho'}$$

or

$$\Delta G^{\ddagger} - \frac{\rho}{\rho'} \Delta G'^{\ddagger} = \text{constant} \tag{9.137}$$

There is thus a linear relationship between the Gibbs energies of activation for one homologous series of reactions and those for another. An equivalent relationship is easily derived for the Gibbs energies of overall reactions.

9.11 Molecular Dynamics*

During recent years there has been increasing interest in what might be called the "fine structure" of reaction rates. Conventional kinetic studies, such as we have considered up to now, provide information relating to the overall rates of formation

*For further details, see R. D. Levine and R. B. Bernstein, *Molecular Reaction Dynamics*, New York: Oxford University Press; 1974, K. J. Laidler, *Theories of Chemical Reaction Rates*. New York: McGraw-Hill, Chapter 7, 1969.

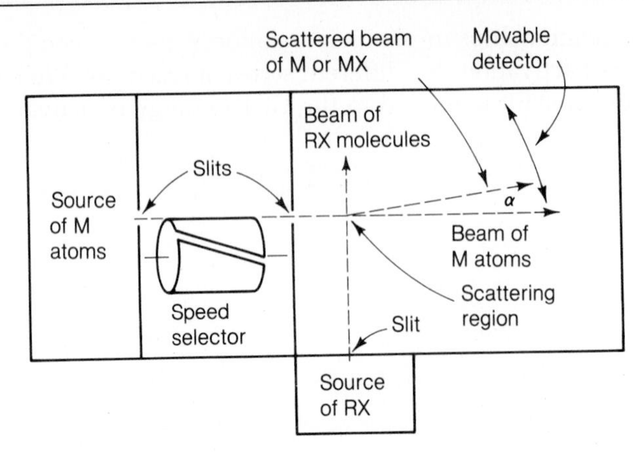

FIGURE 9.20 Schematic diagram of a molecular beam apparatus for the study of a reaction of the type $M + RX \rightarrow MX + R$.

of products of reaction. They cannot reveal the details of how the system progresses over the potential-energy surface or of how the energy released in the reaction is distributed among the product molecules. There have been developed two important experimental techniques, concerned with *crossed molecular beams* and with *chemiluminescence,* which do provide evidence about energy distributions in reaction products. In order to interpret the results of such experiments it is necessary to carry out calculations of the actual passage of systems over potential-energy surfaces. These experimental and theoretical studies are conveniently referred to as being in the area of **molecular dynamics.**

Molecular Beams

The essential feature of a molecular beam experiment, illustrated in Figure 9.20, is that narrow beams of atoms or molecules are caused to cross one another. Movable detectors determine the direction taken by the product molecules and by the unreacted molecules. Analysis of the experimental results yields information about the distribution of energy and angular momentum among the reaction products, the dependence of the total reaction probability on the molecular energies, the lifetime of the collision complex, the quantum states of the products, and a number of other important features.

One important result that has emerged from molecular beam experiments is that there are two types of reaction mechanisms, **stripping mechanisms** and **rebound mechanisms.** Figure 9.21a shows what occurs in a stripping mechanism of the type $A + BC \longrightarrow AB + C$. The reactant atom A, after it has undergone reaction and has combined with B, travels in much the same direction as previously; similarly the direction of motion of C is not greatly altered. In other words, A *strips* B away from C, which is hardly affected and is said to act as a "spectator"; the term *spectator stripping* is also applied to these mechanisms. An example of a reaction of this type is

$$K + HBr \longrightarrow KBr + H$$

Stripping reactions are all characterized by having large collision cross sections; that

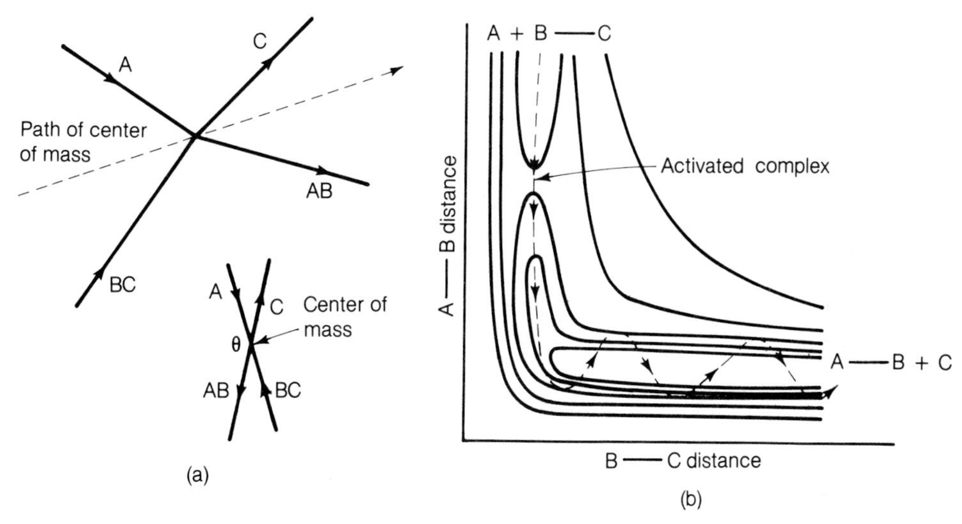

FIGURE 9.21 (a) Directions of motion of reactants and products, for a stripping mechanism. (b) An attractive potential-energy surface, which leads to high reaction cross sections, stripping, and a large proportion of vibrational energy in the reaction product.

is to say, reaction can occur when the reactants are quite far apart. For example, the K + HBr reaction has a collision cross section of 0.34 nm^2, which means that reaction can occur when the reactants are separated by a distance that is much larger than the sum of the kinetic theory radii. This reaction has an activation energy of about 4.2 kJ mol^{-1}. In reactions of this type there is a considerable amount of vibrational energy in the product of reaction (e.g., in KBr in the example just given).

When dynamical calculations are made with various shapes of potential-energy surfaces, it becomes clear that the type of surface that gives rise to a stripping mechanism is the type shown in Figure 9.21b. The characteristic feature of this surface is that the activated complex corresponds to a rather large A—B distance, a feature that accounts for the large reaction cross sections. After the activated complex is formed, the atom A continues to approach B, with at first little change in the B—C distance; after the activated complex is formed, the potential energy falls so that there is attraction between A and BC. Because of this feature the surface is known as an **attractive surface** or as an *early-downhill surface.* If we constructed such a surface and projected a ball along the upper valley, as indicated by the dashed line, it would travel along the lower, right-hand valley with considerable lateral motion. This lateral motion corresponds to vibrational energy in the product A—B, and we therefore have an explanation of why these attractive surfaces lead to a large amount of vibrational energy in the products of reaction. Dynamical calculations confirm this general conclusion.

Other reactions studied in molecular beams occur by a rebound mechanism, the features of which are shown in Figure 9.22a. In such reactions the reactant A, after abstracting B from BC, does not continue on its way but rebounds, remaining on the same side of the center of mass of the system. Reactions of this type tend to have small collision cross sections, and little energy goes into the vibration of the product. An example of a reaction occurring by a rebound mechanism is

$$K + CH_3I \longrightarrow KI + CH_3$$

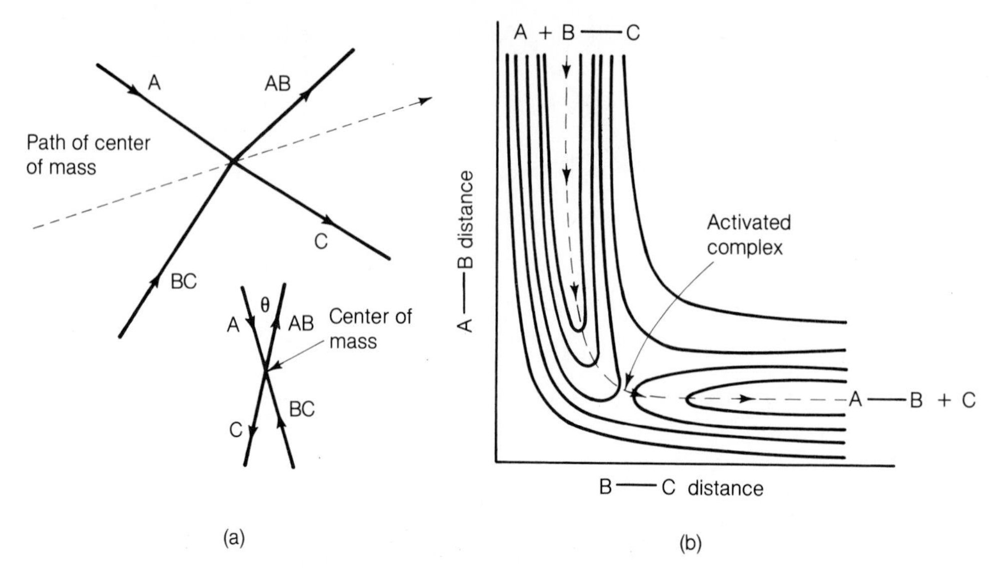

FIGURE 9.22 (a) Directions of motion of reactants and products, for a rebound mechanism. (b) A repulsive potential-energy surface, which leads to low reaction cross sections, a rebound mechanism, and a smaller proportion of vibrational energy in the reaction product.

for which the cross section is ~0.10 nm^2, which means that reaction requires the approach of the reactants to $\sqrt{0.10}$ nm ≈ 0.3 nm.

Dynamical calculations show that the type of potential-energy surface leading to a rebound mechanism is as shown in Figure 9.22b. This is known as a **repulsive surface.** On such a surface the system cannot reach the activated state without there being a significant extension of the B—C bond. A ball projected on a repulsive surface would follow a path such as that shown in Figure 9.22b; it would leave the lower valley without any lateral motion corresponding to the vibration of the A—B bond. Dynamical calculations on a number of repulsive surfaces have confirmed this general conclusion.

Chemiluminescence

The vibrational states of reaction products can be studied by making measurements of the radiation they emit. The chemiluminescence produced in a reaction may be in various regions of the spectrum, including the visible. Many studies have been made of infrared chemiluminescence, which provides information about the vibrational and rotational states of the product molecules.

The pioneer in this field of investigation was Michael Polanyi, who in about 1930 studied the emission of the sodium D radiation in systems in which sodium was reacting with halogens.* For the reaction

$$Na_2 + Cl \longrightarrow Na + NaCl$$

he found that a substantial fraction of the energy released remained as vibrational energy in the NaCl molecule, which in a subsequent collision with a sodium atom gave rise to electronically excited Na, which produced the sodium D line.

*M. Polanyi, *Atomic Reactions*, London: Ernest Benn, 1932.

More recently the infrared emissions from the products of a number of reactions have been measured, and the results correlated with the possible shapes of the potential-energy surfaces.

Dynamical Calculations

A number of calculations have been made of the passage of systems over potential-energy surfaces. Such calculations are useful for testing the validity of transition state theory, which is based on the assumption of equilibrium between reactants and activated complexes. They are also important in leading to certain information, such as the vibrational states of the reaction products and whether a stripping or a rebound mechanism is involved, which is not revealed by transition state theory.

Many of the dynamical calculations have been made for the $H + H_2$ reaction, since reliable potential-energy surfaces have not been constructed for other systems. Some of the dynamical calculations employ quantum-mechanical methods; others use classical mechanics. A particularly important investigation of the $H + H_2$ system, employing classical mechanics, was made by M. Karplus, R. N. Porter, and R. D. Sharma.* One of the conclusions they arrived at is that the rates calculated by the dynamical treatment agree very satisfactorily with those given by transition state theory. They also found—and this is again consistent with the theory—that the reaction system usually does not linger at the activated state and perform a number of vibrations; instead it passes directly through the activated state. Calculations with other potential-energy surfaces have shown that systems will perform vibrations at the activated state only if there is a depression (a "basin") at the intersection of the two valleys in the potential-energy surface.

Key Equations

Definition of *rate of reaction:*

$$v = \frac{1}{V}\frac{d\xi}{dt}$$

where ξ = extent of reaction (Section 2.6).

For an *elementary reaction*

$$v = k[A]^{\alpha}[B]^{\beta}$$

where α and β are partial orders, $\alpha + \beta = n$ is the overall order, and k is the rate constant.

Influence of temperature (Arrhenius law):

$$k = Ae^{-E/RT}$$

*M. Karplus, R. N. Porter, and R. D. Sharma, *J. Chem. Phys.*, *34*, 3259 (1965).

Transition state theory:

$$k = \frac{\mathbf{k}T}{h}K^{\ddagger} = \frac{\mathbf{k}T}{h}e^{-\Delta G^{\ddagger}/RT} = \frac{\mathbf{k}T}{h}e^{\Delta S^{\ddagger}/R}\,e^{-\Delta H^{\ddagger}/RT}$$

$$= e\frac{\mathbf{k}T}{h}e^{\Delta S^{\ddagger}/R}\,e^{-E/RT} \text{ for reactions in solution.}$$

Influence of *ionic strength I* for reaction in aqueous solution at 25°C:

$$\log_{10} k = \log_{10} k_0 + 1.02\, z_A z_B \sqrt{I/\text{mol dm}^{-3}}$$

Influence of *hydrostatic pressure P:*

$$\left(\frac{\partial \ln k}{\partial P}\right)_T = -\frac{\Delta V^{\ddagger}}{RT}$$

Problems

A Rate Constants and Order of Reaction

1. The stoichiometric equation for the oxidation of bromide ions by hydrogen peroxide in acid solution is

$$2Br^- + H_2O_2 + 2H^+ \longrightarrow Br_2 + 2H_2O$$

Since the reaction does not occur in one stage, the rate equation does not correspond to this stoichiometric equation but is

$$v = k[H_2O_2][H^+][Br^-]$$

a. If the concentration of H_2O_2 is increased by a factor of 3, by what factor is the rate of consumption of Br^- ions increased?

b. If, under certain conditions, the rate of consumption of Br^- ions is 7.2×10^{-3} mol dm^{-3} s^{-1}, what is the rate of consumption of hydrogen peroxide? What is the rate of consumption of bromine?

c. What is the effect on the rate constant k of increasing the concentration of bromide ions?

d. If by the addition of water to the reaction mixture the total volume were doubled, what would be the effect on the rate of change of the concentration of Br^-? What would be the effect on the rate constant k?

2. A reaction obeys the stoichiometric equation

$$A + 2B \longrightarrow 2Z$$

Rates of formation of Z at various concentrations of A and B are given in the following table:

[A]/mol dm^{-3}	[B]/mol dm^{-3}	Rate/mol dm^{-3} s^{-1}
3.5×10^{-2}	2.3×10^{-2}	5.0×10^{-7}
7.0×10^{-2}	4.6×10^{-2}	2.0×10^{-6}
7.0×10^{-2}	9.2×10^{-2}	4.0×10^{-6}

What are α and β in the rate equation

$$v = k[A]^{\alpha}[B]^{\beta}$$

and what is the rate constant k?

3. Some results for the rate of a reaction between two substances A and B are shown here. Deduce the order α with respect to A, the order β with respect to B, and the rate constant.

[A]/mol dm^{-3}	[B]/mol dm^{-3}	Rate/mol dm^{-3} s^{-1}
1.4×10^{-2}	2.3×10^{-2}	7.4×10^{-9}
2.8×10^{-2}	4.6×10^{-2}	5.92×10^{-8}
2.8×10^{-1}	4.6×10^{-2}	5.92×10^{-6}

4. A substance decomposes at 600 K with a rate constant of 3.72×10^{-5} s^{-1}.

a. Calculate the half-life of the reaction.

b. What fraction will remain undecomposed if the substance is heated for 3 h at 600 K?

5. How does the time required for a first-order reaction to go to 99% completion relate to the half-life of the reaction?

6. The rate constant for the reaction $H^+ + OH^- \longrightarrow H_2O$ is 1.3×10^{11} dm^3 mol^{-1} s^{-1}. Calculate the half-life for the neutralization process if (a) $[H^+] = [OH^-] = 10^{-1}M$ and (b) $[H^+] = [OH^-] = 10^{-4}M$.

7. The isotope ^{90}Sr emits radiation by a first-order process (as is always the case with radioactive decay) and has a half-life of 28.1 years. When ingested by mammals it becomes permanently incorporated in bone tissue. If 1 μg is absorbed at birth, how much of this isotope remains after (a) 25 years, (b) 50 years, (c) 70 years?

8. The reaction

$$2NO(g) + Cl_2(g) \longrightarrow 2NOCl(g)$$

is second order in NO and first order in Cl_2. In a volume of 2 dm^3, 5 mol of nitric oxide and 2 mol of Cl_2 were brought together, and the initial rate was 2.4×10^{-3} mol dm^{-3} s^{-1}. What will be the rate when one-half of the chlorine has reacted?

***9.** The following results were obtained for the rate of decomposition of acetaldehyde:

Percent decomposed	0	5	10	15	20	25	30	35	40	45	50
Rate/mmHg min^{-1}	8.53	7.49	6.74	5.90	5.14	4.69	4.31	3.75	3.11	2.67	2.29

Employ van't Hoff's differential method to obtain the order of reaction.

10. The isotope $^{32}_{15}$P emits β radiation and has a half-life of 14.3 days. Calculate the decay constant in s^{-1}. What percentage of the initial activity remains after (a) 10 days, (b) 20 days, (c) 100 days?

11. The following counts per minute were recorded on a counter for the isotope $^{35}_{16}$S at various times:

Time/Days	Counts per Minute
0	4280
1	4245
2	4212
3	4179
4	4146
5	4113
10	3952
15	3798

Determine the half-life in days and the decay constant in s^{-1}. How many counts per minute would be expected after (a) 60 days and (b) 365 days?

***12.** The reaction

$$cis\text{-Cr(en)}_2\text{(OH)}_2^+ \underset{k_{-1}}{\overset{k_1}{\rightleftharpoons}} trans\text{-Cr(en)}_2\text{(OH)}_2^+$$

is first order in both directions. At 25°C the equilibrium constant is 0.16 and the rate constant k_1 is 3.3×10^{-4} s^{-1}. In an experiment starting with the pure *cis* form, how long would it take for half the equilibrium amount of the trans isomer to be formed?

B Temperature Dependence

13. The rate constant for a reaction at 30°C is found to be exactly twice the value at 20°C. Calculate the activation energy.

14. The rate constant for a reaction at 230°C is found to be exactly twice the value at 220°C. Calculate the activation energy.

15. Two second-order reactions have identical pre-exponential factors and activation energies differing by 20.0 kJ mol^{-1}. Calculate the ratio of their rate constants (a) at 0°C and (b) at 1000°C.

16. The water flea *Daphnia* performs a constant number of heartbeats and then dies. The flea lives twice as long at 15°C as at 25°C. Calculate the activation energy for the reaction that controls the rate of its heartbeat.

17. A sample of milk kept at 25°C is found to sour 40 times as rapidly as when it is kept at 4°C. Estimate the activation energy for the souring process.

***18.** The gas-phase reaction between nitric oxide and oxygen is third order. The following rate constants have been measured:

T/K	80.0	143.0	228.0	300.0	413.0	564.0
$k \times 10^{-9}$/cm^6 mol^{-2} s^{-1}	41.8	20.2	10.1	7.1	4.0	2.8

The behavior is interpreted in terms of a temperature-dependent preexponential factor; the rate equation is of the form

$$k = aT^n e^{-E/RT}$$

where a and n are constants. Assume the activation energy to be zero and determine n to the nearest half-integer.

C Collision Theory and Transition-State Theory

19. Two reactions of the same order have identical activation energies and their entropies of activation differ by 50 J K^{-1} mol^{-1}. Calculate the ratio of their rate constants at any temperature.

20. The gas-phase reaction

$$H_2 + I_2 \longrightarrow 2HI$$

is second order. Its rate constant at 400.0°C is 2.34×10^{-2} dm^3 mol^{-1} s^{-1}, and its activation energy is 150 kJ mol^{-1}. Calculate $\Delta H^‡$, $\Delta S^‡$, and $\Delta G^‡$ at 400°C and the preexponential factor.

21. A substance decomposes according to first-order kinetics; the rate constants at various temperatures are as follows:

Temperature, °C	Rate Constant, k/s^{-1}
15.0	4.18×10^{-6}
20.0	7.62×10^{-6}
25.0	1.37×10^{-5}
30.0	2.41×10^{-5}
37.0	5.15×10^{-5}

Calculate the activation energy. Calculate also, at 25.0°C, the enthalpy of activation, the Gibbs energy of activation, the preexponential factor, and the entropy of activation.

22. The following data have been obtained for the hydrolysis of adenosine triphosphate, catalyzed by hydrogen ions:

Temperature, °C	k/s^{-1}
39.9	4.67×10^{-6}
43.8	7.22×10^{-6}
47.1	10.0×10^{-6}
50.2	13.9×10^{-6}

Calculate, at 40.0°C, the Gibbs energy of activation, the activation energy, the enthalpy of activation, the preexponential factor, and the entropy of activation.

23. The half-life of the thermal denaturation of hemoglobin, a first-order process, has been found to be 3460 s at 60.0°C and 530 s at 65.0°C. Calculate the enthalpy of activation and entropy of activation at 60.0°C, assuming the Arrhenius law to apply.

***24.** (a) Using Eq. 9.73, calculate the collision number for 6.022×10^{23} molecules of hydrogen iodide present in a volume of 1 m^3 at 300 K. Take d_{AA} = 0.35 nm. (b) If the activation energy for the decomposition of HI is 184 kJ mol^{-1}, what rate constant does kinetic theory predict at 300°C? To what entropy of activation does this result correspond?

D Ionic-Strength Effects

25. The rate constant k for the reaction between persulfate ions and iodide ions varies with ionic strength I as follows:

$I/10^{-3}$ mol dm^{-3}	2.45	3.65	4.45	6.45	8.45	12.45
k/dm^3 mol^{-1} s^{-1}	1.05	1.12	1.16	1.18	1.26	1.39

Estimate the value of $z_A z_B$.

26. The following constants were obtained* for the reaction

$$[CoBr(NH_3)_5]^{2+} + OH^- \longrightarrow [Co(NH_3)_5OH]^{2+} + Br^-$$

under the following conditions:

Concentration/mol dm^{-3}			
$[CoBr(NH_3)_5]^{2+}$	NaOH	NaCl	k/dm^3 mol^{-1} s^{-1}
5.0×10^{-4}	7.95×10^{-4}	0	1.52
5.96×10^{-4}	1.004×10^{-3}	0	1.45
6.00×10^{-4}	0.696×10^{-3}	0.005	1.23
6.00×10^{-4}	0.696×10^{-3}	0.020	0.97
6.00×10^{-4}	0.691×10^{-3}	0.030	0.91

Make an estimate of the rate constant of the reaction at zero ionic strength. Are the results consistent with $z_A z_B = -2$?

***27.** Suppose that the rates of ionic reactions in solution were proportional to the *activity* rather than the concentration of activated complexes. Derive an equation relating the logarithm of the rate constant to the ionic strength and the charge numbers of the ions and contrast it with Eq. 9.123. Can the results in Figure 9.18 be reconciled with the equation you have derived?

E Pressure Effects

28. The rate of a reaction at 300 K is doubled when the pressure is increased from 1 atm to 2000 atm. Calculate $\Delta V^‡$, assuming it to be independent of pressure.

*J. N. Brønsted and R. S. Livingston, *J. Amer. Chem. Soc.*, *49*, 435 (1927).

29. The following results were obtained for the solvolysis of benzyl chloride in an acetone-water solution at 25.0°C:

Pressure/10^2 kPa	1.00	345	689	1033
$k/10^{-6}$ s^{-1}	7.18	9.58	12.2	15.8

Make an appropriate plot and estimate $\Delta V^{\ddagger}$.

*30. The fading of bromphenol blue in alkaline solution is a second-order reaction between hydroxide ions and the quinoid form of the dye:

$$\text{quinoid form(blue)} + OH^- \longrightarrow \text{carbinol form(colorless)}$$

The following results† show the variation of the second-order rate constant k with the hydrostatic pressure P at 25°C:

P/kPa	101.3	2.76×10^4	5.51×10^4	8.27×10^4	11.02×10^4
$k/10^{-4}$ M^{-1} s^{-1}	9.30	11.13	13.1	15.3	17.9

Estimate $\Delta V^{\ddagger}$.

F Supplementary Questions

31. Derive the following relationship for the half-life $t_{1/2}$ of a reaction of order n, with all reactants having an initial concentration a_0:

$$t_{1/2} = \frac{2^{n-1} - 1}{k a_0^{n-1}(n - 1)}$$

32. A drug administered to a patient is usually consumed by a first-order process. Suppose that a drug is administered in equal amounts at regular intervals and that the interval between successive doses is equal to the $(1/n)$-life for the disappearance process (i.e., to the time that it takes for the fraction $1/n$ to disappear). Prove that the limiting concentration of the drug in the patient's body is equal to n times the concentration produced by an individual dose.

33. Equation 9.35 applies to a second-order reaction of stoichiometry $A + B \longrightarrow Z$. Derive the corresponding equation for a second-order reaction of stoichiometry $2A + B \longrightarrow Z$.

*K. J. Laidler and R. Martin, *Int. J. Chem. Kinetics, 1,* 113 (1969).

†D. T. Y. Chen and K. J. Laidler, *Can. J. Chem., 37,* 599 (1959).

34. Derive the integrated rate equation for an irreversible reaction of stoichiometry $2A + B \longrightarrow Z$, the rate being proportional to $[A]^2[B]$ and the reactants present in stoichiometric proportions; take the initial concentration of A as $2a_0$ and that of B as a_0. Obtain an expression for the half-life of the reaction.

35. Prove that for two simultaneous (parallel) reactions

$$A \xrightarrow{k_1} Y \qquad A \xrightarrow{k_2} Z$$

$$\frac{[Y]}{[Z]} = \frac{k_1}{k_2} \text{ at all times.}$$

36. Prove that for two consecutive first-order reactions

$$A \xrightarrow{k_1} B \xrightarrow{k_2} C$$

the rate of formation of C is given by

$$[C] = [A]_0 \left(1 + \frac{k_2 e^{-k_1 t} - k_1 e^{-k_2 t}}{k_1 - k_2}\right)$$

where $[A]_0$ is the initial concentration of A. *Hint:* The solution of the differential equation

$$\frac{dx}{dt} = abe^{-bt} - cx$$

where a, b, and c are constants, is

$$x = \frac{ab}{c - b}(e^{-bt} - e^{-ct}) + I$$

37. **a.** Derive the integrated rate equation for a reversible reaction of stoichiometry

$$A \underset{k_{-1}}{\overset{k_1}{\rightleftharpoons}} Y + Z$$

The reaction is first order from left to right and second order from right to left. Take the initial concentration of A as a_0 and the concentration at time t as $a_0 - x$.

b. Obtain the integrated equation in terms of k_1 and the equilibrium constant $K = k_1/k_{-1}$.

c. A reaction to which this rate equation applies is the hydrolysis of methyl acetate. Newling and Hinshelwood, *J. Chem. Soc., 1936,* 1357 (1936), obtained the following results for the hydrolysis of 0.05 M ester at 80.2°C in the presence of 0.05 M HCl, which catalyzes the reaction:

Time/s	1350	2070	3060	5340	7740	∞
Percent hydrolysis	21.2	30.7	43.4	59.5	73.45	90.0

Obtain values for the rate constants k_1 and k_{-1}.

38. The dissociation of a weak acid

$$HA + H_2O \rightleftharpoons H_3O^+ + A^-$$

can be represented as

$$A \underset{k_{-1}}{\overset{k_1}{\rightleftharpoons}} Y + Z$$

The rate constants k_1 and k_{-1} cannot be measured by conventional methods but can be measured by the T-jump technique (Section 9.5). Prove that the relaxation time is given by

$$t^* = \frac{1}{k_1 + 2k_{-1}x_e}$$

where x_e is the concentration of the ions (Y and Z) at equilibrium.

G Essay Questions

1. Explain clearly the difference between the *order* and the *molecularity* of a reaction.

2. Give an account of experimental methods that might be used to study the kinetics of (a) a reaction having a half-life of about 10^{-1} s and (b) a reaction having a half-life of about 10^{-7} s.

3. Predict the effects of (a) increasing the dielectric constant of the solvent, (b) increasing the ionic strength, and (c) increasing the pressure on the reactions of the following types:

$$A^{2+} + B^- \longrightarrow X^+$$

$$A^+ + B^{2+} \longrightarrow X^{3+}$$

$$A + B \longrightarrow A^+B^-$$

Give a clear explanation in each case. What can you say about the entropy of activation to be expected in each case?

4. Van't Hoff's differential method can be applied to kinetic data in two different ways:

1. Rates can be determined at various stages of a single reaction.
2. Initial rates can be measured at a variety of initial concentrations, the reaction being run a number of times.

In each case $\log_{10}$ (rate) can be plotted against $\log_{10}$ (concentration of a reactant). Can you suggest why a different order of reaction might be obtained when these two different procedures are used?

Suggested Reading

M. H. Back and K. J. Laidler, *Selected Readings in Chemical Kinetics,* Oxford: Pergamon Press, 1967.

S. W. Benson, *Thermochemical Kinetics* (2nd ed.), New York: Wiley, 1976.

S. W. Benson, *The Foundations of Chemical Kinetics,* New York: McGraw-Hill, 1965.

F. F. Bernasconi, *Relaxation Kinetics,* New York: Academic Press, 1976.

E. F. Caldin, *Fast Reactions in Solution,* Oxford: Blackwell Scientific Pub., 1964.

A. A. Frost and R. G. Pearson, *Kinetics and Mechanism* (2nd ed.), New York: Wiley, 1961.

P. C. Jordan, *Chemical Kinetics and Transport,* New York: Plenum, 1979.

K. J. Laidler, *Reaction Kinetics* (Vol. I and II), Oxford: Pergamon Press, 1963.

K. J. Laidler, *Chemical Kinetics* (2nd ed.), New York: McGraw-Hill, 1965.

K. J. Laidler, *Theories of Chemical Reaction Rates,* New York: McGraw-Hill, 1969.

R. D. Levine and R. B. Bernstein, *Molecular Reaction Dynamics,* Oxford: Clarendon Press, 1974.

P. J. Robinson and K. A. Holbrook, *Unimolecular Reactions,* New York: Wiley-Interscience, 1972.

J. W. T. Spinks and R. J. Woods, *An Introduction to Radiation Chemistry* (2nd ed.), New York: Wiley, 1976.

D. H. Truhlar and R. E. Wyatt, "History of H_3 Kinetics," *Ann. Rev. Phys. Chem.,* 27, 1 (1976).

P. R. Wells, *Linear Free Energy Relationships,* London: Academic Press, 1968.

Composite Reaction Mechanisms

Preview

Composite, or *complex,* reactions are reactions that occur in more than one elementary step. There are several different types of composite reactions, which may involve *simultaneous* elementary reactions and *consecutive* elementary reactions.

The analysis of a composite mechanism involves explaining the overall behavior in terms of the kinetics of the elementary steps. This can be difficult mathematically, but it is greatly simplified if reaction intermediates are present in much smaller amounts than the reactants themselves. When this is the case, we can apply the *steady-state treatment,* according to which the rate of change of concentration of the reaction intermediate can, to a good approximation, be set equal to zero. This treatment allows the rate expression for the overall reaction to be obtained in terms of the reactant concentrations and the rate constants for the elementary steps.

A particularly important class of composite reactions comprises the *chain reactions.* An essential feature of a chain reaction is that there must be a *closed sequence,* or *cycle,* of reactions such that certain active intermediates are consumed in one step and are regenerated in another. It is also an essential feature that on the average the cycle is repeated more than once. Elementary reactions that are involved in such cycles are known as *chain-propagating steps.*

Photochemical and *radiation-chemical reactions* usually occur by chain mechanisms. In a particular subgroup of photochemical reactions, *photosensitized reactions,* light energy is absorbed by an atom or molecule that transfers the energy to the substance undergoing reaction.

Many *explosive* reactions occur by a special type of chain reaction in which there are *branching chains.* These are chain-propagation steps in which one active intermediate, usually a free radical, gives rise to two or more active intermediates.

Catalysis can occur by a number of types of composite mechanisms. Catalysis by acids and bases and by enzymes occurs by nonchain mechanisms in which a reaction intermediate is first formed from the catalyst and the substance undergoing reaction (the *substrate*); this intermediate subsequently breaks down into the reaction products and the catalyst. Certain other catalyzed reactions occur by chain mechanisms.

When composite reactions occur in solution, there is an important distinction between *collisions* and *encounters* that occur between reactant molecules. An *encounter* is a *group of collisions,* brought about by the fact that the reactant molecules are *caged in* by the surrounding solvent molecules.

10 Composite Reaction Mechanisms

In Chapter 9 we have been concerned with *elementary* reactions, which occur in a single step; the reactant molecules form an activated complex, which passes directly into products. However, the majority of chemical reactions are not elementary; instead they involve two or more elementary steps and then are said to be **composite,** *stepwise,* or *complex.** This chapter is concerned with how the rates of composite reactions are related to the characteristics of the elementary steps.

There are various types of composite reactions. In some of them, relatively stable molecules occur as intermediates. A simple example is the reaction between hydrogen and iodine monochloride,

$$H_2 + 2ICl \longrightarrow I_2 + 2HCl$$

If this reaction involved a single elementary step, it would be third order—first order in hydrogen and second order in ICl—since a molecule of hydrogen and two molecules of ICl would come together to form an activated complex. In fact, the reaction is second order, being first-order in hydrogen and first order in iodine monochloride:

$$v = k[H_2][ICl] \tag{10.1}$$

This can be explained if there is initially a slow reaction between one molecule of H_2 and one of ICl,

$$H_2 + ICl \xrightarrow{\text{slow}} HI + HCl$$

followed by a rapid reaction between the HI formed in this step and an additional molecule of ICl,

$$HI + ICl \xrightarrow{\text{rapid}} HCl + I_2$$

Addition of these two reactions gives the overall equation. The HI produced in the first reaction is removed as rapidly as it is formed. The rate of the second process has no effect on the overall rate, which is therefore that of the first step; the kinetic behavior is therefore explained. In this scheme the first step is said to be the *rate-determining* or *rate-controlling step.*

Another reaction that involves a fairly stable intermediate and that has a rate-determining step is the oxidation of bromide ions by hydrogen peroxide in aqueous acid solution,

$$2Br^- + H_2O_2 + 2H^+ \longrightarrow Br_2 + 2H_2O$$

The rate is given by the expression

$$v = k[H_2O_2][H^+][Br^-] \tag{10.2}$$

This reaction occurs in the following two stages:

1. $H^+ + Br^- + H_2O_2 \xrightarrow{\text{slow}} HOBr + H_2O$
2. $HOBr + H^+ + Br^- \xrightarrow{\text{fast}} Br_2 + H_2O$

---

*The word *composite* will be employed in this book.

The fact that the rate is proportional to $[H_2O_2][H^+][Br^-]$ suggests at once that reaction 1 is the rate-determining step. It is reasonable to conclude that HOBr is formed in this reaction, since HOBr is a known chemical substance, although it is not very stable. The slow step of the process can be represented as

1. 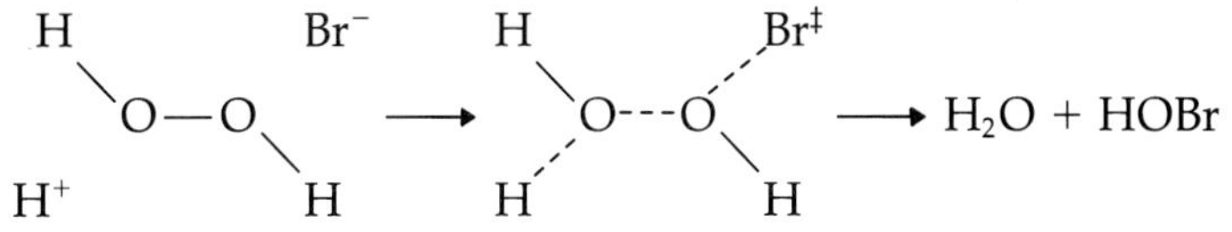 H^+, H–O–O–H, Br^- $\longrightarrow$ [H–O---O(H)---Br, with H---H]‡ $\longrightarrow$ $H_2O + HOBr$

activated complex

The activated complex can be formed quite readily as far as the molecular geometry is concerned; thus there is no reason to doubt that reaction 1 may occur. The hypothesis that the reaction does occur by reactions 1 and 2 is supported by the fact that if solutions of HOBr and Br^- are mixed, and the solution is acidified, bromine is produced very rapidly. On the basis of arguments of this kind, we draw conclusions about the elementary reactions that occur in composite mechanisms.

10.1 Evidence for a Composite Mechanism

There are various indications that a reaction is occurring by a composite mechanism. An obvious piece of evidence is when the kinetic law does not correspond to the stoichiometric equation; two examples of this have just been noted. In other cases the kinetic equation is more complicated, sometimes involving concentrations raised to nonintegral powers and reactant concentrations in the denominator. For example, the gas-phase reaction between hydrogen and bromine

$$H_2 + Br_2 \longrightarrow 2HBr$$

would be first order in hydrogen and first order in bromine if it were an elementary reaction. In fact, the reaction follows an equation of the form

$$v = \frac{k[H_2][Br_2]^{1/2}}{1 + [HBr]/m[Br_2]} \tag{10.3}$$

where k and m are constants. We shall consider the mechanism of this reaction in Section 10.5 and see how this rate equation arises.

Another indication that a mechanism is composite is provided when intermediates can be detected by chemical or other means during the course of reaction. When this can be done, a kinetic scheme must be developed that will account for the existence of these intermediates. Sometimes these intermediates are relatively stable substances; in other cases they are labile substances such as atoms and free radicals. Free radicals can sometimes be observed by spectroscopic methods, and evidence for their existence may be obtained by causing them to undergo specific reactions that less active substances cannot bring about.

When the nature of the reaction intermediates has been determined by methods such as those outlined previously, the next step is to devise a reaction scheme that will involve these intermediates and account for the kinetic features of the reaction. If such a scheme fits the data satisfactorily, it can be tentatively assumed that the mechanism is the correct one. It should be emphasized, however, that additional kinetic work frequently leads to the overthrow of schemes that had previously been supposed to be established firmly.

10.2 Types of Composite Reactions

Composite reaction mechanisms can have a number of features. Reactions occurring in parallel, such as*

$$A \xrightarrow{1} Y$$

$$A \xrightarrow{2} Z$$

are called *simultaneous* reactions. When there are simultaneous reactions, there is sometimes *competition*, as in the scheme

$$A + B \xrightarrow{1} Y$$

$$A + C \xrightarrow{2} Z$$

where B and C compete with one another for A.

Reactions occurring in forward and reverse directions are called *opposing*, for example;

$$A + B \underset{-1}{\overset{1}{\rightleftharpoons}} Z$$

Reactions occurring in sequence, such as

$$A \xrightarrow{1} X \xrightarrow{2} Y \xrightarrow{3} Z$$

are known as **consecutive** reactions, and the overall process is said to occur by *consecutive steps*. Reactions are said to exhibit **feedback** if a substance formed in one step affects the rate of a previous step. For example, in the scheme

$$A \xrightarrow{1} X \xrightarrow{2} Y \longrightarrow Z$$

(with an arrow from Y back to reaction 1)

the intermediate Y may catalyze reaction 1 (*positive feedback*) or inhibit reaction 1 (*negative feedback*). A final product as well as an intermediate may bring about feedback.

10.3 Rate Equations for Composite Mechanisms

Chemical Flux, or Chemiflux

In Section 9.3 we briefly introduced the concept of *chemical flux* or *chemiflux*, with special reference to opposing reactions. The concept is also extremely useful in dealing with composite mechanisms. The rate of an elementary reaction occurring

---

* The filled-in arrow $\longrightarrow$ is used to emphasize that a reaction is elementary.

in isolation is satisfactorily defined in terms of concentration changes, but an alternative treatment is required for elementary reactions that are components of a composite mechanism and for systems at equilibrium or close to equilibrium. In such cases it is necessary to consider the rates of elementary reactions *if they occurred in isolation in one direction*; such a rate has been called the **chemical flux** or **chemiflux.**

For example, for the composite mechanism

$$A \underset{k_{-1}}{\overset{k_1}{\rightleftharpoons}} X \underset{k_{-2}}{\overset{k_2}{\rightleftharpoons}} Z$$

there are four elementary reactions. The chemiflux for reaction 1 can be denoted by ϕ_1 or by $\phi_{-A,X}$ and is the rate of $A \longrightarrow X$ if no other reaction were occurring; it is given by

$$\phi_1 = \phi_{-A,X} = k_1[A] \tag{10.4}$$

Similarly,

$$\phi_{-1} = \phi_{-X,A} = k_{-1}[X] \tag{10.5}$$

$$\phi_2 = \phi_{-X,Z} = k_2[X] \tag{10.6}$$

$$\phi_{-2} = \phi_{-Z,X} = k_{-2}[Z] \tag{10.7}$$

For any species X the *total chemiflux into X*, $\Sigma\phi_X$, is the sum of the chemifluxes of all reactions that produce X. The total chemiflux out of X, $\Sigma\phi_{-X}$, is the sum of the chemifluxes of all the reactions that remove X. For example, for the scheme just considered the total chemiflux into X is

$$\Sigma\phi_X = \phi_1 + \phi_{-2} = k_1[A] + k_{-2}[Z] \tag{10.8}$$

and the total chemiflux out of X is

$$\Sigma\phi_{-X} = \phi_{-1} + \phi_2 = (k_{-1} + k_2)[X] \tag{10.9}$$

For a system at equilibrium the total chemiflux into each species is equal to the total chemiflux out of it.

Consecutive Reactions

The simplest consecutive mechanism is

$$A \xrightarrow{k_1} X \xrightarrow{k_2} Z$$

Equations for the rates of change of the concentrations of A, X, and Z were first given by Harcourt and Esson.* If the initial concentration of A is $[A]_0$ and its concentration at any time t is $[A]$, the rate equation for A is

$$\phi_1 = -\frac{d[A]}{dt} = k_1[A] \tag{10.10}$$

* A. V. Harcourt and W. Esson, *Proc. Roy. Soc.*, *14*, 470 (1865); *Phil. Trans.* *156*, 193 (1866); *157*, 117 (1867).

Integration of this equation, subject to the boundary condition that [A] = $[A]_0$ when $t = 0$, gives (see Eq. 9.30)

$$[A] = [A]_0 e^{-k_1 t} \tag{10.11}$$

The net rate of formation of X is

$$\frac{d[X]}{dt} = k_1[A] - k_2[X] \tag{10.12}$$

which, with Eq. 10.11, is

$$\frac{d[X]}{dt} = k_1[A]_0 e^{-k_1 t} - k_2[X] \tag{10.13}$$

This contains only the variables [X] and t, and integration gives

$$[X] = [A]_0 \frac{k_1}{k_2 - k_1}\left(e^{-k_1 t} - e^{-k_2 t}\right) \tag{10.14}$$

The equation for the variation of [Z] is most easily obtained by noting that

$$[A] + [X] + [Z] = [A]_0 \tag{10.15}$$

so that

$$[Z] = [A]_0 - [A] - [X] \tag{10.16}$$

Insertion of the expressions for [A] and [X] into Eq. 10.16 leads to

$$[Z] = \frac{[A]_0}{k_2 - k_1}\left[k_2\left(1 - e^{-k_1 t}\right) - k_1\left(1 - e^{-k_2 t}\right)\right] \tag{10.17}$$

Figure 10.1a shows the time variations in the concentrations of A, X, and Z as given by these equations. We see that [A] falls exponentially, while [X] goes through a maximum. Since the rate of formation of Z is proportional to the concentration of X, the rate is initially zero and is a maximum when [X] reaches its maximum value. For an initial period of time it may be impossible to detect any of the product Z, and the reaction is said to have an *induction period.* Such induction periods are commonly observed for reactions occurring by composite mechanisms.

Kinetic equations like Eqs. 10.14 and 10.17 are frequently obeyed by nuclides undergoing radioactive decay* (see also Section 9.4), but there are not many examples of chemical reactions that show consecutive first-order behavior. Two good examples are the thermal isomerizations of 1,1-dicyclopropylene and 1-cyclopropylcyclopentene.†

Two limiting cases are of special interest. Suppose, first, that the rate constant k_1 is very large and that k_2 is very small. The reactant A is then rapidly converted into the intermediate X, which slowly forms Z. Figure 10.1b shows plots of the exponentials $e^{-k_1 t}$ and $e^{-k_2 t}$ and of their difference. Since k_2 is small, the exponential $e^{-k_2 t}$ shows a very slow decay, while $e^{-k_1 t}$ shows a rapid fall. The difference

$$e^{-k_2 t} - e^{-k_1 t}$$

---

*E. Rutherford and F. Soddy, *J. Chem. Soc., 81,* 321, 837 (1902).

†G. R. Branton and H. M. Frey, *J. Chem. Soc.* A, 1342 (1966).

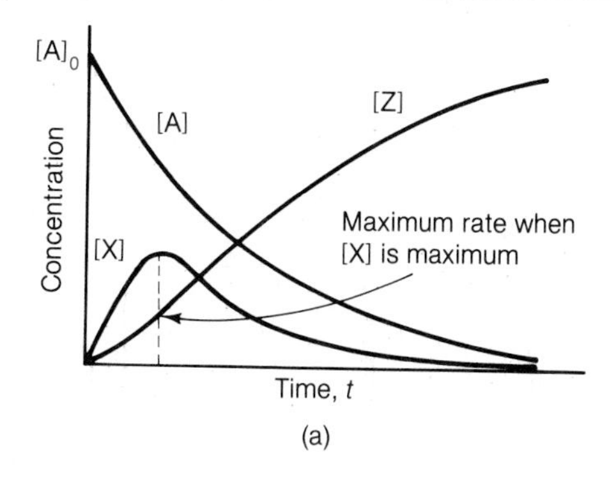

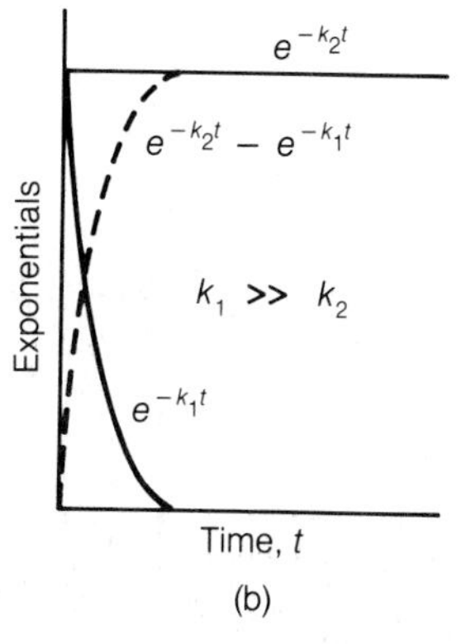

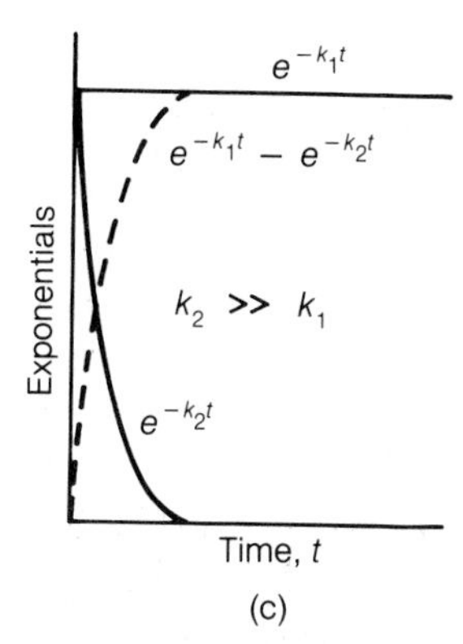

FIGURE 10.1 (a) Variations in the concentrations of A, X, and Z, for a reaction occurring by the mechanism $A \rightarrow X \rightarrow Z$. (b) Variations with time of the exponentials, when $k_1 >> k_2$. (c) Variations of the exponentials when $k_2 >> k_1$.

is shown by the dashed line in Figure 10.1b. The rate of change of the concentration of X is, by Eq. 10.14, equal to this difference multiplied by $[A]_0$ (since $k_1 >> k_2$), and [X] therefore rises rapidly to the value $[A]_0$ and then slowly declines. The rise in [Z] then follows approximately the simple first-order law.

The converse case, when $k_2 >> k_1$, is a particularly interesting one, since it leads us to the concept of the **steady state** and will now be discussed.

Steady-State Treatment

When k_1 is small and k_2 is large, the exponentials change with time in the manner shown in Figure 10.1c. The situation is the same as in Figure 10.1b, with k_1 and k_2 interchanged. Since $k_2 >> k_1$, the concentration of X is now given by (see Eq. 10.14)

$$[X] = [A]_0 \frac{k_1}{k_2}\left(e^{-k_1t} - e^{-k_2t}\right) \tag{10.18}$$

At $t = 0$, $[X] = 0$, but after a very short time, relative to the duration of the reaction, the difference

$$e^{-k_1t} - e^{-k_2t}$$

has attained the value of unity, and the concentration of X is then $[A]_0k_1/k_2$, which is much less than $[A]_0$. After this short induction period the concentration of X remains practically constant, so that to a good approximation

$$\frac{d[X]}{dt} = 0 \tag{10.19}$$

This is the basis of the **steady-state treatment,** which is very commonly applied to reaction mechanisms. What we have proved for the very simple scheme of two consecutive first-order reactions is that if the conditions are such that the concentration of the intermediate X is always much smaller than the reactant concentration, the concentration of X rapidly reaches a value that remains practically constant during the course of the reaction. It is not possible to give a formal proof of this hypothesis, applicable to any reaction mechanism, because the rate equations for more complicated mechanisms are often impossible to solve. However, the derivation we have given for the two-stage system of first-order reactions leads us to an important general conclusion. The rate of change of the concentration of an intermediate can, to a good approximation, be set equal to zero whenever the intermediate is formed slowly and disappears rapidly. In other words, whenever an intermediate X is such that it is always present in amounts much smaller than those of the reactants, the total chemiflux into X, $\Sigma\phi_X$, is nearly the same as the total chemiflux out of X, $\Sigma\phi_{-X}$:

$$\frac{d[X]}{dt} = \Sigma\phi_X - \Sigma\phi_{-X} = 0 \tag{10.20}$$

The steady-state treatment is of great importance in the analysis of composite mechanisms, since often there are mathematical difficulties that make it impossible to obtain an exact solution of the rate equations for the reaction. Consider, for example, the mechanism

$$A + B \underset{k_{-1}}{\overset{k_1}{\rightleftharpoons}} X$$

$$X \xrightarrow{k_2} Z$$

The differential rate equations that apply to this set of reactions are

$$-\frac{d[A]}{dt} = -\frac{d[B]}{dt} = k_1[A][B] - k_{-1}[X] \tag{10.21}$$

$$\frac{d[X]}{dt} = k_1[A][B] - k_{-1}[X] - k_2[X] \tag{10.22}$$

$$\frac{d[Z]}{dt} = k_2[X] \tag{10.23}$$

In order to treat this problem exactly it would be necessary to eliminate [X] and to solve the resulting differential equation to find [Z] as a function of t. Unfortunately, however, in spite of the simplicity of the kinetic scheme, it is not possible to obtain an explicit solution.

The steady-state treatment, which is valid provided that the concentration of X is always small, involves using Eq. 10.19 so that, from Eq. 10.22,

$$k_1[A][B] - k_{-1}[X] - k_2[X] = 0 \tag{10.24}$$

The concentration of [X] is therefore given by

$$[X] = \frac{k_1[A][B]}{k_{-1} + k_2} \tag{10.25}$$

and insertion of this into Eq. 10.23 gives

$$v = v_Z = \frac{d[Z]}{dt} = \frac{k_1 k_2[A][B]}{k_{-1} + k_2} \tag{10.26}$$

Rate-Determining (Rate-Controlling) Steps

Suppose that in the reaction scheme just considered the intermediate X is converted very rapidly into Z, much more rapidly than it can go back into A + B. In that case the rate of the reaction will be the rate of formation of X from A + B; that is,

$$v = k_1[A][B] \tag{10.27}$$

since as soon as X is formed, it is transformed into Z. The initial step is therefore the **rate-determining step.** The exact condition, for this two-step mechanism, is

$$k_2 >> k_{-1}$$

and we see that the steady-state rate equation (Eq. 10.26) becomes Eq. 10.27 if this inequality is satisfied.

Alternatively, suppose that the rate constant for the second reaction, $X \xrightarrow{k_2} Z$, is very small compared to that for the reverse of the first reaction; that is,

$$k_2 << k_{-1}$$

The overall rate is

$$v = k_2[X] \tag{10.28}$$

and since reaction 2 is too slow to disturb the equilibrium $A + B \underset{k_{-1}}{\overset{k_1}{\rightleftharpoons}} X$,

$$[X] = \frac{k_1}{k_{-1}}[A][B] \tag{10.29}$$

Insertion of Eq. 10.29 into Eq. 10.28 gives

$$v = \frac{k_1 k_2}{k_{-1}}[A][B] \tag{10.30}$$

Again, this is the expression to which Eq. 10.26 reduces if the inequality $k_{-1} >> k_2$ is satisfied. Reaction 2 is now the *rate-determining step* or the *rate-controlling step.**

---

* Physical-organic chemists sometimes make a distinction between rate-determining and rate-controlling steps; the former is used when the overall rate is equal to the rate of the *first* step in a consecutive mechanism, the latter when a later reaction is involved, the overall rate also involving equilibrium constants for pre-equilibria (as in Eq. 10.30). However, this distinction is not universally recognized and is not favored by most kineticists; those who make the distinction should therefore emphasize that they are doing so.

It is important to work out each kinetic scheme separately, as there are some pitfalls. Note in particular that when there is a chain reaction (see Section 10.5), there is usually no rate-determining step; the rate is *not* equal to that of any particular step.

10.4 Rate Constants, Rate Coefficients, and Equilibrium Constants

For an *elementary* reaction it is easy to show that the equilibrium constant must be the ratio of the rate constants in forward and reverse directions. Thus, consider the process

$$A + B \underset{k_{-1}}{\overset{k_1}{\rightleftharpoons}} Y + Z$$

in which the reactions in forward and reverse directions, as indicated by the filled-in arrows, are elementary. Their rates are

$$v_1 = k_1[A][B] \tag{10.31}$$

$$v_{-1} = k_{-1}[Y][Z] \tag{10.32}$$

If the system is at equilibrium, these rates are equal; hence

$$\frac{k_1}{k_{-1}} = \left(\frac{[Y]\,[Z]}{[A]\,[B]}\right)_{eq} = K_c \tag{10.33}$$

where K_c is the equilibrium constant.

This argument can be extended to a reaction that occurs in two or more stages. Consider, for example, the reaction

$$H_2 + 2ICl \longrightarrow I_2 + 2HCl$$

which we saw at the beginning of this chapter to occur in two stages, which at equilibrium will be occurring in both directions:

$$H_2 + ICl \underset{k_{-1}}{\overset{k_1}{\rightleftharpoons}} HI + HCl$$

$$HI + ICl \underset{k_{-2}}{\overset{k_2}{\rightleftharpoons}} HCl + I_2$$

At equilibrium the rate of each elementary reaction and its reverse must be the same:

$$k_1[H_2][ICl] = k_{-1}[HI][HCl] \tag{10.34}$$

$$k_2[HI][ICl] = k_{-2}[HCl][I_2] \tag{10.35}$$

The equilibrium constant for each reaction is thus

$$K_1 = \frac{k_1}{k_{-1}} = \left(\frac{[HI]\,[HCl]}{[H_2]\,[ICl]}\right)_{eq} \tag{10.36}$$

$$K_2 = \frac{k_2}{k_{-2}} = \left(\frac{[HCl]\,[I_2]}{[HI]\,[ICl]}\right)_{eq} \tag{10.37}$$

The product of these two equilibrium constants is

$$K_1K_2 = \frac{k_1k_2}{k_{-1}k_{-2}} = \left(\frac{[I_2]\,[HCl]^2}{[H_2]\,[ICl]^2}\right) = K_c \tag{10.38}$$

where K_c is the equilibrium constant for the overall reaction. It is easy to prove that for any mechanism, involving any number of elementary steps, the overall equilibrium constant is the product of the equilibrium constants for the individual steps and is the product of the rate constants for the reactions in the forward direction divided by the product of those for the reverse reactions:

$$K_c = K_1K_2K_3 \cdots = \frac{k_1k_2k_2 \cdots}{k_{-1}k_{-2}k_{-3} \cdots} \tag{10.39}$$

The principle that, at equilibrium, each elementary process is exactly balanced by its reverse reaction is known as the **Principle of Detailed Balancing.**

It is important to realize that if a reaction occurs by a composite mechanism, and we measure a rate coefficient* k_1 for the overall reaction from left to right and also measure a rate coefficient k_{-1} for the overall reaction from right to left, at the same temperature, the ratio k_1/k_{-1} is not necessarily the equilibrium constant for the overall reaction. The reason for this is that rate laws for composite reactions change with the experimental conditions, such as reactant concentrations, and the rate coefficients also change. The ratio of the rate coefficients k_1 and k_{-1} that apply *when the system is at equilibrium* is equal to the equilibrium constant, but rate coefficients determined away from equilibrium are not necessarily the same as those at equilibrium, and their ratio is not necessarily equal to K_c. Great caution should therefore be used in deducing rate coefficients and rate laws for reactions from the equilibrium constant and the rate coefficient for the reverse reaction.

10.5 Free-Radical Reactions

Reactions frequently occur by a series of reactions in which free radicals play a part. The important distinction between a free radical and an ion may be illustrated by comparison of the hydroxyl radical and the hydroxide ion. The oxygen-hydrogen bond in the water molecule may be split *homolytically;* that is, one of the electrons goes with one fragment and the other with the other:

```
   H                   H
   ..                  ..
  :O:H      ——→       :O·     +     .H
   ..                  ..
  water             hydroxyl     hydrogen
 molecule            radical       atom
```

In this process, two electrically neutral free radicals are produced; an atom is a special case of a radical. The hydrogen atom consists of a proton and an electron, and the hydroxyl radical consists of nine protons (one in the nucleus of the hydrogen atom and eight protons in the oxygen nucleus) and nine electrons (seven

*We use the term *coefficient* rather than *constant* when the reaction is not elementary.

valence electrons plus two 1*s* electrons). The hydrogen atom and the hydroxyl radical are both one electron short of the noble gas structures and, therefore, are very reactive species. Radicals combine with one another with very low or zero activation energies, and their reactions with stable molecules occur with quite low activation energies.

In the ionization of water, on the other hand, a bond is split *heterolytically* and the electron pair remains with the oxygen atom:

$$\underset{\text{water molecule}}{\overset{\text{H}}{:\ddot{\underset{\cdot\cdot}{\text{O}}}:\text{H}}} \longrightarrow \underset{\text{hydroxide ion}}{\left[\overset{\text{H}}{:\ddot{\underset{\cdot\cdot}{\text{O}}}:}\right]^-} + \underset{\text{hydrogen ion (proton)}}{[\text{H}]^+}$$

The hydroxide ion is negatively charged; it has nine protons and ten electrons. It has the same electronic configuration as neon and therefore is chemically stable and unreactive. Whereas hydroxyl radicals cannot be stored, solutions of hydroxide ions will remain intact for long periods of time.

Chain Reactions

Ions play little part in ordinary gas-phase reactions, owing to the difficulty with which they are formed in the absence of an ionizing solvent. They do play a role in radiolytic reactions (Section 10.7), where high energies are involved. Atoms and free radicals are produced more easily in the gas phase and, because they enter readily into further reaction, they are important intermediates in reactions. For example, consider the reaction between hydrogen and bromine, for which Eq. 10.3 is the rate equation. This rate equation can be explained by the mechanism

1.	$Br_2 \xrightarrow{k_1} 2Br$	initiation
2.	$Br + H_2 \xrightarrow{k_2} HBr + H$	chain propagation
3.	$H + Br_2 \xrightarrow{k_3} HBr + Br$	
4.	$H + HBr \xrightarrow{k_4} H_2 + Br$	
−1.	$2Br \xrightarrow{k_{-1}} Br_2$	termination

The first reaction, the production of bromine atoms from a bromine molecule, is known as the **initiation reaction,** since it starts the whole process. Reactions 2 and 3, the so-called **chain-propagation steps,** play a very important role in reactions of this type. Bromine atoms disappear in reaction 2 and reappear in reaction 3; hydrogen atoms disappear in reaction 3 and come back again in reaction 2. Because of this feature a small number of Br atoms, produced in reaction 1, can bring about a considerable amount of reaction, since after producing two molecules of hydrogen bromide, one in reaction 2 and one in reaction 3, a bromine atom is regenerated. Reaction 4 accounts for the fact that in the rate equation (Eq. 10.3), HBr appears in the denominator; HBr reduces the rate by removing H atoms. If it were not for reaction -1, a single pair of bromine atoms could bring about the reaction of all the H_2 and Br_2 present. Because of the **termination** reaction -1, however, only a limited amount of reaction is brought about each time a pair of bromine atoms is produced. Bromine atoms are continuously formed by reaction 1, and this keeps the reaction going.

A reaction of this type is known as a **chain reaction.** One essential feature of a chain reaction is that there must be a *closed sequence,* or *cycle,* of reactions such that certain active intermediates are consumed in one step and are regenerated in another; these active intermediates may be atoms, free radicals, or ions. It is also an essential feature that the sequence is, on the average, repeated more than once.

The way in which the chain-reaction mechanism for the hydrogen-bromine reaction explains the experimental rate equation 10.3 can be shown as follows. The net rate of increase of the concentration of hydrogen bromide is equal to

$$v_{HBr} = k_2[Br][H_2] + k_3[H][Br_2] - k_4[H][HBr] \tag{10.40}$$

The concentration of bromine atoms can be obtained by use of the steady-state method, which must now be applied to the two unstable intermediates H and Br. The steady-state equation for H is

$$\frac{d[H]}{dt} = k_2[Br][H_2] - k_3[H][Br_2] - k_4[H][HBr] = 0 \tag{10.41}$$

and that for Br is

$$\frac{d[Br]}{dt} = k_1[Br_2] - k_2[Br][H_2] + k_3[H][Br_2] + k_4[H][HBr] - k_{-1}[Br]^2 = 0 \tag{10.42}$$

We thus have two equations in the two unknowns [H] and [Br], and we can solve for both of these concentrations. A solution for [Br] is quickly obtained if we add Eqs. 10.41 and 10.42:

$$k_1[Br_2] - k_{-1}[Br]^2 = 0 \tag{10.43}$$

and thus

$$[Br] = \left(\frac{k_1}{k_{-1}}\right)^{1/2}[Br_2]^{1/2} \tag{10.44}$$

Note that this is the equilibrium concentration of Br atoms. It is by no means always the case that atoms and free radicals in reaction systems are present at their equilibrium concentrations; indeed in the present example the H atoms are present at much higher concentrations than their equilibrium concentrations.

We can insert this expression for [Br] into either Eq. 10.41 or Eq. 10.42 and obtain an expression for [H]. Insertion into Eq. 10.41 leads to

$$k_2\left(\frac{k_1}{k_{-1}}\right)^{1/2}[Br_2]^{1/2}[H_2] - k_3[H][Br_2] - k_4[H][HBr] = 0 \tag{10.45}$$

so that

$$[H] = \frac{k_2(k_1/k_{-1})^{1/2}[H_2][Br_2]^{1/2}}{k_3[Br_2] + k_4[HBr]} \tag{10.46}$$

If we subtract Eq. 10.41 from Eq. 10.40, we obtain

$$v_{HBr} = 2k_3[H][Br_2] \tag{10.47}$$

and insertion of the expression for [H] leads to

$$v_{HBr} = \frac{2k_2k_3(k_1/k_{-1})^{1/2}[H_2][Br_2]^{3/2}}{k_3[Br_2] + k_4[HBr]} \tag{10.48}$$

$$= \frac{2k_2(k_1/k_{-1})^{1/2}[H_2][Br_2]^{1/2}}{1 + (k_4/k_3)([HBr]/[Br_2])} \tag{10.49}$$

This equation is of the same form as the empirical Eq. 10.3, and we note that the empirical k is equal to $2k_2(k_1/k_{-1})^{1/2}$ and that the empirical m is equal to k_3/k_4.

We can see that the reason the term in $[HBr]/[Br_2]$ appears in the denominator of the rate equation is that HBr inhibits the reaction, by undergoing reaction 4, and that Br_2 reduces the amount of inhibition, since, in reactions 3 and 4, Br_2 and HBr are competing with one another for H atoms.

Organic Decompositions

A typical organic free-radical chain reaction is the decomposition of ethane,

$$C_2H_6 \longrightarrow C_2H_4 + H_2$$

Under most conditions this is a simple first-order reaction, and originally it was thought that it occurred in one stage by what is called a *molecular* reaction; that is, that a small fraction of the molecules have sufficient energy for two C—H bonds to be ruptured, so that a hydrogen molecule is liberated:

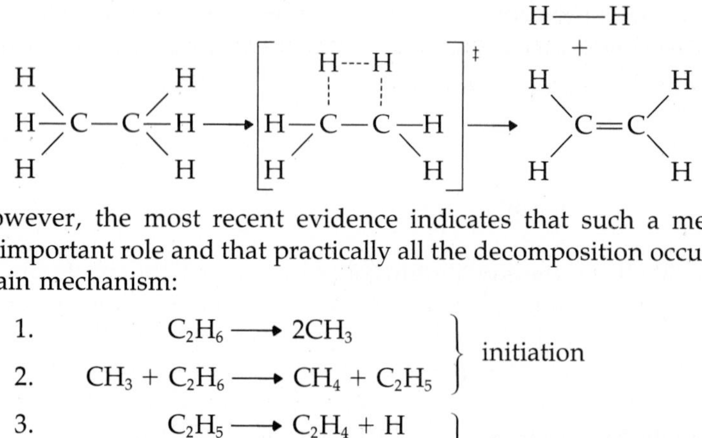

$$H_3C\text{—}CH_3 \longrightarrow \left[\begin{array}{c} H\text{---}H \\ \vdots \quad \vdots \\ H_2C\text{—}CH_2 \end{array}\right]^{\ddagger} \longrightarrow H_2C{=}CH_2 + H\text{—}H$$

However, the most recent evidence indicates that such a mechanism plays an unimportant role and that practically all the decomposition occurs by the following chain mechanism:

$$\begin{array}{lrcll} 1. & C_2H_6 & \longrightarrow & 2CH_3 & \\ 2. & CH_3 + C_2H_6 & \longrightarrow & CH_4 + C_2H_5 & \Big\} \text{ initiation} \\ 3. & C_2H_5 & \longrightarrow & C_2H_4 + H & \\ 4. & H + C_2H_6 & \longrightarrow & H_2 + C_2H_5 & \Big\} \text{ chain propagation} \\ 5. & 2C_2H_5 & \longrightarrow & C_4H_{10} & \text{termination} \end{array}$$

The initiation process involves the breaking of a C—C bond, which is the weakest bond in the molecule. Reaction 2 is in a sense part of the initiation reaction; it converts CH_3 into a radical C_2H_5, which can be involved in propagation; reaction 2 is not a propagation reaction since CH_3 is not regenerated from C_2H_5. Reactions 3 and 4 are chain-propagating steps: C_2H_5 disappears in reaction 3 and appears in reaction 4, while H disappears in reaction 4 and appears in reaction 3. The main products of the reaction, C_2H_4 and H_2, are formed in these propagation steps. The termination step, reaction 5, forms butane, C_4H_{10}, which can be detected as a minor product of the reaction. Methane, formed in reaction 2, has also been observed as a minor product of the reaction.

The thermal decomposition of acetaldehyde occurs to a large extent by the mechanism*

$$CH_3CHO \xrightarrow{k_1} CH_3 + CHO$$
$$CH_3 + CH_3CHO \xrightarrow{k_2} CH_4 + CH_3CO$$
$$CH_3CO \xrightarrow{k_3} CH_3 + CO$$
$$CH_3 + CH_3 \xrightarrow{k_4} C_2H_6$$

To simplify the steady-state treatment we will neglect the subsequent reactions of CHO. The steady-state equation for CH_3 is

$$k_1[CH_3CHO] - k_2[CH_3][CH_3CHO] + k_3[CH_3CO] - k_4[CH_3]^2 = 0 \qquad (10.50)$$

and the steady-state equation for CH_3CO is

$$k_2[CH_3][CH_3CHO] - k_3[CH_3CO] = 0 \qquad (10.51)$$

Addition of these two equations gives

$$k_1[CH_3CHO] - k_4[CH_3]^2 = 0 \qquad (10.52)$$

and therefore

$$[CH_3] = \left(\frac{k_1}{k_4}\right)^{1/2}[CH_3CHO]^{1/2} \qquad (10.53)$$

The rate of change of the concentration of methane, which is approximately the rate of change of the concentration of acetaldehyde, is

$$v = k_2[CH_3][CH_3CHO] \qquad (10.54)$$

and insertion of Eq. 10.53 gives

$$v = k_2\left(\frac{k_1}{k_4}\right)^{1/2}[CH_3CHO]^{3/2} \qquad (10.55)$$

This agrees with the experimental fact that the order of the acetaldehyde decomposition is three-halves.

10.6 Photochemical Reactions

The reactions we have considered in Section 10.5 are known as **thermal reactions;** the energy needed for the activation barriers to be surmounted is provided by the thermal motions of the molecules and radicals. It is also possible for reactions to be

*F. O. Rice and K. F. Herzfeld, *J. Am. Chem. Soc.*, *56*, 284 (1934); a modified mechanism, which better accounts for the minor products, has been given by K. J. Laidler and M. T.H. Liu, *Proc. Roy. Soc., A* *297*, 365 (1967).

brought about by electromagnetic radiation. For example, if a mixture of dry hydrogen and chlorine is irradiated by visible light, the reaction

$$H_2 + Cl_2 \longrightarrow 2HCl$$

occurs with explosive violence. The reaction between hydrogen and bromine can also be brought about by irradiating the mixture with light of suitable wavelength, and the formation of hydrogen bromide is found to follow the rate equation

$$v_{HBr} = \frac{k'[H_2]I^{1/2}}{1 + [HBr]/m'[Br_2]} \tag{10.56}$$

where k' and m' are constants (compare Eq. 10.3) and I is the intensity of the light that is absorbed.

We will later discuss the mechanism of this reaction, but first we must consider some important principles that apply to the interaction between molecules and radiation. Radiation is of two kinds, electromagnetic and particle, and examples of each are given below:

Electromagnetic Radiation	*Particle Radiation*
Infrared radiation	α particles (He nuclei)
Visible light	β particles (electrons)
Ultraviolet radiation	Cathode rays (electrons)
X rays	Beams of protons, deuterons, etc., produced in a cyclotron
γ rays	

Chemical reactions brought about by any of these radiations are known as either *photochemical* or *radiation-chemical reactions.* The distinction between these two classes of reactions is not a sharp one, but it has been found useful. In the case of electromagnetic radiation from the mid-ultraviolet range and beyond, and with high-energy particle radiation, ions can be detected in the reaction system, and in that case we speak of a **radiation-chemical reaction;** such reactions are considered in Section 10.7. Electromagnetic energy of higher wavelengths, such as visible and near-ultraviolet radiation, does not possess enough energy to bring about ionization and we then speak of a **photochemical** reaction.

Various special kinds of photochemical reactions may be identified by the prefix *photo.* For example, a *photolysis* is a photochemical reaction in which there is molecular dissociation, while a *photoisomerization* is a photochemical isomerization.

A number of important principles are associated with the photochemical production of excited molecules and of atoms and radicals; some of these will be discussed in Chapter 13. For a gas there is often a fairly sharp transition from a spectral region where there is no absorption and no chemical reaction to one in which a considerable amount of chemical reaction occurs. The frequency or wavelength at this transition is known as the *photochemical threshold.* Reaction occurs on the higher-frequency, or shorter-wavelength, side of the threshold.

Since the lifetime of an electronically excited species is very short, often about 10^{-8} s, it is very unlikely for a molecule that has absorbed one photon to absorb another before it has become deactivated.* There is usually, therefore, a one-to-one

*Exceptions to this are found when lasers are used; the intensity of the radiation can then be so great that two or more photons can be absorbed.

relationship between the number of photons absorbed by the system and the number of excited molecules produced. If, for example, light is absorbed with the production of an excited species that dissociates into two radicals, as in the process

$$CH_3COCH_3 + h\nu \longrightarrow CH_3COCH_3^* \longrightarrow CO + 2CH_3$$

the rate of formation of methyl radicals is twice the rate of absorption of photons. This principle, which is due to Einstein and is known as the **law of photochemical equivalence,** has been of great value in photochemical studies, since it enables the rates of formation of radicals to be calculated from optical measurements. It is convenient to speak of a mole of photons as an *einstein;* the rate of production of methyl radicals in the photolysis of acetone would then be, in mol dm^{-3} s^{-1}, twice the number of einsteins absorbed per cubic decimeter per second.

In photochemical experiments it is often found that the number of molecules that are transformed chemically differs markedly from the number of photons absorbed. In such cases it is customary to state that Einstein's law is not obeyed, although the violation is only an apparent one. There are two main reasons for failure of the law. In the first place, radicals that are produced initially may recombine before they can undergo reaction; as will be seen, this very commonly occurs in solution. In this case the rate of reaction is less than the Einstein law predicts. In other systems the radicals produced may initiate chain reactions, in which case the rate of reaction may be much larger than expected. Deviations from the Einstein law obviously provide valuable information about reaction mechanisms. The ratio of the number of molecules undergoing reaction in a given time to the photons absorbed is known as the **quantum yield,** Φ.

Photosensitization

Many molecules do not absorb radiation at a convenient wavelength. The production of hydrogen atoms from a hydrogen molecule requires 431.8 kJ mol^{-1}, which corresponds to a wavelength of 277.6 nm. However, hydrogen gas does not absorb at this wavelength and it is necessary to go to very much lower wavelengths in order to produce hydrogen atoms. A very convenient procedure for avoiding this difficulty is to introduce mercury vapor into the hydrogen. Mercury atoms are normally in a 6^1S_0 state, and they absorb 253.7-nm radiation to give the 6^3P_1 state, which is 468.6 kJ mol^{-1} higher:

$$Hg(6^1S_0) + h\nu(253.7\ nm) \longrightarrow Hg^*(6^3P_1)$$

(See Section 13.2 for a discussion of state symbols.) The excited mercury atom has more than enough energy to dissociate a hydrogen molecule, and it does so with high efficiency:

$$Hg^*(6^3P_1) + H_2 \longrightarrow Hg(6^1S_0) + 2H$$

Many other bonds having strengths of less than 468.6 kJ mol^{-1} can be split in this way.

Processes of this kind, in which the radiation is absorbed by one species and the energy is passed on to another, are known as **photosensitized processes.** Zinc and cadmium atoms have also been used in photosensitization experiments. Sometimes the photosensitizer produces an excited state rather than a dissociated state. For example, with ethylene, excited mercury induces the reaction

$$Hg^* + C_2H_4 \longrightarrow Hg + C_2H_4^*$$

as well as

$$Hg^* + C_2H_4 \longrightarrow Hg + H + C_2H_3$$

The Photochemical Hydrogen-Bromine Reaction

When light of sufficiently short wavelength is passed through a mixture of hydrogen and bromine, at room temperature, reaction occurs with the formation of hydrogen bromide, and the rate equation is Eq. 10.56. The similarity of this equation to that for the thermal reaction, Eq. 10.3, suggests that the difference is only in the initiation step. In the thermal reaction, which requires higher temperatures than the photochemical reaction, the initiation reaction is the thermal dissociation of bromine molecules. In the photochemical reaction the initiation process is the photochemical dissociation of Br_2 into bromine atoms. The mechanism is therefore

1. $Br_2 + h\nu \longrightarrow 2Br$ — initiation
2. $Br + H_2 \xrightarrow{k_2} HBr + H$ — chain propagation
3. $H + Br_2 \xrightarrow{k_3} HBr + Br$ — chain propagation
4. $H + HBr \xrightarrow{k_4} H_2 + Br$
5. $2Br \xrightarrow{k_5} Br_2$ — termination

If I_a is the intensity of light absorbed, expressed as einstein dm^{-3} s^{-1}, the rate of change of concentration of Br atoms by reaction 1 is $2I_a$. The steady-state equation for Br atoms is therefore

$$2I_a - k_2[Br][H_2] + k_3[H][Br_2] - k_4[H][HBr] - k_5[Br]^2 = 0 \tag{10.57}$$

The steady-state equation for H atoms is

$$k_2[Br][H_2] - k_3[H][Br_2] - k_4[H][HBr] = 0 \tag{10.58}$$

These equations are very similar to Eqs. 10.42 and 10.41, respectively; the only differences are that $k_1[Br_2]$ has been replaced by $2I_a$ and k_{-1} has been replaced by k_5. If we make these replacements in Eq. 10.49, we see that the rate law for the photochemical reaction is

$$v_{HBr} = \frac{2k_2(2/k_5)^{1/2}[H]I_a^{1/2}}{1 + k_4[HBr]/k_3[Br_2]} \tag{10.59}$$

This is of the same form as Eq. 10.56.

10.7 Radiation-Chemical Reactions*

We saw in Section 10.6 that an arbitrary but useful distinction is often made between photochemical and radiation-chemical reactions; the latter term is usually only applied when ions can be detected in the reaction system. When molecular dissociation occurs in a radiation-chemical reaction, the process is called a *radiolysis*.

* See, for example, J. W. T. Spinks and R. J. Woods, *An Introduction to Radiation Chemistry* (2nd ed.), New York: Wiley, 1977.

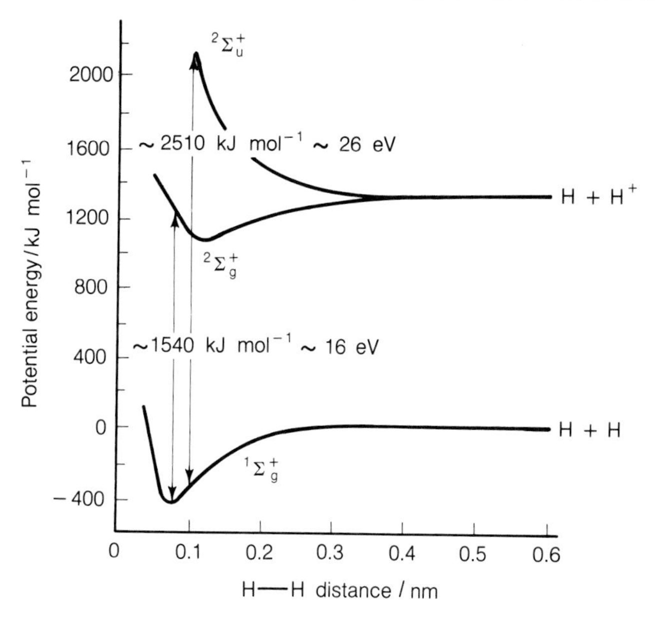

FIGURE 10.2 Potential-energy curves for H_2 and H_2^+, showing how the positive H_2^+ ion is formed by electron impact.

Whereas photochemical initiation reactions are fairly clear-cut, radiolytic initiation reactions frequently occur in a number of stages, and there are marked differences between different kinds of radiation. As a fairly simple example we will consider the effect of a high-energy electron beam on hydrogen molecules. Figure 10.2 shows the ground state of the hydrogen molecule and also two states of H_2^+ which dissociate into $H + H^+$. Electrons of energy 2.6×10^{-18} J (= 1540 kJ mol^{-1} = 16 eV) will eject an electron from H_2; the process is

$$e^- + H_2 \longrightarrow H_2^+ + 2e^-$$

When H_2^+ encounters an electron, neutralization occurs with the liberation of a considerable amount of energy which brings about dissociation into two H atoms:

$$e^- + H_2^+ \longrightarrow H_2^* \longrightarrow 2H$$

Electrons of higher energy, about 4.2×10^{-18} J (= 2510 kJ mol^{-1} = 26 eV) can also produce H_2^+ in its repulsive $^2\Sigma_u^+$ state, which at once dissociates into $H + H^+$:

$$e^- + H_2 \longrightarrow 2e^- + H_2^+\ (^2\Sigma_u^+)$$

$$H_2^+\ (^2\Sigma_u^+) \longrightarrow H + H^+$$

When H^+ encounters an electron, it forms H:

$$H^+ + e^- \longrightarrow H$$

The net result of these several processes is the formation of two hydrogen atoms from the hydrogen molecule. These various processes are usually referred to collectively as the *primary radiolytic process*, which is written as

$$H_2 \overset{e^-}{\rightsquigarrow} 2H$$

The primary process in the irradiation of hydrogen by α particles is written as

$$H_2 \overset{\alpha}{\rightsquigarrow} 2H$$

Again, this primary process occurs in a number of stages, involving the formation of ions. The first step is usually the ejection of an electron:

$$\alpha + M \longrightarrow M^+ + e^- + \alpha$$

The α particle continues on its way with little deflection and ionizes other molecules. The ejected electron frequently has sufficient energy to bring about ionization of additional molecules, by the mechanism just described.

Photons of very high energy (very low wavelengths) frequently eject electrons with the formation of the positive ion:

$$M + h\nu \longrightarrow M^+ + e^-$$

Although the details of primary radiolytic processes are often complex, the overall result is usually fairly simple. With hydrogen, for example, the initiation process with most types of radiation is largely the production of atoms:

$$H_2 \rightsquigarrow 2H$$

Similarly, with HI the main process is

$$HI \rightsquigarrow H + I$$

In addition, the formation of ions must be taken into account.

In investigations of overall radiation-chemical reactions two quantities are frequently quoted. The first is the *ion-pair yield*, or *ionic yield*, which is denoted as M/N and is defined by

$$\frac{M}{N} = \frac{\text{rate of formation of product molecules}}{\text{rate of formation of ion pairs}} \tag{10.60}$$

The rate of formation of ion pairs is determined by applying to the system an electric potential sufficient to produce a limiting current (the *saturation current*); when this is done, the ions are swept out of the system as rapidly as they are produced by the radiation, and the rate of formation of the ions can be calculated from the saturation current.

The second quantity that is frequently quoted is the *G value*, which is the number of product molecules formed per 100 eV input (1 eV = 1.602×10^{-19} J = 96.47 kJ mol^{-1}). Both the ion-pair yield and the G value give some idea of the chain length.

When a gas is passed through an electric discharge, the stream of electrons usually brings about dissociation. It was first demonstrated in 1920 by the American physicist Robert W. Wood (1868–1955) that hydrogen atoms are conveniently produced in this way.* Later the German physical chemist Karl Friedrich Bonhoeffer (1899–1957) modified Wood's apparatus and made a number of studies of the reactions of hydrogen atoms.† This method of preparing hydrogen atoms is usually known as the *Wood-Bonhoeffer method*.

---

*R. W. Wood, *Proc. Roy. Soc., A97*, 455 (1920); *A102*, 1 (1922); *Phil. Mag., 42*, 729 (1921); *44*, 538 (1922).

†K. F. Bonhoeffer, *Z. Phys. Chem., 113*, 119, 492 (1924); *116*, 391 (1925); *Z. Electrochem. 31*, 521 (1925); *Engeb. Exact. Naturw., 6*, 201 (1927).

10.8 Explosions

Explosions are chemical reactions that occur extremely rapidly with the release of a considerable amount of energy. The release of this energy causes a very rapid rise in pressure in the gaseous products of the reaction.

The reaction between hydrogen and oxygen in the gas phase is a typical explosive reaction. At ordinary temperatures, in the absence of catalysts, the reaction between hydrogen and oxygen is extremely slow (the half-life at 25°C is longer than the age of the solar system). If, however, a catalyst such as platinum is introduced into the system, the mixture explodes. Explosion also results if a lighted match is applied to the system. Catalysts such as platinum produce hydrogen atoms from hydrogen molecules (Section 17.4), and these atoms initiate a rapid chain reaction in the mixture. Similarly, the flame of a lighted match is a good source of free radicals.

Gas-phase reactions that occur explosively occur by mechanisms that have a special feature. Some of the free radicals involved in the mechanism are capable of undergoing a **branching reaction,** in which one radical produces more than one radical. For example, a hydrogen atom can react with O_2 to form OH and O,

$$H + O_2 \longrightarrow HO + O$$

so that two radicals have been formed from one. Another example of a branching reaction is

$$O + H_2 \longrightarrow OH + H$$

When reactions of this type occur to a significant extent, the total number of free radicals in the system may increase rapidly, so that steady-state conditions no longer hold. The reaction therefore occurs with very high velocity and, since energy is released, the result is an explosion.

10.9 Catalysis*

The rates of many reactions are influenced by the presence of a substance that remains unchanged at the end of the process. Examples are the conversion of starch into sugars, the rate of which is influenced by acids; the decomposition of hydrogen peroxide, influenced by ferric ions; and the formation of ammonia in the presence of spongy platinum. In 1836 such reactions were classified by the Swedish chemist Jons Jakob Berzelius (1779–1848) under the title of *catalyzed processes.* A substance that decreases the rate of reaction is referred to as an **inhibitor.** It is convenient to classify catalyzed reactions according to whether they occur homogeneously (in a single phase) or heterogeneously (at an interface between two phases). The present section is concerned mainly with homogeneous catalysis; heterogeneous catalysis is considered in Section 17.4.

Various definitions of catalysis have been proposed. An early definition, suggested in 1895 by Friedrich Wilhelm Ostwald, was that a **catalyst** is "any substance

*This word comes from the Greek *kata,* wholly; *lyein,* to loosen.

that alters the velocity of a chemical reaction without modification of the energy factors of the reaction." Another definition is that "a catalyst alters the velocity of a chemical reaction and is both a reactant and a product of the reaction." These definitions were intended to exclude from the category of catalysts substances that accelerated the rate of a reaction by entering into reaction, thus disturbing the position of equilibrium; such substances are reactants in the ordinary sense. Perhaps the most satisfactory definition of a catalyst is that it is *a substance that increases the rate of a reaction without modifying the overall standard Gibbs energy change in the reaction*. In these definitions of catalysis there is no reference to the fact that a small amount of a catalyst has a large effect on the rate; this is frequently the case, but it is not an essential characteristic of a catalyst.

Although by definition the amount of a catalyst must be unchanged at the end of the reaction, the catalyst is invariably involved in the chemical process. In the case of a single reacting substance, a complex may be formed between this reactant (the substrate) and the catalyst. If there is more than one substrate, the complex may involve one or more molecules of substrate combined with the catalyst. These complexes are formed only as intermediates and decompose to give the products of the reaction, with the regeneration of the catalyst molecule. For example, when a reaction is catalyzed by hydrogen ions, an intermediate complex, involving the substrate and a hydrogen ion, is formed, and this later reacts further with the liberation of the ion and the formation of the products of the reaction.

The catalyst is unchanged at the end of the reaction and, therefore, gives no energy to the system; according to thermodynamics it can have no influence on the position of equilibrium. It follows that since the equilibrium constant K_c is, at equilibrium, the ratio of the rate constants in the forward and reverse directions (i.e., $K_c = k_1/k_{-1}$), a catalyst must influence the forward and reverse rates in the same proportion. This conclusion has been verified experimentally in a number of instances.

An extremely small amount of a catalyst frequently causes a considerable increase in the rate of a reaction. Colloidal palladium at a concentration of 10^{-8} mol dm^{-3} has a significant catalytic effect on the decomposition of hydrogen peroxide. The effectiveness of a catalyst is sometimes expressed in terms of its *turnover number*, which is the number of molecules of substrate decomposed per minute by one molecule of the catalyst. For example, the enzyme catalase has, under certain conditions, a turnover number of 5 million for the decomposition of hydrogen peroxide ($2H_2O_2 \rightarrow 2H_2O + O_2$). Since the turnover number generally varies with the temperature and with the concentration of substrate, it is not a particularly useful quantity in kinetic work; the effectiveness of a catalyst is best measured in terms of a rate coefficient.

The rate of a catalyzed reaction is often proportional to the concentration of the catalyst,

$$v = k[\mathrm{C}] \tag{10.61}$$

where [C] represents the concentration of the catalyst, and k is a function of the concentration of the substrate. If Eq. 10.61 were obeyed exactly, the rate of reaction in the absence of the catalyst would be zero. Many examples are known for which it is necessary to introduce an additional term that is independent of the catalyst concentration,

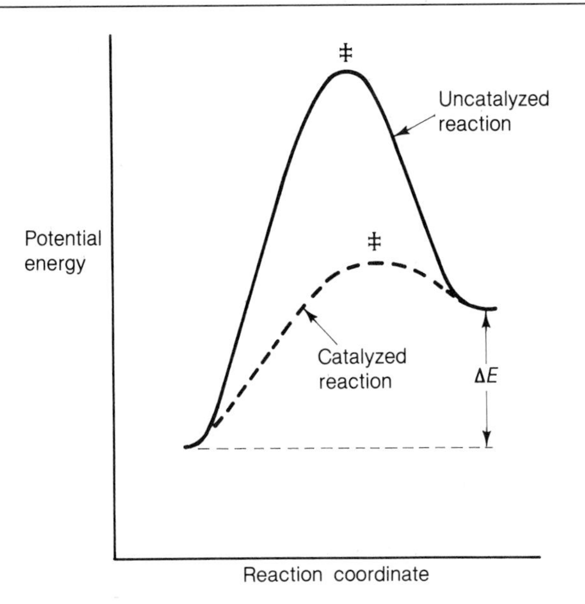

FIGURE 10.3 The lowering of the energy barrier brought about by a catalyst. Since the catalyst is not used up during the reaction, ΔE is the same for the catalyzed reaction as for the uncatalyzed, but the activation energies in both directions are lower for the catalyzed reaction.

$$v = k[\mathrm{C}] + v_0 \tag{10.62}$$

At zero concentration of the catalyst the reaction occurs with the velocity v_0.

It is usually found that the activation energy of a catalyzed reaction is lower than that of the same reaction when it is uncatalyzed. In other words, catalysts generally work by permitting the reaction to occur by another reaction path that has a lower energy barrier. This is shown schematically in Figure 10.3. It is important to note that inhibitors do *not* work by introducing a higher reaction path; this would not reduce the rate, since the reaction would continue to occur by the alternative mechanism. Inhibitors act either by destroying catalysts already present or by removing reaction intermediates such as free radicals.

Acid-Base Catalysis

The study of catalysis by acids and bases played a very important part in the development of chemical kinetics, since many of the reactions studied in the early days of the subject were of this type. The early investigations of the kinetics of reactions catalyzed by acids and bases were carried out at the same time that the electrolytic dissociation theory was being developed, and the kinetic studies contributed considerably to the development of that theory. The reactions considered from this point of view were chiefly the inversion of cane sugar and the hydrolysis of esters. It was first realized in 1884 by Ostwald and later by Arrhenius that the ability of an acid to catalyze these reactions is independent of the nature of the anion but is approximately proportional to its electrical conductivity, which is a measure of an acid's strength. It was originally assumed that the hydrogen ions were the sole effective acid catalysts. Similarly, in catalysis by alkalis the rate is proportional to the concentration of the alkali but independent of the nature of the cation, suggesting that the active species is the hydroxide ion.

The idea that the only catalyzing species in reactions of this type are the hydrogen and hydroxide ions has been found to require modification in a number of instances. Many reactions do exist, however, for which only these two ions are effective catalysts; and we then say that the catalysis is **specific.**

If such reactions are carried out in a sufficiently strong acid solution, the concentration of hydroxide ions may be reduced to such an extent that these ions do not have any appreciable catalytic action. The hydrogen ions are then the only effective catalysts, and the rate (at least at concentrations of catalyst and substrate that are not too high) is given by an expression of the type

$$v = k_{H^+}[H^+][S] \tag{10.63}$$

where k_{H^+} is the rate constant for the hydrogen-ion-catalyzed reaction. Such a reaction would be overall of the second order with respect to concentration. If catalysis is effected simultaneously by hydrogen and hydroxide ions, and reaction may also occur spontaneously (i.e., without a catalyst), the rate is

$$v = k_0[S] + k_{H^+}[H^+][S] + k_{OH^-}[OH^-][S] \tag{10.64}$$

The first-order rate coefficient is therefore given by

$$k = k_0 + k_{H^+}[H^+] + k_{OH^-}[OH^-] \tag{10.65}$$

In these equations, k_0 is the rate constant of the spontaneous reaction, and k_{H^+} and k_{OH^-} are known as the **catalytic constants** for H^+ and OH^-, respectively.

The rate equation may be expressed in a different form by making use of the fact that $[H^+][OH^-] = K_w$, where K_w is the ionic product of water. Elimination of $[OH^-]$ gives

$$k = k_0 + k_{H^+}[H^+] + \frac{k_{OH^-}K_w}{[H^+]} \tag{10.66}$$

while elimination of $[H^+]$ gives

$$k = k_0 + \frac{k_{H^+}K_w}{[OH^-]} + k_{OH^-}[OH^-] \tag{10.67}$$

In many cases one of these terms containing concentration is negligibly small compared with the other. If work is carried out with 0.1 *M* hydrochloric acid, for example, the second term in Eq. 10.66 is $k_{H^+} \times 10^{-1}$ while the third term is $k_{OH^-} \times 10^{-13}$ (since $K_w = 10^{-14}$); consequently, unless k_{OH^-} is at least 10^9 greater than k_{H^+}, the third term will be negligible compared with the second; at this acid concentration, therefore, catalysis by hydroxide ions will be negligible compared with that by hydrogen ions. Similarly, in 0.1 *M* sodium hydroxide solution catalysis by hydrogen ions will usually be unimportant compared with that by hydroxide ions. In general, there will be an upper range of hydrogen-ion concentrations at which catalysis by hydroxide ions will be unimportant and a lower range at which catalysis by hydroxide ions will predominate and catalysis by hydrogen ions will be unimportant. Within each of these ranges the rate will be a linear function of $[H^+]$ and of $[OH^-]$, respectively. In the upper range the value of the catalytic constant k_{H^+} can readily be determined from the experimental data; in the lower range k_{OH^-} can be

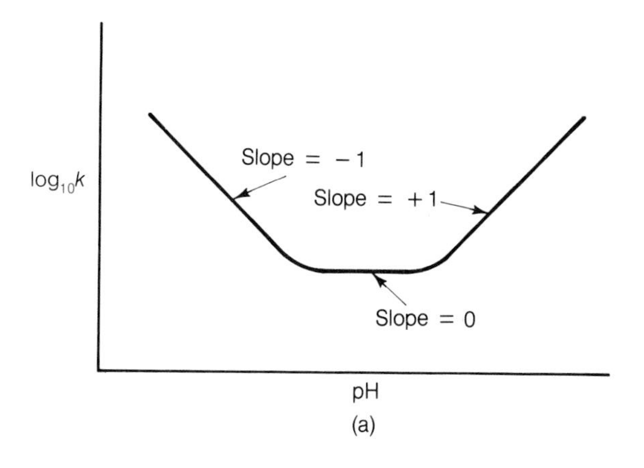

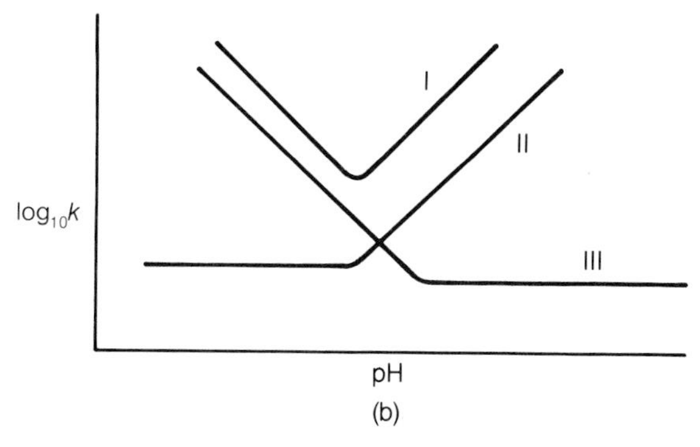

FIGURE 10.4 (a) A schematic plot of $\log_{10}k$ against pH for a reaction that is catalyzed by both hydrogen and hydroxide ions and for which the uncatalyzed reaction occurs at an appreciable rate. (b) Variants of the plot shown in (a).

so determined. The constants for the hydrolysis of ethyl acetate were measured in this manner by J. J. A. Wijs* in 1893, and he also obtained a value for K_w by making use of the fact that the velocity is a minimum when the second and third terms in Eq. 10.66 are equal; this gives rise to

$$[H^+]_{min} = \left(\frac{k_{OH^-}K_w}{k_{H^+}}\right)^{1/2} \tag{10.68}$$

Thus, from the values of $[H^+]_{min}$, k_{H^+}, and k_{OH^-}, the ionic product K_w can be obtained.

A plot of the logarithm of the rate constant against the pH of the solution is shown in Figure 10.4a. There are regions of catalysis by hydrogen and hydroxide ions, separated by a region in which the amount of catalysis is unimportant in comparison with the spontaneous reaction. When the catalysis is largely by hydrogen ions, $k = k_{H^+}[H^+]$, so that

$$\log k = \log k_{H^+} + \log[H^+] \tag{10.69}$$

*J. J. A. Wijs, *Z. Physik. Chem.*, *11*, 492 (1893); *12*, 415 (1893).

or

$$\log k = \log k_{H^+} - \text{pH} \tag{10.70}$$

The slope is therefore −1, which is the slope of the left-hand limb. The slope of the right-hand limb is similarly +1. The velocity in the intermediate region is equal to $k_0[S]$, and k_0 can thus be determined directly from the rate in this region. If the rate of the spontaneous reaction is sufficiently small, the horizontal part of the curve is not found, and the two limbs intersect fairly sharply (Figure 10.4b, curve I). If either k_{H^+} or k_{OH^-} is negligibly small, the corresponding sloping limb of the curve is not found (Figure 10.4b, curves II and III).

The evidence is that acid-base catalysis involves the transfer of a proton to or from the substrate molecule; it is therefore to be expected that catalysis may be effected by acids and bases other than H^+ and OH^-. This has been found in a number of instances, and **general** acid-base catalysis is then said to occur. General acid-base catalysis is to be contrasted with specific acid-base catalysis, in which one can only detect catalysis by H^+ and OH^- ions.

An example of acid-base catalysis is the reaction between acetone and iodine in aqueous solution:

$$CH_3COCH_3 + I_2 \longrightarrow CH_3COCH_2I + HI$$

The rate of this reaction is linear in the acetone concentration and in any acid species present in solution, but it is independent of the concentration of iodine; indeed, if the iodine is replaced by bromine, the corresponding bromination reaction proceeds at the same rate. This suggests that the iodine or bromine is involved in a rapid step that has no effect on the overall reaction rate. The evidence is that the rate-determining step is the conversion of the ordinary keto form of acetone into its enol form:

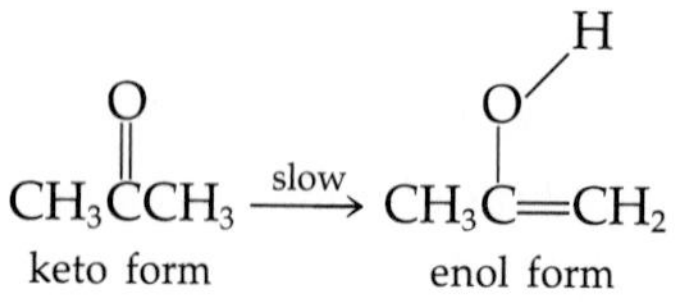

$$\underset{\text{keto form}}{CH_3\overset{\overset{\displaystyle O}{\|}}{C}CH_3} \xrightarrow{\text{slow}} \underset{\text{enol form}}{CH_3\overset{\overset{\displaystyle O-H}{|}}{C}{=}CH_2}$$

The enol form is then rapidly iodinated or brominated:

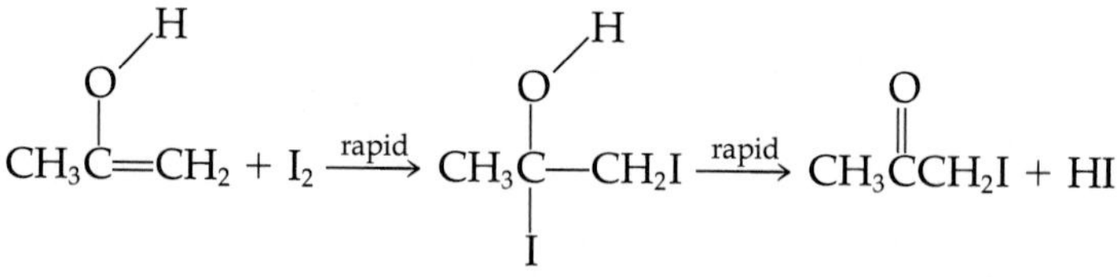

$$CH_3\overset{\overset{\displaystyle O-H}{|}}{C}{=}CH_2 + I_2 \xrightarrow{\text{rapid}} CH_3\underset{\underset{\displaystyle I}{|}}{\overset{\overset{\displaystyle O-H}{|}}{C}}{-}CH_2I \xrightarrow{\text{rapid}} CH_3\overset{\overset{\displaystyle O}{\|}}{C}CH_2I + HI$$

The way in which acids catalyze the conversion of the keto form into the enol form is as follows. First, the acidic species HA transfers a proton to the oxygen atom on the acetone molecule:

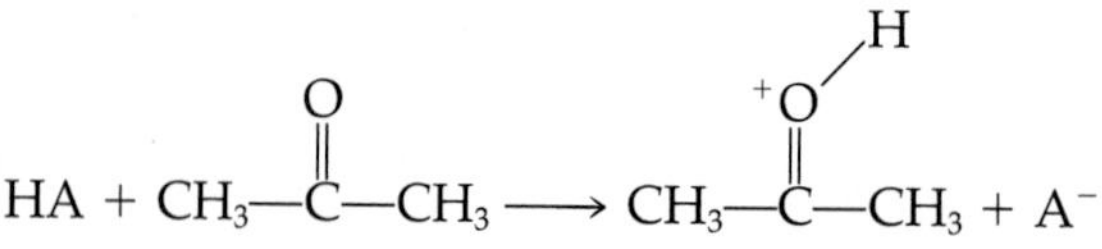

$$HA + CH_3{-}\overset{\overset{\displaystyle O}{\|}}{C}{-}CH_3 \longrightarrow CH_3{-}\overset{\overset{\displaystyle ^{+}O-H}{\|}}{C}{-}CH_3 + A^-$$

The transferred proton is bound to the oxygen atom by one of the oxygen's lone pairs of electrons. The protonated acetone then gives up one of its other hydrogen atoms to some base B present in solution (which may be water), at the same time forming the enol form of acetone:

$$\mathrm{CH_3{-}\overset{\overset{\displaystyle ^{+}O{-}H}{\|}}{C}{-}CH_3 + B \longrightarrow CH_3{-}\overset{\overset{\displaystyle O{-}H}{|}}{C}{=}CH_2 + BH^+}$$

It has been demonstrated that this process is catalyzed not only by H^+ ions but by other species present in solution (i.e., there is *general* acid catalysis). There is also general basic catalysis of this reaction.

Brønsted Relationships

Since acid-base catalysis always involves the transfer of a proton from an acid catalyst or to a basic catalyst, it is natural to seek a correlation between the effectiveness of a catalyst and its strength as an acid or base; this strength is a measure of the ease with which the catalyst transfers a proton to or from a water molecule. The most satisfactory relationship between the rate constant k_a and the acid dissociation constant K_a of a monobasic acid is the equation

$$k_a = G_a K_a^\alpha \tag{10.71}$$

which was proposed in 1924 by J. N. Brønsted and K. J. Pederson.* G_a and α are constants, and the latter is less than unity. The analogous equation for basic catalysis is

$$k_b = G_b K_b^\beta \tag{10.72}$$

Similarly, the relationship between the catalytic constant of a base and the acid strength of the conjugate acid may be expressed as

$$k_b = G_b'\left(\frac{1}{K_a}\right)^\beta \tag{10.73}$$

β is again always less than unity. These equations are commonly spoken of as *Brønsted relationships*.

The equations require a modification if they are to be applied to an acid that has more than one ionizable proton or to a base that can accept more than one proton. The conclusions may be generalized by means of the following relationships, given by Brønsted:†

$$\frac{k_a}{p} = G_a\left(\frac{qK_a}{p}\right)^\alpha \tag{10.74}$$

*J. N. Brønsted and K. J. Pederson, *Z. Physik. Chem.*, *108*, 185 (1924).
†J. N. Brønsted, *Chem. Rev.*, *5*, 322 (1928).

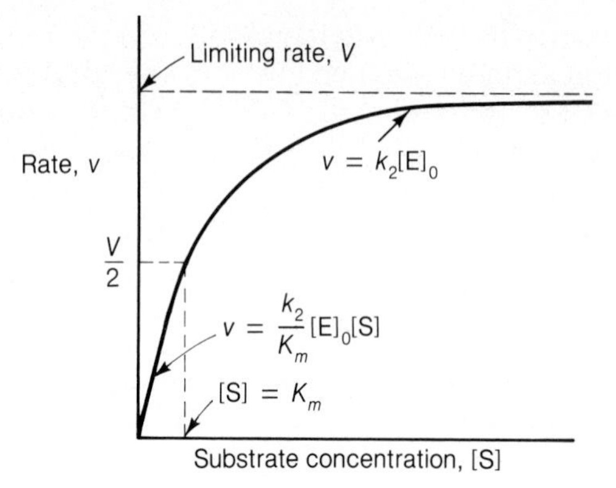

FIGURE 10.5 The variation of rate with substrate concentration for an enzyme-catalyzed reaction obeying the Michaelis-Menten equation.

and

$$\frac{k_b}{q} = G_b\left(\frac{p}{qK_a}\right)^{\beta} \tag{10.75}$$

In Eq. 10.74, p is the number of dissociable protons bound equally strongly in the acid, whereas q is the number of equivalent positions in the conjugate base to which a proton may be attached. Similarly, in Eq. 10.75, q is the number of positions in the catalyzing base to which a proton may be attached, and p is the number of equivalent dissociable protons in the conjugate acid. Very satisfactory agreement has been obtained in all the cases to which these equations have been applied.

The Brønsted relationships are special cases of linear Gibbs energy relationships (see Problem 19).

Enzyme Catalysis

The enzymes, which are proteins, are the biological catalysts. Their action shows some resemblance to the catalytic action of acids and bases but is considerably more complicated. The details of the mechanisms of action of enzymes are still being worked out, and much research remains to be done. Only a brief account can be given here.

The simplest case is that of an enzyme-catalyzed reaction where there is a single substrate; an example is the hydrolysis of an ester. The dependence on substrate concentration in such cases is frequently as shown in Figure 10.5. The rate varies linearly with the substrate concentration at low concentrations (first-order kinetics) and becomes independent of substrate concentration (zero-order kinetics) at high concentrations. This type of behavior was first explained in 1913 by the German-American chemist Leonor Michaelis (1875–1949) and his Canadian colleague Maud L. Menten (1879–1960) in terms of the mechanism

$$\mathrm{E + S \underset{k_{-1}}{\overset{k_1}{\rightleftharpoons}} ES}$$

$$\mathrm{ES \xrightarrow{k_2} E + Z}$$

Here E and S are the enzyme and substrate, Z is the product, and ES is an addition complex.

Usually the concentration of the substrate is much greater than that of the enzyme, and it is then permissible to apply the steady-state treatment in order to obtain the rate equation. The reason that this can be done is that under these conditions the concentration of the enzyme-substrate complex ES must be very much less than that of the substrate; the rate of change of its concentration is therefore much less than that of the substrate. Thus

$$\frac{d[\mathrm{ES}]}{dt} = k_1[\mathrm{E}][\mathrm{S}] - k_{-1}[\mathrm{ES}] - k_2[\mathrm{ES}] = 0 \tag{10.76}$$

The concentration [E] that appears in this equation is the concentration of the *free* enzyme, and it may be very much less than the *total* concentration $[\mathrm{E}]_0$ of enzyme, since much of the enzyme may be in the form of ES. The total concentration is given by

$$[\mathrm{E}]_0 = [\mathrm{E}] + [\mathrm{ES}] \tag{10.77}$$

Elimination of [E] between Eqs. 10.76 and 10.77 gives

$$k_1([\mathrm{E}]_0 - [\mathrm{ES}])[\mathrm{S}] - (k_{-1} + k_2)[\mathrm{ES}] = 0 \tag{10.78}$$

and therefore

$$[\mathrm{ES}] = \frac{k_1[\mathrm{E}]_0[\mathrm{S}]}{k_{-1} + k_2 + k_1[\mathrm{S}]} \tag{10.79}$$

The rate is

$$v = k_2[\mathrm{S}] = \frac{k_1k_2[\mathrm{E}]_0[\mathrm{S}]}{k_{-1} + k_2 + k_1[\mathrm{S}]} \tag{10.80}$$

$$= \frac{k_2[\mathrm{E}]_0[\mathrm{S}]}{[(k_{-1} + k_2)/k_1] + [\mathrm{S}]} \tag{10.81}$$

This equation is conveniently written as

$$v = \frac{V[\mathrm{S}]}{K_m + [\mathrm{S}]} \tag{10.82}$$

where $V = k_2[\mathrm{E}]_0$ and K_m, known as the **Michaelis constant,** is equal to $(k_{-1} + k_2)/k_1$.

Two limiting cases of Eq. 10.82 are of interest. If $[\mathrm{S}] >> K_m$, the rate becomes V, which is therefore called the **limiting rate.** If $[\mathrm{S}] << K_m$,

$$v = \frac{V[\mathrm{S}]}{K_m} = \frac{k_2}{K_m}[\mathrm{E}]_0[\mathrm{S}] \tag{10.83}$$

and the kinetics are therefore first order in substrate. Also, when $[\mathrm{S}] = K_m$, Eq. 10.82 becomes

$$v = \frac{V[\mathrm{S}]}{[\mathrm{S}] + [\mathrm{S}]} = \frac{V}{2} \tag{10.84}$$

These relationships are illustrated in Figure 10.5.

It is important to note that adherence to the Michaelis-Menten Eq. 10.82 does not necessarily mean that the simple Michaelis-Menten mechanism applies. The mechanism

$$E + S \underset{k_{-1}}{\overset{k_1}{\rightleftharpoons}} ES \xrightarrow{k_2} ES' \xrightarrow{k_3} E + Z$$
$$\qquad\qquad\qquad\qquad \searrow Y$$

also gives the same type of rate equation (see Problem 29); in fact there can be any number of intermediates ES, ES′, ES″, etc., and the same dependence of v on [S] is obtained. In view of this, if the Michaelis equation applies but the mechanism is unknown, it is usual to write the equation either as Eq. 10.82 or as

$$v = \frac{k_c[E]_0[S]}{K_m + [S]} \tag{10.85}$$

In this more general equation, k_2 in Eq. 10.81 is replaced by k_c, which is known as the *catalytic constant*.

Similar treatments have been given for enzyme reactions in which two or more substrates are involved (see Problem 27) and for the effects of inhibitors (Problem 28), pH, and temperature. A remarkable feature of enzymes is that they are exceedingly effective as catalysts—much more so than other types of catalysts. Their catalytic action usually, but not always, arises from the fact that they reduce the activation energy very substantially.

10.10 Reactions in Solution

Most of the general principles we have discussed in this chapter apply equally well to gas reactions and to solution reactions. In Section 9.10 we have considered some aspects of solvent effects, and in the present section we consider some additional matters.

One important difference between gas-phase and solution reactions is that solvents frequently favor mechanisms involving ions. We have seen that ordinary gas-phase reactions rarely involve ionic intermediates and usually occur by mechanisms in which atoms and free radicals are intermediates. Exceptions are reactions induced by high-energy radiation (Section 10.7), where the energy required to produce ions is provided by the radiation. For reactions in solution, on the other hand, ionic intermediates are much more common, since the ions may be stabilized by the solvent.

Collisions and Encounters

In Section 9.10 we noted that for a number of reactions the rates and the activation parameters (activation energy and preexponential factor) are sometimes the same in the gas phase and in a variety of solvents. This tends to be true if ionic effects are not involved. The similarity between preexponential factors for gas-phase and solution reactions suggests that the collision frequencies are much the same.

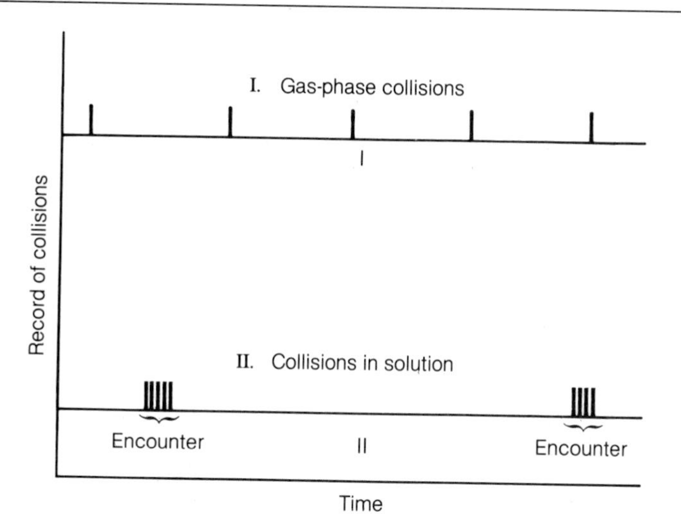

FIGURE 10.6 The distribution of collisions between solute molecules, as shown by the demonstration of Rabinowitch and Wood. Curve I shows schematically the gas-phase collisions; curve II shows the distribution of collisions when a solvent is present.

Further investigation of this problem reveals that there are important differences between collisions in solution and collisions in the gas phase and that these differences have important implications for certain composite mechanisms. In a gas the molecules are relatively far apart and move freely between collisions; after two molecules A and B collide, they separate and may not meet again for some time. In solution, on the other hand, when two solute molecules come together, they are "caged" in by surrounding solvent molecules, which hold the solute molecules together and cause them to collide a number of times before they finally separate. This effect was demonstrated in a very interesting way by Eugene Rabinowitch and W. C. Wood,* who employed a tray on which spheres were allowed to roll. Agitation of the tray caused the spheres to move around, and an electrical device was employed to record collisions between a given pair of spheres. When very few spheres were present on the tray, the collisions between a given pair occurred individually, with relatively long intervals between successive collisions (see Figure 10.6, curve I). This represents the behavior in the gas phase. The effect of solvents was studied by adding additional spheres to the tray, so that there was fairly close packing. It was then found that the distribution of collisions between a given pair was quite different. After one collision occurred, the pair tended to stay together and undergo a succession of collisions, as shown in curve II of Figure 10.6. A group of such collisions is known as an **encounter.** The effect of solvents of caging in the solute molecules is known as the **cage effect** or as the **Franck-Rabinowitch** effect.†

The important result obtained in this demonstration of Rabinowitch and Wood is that although the *distribution* of collisions is quite different in the two cases, the average *frequency* of collisions is much the same. When the tray was packed with

*E. Rabinowitch and W. C. Wood, *Trans. Faraday Soc.*, *32*, 1381 (1936); E. Rabinowitch, *ibid.*, *33*, 1225 (1937).

†J. Franck and E. Rabinowitch, *Trans. Faraday Soc.*, *30*, 120 (1934).

spheres, so as to represent collisions in solution, the collisions occurred in groups within an encounter, but the time between successive encounters was correspondingly long, so that the frequency of collisions was very similar.

For ordinary chemical reactions, involving an activation energy, the distribution of collisions is irrelevant; reaction may occur at any collision within the encounter, so that the rate is controlled by the frequency of collisions and not by their distribution.

On the other hand, reactions involved in composite mechanisms frequently have zero activation energies and then the distribution of collisions becomes important. Free-radical combinations, for example, do not require an activation energy, and the process will therefore occur at the first collision within the encounter; the remaining collisions within the encounter are of no significance. The pre-exponential factor of such reactions therefore depends on the frequency of *encounters* and not on the frequency of collisions.

A related example is provided by certain photochemical reactions, where a pair of radicals may be produced initially. In the gas phase they will separate rapidly, but in solution they will be caged in by the surrounding solvent molecules and may combine with one another before they can separate. This effect is known as *primary combination*, as opposed to *secondary combination* which occurs after the free radicals have separated from each other.

Problems

A Composite Mechanisms and Rate Equations

1. A reaction of stoichiometry A + 2B = Y + Z occurs by the mechanism

$$A + B \xrightarrow{k_1} X \qquad \text{(very slow)}$$

$$X + B \xrightarrow{k_2} Y + Z \qquad \text{(very fast)}$$

where X is an intermediate. Write the expression for the rate of formation of Y.

2. A reaction A + 2B = 2Y + 2Z occurs according to the mechanism

$$A \underset{k_{-1}}{\overset{k_1}{\rightleftharpoons}} 2X \qquad \text{(very rapid equilibrium)}$$

$$X + B \xrightarrow{k_2} Y + Z \qquad \text{(slow)}$$

Obtain an expression for the rate of formation of the product Y.

3. A reaction A + B = Y + Z occurs according to the mechanism

$$A \underset{k_{-1}}{\overset{k_1}{\rightleftharpoons}} X$$

$$X + B \xrightarrow{k_2} Z$$

Apply the steady-state treatment and obtain an expression for the rate. To what expressions does the general rate equation reduce if

a. The second reaction is slow, the initial equilibrium being established very rapidly?

b. The second reaction is very rapid compared with the first reaction in either direction?

4. A reaction of stoichiometry

$$A + B \longrightarrow Y + Z$$

is found to be second order in A and zero order in B. Suggest a mechanism that is consistent with this behavior.

***5.** The rate of formation of the product of a reaction is found to give a nonlinear Arrhenius plot, the line being convex to the $1/T$ axis (i.e., the activation energy is higher at higher temperatures). Suggest a reason for this type of behavior.

***6.** An Arrhenius plot is concave to the $1/T$ axis (i.e., a lower activation energy at higher temperatures). Suggest a reason for this type of behavior.

*7. Nitrogen pentoxide reacts with nitric oxide in the gas phase according to the stoichiometric equation

$$N_2O_5 + NO \longrightarrow 3NO_2$$

The following mechanism has been proposed:

$$N_2O_5 \xrightarrow{k_1} NO_2 + NO_3$$
$$NO_2 + NO_3 \xrightarrow{k_{-1}} N_2O_5$$
$$NO + NO_3 \xrightarrow{k_2} 2NO_2$$

Assuming that the steady-state treatment can be applied to NO_3, derive an equation for the rate of consumption of N_2O_5.

*8. The reaction $2NO + O_2 \longrightarrow 2NO_2$ is believed to occur by the mechanism

$$2NO \xrightarrow{k_1} N_2O_2$$
$$N_2O_2 \xrightarrow{k_{-1}} 2NO$$
$$N_2O_2 + O_2 \xrightarrow{k_2} 2NO_2$$

Assume N_2O_2 to be in a steady state and derive the rate equation. Under what conditions does the rate equation reduce to second-order kinetics in NO and first-order kinetics in O_2?

*9. The gas-phase reaction

$$Cl_2 + CH_4 \longrightarrow CH_3Cl + HCl$$

proceeds by a free-radical chain reaction in which the chain propagators are Cl and CH_3 (but not H), and the chain-ending step is $2Cl \rightarrow Cl_2$. Write the mechanism, identify the initiation reaction and the chain-propagating steps, and obtain an expression for the rate of the overall reaction.

10. The following mechanism has been proposed for the thermal decomposition of pure ozone in the gas phase:

$$2O_3 \underset{k_{-1}}{\overset{k_1}{\rightleftharpoons}} O_3 + O_2 + O$$

$$O + O_3 \xrightarrow{k_2} 2O_2$$

Derive the rate equation.

B Photochemistry and Radiation Chemistry

11. Calculate the maximum wavelength of the radiation that will bring about dissociation of a diatomic molecule having a dissociation energy of 390.4 kJ mol^{-1}.

*12. Hydrogen iodide undergoes decomposition into $H_2 + I_2$ when irradiated with radiation having a wavelength of 207 nm. It is found that when 1 J of energy is absorbed, 440 μg of HI is decomposed. How many molecules of HI are decomposed by one photon of radiation of this wavelength? Suggest a mechanism that is consistent with this result.

*13. A 100-W mercury-vapor lamp emits radiation of 253.7-nm wavelength and may be assumed to operate with 100% efficiency. If all the light emitted is absorbed by a substance that is decomposed with a quantum yield of unity, how long will it take for 0.01 mol to be decomposed?

*14. Suppose that the radiation emitted by the lamp in Problem 13 is all absorbed by ethylene, which decomposes into $C_2H_2 + H_2$ with a quantum yield of unity. How much acetylene will be produced per hour?

*15. The photochemical reaction between chlorine and chloroform in the gas phase follows the stoichiometric equation

$$CHCl_3 + Cl_2 \longrightarrow CCl_4 + HCl$$

It is believed to occur by the mechanism

$$Cl_2 + h\nu \longrightarrow 2Cl$$
$$Cl + CHCl_3 \xrightarrow{k_2} HCl + CCl_3$$
$$CCl_3 + Cl_2 \xrightarrow{k_3} CCl_4 + Cl$$
$$2Cl \xrightarrow{k_4} Cl_2$$

Assume the rate of the initiation reaction to be $2I_a$, where I_a is the intensity of light absorbed, and obtain an expression for the overall rate in terms of I_a and $[CHCl_3]$.

*16. When water vapor is irradiated with a beam of high-energy electrons, various ions such as H^+ and O^- ions appear. Calculate the minimum energies required for the formation of these ions, given the following thermochemical data:

$$H_2O(g) \longrightarrow H(g) + OH(g) \qquad \Delta H° = 498.7 \text{ kJ mol}^{-1}$$

$$OH(g) \longrightarrow H(g) + O(g) \qquad \Delta H° = 428.2 \text{ kJ mol}^{-1}$$

$$H(g) \longrightarrow H^+(g) + e^-(g) \qquad \Delta H° = 1312.2 \text{ kJ mol}^{-1}$$

$$O(g) + e^-(g) \longrightarrow O^-(g) \qquad \Delta H° = -213.4 \text{ kJ mol}^{-1}$$

Are the results you obtain consistent with the experimental appearance potentials of 19.5 eV for H^+ and 7.5 eV for O^-?

C Catalysis

17. A reaction is specifically catalyzed by hydrogen ions; the rate constant is given by

$$k/\text{dm}^3\ \text{mol}^{-1}\ \text{s}^{-1} = 4.7 \times 10^{-2}([\text{H}^+]/\text{mol dm}^{-3})$$

When the substance was dissolved in a 10^{-3} *M* solution of an acid HA, the rate constant was 3.2×10^{-5} dm^3 mol^{-1} s^{-1}. Calculate the dissociation constant of HA.

***18.** The following is a slightly simplified version of the mechanism proposed in 1937 by G. K. Rollefson and R. F. Faull to explain the iodine-catalyzed decomposition of acetaldehyde:

$$I_2 \xrightarrow{k_1} 2I$$

$$I + CH_3CHO \xrightarrow{k_2} HI + CH_3CO$$

$$CH_3CO \xrightarrow{k_3} CH_3 + CO$$

$$CH_3 + HI \xrightarrow{k_4} CH_4 + I$$

$$2I \xrightarrow{k_{-1}} I_2$$

Apply the steady-state treatment to I, CH_3CO, and CH_3, and obtain an expression for the rate.

***19.** Suppose that a reaction is catalyzed by a series of homologous acids and that the Hammett equation applies:

$$\log_{10} k = \log_{10} k_0 + \sigma\rho$$

where σ is the substituent constant and ρ is the reaction constant. Suppose that the corresponding equation for the dissociation of the acid is

$$\log_{10} K = \log_{10} K_0 + \sigma\rho'$$

where ρ' is the reaction constant for the dissociation; the substituent constants are the same in both equations. Prove that the Brønsted equation

$$k = GK^\alpha$$

applies. How does α relate to the reaction constants ρ and ρ'?

20. The hydrolysis of ethyl acetate catalyzed by hydrochloric acid obeys the rate equation

$$v = k[\text{ester}][\text{HCl}]$$

and the reaction essentially goes to completion. At 25.0°C the rate constant is 2.80×10^{-5} dm^3 mol^{-1} s^{-1}. What is the half-life of the reaction if [ester] = 0.1 *M* and [HCl] = 0.01 *M*?

21. The following mechanism has been proposed for the alkaline hydrolysis of $Co(NH_3)_5Cl^{2+}$:

$$Co(NH_3)_5Cl^{2+} + OH^- \underset{k_{-1}}{\overset{k_1}{\rightleftharpoons}} Co(NH_3)_4(NH_2)Cl^+ + H_2O$$

$$Co(NH_3)_4(NH_2)Cl^+ \xrightarrow{k_2} Co(NH_3)_4(NH_2)^{2+} + Cl^-$$

$$Co(NH_3)_4(NH_2)^{2+} + H_2O \xrightarrow{k_3} Co(NH_3)_5(OH)^{2+}$$

Assume $Co(NH_3)_4(NH_2)Cl^+$ and $Co(NH_3)_4(NH_2)^{2+}$ to be in the steady state and derive an expression for the rate of reaction.

Experimentally the rate is proportional to $[Co(NH)_5Cl^{2+}][OH^-]$; does this fact tell us anything about the relative magnitudes of the rate constants?

D Enzyme-Catalyzed Reactions

22. The following rates have been obtained for an enzyme-catalyzed reaction at various substrate concentrations:

$10^3[S]$ / mol dm^{-3}	Rate, v (arbitrary units)
0.4	2.41
0.6	3.33
1.0	4.78
1.5	6.17
2.0	7.41
3.0	8.70
4.0	9.52
5.0	10.5
10.0	12.5

Plot v against [S], $1/v$ against $1/[S]$, and $v/[S]$ against v, and from each plot estimate the Michaelis constant. Which plot appears to give the most reliable value?

23. The following data have been obtained for the myosin-catalyzed hydrolysis of ATP, at 25.0°C and pH 7.0:

$10^6[\text{ATP}]$ / mol dm^{-3}	$10^6 v$ / mol dm^{-3} s^{-1}
7.5	0.067
12.5	0.095
20.0	0.119
32.5	0.149
62.5	0.185
155.0	0.191
320.0	0.195

Plot v against [S], $1/v$ against $1/[S]$, and $v/[S]$ against v, and from each plot calculate the Michaelis constant K_m and the limiting rate V.

***24.** The following values of V (limiting rate at high substrate concentrations) and K_m have been obtained at various temperatures for the hydrolysis of acetylcholine bromide, catalyzed by acetylcholinesterase:

$\frac{T}{°C}$	$\frac{10^6 V}{\text{mol dm}^{-3}\text{ s}^{-1}}$	$\frac{K_m \times 10^4}{\text{mol dm}^{-3}}$
20.0	1.84	4.03
25.0	1.93	3.75
30.0	2.04	3.35
35.0	2.17	3.05

a. Assuming the enzyme concentration to be 1.00×10^{-11} mol dm^{-3}, calculate the energy of activation, the enthalpy of activation, the Gibbs energy of activation, and the entropy of activation for the breakdown of the enzyme-substrate complex at 25.0°C.

b. Assuming K_m to be the *dissociation* constant k_{-1}/k_1 for the enzyme-substrate complex ($\text{ES} \underset{k_1}{\overset{k_{-1}}{\rightleftharpoons}} \text{E} + \text{S}$), determine the following thermodynamic quantities for the *formation* of the enzyme-substrate complex at 25.0°C: $\Delta G°$, $\Delta H°$, $\Delta S°$.

c. From the results obtained in (a) and (b), sketch a Gibbs energy diagram and an enthalpy diagram for the reaction.

25. The following data relate to an enzyme reaction:

$\frac{10^3[\text{S}]}{\text{mol dm}^{-3}}$	$\frac{10^5 V}{\text{mol dm}^{-3}\text{ s}^{-1}}$
2.0	13
4.0	20
8.0	29
12.0	33
16.0	36
20.0	38

The concentration of the enzyme is 2.0 g dm^{-3}, and its relative molecular mass is 50 000. Calculate K_m, the maximum rate V, and k_c.

26. The following data have been obtained for the myosin-catalyzed hydrolysis of ATP:

$\frac{\text{Temperature}}{°C}$	$\frac{k_c \times 10^6}{\text{s}^{-1}}$
39.9	4.67
43.8	7.22
47.1	10.0
50.2	13.9

Calculate, at 40.0°C, the energy of activation, the enthalpy of activation, the Gibbs energy of activation, and the entropy of activation.

***27.** The following mechanism has been proposed by H. Theorell and B. Chance for certain enzyme reactions involving two substrates A and B:

$$\text{E} + \text{A} \underset{k_{-1}}{\overset{k_1}{\rightleftharpoons}} \text{EA}$$

$$\text{EA} + \text{B} \xrightarrow{k_2} \text{EZ} + \text{Y}$$

$$\text{EZ} \xrightarrow{k_3} \text{E} + \text{Z}$$

Assume that the substrates A and B are in excess of E so that the steady-state treatment can be applied to EA and EZ, and obtain an expression for the rate.

***28.** When an inhibitor I is added to a single substrate enzyme system, the mechanism is sometimes

$$\begin{array}{ccc} & \text{S} & \\ & + & \\ \text{I} + & \text{E} & \underset{k_{-i}}{\overset{k_i}{\rightleftharpoons}} \text{EI} \\ & k_{-1} \updownarrow k_1 & \\ & \text{ES} & \\ & \downarrow k_2 & \\ & \text{E} + \text{Y} & \end{array}$$

This is known as a *competitive* mechanism, since S and I compete for sites on the enzyme.

a. Assume that the substrate and inhibitor are present in great excess of the enzyme, apply the steady-state treatment, and obtain the rate equation.

b. Obtain an expression for the degree of inhibition ϵ, defined as

$$\epsilon = \frac{v_0 - v}{v_0}$$

where v is the rate in the presence of inhibitor and v_0 is the rate in its absence.

***29.** Obtain the rate equation corresponding to the mechanism

$$\text{E} + \text{S} \underset{k_{-1}}{\overset{k_1}{\rightleftharpoons}} \text{ES} \xrightarrow{k_2} \text{ES}' \xrightarrow{k_3} \text{E} + \text{Z}$$
$$\qquad\qquad\qquad\quad \searrow \text{Y}$$

assuming ES and ES′ to be in the steady state and the substrate concentration to be much higher than the enzyme concentration. Express the catalytic constant k_c and the Michaelis constant K_m in terms of k_1, k_{-1}, k_2, and k_3.

E Supplementary Problems

30. A reaction occurs by the mechanism

$$\mathrm{A + B \underset{k_{-1}}{\overset{k_1}{\rightleftharpoons}} X \xrightarrow{k_2} Z}$$

and the concentration of X is sufficiently small compared with the concentrations of A and B that the steady-state treatment applies. Prove that the activation energy E at any temperature is given by

$$E = \frac{k_{-1}(E_1 + E_2 - E_{-1}) + k_2 E_1}{k_{-1} + k_2}$$

that is, is the weighted mean of the values $E_1 + E_2 - E_{-1}$, and E_1, which apply, respectively, to the limiting cases of $k_{-1} >> k_2$ and $k_2 >> k_{-1}$.

31. Enzyme-catalyzed reactions frequently follow an equation of the form of Eq. 10.85. Suppose that k_c and K_m show the following temperature dependence:

$$k_c = A_c e^{-E_c/RT} \quad \text{and} \quad K_m = Be^{-\Delta H_m/RT}$$

where A_c, B, E_c, and ΔH_m are temperature-independent parameters. Explain under what conditions, with [S] held constant, the rate may pass through a maximum as the temperature is raised.

32. F. A. Lindemann, *Trans. Faraday Soc.*, *17*, 598 (1922), proposed the following mechanism for a unimolecular gas reaction

$$\mathrm{A + A \underset{k_{-1}}{\overset{k_1}{\rightleftharpoons}} A^* + A}$$

$$\mathrm{A^* \xrightarrow{k_2} Y + Z}$$

The species A* is an energized molecule that is present in low concentrations. Apply the steady-state treatment to A* and obtain an expression for the rate in terms of [A], k_1, k_{-1}, and k_2. Show that the mechanism predicts first-order kinetics at higher A concentrations and second-order kinetics at lower ones.

33. Some enzyme reactions involving two substrates A and B occur by the following mechanism:

$$\mathrm{E + A \underset{k_{-1}}{\overset{k_1}{\rightleftharpoons}} EA}$$

$$\mathrm{EA + B \underset{k_{-2}}{\overset{k_2}{\rightleftharpoons}} EAB \xrightarrow{k_3} E + Y + Z}$$

(This is known as the *ordered ternary-complex mechanism;* A must add first to E, and the resulting complex EA reacts with B; the complex EB is not formed.) The concentrations of A and B are much greater than the concentration of E. Apply the steady-state treatment and obtain an expression for the rate.

34. Certain polymerizations involve esterification reactions between —COOH groups on one molecule and —OH groups on another. Suppose that the concentration of such functional groups is c and that the rate of their removal by esterification obeys the equation

$$-\frac{dc}{dt} = kc^2$$

Obtain an equation relating the time t to the fraction f of functional groups remaining and to the initial concentration c_0 of functional groups.

F Essay Questions

1. Explain the essential features of a chain reaction.

2. Give an account of catalysis by acids and bases, distinguishing between specific and general catalysts.

3. Will the rate of an enzyme-catalyzed reaction usually be more sensitive to temperature than that of the same reaction when it is uncatalyzed? Discuss.

4. Explain how you would determine the parameters K_m and k_c for an enzyme reaction involving a single substrate.

5. Explain clearly the difference between collisions and encounters. What significance does this distinction have in chemical kinetics?

Suggested Reading

Also see Suggested Reading for Chapter 13.

P. G. Ashmore, *Catalysis and Inhibition of Chemical Reactions*, London: Butterworth, 1963.

R. P. Bell, *Acid-Base Catalysis*, Oxford: Clarendon Press, 1941.

W. P. Jencks, *Catalysis in Chemistry and Enzymology*, New York: McGraw-Hill, 1969.

K. J. Laidler and P. S. Bunting, *The Chemical Kinetics of Enzyme Action* (2nd ed.), Oxford: Clarendon Press, 1973.

G. J. Minkoff and C. F. H. Tipper, *Chemistry of Combustion Reactions*, London: Butterworth, 1962.

The many volumes of *Comprehensive Chemical Kinetics* (Eds. C. H. Bamford and C. F. H. Tipper), Amsterdam: Elsevier, cover most aspects of kinetics in considerable detail.

Quantum Mechanics and Atomic Structure

Preview

This chapter introduces a new phase of physical chemistry. Up to now we have made some mention of atoms and molecules but have not concerned ourselves much with their structures and properties. Now we are going to do so and will begin by examining the principles of *quantum mechanics,* a subject that is essential to an understanding of atoms and molecules. We will apply these principles to some simple systems, such as the hydrogen atom and other atoms. Quantum mechanics will also provide us with a basis for understanding chemical bonding and the spectra of atoms and molecules and for understanding the subject of statistical mechanics in its many applications; these topics are dealt with in later chapters.

We begin this chapter by outlining what is known about wave motion. In 1924 de Broglie predicted that electrons and other particles exhibit *wave properties,* a prediction that was soon confirmed experimentally. This discovery led Schrödinger and others in the 1930s to develop *wave mechanics* or *quantum mechanics.* Schrödinger obtained an equation for particles by analogy with the wave equation for electromagnetic radiation. His equation was successful for calculating energies, but his approach was made more versatile by formulating quantum mechanics on the basis of a number of *postulates,* which are a set of rules for arriving at an appropriate *quantum-mechanical operator.* For an energy problem this operator is known as the *Hamiltonian operator* and given the symbol $\hat{H}$. The energy E is then obtained by solving the differential equation

$$\hat{H}\psi = E\psi$$

This equation can only be solved for certain values of the energy E, and energy is therefore *quantized;* this matter is dealt with by introducing *quantum numbers.* Only certain functions ψ, known as *eigenfunctions* or *wave functions,* are solutions of the equation. These eigenfunctions have a special significance, since their square (or, if they are complex, the product of them and their complex conjugate) gives the *probability density* that an electron is present in a particular small region of space. To use this procedure we have to *normalize* the wave function, a device that ensures that the probability that the electron is anywhere in space is unity.

The principles of quantum mechanics can be applied to a number of important problems. One of them is the *particle in the box* and another is the *harmonic oscillator,* an example of which is a *vibrating molecule.* The solution of the wave equations for these systems requires the introduction of *quantum numbers,* which can have only integral values.

The latter half of this chapter is mainly concerned with the quantum mechanics of hydrogenlike atoms, or ions that have a nucleus and a single electron. Solution of the Schrödinger equation for such systems leads to the conclusion that the description of the orbital motion of the electron requires the use of three quantum numbers:

1. The principal quantum number n, which can have values 1, 2, 3, ...
2. The azimuthal quantum number l, which can have the values 0, 1, 2, . . . , $n - 1$.
3. The magnetic quantum number m_l, which can have the values $-l, -l + 1, \ldots, 0, \ldots, l - 1, l$.

Some of the wave functions, or *orbitals*, for the hydrogenlike atoms are spherically symmetrical, while others are directed in space. For hydrogenlike atoms the energy of the electron is determined only by the principal quantum number n.

For a complete description of atoms, a fourth quantum number, the *spin quantum number s*, is introduced. This can have only two values, $+\frac{1}{2}$ and $-\frac{1}{2}$. Associated with electron spin there is an *angular momentum* and a *magnetic moment*. These two properties can also arise from orbital motion.

The latter part of this chapter deals briefly with atoms having more than one electron. The orbitals into which the electrons go depend in part on the order of orbital energy levels. There is also an important restriction imposed by the *Pauli principle*, according to which no two electrons in an atom can have the same set of four quantum numbers.

11 Quantum Mechanics and Atomic Structure

So far we have developed physical chemistry with little regard to the existence of the fundamental particles that comprise all matter. This approach reveals a great deal about the properties of matter, but much more can be accomplished on the basis of the laws that govern the behavior of these fundamental particles. During recent years it has become apparent that most progress is made by considering the *wave properties* of particles, and we will therefore begin by summarizing what is known about wave motion.

11.1 The Nature of Electromagnetic Radiation

Visible light and many other apparently different types of radiation are all forms of *electromagnetic radiation*, and in a vacuum they all travel with the same velocity, namely 2.998×10^8 metres per second (m s^{-1}). Electromagnetic radiation is characterized by a **wavelength** λ and a **frequency** ν. These two physical quantities are related to the velocity of light c by the equation

$$\lambda\nu = c \tag{11.1}$$

The SI unit of wavelength is the metre, although multiples such as nanometres (nm) are frequently used, especially in spectroscopy (Chapter 13). The SI unit of frequency is the reciprocal second (s^{-1}), which is also called the *hertz* (Hz).

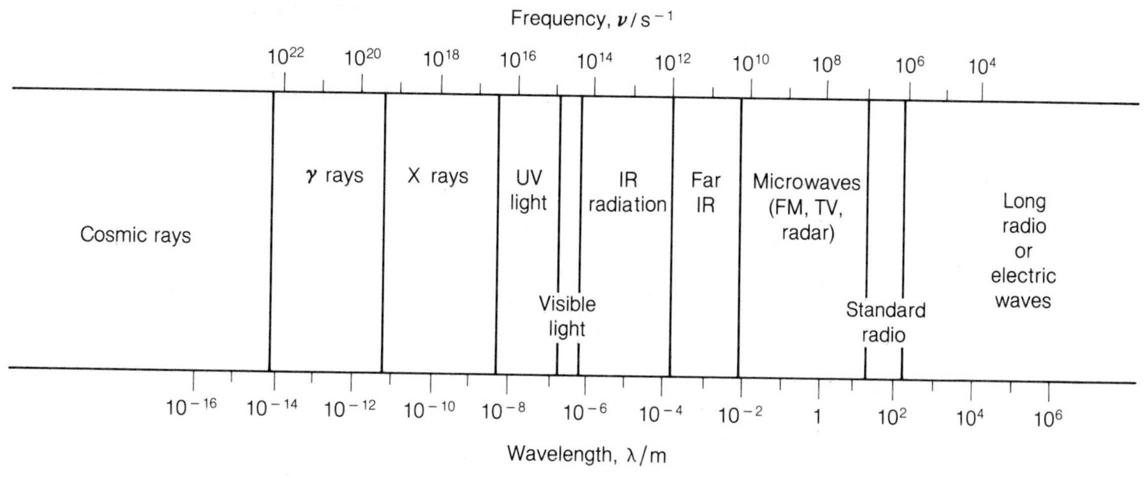

FIGURE 11.1 The electromagnetic spectrum.

Because of this relationship, all electromagnetic radiation can be classified in terms of either its wavelength or its frequency, as is done in Figure 11.1. We see that long radio waves and electric waves may have wavelengths of many kilometres and very low frequencies, whereas γ (Greek lowercase letter gamma) rays have very high frequencies and extremely short wavelengths.

Electromagnetic radiation differs from certain other types of waves in a number of important respects. Sound waves, water waves, seismic waves, and waves in a plucked guitar string exist only by virtue of the medium in which they exist, whereas electromagnetic waves can travel in a vacuum. Sound waves traveling through a gas, for example, consist of alternating zones of compression and rarefaction, and the molecular displacements that occur are in the direction in which the wave travels. As the wave passes a certain region, the gas molecules undergo changes in energy and momentum which then pass on to the next region. This type of wave propagation is also found with a seismic wave traveling through the earth.

An electromagnetic wave is essentially different, since it can travel through a vacuum, and the medium is not essential. However, when such a wave comes in contact with matter, there are important interactions that affect the wave and the material. There is a coupling of the radiation with the medium, and how this occurs is best considered with reference to Figure 11.2, which shows that the wave has two components, one an electric field and the other a magnetic field. These components are in two planes at right angles to each other. A given point in space experiences a periodic disturbance in electric and magnetic field as the wave passes by. A charged particle such as an electron couples its charge with these field fluctuations and oscillates with the frequency of the wave. A useful analogy is provided by a cork floating on still water. If a ripple passes, the cork bobs up and down; its motion is perpendicular to the direction of the wave and it has the same frequency. Similarly an electromagnetic wave causes the particle to move *transverse* (i.e., at right angles) to the direction of propagation, and it is therefore called a *transverse* wave. By contrast, a sound wave is a *longitudinal* wave since the motion of the medium is in the direction of propagation. Both longitudinal and transverse waves are possible in a given medium, a fact that is used to locate the epicenters of earthquakes.

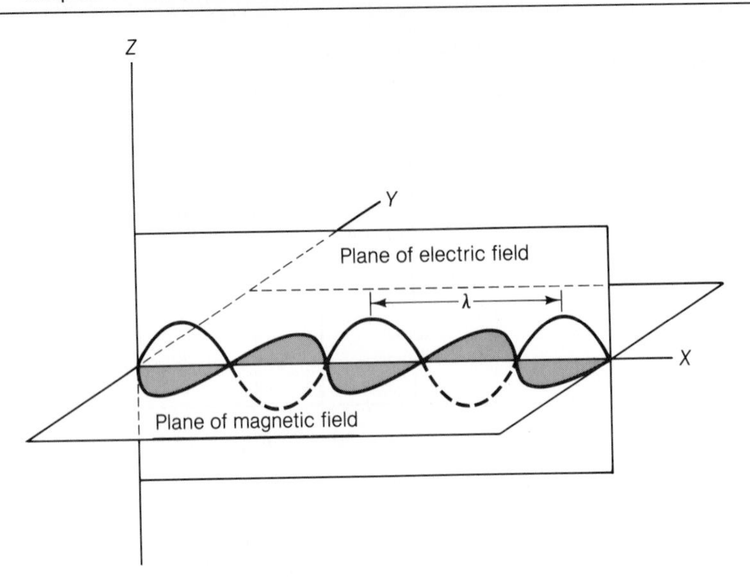

FIGURE 11.2 The propagation of an electromagnetic wave, showing the oscillation of the electric field and the magnetic field in planes at right angles to one another, at a given time.

The analogy of a cork on water may be carried further. If the cork is made to oscillate up and down on the surface, a ripple wave is generated. Similarly, an oscillating electron induces electric and magnetic fields and generates an electromagnetic wave.

Simple Harmonic Motion

It is important to understand the physics and mathematics of wave motion, since this is essential to an understanding of atomic and molecular structure. Some fundamental principles are illustrated in Figure 11.3. Suppose that point C in Figure 11.3a moves counterclockwise on the circumference of the circle; we will call this the *positive* direction. We will be particularly interested in the motion of point P, the projection of point C on the Y axis, which is plotted in Figure 11.3b. Suppose that point C is at B at zero time and moves with constant speed. During a complete revolution with the point C starting at B, the distance y between P and the origin starts from zero, becomes equal to A after a quarter revolution ($\pi/2$ rad), is zero again after half a revolution (π rad), is $-A$ at $3\pi/2$ rad, and is zero again after a complete revolution. This variation of y is shown in Figure 11.3b, the same pattern then continuing indefinitely. The maximum displacement A is known as the **amplitude.** The **period** τ is the time for one revolution, and since the angular path for one revolution is 2π rad, we have*

$$\tau = \frac{2\pi \text{ rad}}{\omega} \tag{11.2}$$

* A complete revolution is 2π rad. Note that the unit, the radian, must be included in Eqs. 11.2–11.4 to balance the units.

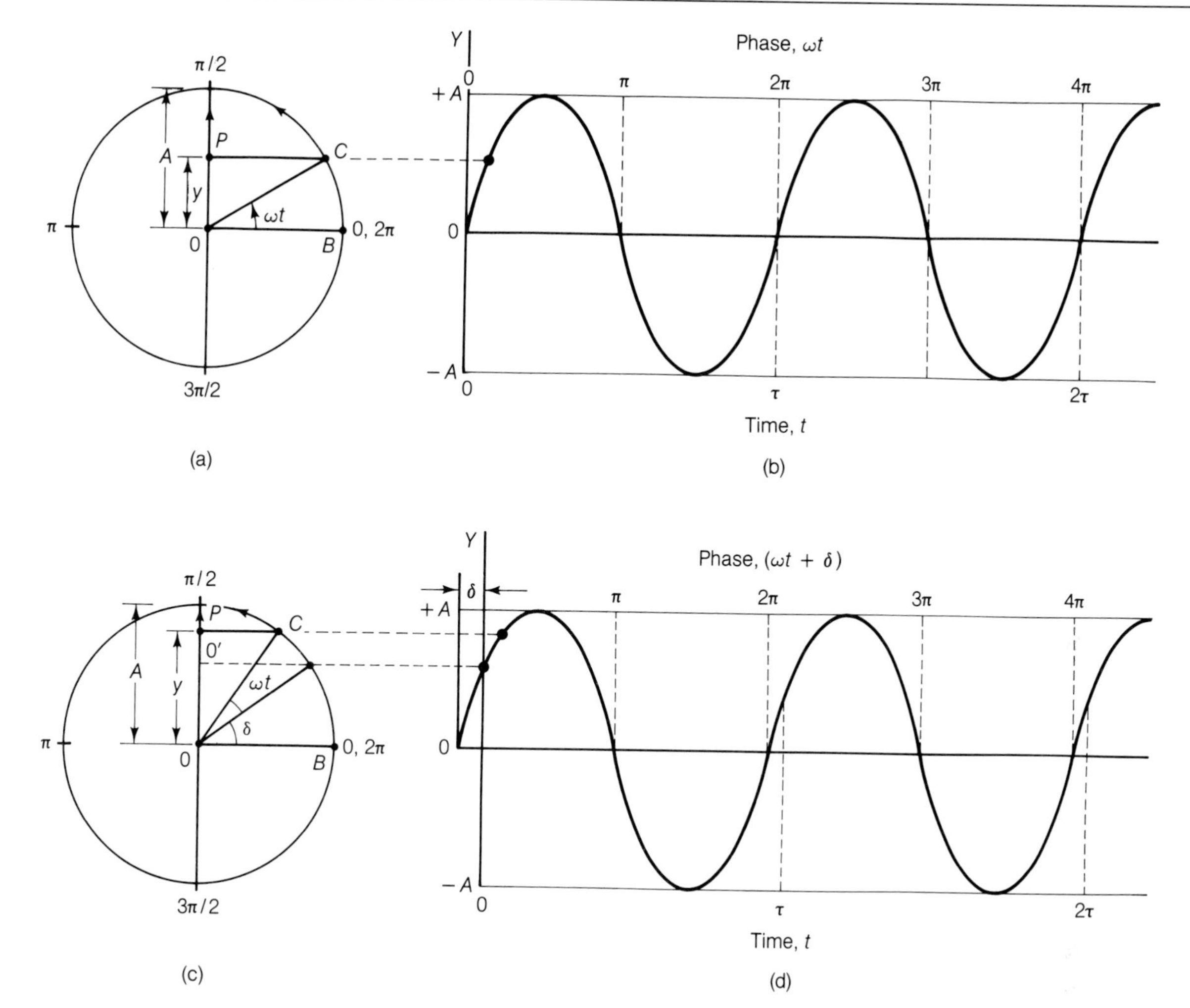

FIGURE 11.3 The generation of simple harmonic motion. (a) A point C undergoes circular motion in the anticlockwise direction. (b) The variation with time of the displacement y of point P; the projection of C on the y axis. (c) and (d) The same as (a) and (b) but with an initial phase displacement δ.

where ω is the angular velocity; its SI unit is rad s^{-1}. The **frequency** is the reciprocal of the period,

$$\nu = \frac{1}{\tau} = \frac{\omega}{2\pi \text{ rad}} \tag{11.3}$$

and therefore

$$\omega = (2\pi \text{ rad})\nu \tag{11.4}$$

The mathematical form of the displacement y, shown in Figure 11.3b, is

$$y = A \sin(\omega t) \tag{11.5}$$

where the quantity ωt (unit: radian) is called the **phase.** More generally, we can allow a displacement by the angle δ before simple harmonic motion begins, as shown in Figure 11.3c. This displacement δ is known as the **phase constant** and this modification requires that

$$y = A \sin (\omega t + \delta) \tag{11.6}$$

The corresponding variation of y with t is shown in Figure 11.3d. Since the curves are not superimposable without a *phase shift*, they are said to be *out of phase*.

To obtain the *acceleration* of point P we differentiate this equation twice:

$$\frac{d^2y}{dt^2} = -A\omega^2 \sin (\omega t + \delta) \tag{11.7}$$

Elimination of $A \sin (\omega t + \delta)$ between these last two equations gives the equation

$$\frac{d^2y}{dt^2} = -\omega^2 y \tag{11.8}$$

This is the equation for **simple harmonic motion.**

Any motion obeying Eq. 11.8 is referred to as simple harmonic motion. Such motion is also found with a mass attached to the end of a spring in which the restoring force obeys Hooke's law, which means that it is proportional to the displacement y:

$$F = -k_h y \tag{11.9}$$

where k_h is the *force constant*. According to Newton's second law of motion this force is the mass m times the acceleration d^2y/dt^2:

$$F = m\frac{d^2y}{dt^2} \tag{11.10}$$

Equating these two expressions for F gives

$$\frac{d^2y}{dt^2} = -\frac{k_h}{m}y \tag{11.11}$$

To solve this we need a function y that when differentiated twice gives the function back again. This condition is satisfied by the sine and cosine functions or any simple combination of them. For simplicity we choose the sine function,

$$y = a \sin \left(\sqrt{\frac{k_h}{m}}\, t + b\right) \tag{11.12}$$

Double differentiation gives

$$\frac{d^2y}{dt^2} = -\frac{k_h}{m} a \sin \left(\sqrt{\frac{k_h}{m}}\, t + b\right) = -\frac{k_h}{m}y \tag{11.13}$$

Our chosen function has therefore satisfied Eq. 11.11. The choices of a and b in Eq. 11.12 are arbitrary and can be determined from initial conditions. Equation 11.12 becomes identical with Eq. 11.6 if a is the amplitude A, if b is the initial phase angle δ, and if

$$\sqrt{\frac{k_h}{m}} = \omega/\text{rad} \tag{11.14}$$

Substitution into this expression of the value of ω from Eq. 11.4 gives

$$\nu = \frac{1}{2\pi}\sqrt{\frac{k_h}{m}} \tag{11.15}$$

This frequency, known as the **natural frequency** of the simple harmonic motion, thus varies inversely with the square root of the mass.

EXAMPLE

Suppose that a hydrogen atom (mass = 1.67×10^{-27} kg) is attached to the surface of a solid by a bond having a force constant of 5.0 kg s^{-2}. Calculate the frequency of its vibration.

SOLUTION

From Eq. 11.15 it follows that

$$\nu = \frac{1}{2\pi}\sqrt{\frac{k_h}{m}} = \frac{1}{2\pi}\sqrt{\frac{5.0\ (\text{kg s}^{-2})}{1.67 \times 10^{-27}(\text{kg})}}$$

$$= 8.70 \times 10^{12}\ \text{s}^{-1}$$

■

When a body is oscillating, its kinetic energy E_k and its potential energy E_p are continuously varying, but their sum E_{total} is a constant:

$$E_{\text{total}} = E_k + E_p \tag{11.16}$$

We have seen (Eq. 1.12) that the potential energy is the work done in moving a mass from its equilibrium position to a new position; thus for the displacement y from the equilibrium position

$$E_p = \int_0^y (-F)\, dy = \int_0^y k_h y\, dy = \frac{k_h y^2}{2} \tag{11.17}$$

The kinetic energy is $\frac{1}{2}mu^2$ where u is the velocity. We can evaluate the total energy by calculating its value when the oscillator reverses its direction; then $u = 0$ and $y = A$, the maximum amplitude. The potential energy is then $\frac{1}{2}k_hA^2$ (Eq. 11.17) and this is the total energy:

$$E_{\text{total}} = \tfrac{1}{2}k_hA^2 \tag{11.18}$$

This result, that the energy is proportional to A^2, is true for any type of simple harmonic motion. For electromagnetic wave motion in a vacuum, for example, the electrical energy is proportional to the square of the electrical field vector (see Figure 11.2), and the magnetic energy is proportional to the square of the magnetic field vector.

Plane Waves and Standing Waves

So far we have considered oscillations at right angles to the direction of propagation of the wave. Energy must also be transported along the wave. A simple type of wave is the *plane wave,* which consists of parallel waves of constant amplitude and has a planar wave front; a useful model of a plane wave is provided by a string stretched horizontally. If it is given a single vertical pulse (Figure 11.4a), the pulse

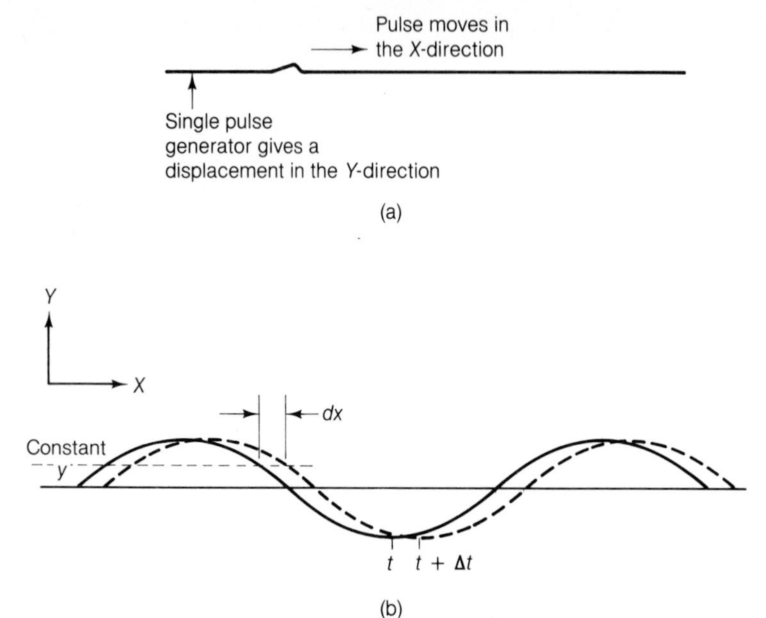

FIGURE 11.4 (a) A single wave pulse traveling along a stretched string. (b) Two profiles of a traveling wave. The solid line shows the transverse displacement y at a position x at time t. The dashed line shows the same curve at a later time $t + \Delta t$.

travels the length of the string. If a fixed-frequency vertical oscillation is applied, a *traveling wave* is generated. This is a wave whose amplitude at a particular position along the string changes periodically with time. Figure 11.4b shows two wave patterns captured at two different instants; these two *profiles* illustrate the fact that the transverse displacement is moving along the string.

Analysis of this type of wave motion leads to the result that the displacement y is given by the equation

$$y = A \sin \frac{2\pi}{\lambda}(x \pm ut) \tag{11.19}$$

The quantity u is called the **phase velocity** because it is the velocity with which a given phase of the wave travels along the X axis. If the positive sign is taken, the wave is traveling to the right with velocity u (see Figure 11.4b); the negative sign means that it is traveling to the left. For electromagnetic radiation traveling in a vacuum the phase velocity is 2.998×10^8 m s^{-1}.

So far we have considered the length of the string to be infinite. A finite string vibrates in a pattern having evenly spaced **nodes** (points of zero displacement) and **antinodes** (points of maximum displacement). An example is shown in Figure 11.5, in which individual profiles are superimposed to show the patterns at various times. The fixed ends are themselves nodes and may be the only nodes; there may be any integral number of nodes as shown in Figure 11.5. This wave form is called a **standing wave** or a *stationary wave.* This situation is dealt with by applying the **principle of superposition,** according to which when two or more waves are involved, the displacement at any position is the sum of the displacements at that

The solution to this dilemma was obtained by the German physicist Max Karl Ernst Ludwig Planck* (1858–1947), who developed the **quantum theory.** According to this theory, energy is not continuous but consists of discrete units that Planck called **quanta** (Latin *quantum,* how much?). He suggested that the energy of a quantum is proportional to the frequency ν of the oscillator:

$$E = h\nu \tag{11.23}$$

The proportionality factor h is known as the **Planck constant;** its modern value is $6.626\,196 \times 10^{-34}$ J s.

On the basis of this theory Planck obtained an equation for the variation with frequency of the energy emitted by an oscillator. The treatment is quite straightforward for a collection of oscillators all having the same fundamental vibrational frequency ν. The energy comes in quanta of $h\nu$, so that there can be

N_0 oscillators having zero energy

N_1 oscillators having energy $h\nu$

N_2 oscillators having energy $2h\nu$, and so on.

It is further assumed that the distribution is governed by the Boltzmann equation (Section 14.2), according to which the number of oscillators N_i having energy ϵ_i is

$$N_i = N_0 e^{-\epsilon_i/\mathbf{k}T} = N_0 e^{-nh\nu/\mathbf{k}T} \tag{11.24}$$

where N_0 is the number with zero energy and $\mathbf{k}$, the Boltzmann constant, is R/L, the gas constant per molecule. The total number of oscillators in all states is thus

$$N = N_0 + N_0 e^{-h\nu/\mathbf{k}T} + N_0 e^{-2h\nu/\mathbf{k}T} + \cdots \tag{11.25}$$

$$= N_0 \sum_{n=0}^{\infty} e^{-nh\nu/\mathbf{k}T} \tag{11.26}$$

This series is of the form

$$N = N_0(1 + x + x^2 + x^3 + \cdots) \tag{11.27}$$

where x is $e^{-h\nu/\mathbf{k}T}$. The series is an expansion of $1/(1 - x)$, and Eq. 11.26 thus becomes

$$N = \frac{N_0}{1 - x} = \frac{N_0}{1 - e^{-h\nu/\mathbf{k}T}} \tag{11.28}$$

The total energy E of all the oscillators in all the states is the number in each state multiplied by the energy of that state:

$$E = N_0(0) + N_1(h\nu) + N_2(2h\nu) + \cdots \tag{11.29}$$

With the use of Eq. 11.24 this becomes

$$E = N_0 h\nu e^{-h\nu/\mathbf{k}T} + N_0 2h\nu e^{-2h\nu/\mathbf{k}T} + \cdots \tag{11.30}$$

$$= N_0 h\nu e^{-h\nu/\mathbf{k}T}(1 + 2e^{-h\nu/\mathbf{k}T} + 3e^{-2h\nu/\mathbf{k}T} + \cdots) \tag{11.31}$$

*M. Planck, *Ann. Physik.,* 4, 553 (1901).

The series in parentheses is of the form $(1 + 2x + 3x^2 + \cdots)$, which is an expansion of $1/(1 - x)^2$. Equation 11.31 may thus be written as

$$E = \frac{N_0 h\nu e^{-h\nu/\mathbf{k}T}}{(1 - e^{-h\nu/\mathbf{k}T})^2} \tag{11.32}$$

However, by Eq. 11.28, N_0 is equal to $N(1 - e^{-h\nu/\mathbf{k}T})$ and therefore the average energy of an oscillator is

$$\bar{\epsilon} = \frac{E}{N} = \frac{h\nu e^{-h\nu/\mathbf{k}T}}{1 - e^{-h\nu/\mathbf{k}T}} \tag{11.33}$$

$$\bar{\epsilon} = \frac{h\nu}{e^{h\nu/\mathbf{k}T} - 1} \tag{11.34}$$

This is Planck's expression for the average energy. When $h\nu$ is much less than $\mathbf{k}T$, the term in the denominator is $h\nu/\mathbf{k}T$, and the average energy therefore becomes

$$\lim_{\nu \to 0} \bar{\epsilon} = \mathbf{k}T \tag{11.35}$$

which is the classical value. The quantum theory thus reduces to classical theory when the quanta are very small.

Equation 11.34 leads to the expression

$$M_e(\lambda, T) = \frac{2\pi hc^2}{\lambda^5}\left(\frac{1}{e^{hc/\mathbf{k}T\lambda} - 1}\right) \tag{11.36}$$

for the spectral radiant excitance. This expression gives extremely close agreement to the experimental curves, such as those shown in Figure 11.7. This was the first outstanding success of the quantum theory, which soon became very firmly established.

The Photoelectric Effect

Another important piece of evidence for the quantum theory was provided by the **photoelectric effect,** which was first discovered in 1887 by Heinrich Rudolf Hertz (1857–1894). He found that when suitable radiation (usually visible or ultraviolet) falls on a metal surface, electrons are emitted from the surface. Later experiments* showed that whether or not electrons are emitted depends only on the frequency of the light and not on its intensity but that the *number* of electrons emitted is proportional to the intensity. Another important result† was that the electron emission occurred the instant the radiation struck the metal surface.

It is impossible to explain these results on the basis of classical physics, according to which the energy of the radiation should depend only on its intensity. Whether or not electrons are emitted should therefore depend on the energy of the radiation, and even weak intensities should be effective if long periods of time were allowed for their absorption.

The correct explanation was given in 1905 by Albert Einstein (1879–1955) who extended Planck's quantum theory to radiation itself. Planck had realized that the

---

*J. Elster and H. Geitel, *Ann. Physik, 38*, 40 (1889).

†P. Lenard, *Ann. Physik., 2*, 359 (1900).

energy of an oscillator is quantized, but Einstein took the argument a step further and postulated that light itself is a beam of particles each of which has energy equal to $h\nu$. He envisaged the photoelectric effect as a process in which the quantum of energy $h\nu$ is given to an individual electron, which may then have sufficient energy to leave the metal surface. If w is the potential energy (called the **work function**) that is required to remove the electron from the surface, then

$$h\nu = \tfrac{1}{2}mu^2 + w \tag{11.37}$$

where $\frac{1}{2}mu^2$ is the kinetic energy of the electron once it has left the metal. If the energy $h\nu$ is less than w, the electron cannot leave the surface; but if $h\nu$ is greater than w, it can leave, and the excess, $h\nu - w$, will appear as kinetic energy of the electron. There is therefore a frequency ν_0, equal to w/h, which is just sufficient to remove the electron, and this is known as the **threshold frequency.**

This theory clearly explains why whether or not electrons are emitted depends on the frequency of the radiation and not on the intensity. The intensity affects not whether they are emitted but how many are emitted. The theory also accounts for the fact that the emission is instantaneous; the energy $h\nu$ interacts with a single electron and cannot be stored in the metal.

In 1926 the American chemist Gilbert Newton Lewis (1875–1946) suggested the name **photon** (Greek *photos*, light) for these particles of radiation of energy $h\nu$. The essential conclusion from the photoelectric experiments was that there is a one-to-one relationship between a photon absorbed and an electron emitted. When a photon interacts with an electron, the electron gains the energy $h\nu$ and the photon disappears. We shall see later, especially in Chapter 13, that in spectroscopy there is usually the same kind of one-to-one relationship between a photon and an electron.

11.2 Atomic Spectra and the Old Quantum Theory

The first two great triumphs of Planck's quantum theory were his explanation of blackbody radiation and Einstein's explanation of the photoelectric effect. A third triumph was the interpretation of spectra. Whereas a blackbody emits radiation spread continuously over a range of wavelengths, the emission spectrum of hydrogen consists of sharp lines, as shown in Figure 11.8a. It proved impossible to predict from classical physics the positions of these and other atomic spectral lines.

Some useful empirical relationship had, however, been obtained for the positions of hydrogen atom spectral lines. In 1885 the Swiss mathematician and physicist Johann Jakob Balmer (1825–1898) developed an empirical equation to locate the lines now known as the Balmer series (Figure 11.8). Other series were similarly predicted by empirical equations developed by the American physicist Theodore Lyman (1874–1954) and the German physicist Friedrich Paschen (1865–1940). These equations were all generalized by the Swedish physicist Johannes Robert Rydberg (1854–1919) and by the Swiss physicist Walter Ritz (1878–1909). The modern form of the Balmer-Rydberg-Ritz formula involves two integers, n_1 and n_2, and is

$$\bar{\nu} = \frac{1}{\lambda} = R\left(\frac{1}{n_1^2} - \frac{1}{n_2^2}\right) \tag{11.38}$$

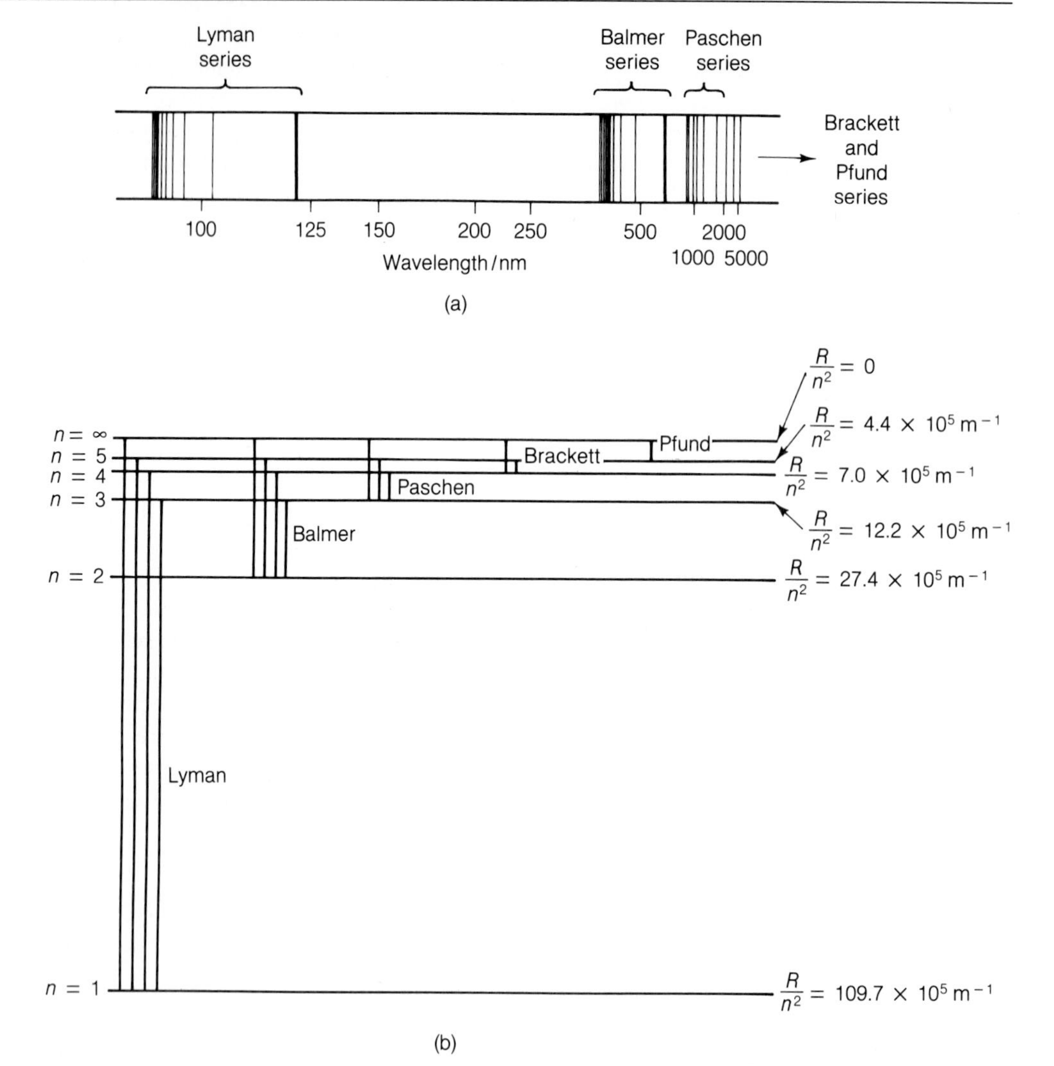

FIGURE 11.8 (a) The hydrogen spectrum in the visible and the near-ultraviolet and near-infrared regions, showing the Lyman, Balmer, and Paschen series. (b) The interpretation of the series as given by the Balmer-Rydberg-Ritz formula. The horizontal lines show R/n^2 values (the spectral terms) for various values of n. The transitions involved in the various series are shown.

where $\bar{\nu}$, the reciprocal of the wavelength λ, is known as the **wave number** and R is a constant known as the **Rydberg constant;** its modern value is 1.0968×10^7 m^{-1}. The wave numbers and frequencies are thus the differences between certain *spectral terms,* taken in pairs. The value of n_2 must be greater than n_1, and all known spectral lines in the hydrogen spectrum may be obtained by appropriate integral values of n_1 and n_2. Figure 11.8b shows the n values and the particular combinations that are involved in the various series found in the hydrogen spectrum.

However, these empirical relationships provided no real explanation for the spectral lines. Moreover, as spectroscopic instruments improved, it was found that what appeared to be single lines at low resolution were actually two or more lines. This *fine structure* of the lines could not be explained by classical theories.

The first satisfactory explanation of spectra was given in 1913 by the Danish physicist Niels Henrik David Bohr (1885–1962), whose theory was based on the nuclear model of the atom proposed in 1911 by Ernest Rutherford (1871–1937). According to Rutherford the nucleus of an atom was very much smaller than the atom itself. Around this dense positively charged nucleus revolved the electrons, and the size of the atom was determined by the size of the electronic orbits. The simplest atom, hydrogen, consists of a nucleus having a single positive charge and a single orbiting electron. This model presented a serious theoretical problem, however, since according to classical electrodynamics an orbiting charged particle, having an acceleration toward the center, must continuously lose energy. It would therefore fall into the nucleus, and hence the atom would not survive.

Niels Bohr overcame this dilemma by combining Rutherford's concept of the nuclear atom with the new quantum theory and with some important new ideas of his own. He followed the suggestion of the Irish physicist Arthur William Conway (1875–1950) that spectral lines are produced by the transitions of single electrons. He also accepted a postulate made by the English mathematician John William Nicholson (1881–1955) that electrons produce spectra because of jumps between different energy states characterized by discrete (quantized) values of the **angular momentum** L. Bohr's special contribution was to realize that in an electronic transition two different states, known as **stationary states,** are involved and that these correspond to the spectral terms that appear in the Balmer-Rydberg-Ritz formula (Eq. 11.38). Bohr further postulated that when an electronic transition occurs between two states of energies E_1 and E_2, the frequency ν of the spectral line is given by

$$h\nu = E_1 - E_2 \tag{11.39}$$

Angular momentum L is (mass) $\times$ (velocity) $\times$ (orbital radius) and therefore has the dimensions (mass)(distance)^2(time)$^{-1}$, which is energy multiplied by time, and the same is true of the Planck constant (6.626×10^{-34} J s). Bohr therefore proposed that the angular momentum is quantized. An electron of mass m moving with a velocity u in a circle of radius r has an angular momentum L of mur, and Bohr's suggestion was that this quantity must be an integer n multiplied by the Planck constant divided by 2π:

$$L = mur = n\frac{h}{2\pi} \tag{11.40}$$

For an electron in an atomic orbital the integer n is called the **principal quantum number.**

We will consider the application of these ideas to a hydrogenlike atom; this is a species having a nucleus of charge Ze and a single orbiting electron (when $Z = 1$, we have the hydrogen atom itself). The electron actually moves around the center of mass of the nucleus-electron system and it is its reduced mass μ that strictly speaking we should consider; this is very close to its true mass. For the circular orbit to be stable, centripetal force $\mu u^2/r$ must be supplied by the coulombic force $Ze^2/4\pi\epsilon_0 r^2$. Thus

$$\frac{\mu u^2}{r} = \frac{Ze^2}{4\pi\epsilon_0 r^2} \tag{11.41}$$

By Eq. 11.40, u is equal to $nh/2\pi\mu r$, and insertion of this expression into Eq. 11.41 gives

$$\frac{n^2h^2}{4\pi^2\mu r^3} = \frac{Ze^2}{4\pi\epsilon_0 r^2} \tag{11.42}$$

Thus

$$r = \frac{n^2h^2\epsilon_0}{\pi\mu Ze^2} \tag{11.43}$$

It is convenient to write this as

$$r = \frac{n^2}{Z}a_0 \tag{11.44}$$

where a_0 is given by

$$a_0 = \frac{h^2\epsilon_0}{\pi\mu e^2} \tag{11.45}$$

This quantity a_0 is a length and is the radius of the orbit for $n = 1$ for the hydrogen atom itself ($Z = 1$). The length a_0 is known as the **first Bohr radius** and has a value of 52.92 pm (1 picometre = 10^{-12}m.).

The energy corresponding to the various electronic levels is the sum of the kinetic and potential energies. The kinetic energy is $\frac{1}{2}\mu u^2$ and, using Eq. 11.41, is given by

$$E_k = \tfrac{1}{2}\mu u^2 = \frac{Ze^2}{8\pi\epsilon_0 r} \tag{11.46}$$

With Eq. 11.44 this gives

$$E_k = \frac{Z^2e^2}{8\pi\epsilon_0 n^2 a_0} \tag{11.47}$$

The potential energy of the electron is conveniently defined with respect to infinite separation from the nucleus and is the negative of the work required to remove an electron from the distance r to infinity. It is therefore

$$E_p = -\int_r^\infty \frac{Ze^2}{4\pi\epsilon_0 r^2}\,dr = -\frac{Ze^2}{4\pi\epsilon_0 r} \tag{11.48}$$

With Eq. 11.44 this gives

$$E_p = -\frac{Z^2e^2}{4\pi\epsilon_0 n^2 a_0} \tag{11.49}$$

Note that the potential energy is twice the kinetic energy with the sign changed. The total energy is

$$E = E_k + E_p = -\frac{Z^2e^2}{8\pi\epsilon_0 n^2 a_0} \tag{11.50}$$

The lowest possible state for the hydrogen atom is when $n = 1$, and the value of this ground-state energy is

-2.1798×10^{-18} J or -13.605 eV

Bohr then derived the Balmer-Rydberg-Ritz formula. He assumed that the atom can exist in various *stationary states* characterized by the value of the quantum number n; no radiation is emitted or absorbed as long as the electron remains in a stationary state. Equation 11.39 gives the value of the frequency ν as $(E_1 - E_2)/h$, and the wave number $\bar{\nu}$ is therefore $(E_1 - E_2)/hc$. If the quantum numbers associated with E_1 and E_2 are n_1 and n_2, Eq. 11.50 therefore leads to

$$\bar{\nu} = \frac{Z^2 e^2}{8\pi\epsilon_0 a_0 hc}\left(\frac{1}{n_2^2} - \frac{1}{n_1^2}\right) \tag{11.51}$$

This is the same form as Eq. 11.38, and we see that when $Z = 1$, the Rydberg constant is given by

$$R = \frac{e^2}{8\pi\epsilon_0 a_0 hc} \tag{11.52}$$

The value calculated by Bohr from this relationship was extremely close to the experimental value.

This good agreement was very encouraging, but serious problems still remained. The Bohr theory provided no explanation for the fine structure of the lines, and it was not found possible to extend it satisfactorily to atoms containing more than one electron. Certain attempts to modify the Bohr theory introduced some improvement, but difficulties remained. The German physicist Arnold Johannes Wilhelm Sommerfeld* (1868–1951) gave a treatment of elliptical orbits and introduced a second quantum number. The German physicist Werner Karl Heisenberg (1901–1976), in a paper published when he was 20 years old, suggested that quantum numbers could be half-integral as well as integral. However, these treatments did not prove satisfactory, and it became clear that the Bohr theory was inadequate and that a completely new approach was needed. Bohr himself again helped to point the way through his **correspondence principle,** which states that any quantum theory must reduce to the familiar classical laws under the conditions for which classical behavior is expected.

11.3 The Wave Nature of Electrons

The new developments were based on the realization that elementary particles such as electrons have wave properties, just as radiation has particle properties. We have seen that Planck's treatment of blackbody radiation and Einstein's interpretation of the photoelectric effect requires us to regard radiation as consisting of photons, and a third phenomenon, the **Compton effect,** leads to the same conclusion. The American physicist Arthur Holly Compton† (1892–1962) investigated the scattering of monochromatic (Greek *mono,* one; *chroma,* color) X rays by a target such as a piece of graphite. He found that the scattered beam consisted of radiation of two different wavelengths; one wavelength, λ, was the same as for the original beam,

* A. Sommerfeld, *Ann. Physik, 51,* 1 (1916).
† A. H. Compton, *Phys. Rev., 22,* 409 (1923).

whereas the other, λ', was slightly longer. The results could only be interpreted on the hypothesis that photons interacted with the material and obeyed the laws of conservation of energy and momentum. The results could not be explained by treating the radiation as having wave properties. Other properties of radiation, however, such as refraction, can *only* be explained on the basis of wave properties. It follows that we need a **dual theory of radiation,** which in some experiments exhibits its wave properties and in others its particle properties.

This realization led the French physicist Louis Victor, Prince de Broglie (b. 1892), to suggest in 1924* the converse hypothesis, that particles such as electrons can also have wave properties. To obtain the wavelength of the wave associated with a particle of mass m moving with velocity u, de Broglie reasoned as follows. According to Einstein's special theory of relativity the energy E of a particle of rest mass m and momentum p are related by the equation

$$E^2 = p^2c^2 + m^2c^4 \tag{11.53}$$

where c is the velocity of light. The relativistic rest mass of the photon is zero and therefore

$$E = pc = p\lambda\nu \tag{11.54}$$

However, since $E = h\nu$, it follows that

$$\lambda = \frac{h}{p} \tag{11.55}$$

Since this expression is for electromagnetic radiation, de Broglie suggested that a parallel expression would apply to a beam of particles of momentum p and velocity u:

$$\boldsymbol{\lambda = \frac{h}{p} = \frac{h}{mu}} \tag{11.56}$$

where m is the mass of the particle.

EXAMPLE

An electron has a mass of 9.11×10^{-31} kg and a charge of 1.602×10^{-19} C. Calculate the de Broglie wavelength of an electron that has been accelerated by a potential of 100 V.

SOLUTION

The kinetic energy of the electron is

$$100 \times 1.602 \times 10^{-19} \text{ V C} = 1.602 \times 10^{-17} \text{ J}$$

The kinetic energy is $\frac{1}{2}mu^2$ and therefore

$$u = \sqrt{\frac{2 \times 1.602 \times 10^{-17}(\text{J})}{9.11 \times 10^{-31}(\text{kg})}} = 5.93 \times 10^6 \text{ m s}^{-1}$$

*L. V. de Broglie, "Researches on the Quantum Theory," Thesis, University of Paris, 1924.

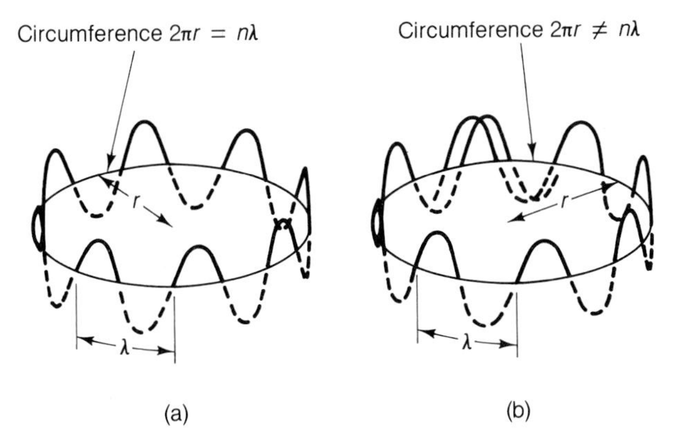

FIGURE 11.9 The de Broglie wave associated with an electron in a Bohr orbit of radius r. (a) Constructive interference: the wave fits into the orbit. (b) Destructive interference: the wave does not fit an integral number of times into the orbit. If in this diagram we continued the waves indefinitely, they would completely obliterate one another, the net amplitude becoming zero.

The de Broglie wavelength λ is therefore, from Eq. 11.56,

$$\frac{6.626 \times 10^{-34}(\text{J s})}{9.11 \times 10^{-31}(\text{kg}) \times 5.93 \times 10^{6}(\text{m s}^{-1})}$$
$$= 1.23 \times 10^{-10} \text{ m}$$
$$= 123 \text{ pm}$$ ■

This wavelength is of a magnitude similar to the distance between neighboring atoms or ions in a crystal, and it was therefore suggested that a beam of electrons could be diffracted by using a crystal as a diffraction grating. This prediction was confirmed experimentally in 1927 by the English physicists Sir George Paget Thomson (1892–1975) and A. Reid and by the American physicists Clinton Joseph Davisson (1881–1958) and Lester Halbert Germer* (b. 1896). The diffraction of electrons is now commonly employed as a technique for investigating molecular structure.

The realization that electrons have wave properties leads at once to an interpretation of Bohr's hypothesis of stationary orbits. Figure 11.9 shows the de Broglie wave associated with an electron in a Bohr orbit or radius r. If the circumference $2\pi r$ is equal to an integer n multiplied by the wavelength λ, the wave will fit into the orbit an integral number of times, and *constructive interference* will occur. The condition for this is

$$2\pi r = n\lambda \qquad (11.57)$$

Destructive interference occurs with wavelengths that do not satisfy this condition, and the orbit is not a stationary one. The requirement that n must be an integer for

*C. Davisson and L. H. Germer, *Phys. Rev.*, *31*, 705 (1927).

constructive interference thus leads to a quantization condition. Substitution of λ as given by Eq. 11.56 into Eq. 11.57 gives

$$2\pi r = \frac{nh}{mu} \tag{11.58}$$

or

$$mur = n\frac{h}{2\pi} \tag{11.59}$$

The quantity mur is the orbital angular momentum, which Bohr has postulated to be an integral number of $h/2\pi$. We can now see that the de Broglie relationship leads at once to this quantization condition.

The Uncertainty Principle

Scientists have often found it useful to carry out what are called *gedanken* (German *Gedanke,* thought) experiments. These are experiments that would be very difficult if not impossible to perform but that can easily be carried out in imagination with results that are completely reliable. A very significant gedanken experiment was carried out in 1926 by Werner Heisenberg, who considered the procedures that would have to be followed in order to make a simultaneous measurement of the position and momentum of a small particle such as an electron. He realized that it would be impossible to make accurate measurements of both properties, since any technique for measuring one of them will necessarily disturb the system and will cause the measurement of the other one to be imprecise. Suppose, for example, that radiation is used to make an accurate determination of the position of an electron. To do so we must use radiation of short wavelength, since otherwise the position is not well defined. Radiation of short wavelength, however, involves high-energy photons, and we have seen in the Compton effect that these will change the particle's momentum. If we avoid the momentum change by using radiation of long wavelength, the position of the particle will not be precisely defined. These conclusions are expressed in the **Heisenberg uncertainty principle,** or principle of indeterminancy, which may be written as

$$\Delta q \, \Delta p \approx \frac{h}{4\pi} \tag{11.60}$$

That is, in a given direction the product of the uncertainty in the position Δq and the uncertainty in the momentum Δp is approximately equal to the Planck constant divided by 4π. Since the momentum p is equal to mu, the mass times the velocity, the relationship can also be written as

$$\Delta q \, \Delta u \approx \frac{h}{4\pi m} \tag{11.61}$$

Thus the product of the uncertainties decreases as the mass of the particle increases. The uncertainty principle is thus particularly important for light particles such as electrons; it does not apply to any significant extent to massive particles (see Problem 10).

Another form of the principle, involving energy and time, is also of importance. If a particle has a velocity u, the uncertainty in the time at which it exists in a given position is related to Δq by

$$\Delta t = \frac{\Delta q}{u} \tag{11.62}$$

Since energy is $\frac{1}{2}mu^2$ and momentum is mu, the uncertainty in the energy is given by

$$\Delta E = u \, \Delta p \tag{11.63}$$

With Eq. 11.60 these equations give

$$\Delta E \, \Delta t \approx \frac{h}{4\pi} \tag{11.64}$$

Two examples of this relationship may be mentioned. The first relates to α particles emitted in the nuclear decay of radioactive substances. The energy of such particles can be measured rather precisely, which means that the time at which the emission occurs is very uncertain. Another application, considered further in Section 13.9, relates to the widths of spectral lines. If an electronic transition occurs from the ground state of an atom or molecule, the spectral line is sharp; that is, it covers only a narrow range of wavelengths. However, if an atom or molecule is in an electronically excited state, its lifetime may be very short; Δt is therefore small, which means that ΔE is large. The energy of the transition therefore covers a range of values, and hence the spectral line is broadened.

11.4 Schrödinger's Wave Equation

It was well recognized in the early 1920s that there was a need for a new theoretical approach to the problem of atomic structure, one that avoided the arbitrary introduction of quantization and that had a wider application than Bohr's theory. As sometimes happens, several theories were proposed by different people at about the same time. The most successful of these were the theories of Werner Heisenberg and of the Austrian theoretical physicist Erwin Schrödinger (1887–1961). Heisenberg's theory* was presented in matrix algebra and its somewhat unfamiliar mathematical form delayed its acceptance, particularly by chemists. A more easily understandable theory was put forward in 1926 by Schrödinger.† These theories are usually referred to as **quantum mechanics** or as **wave mechanics.** The latter expression is particularly appropriate for Schrödinger's theory, which is based on the application of wave theory to electrons.

Schrödinger's equation cannot be derived but is based on intuitive arguments. He took as his starting point a fundamental equation for a wave traveling in the

---

*W. Heisenberg, *Z. Physik, 33*, 879 (1925).

†E. Schrödinger, *Ann. Physik, 79*, 361, 489, 734; *81*, 109 (1926); *Phys. Rev., 28*, 1049 (1926).

X-direction. For an electromagnetic wave traveling in a vacuum with velocity c, this equation is

$$\frac{\partial^2 y}{\partial x^2} = \frac{1}{c^2}\frac{\partial^2 y}{\partial t^2} \tag{11.65}$$

where y is the displacement in the Y-direction. Schrödinger replaced this equation by

$$\frac{\partial^2 \Psi}{\partial x^2} = \frac{1}{c^2}\frac{\partial^2 \Psi}{\partial t^2} \tag{11.66}$$

where $\Psi(x, t)$ is a function that describes the behavior of the electron and is a function of the distance x and the time t. A solution of Eq. 11.66 is

$$\Psi(x, t) = Ce^{i\alpha} \tag{11.67}$$

where C is a constant, i is $\sqrt{-1}$, and α is the phase, given by

$$\alpha = 2\pi\left(\frac{x}{\lambda} - \nu t\right) \tag{11.68}$$

This function $\Psi(x, t)$ is known as a **wave function,** and it can be split into two factors as follows:

$$\Psi(x, t) = Ce^{2\pi i x/\lambda}e^{-2\pi i \nu t} \tag{11.69}$$

This may be written as

$$\Psi(x, t) = \psi(x)e^{-2\pi i \nu t} \tag{11.70}$$

where $\psi(x)$ is a function of x but not of t and $e^{-2\pi i \nu t}$ is a function of t but not of x.

The next stage in Schrödinger's argument was to replace ν by E/h (Eq. 11.23) and λ by h/p_x (Eq. 11.56). Equation 11.69 therefore becomes

$$\Psi(x, t) = Ce^{2\pi i x p_x/h}e^{-2\pi i E t/h} \tag{11.71}$$

Partial differentiation with respect to t then gives

$$\frac{\partial \Psi(x, t)}{\partial t} = -\frac{2\pi i E C}{h}e^{2\pi i x p_x/h}e^{-2\pi i E t/h} \tag{11.72}$$

$$= -\frac{2\pi i E}{h}\Psi(x, t) \tag{11.73}$$

This equation rearranges to

$$-\frac{h}{2\pi i}\frac{\partial \Psi}{\partial t} = E\Psi \tag{11.74}$$

This is an *operator equation*. The operator, written as $\hat{E}$, is

$$-\frac{h}{2\pi i}\frac{\partial}{\partial t}$$

which means that we take the partial derivative of the function Ψ and multiply the result by $-h/2\pi i$. Equation 11.74 thus tells us that if we perform this operation on the function Ψ, we obtain the energy E multiplied by the function.

We can also take the partial derivative of $\Psi(x, t)$ with respect to x. We obtain, from Eq. 11.71,

$$\frac{\partial \Psi(x, t)}{\partial x} = \frac{2\pi i p_x C}{h} e^{2\pi i x p_x/h} e^{-2\pi i E t/h} \tag{11.75}$$

$$= \frac{2\pi i p_x}{h} \Psi(x, t) \tag{11.76}$$

This rearranges to

$$\frac{h}{2\pi i} \frac{\partial \Psi}{\partial x} = p_x \Psi \tag{11.77}$$

The operator $\hat{p}_x$, which is

$$\frac{h}{2\pi i} \frac{\partial}{\partial x}$$

is called a **linear momentum operator,** since when it operates on the function Ψ it gives the linear momentum multiplied by the function.

The next step in the argument is to introduce the total energy E of the system. This is the sum of the kinetic energy E_k and the potential energy E_p, and for this one-dimensional system the kinetic energy is $p_x^2/2m$. Thus

$$E = \frac{p_x^2}{2m} + E_p(x, t) \tag{11.78}$$

When energy is expressed in terms of momentum, it is said to be a *Hamiltonian,* after the Irish mathematician Sir William Rowan Hamilton (1805–1865). We can now replace p_x in this equation by the linear momentum operator (Eq. 11.77), and we obtain the operator

$$\hat{H} = \frac{-h^2}{8\pi^2 m} \frac{\partial^2}{\partial x^2} + E_p(x, t) \tag{11.79}$$

This is known as the **Hamiltonian operator.**

If we replace the energy E in Eq. 11.74 by this operator $\hat{H}$, we obtain

$$\left[-\frac{h^2}{8\pi^2 m} \frac{\partial^2}{\partial x^2} + E_p(x, t) \right] \Psi = -\frac{h}{2\pi i} \frac{\partial \Psi}{\partial t} \tag{11.80}$$

This equation is easily extended to three dimensions. The operator $\partial^2/\partial x^2$ is now replaced by the sum

$$\nabla^2 = \frac{\partial^2}{\partial x^2} + \frac{\partial^2}{\partial y^2} + \frac{\partial^2}{\partial z^2} \tag{11.81}$$

This is the **del-squared operator** or the **Laplacian operator,** named for the French astronomer and mathematician Pierre Simon, Marquis de Laplace (1749–1827). The functions $\Psi(x, t)$ and $E_p(x, t)$ are now replaced by $\Psi(x, y, z, t)$ and $E_p(x, y, z, t)$, and Eq. 11.80 becomes

$$\left[-\frac{h^2}{8\pi^2 m} \nabla^2 + E_p(x, y, z, t) \right] \Psi = -\frac{h}{2\pi i} \frac{\partial \Psi}{\partial t} \tag{11.82}$$

This is the **time-dependent Schrödinger equation.**

For many problems, such as those concerned with the structures of atoms and molecules, we are not concerned with time-dependent wave functions and energies but with *stationary states* which imply that the potential function is time independent. The time-dependent function $\Psi(x, y, z, t)$ may be expressed as

$$\Psi(x, y, z, t) = \psi(x, y, z)e^{-2\pi iEt/h} \tag{11.83}$$

where $\psi(x, y, z)$ is the time-independent wave function. (Compare Eq. 11.71 for the one-dimensional form.) Partial differentiation of $\Psi(x, y, z, t)$ with respect to t gives

$$\frac{\partial\Psi}{\partial t} = -\frac{2\pi iE}{h}\psi(x, y, z)e^{-2\pi iEt/h} = -\frac{2\pi i}{h}E\Psi \tag{11.84}$$

For stationary states where E_p is independent of time, Eq. 11.84 is substituted into Eq. 11.82; and, since there is no operator in the Hamiltonian involving time, the exponential term involving time in Ψ is factored from both sides. Equation 11.82 thus becomes

$$\left[-\frac{h^2}{8\pi^2 m}\nabla^2 + E_p(x, y, z)\right]\psi(x, y, z) = E\psi(x, y, z) \tag{11.85}$$

This is the **time-independent Schrödinger equation,** and it is frequently written in the very compact form

$$\hat{H}\psi = E\psi \tag{11.86}$$

where $\hat{H}$ represents the Hamiltonian operator that appears in the brackets in Eq. 11.85.

Solution of Eq. 11.86, with an appropriate expression for the potential energy E_p, thus leads to stationary-state solutions, which are analogous to the stationary-state or standing-wave solutions obtained for a vibrating spring (Figure 11.6). Certain ψ functions are possible solutions of the equation, and they are called **eigenfunctions,** *characteristic functions,* or *wave functions.* They are analogous to the amplitude functions y of Eq. 11.20. Corresponding to each of these wave functions is an allowed energy level, and these energy levels are called the *characteristic values* or the **eigenvalues** for the energy of the system.

In most of the rest of this book we will be concerned with time-independent systems, which are often called **conservative** systems.

Eigenfunctions and Normalization

A further, and very significant, interpretation of the eigenfunction ψ was suggested by the German-British physicist Max Born* (1882–1970). His idea was that if the eigenfunction is real (i.e., does not involve $\sqrt{-1}$,) its square ψ^2 multiplied by a small volume element $dx\, dy\, dz$ is proportional to the *probability* that the electron is present in that element.

This concept is again analogous to that in classical physics, in which the intensity of radiation at any point is proportional to the square of the amplitude of the

*M. Born, Z. *Physik,* 37, 863; 38, 803 (1926).

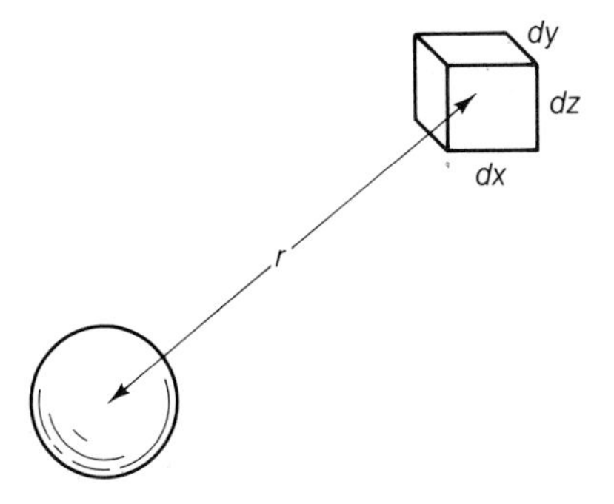

FIGURE 11.10 A volume element $dx\ dy\ dz$ at a distance r from an atomic nucleus. The probability that the electron is in the volume element is proportional to $\psi^2\ dx\ dy\ dz$ if ψ is real and to $\psi\psi^*\ dx\ dy\ dz$ if ψ is complex. If the function ψ is normalized (Eq. 11.89), the probability is *equal* to $\psi\psi^*\ dx\ dy\ dz$.

wave at that point. In other words, the greater the amplitude of the light wave in a particular region, the greater is the probability that a photon is present in that region. For an electron, Born's proposal, which is now accepted as correct, is that the probability of finding an electron within an element of volume $dx\ dy\ dz$ is proportional to

$$\psi^2(x, y, z,)\ dx\ dy\ dz$$

This is illustrated in Figure 11.10. In other words, the eigenfunction has provided information about the **probability density** or *probability distribution* of the electron.

For an eigenfunction to be an acceptable one, several conditions must be satisfied. One is that it must be *single valued,* since the probability in any region can have only one value. Another is that it must be *finite* in all regions of space; otherwise the Schrödinger equation would not apply since the electron would be located at a region where ψ was infinite and would not be acting as a wave. Third, the eigenfunction must be *continuous,* since discontinuities are not possible in nature. Finally, the eigenfunction must be *square integrable,* by which is meant that integrals such as

$$\int \psi^2(x, y, z)\ dx\ dy\ dz$$

over all space must be finite.

An eigenfunction may, however, be *complex;* that is, it may contain $i = \sqrt{-1}$. A complex function contains a real part and an imaginary part and may be written as $a + ib$. The square of this is $a^2 + 2iab - b^2$, which is also complex and cannot therefore represent a probability. To avoid this difficulty the probability is considered to be $\psi\psi^*$, where ψ^* is the complex conjugate of the function. If the function ψ is written as $a + ib$, the complex conjugate ψ^* is $a - ib$ and the product is $a^2 + b^2$, which is real.

We have said that the probability of finding an electron within the element $dx\ dy\ dz$ is *proportional* to $\psi\psi^*\ dx\ dy\ dz$. However it is possible to *normalize* an eigenfunction so that the probability is *equal* to this function. First we should note that there is an arbitrariness about a function ψ obtained by solving the Schrödinger equation, Eq. 11.86. Thus if ψ is a solution and we multiply ψ by any *number a*, the function $a\psi$ is still an eigenfunction, because

$$\hat{H}(a\psi) = E(a\psi) \tag{11.87}$$

Thus, for a given Schrödinger equation there is an infinite set of eigenfunctions differing from each other by numerical factors. We need to know which particular eigenfunction is such that

$$\psi\psi^* \, dx \, dy \, dz$$

is *equal* to the probability that the electron is in the volume element of dimensions $dx\,dy\,dz$. To determine this, we note that the total probability of finding the electron anywhere at all in space must be unity. The integral over all space is

$$\int_{-\infty}^{\infty}\int_{-\infty}^{\infty}\int_{-\infty}^{\infty} \psi\psi^* \, dx \, dy \, dz = 1 \tag{11.88}$$

This may conveniently be written as

$$\int \psi\psi^* \, d\tau = 1 \tag{11.89}$$

where $d\tau$ represents $dx\,dy\,dz$ and it is understood that the integral is over all space. This is known as the **normalization condition.**

Suppose that when Eq. 11.89 is integrated, the eigenfunction ϕ obtained by solving the Schrödinger equation gives us not unity but a number b:

$$\int \phi\phi^* \, d\tau = b \tag{11.90}$$

The eigenfunction therefore requires adjustment to make the integral unity. This is done by dividing both ϕ and ϕ^* by $\sqrt{b}$, and then

$$\int \frac{\phi}{\sqrt{b}} \frac{\phi^*}{\sqrt{b}} \, d\tau = 1 \tag{11.91}$$

The new function $\psi = \phi/\sqrt{b}$ is therefore such that $\psi\psi^* \, d\tau$ correctly represents the probability in any region. This process of adjusting an eigenfunction to satisfy Eq. 11.89 is known as **normalization,** and the resulting function is a *normalized eigenfunction.*

11.5 Quantum-Mechanical Postulates

In Section 11.4 we have seen how quantum mechanics was developed by analogy with the wave theory of electromagnetic radiation. This approach has been very fruitful, but for many purposes it is more useful to start with a series of postulates that can be shown to lead to conclusions that are verified experimentally. This approach leads to the treatment in Section 11.4, but it also leads to more general procedures.

Postulate I states that

The physical state of a particle is described as fully as possible by an appropriate wave function Ψ (x, y, z, t).

We have already considered this, and its implications, in Section 11.4.

Postulate II states that

The possible wave functions Ψ (x, y, z, t) are obtained by solving the appropriate Schrödinger equation.

More explicitly, the Schrödinger equation is written as Eq. 11.80, where $E_p(x, y, z, t)$ is the potential-energy function that relates to the system being considered. If E_p is independent of time, the equation reduces to

$$\left[-\frac{h^2}{8\pi^2 m}\nabla^2 + E_p(x, y, z)\right]\psi = E\psi \tag{11.92}$$

Postulate III states that

Every dynamical variable, corresponding to a physically observable quantity, can be expressed as a linear operator.

We will represent an operator by a circumflex accent over the symbol; for example, $\hat{o}$ is an operator, and $\hat{o}\psi$ means that $\hat{o}$ is operating on the function ψ. An operator is *linear* when for any two functions ψ and ϕ

$$\hat{o}(a\psi + b\phi) = a\hat{o}\psi + b\hat{o}\phi \tag{11.93}$$

where a and b are arbitrary numbers. Multiplication and taking a derivative are linear operators, but taking a square root is not; $\sqrt{\psi + \phi}$ is not in general equal to $\sqrt{\psi} + \sqrt{\phi}$. If operators were not linear, the wave functions obtained from a solution of the Schrödinger equation could not be superimposed.

The operator is always written to the left of the function on which it is to operate. The symbol $\hat{o}_2\hat{o}_1\psi$ means that we perform the operation $\hat{o}_1$ first and then perform the operation $\hat{o}_2$ on the result. Two operators are said to **commute** if for any arbitrary function

$$\hat{o}_1\hat{o}_2\psi = \hat{o}_2\hat{o}_1\psi \tag{11.94}$$

In other words, the order of operation is irrelevant. We will now show that *if two operators commute, any eigenfunction for one of them is also an eigenfunction for the other.* Suppose that the function ψ is an eigenfunction of $\hat{o}_1$; this means that

$$\hat{o}_1\psi = o_1\psi \tag{11.95}$$

where o_1 is the eigenvalue, which simply multiplies ψ. If we operate on both sides of Eq. 11.95 by $\hat{o}_2$, we obtain

$$\hat{o}_2\hat{o}_1\psi = \hat{o}_2 o_1\psi \tag{11.96}$$

The right-hand side can also be written as $o_1\hat{o}_2\psi$, since multiplication commutes with any operator. However, since the operators commute, the left-hand side of Eq. 11.96 is $\hat{o}_1\hat{o}_2\psi$ and therefore

$$\hat{o}_1(\hat{o}_2\psi) = o_1(\hat{o}_2\psi) \tag{11.97}$$

As indicated by the parentheses in this equation, $\hat{o}_2\psi$ is an eigenfunction for the operator $\hat{o}_1$. However, since ψ is an eigenfunction for $\hat{o}_1$, so also is $o_2\psi$; in other words

$$\hat{o}_1(o_2\psi) = o_1 o_2\psi = o_1(\hat{o}_2\psi) \tag{11.98}$$

and therefore

$$\hat{o}_2\psi = o_2\psi \tag{11.99}$$

The function ψ, an eigenfunction for $\hat{o}_1$, is therefore also an eigenfunction for $\hat{o}_2$.

EXAMPLE

The operator x is the one-dimensional position operator, and $(h/2\pi i)\,(\partial/\partial x)$ is the one-dimensional momentum operator. Do these operators commute?

SOLUTION

One way of dealing with this problem is to choose a function that is an eigenvalue of one of the operators and see if it is also an eigenvalue of the other. The function $2x$ is obviously an eigenvalue of x. Operating by the momentum operator gives

$$\frac{h}{2\pi i}\frac{\partial}{\partial x}(2x) = \frac{h}{\pi i}$$

Since the result is not a constant times $2x$, the function $2x$ is not an eigenfunction for the momentum operator.

Alternatively, we can first operate on the function $2x$ by the operator x and then by the momentum operator. Then we can reverse the order of operation and see if the result is the same. Thus

$$\frac{h}{2\pi i}\frac{\partial}{\partial x}(x)(2x) = \frac{h}{2\pi i}\frac{\partial}{\partial x}(2x^2) = \frac{2hx}{\pi i}$$

and

$$(x)\frac{h}{2\pi i}\frac{\partial}{\partial x}(2x) = (x)\frac{h}{\pi i} = \frac{hx}{\pi i}$$

Since the result is different in the two cases, the operators do not commute. ■

This result, that the position and momentum operators do not commute, is consistent with the Heisenberg uncertainty principle. If two operators commute, there is an eigenfunction for both operators, which means that the corresponding physical properties both have precise values. When two operators do not commute, the corresponding physical properties cannot both be measured simultaneously and precisely, and this is the case with position and momentum (compare Eq. 11.60).

Postulate IV states that

> **Quantum mechanical operators corresponding to physical properties are obtained from the classical expressions for these quantities, using certain established procedures.**

These procedures are summarized in Table 11.1. As an example of the application of the method, if we are concerned with calculating the energy for a conservative (time-independent) system, we start with the equation

$$\hat{H}\psi = E\psi \tag{11.100}$$

For the operator $\hat{H}$ we use

$$\hat{H} = -\frac{h^2}{8\pi^2 m}\nabla^2 + E_p \tag{11.101}$$

and at once obtain

$$\left(-\frac{h^2}{8\pi^2 m}\nabla^2 + E_p\right)\psi = E\psi \tag{11.102}$$

which is Eq. 11.85. This Schrödinger equation is a special case of many quantum-mechanical equations, for a variety of physical properties, that can be set up. The postulatory approach that we are using thus has a much wider applicability than the original Schrödinger equation.

TABLE 11.1 Common Quantum-Mechanical Operators as Derived from their Classical Expressions

Classical Variable	Quantum Mechanical Operator	Operation
Position		
x	$\hat{x}$ (similarly for the Y- and Z-directions)	x (multiplication)
Linear momentum		
p_x (X-direction)	$\hat{p}_x$ (similarly for the Y- and Z- directions)	$\frac{h}{2\pi i}\left(\frac{\partial}{\partial x}\right)$
Angular momentum		
L_z (rotation about the Z axis)	$\hat{L}_z$	$\frac{h}{2\pi i}\left(\frac{\partial}{\partial \phi}\right)$
Time		
t	$\hat{t}$	t (multiplication)
Kinetic energy		
E_k	$\hat{E}_k$	$-\frac{h^2}{8\pi^2 m}\nabla^2$
Potential energy		
$E_p(x, y, z)$	$\hat{E}_p$	E_p (multiplication)
Total energy (time-dependent system)		
$E = E_k + E_p$	$\hat{E}$	$-\frac{h}{2\pi i}\left(\frac{\partial}{\partial t}\right)$
Total energy (conservative system)		
H (Hamiltonian)	$\hat{H} = \hat{E}_k + \hat{E}_p$	$-\frac{h^2}{8\pi^2 m}\nabla^2 + E_p$

Postulate V states that

> **When the equation $\hat{F}\psi = F\psi$ has been set up, with $\hat{F}$ corresponding to a physical quantity (as in Table 11.1), the eigenvalues F that are obtained by solving the equation represent all possible values of an individual measurement of that quantity.**

For example, solution of the energy equation

$$\hat{H}\psi = E\psi \tag{11.103}$$

gives us all the possible values of the energy E for a particular system.

Postulate VI states that

> **The mean value $\bar{F}$ of a physical quantity, or the expectation value $\langle F\rangle$ when measurements are made for a large number of particles, is given by the equation**

$$\bar{F} \equiv \langle F\rangle = \frac{\int_{-\infty}^{\infty} \Psi^*\hat{F}\Psi \, d\tau}{\int_{-\infty}^{\infty} \Psi^*\Psi \, d\tau} \tag{11.104}$$

Here $d\tau$ stands for $dx\,dy\,dz$; the integration from minus infinity to infinity means that we are integrating over all space. The denominator in this expression allows for the possibility that the wave functions are not normalized; if they are normalized, only the numerator is needed.

This postulate allows us to calculate average values for quantities for which stationary-state values cannot be measured exactly. If Ψ is an eigenfunction of $\hat{F}$, that is, if

$$\hat{F}\Psi = F\Psi \tag{11.105}$$

$\bar{F}$ is equal to F; that is, the eigenvalue is the average value.

Measurable physical properties must be real and not complex. The operators in Table 11.1 all belong to the class of **Hermitian operators**, which means that the following relationship applies to them:

$$\int \psi_1^*\hat{F}\psi_2 \, d\tau = \int \psi_2\hat{F}^*\psi_1^* \, d\tau \tag{11.106}$$

We now show that all measurable quantities (i.e., eigenvalues corresponding to operators such as those in Table 11.1) are real. We take complex conjugates of everything in Eq. 11.104 and obtain

$$\bar{F}^* = \frac{\int_{-\infty}^{\infty} \Psi\hat{F}^*\Psi^* \, d\tau}{\int_{-\infty}^{\infty} \Psi\Psi^* \, d\tau} \tag{11.107}$$

The denominators in Eqs. 11.104 and 11.107 are identical, and the numerators are equal because of the Hermitian condition. Thus

$$\bar{F}^* = \bar{F} \tag{11.108}$$

which means that $\bar{F}$ is real.

Orthogonality of Wave Functions

Eigenfunctions for a given Hermitian operator and corresponding to different eigenvalues are necessarily orthogonal to one another. Suppose, for example, that ψ_1 and ψ_2 are two different eigenfunctions for the Hermitian operator $\hat{F}$ and that the eigenvalues f_1 and f_2 are different from one another:

$$\hat{F}\psi_1 = f_1\psi_1 \tag{11.109}$$

and

$$\hat{F}\psi_2 = f_2\psi_2 \tag{11.110}$$

We will now show that ψ_1 and ψ_2 are orthogonal to each other, which means that the relationship

$$\int \psi_1^*\psi_2 \, d\tau = 0 \tag{11.111}$$

is true. When we use the simplified notation of Eq. 11.111, it is to be understood that the integration is over all space.

The proof is as follows. Multiplication of both sides of Eq. 11.109 by ψ_2^* and integration over all space gives

$$\int \psi_2^*\hat{F}\psi_1 \, d\tau = f_1 \int \psi_2^*\psi_1 \, d\tau \tag{11.112}$$

If we take complex conjugates of both sides of Eq. 11.110, multiply by ψ_1, and integrate, we obtain

$$\int \psi_1\hat{F}^*\psi_2^* \, d\tau = f_2^* \int \psi_1\psi_2^* \, d\tau \tag{11.113}$$

The Hermitian condition requires that the left-hand sides of Eqs. 11.112 and 11.113 are equal so that, since $f_2^* = f_2$ (the eigenvalues must be real), we have

$$f_1 \int \psi_2^*\psi_1 \, d\tau = f_2 \int \psi_1\psi_2^* \, d\tau \tag{11.114}$$

These two integrals are identical, but since by hypothesis f_1 is not equal to f_2, the integrals must be zero:

$$\boldsymbol{\int \psi_1\psi_2^* \, d\tau = 0} \tag{11.115}$$

This is the **orthogonality** condition.

The situation is somewhat different if the eigenvalues for two eigenfunctions are equal to one another, in which case the eigenvalue is said to be **degenerate.** Suppose, for example, that two eigenfunctions ψ_1 and ψ_2 have the same eigenvalue f; in other words $\hat{F}\psi_1 = f\psi_1$ and $\hat{F}\psi_2 = f\psi_2$. It can then be seen at once that *any linear combination of ψ_1 and ψ_2 is also an eigenfunction*, because

$$\hat{F}(c_1\psi_1 + c_2\psi_2) = f(c_1\psi_1 + c_2\psi_2) \tag{11.116}$$

where c_1 and c_2 can be constants. It is not necessarily the case that ψ_1 and ψ_2 are orthogonal to each other, but it is always possible to take linear combinations of them and to obtain orthogonal functions. The way this is done is explained in books on quantum mechanics (see Selected Reading, p. 513).

A special case of Eq. 11.116 is when the operator is the Hamiltonian operator $\hat{H}$, in which case the eigenvalue is the energy E:

$$\hat{H}(c_1\psi_1 + c_2\psi_2) = E(c_1\psi_1 + c_2\psi_2) \tag{11.117}$$

If there are two wave functions ψ_1 and ψ_2 for an energy E, the energy level is said to have *twofold degeneracy*. Again, it is often convenient to take linear combinations of ψ_1 and ψ_2 and convert them into two wave functions that are orthogonal to each other. If there are n wave functions corresponding to an energy E, the energy level is said to have *n-fold degeneracy*.

It is often useful (for example, in connection with the theory of *resonance* and when using the *variation method*) to work with functions that are linear combinations of wave functions that correspond to different energies. For example, suppose that ψ_1 and ψ_2 are normalized wave functions for a given Hamiltonian operator; we have seen that they must be orthogonal . We can construct the normalized function

$$\psi_3 = c_1\psi_1 + c_2\psi_2 \tag{11.118}$$

where c_1 and c_2 are constants. Then

$$\int \psi_3^*\psi_3 \, d\tau = \int (c_1^*\psi_1^* + c_2^*\psi_2^*)(c_1\psi_1 + c_2\psi_2) \, d\tau \tag{11.119}$$

$$= c_1^*c_1 \int \psi_1^*\psi_1 \, d\tau + c_2^*c_2 \int \psi_2^*\psi_2 \, d\tau$$

$$+ c_1^*c_2 \int \psi_1^*\psi_2 \, d\tau + c_2^*c_1 \int \psi_2^*\psi_1 \, d\tau \tag{11.120}$$

The integral on the left-hand side is unity because of the normalization condition, and the same is true of the first two integrals on the right-hand side. The remaining two integrals are zero because of the orthogonality condition. It therefore follows that

$$c_1^*c_1 + c_2^*c_2 = 1 \tag{11.121}$$

This relationship is useful for the construction of suitable linear combinations of wave functions. It is easily extended to more than two wave functions.

11.6 Quantum Mechanics of Some Simple Systems

We will now apply the postulates of Section 11.5 to some simple systems that are related to some important chemical problems.

The Free Particle

The simplest application is to a particle of mass m moving freely in one dimension, which we will take to be the X-direction. This is known as the *free particle problem in one dimension*. Since the particle is moving freely with no forces acting on it, the potential energy E_p is constant throughout its motion and we take it to be zero. The total energy is therefore the kinetic energy $\frac{1}{2}mu_x^2$, which is $p_x^2/2m$ where p_x is the momentum. The total energy in Hamilton's form is thus

$$H = \frac{p_x^2}{2m} \tag{11.122}$$

We see from Table 11.1 that p_x is to be replaced by the operator $(h/2\pi i)\,(\partial/\partial x)$, so that the quantum-mechanical operator is

$$\hat{H} = \frac{1}{2m}\left(\frac{h}{2\pi i}\right)^2 \frac{\partial^2}{\partial x^2} = -\frac{h^2}{8\pi^2 m}\frac{\partial^2}{\partial x^2} \tag{11.123}$$

The Schrödinger equation $\hat{H}\psi = E\psi$ thus takes the form

$$-\frac{h^2}{8\pi^2 m}\frac{\partial^2\psi}{\partial x^2} = E\psi \tag{11.124}$$

or

$$\frac{\partial^2\psi}{\partial x^2} + \frac{8\pi^2 mE\psi}{h^2} = 0 \tag{11.125}$$

There are two solutions to this differential equation, namely

$$\psi_1 = Ae^{i\sqrt{8\pi^2 mE}x/h} \tag{11.126}$$

and

$$\psi_2 = Be^{-i\sqrt{8\pi^2 mE}x/h} \tag{11.127}$$

where A and B are constants. The fact that these are solutions may easily be verified by substituting them into Eq. 11.125.

The relative probability density of the particle is $|\psi|^2 = \psi^*\psi$, and for the function ψ_1 this is

$$|\psi_1|^2 = \psi^*\psi = A^*e^{-i\sqrt{8\pi^2 mE}x/h}Ae^{i\sqrt{8\pi^2 mE}x/h} \tag{11.128}$$

$$= A^*A = |A|^2 \tag{11.129}$$

Similarly for the function ψ the relative probability is $|B|^2$. Both of these quantities are independent of x, so that there is an equal probability of finding the particle at any distance along the X axis. The particle is therefore *nonlocalized.*

We can also use Postulate V to obtain allowed values of the momentum p_x. The momentum operator $\hat{p}_x$ is $(h/2\pi i)\,(\partial/\partial x)$, and the wave equation is therefore

$$\frac{h}{2\pi i}\frac{\partial\psi}{\partial x} = p_x\psi \tag{11.130}$$

If $\psi = \psi_1$ from Eq. 11.126, we obtain

$$\frac{h}{2\pi i}\frac{\partial}{\partial x}(Ae^{i\sqrt{8\pi^2 mE}x/h}) = p_xAe^{i\sqrt{8\pi^2 mE}x/h} \tag{11.131}$$

This reduces to

$$\sqrt{2mE}Ae^{i\sqrt{8\pi^2 mE}x/h} = p_xAe^{i\sqrt{8\pi^2 mE}x/h} \tag{11.132}$$

and therefore

$$p_x = \sqrt{2mE} \qquad (11.133)$$

Similarly with ψ_2 it is found that $p_x = -\sqrt{2mE}$, so that the two possible solutions are

$$p_x = \pm\sqrt{2mE} \qquad (11.134)$$

These solutions correspond to the classical values; p_x is positive if the particle is moving in one direction and negative if it is moving in the other. For this particular case there is therefore no quantization; we can give the particle any energy we like.

An example of a free particle is an electron that has been separated from an atom; its energy is no longer quantized.

The Particle in a Box

Quantization does, however, appear if the particle is not permitted to travel an infinite distance but is confined to a certain region of space. In three dimensions this problem is referred to as the *particle in a box*. We will first consider the one-dimensional problem, with the particle moving along the X axis over a distance from 0 to a, as shown in Figure 11.11a. The potential energy E_p is taken to be zero within the box ($0 < x < a$) and infinity for $x < 0$ and $x > a$.

Within the box the wave equation is Eq. 11.124, as for the free particle, and the solutions are given by Eqs. 11.126 and 11.127. We now, however, have the additional conditions that ψ must be zero when $x = 0$ and $x = a$ and must be zero for $0 > x > a$. Neither ψ_1 nor ψ_2 as given by these equations can *individually* satisfy these *boundary conditions*, because the condition that ψ must be zero when $x = a$ requires both A and B to be zero. However, this dilemma is avoided by taking a linear combination of ψ_1 and ψ_2, which we know is also a solution of the wave equation (compare Eq. 11.116). It is perfectly general to take $\psi_1 + \psi_2$ as the linear combination, since A and B are in any case adjustable. Our selected wave function is therefore

$$\psi = Ae^{i\sqrt{8\pi^2 mE}x/h} + Be^{-i\sqrt{8\pi^2 mE}x/h} \qquad (11.135)$$

We will now see if this function is consistent with the boundary conditions, and if so with what restrictions. The boundary condition that $\psi = 0$ when $x = 0$ requires that $A + B = 0$, so that $B = -A$. Equation 11.135 can thus be rewritten as

$$\psi = A(e^{i\sqrt{8\pi^2 mE}x/h} - e^{-i\sqrt{8\pi^2 mE}x/h}) \qquad (11.136)$$

It is now convenient to apply Euler's theorem* and rewrite this equation as

$$\psi = 2iA \sin(\frac{\sqrt{8\pi^2 mE}}{h}x) \qquad (11.137)$$

The second boundary condition, that $\psi = 0$ when $x = a$, gives

$$0 = 2iA \sin(\frac{\sqrt{8\pi^2 mE}}{h}a) \qquad (11.138)$$

*According to Euler's theorem, $e^{iy} = \cos y - i \sin y$, and therefore $e^{iy} - e^{-iy} = 2i \sin y$.

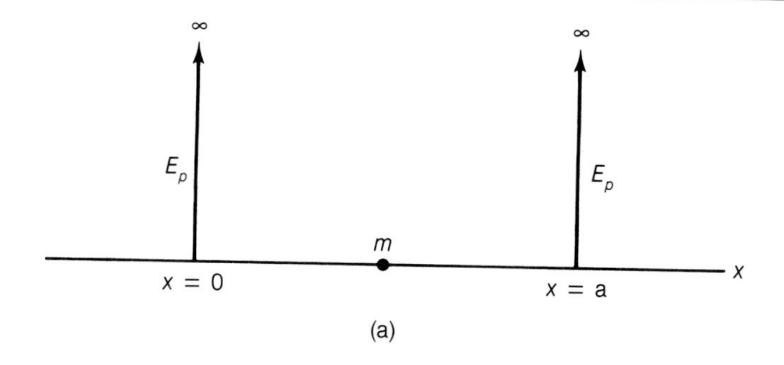

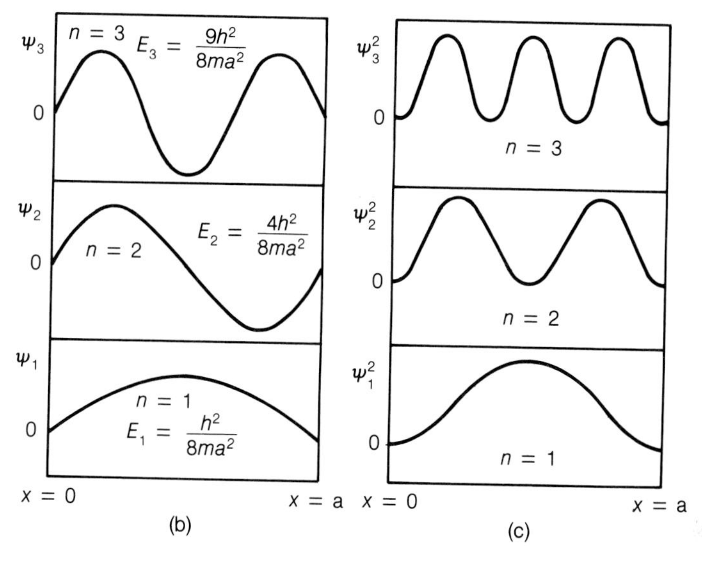

FIGURE 11.11 (a) A particle confined to a one-dimensional box of length a. The potential energy is infinite for $x < 0$ and $x > a$. (b) The form of some of the wave functions for the particle in a one-dimensinal box and the corresponding energies. (c) The form of the probability densities ψ^2.

The factor $2iA$ cannot be zero, but the sine of an angle is zero when the angle is an integral multiple of π. Thus

$$\frac{\sqrt{8\pi^2 mE_n}}{h}a = \pm n\pi \tag{11.139}$$

where $n = 0, 1, 2, \ldots, \infty$. We have now written E_n instead of E since there is a different energy E_n for each value of n. The wave function (Eq. 11.137) therefore becomes

$$\psi_n = \pm 2iA \sin \frac{n\pi x}{a} \tag{11.140}$$

Note that the $\pm$ sign can go outside the sine function since $\sin(-n\pi x/a) = -\sin(n\pi x/a)$.

To determine the value of $2iA$ we use the normalization condition that the total probability $\psi_n^*\psi_n$ of finding the particle in the box must be unity. Thus

$$\int_0^a \psi_n^*\psi_n \, dx = \pm 4A^2 \int_0^a \sin^2\left(\frac{n\pi x}{a}\right) dx = 1 \tag{11.141}$$

The value of the integral is $a/2$ and therefore

$$\pm 4A^2 \frac{a}{2} = 1 \tag{11.142}$$

or

$$A = \pm\sqrt{\frac{\pm 1}{2a}} \tag{11.143}$$

To make the wave function real, we select the negative sign within the square root so that $A = \pm i\sqrt{1/2a}$. The wave function (Eq. 11.140) thus becomes

$$\psi_n = \pm\sqrt{\frac{2}{a}} \sin\left(\frac{n\pi x}{a}\right) \tag{11.144}$$

The quantum numbers can now be given as $n = 1, 2, 3, \ldots, \infty$, since $n = 0$ would mean that no particle is present.

The possible energy values are obtained by rearrangement of Eq. 11.139:

$$E_n = \frac{n^2h^2}{8ma^2} \tag{11.145}$$

These are therefore the possible stationary-state energies for a particle in a one-dimensional box.

The wave functions for various values of n and the corresponding energies are shown in Figure 11.11b, and the probability densities $\psi^*\psi$ are shown in Figure 11.11c. Note that the energy increases in proportion to n^2 and that the number of antinodes is equal to n.

The energy levels and, therefore, the separations between them are inversely proportional to ma^2. Thus if the particles are heavy or if the box is very large, the energy levels are very close together and the system behaves classically, with no observable quantization. This result is consistent with Bohr's correspondence principle, according to which at certain limits quantum-mechanical behavior becomes classical behavior. At the other extreme, if m and a are of atomic dimensions, the quantization is important and can easily be detected experimentally. The following two examples illustrate the effect of mass and box size.

EXAMPLE

Calculate the energy difference between the $n = 1$ and $n = 2$ levels for an electron ($m = 9.1 \times 10^{-31}$ kg) confined to a one-dimensional box having a length of 4.0×10^{-10} m (this is the order of magnitude of an atomic diameter). What wavelength corresponds to a spectral transition between these levels?

SOLUTION

The ground-state energy is obtained from Eq. 11.145 with $n = 1$:

$$E_1 = \frac{[6.626 \times 10^{-34}\ (\text{J s})]^2}{8[9.1 \times 10^{-31}(\text{kg})][4.0 \times 10^{-10}(\text{m})]^2}$$

$$= 3.77 \times 10^{-19}\ \text{J}$$

The energy for $n = 2$ is just four times this, and the energy difference is therefore three times this value:

$$\Delta E = E_2 - E_1 = 3 \times 3.77 \times 10^{-19}\ \text{J} = 1.33 \times 10^{-18}\ \text{J}$$

The frequency is this energy divided by the Planck constant, and the wavelength is the velocity of light divided by the frequency:

$$\lambda = \frac{c}{\nu} = \frac{hc}{\Delta E} = \frac{6.626 \times 10^{-34}(\text{J s})\ 2.998 \times 10^{8}(\text{m s}^{-1})}{1.13 \times 10^{-18}(\text{J})}$$

$$= 1.76 \times 10^{-7}\ \text{m} = 176\ \text{nm}$$

This wavelength is in the ultraviolet. ■

EXAMPLE

Calculate the energy difference between the $n = 1$ and $n = 2$ levels for a marble of mass 1 g confined in a one-dimensional box of length 0.10 m. What wavelength corresponds to a spectral transition between these levels?

SOLUTION

The ground-state ($n = 1$) energy is

$$E_1 = \frac{[6.626 \times 10^{-34}(\text{J s})]^2}{8[0.001(\text{kg})][0.10(\text{m})]^2}$$

$$= 5.48 \times 10^{-63}\ \text{J}$$

The energy difference ΔE is three times this:

$$E_2 - E_1 = 3 \times 5.48 \times 10^{-63}\ \text{J} = 1.65 \times 10^{-62}\ \text{J}$$

Such an energy is much too small to be measurable; it is impossible to measure an energy of less than about 10^{-25} J. The corresponding wavelength is

$$\lambda = \frac{hc}{\Delta E} = 1.21 \times 10^{87}\ \text{m}$$

This wavelength is quite impossible to measure, as it is greater than the distance to any observed star! No quantization can be detected in such a system. ■

So far we have considered only a one-dimensional box. The extension to a three-dimensional box is quite straightforward, and we will simply state the result. For a particle confined in a box of sides a, b, and c, the allowed energy levels are

$$E = \frac{h^2}{8m}\left(\frac{n_1^2}{a^2} + \frac{n_2^2}{b^2} + \frac{n_3^2}{c^2}\right) \tag{11.146}$$

There are now three distinct quantum numbers, n_1, n_2, and n_3. If the box is a cube of sides a, the expression becomes

$$E = \frac{h^2}{8ma^2}(n_1^2 + n_2^2 + n_3^2) \tag{11.147}$$

It is now possible for three different combinations of quantum numbers to give the same energy. For example, the same energy is found for

$$n_1 = 2, \qquad n_2 = 1, \qquad n_3 = 1$$

as for

$$n_1 = 1, \qquad n_2 = 2, \qquad n_3 = 1$$

and for

$$n_1 = 1, \qquad n_2 = 1, \qquad n_3 = 2$$

The energy level is thus *degenerate*. In this case we speak of *threefold degeneracy*, since there are three combinations of quantum numbers that give the same energy.

EXAMPLE

What is the degree of the degeneracy if the three quantum numbers n_1, n_2, and n_3 can have the values 1, 2, and 3?

SOLUTION

The following combinations are possible:

123 132 213

231 312 321

There is therefore *sixfold* degeneracy. ■

The Harmonic Oscillator

Another quantum-mechanical problem of great importance is that of the vibrating molecule. The simplest case to consider is a diatomic molecule in which the two atomic masses are m_1 and m_2. To a good approximation the vibration of such a molecule obeys the equations of simple harmonic motion (Eqs. 11.8 to 11.18). However, as we will discuss in more detail in Section 13.4, we must replace the mass m in those equations by a *reduced mass* μ, given by

$$\mu = \frac{m_1 m_2}{m_1 + m_2} \tag{11.148}$$

If the bond is extended by a distance x, the potential energy is given by

$$E_p = \frac{1}{2} k_h x^2 \tag{11.149}$$

(compare Eq. 11.17), where k_h is the force constant. The kinetic energy is

$$E_k = \frac{p_x^2}{2\mu} \tag{11.150}$$

and the total energy in Hamiltonian form is therefore

$$H = \frac{p_x^2}{2\mu} + \frac{1}{2}k_h x^2 \tag{11.151}$$

The Hamiltonian operator is then obtained by replacing p_x by $(h/2\pi i)(\partial/\partial x)$:

$$\hat{H} = -\frac{h^2}{8\pi^2\mu}\frac{\partial^2}{\partial x^2} + \frac{1}{2}k_h x^2 \tag{11.152}$$

The Schrödinger equation for the system is therefore

$$\left(-\frac{h^2}{8\pi^2\mu}\frac{\partial^2}{\partial x^2} + \frac{1}{2}k_h x^2\right)\psi = E\psi \tag{11.153}$$

The solution of this equation involves the use of *Hermite polynomials.* We will not give details but will merely state the main conclusions. Solutions can only be obtained provided that the total energy E has a value given by the equation

$$E_v = \frac{h}{2\pi}\sqrt{\frac{k_h}{\mu}}\left(v + \frac{1}{2}\right) \tag{11.154}$$

Here v is the *vibrational quantum number,* which can have only integral values:

$$v = 0, 1, 2, 3, \ldots \tag{11.155}$$

Equation 11.154 can also be written as

$$\mathbf{E_v = h\nu_0\left(v + \frac{1}{2}\right)} \tag{11.156}$$

where ν_0, equal to $(\frac{1}{2}\pi)\sqrt{k_h/\mu}$, is the natural frequency of the harmonic oscillator (compare Eq. 11.15).

We can see from Eq. 11.156 that the quantized energy levels for the harmonic oscillator are equidistant from one another. A very important feature is that the lowest possible energy, corresponding to $v = 0$, is equal to

$$E_0 = \tfrac{1}{2}h\nu_0 \tag{11.157}$$

In classical theory a bond could be undergoing no vibration, but in quantum mechanics this is not allowed; even at the absolute zero of temperature vibration will still be occurring with this energy $\frac{1}{2}h\nu_0$. This energy is known as the *zero-point energy*. The fact that there is a zero-point energy is a consequence of Heisenberg's principle of indeterminancy. If no vibration occurred, the position and momentum of the atoms would both have precise values, and this is impossible.

The wave functions corresponding to the various energy levels involve the Hermite polynomials, and the shapes of some of them are shown in Figure 11.12a. The curves for even values of v are symmetric with respect to $x = 0$, while with v odd the curves are antisymmetric. The probability density functions $\psi^*\psi$ for several values of v are shown in Figure 11.12b. Also shown on this diagram are the corresponding classical probability functions. The differences between the classical and the quantum-mechanical behavior are very striking. For $v = 0$, for example, we see that the highest probability according to the quantum-mechanical treatment is when the system is passing through its equilibrium position ($x = 0$). In classical mechanics, on the other hand, the probability is *lowest* at the equilibrium position,

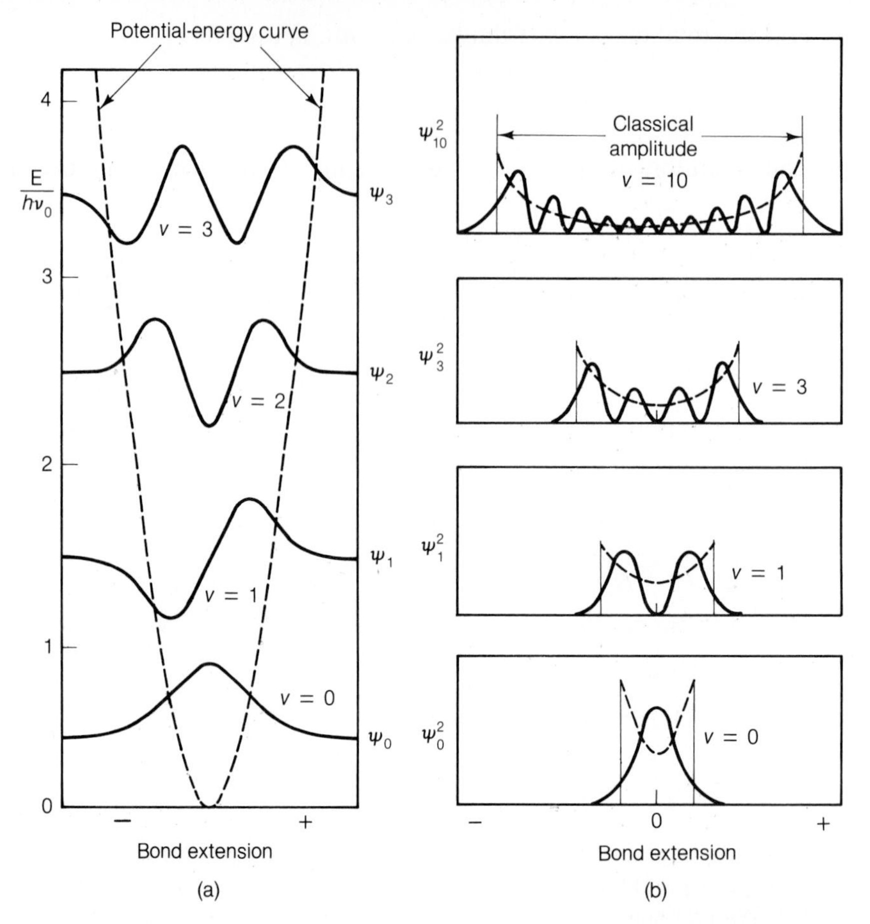

FIGURE 11.12 (a) The form of some wave functions for a harmonic oscillator. (b) Some probability density functions for a harmonic oscillator. The corresponding classical functions are shown as dashed lines.

since the system has its highest velocity at this position and therefore passes through it rapidly; the vibrational motion is slowest at the extremities of the vibration, and the probability is therefore the highest at these extremities. As the vibrational quantum number increases, however, the quantum-mechanical probabilities become closer to the classical ones, and the highest probabilities are close to the turning points of the vibrations.

Another important difference between the quantum-mechanical and classical probabilities is that the quantum-mechanical treatment gives a finite probability that the bond is extended to a greater extent than the value permitted by classical theory. In a classical vibration the extension is restricted to a certain fixed value, but the quantum-mechanical curves in Figure 11.12b show that there is a certain probability of an even greater extension. There are a number of situations where quantum mechanics permits a system to penetrate into regions that are forbidden in classical mechanics, and the effect is known as the **tunnel effect.**

11.7 Quantum Mechanics of Hydrogenlike Atoms

Bohr's theory of the hydrogen atom was a great step forward, but it failed to provide a satisfactory basis for the understanding of more complex atoms. Much greater success has been achieved through quantum mechanics. We will consider treatment of a hydrogenlike atom, having a nucleus of charge Ze and a single electron. Since the mass of the electron is very much smaller than that of the nucleus, the reduced mass μ, given by Eq. 11.148, is almost exactly equal to the mass of the electron.

The potential energy of the electron at a distance r from the nucleus arises entirely from the Coulombic attraction and is given by

$$E_p = -\frac{Ze^2}{4\pi\epsilon_0 r} \tag{11.158}$$

where ϵ_0 is the permittivity of a vacuum. The energy is independent of the direction, so that we have a *symmetrical field* or a *central field*.

The Hamiltonian for the system involves the components of momentum along the three axes and is

$$H = \frac{1}{2\mu}(p_x^2 + p_y^2 + p_z^2) - \frac{Ze^2}{4\pi\epsilon_0 r} \tag{11.159}$$

The Hamiltonian operator is obtained by making the substitutions of Table 11.1 and is

$$\hat{H} = -\frac{h^2}{8\pi^2\mu}\nabla^2 - \frac{Ze^2}{4\pi\epsilon_0 r} \tag{11.160}$$

The time-independent Schrödinger equation $\hat{H}\psi = E\psi$ is therefore

$$-\frac{h^2}{8\pi^2\mu}\nabla^2\psi - \frac{Ze^2}{4\pi\epsilon_0 r}\psi = E\psi \tag{11.161}$$

Since the system is spherically symmetrical, it is most convenient to use spherical polar coordinates, which are related to Cartesian coordinates as shown in Figure 11.13. When the Laplacian operator ∇^2 is converted to polar coordinates, the Schrödinger equation takes the form

$$\frac{1}{r^2}\frac{\partial}{\partial r}\left(r^2\frac{\partial\psi}{\partial r}\right) + \frac{1}{r^2 \sin\theta}\frac{\partial}{\partial\theta}\left(\sin\theta\frac{\partial\psi}{\partial\theta}\right) + \frac{1}{r^2\sin^2\theta}\frac{\partial^2\psi}{\partial\phi^2} + \frac{8\pi^2\mu}{h^2}\left(E + \frac{Ze^2}{4\pi\epsilon_0 r}\right)\psi = 0 \tag{11.162}$$

This partial differential equation may be separated into three ordinary differential equations, in each of which we have only one of the variables r, θ, and ϕ. To do this we write the wave function ψ, which is a function of r, θ, and ϕ, as the product of three functions $R(r)$, $\Theta(\theta)$, and $\Phi(\phi)$:

$$\psi(r, \theta, \phi) = R(r)\Theta(\theta)\Phi(\phi) \tag{11.163}$$

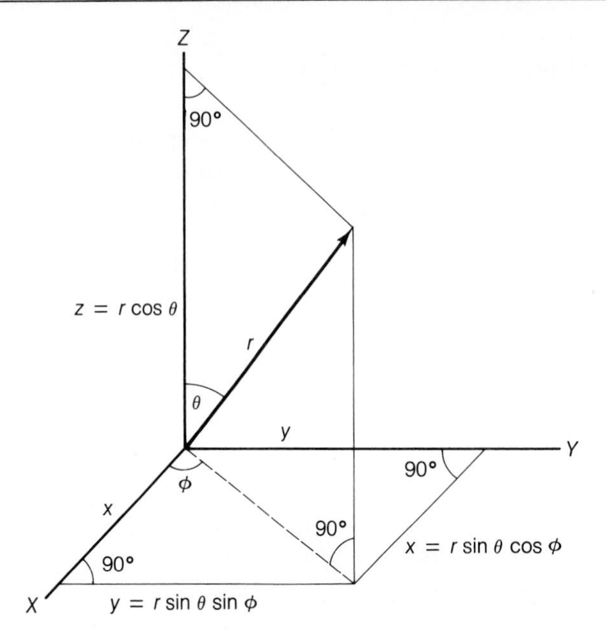

FIGURE 11.13 The relation between Cartesian and spherical polar coordinates.

Substitution of this expression into Eq. 11.162 and division by $R\Theta\Phi$ give

$$\frac{1}{r^2R}\frac{d}{dr}\left(r^2\frac{dR}{dr}\right) + \frac{1}{r^2(\sin\theta)\Theta}\frac{d}{d\theta}\left(\sin\theta\frac{d\Theta}{d\theta}\right) + \frac{1}{r^2\sin^2\theta}\cdot\frac{1}{\Phi}\frac{d^2\Phi}{d\phi^2} + \frac{8\pi^2\mu}{h^2}\left(E + \frac{Ze^2}{4\pi\epsilon_0 r}\right) = 0 \quad (11.164)$$

The partial derivatives have been replaced by ordinary derivatives since each function now depends on a single variable only.

The third term in this expression contains

$$\frac{1}{\Phi}\cdot\frac{d^2\Phi}{d\phi^2}$$

and this is the only term in which Φ and ϕ appear. Since the three polar coordinates are quite independent of one another, this term must therefore be constant. We will write it as

$$\frac{1}{\Phi}\cdot\frac{d^2\Phi}{d\phi^2} = -m_l^2 \quad (11.165)$$

and we will see later that m_l is the **magnetic quantum number.** We will call Eq. 11.165 the Φ *equation.*

If we make this substitution (Eq. 11.165) into Eq. 11.164 and then multiply by r^2, we obtain, after some rearrangement,

$$\frac{1}{R}\frac{d}{dr}\left(r^2\frac{dR}{dr}\right) + \frac{8\pi^2\mu}{h^2}\left(E + \frac{Ze^2}{4\pi\epsilon_0 r}\right)r^2 = \frac{m_l^2}{\sin^2\theta} - \frac{1}{\sin\theta}\frac{1}{\Theta}\frac{d}{d\theta}\left(\sin\theta\frac{d\Theta}{d\theta}\right) \quad (11.166)$$

The left-hand side involves R and r but not Θ and θ, while the right-hand side involves Θ and θ but not R and r. Both sides must therefore be equal to a *separation constant,* which we will write as $l(l + 1)$ in anticipation of the fact that l is the **azimuthal quantum number.** Equating the left-hand side of Eq. 11.166 to $l(l + 1)$ gives the *radial equation,* or *R equation:*

$$\frac{1}{R}\frac{d}{dr}\left(r^2\frac{dR}{dr}\right) + \frac{8\pi^2\mu}{h^2}\left(E + \frac{Ze^2}{4\pi\epsilon_0 r}\right)r^2 = l(l + 1) \tag{11.167}$$

Similarly, from the right-hand side, with a little rearrangement, we obtain the *angular equation,* or Θ *equation:*

$$\frac{1}{\sin\theta}\frac{d}{d\theta}\left(\sin\theta\frac{d\Theta}{d\theta}\right) - \frac{m_l^2\Theta}{\sin^2\theta} + l(l + 1)\Theta = 0 \tag{11.168}$$

We have thus split the Schrödinger equation (Eq. 11.164) into three equations, one involving Φ and ϕ (Eq. 11.165), one involving R and r (Eq. 11.167), and one involving Θ and θ (Eq. 11.168). These equations must now be solved so as to eliminate the differentials. We will first solve the Φ equation so as to obtain the allowed values of m_l. These will then be used to solve the Θ equation so as to obtain the allowed l values. Finally, the l values will be used to solve the R equation.

Solution of the Φ Equation

Equation 11.165 is of a familiar form (compare Eq. 11.125) and its solution is

$$\Phi = Ae^{im_l\phi} \tag{11.169}$$

where A is a normalization constant. The function Φ must have the same value at $\phi = 0$, $\phi = 2\pi$, $\phi = 4\pi$, . . . , because these angles correspond to the same position; this requires that

$$m_l = 0, \pm1, \pm2, \pm3, \ldots \tag{11.170}$$

The positive and negative values relate to distinct solutions. The quantity m_l has thus become a quantum number because of the mathematical constraints on the system and not in any arbitrary way.

The value of A is obtained by applying the normalization condition, the range of ϕ being 0 to 2π:

$$\int_0^{2\pi}\Phi_{m_l}\Phi_{m_l}^*\,d\phi = A^2\int_0^{2\pi}e^{im_l\phi}e^{-im_l\phi}\,d\phi = A^2 2\pi = 1 \tag{11.171}$$

Therefore $A = 1/\sqrt{2\pi}$ and the solution becomes

$$\Phi = \frac{1}{\sqrt{2\pi}}e^{im_l\phi} \tag{11.172}$$

When $m_l = 0$, the value Φ_0 is $1/\sqrt{2\pi}$, but for other values of m_l the solutions involve imaginary exponents. This is awkward, and it is more usual to employ a linear combination of the functions, $\Phi_{m_l} \pm \Phi_{m_l'}$; as we have seen (Eq. 11.116), linear combinations of wave functions having the same eigenvalue are also solutions of wave equations. For example,

$$\text{for } m_l = 1, \qquad \Phi_1 = \frac{1}{\sqrt{2\pi}} e^{i\phi} \tag{11.173}$$

$$\text{for } m_l = -1, \qquad \Phi_{-1} = \frac{1}{\sqrt{2\pi}} e^{-i\phi} \tag{11.174}$$

We can take the sum of these and divide by $\sqrt{2}$ to preserve the normalization,

$$\Phi_x = \frac{1}{\sqrt{2}}(\Phi_1 + \Phi_{-1}) = \frac{1}{2\sqrt{\pi}}(e^{i\phi} + e^{-i\phi}) = \frac{\cos \phi}{\sqrt{\pi}} \tag{11.175}$$

We designate this orbital Φ_x since $\cos \phi$ has its maximum value when $\phi = 0$, which (as we see from Figure 11.13) corresponds to the X axis. Similarly we can take the difference $\Phi_1 - \Phi_{-1}$ divided by $\sqrt{2}$:

$$\Phi_y = \frac{1}{\sqrt{2}}(\Phi_1 - \Phi_{-1}) = \frac{1}{2\sqrt{\pi}}(e^{i\phi} - e^{-i\phi}) = \frac{i \sin \phi}{\sqrt{\pi}} \tag{11.176}$$

Since we are usually interested in probability densities $\Phi\Phi^*$, is it common to drop the i in this expression, since it disappears when we take the complex conjugate. The function is therefore written as

$$\Phi_y = \frac{\sin \phi}{\sqrt{\pi}} \tag{11.177}$$

The value of this function is a maximum when $\phi = \pi/2$, which is along the Y axis.

Table 11.2 lists Φ functions for the first three values of $\pm m_l$.

Solution of the Θ Equation

The solution of the Θ equation (Eq. 11.168) is mathematically more difficult, and we will present only a very brief outline, with emphasis on the main results. For details the reader is referred to textbooks of quantum mechanics.*

We may introduce a transformation into Eq. 11.168 by putting

$$\xi = \cos \theta \quad \text{and} \quad P_l(\xi) = \Theta \tag{11.178}$$

and obtain

$$(1 - \xi)^2 \frac{d^2 P_l(\xi)}{d\xi^2} - 2\xi \frac{dP_l(\xi)}{d\xi} + \left[l(l+1) - \frac{m_l^2}{(1 - \xi^2)} \right] P_l(\xi) = 0 \tag{11.179}$$

When m_l is zero, this equation is the *Legendre equation*, named after the French mathematician Adrien Marie Legendre (1752–1833). Solutions for the Legendre equation are possible only *when l is zero or has positive integral values*. These solutions are known as the *Legendre polynomials of degree l*. When m_l is not zero, a solution can only be obtained if m_l has one of the integral values $-l, -l + 1, \ldots, 0, \ldots, l - 1, l$. The solutions are then the *associated Legendre functions* and are usually expressed by means of a *recursion formula*.

---

*A particularly explicit treatment is given by L. Pauling and E. B. Wilson, *Introduction to Quantum Mechanics*, New York: McGraw-Hill, 1935.

TABLE 11.2 Solutions of the Φ Equation

Value of m_l	Solution in Complex Form	Real Form
0	—	$\Phi_0 = \dfrac{1}{\sqrt{2\pi}}$
1	$\Phi_1 = \dfrac{1}{\sqrt{2\pi}} e^{i\phi}$	$\Phi_x = \dfrac{1}{\sqrt{2}}(\Phi_1 + \Phi_{-1}) = \dfrac{\cos \phi}{\sqrt{\pi}}$
−1	$\Phi_{-1} = \dfrac{1}{\sqrt{2\pi}} e^{-i\phi}$	$\Phi_y = \dfrac{1}{\sqrt{2}}(\Phi_1 - \Phi_{-1}) = \dfrac{\sin \phi}{\sqrt{\pi}}$
2	$\Phi_2 = \dfrac{1}{\sqrt{2\pi}} e^{i2\phi}$	$\Phi_{x^2-y^2} = \dfrac{1}{\sqrt{2}}(\Phi_2 + \Phi_{-2}) = \dfrac{\cos 2\phi}{\sqrt{\pi}}$
−2	$\Phi_{-2} = \dfrac{1}{\sqrt{2\pi}} e^{-i2\phi}$	$\Phi_{xy} = \dfrac{1}{\sqrt{2}}(\Phi_2 - \Phi_{-2}) = \dfrac{\sin 2\phi}{\sqrt{\pi}}$

The reasons for the $x^2 - y^2$ and xy subscripts are as follows:
1. $\cos 2\phi = \cos^2 \phi - \sin^2 \phi$; from Figure 11.13, $\cos^2 \phi = x^2/r^2 \sin^2 \theta$ and $\sin^2 \phi = y^2/r^2 \sin^2 \theta$; thus $\cos 2\phi$ has the same dependence on ϕ as $x^2 - y^2$.
2. $\sin 2\phi = 2 \sin \phi \cos \phi = xy/r^2 \sin^2 \theta$; thus $\sin 2\phi$ has the same dependence on ϕ as xy.

The conclusion is therefore that l can only be zero or have a positive value and that the m_l values are determined by the value of l:

$$l = 0, 1, 2, 3, \ldots \tag{11.180}$$

$$m_l = -l, -l + 1, \ldots, -1, 0, 1, \ldots, l - 1, l \tag{11.181}$$

We shall see that the solution of the R equation imposes an upper limit on the value of l.

Table 11.3 gives solutions of the Θ equation for l values of 0, 1, and 2 and the corresponding permitted m_l values. Because the sine and cosine functions can have

TABLE 11.3 Solutions of the Θ Equation

l	m_l	Function
0	0	$\Theta_{00} = \dfrac{\sqrt{2}}{2}$
1	0	$\Theta_{10} = \dfrac{\sqrt{6}}{2} \cos \theta$
1	+1, −1	$\Theta_{1\pm1} = \dfrac{\sqrt{3}}{2} \sin \theta$
2	0	$\Theta_{20} = \dfrac{\sqrt{10}}{4}(3 \cos^2 \theta - 1)$
2	+1, −1	$\Theta_{2\pm1} = \dfrac{\sqrt{15}}{2} \sin \theta \cos \theta$
2	+2, −2	$\Theta_{2\pm2} = \dfrac{\sqrt{15}}{4} \sin^2 \theta$

TABLE 11.4 Solutions of the R Equation

Quantum Numbers n	l	Function
1	0	$R_{10} = 2\left(\frac{Z}{a_0}\right)^{3/2} e^{-Zr/a_0}$
2	0	$R_{20} = \frac{1}{2\sqrt{2}}\left(\frac{Z}{a_0}\right)^{3/2}\left(2 - \frac{Zr}{a_0}\right)e^{-Zr/2a_0}$
2	1	$R_{21} = \frac{1}{2\sqrt{6}}\left(\frac{Z}{a_0}\right)^{3/2}\frac{Zr}{a_0}e^{-Zr/2a_0}$
3	0	$R_{30} = \frac{1}{9\sqrt{3}}\left(\frac{Z}{a_0}\right)^{3/2}\left(6 - \frac{4Zr}{a_0} + \frac{4Z^2r^2}{9a_0^2}\right)e^{-Zr/3a_0}$
3	1	$R_{31} = \frac{1}{9\sqrt{6}}\left(\frac{Z}{a_0}\right)^{3/2}\left(4 - \frac{2Zr}{3a_0}\right)\frac{2Zr}{3a_0}e^{-Zr/3a_0}$
3	2	$R_{32} = \frac{1}{9\sqrt{30}}\left(\frac{Z}{a_0}\right)^{3/2}\left(\frac{2Zr}{3a_0}\right)^2 e^{-Zr/3a_0}$

The quantity a_0 is the same as the first Bohr radius, defined by Eq. 11.45.

positive and negative values, there are positive and negative regions of the wave functions. The functions in Table 11.3 are orthogonal to one another (as required for eigenfunctions of a Hermitian operator) and have been normalized.

Solution of the *R* Equation

The R equation, Eq. 11.167, may be cast into the form of a type of equation studied in the nineteenth century by the French mathematician Edmond Laguerre (1834–1886). Its solution leads to the conclusion that there is a quantum number n that can have positive integral values starting with unity:

$$n = 1, 2, 3, \ldots \tag{11.182}$$

The relationship between n and l is that the maximum value that l can have is one less than the value of n:

$$l = 0, 1, \ldots, n - 1 \tag{11.183}$$

The solutions of the R equation, under these restrictions, are the *Laguerre polynomials.* The wave functions for the first three values of n and the possible values of l are given in Table 11.4.

Complete Wave Functions

The complete wave functions, known also as **orbitals,** are obtained by multiplying together the appropriate functions that are given in Tables 11.2, 11.3, and 11.4. Some examples are given in Table 11.5. The way they are constructed is shown by the following example.

TABLE 11.5 Selected Complete Hydrogen Atom Wave Functions, ψ_n, l, m, for the Hydrogen Atom ($Z = 1$)

Quantum Numbers n	l	m_l	Function
1	0	0	$\psi_{1s} = \psi_{100} = \dfrac{1}{\sqrt{\pi}}\left(\dfrac{1}{a_0}\right)^{3/2} e^{-r/a_0}$
2	0	0	$\psi_{2s} = \psi_{200} = \dfrac{1}{4\sqrt{2\pi}}\left(\dfrac{1}{a_0}\right)^{3/2}\left(2 - \dfrac{r}{a_0}\right)e^{-r/2a_0}$
2	1	0	$\psi_{2p_z} = \psi_{210} = \dfrac{1}{4\sqrt{2\pi}}\left(\dfrac{1}{a_0}\right)^{3/2}\dfrac{r}{a_0}e^{-r/2a_0}\cos\theta$
2	1	+1	$\psi_{2p_x} = \dfrac{1}{4\sqrt{2\pi}}\left(\dfrac{1}{a_0}\right)^{3/2}\dfrac{r}{a_0}e^{-r/2a_0}\sin\theta\cos\phi$
2	1	−1	$\psi_{2p_y} = \dfrac{1}{4\sqrt{2\pi}}\left(\dfrac{1}{a_0}\right)^{3/2}\dfrac{r}{a_0}e^{-r/2a_0}\sin\theta\sin\phi$
3	0	0	$\psi_{3s} = \psi_{300} = \dfrac{1}{81\sqrt{3\pi}}\left(\dfrac{1}{a_0}\right)^{3/2}\left[6 - 6\left(\dfrac{2r}{3a_0}\right) + \left(\dfrac{2r}{3a_0}\right)^2\right]e^{-r/3a_0}$

The significance of the notation 1s, 2s, $2p_z$, etc., is considered later.

EXAMPLE

Obtain a complete wave function for an electron having $n = 3$, $l = 1$, $m_l = 0$ (a 3p orbital).

SOLUTION

In order to determine an expression for Ψ_{nlm_l}, we can multiply the radial wave function for an $n = 3$, $l = 1$ orbital from Table 11.4 (R_{31}) by the Θ function for $l = 1$, $m_l = 0$ from Table 11.3 (Θ_{10}) and then by the Φ function for $m_l = 0$ from Table 11.2 (Φ_0). The final expression is

$$\psi_{310} = \frac{1}{27\sqrt{2\pi}}\left(\frac{Z}{a_0}\right)^{3/2}\left(6 - \frac{Zr}{a_0}\right)\left(\frac{Zr}{a_0}\right)e^{-Zr/3a_0}\cos\theta$$

Note that two other equally acceptable solutions for the 3p orbitals could have been obtained by taking the products $R_{31}\Theta_{11}\Phi_x$ or $R_{31}\Theta_{1-1}\Phi_y$. These give identical orbitals to the one described except for orientation. ■

11.8 Physical Significance of the Orbital Quantum Numbers

The mathematical solution of the Schrödinger equation for hydrogenlike atoms has thus revealed that there are three orbital quantum numbers n, l, and m_l; their magnitudes are related to one another by Eqs. 11.180, 11.183, and 11.181, respectively. A special notation has been introduced to designate the quantum numbers n and l and the orbitals to which they correspond. The principal quantum number n is given first, followed by a letter that indicates the quantum number l, as follows:

$l = 0$ s

$l = 1$ p

$l = 2$ d

$l = 3$ f

From then on we follow the letters in alphabetical order (g, h, etc.). The letters s, p, d, and f relate to the descriptions "sharp," "principal," "diffuse," and "fundamental" that the early spectroscopists had given to series of lines in atomic spectra. As an example of the use of this notation, a 3p orbital is one for which $n = 3$ and $l = 1$. We will see later that m_l values can be indicated by the addition of subscripts to the letters, but this is frequently unnecessary.

The Principal Quantum Number *n*

Inspection of Eqs. 11.165, 11.167, and 11.168 shows that the only one that contains the energy E is Eq. 11.167, the R equation. It therefore follows that the energies for a hydrogenlike atom depend only on the solutions of the R equation and not at all on the solutions of the Θ and Φ equations. In other words, the energy depends only on the principal quantum number n, which comes from the R equation, and not on l and m_l. This is only true for hydrogenlike atoms.

We can obtain the energies by substituting the various solutions for R, listed in Table 11.4, into the R equation and solving for E. When this is done, the permitted energies are given by the expression

$$E_n = -\frac{Z^2e^2}{8\pi\epsilon_0 a_0 n^2} = -\frac{\mu Z^2 e^4}{8\epsilon_0^2 h^2 n^2} \tag{11.184}$$

It is of interest that this is exactly the same expression that was given by Bohr's theory (Eq. 11.50). The lowest (i.e., most negative) energy occurs when $n = 1$, and this is therefore the most stable state, or *ground* state, of the atom. Since when $n = 1$, l must be 0, this is the 1s state.

The R equation, and therefore the value of n, determines the expectation value of the distance between the nucleus and the electron. Thus if $n = 1$, the electron is more likely to be close to the nucleus than if $n = 2$, and so on. This may be represented quantitatively by diagrams in which the function R is plotted against the distance r. Some such diagrams are shown in Figure 11.14. Note that because of the exponential factor appearing in the expression for R (Table 11.4), the function approaches zero as r becomes large. For l equal to $n - 1$, this approach to zero is monatonic, with no maxima and minima, but for other values of l the function passes through zero at various distances and there are maxima and minima. The number of *nodes*, where $R = 0$, is equal to $n - l - 1$.

We have seen (Figure 11.10) that the probability that an electron is in a volume element $d\tau = dx\, dy\, dz$ is given by $\psi\psi^*\, d\tau$, or $\psi^2\, d\tau$ if the wavefunction is real. If we divide by $d\tau$, the result ψ^2 is the probability per unit volume, or **probability density.**

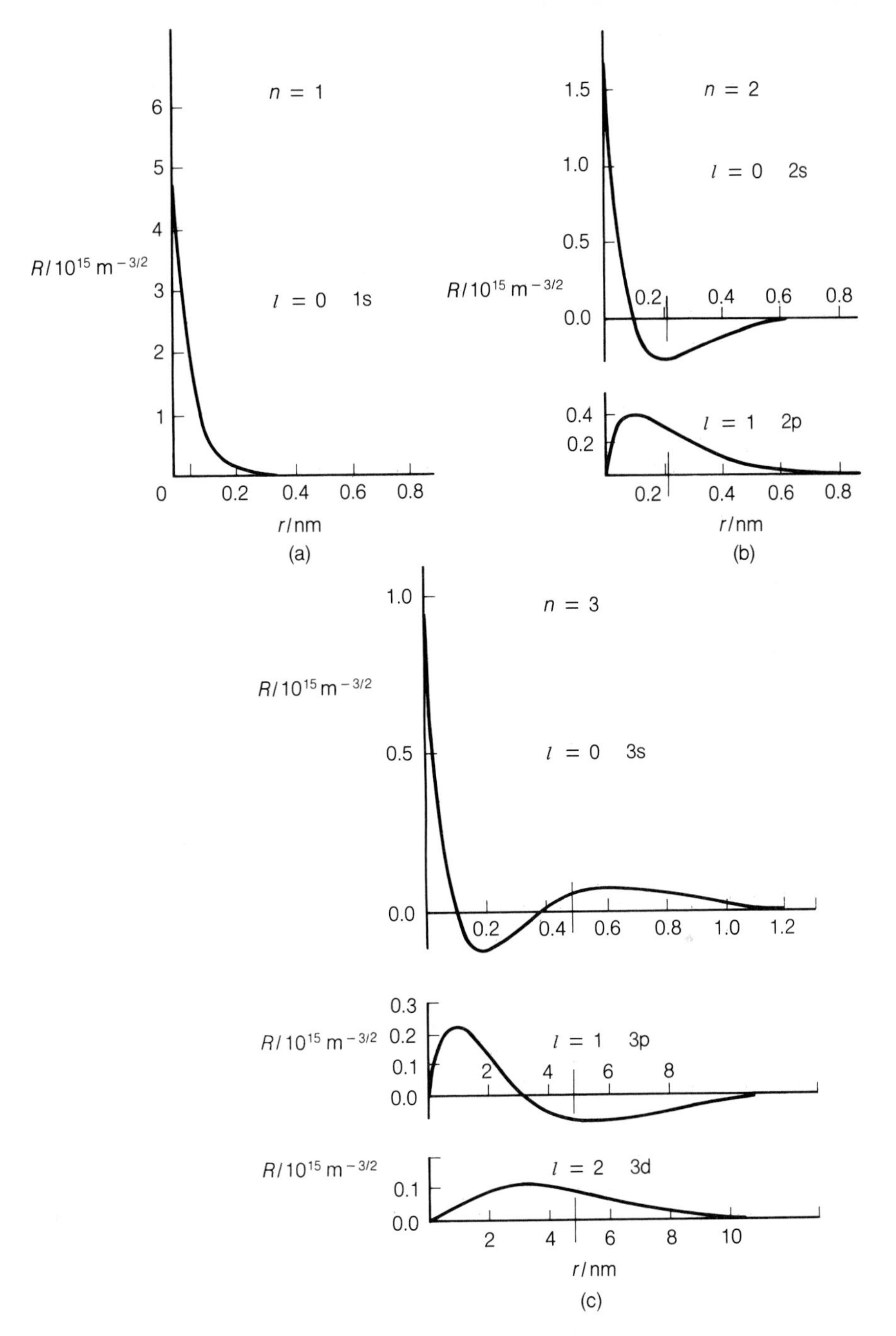

FIGURE 11.14 Radial wave functions $R\ (r)$ for the hydrogen atom, plotted against distance r from the nucleus. (a) The curve for $n = 1$; l must be 0. (b) The curves for $n = 2$; l can be 0 or 1. (c) The curves for $n = 3$; l can be 0, 1, or 2.

For an electron in the 1s state the probability density, obtained from the first item in Table 11.5, is therefore

$$\psi_{1s}^2 = \frac{e^{-2r/a_0}}{\pi a_0^3} \tag{11.185}$$

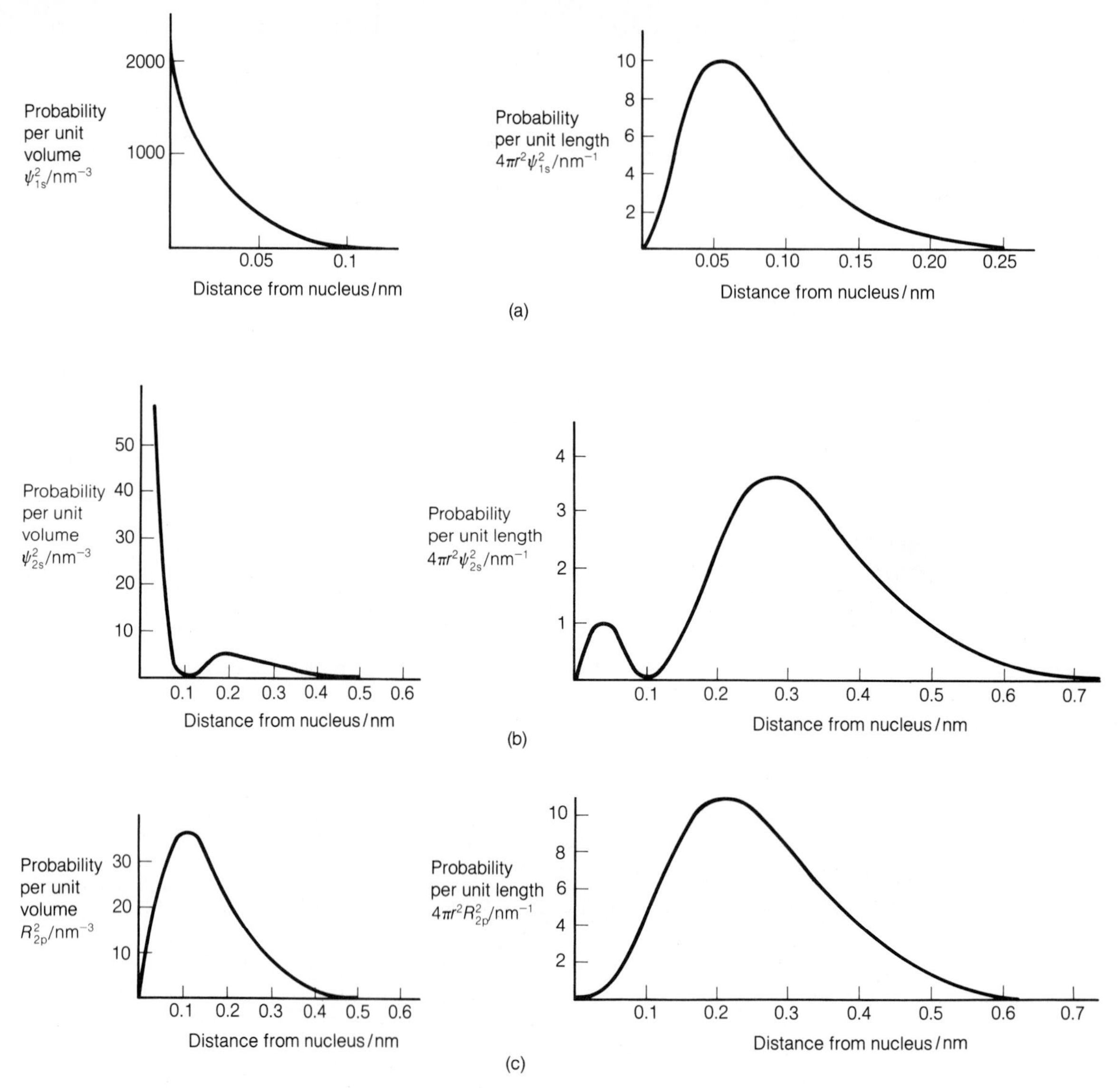

FIGURE 11.15 a) Plots of ψ_{1s}^2 and $4\pi r^2\psi_{1s}^2$ against the distance r from the nucleus, for a 1s electron in the hydrogen atom. (b) Plots of ψ_{2s}^2 and $4\pi r^2\psi_{2s}^2$ against r, for a 2s electron. (c) Plots of R_{2p}^2 and $4\pi r^2 R_{2p}^2$, for a 2p electron.

A plot of this function against r is shown in Figure 11.15a, and we see that the probability density has a maximum value at $r = 0$ and falls toward zero as r increases. It is of greater interest, however, to consider the probability that the electron is present at a distance between r and $r + dr$ from the nucleus—in other words, the probability that the electron is present between two concentric spherical surfaces; one has a radius of r and the other a radius of $r + dr$. The volume between two such concentric surfaces is $4\pi r^2\, dr$, and the probability that the 1s electron lies within the distances r and $r + dr$ is thus

$$4\pi r^2 \psi_{1s}^2\, dr$$

The corresponding probability density (which is now per unit radial distance r) is obtained by dividing by dr and is therefore

$$4\pi r^2 \psi_{1s}^2$$

A plot of this function against r is shown in Figure 11.15a. The curve passes through a maximum when $r = a_0 = 52.92$ pm, and this distance is therefore the most probable distance between the nucleus and the electron. This emphasizes a very important similarity, and also a very important difference, between the Bohr theory and the wave-mechanical theory. In the Bohr theory the electron in the ground (1s) state of the hydrogen atom moves in a *precise orbit* of radius a_0, while in quantum mechanics this distance is only the *most probable distance;* the electron can be at other distances from the nucleus.

Figure 11.15b shows plots of ψ_{2s}^2 and of $4\pi r^2 \psi_{2s}^2$ for an electron in the 2s state. The latter shows two maxima. Similar plots can be given for 3s, 4s, etc., states, and the number of maxima in the $4\pi r^2 \psi^2$ plots is always equal to the principal quantum number n.

For states where l is not zero (e.g., 2p, 3d, . . . states), we have to proceed differently because there is now an angular dependence (see Table 11.5). We could make plots of ψ^2 with $\cos\theta$ and $\sin\theta$ taken to be unity, their maximum values. Such plots would represent probability densities *along the axis for which these have their maximum values.* However, a more usual practice is instead to make plots of R^2 against r. This is not quite the same thing, since a comparison of Tables 11.4 and 11.5 shows us that even with $Z = 1$ the numerical factors are not the same. In spite of this, however, the plots of R^2 against r do give us the *general form* of the probability density function. To obtain the function along the axis for which it is a maximum we would have to multiply by an appropriate factor. A plot of R^2 against r for the 2p state of the hydrogen atom is shown in Figure 11.15c.

We can also make plots of $4\pi r^2 R^2$ against r, and Figure 11.15c shows such a plot for the 2p state. There are now two points to note about the significance of such plots. One is that the numerical factor in R is not the same as that in ψ. But aside from this, the plot of $4\pi r^2 R^2$ would only represent the probability that the electron is in a spherical element at a distance r from the nucleus *if the orbital had spherical symmetry.* In reality, orbitals for which l is other than zero have maximum values in certain directions. The function $4\pi r^2 R^2$ therefore does not have any simple significance in such cases, but plots of $4\pi r^2 R^2$ against r are informative as long as one recognizes this limitation.

We have seen that an electron in the 1s orbital is *most likely* to be at a distance a_0 from the nucleus. This distance, however, because of the skewed distribution of the orbitals, is not the same as the *expectation value* of the distance between the nucleus and the electron (i.e., the distance the electron is expected to be from the nucleus). We can calculate this expectation value by using Postulate VI. According to Eq. 11.104, with the wave functions real and normalized,

$$\langle r_{1s} \rangle = \int \psi_{1s} r \psi_{1s}\, d\tau = \int_0^\infty \psi_{1s} r \psi_{1s} 4\pi r^2\, dr \qquad (11.186)$$

Introduction of the expression for ψ_{1s} in Table 11.5 then gives

$$\langle r_{1s} \rangle = \frac{1}{\pi a_0^3} \int_0^\infty re^{-2r/a_0}\, 4\pi r^2\, dr \tag{11.187}$$

$$= \frac{3}{2} a_0 \tag{11.188}$$

The reason that this is greater than a_0, the most probable distance, is that the wave functions do not give a symmetrical curve when plotted against r (see Figure 11.14); they are skewed in favor of larger r values. For s states with other values of n, the expectation values are given by

$$\langle r_{ns} \rangle = \tfrac{3}{2} n^2 a_0 \tag{11.189}$$

Angular Dependence of the Wave Function: The Quantum Numbers l and m_l

The Θ and Φ wave functions are best considered together, since both are concerned with the angular dependence of the orbitals. Plots of $[\Theta(\theta)\Phi(\phi)]^2$ for various l and m_l values are shown in Figure 11.16. It is to be emphasized that these plots are in no way related to distance from the nucleus but only to the variation of the wave function with the angles θ and ϕ. This is illustrated explicitly for the p_z orbital, for which the angles θ and ϕ are shown and in which the plot is shown in its three-dimensional form. For given values of θ and ϕ, the length of the line joining the origin to the surface of the solid figure is the relative probability that the electron is to be found in that direction. For this p_z orbital the *maximum* probability is along the Z axis. The other diagrams are given in a simpler form, but they are to be interpreted in the same way.

All s orbitals ($l = 0$) are of spherical symmetry, as shown in Figure 11.16a. When $l = 1$ (a p orbital), the quantum number m_l can have the value -1, 0, or $+1$, and there are orbitals that have their maximum values along the X, Y, and Z axes. For the d orbitals ($l = 2$) there are five equivalent orbitals corresponding to $m_l = -2, -1, 0, 1, 2$. The subscripts for these orbitals relate to the symmetry properties. For example, with $l = 2$ and $m_l = -1$ we obtain, from Tables 11.2 and 11.3, the product

$$\Phi_y \Theta_{2\pm1} = \frac{\sin\phi}{\sqrt{\pi}} \cdot \frac{\sqrt{15}}{2} \sin\theta\cos\theta = \frac{1}{2}\sqrt{\frac{15}{\pi}} \sin\theta\cos\theta\sin\phi \tag{11.190}$$

From Figure 11.13 we see that $y = r\sin\theta\sin\phi$ and that $z = r\cos\theta$. The product yz therefore has the same angular dependence as the function in Eq. 11.190, which is therefore designated d_{yz}.

The f orbitals ($l = 3$) are difficult to represent, but they bear some resemblance to the d orbitals. There are seven such orbitals, corresponding to $m_l = -3, -2, -1, 0, 1, 2, 3$. Three of them, f_{z^3}, f_{x^3}, and f_{y^3} have *two* doughnut-shaped rings perpendicular to the appropriate axis, instead of the one found in the d_{z^2} orbital. The other four resemble the $d_{x^2-y^2}$ orbitals except that there are now eight lobes instead of four.

From what has been said, it might be thought that the various orbitals corresponding to a given value of l would mean that a complete orbital diagram for an atom would look like a pincushion, with orbitals sticking out in various directions. However, a theorem due to the German physicist Albrecht Otto Johannes Unsöld

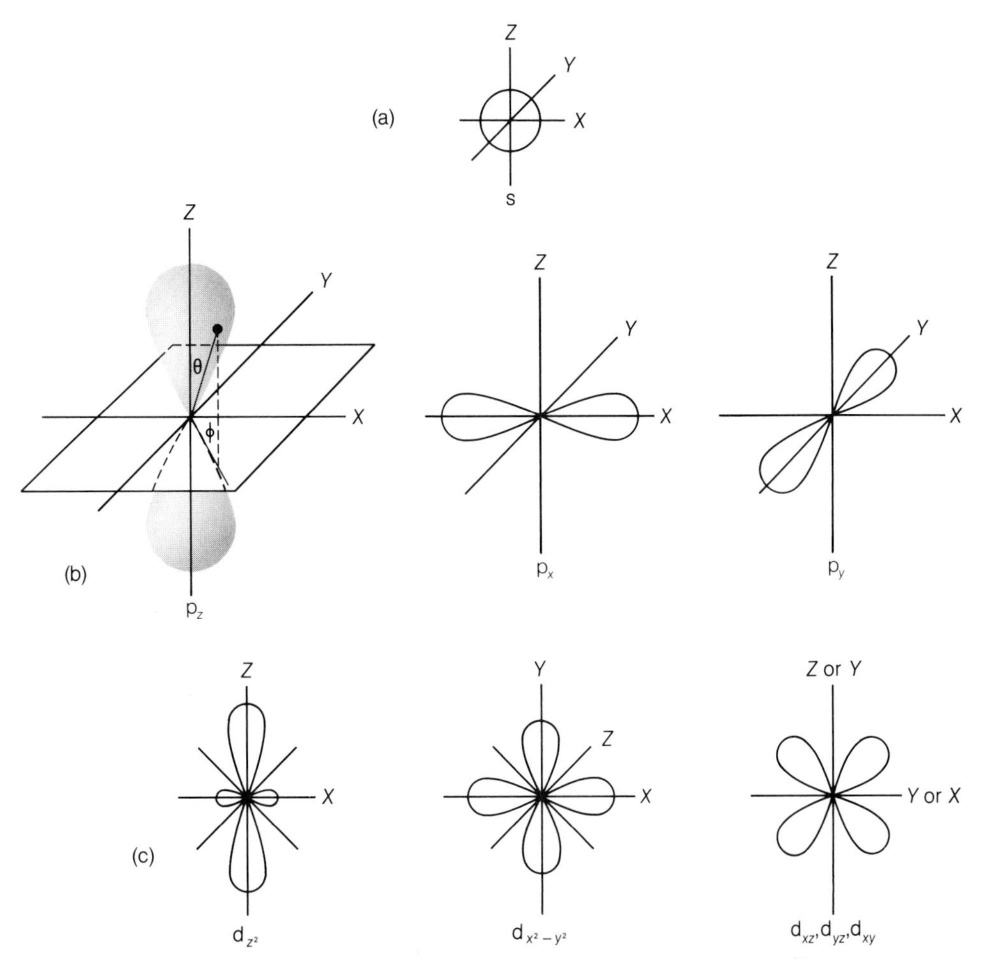

FIGURE 11.16 Plots of $[\Theta(\theta)\Phi(\phi)]^2$. (a) An s orbital ($l = 0$). (b) Three p orbitals; the p_z plot is shown enlarged for clarity in three-dimensional form. (c) The d_{z^2}, $d_{x^2-y^2}$, d_{xz}, d_{yz}, and d_{xy} orbitals.

(b. 1905) states that for the orbitals corresponding to a particular value of l the sum of the values of $[\Theta(\theta)\Phi(\phi)]^2$ is a constant. Such a sum therefore corresponds to completely spherical symmetry for the atom. It follows from this that the atoms of the noble gases, in which all orbitals corresponding to a given l value are filled, have complete spherical symmetry.

11.9 Angular Momentum and Magnetic Moment

Besides determining the shape of an orbital and its orientation in space, the quantum numbers l and m_l have another important significance. They relate to the angular momentum of the electron in its orbital and also to the magnetic moment of the orbital. In this section we will see that the quantum number l determines the *magnitude* of the angular momentum and magnetic moment, and that the quantum number m_l determines the *direction in space* of the angular momentum and magnetic moment.

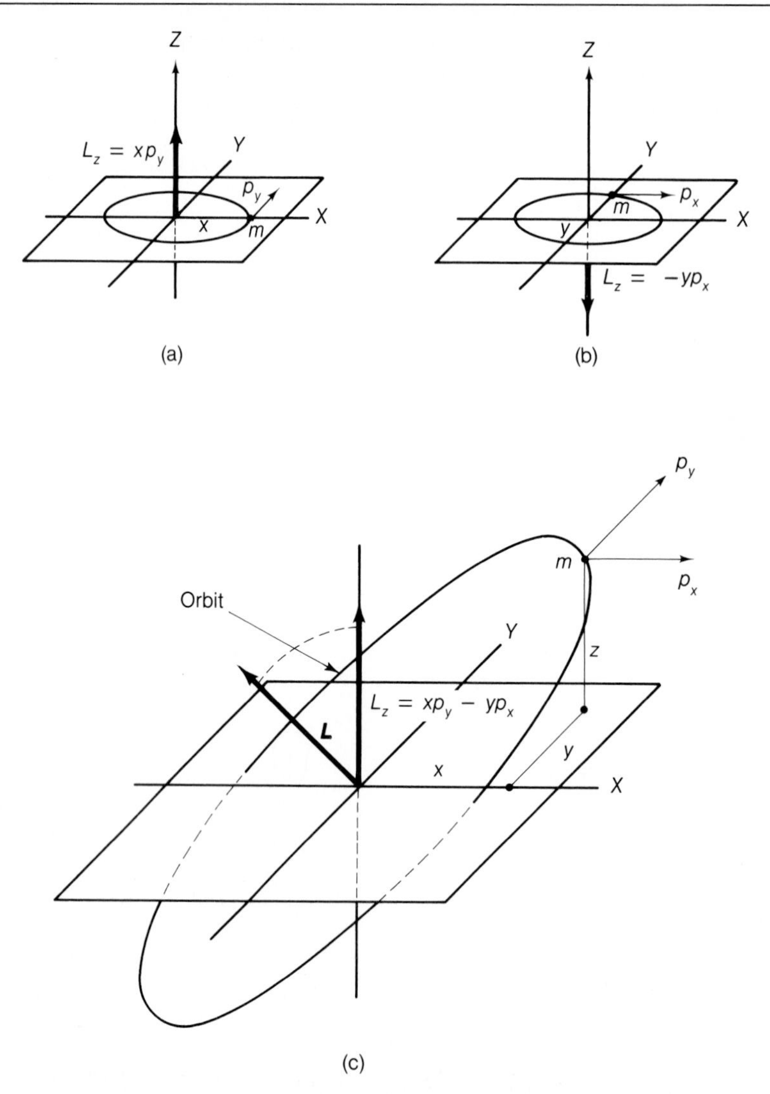

FIGURE 11.17 (a) A particle of mass m undergoing circular motion in the XY plane in an anticlockwise direction. (b) Clockwise motion in the XY plane. (c) The general case, showing the components of velocity and of linear momentun, and the component of angular momentum along the Z axis.

Angular Momentum

The classical relationship between angular momentum and linear velocity is illustrated in Figure 11.17. Suppose that a particle of mass m is undergoing circular motion in the XY plane, in an anticlockwise direction if one is looking down along the Z axis (Fig. 11.17a). When the particle is on the X axis, it has a positive velocity u_y in the Y-direction, and its linear momentum p_y is mu_y. Its angular momentum is then, by definition, positive along the Z axis and is given by

$$L_z = p_y x = mu_y x \tag{11.191}$$

The sign of the angular momentum is determined by the *right-hand rule;* the curved fingers of the right hand are caused to point in the direction of the linear momentum vector, and the thumb then points in the direction of the angular momentum vector. Figure 11.17b shows the situation when the particle is on the Y axis; a positive momentum p_x then means that it is moving in the clockwise direction, so that the angular momentum L_z is now $-yp_x$. We can generalize these results for a particle moving in any direction (Figure 11.17c). If its momentum components are p_x and p_y and its position coordinates are x and y, the resultant component of angular momentum along the Z axis is given by

$$L_z = xp_y - yp_x \tag{11.192}$$

The corresponding quantum-mechanical operator, as given in Table 11.1, is

$$\hat{L}_z = \frac{h}{2\pi i}\left(x\frac{\partial}{\partial y} - y\frac{\partial}{\partial x}\right) \tag{11.193}$$

In polar coordinates (Figure 11.13) this becomes

$$\hat{L}_z = \frac{h}{2\pi i}\frac{\partial}{\partial \phi} \tag{11.194}$$

Exactly the same type of treatment can be applied to the components of angular momentum along the X and Y axes, leading to expressions for the operators $\hat{L}_x$ and $\hat{L}_y$. Of particular significance is the square of the *total* angular momentum operator, defined by

$$\hat{L}^2 = \hat{L}_x^2 + \hat{L}_y^2 + \hat{L}_z^2 \tag{11.195}$$

This is given by

$$\hat{L}^2 = -\frac{h^2}{4\pi^2}\left[\frac{1}{\sin\theta}\frac{\partial}{\partial\theta}\left(\sin\theta\frac{\partial}{\partial\theta}\right) + \frac{1}{\sin^2\theta}\frac{\partial^2}{\partial\phi^2}\right] \tag{11.196}$$

We saw in Section 11.5, in our discussion of Postulate III, that if two operators are such that they have the same eigenfunctions, the corresponding physical properties can be measured simultaneously and precisely; in other words, the Heisenberg uncertainty principle places no restriction on their measurement. We will now show that the operators $\hat{H}$, $\hat{L}_z$, and $\hat{L}^2$ have the same set of eigenfunctions; this means that the energy, the component of angular momentum along the Z axis, and the square of the total angular momentum can all be measured simultaneously.

The operator $\hat{L}_z$ involves only the angle ϕ (Eq. 11.194), and the function

$$\Phi = \frac{1}{\sqrt{2\pi}}e^{im_l\phi} \tag{11.197}$$

is the solution of the $\hat{H}$ equation that involves ϕ (Eq. 11.172). This function is therefore an eigenfunction for $\hat{H}$, and we will now see if it is an eigenfunction for $\hat{L}_z$:

$$\hat{L}_z\Phi = \frac{h}{2\pi i}\frac{\partial}{\partial\phi}\left(\frac{1}{\sqrt{2\pi}}e^{im_l\phi}\right) = \frac{hm_l}{2\pi}\cdot\frac{1}{\sqrt{2\pi}}e^{im_l\phi} \tag{11.198}$$

$$= m_l\frac{h}{2\pi}\Phi \tag{11.199}$$

The function Φ is therefore an eigenfunction for $\hat{L}_z$ as well as for $\hat{H}$, and the eigenvalue for $\hat{L}_z$ is

$$L_z = m_l \frac{h}{2\pi} \tag{11.200}$$

This expression therefore gives the quantized values of the Z component of the angular momentum.

The operator $\hat{L}^2$ involves only θ and ϕ, and we know that the function $\Theta\Phi$ is an eigenfunction for the Hamiltonian operator $\hat{H}$; we may therefore see whether it is also an eigenfunction for $\hat{L}^2$:

$$\hat{L}^2\Theta\Phi = -\frac{h^2}{4\pi^2}\left[\frac{\Phi}{\sin\theta}\frac{\partial}{\partial\theta}\left(\sin\theta\frac{\partial\Theta}{\partial\theta}\right) + \frac{\Theta}{\sin^2\theta}\frac{\partial^2\Phi}{\partial\phi^2}\right] \tag{11.201}$$

Comparison of this equation with the Φ equation (Eq. 11.165) and the Θ equation (Eq. 11.168) shows that the function in brackets in Eq. 11.201 has the eigenvalues $l(l+1)$ and may therefore by written as $l(l+1)\Theta\Phi$; thus

$$L^2\Theta\Phi = l(l+1)\frac{h^2}{4\pi^2}\Theta\Phi \tag{11.202}$$

The eigenfunction for $\hat{H}$ is therefore also an eigenfunction for $\hat{L}^2$, with the eigenvalue

$$L^2 = l(l+1)\frac{h^2}{4\pi^2} \tag{11.203}$$

We have thus proved that it is possible to make simultaneous and precise measurements of the total energy, the total angular momentum, and the component of the angular momentum along the Z axis. This arises because the operators $\hat{H}$, $\hat{L}^2$, and $\hat{L}_z$ commute (Postulate III). Of course, our choice of Z axis is entirely arbitrary; we can equally well show that $\hat{H}$, $\hat{L}^2$, and $\hat{L}_x$ commute and that $\hat{H}$, $\hat{L}^2$, and $\hat{L}_y$ commute. It is therefore possible to make simultaneous and precise measurements of the total energy, the total angular momentum, and the component of the angular momentum along the X axis *or* along the Y axis *or* along the Z axis. However, $\hat{L}_x$, $\hat{L}_y$, and $\hat{L}_z$ do *not* commute with each other, and we therefore cannot make simultaneous and precise measurements of the angular momentum along the X, Y, and Z axes.

We therefore have to choose one axis, and the convention is to choose the Z axis. For example, when an atom is placed in an electric or magnetic field, the field is taken to be along the Z axis. Similarly, when an atom is present in a diatomic molecule, we take the axis of the molecule to be the Z axis.

Magnetic Moment

An orbiting electron, being charged, generates a magnetic field. Since the electron is negatively charged, the direction of the magnetic field is exactly opposite to that of the angular momentum, as shown in Figure 11.18a. The direction is conveniently obtained by using the *left-hand* rule; the curved fingers of the left hand point in the direction of motion of the orbiting electron, and the thumb then points toward the north pole of the electromagnet. The magnitude of the magnetic moment is propor-

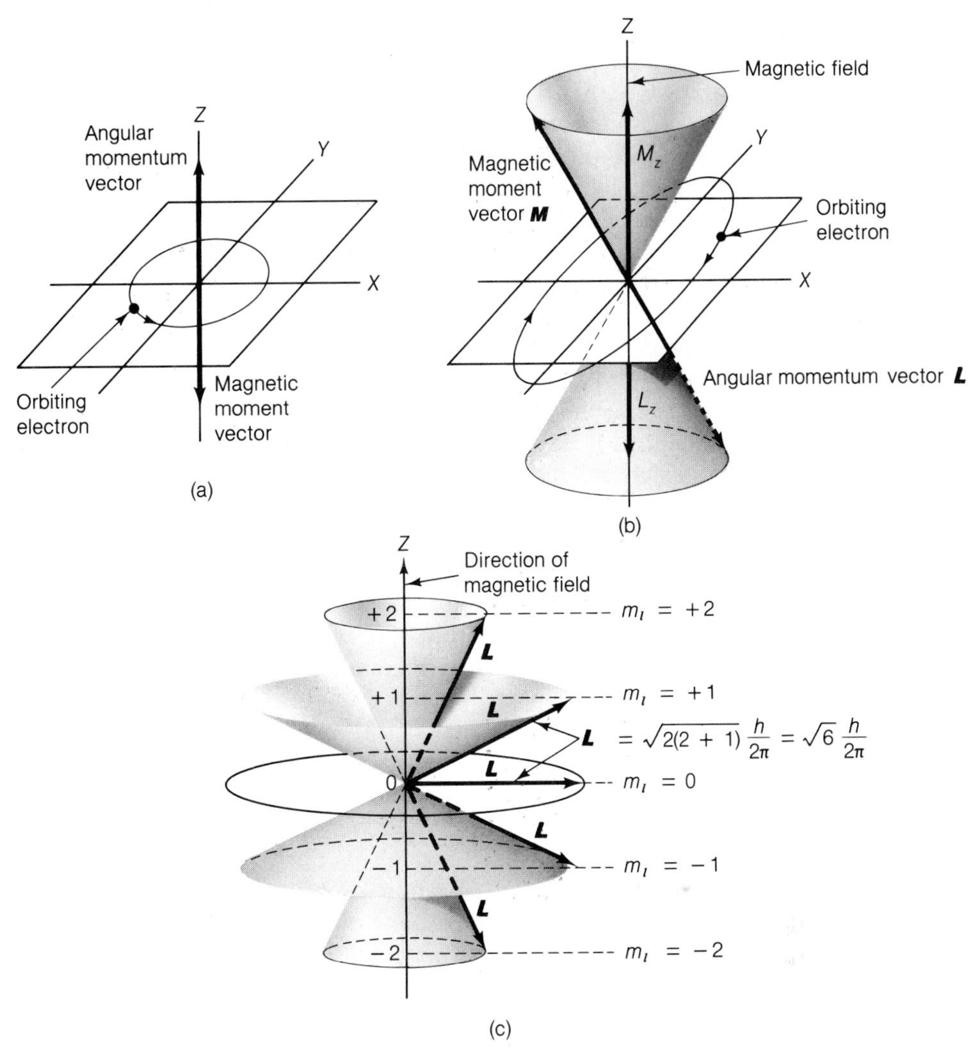

FIGURE 11.18 (a) An electron moving in an orbit in the XY plane, showing the angular momentum and magnetic moment vectors along the Z axis. (b) An orbiting electron in a magnetic field that lies along the Z axis, for $l = 1$. The Z component of the magnetic moment is shown aligned with the field and the angular momentum against it. The cones represent precession about the Z axis. (c) The five possible orientations of the orbital angular momentum vector, for $l = 2$, in a magnetic field. The magnetic moment lies in the opposite direction.

tional to that of the angular momentum, and aside from the difference of orientation the treatment of the magnetic moment is very similar to that of the angular momentum. Again there is quantization, and precise and simultaneous measurements can be made of the energy, the component M_z of magnetic moment along the Z axis, and the square of the total magnetic moment. This arises because the magnetic moment operators $\hat{M}^2$ and $\hat{M}_z$ commute with $\hat{H}$, $\hat{L}^2$, and $\hat{L}_z$ and, therefore, have the same set of eigenfunctions. The same quantum numbers l and m_l therefore relate to the magnetic moment.

The magnetic moment differs in the *units* used for its measurement. We will discuss this further in Section 13.2 in connection with the Zeeman effect. Here we will simply state that the ratio of the magnetic moment to the angular momentum is known as the **gyromagnetic ratio,** or the **magnetogyric ratio.** For an electron of charge e and mass m_e this ratio is $-e/m_e$. The SI unit of magnetic moment is ampere metre^2 (A m^2), but it is usual to employ the **Bohr magneton,** which is equal to 9.274×10^{-24} A m^2.

The importance of the magnetic moment is that it dictates the orientation of the orbital in a magnetic field, which may be that due to neighboring atoms. Figure 11.18b shows one arrangement that can arise when a p orbital ($l = 1$) is in a magnetic field along the Z axis. The Z components of the angular momentum and of the magnetic moment, opposite to one another, must have quantized values, corresponding to the magnetic quantum number m_l, along the Z axis. In the case shown in Figure 11.18b the Z component of the magnetic moment is aligned with the field, which means that the Z component of the angular momentum is aligned against the field. This is the case of $m_l = -1$, and since the magnetic moment is in the same general direction of the field, the energy has its lowest value. Another possibility is that the component L_z of angular momentum is aligned with the field ($m_l = +1$) and the magnetic moment against it; this corresponds to the highest energy. A third possibility is that L_z is at right angles to the field; this corresponds to $m_l = 0$ and to zero energy of interaction between the magnetic moment and the field.

The L_x and L_y components of the angular momentum and the M_x and M_y components of the magnetic moment do not have precise values, and the angular momentum and the magnetic moment therefore precess about the Z axis, as shown in Figure 11.18b. This motion, which corresponds to the curved surface of a cone, keeps L, L_z, M, and M_z fixed but leaves L_x, L_y, M_x, and M_y indeterminate.

Figure 11.18c shows the situation for a d orbital, for which $l = 2$ and m_l can have the values $-2, -1, 0, +1$, and $+2$. The diagram shows the *angular momentum vectors;* the magnetic moment vectors are not shown to avoid complicating the diagram, but they lie in the opposite direction. The case of $m_l = -2$ means that the angular momentum has its maximum value in the direction opposed to the field, and the magnetic moment then has its maximum value in the direction of the field. This corresponds to the lowest energy for the system. The highest energy is for $m_l = +2$, and $m_l = 0$ gives zero interaction energy.

11.10 The Rigid Linear Rotor

It is convenient to make a brief digression at this point and consider the rotation of a rigid linear molecule. We will see that the angular momentum of such a system obeys the same types of equations that apply to orbitals.

The angular momentum L of a body having a moment of inertia I and an angular velocity of ω is given by

$$L = I\omega \tag{11.204}$$

Its kinetic energy E_k is given by

$$E_k = \tfrac{1}{2}I\omega^2 \tag{11.205}$$

For free rotation the potential energy is zero, and the total energy E is therefore $\frac{1}{2}I\omega^2$. Elimination of ω between these two equations gives

$$E = E_k = \frac{L^2}{2I} \tag{11.206}$$

To obtain the angular momentum values for the system we first convert L^2 into its operator (Eq. 11.196):

$$\hat{L}^2\Theta\Phi = -\frac{h^2}{4\pi^2}\left[\frac{\Phi}{\sin\theta}\frac{\partial}{\partial\theta}\left(\sin\theta\frac{\partial\Theta}{\partial\theta}\right) + \frac{\Theta}{\sin^2\theta}\frac{\partial^2\Phi}{\partial\phi^2}\right] \tag{11.207}$$

This is the same equation as Eq. 11.201, and its eigenvalues are given by Eq. 11.203. It is customary, however, to use J for the rotational quantum number, and therefore

$$L^2 = J(J+1)\frac{h^2}{4\pi^2} \tag{11.208}$$

where J can have the values 0, 1, 2, ... The permitted energy values are then obtained by use of Eq. 11.206:

$$E = J(J+1)\frac{h^2}{8\pi^2 I} \tag{11.209}$$

It is to be noted that when $J = 0$, the energy is zero; there is no zero-point energy for rotation. In this treatment the angles θ and ϕ are completely unspecified, so that one cannot assign values to the angular momenta about the X and Y axes. For a linear molecule the angular momentum about the Z axis, which is the axis of the molecule, is zero.

We shall see applications of this treatment in Chapter 13.

11.11 Spin Quantum Numbers

An important experiment carried out in 1921 by the physicists Otto Stern (b. 1888) and Walter Gerlach* (b. 1889) showed that the three quantum numbers n, l, and m_l were insufficient to explain atomic behavior. They found that a beam of silver atoms, each one of which has an odd electron, is split into two beams when passed through an inhomogeneous magnetic field. This result was later explained by the Austrian physicist Wolfgang Pauli† (1900–1958), by the Dutch-American physicists George Eugene Uhlenbeck (b. 1900) and Samuel Abraham Goudsmit‡ (b. 1902), and by the British physicist Paul Adrien Maurice Dirac§ (b. 1902) in terms of a spin angular momentum of the electron.

Dirac's theory was an extension of the theory of angular momentum in which relativistic effects were introduced, and it led to the conclusion that for an electron there can be two possible components of spin angular momentum about an axis.

---

* O. Stern, *Z. Physik, 7*, 249 (1921); W. Gerlach and O. Stern, *Z. Physik, 8*, 110 (1922); *9*, 349, 353 (1922).
† W. Pauli, *Z. Physik, 31*, 765 (1926).
‡ G. E. Uhlenbeck and S. Goudsmit, *Naturwiss., 13*, 953 (1925); *Nature, 117*, 264 (1926).
§ P. A. M. Dirac, *Proc. Roy. Soc.* (London), *A117*, 610 (1928); *A118*, 351 (1928).

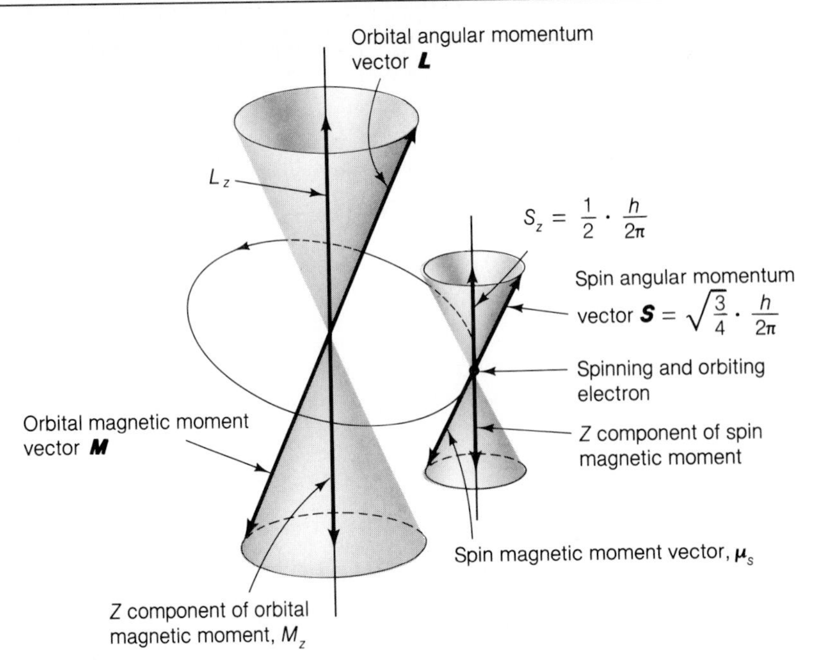

FIGURE 11.19 The spin angular momentum vector $\boldsymbol{S}$ and its relationship to the orbital angular momentum vector $\boldsymbol{L}$. The vectors are shown oriented in the same direction, which leads to a somewhat higher energy than when they are opposed.

The *spin angular momentum vector* $\boldsymbol{S}$ is exactly analogous to the orbital angular momentum $\boldsymbol{L}$ and is quantized in a similar way (compare Eqs. 11.200 and 11.203); thus

$$S = \sqrt{s(s+1)}\frac{h}{2\pi} \tag{11.209a}$$

where the quantum number s can have only the value $+\frac{1}{2}$. The component S_z of spin angular momentum along the Z axis is quantized in a similar way to L_z,

$$S_z = m_s \frac{h}{2\pi} \tag{11.210}$$

but the **spin quantum number** m_s can have only the values $+\frac{1}{2}$ and $-\frac{1}{2}$.

The orbital angular momentum of an orbital electron and also the spin angular momentum are shown in Figure 11.19. The orbiting electron is producing a magnetic moment vector in the direction indicated, and the electron may be considered to spin on its own axis in the magnetic field produced by the orbital motion. The magnetic moment due to the spin may be oriented in the same direction as the magnetic field produced by the orbiting electron (as shown in the figure) or in the opposite direction. The former gives a somewhat higher energy than the latter. We shall see in Section 13.2 that the resulting splitting of energy levels gives closely spaced lines in atomic spectra.

We are now in a position to add two additional postulates to the six that we introduced in Section 11.5: *Postulate VII* states that

> **There is a spin angular momentum operator $\hat{S}$ and a Z-component spin operator $\hat{S}_z$ that correspond to the angular momentum operators $\hat{L}$ and $\hat{L}_z$.**

Postulate VIII states that

> **In contrast to $\hat{L}^2$ and $\hat{L}_z$, the spin operators $\hat{S}^2$ and $\hat{S}_z$ have only two eigenfunctions α and β.**

The eigenvalue equations for $\hat{S}_z$ are

$$\hat{S}_z\alpha = \frac{1}{2}\frac{h}{2\pi}\alpha \tag{11.211}$$

and

$$\hat{S}_z\beta = -\frac{1}{2}\frac{h}{2\pi}\beta \tag{11.212}$$

The eigenvalue equations for $\hat{S}^2$ are

$$\hat{S}^2\alpha = \frac{1}{2}\left(\frac{1}{2}+1\right)\frac{h^2}{4\pi^2}\alpha \tag{11.213}$$

and

$$\hat{S}^2\beta = \frac{1}{2}\left(\frac{1}{2}+1\right)\frac{h^2}{4\pi^2}\beta \tag{11.214}$$

It should again be emphasized that the Z axis is entirely arbitrary and only has significance if there is an external magnetic field.

11.12 Many-Electron Atoms

For atoms containing more than one electron the form of the Schrödinger equation is more complicated, because the energy expression involves terms for the repulsive interactions between the different electrons. Even for helium, with two electrons, the Schrödinger equation is too complex to be solved in explicit form, although numerical solutions can be obtained fairly easily. For larger atoms the procedures are extremely complicated and require the use of approximation methods. We will consider some of these methods of solving the equations in Section 11.13. In this section we will outline a simple qualitative approach based on the orbitals for the hydrogen atom.

The Aufbau Principle

Only a brief account of this topic will be given, as it is assumed that the reader will already be familiar with the periodic table and with how it is interpreted in terms of adding electrons successively into the various orbitals. Use is made of the Aufbau (German *building up*) principle, originally suggested by Pauli. The basis of the method is that it is assumed that the orbitals obtained from the solution of the Schrödinger equation for the hydrogen atom can also be occupied in multielectron atoms. There are two important aspects to the problem: the *order* in which the different orbitals are filled and the *number of electrons* that can go into each orbital.

The first of these aspects is illustrated in Figure 11.20, which shows the order of filling of the orbitals. This order can be deduced from spectroscopic data on the

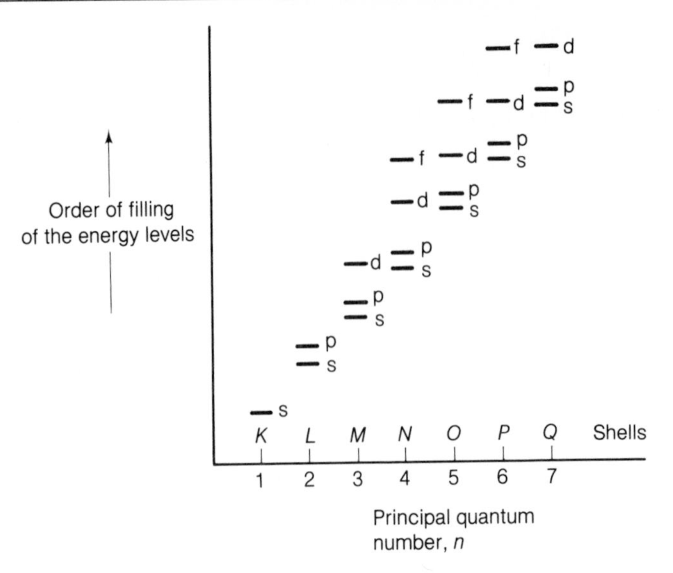

FIGURE 11.20 The order of filling of the various energy levels. This is very close to the order of the energy levels in an atom having a single electron, but there are some exceptions as discussed in the text.

various atoms and also on numerical calculations based on some of the approximation methods. There are several important points to note about this order of filling. The first relates to the influence of the azimuthal quantum number l on the energy levels. We have seen that for the hydrogen atom the energy is determined only by the principal quantum number n and not at all by l, so that, for example, the energy levels are the same for 2s and 2p orbitals. However, when there is more than one electron, the effect of the electron repulsions is to cause the 2p level to be somewhat higher than the 2s level. Similarly, when $n = 3$, the 3d level is higher than the 3p, which is higher than the 3s.

To a first approximation the order of filling of the levels, shown in Figure 11.20, is the order of the energy levels themselves. This is certainly true for a given value of n, but there are some discrepancies for different n values. For example, the diagram shows that the 4s level is filled before the 3d level, and this is often explained by saying that the 3d orbital is higher in energy than the 4s. This statement, however, is an oversimplification, as the order of energy levels depends on what electrons are present in the atom. Calculations for a *single*-electron system* show that the energy of the electron in the 3d orbital is below the energy when it is in the 4s orbital. The situation is, however, different when a number of electrons are present, because of the electron repulsions. The 3d orbital is more concentrated than the 4s, which has spherical symmetry, and there is a greater repulsion between a 3d electron and the other electrons than between a 4s electron and the other electrons. As a result, the 4s level fills before the 3d level.

*F. L. Pilar, *J. Chem. Ed.*, *55*, 2 (1978).

The Pauli Exclusion Principle

The energy factor is not the only one that determines the building up of the periodic table; if it were, all the electrons would go into the lowest orbitals. An important limitation is imposed by a principle due to Pauli, and it is known as the **Pauli exclusion principle.** One statement of it is

> **In an atom no two electrons can have all four quantum numbers (n, l, m_l, m_s) the same.**

Another formulation, which follows from the first, is

> **If, in an atom, two electrons have the same three orbital quantum numbers (n, l, and m_l), their spins must be opposed (i.e., the spin quantum number of one must be $+\frac{1}{2}$ and of the other $-\frac{1}{2}$).**

It follows at once from this principle that for a given principal quantum number n the maximum number of electrons that can be in the various subshells is as follows:

s	2
p	6
d	10
f	14

In general, the maximum number is $2(2l + 1)$.

Hund's Rule

The carbon atom, with six electrons, provides a simple example of the principles we have just considered. The first two electrons will go into the 1s orbital and have opposite spins. This shell is now filled and the next two go into the 2s orbital. The remaining two go into the 2p orbital, and now there are three possibilities which are conveniently represented as follows; the boxes indicate different m_l values and the arrow directions indicate the two different m_s values:

(a) [↑↓][ ][ ]   (b) [↑][↑][ ]   (c) [↑][↓][ ]

An important rule, formulated by the German physicist Friedrich Hund (b. 1890) enables us to decide which of these possibilities corresponds to the lowest energy. According to **Hund's rule,** the favored configuration is the one in which the electrons have different magnetic quantum numbers m_l and the same spin quantum numbers. Thus the arrangement b is found in the lowest energy state of carbon. The other arrangements are permitted but correspond to slightly higher energies.

The reason for the effect is rather subtle as two opposing factors are at work. When electrons have different magnetic quantum numbers, they tend to be further apart, and they therefore repel each other less than if they had the same m_l value. This factor alone favors arrangements b and c over arrangement a. When two electrons have the same spin quantum numbers, their magnetic moments are aligned in the same direction and they therefore repel one another more than if their spins are opposed. This factor alone favors arrangements a and c over arrangement

b. It might then appear that arrangement c would lead to the lowest energy. In this arrangement, however, there is not enough repulsion to keep the electrons in separate orbitals, and on balance arrangement b is preferred. This tendency for electrons with the same spin to avoid each other is called **spin correlation.**

The best evidence for Hund's rule and spin correlation comes from spectroscopic measurements of the magnetic properties of atoms. The magnetism of an atom is related to the number of unpaired electrons in orbitals where there is orbital angular momentum (i.e., all orbitals except s orbitals). If carbon has two unpaired p electrons, as predicted by Hund's rule, the magnetism of the carbon atom should be roughly twice that of the boron atom, which has one p electron. In fact, this is the case. If the two 2p electrons in the carbon atom were paired with each other (i.e., had opposite spins as in arrangements a and c), the atom would exhibit zero magnetism.

11.13 Approximate Methods in Quantum Mechanics

It is a straightforward matter to write the Schrödinger equation for any atom, but an explicit solution is possible only for the hydrogenlike atoms, which have a single electron. This is because of the terms for electron-electron repulsion, which make it impossible to separate the equation into simpler forms to which solutions can be obtained.

For many-electron atoms it is therefore necessary to employ approximate methods and even then it is usually impossible to obtain explicit solutions. However, these approximate methods allow numerical solutions to be obtained in a reasonable time with the use of a modern computer.

Many of these approximate methods make use of the **central-field approximation.** The outermost electron in the atom is considered to be in the electric field that is created by the nucleus and the remaining electrons. The effect of these electrons is to shield the outermost electron from the nucleus, which is considered to have an *effective charge* $Z_{eff}e$, which is less than the true charge Ze on the nucleus. The shielded nucleus is assumed to produce a spherically symmetric field, so that the atom can in this approximate manner be treated quantum mechanically as a one-electron atom.

A simple application of this procedure is in the estimation of **ionization energies,** which are the energies required to remove an electron from its orbital to infinity. According to Bohr's theory of the hydrogenlike atom, the potential energy of the electron, relative to infinite separation, is given by Eq. 11.49:

$$E_p = -\frac{Z^2e^2}{4\pi\epsilon_0 a_0 n^2} \tag{11.215}$$

Exactly the same expression for E_p is given by the quantum-mechanical treatment. The value of this energy, with the sign dropped and with $Z = 1$ and $n = 1$, is 4.36×10^{-18} J or 27.2 eV, and it is a common practice to refer to this quantity as one **atomic unit of energy** (au). It is also known as the *hartree,* after the British physicist Douglas Rayner Hartree (1897–1958). The Bohr radius a_0, which is equal to 52.92 pm, is known as the **atomic unit of length.**

The total energy of the electron, according to Bohr's theory and to quantum mechanics, is given by Eqs. 11.50 and 11.184:

$$E_n = -\frac{Z^2e^2}{8\pi\epsilon_0 a_0 n^2} \tag{11.216}$$

This is just one-half the potential energy, and the negative of its value is the ionization energy I for an electron of quantum number n. The ionization energy is therefore given by

$$I = \frac{Z^2e^2}{8\pi\epsilon_0 a_0 n^2} = \frac{Z^2}{2n^2} \times 4.36 \times 10^{-18}\ \text{J} \tag{11.217}$$

$$= \frac{Z^2}{2n^2}\ \text{atomic units (au)} \tag{11.218}$$

To apply the central-field approximation we simply replace Z by Z_{eff}:

$$I = \frac{Z_{eff}^2}{2n^2}\ \text{au} \tag{11.219}$$

The following example shows how the value of Z_{eff} can be estimated from an experimental value of the ionization energy.

EXAMPLE
Estimate Z_{eff} for a 1s electron in He, if the first ionization energy of helium is 24.6 eV.

SOLUTION
In atomic units this ionization energy is

$$\frac{24.6}{27.2} = 0.904\ \text{au}$$

For the helium atom $n = 1$ and therefore, from Eq. 11.219,

$$0.904 = \frac{Z_{eff}^2}{2} \quad \text{or} \quad Z_{eff} = 1.34$$

This is significantly less than the Z value of He and shows that there is substantial screening by the other 1s electron. Note that the Z_{eff} value obtained relates to He^+ (i.e., to the nucleus screened by *one* electron). ■

The Variation Method

There is a very powerful quantum-mechanical procedure that forms the basis of most calculations that have been made on atoms and molecules. This procedure, known as the **variation method,** not only allows reliable calculations to be made on complex atoms (and, as we shall see in Chapter 12, on molecules), but it also provides considerable insight into the nature of atoms and of chemical bonds. It is most useful for obtaining the energies of atoms and molecules in their most stable states.

Suppose, first, that the Schrödinger equation can be solved exactly and that we have obtained an eigenfunction ψ, which need not be normalized. In other words, for a particular system we have set up the equation

$$\hat{H}\psi = E\psi \tag{11.220}$$

and have obtained the function ψ. We can now multiply both sides by the complex conjugate ψ^* of the eigenfunction:

$$\psi^*\hat{H}\psi = \psi^*E\psi = E\psi\psi^* \tag{11.221}$$

The transformation of $\psi^*E\psi$ into $E\psi\psi^*$ is permissible because E is a constant and ψ, ψ^*, and E all commute. Integration of both sides of Eq. 11.221 over all space gives

$$\int \psi^*\hat{H}\psi \, d\tau = E \int \psi\psi^* \, d\tau \tag{11.222}$$

and therefore

$$E = \frac{\int \psi^*\hat{H}\psi \, d\tau}{\int \psi\psi^* \, d\tau} \tag{11.223}$$

The denominator of this equation is unity if the eigenfunction has been normalized.

On the other hand, suppose that ψ was not an eigenfunction of the operator $\hat{H}$ but was instead a trial function obtained in some way. We could then calculate the values of the integrals on the right-hand side of Eq. 11.223 and calculate an energy, if necessary by using numerical methods. The variation principle now tells us that the energy calculated from Eq. 11.223, with any trial function, *cannot be below* the true energy for the ground state of the system. If the function chosen happened to be the true eigenfunction, the exact energy would be obtained. Otherwise, the calculated energy is *higher* than the true energy.

In order to calculate reliable energies by the variation method, it is best to choose trial wave functions, containing adjustable parameters, that relate as closely as possible to the system that is being studied. For a many-electron atom, for example, one might choose a linear combination of hydrogenlike wave functions and vary the coefficients, until a minimum energy is obtained. For more reliable results, a number of trial wave functions of different types may be employed.

Another method that has been used for the same purpose is the *perturbation method*. The procedure is to start with an exact solution of the Schrödinger equation for a known system and then to add additional small terms to deal with the actual system. In first-order perturbation theory a single term is added to the Hamiltonian for the unperturbed system. For details the reader is referred to texts on quantum mechanics.

The Self-Consistent Field (SCF) Method

A very important procedure for dealing with complex atoms is the **self-consistent field (SCF) method,** which is a development of the central-field approach and which uses the variational principle. The method was first suggested in 1933 by

Douglas Rayner Hartree* (1897–1958) and it was improved by the introduction of methods earlier suggested by the Russian physicist Vladimir Alexandrovitch Fock† (b. 1898). The method is therefore often known as the **Hartree-Fock method.** It is based on the construction of an approximate electronic distribution for the atom, by first considering the individual shells. Closed shells have spherical symmetry, by Unsöld's theorem, and for nonfilled shells the distribution is averaged over all angles.

The next step is to consider one particular electron in the potential field created by the nucleus and the remaining electrons, averaged as indicated previously. Since the averaged potential is symmetrical, the Schrödinger equation can be separated into R, Φ, and Θ equations that can be solved. This procedure can be applied to all the electrons. On the basis of the wave functions obtained in this way one can then calculate a new potential-energy field that will be closer to the truth than the one originally assumed. The whole procedure is then repeated and a second set of solutions obtained. This *iterative* procedure is repeated until there is no further change in the individual electronic distributions or their potential-energy fields, which are then said to be self-consistent. The electron densities calculated by this method have been found to give good agreement with the results of X-ray and electron-diffraction experiments.

Slater Orbitals

The SCF method is rather laborious, and a very useful approximate procedure (involving analytical expressions for atomic orbitals) has been suggested by the American physicist John Clarke Slater‡ (b. 1900). These orbitals are calculated using the effective nuclear charge that influences each electron. This effective nuclear charge is calculated from the equation

$$Z_{\text{eff}} = Z - \sigma \tag{11.224}$$

where Z is the actual charge number of the nucleus and σ is the amount of shielding, which is obtained from some semiempirical rules:

1. There is no contribution to σ from any electron with n greater than that of the electron under consideration.
2. For an s electron, the shielding σ of the other s electron is 0.30. For all other electrons that have the same value of n as the electron under consideration, the contribution is 0.35 for each electron.
3. If the value of n is 1 less than that of the electron of interest, σ is 0.85 for each s or p electron and is 1.00 for each d or f orbital.
4. If n is 2 or more less than the quantum number of the electron of interest, σ is 1.00 for each electron.

---

*D. R. Hartree, *Proc. Roy. Soc.*, *A141*, 282 (1933); *A143*, 506 (1933).
†V. Fock, *Z. Physik*, *61*, 126 (1930).
‡J. C. Slater, *Phys. Rev.*, *42*, 33 (1932).

EXAMPLE

Calculate the effective nuclear charge for one of the 1s electrons in the helium atom.

SOLUTION

Of the two electrons in the ground state of helium, one electron contributes to the shielding of the other. Therefore, from rule 2, $\sigma = 0.30$, and $Z_{\text{eff}} = Z - \sigma = 2.00 - 0.30 = 1.70$. Note that this is somewhat higher than the value (1.34) estimated from the ionization energy. ■

EXAMPLE

Calculate the effective nuclear charge for a 2p electron in nitrogen.

SOLUTION

The electron configuration for nitrogen is $1s^2 2s^2 2p^3$ and it has a nuclear charge of 7. The shielding is $2(0.85) = 1.70$ for the 1s electrons and $(4 \times 0.35) = 1.40$ for the two 2s and the two remaining 2p electrons. The effective nuclear charge for any 2p electron is thus $Z - \sigma = 7 - (1.70 + 1.40) = 3.90$. ■

The values of Z_{eff} as calculated are used to calculate *Slater orbitals,* which are of a slightly different form than the hydrogenlike functions. The Slater ns orbitals are defined as

$$\psi(n\text{s}) = r^{n-1} e^{-Z_{\text{eff}} r/n} \tag{11.225}$$

and the np orbitals as

$$\psi(n\text{p}_x) = x r^{n-2} e^{-Z_{\text{eff}} r/n} \tag{11.226}$$

where x is the contribution of $\text{p}(r)$ (a function of the variable r only) to $n\text{p}_x$. The orbitals generated in this way are useful for order-of-magnitude calculations and can be used as a starting point in SCF calculations.

Key Equations

Electromagnetic radiation:

$$\lambda \nu = c$$

Natural frequency ν of simple harmonic motion:

$$\nu = \frac{1}{2\pi}\sqrt{\frac{k_h}{m}}$$

Energy of a *photon:*

$$E = h\nu$$

Average energy of an *oscillator* of frequency ν:

$$\bar{\boldsymbol{\epsilon}} = \frac{h\nu}{e^{h\nu/\mathbf{k}T} - 1}$$

Wavenumbers $\bar{\nu}$ of hydrogen atom spectral lines:

$$\bar{\nu} \equiv \frac{1}{\lambda} = R\left(\frac{1}{n_1^2} - \frac{1}{n_2^2}\right) \qquad (R = \text{Rydberg constant})$$

Uncertainty principle:

$$\Delta q\, \Delta p \approx \frac{h}{4\pi} \quad \text{or} \quad \Delta q\, \Delta u \approx \frac{h}{4\pi m}$$

Schrödinger equation:

Time dependent: $$\left[-\frac{h^2}{8\pi^2 m}\nabla^2 + E_p(x, y, z, t)\right]\Psi = -\frac{h}{2\pi i}\frac{\partial \Psi}{\partial t}$$

Time independent: $$\left[-\frac{h^2}{8\pi^2 m}\nabla^2 + E_p(x, y, z)\right]\psi = E\psi$$

or

$$\hat{H}\psi = E\psi$$

Normalization condition:

$$\int \psi\psi^*\, d\tau = 1$$

Orthogonality condition:

$$\int \psi_1\psi_2^*\, d\tau = 0$$

Average value of a quantity F:

$$\bar{F} = \langle F \rangle = \frac{\int \psi^* F\psi\, d\tau}{\int \psi^*\psi\, d\tau}$$

Energy of a *particle in a box* of sides a, b, and c:

$$E = \frac{h^2}{8m}\left(\frac{n_1^2}{a^2} + \frac{n_2^2}{b^2} + \frac{n_3^2}{c^2}\right)$$

Energy of a *harmonic oscillator* of reduced mass μ:

$$E_v = \frac{h}{2\pi}\sqrt{\frac{k_h}{\mu}}\left(v + \frac{1}{2}\right) = h\nu_0\left(v + \frac{1}{2}\right)$$

Quantum numbers:

$$n = 1, 2, 3, \ldots$$

$$l = 0, 1, 2, \ldots, n - 1$$

$$m_l = -l, -l + 1, -l + 2, \ldots, -1, 0, 1, \ldots, l - 1, l$$

$$m_s = +\tfrac{1}{2}, -\tfrac{1}{2}$$

Angular momentum:

$$L_z = m_l \frac{h}{2\pi}$$

$$L^2 = l(l + 1)\frac{h^2}{4\pi^2}$$

Variation method:

$$E = \frac{\int \psi^* \hat{H} \psi \, d\tau}{\int \psi \psi^* \, d\tau}$$

Problems

A Electromagnetic Radiation and Wave Motion

1. Calculate, for light of 325-nm wavelength,
  a. the frequency;
  b. the wave number;
  c. the photon energy in J, eV, and kJ mol^{-1}; and
  d. the momentum of the photon.

2. A pulsar in the Crab Nebula, NP 0532, emits both radio pulses and optical pulses. A radio pulse is observed at 196.5 MHz. Calculate
  a. the corresponding wavelength;
  b. the energy of the photon in J, eV, and J mol^{-1}; and
  c. the momentum of the photon.

3. The potassium spectrum has an intense doublet with lines at 766.494 nm and 769.901 nm. Calculate the frequency difference between these two lines.

4. Suppose that the position y of a particle that travels along the Y axis of a coordinate system is given by

$$y(t) = y_0 \sin\left[\frac{3\pi \text{ rad}}{5}\left(\frac{t}{s}\right) + C\right]$$

What is the frequency of the wave motion?

5. A mass of 0.2 kg attached to a spring has a period of vibration of 3.0 s.
  a. What is the force constant of the spring?
  b. If the amplitude of vibration is 0.010 m, what is the maximum velocity?

***6.** If the average energy associated with a standing wave of frequency ν in a cavity is

$$\overline{E} = \frac{h\nu}{e^{h\nu/kT} - 1}$$

deduce the limiting low-frequency value of the average energy associated with the standing wave.

B Particles and Waves

7. A sodium lamp of 50-W power emits yellow light at 550 nm. How many photons does it emit each second? What is the momentum of each photon?

8. The threshold frequency ν_0 for emission of photoelectrons from metallic sodium was found by Millikan, *Phys. Rev.*, *7*, 362 (1916), to be 43.9×10^{13} s^{-1}.

Calculate the work function for sodium. A more recent value, for a carefully outgassed sample of sodium, is 5.5×10^{13} s^{-1}. What work function corresponds to that value?

9. Calculate the value of the de Broglie wavelength associated with

a. An electron (mass = 9.11×10^{-31} kg) moving with a speed of 6.0×10^7 m s^{-1} (this is the approximate velocity produced by a potential difference of 10 kV).

b. An oxygen molecule moving with a speed of 425 m s^{-1} at 0°C.

c. An α particle emitted by the disintegration of radium, moving at a speed of 1.5×10^7 m s^{-1}.

d. An electron having a speed of 2.818×10^8 m s^{-1}.

10. Consider a colloidal particle with a mass of 6×10^{-16} kg. Suppose that we measure the position to within 1.0 nm, which is about the resolving power of an electron microscope. Calculate the uncertainty in the velocity and comment on the significance of the result.

11. Calculate the velocity and the de Broglie wavelength of an electron accelerated by a potential of

a. 10 V,

b. 1 kV, and

c. 1 MV.

***12.** The group velocity of a wave is given by the equation

$$v_g = \frac{d\nu}{d(1/\lambda)}$$

Prove that the group velocity of a de Broglie particle wave is equal to the ordinary velocity of the particle.

C Quantum-Mechanical Principles

13. Assume that the three real functions ψ_1, ψ_2, and ψ_3 are normalized and orthogonal. Normalize the following functions:

a. $\psi_1 + \psi_2$

b. $\psi_1 - \psi_2$

c. $\psi_1 + \psi_2 + \psi_3$

d. $\psi_1 - \frac{1}{\sqrt{2}}\psi_2 + \frac{\sqrt{3}}{\sqrt{2}}\psi_3$

14. Is the function Ae^{-ax} an eigenfunction of the operator d^2/dx^2? If so, what is the eigenvalue?

15. Prove that m_l must be integral in order for the function

$$\Phi = \sin m_l \phi$$

to be an acceptable wave function.

16. The energy operator for a time-dependent system (Table 11.1) is

$$-\frac{h}{2\pi i}\frac{\partial}{\partial t}$$

A possible eigenfunction for the system is $\Psi(x, y, z, t) = \psi(x, y, z)e^{2\pi iEt/h}$. Show that $\Psi^*\Psi$, the probability density, is independent of time.

***17.** Prove that the momentum operator corresponding to p_x is a Hermitian operator.

18. Which of the following functions is an eigenfunction of the operator d/dx?

a. k

b. kx^2

c. $\sin kx$

d. e^{kx}

e. e^{kx^2}

f. e^{ikx}

(k is a constant.) Give the eigenvalue where appropriate.

D Particle in a Box

19. Calculate the lowest possible energy for an electron confined in a cube of sides equal to

a. 10 pm and

b. 1 fm (1 femtometre = 10^{-15} m).

The latter cube is the order of magnitude of an atomic nucleus; what do you conclude from the energy you calculate about the probability of a free electron being present in a nucleus?

20. A particle is moving in one dimension between $x = a$ and $x = b$. The potential energy is such that the particle cannot be outside these limits and that the wave function in between is

$$\psi = \frac{A}{x}$$

a. Determine the normalization constant A.

b. Calculate the average value of x.

21. An electron is confined in a one-dimensional box 1 nm long. How many energy levels are there with energy less than 10 eV? How many levels are there with energy between 10 and 100 eV?

22. Determine whether the eigenfunctions obtained in Section 11.6 for a particle in a one-dimensional box are eigenfunctions for the momentum operator. If they are, obtain the eigenvalues; if they are not, explain why.

E Vibration and Rotation

23. The vibration frequency of the N_2 molecule corresponds to a wave number of 2360 cm^{-1}. Calculate the zero-point energy and the energy corresponding to $v = 1$.

***24.** If a rigid body rotates in the XY plane, about the Z axis, the angular momentum operator is

$$\hat{L} = \frac{h}{2\pi i}\frac{\partial}{\partial \phi}$$

(see Figure 11.13). If the moment of inertia is I, what is the energy operator?

(For additional problems dealing with molecular vibrations and rotations, see Chapter 13.)

F The Atom

25. Calculate the ionization energy of the hydrogen atom on the basis of the Bohr theory.

26. Calculate, on the basis of the Bohr theory, the linear velocity of an electron (mass = 9.11×10^{-31} kg) in the ground state of the hydrogen atom. To what de Broglie wavelength does this velocity correspond? Deduce an equation for the de Broglie wavelength, in a Bohr orbit of quantum number n, with $Z = 1$, in terms of a_0 and n. What is the ratio of the circumference of a Bohr orbit of quantum number n to the de Broglie wavelength?

27. Calculate the wavelength and energy corresponding to the $n = 4$ to $n = 5$ transition in the hydrogen atom.

28. Calculate, in joules and in atomic units, the potential energy of an electron in the $n = 2$ orbit of the hydrogen atom.

29. The first ionization energy of the Li atom is 5.39 eV. Estimate an effective nuclear charge Z_{eff} for the valence electron in the Li atom.

30. The first ionization energy of the Na atom is 5.14 eV. Estimate the effective nuclear charge Z_{eff} for the valence electron in the Na atom.

31. Use Slater's method (Section 11.13) to determine the effective nuclear charge for

- **a.** a 3s electron in the chlorine atom,
- **b.** a 3p electron in the phosphorus atom, and
- **c.** the 4s electron in the potassium atom.

G Supplementary Problems

32. Treat the three-dimensional particle in a box of sides a, b, and c, by analogy with the treatment in Section 11.6. Assume the potential to be zero inside the box and infinite outside, and proceed by the following steps:

- **a.** Write the basic differential equation that must be solved for the three-dimensional problem.
- **b.** Separate the equation from (a) into terms involving $X(x)$, $Y(y)$, and $Z(z)$.
- **c.** Determine the expressions for X, Y, and Z.
- **d.** Obtain the expression (Eq. 11.146) for the total energy.

33. Calculate the reduced masses of the hydrogen and deuterium atoms, using the following masses for the particles:

Electron:	9.1095×10^{-31} kg
Proton:	1.6727×10^{-27} kg
D nucleus:	3.3434×10^{-27} kg

- **a.** Explain qualitatively what effect the different reduced masses will have on the Bohr radii and therefore on the positions of the lines in the atomic spectra.
- **b.** The Balmer spectrum of hydrogen has a line of wavelength 656.47 nm. Deduce the wavelength of the corresponding line in the spectrum of D.

34. Use the wave function for the 1s orbital of the hydrogen atom, given in Table 11.5, to obtain an expression for the probability that the electron lies between the distances r and $r + dr$ from the nucleus. (Use spherical polar coordinates, for which the volume element is $r^2\, dr \sin\theta\, d\theta\, d\phi$.)

35. Unsöld's theorem (Section 11.8) states that, for a given value of l, the sum of all the functions

$$\Theta \quad \Phi \quad \Theta^* \quad \Phi^*$$

is independent of θ and ϕ. Write all these functions for the 2p orbitals (see Tables 11.2 and 11.3), and show that their sum shows no angular dependence.

36. Problem 19 was concerned with the calculation of the minimum energy for an electron confined in a cube. Another approach to the problem is to consider, on the basis of the Uncertainty Principle (Eq. 11.60), the uncertainty in the energy if the uncertainty in the position is equal to the length of the side of the cube. Calculate ΔE for a cube of sides equal to

- **a.** 10 pm and
- **b.** 1 fm (10^{-15} m),

and compare the results with the minimum energies found for Problem 19.

37. Prove that any two wave functions for a particle in a one-dimensional box of length a are orthogonal to each other; that is, they obey the relationship

$$\int_0^a \psi_m \psi_n \, dx = 0$$

H Essay Questions

1. With emphasis on the physical significance, explain precisely what is meant by a *normalized* wave function.

2. Explain clearly the relationship between the Heisenberg uncertainty principle and the question of whether two operators commute.

3. Give an account of the main principles underlying the variation method in quantum mechanics.

4. Discuss the reasons for abandoning the Bohr theory of the atom.

Suggested Reading

P. W. Atkins, *Quanta: A Handbook of Concepts*, Oxford: Clarendon Press, 1974.

P. W. Atkins, *Physical Chemistry*, chap. 13 and 14, San Francisco: W. H. Freeman, 1978.

W. Heitler, *Elementary Wave Mechanics*, Oxford: Clarendon Press, 1956.

M. Karplus and R. N. Porter, *Atoms and Molecules: An Introduction for Students of Physical Chemistry*, New York: Benjamin, 1970.

J. W. Linnett, *Wave Mechanics and Valency*, London: Methuen, 1960.

W. J. Moore, *Physical Chemistry* (4th ed.) chap. 13, 14, and 15, Englewood Cliffs, N.J.: Prentice-Hall, 1972.

Chapter 12

The Chemical Bond

Preview

In this chapter we continue our study of quantum mechanics, and we will see how it is applied to *molecules*. We will be concerned almost entirely with *covalent bonds,* which arise from the sharing of electrons between atoms. The other important type of bond, the *ionic bond,* which is purely electrostatic in character, is dealt with in Chapter 15 where we will be concerned with the structure of crystals.

The simplest of all molecules is the *hydrogen molecule ion* H_2^+ in which two protons are bound together by a single electron. The quantum-mechanical equation for this system can be solved exactly, but to gain an insight into the chemical bond it is more useful to explore approximate methods in which wave functions for the molecule are constructed from the orbitals for the isolated atoms. One way of doing this is to take a *linear combination of atomic orbitals* (LCAO). Such a combination gives what is called a *molecular orbital* (MO); the variation method, which we considered in Chapter 11, can be used to obtain the energy of the molecule.

The *hydrogen molecule* H_2 is the simplest molecule having the most usual type of chemical bond, the *two-electron* or *electron-pair bond.* There are two essentially different quantum-mechanical methods for dealing with such bonds. Closely related to the familiar chemical concept of a chemical bond is the *valence-bond method,* in which a wave function for the electron-pair bond is constructed by first taking the product of two atomic wave functions, one for each electron. The other method is the *molecular-orbital method,* which focuses attention not on the electron pair but on the individual electrons. Often MOs are constructed as linear combinations of atomic orbitals.

An important contribution to the valence-bond method was made by *Heitler and London,* who recognized that in constructing an orbital for the molecule one must not associate a particular electron with a particular nucleus. They avoided this difficulty by adding or subtracting functions that were related to one another simply by an exchange of electrons. They found that this device leads to an important contribution to the binding energy. This *exchange* or *resonance energy* usually accounts for most of the binding, and it is an energy that is not explained at all by classical mechanics.

The nature of a chemical bond is conveniently treated in terms of the concept of *electronegativity.* A bond may have a *dipole moment,* which is related to the difference between the electronegativities of the atoms that the bond connects. Certain bonds are *hybrids* of different kinds of orbitals, and this concept leads to an interpretation of the *shapes* of molecules.

A very important contribution to the study of molecules is provided by the subject of *symmetry*. A consideration of planes of symmetry, centers of symmetry, and other symmetry elements allows decisions to be made as to what types of wave functions are possible for particular molecules. The systematic study of symmetry is known as *group theory*, which leads to useful rules for designating the different types of molecular wave functions.

12 The Chemical Bond

It has long been recognized that many chemical bonds can be classified as either **ionic bonds** or **covalent bonds.** The former can be treated satisfactorily by classical electrostatic theory, but for an understanding of the covalent bond it is necessary to use the methods of quantum mechanics. In almost all cases it is necessary to employ approximate methods, such as the variation method (Section 11.13).

Other types of bonding occur, and Table 12.1 gives information about the more important forces that hold atoms together. The ionic and covalent bonds are the strongest, but ion-dipole and dipole-dipole forces also play an important role in chemical structures. The best known bond formed largely as a result of dipole-

TABLE 12.1 The Main Types of Chemical Bonds

Type of Force or Chemical Bond	Example	Equilibrium Separation/nm	Dissociation Energy*/ $kJ\ mol^{-1}$
Ionic bond (ion-ion force)	$Na^+ \cdots F^-$	0.23	670.0
Covalent bond	H—H	0.074	435.0
Ion-dipole force	$Na^+ \cdots OH_2$ (O bonded to two H)	0.24	84.0
Hydrogen bond (dipole-dipole bond)	$H_2O \cdots H—OH$	0.28	20.0
Hydrophobic bond	$>CH_2 \cdots H_2C<$	≈0.30	≈4.0
Van der Waals (dispersion forces)	Ne ··· Ne	≈0.33	≈0.25

*This is the energy that would be required per mole to dissociate the species into its units (e.g., H—H into H + H or $Na^+ \cdots OH_2$ into $Na^+ + H_2O$) in a vacuum.

dipole attractions is the **hydrogen bond,** which is very important in the structure of liquid water (see Chapter 16). **Hydrophobic bonds** are of an indirect kind; they occur when nonpolar groups are present in aqueous solution, and they result from the fact that such groups have an effect on the hydrogen-bonded structure of water.

Ionic bonds are described in terms of the Coulombic forces of attraction, which will be further discussed in Section 15.9. The remainder of this chapter will be concerned with the application of quantum mechanics to the covalent bond.

12.1 The Hydrogen Molecule-Ion, H_2^+

The simplest molecular system is H_2^+, which consists of two protons and a single electron. This molecule-ion forms no stable salts, but there is evidence for its existence in electrical discharges passed through hydrogen gas, and some of its properties have been studied by spectroscopic methods. The experimental potential-energy curve for H_2^+ is shown in Figure 12.1a. The minimum energy occurs at an internuclear separation of 106 pm and corresponds to an energy 269.3 kJ mol^{-1} (= 2.791 eV) below that of the separate particles H + H$^+$.

It is possible to obtain an exact quantum-mechanical solution for the H_2^+ ion. The system of two protons and an electron is shown in Figure 12.1b, and we make use of confocal-elliptical coordinates in order to effect a separation of variables. The problem is simplified by the fact that the electron moves very much more rapidly than the nuclei. The assumption may therefore be made that the nuclei are held fixed in position during the period of the electron's motion. This is known as the **Born-Oppenheimer approximation,** and it permits the calculation of both the stationary-state wave functions and the energy of the electron. The energy is calculated with a fixed r_{AB}; then new fixed values of r_{AB} are chosen and the corresponding energies are determined. The potential energy E_p at different values of r_{AB} is then plotted against r_{AB} to obtain the **potential-energy curve** of the system. It is possible to obtain an exact quantum-mechanical solution for this system, but instead we will consider a much simpler solution based on the variation principle.

The system shown in Figure 12.1b consists of two electron-proton attractions and a proton-proton repulsion, and its potential energy is

$$E_p = \frac{e^2}{4\pi\epsilon_0}\left(\frac{1}{r_{AB}} - \frac{1}{r_A} - \frac{1}{r_B}\right) \tag{12.1}$$

The Hamiltonian operator for the system is therefore, from Table 11.1,

$$\hat{H} = -\frac{h^2}{8\pi^2\mu}\nabla^2 + \frac{e^2}{4\pi\epsilon_0}\left(\frac{1}{r_{AB}} - \frac{1}{r_A} - \frac{1}{r_B}\right) \tag{12.2}$$

where μ is the reduced mass. The Schrödinger equation to be solved is therefore

$$\left[-\frac{h^2}{8\pi^2\mu}\nabla^2 + \frac{e^2}{4\pi\epsilon_0}\left(\frac{1}{r_{AB}} - \frac{1}{r_A} - \frac{1}{r_B}\right)\right]\psi = E\psi \tag{12.3}$$

To obtain the best energy we choose some suitable trial eigenfunction ψ that displays the generally expected shape of the true function. One function used was*

*H. M. James, *J. Chem. Phys.*, 3, 9 (1935).

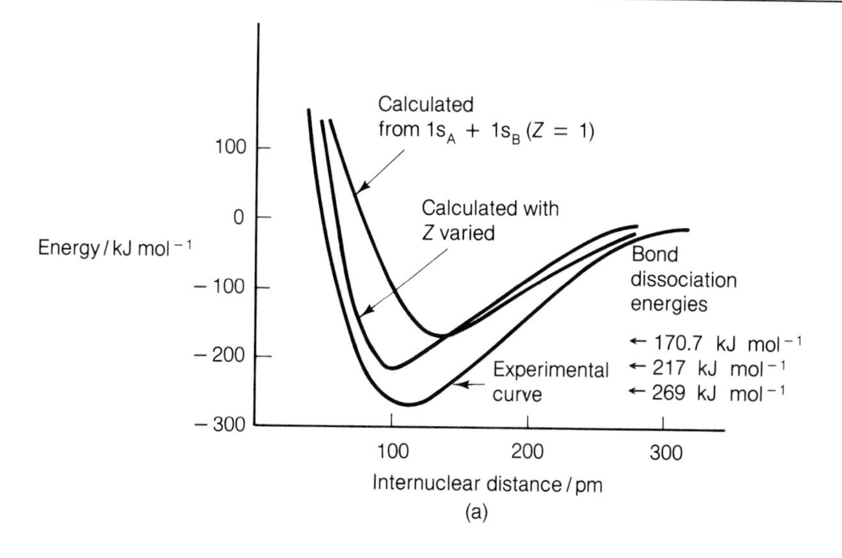

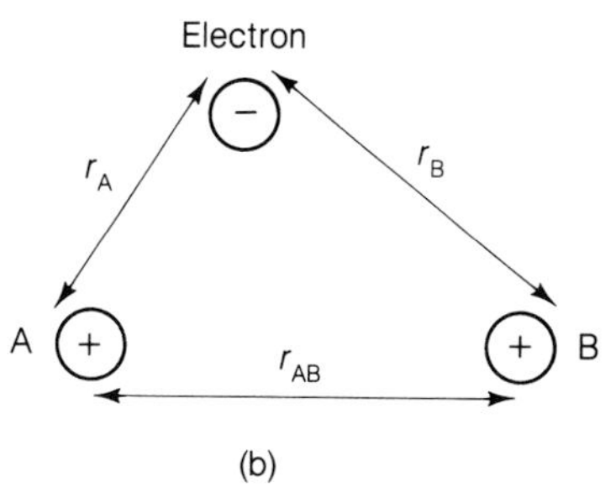

FIGURE 12.1 The hydrogen molecule-ion H_2^+. (a) Experimental and theoretical potential-energy curves. (b) The system of two protons and an electron.

$$\psi = e^{-b\xi}(1 + c\eta^2) \tag{12.4}$$

where ξ and η are the confocal-elliptic variables and b and c are constants adjusted to minimize the energy. The energy is calculated from Eq. 11.223 and according to the variation principle the calculated energy cannot be below the experimental value. This function gave a dissociation energy of 268.8 kJ mol^{-1} (2.786 eV), very close to the experimental value, but the internuclear distance is 206 pm, which is in considerable error.

Not many problems are simple enough for arbitrary trial variation functions to be used. A more common procedure is to construct trial functions for the molecule from the exact wave functions that apply to the atoms from which the molecule is formed. This approach is an approximation, but the atomic wave functions provide a useful starting point.

To deal with H_2^+ in this way we start with the 1s wave function for the hydrogen atom, given in Table 11.5. We will write the wave function for the electron bound to proton A as $1s_A$, and the corresponding wave function for the electron associated with nucleus B as $1s_B$:

$$1s_A = \frac{1}{\sqrt{\pi}}\left(\frac{1}{a_0}\right)^{3/2} e^{-r_A/a_0} \quad \text{and} \quad 1s_B = \frac{1}{\sqrt{\pi}}\left(\frac{1}{a_0}\right)^{3/2} e^{-r_B/a_0} \tag{12.5}$$

These 1s functions are unsatisfactory in themselves because they do not properly describe the bonding of the molecule-ion.

We then form a linear combination of the atomic orbitals (LCAO), which when normalized gives the wave function

$$\psi = \frac{1}{\sqrt{2}}(1s_A + 1s_B) \tag{12.6}$$

Such a wave function, which is spread over the entire molecule, is called a **molecular orbital** (MO). The distribution of the electron throughout the molecule is proportional to $\int \psi^2 \, d\tau$ and therefore to

$$\int (1s_A + 1s_B)^2 \, d\tau = \int (1s_A)^2 \, d\tau + \int (1s_B)^2 \, d\tau + 2\int (1s_A)(1s_B) \, d\tau \tag{12.7}$$

When the electron is near proton A, the wave function resembles $1s_A$, the wave function on the isolated hydrogen atom, and the contribution from $1s_B$ is small; when the electron is near proton B, the wave function is approximately $1s_B$. Therefore, when the electron is near A or B, the distribution of electron density is similar to that in the isolated atom. In the region between the nuclei the probability of finding the electron is not simply the sum of $(1s_A)^2$ and $(1s_B)^2$ but is enhanced by the term $2(1s_A)(1s_B)$, and this accounts for the binding.

It is often stated that the accumulation of electron density between nuclei lowers the energy of the molecule and therefore accounts for its stability. This may be the case for more complicated molecules, but for the H_2^+ species detailed calculations have shown that the shifting of the electron away from its position on either nucleus into the internuclear region *raises* its potential energy. A more satisfactory explanation of the bonding is that, coincident with the electron shift into the internuclear region, the atomic orbitals shrink closer to their respective nuclei. This causes an increase in the electron-nuclear attractions and thus a lowering of the potential energy, which more than makes up for the loss in space-filling character of the orbitals. Although the kinetic energy is also changed, the overall effect is that the electron-nuclear attractions dominate.

The wave function of the bonding and lowest-energy MO in Eq. 12.6 is not spherically symmetrical like the atomic s orbital; instead it has a cylindrical symmetry about the internuclear axis. Such an MO is called a **σ (sigma) orbital** because of its similarity to the symmetry of the s atomic orbital. The bond formed by such a σ orbital is called a **σ bond.**

The variation method is used to calculate the energy of this orbital σ from Eq. 11.223:

$$E = \frac{\int \psi \hat{H} \psi \, d\tau}{\int \psi^2 \, d\tau} \tag{12.8}$$

When the integrals are evaluated and the energies calculated for various internuclear distances, the results are as shown in Figure 12.1a. As expected from the variation principle, the calculated energies are all higher than the experimental energies. The theory certainly gives a curve of the right form, but the calculated energies are not very accurate. For example, the calculated dissociation energy is 170.7 kJ mol^{-1} as compared with the experimental value of 269.3 kJ mol^{-1}. The calculated internuclear distance for this minimum is 132 pm.

Since the selected wave function ψ does give some success in predicting the proper curve of energy, the next step is to vary the function in Eq. 12.6 to obtain better agreement with experiment. The nuclear charge may be used as a variable in order to obtain the lowest possible energy at each internuclear separation. In other words, instead of Eq. 12.6, the new trial eigenfunction is

$$\psi = \frac{1}{\sqrt{\pi}}\left(\frac{Z}{a_0}\right)^{3/2} e^{-Zr/a_0} + \frac{1}{\sqrt{\pi}}\left(\frac{Z}{a_0}\right)^{3/2} e^{-Zr/a_0} \tag{12.9}$$

and Z is varied after each new energy is calculated from Eq. 12.8. The best value for Z is 1.23, and with this value the calculated potential-energy curve is much closer to the experimental curve, as shown in Figure 12.1a. The calculated dissociation energy now is 217.0 kJ mol^{-1}, much closer to the experimental value of 269.3 kJ mol^{-1}.

With the advent of large computer facilities, it is now possible to include contributions from the complex set of atomic orbitals; each orbital has the proper weighting factor determined by the variation principle.

Although the results of this simple LCAO method are not in good agreement with the experimental energy, it is possible to improve the agreement and to give better values for molecular properties; we will later see how this can be done.

12.2 The Hydrogen Molecule

Most chemical bonds consist of a pair of electrons which hold two nuclei together. The hydrogen molecule H_2 is the simplest molecule in which there is an *electron-pair bond*. Many calculations have been made for this molecule, which is a prototype for all covalent bonds. There are two basic quantum-mechanical treatments of the hydrogen molecule. One of them, the **valence-bond method,** considers the *two-electron* (*electron-pair*) bond and constructs a wave function for it by first taking the product of two atomic wave functions, one for each electron. The other treatment, the **molecular-orbital (MO) method,** is primarily concerned with the orbitals for single electrons and is based on the type of treatment we have given for the H_2^+ ion. Molecular orbitals are often constructed as linear combinations of atomic orbitals (LCAO), and electrons are fed into the corresponding orbitals with due regard to the Pauli principle.

The Valence-Bond Method

The valence-bond method is based on the familiar chemical ideas of **resonance** and *resonance structures*. This method is favored by some chemists because it gives a more pictorial view of bonding and is closely related to the classical structural theory of organic chemistry.

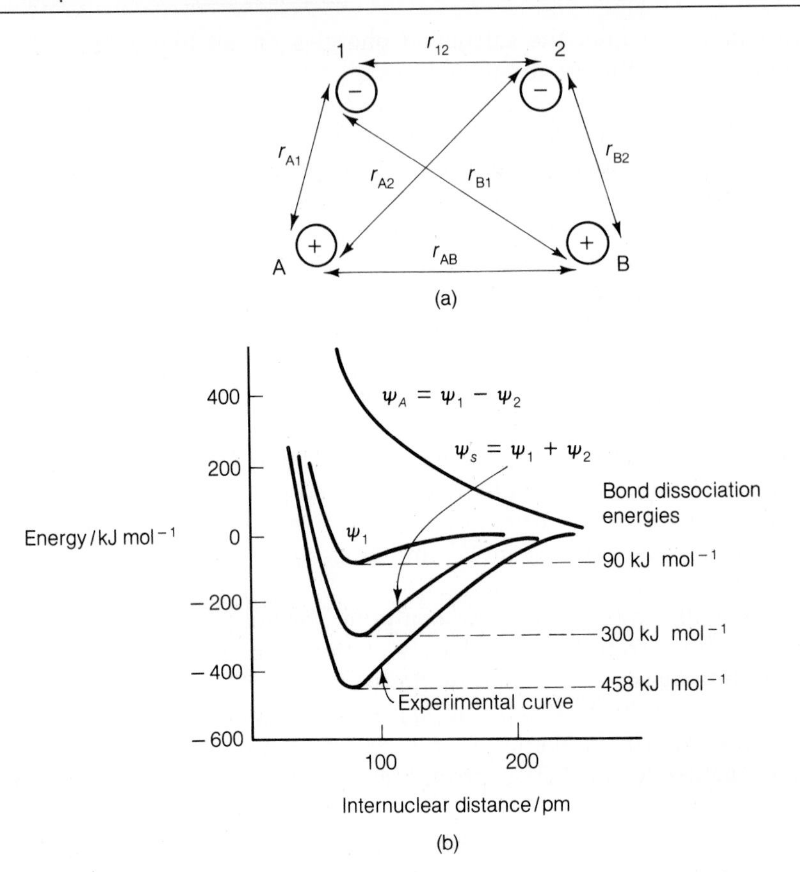

FIGURE 12.2 (a) The hydrogen molecule H_2, consisting of two protons and two electrons. (b) Experimental and theoretical potential-energy curves for the hydrogen molecule.

The system is represented in Figure 12.2a, where the protons are labeled A and B and the electrons are labeled 1 and 2. The potential energy is

$$E_p = \frac{e^2}{4\pi\epsilon_0}\left(\frac{1}{r_{AB}} + \frac{1}{r_{12}} - \frac{1}{r_{A1}} - \frac{1}{r_{A2}} - \frac{1}{r_{B1}} - \frac{1}{r_{B2}}\right) \tag{12.10}$$

and the Hamiltonian operator is therefore

$$\hat{H} = -\frac{h^2}{8\pi^2\mu}\nabla^2 + \frac{e^2}{4\pi\epsilon_0}\left(\frac{1}{r_{AB}} + \frac{1}{r_{12}} - \frac{1}{r_{A1}} - \frac{1}{r_{A2}} - \frac{1}{r_{B1}} - \frac{1}{r_{B2}}\right) \tag{12.11}$$

We initially simplify the problem by ignoring the electrical repulsion between the electrons, which is the case if the two hydrogen atoms are infinitely far apart. The state of electron 1 on nucleus A is described by the wave function $1s_A(1)$, and electron 2 on nucleus B is described similarly by the wave function $1s_B(2)$. When the two atoms are infinitely far apart, their energy is simply the sum of the two atomic energies, $E_A + E_B$. To meet this requirement, the Schrödinger wave function for this system must be the *product* of the individual wave functions. Thus, a possible wave function is

$$\psi_1 = 1s_A(1)1s_B(2) \tag{12.12}$$

An equally acceptable solution is obtained if electron 2 is associated with nucleus A and electron 1 with nucleus B:

$$\psi_2 = 1s_A(2)1s_B(1) \tag{12.13}$$

However, the energy calculated from either of these functions is in poor agreement with the experimental results, as shown in Figure 12.2b. The fault with each of these trial functions was recognized by the German physicists Walter Heinrich Heitler (b. 1904) and Fritz London, who carried out in 1927 (just after Schrödinger's equation appeared in 1926) the first calculations of molecular energies. What is wrong with both functions is that, although electrons are indistinguishable particles, Eq. 12.12 implies that one electron can be designated as electron 1 and is particularly associated with a nucleus designated nucleus A; a similar objection applies to the function in Eq. 12.13. Heitler and London realized that this difficulty is avoided if one uses a linear combination of these wave functions. In effect, this accounts for the electron repulsion. Two possible combinations exist:

$$\psi_S = \frac{1}{\sqrt{2}}(\psi_1 + \psi_2) = \frac{1}{\sqrt{2}}[1s_A(1)1s_B(2) + 1s_A(2)1s_B(1)] \tag{12.14}$$

and

$$\psi_A = \frac{1}{\sqrt{2}}(\psi_1 - \psi_2) = \frac{1}{\sqrt{2}}[1s_A(1)1s_B(2) - 1s_A(2)1s_B(1)] \tag{12.15}$$

where the $1/\sqrt{2}$ is a normalization factor.

For a stable bond the distribution of electrons must reduce the repulsion between the two protons. This is accomplished if the electrons are between the nuclei, which implies that the electrons must be close to one another. Consequently,

$$1s_A(1) \approx 1s_A(2) \quad \text{and} \quad 1s_B(1) \approx 1s_B(2)$$

Substitution of these relations into Eqs. 12.14 and 12.15 leads to

$$\psi_S \approx \frac{2}{\sqrt{2}}1s_A(1)1s_B(2) = \frac{2}{\sqrt{2}}\psi_1 \quad \text{and} \quad \psi_A \approx 0 \tag{12.16}$$

The description of the system is given by the probabilities ψ_S^2 and ψ_A^2. The probability of finding the electrons close together is, therefore, very small with the ψ_A wave function and is larger with ψ_S. Plots of ψ_S and ψ_A are made in Figure 12.2b, where the energy is calculated as a function of r_{AB}, the internuclear distance. The existence of a minimum in the ψ_S curve indicates that a stable molecule is formed; the state is *attractive.* The difference between the zero of energy and the minimum is the *classical binding energy,* or the *classical dissociation energy,* of the molecule.* The energy of the state ψ_A is always above the energy of two separated atoms and is therefore called an **antibonding** state; it corresponds to repulsion between the atoms. The energy difference between ψ_1 and ψ_S at the minimum in the curves is called the **resonance energy** or *resonance stabilization energy.* The energy of the electrons is decreased by their being spread out over both nuclei. There is therefore a buildup of electron density between the nuclei, and this may be thought of as an

*We have seen in Section 11.6 that there is a *zero-point energy* and that the true dissociation energy is somewhat less than the classical value.

overlapping of the electron clouds originally based on the atoms. This rather semi-quantitative rationale is the basis of the **principle of maximum overlap** first proposed by the American physical chemist Linus Pauling (b. 1901).

When ψ_S is used in the variation technique to calculate the energy, the agreement with experiment is more satisfactory than when ψ_1 or ψ_2 is used, as seen in Figure 12.2b. There is still much room for improvement, however; the calculated dissociation energy is only 66% of the experimental value. Evaluation of the integrals involving ψ_S, which occur in the variation treatment, leads to two contributions to the energy, the **Coulombic energy J** and the **exchange energy K.** At the normal internuclear separation both of these energies are negative in value with respect to the energy of the separated atoms. The Coulombic energy accounts for only about 10% of the binding; the remaining 90% is exchange energy. The Coulombic energy is approximately the energy that would be calculated on the basis of electrostatic effects in a purely classical treatment, and such a treatment is therefore completely inadequate. The exchange energy, which accounts for most of the binding, is a purely quantum-mechanical contribution, arising from the interchange of electrons allowed for in the Heitler-London wave function.

The Heitler-London treatment can be improved in various ways. For example, we can add terms corresponding to ionic states, which were completely neglected in the preceding treatment. If both electrons are associated with nucleus A, we have the function

$$\psi_3 = 1s_A(1)1s_A(2) \tag{12.17}$$

but if both are associated with B, we have

$$\psi_4 = 1s_B(1)1s_B(2) \tag{12.18}$$

Therefore, a reasonable trial function is

$$\psi = c_1(\psi_1 + \psi_2) + c_2(\psi_3 + \psi_4) \tag{12.19}$$

where c_1 and c_2 are numbers that can be varied, after the integrals in Eq. 12.8 have been evaluated, to obtain the lowest energy. The number c_1 multiplies both ψ_1 and ψ_2 because these two functions are equivalent. Similarly, ψ_3 and ψ_4 have both been multiplied by c_2, because in the symmetrical H_2 molecule one ionic state cannot be favored over the other. After the variation method is applied, the function ψ leads to better energies than those obtained when only the first two terms are used. Of course, other refinements are possible to give still better agreement with experiment, but when they are made, we lose some of the clarity of the chemical description.

Electron Spin

An important feature of ψ_S and ψ_A must be noted. If the coordinates of the electrons in Eq. 12.14 are interchanged, that is,

$$1s_A(1)1s_B(2) + 1s_A(2)1s_B(1) \rightarrow 1s_A(2)1s_B(1) + 1s_A(1)1s_B(2)$$

the wave function ψ_S is *left unchanged;* it is said to be **symmetric** with respect to the interchange of electrons. The function ψ_A, however, *changes sign* upon interchange of the electrons:

$$1s_A(1)1s_B(2) - 1s_A(2)1s_B(1) \rightarrow 1s_A(2)1s_B(1) - 1s_A(1)1s_B(2)$$

This operation causes the first term on the left-hand side to be the same as the second term on the right-hand side, *with the sign changed*. Similarly, the second term on the left becomes the first term on the right, *with the sign changed*. Such behavior under this operation is said to be **antisymmetric**.

We saw in Section 11.12 that the Pauli exclusion principle places an important restriction on the possible electron spins when more than one electron is present. The most general statement of the principle is that

> **The total wave function of a system (*i.e.*, the product of the orbital and spin wave functions) must be antisymmetric with respect to an interchange of electrons.**

This means that the symmetric orbital function ψ_S must be multiplied by an antisymmetric spin function and that the antisymmetric orbital function ψ_A must be multiplied by a symmetric spin function.

We saw in Section 11.11 that electron spin is determined by the quantum number m_s, which can have two possible values, $+\frac{1}{2}$ and $-\frac{1}{2}$. The corresponding spin functions may be written as α and β, and the eigenvalue equations are Eqs. 11.211 and 11.212. For the pair of electrons in the hydrogen molecule we can set up four possible products of the spin functions, namely

$$\alpha(1)\alpha(2), \qquad \beta(1)\beta(2), \qquad \alpha(1)\beta(2), \quad \text{and} \quad \beta(1)\alpha(2)$$

The first two of these remain unchanged when we interchange electrons and are *symmetric;* in both of them the spins are in the same direction and are said to be *parallel*. These two functions may therefore be combined with the antisymmetric orbital function ψ_A to give an acceptable function:

$$\Psi_1 = \psi_A\alpha(1)\alpha(2) \tag{12.20}$$

$$\Psi_2 = \psi_A\beta(1)\beta(2) \tag{12.21}$$

The other two spin functions $\alpha(1)\beta(2)$ and $\beta(1)\alpha(2)$ are, however, neither symmetric nor antisymmetric with respect to interchanging the electrons; each of them changes into the other. These functions therefore suffer from the same fault as the orbital wave functions in Eqs. 12.12 and 12.13; they are inconsistent with the fact that electrons are indistinguishable. This difficulty is overcome by a device equivalent to that used by Heitler and London; we take linear combinations of the functions:

$$\alpha(1)\beta(2) + \beta(1)\alpha(2) \quad \text{and} \quad \alpha(1)\beta(2) - \beta(1)\alpha(2)$$

The first of these is symmetric and can therefore be combined with the antisymmetric orbital ψ_A:

$$\Psi_3 = \psi_A[\alpha(1)\beta(2) + \beta(1)\alpha(2)] \tag{12.22}$$

The second is antisymmetric and can be combined with the symmetric orbital ψ_S:

$$\Psi_4 = \psi_S[\alpha(1)\beta(2) - \beta(1)\alpha(2)] \tag{12.23}$$

We have thus obtained four combined orbitals, all of them antisymmetric and therefore consistent with the Pauli principle. Three of them, Ψ_1, Ψ_2, and Ψ_3, involve the antisymmetric orbital ψ_A, which we have seen to lead to repulsion. Since there are three functions, corresponding to different spin functions, we refer to this

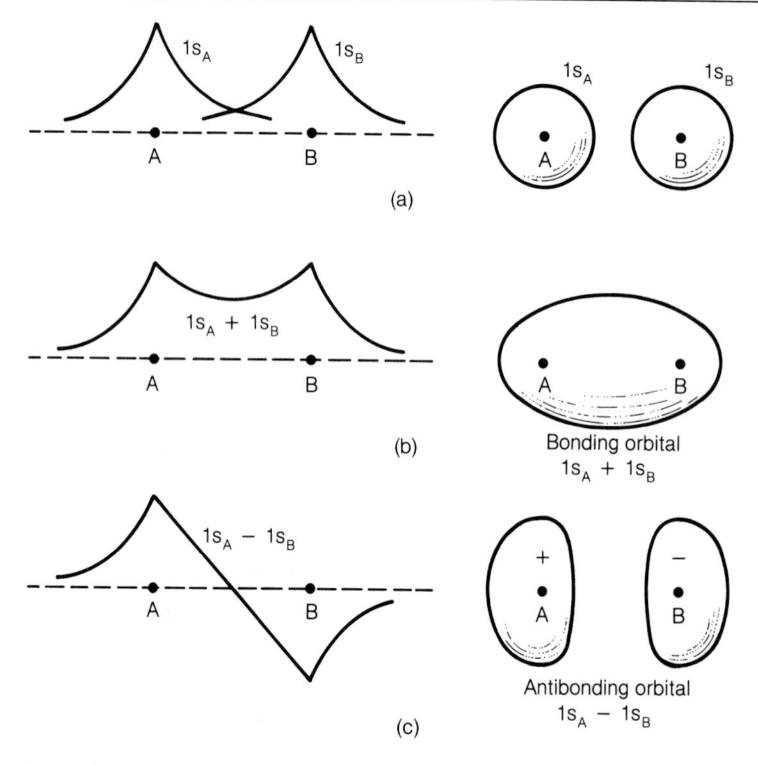

FIGURE 12.3 Wave function plots: To the right are shown contours that represent the general shapes of the orbitals. (a) The individual atomic orbitals $1s_A + 1s_B$ for the hydrogen atoms. (b) The bonding orbital $1s_A + 1s_B$. (c) The antibonding orbital $1s_A - 1s_B$.

repulsive state as a **triplet state.** The single state Ψ_4 giving attraction is referred to as a **singlet state.** More is said about these states in Section 13.2.

The Molecular-Orbital Method

The valence-bond calculations necessary to make numerical predictions for large molecules become rather tedious. It was for that reason that Hund and Mulliken developed the *molecular-orbital method,* which deals with the molecule as a whole and generates a system of molecular energy levels. These levels are formed from atomic orbitals, and the available electrons are placed in them with due regard to the Pauli principle. With all the current refinements of the two methods there is practically no difference in the results obtained by the valence-bond and molecular-orbital methods.

The molecular orbitals are formed from a linear combination of atomic orbitals, without consideration of the electrons they have to accommodate, and the MO's for the hydrogen molecule are therefore identical to those for the hydrogen molecule-ion. The form of the bonding orbital from Eq. 12.7 is thus $1s_A + 1s_B$, and this is the MO of lowest energy. Figure 12.3a shows the individual atomic orbitals, and Figure 12.3b shows their sum; the value of the MO is plotted along the internuclear axis A-B. This MO corresponds to a high electron density between the nuclei, and the effect is to hold the nuclei together; there is therefore *bonding.*

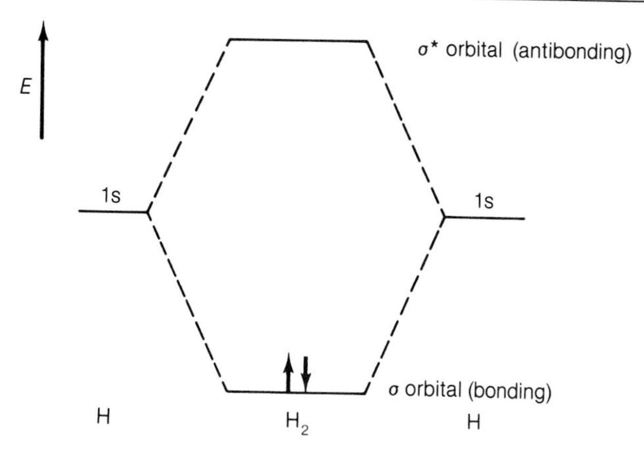

FIGURE 12.4 The energies of bonding and antibonding orbitals for the H_2 molecule. Two electrons, with opposite spins, are shown in the bonding orbital.

To describe the molecular orbitals, we again use the quantum numbers that relate to the eigenvalue L_z for the angular momentum along the internuclear axis. The notation is similar to that for atomic orbitals except that Greek letters are now used, as follows:

	$\lambda = 0$	1	2
orbital designation:	σ	π	δ
	sigma	pi	delta

The bonding orbital is thus

$$\sigma = 1s_A + 1s_B \tag{12.24}$$

When one electron is in this orbital, we have the H_2^+ ion in which there is bonding; the energy of the H_2^+ is below that of $H + H^+$. Since the bond is due to one bonding electron, the bond order is 0.5. A second electron, with opposite spin, can also go into this orbital, with a consequent increase in bonding; the bond order is now 1.

A second MO can be obtained by subtracting the two atomic orbitals:

$$\sigma^* = 1s_A - 1s_B \tag{12.25}$$

This corresponds to a low electron density between the nuclei, as shown in Figure 12.3c. It corresponds to $\lambda = 0$, and it is also a σ orbital; the fact that it is antibonding is shown by the asterisk. The two wave functions σ and σ^* are orthogonal to each other (see Problem 8).

Figure 12.4 is an energy diagram showing the separated atoms and the bonding and antibonding levels. The ground state of H_2 is shown, with two electrons in the bonding orbital. If one of these is promoted into the antibonding σ^* orbital, the antibonding effect more than overcomes the bonding, and there is a net repulsion between the nuclei; the resulting molecule would at once dissociate into two hydrogen atoms.

The molecular orbital for the electron-pair bond in the hydrogen molecule is obtained by taking the product of the molecular orbitals for each of the two electrons, as given by Eq. 12.24. This description allows the electrons to move freely throughout the molecule. The wave function for electron 1 is written as

$$\sigma(1) = N_b[1s_A(1) + 1s_B(1)] \tag{12.26}$$

and that for electron 2 as

$$\sigma(2) = N_b[1s_A(2) + 1s_B(2)] \tag{12.27}$$

where N_b is the normalization factor. The trial molecular orbital is, therefore,

$$\sigma = \sigma(1)\sigma(2) = N_b^2[1s_A(1) + 1s_B(1)][1s_A(2) + 1s_B(2)] \tag{12.28}$$

$$= N_b^2[1s_A(1)1s_B(2) + 1s_B(1)1s_A(2) + 1s_A(1)1s_A(2) + 1s_B(1)1s_B(2)] \tag{12.29}$$

The first two terms correspond to the Heitler-London wave function; the other two terms describe ionic states. These ionic structures contribute slightly, ≈3%, to the total energy of the hydrogen molecule. In this simple treatment these forms are weighted equally with the covalent structures. In more advanced treatments a weighting factor is used to cut the contribution to the proper level. The valence-bond and MO methods are thus quite similar.

In order to determine the energy of the system, absolute values of N_b and N_a must be known. These are found from the normalization (Eq. 11.89). The value of N_b is determined as follows:

$$\int \sigma(1)^2 \, d\tau = 1 = N_b^2 \int [1s_A(1) + 1s_B(1)]^2 \, d\tau$$

$$= N_b^2\left[\int 1s_A(1)^2 \, d\tau + \int 1s_B(1)^2 \, d\tau + 2\int 1s_A(1)1s_B(1) \, d\tau\right. \tag{12.30}$$

Since $1s_A(1)$ and $1s_B(1)$ are separately normalized, each of the first two integrals is equal to 1. The integral $\int 1s_A(1)1s_B(1) \, d\tau$ is abbreviated S and is called the **overlap integral;** it represents the *degree of overlap* of the two orbitals. This is evident if we consider the definition

$$S = \int 1s_A 1s_B \, d\tau \tag{12.31}$$

The wave functions extending from nucleus A and nucleus B each have a particular value at any location. If the nuclei are very far apart, $1s_A$ is large at a point near A, but $1s_B$ is so small that their product is extremely small. At a point near B, the reverse is true for the wave functions, so that their product is again extremely small. As the nuclei approach the equilibrium distance, S becomes larger. When S is not zero, that is, when r_{AB} is not infinite, $1s_A$ and $1s_B$ are not orthogonal and consequently are only approximations to the proper wave functions of the hydrogen molecule. The integral S is thus a measure of the interpenetration or overlapping of the electron clouds on the nuclei. Equation 12.30 with these substitutions becomes

$$N_b^2(1 + 1 + 2S) = 1 \quad \text{and} \quad N_b = \frac{1}{\sqrt{2(1 + S)}} \tag{12.32}$$

The energy of the orbitals may now be calculated for the bonding orbitals of

Eqs. 12.26 and 12.27 using the variation principle, where

$$E = \frac{\int \sigma(1)\hat{H}\sigma(1)\,d\tau}{\int \sigma(1)\sigma(1)\,d\tau} \tag{12.33}$$

The Hamiltonian in Eq. 12.11 is used in the formal calculation.

Since $\sigma(1)$ is already normalized, we may drop the denominator in Eq. 12.33, and the energy becomes

$$E = \int \sigma(1)\hat{H}\sigma(1)\,d\tau \tag{12.34}$$

Substitution of Eq. 12.26 into Eq. 12.34 gives, for the bonding energy,

$$E_b = N_b^2 \int [1s_A(1) + 1s_B(1)]\hat{H}[1s_A(1) + 1s_B(1)]\,d\tau \tag{12.35}$$

$$= N_b^2 \int 1s_A(1)\hat{H}1s_A(1)\,d\tau + N_b^2 \int 1s_B(1)\hat{H}1s_B(1)\,d\tau + 2N_b^2 \int 1s_A(1)\hat{H}1s_B(1)\,d\tau \tag{12.36}$$

The first two integrals are equal since $1s_A$ and $1s_B$ are identical orbitals. They are the *Coulomb integrals,* equal to the **electrostatic energy** J contributing to the bond. The last integral is the *resonance* or *exchange integral,* which is the **exchange energy** K. Substitution of J and K for the integrals and the value of N_b from Eq. 12.32 gives

$$E_b = \frac{1}{2(1 + S)}(2J + 2K) = \frac{J + K}{1 + S} \tag{12.37}$$

A further simple approximation is to neglect the overlap integral S; then the energy of the bonding orbital becomes

$$E_b = J + K \tag{12.38}$$

Since K is a negative quantity, the energy of the molecular orbital is *below* the energy of the individual atoms, as shown in Figure 12.4.

The requirement of symmetry is that the molecular wave function is symmetric or antisymmetric under the interchange of nuclei. For the hydrogen molecule this is equivalent to an *inversion* of spatial coordinates through a center of symmetry. Inversion means that a straight line is drawn from every element of space through the center of symmetry and is continued for the same distance beyond the center. A wave function that retains its sign after inversion is said to be **symmetric** and is labeled g (German *gerade,* even). One that changes its sign on inversion is said to be **antisymmetric** and is labeled u (German *ungerade,* odd). The bonding orbital may then be written as

$$\sigma_g(1) = N_b[1s_A(1) + 1s_B(1)] \tag{12.39}$$

For the antibonding orbital we have

$$\sigma_u^*(1) = N_a[1s_A(1) - 1s_B(1)] \tag{12.40}$$

where $N_a = 1/\sqrt{2(1 - S)}$. The energy E_a of the antibonding orbital is thus

$$E_a = \frac{J - K}{1 - S} \tag{12.41}$$

Its approximate value is

$$E_a = J - K \tag{12.42}$$

Again K is a negative quantity, which explains why the energy of σ_u^* lies above the value of J. As long as some overlap S exists, the difference between the energy J is greater than the difference between J and the energy of the bond state. We will later see that this accounts for the nonexistence of molecules such as He_2.

12.3 Valence-Bond Theory

For molecules larger than hydrogen, valence-bond theory has not been applied as much as MO theory because of the difficulty of including the contribution to bonding from all the possible ionic structures as well as weakly bonded structures for organic molecules. Nevertheless, several important concepts that have been developed from valence-bond theory have profoundly influenced our chemical thinking.

The Covalent Bond

In general, the covalent bond between two atoms, such as H and Cl, can be described by a wave function that is similar to that for hydrogen (Eq. 12.12). Thus, for the structures marked 1 and 2,

1. $H^{\cdot 1} \quad {}^{2\cdot}Cl \qquad \psi_1 = \psi_H(1)\psi_{Cl}(2)$ (12.43)

2. $H^{\cdot 2} \quad {}^{1\cdot}Cl \qquad \psi_2 = \psi_H(2)\psi_{Cl}(1)$ (12.44)

ψ_1 and ψ_2 are the valence-bond orbitals made from the appropriate H and Cl orbitals. The symmetric, bonding orbital is then

$$\psi_S = \frac{1}{\sqrt{2(1 + S)}}(\psi_1 + \psi_2) \tag{12.45}$$

This wave function ψ_S leads to a minimum in the potential-energy curve and the resonance energy is calculated as the difference between that computed for ψ_S and that computed for ψ_1 or ψ_2. Since the resonance energy is the prime source of stability for the bond, it is important to know the factors that influence this resonance stability. In writing structures that can contribute to the resonance stability, two factors have been recognized. One is that atomic positions must be the same for all contributing forms. The second is that the forms must have the same number of unpaired electrons. In Eq. 12.45 both ψ_1 and ψ_2 have exactly the same energy since they differ only by the exchange of electron coordinates, and they therefore contribute greatly to the resonance stabilization.

When ionic structures are considered, two possibilities exist:

$$\begin{array}{ll} 3. & H^+ \quad {}^1_2\!:\!Cl^- \\ 4. & H{}^{1:-}_{2\cdot} \quad Cl^+ \end{array} \tag{12.46}$$

In the general case, both forms contribute to the overall structure of the molecule. For heteronuclear diatomic molecules, however, one structure is normally much lower in energy than the other. In larger molecules many ionic structures may contribute to the overall structure of the molecule. In our example, structure 3 is more important than structure 4, so that the structure of HCl may best be described by the wave function

$$\psi = [\psi_1 + \psi_2]_{\text{covalent}} + \lambda[\psi_3]_{\text{ionic}} \tag{12.47}$$

which corresponds to a resonance hybrid between covalent and ionic forms. The coefficient λ indicates that the contribution of ψ_3 is different from that of ψ_1 and ψ_2. Calculations show that these contributions are 26% each for the covalent structures ψ_1 and ψ_2 and 48% for the ionic structure ψ_3.

Even in homonuclear molecules there is a small contribution from ionic forms. For hydrogen this contribution is approximately 3%.

Electronegativity

As with the hydrogen molecule, the calculations for hydrogen chloride lead to the conclusion that there is a piling-up of electron density between the nuclei. The H_2 molecule is symmetrical, so that the electron cloud lies symmetrically between the nuclei. The quantum-mechanical calculations for hydrogen chloride, on the other hand, show that the electron cloud lies more toward the chlorine atom.

The consequence of this asymmetry is that the molecule has a **dipole moment.** The dipole moment of a diatomic molecule is equal to the effective charge q at the positive and negative ends multiplied by the distance between them:

$$\mu = qd \tag{12.48}$$

This is shown in Figure 12.5. It has been common practice to express the charge in electrostatic units (esu) and the distance in angstroms. If an electronic charge of 4.8×10^{-10} esu were separated by a distance of 1 Å (10^{-8} cm or 0.1 nm) from an equal

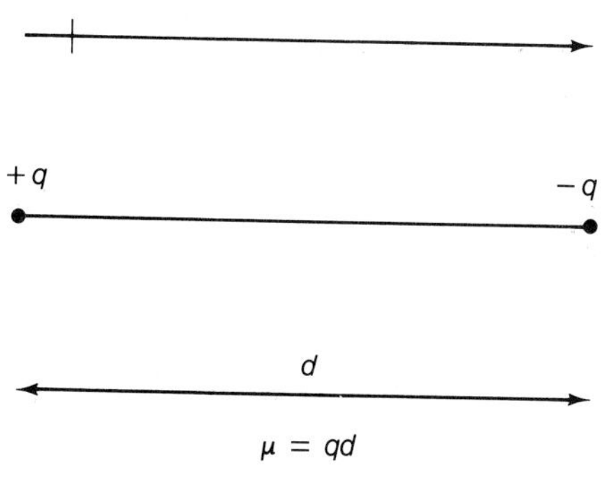

FIGURE 12.5 Two charges $+q$ and $-q$, separted by a distance d; the dipole moment is qd. The direction of the moment is often represented by an arrow $\leftarrow\!\!\!\!+\!\!\!\rightarrow$, as shown.

charge of opposite sign, the dipole moment would be 4.8×10^{-18} esu cm or 4.8 debyes (D); one debye equals 10^{-18} esu cm.

The SI unit of dipole moment is coulomb metre, C m. The elementary charge is 1.602×10^{-19} C; and if two such charges, one positive and one negative, are separated by 1 Å (1×10^{-10} m), the dipole moment is

$$1.602 \times 10^{-19} \text{ C} \times 1 \times 10^{-10} \text{ m} = 1.602 \times 10^{-29} \text{ C m}$$

Since this is 4.8 D, it follows that

$$1 \text{ D} = \frac{1.602 \times 10^{-29} \text{ C m}}{4.8} = 3.336 \times 10^{-30} \text{ C m}$$

If we know the distance between two atoms in a diatomic molecule, we can calculate a dipole moment μ_{ionic} on the assumption that the atoms bear a full elementary charge. The percentage ionic character of the bond can then be calculated as

$$\% \text{ ionic character} = \frac{\mu_{\text{exp}}}{\mu_{\text{ionic}}} \times 100 \tag{12.49}$$

where μ_{exp} is the experimental dipole moment. The percent ionic character can also be calculated from valence-bond theory. Equation 12.47 is constructed as a linear combination of covalent and ionic wave functions. The lowest-energy ionic function is weighted by λ, as compared with unity for the covalent function. When the energy is calculated using the variation Eq. 12.8, the ratio of the contribution to the energy of the ionic wave function to that of the covalent wave function is $\lambda^2/1$. The theoretical percent ionic character is therefore

$$\% \text{ ionic character} = \frac{\lambda^2}{1 + \lambda^2} \times 100 \tag{12.50}$$

The American chemist Linus Pauling has made important contributions to our understanding of electronegativity and the ionic character of bonds by using both quantum-mechanical theory and experimental results. He considered a reaction such as

$$\text{AA} + \text{BB} \longrightarrow 2\text{AB}$$

in which two homonuclear molecules form two heteronuclear molecules. He regarded the molecules AA and BB as purely covalent in the sense that the molecules are symmetrical and cannot have dipole moments. However, the unsymmetrical molecule AB can have a dipole moment, and there will be an ionic contribution to its energy, which will make the molecule more stable. Pauling concluded empirically that the purely covalent bond dissociation energy of AB is the geometric mean of the values for AA and BB:

$$E_{\text{covalent}} = [D(\text{AA})D(\text{BB})]^{1/2} \tag{12.51}$$

The experimental dissociation energy $D(\text{AB})$ will, in general, be greater than this, and the difference is taken to be the ionic energy of the bond:

$$E_{\text{ionic}} = D(\text{AB}) - [D(\text{AA})D(\text{BB})]^{1/2} \tag{12.52}$$

This quantity, therefore, can be calculated from the dissociation energies, which are usually known.

Pauling found empirically that the square roots of these ionic energies, $(E_{\text{ionic}})^{1/2}$, were additive with respect to the atoms A and B. In other words, the $(E_{\text{ionic}})^{1/2}$ values were proportional to the difference between certain numbers χ assigned to each atom:

$$(E_{\text{ionic}})^{1/2} = K|\chi_{\text{A}} - \chi_{\text{B}}| \tag{12.53}$$

Pauling chose his proportionality constant K in such a way that the difference $\chi_{\text{A}} - \chi_{\text{B}}$ also gave a reliable estimate of the dipole moment of AB measured in debyes. For energies in kJ, $K = 10$. In this way, he was able to construct a table of χ values, or **electronegativities.** A few such values are given in Table 12.2. These values are very useful for making rough estimates of dipole moments. For example, hydrogen has an electronegativity of 2.1 and chlorine of 2.8; the estimated dipole moment of HCl is thus $2.8 - 2.1 = 0.7$ D (debye) with the chlorine atom being at the negative end of the dipole.

A second but limited method of expressing electronegativities is due to the American physicist Robert S. Mulliken. Mulliken considered that the attraction of an atom in a molecule for a pair of electrons in the bond is an average of the attraction of the free ion for an electron (the ionization potential I) and the attraction of the neutral atom for an electron (the electron affinity A). A scale factor of 5.6 is used to make coincident the values of I and A with χ from the Pauling scale. Thus, if I and A are expressed in electron volts,

$$\chi_M = \frac{(I + A)/\text{eV}}{5.6} \tag{12.54}$$

Values of the Mulliken χ_M are listed in parentheses in Table 12.2.

Orbital Overlap

We think of the covalent bond as a pair of electrons united with their spins opposed in a stable orbital based on two adjacent atoms in a molecule. The strength of the bond depends on the extent of overlap of the charge clouds on the two atoms. We have already seen that the overlap integral (Eq. 12.31) is a measure of this overlap and that its value depends on the orientation of the orbitals. In addition, overlap of orbitals, and hence bond formation, can occur only in regions of like sign. Associated with the orbitals s, p, and d are + or − regions that give the algebraic

TABLE 12.2 Atomic Electronegativities on the Pauling Scale*

H 2.1 (2.5)						
Li 1.0	Be 1.5	B 2.0	C 2.5	N 3.1	O 3.5	F 4.1 (3.8)
Na 1.0	Mg 1.23	Al 1.5	Si 1.7	P 2.1	S 2.4	Cl 2.8 (3.0)
						Br 3.1 (2.7)

*Selected values from the Mulliken scale are in parentheses

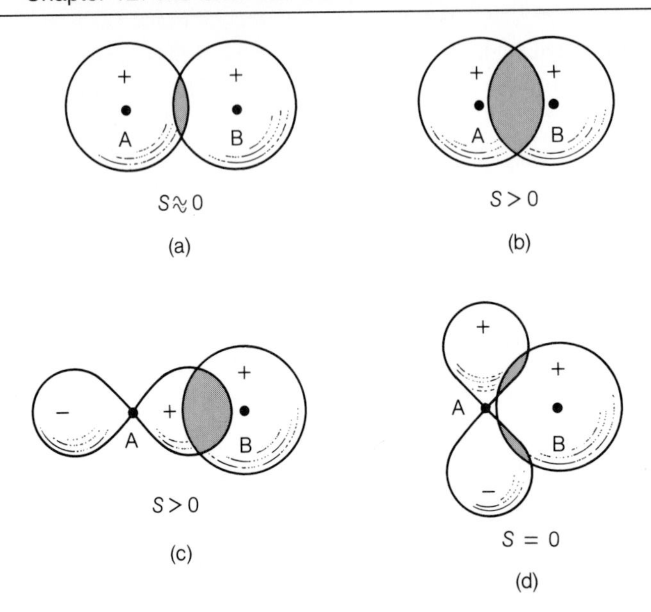

FIGURE 12.6 Orbital overlap involving s and p orbitals. (a) Slight overlap between two s orbitals. (b) Larger overlap between two s orbitals. (c) Overlap between an s orbital and a p orbital, with the ss orbital interacting with the positive lobe of the p orbital; this gives significant orbital overlap and hence bonding. (d) Lateral overlap between an s orbital and a p orbital. The overlap involving the positive lobe of the p orbital is exactly canceled by that involving the negative lobe, and there is no bonding.

sign of the wave function in its different regions. Bonding can only occur if the signs are the same for two orbitals in an overlap region.

Since the s orbital has a positive value everywhere, the overlap of s orbitals on two different atoms is independent of the direction of approach, as shown in Figure 12.6a and b. There can be more or less overlap. Bonds of this type are called σ (sigma) bonds.

A p orbital has a positive lobe and a negative lobe, and orientation is important. If the positive lobe is oriented toward an s orbital, which is positive, significant overlap may occur (Figure 12.6c); this is a σ bond. The s orbital can also be normal to the longitudinal axis of the p orbital, as shown in Figure 12.6d. In this case the integral formed from the s orbital and the negative region of the p orbital exactly counteracts that from the s orbital and the positive lobe of the p orbital; the integral S is therefore zero and there is no bonding. Orientations intermediate to these are also possible.

Two p orbitals can also overlap, and three cases are shown in Figure 12.7. The bond shown in Figure 12.7a is a σ bond, whereas in Figure 12.7b there is no bonding. Figure 12.7c shows lateral overlapping of p orbitals with the positive lobes coming together and the negative lobes coming together; the bond formed from such an arrangement is called a π (pi) bond.

The value of S depends on the orientation of the orbitals. Values in the range S = 0.2–0.3 normally indicate an effective bond.

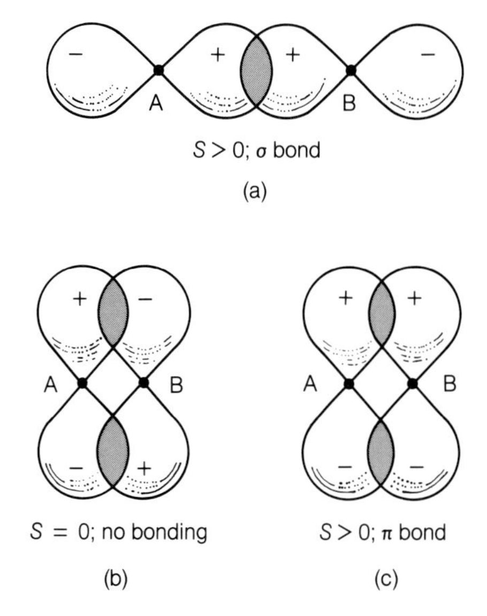

FIGURE 12.7 Orbital overlap involving two p orbitals. (a) Overlap along the internuclear axis, with lobes of the same sign coming together; this is a σ bond. (b) Lateral overlap in which lobes of opposite sign come together, giving no net bonding. (c) Lateral overlap in which lobes of the same sign come together, giving a π bond.

Orbital Hybridization

In the description of H_2 by the valence-bond method, one atomic orbital (1s) was used from each of the hydrogen atoms. However, sometimes this is not satisfactory and we must use *two or more orbitals from a given atom.*

The simplest **hybrid** to imagine is one made from one s orbital and one p orbital. The hybrid is formed by adding the wave functions for the s and p orbitals and dividing by $\sqrt{2}$. Figure 12.8 shows how these wave functions constructively and destructively interfere to form the sp hybrids.

A simple example of a more complicated hybridization is provided by the molecule methane, CH_4. In its ground state, carbon has two unpaired 2 p electrons with the same spin. Although it might appear that carbon would form only two single bonds, carbon usually shows a valency of 4. To obtain four unpaired electrons, one of the 2s electrons is "promoted" to the 2p state to give the configuration

$$C^* \quad 1s^2 2s 2p^3$$

for the excited C* carbon atom.

Initially it would appear that, of the four bonds formed by a carbon atom, one would involve the 2s orbital and the other three would involve the three 2p orbitals. This, however, implies that one bond is different from the other three, whereas experimentally the CH_4 molecule is perfectly symmetrical; all four bonds are identical. The solution to this dilemma was given by Pauling, who suggested that we should use a linear combination of orbitals instead of the pure s and p orbitals of the carbon atom. On the basis of the required geometry, Pauling concluded that,

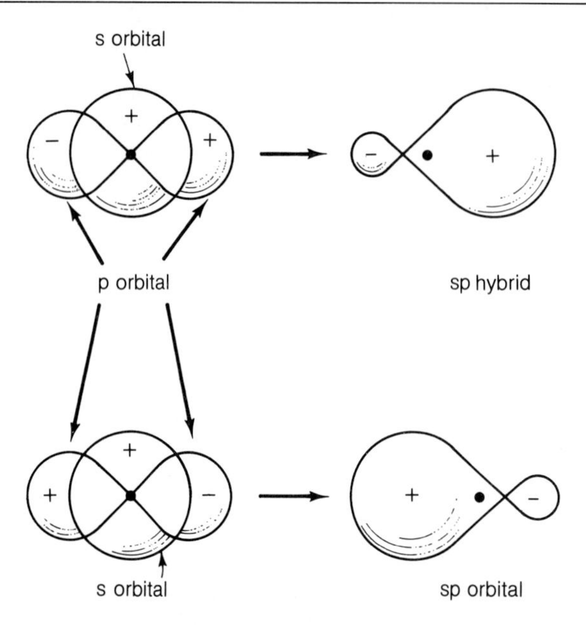

FIGURE 12.8 The formation of two sp hybrids from an s orbital and a p orbital.

from one s and three p orbitals, four *hybridized* orbitals, labeled t_1 to t_4, can be constructed:

$$t_1 = \tfrac{1}{2}(\mathrm{s} + \mathrm{p}_x + \mathrm{p}_y + \mathrm{p}_z) \tag{12.55}$$

$$t_2 = \tfrac{1}{2}(\mathrm{s} + \mathrm{p}_x - \mathrm{p}_y - \mathrm{p}_z) \tag{12.56}$$

$$t_3 = \tfrac{1}{2}(\mathrm{s} - \mathrm{p}_x + \mathrm{p}_y - \mathrm{p}_z) \tag{12.57}$$

$$t_4 = \tfrac{1}{2}(\mathrm{s} - \mathrm{p}_x - \mathrm{p}_y + \mathrm{p}_z) \tag{12.58}$$

These orbitals are normalized and are mutually orthogonal (see Problem 13). Their extension in space is as large as possible, so that when bonds are formed, the overlap is a maximum. This condition embodies *Pauling's principle of maximum overlap*. These four new orbitals are directed to the apices of a tetrahedron, and the individual bonds are formed by combining one of these hybrid orbitals and a 1s orbital from the hydrogen atom. This can be visualized as the overlapping of the electron clouds, as shown in Figure 12.9a. Obviously the maximum overlapping is obtained when the hydrogen atoms are located along the axes of the orbitals. This principle of maximum overlap is of great value in predicting the shapes of molecules, which follow at once from the shapes of the orbitals.

Hybridization also arises with the bonding in water and ammonia, although matters are not so clear as with methane. The ground-state oxygen atom configuration is $1s^2 2s^2 2p^4$. In accordance with Hund's rule, there are two unpaired 2p electrons in orbitals at right angles to each other. If the H_2O molecule were made up directly from these 2p orbitals, a bond angle of 90° is predicted from the principle of maximum overlap. Experimentally, however, the angle is about 104.5°, as shown in Figure 12.9b.

Alternatively, hybridization of the one 2s and the three 2p orbitals gives a tetrahedral sp^3 arrangement, in which the angle between the orbitals is 109.47°, the angle between bonds shown for methane in Figure 12.9a. Detailed quantum-

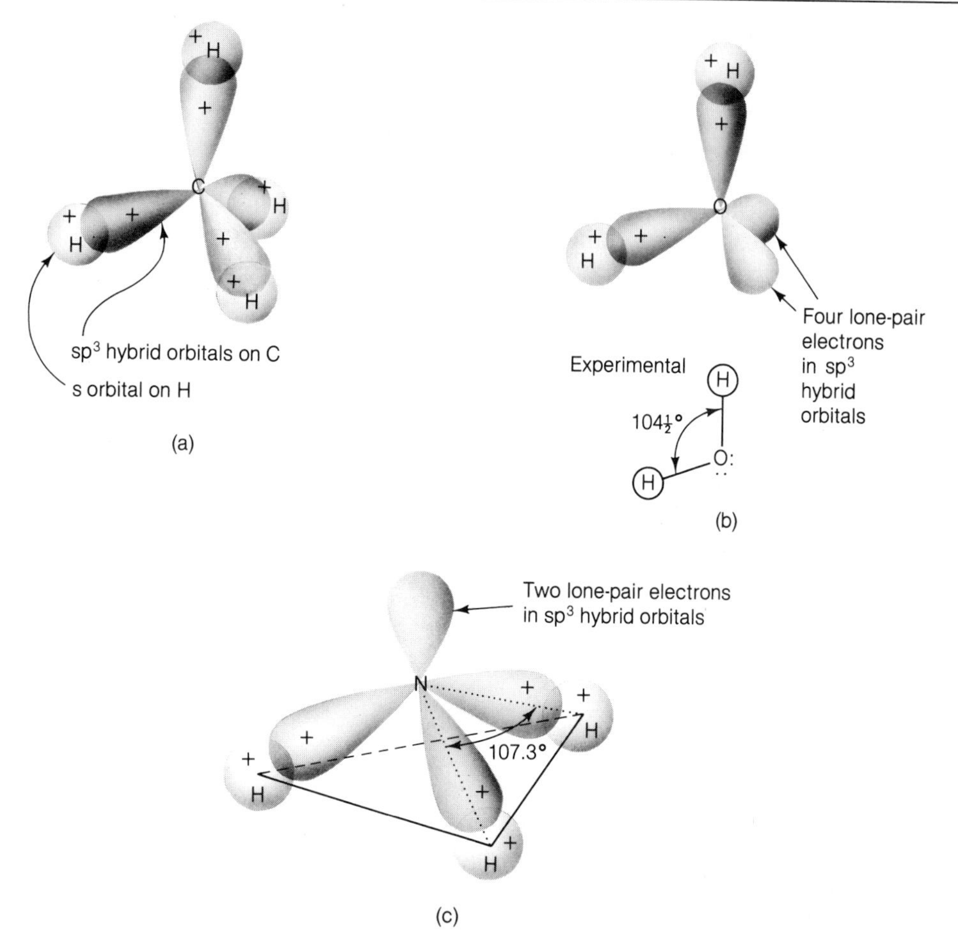

FIGURE 12.9 Hybrid orbitals in the bonding of (a) methane, (b) water, and (c) ammonia. Small negative lobes of the hybrids are not shown for clarity.

mechanical treatments suggest that there is only partial hybridization of the orbitals.

A similar case is found with ammonia. If there were no hybridization, the bond angles would be 90°. The experimental bond angle is 107.3° (Figure 12.9c), and the fact that this is only slightly below the tetrahedral angle suggests that the sp^3 hybridization is almost complete.

A theory that is particularly useful in making predictions about the shapes is the **valence-shell electron-pair repulsion (VSEPR) theory.** The ideas behind this theory were first suggested in 1940 by the British chemists Nevil Vincent Sidgwick (1873–1952) and H. E. Powell. These concepts were later developed further by Sir Ronald Nyholm (1917–1971) and, more particularly, by the Canadian chemist Ronald James Gillespie (b. 1924).

The basis of the theory is the following fundamental rule:

The pairs of electrons in a valence shell adopt an arrangement that maximizes their distance apart; that is, the electron pairs behave as if they repel each other.

For an application of this rule, consider the water molecule. This molecule has two pairs of bonding electrons and two lone pairs:

$$\begin{array}{l} :\ddot{\underset{..}{O}}:H \\ H \end{array}$$

Application of the VSEPR theory suggests that the four pairs will be arranged in an approximately tetrahedral manner, as seen in Figure 12.9b. The lone-pair orbitals lie closer to the oxygen outermost shell and there will be a greater repulsion between them than between a lone-pair orbital and a bonding-pair orbital containing the H—O bond. Consequently, to minimize this repulsion, the H—O—H angle is less than the tetrahedral angle, in agreement with experiment.

Rules have been worked out for more complicated cases and are very valuable in predicting the geometries of more complicated molecules.

Multiple Bonds

In addition to the sp^3 hybridization that carbon exhibits, two other types must be invoked to explain molecules having double and triple bonds. Simple examples of these compounds are

$H_2C{=}CH_2$ (ethylene) $\quad H_2C{=}O$ (formaldehyde) $\quad H{-}C{\equiv}C{-}H$ (acetylene)

The existence of such molecules is explained in valence-bond theory in terms of two different kinds of hybridization, sp^2 and sp, which are illustrated in Figure 12.10b and c. In sp^2 hybridization, the bonding orbitals involve a linear combination of an s orbital and two p orbitals (e.g., p_x and p_y) as follows:

$$\psi_1 = \frac{1}{\sqrt{3}}\psi_{2s} + \frac{\sqrt{2}}{\sqrt{3}}\psi_{2p_x} \tag{12.59}$$

$$\psi_2 = \frac{1}{\sqrt{3}}\psi_{2s} - \frac{1}{\sqrt{6}}\psi_{2p_x} + \frac{1}{\sqrt{2}}\psi_{2p_y} \tag{12.60}$$

$$\psi_3 = \frac{1}{\sqrt{3}}\psi_{2s} - \frac{1}{\sqrt{6}}\psi_{2p_x} - \frac{1}{\sqrt{2}}\psi_{2p_y} \tag{12.61}$$

As shown in Figure 12.11b, these orbitals lie symmetrically in the XY plane; the angle between orbitals is 120°. In sp hybridization, we combine the s orbital with one p orbital (e.g., p_x) as follows:

$$\psi_1 = \psi_{2s} + \psi_{p_x} \tag{12.62}$$

$$\psi_2 = \psi_{2s} - \psi_{p_x} \tag{12.63}$$

The hybrid sp orbital is linear as shown in Figure 12.10c.

In sp^2 hybridization, the p orbital that is not involved in hybridization lies above and below the plane of the hybrid bonds and is thus capable of overlapping with a similar p orbital on another atom. This accounts for one rather weak bond, called a π bond. The second bond, the σ bond, is formed by overlap of two of the sp^2

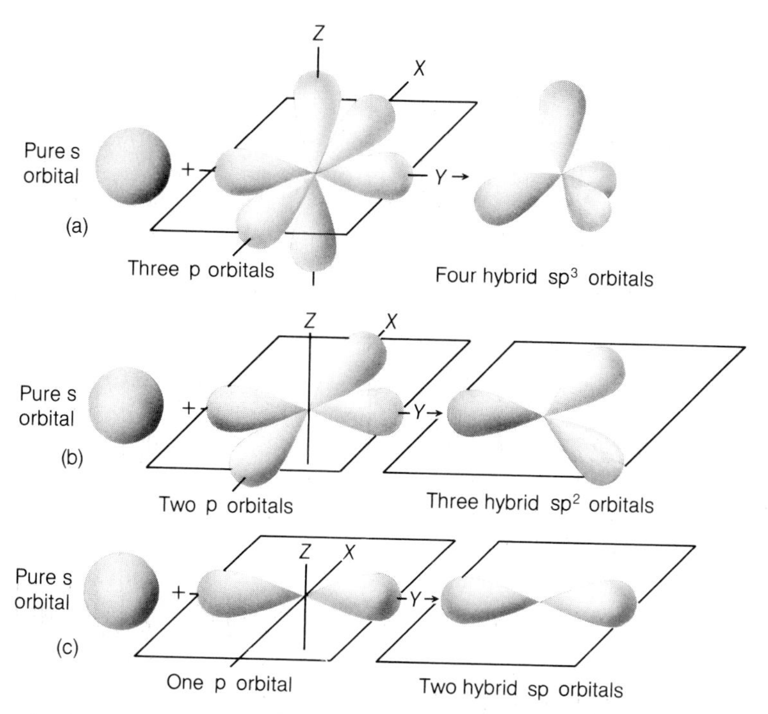

FIGURE 12.10 The three sets of hybrids formed from s and p orbitals.

hybrid orbitals. The σ bond is much stronger than the π bond. In ethylene, all the atoms lie in one plane.

An interesting feature of the double bond is the *torsional rigidity* (absence of free rotation) which occurs because of the overlap of the two p orbitals. The rotation of one CH_2 group out of the plane relative to the other CH_2 group decreases overlap. This can only be accomplished by the performance of work on the molecule to distort the planar arrangement. If this happens, the π bond becomes weaker because the energy of the molecules rises. Only a small amount of energy is required to do this. Thus the reactivity of the double bond is easy to understand since replacement of the π bond by a σ bond is rather easily accomplished; the energy of the system is lowered when this occurs.

12.4 Symmetry in Chemistry

Although the valence-bond method gives good pictorial representations of molecules, it is difficult to apply it to complex molecules, particularly when hybridized orbitals are involved. As a result, there is now more emphasis on the molecular-orbital method. However, even this method usually requires very complicated calculations, and any device that simplifies the computational procedures is very welcome. Considerable aid is provided by the study of *symmetry*.

Symmetry Elements and Symmetry Operations

A molecule, like any other geometrical figure or object, may have one or more **symmetry elements.** For example, a molecule may have an *axis of symmetry* which is such that a rotation about it, through a specified angle, leads to a configuration that is superimposable on the original molecule and indistinguishable from it. Such a rotation is an example of a *symmetry operation*. Other symmetry elements are a *plane of symmetry* and a *center of symmetry*. Associated with these three symmetry elements are four different symmetry operations, each one of which leaves the center of gravity unchanged but transforms the molecule into a configuration that is indistinguishable from the original one.

1. *Rotation About a Symmetry Axis.* A molecule may have one or more axes about which a rotation leads to a configuration that is indistinguishable from the original one. Such an axis is a symmetry element and is called a **symmetry axis** or a **rotational axis.** A rotation about a symmetry axis is a symmetry operation. A simple example is provided by the water molecule, shown in Figure 12.11a and b. We can draw the Z axis through the oxygen atom, bisecting the angle between the two O—H bonds. The X axis can be drawn through the center of gravity of the molecule and in the plane of the molecule, and the Y axis is at right angles to these two axes. The X and Y axes are not symmetry axes, but the Z axis is a symmetry axis because a 180° rotation about this axis leads to an indistinguishable configuration. To show this we have labeled the two hydrogen atoms A and B. If Figure 12.11a represents the original configuration, Figure 12.11b represents the configuration after rotation through 180°. Since the labels are only mental constructs, the configuration in Figure 12.11b is indistinguishable from that in Figure 12.11a.

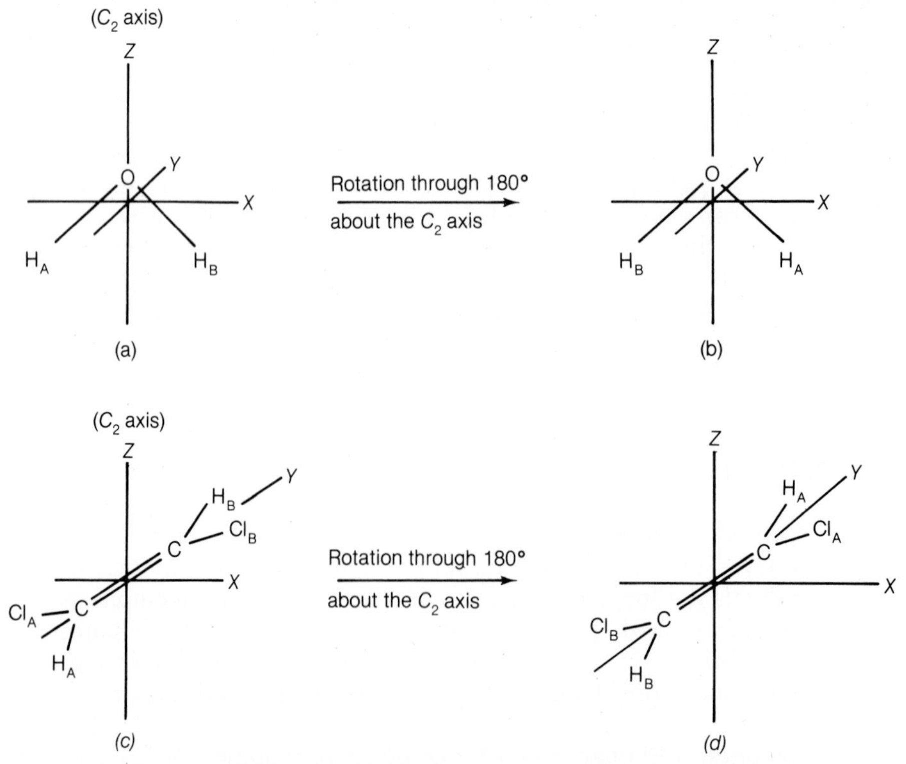

FIGURE 12.11 Two molecules having a C_2 rotational axis. (a) and (b) The water molecule. (c) and (d) *Trans*-dichloroethylene. In both cases rotation through 180° about the Z axis leads to an indistinguishable configuration.

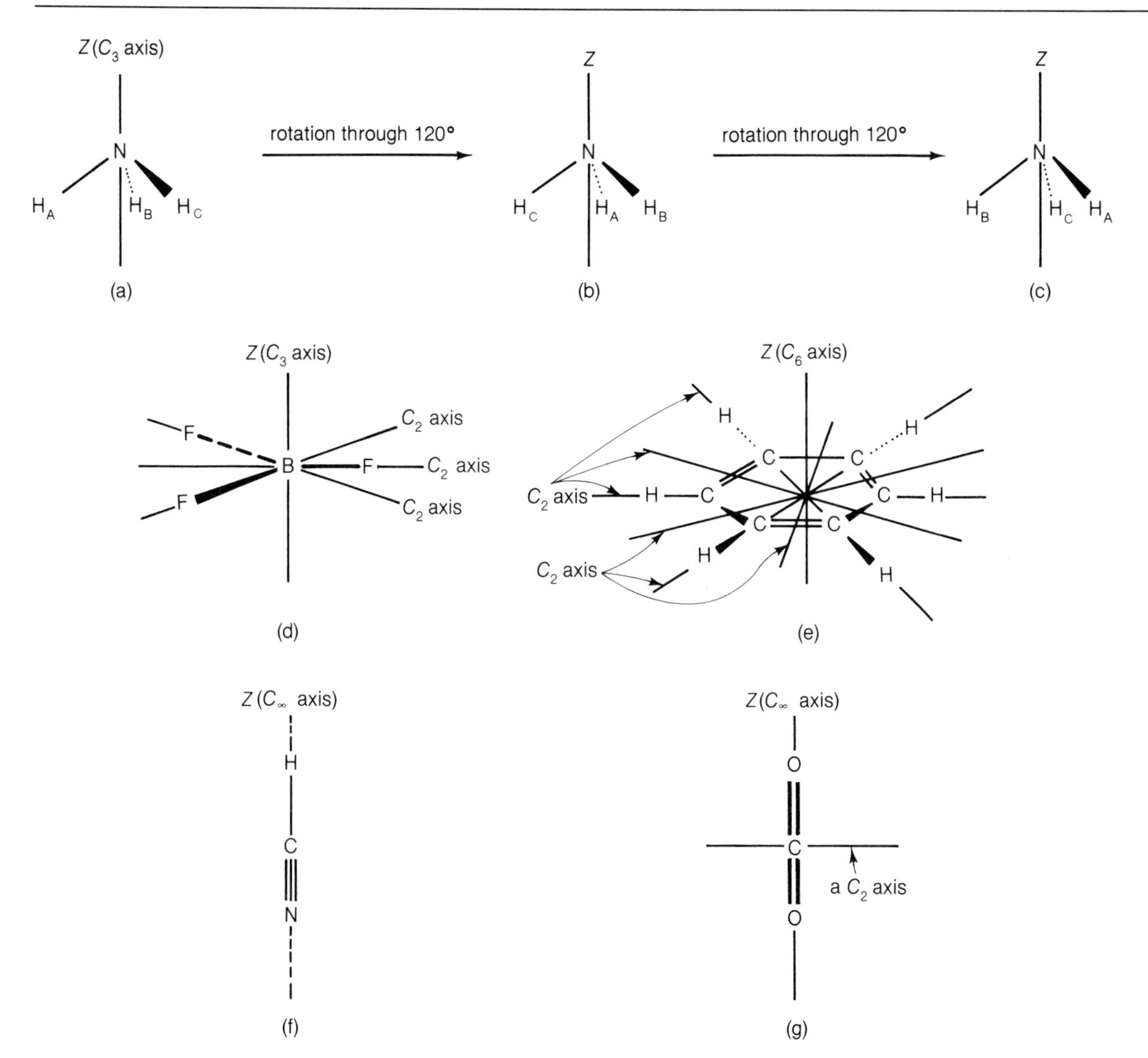

FIGURE 12.12 (a), (b), and (c) The nonplanar ammonia molecule, having a C_3 axis. (d) The planar BF_3 molecule, which has a C_3 axis and three C_2 axes. (e) Benzene, having a C_6 axis and six C_2 axes. (f) The linear molecule HCN, having a C_∞ axis. (g) The linear molecule CO_2, which has a C_∞ axis and an infinite number of C_2 axes.

If a molecule is rotated through an angle θ in order to achieve the indistinguishable configuration, the axis is said to be a $360°/\theta$-fold rotational axis and to have an **order p** of $360°/\theta$; such an axis is designated C_p. In the example of water, $p = 360°/180° = 2$ and the designation is therefore C_2; the axis is said to be a *twofold* rotational axis. The planar molecule *trans*-dichloroethylene also has a twofold (C_2) axis, as shown in Figure 12.11c. All molecules remain the same if they are rotated through 360°, so that all molecules have C_1 axes.

The nonplanar ammonia molecule has a threefold (C_3) axis; its order is $360°/120° = 3$. If we label the hydrogen atoms A, B, and C, Figure 12.12a shows one configuration, and Figure 12.12b shows the configuration after a rotation though 120°; Figure 12.12c shows the result of a further rotation through 120°.

The planar molecule BF_3 (Figure 12.12d) has a threefold axis at right angles to the plane of the molecule and passing through the boron atom. In addition, each of the bonds lies along a twofold rotational axis; there are thus three C_2 axes as well as the one C_3 axis. When a molecule has symmetry axes of different orders, a useful convention is to take an axis of the highest order as the Z axis and to draw it vertically. The axis of highest order is often called the **principal axis.** Thus in the example of BF_3, shown in Figure 12.12d, the C_3 axis is taken to be the vertical Z axis, so that the molecule lies in a horizontal plane.

Benzene, shown in Figure 12.12e, is a molecule having a C_6 axis, and it also has six C_2 axes at right angles to the C_6 axis. The C_6 axis is thus the principal axis. All linear molecules have a C_∞ rotational axis, since an infinitesimal rotation about the axis of the molecule leaves the molecule unchanged; thus $\theta \rightarrow 0$ and $p = 360°/\theta \rightarrow \infty$. An example is H—C≡N, shown in Figure 12.12f. The linear symmetrical molecule CO_2 has, in addition to the C_∞ axis, which is the principal axis, an infinite number of C_2 axes, since any axis through the carbon atom and at right angles to the C_∞ axis is a C_2 axis (see Figure 12.12g).

2. *Inversion About a Center of Symmetry.* A molecule has a **center of symmetry,** designated i (for *inversion*), if a straight line drawn from every atom through the center and extended in the same direction encounters an equivalent atom equidistant from the center. Such a center of symmetry, which can also be called a **center of inversion,** is another example of a symmetry element. Figure 12.13 shows some examples of molecules having a center of symmetry. In each case there are pairs of atoms equidistant from the center and situated in opposite directions from the center.

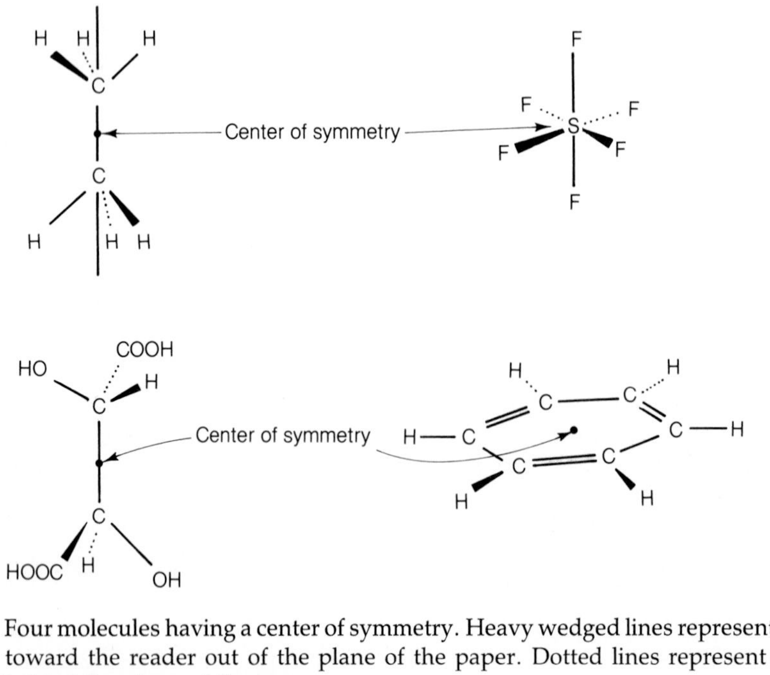

FIGURE 12.13 Four molecules having a center of symmetry. Heavy wedged lines represent bonds projecting toward the reader out of the plane of the paper. Dotted lines represent bonds projecting behind the plane of the paper.

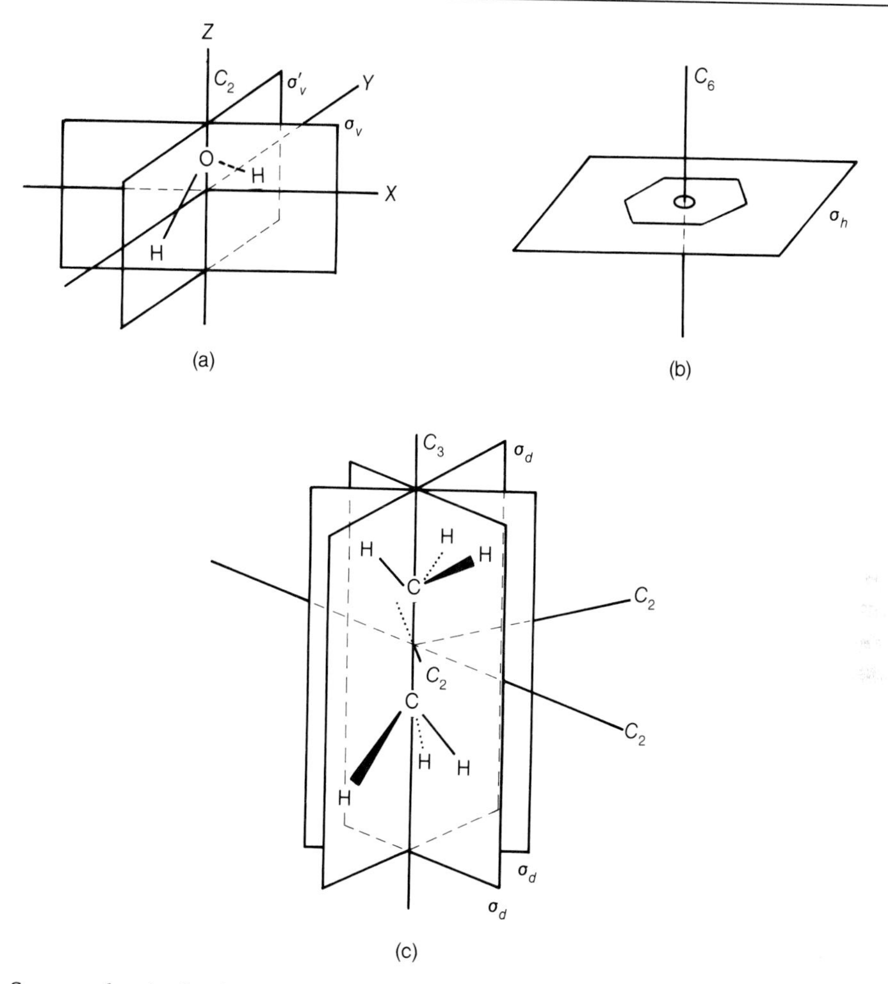

FIGURE 12.14 Some molecules having planes of symmetry. (a) H_2O: The σ'_v plane is the plane of the molecule: the σ_v plane is at right angles to the σ'_v plane. (b) Benzene, C_6H_6, which has a σ_h plane. (c) Staggered ethane, which has three dihedral (σ_d) planes; these bisect the angles between the three C_2 axes.

3. *Reflection Through a Plane of Symmetry.* For some molecules there is a plane such that if the molecule is reflected in the plane, it is indistinguishable from the original. Such a plane, which is another symmetry element, is called a **plane of symmetry** and is denoted by the symbol σ (sigma); it is also often called a **mirror plane.** Several types of mirror planes may be distinguished according to their orientation. Consider, for example, the water molecule, shown again in Figure 12.14a. We have seen that there is one axis of symmetry, a C_2 axis, and that this is conventionally taken as the Z axis and drawn vertically. Both planes of symmetry pass through this axis and are therefore vertical planes. One of them is the plane of the molecule itself; we take it to be the YZ plane and denote it by the symbol σ'_v. The other plane of symmetry is at right angles to this σ'_v plane and is the XZ plane; we denote it by the symbol σ_v. In this example of water there are only two planes and both pass through the axis of symmetry. Some molecules have a plane that is

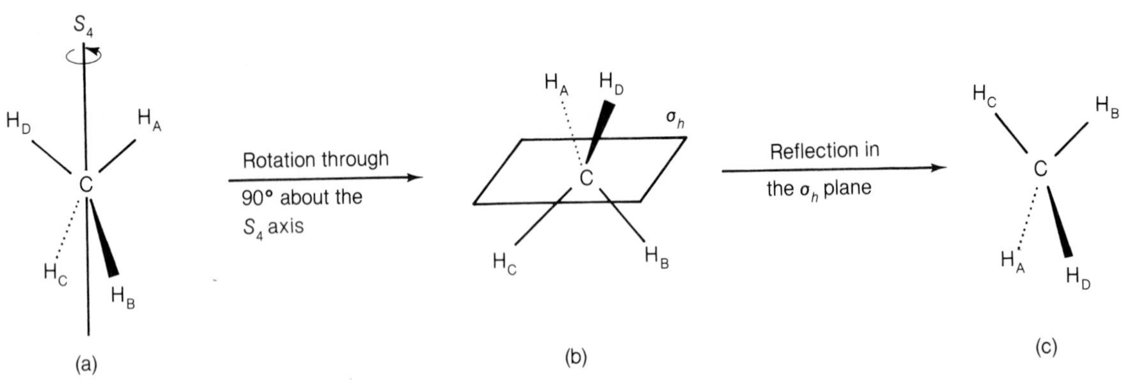

FIGURE 12.15 Methane, which has an S_4 rotation reflection axis. (a) The original form. (b) After rotation through 90° (i.e., a C_4 rotation). (c) After reflection in the horizontal plane perpendicular to the S_4 axis.

perpendicular to the principal axis of symmetry. Since this axis is conventionally taken to be vertical, this type of plane is horizontal and is designated σ_h. An example of such a plane is found in benzene, as shown in Figure 12.14b.

Another type of plane, called a **dihedral** plane and given the symbol σ_d, is found in ethane in the staggered conformation (Figure 12.14c). In this molecule there is a principal C_3 axis, and there are also three C_2 axes at right angles to the C_3 axis. The three dihedral planes contain the principal axis and also bisect the angles between the three C_2 axes.

4. *Rotation About an Axis Followed by Reflection in a Plane.* The symmetry of some molecules is such that if there is first a rotation about an axis and then a reflection in a plane perpendicular to this axis, the result is superimposable on the original. Such an axis is known as an **axis of improper rotation,** or a **rotation-reflection axis.** The designation used for such an axis is S_p; p is equal to $360°/\theta$, where θ is the angle through which the molecule is rotated in order for the reflection to give a superimposable form. Methane has an S_4 rotation-reflection axis, as shown in Figure 12.15.

5. *The Identity Operation.* We have seen that all molecules have a C_1 axis of symmetry, since rotation through 360° restores anything to its original condition. The same is, of course, true if we perform no operation at all; we then say that we are performing the **identity operation.** This operation is included for mathematical completeness, and the symbol E is used for it.

If an operation $\hat{A}$ is performed and is followed by an operation $\hat{B}$, we write the combined operation as $\hat{B}\hat{A}$. If a C_2 operation is performed, and then another C_2 operation, the combined operation is written as C_2^2. Since C_2 rotates about 180°, C_2^2 rotates about 360° and is therefore equal to C_1 and to E.

Point Groups and Multiplication Tables

None of the operations we have considered displaces the molecule to another position. There is always one point in the molecule (at which an atom is not necessarily present) that remains unmoved by the operation. Since this is the case, we call the group of all possible symmetry operations for a given molecule a **point group.** This is in contrast to a *space group,* where we are concerned with operations that move the molecule to another position in space.

The number of elements in a group is known as its **order**. For a symmetry group the order is therefore the number of operations that leave the molecule unchanged. An important property of a mathematical group is that the product of any pair of operations in the group results in another operation that is also a member of the group. This result is helpful in that by taking products of known symmetry operations for a molecule we may discover other symmetry operations that were not at once apparent.

Point groups are defined with respect to the particular symmetry elements involved. Table 12.3 lists the more important point groups and gives some examples. A molecule having no symmetry at all, such as CHClBrI (Figure 12.16a), is said to belong to the C_1 point group. The hydrogen peroxide molecule, in which the two O—H bonds do not lie in the same plane (Figure 12.16b), has a twofold (C_2) axis of symmetry. Its symmetry operations are therefore C_2 and E, and it said to belong to the C_2 point group. Since the identity operation E does nothing, EC_2 and C_2E are both equal to C_2, and the C_2 operation performed twice (i.e., C_2^2) is equal to E. A *multiplication table* for a point group shows all possible products of the symmetry operations and such a table is included in Figure 12.16b.

The water molecule contains the symmetry elements E, C_2, σ_v, and σ_v' and is said to belong to the point group C_{2v}. Figure 12.16c gives the multiplication table. The reader should verify the individual items in this table. Consider, for example, the item

$$\sigma_v' C_2 = \sigma_v \tag{12.64}$$

This means that rotation by 180° about the C_2 axis followed by reflection in the σ_v' plane is equivalent to reflection in the σ_v plane.

TABLE 12.3 The More Common Point Groups, with Some Examples

Point Group	Symmetry Elements (besides E)	Examples
C_1	None	CHFClBr
C_2	C_2	H_2O_2
C_{2v}	C_2, $2\sigma_v$	H_2O, H_2CO
C_{3v}	C_3, $3\sigma_v$	NH_3, CH_3Cl
$C_{\infty v}$	C_∞, $\infty\sigma_v$	HCN
C_{2h}	C_2, σ_h, i	*trans*-$C_2H_2Cl_2$
D_{2h}	C_2, $2C_2$, 3σ, i	C_2H_4
D_{3h}	C_3, $3C_2$, $3\sigma_v$, σ_h	BF_3
$D_{\infty h}$	C_∞, ∞C_2, $\infty\sigma_v$, σ_h, i	H_2, O_2, CO_2
T_d	$3C_2$, $4C_3$, 6σ, $3S_4$	CH_4
O_h	$3C_4$, $4C_3$, i, $3S_4$, $8C_2$, 9σ, $4S_6$	SF_6

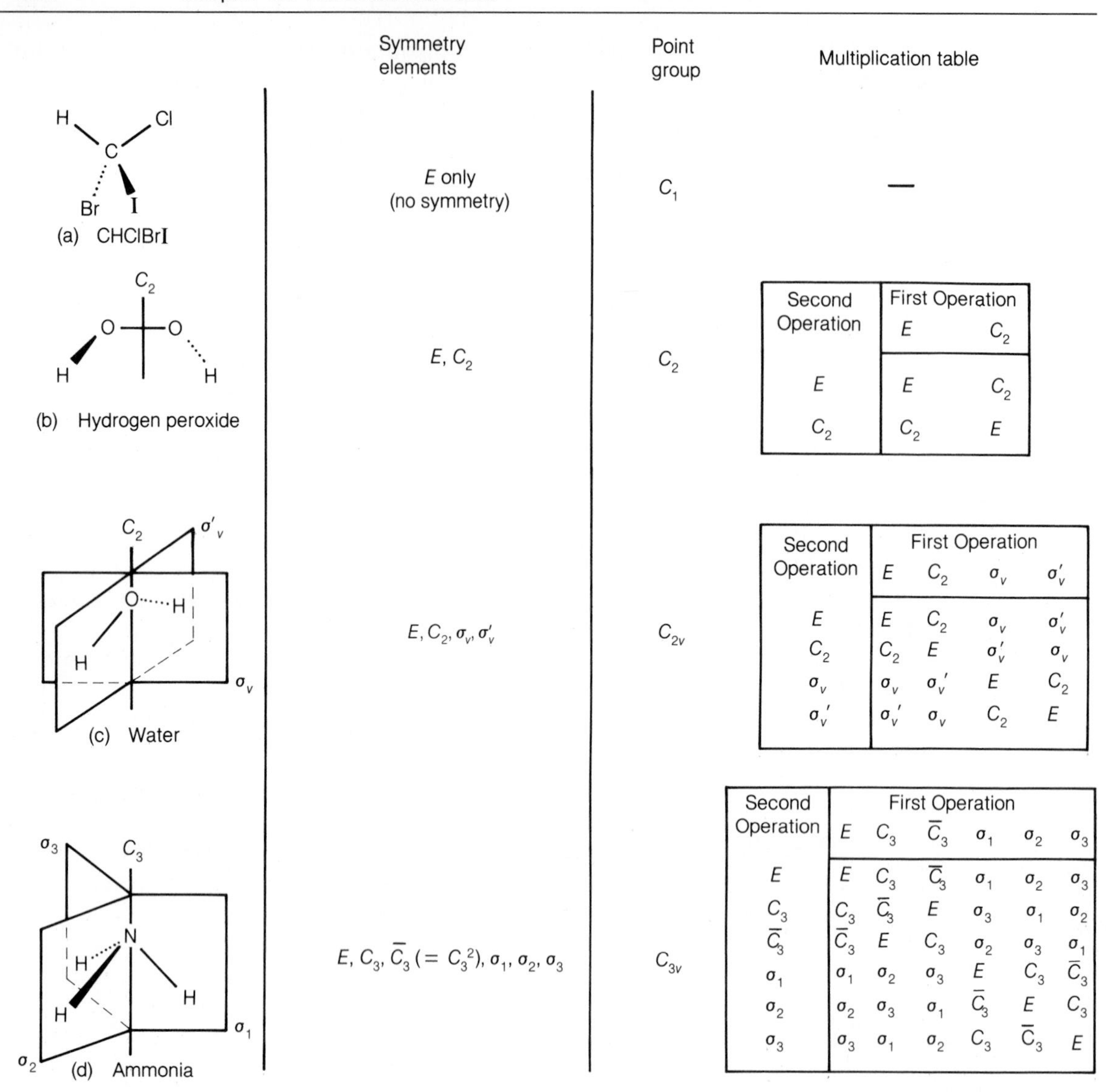

Second Operation	First Operation: E	C_2
E	E	C_2
C_2	C_2	E

Second Operation	First Operation: E	C_2	σ_v	σ'_v
E	E	C_2	σ_v	σ'_v
C_2	C_2	E	σ'_v	σ_v
σ_v	σ_v	σ'_v	E	C_2
σ'_v	σ'_v	σ_v	C_2	E

Second Operation	First Operation: E	C_3	$\bar{C}_3$	σ_1	σ_2	σ_3
E	E	C_3	$\bar{C}_3$	σ_1	σ_2	σ_3
C_3	C_3	$\bar{C}_3$	E	σ_3	σ_1	σ_2
$\bar{C}_3$	$\bar{C}_3$	E	C_3	σ_2	σ_3	σ_1
σ_1	σ_1	σ_2	σ_3	E	C_3	$\bar{C}_3$
σ_2	σ_2	σ_3	σ_1	$\bar{C}_3$	E	C_3
σ_3	σ_3	σ_1	σ_2	C_3	$\bar{C}_3$	E

FIGURE 12.16 Examples of the point groups C_1, C_2, C_{2v}, and C_{3v}, with their multiplication tables.

Ammonia (Figure 12.16d) has the symmetry elements E and C_3 and it has three vertical planes that we may designate σ_1, σ_2, and σ_3. Its point group is called C_{3v}, and its multiplication table is shown in Figure 12.16d. One special point that arises here is the direction of rotation. If we take C_3 to mean a clockwise rotation through 120°, then C_3^2 means a clockwise rotation through 240°; this however is equivalent to an anticlockwise rotation through 120°, an operation that can be written as $\bar{C}_3$. Thus $C_3^2 = \bar{C}_3$. In the multiplication table it is therefore necessary to include both C_3 and $\bar{C}_3$.

Group Theory

The branch of mathematics that deals with symmetry operations is known as **group theory.** It is outside the scope of this book to give a detailed account of this subject. Instead we will be content with presenting some of the more important aspects, which are sufficient to allow us to appreciate the symbolism used in designating molecular states. We will also be able to understand some of the spectroscopic selection rules, discussed in Chapter 13.

The essential feature of group theory is that the symmetry operations are replaced by numbers or matrices that multiply in the same way as the operations themselves. These sets of numbers and matrices are known as *representations* or as *symmetry species*, and we will first give some examples for the point group C_{2v}.

The multiplication table for this point group is shown in Figure 12.16c. There are four operations, E, C_2, σ_v, and σ_v', and we ask what sets of numbers multiply in the same way as the actual operations. One obvious possibility is to replace all the operations by unity:

$$E \longrightarrow 1, \quad C_2 \longrightarrow 1, \quad \sigma_v \longrightarrow 1, \quad \sigma_v' \longrightarrow 1$$

Another is to replace E and C_2 by unity, and σ_v and σ_v' by -1:

$$E \longrightarrow 1; \quad C_2 \longrightarrow 1, \quad \sigma_v \longrightarrow -1, \quad \sigma_v' \longrightarrow -1$$

We can easily verify that this representation also leads to the same multiplication table; for example, $\sigma_v\sigma_v' = C_2$ and $(-1)(-1) = 1$. Two other representations are also shown in Table 12.4. The four representations have been designated A_1, A_2, B_1, and B_2. The A designation means that there is no change on rotation about the principal axis; B means that there is a change.

It turns out that there are certain representations that are the most fundamental and useful, and these are known as **irreducible representations.** The symmetry species shown in Table 12.4 are in fact all the irreducible representations for this point group.

For the point group C_{2v} there are four different types of operation: the identity operation (E), rotation (C_2), and two reflections (σ_v and σ_v'). We refer to these types of operation as **classes.** Sometimes (although not in the present example) there is

TABLE 12.4 Representations (Symmetry Species) for the Point Group C_{2v}

Designation of Symmetry Species	E	C_2	$\sigma_v(xz)$	$\sigma_v'(yz)$
A_1	+1	+1	+1	+1
A_2	+1	+1	−1	−1
B_1	+1	−1	+1	−1
B_2	+1	−1	−1	+1

The modern convention is to take the $\sigma_v'(yz)$ plane to be the plane of the molecule. For the B_1 species there is therefore antisymmetry with respect to a reflection in the plane of the molecule.

more than one operation in a class, and we then speak of the *degeneracy* of the class. An important theorem in group theory is that the number of irreducible representations is equal to the number of classes. Thus, for the C_{2v} point group there are four classes, E, C_2, σ_v, and σ_v' and, therefore, four irreducible representations as listed in Table 12.4.

Tables such as Table 12.4 are known as **character tables.** The numbers +1, −1, etc., that appear in these tables are known as **characters.*** In the appendix to this chapter we give character tables for some of the commoner point groups. In these tables we have made use of symbols such as A_1, B_2, and Σ_g^+, and these are part of a scheme proposed by the American chemical physicist Robert S. Mulliken (b. 1896).

The terms **symmetric** and **antisymmetric** are used to refer to the characters +1 and −1, respectively, for a particular operation. For example, suppose that a molecular orbital is such that it retains its sign when a rotation is carried out; the character is +1 and the orbital is said to be symmetric with respect to that rotation. If it changes sign, the character is −1 and the function is said to be antisymmetric. A particular notation is used if the molecule (e.g., a homonuclear diatomic molecule) has a center of symmetry, in which case one of the symmetry elements is inversion. If the symmetry species is such that inversion brings about no change of sign, the species is *symmetric with respect to inversion* and the subscript g (German *gerade*) is added to the symbol. An example is the Σ_g^+ species shown in the appendix to this chapter for point group $D_{\infty h}$. If, on the other hand, inversion brings about a change of sign, the subscript is u (German *ungerade*). An example is the symmetry species Σ_u^+ found in the point group $D_{\infty h}$. Other features of the notation for the symmetry species of diatomic molecules are considered in Section 12.5.

Many of the applications of group theory to molecular problems depend on the fact that certain simple functions behave, with respect to the symmetry operations, in the same way as certain symmetry species. Such a function is said to be a **basis** for the particular species.

Consider, for example, the point group C_{2v}, which is shown in Figure 12.16c. A z coordinate is unchanged by a rotation about the C_2 axis and by reflections in the σ_v and σ_v' planes. It therefore behaves in the same way as the species A_1;

E	C_2	σ_v	σ_v'
+1	+1	+1	+1

A z coordinate is therefore a basis for the A_1 representation. An x coordinate, however, changes its sign on a C_2 rotation and on reflection in the σ_v' plane; reflection in the σ_v plane, however, does not affect it. It therefore transforms according to the scheme

E	C_2	σ_v	σ_v'
+1	−1	+1	−1

and it is therefore a basis for the B_1 representation (Table 12.4). A y coordinate can similarly be shown to be a basis for the B_2 representation.

---

*They are also known as *traces*. The word *character* is then reserved for the *set* of traces that are given in a character table.

Rotations can be considered from the same point of view. Suppose that the water molecule shown in Figure 12.14a is rotating about the Z axis. The direction of rotation is unchanged by the C_2 operation, whereas it is reversed by reflections in the σ_v and σ_v' planes. The rotation R_z therefore transforms according to the scheme

E	C_2	σ_v	σ_v'
$+1$	$+1$	-1	-1

and the function R_z therefore belongs to the A_2 symmetry species. Rotation about the X axis is unchanged by the σ_v' operation but is reversed by C_2 and by σ_v, and R_x therefore transforms according to

E	C_2	σ_v	σ_v'
$+1$	-1	-1	$+1$

The function R_x therefore belongs to the B_2 symmetry species. Similarly, R_y belongs to the B_1 species.

These and other relationships are shown in the character tables given in the appendix to this chapter. We shall see in Chapter 13 that the x, y, and z relationships are relevant to infrared vibrational spectra; R_x, R_y, and R_z to infrared rotational spectra; and x^2, y^2, z^2, xy, xz, and yz to Raman spectra.

12.5 Molecular Orbitals

The reason that the study of symmetry, particularly by the use of group theory, is of such importance in the quantum-mechanical treatment of molecules is that it places restrictions on the possible orbitals. The pattern of electron density around a molecule must have the same symmetry as the molecule itself. For example, the plane of the water molecule (Figure 12.14a) is a plane of symmetry, and there cannot be a different pattern of electron density on one side of this plane than on the other. The same is true for every symmetry operation.

Since the electron density is given by ψ^2, it follows that a wave function ψ must be such that if a symmetry operation is performed, ψ either remains unchanged or simply changes its sign. For the water molecule, for example, the wave function may belong to any one of the symmetry species A_1, A_2, B_1, and B_2, but it cannot show any other behavior. This conclusion is of great help in constructing molecular orbitals from atomic orbitals, as we will now examine in more detail.

Homonuclear Diatomic Molecules

Every homonuclear diatomic molecule (Figure 12.17a) is of symmetry $D_{\infty h}$. Its axis is a C_∞ axis of rotation, and there is an infinite number of axes at right angles to the C_∞ axis and passing through the center of gravity; these are C_2 axes. Any plane passing through the molecule is a σ_v plane, and there is also a σ_h plane passing through the center of gravity and at right angles to the principal C_∞ axis. There is also a center of inversion i. Examples of these symmetry elements are shown in Figure 12.17a.

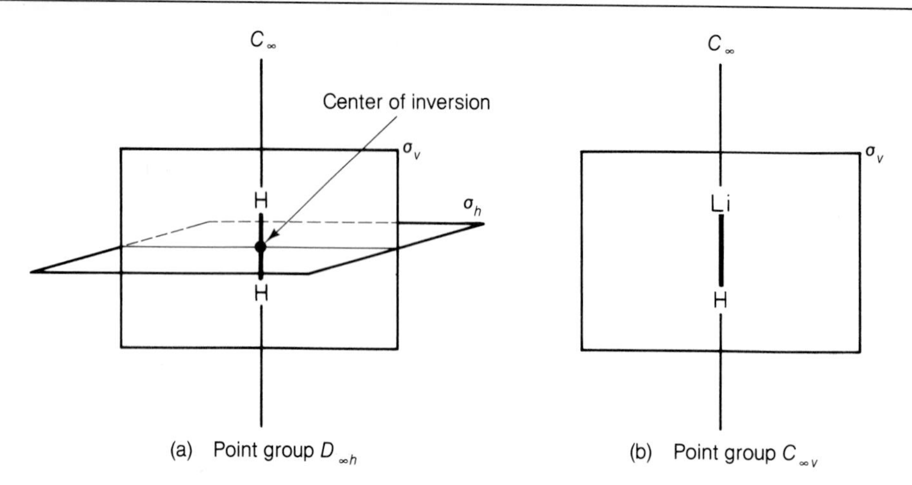

FIGURE 12.17 Examples of the symmetry elements of a homonuclear (a) and a heteronuclear (b) molecule.

The complete wave functions for homonuclear diatomic molecules must be consistent with this $D_{\infty h}$ symmetry and must therefore correspond to one of the symmetry species listed in the character table (see appendix to this chapter).

For example, a molecular orbital for the H_2 molecule can be constructed by adding the 1s orbitals for the individual atoms. The resulting MO, $1s_A + 1s_B$, belongs to the Σ_g^+ symmetry species for this point group. The convention used for individual molecular orbitals, as opposed to the complete wave functions for the molecule, is to use lowercase Greek letters. Thus, this orbital would be represented as σ_g (the plus sign is not usually added), and to indicate that the orbital is constructed from 1s atomic orbitals we can write it as $1s\sigma_g$.

The molecular orbital obtained by subtracting the 1s atomic orbitals, namely $1s_A - 1s_B$, is of symmetry species Σ_u^+, and we write it as $1s\sigma_u^*$; the asterisk indicates that it corresponds to a repulsive state.

Earlier in this chapter we discussed the simplest homonuclear molecule H_2 from both the valence-bond and molecular-orbital points of view. The next simplest homonuclear molecule He_2 does not exist, but if it did, it would have the same MO diagram as hydrogen (Figure 12.4), since only two molecular orbitals are needed to accommodate four electrons. The first two electrons enter the lowest-lying orbital $1s\sigma_g$ with spins paired as shown in Figure 12.18. The Pauli exclusion principle forces the next electron to be placed in the antibonding $1s\sigma_u^*$ orbital. The two electrons in the $1s\sigma_g$ orbital produce a stabilization of the molecule, but the electron in the antibonding orbital tends to destabilize the structure since the energy of the antibonding electron is higher than that of the original AO. The fourth electron must also be placed into the antibonding orbital, with opposite spin. The net effect is a complete destabilization of the molecule, which now has a slightly higher energy than the separate atoms. It is convenient to refer to one-half the difference between the number of bonding electrons and the number of antibonding electrons as the **bond order.** Thus, He_2 has two bonding electrons and two antibonding electrons, and its bond order is therefore zero.

Lithium can exist as dilithium, Li_2, and alternative molecular-orbital diagrams are shown in Figure 12.19. In Figure 12.19a the 1s orbitals remain intact, whereas

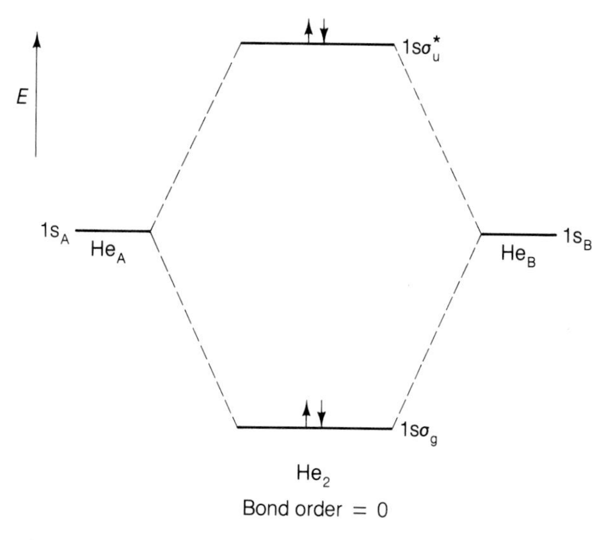

FIGURE 12.18 The energies of bonding and antibonding orbitals for the nonexistent He_2 molecule. The increased energy of the antibonding orbital cancels the stability of the bonding orbital; therefore, no bond exists.

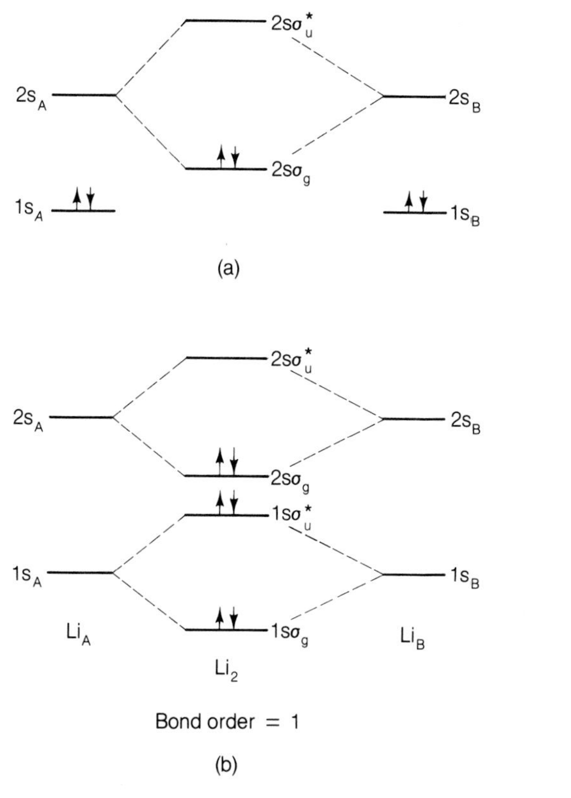

FIGURE 12.19 Molecular-orbital descriptions for Li_2. In (a) the 1s electrons are regarded as remaining in atomic orbitals. In (b) they are in σ_g and σ_u^* molecular orbitals.

in Figure 12.19b they combine to give $1s\sigma_g$ and $1s\sigma_u$ orbitals. In both representations the two 2s orbitals combine to form two molecular orbitals $2s\sigma_g$ and $2s\sigma_u^*$. Six electrons are to be accommodated, and in the second representation (Figure 12.19b) the first two, with spins paired, occupy the lowest-energy $1s\sigma_g$ orbital; the next two occupy the $1s\sigma_u^*$ orbital. As a first approximation we may assume that the energy of the bonding MO is as much below the atomic 1s orbital as the energy of the antibonding orbital is above it. This is equivalent to setting the overlap integral S_{AB} equal to zero. In other words, this arrangement of four electrons is no more stable than the separated atoms (Figure 12.19b). If S_{AB} is included, the antibonding electrons destabilize the bond, an effect known as **inner-shell repulsion;** however, S_{AB} is small because of the small size of the 1s orbitals. The stability of Li_2 therefore comes from the final two electrons that are placed in the bonding $2s\sigma_g$ molecular orbital.

In Be_2 there are two additional electrons, which must go into the $2s\sigma_u^*$ orbital. There are therefore as many antibonding electrons as bonding electrons, and the bond order is zero; the molecule does not exist.

To deal with diatomic molecules having more electrons than Be_2, we must construct molecular orbitals from 2p atomic orbitals. If we start with p orbitals that lie along the axis of the diatomic molecule (i.e., p_z orbitals), these combine as follows:

$$2p\sigma_g = 2p_{zA} + 2p_{zB} \tag{12.65}$$

$$2p\sigma_u^* = 2p_{zA} - 2p_{zB} \tag{12.66}$$

The former is bonding, and the latter is antibonding, as shown in Figure 12.20a.

We can also add and subtract $2p_y$ orbitals, which lie perpendicular to the axis of the molecule, and obtain bonding and antibonding MOs:

$$2p\pi_u = 2p_{yA} + 2p_{yB} \tag{12.67}$$

$$2p\pi_g^* = 2p_{yA} - 2p_{yB} \tag{12.68}$$

These orbitals are represented in Figure 12.20b. Note the analogy with the π bonds formed in valence-bond theory by the sideways overlapping of p orbitals. We see from the figure, in which the + and − signs on the orbitals are shown, that the bonding orbital is now of u symmetry, whereas the antibonding orbital is of g symmetry. We can also form an equivalent set of MOs from the $2p_x$ orbitals, as shown in Figure 12.20c:

$$2p\pi_u = 2p_{xA} + 2p_{xB} \tag{12.69}$$

$$2p\pi_g^* = 2p_{xA} - 2p_{xB} \tag{12.70}$$

Again, the bonding orbital is u and the antibonding is g. The π levels and the π^* levels are therefore *degenerate,* and each can be occupied by two pairs of electrons. Note that we cannot combine, for example, p_x and p_z orbitals, since the resulting MO would be inconsistent with the symmetry of the molecule.

The electronic configurations in molecules like N_2, O_2, and F_2 depend on the order of the energy levels. Calculations by the self-consistent field method have shown that the $2p\sigma_g$ and $2p\pi_u$ levels have much the same energy, but photoelectron spectroscopy has shown that for N_2 and F_2 the order is as indicated in Figure 12.21a, which gives the assignment of electrons in N_2. The first four electrons go into the

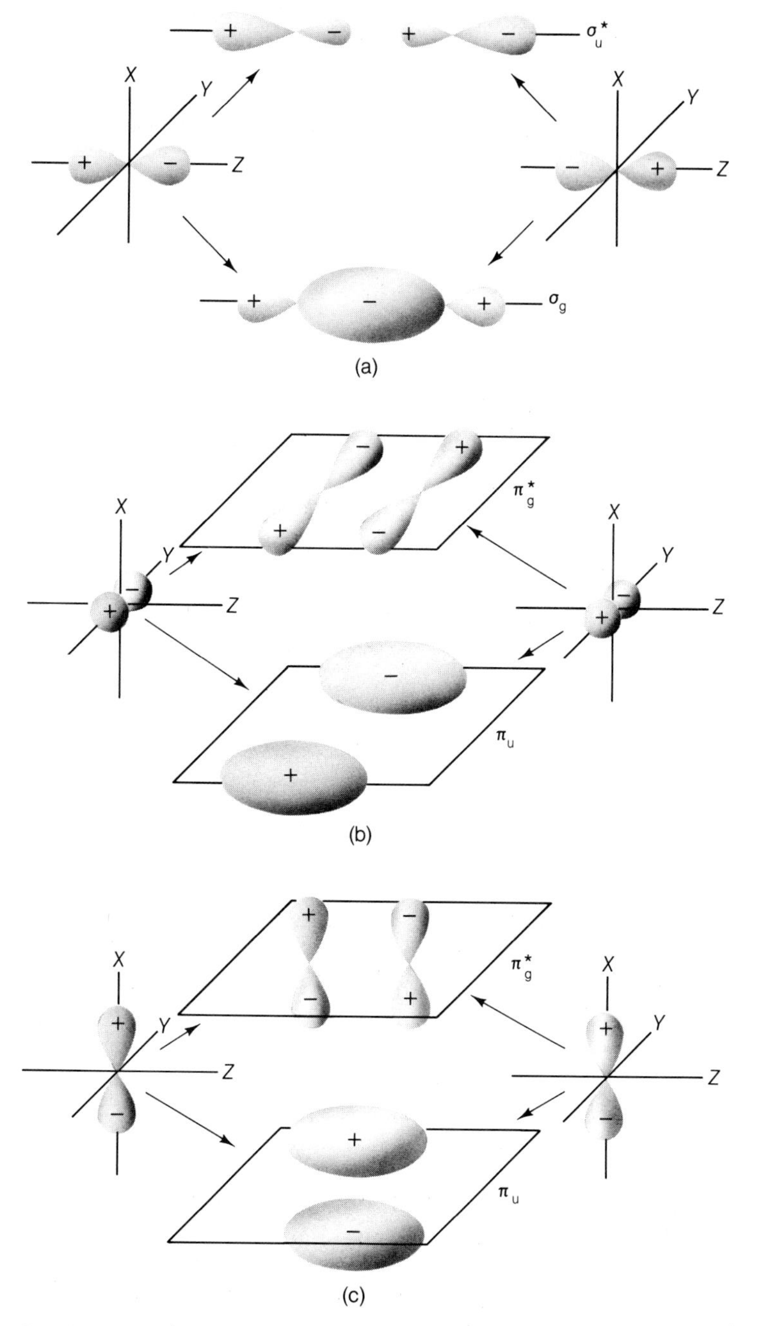

FIGURE 12.20 Combination of p atomic orbitals to give molecular orbitals. (a) σ orbitals formed from two p_z orbitals, which lie along the axis of the molecule (the Z axis). (b) π orbitals formed from p_y orbitals. (c) π orbitals formed from p_x orbitals.

$1s\sigma_g$ and $1s\sigma_u^*$ orbitals. The high nuclear charge draws these electrons close to the nuclei, and there is virtually no overlap for the 1s AO's. The next four electrons fill the $2s\sigma_g$ and $2s\sigma_u^*$ MOs. The next level, $2p\pi_u$, shown as two lines, is degenerate and

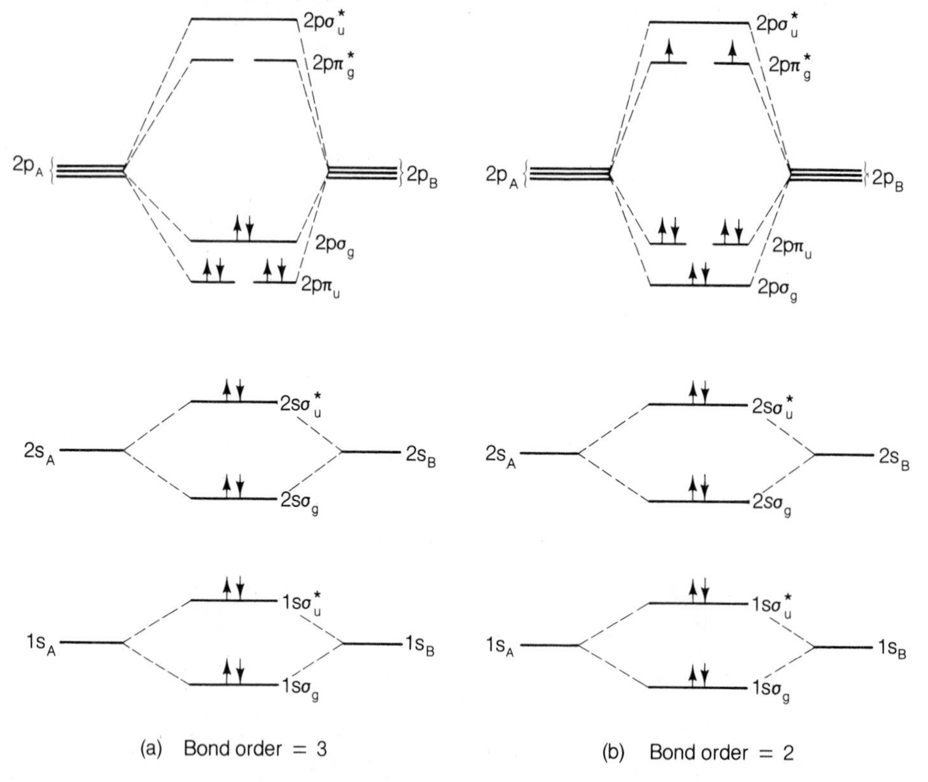

FIGURE 12.21 Molecular-orbital energy diagrams for homonuclear diatomic molecules. (a) The order of energy levels for N_2, showing the assignment of the 14 electrons; the order is the same for F_2. (b) The order of levels for O_2, showing the electrons.

can accommodate four electrons. The remaining two electrons go into the $2p\sigma_g$ orbital. There are therefore six bonding electrons, and the bond order is three. The configuration of N_2 can be written as

$$1s\sigma_g^2 1s\sigma_u^{*2} 2s\sigma_g^2 2s\sigma_u^{*2} 2p\pi_u^4 2p\sigma_g^2$$

For O_2, however, the order of filling is different, as shown in Figure 12.21b.

A valuable feature of MO theory is that it provides a very simple interpretation of molecular magnetism. In Section 11.9 we saw that an electron in any orbital has a magnetic moment. If all such electrons in an atom or molecule are paired, there is no resultant magnetic moment, and there is no interaction with a magnetic field; the atom or molecule is then said to be **diamagnetic.** If, on the other hand, there are p, d, etc., electrons that are not paired and have the same spin (i.e., *odd* electrons), there is a resultant magnetic moment. The atom or molecule then interacts with an extenal magnetic field, and is said to be **paramagnetic.** The degree of paramagnetism depends on the number of unpaired electrons.

In the MO diagram of Figure 12.21b we see that 14 of the 16 electrons in O_2 are paired with one another. The remaining 2 electrons, on the other hand, are in the $2p\pi_g^*$ orbital, which could accommodate a total of 4 electrons, and by Hund's rule they have parallel spins. The oxygen molecule is therefore predicted to have a resultant magnetic moment (i.e., to be paramagnetic), and this is in agreement with experiment.

The MO energy diagram of F_2 is similar to that for N_2, but four additional electrons must be accommodated. These electrons go into the $2p\pi_g^*$ orbitals in Figure 12.21a. These electrons in the two antibonding orbitals cancel the bonding from the $2p\pi_u$ orbitals, leaving only one bonding orbital having its pair of electrons; the bond order is therefore one. Because the antibonding cancels the bonding, there are three lone pairs on each atom, corresponding to the conventional valence-bond structure $:\ddot{\underset{..}{F}}:\ddot{\underset{..}{F}}:$.

In the series N, O, F there is increasing nuclear charge but decreasing bond order; this is partly responsible for the decreased bonding in the order N_2, O_2, and F_2. However, the importance of the antibonding orbitals is illustrated by the decreasing bond energies of N≡N, O=O, and F—F, which are 941.4, 493.7, and 153.0 kJ mol^{-1}, respectively, and by the increasing bond distances, which are 110 pm for N_2, 121 pm for O_2, 142 pm for F_2.

Heteronuclear Diatomic Molecules

Every heteronuclear diatomic molecule belongs to the point group $C_{\infty v}$. The axis of the molecule is a C_∞ axis, and any plane passing through the molecule is a plane of symmetry, designated σ_v (see Figure 12.17b). These are the only symmetry elements. In particular, a heteronuclear molecule lacks a center of symmetry and in this respect differs from a homonuclear diatomic molecule.

The wave functions for heteronuclear molecules must therefore correspond to one of the symmetry species for this point group, which are given in the appendix to this chapter. Subscripts g and u are now inappropriate, since there is no center of inversion, but for some of the representations the superscripts + and − are used, the former to indicate no change of sign on reflection in a σ_v plane and the latter to indicate a change of sign.

Lowercase Greek letters are again used to indicate the symmetry species of the individual molecular orbitals. For example, the MO formed by adding two 1s atomic orbitals is called a $1s\sigma$ orbital, while that obtained by subtracting them is a $1s\sigma^*$ orbital. Since the gerade-ungerade notation no longer applies, these two orbitals are distinguished simply by using the asterisk to indicate antibonding. If a molecular orbital is formed by adding or subtracting two different atomic orbitals, such as $1s_A + 2s_B$, $1s_A - 2s_B$, the symbol used is simply σ or σ^*.

On this basis we may consider LiH as follows. There is a total of four electrons, and two of these may be considered to remain in the 1s atomic orbital of Li. The bond is formed from the valence (2s) electron of Li and the 1s electron of H. Addition of the $1s_H$ and $2s_{Li}$ orbitals will give a σ molecular orbital that will accommodate the two electrons.

The two modifications required to deal with heteronuclear diatomic molecules are

1. The g and u suffixes are dropped, since the molecules have no center of symmetry.

2. The diagrams are no longer symmetrical.

Figure 12.22 shows a schematic MO diagram for a heteronuclear diatomic molecule. As an example, carbon monoxide has 14 electrons, and on the basis of the Aufbau Principle we have

$$(1s\sigma)^2(1s\sigma^*)^2(2s\sigma)^2(2s\sigma^*)^2(2p\pi)^4(2p\sigma)^2$$

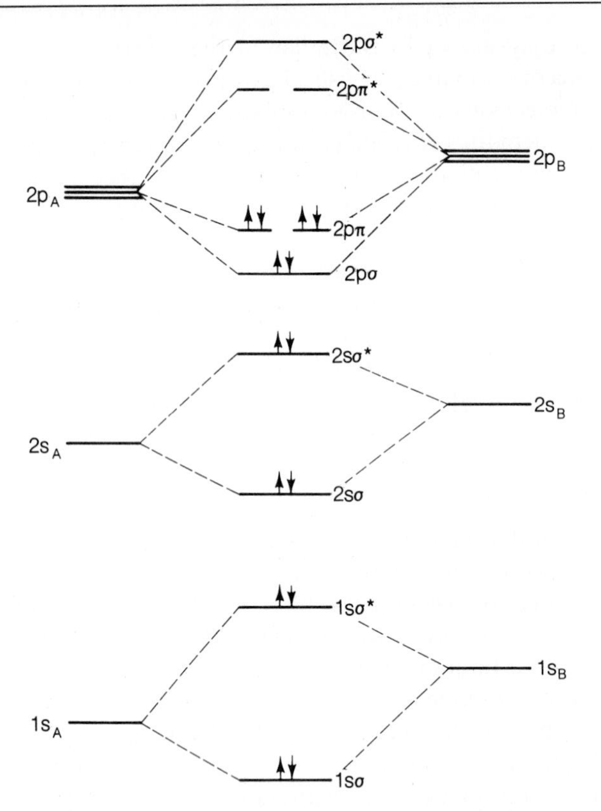

FIGURE 12.22 Molecular-orbital diagram for a heteronuclear diatomic molecule. As an example, the 14 electrons of carbon monoxide are shown as arrows.

The two pairs of σ—σ^* orbitals will cancel and give no bonding. Bonding, however, results from the six electrons in the 2pπ and 2pσ orbitals, and the bond order is three.

The Water Molecule

The procedures for constructing molecular orbitals for more complicated molecules may be exemplified by a brief treatment of the water molecule, the symmetry elements for which are shown in Figure 12.14a. The oxygen atom lies on all three of these symmetry elements, and in Figure 12.23a we show orbitals of the oxygen atom in relation to these elements. Since the 1s and 2s orbitals are spherically symmetrical, none of the symmetry operations brings about any change in them, and both are therefore of A_1 symmetry. The 2p$_z$ orbital has a positive lobe in one direction along the Z axis and a negative lobe in the other. The C_2, σ_v, and σ_v' operations bring about no change in this orbital, which is therefore also of A_1 symmetry.

The 2p$_x$ and 2p$_y$ orbitals, however, behave differently. Reflection of the 2p$_x$ orbital in the σ_v' plane interchanges the + and − lobes and therefore changes the sign, and the same happens with a C_2 rotation. Reflection in the σ_v plane brings about no change. The 2p$_x$ orbital is therefore of B_1 symmetry. By similar arguments we find that the 2p$_y$ orbital is of B_2 symmetry.

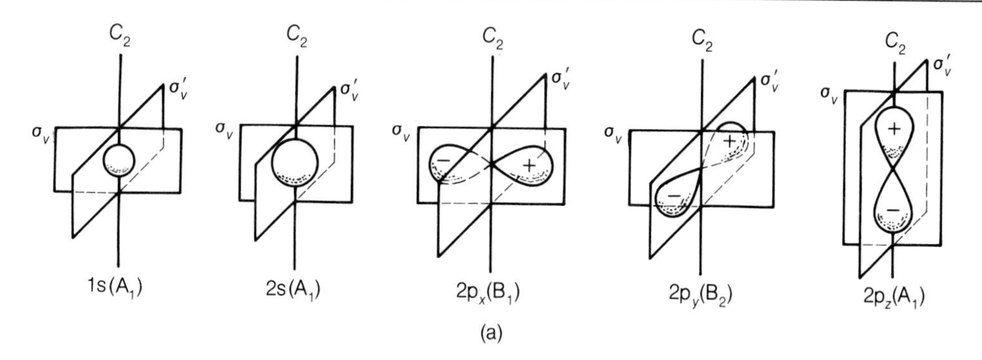

	E	C_2	σ_v	σ_v'	Atomic Orbitals
A_1	+1	+1	+1	+1	$1s(O)$, $2s(O)$, $2p_z(O)$, $1s_A + 1s_B$
A_2	+1	+1	−1	−1	—
B_1	+1	−1	+1	−1	$2p_x(O)$
B_2	+1	−1	−1	+1	$2p_y(O)$, $1s_A - 1s_B$

(b)

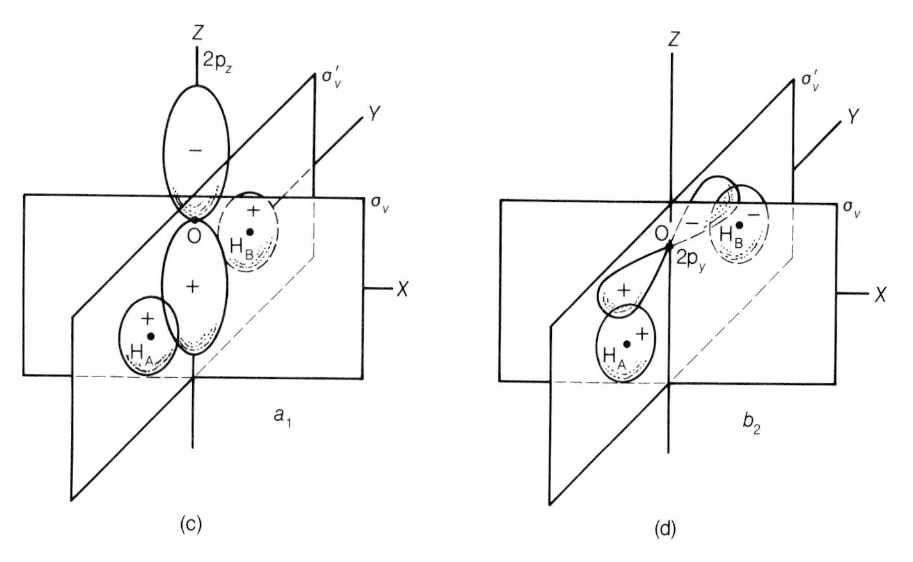

FIGURE 12.23 Orbitals for the water molecule. (a) The atomic orbitals and their symmetry species. The σ_v' plane is the plane of the molecule. (b) The C_{2v} character table showing the symmetries of the atomic orbitals and the symmetry-adapted orbitals. (c) and (d) The bonding molecular orbitals a_1 and b_2.

The situation with the orbitals on the hydrogen atoms in the water molecule is a little more complicated, since these atoms do not lie on all the symmetry elements of the molecule; they only lie in the σ_v' plane. The 1s orbital of the H_A atom, which we will write as $1s_A$, therefore does not correspond to the symmetry of the molecule, and the same is true of the $1s_B$ orbital. However, from these two orbitals we can construct orbitals that do have the right symmetry. Thus the sum of the atomic orbitals, $1s_A + 1s_B$, is unchanged when any of the symmetry operations is performed, and it is therefore of A_1 symmetry. The difference, $1s_A - 1s_B$, changes its sign when the C_2 operation is performed and when the σ_v operation is performed, but it remains unchanged when σ_v' is performed. It is therefore of B_2 symmetry.

The symmetry properties of the various orbitals on the H and O atoms are summarized in Figure 12.23b. The complete molecular orbital must correspond in symmetry to one of the symmetry species of the molecule, and this is impossible if we combine atomic orbitals of different symmetries. We can, however, obtain a molecular orbital of symmetry A_1 if we make a linear combination of 2s, $2p_z$, and $1s_A + 1s_B$ orbitals, all of which are of A_1 symmetry. Such an orbital is designated a_1 (the lowercase letter is used) and is of the form

$$a_1 = 1s_A + 1s_B + \lambda_1 2s + \lambda_2 2p_z \tag{12.71}$$

where λ_1 and λ_2 are coefficients which could be determined by a variation procedure. This orbital is represented schematically in Figure 12.23c.

We can also construct the orbital

$$b_2 = 1s_A - 1s_B + \lambda 2p_y \tag{12.72}$$

in which orbitals of B_2 symmetry are combined. This orbital is represented in Figure 12.22d.

Both of these orbitals are bonding, since the atomic orbitals have been added together, and there is a piling up of charge between the nuclei. For each bonding molecular orbital there is an antibonding one, of the same symmetry, formed by subtracting the atomic orbitals. These are denoted by an asterisk. The antibonding molecular orbital corresponding to a_1 is

$$a_1^* = 1s_A + 1s_B - \lambda_1' 2s - \lambda_2' 2p_z \tag{12.73}$$

That corresponding to b_2 is

$$b_2^* = 1s_A - 1s_B - \lambda' 2p_y \tag{12.74}$$

There is also an antibonding orbital of B_1 symmetry, which consists only of $2p_x$(O) itself; this cannot combine in any way with the hydrogen atom orbitals and, therefore, gives no bonding.

These conclusions are shown in Figure 12.24 in the form of an energy diagram. The order of the energy levels is that predicted by self-consistent field calculations. The 1s atomic orbital of the oxygen atom has a much lower energy than the other atomic orbitals, and it is therefore carried over as the lowest MO; electrons in this orbital are nonbonding. The 2s atomic orbital of the oxygen atom also forms an MO, designated a_1', and electrons in this level are also nonbonding.

The assignment of the 10 electrons in the water molecule is shown in Figure 12.24. The 2 electrons in the 1s orbital of oxygen make no contribution to bonding.

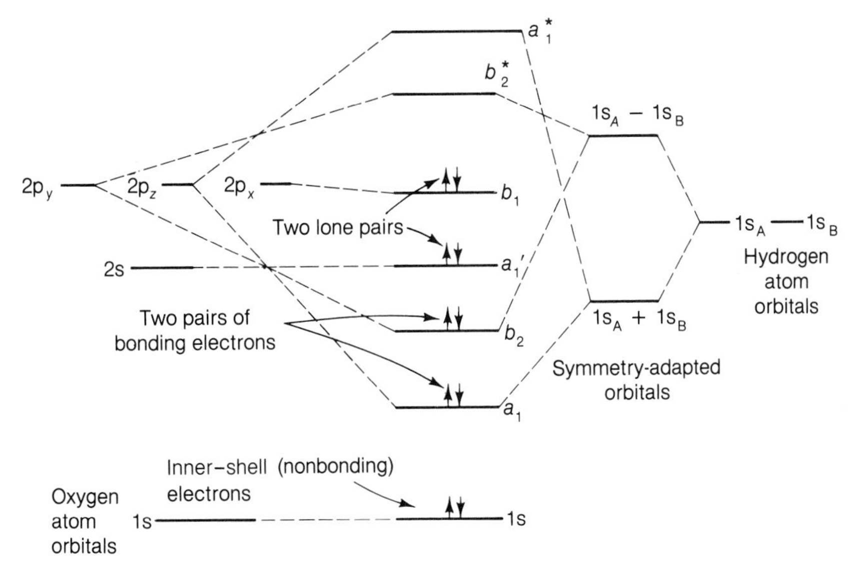

FIGURE 12.24 The energy levels corresponding to the molecular orbitals for water, showing how they are derived from the atomic orbitals of oxygen and hydrogen (not to scale).

Two electrons are in each of the a_1 and b_2 orbitals, and these four electrons are responsible for the two O—H bonds. The remaining four electrons are in the a_1' and b_1 orbitals and constitute the two lone pairs.

Appendix to Chapter 12: Character Tables

C_{2v}	E	C_2	$\sigma_v(xz)$	$\sigma_v'(yz)$*					
A_1	+1	+1	+1	+1	z		x^2	y^2	z^2
A_2	+1	+1	−1	−1		R_z		xy	
B_1	+1	−1	+1	−1	x	R_y		xz	
B_2	+1	−1	−1	+1	y	R_x		yz	

C_{3v}	E	$2C_3$	$3\sigma_v$						
A_1	+1	+1	+1	z		$x^2 + y^2$	z^2		
A_2	+1	+1	−1		R_z				
E	+2	−1	0	x y	R_x R_y	$x^2 - y^2$	xy	xz	yz

*The modern convention is to take the σ_v' (yz) plane to be the plane of the molecule.

$C_{\infty v}$	E	$\infty\sigma_v$				
Σ^+	+1	+1	z		$x^2 + y^2$	z^2
Σ^-	+1	−1		R_z		
π	+2	0	$x\ y$	$R_x\ \ R_y$	xz	yz
Δ	+2	0			$x^2 - y^2$	xy
Φ	+2	0				

$D_{\infty h}$	E	σ_h	σ_v	i				
Σ_g^+	+1	+1	+1	+1			$x^2 + y^2$	z^2
Σ_u^+	+1	−1	+1	−1	z			
Σ_g^-	+1	+1	−1	+1		R_z		
Σ_u^-	+1	−1	−1	−1				
π_g	+2	−2	0	+2		$R_x\ \ R_y$	xz	yz
π_u	+2	+2	0	−2	$x\ y$			
Δ_g	+2	+2	0	+2			$x^2 + y^2$	xy
Δ_u	+2	−2	0	−2				
Φ_g	+2	−2	0	+2				
Φ_u	+2	+2	0	−2				

D_{3h}	E	$2C_3$	$3C_2$	σ_h	$2S_3$	σ_v			
A_1'	+1	+1	+1	+1	+1	+1			z^2
A_1''	+1	+1	+1	−1	−1	−1			
A_2'	+1	+1	−1	+1	+1	−1		R_z	
A_2''	+1	+1	−1	−1	−1	+1	z		
E'	+2	−1	0	+2	−1	0	$x\ y$		xy
E''	+2	−1	0	−2	+1	0		$R_x\ \ R_y$	$xz\ \ yz$

C_{2h}	E	C_2	σ_h	i			
A_g	+1	+1	+1	+1		R_z	x^2 y^2 z^2 xy
A_u	+1	+1	−1	−1	z		
B_g	+1	−1	−1	+1		R_x R_y	xz yz
B_u	+1	−1	+1	−1	x y		

D_{6h}	E	$3C_2$	$2C_3$	$2C_6$	σ_h	σ_v	i			
A_{1g}	+1	+1	+1	+1	+1	+1	+1			x^2 y^2 z^2
A_{1u}	+1	+1	+1	+1	−1	−1	−1			
A_{2g}	+1	−1	+1	+1	+1	−1	+1		R_z	
A_{2u}	+1	−1	+1	+1	−1	+1	−1	z		
B_{1g}	+1	+1	+1	−1	−1	−1	+1			
B_{1u}	+1	+1	+1	−1	+1	+1	−1			
B_{2g}	+1	−1	+1	−1	−1	−1	+1			
B_{2u}	+1	−1	+1	−1	+1	+1	−1			
E_{1g}	+2	0	−1	+1	−2	0	+2		R_x R_y	xz yz
E_{1u}	+2	0	−1	+1	+2	0	−2	x y		
E_{2g}	+2	0	−1	−1	+2	0	+2			x^2 y^2 xy
E_{2u}	+2	0	−1	−1	−2	0	−2			

Key Equations

Heitler-London wave functions for the H_2 molecule:

$$\psi_S = \frac{1}{\sqrt{2}}[1s_A(1)1s_B(2) + 1s_A(2)1s_B(1)]$$

$$\psi_A = \frac{1}{\sqrt{2}}[1s_A(1)1s_B(2) - 1s_A(2)1s_B(1)]$$

Molecular orbitals for the H_2 molecule:

$$\sigma = 1s_A + 1s_B$$

$$\sigma^* = 1s_A - 1s_B$$

Problems

A Bond Energies, Shapes of Molecules, and Dipole Moments

1. The attractive energy between two univalent ions M^+ and A^- separated by a distance r is

$$\frac{137.2}{r/\text{nm}} \text{ kJ mol}^{-1}$$

Suppose that there is also a repulsive energy given by

$$\frac{0.0975}{(r/\text{nm})^6} \text{ kJ mol}^{-1}$$

Plot the attractive and repulsive energies against r, and also plot the resultant energy, all on the same graph. By differentiating the equation for the resultant energy, calculate the equilibrium interionic distance and the net energy at that distance.

2. The equilibrium internuclear distance in gaseous LiI is 239 pm, and the dipole moment is 2.09×10^{-29} C m. Estimate the percentage ionic character of the bond.

3. The following are bond dissociation energies:

$$D(Li_2) = 113 \text{ kJ mol}^{-1}$$
$$D(H_2) = 435 \text{ kJ mol}^{-1}$$
$$D(LiH) = 243 \text{ kJ mol}^{-1}$$

a. Use Pauling's relationship (Eq. 12.53) to estimate the electronegativity difference between Li and H.
b. Estimate the percentage ionic character of the Li-H bond, given the following covalent radii:

Li: 126 pm
H: 36 pm

(*Note:* Since the Pauling relationship leads to dipole moments in debyes, it is most convenient to work problems of this type taking the electronic charge as 4.8×10^{-10} esu.)

4. Deduce the shapes of the following, on the basis of valence-shell electron-pair repulsion (VSEPR) theory:

$BeCl_2$, SF_6, H_3O^+, NH_4^+, PCl_6^-, AlF_6^{3-}, PO_4^{3-}, CO_2, SO_2, NH_3^{2+}, CO_3^{2-}, NO_3^-

5. Calculate the percentage ionic character of the HCl, HBr, HI, and CO bonds from the following data:

	HCl	HBr	HI	CO
Internuclear distance/pm	127	141	160	113
Dipole moment/10^{-30} C m	3.60	2.67	1.40	0.33

B Molecular Orbitals

6. Use Figures 12.21 and 12.22 to construct molecular-orbital diagrams for the following:

B_2, CO, BN, BN^{2-}, BO, BF, OF, OF^-, OF^+

Deduce the bond order and paramagnetism in each case.

7. Sketch the molecular-orbital diagrams for the following:

N_2, O_2, C_2, F_2, CN, NO

Which of these species would you expect to become more stable if (a) an electron is added and (b) an electron is removed?

8. The hydrogen atom wave functions $1s_A$ and $1s_B$ are normalized. Prove that the molecular orbitals σ and σ^* given in Eqs. 12.24 and 12.25 are mutually orthogonal.

C Group Theory

9. Give the point groups to which the following belong: (a) an equilateral triangle, (b) an isosceles triangle, and (c) a cylinder.

10. List the symmetry elements for each of the following molecules, and give the point group:

$CHCl_3$, CH_2Cl_2, naphthalene, chlorobenzene, NO_2 (bent), cyclopropane, CO_3^{2-}, C_2H_2

11. The condition for optical activity, and for a molecule to exist in two enantiomeric forms, is that the molecule has neither a plane of symmetry nor a center of inversion. Can H_2O_2 exist in two mirror-image forms?

***12.** Deduce the symmetry species of the following vibrations:

a. H_2:

b. H_2O:

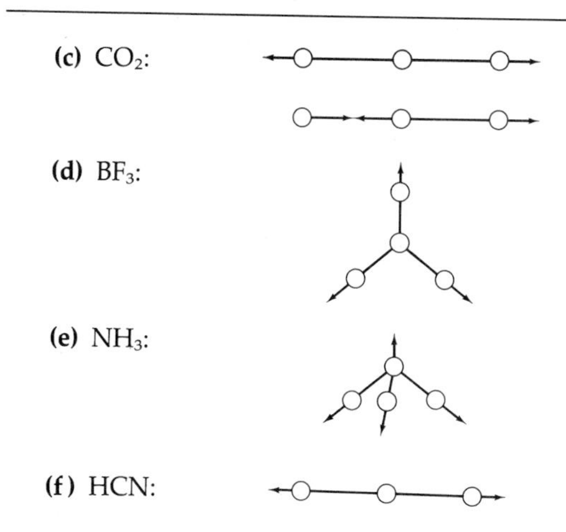

D Supplementary Problems

13. The four sp^3 hydrid orbitals are given in Eqs. 12.55–12.58. They are constructed from atomic orbitals, s, p_x, p_y, and p_z, which are normalized and mutually orthogonal.

a. The orbitals t_1, t_2, t_3, and t_4 are normalized. Prove that this is the case for t_3.

b. The orbitals t_1, t_2, t_3, and t_4 are mutually orthogonal. Prove that this is the case for t_2 and t_4.

14. Eigenfunctions for sp^2 hybridization, with maxima at 120° to one another, can be constructed with reference to the following diagram:

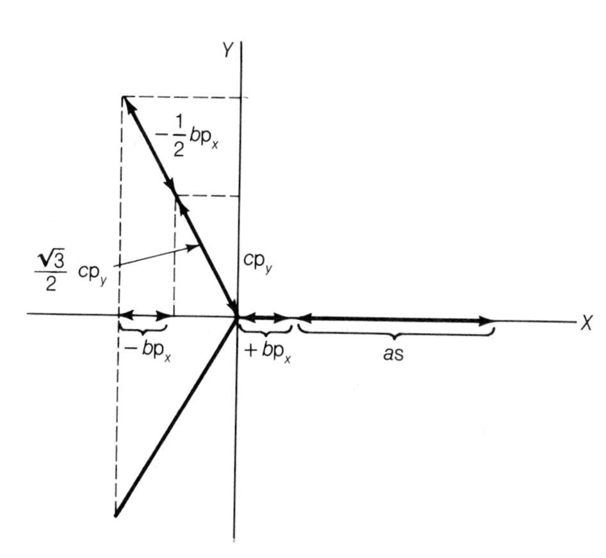

The plane is chosen as the XY plane, and there will be no contribution from the p_z orbital. An orbital along the X axis can be constructed as a linear combination of the s and p_x orbitals:

$$\psi_1 = a\mathrm{s} + b\mathrm{p}_x$$

where a and b are numbers to be determined. The orbitals s and p_x are normalized and orthogonal. To obtain the other two orbitals, the p_x and p_y orbitals are resolved along the two directions 120° from the X axis and are combined with the s orbital:

$$\psi_2 = a\mathrm{s} - \frac{1}{2}b\mathrm{p}_x + \frac{\sqrt{3}}{2}c\mathrm{p}_y$$

$$\psi_3 = a\mathrm{s} - \frac{1}{2}b\mathrm{p}_x - \frac{\sqrt{3}}{2}c\mathrm{p}_y$$

Make use of the normalization and orthogonality conditions to determine the numbers a, b, and c.

15. Use the procedure of Problem 14 to construct two normalized and orthogonal wave functions for sp hydridization, at an angle of 180° to one another.

16. A molecule having a center of symmetry (i) or having an axis of improper rotation (S) cannot have a dipole moment; all other molecules can. Note, however, that in tables such as Table 12.3 the symmetry operation S is not always included, since it follows from other operations (for example, as seen in Figure 12.15, S_4 is the product $C_4\sigma_h$). The following are molecules that have no center of symmetry but have no dipole moment:

cyclopropane   cyclopentane

In each case, identify the axis of improper rotation, designate the operation, and relate the operation to the appropriate rotations and reflections. Which of the point groups in Table 12.3 are such that there can be a dipole moment?

E Essay Questions

1. Give an account of the valence-shell electron-pair repulsion (VSEPR) theory of the shapes of molecules.

2. Explain orbital hydridization, with special reference to sp, sp^2, and sp^3 hydridization.

3. Explain the principles underlying the construction of trial wave functions (a) in the valence-bond method and (b) in the molecular-orbital method.

4. Explain the theory underlying the estimation of bond dipole moments from electronegativities.

Suggested Reading

See also Suggested Reading for Chapter 11 (p. 513).

C. J. Ballhausen and H. B. Gray, *Molecular Electronic Structures, An Introduction,* Reading; Mass.: Benjamin/Cummings, 1980.

G. M. Barrow, *The Structure of Molecules,* New York: Benjamin, 1963.

D. M. Bishop, *Group Theory and Chemistry,* Oxford: Clarendon Press, 1973.

E. Cartmell and G. W. A. Fowles, *Valency and Molecular Structure,* London: Butterworth, 1961.

C. A. Coulson, *The Shape and Structure of Molecules,* Oxford: Clarendon Press, 1973.

C. A. Coulson, *Valence,* Oxford: Clarendon Press, 1961.

R. J. Gillespie, *Molecular Geometry,* London: Van Nostrand Reinhold, 1972.

H. B. Gray, *Electrons and Chemical Bonding,* New York: Benjamin, 1964.

J. W. Linnett, *The Electronic Structure of Molecules: A New Approach,* London: Methuen, 1964.

J. N. Murrell, S. F. A. Kettle, and J. M. Tedder, *Valence Theory,* New York: Wiley, 1965.

L. Pauling, *The Nature of the Chemical Bond* (3rd ed.), Ithaca, N.Y. Cornell University Press, 1960.

Chemical Spectroscopy

Preview

Chapters 11 and 12 have dealt with atoms and molecules, from the point of view of quantum mechanics. Very important experimental information about atoms and molecules is provided by *spectroscopy,* which is concerned with the interaction between electromagnetic radiation and matter. A fundamental relationship is the *Lambert-Beer Law,* which relates the amount of radiation absorbed to the concentration of absorbing material and the distance through which the light passes.

Atomic spectra involve electronic transitions only. The simple Bohr treatment in Chapter 11 of the atomic spectrum of the hydrogen atom can be improved by the inclusion of some additional effects; in particular, *Coulombic interactions* between the different electrons in an atom, *exchange interactions,* and *spin-orbit interactions* which result from the magnetic fields associated with the spinning electron and the orbiting electron. The *Zeeman effect* and the *anomalous Zeeman effect* relate to the effects of external magnetic fields on spectra.

Molecular spectra are much more complicated, because there is the possibility of rotational and vibrational transitions as well as electronic transitions. When the light quanta are of low energy (in the far infrared or microwave regions of the spectrum), only rotational transitions can occur. At higher energies (the near infrared) both rotational and vibrational transitions occur. At still higher energies (the visible and ultraviolet) electronic transitions are possible, and vibrational and rotational transitions occur at the same time. Analysis of the rotational transitions allows *interatomic distances* to be obtained, and analysis of the vibrational transitions leads to *force constants* of bonds. There are certain restrictions to transitions that are described by *selection rules.*

In *Raman spectroscopy* the light is not absorbed but its frequency is changed by interaction with the material through which it passes. A *laser beam* is very intense and covers a very narrow range of wavelengths, and *laser spectroscopy* provides some very important information. *Spectral line widths* are related to the *Doppler effect* and to the *lifetimes* of excited species.

Three types of spectroscopy depend on *resonance effects:* These are electron spin resonance spectroscopy, nuclear magnetic resonance spectroscopy, and Mössbauer spectroscopy. *Electron spin resonance* is obtained with a molecule that has an odd electron, and it arises from the alignment of the electron spin in a magnetic field; special features (the *hyperfine structure*) result from the interaction between electron spin and nuclear spin. *Nuclear magnetic resonance* arises from the alignment of nuclear spins in a magnetic field; an important effect is the *chemical shift*, which depends on the interaction between the nuclear spin and the orbiting electrons. *Mössbauer spectra* are quite different, and they arise from the absorption of γ rays by nuclei. They are made possible by Mössbauer's discovery of a way of generating highly homogeneous beams of γ rays and of controlling their frequency by use of the Doppler effect.

13 Chemical Spectroscopy

The subject of *spectroscopy* is concerned with the interaction of matter with electromagnetic radiation. Spectroscopic measurements, combined with theoretical interpretations, provide very detailed information about chemical structure and the arrangement of electrons in atoms and molecules. Atomic spectra, for example, have been valuable in providing information about the energy levels that are available to the electrons, and the results can be compared with those predicted by quantum mechanics. Molecules are much more complex because of the variety of ways in which energy transformations can occur, and their spectra are correspondingly more complicated. Molecular spectra provide very valuable information about molecular vibrations, bond energies, and many other properties.

Spectroscopy also plays a very useful role in the determination of concentrations of substances. Since some of the experimental methods are quite simple, they are very widely used.

Other areas of spectroscopy are concerned with the behavior of substances in a magnetic field. *Nuclear magnetic resonance* (nmr) spectroscopy deals with effects arising from nuclear spins, while *electron spin resonance* (esr) spectroscopy is concerned with electron spin. These areas are briefly considered in Sections 13.11 and 13.10.

13.1 Emission and Absorption Spectra

The foundation stone of spectroscopy was laid in 1665–1666 when the English scientist Sir Isaac Newton (1642–1727) demonstrated that white light passed through a prism is split into a spectrum of colors ranging from red to violet. The red end of the visible spectrum corresponds to the longer wavelengths (≈700 nm or 7000 Å) and lower frequencies. The violet end of the spectrum has shorter wavelengths (≈400 nm or 4000 Å) and higher frequencies. The violet end corresponds to higher photon energies, since photon energies are equal to the frequency multiplied by the Planck constant. Violet light is in general more effective than red light in bringing about chemical and biological changes because of the higher energy it carries.

Units of Wavelength and Frequency

In Section 11.1 we considered the electromagnetic spectrum, which is represented in Figure 11.2, and we saw that there are vast spectral regions to which the human eye is not sensitive. Electromagnetic radiation can be characterized by either a frequency ν or a wavelength λ, the product of these being c, the velocity of light (see Eq. 11.1). Although frequency can be expressed in the SI unit of reciprocal seconds ($s^{-1} \equiv Hz$), for practical applications of spectroscopy this unit is often inconvenient, because the numbers involved are usually large. A common practice is to use the reciprocal of the wavelength instead of the frequency; that is, $1/\lambda$ instead of c/λ. The term *wave number* is then employed, and the usual symbol is $\bar{\nu}$. If λ is expressed in metres, $\bar{\nu} = 1/\lambda$ will be in reciprocal metres (m^{-1}), and multiplication by the velocity of light (2.998×10^8 m s^{-1}) gives ν in reciprocal seconds (s^{-1}). Wave numbers are more often expressed in the units of reciprocal centimetres; 1 cm^{-1} = 100 m^{-1}. Multiplication of $\bar{\nu}$/cm^{-1} by 2.998×10^{10} cm s^{-1} gives the frequency ν/s^{-1}.

Various units have been used for wavelength. The most important of these, along with their SI equivalences are as follows:

	SI Designation
1 angstrom (Å) = 10^{-8} cm = 10^{-10} m	Accepted but not recommended
1 micron (μ) = 10^{-4} cm = 10^{-6} m	Micrometre (μm)
1 millimicron (mμ) = 10^{-7} cm = 10^{-9} m	Nanometre (nm)

The use of the terms *micron* and *millimicron* is now discouraged.

The Energy of Radiation in Emission and Absorption

Spectra can result from either emission or absorption of energy. In order to observe an emission spectrum, the substance must be excited in some way. This may be done for an atomic species by introducing it into a flame or by passing an electric discharge through it. If the emitted light is observed through a spectrometer, the characteristic *emission spectrum* of the substance will be seen. In Section 11.2 we considered the emission spectrum of hydrogen.

The procedure for obtaining an *absorption spectrum* is quite different. An absorption spectrum can be observed by passing continuous radiation, such as white light, through a substance and observing the spectrum with a spectrometer. When this is done, certain wavelengths are missing from the spectrum. For example, if continuous radiation of white light is used, its normal spectrum consists of colors from red to violet, which blend smoothly into one another. After the white light has passed through the substance, however, the absorption spectrum consists of black lines superimposed on the continuous spectrum. These black lines occur because the substance through which the light passed has removed light corresponding to certain wavelengths. The energy of a transition causing such a line is the frequency of the line multiplied by the Planck constant (Eq. 11.23).

It is often important to evaluate the energy associated with absorption in a specific region of the spectrum. Since the energy per molecule ϵ is equal to $h\nu$, we can calculate an energy per mole as follows:

$$E = L\epsilon = Lh\nu = \frac{6.022 \times 10^{23}(\text{mol}^{-1}) \times 6.626 \times 10^{-34}(\text{J s}) \times 2.998 \times 10^{8}(\text{m s}^{-1}) \times 10^{-3}(\text{kJ J}^{-1})}{10^{-9}(\text{m nm}^{-1}) \times \lambda}$$

$$= \frac{1.196 \times 10^5}{\lambda/\text{nm}} \text{ kJ mol}^{-1} \qquad (13.1)$$

EXAMPLE
Calculate the energy in kJ mol^{-1} corresponding to absorption at 400 nm.

SOLUTION
Substitution of $\lambda = 400$ nm into Eq. 13.1 gives 2.99×10^2 kJ mol^{-1}. ■

The amount of energy absorbed by 1 mol of a compound at a particular wavelength has been called an *einstein* and is the energy of 1 mol of photons at that wavelength. In the previous example 1 einstein equals 2.99×10^2 kJ.

The Laws of Lambert and Beer

When spectroscopic studies of substances are combined with measurements of light absorbed and transmitted at various wavelengths, the term **spectrophotometry** is employed. This type of investigation is based on laws propounded by the German mathematician Johann Heinrich Lambert (1728–1777) and the German astronomer Wilhelm Beer (1797–1850). These laws are concerned with the intensities of light absorbed or transmitted when incident light is passed through some material.

Lambert's law states that the proportion of radiation absorbed by a substance is independent of the intensity of the incident radiation. This means that each successive layer of thickness dx of the medium absorbs an equal fraction $-dI/I$ of the radiant intensity I incident upon it. In other words,

$$-\frac{dI}{I} = b\,dx \qquad (13.2)$$

where b is a constant.* Integration of this equation, for passage of light through a distance l, proceeds as follows:

$$\int \frac{dI}{I} = -b \int_0^l dx \qquad (13.3)$$

or

$$\ln I = -bl + g \qquad (13.4)$$

where g is the constant of integration. This constant can be evaluated using the

* The IUPAC Manual gives special names to a number of the parameters that relate to the absorption of radiation. The names that relate to Napierian (natural) logarithms are not much used, but we include them in this footnote for reference. The parameter b that appears in Eq. 13.2 and 13.8 is known as the *Napierian absorption coefficient*. The product bl which appears in Eq. 13.7 and other equations can be written as B and is known as the *Napierian absorbance*.

boundary condition that $I = I_0$ when $l = 0$, where I_0 is the intensity of radiation before passage through the medium. Thus,

$$g = \ln I_0 \tag{13.5}$$

and therefore

$$\ln I = -bl + \ln I_0 \tag{13.6}$$

or

$$\ln \frac{I_0}{I} = bl \tag{13.7}$$

or

$$I = I_0 e^{-bl} \tag{13.8}$$

It is more usual to employ common logarithms, and instead of Eq. 13.7 we have

$$\mathbf{\log_{10} \frac{I_0}{I} = \frac{bl}{2.303} = A} \tag{13.9}$$

The quantity A is known as the *decadic absorbance,* or simply as the **absorbance**. It was formerly called the *extinction* or the *optical density,* but IUPAC now discourages these terms.

The **transmittance** T is the ratio of the intensities of transmitted to incident light:

$$\mathbf{T \equiv \frac{I}{I_0}} \tag{13.10}$$

Thus, from Eq. 13.9,

$$\log_{10} \frac{1}{T} = A \tag{13.11}$$

The **percentage transmittance** $T\%$ is the transmittance multiplied by 100:

$$T\% = 100\, T = \frac{100\, I}{I_0} \tag{13.12}$$

Taking logarithms gives

$$\log_{10} T\% = \log_{10} 100 + \log_{10} T \tag{13.13}$$

$$\mathbf{\log_{10} T\% = 2 - A} \tag{13.14}$$

Beer studied the influence of the concentration of a substance in solution on the absorbance, and he found the same linear relationship between absorbance and concentration as Lambert had found between absorbance and thickness (see Eq. 13.9). Thus, for a substance in solution at a concentration c, **Beer's law** states that

$$\log_{10} \frac{I_0}{I} = A = \text{constant} \times c \tag{13.15}$$

The **Lambert-Beer law** combines Eqs. 13.9 and 13.15 into a single equation

$$A = \log_{10} \frac{I_0}{I} = \epsilon cl \tag{13.16}$$

where ϵ is a constant known as the *absorption coefficient* and l is the light path. The absorption coefficient is usually written as $\epsilon_l^c(\lambda)$ so that if the concentration units are moles per cubic decimeter, the light path is 1 cm, and the wavelength is 430 nm, the coefficient would be written as

$$\epsilon_{1\ \mathrm{cm}}^{1\ M}\ (430\ \mathrm{nm})$$

and would then be called the *molar absorption coefficient*. Since ϵcl is dimensionless, the molar absorption coefficient then has the units $\mathrm{dm^3\ mol^{-1}\ cm^{-1}}$.

EXAMPLE

An aqueous solution of iridine-5′-triphosphate, at a concentration of 57.8 mg $\mathrm{dm^{-3}}$ of trisodium dihydrate (molar mass 586 g $\mathrm{mol^{-1}}$), gave an absorbance of 1.014 with a light path of 1 cm. Calculate the molar absorption coefficient. What would be the absorbance of a 10 μM solution, and what would be the percentage of light transmittance?

SOLUTION

From Eq. 13.16

$$A = 1.014 = \epsilon cl$$

$$c = \frac{57.8 \times 10^{-3}(\mathrm{g\ dm^{-3}})}{586(\mathrm{g\ mol^{-1}})} = 9.86 \times 10^{-5}\ M$$

Thus

$$\epsilon = \frac{1.014}{9.86 \times 10^{-5}(\mathrm{mol\ dm^{-3}}) \times 1(\mathrm{cm})} = 1.028 \times 10^4\ \mathrm{dm^3\ mol^{-1}\ cm^{-1}}$$

For a 10 μM solution,

$$A = \epsilon cl = 1.028 \times 10^4(\mathrm{dm^3\ mol^{-1}\ cm^{-1}}) \times 10 \times 10^{-6}(\mathrm{mol\ dm^{-3}}) \times 1(\mathrm{cm}) = 0.1028$$

The percentage light transmittance $T\%$ is given by Eq. 13.14,

$$\log_{10} T\% = 2 - A = 1.8972$$

and thus

$$T\% = 78.9\%$$

■

The Lambert-Beer equation is always considered to be obeyed exactly. However, apparent deviations are sometimes observed, and it is then necessary to find an explanation in terms of extraneous effects such as dissociation and complex formation.

13.2 Atomic Spectra

Atomic spectra may be observed in different regions of the spectrum, depending on the energies involved in the electronic transitions. The hydrogen spectrum, which we considered in Section 11.2, occurs in the near-ultraviolet, visible, and infrared regions. Sometimes, on the other hand, the differences between atomic energy levels are very small, and lines may only be observed in the microwave region.

The characteristics of atomic spectra depend to a considerable extent on the type of spectrometer used. Simple instruments give what are called *low-resolution* spectra. If more advanced instruments are employed, a *high-resolution* spectrum is observed; what appeared on the simpler instrument to be single lines are split into two or more lines. Several effects are responsible for this splitting; in order of decreasing strength they are

1. Coulombic interactions.
2. Exchange interactions.
3. Spin-orbit interactions.

In addition, splitting can be brought about by an external magnetic field. We shall now consider these effects in a little detail.

Coulombic Interaction and Term Symbols

For a hydrogen atom the energy of the electron depends to a very close approximation only on the principal quantum number, as seen from Eq. 11.184. Thus, an electron has the same energy whether it is in the 2s or the 2p state; the energy is the same for 3s, 3p, and 3d electrons, and so on. This is true for all hydrogenlike atoms (i.e., for atoms or ions having a single electron). When there is more than one electron, however, there is a significant difference in energy for electrons having different second quantum numbers. Thus, the 2p level lies above the 2s level, and 3d above the 3p level, and so on (see Figure 11.20). These differences are due to **Coulombic interactions** between the electrons in the atom.

To take account of the effect of Coulombic interactions the atomic energy levels are described by capital letters, called **term symbols**, which indicate the magnitude of $\boldsymbol{L}$, the total angular momentum. For a single electron the orbital angular momentum is given by

$$\boldsymbol{L} = \sqrt{l(l+1)}\frac{h}{2\pi} \tag{13.17}$$

(see Eq. 11.203 and Section 11.9). For more than one electron in an atom the resultant total angular momentum is

$$\boldsymbol{L} = \Sigma \boldsymbol{l}_i = \sqrt{L(L+1)}\frac{h}{2\pi} \tag{13.18}$$

where $\boldsymbol{l}_i$ refers to the angular momentum vector of each electron in an atom and L, which has only integral values, is the total angular momentum quantum number of the atom. The significance of Eq. 13.18 is that the vector sum of the $\boldsymbol{l}_i$ is constrained to certain values. When $L = 0, 1, 2, 3 \ldots$, the corresponding term symbols are S, P, D, F, . . . by analogy to the one-electron angular-momentum terms.

For hydrogen, with only one electron, $L = l$. The term symbols for hydrogen in its ground and first two excited states are therefore

$$\begin{array}{lll} n = 1 & l = 0 & \text{S} \\ n = 2 & l = 0, 1 & \text{S, P} \\ n = 3 & l = 0, 1, 2 & \text{S, P, D} \end{array} \tag{13.19}$$

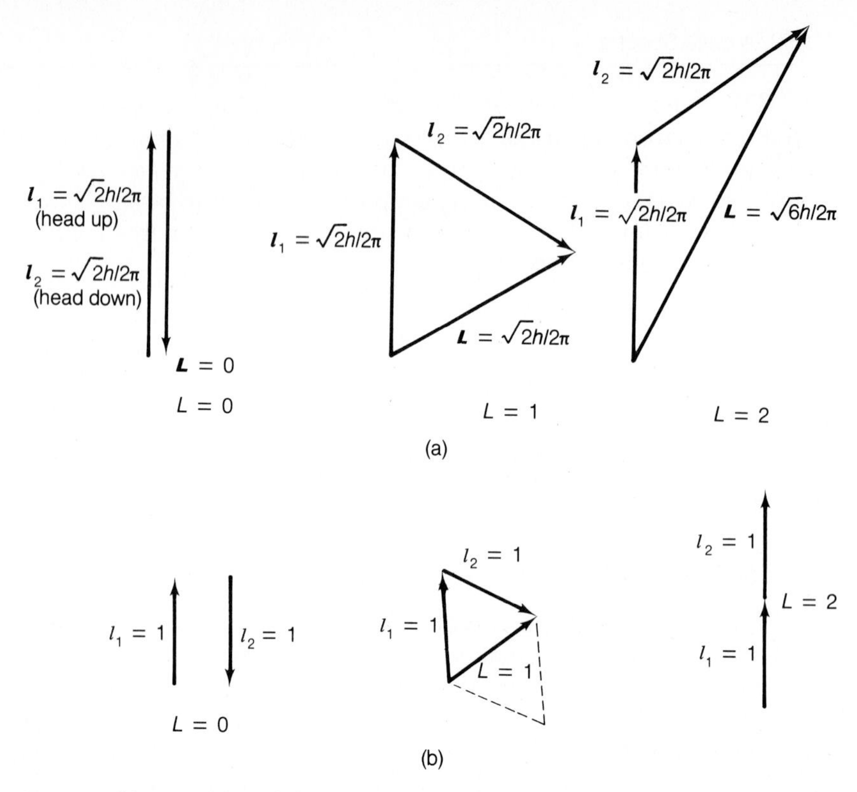

FIGURE 13.1 Vector addition of l_1 and l_2 to give the resultant values of L. (a) In this case $l_1 = l_2 = 1$ and the values of L are 0, 1, and 2. (b) Simplified vector additions for $l_1 = l_2 = 1$ that appear in (a).

For a multielectron system, vector addition is used to find the resulting value of L. We will illustrate this process for two electrons with the same value of l but different values of n. The orbital angular momenta for the individual electrons are assumed to be coupled strongly and the vectors representing the angular momentum of the two electrons, l_1 and l_2, are placed head to tail as shown in Figure 13.1a. Only those angles between l_1 and l_2 are allowed that result in integral values of L. These vectors are cumbersome to use because of the square-root terms, and it is simpler to use schematic vectors that are in direct proportion to the quantum numbers. The simplified technique is demonstrated in Figure 13.1b and by the following example.

EXAMPLE

Determine the allowed values of L for two electrons in an atom where $l_1 = 2$ and $l_2 = 1$.

SOLUTION

Using the same method as in Figure 13.1b, we have

$l_2 = 1$ $l_1 = 2$ $L = 3$ $l_2 = 1$ $l_1 = 2$ $L = 2$ $l_1 = 2$ $l_2 = 1$ $L = 1$

The three values allowed for L are thus 1, 2, and 3. ■

Exchange Interaction: Multiplicity of States

After the Coulombic interaction, the greatest splitting of spectral lines arises from what is known as **exchange interaction** or, perhaps better, as **spin correlation**. This effect arises from the fact that the electrons may be thought of as spinning (Section 11.11) and that electrons with the same spin interact with one another differently from electrons with opposite spins. Consider, for example, the helium atom, in which there are two electrons. If these are both in the 1s orbital, the Pauli principle requires them to have opposite spins. Suppose, however, that one is in the 1s orbital and the other in the 2s orbital. There are then two possibilities:

1. They may both have the same spin
2. The spins may be opposed.

According to Hund's rule (Section 11.12) for equivalent electrons (same n and l), the state in which the spins are the same is lower in energy than that in which the spins are opposed. Two different energy states therefore arise, according to whether the spins are the same or are opposed.

The resultant spin quantum number for more than one electron is represented by the symbol S (not to be confused with the term symbol). The individual spin quantum numbers are $+\frac{1}{2}$ and $-\frac{1}{2}$ (see Section 11.11, especially Eq. 11.210), and we obtain the resultant S in much the same way as we obtained the resultant orbital quantum number L from the l values for the individual electrons (Figure 13.1). Figure 13.2 shows a few examples. The case of a single electron is shown in Figure 13.2a, and of course the resultant spin quantum number is the spin quantum number of the individual electron; thus $m_s = \frac{1}{2}$ and $S = \frac{1}{2}$. If there are two electrons, there are two possibilities, shown in Figure 13.2b. If the spins are opposed, the resultant spin S is zero; whereas if the electrons have the same value of m_s, the resultant spin S is $\frac{1}{2} + \frac{1}{2} = 1$. The case of two electrons of opposite spins is of particular interest, because this is what arises when two electrons in an atom have the same three orbital quantum numbers, n, l, and m. By the Pauli principle (Section 11.11) they must have opposite spins, and it follows that electrons in a completed shell or in completed subshells do not contribute to the resultant spin quantum number S.

The case of two electrons that are not restricted in this way is also of special interest. We considered this problem for molecules in Section 12.2, where we saw that there are four possible wave functions, three of them symmetric and one antisymmetric:

$$\left.\begin{array}{lll} 1.\ \alpha(1)\alpha(2) & m_{s1} + m_{s2} = +1 \\ 2.\ \alpha(1)\beta(2) + \alpha(2)\beta(1) & m_{s1} + m_{s2} = 0 \\ 3.\ \beta(1)\beta(2) & m_{s1} + m_{s2} = -1 \end{array}\right\} \quad S = 1$$

$$4.\ \alpha(1)\beta(2) - \alpha(2)\beta(1) \qquad m_{s1} + m_{s2} = 0 \qquad S = 0$$

Two electrons with spins combine to give a resultant spin S that can be 1 or 0, and the components $m_{s1} + m_{s2}$ can be 1, 0, or -1. It is obvious that the spin functions 1 and 3 must belong to $S = 1$, because only $S = 1$ can have components $+1$ or -1. Both of these functions are symmetric with respect to an interchange of electrons, as is also function 2. This function, having a spin component 0, also belongs to the

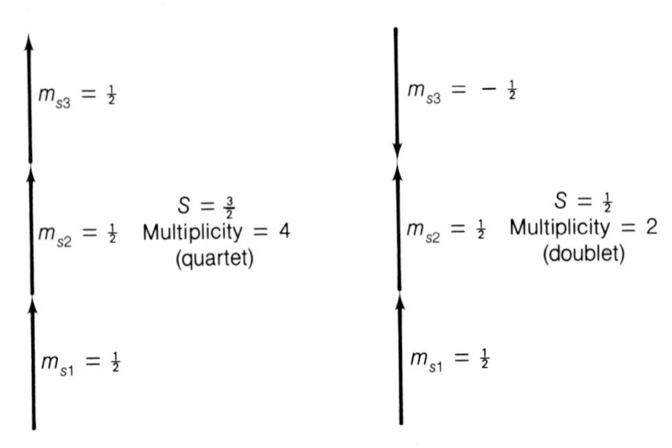

FIGURE 13.2 Vector diagrams showing how the resultant spin quantum number S is obtained from the spin quantum numbers for the individual electrons.

group of three functions having $S = 1$. The remaining function 4, which is antisymmetric, has $S = 0$.

Since there are three wave functions corresponding to two electrons having $S = 1$, we say that the state has a **multiplicity** of 3 or that it is a *triplet* state. In general the multiplicity is related to the resultant spin by the formula

$$\text{multiplicity} = 2S + 1 \tag{13.20}$$

In this particular example $S = 1$ and the multiplicity is 3; for $S = 0$ the multiplicity is 1, and we say that the state is a *singlet* state.

Other situations are also represented in Figure 13.2. If we have three electrons, there can be two resultant spin quantum numbers, $\frac{3}{2}$ and $\frac{1}{2}$, as shown in Figure 13.2c. If $S = \frac{3}{2}$, the multiplicity is 4 (a *quartet*), while if $S = \frac{1}{2}$, the multiplicity is 2 (a *doublet*).

The multiplicity $2S + 1$ is indicated in the term symbol as a left-hand superscript. Thus for the ground state of hydrogen with $S = \frac{1}{2}$, the multiplicity is $2(\frac{1}{2}) + 1 = 2$. The term symbol is thus ^2S and is read *doublet S*.

EXAMPLE

Determine the multiplicities possible for an atom in the configuration $1s^22s^22p^2$.

SOLUTION

Electrons in a complete shell or in completed subshells do not contribute to the total spin angular momentum. Consequently, only the interaction of the spins of the two

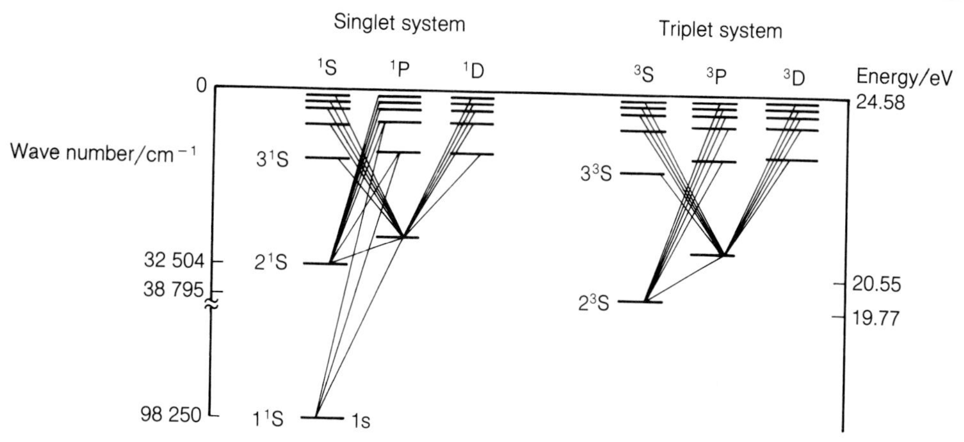

FIGURE 13.3 Simplified energy-level diagram for helium showing the singlet and triplet term systems.

p electrons need be considered. The resultant values of S are 1 and 0, giving rise to two states: $2(1) + 1 = 3$, a triplet state; and $2(0) + 1 = 1$, a singlet state. ■

The way in which exchange interaction (spin correlation) brings about the splitting of energy levels can be illustrated by the simplest atom that shows this effect, namely helium. The term diagram of helium is shown in Figure 13.3, in which the terms are divided into two sets. At one time these were thought to arise from two distinct forms of helium, *parahelium* and *orthohelium*, but this is now known not to be the case. The first set consists of *singlets* (multiplicity = 1) and the other of *triplets* (multiplicity = 3). The ground state of helium is $1s^2$ with $L = 0$ ($l_1 = 0$, $l_2 = 0$) and $S = 0$ since the electrons must have opposite spins. Hence its ground-state term symbol is ^{1}S.

If one electron is promoted to $n = 2$, two possibilities exist, $1s^12s^1$ and $1s^12p^1$. The first of these has $L = 0$ and is an S state. The second has $L = 1$ and is a P state. Since in both states the electrons may have either $m_s = \pm\frac{1}{2}$, the value of S is 0 or 1, and ^{1}S, ^{3}S, ^{1}P, and ^{3}P states are formed. The S terms lie below the P terms of the same principal quantum number, and by Hund's rule the triplet states lie below the singlets of the corresponding n and l values. The electrostatic energies of two such states may in fact be written as

$$\begin{array}{ll} {}^1\text{S} & E_1 = J + K \\ {}^3\text{S} & E_3 = J - K \end{array} \tag{13.21}$$

where J is the Coulombic integral and K is the exchange integral. These integrals are analogous to those we encountered for molecules in Section 12.3. These energy integrals are usually difficult to calculate from quantum-mechanical theory, but spectroscopic measurements can provide very accurate values for them.

Spin-Orbit Interactions

The effects that we have just considered, Coulombic interactions between electrons and exchange interactions between electrons, do not apply to the hydrogen atom, in which there is only one electron. We have seen that to a very good approximation

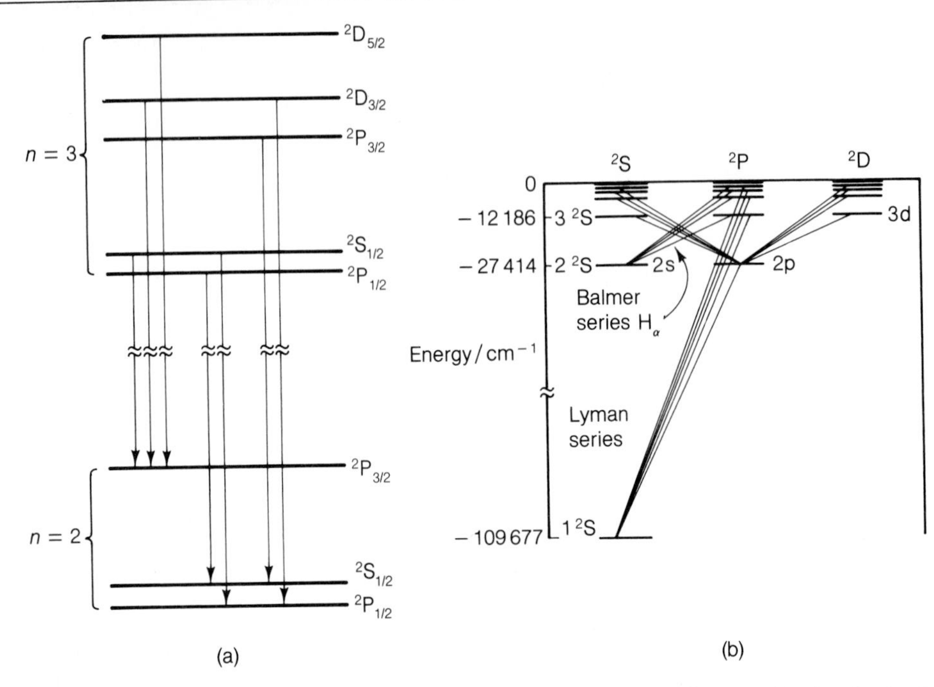

FIGURE 13.4 (a) States associated with the H_α line ($n_1 = 2 \leftrightarrow n_2 = 3$) of the Balmer series for the hydrogen atom spectrum, as detected by high-resolution spectroscopy. (b) A simplified energy diagram for the hydrogen atom, showing various transitions. This type of diagram is known as a Grotrian diagram.

the energy levels of the hydrogen atom depend only on the principal quantum number n and not on the second quantum number l. However, if measurements are made with a spectrograph of very high resolving power, the situation is not quite so simple. Some of the lines found in a low-resolution instrument are then split into component lines. For example, the H_α line of the Balmer series (Section 11.2 and Figure 11.8) corresponds to transitions between $n = 2$ and $n = 3$. In an ordinary spectrometer it appears as a single line, but in a high-resolution instrument it is split into seven lines, corresponding to transitions between the states shown in Figure 13.4a. Lines in other series, such as the Lyman series, are also split, and some of the transitions are shown in Figure 13.4b.

These splittings are smaller than those resulting from Coulombic and exchange interactions between electrons. They are due to a magnetic coupling between the magnetic moment associated with the spin of the electron and the magnetic moment associated with the orbital motion of the electron. This coupling is referred to as a **spin-orbit interaction** and it can arise in two different ways. In hydrogen, $\boldsymbol{l}$ and $\boldsymbol{s}$ are coupled in such a way that there is formed a new *inner quantum number* j with the values

$$j = l \pm s \tag{13.22}$$

This quantum number j describes the allowed energy levels associated with the total angular momentum, both spin and orbital. Since m_s can only have the values $\pm\frac{1}{2}$, for a P state ($l = 1$) only two possibilities exist, namely

$$j = l + s = \tfrac{3}{2} \quad \text{and} \quad j = l - s = \tfrac{1}{2} \tag{13.23}$$

The value of j is written as a right-hand subscript on the term symbol. Corresponding to Eq. 13.19, we have

$$n = 1: \quad {}^2S_{1/2}$$
$$n = 2: \quad {}^2S_{1/2},\ {}^2P_{1/2},\ {}^2P_{3/2} \tag{13.24}$$
$$n = 3: \quad {}^2S_{1/2},\ {}^2P_{1/2},\ {}^2P_{3/2},\ {}^2D_{3/2},\ {}^2D_{5/2}$$

Only certain transitions occur and these are given by a set of **selection rules,** namely

$$\Delta n = \pm 1, \pm 2, \ldots, \quad \Delta l = \pm 1 \tag{13.25}$$

and

$$\Delta j = 0, \pm 1 \tag{13.26}$$

Transitions are allowed between S levels and P levels. The ${}^2D_{3/2}$–${}^2P_{3/2}$ ($\Delta j = 0$) transition* between the closest pair of these levels is lowest in frequency and the ${}^2D_{5/2} \leftarrow {}^2P_{3/2}$ ($\Delta j = +1$) transition is next lowest. The lines for these are close together because the separation between the ${}^2D$ states is very small. The next line, farther away, is for the transition ${}^2D_{3/2}$–${}^2P_{1/2}$ ($\Delta j = \pm 1$) but the transition ${}^2D_{5/2}$–${}^2P_{1/2}$ is not allowed since for this case $\Delta j = \pm 2$. When three lines arise from transitions between doublet levels, we speak of a *compound doublet* spectrum. Thus in the ground-state spectrum of hydrogen, the Lyman spectrum, practically every line is a doublet. This is difficult to observe experimentally with hydrogen, but it is rather easily seen in sodium where the yellow line, the strongest line in the spectrum, has two maxima, one at 589.0 nm and the other at 589.6 nm. This splitting of the yellow sodium line into two separate lines was in fact the first spectroscopic proof of the existence of electron spin.

The Vector Model of the Atom

The vector additions of angular momenta that we have considered earlier in this section are examples of a more general method for dealing with the energy levels of molecules and, therefore, of interpreting spectra. We have seen that individual electrons have orbital angular momenta, which couple together to give a resultant orbital angular momentum $\boldsymbol{L}$. Each individual l_i precesses around the resultant vector, much as the angular momentum vector of Figure 11.18 interacts with an external magnetic field. Similar considerations apply to the spin angular momenta. These couple together to form a resultant $\boldsymbol{S}$, which interacts with the resultant orbital angular momentum $\boldsymbol{L}$. From the standpoint of an orbital electron the nucleus is rotating around it, just as the sun appears to us to rotate around the earth. The magnetic field generated by the nucleus causes an electron to align in one of two possible directions, and there is a coupling of the resultant spin angular momentum with the resultant orbital angular momentum.

The details of the way in which coupling occurs vary somewhat, and it is possible to distinguish two extreme situations, according to whether the nuclear charge is small (i.e., the atoms are light) or is large (the atoms are heavy). The way

---

*The convention employed in spectroscopy is always to write the *upper* state first, with the arrow pointing in the appropriate direction; ← represents absorption and → represents emission. If either emission or absorption is referred to, a simple dash (–) is used.

in which the coupling occurs for light atoms is described by **Russell-Saunders coupling,** named in recognition of the pioneer work in this area by the American astronomer Henry Norris Russell (1877–1957) and physicist Frederick Albert Saunders (1875–1963).* In this scheme, the individual spin vectors of each electron $\boldsymbol{s}_i$ form a resultant spin vector $\boldsymbol{S}$. In the same way that we treated the vector addition of the $\boldsymbol{l}_i$ in Figure 13.1b, we may simplify the analysis by requiring that the spin quantum numbers add to give S, the total spin quantum number, with integral or half-integral values:

$$\sum_i m_{s_i} = S \tag{13.27}$$

where vector addition is implied. For example, two spins of $+\frac{1}{2}$ can couple to yield $S = 0$ or 1, whereas three spins could give $S = \frac{1}{2}$ or $\frac{3}{2}$ as shown in Figure 13.2.

The individual orbital angular momentum quantum numbers l_i couple to form a resultant L which is restricted to integral values as seen earlier in Eq. 13.18.

The vectors represented by $\boldsymbol{L}$ and $\boldsymbol{S}$ exert magnetic forces on each other, and they couple to form a resultant total angular momentum vector $\boldsymbol{J}$ for the atom. Thus

$$\boldsymbol{J} = \boldsymbol{L} + \boldsymbol{S} \tag{13.28}$$

where the values of $\boldsymbol{J}$ are quantized through the expression

$$\boldsymbol{J} = \sqrt{J(J+1)}\frac{h}{2\pi} \tag{13.29}$$

and the inner quantum number J may have integral or half-integral values. In the absence of an external field, $\boldsymbol{L}$ and $\boldsymbol{S}$ precess around their resultant $\boldsymbol{J}$ in order to keep the total angular momentum a constant as shown in Figure 13.5. Because of the coupling, $\boldsymbol{J}$ precesses relatively more slowly around the field direction.

In Russell-Saunders coupling, which applies to light atoms, the coupling to form $\boldsymbol{L}$ and to form $\boldsymbol{S}$ is strong. This means that the different values of L and S

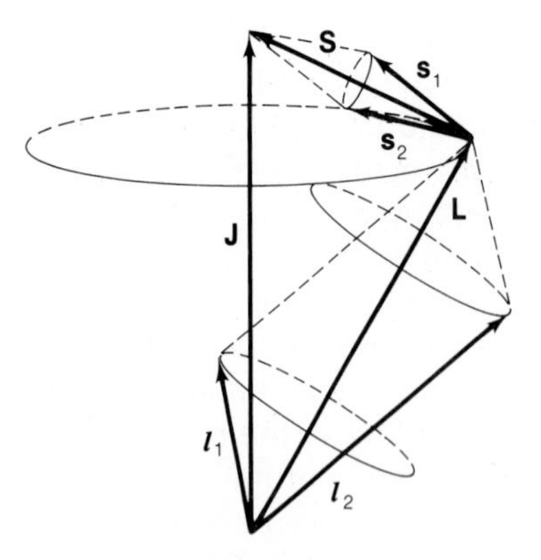

FIGURE 13.5 Precession of $\boldsymbol{L}$ and $\boldsymbol{S}$ about $\boldsymbol{J}$ for Russell-Saunders coupling. The individual $\boldsymbol{l}$ couple and $\boldsymbol{s}$ couple to form resultant $\boldsymbol{L}$ and $\boldsymbol{S}$ *before* coupling to form $\boldsymbol{J}$.

*H. N. Russell and F. A. Saunders, *Astrophys. J.*, *61*, 38 (1925).

represent states of considerably different energies. On the other hand, the coupling of $\boldsymbol{L}$ and $\boldsymbol{S}$ to form $\boldsymbol{J}$ is weak, resulting in $\boldsymbol{J}$ values that do not differ much in energy.

The selection rules that apply to transitions in Russell-Saunders coupling can be summarized as

$$\Delta S = 0 \tag{13.30}$$

$$\Delta L = 0, \pm 1 \tag{13.31}$$

$$\Delta J = 0, \pm 1 \tag{13.32}$$

There is a further restriction, known as the *Laporte rule,* that even terms only combine with odd terms:

$$\mathrm{g} \not\leftrightarrow \mathrm{g} \qquad \mathrm{u} \not\leftrightarrow \mathrm{u} \qquad \mathrm{g} \leftrightarrow \mathrm{u} \tag{13.33}$$

In addition

$$J = 0 \not\leftrightarrow J = 0 \tag{13.34}$$

The underlying reason behind these selection rules is that the photon carries one unit of spin angular momentum; there therefore must be a change of angular momentum when a photon is absorbed or emitted.

For heavy atoms (i.e., for atoms of high atomic number) there is another type of coupling, known as *j-j* coupling. The spin-orbit interactions are now stronger and cause the individual $\boldsymbol{l}$ and $\boldsymbol{s}$ vectors for each electron to couple and to form resultant $\boldsymbol{j}$ vectors. These in turn couple to form a total $\boldsymbol{J}$ for the atom. This *j-j* coupling may be explained in terms of enhanced spin-orbit interactions caused by the shielding of the outermost electrons by all the remaining electrons. In effect, the electrostatic potential changes very rapidly as a function of distance from the nucleus as a valence electron penetrates deeply into the shielding electron cloud. As a result of an increase in the number of electrons, there is an increase in the electron velocity and in the average magnetic field felt by an electron in a heavy atom. This type of spin-orbit interaction is strong even in halogens since their valence electrons may also contribute to the increased magnetic field. Thus for heavy atoms the quantum numbers S and L lose their significance, and the selection rules for Russell-Saunders coupling fail. Instead, singlet-triplet transitions ($\mathrm{S} = 0 \leftrightarrow \mathrm{S} = 1$) are permitted in heavier atoms, whereas they are forbidden in light atoms.

The value of J is written as a right-hand subscript to the term symbol. Since a closed shell is spherically symmetric, its contribution to the total angular momentum is zero; for a closed shell, therefore, $L = 0$, $S = 0$, and $J = 0$.

So far the discussion has been concerned with nonequivalent electrons. For two such electrons, with $L > \mathrm{S}$, the permitted values, which must be positive, are

$$L = (l_1 + l_2), (l_1 + l_2 - 1), \ldots, (l_1 - l_2) \tag{13.35}$$

$$S = (s_1 + s_2), (s_1 - s_2) = 1, 0 \tag{13.36}$$

$$J = (L + S), (L + S - 1), \ldots, (L - S) \tag{13.37}$$

A procedure can be worked out for equivalent electrons (having the same value of n and l), but it is rather lengthy and outside the scope of this book.

The Effect of an External Magnetic Field*

In 1896 the Dutch physicist Pieter Zeeman† (1865–1943) discovered that spectral lines were split into component lines by a magnetic field. This effect is sometimes due to the interaction of the orbital magnetic moment of the atom with the magnetic field and is then called the **Zeeman effect.** A more common effect in a magnetic field is the *anomalous Zeeman effect,* in which the electron spin is also involved.

We have seen in Section 11.9 that an orbiting electron has an angular momentum and also a magnetic moment which is in the opposite direction to the angular momentum. We can arrive at a simple quantitative interpretation of the normal Zeeman effect by considering an electron moving in a circular orbit of radius r. The electric current that travels around this orbit is the charge that passes a given point per unit time; it is therefore the product of the charge $-e$ and the frequency with which the electron passes the given point. The circumference of the orbit is $2\pi r$, and if the electron moves with velocity u, it performs $u/2\pi r$ revolutions in unit time. The current is, therefore,

$$I = -\frac{eu}{2\pi r} \tag{13.38}$$

The magnetic moment μ_l produced by this current is the product of the current and the area πr^2 of the orbit:

$$\mu_l = -\frac{eu}{2\pi r} \cdot \pi r^2 = -\frac{eur}{2} \tag{13.39}$$

The direction of this magnetic moment is given by the left-hand rule (Section 11.9, and especially Figure 11.18a).

The angular momentum is in the opposite direction, and its direction is given by the right-hand rule (Figure 11.18a). Its magnitude is

$$L = m_e ur \tag{13.40}$$

where m_e is the mass of the electron. The ratio of the magnetic moment to the angular momentum is known as the **gyromagnetic ratio** or the **magnetogyric ratio** and is given the symbol γ:

$$\gamma \equiv \frac{\boldsymbol{\mu}_l}{\boldsymbol{L}} = \frac{-\boldsymbol{eur}/2}{\boldsymbol{m}_e\boldsymbol{ur}} = -\frac{e}{2\boldsymbol{m}_e} \tag{13.41}$$

In Section 11.9 we considered the orientation of the orbital angular momentum in a magnetic field, and we have seen (Eq. 11.200) that the component L_z of the angular momentum along the direction of the field (the Z axis) is

$$L_z = m_l \frac{h}{2\pi} \tag{13.42}$$

---

*This section could be omitted on first reading. However, some of the results are used in Sections 13.10 and 13.11.

†P. Zeeman, *Proc. Acad. Sci. Amsterdam, 181,* 242 (1896); *Phil. Mag., 43,* (1897); *44,* 55, 255 (1897).

where m_l is the magnetic quantum number. It then follows that the corresponding component $\mu_{l,z}$ of the magnetic moment along the Z axis is

$$\mu_{l,z} = m_l \frac{h}{2\pi} \cdot \frac{e}{2m_e} = m_l \frac{eh}{4\pi m_e} \tag{13.43}$$

(We can drop the negative sign since m_l can have positive and negative values, namely $l, l - 1, \ldots, 1, 0, -1, \ldots, -l$, where l is the azimuthal quantum number.) The quantity $eh/4\pi m_e$, which is the product of $h/2\pi$ and the gyromagnetic ratio, plays an important part in the theory; it is called the **Bohr magneton** and given the symbol μ_B. Insertion of the values of e, h, and m_e leads to

$$\boldsymbol{\mu_B = \frac{eh}{4\pi m_e} = 9.274 \times 10^{-24}\ \mathrm{A\ m^2}}$$

The Z component of the magnetic moment is thus

$$\mu_{l,z} = m_l \mu_B \tag{13.44}$$

In order to treat the energy of a magnetic dipole in a magnetic field it is simplest to work not with the magnetic field strength H (SI unit: ampere/metre, $\mathrm{A\ m^{-1}}$) but with the **magnetic flux density*** B. The SI unit of magnetic flux density is $\mathrm{kg\ s^{-1}\ A^{-1}}$ ($\equiv \mathrm{V\ s\ m^{-2}}$), and this unit is known as the **tesla,**† for which the symbol is T. Use of the magnetic flux density produces a considerable simplification, since the energy of a magnetic moment $\mu_{l,z}$ in a field of magnetic flux density B is the product $\mu_{l,z}B$:

$$E = \mu_{l,z} B \tag{13.45}$$

Insertion of the expression for $\mu_{l,z}$(Eq. 13.44) thus gives

$$E = m_l \frac{eh}{4\pi m_e} B = m_l \mu_B B \tag{13.46}$$

Since the unit of E is the joule (J) and that of B is the tesla (T), the unit of μ_B is most conveniently written as $\mathrm{J\ T^{-1}}$, which is equivalent to $\mathrm{A\ m^2}$.

Since there are $2l + 1$ possible values of m_l, it follows from Eq. 13.46 that the magnetic field splits each energy level of the atom into $2l + 1$ components. The energy separation between adjacent components is

$$\Delta E = \mu_B B \tag{13.47}$$

and is independent of m_l.

EXAMPLE

Calculate the energy splitting when the hydrogen atom is placed in a magnetic field of 1 T (10^4 gauss). Calculate the wavelength splitting expected for the $n = 2$ to $n = 3$ transition in the Balmer series (the 656.2-nm line).

---

* The magnetic flux density B is the field strength H multiplied by the permeability μ; see Appendix A.

† The old unit, the gauss (G), is now discouraged; $1\ \mathrm{T} \equiv 10^4\ \mathrm{G}$.

SOLUTION
For such a field, we have from Eq. 13.47,

$$\Delta E = \mu_B B = \frac{ehB}{4\pi m_e} = \frac{1.60 \times 10^{-19}(\text{C})\ 6.626 \times 10^{-34}(\text{J s})\ 1(\text{T})}{4\pi(9.11 \times 10^{-31})(\text{kg})}$$

$$= 9.274 \times 10^{-24}\ \text{J} \quad \text{or} \quad 5.79 \times 10^{-5}\ \text{eV}$$

The energy of the $n = 2$ level is $(1/2^2)(13.6\ \text{eV}) = 3.40\ \text{eV}$, and that for the $n = 3$ level, $(1/3^2)(13.6\ \text{eV}) = 1.51\ \text{eV}$, a difference of about 1.9 eV. The fractional change in energy is

$$\frac{5.8 \times 10^{-5}}{1.9} \approx 3 \times 10^{-5}$$

Thus, for the 656.2-nm line, we expect a splitting of

$$\Delta\lambda = (656.2\ \text{nm})(3 \times 10^{-5}) = 0.02\ \text{nm}$$

This is easily observable with modern grating spectrometers. ■

For a multielectron atom the treatment is similar, but l and Δm_l are now replaced by L and ΔM_L, the resultant values. The normal Zeeman effect consists of lines arising from transitions between the $2L + 1$ new energy levels produced by the magnetic field. However, the normal Zeeman effect applies only to transitions between singlet states, for which $S = 0$, and as a consequence $L = J$. A $^1S_0$ term with $L = J = 0$ remains unsplit, but a $^1P_1$ term with $L = J = 1$ has $M_L = +1, 0, -1$ and splits into three levels. The transition $^1P_1 \rightarrow {}^1S_0$ thus gives a splitting into three lines, as shown in Figure 13.6a. Because the energy levels are equally spaced and the selection rule $\Delta M_L = 0, \pm 1$ applies, the transition $^1D_2 \rightarrow {}^1P_1$ also gives three lines, as shown in Figure 13.6b.

In a magnetic field most atoms behave in a more complicated way than predicted on the basis of the normal Zeeman effect. The behavior, known as the **anomalous Zeeman effect,** arises when $S > 0$; the Zeeman splitting is then no longer the same in different terms of the atom. The electron generates a spin magnetic momentum vector, as shown in Figure 11.19. On account of this vector and the coupling of $\boldsymbol{L}$ and $\boldsymbol{S}$ to give a resultant $\boldsymbol{J}$, Eq. 13.44 is modified to

$$\mu_z = g_J M_J \mu_B \tag{13.48}$$

In this equation the orbital quantum number m_l has become M_J, and the factor g_J has been introduced. This is known as the **Landé-g-factor,** after the German-American physicist Alfred Landé (b. 1888).* This factor g_J allows for the interactions between $\boldsymbol{L}$ and $\boldsymbol{S}$, and for Russell-Saunders coupling is given by

$$g_J = 1 + \frac{J(J + 1) + S(S + 1) - L(L + 1)}{2J(J + 1)} \tag{13.49}$$

The treatment is more complicated for j-j splitting, but g_J normally has a value between 0 and 2.

---

* A. Landé, *Z. Physik*, *5*, 234 (1921); *15*, 189 (1923).

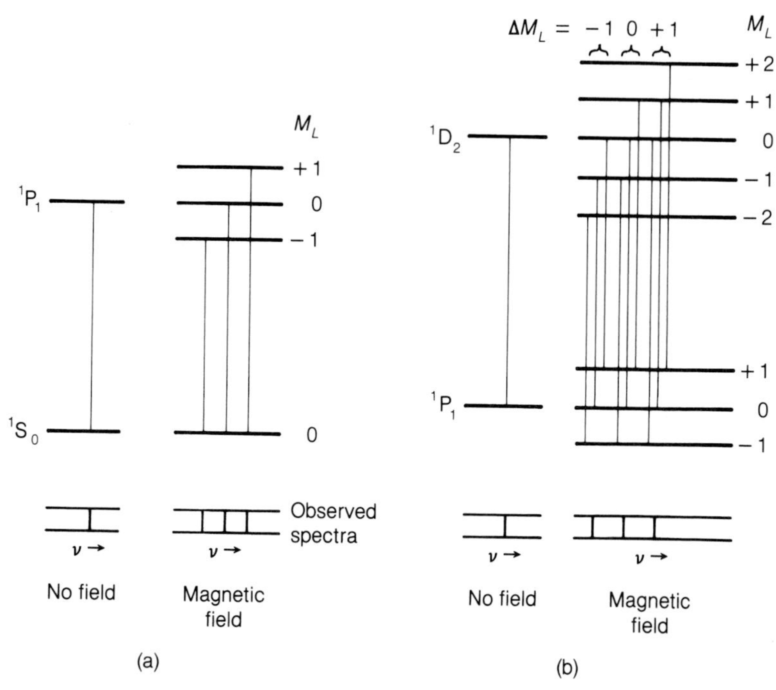

FIGURE 13.6 The normal Zeeman effect, arising from the interaction of the orbital angular momentum with a magnetic field. (a) The $^1P_1$—$^1S_0$ transition, where the $^1P_1$ state is split into three levels so that three Zeeman lines are observed. (b) The $^1D_2$—$^1P_1$ transition. Because the split levels are equally spaced, there are still only three Zeeman lines.

The energy of a magnetic dipole μ_z in a magnetic field is

$$E = \mu_z B \tag{13.50}$$

and therefore each energy level in the atom is split into the values

$$E = g_J M_J \mu_B B \tag{13.51}$$

For a ^3S state the experimental value of g_J is 2.0023, and the splitting of the energy levels in the anomalous Zeeman effect is therefore about twice the value for the normal Zeeman effect. For other terms the g_J values are as follows:

$^3P_0$: 0

$^3P_1$ and $^3P_2$: $\frac{3}{2}$

Figure 13.7 shows the anomalous Zeeman effect for a $^3P_2$–$^3S_1$ transition.

Matters become more complicated at high magnetic fields. If the separations between Zeeman components exceed the separation arising because of the fine structure, the magnetic coupling between $\boldsymbol{J}$ and the field exceeds that between $\boldsymbol{L}$ and $\boldsymbol{S}$. In this case, $\boldsymbol{L}$ and $\boldsymbol{S}$ are uncoupled and precess independently about the field direction. The spectral lines now revert back to the normal triplet but are split

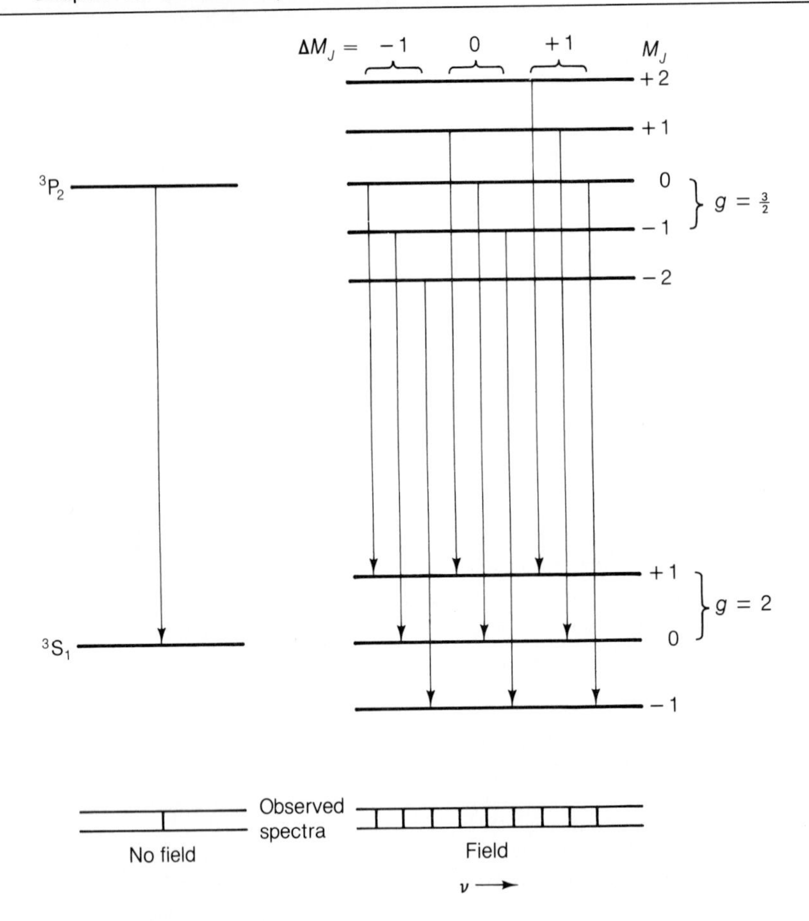

FIGURE 13.7 The anomalous Zeeman effect for the transition $^3P_2$—$^3S_1$.

into closely spaced components. This is known as the **Paschen-Back effect**, named after the German physicists Friedrich Paschen (1865–1940) and E. Back.*

EXAMPLE
Into how many levels will a $^3D_2$ level split in a magnetic field? If the field is 4.0 T, calculate the separation between neighboring energy levels.

SOLUTION
For the $^3D_2$ level,

$$J = 2, \quad L = 2, \quad \text{and} \quad S = 1$$

and the M_J values are 2, 1, 0, −1, and −2; there is therefore a splitting into five levels.

The Landé g factor is, according to Eq. 13.49,

$$g_J = 1 + \frac{(2 \times 3) + (1 \times 2) - (2 \times 3)}{2 \times 2 \times 3} = 1.167$$

*F. Paschen and E. Back, *Ann. Physik, 39,* 929 (1912); *40,* 960 (1913); *Physica, 1,* 261 (1921).

The energy levels are given by Eq. 13.51:

$$E = 1.167 \times 9.273 \times 10^{-24}(\text{J T}^{-1}) \times 4.0(\text{T})M_J$$

where M_J can have the aforementioned values. The separation between neighboring levels is thus

$$E = 1.167 \times 9.273 \times 10^{-24} \times 4.0 \text{ J} = 4.329 \times 10^{-23} \text{ J}$$

The separation in cm^{-1} is obtained by dividing by hc with $c = 2.998 \times 10^{10}$ cm s^{-1}:

$$\frac{E}{hc} = \frac{4.329 \times 10^{-23} \text{ (J)}}{6.626 \times 10^{-34} \text{ (J s)} \times 2.998 \times 10^{10} \text{ (cm s}^{-1})}$$

$$= 2.18 \text{ cm}^{-1}$$

■

13.3 Pure Rotational Spectra of Molecules

The energy of an atom can only be electronic energy, and atomic spectra are due to transitions between different electronic states. Molecular spectra may also involve transitions between electronic states, but energy can also reside in molecules in the form of rotational and vibrational energy. As a result, molecular spectra are considerably more complicated than atomic spectra, because in a molecule a transition may involve simultaneous changes in electronic, vibrational, and rotational energy. However, some simplification is brought about by the fact that the amount of energy involved in a simple rotational transition is considerably less than that in a change of vibrational state, which in turn involves less energy than a change of electronic state. An indication of the different energies and of the spectral regions where each type of transition occurs is given in Table 13.1 The relationships between the different types of motion are illustrated in Figure 13.8, in which rotational levels are superimposed on vibrational levels, and vibrational levels are superimposed on electronic levels.

In representing the situation as we do in Figure 13.8 we are making use of the **Born-Oppenheimer approximation** (see also Section 12.1). When we try to solve the exact Schrödinger equation, which takes into account different forms of energy,

TABLE 13.1 Types of Optical Spectra

Spectroscopic Region	Approximate Range: Frequency s^{-1}	Approximate Range: Wave Number cm^{-1}	Approximate Range: Energy kJ mol^{-1}	Types of Molecular Energy	Information Obtained
Microwave and far infrared	10^9–10^{12}	0.03–30	4×10^{-4}–0.4	Rotation	Interatomic distances
Infrared	10^{12}–10^{14}	30–3000	0.4–40	Vibration and rotation	Interatomic distances and force constants of bonds
Visible and ultraviolet	10^{14}–10^{16}	3×10^3–3×10^5	40–4000	Electronic, vibration, and rotation	Electronic energy levels, bond dissociation energies, force constants of bonds, and interatomic distances

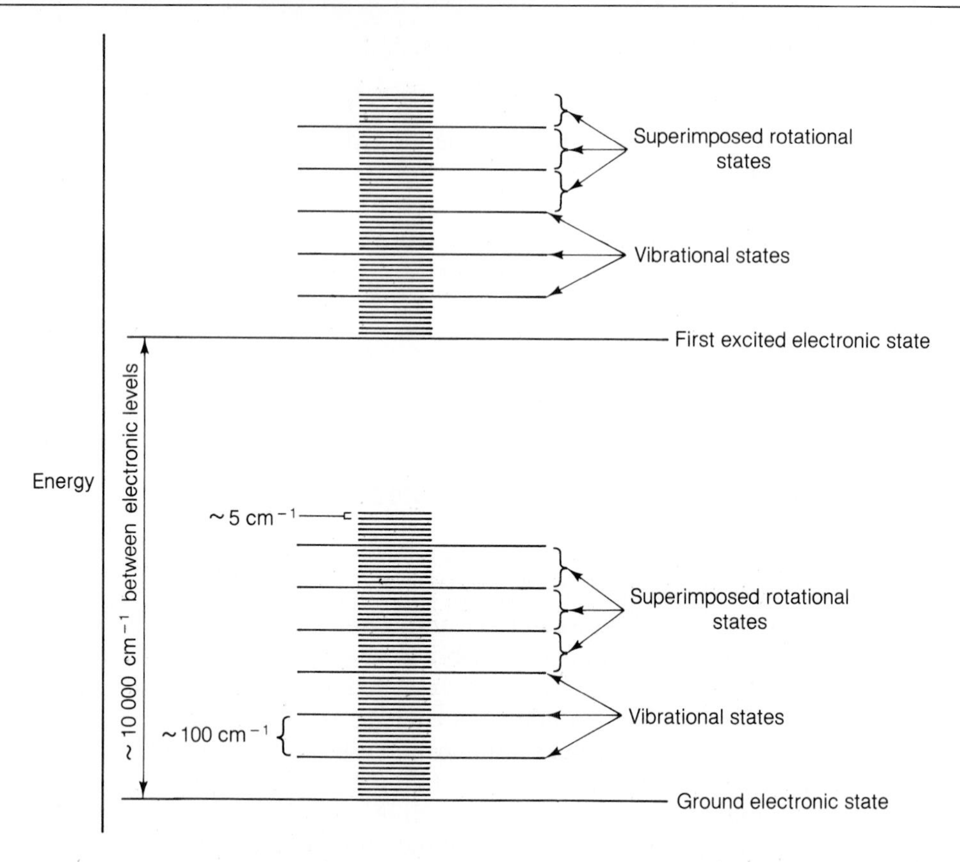

FIGURE 13.8 Typical energy separations for rotational, vibrational and electronic levels. As permitted by the Born-Oppenheimer approximation, the energy is separated into the three types, so that rotational levels are superimposed on vibrational levels and vibrational levels on electronic levels.

the energy cannot be separated into rotational, vibrational, and electronic contributions. However, Max Born and J. R. Oppenheimer were able to show that by neglecting certain very small terms in the equation the separation can be made, so that the total energy E is the sum of the electronic energy E_e, the vibrational energy E_v, and the rotational energy E_r. This means, for example, that in dealing with electronic energy we can take the distance between the nuclei to be the equilibrium distance, as if the molecule were not vibrating. Similarly, in dealing with rotations we can consider only the equilibrium internuclear distances when we calculate moments of inertia.

If we make spectroscopic observations in the microwave and far-infrared regions, the only transitions that occur are in the rotational state of the molecule. *Pure rotational spectra* observed in those regions of the spectrum are therefore relatively simple. If we work in the near infrared, vibrational transitions occur and superimposed are rotational transitions since there is ample energy to bring these about; a near-infrared spectrum is therefore a *vibrational-rotational spectrum* and is a good deal more complicated than a pure rotational spectrum. In the visible and ultraviolet regions electronic transitions occur, and at the same time there are rotational and vibrational transitions. *Electronic spectra* are therefore even more complex than the vibrational-rotational spectra.

We will deal with pure rotational spectra first. These involve transitions between different rotational states, but there is an important restriction; rotational transitions are only observed in the spectrum if the molecule has a permanent dipole moment. The reason for this is that the rotational motion must involve an oscillating dipole, which can interact with an electromagnetic field; otherwise there can be no absorption or emission of radiation. This means, as far as *linear* molecules are concerned, that molecules with a center of symmetry (point group $D_{\infty h}$, see Figure 12.17a), such as N_2 and C_2H_2, do not have a pure rotational spectrum, whereas those without a center of symmetry (point group $C_{\infty v}$, see Figure 12.17b), such as HF and HCN, do have a pure rotational spectrum.

Diatomic Molecules

The rotational spectrum of a diatomic molecule can be treated by considering the molecule to be a rigid rotor. The solution of the Schrödinger equation for this problem (Section 11.10; Eq. 11.208) shows that the magnitude of the angular momentum is given by the equation

$$L = \sqrt{J(J+1)}\frac{h}{2\pi} \tag{13.52}$$

where J is the rotational quantum number. The energy is given by Eq. 11.209,

$$E_J = J(J+1)\frac{h^2}{8\pi^2 I} \tag{13.53}$$

where I is the moment of inertia and J can have the values 0, 1, 2, ...

Each rotational level with quantum number J has a $2J + 1$ degeneracy which arises in the same way that the m_l values come from the degeneracy of the l states in the hydrogen atom. That is, the angular momentum vector $\boldsymbol{J}$ with magnitude $Jh/2\pi$ along the Z axis may have $2J + 1$ different values according to its orientation in space with respect to a fixed axis. Each of the states has the same energy, and all are equally probable. A **statistical weight** factor of $2J + 1$ is therefore assigned to each rotational level. The relative population of states depends on this statistical factor and also on the Boltzmann factor (Eq. 14.35), according to which the probability decreases as the energy increases. The combination of these two factors causes the intensity of the rotational states to increase and later decrease as the energy increases.

The moment of inertia of a diatomic molecule is calculated in the following way (see Figure 13.9a). The moment of inertia about the axis of the molecule (the Z axis) is zero. The moments about any axes at right angles to the Z axis are equal and are given by

$$I = m_1 r_1^2 + m_2 r_2^2 \tag{13.54}$$

where r_1 and r_2 represent the distances of the two atomic masses, m_1 and m_2, respectively, from the center of gravity of the molecule. If r_0 is the internuclear distance, the values of r_1 and r_2 are given by

$$r_1 = \frac{m_2}{m_1 + m_2} r_0 \qquad r_2 = \frac{m_1}{m_1 + m_2} r_0 \tag{13.55}$$

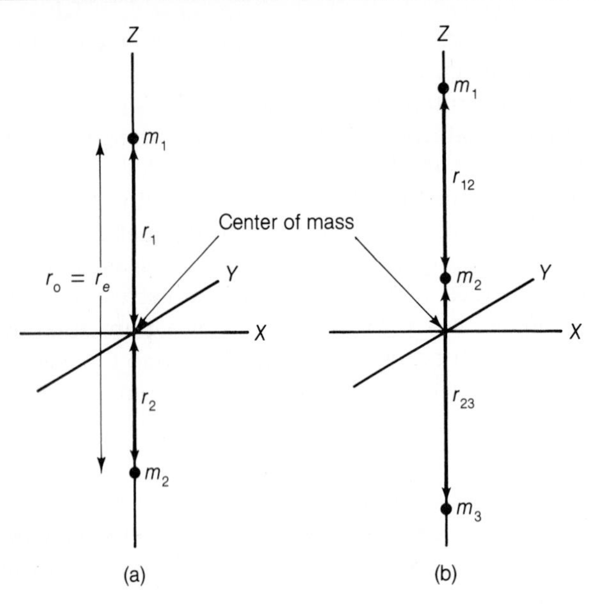

FIGURE 13.9 (a) A diatomic molecule. (b) A linear triatomic molecule.

Substitution of these expressions into Eq. 13.54 gives, after some reduction,

$$I = \frac{m_1 m_2}{m_1 + m_2} r_0^2 = \mu r_0^2 \tag{13.56}$$

where μ is known as the **reduced mass** and for a diatomic molecule is given by

$$\mu = \frac{m_1 m_2}{m_1 + m_2} \tag{13.57}$$

As we have seen, by the Born-Oppenheimer approximation we take the distance r_0 to be the equilibrium distance r_e.

Pure rotational spectra are simple in appearance, consisting of equally spaced lines. The energy difference for the absorption $J + 1 \leftarrow J$ between two neighboring rotational states is found from Eq. 13.53 to be

$$\Delta E = E_{J+1} - E_J = \left[(J + 1)(J + 2) - J(J + 1) \right] \frac{h^2}{8\pi^2 I} \tag{13.58}$$

$$= 2(J + 1) \frac{h^2}{8\pi^2 I} \tag{13.59}$$

The frequency ν_J associated with the transitions is therefore

$$\nu_J = 2(J + 1) \frac{h}{8\pi^2 I} \tag{13.60}$$

This equation is conveniently written as

$$\boldsymbol{\nu_J = 2(J + 1)B} \tag{13.61}$$

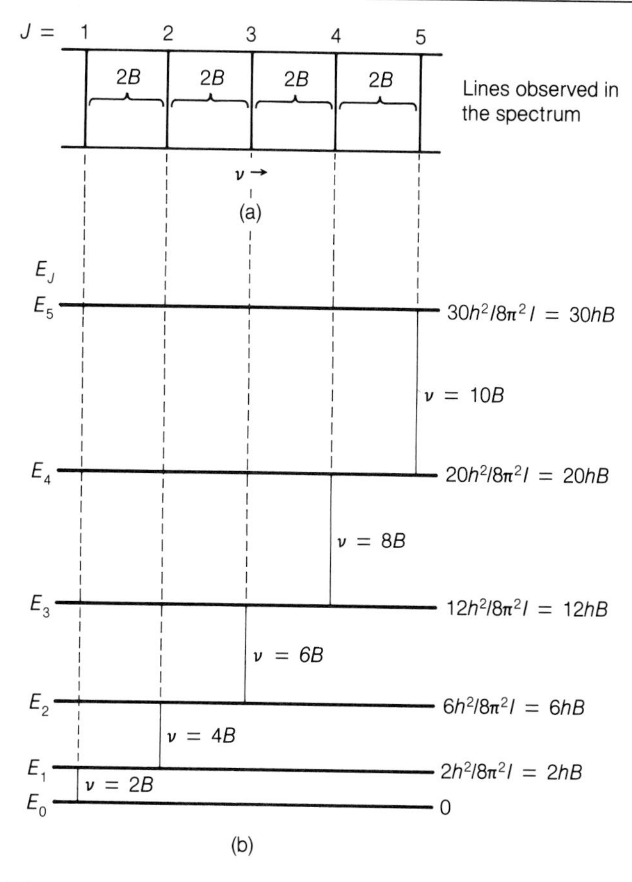

FIGURE 13.10 The constant $2B$ spacing of a pure rotational spectrum is shown in (a). The linearly increasing energy differences between the corresponding transitions are shown in (b).

where B is given by

$$B = \frac{h}{8\pi^2 I} \tag{13.62}$$

and is known as the **rotational constant**. It follows from Eq. 13.61 that the rotational spectrum will consist of equally spaced lines, the separation between neighboring lines being $2B$. This is shown in Figure 13.10, and it is seen to result from the fact that the spacing between neighboring energy levels increases linearly with J. It is usual to express the lines in wave numbers, usually cm^{-1}; Eq. 13.61 then becomes

$$\bar{\nu}_J = \frac{\nu_J}{c} = 2(J+1)\frac{B}{c} \tag{13.63}$$

The selection rule for a rotational transition is

$$\Delta J = \pm 1 \tag{13.64}$$

A simple interpretation of this selection rule is that photons have unit spin and therefore have a spin angular momentum of $h/2\pi$; thus, since angular momentum is conserved in a transition, the rotational quantum number must change by $\Delta J = \pm 1$.

It follows that if the spacing between rotational lines is measured, the moment of inertia of the molecule may be determined. This is illustrated by the following example.

EXAMPLE

Absorption by $H^{35}Cl$ occurs in the far infrared near $\bar{\nu} = 200$ cm^{-1}, and the spacing between the neighboring lines is 20.89 cm^{-1}. Find the moment of inertia and the internuclear distance in $H^{35}Cl$.

SOLUTION

From Eq. 13.63, with $J = 0$ (which gives the spacing),

$$B = \frac{c\Delta\bar{\nu}}{2} = \frac{2.998 \times 10^{10}\ (\text{cm s}^{-1})\ 20.89\ (\text{cm}^{-1})}{2}$$

$$= 3.13 \times 10^{11}\ \text{s}^{-1}$$

From Eq. 13.62 the moment of inertia is

$$I = \frac{h}{8\pi^2 B} = \frac{6.626 \times 10^{-34}\ (\text{J s})}{8\pi^2 \times 3.13 \times 10^{11}\ (\text{s}^{-1})}$$

$$= 2.68 \times 10^{-47}\ \text{kg m}^2$$

The reduced mass μ for a diatomic molecule is given by Eq. 13.57:

$$\mu = \frac{M_r(\text{Cl})M_r(\text{H})}{M_r(\text{Cl}) + M_r(\text{H})} \times \frac{1}{6.022 \times 10^{23}}$$

$$= \frac{(35.0)(1)}{35.0 + 1} \times \frac{1}{6.022 \times 10^{23}} = 1.614 \times 10^{-24}\ \text{g}$$

$$= 1.614 \times 10^{-27}\ \text{kg}$$

The equilibrium distance r_e is found from $I = \mu r_e^2$:

$$r_e = \sqrt{\frac{I}{\mu}} = \left(\frac{2.68 \times 10^{-47}}{1.614 \times 10^{-27}}\right)^{1/2} = 1.29 \times 10^{-10}\ \text{m}$$

$$= 0.129\ \text{nm}$$

■

Linear Triatomic Molecules

The moment of inertia of a linear triatomic molecule is given by

$$I = m_1 r_{12}^2 + m_3 r_{23}^2 - \frac{(m_1 r_{12} - m_3 r_{23})^2}{m} \tag{13.65}$$

where $m = m_1 + m_2 + m_3$ and the individual atomic masses and positions are shown in Figure 13.9b. There are now two distances, r_{12} and r_{23}, to determine, and this may be done by studying two molecules that are isotopically different. For example, deuterium may be substituted for hydrogen in HCN, and the pure rotational spectra of HCN and DCN studied. It is assumed that the bond lengths do not change on isotopic substitution. The method of determining the internuclear distances is illustrated by the following example.

EXAMPLE

The B values for HCN and DCN are as follows:

$$B(\text{HCN}) = 1.478 \text{ cm}^{-1} = 4.43 \times 10^{10} \text{ s}^{-1}$$

$$B(\text{DCN}) = 1.209 \text{ cm}^{-1} = 3.62 \times 10^{10} \text{ s}^{-1}$$

Calculate the bond lengths H—C (= D—C) and C—N.

SOLUTION

The moments of inertia are equal to $h/8\pi^2 B$ and from the B values are found to be

$$I(\text{HCN}) = 1.89 \times 10^{-46} \text{ kg m}^2$$

$$I(\text{DCN}) = 2.32 \times 10^{-46} \text{ kg m}^2$$

The atomic masses are

$$\text{H:} \quad m_1 = 1.674 \times 10^{-27} \text{ kg} \qquad \text{D:} \quad m_1' = 3.346 \times 10^{-27} \text{ kg}$$

$$\text{C:} \quad m_2 = 1.99 \times 10^{-26} \text{ kg} \qquad \text{N:} \quad m_3 = 2.33 \times 10^{-26} \text{ kg}$$

For HCN,

$$\text{total mass } m = 4.49 \times 10^{-26} \text{ kg}$$

For DCN,

$$\text{total mass } m' = 4.54 \times 10^{-26} \text{ kg}$$

Substitution of these two sets of values into Eq. 13.65 gives two simultaneous quadratic equations, and their solution is

$$r_{12} = r(\text{H—C}) = r(\text{D—C}) = 0.106 \text{ nm}$$

$$r_{23} = r(\text{C—N}) = 0.116 \text{ nm}$$

■

Microwave Spectroscopy

The $H^{35}Cl$ molecule has a moment of inertia of 2.68×10^{-47} kg m^2 and this gives a B value of 10.455 cm^{-1}; there are therefore lines at 20.89 cm^{-1}, 41.78 cm^{-1}, and so on. Measurements in this region of the spectrum can be made by the methods of infrared spectroscopy, in which the radiation is passed through a cell and then a diffraction grating, and the transmitted radiation detected by a heat-sensitive device. Heavier and more complicated molecules, however, have larger moments of inertia and consequently the rotational frequencies are much lower. For example, $^{16}O^{12}C^{32}S$ has a moment of inertia of 1.38×10^{-45} kg m^2, which is about 50 times that of HCl, and its B value is 0.203 cm^{-1}. The lines are therefore at 0.406 cm^{-1}, 0.812 cm^{-1}, and so on. It is extremely difficult to make measurements at these frequencies by infrared techniques. However, the wavelengths of *microwaves* are in the right range, and the larger molecules are therefore usually studied by the techniques of microwave spectroscopy.

In microwave spectroscopy the radiation is generated by an electronically controlled oscillator, and it is possible for it to be highly monochromatic, covering an extremely narrow band of wavelengths. The radiation is passed through a metal wave guide of suitable dimensions and the substance under investigation is present in this wave guide. The microwave beam that emerges from the wave guide impinges on a crystal detector, the signal from which is amplified and then recorded on a cathode-ray oscilloscope. Because the radiation is so highly monochromatic and the electronic recording is so precise, the sensitivity of the microwave technique is about 10^5 times that of an infrared grating spectrometer. Frequency measurements can in fact be made to more than seven significant figures. The precision with which interatomic distances can be obtained by the use of this technique was until recently limited not by the precision of the frequency measurements but by the precision with which the Planck constant had been determined. However, the Planck constant is now known to eight significant figures [$6.626\,176\ (36) \times 10^{-34}$ J s^{-1}].

The results of microwave experiments are usually quoted not in wave numbers but in frequencies. For example, for the $^{12}C^{16}O$ spectrum a line has been reported at $1.152\,712\,04 \times 10^{11}$ s^{-1}, which can be written as 115 271.204 MHz or as 115.271 204 GHz. This corresponds to 3.845 033 49 cm^{-1}.

Nonlinear Molecules

Linear molecules have a single moment of inertia, but nonlinear molecules have three moments of inertia, two or all three of which may be equal. By a simple theorem of mechanics, there must be at least one axis passing through the center of gravity about which the moment of inertia is a *maximum*. There must also be at least one axis about which the moment of inertia is a *minimum*. These two axes, and a third axis at right angles to both of them, are known as the **principal axes.** If the molecule has an axis of symmetry, this is bound to be a principal axis. Moments of inertia must be calculated about principal axes.

If a molecule has three moments of inertia that are different, it is referred to as an **asymmetric top;** an example is the water molecule, which is illustrated in Figure 13.11a. If two of the three moments of inertia are equal (i.e., if there are two different moments of inertia), the molecule is known as a **symmetric top.** Examples are molecules belonging to the point group C_{3v}, such as ammonia and methyl chloride (Figure 13.11b). If all three moments of inertia are equal, the molecule is known as a **spherical top;** an example is methane.

The pure rotational spectra of nonlinear molecules are more complex than those of linear molecules, and their interpretation requires the use of two or three quantum numbers. For a treatment of such spectra the reader is referred to more advanced texts.

The Stark Effect

Since a molecule that has a rotational spectrum also has an electric dipole moment, an electric field will cause an interaction. Known as the *Stark effect,** in honor of the German physicist Johannes Stark (1874–1957), the application of the field causes the

*J. Stark, *Sitzber, Deut. Akad. Wiss. Berlin, 40,* 932 (1913).

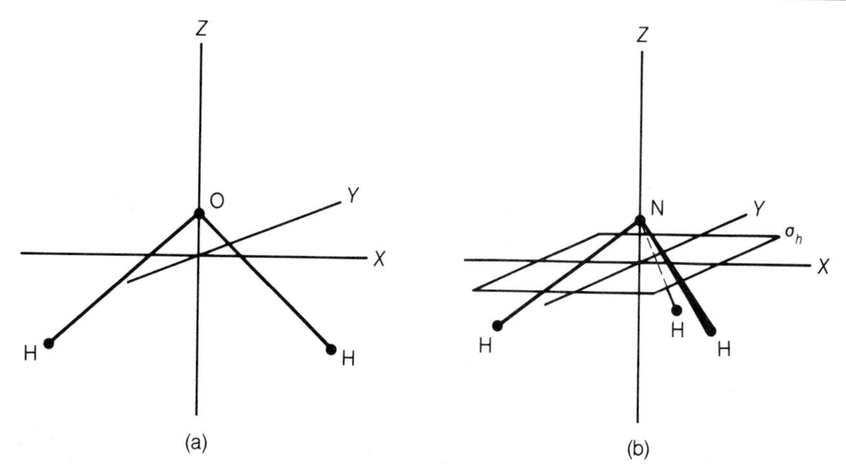

FIGURE 13.11 (a) The water molecule, an example of an asymmetric top. The principal moments of inertia, about the X, Y, and Z axes, are all different. (b) The ammonia molecule, an example of a symmetric top. The moments of inertia about any axis at right angles to the Z axis are equal.

$2J + 1$ degenerate rotational energy levels to be split into $2J + 1$ lines, and multiplet structure is observed for all lines with $J > 0$. This allows J values to be assigned to particular observed spectral lines, since the lowest frequency line observed need not be the one for $J = 0$. Since the number of Stark components depends on J, unambiguous assignments can be made.

A second advantage of the Stark effect is that the dipole moment may easily be found because the absorption line shift in an electric field depends both on the electric field and on the dipole moment.

13.4 Vibrational-Rotational Spectra of Molecules

The spectra observed in the infrared region involve vibrational transitions accompanied by rotational transitions. Infrared spectra thus consist of series of *bands*, each band corresponding to changes in the vibrational state of the molecule, and each line in the band corresponding to a superimposed change in the rotational state.

The infrared bands obtained with substances in the gas phase consist of fairly sharp lines, since the molecules can rotate freely. When the substance is in the liquid phase or in solution, however, the molecules cannot rotate freely, and the bands are blurred. It is not possible to make infrared measurements on substances in aqueous solution, since water absorbs intensely in this region and completely masks the spectrum of the solute. Measurements in the infrared must therefore be made with the substance in the gas phase, or in a medium that does not absorb. A common modern technique is to disperse the sample in a suitable inorganic salt, usually potassium bromide. The sample is mixed with the powdered crystalline salt, which is then pressed into a transparent disk; this is mounted in a holder that is supported in the beam of the infrared instrument.

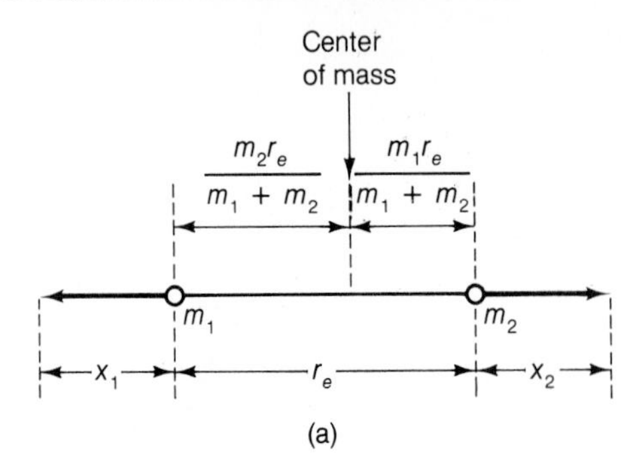

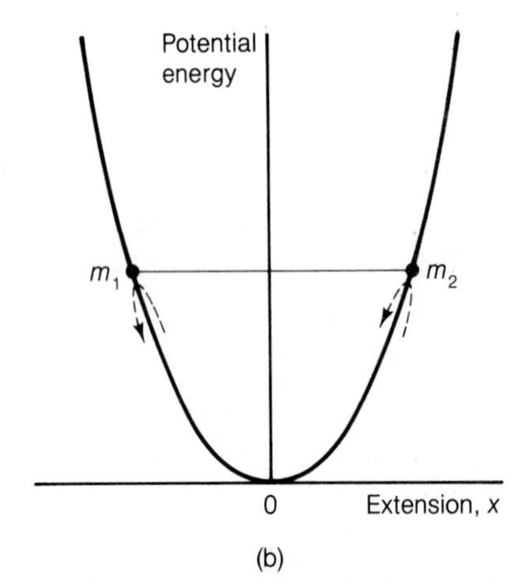

FIGURE 13.12 (a) The displacements of the masses m_1 and m_2 during the vibration of a diatomic molecule. (b) The potential-energy curve (a parabola) for a harmonic oscillator.

Diatomic Molecules

The simplest infrared spectra are those of diatomic molecules, which are normally studied in the gas phase. In Section 11.6 we treated the quantum mechanics of the harmonic oscillator, and the energy (Eq. 11.156) was found to be

$$E_v = (v + \tfrac{1}{2})h\nu_0 \tag{13.66}$$

where v is the *vibrational quantum number*. The frequency of vibration ν_0 can be related to the *force constant* k in the following way. Figure 13.12a shows two masses m_1 and m_2 connected by a spring. The center of mass is taken as the origin, and if the distance between the masses when the system is at rest is r_e, the center of mass is at distances

$$\frac{m_2}{m_1 + m_2} r_e \quad \text{and} \quad \frac{m_1}{m_1 + m_2} r_e$$

from the masses m_1 and m_2. Suppose that at a particular time during the vibration the masses are displaced by distances x_1 and x_2, as shown in Figure 13.12. If moments are taken about the center of mass

$$m_1 \left(\frac{m_2}{m_1 + m_2} r_e + x_1 \right) = m_2 \left(\frac{m_1}{m_1 + m_2} r_e + x_2 \right) \tag{13.67}$$

from which it follows that

$$m_1 x_1 = m_2 x_2 \tag{13.68}$$

If x is the extension,

$$x = x_1 + x_2 \tag{13.69}$$

so that

$$x_1 = \frac{m_2}{m_1 + m_2} x \tag{13.70}$$

$$x_2 = \frac{m_1}{m_1 + m_2} x \tag{13.71}$$

If the motion is harmonic, the restoring force on each mass is proportional to the extension x; the proportionality constant k is known as the **force constant** for the bond. This force is mass times acceleration, so that the equation of motion for particle 1 is

$$m_1 \ddot{x}_1 = -kx \tag{13.72}$$

Double differentiation of Eq. 13.70 gives

$$\ddot{x}_1 = \frac{m_2}{m_1 + m_2} \ddot{x} \tag{13.73}$$

and therefore

$$\frac{m_1 m_2}{m_1 + m_2} \ddot{x} = -kx \tag{13.74}$$

This may be written as

$$\mu \ddot{x} = -kx \tag{13.75}$$

where μ, the reduced mass, is given by Eq. 13.57:

$$\mu = \frac{m_1 m_2}{m_1 + m_2} \tag{13.76}$$

A solution of Eq. 13.75 is

$$x = A \cos 2\pi \nu_0 t \tag{13.77}$$

double differentiation of which leads to

$$\ddot{x} = -4\pi^2 \nu_0^2 A \cos 2\pi \nu_0 t \tag{13.78}$$

$$= -4\pi^2 \nu_0^2 x \tag{13.79}$$

Comparison of this with Eq. 13.75 shows that

$$k = 4\pi^2 \nu_0^2 \mu \tag{13.80}$$

and the frequency of motion is

$$\nu_0 = \frac{1}{2\pi}\sqrt{\frac{k}{\mu}} \tag{13.81}$$

This is of the same form as Eq. 11.15, which was for a single mass m attached by a spring to an infinite mass. The only difference is that the mass is now replaced by the reduced mass μ.

Since the restoring force F is proportional to the extension x,

$$F = -kx \tag{13.82}$$

and the force is the negative of the derivative of the potential energy E_p,

$$F = -\frac{dE_p}{dx} \tag{13.83}$$

it follows that the potential energy is given by

$$E_p = -\int F\,\mathrm{d}x = k\int x\,dx \tag{13.84}$$

$$= \tfrac{1}{2}kx^2 \tag{13.85}$$

This is the equation of a parabola (Figure 13.12b). Simple harmonic motion therefore corresponds to the movement of a particle back and forth on the parabola. At the extremities of the motion (points A and B in Figure 13.12b) the system has no kinetic energy and maximum potential energy. As it approaches the equilibrium position, the kinetic energy increases to a maximum and the potential energy decreases; the sum of the two remains constant. The kinetic energy is then reconverted into potential energy as the system continues its motion.

The actual shape of a potential-energy curve is shown in Figure 13.13. The lower regions of this curve are satisfactorily represented by a parabola, shown as

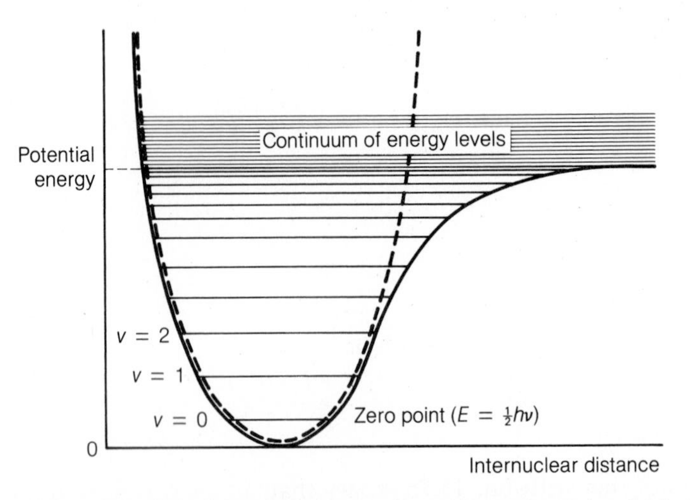

FIGURE 13.13 The actual potential-energy curve for a diatomic molecule (solid curve) with a parabola imposed on it (dashed line). The lowest horizontal lines, representing the quantized vibrational energy levels, are approximately equally spaced. However, as the curve deviates from the parabola, the levels become closer together.

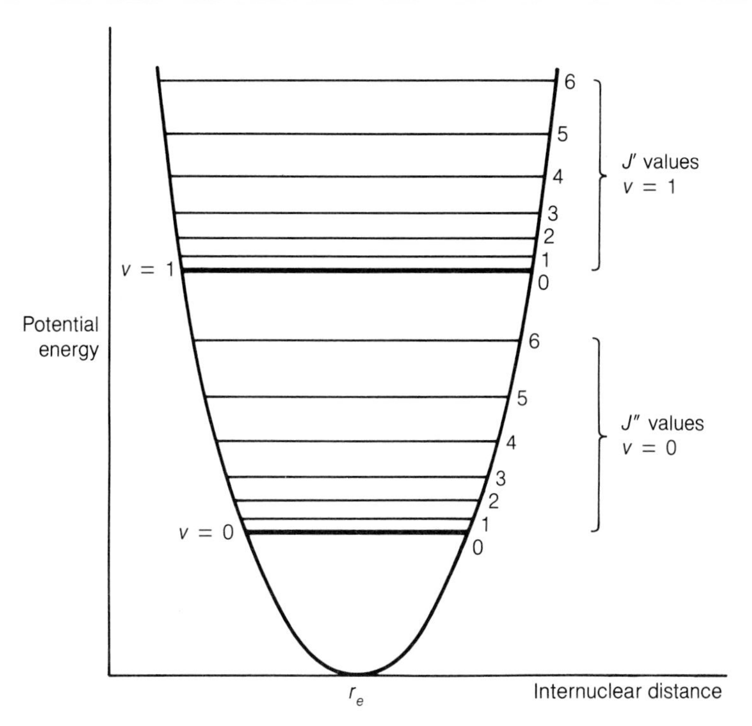

FIGURE 13.14 A potential-energy curve, showing the two lowest vibrational states ($v = 0$ and $v = 1$) with rotational states superimposed.

the dashed curve. The parabolic curve, corresponding to the harmonic oscillator, therefore represents a good starting point for the treatment of molecular vibrations, at any rate at low vibrational energies. We will consider this situation first and later see what modifications are required to give a more realistic representation of the vibration of an actual molecule.

To interpret vibrational-rotational spectra on this model we use the assumption (the Born-Oppenheimer approximation) that the vibrational and rotational energies are separable and that the energy can be expressed as the sum of them. Thus, if we combine Eqs. 13.66 and 13.53 (with $B = h/8\pi^2 I$), we obtain

$$E_{n,J} = (v + \tfrac{1}{2})h\nu_0 + BJ(J + 1)h \tag{13.86}$$

If $\Delta v = 0$, we have the pure rotational spectrum; otherwise we have a vibrational spectrum with rotational changes superimposed. The situation is represented schematically in Figure 13.14, which shows two vibrational states corresponding to $v = 0$ and $v = 1$. The rotational quantum numbers in the $v = 0$ state are designated J'' and those in the $v = 1$ state are J'. To the extent that the behavior is harmonic, the average internuclear distance is r_e for all vibrational states, so that the rotational constant B may be assumed to be the same for all values of v and J.

For strict harmonic motion the selection rules for vibrational-rotational transitions are usually

$$\Delta v = \pm 1 \quad \text{and} \quad \Delta J = \pm 1 \tag{13.87}$$

In other words, the vibrational state can only change to a neighboring one, and at the same time there must be a change to a neighboring rotational level. A transition between an upper level having quantum numbers v' and J' and a lower level with v'' and J'' involves an energy change of

$$\Delta E_{v,J} = E_{v',J'} - E_{v'',J''} \tag{13.88}$$

$$= (v' - v'')h\nu_0 + B[J'(J' + 1) - J''(J'' + 1)]h \tag{13.89}$$

If $v' - v'' = 1$,

$$\Delta E_{v,J} = h\nu_0 + \text{B}\,[J'(J' + 1) - J''(J'' + 1)]h \tag{13.90}$$

Associated with this change of v there are two allowed changes of J. If $J' = J'' + 1$, the allowed frequencies are given by

$$\nu_R = \frac{\Delta E_{n,J}}{h} = \nu_0 + B[(J'' + 1)(J'' + 2) - J''(J'' + 1)] \tag{13.91}$$

$$= \nu_0 + 2J''B \tag{13.92}$$

Since J'' can have the values 1, 2, . . . , this corresponds to a series of equally spaced lines. This particular series is known as the *R-branch*. If, on the other hand, $J'' = J'' - 1$, the allowed frequencies are given by

$$\nu_p = \nu_0 + B[J''(J'' - 1) - J''(J'' + 1)] \tag{13.93}$$

$$= \nu_0 - 2J''B \tag{13.94}$$

This corresponds to a series of equally spaced lines on the low-frequency side of ν_0 and is known as the *P-branch*. In some cases $\Delta J = 0$ is permitted, and the resulting line, of frequency ν_0, is known as the *Q-branch*.

Figure 13.15a shows a representation of the infrared spectrum of HCl. The absorption lines constituting the P-branch are to be seen on the low-frequency side, and the R-branch is on the high-frequency side. In between them is a gap, because the transition $\Delta J = 0$, in this case, is not permitted by the selection rules. Figure 13.15b shows the rotational levels for the two vibrational states and indicates the transitions that give rise to the absorption lines shown in Figure 13.15a. By measuring the separation between lines in the P- and R-branches, the value of B can be obtained using Eqs. 13.92 and 13.94. It is therefore possible to calculate the intermolecular distance in a diatomic molecule.

Deviations from the simple type of behavior that we have just described arise because of the **anharmonicity** of molecular vibrations. If a potential-energy curve had a parabolic shape, the bond could never be broken. In reality, as shown in Figure 13.13, at higher internuclear separations the potential energy is lower than represented by the parabola; the deviation becomes greater and greater as the bond length increases. Eventually the curve is horizontal as the molecule dissociates into atoms.

Because at higher internuclear separations the true potential is less confining than the parabolic approximation, the actual quantized vibrational energy levels become more and more closely spaced at higher energies than would be the case for a parabolic energy curve. Eventually, when nothing is holding the atoms together, there is no separation between the energy levels (as with a particle in a very large box), and we say that there is a *continuum* of energy levels.

When there are significant deviations from the parabolic curve, the vibrations

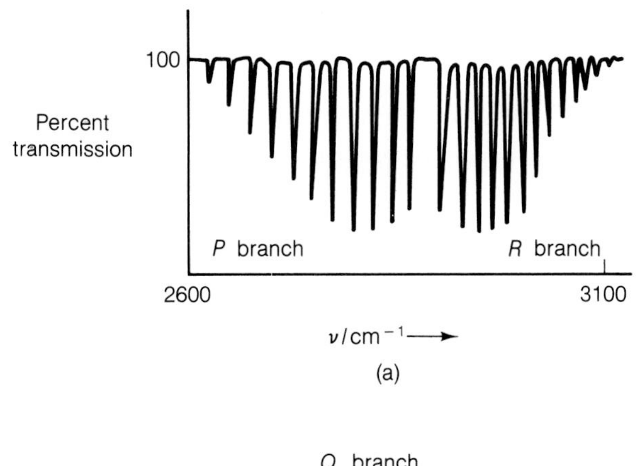

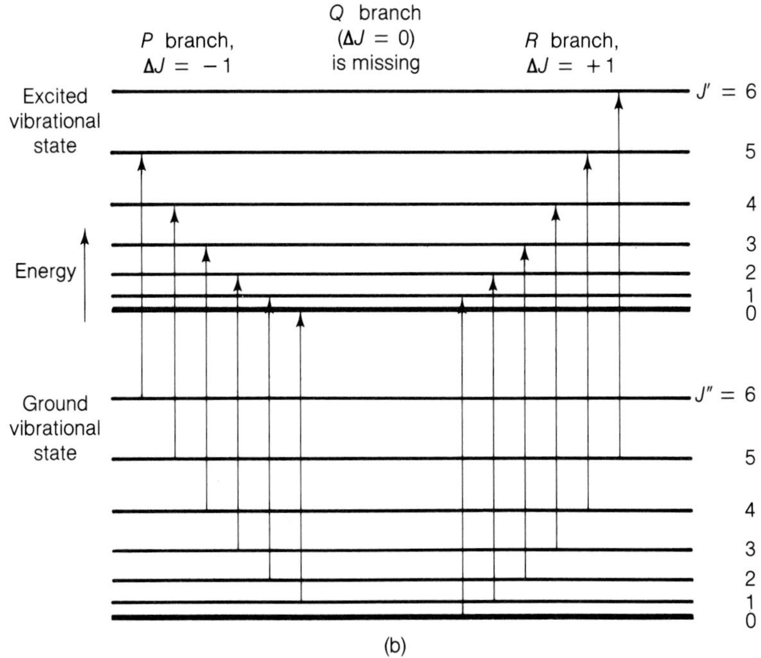

FIGURE 13.15 (a) The vibrational-rotational spectrum of HCl, showing the P and R branches. There is a missing line in between them, because $\Delta J = 0$ is forbidden. (b) Rotational levels corresponding to two vibrational states, $v = 0$ and $v = 1$, showing the transitions that give rise to the P and R branches.

are no longer harmonic and are said to be *anharmonic*. In addition, overtones and combination terms are observed in the spectrum.

There are various ways of improving parabolic potential-energy curves in order to make them represent the experimental behavior more closely. One procedure is to express the potential energy as a power series in $(v + \frac{1}{2})$. When this is done, the energy levels for the anharmonic oscillator are given by the equation

$$E_n = h\nu_0[(v + \tfrac{1}{2}) - x_e(v + \tfrac{1}{2})^2 + y_e(v + \tfrac{1}{2})^3 - \cdots] \tag{13.95}$$

where x_e, y_e . . . are dimensionless constants. It is usually sufficient to include only the first anharmonic term:

$$E_n = h\nu_0[(v + \tfrac{1}{2}) - x_e(v + \tfrac{1}{2})^2] \tag{13.96}$$

and x_e is then called the *anharmonicity constant*. A typical value for it is 0.01. This expression brings the higher levels closer together, as observed experimentally. We can write this expression as

$$E_n = h\nu_0[1 - x_e(v + \tfrac{1}{2})](v + \tfrac{1}{2}) \tag{13.97}$$

from which we see that the apparent oscillation frequency is $\nu_0[1 - x_e(v + \frac{1}{2})]$. This frequency therefore decreases as v increases. The zero-point energy is found by substituting $v = 0$ into this equation:

$$E_0 = \tfrac{1}{2}h_0(1 - \tfrac{1}{2}x_e)\nu_0 \tag{13.98}$$

If $x_e = 0.01$, this zero-point energy is very close to that for the harmonic oscillator.

Whereas for the harmonic oscillator the selection rule is $\Delta v = \pm 1$, that for the anharmonic oscillator is

$$\Delta v = \pm 1, \pm 2, \pm 3, \text{etc.} \tag{13.99}$$

Multiple jumps can thus occur, but the probability is less and the line intensity less, when v changes by more than 1. At ordinary temperatures most molecules are in the lowest vibrational state, and we can see from Eq. 13.97 that the following are the energy changes in a transition from the lowest level to $v = 1$, $v = 2$, and $v = 3$, with no change of J:

$$\Delta E_1 = E_1 - E_0 = h\nu_0(1 - 2x_e) \tag{13.100}$$

$$\Delta E_2 = E_2 - E_0 = 2h\nu_0(1 - 3x_e) \tag{13.101}$$

$$\Delta E_3 = E_3 - E_0 = 3h\nu_0(1 - 4x_e) \tag{13.102}$$

The frequency of the transition corresponding to ΔE_1 lies very close to ν_0 and is known as the *fundamental* absorption. The transitions corresponding to ΔE_2 and ΔE_3 occur at much reduced intensity and are known as the *first overtone* and the *second overtone*, respectively. These transitions are, of course, accompanied by a rotational fine structure.

At higher temperatures (over 500 K for a typical molecule) there may be enough molecules in the $v = 1$ state to give a weak absorption corresponding to a transition to higher states. For the transition to $v = 2$,

$$E_{2\leftarrow 1} = E_2 - E_1 = h\nu_0(1 - 4x_e) \tag{13.103}$$

This weak absorption will thus occur at a slightly lower frequency than the fundamental and will increase in intensity as the temperature is raised. Because of this, such bands are known as *hot bands*.

An equation that closely approximates the experimental potential-energy curves for diatomic molecules was suggested in 1929 by the American physicist Philip M. Morse (b. 1903). This equation, known as the **Morse potential function,** is

$$\mathbf{E_p = D_e(1 - e^{-ax})^2} \tag{13.104}$$

where $x = r - r_e$ (i.e., the extension of the bond from its equilibrium distance), and

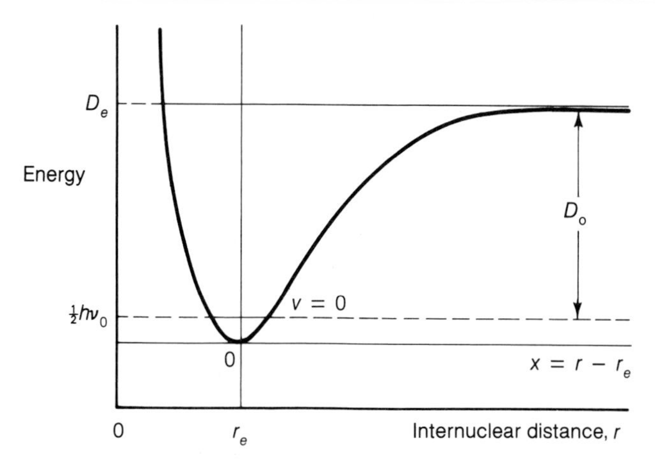

FIGURE 13.16 The potential-energy curve predicted by the Morse equation, Eq. 13.104.

D_e and a are constants. The curve represented by this equation is shown in Figure 13.16. When $x = 0$, $E_p = 0$; when x approaches infinity, E_p approaches D_e. The quantity D_e is therefore the *classical* dissociation energy and is equal to the experimental dissociation energy plus the zero-point energy. One advantage of the equation is that if the form of the potential-energy curve can be deduced from spectroscopic data, the curve can be fitted to the Morse function; the energy D_e is then known, and the dissociation energy D_0 can easily be calculated.

Normal Modes of Vibration

A diatomic molecule vibrates according to a simple pattern and has only one vibrational frequency. The vibrational motions of molecules having more than two atoms are a superposition of a number of basic vibrations, known as **normal modes of vibration.** In order to bring out this concept, we will work through the dynamics for a linear triatomic molecule.

The linear triatomic system is shown schematically in Figure 13.17a. In this treatment we will ignore bending vibrations and will consider only displacements along the axis of the molecule. The center of mass is again taken as the origin, and if moments are taken about the origin when the molecule is at rest,

$$m_1(r_{12} - y) = m_2 y + m_3(r_{23} + y) \tag{13.105}$$

where y is the distance between the center of mass and the atom B. If the three masses are displaced to the right by the distances x_1, x_2, and x_3,

$$m_1(r_{12} - y - x_1) = m_2(y + x_2) + m_3(r_{23} + y + x_3) \tag{13.106}$$

and therefore

$$m_1 x_1 + m_2 x_2 + m_3 x_3 = 0 \tag{13.107}$$

The displacements cause the distance between particles A and B to increase by $x_2 - x_1$; the restoring force is therefore $k_{12}(x_2 - x_1)$, where k_{12} is the force constant for the bond. The equation of motion for particle A is therefore

$$k_{12}(x_2 - x_1) = m_1 \ddot{x}_1 \tag{13.108}$$

(a)

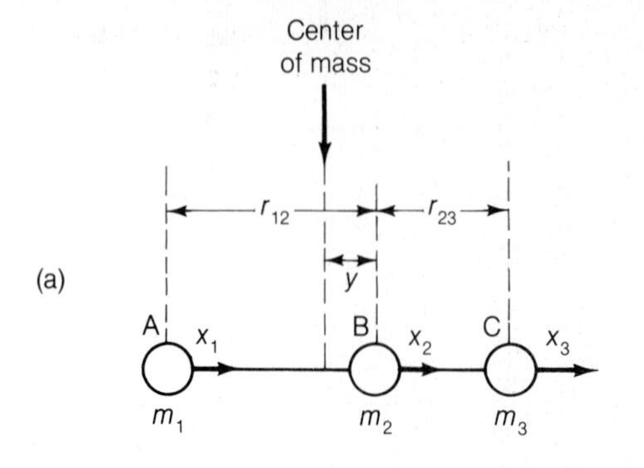

(b)

(c)

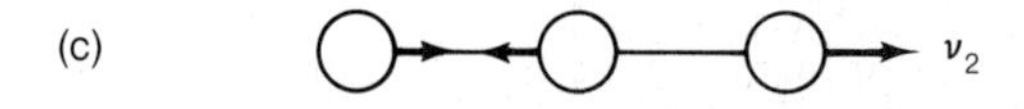

(d)

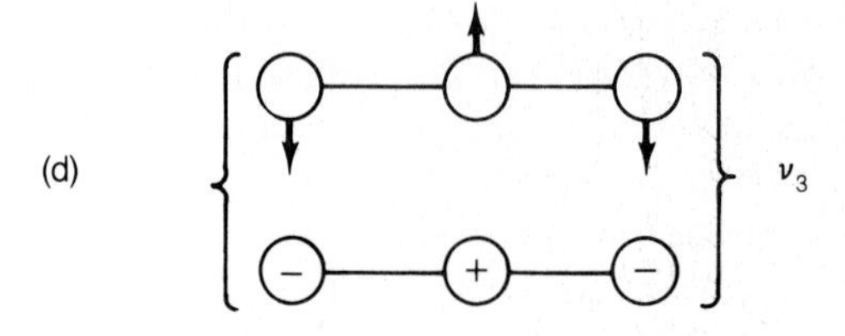

FIGURE 13.17 (a) Stretching vibrations of a linear triatomic molecule. (b) The symmetric stretching (breathing) vibration of a linear triatomic molecule. (c) The antisymmetric stretching vibration. (d) The two degenerate bending vibrations.

Similarly for particle C

$$k_{23}(x_3 - x_2) = -m_3\ddot{x}_3 \tag{13.109}$$

The equation for particle B need not be used since it is simply a linear combination of Eqs. 13.107 to 13.109.

Elimination of x_2 in Eqs. 13.108 and 13.109 by use of Eq. 13.107 leads to

$$m_1m_2\ddot{x}_1 + k_{12}[(m_1 + m_2)x_1 + m_3x_3] = 0 \tag{13.110}$$

and

$$m_2m_3\ddot{x}_3 + k_{23}[m_1x_1 + (m_2 + m_3)x_3] = 0 \tag{13.111}$$

To solve these equations, we look for solutions of the form

$$x_1 = A_1 \cos 2\pi\nu t \tag{13.112}$$

$$x_3 = A_3 \cos 2\pi\nu t \tag{13.113}$$

with the same frequency ν for both displacements. Double differentiation gives

$$\ddot{x}_1 = -4\pi^2\nu^2x_1 = -\lambda x_1 \tag{13.114}$$

$$\ddot{x}_3 = -4\pi^2\nu^2x_3 = -\lambda x_3 \tag{13.115}$$

where λ has been written for $4\pi^2\nu^2$. Insertion of these expressions into Eqs. 13.110 and 13.111 leads to two simultaneous equations in x_1 and x_3:

$$[-\lambda m_1 m_2 + k_{12}(m_1 + m_2)]x_1 + k_{12}m_3x_3 = 0 \tag{13.116}$$

$$m_1k_{23}x_1 + [-\lambda m_2m_3 + k_{23}(m_2 + m_3)]x_3 = 0 \tag{13.117}$$

Equations of this type are only consistent with one another provided that

$$\begin{vmatrix} -\lambda m_1m_2 + k_{12}(m_1 + m_2) & k_{12}m_3 \\ m_1k_{23} & -\lambda m_2m_3 + (m_2 + m_3)k_{23} \end{vmatrix} = 0 \tag{13.118}$$

This equation is a quadratic equation in λ:

$$\lambda^2 - \left(\frac{m_1 + m_2}{m_1m_2}k_{12} + \frac{m_2 + m_3}{m_2m_3}k_{23}\right)\lambda + \frac{k_{12}k_{23}(m_1 + m_2 + m_3)}{m_1m_2m_3} = 0 \tag{13.119}$$

Its solution gives two values of λ and therefore two frequencies ν_1 and ν_2. This equation may be written as

$$\lambda^2 - b\lambda + c = 0 \tag{13.120}$$

and the roots are

$$\lambda_1 = \frac{b - \sqrt{b^2 - 4c}}{2} \tag{13.121}$$

and

$$\lambda_2 = \frac{b + \sqrt{b^2 - 4c}}{2} \tag{13.122}$$

Since $\lambda_2 > \lambda_1$, the frequency ν_2 is the larger of the two frequencies. If Eq. 13.122 for λ is inserted into either Eq. 13.116 or Eq. 13.117, there results an equation from which it is possible to evaluate the ratio x_1/x_3; when this is done, x_1/x_3 is found to have a positive value. If, on the other hand, the lower root λ_1 is inserted into Eq. 13.116 or Eq. 13.117, the ratio x_1/x_3 is found to be negative.

The significance of this result is that the lower frequency ν_1, which relates to λ_1, is the frequency of a symmetric vibration in which the bonds stretch and shorten in unison, as shown in Figure 13.17b. The higher frequency ν_2 is the frequency of an asymmetric vibration in which one bond shortens while the other is stretching, as shown in Figure 13.17c.

A similar treatment of the bending vibrations, in terms of a force constant for the bending, leads to the result that there are two *degenerate* (i.e., equivalent) bending motions in planes at right angles to one another. In the upper representation in Figure 13.17d, the bending is occurring in the plane of the paper, while in the lower representation it is at right angles to the plane of the paper. A bending motion in any plane can be treated as a superposition of the bending motions in these two planes; expressed differently, any bending motion can be resolved into the bending motions in these two planes. In the same way, any stretching motion that the molecule can undergo is the resultant of the two basic motions represented in Figure 13.17b and c.

The stretching and bending motions shown in Figure 13.17b–d are known as *normal modes of vibration*, and they play a similar role in vibrational theory as do motions along the X, Y, and Z axes in translational theory. For any linear triatomic molecule there are four normal modes of vibration; the two bending modes are degenerate. In each of the modes all three atoms are either completely in phase or completely out of phase with each other and move with the same frequency. The motion of each atom, with reference to the center of mass, corresponds to simple harmonic motion.

The question of the number of normal modes in a molecule is discussed in Section 14.1. For a linear molecule the number is $3N - 5$, where N is the number of atoms in the molecule; for the linear triatomic molecule that we have just considered the number is $3(3) - 5 = 4$, and these four modes are represented in Figure 13.17. For a nonlinear molecule the number is $3N - 6$; for benzene, for example, there are $3(12) - 6 = 30$ normal modes. Each one of these has a characteristic vibrational frequency, some of them degenerate.

A detailed treatment of molecular vibrations is outside the scope of the present book, and the reader is referred to more specialized texts such as some of those listed in Suggested Reading (p. 642). Mention should, however, be made of the relationship between the vibrational modes and the symmetry of the molecule as indicated by its point group. Each vibrational mode must transform according to one of the symmetry species of the molecule. We may illustrate this for the normal modes of the symmetrical linear carbon dioxide molecule, shown in Figure 13.18a. The point group is $D_{\infty h}$, an abbreviated character table for which is given in the appendix to Chapter 12 (p. 558). The symmetric vibration 1 belongs to the Σ_g^+ symmetry species, since all symmetry operations leave the vibration unchanged. It is usual to denote the vibration as σ_g^+, using the lowercase Greek letter. The antisymmetric vibration 2 is antisymmetric with respect to reflection in the σ_h plane and with respect to inversion, but symmetric with respect to a σ_v reflection; it therefore belongs to the Σ_u^+ symmetry species and is denoted σ_u^+. The reader should confirm that the two bending vibrations are π_u. These assignments are important in connection with the selection rules, as will be discussed later.

Infrared Spectra of Complex Molecules

For a diatomic molecule there is only one set of bands in the infrared, since there is only a single vibrational frequency. With more complicated molecules there may be a considerable number of sets of bands corresponding to vibrational-rotational transitions, because of the various vibrational frequencies corresponding to the different normal modes of vibration. Each one of these bands has a P- and an R-branch corresponding to whether the rotational quantum number in the upper vibrational level is lower or higher, respectively, than in the ground state. In some cases $\Delta J = 0$ is also allowed, and then there is a Q-branch in addition.

The presence of infrared bands is restricted by another factor. We saw earlier that a *diatomic* molecule can give a vibrational-rotational spectrum only if it has a permanent dipole moment. This rule is a special case of a more general selection rule for vibrational-rotational transitions, namely that the normal-mode vibration must give rise to an *oscillating dipole moment*. For a diatomic molecule this requires the molecule to have a permanent dipole moment, because if a diatomic molecule has no dipole moment, the moment remains zero throughout the vibration. It is, however, possible for a more complex molecule having no dipole moment to give

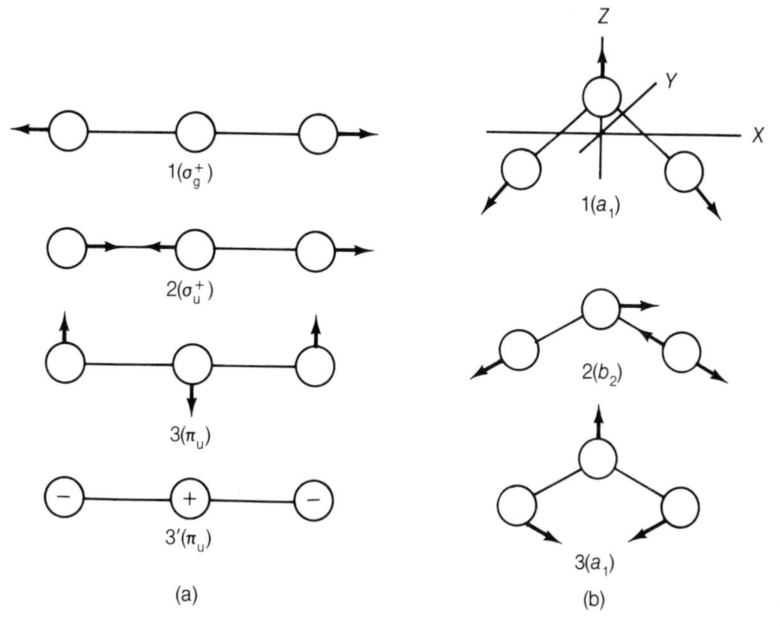

FIGURE 13.18 Normal modes of vibration. (a) The carbon dioxide molecule, which belongs to the point group $D_{\infty h}$. (b) The water molecule, which belongs to the point group C_{2v}. The character tables for these point groups are to be found in the appendix to Chapter 12 (pp. 557–558).

rise to an oscillating moment during the course of its vibration. Consider, for example, the normal modes of vibration of the carbon dioxide molecule, shown in Figure 13.18a. In the purely symmetric mode the dipole moment remains zero during the course of the vibration, and this vibration does not give rise to an infrared spectrum. The bending vibrations give rise to a small moment, and therefore they give rise to an infrared spectrum, although it may be of weak intensity. The antisymmetric vibration, on the other hand, produces an oscillating dipole moment, and there is a spectrum arising from this motion. The carbon dioxide spectrum is an example of one in which $\Delta J = 0$ is allowed, so that there is a Q-branch.

These symmetry restrictions may be formulated in terms of group theory. We have seen that the normal modes of vibration must correspond to one of the symmetry species of the point group to which the molecule belongs. Let us suppose that a molecule is vibrating in a normal mode and that at a particular instant it has a dipole moment μ. This dipole moment, being a vector, has components along the X, Y, and Z axes, and we can write these components as μ_x, μ_y, and μ_z. Suppose that one of these components, μ_x, is affected by all the symmetry operations in the same way as the normal mode of vibration itself. If this is the case, the component will vary in magnitude as the vibration occurs (i.e., will be an oscillating dipole). If, on the other hand, the component μ_x is of a different symmetry species from the normal mode, its magnitude will remain unchanged as the vibration occurs. It follows that if all the components μ_x, μ_y, and μ_z are of different species from the normal mode, there can be no oscillation of the dipole moment and hence no vibrational-rotational spectrum corresponding to that normal mode of vibration.

However, if one or more of the three components is of the same species as the normal mode, there is an infrared spectrum.

Since the components μ_x, μ_y, and μ_z are along the X, Y, and Z axes, their symmetry properties are the same as those of x, y, and z coordinates, respectively. The symmetry restriction can therefore be stated as follows:

> **There can be a vibrational-rotational spectrum corresponding to a normal mode of vibration only if that mode belongs to the same symmetry species as one or more of the three coordinates x, y, and z.**

We may illustrate this rule by some examples. Consider first the normal modes of vibration for carbon dioxide, shown in Figure 13.18a. On page 558 is an abbreviated character table for $D_{\infty h}$, which shows how the x, y, and z coordinates transform. The purely symmetric vibration is of symmetry species Σ_g^+ and is designated σ_g^+ (the lowercase letters are used); however, none of the coordinates belongs to this species, and this normal mode therefore does not give rise to a spectrum. The antisymmetric vibration, however, is σ_u^+, to which z belongs; there is therefore a spectrum. The two degenerate bending vibrations are π_u, and x and y are also of this species. There can therefore be a spectrum, although it will be weak since the bending vibrations give rise to only a very small dipole moment.

The case of water is shown in Figure 13.18b, and the reader should verify the assignment of symmetry species with reference to the character table. All the three normal modes are either a_1 or b_2, and the coordinate z is A_1 and y is B_2. All three modes therefore give rise to an infrared spectrum.

Two more complicated cases are illustrated in Figure 13.19. In Figure 13.19a are shown the six normal modes of vibration of a planar symmetrical molecule such as BF_3 (point group D_{3h}), and the reader should verify the assignment of symmetry species for the vibrations and for the axes. Vibration 2 is of species a_2'', which is the same as z; it therefore gives rise to infrared bands. Vibrations $3a$, $3b$, $4a$, and $4b$ are e', as are x and y, so that these vibrations also give a spectrum. Vibration 1, however, is a_1', which does not correspond to x, y, or z; it therefore gives no spectrum.

Figure 13.19b is for a molecule like CH_2O, which belongs to the point group C_{2v}. The reader should verify that from the character table it can be deduced that all the normal modes give rise to an infrared spectrum.

Some of the selection rules for infrared and Raman spectra (Section 13.5) are conveniently summarized in the appendix to this chapter (p. 638). It should be noted that intuitive arguments, such as we used earlier for the vibrations of CO_2 (Figure 13.18a), are not reliable for more complicated systems.

Characteristic Group Frequencies

The complete analysis of the infrared spectrum of a molecule containing more than a dozen atoms is very difficult, and it is sometimes impossible. However, very useful information can be obtained from the spectra of quite large molecules if one makes use of simple theory and empirical correlations. In particular, individual functional groups in molecules absorb in characteristic regions of the infrared spectrum. For example, a molecule containing the alcohol group, O—H, always shows an absorption in the 3100–3500 cm^{-1} region of the spectrum, while molecules containing the carbonyl group C═O characteristically absorb in the 1600–1800 cm^{-1}

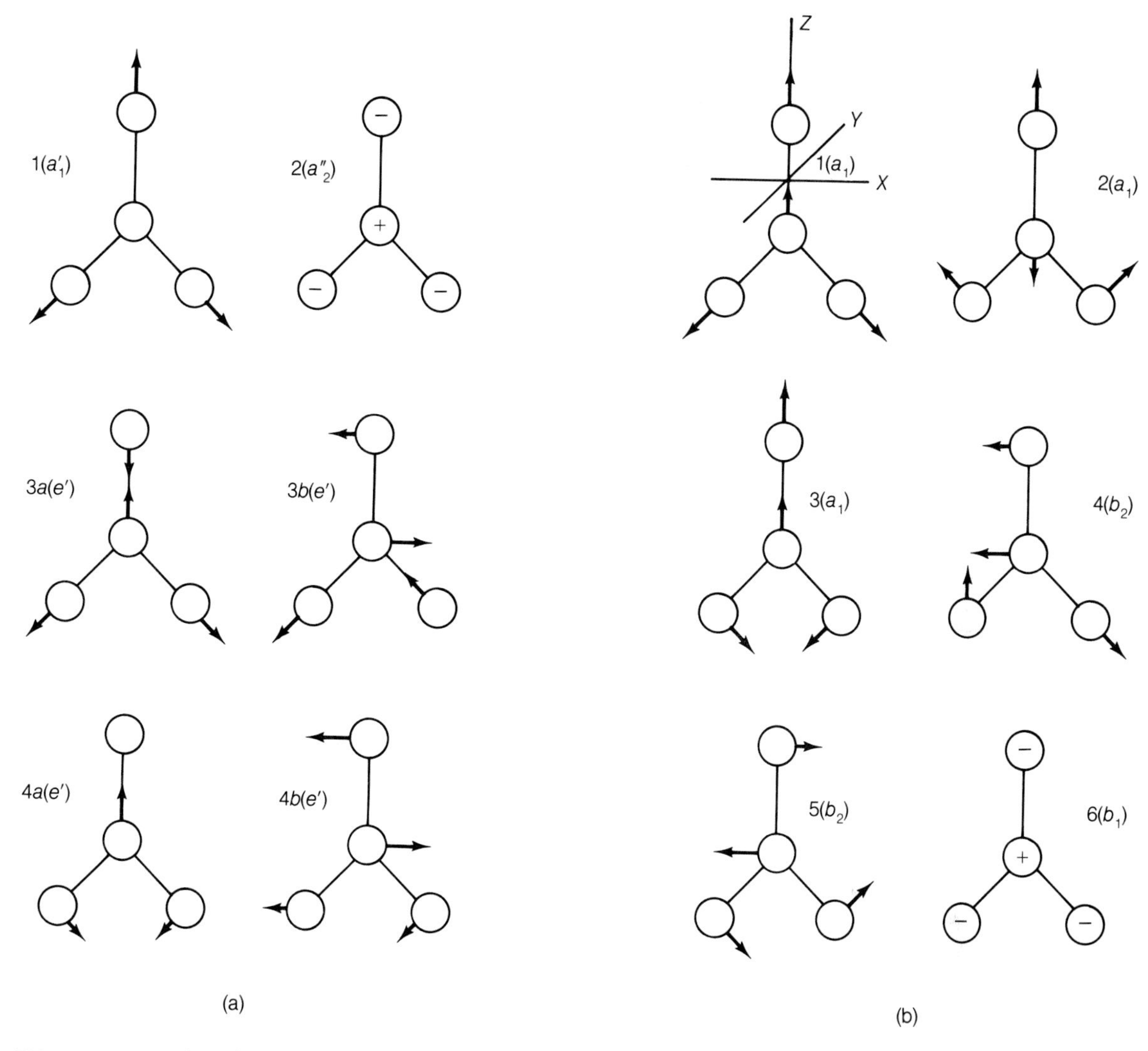

FIGURE 13.19 (a) The normal modes of vibration of a planar symmetrical molecule such as BF_3, which belongs to the point group D_{3h}. (b) The normal modes of vibration of a molecule such as CH_2O, which belongs to the point group C_{2v}. The character tables are in the appendix to Chapter 12 (pp. 557–558).

region. Table 13.2 gives a few wave numbers and wavelengths for different kinds of bonds.

There are two main reasons why these convenient simplifications occur. If a bond involves a hydrogen atom attached to a much heavier atom such as an oxygen atom, some of the normal modes of vibration involve the very light atom moving with respect to the rest of the molecule, which does not move very much in relation to the center of gravity. A very simple example is provided by the spectrum of HCN. One of the normal-mode wave numbers is 2089 cm^{-1}, and this relates to the symmetric normal mode of vibration:

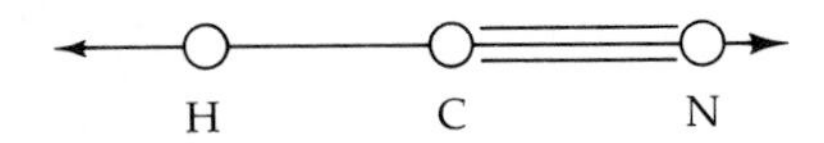

When this vibration occurs, the C—H bond stretches very considerably compared with the C≡N bond, because of the relative lightness of the H atom, and the C and N atoms do not move much relative to the center of gravity. This normal vibration is therefore very close to being simply a stretching of the C—H bond. Similarly, with an alcohol, some of the normal modes of vibration are very close to being the simple stretching of the O—H bond.

Another factor that tends to favor characteristic frequencies is the presence of multiple bonds. Here the effect is due to the wide disparity between the force constant of the multiple bond and the force constants for other bonds in the molecule. A simple example is again provided by HCN, which has a normal-mode wave number of 3312 cm^{-1} for the antisymmetric vibration:

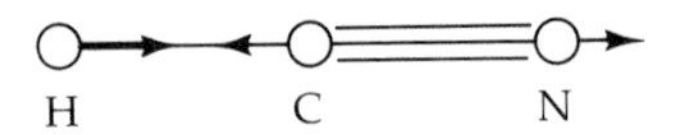

It happens that the force constant for the C≡N bond is over three times that for the H—C bond, and the normal-mode analysis shows that this antisymmetric vibration is very largely the stiff vibration of the C≡N bond.

Bonds involving hydrogen atoms and multiple bonds therefore tend to lead to characteristic group frequencies. There are also some effects that work against characteristic group frequencies; the most important is when there is considerable symmetry so that the molecule contains two or more equivalent bonds. A simple example is provided by the water molecule (Figure 13.18b), which has normal-mode frequencies of 3756, 3652, and 1595 cm^{-1}. The lowest of these may be attributed to the bending vibration. However, neither of the two higher frequencies can be ascribed to the stretching of one bond, since the two O—H bonds are equivalent. The stretching motions of the two bonds will therefore be strongly coupled, and in these two normal modes the two bonds stretch either in phase or out of phase with one another.

These limitations, however, arise largely with fairly small molecules. With molecules of any complexity the characteristic group frequencies can provide useful supporting evidence for the presence of specific functional groups. Moreover,

TABLE 13.2 Characteristic Group Frequencies

Bond	Wave Number, $\bar{\nu}$ cm^{-1}	Wavelength, λ μm
O—H, N—H	3100–3500	3.2–2.85
C—H	2800–3100	3.6–3.25
C—O, C—N	800–1300	12.5–7.7
C=C, C=O	1600–1800	6.25–5.55
P—O	1200–1300	8.3–7.7

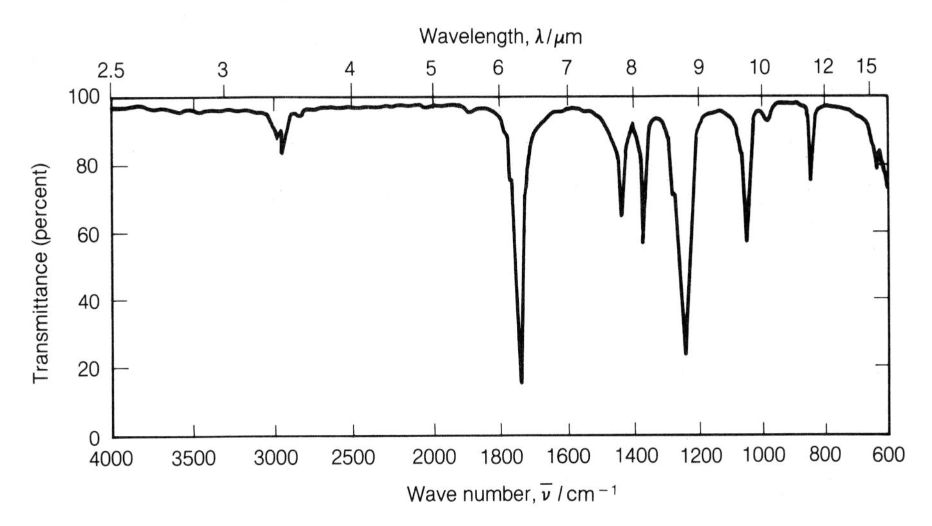

FIGURE 13.20 The infrared spectrum of methyl acetate. This spectrum was obtained with the material in the liquid phase.

because of the large number of vibrational modes in a molecule of any size, the infrared spectrum is very complicated and is unique to the molecule. Figure 13.20 shows a typical infrared spectrum; since this was obtained with the substance in the liquid phase, the rotational fine structure is blurred, but the features of this spectrum will not be found with any other molecule and will allow a very convincing identification of the substance to be made. Because the infrared spectrum of a substance is so uniquely characteristic, it is often referred to as a *fingerprint spectrum*.

13.5 Raman Spectra

When a beam of light passes through a medium, a certain amount of the light is scattered and can be detected by making observations perpendicular to the incident beam. Most of the light is scattered without a change in wavelength, an effect known as *Rayleigh scattering*. However, if the incident light is monochromatic (i.e., if only a narrow range of wavelength is represented), a small amount of the scattered light may have either higher or lower wavelengths than the original light. The spectrum consists then of lines of both longer and shorter wavelengths than the incident wavelength. This effect was first observed in 1928 by the Indian physicist Sir Chandrasekhara Venkata Raman (1888–1970) and his co-worker Sir Kariamanikkam Srinivasa Krishnan (1898–1961) and is known as the **Raman effect.**

The Raman effect is a result of the interaction that occurs between a molecule and a photon. If a photon had a perfectly elastic collision with the molecule, it would be scattered with no change of frequency. The radiation would then be emitted in all directions, with an intensity inversely proportional to the fourth power of the wavelength. Because of this dependency, shorter wavelengths are more prominent in scattered light and particles suspended in a gas are rendered visible by a light beam.

On the other hand, in an inelastic collision, energy is exchanged between the molecule and the incident photon, according to quantum rules. The selection rule for rotational transitions in Raman spectroscopy is

$$\Delta J = 0, \pm 2 \tag{13.123}$$

The reason for this is that Raman scattering is a two-photon process, with one photon going in and one coming out, and as a result the angular momentum of the photons can remain unchanged or change by two units. If enough energy is available, a vibrational transition may occur, and the selection rule is then the same as for the infrared spectra of anharmonic oscillators; namely $\Delta v = \pm 1, \pm 2, \ldots$. Generally only the lowest vibrational level is occupied to any extent and the transition $v = 1 \leftarrow v = 0$ with rotational states superimposed gives the strongest Raman band.

If the molecule gains energy ΔE, the photons will be scattered with reduced energy $h\nu_0 - \Delta E$ and result in **Stokes lines** of lower frequency than the frequency ν_0 of the incident beam; their frequency is

$$\nu = \nu_0 - \frac{\Delta E}{h} \tag{13.124}$$

Anti-Stokes lines correspond to a decrease in molecular energy, which can occur if the molecule is initially in an excited state. If the molecule loses energy, the frequency of these lines is

$$\nu = \nu_0 + \frac{\Delta E}{h} \tag{13.125}$$

Generally these lines are much weaker than the Stokes lines.

Figure 13.21a is a schematic representation of a pure rotational Raman spectrum. The line corresponding to the incident beam, of frequency ν_0, is known as the **Rayleigh line.** For pure rotational transitions the energy is given in Eq. 13.59; but since for a Raman line $\Delta J = \pm 2$, the energy difference between two levels is

$$\Delta E = E_{J+2} - E_J = Bh(4J + 6) \tag{13.126}$$

where for anti-Stokes lines J is the rotational quantum number for the lower state. The frequency of the lines is thus

$$\nu = B(4J + 6) \tag{13.127}$$

The first line is thus $6B$ from the Rayleigh line and the subsequent spacing is $4B$ (see Figure 13.21). The anti-Stokes lines are referred to as the S-branch. For Stokes lines (the O-branch) the spacings are the same.

An important advantage of Raman spectroscopy is that it is not necessary to work in the infrared and microwave regions of the spectrum. The region is determined by the choice of frequency of the incident light and can be the visible region. Also, with Raman spectra it is possible to work with aqueous solutions since one can choose an incident frequency where there is no absorption by water.

Another advantage of Raman spectroscopy is that certain lines appear in the Raman spectrum that do not appear in the infrared. The reason is that to be active in the Raman spectrum, a molecule must produce an *oscillating polarizability* α. This

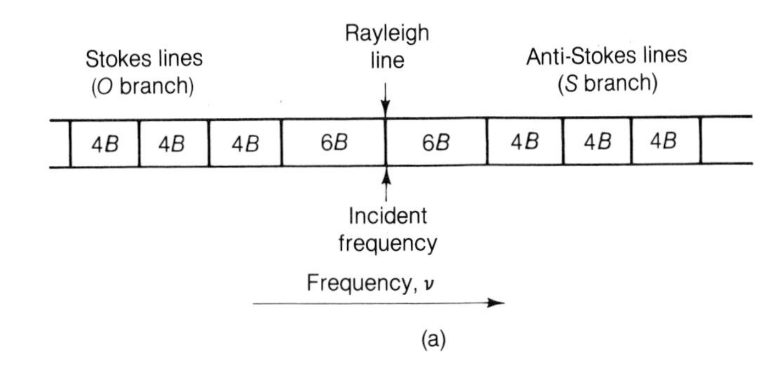

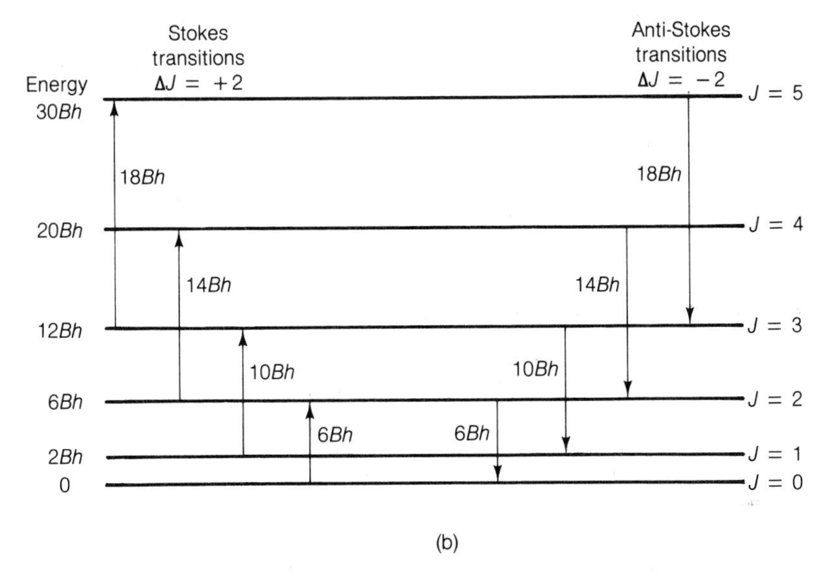

FIGURE 13.21 (a) Schematic representation of a Raman spectrum arising from pure rotational transitions. (b) The rotational transitions giving rise to the Raman lines.

is in contrast to being active in the infrared where a vibration must produce an *oscillating dipole.* This polarizability is the result of the oscillating electric field **E** in the electromagnetic radiation; this causes a distortion in the molecule, the positively charged nuclei being attracted to the negative pole of the field and the electrons to the positive pole. The polarizability of a molecule is therefore a measure of the effectiveness of the field *E* in disturbing the electron clouds. A molecule or atom with loosely bound electrons is more polarizable than one in which the electrons are tightly bound to the nucleus. For example, xenon is more polarizable than helium. The polarizability of a molecule increases with an increase in the number of electrons, and as a consequence the intensity of the scattered radiation increases with an increase of molecular mass. In general, a bond is more polarizable when it is lengthened than when it is shortened, because in the extended bond the electrons are less under the control of the nuclei.

When a molecule is distorted, an induced electric dipole moment is produced and the molecule is said to be *polarized*. The induced dipole $\boldsymbol{\mu}$ depends on the strength of the electric field and the polarizablity α of the molecule:

$$\boldsymbol{\mu} = \alpha \boldsymbol{E} \tag{13.128}$$

A rotation or vibration of frequency ν_0 will cause the electric field $\boldsymbol{E}$ to oscillate with this frequency:

$$\boldsymbol{E} = E_0 \sin 2\pi\nu_0 t \tag{13.129}$$

and the induced dipole moment also oscillates with the same frequency:

$$\boldsymbol{\mu} = \alpha \boldsymbol{E} = \alpha E_0 \sin 2\pi\nu_0 t \tag{13.130}$$

Raman lines appear if the molecular rotations or vibrations bring about an oscillating polarizability. For example, if a vibrational frequency ν_{vib} changes the polarizability, the variation of polarizability is given by

$$\alpha = \alpha_0 + R \sin 2\pi\nu_{\text{vib}} t \tag{13.131}$$

where α_0 is the polarizability before vibration and R is a coefficient that relates the change of α to the vibration. From Eq. 13.128 we have

$$\boldsymbol{\mu} = \alpha \boldsymbol{E} = (\alpha_0 + R \sin 2\pi\nu_{\text{vib}} t) E_0 \sin 2\pi\nu_0 t \tag{13.132}$$

This can be cast into the form

$$\boldsymbol{\mu} = \alpha_0 E_0 \sin 2\pi\nu_0 t + \tfrac{1}{2} R E_0 \cos 2\pi(\nu_0 - \nu_{\text{vib}})t - \cos 2\pi(\nu_0 + \nu_{\text{vib}})t \tag{13.133}$$

The components $\nu_0 \pm \nu_{\text{vib}}$ appear in addition to the incident frequency.

Raman spectra are therefore found with homonuclear diatomic molecules like H_2 and O_2,* which have no infrared spectra. The situation with polyatomic molecules is a little more complicated. We saw earlier that the symmetric vibration of carbon dioxide does not give rise to an infrared spectrum because there is no change in dipole moment, although the antisymmetric mode does give an infrared spectrum as a result of the oscillating dipole moment. The situation with Raman spectra is the converse. The symmetric vibration stretches and compresses both bonds, and the polarizability therefore varies. As a result, this vibration gives rise to Raman lines. The antisymmetric vibration, on the other hand, does not lead to a net polarizability variation because as one bond lengthens, the other bond shortens, and the two effects cancel. Thus there are no Raman lines associated with the antisymmetric vibration. To summarize the situation for CO_2: Symmetric vibrations are infrared inactive and Raman active. Antisymmetric vibrations are infrared active and Raman inactive. This can be expressed in a more general way in the form of an exclusion rule: If the molecule has a center of symmetry, vibrations that are infrared inactive are Raman active and vice versa. In view of these differences, it is advantageous to make a parallel study of both infrared and Raman spectra.

The selection rules for vibrational Raman spectra of polyatomic molecules are conveniently formulated with reference to the symmetry species of the vibrations. We have seen in our discussion of infrared spectra that an induced dipole moment

*See J. J. Barett and N. I. Adams, III, *J. Opt. Soc. Am.*, *58*, 311 (1968).

μ may be resolved along the axes X, Y, and Z. In considering Raman spectra we must also consider the inducing field $\boldsymbol{E}$, and this also has components along the X, Y, and Z axes. The dipole moment vector $\boldsymbol{\mu}$ is the product of the polarizability α and the inducing field (Eq. 13.130). Suppose that we consider the dipole moment μ_x in the X-direction induced by a field in the Y-direction:

$$\mu_x = \alpha_{xy} E_y \quad (13.134)$$

The component μ_x behaves like a translation x and it can either remain unchanged or change sign under a symmetry operation. Similarly E_y behaves like a translation y. Suppose that a particular symmetry operation changes the sign of μ_x and also changes the sign of E_y; by Eq. 13.128 this operation leaves the sign of α_{xy} unchanged. The component of the polarizability α_{xy} thus changes in the same way as x/y or as xy.

This conclusion applies to all the polarizability components, which are μ_x^2, μ_y^2, μ_z^2, μ_{xy}, μ_{yz}, and μ_{xz}. Each one is affected by the symmetry operations of the group in the same way as the corresponding product of coordinates, namely x^2, y^2, z^2, xy, yz, and xz. In order for there to be a Raman spectrum, one or more of these components must belong to the same symmetry species as one of the normal modes of vibration. To obtain the selection rules for Raman spectra we must therefore compare the symmetry species of the vibrations with those of the products x^2, y^2, z^2, xy, yz, and xz (see the appendix to this chapter, p. 638).

As an example, consider the carbon dioxide molecule, shown in Figure 13.18a. The character table (p. 558) shows the species of the previous products, and we see that the symmetric σ_g^+ vibration is of the same symmetry species as z^2; it therefore gives a Raman spectrum. The antisymmetric σ_u^+ vibration does not, nor do the bending vibrations. These conclusions are the same as those arrived at earlier.

Similarly we can see that for water (Figure 13.18b) all three vibrations are Raman active. For boron trifluoride (Figure 13.19a) the vibrations $\nu_1(a_1')$, $\nu_{3a}(e')$, $\nu_{3b}(e')$, $\nu_{4a}(e')$, and $\nu_{4b}(e')$ are Raman active, while $\nu_2(a_1)$ is Raman inactive. For formaldehyde (Figure 13.19b) all modes are Raman active.

13.6 Electronic Spectra of Molecules

We have seen (Figure 13.8) that transitions in which there is a change from one electronic state to another generally involve much higher energies than those in which there are vibrational and rotational changes only. Electronic transitions usually give rise to absorption or emission of radiation in the visible and ultraviolet regions of the spectrum. Associated with electronic transitions there are also changes of vibrational energy, which give rise to spectral bands, and changes of rotational energy, which give the fine structure of the bands. If the substance is in the gas phase, the molecular electronic spectrum is sharply defined, but in solution there is considerable blurring since the vibrations and rotations do not occur freely. Even in solution, however, substances give characteristic spectra that are useful for their identification or for determinations of their concentration.

Whereas pure rotational and vibrational-rotational spectra are not always exhibited by molecules, every substance can give an electronic spectrum because there are always excited states to which molecules can be raised by absorption of radiation. There are selection rules for electronic transitions, but before considering these we must deal with the classification of electronic states.

Term Symbols for Linear Molecules

The designations used for electronic states of molecules are somewhat analogous to those for atoms, which we considered in Section 13.2, but there are important differences. Whereas for an atom we deal with the resultant total orbital angular momentum $\boldsymbol{L}$, with quantum number L, for a linear molecule we must consider the component of orbital angular momentum along the axis of the molecule (the Z axis). This component is quantized, and it is related to $h/2\pi$ by a quantum number Λ, which can have the values 0, 1, 2 . . . :

$$\text{axial component of orbital angular momentum} = \frac{\Lambda h}{2\pi} \tag{13.135}$$

The quantum number Λ is therefore not the analog of the atomic quantum number L, which relates to the resultant angular momentum of the atom, but is equivalent to the atomic quantum number $|M_L|$ for an atom when there is an electric field along one axis.

The value of the quantum number Λ is related to the symmetry of the electron cloud with respect to the Z axis. When $\Lambda = 0$, the cloud is cylindrically symmetrical about the internuclear axis, and the state is given the term symbol Σ (Greek capital letter sigma), which is the Greek equivalent of the S state ($L = 0$) for atoms. The term symbols for all Λ values are in fact the Greek equivalents of the corresponding capital letters used with respect to the L values for atoms:

Atoms		*Linear Molecules*	
$L = 0$	S	$\Lambda = 0$	Σ (sigma)
$L = 1$	P	$\Lambda = 1$	Π (pi)
$L = 2$	D	$\Lambda = 2$	Δ (delta)
$L = 3$	F	$\Lambda = 3$	Φ (phi)

A linear molecule belongs either to the point group $C_{\infty v}$ or to the point group $D_{\infty h}$. If it has no center of symmetry, as with LiH, shown in Figure 12.17b, it is $C_{\infty v}$, and in that case a stretch is either Σ^+ or Σ^- depending on whether the state is symmetric or antisymmetric with respect to any of the σ_v planes passing through the axis of the molecule. If the linear molecule has a center of inversion, it is $D_{\infty h}$ (Figure 12.17a) and an additional classification is then possible, according to whether the wave function remains unchanged (g) or changes sign (u) on inversion.

As with atoms, the electrons in a molecule may not all be paired, and the multiplicity $(2S + 1)$ is shown as a prefix to the term symbol; for example, $^3\Sigma$.

For example, in the ground state of H_2 the spins are paired ($S = 0$), there is no angular momentum along the axis ($\Lambda = 0$), and there is no change of sign on reflection in a σ_v axis (+) or on inversion (g); the term symbol is therefore $^1\Sigma_g^+$. This is easily understood from our discussion of the hydrogen molecule in Section 12.2; in particular, the molecular orbitals are made up of the sum of 1s atomic orbitals, which are spherically symmetrical. The situation with O_2, however, is different because π orbitals are involved and two of the electrons are unpaired. We see from the molecular orbital diagram in Figure 12.21b that these two unpaired electrons are of g symmetry; this is also seen from Figure 12.20b, which in addition shows that the sign changes (−) on reflection in the XZ plane. Since these π electrons are in orbitals at right angles to the molecular axis, they do not contribute to the angular momentum along that axis; thus $\Lambda = 0$ and the state is a Σ state. The term symbol for ground-state O_2 is therefore ${}^3\Sigma_g^-$.

The term symbols for *nonlinear* molecules are again based on the symmetry species for the appropriate point groups. For further details the reader should consult the texts listed in Suggested Reading (p. 642).

Selection Rules

The spin selection rule for molecules is the same as for atoms. For light molecules the rule

$$\Delta S = 0 \tag{13.136}$$

is obeyed quite strictly, but for heavier molecules changes of multiplicity are more likely. Just as for atoms the selection rule is $\Delta L = 0, \pm 1$, that for molecules is

$$\Delta\Lambda = 0, \pm 1 \tag{13.137}$$

For Σ–Σ transitions the selection rules are

$$\begin{aligned} &+ \longleftrightarrow + \\ &- \longleftrightarrow - \\ &+ \not\longleftrightarrow - \end{aligned} \tag{13.138}$$

If the molecule has a center of inversion (point group $D_{\infty h}$), an even state tends to transform into an odd state:

$$\begin{aligned} &\mathrm{g} \longleftrightarrow \mathrm{u} \\ &\mathrm{g} \not\longleftrightarrow \mathrm{g} \\ &\mathrm{u} \not\longleftrightarrow \mathrm{u} \end{aligned} \tag{13.139}$$

However, exceptions to these restrictions are not uncommon, especially with heavier molecules.

For example, the strongest bands in the absorption spectrum of the oxygen molecule, the so-called Schumann-Runge bands, at about 200 nm, arise from the transition ${}^3\Sigma_u^- \leftarrow {}^3\Sigma_g^-$. This transition obeys the selection rules previously given. The oxygen spectrum also includes some weak bands that violate the selection rules. For example, at about 760 nm there are bands due to a ${}^1\Sigma_g^+ \leftarrow {}^3\Sigma_g^-$ transition, and there are very weak bands at about 1300 nm that arise from a ${}^1\Delta_g \leftarrow {}^3\Sigma_g^-$ transition; in both of these transitions there is a change of multiplicity, and both violate the $\mathrm{g} \not\leftrightarrow \mathrm{g}$ prohibition.

The Structure of Electronic Band Systems

Since electronic spectra also involve vibrational changes, they are often called **vibronic** spectra. Each band in a vibronic spectrum is associated with a change in vibrational energy and consists of many closely spaced rotational lines that usually have on one side a sharp edge, or *band head*. Beyond this band head the intensity falls sharply to zero, while on the other side of the band the intensity falls off more gradually and is said to be *shaded* or *degraded.*

In the Born-Oppenheimer approximation each electronic state of a diatomic molecule is represented by a curve showing potential energy plotted against internuclear distance. On this curve the various vibrational states can be indicated by horizontal lines. For a molecule containing more than two atoms the energy must in principle be represented by a multidimensional surface in which the potential energy is plotted against a number of distances and angles that specify the configuration of the molecule. Although polyatomic molecules are more difficult to deal with in practice, the basic ideas are the same as for diatomic molecules, to which our discussion will initially be confined.

The equilibrium bond distances and the shapes of potential-energy curves are in general different for different electronic states of a molecule, and the spacing of the vibrational levels is therefore different. Figure 13.22 shows two electronic states of a diatomic molecule. At ordinary temperatures the vast majority of the molecules will be in the ground electronic state and be at the lowest ($v'' = 0$) vibrational level. The absorption spectrum of such a molecule will therefore consist of a series of bands corresponding to the following transitions to vibrational levels of the upper electronic state:

$$v' = 0 \longleftarrow v'' = 0$$

$$v' = 1 \longleftarrow v'' = 0$$

$$v' = 2 \longleftarrow v'' = 0, \text{ etc.}$$

The transitions corresponding to these bands are shown in Figure 13.22. Such a series of bands is called a *progression,* and each band is usually labeled v'–v'' with the *upper* state given first; the absorption bands indicated here would thus be labeled, 0–0, 1–0, 2–0. For an emission spectra to be studied, however, energy must be supplied to the system in some way, with the production of electronically excited species in which higher vibrational levels may be populated. Transitions from these levels can occur to vibrational levels in the electronic ground state, and some of these are represented in Figure 13.22. It is evident that there may be many more bands in emission than in absorption. Some of the absorption and emission bands will be of the same frequency, but there will be emission bands (for example, the 3–1 and 1–3 bands shown in the figure) that do not appear in absorption. The detailed analysis of electronic spectra of molecules is often a matter of considerable difficulty.

There can be large differences between the intensities of different bands, and the intensity pattern is very helpful in the analysis of spectra. An important principle that governs the probabilities of transitions from one vibrational level to another is the **Franck-Condon principle.** This was first expressed qualitatively by the German-American physicist James Franck (1882–1964) who based his conclusions on the relative speeds of electronic transitions and vibrational motions. An electronic transition may take place in 10^{-15}–10^{-18} s and is so rapid compared with vibrational motion ($\approx 10^{-13}$ s) that immediately after an electronic transition has occurred, the internuclear distance is much the same as before. Franck therefore argued that the most probable electronic transitions are those that can be represented as more or less vertical transitions in diagrams such as Figure 13.22.

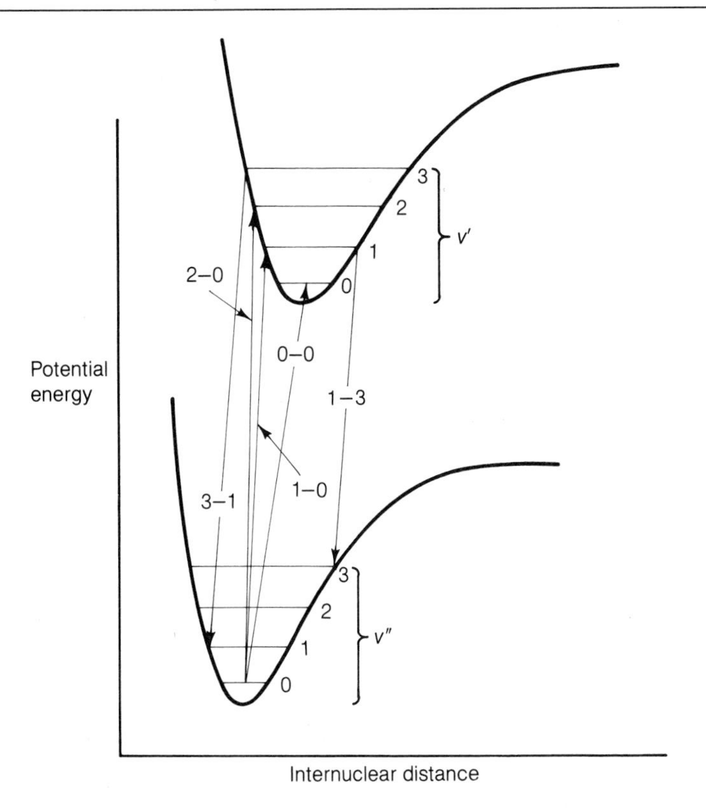

FIGURE 13.22 Two electronic states of a diatomic molecule, showing a number of vibrational levels. Transitions giving rise to absorption bands (↑) and to emission bands (↓) are shown.

This idea was put on a more rigorous quantum-mechanical basis by the American physicist Edward Uhler Condon (b. 1902). An important point in considering transitions of this kind relates to the most probable internuclear distances in vibrating molecules. According to classical mechanics, this distance is changing more slowly near the turning point of the motion; thus in Figure 13.12b the molecule is vibrating more slowly near points 1 and 2, where the motions are reversing, than when it is passing through the equilibrium position. If classical mechanics applied, we would therefore represent transitions as occurring to and from the extremities of the lines representing the vibrational levels. In quantum mechanics, on the other hand, the situation is somewhat different, as may be seen by reference to Figure 11.12b. There we see that for the *lowest* vibrational state ($v = 0$) the probability is a maximum at the *center* of the vibration. For all other vibrational states, there is a maximum slightly inside each end of the classical vibrational amplitude, and there are smaller maxima in between. In representing the probabilities of transitions we should therefore draw the connecting arrows between the *center* of the amplitude for $v = 0$ and a point close to one of the two turning points for all other vibrational states. This is the procedure used in Figure 13.22. This is of course an oversimplification, because there are smaller probability peaks at other positions, so that weaker transitions can occur from those positions.

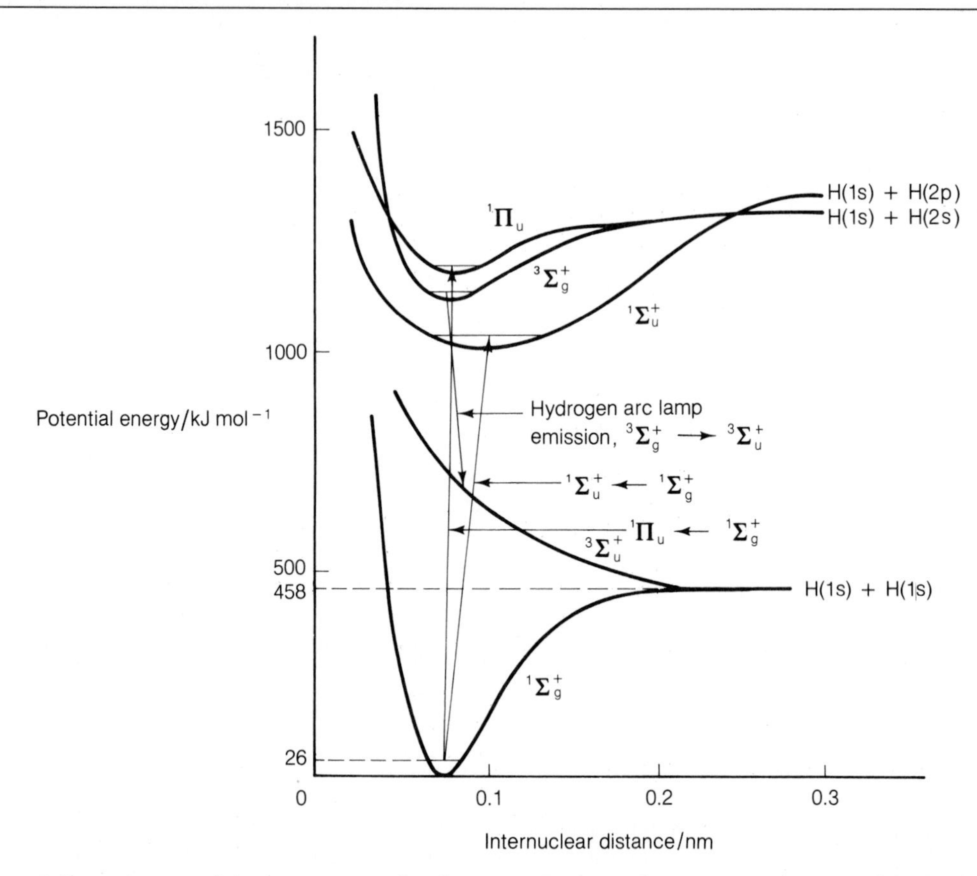

FIGURE 13.23 Schematic potential-energy curves for the ground state and some excited states of the hydrogen molecule.

Excited Electronic States

Various types of electronic transitions may be involved in the formation of electronically excited molecules. The simplest situation arises with diatomic molecules, where an electron in a bonding orbital may be transferred to a higher orbital, which may be an antibonding orbital. As an example, Figure 13.23 shows potential-energy curves for the ground state of the hydrogen molecule and also for a number of excited states. In the ground state of hydrogen both electrons are in the $1s\sigma_g$ orbital (Figure 12.4), and this electronic state is of symmetry $^1\Sigma_g^+$. Some of the higher orbitals available to electrons are shown in Figure 12.21, and one or both of the $1s\sigma_g$ electrons can be promoted to these orbitals. As a result there are a considerable number of excited states, some of which are shown in Figure 13.23. The lowest excited state is the one in which an electron has been promoted to the $1s\sigma_g^*$ orbital, and since this is antibonding, the state is repulsive; this is the $^1\Sigma_u^+$ state shown in the figure. Promotion to higher orbitals leads to states of higher energy that dissociate not into two ordinary hydrogen atoms but into atoms in excited states. States in which both electrons have opposite spins ($S = 0$) are singlet states, while those in which the spins are the same ($S = 1$) are triplets. Singlet-triplet transitions are forbidden by the selection rules, which apply rather rigorously to a molecule as small as hydrogen. The absorption spectrum of H_2 shows bands corresponding to the $^1\Sigma_u^+ \leftarrow {}^1\Sigma_g^+$ and $^1\Pi_u \leftarrow {}^1\Sigma_g^+$ transitions. In a hydrogen arc lamp, molecules are produced in the $^3\Sigma_g^+$ state, and the characteristic emission is due to the transition $^3\Sigma_g^+ \rightarrow {}^3\Sigma_u^+$.

A somewhat different situation may be found with more complex molecules containing groups known as **chromophores** (Greek *chroma,* color; *phoros,* carrying). For example, substances such as acetone that contain the carbonyl group, C=O, all exhibit a weak absorption band having a maximum at about 285 nm and a stronger absorption band near 200 nm. The absorption coefficient for the band at 285 nm is usually around 10 $dm^3\ mol^{-1}\ cm^{-1}$. At 200 nm the absorption coefficient is usually between 10^3 and 10^4 $dm^3\ mol^{-1}\ cm^{-1}$. These two absorption bands are easily understood with reference to the nature of the carbonyl bond.

Some of the electrons relating to the carbonyl group are nonbonding and are given the symbol n. Others are in the π bond (valence-bond theory) or π orbital (molecular-orbital theory). Figure 13.24 shows the types of transitions that can occur. A nonbonding (n) electron can be promoted into an empty antibonding π orbital based on carbon, and the transition marked $\pi^* \leftarrow n$ at ~285 nm is known as a *$\pi^* \leftarrow n$ transition,* read "n-to-π-star transition." Since $\pi^* \leftarrow n$ transitions are often symmetry forbidden, the absorption is weak. The more intense absorption at about 200 nm is caused by the promotion of a π electron into the antibonding π^* orbital in the electronically excited state. This is known as a $\pi^* \leftarrow \pi$ transition. This is the same mechanism responsible for the absorption found in other double bonds such as the carbon-carbon double bond, C=C.

The presence of chromophores such as the carbonyl group often causes the molecule to absorb in the visible part of the spectrum and therefore to be colored. The contribution of a particular chromophore to the absorption spectrum is considerably altered by conjugation with other chromophores. Thus, an isolated carbon-carbon double bond exhibits a $\pi^* \leftarrow \pi$ transition at about 180 nm. However, if there

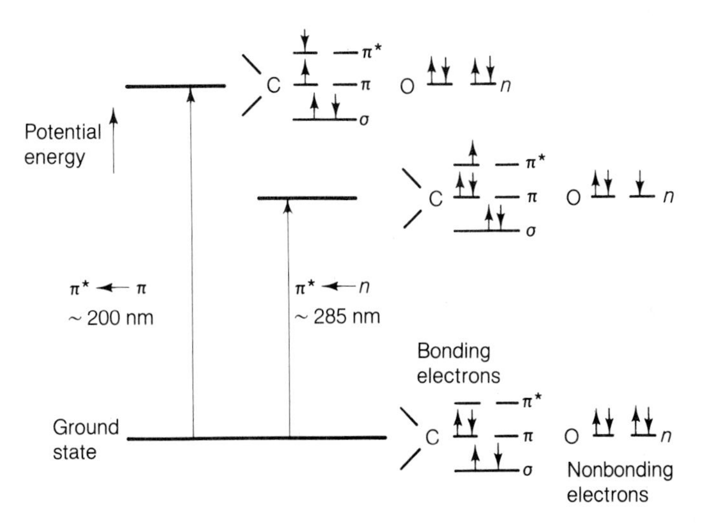

FIGURE 13.24 Two types of electronic transitions that can occur with a chromophone such as the carbonyl group.

is conjugation (i.e., alternating double and single bonds), the absorption is shifted to much higher wavelengths of around 450–500 nm.

Numerous electronic-spectrum assay procedures for determining concentrations are based on the absorption by chromophores. For example, the reaction of ninhydrin with most amino acids yields a purple product having a maximum absorption at 570 nm. Ninhydrin has a molar absorption coefficient $\epsilon_{1\,\mathrm{cm}}^{1\,M}$ of $\approx 10^{11}$ dm^3 mol^{-1} cm^{-1}, which is extremely large, and this means that amino acids in amounts of the order of 10 nanomoles (nmol) can be estimated. This reagent is used in forensic science (e.g., to obtain the fingerprints of a person who has handled a piece of paper).

The Fate of Electronically Excited Species

After an electronically excited molecule or radical has been produced (e.g., by absorption of radiation or in an electric discharge), a number of things can happen to it.

In the first place, the species can emit radiation and pass into a lower state, which may be the ground state or a lower excited state. This emission of radiation is referred to as **fluorescence.*** This term is generally restricted to processes in which there is no change of multiplicity, in contrast to **phosphorescence** where there is a multiplicity change and in which the emission is slower. Transitions giving rise to fluorescence are included in Figure 13.22. Whereas absorption generally occurs from the ground vibrational state, fluorescent emissions often give excited vibrational states, as shown in Figure 13.22, and therefore correspond to longer wavelengths than observed in absorption. This shift to longer wavelengths is called the *Stokes shift*. Fluorescence often occurs after certain radiationless processes have occurred, and this is considered later.

The formation of an excited species can also lead to *molecular dissociation*, which can occur by different mechanisms according to the relative positions of the potential-energy curves for ground and excited states. Figures 13.25a and b relate to a diatomic molecule AB and show two situations that are of particular importance. In Figure 13.25a a more-or-less vertical transition from the lowest vibrational level ($v'' = 0$) of the ground state takes the molecule to point *a* on the diagram. Here the energy level is higher than for A + B, and as a result the molecule dissociates in its first vibration. The spectrum corresponding to this transition shows no rotational fine structure and appears *diffuse*, because the excited molecule AB* does not live long enough for any rotations to occur. Other transitions from $v'' = 0$ may take the system to lower vibrational levels, such as to point *b* for which $v' = 1$, and then no dissociation can occur. The band corresponding to this $1 \leftarrow 0$ transition will have a rotational fine structure, as will the $1 \rightarrow 0$ emission band.

Another situation, corresponding to what is known as **predissociation,** is illustrated in Figure 13.25b. A transition from $v'' = 0$ to $v' = 2$, for example, will produce an AB* molecule that has insufficient energy to dissociate, and a rotational fine structure will therefore be observed. However, the diagram shows a *repulsive* state of AB* that crosses the upper potential-energy curve at a level corresponding to $v' = 4$, and $4 \leftarrow 0$ transition therefore produces a species that, in its first vibration, reaches point *a* where it undergoes a transition to the repulsive state and dis-

* This word is derived from the mineral fluorspar (Latin *fluor*, flowing; German *Spat*, crystalline material).

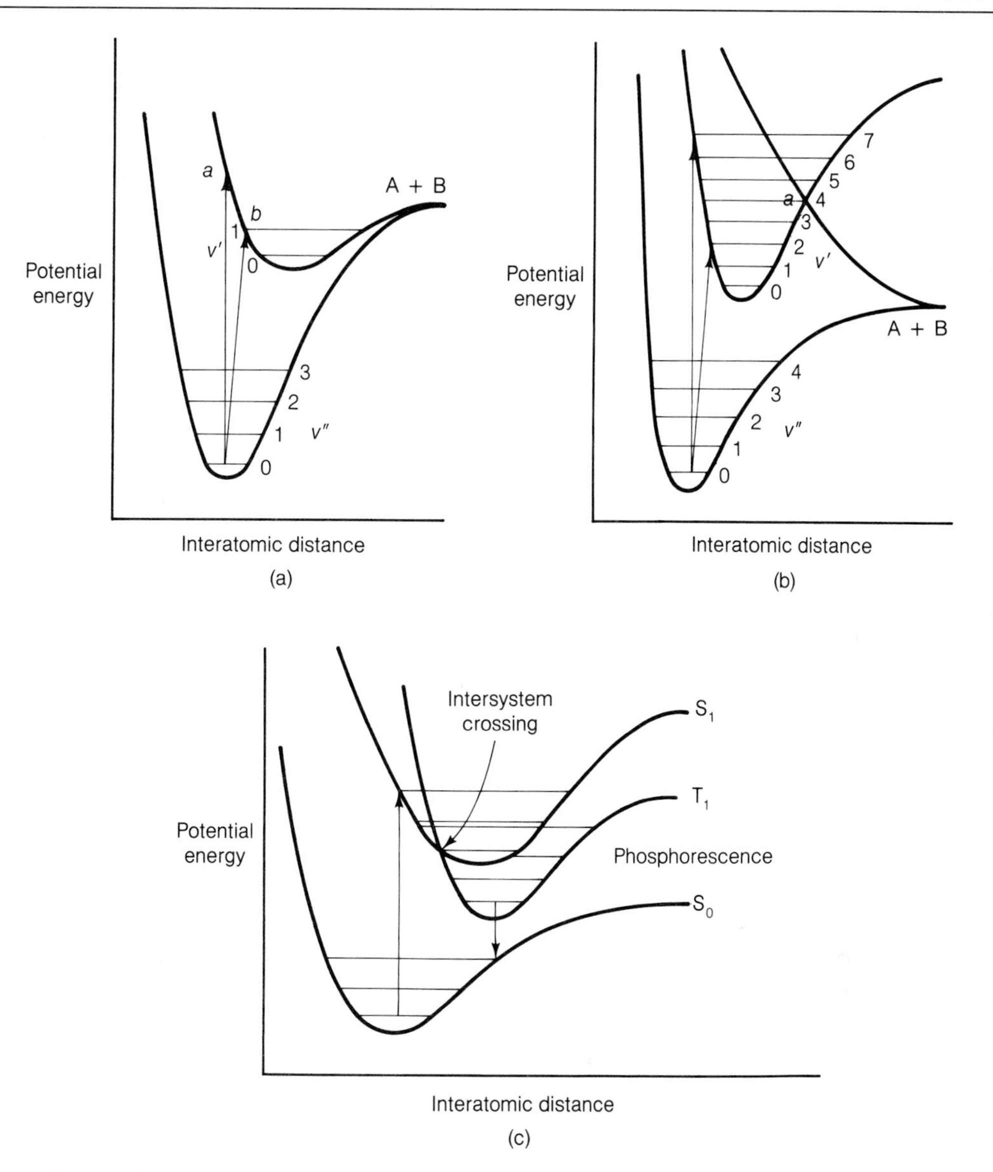

FIGURE 13.25 (a) Potential-energy curves for a diatomic molecule. The transition to point *a* leads to immediate dissociation. (b) Potential-energy curves for the case of predissociation. (c) Potential-energy curves that will lead to phosphorescence. Absorption from the singlet ground state S_0 gives the singlet excited state S_1; after a transition to the triplet state T_1, the molecule phosphoresces.

sociates. The $4 \leftarrow 0$ band in the absorption spectrum therefore appears diffuse, as no rotations can occur. Transitions to a vibrational level such as $v' = 7$ may, on the other hand, produce a species that does not easily dissociate, since when it vibrates, it moves rather rapidly past point *a* and as a result the transition probability is reduced. The $v' = 7$ vibrational state of AB* may therefore have a long enough life for rotations to occur, and there will be a rotational fine structure corresponding to this $7 \leftarrow 0$ transition.

A characteristic feature of a predissociation spectrum is therefore that there is a diffuse region, corresponding to transitions to the repulsive state followed by dissociation, and that at longer and shorter wavelengths the spectrum shows a rotational fine structure. This is a different situation from that found with ordinary dissociation (Figure 13.25a), where the rotational fine structure can only appear on the lower-energy (longer-wavelength) side of the diffuse bands.

These diagrams (Figure 13.25a and b) relate to a diatomic molecule AB, but the principles are somewhat the same for more complex molecules. However, some additional matters must be taken into consideration for molecules having more than two atoms. If a *diatomic* molecule has enough energy to undergo dissociation, it is bound to do so in the period of its first vibration (i.e., within 10^{-13}–10^{-14} s). However it is possible for a *polyatomic* molecule to have sufficient energy to dissociate without dissociating in the first vibration. This is because the energy may not at first be suitably distributed among the normal modes of vibration for bond rupture to occur. As the molecule vibrates, the energy is constantly redistributed between the normal modes, and eventually a vibration may lead to dissociation. As a result the lifetime of a complex molecule with enough energy to dissociate may be 10^{-8} s or more, which is several orders of magnitude greater than that of a diatomic molecule. During this period of time a number of alternative processes may occur to a polyatomic molecule, such as the following:

1. **Vibrational Relaxation.** During this process the molecule undergoes collisions with other molecules and drops to lower vibrational levels within the same electronic state.

2. **Collisional Quenching,** or **External Conversion.** On collision with other molecules the electronically excited molecule may pass into a lower electronic state, which is often the ground state. The electronic energy is converted in this process into translational and internal energy and is eventually dissipated as heat.

Since these two processes *involve collisions,* their rates depend on the frequency of collisions and therefore on the availability of the colliding molecules. In the gas phase these processes may occur within 10^{-8} s at ordinary pressures, but in the liquid phase they are much faster and may occur in 10^{-12}–10^{-14} s.

The following processes that a molecule may undergo *do not involve collisions:*

3. **Internal Quenching,** or **Internal Conversion.** In this process the electronic energy is converted into vibrational energy of the molecule itself, which therefore acts as a self-quencher.

The following processes are *radiative:*

4. **Fluorescence.** We have already considered fluorescence, which involves emission of radiation (generally with no change of multiplicity) and occurs in 10^{-9}–10^{-8} s. This is usually a much slower process than vibrational relaxation, quenching, or internal conversion.

5. **Phosphorescence.** The term *fluorescence* is usually restricted to processes in which there is no change of multiplicity. The term *phosphorescence,* on the other hand, is applied to processes in which there is a change of multiplicity. Many

examples of phosphorescence involve triplet → singlet transitions and are preceded by processes in which an excited singlet state is converted into an excited triplet state. Figure 13.25c shows a typical system of potential-energy curves that lead to phosphorescence. The molecule in a singlet ground electronic state S_0 absorbs radiation that excites it to the state S_1, the potential-energy curve for which crosses that for an excited triplet state T_1. The species S_1 can therefore make an *intersystem crossing* to the triplet state at the point where the two curves cross. This is a *radiationless transition* and it often occurs only with difficulty. However, if the molecule is not too light, there is a considerable amount of spin-orbit coupling, in which states with different spin and orbital angular momenta mix because they have the same total angular momentum. This triplet state may be vibrationally excited and can undergo vibrational relaxation on collision with other molecules. The triplet species is then in its ground vibrational state where it cannot readily lose radiation since the transition involves a change of multiplicity (triplet → singlet). However, because of spin-orbit coupling the emission does occur, but very slowly. This emission sometimes occurs over periods of seconds or minutes. The process is particularly slow with certain solids at low temperatures.

Figure 13.26, known as a **Jablonski diagram**, summarizes some of the processes that can occur after absorption of radiation. Solid vertical lines represent

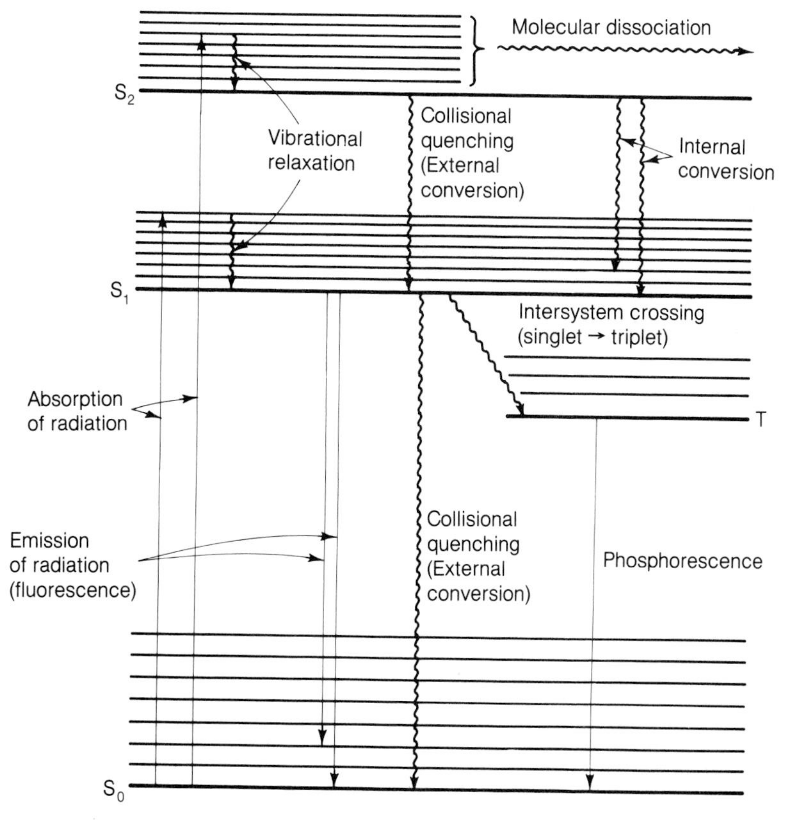

FIGURE 13.26 A Jablonski diagram, which shows the ground singlet state, S_0, two excited singlet states S_1 and S_2, and a triplet state, T. Various possible processes are represented.

processes accompanied by the emission of radiation, while radiationless transitions are indicated by wavy lines. More processes are possible than shown, and further details are to be found in some of the books listed in Suggested Reading (p. 642).

13.7 Laser Spectroscopy

During recent years, the techniques of spectroscopy have been greatly enhanced by the introduction of lasers. The word *laser* is an acronym for "*l*ight *a*mplification by *s*timulated *e*mission of *r*adiation." The development of this field began in 1953 with the introduction by the American physicist Charles H. Townes of the *maser*, which stands for "*m*icrowave *a*mplification by *s*timulated *e*mission of *r*adiation." Work in the microwave region was soon extended to other regions of the spectrum, including the visible and ultraviolet. Today, laser spectroscopy is carried out in many laboratories and a variety of applications have been developed.

In a laser beam, excitation energy is supplied to a substance in such a way as to produce a *population inversion*, in which more atoms are in a specific excited level than in the ground-state level. Normally there are more molecules in ground states than in any of the excited states, but if energy is supplied (a process called *pumping*), this situation can be reversed. Such an excited state can release energy spontaneously, but the unique feature of a laser is that the release may be accomplished by a process known as *stimulated emission*. During this process, a photon released by one atom or molecule will interact with an excited species having a population inversion and will stimulate the species to release a photon of the same wavelength and traveling in the same direction. Thus, as a beam progresses through the excited laser medium, the intensity is greatly increased. A common device is to reflect the initial beam by means of a mirror, with the result that it passes again through the medium and is further amplified. When the beam reaches a partially reflecting mirror, a portion of it escapes, and this is the active emission from the laser.

A commonly used laser is the *ruby laser*, in which the medium is a rod of ruby; this is sapphire, Al_2O_3, in which one out of every 10^2 to 10^3 Al^{3+} ions has been replaced by a Cr^{3+} ion; this gives the resulting crystal its characteristic red color. The population inversion is usually brought about by means of xenon flashlamps that are placed in a highly reflecting housing in order to focus the lamp emission on the laser rod. Some lasers operate continuously and we speak of *continuous-wave* (cw) *lasers*. However, a common technique is to pulse the exciting light for short periods of time (e.g., a few milliseconds), and the power developed in a pulsed laser may be as high as a gigawatt (GW = 10^9 W). The ruby laser produces light of 694.3-nm wavelength, which lies toward the red end of the visible spectrum. Of particular importance is the fact that the light is highly monochromatic (i.e., it covers a very narrow range of wavelengths). Also, if the laser medium (e.g., the ruby) has a suitable geometry, the beam divergence can be extremely small, perhaps of the order of a milliradian, and is said to be *coherent*. Although most lasers have a specific wavelength, a recent development is the introduction of tunable chemical lasers that within limits can be adjusted to a desired wavelength of emission.

Because of these special features, laser investigations have many advantages over conventional spectroscopic studies. For instance, for many years Raman spectroscopy presented serious experimental difficulties, but laser Raman spectroscopy has largely eliminated the main problems. Furthermore, the fact that very narrow beams of coherent light can be produced leads to new techniques, such as measurements on individual biological cells. New methods using laser beams are still being developed and some are carried out routinely for analytical determinations on chemical and biological materials.

13.8 Transition Probabilities

If atoms or molecules are excited from a lower state m to a higher state n by the absorption of light of frequency ν_{mn}, the number of transitions N_{mn} to the higher level is proportional to the number of molecules N_m in the lower level and also to the *spectral energy density*, $w(\nu_{mn})$, which is the density of radiation at the frequency ν_{mn}. The relationship is therefore

$$N_{mn} = B_{mn} N_m w(\nu_{mn}) \tag{13.140}$$

where the proportionality factor B_{mn} is known as the **Einstein transition probability for absorption.**

It was Einstein who pointed out that there are two processes to consider in emission, *radiation-induced emission* and *spontaneous emission*. The number of transitions N_{nm} brought about by the radiation is

$$N_{nm}(\text{induced}) = B_{nm} N_n w(\nu_{nm}) \tag{13.141}$$

where B_{nm} is the **Einstein coefficient of stimulated emission.** These two coefficients B_{nm} and B_{mn} are equal to one another. The spontaneous emission does not depend on the radiation and is given by

$$N_{nm}(\text{spontaneous}) = A_{nm} N_n \tag{13.142}$$

where A_{nm} is the **Einstein coefficient of spontaneous emission.**

The total number of transitions from the higher to the lower state is thus

$$N_{nm}(\text{induced and spontaneous}) = N_n[A_{nm} + B_{nm} w(\nu_{nm})] \tag{13.143}$$

A steady state will be established in which the number of molecules passing from m to n is equal to the number passing from n to m:

$$N_m B_{mn} w(\nu_{mn}) = N_n[A_{nm} + B_{nm} w(\nu_{nm})] \tag{13.144}$$

The value of A_{nm} cannot be calculated from radiation theory, but it can be deduced in the following way on the basis of the theory of the Boltzmann distribution. According to this theory, which is considered in Section 14.2, at equilibrium the ratio of populations of molecules in the two states n and m is

$$N_n/N_m = e^{-h\nu_{mn}/\mathbf{k}T} \tag{13.145}$$

Equations 13.144 and 13.145 give, since $B_{nm} = B_{mn}$,

$$e^{-h\nu_{mn}/\mathbf{k}T} = \frac{B_{nm} w(\nu_{mn})}{A_{nm} + B_{nm} w(\nu_{nm})} \tag{13.146}$$

The spectral energy density $w(\nu_{nm})$ is given by the Planck radiation law (compare Eq. 11.36):

$$w(\nu_{nm}) = \frac{8\pi h \nu_{nm}^3}{c^3} \frac{1}{1 - e^{-h\nu_{nm}/\mathbf{k}T}} \tag{13.147}$$

Since Eq. 13.146 holds at equilibrium, we may substitute Eq. 13.147 into it and obtain, after simplification,

$$\frac{A_{nm}}{B_{nm}} = \frac{8\pi h \nu_{nm}^3}{c^3} \tag{13.148}$$

Note the frequency dependence of this ratio. In optical spectroscopy, spontaneous emission is usually much larger than stimulated emission, but when the frequencies are much smaller, as in magnetic resonance (Section 13.10), the stimulated emission is predominant.

13.9 Spectral Line Widths

Transitions between different sets of well-defined energy levels result in spectral lines of different line widths. There are several reasons why the lines spread over a range of frequencies. In molecular spectra in the gaseous state, superposition of a number of energy levels that are close together may cause broadening, but we will consider two more fundamental reasons.

Doppler Broadening

One mechanism that leads to a broadening of a spectral line is the *Doppler effect*. An object that approaches a static observer with a speed u and at the same time emits radiation of wavelength λ appears to be emitting radiation at a wavelength $[1 - (u/c)]\lambda$. If the object recedes from the observer, its radiation appears to have a wavelength of $[1 + (u/c)]\lambda$. Since molecules in a gas travel at very high speeds in all directions, there is a range of Doppler shifts and the line is therefore broadened. The magnitude of this broadening can be predicted from the distribution of velocities as given by the Maxwell-Boltzmann distribution (Eq. 14.47). We thus find that the predicted shape of this distribution in a single direction is a bell-shaped Gaussian curve, as shown in Figure 13.27. The distribution is temperature dependent, broadening as the temperature rises. (Further discussion of this is given in Section 14.3.) The width $\Delta\lambda$ at half-height of any line may be found from

$$\Delta\lambda = 2\left(\frac{\lambda}{c}\right)(2\mathbf{k}T/m)^{1/2} \tag{13.149}$$

where m is the mass of the species involved in the transition. To obtain the maximum resolution of spectral lines, the spectra should be taken at the lowest possible temperature. High temperatures (such as the temperature of the sun) may be determined by measuring the broadening of a particular spectral line.

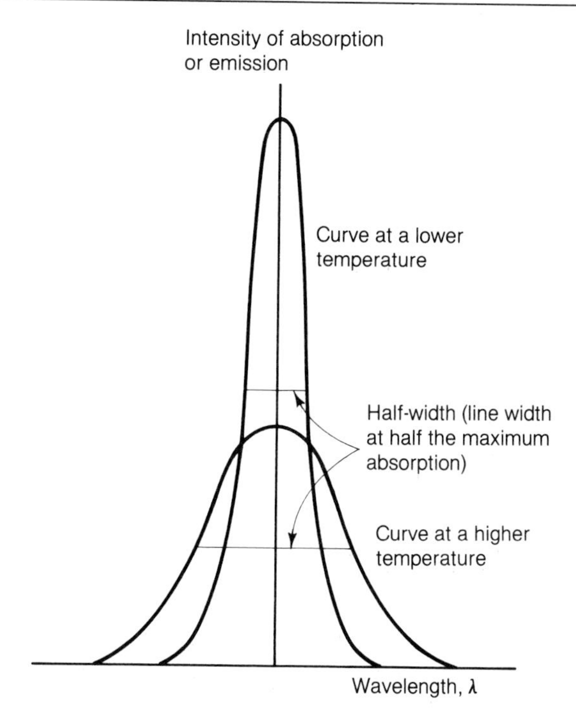

FIGURE 13.27 Line widths at two different temperatures as determined by the Gaussian distribution.

Lifetime Broadening

A second mechanism which is important even at very low temperatures is a quantum-mechanical effect called *lifetime broadening*. Because of the Heisenberg uncertainty relation it is impossible to specify exactly the energies of the levels in a transition. Only if the state has an infinite lifetime can its energy be specified exactly; in reality, since no excited state has an infinite lifetime, no excited state has a precisely defined energy. The shorter the lifetime of an excited state, the broader the expected spectral lines.

The uncertainty principle may be used to estimate the extent of broadening. From Eq. 11.64,

$$\Delta E \, \Delta\tau \approx \frac{h}{4\pi} \tag{13.150}$$

The energy spread ΔE is therefore inversely proportional to the lifetime $\Delta\tau$, and it follows that the corresponding broadening of the line is

$$\Delta\bar{\nu}/\text{cm}^{-1} \approx \frac{2.7 \times 10^{-12}}{\Delta\tau/\text{s}} \tag{13.151}$$

The radiative lifetime of an electronically excited species may be as long as 10^{-7} s, which leads to a line width of $\approx 2.75 \times 10^{-5}$ cm^{-1}; this is too small to be measured. When there can be molecular dissociation or internal conversion, however, the lifetime may be 10^{-11} s or less; this value gives a broadening of about 0.27 cm^{-1}, which can easily be measured with a high-resolution instrument.

13.10 Electron Spin Resonance Spectroscopy

The phenomenon of *resonance* has many familiar examples. If two pendulums of the same frequency are attached to a common support, the oscillation of one induces oscillation in the other. The oscillations are said to be *coupled*, and energy flows back and forth between the two. The tuning of a radio to the electromagnetic field produced by a transmitter is another example of resonance.

There are some important spectroscopic applications of resonance. They are based on observing the strong absorption of radiation that occurs when there is perfect matching of energy. In *electron spin resonance* (esr) *spectroscopy*, also known as *electron paramagnetic resonance* (epr) *spectroscopy*, the energy difference between levels is created by applying a powerful magnetic field to species containing unpaired electrons. There are two ways in which the electron spins can be aligned in the field, and there are therefore two energies. Resonance occurs when the frequency of the radiation corresponds to a photon energy that matches the difference between the energy levels. The radiation required in experiments of this kind is usually in the microwave region of the spectrum, since the energy difference is very small.

The apparatus used in esr spectroscopy is shown schematically in Figure 13.28a. Microwaves are generated in a *klystron tube* and are directed by means of a metal *wave guide* into the sample under investigation, which is maintained in a glass or quartz container supported between the poles of a powerful electromagnet. It is usual to operate the klystron tube at a fixed frequency and to vary the magnetic field until resonance is achieved. Some commercial esr instruments operate in the *X band* of the microwave region (8 to 12 GHz, or 3.75 to 2.5 cm); others operate in the *K band* (27 to 35 GHz, or 1.1 to 0.85 cm).

To give an esr spectrum the species under investigation must have at least one unpaired electron. This excludes most molecules; nitric oxide (NO), which has 15 electrons, is one example of a molecule with an odd electron. Free radicals, such as CH_3, also have an odd electron, and certain diradicals have two unpaired electrons; the same is true of certain molecules, such as O_2, that occur in triplet states.

The theory of esr is similar to that of the Zeeman effect, which we considered in Section 13.2. Equation 13.46 gives the energy of an *orbiting* electron in a magnetic field:

$$E = \mu_B m_l B \tag{13.152}$$

In electron spin resonance we are instead concerned with the magnetic moment arising from the electron *spin*, for which the spin quantum number has the possible values $+\frac{1}{2}$ and $-\frac{1}{2}$. The corresponding energy equation is now

$$E = g\mu_B m_s \mathrm{B} \tag{13.153}$$

where g is known as the *g factor*. For a completely free electron, g has the value of 2.0023. The g value for an electron depends somewhat on its environment and ranges from about 1.9 to 2.1. As a result, precise measurements of g values can be used in the identification of radicals.

Since m_s can have a value of $+\frac{1}{2}$ or $-\frac{1}{2}$, the variation of energy with the magnetic flux density is as shown in Figure 13.28b. At a given B the difference between the

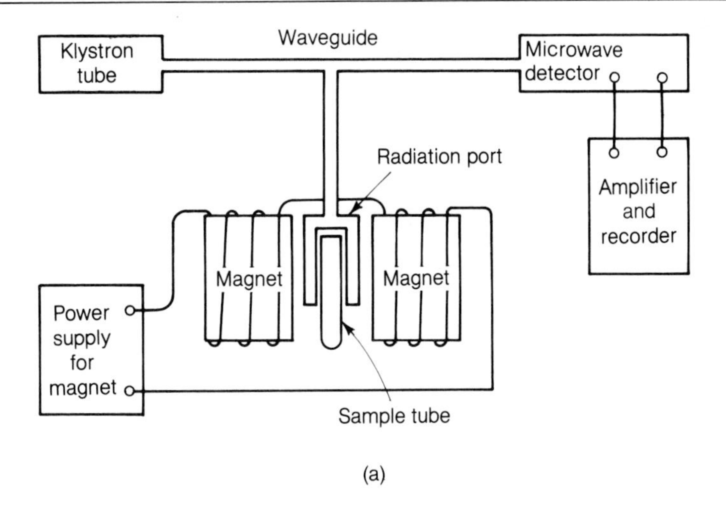

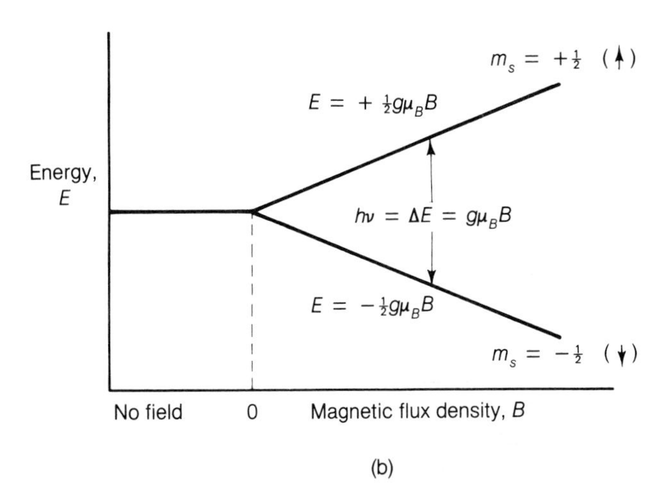

FIGURE 13.28 Schematic diagram of an electron spin resonance spectrometer. (b) The variation of the electron spin energy with the magnetic flux density.

energy states for the two spins is

$$\Delta E = \tfrac{1}{2}g\mu_B B - (-\tfrac{1}{2})g\mu_B B = g\mu_B B \tag{13.154}$$

Resonance therefore occurs at a frequency given by

$$\nu = \frac{g\mu_B B}{h} \tag{13.155}$$

EXAMPLE

An esr instrument is operating at a frequency of 9.10 GHz, and measurements are made with atomic hydrogen. Resonance is observed at a magnetic flux density of 0.3247 T. Calculate the g value for an electron in the hydrogen atom.

SOLUTION

From Eq. 13.155 the g value is given by

$$g = \frac{h\nu}{\mu_B B}$$

$$= \frac{6.626 \times 10^{-34}(\text{J s}) \times 9.10 \times 10^{9}(\text{s}^{-1})}{9.274 \times 10^{-24}(\text{J T}^{-1}) \times 0.3247(\text{T})}$$

$$= 2.0024$$

■

Hyperfine Structure

The type of esr spectrum that is commonly observed is shown in Figure 13.29. We usually see more than one peak, and we then speak of the *hyperfine structure*. The most important contribution to this splitting of lines is the interaction between the electron spin magnetic moment and the nuclear spin magnetic moment. The magnitude of the nuclear spin angular momentum is $\sqrt{I(I+1)}h/2\pi$ where I is the nuclear spin quantum number. This expression is analogous to that for electron spin (Eq. 11.210), with the spin quantum number s replaced by the nuclear spin quantum number I. Again, by analogy with the electron spin, the nuclear spin angular momentum must be quantized along the direction of a magnetic field and can have the values

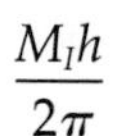

$$\frac{M_I h}{2\pi}$$

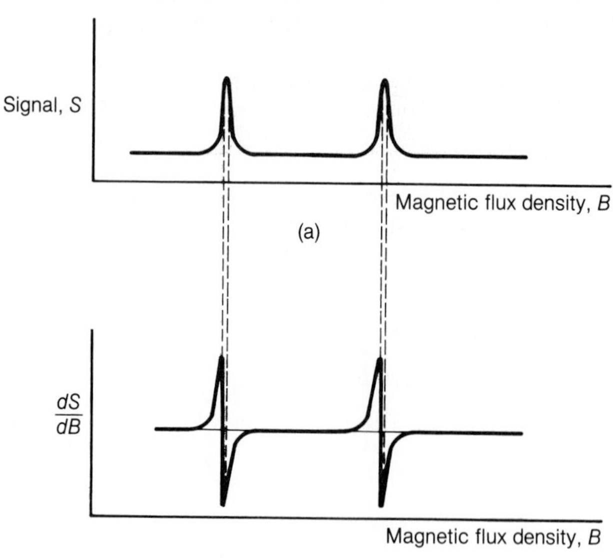

FIGURE 13.29 (a) An esr absorption curve, in which the signal S is shown as a function of the magnetic field B. (b) The first derivative of the signal, dS/dB, shown as a function of B. A common experimental technique produces the results in this form.

where M_I is a quantum number; its permitted values are

$$M_I = I, I-1, I-2, \ldots, 0, \ldots, -I \tag{13.156}$$

The simplest behavior is found with the hydrogen atom, the nucleus of which is a single proton; I is now $\frac{1}{2}$, and the permitted values of M_I are $+\frac{1}{2}$ and $-\frac{1}{2}$. The situation is thus analogous to that of the spinning electron. With composite nuclei, however, other values of I are possible, and some values are given in Table 13.3. The value of I must be integral or half-integral.

Again, the magnetic moment of the nucleus is proportional to the angular momentum, and it is in the opposite direction. In a magnetic field the nuclear magnetic dipole precesses about the direction of the field, in the same manner as for the spinning electron (Figure 11.19). The component of the magnetic moment in the direction of the field (the Z-direction) is

$$\mu_{I,z} = g_N M_I \mu_N \tag{13.157}$$

This expression is to be compared with Eq. 13.44 for the orbiting electron. The Bohr magneton μ_B has been replaced by the **nuclear magneton μ_N**, defined by

$$\mu_N = \frac{eh}{4\pi m_p} \tag{13.158}$$

where m_p is the mass of the proton. The quantum number m_l has been replaced by M_I, which relates to nuclear spin, and Eq. 13.157 also involves the **nuclear g factor** g_N. Some values of this factor are given in Table 13.3. An important point to note is that the nuclear magnetic moments are much smaller than those due to orbital motion or to electron spin. The reason is that the nuclear magneton μ_N is very much smaller than the Bohr magneton μ_B, because the much greater nuclear mass appears in the denominator of the expression. The value of the nuclear magneton is

$$\mu_N = \frac{1.602\ 19 \times 10^{-19}(\text{C})6.6262 \times 10^{-34}(\text{J s})}{4\pi(1.672\ 65 \times 10^{-27})(\text{kg})} = 5.0508 \times 10^{-27}\ \text{J T}^{-1}$$

to be compared with 9.274×10^{-24} J T^{-1} for the Bohr magneton.

The electron-nuclear magnetic interaction energy is small in comparison with the electron spin resonance energy and is independent of the strength of the magnetic field. The situation with the hydrogen atom is particularly simple, since I is $\frac{1}{2}$ and

TABLE 13.3 Some Spin Properties of Nuclei

Isotope	Spin, I	g_N
^{1}H	$\frac{1}{2}$	5.5856
^{2}H	1	0.8574
^{13}C	$\frac{1}{2}$	1.4048
^{19}F	$\frac{1}{2}$	5.257
^{35}Cl	$\frac{3}{2}$	0.5479
^{37}Cl	$\frac{3}{2}$	0.4560

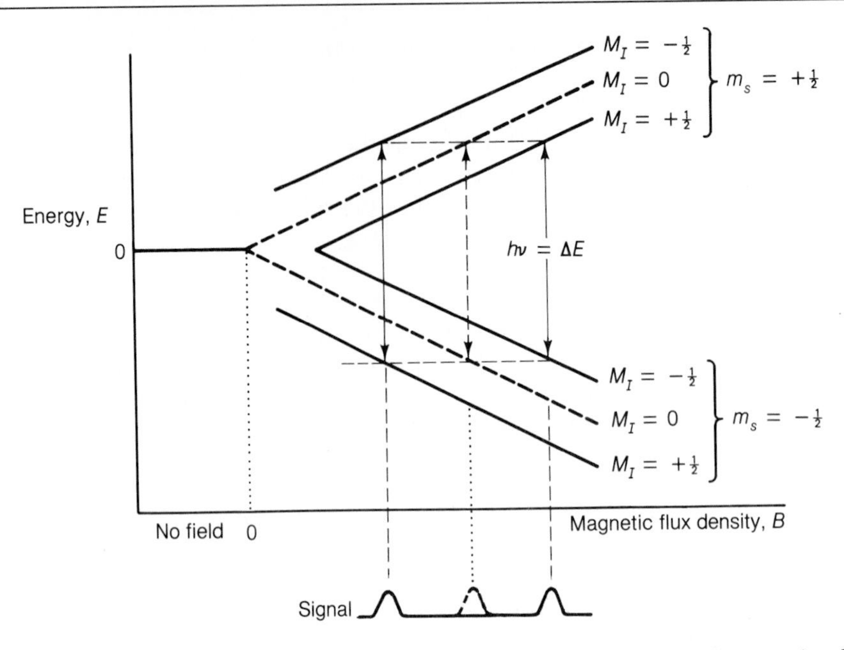

FIGURE 13.30 The energy splitting obtained with a hydrogen atom, for which the permitted nuclear spin quantum numbers are $+\frac{1}{2}$ and $-\frac{1}{2}$. The dashed lines correspond to the situation in Figure 13.28, where there is no interaction with the nucleus.

M_I can be only $+\frac{1}{2}$ or $-\frac{1}{2}$. Each one of these can combine in two ways with the two electron spins, and four levels are therefore generated, as shown in Figure 13.30. The vertical dashed line corresponds to the magnetic field that would give resonance if there were no interaction with the nuclear magnetic moment. Because of the interaction there is resonance at a lower field and at a higher field; there are therefore two lines. The separation between the lines depends on the interaction between the nucleus and the electron.

In general, a nuclear quantum number of I gives rise to $2I + 1$ values of M_I (Eq. 13.156) and therefore to $2I + 1$ peaks in the spectrum. The number of lines observed is thus useful in identifying the nucleus. For example, ^{35}Cl has a spin quantum number I of $\frac{3}{2}$, and the M_I values are therefore $+\frac{3}{2}$, $+\frac{1}{2}$, $-\frac{1}{2}$, $-\frac{3}{2}$. The hyperfine structure in the esr spectrum of the ^{35}Cl atom therefore consists of four lines. The situation is more complicated for species other than atoms, where there is more than one nucleus.

13.11 Nuclear Magnetic Resonance Spectroscopy

Whereas electron spin resonance arises from the spin of the electron, nuclear magnetic resonance (nmr) results from nuclear spin. We have seen that the nuclear magneton is three orders of magnitude smaller than the Bohr magneton, so that the interaction between a nucleus and a magnetic field is very much smaller than that of an electron. The nuclear energy splitting is therefore much smaller, and the frequencies required to achieve resonance are much smaller. For esr the frequencies are microwave frequencies in the gigahertz (GHz $= 10^9\ s^{-1}$) region. By contrast, for

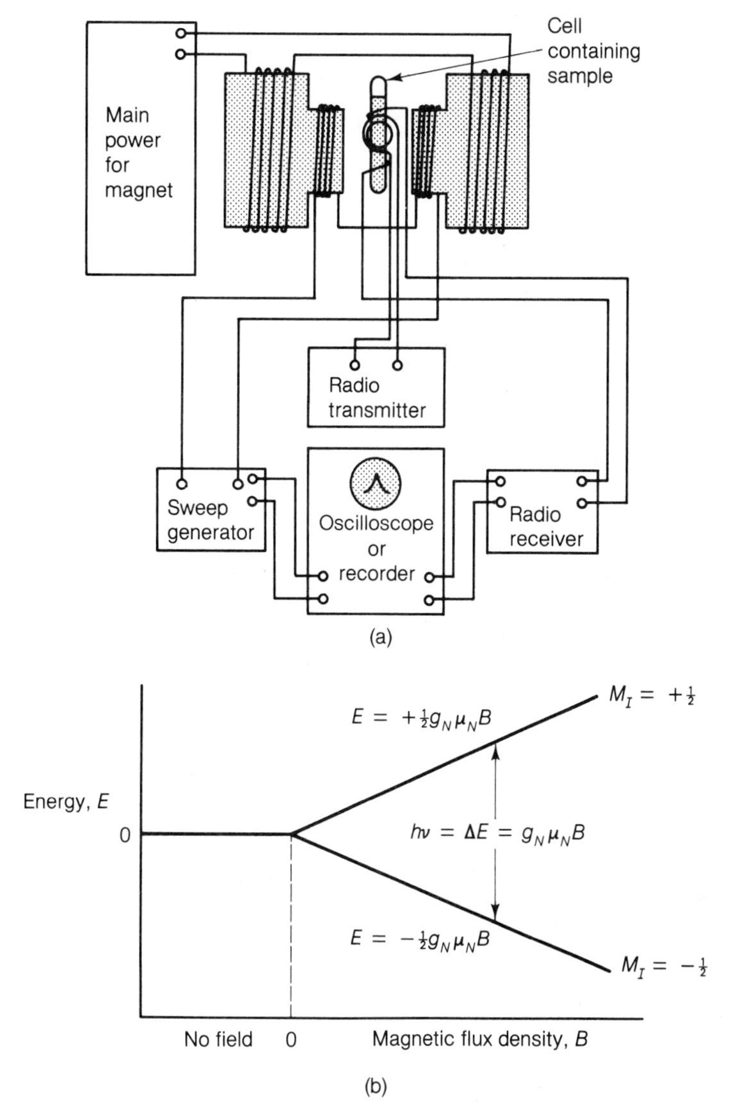

FIGURE 13.31 (a) Schematic diagram of a nuclear magnetic resonance spectrometer. (b) The splitting by a magnetic field of the energy levels for the spin of the proton.

comparable magnetic fields the frequencies required for nmr are in the megahertz (10^6 s^{-1}) region; these are shortwave radio frequencies. In fact, for a magnetic field of 1.5 T almost all nmr spectra fall between 10 and 65 MHz, which means wavelengths of about 30 to 5 m.

Nuclear magnetic resonance was discovered in 1946 independently by Edward Mills Purcell* (b. 1912) and Felix Bloch† (b. 1905). The type of experimental arrangement commonly used today is shown in Figure 13.31a. It is most convenient, as

---

* E. M. Purcell, H. C. Torrey, and R. V. Pound, *Phys. Rev.*, *69*, 37 (1946).

† F. Bloch, W. W. Hansen, and M. E. Packard, *Phys. Rev.*, *69*, 127 (1946).

with esr spectroscopy, to hold the frequency constant and to vary the magnetic field until resonance is achieved. The magnetic field is usually varied by superimposing a small *sweep field* on the main field, which is produced by the large magnet. When resonance occurs, there is a voltage oscillation that is detected by the radio receiver and is amplified and displayed on an oscilloscope screen or otherwise recorded. Typical modern instruments use a magnet producing a field of about 1 T, which means that they work at a radio frequency of about 60 MHz. However, some instruments use superconducting magnets that produce fields of about 7 T, and resonance is then achieved at frequencies of about 300 MHz.

The basic theory of nmr has been covered in our discussion of hyperfine structure in esr spectroscopy. Equation 13.157 gives the Z component of the nuclear magnetic moment. The possible values of M_I relate to the I values, some of which are given in Table 13.3. In a magnetic field of flux density B the corresponding energy is the magnetic moment $\mu_{I,z}$ multiplied by B:

$$E = \mu_{I,z}B = g_N M_I \mu_N B \tag{13.159}$$

For the proton $I = \frac{1}{2}$, and the allowed values of M_I are $+\frac{1}{2}$ and $-\frac{1}{2}$; the energy is thus

$$E = \pm\tfrac{1}{2}g_N\mu_N B \tag{13.160}$$

The relationship between E and B for the proton is shown in Figure 13.31b. Since g_N for the proton is 5.586, the magnetic moment is $\frac{1}{2} \times 5.586 = 2.793$ nuclear magnetons. The resonance condition arises when the frequency of the radiation corresponds to the difference between two nuclear spin states, and for the proton it is therefore given by

$$\Delta E = h\nu = \tfrac{1}{2}g_N\mu_N B - (-\tfrac{1}{2})g_N\mu_N B = g_N\mu_N B \tag{13.161}$$

EXAMPLE

Calculate the energy difference between the proton spin states and the corresponding frequency for a magnetic field of 1.50 T.

SOLUTION

From Eq. 13.161 with $g_N = 5.5856$,

$$\Delta E = g_N\mu_N B = 5.5856(5.0508 \times 10^{-27}\ \text{J T}^{-1})(1.50\ \text{T}) = 4.23 \times 10^{-26}\ \text{J}$$

The frequency is therefore

$$\nu = \frac{\Delta E}{h} = \frac{4.23 \times 10^{-26}(\text{J})}{6.626 \times 10^{-34}(\text{J s}^{-1})}$$

$$= 63.8 \times 10^{6}\ \text{s}^{-1} = 63.8\ \text{MHz}$$

■

Equation 13.161 also applies to other nuclei, since the selection rule is $\Delta M_I = \pm 1$.

Chemical Shifts

One reason for the importance of nmr spectroscopy is that several absorption lines are obtained, corresponding to effects of the chemical environment on the various atomic nuclei. What happens is that a secondary magnetic field is generated by the original external magnetic field. This secondary field opposes the applied field but is proportional to it. The *effective magnetic field* B_{eff} at the nucleus is written as

$$B_{\text{eff}} = (1 - \sigma)B \tag{13.162}$$

where σ is called the *shielding constant* and has a value of approximately 10^{-5}–10^{-6}. The effect of the field σB on the nmr spectrum is referred to as the **chemical shift.**

This technique has been of enormous value in structural determinations, particularly with organic molecules, and in providing more detailed information about the nature of the chemical bond. The principle underlying such studies is that the magnetic field that penetrates to the nucleus is highly dependent on the nature of the electron clouds surrounding the nucleus. As an example, the nuclear magnetic spectrum of ethanol

```
    H  H
    |  |
H—C—C—O
    |  |    \
    H  H     H
```

is shown in Figure 13.32. Figure 13.32a shows the results obtained with a low-resolution apparatus, and Figure 13.32b shows the results with a high-resolution apparatus capable of splitting the main peaks into a number of smaller peaks.

The characteristic feature of the spectrum in Figure 13.32a is that there are three peaks, A, B, and C. The areas below the peaks are in the ratio 1/2/3. These peaks correspond to three groups of protons in three different magnetic environments. One group, A, consists of the single proton bonded to the oxygen atom; the second group, B, consists of the two protons of the methylene group; and the third group, C, consists of the three protons of the methyl group. These are easily identified

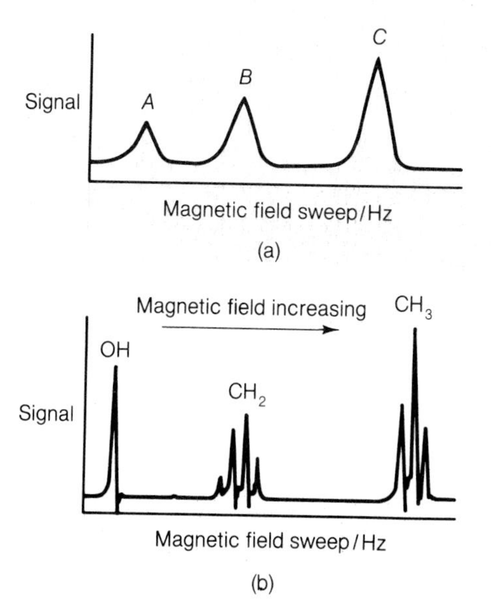

FIGURE 13.32 The nuclear magnetic resonance spectrum of ethanol. (a) The results with a low-resolution instrument. (b) The spin-spin splitting obtained with a high-resolution instrument.

because the areas under the peaks are proportional to the numbers of protons in the groups.

To find how the spacing between these groups varies, we can compare the value of B needed to produce resonance in a standard compound with the value needed to produce the same effect in the compound under investigation. Equating the right-hand side of Eq. 13.162 for a compound with that for the reference standard, we have

$$B_{ref}(1 - \sigma_{ref}) = B(1 - \sigma) \tag{13.163}$$

This gives, on rearrangement,

$$\frac{B_{ref} - B}{B_{ref}} = \frac{\sigma_{ref} - \sigma}{1 - \sigma} \approx \sigma_{ref} - \sigma \tag{13.164}$$

since $1 - \sigma \approx 1$. It is usual to express this in terms of a field-independent chemical shift δ, defined in parts per million (ppm) as

$$\delta = (\sigma_{ref} - \sigma) \times 10^6 \tag{13.165}$$

Since the free nucleus is not a convenient reference, the field at which a single sharp resonance occurs for some other compound may be taken as the reference standard. Tetramethylsilane, $(CH_3)_4Si$, which has a large chemical shift, is often used for this purpose.

A second method of expressing the chemical shift is to add 10 ppm to the δ values based on tetramethylsilane. This is called the τ scale, and most protons then have positive values. With ethanol, for example, the methyl protons lie at $\tau = 8.83$, and the methylene protons at 6.41; with aromatic compounds, the values are still lower.

Spin-Spin Splitting

When the experiments are carried out at higher resolution, as shown in Figure 13.32b, the peak for CH_2 is split into four peaks, and the peak for CH_3 into three peaks. The extent of this splitting (in contrast to the splitting into the three main peaks) does not depend on the strength of the magnetic field and therefore cannot be a chemical shift due to shielding of the nuclei by the surrounding electrons. The effect is caused by the interaction of the nuclear spins of one set of protons with those of another (*spin-spin coupling*) and is known as *spin-spin splitting*. The reasons for the coupling are the same as involved in spin-spin coupling discussed in Section 13.2. In each case the effect is transmitted from one nucleus through the intervening electrons to the neighboring nucleus. The greatest effect involves neighboring nuclei, although next-nearest-neighbor effects are also observed.

In order to predict the number of lines, we note that each proton in a group feels a slightly different magnetic field depending on its relation to the spins in the neighboring nuclei. As an example, the methyl group protons in ethanol may have the four following sets of spins:

1. $+\frac{1}{2}, +\frac{1}{2}, +\frac{1}{2}$ (1 spin state) $\quad I_{total} = +\frac{3}{2}$
2. $+\frac{1}{2}, +\frac{1}{2}, -\frac{1}{2}; +\frac{1}{2}, -\frac{1}{2}, +\frac{1}{2}; -\frac{1}{2}, +\frac{1}{2}, +\frac{1}{2}$ (3 spin states) $\quad I_{total} = +\frac{1}{2}$

3. $+\frac{1}{2}, -\frac{1}{2}, -\frac{1}{2}; -\frac{1}{2}, +\frac{1}{2}, -\frac{1}{2}; -\frac{1}{2}, -\frac{1}{2}, +\frac{1}{2}$ (3 spin states) $I_{\text{total}} = -\frac{1}{2}$
4. $-\frac{1}{2}, -\frac{1}{2}, -\frac{1}{2}$ (1 spin state) $I_{\text{total}} = -\frac{3}{2}$

There are therefore four possible values of M_I, and each proton in the methylene group experiences slightly different magnetic fields arising from these four values. The signals for the methylene protons are therefore split into four lines, with intensities given by the ratio of the possible ways in which the spin states may combine, namely 1/3/3/1. In the same way, the methylene protons may have the three following sets of spins:

1. $+\frac{1}{2}, +\frac{1}{2}$ (1 spin state) $I_{\text{total}} = +1$
2. $+\frac{1}{2}, -\frac{1}{2}; -\frac{1}{2}, +\frac{1}{2}$ (2 spin states) $I_{\text{total}} = 0$
3. $-\frac{1}{2}, -\frac{1}{2}$ (1 spin state) $I_{\text{total}} = -1$

Each proton in the methyl group thus feels three slightly different magnetic fields; there are therefore three lines with intensities in the ratio of the number of spin states for each energy, namely 1/2/1.

On this basis we would expect the hydroxyl proton resonance to be split into three lines because of the neighboring methylene group. However, Figure 13.32b shows only one line for OH. The reason for this is the presence of small amounts of protonic acids that catalyze the rapid exchange of protons between the OH groups of different ethanol molecules. The time spent by a proton in a particular conformation is therefore too short to permit spin-spin interaction with the proton of the methylene and methyl groups. However, if the sample is carefully prepared, with no acid present, the OH proton resonance is split into three peaks, and the four methylene resonances are split into eight. Similar effects are also seen with the methyl group.

Nuclear magnetic resonance has been very important in the determination of the structures of organic molecules and for analyses. It has also been used to give information about equilibrium and rates of rapid chemical reactions in which there is a change in the magnetic environment of a nucleus.

13.12 Mössbauer Spectroscopy

In 1958 the German physicist Rudolf Ludwig Mössbauer (b. 1929) discovered what has come to be known as the **Mössbauer effect.** It is concerned with the emission of γ (Greek *gamma*) rays, which under ordinary conditions are emitted with a considerable spread of wavelength. This is because of the recoil of the emitting nuclei. However, in Mössbauer's experiments the emitting nuclei were part of a crystal, the recoil of which is very small, and the γ rays are therefore emitted with an exceedingly narrow spread of wavelength. In some experiments the spread is as low as 1 part in 10^{13}. An important application of these highly homogeneous γ rays was to provide a very crucial test of Einstein's theory of relativity. They have also found useful spectroscopic applications.

To take a specific example, the primary source of the γ rays is sometimes the ^{57}Co nucleus. This decays slowly (with a half-life of 270 days) to give an excited ^{57}Fe

nucleus, which emits γ rays with a half-life of 2×10^{-7} s:

$$^{57}\text{Co} \xrightarrow{t_{1/2} = 270 \text{ days}} {}^{57}\text{Fe}^* \xrightarrow{t_{1/2} = 2 \times 10^{-7} \text{ s}} {}^{57}\text{Fe} \quad (\searrow \gamma \text{ rays})$$

The frequency of the emitted γ rays is about 3.5×10^{18} Hz, which corresponds to an energy of 2.32×10^{-15} J or 14.5 keV. The half-life of 2×10^{-7} s corresponds to a frequency uncertainty of about 4×10^5 Hz (see Problem 35). If this were the only factor, the ratio of the frequency uncertainty to the frequency would be $(4 \times 10^5$ Hz)/(3.5×10^{18} Hz), which is about 10^{-13}. There would thus be a very high degree of homogeneity. However, under ordinary conditions the nucleus recoils and the homogeneity is much smaller. Mössbauer's important contribution was to overcome this difficulty by holding the nuclei rigidly in a crystal.

EXAMPLE

Calculate the frequency range of γ rays of frequency 3.5×10^{18} Hz emitted by (a) a free ^{57}Fe atom and (b) a ^{57}Fe atom held in a crystal of mass 1 g.

SOLUTION

The momentum of the photon is $h\nu/c$ (from Eq. 11.55), and if the recoiling entity is of mass m and recoils with a velocity u,

$$\frac{h\nu}{c} = mu$$

The energy $h\nu$ is $6.626 \times 10^{-34} \times 3.5 \times 10^{18}$ J = 2.32×10^{-15} J.

a. The mass of the ^{57}Fe atom is 57×10^{-3} kg/6.022×10^{23} = 9.47×10^{-26} kg. The speed of recoil is therefore

$$u = \frac{h\nu}{cm} = \frac{2.32 \times 10^{-15}(\text{J})}{2.998 \times 10^8(\text{m s}^{-1}) \times 9.47 \times 10^{-26}(\text{kg})}$$
$$= 81.7 \text{ m s}^{-1}$$

We saw in Section 13.9 that the change of wavelength due to the Doppler effect is $u\lambda/c$, and the corresponding change of frequency is thus $u\nu/c$. The frequency range is therefore

$$\Delta\nu = \frac{81.7(\text{m s}^{-1}) \times 3.5 \times 10^{18}(\text{Hz})}{2.998 \times 10^8(\text{m s}^{-1})}$$
$$= 9.5 \times 10^{11} \text{ Hz}$$

This is very much larger than the frequency uncertainty of 4×10^5 Hz that arises from the half-life of the parent nucleus.

b. If the mass is 1 g, the speed of recoil is

$$u = \frac{2.32 \times 10^{-15}(\text{J})}{2.998 \times 10^8(\text{m s}^{-1}) \times 10^{-3}(\text{kg})} = 7.74 \times 10^{-21} \text{ m s}^{-1}$$

The frequency range is now

$$\Delta\nu = \frac{7.74 \times 10^{-21}(\text{m s}^{-1}) \times 3.5 \times 10^{18}(\text{Hz})}{2.998 \times 10^{8}(\text{m s}^{-1})}$$

$$= 9.04 \times 10^{-11}\ \text{Hz}$$

Since this is negligible compared with 4×10^5 Hz, the latter is the uncertainty in the frequency. ■

Mössbauer spectroscopy involves the *resonant absorption of* γ *rays by atomic nuclei*. For example, the γ rays emitted by a $^{57}Fe^*$ nucleus, of frequency 3.5×10^{18} Hz, may be absorbed by another ^{57}Fe nucleus. The frequency at which resonant absorption occurs is affected to a small extent by the valence state of the atom containing the absorbing nucleus. For example, ^{57}Fe in Fe_2O_3 absorbs at a slightly different frequency from ^{57}Fe in $Fe(CN)_6^{4-}$ or in hemoglobin. Similarly, ^{119}Sn in Sn(II) covalent molecules absorbs at a different frequency from ^{119}Sn in Sn(IV) compounds. The differences may be as large as 100 MHz; this is exceedingly small compared with the frequency of the γ radiation itself, but it is much larger than the uncertainty in the frequency. Consequently, the frequency changes brought about by the different valence states can be measured satisfactorily.

To do this it is necessary to be able to control the frequency of the γ radiation, and this is done by taking advantage of the Doppler effect. The source of radiation is mounted on a sliding support, which is moved at a carefully controlled rate. The following example shows that quite small speeds are sufficient to produce frequency shifts of 100 MHz or less, which are required for resonance to be achieved.

EXAMPLE

Calculate the frequency shift brought about by moving a ^{57}Co source at a speed of 5 mm s^{-1}.

SOLUTION

The Doppler shift is of magnitude $\nu u/c$, where u is the speed of the source. The frequency of the radiation is 3.5×10^{18} Hz, and the shift is therefore

$$\Delta\nu = \frac{3.5 \times 10^{18}(\text{Hz}) \times 5 \times 10^{-3}(\text{m s}^{-1})}{2.998 \times 10^{8}(\text{m s}^{-1})} = 5.84 \times 10^{7}\ \text{Hz}$$

$$= 58.4\ \text{MHz}$$

■

The remarkable feature of Mössbauer spectroscopy is that although the energies of the γ rays are extremely large, very much smaller energy differences can be measured. This was only made possible by Mössbauer's device of limiting the range of frequencies by holding the emitting source in a crystal lattice.

Appendix to Chapter 13: Symmetry Species Corresponding to Infrared and Raman Spectra

Symmetry Species	Infrared Spectra			Raman Spectra					
	x	y	z	x^2	y^2	z^2	xy	xz	yz
C_{2v}	b_1	b_2	a_1	a_1	a_1	a_1	a_2	b_1	b_2
C_{3v}	e	e	a_1	a_1, e	a_1, e	a_1	e	e	e
$C_{\infty v}$	π	π	σ^+	$\sigma^{\cdot}\ \delta$	σ^+, δ	σ^+	δ	π	π
$D_{\infty h}$	π_u	π_u	σ_u^+	σ_g^+, δ	σ_g^+, δ_g	σ_g^+	δ_g	π_g	π_g
D_{3h}	e'	e'	a_2''	a_1', e'	a_1', e'	a_1'	e'	e''	e''
C_{2h}	b_u	b_u	a_u	a_g	a_g	a_g	a_g	b_g	b_g
D_{6h}	e_{1u}	e_{1u}	a_{2u}	a_{1g}, e_{2g}	a_{1g}, e_{2g}	a_{1g}	e_{2g}	e_{1g}	e_{1g}

A more complete table is given in G. Herzberg, *Infrared and Raman Spectra*, New York: D. Van Nostrand Co., 1945, p. 252.

Key Equations

Lambert-Beer law:

$$A = \log_{10} \frac{I_0}{I} = \epsilon c l$$

where A = absorbance; ϵ = absorption coefficient.

$$\log_{10} T\% = 2 - A$$

where $T\%$ = percentage transmittance.

The *gyromagnetic* (magnetogyric) *ratio* is the ratio of the magnetic moment and the angular momentum:

$$\gamma \equiv \frac{\mu_l}{L} = -\frac{e}{2m_e}$$

for an orbiting electron.

The Bohr magneton:

$$\mu_B = \frac{eh}{4\pi m_e}$$

Frequencies of rotational transitions:

$$\nu_J = 2(J + 1)B$$

where $B = h/8\pi^2 I$ and I = moment of inertia.

Morse potential function for the dissociation energy D:

$$E_p - D_e(1 - e^{-ax})^2$$

where $x = r - r_e$.

Problems

A Absorption of Radiation

1. A spectrophotometer has a meter that gives a reading directly proportional to the amount of light reaching the detector. When the light source is off, the reading is zero. With pure solvent in the light path, the meter reading is 78; with an 0.1 M solution of a solute in the same solvent, the meter reading is 55. The light path is 0.5 cm. Calculate the absorbance, the transmittance, and the molar absorption coefficient.

2. An aqueous solution containing 0.95 g of oxygenated myoglobin (M_r = 18 800) in 100 cm^3 gave a transmittance of 0.87 at 580 nm, with a path length of 10.0 cm. Calculate the molar absorption coefficient.

3. The molar absorption coefficient of hemoglobin at 430 nm is 532 dm^3 mol^{-1} cm^{-1}. A solution of hemoglobin was found to have an absorbance of 0.155 at 430 nm, with a light path of 1.00 cm. Calculate the concentration.

4. Two substances of biological importance, NAD^+ and NADH, have equal absorption coefficients, 1.8×10^4 dm^3 mol^{-1} cm^{-1}, at 260 nm (a wavelength at which absorption coefficients are equal is known as the *isosbestic** point). At 340 nm, NAD^+ does not absorb at all, but NADH has an absorption coefficient of 6.22×10^3 dm^3 mol^{-1} cm^{-1}. A solution containing both substances had an absorbance of 0.215 at 340 nm and of 0.850 at 260 nm. Calculate the concentration of each substance.

*From the Greek prefix *iso-*, the same, and *sbestos*, quench. This word is sometimes incorrectly written as "isobestic."

5. The transmittance of a 0.01 M solution of bromine in carbon tetrachloride, with a path length of 2 mm, is 28%. Calculate the molar absorption coefficient of bromine at that wavelength. What would the percentage transmittance be in a cell 1 cm thick?

B Atomic Spectra

6. The ground state of the Li atom has the electronic configuration $1s^2 2s^1$. What is its spectroscopic term? If the 2s electron is excited to the 2p state, what terms are then possible?

7. Suppose that an excited state of the carbon atom has the electronic configuration $1s^2 2s^2 2p^1 3p^1$. What are the possible spectroscopic terms?

8. What are the terms for the following electronic configurations?

a. Na ($1s^2 2s^2 2p^6 3p^1$)
b. Sc ($1s^2 2s^2 2p^6 3s^2 3p^6 4s^2 3d^1$)

9. What values of J may arise in the following terms?

$$^1P, \ ^3P, \ ^4P, \ ^1D, \ ^2D, \ ^3D, \ ^4D$$

***10.** Calculate the Landé g factor for a $^2P_{1/2}$ level. What would be the anomalous Zeeman splitting for this level in a magnetic field of 4.0 T?

***11.** Calculate the spacing between the lines for a $^3D_1 \rightarrow {}^3P_0$ transition, in an anomalous Zeeman experiment with a magnetic field of 4.0 T.

C Rotational and Microwave Spectra

12. The separation between neighboring lines in the pure rotational spectrum of $^{35}Cl^{19}F$ is found to be 1.023 cm^{-1}. Calculate the interatomic distance.

13. The lines in the pure rotational spectrum of HF are 41.9 cm^{-1} apart. Calculate the interatomic distance. Predict the separation between the lines for DF and TF.

14. In the microwave spectrum of $^{12}C^{16}O$ the separation between lines has been measured to be 115 270 MHz. Calculate the interatomic distance.

D Vibrational-Rotational and Raman Spectra

15. Consider the following molecules H_2, HCl, CO_2, CH_4, H_2O, CH_3Cl, CH_2Cl_2, H_2O_2, NH_3, SF_6. Which of them will give

- **a.** a pure rotational spectrum,
- **b.** a vibrational-rotational spectrum,
- **c.** a pure rotational Raman spectrum,
- **d.** a vibrational Raman spectrum?

16. Analysis of the vibrational-rotational spectrum of the $H^{35}Cl$ molecule shows that its fundamental vibrational frequency ν_0 is 2988.9 cm^{-1}. Calculate the force constant of the H—Cl bond.

17. The vibrational Raman spectrum of $^{35}Cl_2$ shows series of Stokes and anti-Stokes lines; the separation between the lines in each of the two series is 0.9752 cm^{-1}. Estimate the bond length in Cl_2.

18. The dissociation energy of H_2 is 432.0 kJ mol^{-1} and the fundamental vibrational frequency of the molecule is 1.257×10^{14} s^{-1}. Calculate the classical dissociation energy. Estimate the zero-point energies of HD and D_2 and their dissociation energies.

19. A molecule AB_2 is known to be linear but it is not known whether it is B—A—B or A—B—B. Its infrared spectrum is found to show bands corresponding to three normal modes of vibration. Which is the structure?

20. The frequency of the O—H stretching vibration in CH_3OH is 3300 cm^{-1}. Estimate the frequency of the O—D stretching vibration in CH_3OD.

21. Irradiation of acetylene with mercury radiation at 435.83 nm gives rise to a Raman line at 476.85 nm. Calculate the vibrational frequency that corresponds to this shift.

22. The fundamental vibrational frequency of $H^{127}I$ is 2309.5 cm^{-1}. Calculate the force constant of the bond.

E Electronic Spectra and Line Widths

23. Sketch-potential energy curves for a diatomic molecule in its ground electronic state and in an excited state, consistent with the following observations:

- **a.** There is a strong $0 \leftarrow 0$ absorption band, and strong $0 \rightarrow 0$, $1 \rightarrow 1$, and $2 \rightarrow 2$ emission bands.
- **b.** The strongest absorption band is $4 \leftarrow 0$, and the strongest emission band is $0 \rightarrow 2$.
- **c.** There is no sharp rotational fine structure in absorption, but there is a sharp emission spectrum.
- **d.** The absorption spectrum shows a well-defined fine structure for the $0 \leftarrow 0$, $1 \leftarrow 0$, $2 \leftarrow 0$, $3 \leftarrow 0$, and $4 \leftarrow 0$ transitions and for the $6 \leftarrow 0$ and $7 \leftarrow 0$, but not in between.

24. The sun emits a spectral line at 677.4 nm and it has been identified as due to an ionized ^{57}Fe atom, which has a molar mass of 56.94 g mol^{-1}. Its width is 0.053 nm. Estimate the temperature of the surface of the sun.

25. Estimate the lifetime of a state that, because of lifetime broadening, gives rise to a line of width

- **a.** 0.01 cm^{-1},
- **b.** 0.1 cm^{-1},
- **c.** 1.0 cm^{-1},
- **d.** 200 MHz.

26. The dissociation energy (from the zero-point level) of the ground state O_2 ($^3\Sigma_g^-$) molecule is 5.09 eV. There exists an electronically excited $^3\Sigma_u^-$ state of O_2, whose zero-point level lies 6.21 eV above the zero-point level of the ground state. The ground-state molecule dissociates into two ground-state O (^3P) atoms, while the $^3\Sigma_u^-$ species dissociates into one ground-state O (^3P) atom and an O* (^1D) atom that lies 1.97 eV above the ground state. Sketch the potential-energy curves and calculate the dissociation energy of O_2 into O + O*.

27. Sodium vapor, which consists mainly of Na_2 molecules, has a system of absorption bands in the green, the origin of the 0, 0 band being at 20 302.6 cm^{-1}. From the spacing of the vibrational levels it can be deduced that the dissociation energy of the upper state is 0.35 eV. The dissociation of the excited Na_2 gives a normal atom and an atom that emits the yellow sodium D line at 589.3 nm. Calculate the energy of dissociation of Na_2 in its ground state.

F Resonance Spectroscopy

28. Calculate the magnetic flux density that is required to bring a free electron ($g = 2.0023$) into resonance in an esr spectrometer operating at a wavelength of 8.00 mm.

29. An esr spectrometer is operating at a frequency of 10.42 GHz and study is made of methyl radicals. Resonance is observed at a magnetic flux density of 0.371 75 T. Calculate the g value of the methyl radical.

30. How many hyperfine lines would you expect to find in the esr spectrum of ^{2}H, ^{19}F, ^{35}Cl, ^{37}Cl? Calculate the nuclear magnetic moment in each case (refer to Table 13.3).

31. In a nuclear magnetic resonance instrument operating at a frequency of 60 MHz, at what magnetic fields would you expect to observe resonance with $^{1}H^{35}Cl$?

32. The chemical shift δ of the methyl protons in acetaldehyde is 2.20 ppm, and that of the aldehydic proton is 9.80 ppm. What is the difference in the effective magnetic field for the two types of proton when the applied field is 1.5 T? If resonance is observed at 60 MHz, what is the splitting between the methyl and aldehyde proton resonances?

33. The nuclear spin quantum number I of the ^{39}K nucleus is $\frac{3}{2}$, and the nuclear g factor is 0.2606. How many orientations does the nucleus have in a magnetic field? At what frequency would there be resonance in a field of 1.0 T?

34. The ^{11}B nucleus has a spin I of $\frac{3}{2}$ and a nuclear g factor of 1.7920. At what field would resonance be observed at 60 MHz?

35. The lifetime of $^{57}Fe^{*}$ is 2×10^{-7} s. Calculate the uncertainty in the frequency of the γ radiation emitted and in the wave number.

G Supplementary Problems

36. The following are some normal modes of vibration for several molecules:

cis-ClHC=CHCl : **a.** C=C stretch:

b. C=C twist:

c. Asymmetric C—H and C—Cl stretches:

d. Symmetric C—H and C—Cl stretches:

trans-ClHC=CHCl : **e.** C=C stretch:

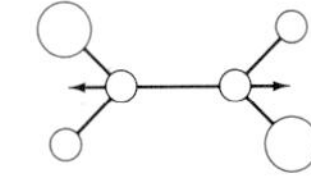

f. C=C twist:

g. Asymmetric C—H and C—Cl stretches:

h. Symmetric C—H and C—Cl stretches:

C_6H_6 (benzene): **i.** "Breathing":

j. "Flexing":

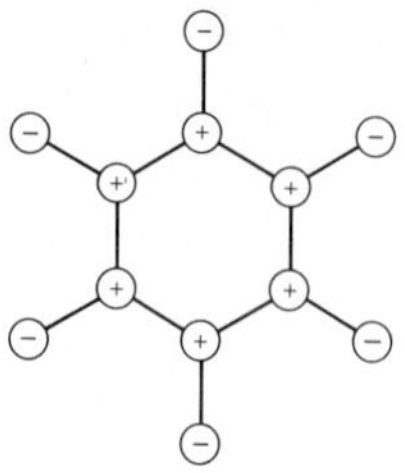

k. Asymmetric C—H stretches:

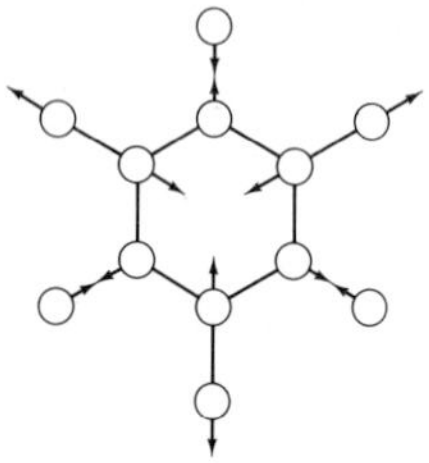

In each case, determine the point group and refer to the appendix to Chapter 12 (p. 557) to determine whether the vibration is active in the infrared and in the Raman spectrum. Then check your conclusions by reference to the appendix to this chapter (p. 638).

37. The microwave spectrum of $^{16}O^{12}C^{32}S$ shows absorption lines separated by 12.163 GHz. That of $^{16}O^{12}C^{34}S$ shows lines separated by 11.865 GHz. The determination of the distances involves solving two simultaneous quadratic equations, which is best done by successive approximations. To avoid all that labor, simply confirm that the results are consistent with r(O—C) = 116 pm and r(C—S) = 156 pm.

38. The free radical CH_3 is found experimentally to be planar. Give an interpretation of this result in terms of orbital hybridization. What microwave spectrum would the radical show? What vibrations would be active in the infrared?

39. Prove that the force constant k corresponding to the Morse potential function (Eq. 13.104) at small bond extensions is

$$k = 2D_e\alpha^2$$

Calculate the vibration frequency ν_0 on the basis of the following Morse parameters for $H^{35}Cl$:

$$D_e = 4.67 \text{ eV}$$

$$\alpha = 1.85 \times 10^8 \text{ cm}^{-1}$$

H Essay Questions

1. Give an account of the fundamental origins of ultraviolet, infrared, and nuclear magnetic resonance spectra.

2. State the laws of Lambert and Beer, and write an equation comprising the two laws.

3. Explain clearly what is meant by absorbance and transmittance, and derive a relationship between them.

4. Explain the selection rules for infrared spectra, with examples.

5. Explain the selection rules for Raman spectra, with examples.

6. Give an account of the fundamental principles underlying Mössbauer spectroscopy.

Suggested Reading

C. N. Banwell, *Fundamentals of Molecular Spectroscopy*, New York: McGraw-Hill, 1966.

G. M. Barrow, *Introduction to Molecular Spectroscopy*, New York: McGraw-Hill, 1962.

A. Carrington and A. D. McLachlan, *Introduction to Magnetic Resonance with Applications to Chemistry and Chemical Physics*, New York: Harper & Row, 1967.

R. N. Dixon, *Spectroscopy and Structure*, London; Methuen, 1965.

H. B. Dunford, *Elements of Diatomic Molecular Spectra*, Reading, Mass.: Addison-Wesley, 1968.

G. Herzberg, *Atomic Spectra and Atomic Structure*, New York: Dover Pub., 1944.

G. Herzberg, *Molecular Structure, I. Spectra of Diatomic Molecules*, New York: D. Van Nostrand Co., 1950.

G. Herzberg, *Molecular Spectra and Molecular Structure, II. Infrared and Raman Spectra of Polyatomic Molecules*, New York: D. Van Nostrand Co., 1945.

G. Herzberg, *Molecular Spectra and Molecular Structures, III. Electronic Spectra and Electronic Structure of Polyatomic Molecules*, New York: D. Van Nostrand Co., 1966.

G. W. King, *Spectroscopy and Molecular Structure,* New York: Holt, Rinehart and Winston, 1964.

K. A. McLauchlan, *Magnetic Resonance,* Oxford: Clarendon Press, 1972.

M. Orchin and H. H. Jaffe, *Symmetry, Orbitals, and Spectra,* New York: Wiley-Interscience, 1971.

J. A. Pople, W. G. Schneider, and H. J. Bernstein, *High-Resolution Nuclear Magnetic Resonance,* New York: McGraw-Hill, 1959.

C. H. Townes and A. L. Schawlow, *Microwave Spectroscopy,* New York: McGraw-Hill, 1955.

D. H. Whiffen, *Spectroscopy,* London: Longman, 1972.

E. B. Wilson, J. C. Decius, and P. C. Cross, *Molecular Vibrations,* New York: McGraw-Hill, 1955.

Chapter 14

Molecular Statistics

Preview

This chapter is concerned with interpreting the macroscopic behavior of matter by applying a statistical treatment to large assemblies of molecules. This subject is known as *statistical mechanics,* and the branch of it that is concerned with thermodynamic properties is known as *statistical thermodynamics*.

The energy relating to the motions of atoms and molecules can be classified as translational, rotational, and vibrational energy. A molecule having N_a atoms has $3N_a$ *degrees of freedom,* and these are divided between the various modes of motion. A simple hypothesis suggests that there is *equipartition of energy* between degrees of freedom, except that vibrational degrees of freedom (normal modes) have twice as much as the others since both potential energy and kinetic energy are involved. However, measurements of heat capacities of gases show that equipartition only occurs in the limit of very high temperatures.

To treat these problems more satisfactorily it is necessary to have a statistical treatment of the way molecules distribute themselves among the various energy levels. The fundamental equation dealing with this is the *Boltzmann distribution law* according to which the number of molecules in a given state is proportional to $e^{-\epsilon_i/\mathbf{k}T}$, where ϵ_i is the energy of the state and $\mathbf{k}$ is the Boltzmann constant. The sum of the $e^{-\epsilon_i/\mathbf{k}T}$ terms over all energy states is known as the *partition function.*

Two applications of the Boltzmann distribution law are of special importance: the distribution of speeds of gas molecules and of their energies. The equation for the distribution of speeds is often known as the *Maxwell distribution law,* while that for the distribution of energies is the *Maxwell-Boltzmann distribution law.*

A partition function can relate to an individual molecule or to a system of N molecules. The relationship between the molecular partition function q and the system partition function Q depends on whether the molecules are distinguishable from one another or are indistinguishable. The system partition function can be related to the various thermodynamic quantities, such as entropy, enthalpy, and Gibbs energy.

To evaluate partition functions for a particular molecular system the translational, rotational, vibrational, electronic, and nuclear forms of energy are considered separately. In most cases there are compact expressions for the partition functions, which are therefore easily evaluated if we have the necessary data about the molecules; in particular we need the mass, the moments of inertia, the vibrational frequencies, and the electronic energy levels.

An important application of molecular statistics is in the calculation of *equilibrium constants.* One useful device for doing this is to define certain functions, such as the *enthalpy function* and the *Gibbs energy function.* These can be calculated from partition functions and have been tabulated for many molecules. From them the equilibrium constant can be calculated.

Equilibrium constants can also be derived directly from partition functions.

14 Molecular Statistics

In the earlier chapters of this book we have been concerned almost exclusively with the *macroscopic* properties of matter. In Chapters 11–13 we have gone to the other extreme and have dealt with the structure and behavior of individual atoms and molecules. In the present chapter we will see how the macroscopic properties can be related to the behavior of the microscopic systems.

In principle, if the positions and velocities of all molecules in a given system were known at any one time, the behavior of the system, at that time and at later times, could be determined by applying the laws of classical and quantum mechanics. There are, however, two reasons why this is impossible. In the first place, because of the Uncertainty Principle the actual positions and velocities of individual molecules cannot be known. Second, because of the enormous numbers of molecules present in the systems with which chemists and physicists normally deal, the calculation of macroscopic properties from the states of the individual molecules presents insuperable computational difficulties.

Instead of attacking this impossible problem, the science of *statistical mechanics* deals with a large collection or *assembly* of molecules and applies the methods of statistics in order to predict the *most probable* behavior of the assembly. The assembly will be characterized by certain properties, such as the total volume, the number of molecules, and the total energy, but the individual molecules are distributed over a range of states; for example, the coordinates and velocities of the individual atoms differ from one molecule to another. Statistical mechanics makes full use of the information derived from quantum mechanics, but it is not concerned with the precise states of individual molecules. Instead it directs its attention to the most probable states and hence deduces the macroscopic properties. Although it only indicates the most probable behavior, the number of molecules in the assembly is usually so large that the most probable behavior is very close to the behavior actually observed. Only if the number of molecules is small will there be observable statistical fluctuations from the most probable behavior.

The subject of statistical mechanics is a very vast one, and here we will deal only with those aspects that are of particular application to chemical problems. We shall be concerned, for example, with the distribution laws for energies and molecular speeds and with the question of average and most probable energies and speeds. We will then see how the thermodynamic properties can be calculated from these distribution laws. This particular aspect is known as *statistical thermodynamics*, which among other things allows us to calculate equilibrium constants from such basic information as the shapes and sizes of individual molecules, the vibrational frequencies, and the zero-point levels of the molecules.

14.1 Forms of Molecular Energy

Statistical mechanics is primarily concerned with the distribution of energies, and we will begin by considering the different forms of molecular energy. One form is *translational energy,* which is the energy arising from the movement of the molecule from one position to another. We have already discussed this in Section 1.9, where we saw that the average kinetic energy per molecule $\overline{\epsilon}_k$, resulting from translational motion, is equal to $\frac{3}{2}\mathbf{k}T$ (Eq. 1.50). This average energy can be split into three equal contributions for the energies corresponding to motion along the three Cartesian axes X, Y, and Z:

$$\overline{\epsilon}_x = \overline{\epsilon}_y = \overline{\epsilon}_z = \tfrac{1}{2}\mathbf{k}T \tag{14.1}$$

This equal division of the total kinetic energy $\frac{3}{2}\mathbf{k}T$ into three contributions of $\frac{1}{2}\mathbf{k}T$ is known as the *equipartition of energy.* It is obvious that there must be an *equal* splitting, since there is no reason why any direction should be favored over the others. We refer to the components of speed and energy along the three Cartesian axes as *three degrees of translational freedom.*

In Section 13.3, the rotational energies of molecules were discussed. For a *linear* molecule (Figure 13.9) rotation about two axes at right angles to the axis of the molecule and to one another displaces the atoms in the molecule; there are therefore *two degrees of rotational freedom* for a linear molecule. For a *nonlinear* molecule (Figure 13.11) there are *three degrees of rotational freedom,* since rotation about all three Cartesian axes causes a displacement of the atoms in the molecule.

Vibrational energy was discussed in Section 13.4, where we saw that the vibrational motion can be resolved into a certain number of normal modes of vibration, which are the degrees of freedom for vibration.

To describe the position of an atom in space we need to specify three position coordinates x, y, and z. If, therefore, a molecule contains N_a atoms, $3N_a$ position coordinates must be specified. It is much more convenient, however, to work with *degrees of freedom,* which give us much more insight into the atomic motions. If a molecule contains N_a atoms, it must have $3N_a$ degrees of freedom, and it is easy to see how these are distributed among the different kinds of motion. Let us consider some simple examples of molecules in the gas phase:

1. *A monatomic molecule,* such as He. There are 3 degrees of freedom, all corresponding to translational motion.

2. *A diatomic molecule,* such as H_2. There are now $3 \times 2 = 6$ degrees of freedom, 3 of which are for translation. Since the molecule is linear, there are 2 degrees of rotational freedom. There is therefore 1 degree of vibrational freedom. This is expected since in Section 13.4 (Figure 13.12) it was found that there is just one mode of vibrational motion for a diatomic molecule.

3. *A linear triatomic molecule,* such as CO_2. There are $3 \times 3 = 9$ degrees of freedom, of which 3 are for translation and 2 for rotation. That leaves 4 degrees of

vibrational freedom. These four vibrational modes were shown in Figure 13.18a.

4. *A nonlinear triatomic molecule,* such as H_2O. Again there are 9 degrees of freedom, of which 3 are for translation. Now, however, 3 are for rotation, and therefore 3 are for vibration. The three vibrational modes were shown in Figure 13.18b.

If we look at Figure 13.18, we can easily see why there is this difference between the numbers of vibrational and rotational degrees of freedom for linear and nonlinear molecules. The vibrations designated 1, 2, and 3 in Figure 13.18a correspond to the 1, 2, and 3 vibrations in Figure 13.18b. However, vibration 3′ for the linear molecule, if applied to the nonlinear molecule, would become a *rotation* about the X axis.

We can generalize our conclusions and give the following numbers of degrees of freedom for a gaseous polyatomic molecule containing N_a atoms:

	Translational	*Rotational*	*Vibrational*	*Total*
Linear	3	2	$3N_a-5$	$3N_a$
Nonlinear	3	3	$3N_a-6$	$3N_a$

Note that the number of vibrational modes can be very large; for benzene, for example, with $N_a = 12$, there are $3 \times 12 - 6 = 30$ vibrational degrees of freedom.

Molar Heat Capacities of Gases

Important light on energy distributions in molecules is thrown by measurements of the molar heat capacities of gases. It is simplest to consider the molar heat capacities at constant volume $C_{V,m}$, since the gas then performs no work and the heat capacity arises from the change in internal energy U as the temperature is raised (Section 2.5). The situation is most straightforward for a monatomic gas, since the internal energy then resides solely in the translational motion. We have seen that the average internal energy per molecule is $\frac{1}{2}\mathbf{k}T$ for each degree of freedom or $\frac{3}{2}\mathbf{k}T$ for the three translational degrees of freedom. For 1 mol the internal energy is therefore

$$U_m = \tfrac{3}{2}L\mathbf{k}T = \tfrac{3}{2}RT \tag{14.2}$$

where L is the Avogadro constant and R is the gas constant. We therefore predict that $C_{V,m}$ is

$$C_{V,m} = \frac{dU}{dT} = \frac{3}{2}R \tag{14.3}$$

In other words, the value of $C_{V,m}/R$ should be $\frac{3}{2}$ or 1.5. This prediction is confirmed by the experimental values for some monatomic gases shown in Table 14.1.

The situation with other than monatomic molecules is, however, considerably more complicated. Obviously some energy will reside in the other degrees of freedom, and the question is: how much? Let us consider first the *rotational motions.* If the molecule is linear, we can express its rotational energy as

$$\epsilon_{\text{rot}} = \tfrac{1}{2}I\omega_x^2 + \tfrac{1}{2}I\omega_y^2 \tag{14.4}$$

TABLE 14.1 Experimental and Calculated Values of $C_{V,m}/R$ for Gases at 298.15 K

Species	$C_{V,m}/R$ Experimental	Calculated: trans. + rot.	Calculated: trans. + rot. + vib.
He, Ne, Ar, Kr, Xe	1.500	1.500	1.500
Diatomic:			
H_2	2.468	2.50	3.50
N_2, HF, HCl, HBr	2.50	2.50	3.50
O_2	2.531	2.50	3.50
NO	2.591	2.50	3.50
Cl_2	3.08	2.50	3.50
Linear triatomic:			
CO_2	3.466	2.50	6.50
CS_2	4.490	2.50	6.50
Nonlinear triatomic:			
H_2O	3.038	3.00	6.00
NO_2	3.56	3.00	6.00
SO_2	3.79	3.00	6.00
Linear polyatomic:			
C_2H_2	4.283	2.50	9.50
Nonlinear polyatomic:			
NH_3	3.289	3.00	9.00
P_4	7.05	3.00	9.00

where I is the moment of inertia and ω is the angular velocity about the axis indicated. If the molecule is nonlinear, we have

$$\epsilon_{\text{rot}} = \tfrac{1}{2}I_x\omega_x^2 + \tfrac{1}{2}I_y\omega_y^2 + \tfrac{1}{2}I_z\omega_z^2 \tag{14.5}$$

There are now three moments of inertia (two or all of which may be the same). In either case, the energy is proportional to a sum of velocity-squared terms, just as in the translational case. It therefore seems reasonable to postulate that each of the degrees of rotational freedom should carry $\frac{1}{2}\mathbf{k}T$ of energy per molecule. Thus for a linear molecule there is a contribution of $2 \times \frac{1}{2}\mathbf{k}T = \mathbf{k}T$ per molecule, or of RT per mole. For a nonlinear molecule there are $\frac{3}{2}\mathbf{k}T$ per molecule or $\frac{3}{2}RT$ per mole. The corresponding contributions to $C_{V,m}$ are thus R and $\frac{3}{2}R$, respectively.

In the third column of Table 14.1 are the contributions to $C_{V,m}/R$ that arise from the translational and rotational motions. For a linear molecule this contribution is $\frac{3}{2}$ from translation and one from rotation, giving us $\frac{5}{2}$ or 2.50. For a nonlinear molecule we have $\frac{3}{2} + \frac{3}{2} = 3$. We see that the prediction is now closer to the experimental value than if we had considered translation alone. With H_2, N_2, HF, etc., the agreement is in fact excellent, but with the other molecules the theoretical values are too low. With these there must therefore be a contribution from vibrations.

The situation with *vibrational motion* is a little different. In a vibration, the atoms have not only kinetic energy but potential energy, by virtue of being displaced from their equilibrium positions. The energy in a vibrational mode must therefore be written as the sum of two terms, one for kinetic energy and one for potential energy. If there is equipartition of energy, *each* of these terms will contribute $\frac{1}{2}\mathbf{k}T$ to the energy. There is therefore a contribution from each normal mode of $\mathbf{k}T$ per molecule, or RT per mole. The corresponding contribution to $C_{V,m}/R$ is thus one per normal mode.

Consider, for example, the case of CO_2, a linear triatomic molecule. Its predicted $C_{V,m}/R$ value, assuming equipartition of energy, is $\frac{3}{2}$ from translation, 1 from the two rotational degrees of freedom, and 4 from the four vibrational modes; the total is therefore $\frac{3}{2} + 1 + 4 = 6.50$. This value, together with similarly obtained values for other molecules, is included in the last column of Table 14.1.

For the monatomic gases, where there is no rotation or vibration, the value in the last column of course agrees with the experimental value. In all other cases, however, the theoretical values are a good deal too large. The obvious conclusion to draw from these various comparisons is that the vibrational modes are not making their expected contributions to the molar heat capacity. The agreement for the monatomic gases indicates that the translational motions are doing so, and there is no reason to doubt, from the figures in Table 14.1, that the rotations are also behaving as predicted. If we were to look at the heat capacity data at *very high temperatures,* we would find that there was much better agreement with the values in the last column of the table. In other words, there is equipartition of energy at high temperatures. The principle of equipartition of energy therefore breaks down at lower temperatures, particularly for the vibrational modes. In Section 13.3 (Figure 13.8) we have seen that the vibrational levels are much more widely spaced than the rotational and translational levels, and this gives us a clue as to what is happening. Because of this wide spacing, energy cannot move freely into and out of vibrational modes; a transition can only occur if a sufficiently large quantum of energy is involved. These ideas were first put forward in 1906 by Albert Einstein, who applied them quantitatively to the heat capacities of solids, a matter we deal with in Section 15.12.

We will now develop the general principles of statistical mechanics, which are concerned with this problem of how energy is distributed among different modes of motion.

14.2 Statistical Mechanics

We start by considering a large number N of identical atoms or molecules in the gas phase; from now on we will refer to them simply as molecules, which includes atoms and ions. Such a collection is known as an **assembly.** We assume the molecules to be *independent,* by which we mean that the state of each is unaffected by the state of any other. A number of discrete energy levels are available to the molecules, and in Figure 14.1 we represent these levels as compartments, which are labeled, starting with the lowest energy as ϵ_0 and higher levels as ϵ_1, ϵ_2, etc. The numbers of molecules in each level are represented as n_0, n_1, n_2, etc.

In any assembly of a large number N of identical molecules, corresponding to a given total energy, there are very many different ways in which the molecules can be distributed among the energy levels. For a given total energy we have a given

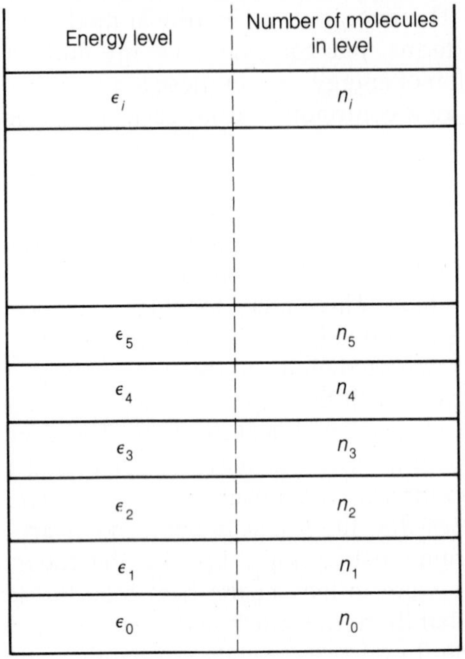

Energy level	Number of molecules in level
ϵ_i	n_i
ϵ_5	n_5
ϵ_4	n_4
ϵ_3	n_3
ϵ_2	n_2
ϵ_1	n_1
ϵ_0	n_0

FIGURE 14.1 A series of energy levels ϵ_0, ϵ_1, ϵ_2, . . . , represented as compartments into which n_0, n_1, n_2, . . . molecules can go.

state of the system, and we may ask how many ways there are of achieving this state. This is equivalent to asking: in how many ways can we put n_0 molecules into ϵ_0, n_1 into ϵ_1, and so on? For example, suppose that for the first three levels we have $n_0 = 3$, $n_1 = 4$, and $n_2 = 2$. If we consider the ϵ_0 level first, with three molecules in it, we have N choices for the first molecule; once that is in the level, we have $N - 1$ choices; then $N - 2$. If the molecules were all distinguishable, there would therefore be $N(N - 1)(N - 2)$ ways of selecting the three molecules to go into the lowest energy level. However, we are considering N identical and therefore indistinguishable molecules, so that the order in which we choose them does not matter. Thus, if the three molecules in $\epsilon_0 = 0$ are labeled a, b, and c, the choices abc, bca, cba, acb, bac, and cab all amount to the same selection. To obtain the correct number of choices we should therefore divide the product $N(N - 1)(N - 2)$ by 6. In general, if n molecules go into a level, we divide by $n!$, which is the number of ways we can select n molecules. Thus for the ϵ_0 level, with $n_0 = 3$, the number of ways is

$$\frac{N(N - 1)(N - 2)}{3!}$$

For the second level, ϵ_1, in which there are four molecules, we can choose the first in $N - 3$ ways, the second in $N - 4$ ways, the third in $N - 5$ ways, and the fourth in $N - 6$ ways. Again, we must divide, in this case by 4!, and the number of ways is therefore

$$\frac{(N-3)(N-4)(N-5)(N-6)}{4!}$$

Similarly, for the ϵ_2 level, with $n_2 = 2$, there are

$$\frac{(N-7)(N-8)}{2!}$$

ways of selecting.

The product of these three expressions gives the total number of ways of choosing three molecules to go into ϵ_0, four to go into ϵ_1, and two to go into ϵ_2. This number of ways is known as the number of **complexions,** and in our example it is given by

$$\frac{N(N-1)(N-2)(N-3)(N-4)(N-5)(N-6)(N-7)(N-8)}{3!4!2!}$$

So far we have only considered the first three levels; in reality we have to distribute all the N molecules among the levels. This means that the numerator in this expression will continue with $(N-9)(N-10)\ldots$ until we have used up all the molecules; the numerator is thus $N!$. The denominator is the product of the $n_i!$ values for all the levels, which we write as $\prod_i n_i!$. The general expression for the number of complexions is therefore

$$\boldsymbol{\Omega = \frac{N!}{\prod_i n_i!}} \qquad (14.6)$$

However, there is an important additional factor, arising from the fact that the total energy corresponding to a given state is fixed. We cannot distribute the molecules in any way we like among the energy levels; we are restricted to distributions that are consistent with the total energy of the system. To understand this in a simple way, suppose that we have three molecules to distribute among four energy levels ϵ_0, ϵ_1, ϵ_2, and ϵ_3. We will suppose that the energy levels are equally spaced, with a difference of ϵ between neighboring levels, so that

$$\epsilon_0 = 0, \qquad \epsilon_1 = \epsilon, \qquad \epsilon_2 = 2\epsilon, \qquad \epsilon_3 = 3\epsilon$$

as shown in Figure 14.2a. We will suppose that the total energy of the system is 3ϵ, and the question we ask is: how many complexions are there in which three molecules are in the various energy levels, such that the total energy is 3ϵ?

These complexions are shown in Figure 14.2b. One distribution, designated Ω_1 in the figure, is the one in which each of the three molecules is in level ϵ_1, so that it has an energy ϵ; the total energy is 3ϵ. Since the molecules are indistinguishable, the order of filling is immaterial (i.e., *abc* is the same as *acb*, etc.), so that there is only one complexion corresponding to this distribution. This conclusion is confirmed by use of Eq. 14.6, since

$$\Omega_1 = \frac{3!}{0!3!0!0!} = 1 \qquad (14.7)$$

(Remember that $0! \equiv 1$.)

A second distribution, designated Ω_2 and shown in the figure, involves having one of the molecules in ϵ_3, with energy 3ϵ, and the other two in ϵ_0, with energy

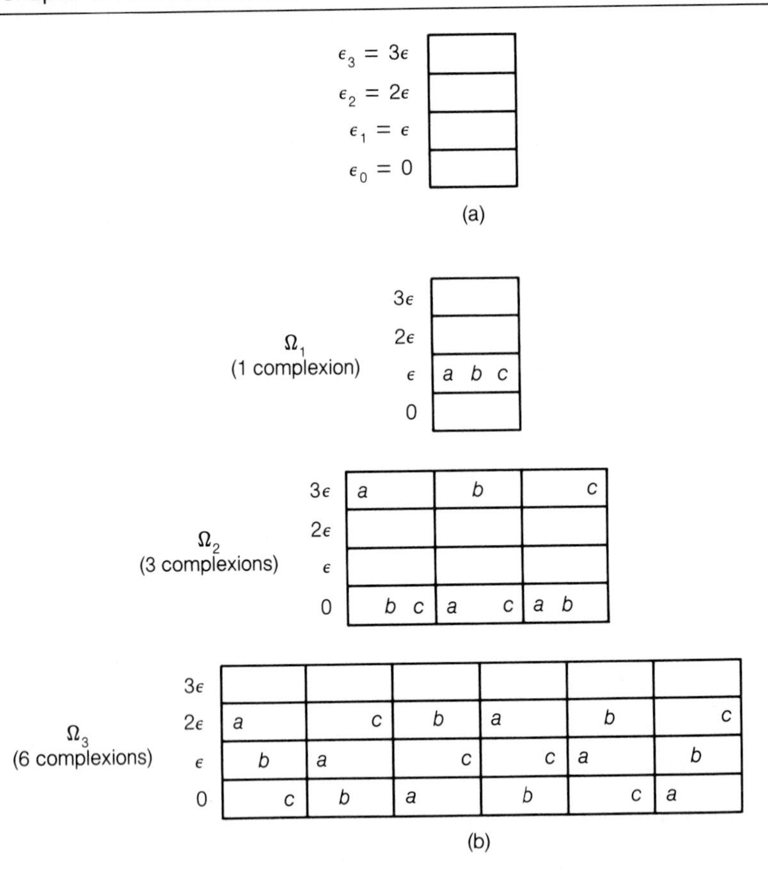

FIGURE 14.2 (a) Four equally spaced energy levels, shown as compartments into which we are going to put three molecules, the total energy of the system being 3ϵ. (b) The various ways (i.e., complexions) in which the three molecules can go into the energy levels. The ways can be grouped into Ω_1, Ω_2, and Ω_3.

zero. This can be done in three different ways, since any one of the molecules can be in ϵ_3, leaving the other two in ϵ_0. Use of Eq. 14.6 again confirms that there are three complexions:

$$\Omega_2 = \frac{3!}{2!0!0!1!} = 3 \tag{14.8}$$

A third possible arrangement is to have one molecule in ϵ_0 ($\epsilon = 0$), one in ϵ_1 ($\epsilon = 1$), and one in ϵ_2 ($\epsilon = 2$). The figure shows that there are six ways of doing this, as confirmed by use of Eq. 14.6:

$$\Omega_3 = \frac{3!}{1!1!1!0!} = 6 \tag{14.9}$$

In statistical mechanics, each type of distribution, such as Ω_1, is referred to as a **state.*** The system of three molecules that we are considering will be constantly

---

* Note that the word *state* is here being used in a more restricted sense than in thermodynamics (Section 2.3).

shifting from one state to another. Thus at a given instant we may have the Ω_1 distribution, with all three molecules in ϵ_1, but collisions may bring about a transition to one of the three Ω_2 complexions; two of the molecules drop down to ϵ_0 and one is promoted to ϵ_3, there being no change of total energy. It is easy to see intuitively that *all the complexions are equally probable.* In our example, the Ω_2 distribution is three times as probable as the Ω_1, because there are three ways of forming it as compared with one way for Ω_1. Similarly Ω_3 is six times as probable as Ω_1 and twice as probable as Ω_2.

We now come to a point of very great importance. In our example of three molecules we see that one state, Ω_3, is more probable than the other two but the difference is not enormous. However, we are usually dealing with very large numbers of molecules, of the order of 10^{23} and more. We then find that one state has a much larger number of complexions than any of the others.

To illustrate this point, suppose that we had an assembly of N molecules, where N is very large. We can imagine a state in which all the N molecules are in different energy levels. The number of complexions is then, from Eq. 14.6,

$$\Omega = \frac{N!}{1!1!\ldots} = N! \tag{14.10}$$

This is the largest possible value of Ω for this particular state. If two of the molecules were in the lowest level and another were promoted, in order to keep the energy the same, the number of complexions drops to

$$\Omega = \frac{N!}{2!1!1!\ldots} = \frac{1}{2}N! \tag{14.11}$$

To go to the limit, if all the molecules except one were in the lowest level, the number of complexions is

$$\Omega = \frac{N!}{(N-1)!1!} = N \tag{14.12}$$

which is vastly smaller than $N!$

The important point that we have made is that when we are dealing with very large numbers of molecules, a certain state will greatly predominate over the others. This is the state corresponding to the broadest distribution of the molecules over all the energy levels.

The Boltzmann Distribution Law

This discussion has provided us with the background to proceed in a more formal way, with the objective of finding a general expression for the distribution of molecules among energy levels.

We start by writing the total number W of complexions; this is the sum of the number of complexions for each state:

$$W = \sum \Omega = \sum \frac{N!}{\prod_i n_i!} \tag{14.13}$$

It is understood that the summation is over all the states. There are two additional restrictions. The first is that the sum of all the n_i values is equal to the total number N of molecules in the assembly:

$$n_1 + n_2 + n_3 + \cdots = \sum_i n_i = N \tag{14.14}$$

The second is that the total energy is fixed; if we call the total energy E, this means that

$$n_1\epsilon_1 + n_2\epsilon_2 + \cdots = \sum_i n_i\epsilon_i = E \tag{14.15}$$

We now apply the conclusion that one of the Ω values is predominant, and we take the largest term in Eq. 14.13 to be the value of W. What we therefore do is to refer to Eq. 14.6, which defined Ω, and find what values of n_1, n_2, etc., will make Ω a maximum. Any small change in these values, namely δn_1, δn_2, etc., will produce a small change $\delta\Omega$, and the condition that Ω has its maximum value is

$$\delta\Omega = \frac{\partial\Omega}{\partial n_1}\,\delta n_1 + \frac{\partial\Omega}{\partial n_2}\,\delta n_2 + \cdots + \frac{\partial\Omega}{\partial n_i}\,\delta n_i + \cdots = 0 \tag{14.16}$$

If there were no restrictions on the n_i values, we could obtain the maximum by setting each $\partial\Omega/\partial n_i$ equal to zero. This, however, is not possible because of Eqs. 14.14 and 14.15, which do not allow the n_i values to be varied independently. Instead we make use of the *Lagrange method of undetermined multipliers,* introduced by the astronomer and mathematician Joseph Louis Lagrange (1736–1813). In this method the two restrictive conditions

$$\delta n_1 + \delta n_2 + \cdots + \delta n_i + \cdots = 0 \tag{14.17}$$

and

$$\epsilon_1\delta n_1 + \epsilon_2\delta n_2 + \cdots + \epsilon_i\delta n_i + \cdots = 0 \tag{14.18}$$

(which came from Eqs. 14.14 and 14.15) are first multiplied by constants, which we write as α and $-\beta$, respectively:

$$\alpha(\delta n_1 + \delta n_2 + \cdots + \delta n_i + \cdots) = 0 \tag{14.19}$$

$$-\beta(\epsilon_1\,\delta n_1 + \epsilon_2\,\delta n_2 + \cdots + \epsilon_i\,\delta n_i + \cdots) = 0 \tag{14.20}$$

We then take the sum of Eqs. 14.16, 14.19, and 14.20 and obtain

$$\Omega = \left(\frac{\partial\Omega}{\partial n_1} + \alpha - \beta\epsilon_1\right)\delta n_1 + \left(\frac{\partial\Omega}{\partial n_2} + \alpha - \beta\epsilon_2\right)\delta n_2 + \cdots + \left(\frac{\partial\Omega}{\partial n_i} + \alpha - \beta\epsilon_i\right)\delta n_i + \cdots = 0 \tag{14.21}$$

The only way that this can hold for any values of the independent small quantities δn_1, δn_2, etc., is if every one of the enclosed terms is equal to zero; in other words, for all values of i,

$$\frac{\partial\Omega}{\partial n_i} + \alpha - \beta\epsilon_i = 0 \tag{14.22}$$

Since $\partial\Omega/\partial n_i$ proves difficult to work with, we replace Ω by $\ln\Omega$, which is permissible since $\ln\Omega$ reaches a maximum when Ω does. The equation we need to solve is therefore

$$\frac{\partial \ln \Omega}{\partial n_i} + \alpha - \beta \epsilon_i = 0 \tag{14.23}$$

To solve this equation, we first take the natural logarithms of both sides of Eq. 14.6 and obtain

$$\ln \Omega = \ln N! - \sum_i \ln n_i! \tag{14.24}$$

We now make use of Stirling's approximation (derived in 1730), which for values of n greater than about 10 reduces satisfactorily to

$$\ln n! = n \ln n - n \quad \text{or} \quad n! = \frac{n^n}{e^n} \tag{14.25}$$

Application of this approximation to Eq. 14.24 gives

$$\ln \Omega = \ln N! - \sum_i (n_i \ln n_i - n_i) \tag{14.26}$$

Then, since $\ln N!$ is constant,

$$\frac{\partial \ln \Omega}{\partial n_i} = -\frac{\partial}{\partial n_i}(n_i \ln n_i - n_i) \tag{14.27}$$

$$= -\ln n_i - 1 + 1 = -\ln n_i \tag{14.28}$$

Insertion of this into Eq. 14.23 gives

$$\ln n_i = \alpha - \beta \epsilon_i \tag{14.29}$$

and therefore

$$n_i = e^{\alpha} e^{-\beta \epsilon_i} \tag{14.30}$$

This is conveniently written as

$$n_i = A e^{-\beta \epsilon_i} \tag{14.31}$$

where A is equal to e^{α}.

In Section 14.3 we will apply this equation to the distribution of molecular speeds and prove that β is equal to $1/\mathbf{k}T$, where T is the absolute temperature and **k** is the Boltzmann constant. In the meantime we will anticipate this result and write Eq. 14.31 as

$$n_i = A e^{-\epsilon_i/\mathbf{k}T} \tag{14.32}$$

The total number of molecules N is

$$N = \sum_i n_i = A \sum_i e^{-\epsilon_i/\mathbf{k}T} \tag{14.33}$$

Therefore

$$\frac{n_i}{N} = \frac{e^{-\epsilon_i/\mathbf{k}T}}{\sum_i e^{-\epsilon_i/\mathbf{k}T}} \tag{14.34}$$

This is the fundamental equation of the **Boltzmann distribution.** It gives us the fraction of the molecules that are in a specified state of energy ϵ_i. If we are interested in two states of energies ϵ_i and ϵ_j, we also have, from Eq. 14.32,

$$\frac{n_i}{n_j} = \frac{e^{-\epsilon_i/\mathbf{k}T}}{e^{-\epsilon_j/\mathbf{k}T}} = \exp\left[\frac{-(\epsilon_i - \epsilon_j)}{\mathbf{k}T}\right] \tag{14.35}$$

That is, the ratio depends on the energy difference $\epsilon_i - \epsilon_j$.

14.3 The Maxwell-Boltzmann Distribution of Molecular Speeds and Translational Energies

The Distribution of Speeds

The equations that we have just obtained are quite general and apply to all forms of energy. What we are now going to do is to apply them to the translational energies and speeds of gas molecules. This topic relates to what we discussed in Section 1.9 about the kinetic theory of gases. In this section we are not going to start with the assumption that $\beta = 1/\mathbf{k}T$; instead we will start with Eq. 14.31, and we will be able to prove, from our development of the distribution of speeds, that β is equal to $1/\mathbf{k}T$.

The treatment that we will give is essentially that given in about 1860 by J. Clerk Maxwell and Ludwig Boltzmann. Their work was done before the development of the quantum theory, and their treatment is therefore a classical one. In the case of translational energy it does not much matter whether we have a classical or a quantum treatment, because even in quantum theory the translational energy levels (Section 11.6) are so close together that the behavior is almost exactly the same as if we had a continuum of energies.

The speed u of a molecule can be resolved into its three components along the X, Y, and Z axes. If u_x is the component along the X axis, the corresponding kinetic energy is $\frac{1}{2}mu_x^2$, where m is the mass of the molecule. We can then ask: what is the probability dP_x that the molecule has a speed component along the X axis between u_x and $u_x + du_x$? We must, of course, consider a range of speeds, since otherwise the probability is zero. This probability is proportional to the range du_x; also, by Eq. 14.31, it is proportional to $e^{-\beta\epsilon_x}$, which in this case is $e^{-mu_x^2\beta/2}$. We can therefore write

$$dP_x = Be^{-mu_x^2\beta/2}\, du_x \tag{14.36}$$

where B is the proportionality constant. We also have similar expressions for the other components:

$$dP_y = Be^{-mu_y^2\beta/2}\, du_y \quad \text{and} \quad dP_z = Be^{-mu_z^2\beta/2}\, du_z \tag{14.37}$$

The product of these three expressions is

$$dP_x dP_y dP_z = B^3 e^{-mu_x^2\beta/2} e^{-mu_y^2\beta/2} e^{-mu_z^2\beta/2}\, du_x\, du_y\, du_z \tag{14.38}$$

Since $u^2 = u_x^2 + u_y^2 + u_z^2$, where u is the speed,

$$dP_x dP_y dP_z = B^3 e^{-mu^2\beta/2}\, du_x\, du_y\, du_z \tag{14.39}$$

This expression is the probability that the three components of speed have values between u_x and $u_x + du_x$, u_y and $u_y + du_y$, and u_z and $u_z + du_z$. However, what we are really interested in is the probability that the actual speed of the molecule lies

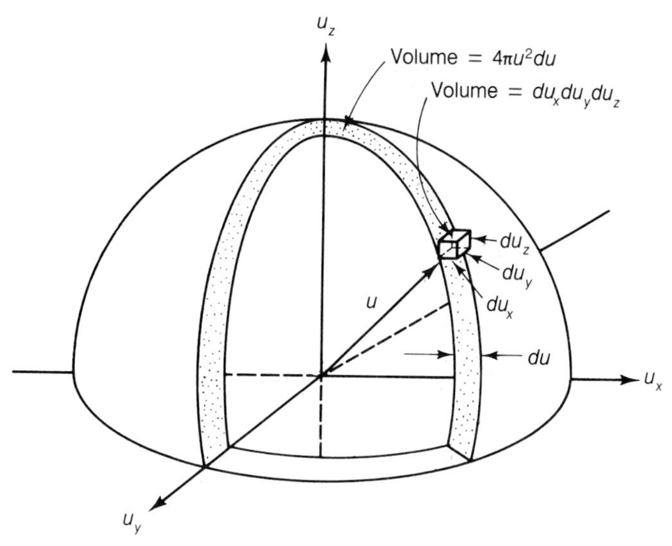

FIGURE 14.3 A diagram with axes u_x, u_y, and u_z. Equation 14.39 gives the probability that the velocity is represented by the cube of volume $du_x\,du_y\,du_z$. We are interested in the speed corresponding to the volume $4\pi u^2\,du$ that lies between the concentric spherical surfaces.

between u and $u + du$, and this can be achieved by a variety of combinations of the three components of speed. This is conveniently considered in a manner that is reminiscent of our treatment in Section 7.4 of the distribution of ions in the ionic atmosphere. We construct a velocity diagram, shown in Figure 14.3, with the components u_x, u_y, and u_z as the three axes. We then construct a spherical shell, of radius u and thickness du. What we want is the probability that systems lie within the shell, but the product $dP_x\,dP_y\,dP_z$ has only given us the probability that they lie within the small cube shown in the diagram. The volume of this cube is $du_x\,du_y\,du_z$, and the volume of the spherical shell is $4\pi u^2\,du$. In order to obtain the probability dP that the speed lies between u and $u + du$, we must therefore replace $du_x\,du_y\,du_z$ in Eq. 14.39 by $4\pi u^2\,du$:

$$dP = 4\pi B^3 e^{-mu^2\beta/2} u^2\,du \qquad (14.40)$$

The fraction dN/N of the N molecules that have speeds between u and $u + du$ is given by the ratio of this expression to its value integrated from $u = 0$ to $u = \infty$, since that is the range of possible speeds. Thus

$$\frac{dN}{N} = \frac{4\pi B^3 e^{-mu^2\beta/2} u^2\,du}{4\pi B^3 \int_0^\infty e^{-mu^2\beta/2} u^2\,du} \qquad (14.41)$$

The integral in the denominator is a standard one,* and when we evaluate it, we obtain

$$\frac{dN}{N} = 4\pi\left(\frac{\beta m}{2\pi}\right)^{3/2} e^{-mu^2\beta/2} u^2\,du \qquad (14.42)$$

*The appendix to this chapter (p. 685) gives a number of the integrals that are useful in distribution problems.

We now consider what the mean square speed is on the basis of this treatment. The point of doing this is that we already know the mean square speed (which is the same as the mean square velocity) from kinetic theory. Thus from Eq. 1.43 we have

$$\overline{u^2} = \frac{3RT}{Lm} = \frac{3\mathbf{k}T}{m} \tag{14.43}$$

This quantity is obtained from Eq. 14.42 by multiplying u^2 by the fraction dN/N and integrating:

$$\overline{u^2} = \int_0^\infty u^2 \frac{dN}{N} = 4\pi \left(\frac{\beta m}{2\pi}\right)^{3/2} \int_0^\infty e^{-mu^2\beta/2} u^4 \, du \tag{14.44}$$

The integral is again a standard one (see appendix to this chapter, p. 685), and evaluating it leads to

$$\overline{u^2} = \frac{3}{m\beta} \tag{14.45}$$

Comparison of the expressions in Eqs. 14.43 and 14.45 then gives the important result that

$$\beta = \frac{1}{\mathbf{k}T} \tag{14.46}$$

Equation 14.42 may thus be written as

$$\frac{dN}{N} = 4\pi \left(\frac{m}{2\pi \mathbf{k}T}\right)^{3/2} e^{-mu^2/2\mathbf{k}T} u^2 \, du \tag{14.47}$$

This is usually known as the **Maxwell distribution law,** although Boltzmann also made important contributions to the theory.

Figure 14.4 shows a plot of $(1/N)(dN/du)$ for oxygen gas at two temperatures. Near the origin the curves are parabolic because of the dominance of the u^2 term in the equation, but at higher speeds the exponential term is more important. It is to be noted that the curve becomes much flatter as the temperature is raised. Indicated on the curve for 300 K are the root-mean-square speed $\sqrt{\overline{u^2}}$, the average speed $\overline{u}$, and the most probable speed u_{mp}. The mean square speed is given in Eq. 14.45, and insertion of $1/\mathbf{k}T$ for β gives the expression in Table 14.2; the root-mean-square speed is its square root. The mean speed or average speed is given by

$$\overline{u} = \int_0^\infty u \frac{dN}{N} \tag{14.48}$$

and insertion of Eq. 14.47 and integration leads to the expression in the table. The most probable speed is the speed at the maximum of the curve and is obtained by differentiating $(1/N)$ (dN/du) with respect to u and setting the result equal to zero. The resulting expression for this quantity is also given in Table 14.2.

The Distribution of Energy

We can convert Eq. 14.47 into an equation for the energy distribution. We start with

$$\epsilon = \tfrac{1}{2}mu^2 \tag{14.49}$$

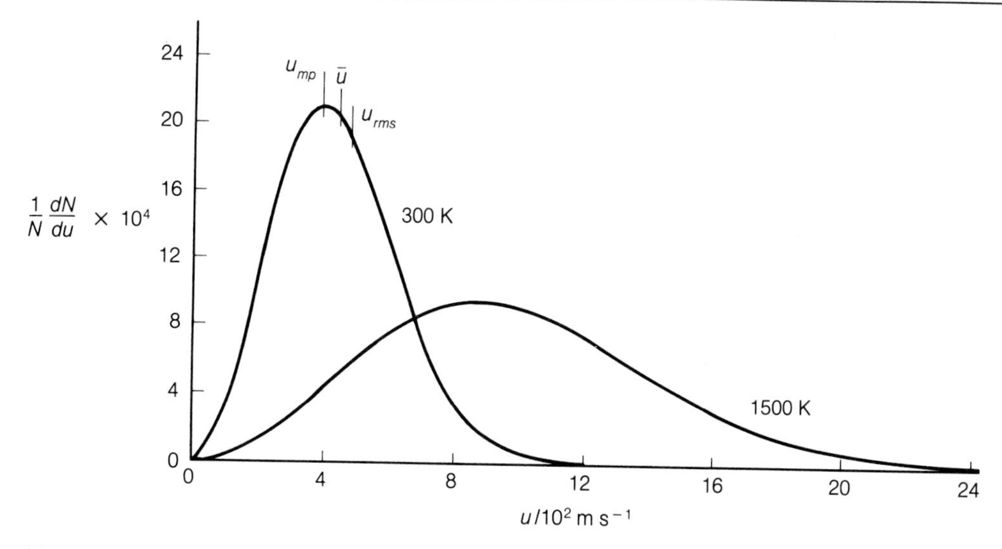

FIGURE 14.4 The Maxwell distribution of molecular speeds for oxygen at 300 K and 1500 K.

and differentiation gives

$$\frac{d\epsilon}{du} = mu = m\left(\frac{2\epsilon}{m}\right)^{1/2} = (2\epsilon m)^{1/2} \tag{14.50}$$

We can therefore replace du in Eq. 14.47 by $d\epsilon/(2\epsilon m)^{1/2}$ and u^2 by $2\epsilon/m$, and we obtain

$$\frac{dN}{N} = 4\pi\left(\frac{m}{2\pi \mathbf{k}T}\right)^{3/2} e^{-\epsilon/\mathbf{k}T}\frac{2\epsilon}{m(2\epsilon m)^{1/2}}\, d\epsilon \tag{14.51}$$

$$= \frac{2\pi}{(\pi \mathbf{k}T)^{3/2}} e^{-\epsilon/\mathbf{k}T}\epsilon^{1/2}\, d\epsilon \tag{14.52}$$

This is often known as the *Boltzmann distribution law* or as the **Maxwell-Boltzmann distribution law.**

An expression for the average energy is obtained by evaluating the integral

TABLE 14.2 Quantities Relating to the Maxwell Distribution of Molecular Speeds

Mean square speed	$\overline{u^2} = \frac{3\mathrm{k}T}{m}$
Square root of the mean square speed	$\sqrt{\overline{u^2}} = \sqrt{\frac{3\mathrm{k}T}{m}}$
Average speed	$\bar{u} = \sqrt{\frac{8\mathrm{k}T}{\pi m}}$
Most probable speed	$u_{mp} = \sqrt{\frac{2\mathrm{k}T}{m}}$
Average translational energy	$\bar{\epsilon} = \frac{3}{2}\mathrm{k}T$

$$\bar{\epsilon} = \int_0^\infty \epsilon \frac{dN}{N} \tag{14.53}$$

The result is $\frac{3}{2}\mathbf{k}T$, which we already found, in another way, in Section 1.9.

14.4 The Partition Function

In Section 14.3 we have been concerned exclusively with the translational motions of gas molecules. This is an important topic in itself, particularly as it led to a simple way of identifying the quantity β with $1/\mathbf{k}T$, but the Boltzmann distribution law has a much wider applicability. We will now apply it to a variety of different kinds of energy.

The law in its most general form is expressed in Eq. 14.34. The denominator in this expression has a special significance, and it is given the symbol q or Q and called the **partition function:**

$$q \equiv \sum_i e^{-\epsilon_i/\mathbf{k}T} \tag{14.54}$$

Its German name *Zustandsumme* (sum over states) indicates its significance more clearly.

The Molecular Partition Function

When the partition function relates to a single molecule, we call it the **molecular partition function** or the *particle partition function* and give it the symbol q. We will consider this type of partition function first. Figure 14.5 shows a number of molecular energy levels, starting with ϵ_0, which is taken to be zero. The terms for each level are indicated, and their sum is

$$q = e^{-0/\mathbf{k}T} + e^{-\epsilon_1/\mathbf{k}T} + e^{-\epsilon_2/\mathbf{k}T} + \cdots \tag{14.55}$$

The first term is unity, the second is a fraction, the third a smaller fraction, and so on. The partition function is thus unity plus a series of fractions that become smaller

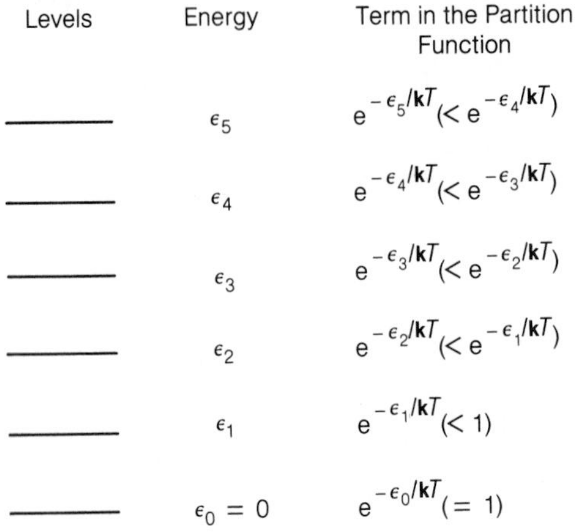

Levels	Energy	Term in the Partition Function
———	ϵ_5	$e^{-\epsilon_5/\mathbf{k}T} (< e^{-\epsilon_4/\mathbf{k}T})$
———	ϵ_4	$e^{-\epsilon_4/\mathbf{k}T} (< e^{-\epsilon_3/\mathbf{k}T})$
———	ϵ_3	$e^{-\epsilon_3/\mathbf{k}T} (< e^{-\epsilon_2/\mathbf{k}T})$
———	ϵ_2	$e^{-\epsilon_2/\mathbf{k}T} (< e^{-\epsilon_1/\mathbf{k}T})$
———	ϵ_1	$e^{-\epsilon_1/\mathbf{k}T} (< 1)$
———	$\epsilon_0 = 0$	$e^{-\epsilon_0/\mathbf{k}T} (= 1)$

FIGURE 14.5 A series of energy levels and the corresponding terms in the partition function.

and smaller. If the energy levels are very close together, the second term is close to unity, and there may be a very large number of terms that are close to unity. This is the situation with translational energy; the levels are very closely spaced, and there are very many terms that are close to unity. The partition function is thus very large; we shall see later that for translational motion in a volume of ordinary dimensions the molecular partition function at 300 K is 10^{30} or more.

At the other extreme, if the energy levels are widely spread, even the second term $\exp(-\epsilon_1/\mathbf{k}T)$ is very small, and subsequent ones are even smaller. The partition function will thus be close to unity (apart from *degeneracy*, which is discussed later). This is often the situation for electronically excited states. For example, the first electronically excited state of the Cl_2 molecule is 18 310.5 cm^{-1} above the ground state. This is 3.64×10^{-19} J, and at 300 K the term is

$$\exp\left(\frac{-3.64 \times 10^{-19}\ \text{J}}{\mathbf{k}T}\right) = e^{-87.8} = 7.4 \times 10^{-39}$$

This is negligible compared to unity, so that the partition function is very close to unity.

The situation with rotational and vibrational energy levels is in between these two extremes. Vibrational partition functions are often close to unity because of the fairly wide separation of the energy levels. The separation is much less for the rotational levels, and the partition functions are usually substantially larger than unity.

The significance of the partition function is that it gives an indication of the number of states that are accessible to the system. At the absolute zero only the ground state is accessible, and the partition function is unity. As the temperature is raised, more and more states become accessible, and as T approaches infinity, all states become accessible. A system that has a large number of accessible states is inherently more likely to exist than one with a more limited accessibility. We will see later in this chapter that this concept is of great significance in connection with the evaluation of equilibrium constants.

It is possible for more than one state to have an energy equal to ϵ_i, and the energy level is then said to be *degenerate*. If the energy level i has g_i states, we can write

$$q = \sum_i g_i e^{-\epsilon_i/\mathbf{k}T} \tag{14.56}$$

where the factor g_i is known as the **degeneracy** or the **statistical weight.** We can now rewrite Eq. 14.34 in a form that includes the degeneracies:

$$\frac{n_i}{N} = \frac{g_i e^{-\epsilon_i/\mathbf{k}T}}{\sum_i g_i e^{-\epsilon_i/\mathbf{k}T}} = \frac{g_i e^{-\epsilon_i/\mathbf{k}T}}{q} \tag{14.57}$$

EXAMPLE

In a C—I bond, iodine has a low frequency of vibration because of its large mass. As a result, energy levels higher than ϵ_0 are occupied to a greater extent than for atoms such as hydrogen bound to carbon. If the frequency of vibration of iodine in a C—I bond is 4.0×10^{12} s^{-1} at 298 K, calculate the value of n_8/n_0, the value of q, and the value of n_8/N.

SOLUTION

The value of the energy difference $\epsilon_i - \epsilon_0$ corresponds to $nh\nu$ since each level is higher than the previous one by $h\nu$. Thus

$$\epsilon_8 - \epsilon_0 = 8h\nu = 8(6.626 \times 10^{-34} \text{ J s})(5.0 \times 10^{12} \text{ s}^{-1}) = 2.65 \times 10^{-20} \text{ J}$$

From Eq. 14.35 we can write

$$\ln \frac{n_8}{n_0} = -\frac{1}{\mathbf{k}T}(\epsilon_8 - \epsilon_0) = -\frac{2.65 \times 10^{-20}(\text{J})}{1.38 \times 10^{-23} \times 298(\text{J})} = -6.44$$

Thus $n_8/n_0 = 1.59 \times 10^{-3}$.

We can calculate the partition function by adding the $e^{-\epsilon_i/\mathbf{k}T}$ terms until the value becomes negligible. The first terms are as follows:

i	0	1	2	3	4	5	6	7	8
$e^{-\epsilon_i/\mathbf{k}T}$	1	0.447	0.200	0.090	0.040	0.018	0.008	0.004	0.002

The sum of these is 1.8, and this is a good approximation to the partition function q. We will later see a simpler way of obtaining such a partition function, by summation of a series. The ratio n_8/N is

$$\frac{n_8}{N} = \frac{e^{-(\epsilon_8-\epsilon_0)/\mathbf{k}T}}{q} = \frac{e^{-6.44}}{1.8} = 8.9 \times 10^{-4}$$ ■

The Partition Function for the System

The molecular partition function q relates to a single molecule, but in thermodynamics we are concerned with assemblies of molecules. We therefore also define a *partition function Q for the system,* which relates to an assembly of a large number N of molecules. If N is the number of molecules in 1 mol (6.022×10^{23}), it is called the *molar partition function.*

The **system partition function** is defined by

$$Q = \sum_i e^{-\epsilon_i/\mathbf{k}T} \tag{14.58}$$

where the summation is now taken over all the energy levels for all the molecules in the assembly. We must now see how the molecular partition function q is related to the partition function Q. An important distinction arises, according to whether the molecules are localized and therefore *distinguishable* or are nonlocalized and *indistinguishable.*

An ideal crystal is a simple example of a system in which the molecules (or atoms or ions) comprising it are localized. If we label the molecules a, b, c, etc., we see that molecule a is *distinguishable* from molecule b simply because it is in a different position. Each molecule, such as a, will have a set of energy levels ϵ_1^a, ϵ_2^a, ϵ_3^a, etc. The system partition function Q can therefore be expressed as

$$Q = \sum_i \exp\left[\frac{-(\epsilon_i^a + \epsilon_i^b + \epsilon_i^c + \cdots)}{\mathbf{k}T}\right] \tag{14.59}$$

This summation can be factorized into

$$Q = \left(\sum_i e^{-\epsilon_i/\mathbf{k}T}\right)_a \left(\sum_i e^{-\epsilon_i/\mathbf{k}T}\right)_b \left(\sum_i e^{-\epsilon_i/\mathbf{k}T}\right)_c \cdots \tag{14.60}$$

where the first term in parentheses is the summation for molecule a, the second for b, and so on. Since all of these terms are the same, the molecules being identical, we can write

$$Q = \left(\sum_i e^{-\epsilon_i/kT}\right)^N \tag{14.61}$$

where N is the number of molecules. We therefore obtain the important result that for this type of system

$$Q = q^N \tag{14.62}$$

In a gas, however, the situation is different, because the molecules or atoms composing it are not localized and are therefore *indistinguishable.* Suppose, for example, that molecule a has the energy ϵ_a; molecule b has the energy ϵ_b; and so on. The total energy is therefore

$$E = \epsilon_a + \epsilon_b + \epsilon_c + \cdots \tag{14.63}$$

However, the state corresponding to this energy is indistinguishable from the state that we would obtain by interchanging molecules a and b or any pairs of molecules. In this case we therefore count too many states if we proceed as we did for the crystal. We must divide by the number of ways we can permute N molecules, that number being $N!$. For systems of this kind we therefore have the relationship

$$Q = \frac{q^N}{N!} = \left(\frac{qe}{N}\right)^N \tag{14.64}$$

In the second expression here we have made use of Stirling's approximation (Eq. 14.25).

14.5 Thermodynamic Quantities from Partition Functions

The reason that partition functions are of such great importance is that all the thermodynamic quantities can be calculated from them. This means that the entire macroscopic behavior of matter at equilibrium can be predicted if we have evaluated the partition function. For example, if we know the partition functions for all the substances in a chemical reaction, we can calculate the Gibbs energy of each substance and can therefore calculate the equilibrium constant at any temperature. We will see in Section 14.8 exactly how this is done.

First we will consider the relationship between the partition function and the internal energy U of the system. When all the molecules are in their lowest possible energy states, the internal energy has its minimum value, which we call U_0. In general the internal energy is U and what we can obtain from the partition function Q for the system is the value of $U - U_0$. The way we do this is to consider a very large number N^* of assemblies, having energies E_1^*, E_2^*, etc., with n_1^*, n_2^*, etc., being their respective distribution numbers. This large collection of assemblies is known as an **ensemble,** and we use the starred quantities for the ensemble. The energy of each ensemble is represented by a term $n_i^* E_i^*$, and the average value of $U - U_0$ for the large collection of assemblies is*

$$U - U_0 = \frac{\sum_i n_i^* E_i^*}{N^*} \tag{14.65}$$

For an ensemble of N^* assemblies the ratio n_i^*/N^* is given by

$$\frac{n_i^*}{N^*} = \frac{g_i e^{-E_i^*/\mathbf{k}T}}{Q} \tag{14.66}$$

(compare Eq. 14.57, which is for a single assembly). The partition function Q for the system is given by

$$Q = \sum_i g_i e^{-E_i^*/\mathbf{k}T} \tag{14.67}$$

where E_i^* is the energy of the ith assembly. The value of $U - U_0$ is therefore

$$U - U_0 = \frac{\sum_i g_i E_i^* e^{-E_i^*/\mathbf{k}T}}{Q} \tag{14.68}$$

The E_i^* values do not depend on temperature and we can differentiate Eq. 14.67 and obtain

$$\left(\frac{\partial Q}{\partial T}\right)_V = \frac{1}{\mathbf{k}T^2} \sum_i g_i E_i^* e^{-E_i^*/\mathbf{k}T} \tag{14.69}$$

The expression for $U - U_0$ can therefore be written as

$$U - U_0 = \frac{\mathbf{k}T^2}{Q}\left(\frac{\partial Q}{\partial T}\right)_V \tag{14.70}$$

or as

$$U - U_0 = \mathbf{k}T^2 \left(\frac{\partial \ln Q}{\partial T}\right)_V \tag{14.71}$$

This expression is given in Table 14.3, along with other expressions for thermodynamic quantities.

To obtain the entropy we can make use of the relationship

$$\int_0^T dS = \int_0^T \frac{C_V}{T}\, dT \tag{14.72}$$

or

$$S - S_0 = \int_0^T \frac{C_V}{T}\, dT \tag{14.73}$$

* More advanced presentations of this topic discuss in detail an assumption that is involved here, namely that the time average of $U - U_0$ is the same as the average over a large number of assemblies. This is known as the *ergodic hypothesis*.

where S_0 is the entropy at the absolute zero. The heat capacity C_V is

$$C_V = \left(\frac{\partial U}{\partial T}\right)_V = \frac{\partial}{\partial T}\left[\mathbf{k}T^2\left(\frac{\partial \ln Q}{\partial T}\right)_V\right] \tag{14.74}$$

$$= \mathbf{k}\left[T^2\left(\frac{\partial^2 \ln Q}{\partial T^2}\right)_V + 2T\left(\frac{\partial \ln Q}{\partial T}\right)_V\right] \tag{14.75}$$

Then

$$S - S_0 = \mathbf{k}\int_0^T \left[T^2\left(\frac{\partial^2 \ln Q}{\partial T^2}\right)_V + 2T\left(\frac{\partial \ln Q}{\partial T}\right)_V\right] dT \tag{14.76}$$

Integration by parts leads to

$$S - S_0 = \mathbf{k}T\left(\frac{\partial \ln Q}{\partial T}\right)_V + \mathbf{k}\ln Q - \mathbf{k}\ln Q_0 \tag{14.77}$$

where Q_0 is the value of Q at the absolute zero. The values at the absolute zero may be equated to one another,

$$S_0 = \mathbf{k}\ln Q_0 \tag{14.78}$$

and therefore

$$\boldsymbol{S = \mathbf{k}T\left(\frac{\partial \ln Q}{\partial T}\right)_V + \mathbf{k}\ln Q} \tag{14.79}$$

$$= \frac{U - U_0}{T} + \mathbf{k}\ln Q \tag{14.80}$$

To obtain the Helmholtz energy we use the relationship $A = U - TS$, and with Eq. 14.80 we obtain

$$\boldsymbol{A - U_0 = -\mathbf{k}T\ln Q} \tag{14.81}$$

(Note that at the absolute zero $A_0 = U_0$.) Since $P = -(\partial A/\partial V)_T$ (Eq. 3.118) we also have

$$P = \mathbf{k}T\left(\frac{\partial \ln Q}{\partial V}\right)_T \tag{14.82}$$

The definition of the enthalpy, $H = U + PV = U + N\mathbf{k}T$, with Eq. 14.71 gives

$$\boldsymbol{H - U_0 = \mathbf{k}T^2\left(\frac{\partial \ln Q}{\partial T}\right)_V + N\mathbf{k}T} \tag{14.83}$$

The Gibbs energy, $G = A + PV = A + N\mathbf{k}T$, is similarly found, with Eq. 14.81, to be

$$\boldsymbol{G - U_0 = -\mathbf{k}T\ln Q + N\mathbf{k}T} \tag{14.84}$$

These various expressions are included in Table 14.3. The corresponding expressions in terms of the molecular partition function q are easily obtained (see Problem 7).

TABLE 14.3 Thermodynamic Quantities in Terms of the Partition Function Q for the System

$$U - U_0 = kT^2 \left(\frac{\partial \ln Q}{\partial T}\right)_V$$

$$S = kT \left(\frac{\partial \ln Q}{\partial T}\right)_V + k \ln Q$$

$$A - U_0 = -kT \ln Q$$

$$H - U_0 = kT^2 \left(\frac{\partial \ln Q}{\partial T}\right)_V + NkT$$

$$G - U_0 = -kT \ln Q + NkT$$

If N is the number of molecules in 1 mol, it can be replaced by the Avogadro constant L ($=6.022 \times 10^{23}$ mol^{-1}), and the thermodynamic quantity is then the molar quantity.

If two substances are present, the procedure is as follows. Suppose that, as in a mixture of two gases, the molecules of each type are indistinguishable. The partition function for the system is then

$$Q = \frac{q_a^{N_a} q_b^{N_b}}{N_a! N_b!} \tag{14.85}$$

We can now make use of the definition of the chemical potential (Section 4.1); thus for the substance of type a, of which there are N_a molecules, the chemical potential per molecule of a is

$$\mu_a = \left(\frac{\partial A}{\partial N_a}\right)_{T,V} = -kT \left(\frac{\partial \ln Q}{\partial N_a}\right)_{T,V} \tag{14.86}$$

From Eq. 14.85 we obtain

$$\ln Q = N_a \ln q_a + N_b \ln q_b - \ln N_a! - \ln N_b! \tag{14.87}$$

$$= N_a \ln q_a - N_a \ln N_a + N_a + N_b \ln q_b - \ln N_b! \tag{14.88}$$

using Stirling's approximation. Thus

$$\left(\frac{\partial \ln Q}{\partial N_a}\right)_{T,V} = \ln q_a - \ln N_a = \ln \frac{q_a}{N_a} \tag{14.89}$$

The **molecular chemical potential** is therefore

$$\boldsymbol{\mu_a = -kT \ln \frac{q_a}{N_a}} \tag{14.90}$$

This type of expression can be used in the treatment of chemical equilibrium.

14.6 The Partition Function for Some Special Cases

We have already seen in Section 14.1 that the energy of a molecule may be translational, ϵ_t; rotational, ϵ_r; and vibrational, ϵ_v. To these must be added electronic energy, ϵ_e, and nuclear energy, ϵ_n, and the total energy is

$$\epsilon_{tot} = \epsilon_t + \epsilon_r + \epsilon_v + \epsilon_e + \epsilon_n \tag{14.91}$$

Because these energies are additive, the partition function is a product of the contributions for these different types of energy. For example, suppose there is a series of vibrational energies ϵ_j^v and another series of rotational energies ϵ_k^r. There is then a set of total energies given by

$$\epsilon_i = \epsilon_j^v + \epsilon_k^r \tag{14.92}$$

Then, from Eq. 14.54, the partition function for these two forms of energy is

$$q = \sum_i e^{-\epsilon_i/\mathbf{k}T} = \sum_j \sum_k \exp\left[\frac{-(\epsilon_j^v + \epsilon_k^r)}{\mathbf{k}T}\right] \tag{14.93}$$

$$= \sum_j e^{-\epsilon_j^v/\mathbf{k}T} \sum_k e^{-\epsilon_k^r/\mathbf{k}T} = q_v q_r \tag{14.94}$$

In general, for the five forms of energy,

$$q = q_t q_r q_v q_e q_n \tag{14.95}$$

Translational Motion

We will first consider a monatomic gas and will assume, as is usually the case at ordinary temperatures, that only the lowest electronic state has to be considered. There is therefore only translational energy, which consists of the energies in the three independent directions,

$$\epsilon_t = \epsilon_x + \epsilon_y + \epsilon_z \tag{14.96}$$

The translational partition function is also a product, for the reason we have just discussed:

$$q_t = q_x q_y q_z \tag{14.97}$$

Solving for q_t simply involves determining the values of the allowed energies. The translational energies are given by the solution of the Schrödinger equation for a particle of mass m confined in a box with sides of length a, b, c in the X-, Y-, and Z-directions, respectively. The solution, given in Section 11.6 (Eq. 11.146), is

$$\epsilon_t = \epsilon_x + \epsilon_y + \epsilon_z = \frac{h^2}{8m}\left(\frac{n_x^2}{a^2} + \frac{n_y^2}{b^2} + \frac{n_z^2}{c^2}\right) \tag{14.98}$$

where the integers n_x, n_y, n_z are *quantum numbers* and specify the energy states available to the molecule.

The partition function can then be written as

$$q_t = \sum_{n_x} \sum_{n_y} \sum_{n_z} \exp\left[-\frac{(\epsilon_x + \epsilon_y + \epsilon_z)}{\mathbf{k}T}\right] = q_x q_y q_z \tag{14.99}$$

It has been split into three factors into which the values of the individual energies are inserted. For q_x the expression is

$$q_x = \sum_{n_x} e^{-n_x^2 h^2/8ma^2\mathbf{k}T} \tag{14.100}$$

the summation being from 0 to ∞. The spacing between translational levels is extremely small, and the sum in Eq. 14.100 can therefore be replaced by an integral:

$$q_x = \int_0^\infty e^{-n_x^2 h^2/8ma^2\mathbf{k}T}\, dn_x \tag{14.101}$$

The value of the integral can be obtained from the appendix to this chapter (p. 685) and the result is

$$q_x = \frac{(2\pi m \mathbf{k}T)^{1/2}a}{h} \tag{14.102}$$

Similar expressions are obtained for q_y and q_z. The result for q_t is

$$q_t = q_x q_y q_z = \frac{(2\pi m \mathbf{k}T)^{3/2}abc}{h^3} = \frac{(2\pi m \mathbf{k}T)^{3/2}V}{h^3} \tag{14.103}$$

since abc is the volume V.

EXAMPLE
Calculate the partition function for a hydrogen atom at 300 K in a container of 1-m^3 volume.

SOLUTION
Since an atom has only a translational motion, the partition function is found from Eq. 14.103. The mass of the hydrogen atom is

$$\frac{1.008 \times 10^{-3}}{6.022 \times 10^{23}} = 1.674 \times 10^{-27} \text{ kg}$$

With $V = 1$ m^3, the partition function is therefore

$$q = q_t = \frac{[2\pi \times 1.674 \times 10^{-27}(\text{kg}) \times 1.381 \times 10^{-23}(\text{J K}^{-1}) \times 300(\text{K})]^{3/2} 1(\text{m}^3)}{[6.626 \times 10^{-34}(\text{J s})]^3}$$

$$= \frac{2.877 \times 10^{-70}}{2.909 \times 10^{-100}} = 9.89 \times 10^{29}$$

■

Note that the complete partition function is dimensionless. Later, in dealing with equilibrium constants, we will use partition functions per unit volume. Translational partition functions per unit volume typically have the magnitude of $\approx 10^{32}$ m^{-3} at 300 K, and we will see that the rotational and vibrational functions are much smaller.

An expression for the entropy of an ideal monatomic gas was first derived in 1911 by O. Sackur and H. Tetrode.* For a system of N indistinguishable molecules, $\ln Q$ is $N \ln (qe/N)$ (Eq. 14.64), and substitution of the expression for q in Eq. 14.103 gives

$$\ln Q = N \ln\left[\frac{V}{Nh^3}(2\pi m \mathbf{k}T)^{3/2}\right] + N \tag{14.104}$$

Then from Eq. 14.80, and making use of the fact that $U - U_0 = \frac{3}{2}N\mathbf{k}T$, we obtain, for the molar entropy,

$$S_m = \frac{U_m - U_{0,m}}{T} + \mathbf{k} \ln Q = \frac{5}{2}R + R \ln\left[\frac{V_m}{Lh^3}(2\pi m \mathbf{k}T)^{3/2}\right] \tag{14.105}$$

*O. Sackur, *Ann. Physik, 36*, 398 (1911); *40*, 67 (1913); H. Tetrode, *Ann. Physik, 38*, 434 (1912).

This is known as the **Sackur-Tetrode equation,** and it gives excellent agreement for monatomic gases. We can replace V_m by $L\mathbf{k}T/P$ and then obtain for an ideal gas at 1 atm pressure

$$S_m/\text{J K}^{-1}\ \text{mol}^{-1} = -9.620 + 12.47 \ln M_r + 12.47 \ln T/K \tag{14.106}$$

where M_r is the relative molecular mass. At 25°C

$$S_m/\text{J K}^{-1}\ \text{mol}^{-1} = 108.74 + 12.47 \ln M_r \tag{14.107}$$

Rotational Motion

The rotational energy of a linear molecule, treated as a rigid rotor, was given in Eq. 11.209:

$$\epsilon_r = J(J+1)\frac{h^2}{8\pi^2 I} \tag{14.108}$$

where I is the moment of inertia and J is the rotational quantum number, having values 0, 1, 2, 3, ... The rotational partition function is therefore

$$q_r = \sum_i g_i e^{-\epsilon_i/\mathbf{k}T} = \sum_J (2J+1) \exp\left[-\frac{J(J+1)h^2}{8\pi^2 I\mathbf{k}T}\right] \tag{14.109}$$

As seen in Section 13.3, the degeneracy of a rotational level is $2J + 1$.

In general the summation in this equation cannot be expressed in closed form. However, if the energy levels are sufficiently close together compared with $\mathbf{k}T$, the summation can be replaced by an integral. This is a satisfactory approximation for most molecules at room temperature and above. Under these conditions, therefore,

$$q_r = \int_0^\infty (2J+1) \exp\left[-\frac{J(J+1)h^2}{8\pi^2 \mathrm{I}\mathbf{k}T}\right] dJ \tag{14.110}$$

This integration is easily performed if we recognize that $d[J(J+1)] = (2J+1)\,dJ$. We then have

$$q_r = \int_0^\infty \exp\left[-\frac{J(J+1)h^2}{8\pi^2 I\mathbf{k}T}\right] d[J(J+1)] = \frac{8\pi^2 I\mathbf{k}T}{h^2} \tag{14.111}$$

There is another important feature of the rotational partition function that has to be recognized. In the treatment we have given it was assumed that all the rotational energy levels are accessible to the molecule. This is indeed true for unsymmetrical linear molecules such as HCl, HCN, and OCS, and the rotational partition function for these molecules is correctly given by Eq. 14.111. For symmetrical linear molecules such as H_2, CO_2, and C_2H_2, however, not all the energy levels given by Eq. 14.108 are allowed. Which ones are allowed depends on the nature of the nuclei, but the result is that only half of the rotational energy levels can be taken up by the molecule. The partition function in Eq. 14.111 must therefore be divided by 2.

This factor of 2 is known as the **symmetry number** and given the symbol σ. The partition function for a linear molecule is therefore given in general by

$$q_r = \frac{8\pi^2 I\mathbf{k}T}{\sigma h^2} \tag{14.112}$$

where σ is unity for an unsymmetrical linear molecule and is 2 for a symmetrical one. A simple way of determining the symmetry number, and one that is applicable to nonlinear molecules, is to imagine the identical atoms to be labeled and to count the number of different but equivalent arrangements that are obtained when the molecule is rotated. Thus for CO_2 there are two such arrangements:

$$O^1{=}C{=}O^2 \quad \text{and} \quad O^2{=}C{=}O^1$$

The value of Q_r for a system of N molecules is simply q_r^N since the indistinguishability of the molecules has already been taken into account in the translational partition function. Application to thermodynamic quantities is straightforward. For example, for the rotational energy we have, for symmetrical or nonsymmetrical diatomic molecules,

$$U_{m,r} = \mathbf{k}LT^2\left(\frac{\partial \ln q_r}{\partial T}\right) = RT^2\left(\frac{8\pi^2 I\mathbf{k}}{\sigma h}\right)\frac{\sigma h}{8\pi^2 I\mathbf{k}T} = RT \tag{14.113}$$

which is the classical result. Note that the symmetry number does not affect the value of this thermodynamic property, since q appears as the derivative of the logarithm.

EXAMPLE

The internuclear separation in the hydrogen molecule is 0.074 nm.
a. Calculate the molecular rotational partition function at 300 K.
b. Calculate also $\ln Q_r$ for a system of $N = 6.022 \times 10^{23}$ molecules.

SOLUTION

First we calculate the moment of inertia of the hydrogen molecule. The distance between each atom and the center of gravity is $0.074 \times 10^{-9}/2$ m, and the moment of inertia is therefore

$$I = 2 \times 1.674 \times 10^{-27}(\text{kg}) \times \left[\frac{0.074 \times 10^{-9}}{2}(\text{m})\right]^2 = 4.583 \times 10^{-48}\ \text{kg m}^2$$

a. The molecular rotational partition function is therefore, from Eq. 14.112, with $\sigma = 2$,

$$q_r = \frac{8\pi^2 \times 4.583 \times 10^{-48}(\text{kg m}^2) \times 1.381 \times 10^{-23}(\text{J K}^{-1}) \times 300(\text{K})}{2 \times [6.626 \times 10^{-34}(\text{J s})]^2} = 1.707$$

This is an unusually small value for a rotational partition function; this arises because of the low mass of the H atom and the small internuclear separation. Rotational partition functions are more typically 10–1000. The values are much smaller than for translational partition functions (see the example on p. 668) because of the much wider separation of the rotational levels.
b. For a system of $N = 6.022 \times 10^{23}$ molecules,

$$\ln Q_r = N \ln q_r = 6.022 \times 10^{23} \ln 1.707 = 3.22 \times 10^{23}$$ ■

The derivation of the rotational partition function for nonlinear molecules is more difficult since the expressions for the allowed energies are more complicated. The partition function must now contain terms involving the *principal moments of inertia,* which are the moments of inertia I_A, I_B, and I_C about three perpendicular axes (Section 13.3). The final result is

$$q_r = \frac{8\pi^2(8\pi^3 I_A I_B I_C)^{1/2}(\mathbf{k}T)^{3/2}}{\sigma h^3} \tag{14.114}$$

Again, we obtain the symmetry number σ by labeling identical atoms and seeing how many equivalent arrangements can be formed by rotation. For example, for water

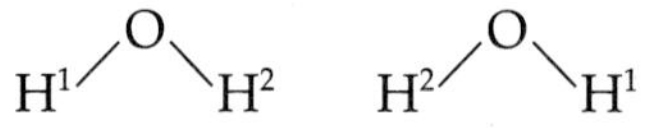

H^1–O–H^2 $\quad$ H^2–O–H^1

the symmetry number is two. For ammonia, which is nonplanar, σ is three. The methyl radical is planar, and the possible arrangements are

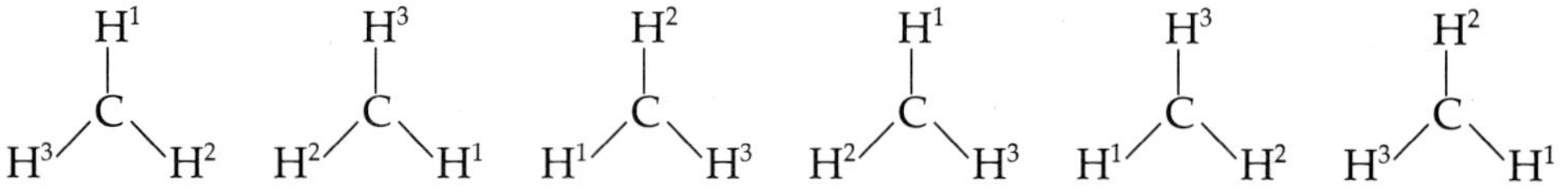

H^1	H^3	H^2	H^1	H^3	H^2
C	C	C	C	C	C
H^3 H^2	H^2 H^1	H^1 H^3	H^2 H^3	H^1 H^2	H^3 H^1

Its symmetry number is therefore six.

Vibrational Motion

The spacings between vibrational levels are appreciable compared to the room-temperature value of $\mathbf{k}T$. We therefore expect that only a few of the terms in the partition function will contribute significantly to the energy. In Section 11.6 (Eq. 11.156) we found that the energy of the harmonic oscillator is

$$\epsilon_v = (v + \tfrac{1}{2})h\nu \qquad v = 0, 1, 2, \ldots \tag{14.115}$$

where v is a quantum number, h is the Planck constant, and ν $(= (1/2\pi)\sqrt{k_h/\mu})$ is the classical frequency of the oscillator. However, in statistical mechanics the zero of energy is most commonly taken to be at $v = 0$ (i.e., at the zero-point level). The energies related to this level are $vh\nu$, and the corresponding partition function is

$$q_v = \sum_{n=0}^{\infty} e^{-vh\nu/\mathbf{k}T} \tag{14.116}$$

The summation is over all levels, but in practice only a few levels contribute substantially to the sum. Because of the wide spacing of the energy levels compared to $\mathbf{k}T$, the sum cannot be replaced by an integral. Thus q_v must be written out as a sum:

$$q_v = \sum e^{-vh\nu/\mathbf{k}T} \tag{14.117}$$

If we substitute x for $e^{-h\nu/\mathbf{k}T}$, the first few terms are

$$q_v = (1 + x + x^2 + \cdots) \tag{14.118}$$

The series $(1 + x + x^2 + \cdots)$, when x is small, is $1/(1 - x)$. Thus

$$q_v = \frac{1}{1 - x} = \frac{1}{1 - e^{-h\nu/\mathbf{k}T}} \tag{14.119}$$

It is convenient to refer to the quantity $h\nu/\mathbf{k}$ as the **characteristic temperature,** with the symbol θ_{vib}. Equation 14.119 may then be rewritten as

$$q_v = \frac{1}{1-e^{-\theta_{vib}/T}} \tag{14.120}$$

When θ_{vib} is large compared to T, the fraction of molecules in *excited states* (levels where $v > 0$) is small. For example, θ_{vib} for HCl is 4140 K, and for the exponential term this gives a value of 1.0×10^{-6} at 300 K. Table 14.4 gives some characteristic temperatures.

Equation 14.120 can be simplified at two extremes of temperature. At very low temperatures, where $\theta_{vib}/T >> 1$, the exponential term becomes negligible compared to unity and therefore

$$q_v \approx 1 \tag{14.121}$$

At high temperature, when $\theta_{vib}/T >> 1$, the exponential in the denominator may be expanded and only the first term accepted. Thus

$$e^{-\theta_{vib}/T} = 1 - \frac{\theta_{vib}}{T} \tag{14.122}$$

and therefore

$$q_v = \frac{T}{\theta_{vib}} \tag{14.123}$$

EXAMPLE

From the data in Table 14.4, calculate, with reference to $v = 0$, the molecular vibrational partition function for H_2 at (a) 300 K and (b) 3000 K.

TABLE 14.4 Characteristic Temperatures,* θ_{vib}

Compound		θ_{vib}/K	Compound		θ_{vib}/K
H_2		6210	CO		3070
N_2		3340	NO		2690
O_2		2230	HCl		4140
Cl_2		810	HBr		3700
Br_2		470	HI		3200
CO_2	θ_1	1890	H_2O	θ_1	5410
	θ_2	3360		θ_2	5250
	$\theta_3 = \theta_4$	954		θ_3	2290

*T. L. Hill, *Introduction to Statistical Thermodynamics*, Reading, Mass.: Addison-Wesley, 1960.

SOLUTION

The value of θ_{vib} is 6210 K, and the molecular vibrational partition function with reference to $v = 0$ is, from Eq. 14.120,

$$q_v = \frac{1}{1 - e^{-6210/T}}$$

a. At $T = 300$ K,

$$q_v = \frac{1}{1 - e^{-6210/300}} = \frac{1}{1 - 1.024 \times 10^{-9}} = 1$$

b. At $T = 3000$ K,

$$q_v = \frac{1}{1 - e^{-6210/3000}} = \frac{1}{1 - 0.1262} = 1.144$$

Note that even at high temperatures these molecular vibrational partition functions do not differ much from unity. ■

From Eq. 14.120 we find for N independent harmonic oscillators that $Q_v = q_v^N$ (indistinguishability is not a factor because it has been taken into account in the translational term). Thus

$$\ln Q_v = N \ln q_v = -N\left[\frac{\theta_{vib}}{2T} + \ln(1 - e^{-\theta_{vib}/T})\right] \quad (14.124)$$

The energy is given in terms of the partition function as

$$U_v = \mathbf{k}T^2\left(\frac{\partial \ln Q_v}{\partial T}\right)_V = \mathbf{k}NT^2\left[\frac{\theta_{vib}}{2T^2} + \frac{\theta_{vib}/T^2}{(e^{\theta_{vib}/T} - 1)}\right] \quad (14.125)$$

If the ground-state energy $U_{0,v}$ is defined as $Nh\nu/2$, then $R\theta_{vib}/2$ is the zero-point energy and the molar energy may be written as

$$U_v - U_{0,v} = RT\frac{\theta_{vib}/T}{e^{\theta_{vib}/T} - 1} \quad (14.126)$$

This is the same result that would have been obtained if we had used the partition function in Eq. 14.119.

When $h\nu_{vib}$ is much less than $\mathbf{k}T$ ($\theta_{vib}/T \rightarrow 0$), Eq. 14.126 reduces to the expected classical value. Since under this condition $e^{\theta_{vib}/T} = 1 + \theta_{vib}/T + \cdots$, we may write for the molar energy

$$(U - U_0)_{v,m} = RT\frac{\theta_{vib}/T}{(1 + \theta_{vib}/T) - 1} = RT \quad (14.127)$$

It is also informative to compare the size of the energy terms expected in the classical and quantum-mechanical expressions. At 298 K the average value of the kinetic and potential energies is RT, or 2480 J mol^{-1}. If 2×10^{-20} J is the energy spacing between typical vibrational levels, then

$$\frac{\theta_{vib}}{T} = \frac{h\nu}{\mathbf{k}T} = \frac{2 \times 10^{-20}(\text{J})}{1.38 \times 10^{-23}(\text{J K}^{-1}) \times 298(\text{K})} = 4.9 \quad (14.128)$$

Thus the energy for this one vibrational frequency is

$$(U - U_0)_v = RT\frac{\theta_{\text{vib}}/T}{e^{\theta_{\text{vib}}/T} - 1} = \frac{8.314(298)(4.9)}{e^{4.9} - 1} = 91 \text{ J mol}^{-1} \tag{14.129}$$

The large difference of approximately 2400 J between the values is the result of the quantum effects. As the temperature increases, the vibrational contribution to the energy increases.

The heat capacity due to the vibrational modes is an important quantity in many theoretical studies since it can be measured accurately. The form of C_V is given by differentiation of $U - U_0$ with respect to T, or from Eq. 14.126. The result is

$$\frac{\overline{C}_V(\text{vib})}{R} = \left(\frac{\theta_{\text{vib}}}{T}\right)^2 \frac{e^{\theta_{\text{vib}}/T}}{(e^{\theta_{\text{vib}}/T} - 1)^2} \tag{14.130}$$

The function on the right-hand side is called an **Einstein function,** and in Figure 14.6 it is plotted as a function of T/θ_{vib}.

The total $\overline{C}_V/R$ for a diatomic molecule is calculated by adding the translational and rotational contributions to Eq. 14.130. Thus

$$\frac{\overline{C}_V}{R} = \frac{3}{2} + \frac{2}{2} + \left(\frac{\theta_{\text{vib}}}{T}\right)^2 \frac{e^{\theta_{\text{vib}}/T}}{(e^{\theta_{\text{vib}}/T} - 1)^2} \tag{14.131}$$

A polyatomic molecule such as CO_2, which has four modes of vibration, two of which are degenerate and thus equal in frequency, has three distinct Einstein functions, one doubly weighted because of the degeneracy. The value $\frac{5}{2}$ arises from the translational and rotational contributions:

$$\frac{\overline{C}_V}{R} = \frac{5}{2} + \left(\frac{\theta_{\text{vib},1}}{T}\right)^2 \frac{e^{\theta_{\text{vib},1}/T}}{(e^{\theta_{\text{vib},1}/T} - 1)^2} + \left(\frac{\theta_{\text{vib},2}}{T}\right)^2 \frac{e^{\theta_{\text{vib},2}/T}}{(e^{\theta_{\text{vib},2}/T} - 1)^2} + 2\left(\frac{\theta_{\text{vib},3}}{T}\right)^2 \frac{e^{\theta_{\text{vib},3}/T}}{(e^{\theta_{\text{vib},3}/T} - 1)^2} \tag{14.132}$$

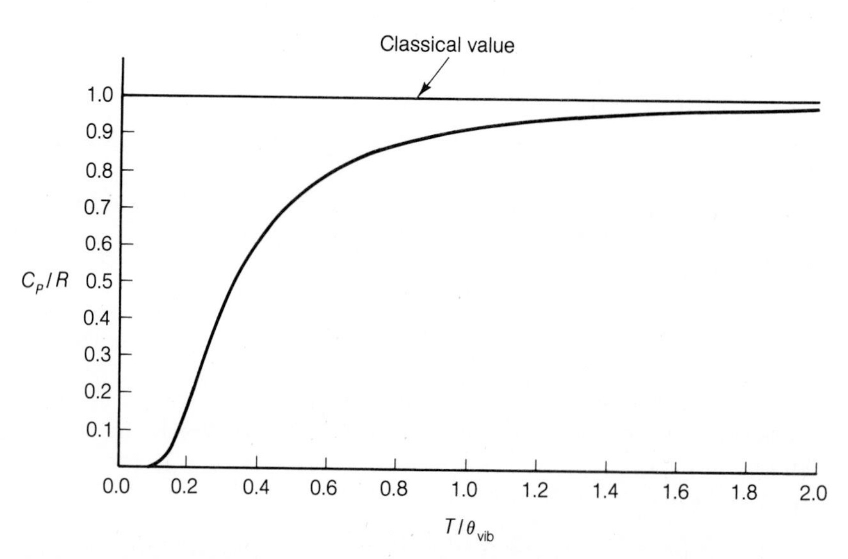

FIGURE 14.6 A plot against T/θ_{vib} of the Einstein function, given by Eq. 14.129.

TABLE 14.5 Partition Functions for Different Types of Molecular Motion

Motion	Degrees of Freedom	Partition Function	Order of Magnitude
Translation	3	$\frac{(2\pi m\mathbf{k}T)^{3/2}}{h^3}$ (per unit volume)	10^{31}–10^{32} m^{-3}
Rotation (linear molecule)	2	$\frac{8\pi^2 I\mathbf{k}T}{\sigma h^2}$	10–10^2
Rotation (nonlinear molecule)	3	$\frac{8\pi^2(8\pi^3 I_A I_B I_C)^{1/2}(\mathbf{k}T)^{3/2}}{\sigma h^3}$	10^{12}–10^3
Vibration (per normal mode)	1	$\frac{1}{1 - e^{-h\nu/\mathbf{k}T}}$	1–10
Restricted rotation	1	$\frac{(8\pi^2 I'\mathbf{k}T)^{1/2}}{h}$	1–10

m = mass of molecule
I = moment of inertia for linear molecule
I_A, I_B, and I_C = moments of inertia for a nonlinear molecule, about three axes at right angles to one another
I' = moment of inertia for restricted rotation
ν = normal-mode vibrational frequency
$\mathbf{k}$ = Boltzmann constant
h = Planck constant
T = absolute temperature
σ = symmetry number

It is useful to remember that the power to which h appears is equal to the number of degrees of freedom.

A listing of partition functions is given in Table 14.5.

The Electronic Partition Function

The electronic levels of an atom or molecule are usually much further apart than the vibrational levels. For the hydrogen atom, for example, the first stable excited electronic state ($^1\Sigma_u^+$; see Figure 13.23) is about 1.6×10^{-18} J (10.2 eV) above the lowest state, and at ordinary temperatures this state makes a negligible contribution to the partition function. As a useful rule we can say that if the energy $\Delta\epsilon$ divided by $\mathbf{k}$ is greater than 5 K, the level can be ignored. At 300 K this means that $\Delta\epsilon/\mathbf{k}$ is greater than 1500 K (i.e., that $\Delta\epsilon$ is greater than about 2×10^{-20} J or about 0.12 eV); this energy corresponds to about 110 cm^{-1}.

Another factor is the degeneracy of an electronic state. The total electronic angular momentum of an atom is determined by the quantum number j, which has positive values given by $l \pm s$. For every j value there are $2j + 1$ possible orientations in a magnetic field, corresponding to identical energies, and the electronic degeneracy is therefore $2j + 1$. For example, the ground state of the hydrogen atom is a $^2S_{1/2}$ state, so that j is $\frac{1}{2}$ and $2j + 1$ is equal to 2. The ground state of the iodine atom is $^2P_{3/2}$, and the degeneracy is therefore 4; in both of these cases the excited states can be ignored at ordinary temperatures.

The oxygen atom is a little more complicated. The lowest state is $^3P_2$, but there are two states very close to it: $^3P_1$, which is 157.4 cm^{-1} above the lowest level, and $^3P_0$, which is 226.1 cm^{-1} above the lowest level. Higher states can be neglected at all but the highest temperatures. The degeneracies of the three levels are 5, 3, and 1, respectively. The electronic partition function is, therefore,

$$q_e = 5e^{-0/\mathbf{k}T} + 3e^{-157.4\ hc/\mathbf{k}T} + e^{-226.1\ hc/\mathbf{k}T}$$

At 300 K this gives $5 + (3 \times e^{-0.0075}) + e^{-0.0108} = 8.97$.

Nearly all stable molecules have $^1\Sigma$ ground terms, and their excited states are high enough to be neglected except at very high temperatures; since $j = 0$, there is no degeneracy and their electronic partition function is unity. Important exceptions are O_2, NO, and most free radicals.

The Nuclear Partition Function

There is also a contribution to the partition function arising from nuclear spins. This has to be taken into account in the calculation of absolute values, such as entropies. For ordinary chemical reactions, however, the nuclear spins do not effect the Gibbs energy change and therefore do not affect the equilibrium constant. The nuclear spin factors will therefore not be considered here.

14.7 The Internal Energy, Enthalpy, and Gibbs Energy Functions

Section 14.6 has shown us how partition functions are evaluated, on the basis of spectroscopic and other information. Earlier, in Section 14.5, we saw how the thermodynamic quantities can be calculated from the partition functions. We are thus provided with a very fundamental way of obtaining the properties of substances and equilibrium constants of reactions from the basic molecular properties as determined by spectroscopy and in other ways.

To do this it has proved convenient to define thermodynamic functions, as follows:

1. The *internal energy function,* or the *thermal energy,* defined as $U - U_0$, where U is the internal energy at any temperature and U_0 is its value at the absolute zero.
2. The *enthalpy function,* defined as $(H - H_0)/T$ where H is the enthalpy at temperature T and H_0 is its value at the absolute zero.
3. The *Gibbs energy function,* defined as $(G - H_0)/T$ where G is the Gibbs energy at temperature T.

A particular advantage of the last of these functions is that it does not vary strongly with temperature, so that tables can give values at widely separated temperatures; the values at intermediate temperatures can be obtained by extrapolation. The internal energy and enthalpy functions are commonly given in tables only at 25°C.

The internal energy function $U - U_0$, also known as the thermal energy, comes at once from the fact that C_V is defined as $(\partial U/\partial T)_V$ (Eq. 2.26); this leads to

$$U_{\text{thermal}} = U - U_0 = \int_0^T C_V\, dT \qquad (14.133)$$

The integration constant U_0 is the value of U when $T = 0$ (i.e., when all the molecules are in their lowest energy states). The thermal energy can be obtained from knowledge of C_V over a range of temperatures and also from the partition function as explained in Section 14.5.

A function similar to the thermal energy, but involving the enthalpy, is obtained as follows:

$$H = U + PV = U_0 + U - U_0 + PV \tag{14.134}$$

and the function $H - U_0$ is therefore

$$H - U_0 = U - U_0 + PV \tag{14.135}$$

For the special case of 1 mol of an ideal gas,

$$H_m - U_{0,m} = U_m - U_{0,m} + RT \tag{14.136}$$

EXAMPLE

Calculate $U - U_0$ and $H - U_0$ for 1 mol of SO_2 at 298 K, assuming that the rotational levels make their classical contribution. The vibrational frequencies are $\bar{\nu}_1 = 1151.4$ cm^{-1}, $\bar{\nu}_2 = 517.7$ cm^{-1}, and $\bar{\nu}_3 = 1361.8$ cm^{-1}. The electronic contribution may be neglected.

SOLUTION

The SO_2 molecule is nonlinear. Its three translational degrees of freedom each contribute $\frac{1}{2}RT = 1243.3$ J mol^{-1} to the internal energy function, as do the three rotational modes. The translational contribution is thus 3729.9 J mol^{-1}, as is the rotational contribution. The vibrational contributions must be calculated separately. Since $h\nu = hc\bar{\nu}$, where c is the velocity of light, the energies of the levels may be calculated as follows:

$$\epsilon_1 = hc\bar{\nu}_1 = 6.62 \times 10^{-34}\,\text{J s} \times 2.998 \times 10^{10}\,\text{cm s}^{-1} \times 1151.4\,\text{cm}^{-1} = 2.28 \times 10^{-20}\,\text{J}$$

$$\epsilon_2 = 6.62 \times 10^{-34} \times 2.998 \times 10^{10} \times 517.7 = 1.03 \times 10^{-20}\,\text{J}$$

$$\epsilon_3 = 6.62 \times 10^{-34} \times 2.998 \times 10^{10} \times 1361.8 = 2.70 \times 10^{-20}\,\text{J}$$

Values for θ_{vib}/T and $(U - U_0)_{vib}$, obtained by the use of Eqs. 14.128 and 14.129, are

Wave Number	θ_{vib}/T	$(U - U_0)_{vib}$
1151.4 cm^{-1}	5.54	54.3 J mol^{-1}
517.7 cm^{-1}	2.50	555.9 J mol^{-1}
1361.8 cm^{-1}	6.57	22.9 J mol^{-1}

The internal energy function is thus the sum of these contributions:

$$U - U_0 = 3729.9 + 3729.9 + 54.3 + 555.9 + 22.9 = 8092.9\,\text{J mol}^{-1}$$

The enthalpy function is given by Eq. 14.136:

$$H - U_0 = U - U_0 + RT = 8092.9 + 2486.5 = 10\,579.4\,\text{J mol}^{-1}$$ ■

These functions are particularly useful for calculating energies and enthalpies of reaction at elevated temperatures, where direct experimental determinations may be difficult. For example, if $H - U_0$ can be calculated for the reactants and products of a reaction, the enthalpy change in the reaction is given by

$$\Delta H = \Sigma(H - U_0)_{\text{products}} - \Sigma(H - U_0)_{\text{reactants}} \tag{14.137}$$

Alternatively, ΔH may have been determined experimentally at one temperature T, say 25°C; then

$$\Delta H = \Delta U_0 + \Delta H - \Delta U_0 = \Delta U_0 + \Delta(H - U_0) \tag{14.138}$$

The value of $\Delta(H - U_0)$ at T_1, calculated from the partition function, therefore allows ΔU_0 to be obtained. From the value of $\Delta(H - U_0)$ at some other temperature T_2, the ΔH at T_2 can therefore be calculated.

For the calculation of equilibrium constants, the Gibbs energy and enthalpy functions are particularly useful. At the absolute zero $U_0 = H_0$ and, therefore, for the standard quantities

$$H^0 - H_0^0 = H^0 - U_0^0 \tag{14.139}$$

$$= (U^0 - U_0^0) + PV \tag{14.140}$$

For 1 mol

$$H_m^0 - H_{0,m}^0 = (U_m^0 - U_{0,m}^0) + RT \tag{14.141}$$

The quantity $H_m^0 - H_{0,m}^0$ can be determined from thermochemical data, making use of the relationship

$$H_m^0 - H_{0,m}^0 = \int_0^T C_{P,m}\, dT \tag{14.142}$$

Alternatively it can be obtained from the partition function, making use of the expression in Table 14.3; the result is

$$H_m^0 - H_{0,m}^0 = \mathbf{k}T^2 \left(\frac{\partial \ln Q}{\partial T}\right)_V + RT \tag{14.143}$$

It is this function divided by T that is referred to as the **standard enthalpy function.**

The Gibbs energy function is arrived at as follows. Since $G^0 = H^0 - TS^0$, we can write

$$\frac{G^0 - H_0^0}{T} = \frac{H^0 - H_0^0}{T} - S^0 \tag{14.144}$$

The quantity is the **standard Gibbs energy function,** and it can be evaluated from the partition function by making use of the expressions in Table 14.3.

Table 14.6 shows a few values of the standard Gibbs energy function, standard enthalpies of formation at 25°C, and $H_{298}^0 - H_0^0$. We shall now see how these values are used to calculate equilibrium constants.

14.8 The Calculation of Equilibrium Constants

For any reaction

$$\Delta H^0 = \Delta H_0^0 + \Delta(H^0 - H_0^0) \tag{14.145}$$

TABLE 14.6 Gibbs Energy and Enthalpy Functions for Substances in the Vapor State (Standard State = 1 atm)

Substance	$-(G^0 - H_0^0)/T$ / J K^{-1}mol^{-1} 298 K	1000 K	$\Delta H^\circ_{f,298}$ / kJ mol^{-1}	$\Delta H^\circ_{298} - H_0^0$ / kJ mol^{-1}
H_2	102.17	136.98	0	8.468
N_2	162.42	197.95	0	8.669
O_2	175.98	212.13	0	8.661
H_2O	155.52	196.69	−241.82	9.908
CO	168.41	204.05	−110.53	8.673
CO_2	182.26	226.40	−393.51	9.364
CH_4	152.55	199.37	− 74.85	10.03
CH_3OH	201.17	257.65	−200.66	11.426
NH_3	158.95	203.47	− 46.11	9.916

For additional values, see G. N. Lewis and M. Randall, *Thermodynamics* (revised by K. S. Pitzer and L. Brewer), New York: McGraw-Hill, 1961, p. 682.

where the symbol Δ represents the sum of the values for the products minus the sum for the reactants. Similarly, for the Gibbs energy change,

$$\frac{\Delta G^0}{T} = \frac{\Delta H_0^0}{T} + \frac{\Delta(G^0 - H_0^0)}{T} \tag{14.146}$$

The quantities $H^0 - H_0^0$ and $G^0 - H_0^0$ can be evaluated for any substance at any temperature, by the methods indicated in Section 14.7. If, therefore, ΔH^0 or ΔG^0 is known at any temperature, the value of ΔH_0^0 can be obtained. This enables ΔG^0 to be obtained at any other temperature, by the use of Eq. 14.146, and using the value of $\Delta(G^0 - H_0^0)/T$ at that temperature.

The equilibrium constant is then obtained through the equation

$$\Delta G^0 = -RT \ln K^u \tag{14.147}$$

Since the standard state for the values given in Table 14.6 is 1 atm, the equilibrium constant obtained is K_P, in units involving atmospheres.

The following example illustrates the application of Gibbs energy and enthalpy functions to equilibrium problems.

EXAMPLE

Use the data in Table 14.6 to calculate, at 1000 K, ΔG^0 and K_P for the reaction

$$CO + 2H_2 \longrightarrow CH_3OH$$

SOLUTION

The standard enthalpy change for the reaction at 298.15 K is

$$\begin{aligned}\Delta H_{298}^0 &= \Delta H_{f,298}^0(CH_3OH) - \Delta H_{f,298}^0(CO) - 2\,\Delta H_{f,298}^0(H_2)\\ &= -200.66 + 110.53 - 0 = -90.13 \text{ kJ mol}^{-1}\end{aligned}$$

Also,

$$\Delta(H^0_{298} - H^0_0) = (H^0_{298} - H^0_0)(CH_3OH) - (H^0_{298} - H^0_0)(CO) - 2(H^0_{298} - H^0_0)(H_2)$$

$$= 11.426 - 8.673 - 2(8.468) = -14.18 \text{ kJ mol}^{-1}$$

From Eq. 14.145 we then have, for ΔH^0 at the absolute zero,

$$\Delta H^0_0 = \Delta H^0_{298} - \Delta(H^0_{298} - H^0_0) = -90.13 + 14.18 = -75.95 \text{ kJ mol}^{-1}$$

The Gibbs energy change at 1000 K is then obtained from Eq. 14.146:

$$\Delta G^0 = \Delta H^0_0 + \Delta(G^0 - H^0_0)$$

$$= -75\,950 - 1000(257.65 - 204.05 - 2 \times 136.98)$$

$$= 144\,410 \text{ J mol}^{-1} = 144.41 \text{ kJ mol}^{-1}$$

Then

$$\ln K^u_p = -\frac{\Delta G^0}{RT} = -17.37$$

and therefore

$$K_p = 2.86 \times 10^{-8} \text{ atm}^{-2}$$ ■

Direct Calculation from Partition Functions

If the thermodynamic functions discussed in Section 14.7 are not available for all the substances in a chemical reaction, the equilibrium constant must be calculated directly from the partition functions.

We will first arrive at the equilibrium equation in an intuitive fashion and then obtain it in a more formal way. Suppose that two substances A and Z are at equilibrium

$$A \rightleftharpoons Z$$

In Figure 14.7 the various energy levels for the two molecules are represented schematically. Although for a given molecule we conventionally take the lowest level as having an energy of zero, we obviously cannot do this when we have molecules present together at equilibrium, since there is, in general, an energy difference between the lowest levels. In the figure we have shown the lowest level of Z as being $\Delta\epsilon_0$ higher than the lowest level of A. If there were no other accessible levels, this would at once give the result as

$$K = \frac{[Z]}{[A]} = e^{-\Delta\epsilon_0/kT} \tag{14.148}$$

However, the other energy levels obviously must be taken into account, and this is conveniently done through the partition function. We saw in Section 14.4 that the partition function is related to the accessibility of the various energy levels and is a measure of the inherent probability of the existence of a particular molecule. Thus, if the zero levels were the same for reactants and products ($\Delta\epsilon_0 = 0$), the

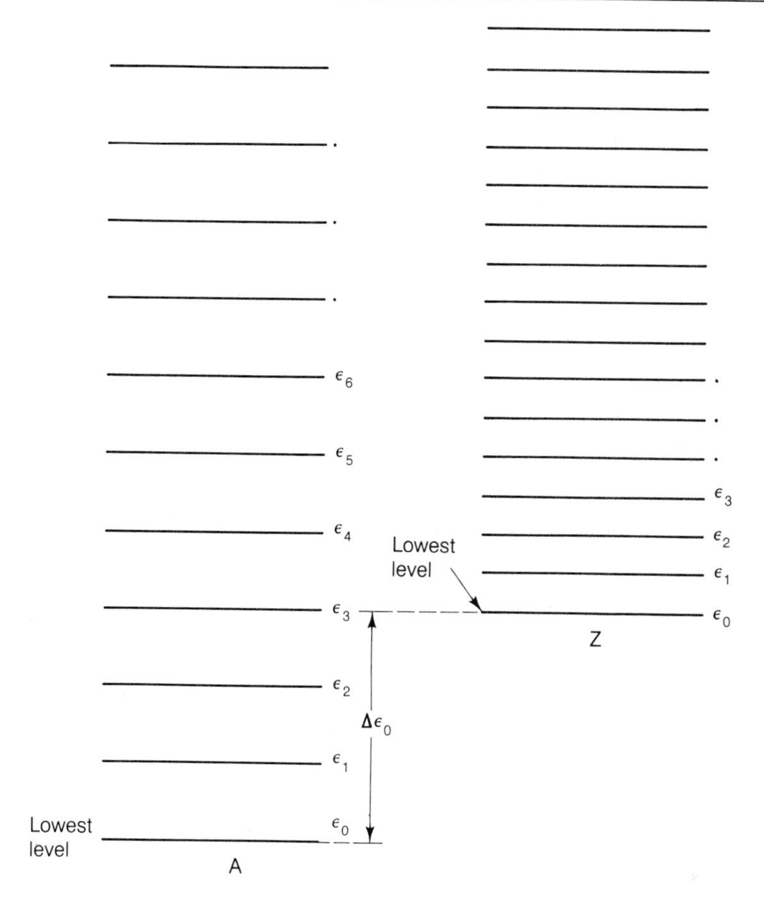

FIGURE 14.7 Two substances A and Z at equilibrium, showing the energy levels and the distance $\Delta\epsilon_0$ between the zero-point levels.

equilibrium constant would be simply the ratio of the molecular partition functions q_Z/q_A. With an energy difference of $\Delta\epsilon_0$ the equilibrium constant is

$$K = \frac{q_Z}{q_A} e^{-\Delta\epsilon_0/\mathbf{k}T} \tag{14.149}$$

For a general chemical reaction

$$aA + bB \rightleftharpoons yY + zZ$$

the corresponding equation is

$$K = \frac{q_Y^y q_Z^z}{q_A^a q_B^b} e^{-\Delta\epsilon_0/\mathbf{k}T} \tag{14.150}$$

The standard state to which K is related depends on how the partition functions are formulated. The simplest procedure is to express the q's as related to unit volume (1 m^3), in which case the equilibrium constant is a molecular equilibrium constant corresponding to the standard state 1 m^{-3}; conversion to other units is easily carried out.

The derivation of the expression for the equilibrium constant is based on Eq. 14.84:

$$G^0 - U_0^0 = -\mathbf{k}T \ln Q + N\mathbf{k}T \tag{14.151}$$

where N is 6.022×10^{23}. The standard molar Gibbs energy change for the reaction we are considering is

$$G^0 = yG_Y^0 + zG_Z^0 - aG_A^0 - bG_B^0 \tag{14.152}$$

(To simplify the notation we omit the subscripts m.) Use of Eq. 14.151 leads to

$$\Delta G^0 = yU_{0,Y}^0 + zU_{0,Z}^0 - aU_{0,A}^0 - bU_{0,B}^0 - RT \ln \left[\frac{\left(\frac{q_Y}{N}\right)^y \left(\frac{q_Z}{N}\right)^z}{\left(\frac{q_A}{N}\right)^a \left(\frac{q_B}{N}\right)^b} \right] \tag{14.153}$$

$$= \Delta U_0^0 - RT \ln \left[\frac{\left(\frac{q_Y}{N}\right)^y \left(\frac{q_Z}{N}\right)^z}{\left(\frac{q_A}{N}\right)^a \left(\frac{q_B}{N}\right)^b} \right] \tag{14.154}$$

$$= \Delta U_0^0 - RT \ln \frac{q_Y^y q_Z^z}{q_A^a q_B^b} N^{-\Sigma\nu} \tag{14.155}$$

where $\Sigma\nu$, the stoichiometric sum, is $y + z - a - b$. Then

$$\Delta G^0 = -RT \ln \left(\frac{q_Y^y q_Z^z}{q_A^a q_B^b} N^{-\Sigma\nu} e^{-\Delta U_0/RT} \right) \tag{14.156}$$

It then follows that

$$K^u = \frac{q_Y^y q_Z^z}{q_A^a q_B^b} N^{-\Sigma\nu} e^{-\Delta U_0/RT} \tag{14.157}$$

This is the equilibrium constant in dimensionless form. Since we have taken the volume to be m^3, the practical equilibrium constant has the same numerical value but has units involving mol m^{-3}. Conversion to other units is easily carried out as illustrated in the following example.

EXAMPLE

Calculate the equilibrium constant K_c for the reaction

$$H_2 \rightleftharpoons 2H$$

at (a) 300 K, and (b) 3000 K, making use of the following information: Ground state of the H atom is $^2S_{1/2}$; higher states can be ignored. H—H dissociation energy = 431.8 kJ mol^{-1}; H—H internuclear distance = 0.074 nm; $\theta_{vib}(H_2)$ = 6210 K.

SOLUTION

According to Eq. 14.157 the equilibrium constant for the reaction is

$$K = \frac{q^2(\text{H})}{q(\text{H}_2)N^{\Sigma\nu}} e^{-\Delta U_0/RT} = \frac{q^2(\text{H})}{q(\text{H}_2)N} e^{-\Delta U_0/RT}$$

since for this reaction $\Sigma\nu = 1$. We have already obtained some of the contributions to the partition functions in earlier examples.

a. At 300 K the translational partition function for H, with $V = 1\ \text{m}^3$, is 9.89×10^{29}, the electronic partition function is 2. The complete partition function for H at 300 K is therefore, for $V = 1\ \text{m}^3$,

$$q(\text{H, 300 K}) = 2 \times 9.89 \times 10^{29} = 1.978 \times 10^{29}$$

The complete partition function for H_2 is the product of the translational, rotational, vibrational, and electronic factors; the last is unity. Since $m^{3/2}$ enters into q_t, the value for H_2 is

$$q_t(\text{H}_2, \text{300 K}) = 2^{3/2} \times 9.89 \times 10^{29} = 2.22 \times 10^{30}$$

The rotational partition function was found (on p. 670) to be 1.707, and the vibrational partition function (on p. 673) to be 1. The total partition function is therefore, for $V = 1\ \text{m}^3$,

$$q_{\text{total}}(\text{H}_2, \text{300 K}) = 2.22 \times 10^{30} \times 1.707 \times 1 \times 1$$
$$= 3.80 \times 10^{30}$$

The dimensionless equilibrium constant K_c^u is therefore

$$K_c^u = \frac{q^2(\text{H})}{q(\text{H}_2)N} e^{-\Delta U_0/RT} = \frac{(1.978 \times 10^{29})^2}{3.80 \times 10^{30} \times 6.022 \times 10^{23}} \exp\left(\frac{-431\ 800}{8.314 \times 300}\right)$$
$$= 1.11 \times 10^{-71}$$

The practical equilibrium constant is thus

$$K_c = 6.00 \times 10^{-68}\ \text{mol m}^{-3} = 6.00 \times 10^{-71}\ \text{mol dm}^{-3}$$

b. Since the translational partition function is proportional to $T^{3/2}$, the values at 3000 K, for $1\ \text{m}^3$, are

$$q(\text{H, 3000 K}) = 1.978 \times 10^{29} \times 10^{3/2} = 1.14 \times 10^{31}$$

$$q_t(\text{H}_2, \text{3000 K}) = 2.22 \times 10^{30} \times 10^{3/2} = 1.28 \times 10^{34}$$

$$q_r(\text{H}_2, \text{3000 K}) = 1.707 \times 10 = 17.07$$

$$q_v(\text{H}_2, \text{3000 K}) = 1.144 \qquad \text{(see p. 673)}$$

The total partition function for H_2 is thus

$$q_{\text{total}}(\text{H}_2, \text{3000 K}) = 1.28 \times 10^{34} \times 17.07 \times 1.144 = 2.50 \times 10^{35}$$

The equilibrium constant is therefore

$$K_c^u = \frac{(1.14 \times 10^{31})^2}{2.50 \times 10^{35} \times 6.022 \times 10^{23}} \exp\left(-\frac{431\ 800}{8.314 \times 3000}\right) = 2590$$

and

$$K_c = 2590 \text{ mol m}^{-3} = 2.59 \text{ mol dm}^{-3}$$

The dissociation is negligible at 300 K but is appreciable at 3000 K. ■

An excellent example of the calculation of an equilibrium constant for a much more complicated reaction, namely

$$C_2H_4 + H_2 \rightleftharpoons C_2H_6$$

has been given by Guggenheim.*

The significance of symmetry numbers in calculations of equilibrium constants is illustrated by the following example.

EXAMPLE

Calculate the equilibrium constant at 300 K for the reaction

$$H_2 + D_2 \rightleftharpoons 2HD$$

The relative atomic masses are H = 1.0078 and D = 2.014 and the vibrational wave numbers are

$$H_2\text{: } 4371 \text{ cm}^{-1}; \qquad HD\text{: } 3786 \text{ cm}^{-1}; \qquad D_2\text{: } 3092 \text{ cm}^{-1}$$

The internuclear separation is 0.074 nm for all three forms.

SOLUTION

The value of ΔU_0 for the reaction can be calculated from the zero-point energies:

$$H_2\text{:} \quad \tfrac{1}{2} h\bar{\nu}c = \tfrac{1}{2} \times 6.626 \times 10^{-34} \times 4371 \times 2.998 \times 10^{10} = 4.341 \times 10^{-20} \text{ J}$$

$$HD\text{:} \quad \tfrac{1}{2} h\bar{\nu}c = \tfrac{1}{2} \times 6.626 \times 10^{-34} \times 3786 \times 2.998 \times 10^{10} = 3.760 \times 10^{-20} \text{ J}$$

$$D_2\text{:} \quad \tfrac{1}{2} h\bar{\nu}c = \tfrac{1}{2} \times 6.626 \times 10^{-34} \times 3092 \times 2.998 \times 10^{10} = 3.071 \times 10^{-20} \text{ J}$$

$$\Delta\epsilon_0 = (2 \times 3.760 - 4.341 - 3.071) \times 10^{-21} = 1.08 \times 10^{-21}$$

$$e^{-\Delta U_0/RT} = e^{-\Delta\epsilon_0/kT} = e^{0.2606} = 1.298$$

The masses and moments of inertia of the three species are

	Mass/kg	I/kg m^2	
H_2	3.345×10^{-27}	4.583×10^{-48}	
HD	5.018×10^{-27}	6.109×10^{-48}	$\left[= \dfrac{1.674 \times 3.345 \times 10^{-27}}{5.018 \times (0.074 \times 10^{-9})^{-2}}\right]$
D_2	6.689×10^{-27}	9.157×10^{-48}	

The equilibrium constant is obtained from Eq. 14.157 with $\Sigma\nu = 0$:

---

* E. A. Guggenheim, *Trans. Faraday Soc.*, *37*, 97 (1941); see also K. S. Pitzer, *J. Chem. Phys.*, *5*, 469 (1937); K. S. Pitzer and W. D. Gwanon, *J. Chem. Phys.*, *10*, 428 (1942); *16*, 303 (1948).

$$K = \frac{q^2(\text{HD})}{q(\text{H}_2)q(\text{D}_2)} e^{-\Delta U_0/RT}$$

$$= \frac{\left[\dfrac{(2\pi m_{\text{HD}}\mathbf{k}T)^3}{h^6}\right] \cdot \left(\dfrac{8\pi^2 I_{\text{HD}}\mathbf{k}T}{\sigma_{\text{HD}}h^2}\right)^2}{\left[\dfrac{(2\pi m_{\text{H}_2}\mathbf{k}T)^{3/2}}{h^3}\right] \cdot \left(\dfrac{8\pi^2 I_{\text{H}_2}\mathbf{k}T}{\sigma_{\text{H}_2}h^2}\right) \cdot \left[\dfrac{(2\pi m_{\text{D}_2}\mathbf{k}T)^{3/2}}{h^3}\right] \cdot \left(\dfrac{(8\pi^2 I_{\text{D}_2}\mathbf{k}T)}{\sigma_{\text{D}_2}h^2}\right)} \times 1.298$$

$$= \frac{m_{\text{HD}}^3}{(m_{\text{H}_2}m_{\text{D}_2})^{3/2}} \cdot \frac{I_{\text{HD}}^2}{I_{\text{H}_2}I_{\text{D}_2}} \cdot \frac{\sigma_{\text{H}_2}\sigma_{\text{D}_2}}{(\sigma_{\text{HD}})^2} = \frac{(5.018)^2}{3.348 \times 6.689} \times \frac{(6.109)^2}{4.583 \times 9.157} \times \frac{2 \times 2}{1} \times 1.298$$

$$= 1.12 \times 0.889 \times 4 \times 1.298 = 5.17$$

■

The equilibrium constant for this reaction is almost exactly the ratio of the symmetry numbers. It is easy to see why this should be so. If we were to start with equal amounts of H_2 and D_2 and allowed the reaction to go to equilibrium, each atom in the molecules present would have one-half a chance of being H and one-half a chance of being D. Thus in a molecule AB, the chance that A is H and that also B is H is $\frac{1}{2} \times \frac{1}{2} = \frac{1}{4}$. The chance that both atoms are D is similarly $\frac{1}{2} \times \frac{1}{2} = \frac{1}{4}$. The molecule HD, however, can be arrived at in two ways. The atom A may be H and atom B may be D, the probability of which is $\frac{1}{2} \times \frac{1}{2} = \frac{1}{4}$. Also, atom A may be D and atom B may be H, with a probability of $\frac{1}{2} \times \frac{1}{2} = \frac{1}{4}$. The probability that a molecule is HD is thus $\frac{1}{4} + \frac{1}{4} = \frac{1}{2}$. The ratio

$$H_2 : HD : D_2$$

is therefore

$$\tfrac{1}{4} : \tfrac{1}{2} : \tfrac{1}{4} \quad \text{or} \quad 1 : 2 : 1$$

and the equilibrium constant $[HD]^2/[H_2][D_2]$ is thus $2^2/1 \times 1 = 4$. The mass effects just about cancel out.

Appendix: Some Definite Integrals Often Used in Statistical Mechanics

$$\int_0^\infty e^{-ax^2}\,dx = \frac{1}{2}\left(\frac{\pi}{a}\right)^{1/2}$$

$$\int_0^\infty e^{-ax^2}x\,dx = \frac{1}{2a}$$

$$\int_0^\infty e^{-ax^2}x^2\,dx = \frac{1}{4}\left(\frac{\pi}{a^3}\right)^{1/2}$$

$$\int_0^\infty e^{-ax^2}x^3\,dx = \frac{1}{2a^2}$$

$$\int_0^\infty e^{-ax^2}x^4\,dx = \frac{3}{8}\left(\frac{\pi}{a^5}\right)^{1/2}$$

$$\int_0^\infty e^{-ax}\,dx = \frac{1}{a}$$

$$\int_0^\infty e^{-ax}x^{1/2}\,dx = \frac{1}{2a}\left(\frac{\pi}{a}\right)^{1/2}$$

$$\int_0^\infty e^{-ax}x\,dx = \frac{1}{a^2}$$

$$\int_0^\infty e^{-ax}x^2\,dx = \frac{2}{a^3}$$

Key Equations

Number of complexions, for N molecules distributed among energy levels:

$$\Omega = \frac{N!}{\prod_i n_i!}$$

Boltzmann distribution law:

$$\frac{n_i}{N} = \frac{g_i e^{-\epsilon_i/\mathbf{k}T}}{\sum_i g_i e^{-\epsilon_i/\mathbf{k}T}} = \frac{g_i e^{-\epsilon_i/\mathbf{k}T}}{q}$$

where q is the *molecular partition function.*

Maxwell distribution law for the distribution of speeds:

$$\frac{dN}{N} = 4\pi\left(\frac{m}{2\pi\mathbf{k}T}\right)^{3/2} e^{-mu^2/2\mathbf{k}T}u^2\,du$$

(see also Table 14.2).

Maxwell-Boltzmann distribution law for the distribution of energies:

$$\frac{dN}{N} = \frac{2\pi}{(\pi\mathbf{k}T)^{3/2}} e^{-\epsilon/\mathbf{k}T}\,\epsilon^{1/2}\,d\epsilon$$

System partition function Q:

Distinguishable molecules: $Q = (\sum_i g_i e^{-\epsilon_i/\mathbf{k}T})^N = q^N$

Indistinguishable molecules: $Q = \dfrac{q^N}{N!} \approx \left(\dfrac{qe}{N}\right)^N$

For *thermodynamic quantities* in terms of Q, see Table 14.3.

Partition functions for some simple cases:

$$\text{Translation: } q_t = \frac{(2\pi m \mathbf{k}T)^{3/2}V}{h^3}$$

$$\text{Rotation: } q_r = \frac{8\pi^2 I\mathbf{k}T}{\sigma h^2} \quad \text{(linear molecule; } \sigma = \text{symmetry number)}$$

$$q_r = \frac{8\pi^2(8\pi^3 I_A I_B I_C)^{1/2}(\mathbf{k}T)^{3/2}}{\sigma h^3} \quad \text{(nonlinear molecule)}$$

$$\text{Vibration: } q_v = \frac{1}{1 - e^{-h\nu/\mathbf{k}T}} = \frac{1}{1 - e^{-\theta_{\text{vib}}/T}}$$

Sackur-Tetrode equation for molar entropy:

$$S_m = \frac{5}{2}R + R\ln\left[\frac{V_m\,(2\pi m\mathbf{k}T)^{3/2}}{Lh^3}\right]$$

Equilibrium constants in terms of molecular partition functions:

$$K^u = \frac{q_Y^y q_Z^z \cdots}{q_A^a q_B^b \cdots} N^{-\Sigma\nu} e^{-\Delta U_0/RT}$$

($\Sigma\nu$ = stoichiometric sum).

Problems

A Distributions of Speeds and Energies (refer to the integrals given in the appendix to this chapter)

1. Refer to Table 14.2 and write expressions and values for (a) the ratio $\sqrt{\overline{u^2}}/\overline{u}$ and (b) the ratio $\overline{u}/u_{mp}$. Note that these ratios are independent of the mass and the temperature. How do the *differences* between them depend on these quantities?

2. The speed that a body of any mass must have to escape from the earth is 1.07×10^4 m s^{-1}. At what temperature would the average speed of (a) an H_2 molecule and (b) an O_2 molecule be equal to this escape speed?

3. **a.** For H_2 gas at 25°C, calculate the ratio of the fraction of molecules that have a speed $2\overline{u}$ to the fraction that have the average speed $\overline{u}$. How does this ratio depend on the mass of the molecules and the temperature?

b. Calculate the ratio of the fraction of the molecules that have the average speed $\overline{u}_{100°C}$ at 100°C to the fraction that have the average speed $\overline{u}_{25°C}$ at 25°C. How does this ratio depend on the mass?

4. Suppose that two ideal gases are heated to different temperatures that are such that their pressures and vapor densities are the same. What is the relationship between their average molecular speeds?

***5.** **a.** If $\overline{u}_{25°C}$ is the average speed of the molecules in a gas at 25°C, calculate the ratio of the fraction that will have the speed $\overline{u}_{25°C}$ at 100°C to the fraction that will have the same speed at 25°C. **b.** Repeat this calculation for a speed of $10\,\overline{u}_{25°C}$.

B Thermodynamic Quantities from Partition Functions

6. Obtain an expression for C_P in terms of the partition function Q for the system.

7. Obtain expressions for each of the functions $U - U_0$, S, $A - U_0$, $H - U_0$, and $G - U_0$, in terms of the molecular partition function q, for (a) distinguishable molecules and (b) indistinguishable molecules. (Use the Stirling approximation $N! = N^N/e^N$).

8. Obtain an expression for the pressure P in terms of the molecular partition function q, for (a) distinguishable molecules and (b) indistinguishable molecules. Express the result in terms of the number of molecules N and also the amount of substance n.

9. Two systems are identical in all respects except that in one the molecules are distinguishable and in the other they are indistinguishable. Calculate the difference between their molar entropies.

10. Calculate the molar energy at 25.0°C of a monatomic gas absorbed on a surface and forming a completely mobile layer. (Such a system may be regarded as a two-dimensional gas.)

11. The partition function for each degree of vibrational freedom is $1/(1 - e^{-h\nu/kT})$ (Eq. 14.119). Obtain from this expression the limiting value of the vibrational C_V as T approaches infinity.

C Partition Functions for Some Special Cases

12. Starting with Eq. 14.103, obtain an expression for the molar internal energy U_m of an ideal monatomic gas.

13. Calculate the molecular translational partition functions q_t for (a) N_2, (b) H_2O, (c) C_6H_6 in a volume of 1 m^3 at 300 K. In each case, calculate also $\ln Q_{tm}$, where Q_{tm} is the molar translational partition function.

14. The internuclear distance for N_2 is 0.1095 nm. Determine the molecular rotational partition function q_r and $\ln Q_{rm}$ for N_2 at 300 K.

15. Use the data in Table 14.4 to calculate, with reference to $v = 0$, the molecular vibrational partition function for CO_2 at (a) 300 K and (b) 3000 K.

16. Expressions such as the Sackur-Tetrode equation for the entropy contain a term ln(constant × T). At temperatures close to the absolute zero this term has large negative values, and the expression therefore leads to a negative value of the entropy. Comment on this.

17. Calculate the entropy of argon gas at 25°C and 1 atm pressure.

18. From the data in Table 14.4, calculate, with reference to $v = 0$, the molecular vibrational partition function for Br_2 at (a) 300 K and (b) 3000 K.

19. Give the symmetry numbers of the following molecules: C_3O_2 (carbon suboxide), CH_4, C_2H_4, C_2H_6 in the staggered conformation, C_2H_6 in the eclipsed conformation, $CHCl_3$, C_3H_6 (cyclopropane), C_6H_6 (benzene), NH_2D, CH_2Cl_2.

20. Show that the rotational partition function for a linear molecule can be expressed as

$$q_r = \frac{\mathbf{k}T}{\sigma Bh}$$

where B is the rotational constant defined by Eq. 13.62.

***21.** Calculate the molar translational entropy of chlorine gas at 25°C and 0.1 atm pressure.

***22.** The carbon monoxide molecule has a moment of inertia of 1.45×10^{-46} kg m^2, and its vibrational frequency is 6.50×10^{13} s^{-1}. Calculate the translational, rotational, and vibrational contributions to the molar entropy of carbon monoxide at 25°C and 1 atm pressure.

***23.** Suppose that a system has equally spaced energy levels, the separation between neighboring levels being $\Delta\epsilon$. Prove that the fraction of the molecules in state i, having energy ϵ_i greater than the lowest level, is

$$(1 - e^{-\Delta\epsilon/\mathbf{k}T})\, e^{-\epsilon_i/\mathbf{k}T}$$

What is the limiting value of this fraction as $T \to \infty$? Explain your answer.

***24.** Deduce the following from the Sackur-Tetrode equation (Eq. 14.105), which applies to an ideal monatomic gas:

- **a.** The dependence of entropy on relative molecular mass M_r; also, obtain an expression for dS_m/dM_r.
- **b.** The dependence of heat capacity C_P on relative molecular mass.
- **c.** The dependence of entropy on temperature; also obtain an expression for dS_m/dT.

D Calculation of Gibbs-Energy Changes and Equilibrium Constants

25. Use the data in Table 14.6 to calculate the molar Gibbs energy of formation of $H_2O(g)$ at 298.15 K.

26. From the data in Table 14.6, calculate K_p at 1000 K for the "water-gas" reaction

$$CO_2(g) + H_2(g) \rightleftharpoons CO(g) + H_2O(g)$$

27. From the data in Table 14.6, calculate K_p for the reaction

$$N_2(g) + 3H_2(g) \rightleftharpoons 2NH_3(g)$$

at (a) 298.15 K and (b) 1000 K.

28. Without making detailed calculations but by using symmetry numbers, estimate the equilibrium constants for the following reactions:

a. $^{35}Cl - ^{35}Cl + ^{37}Cl \rightleftharpoons ^{35}Cl - ^{37}Cl + ^{35}Cl$
b. $^{35}Cl - ^{35}Cl + ^{37}Cl - ^{37}Cl \rightleftharpoons 2\ ^{35}Cl - ^{37}Cl$
c. $C\ ^{35}Cl_4 + ^{37}Cl \rightleftharpoons C\ ^{37}Cl\ ^{35}Cl_3 + ^{35}Cl$
d. $N\ ^{35}Cl_3 + ^{37}Cl \rightleftharpoons N\ ^{37}Cl\ ^{35}Cl_2 + ^{35}Cl$
e. $^{35}Cl_2O + ^{37}Cl \rightleftharpoons ^{37}Cl\ ^{35}ClO + ^{35}Cl$

(Because of the similarity of the masses, these estimates will be quite accurate.)

D. M. Bishop and K. J. Laidler, *J. Chem. Phys.*, *42*, 1688 (1965), have defined a *statistical factor* for a reaction as the number of equivalent ways in which a reaction can occur. Thus for reaction (a) from left to right the statistical factor is 2, since the ^{37}Cl atom can abstract either of the two ^{35}Cl atoms. For the reverse reaction the statistical factor r is 1, since the ^{35}Cl atom can only abstract the ^{35}Cl atom in order to give the desired products. If two identical molecules are involved, the statistical factor must be taken as the number of equivalent products divided by 2; thus for reaction (b) from right to left the statistical factor is $\frac{1}{2}$.

Bishop and Laidler proved that the ratio l/r of statistical factors is always equal to the ratio $\sigma_A\sigma_B/\sigma_Y\sigma_Z$ of symmetry numbers. Verify that this is true for the given reactions.

This statistical factor procedure is useful in providing a simple insight into the factors that appear in equilibrium constants.

E Supplementary Problems

29. On the basis of Eq. 14.36, with $\beta = 1/\mathbf{k}T$, derive an expression for the fraction of molecules in a one-dimensional gas having speeds between u_x and $u_x + du_x$. What is the most probable speed?

30. Derive an expression for the fraction of molecules in a one-dimensional gas having energies between ϵ_x and $\epsilon_x + d\epsilon_x$. Also, obtain an expression for the average energy $\bar{\epsilon}_x$.

31. Derive an expression for the fraction of molecules in a two-dimensional gas having speeds between u and $u + du$. (*Hint:* Proceed by analogy with the derivation of Eq. 14.47.) Then obtain the expression for the fraction having energies between ϵ and $\epsilon + d\epsilon$. What fraction will have energies in excess of ϵ^*?

32. Consider a hypothetical substance for which the molecules can exist only in two states, one of zero energy and one of energy ϵ. Write its molecular partition function q. Assuming the molecules to be distinguishable, obtain expressions for the following, in the limit of very high temperatures:

a. The molar internal energy $U_m - U_{0,m}$.
b. The molar entropy S_m.
c. The molar enthalpy $H_m - U_{0,m}$.
d. The molar Gibbs energy $G_m - U_{0,m}$.

33. Molecules adsorbed on a surface sometimes behave like a two-dimensional gas. Derive an equation, analogous to the Sackur-Tetrode equation 14.105, for the molar entropy of such an adsorbed layer of atoms, in terms of the molecular mass m and the surface area A. What would be the molar entropy if 10^{10} argon atoms were adsorbed on an area of 1 cm^2 at 25°C?

34. Calculate the equilibrium constant at 1000°C for the dissociation

$$I_2 \rightleftharpoons 2I$$

given the following information: moment of inertia of $I_2 = 7.426 \times 10^{-45}$ kg m^2, wave number for I_2 vibration $= 213.67$ cm^{-1}, $\Delta U_0 = 148.45$ kJ mol^{-1}. The I atom is in a $^2P_{3/2}$ state; neglect higher states.

35. Calculate the equilibrium constant K_p for the dissociation

$$Na_2 \rightleftharpoons 2\ Na$$

at 1000 K, using the following data: internuclear separation in $Na_2 = 0.3716$ nm, vibrational wave number, $\bar{\nu} = 159.2$ cm^{-1}, $\Delta U_0 = 70.4$ kJ mol^{-1}. The Na atom is in a $^2S_{1/2}$ state; neglect higher states.

36. Calculate the equilibrium constant K_p at 1200 K for $Cl_2 \rightarrow 2Cl$, from the following data: internuclear separation in $Cl_2 = 199$ pm, wave number for vibration = 565.0 cm^{-1}, ΔU_0 = 240.0 kJ mol^{-1}. The ground state of Cl is a doublet, $^2P_{3/2,1/2,}$ the separation between the states being 881 cm^{-1}.

37. An indication of the spread of energies in the molecules comprising a sample is provided by the *root-mean-square deviation* of the energy. This quantity, more simply called the *fluctuation* in energy, is defined as

$$\delta\epsilon = \sqrt{\langle\epsilon^2\rangle - \langle\epsilon\rangle^2}$$

where $\langle\epsilon^2\rangle$ is the average value of ϵ^2 and $\langle\epsilon\rangle$ is the average ϵ. The probability P_i that a molecule is in a given state is given by Eq. 14.34:

$$P_i = \frac{n_i}{N} = \frac{\epsilon^{-\epsilon_i/\mathbf{k}T}}{\sum_i e^{-\epsilon_i/\mathbf{k}T}} = \frac{e^{-\epsilon_i/\mathbf{k}T}}{q}$$

where q is the molecular partition function. Therefore,

$$\langle\epsilon^2\rangle = \sum_i P_i\,\epsilon_i^2 \quad \text{and} \quad \langle\epsilon\rangle = \sum_i P_i\,\epsilon_i$$

Prove that the square of the fluctuation is given by

$$\delta\epsilon^2 = \frac{1}{q}\left(\frac{\partial^2 q}{\partial\beta^2}\right)_V - \frac{1}{q^2}\left(\frac{\partial q}{\partial\beta}\right)_V^2$$

where $\beta = 1/\mathbf{k}T$. Then use this relationship to prove that for a harmonic oscillator the fluctuation of the vibrational energy is

$$\delta\epsilon_{vib} = \frac{h\nu}{e^{h\nu/2\mathbf{k}T} - e^{-h\nu/2\mathbf{k}T}}$$

Show from this relationship that as $T \rightarrow \infty$,

$$\delta\epsilon_{\text{vib}} = \mathbf{k}T$$

F Essay Questions

1. The molar entropy of a gas increases with the temperature and with the molar mass. Give a physical explanation of these two effects.

2. Explain the factors that influence the magnitudes of partition functions, and comment on the magnitudes of the molecular partition functions for translational, rotational, and vibrational energy. What characteristics of a molecule will lead to a high value of (a) q_t, (b) q_r, and (c) q_v?

3. Give a general account of the Principle of Equipartition of Energy. Emphasize in particular the circumstances under which the principle fails.

Suggested Reading

F. C. Andrews, *Equilibrium Statistical Mechanics*, New York: Wiley, 1963

R. E. Dickerson, *Molecular Thermodynamics*, Menlo Park, Calif.: Benjamin/Cummings, 1969.

J. A. Foy, *Molecular Thermodynamics*, Reading, Mass.: Addison-Wesley, 1960.

T. L. Hill, *Introduction to Statistical Thermodynamics*, Reading, Mass.: Addison-Wesley, 1960.

L. K. Nash, *Elements of Statistical Thermodynamics*, (2nd ed.) Reading, Mass.: Addison-Wesley, 1972.

O. K. Rice, *Statistical Mechanics, Thermodynamics, and Kinetics*, San Francisco: W. H. Freeman, 1967.

G. S. Rushbrooke, *Introduction to Statistical Mechanics*, Oxford: Clarendon Press, 1962.

The Solid State

Preview

We examine here the regularity of the atomic arrangements in crystals. The study of the rules that govern the formation of crystal structures is called *geometrical crystallography*. Symmetry is important in this study. However, only certain of the point groups discussed in Chapter 12 are possible for the whole crystal. The study shows that only 32 arrangements of entities can be found in solids.

Much information is gained concerning the arrangement of atoms in a particular crystal by the use of X rays. When X rays of wavelength between 0.1 and 1.0 nm impinge on a crystalline solid, a *diffraction pattern* is obtained that can be related to the interatomic spacings. Analysis of these patterns allows the crystal structure (placement of atoms or molecules and their orientation) to be determined. In conjunction with X-ray analysis, the *Bragg equation*

$$n\lambda = 2d \sin \theta$$

gives the interplanar spacing d when the angle θ at which the X-ray beam strikes the surface is known.

The intensities of the observed diffraction lines or spots are taken in *reciprocal space*. This is shown through a *Fourier transform* to give us the electron density in real space.

The structure of crystals is discussed in terms of two models: the *bond model*, which considers bonds extending over the entire crystal, and the *band model*, which is analogous to molecular-orbital theory.

As part of the development of the bond model, the *Madelung constant* is introduced. This constant describes the charge interactions in ionic crystals. Thermodynamic information can be obtained from the *Born-Haber* cycle.

The close packing of spheres is discussed along with several crystal structures normally found in metals. An examination of the nature of metals leads us to the *Debye model* for predicting heat capacity and this, in turn, requires a brief look at *quantum statistics*.

We conclude our study of the solid state with a consideration of the band theory and of how the complete order of perfect crystals changes as impurities are added. Metals, semiconductors, and insulators are briefly examined.

15 The Solid State

The characteristic of the solid state is that the substance can maintain itself in a definite shape that is little affected by changes in temperature and pressure. In some solids, such as *glasses* and *amorphous* materials like pitch, there is no orderly arrangement of atoms or molecules at the atomic level. Such materials are really supercooled liquids. Other solids are characterized by a complete regularity of their atomic or molecular structures. Such solids are called *crystals* (Greek *krystallos*, clear ice). Regularity in external form and sharp melting point generally distinguish crystals from amorphous solids.

In the first part of this chapter we focus on the basic feature of crystals, the regularity of their atomic arrangements, and how this is manifested in the periodicity of their structural patterns.

15.1 Crystal Forms

Crystalline minerals and gems have been known for several thousand years. The beauty of crystalline forms took on scientific meaning with the quantitative study of crystalline interfacial angles (the angle between two adjacent faces of a crystal). In 1669 Niels Stensen (Nicolaus Steno) (1638–1686) found that the corresponding interfacial angles in different quartz crystals were always the same. Once the reflection *goniometer* (Greek *gunia metron*, angle measure) was invented in 1780 by the English chemist William Hyde Wollaston (1766–1820), it was rapidly established that other substances also exhibited this same constancy, which is now expressed as the *First Law of Crystallography*. Figure 15.1 shows examples of this first law for three different external shapes or habits.

This constancy of angles may seem surprising because of the large number of shapes a given crystalline material may have. However, the reason for the occurrence of different habits is that atoms or molecules may be deposited on the faces of a crystal in different ways. Suppose, for example, that a perfect cubic crystal is growing in a solution. If one face of the cube preferentially receives more atoms or molecules from the solution than the other sides, the crystal will elongate, and the face receiving more material will not grow as rapidly in area as the other faces. The most rapidly formed rectangular sides are those on which the atoms or molecules deposit most slowly.

In 1678 the Dutch physicist Christiaan Huygens (1629–1695) theorized that the existence of cleavage planes in crystals could be accounted for by the regular packing of spheroidal particles in layers. This idea was extended in 1784 by the French mineralogist René Just Haüy (1743–1822). After accidentally dropping a piece of calcareous spar and noting the similarity between the shapes of the fragments, he suggested that the crystal was formed from little cubes or polyhedra, which he called the *molécules intégrantes*. The orderly internal arrangement of these building blocks produced the regular external faces of the crystals. Haüy is recognized as the founder of crystallography for this geometrical *Law of Crystallization*, and for his formulation of the *Law of Rational Intercepts*, which is explained below.

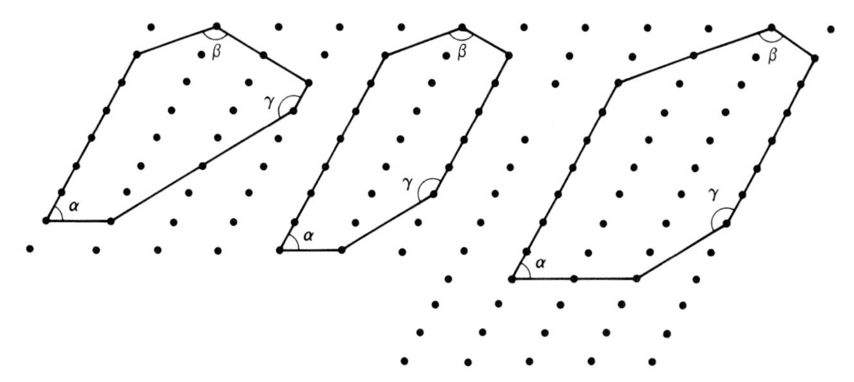

FIGURE 15.1 Constancy of the interfacial angles is shown for three different crystal habits of a material. Each of the angles α, β, and γ has the same value in the three figures.

15.2 Crystal Lattices

Crystals are distinguished from amorphous solids by having a representative unit of structure repeated in space at regular intervals on a periodic array of points called a **lattice**. A lattice is merely a uniform arrangement of points in space. The structural unit of a crystal is called a **basis**, which may consist of an atom, an ion, or a molecule (which may be as large as a protein molecule). In a crystal a basis is present at each lattice point in such a way that the environment of each basis is the same throughout the crystal.

A one-dimensional lattice is generated by the continual repetition of a point after translation through a distance a. This is shown in Figure 15.2a. A linear array is obtained in Figure 15.2b, where the motif 9 is placed on each lattice point. (The motif 9 is used to represent a basis since it has no symmetry of its own.) The *translation operation* takes the point or motif into itself and is, therefore, a symmetry operation. It is normal to describe such a translation by the vector **a**.

If the translation **a** is combined with a translation **b** in another direction, a two-dimensional or *plane lattice* is obtained by repetition of each **a** derived point in the direction of **b**. A two-dimensional lattice generated from the vectors **a** and **b** is shown in Figure 15.2c.

A third translation out of the **a**–**b** plane repeats the entire plane at intervals **c**. This generates a *space lattice*. Placement of our motif on the space lattice produces the *crystal lattice* or three-dimensional lattice array shown in Figure 15.2d. Each point in the crystal lattice has exactly the same environment as any other lattice point.

The Unit Cell

It is usually convenient to divide the lattice into *primitive cells* that completely generate the three-dimensional lattice under the action of suitable translation operations.

A line joining any two lattice points represents a possible translation vector. Indeed, there exists an indefinite number of translations that can be used to gener-

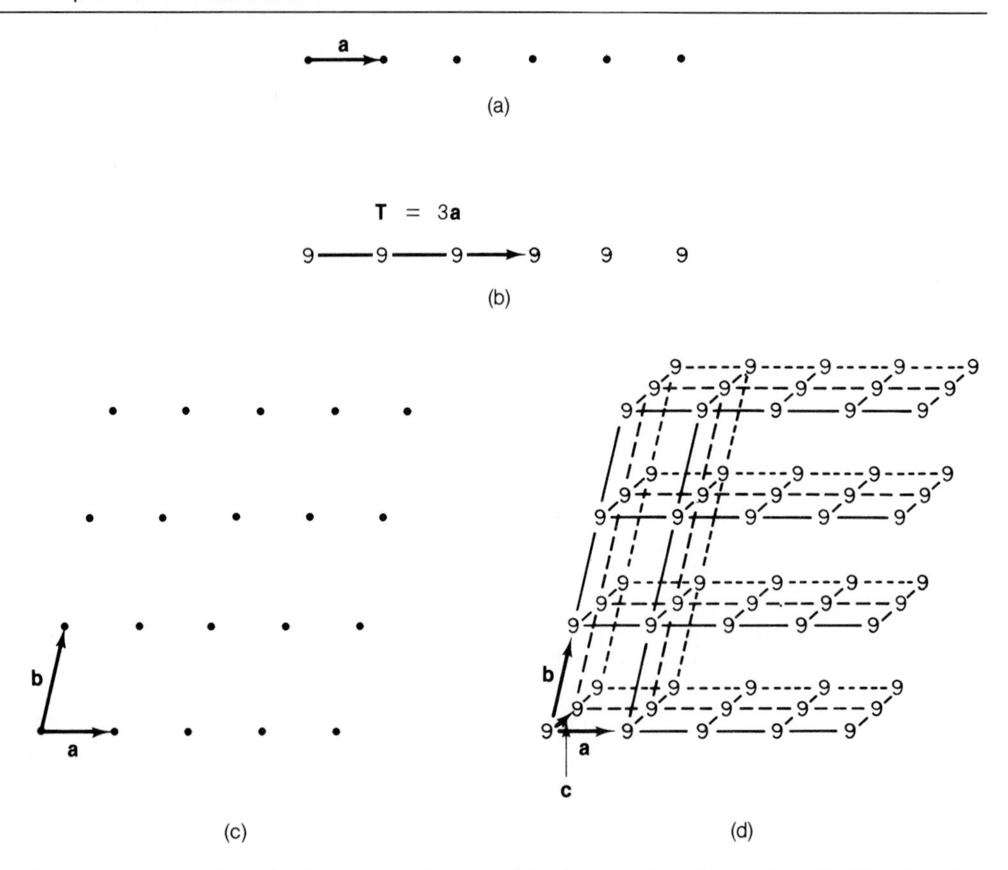

FIGURE 15.2 Patterns representing the formation of a crystal lattice. (a) One-dimensional lattice showing the primitive translation vector **a**. (b) The motif 9 is identically placed on the lattice points of (a). (c) Two-dimensional lattice using the vectors **a** and **b**. (d) Three-dimensional array showing the repetition of the motif by the vectors **a**, **b**, and **c**.

ate the lattice. Figure 15.3 shows a plane lattice and illustrates several possible choices of **a** and **b**.

We will use the primitive translation vectors **a**, **b**, **c** to define the lattice or crystal axes. These axes **a**, **b**, **c** comprise the three adjacent edges of a parallepiped. A translation using these axes can be described in terms of the *lattice translation vector*

$$\mathbf{T} = n_1\mathbf{a} + n_2\mathbf{b} + n_3\mathbf{c} \tag{15.1}$$

where the n's are integers. This vector gives the displacement of the crystal parallel to itself. Note in Figure 15.3 that $\mathbf{b}_3$ and the combination $\mathbf{a}_2$, $\mathbf{b}_2$ in Set 6 are not primitive translation vectors because **T** in this case cannot be formed from integral combinations of $\mathbf{a}_1$ and $\mathbf{b}_3$ or $\mathbf{a}_2$ and $\mathbf{b}_2$.

The primitive cell, therefore, is defined by the primitive crystal axes **a**, **b**, and **c**. It has a volume V_c from elementary vector analysis of

$$V_c = \mathbf{a} \times \mathbf{b} \cdot \mathbf{c} \tag{15.2}$$

and has a lattice point at each corner of the crystal. In Figure 15.3, the choice of pairs of primitive translation vectors in Sets 1 through 4 defines two-dimensional primitive cells. A primitive cell in three dimensions is shown to the left in Figure 15.4.

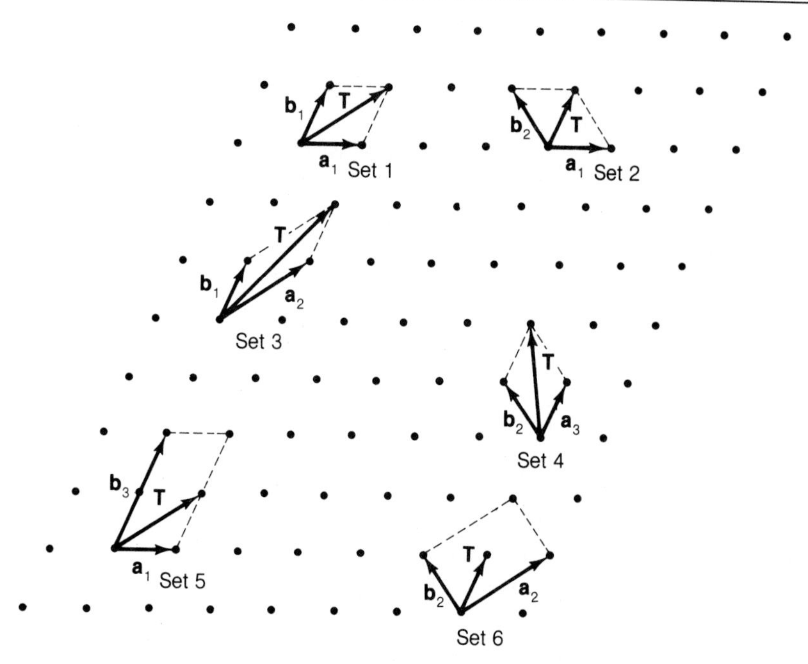

FIGURE 15.3 The translation vectors **a** and **b** may be chosen in several ways to generate a plane lattice. The vectors $\mathbf{a}_1$, $\mathbf{a}_2$, $\mathbf{a}_3$ and $\mathbf{b}_1$, $\mathbf{b}_2$ are primitive translation vectors and form primitive lattice cells in the first four sets. The lattices in Sets 5 and 6 are not primitive because the vector **T** drawn internal to the cell cannot be formed by integral combination of the base vectors. (See Eq. 15.1.) In Sets 1 to 4, **T** is formed by integral combination of the base vectors.

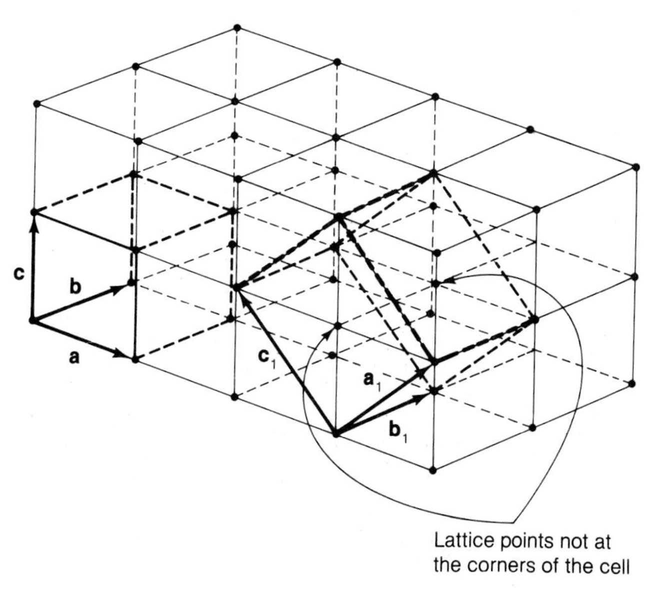

FIGURE 15.4 A unit cell on the left defined by the vectors **a**, **b**, and **c**. The cell to the right is a multiple cell.

The *primitive cell is a minimum-volume cell that has only one lattice point per cell.* Since there are eight corners and only one-eighth of each point belongs to the cell of interest, there is, therefore, $8 \times \frac{1}{8} = 1$ lattice point that belongs to the unit cell. On the other hand, the lattice to the right in Figure 15.4 is called a *multiple cell* since it contains lattice points not at the corners of the cell in addition to the eight shared at the corners.

Either a primitive or a multiple cell may be chosen as a **unit cell** of the lattice. Generally several unit cells are possible. When unit cells larger than the primitive cell are used, it is because they better represent the symmetry of the lattice.

Symmetry Properties

In addition to the translations that reproduce the crystal, crystals also possess rotational axes. Although all the rotational axes studied in Section 12.5 are possible for individual molecules, the necessity of filling all spaces with repeating units places a restriction on the types of rotational axes that a crystal may have.

This is easily demonstrated if we try to cover a plane surface with regular pentagons; we cannot avoid overlapping the pentagons or leaving spaces between them. The C_5 rotational axis is therefore impossible with repeating units, such as we have in a crystal, and similar considerations lead to the conclusion that only C_2, C_3, C_4, and C_6 rotational axes are possible. (The axis C_1 is also possible but is equivalent to doing nothing.) According to the *Hermann-Manguin* or *international* crystallographic notation*, a C_n rotational axis is listed simply as n, and therefore we say that only the rotational axes 2, 3, 4, or 6 are possible. Crystals must be constructed of subunits having one of the four axial symmetries just noted. The observed natural crystal forms must reflect this inner regularity in structure.

From considerations like these it is found that only five different types of two-dimensional lattices are possible. The most general, the *oblique lattice* shown in Figure 15.5a, contains a twofold axis normal to the lattice plane but no other symmetry elements. The placement of a mirror plane, designated m in the international system, would require lattice points to lie on rows parallel and perpendicular to the mirror plane. Two possibilities exist within the framework of a twofold axis: the *rectangular lattice* (Figure 15.5b) and the *centered-rectangular lattice* or *diamond lattice* (Figure 15.5c). The *hexagonal lattice* is a special case of the centered-rectangular lattice. Here the symmetry increases to a threefold or a sixfold axis. This is shown in Figure 15.5d. Figure 15.5e shows a *square lattice* that can accommodate a fourfold axis on each lattice point in addition to mirror planes.

For the sake of completeness, several other designations in the Hermann-Mauguin system should be mentioned. The *rotary-inversion axes* (n-fold rotation followed by inversion in a plane perpendicular to the axis) are represented by $\bar{n}$ (read *n-bar*) where n represents the number of equivalent positions in 360°. Several equivalences are immediately evident. The inversion i is equivalent to a $\bar{1}$ axis. A $\bar{2}$ axis has the same effect as a mirror plane. A special designation n/m is used for a mirror plane perpendicular to an n-fold axis.

---

*Norman F. M. Henry and Kathleen Lonsdale (Eds.), *International Tables for X-Ray Crystallography*, (Vol. I), *Symmetry Groups*, Birmingham, England: Kynock Press, 1952.

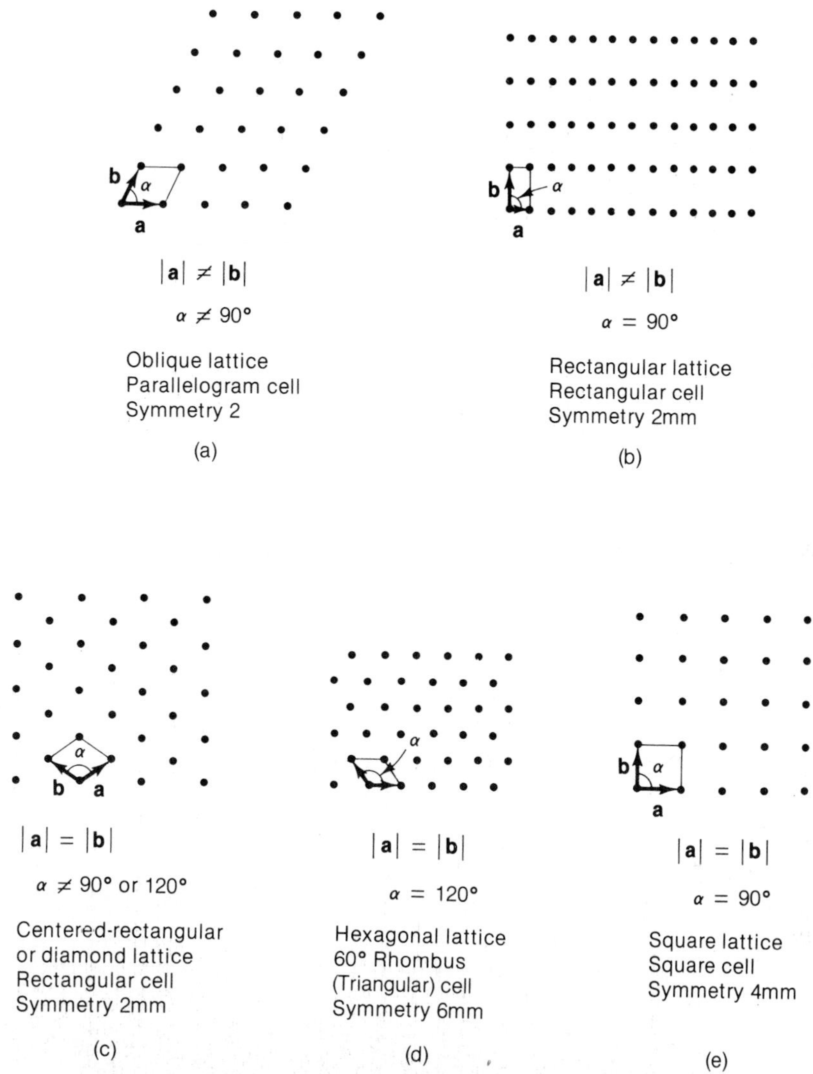

FIGURE 15.5 The five types of plane lattices. The lattice type and its axes are given along with the conventional cell normally chosen. The point group symmetry of the lattice is also given. The notation mm means that two mirror planes are present in addition to the rotational axis.

Two other types of symmetry operations are possible in crystals. The first of these is the *screw axis* formed by a proper rotation axis with a translation parallel to the rotation axis. The action is equivalent to the motion of a point on the thread of a screw, hence the name. The second operation is also a combination of two operations: reflection in a plane followed by a translation parallel to the reflection plane. The symmetry element is known as a *glide plane.* The screw axis and glide plane are not point-symmetry operations. For further discussion the reader is directed to the books listed in Suggested Reading (p. 738).

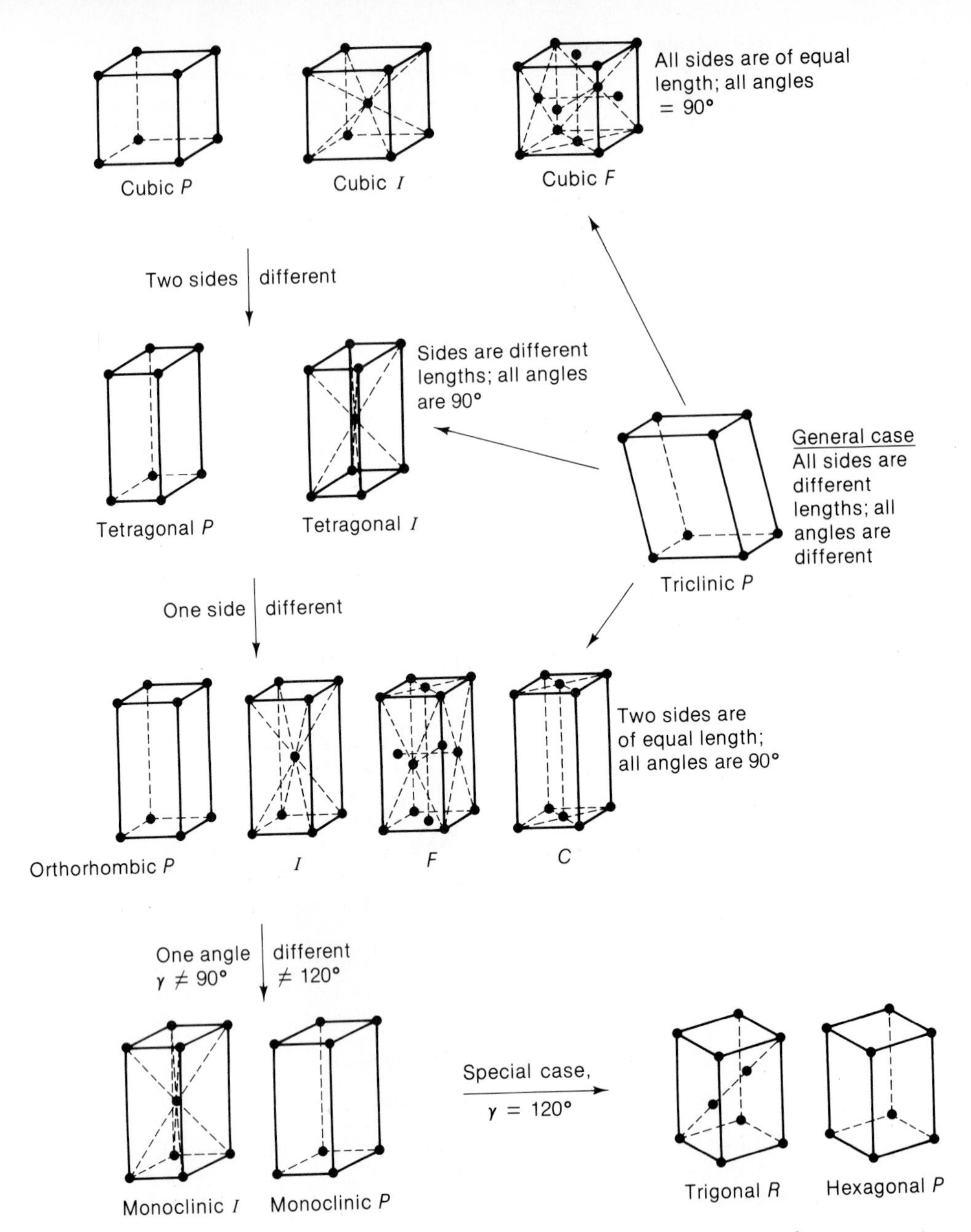

FIGURE 15.6 The 14 Bravais lattices. Conventional cells are shown except for the trigonal case.

Point Groups and Crystal Systems

The original crystallographers had no way to determine the periodicity of the atoms in crystals and had to deduce crystal structures from the external form or *morphology* of the crystal. The symmetry axes are lines, and they and the mirror planes intersect at a common point at the center of the crystal. The sets of symmetry elements found in a particular crystal were therefore designated *point groups* and were used for purposes of classification. Crystals having the same point-group symmetry belong to the same *crystal class*. Because the proper and improper rotational axes are restricted to 1, 2, 3, 4, and 6, only 32 point groups are possible, and these constitute the 32 crystal classes. The effect of screw axes and glide planes cannot be considered here because they are not visible in the external morphology since the translations are too small to be seen.

Space Lattices

We expect that the number of lattices that can fill space is limited, just as in two dimensions only 5 lattice types are permitted. In 1848 the French physicist Auguste Bravais (1811–1863) showed that the point-symmetry groups of the 5 two-dimensional lattices allow only 14 different lattice types in three dimensions. These distinct types of lattices are referred-to as the **Bravais lattices** and are shown in Figure 15.6.

Among the 14 lattices is 1 general lattice type (triclinic), and 13 special types occur because of symmetry restrictions. It is convenient to divide the 14 lattices into seven *systems* based on the seven unit cells conventionally used: *triclinic, monoclinic, orthorhombic, tetragonal, cubic* or *isometric, trigonal,* and *hexagonal.* The conventional cell axes and angles are summarized in Table 15.1, and the axial and angular relationships are defined in Figure 15.7.

The seven systems are further subdivided into *classes* based on whether they are *primitive* (P), *face-centered* (F), *side-* or *end-centered* (C), *body-centered* (I, German *Innenzentriertes*), or *rhombohedral* (R). These classes are shown in the figure along with the symmetry axes.

The cells in Figure 15.6 are conventional cells, and they are not always primitive cells. The reason for using a nonprimitive cell is that it has a more obvious connection with the point-symmetry elements. To illustrate this, the primitive cell for a body-centered cubic lattice is shown in Figure 15.8. Finally we note that it is the

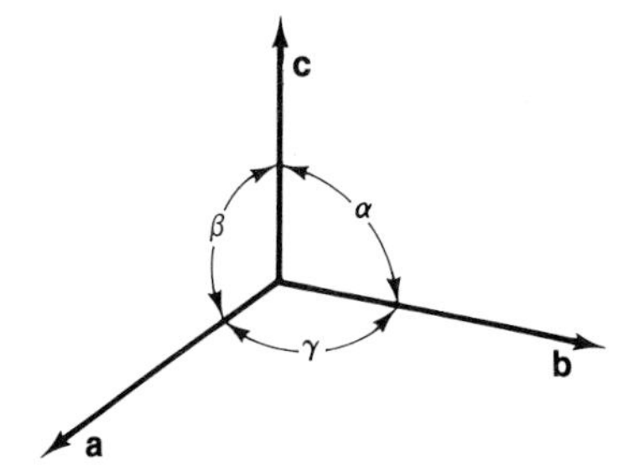

FIGURE 15.7 Crystal axes **a**, **b**, **c**. α is the angle between **b** and **c** (opposite the a axis). β is the angle between **c** and **a** (opposite the b axis). γ is the angle between **a** and **b** (opposite the c axis).

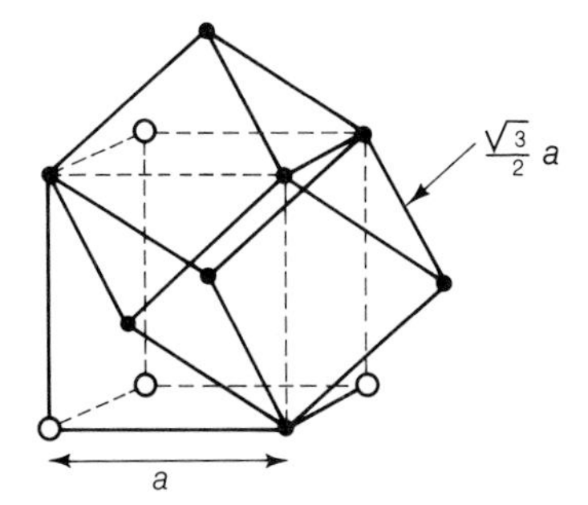

FIGURE 15.8 Demonstration showing how a primitive cell can be formed from a body-centered cubic lattice. The rhombohedron has an edge of $\frac{1}{2}\sqrt{3}a$, where the angle between adjacent edges is the tetrahedral angle 109°28′.

TABLE 15.1 The Seven Crystal Systems and Fourteen Three-Dimensional Bravais Lattice Types

System/Examples	Number of Lattices in System	Lattice Symbols	Restrictions on Conventional Cell Axes and Angles	Number of Space Groups
Triclinic $CuSO_4 \cdot 5H_2O$ $K_2Cr_2O_7$	1	*P*	$a \neq b \neq c$ $\alpha \neq \beta \neq \gamma$	2
Monoclinic $CaCO_3$ (calcite) $NaNO_3$	2	*P*, *C*	$a \neq b \neq c$ $\alpha = \gamma = 90° \neq \beta$	13
Orthorhombic $CaSO_4$ $NaNO_3$	4	*P*, *C*, *I*, *F*	$a \neq b \neq c$ $\alpha = \beta = \gamma = 90°$	59
Tetragonal SnO_2 $LiNH_2$	2	*P*, *I*	$a = b \neq c$ $\alpha = \beta = \gamma = 90°$	68
Isometric or cubic NaCl CaF_2	3	*P* or sc *I* or bcc *F* or fcc	$a = b = c$ $\alpha = \beta = \gamma = 90°$	36
Trigonal (special case of hexagonal system) $KClO_3$ Au_2Te_3	1	*R* (rhombohedral)	$a = b = c$ $\alpha = \beta = \gamma \neq 90°$	25
Hexagonal CdS SiO_2	1	*P*	$a = b \neq c$ $\alpha = \beta = 90°$ $\gamma = 120°$	27
			Total number of space groups = 230	

points that constitute the lattice and not the lines, which are inserted as a matter of convenience.

Space Groups

Just as only certain symmetry operations can be combined with the 5 two-dimensional lattices (resulting in 32 point groups), the combination of the 14 Bravais lattices with translations, screw axes, and glide planes results in 230 different three-dimensional crystal patterns. The determination of the space group that a particular crystal belongs to was not possible until diffraction techniques were available to investigate the internal symmetry of crystals. (We have already pointed out that screw axes and glide planes are not observed in a visual observation of the crystal.) The determination of the structure of a crystal is simplified if the space group is known because, if the asymmetric portion of the unit cell is determined experimentally, the rest of the structure may be obtained through symmetry.

15.3 Crystal Planes and Miller Indices

Haüy noticed that when a crystal was cleaved, the corresponding faces of the different fragments were equivalent. In order to describe those crystal faces Haüy proposed his **Law of Rational Intercepts.** The law states that it is always possible to find a set of axes that can be used to describe a crystal face in terms of intercepts along the axes. The reciprocals of these intercepts are small whole numbers. These numbers, h, k, and l, are called the **Miller indices** in honor of the British mineralogist William Hallowes Miller (1810–1880) who in 1838 first suggested their use for indexing planes.

Crystal planes pass through lattice points and are parallel to crystal faces. They are also described by the three integers h, k, l, either positive or negative. We can view parallel planes as cutting a unit length of each axis of the unit cell into an integral number of equal parts: the a axis into h equal parts, the b axis into k parts, and the c axis into l parts. Since all parallel planes are exactly alike, we generally only consider the plane nearest to the origin passing through the set of points closest to the origin. The next plane parallel to this will pass through the second nearest set of points to the origin, and so forth. The important point is that each plane passes through the dividing points and through the lattice points of the crystal. Such a set of planes will have the interplanar spacing d equal to the distance of the first plane from the origin.

For example, consider Figure 15.9 in which the noncoplanar axes a, b, c are shown with unit length vectors **a**, **b**, **c**. The arrangement of atoms in the lattice is such that a plane can intercept these axes at $\frac{1}{2}$ on the a axis, $\frac{2}{3}$ on the b axis, and $\frac{1}{4}$ on

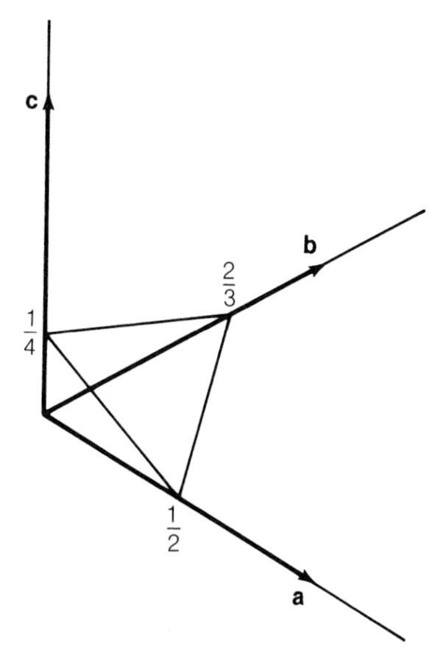

FIGURE 15.9 A plane cutting the axes at $\frac{1}{2}$ **a**, $\frac{2}{3}$ **b**, and $\frac{1}{4}$ **c**. The Miller indices are (438).

the c axis. The reciprocals of these in the order a, b, c is $\frac{2}{1}$, $\frac{3}{2}$, and $\frac{4}{1}$. We clear the fractions by multiplying through by 2. The Miller indices are (438) and are placed in parentheses.

The procedure for finding the Miller indices for any set of planes follows:

1. Determine the intercepts.
2. Find the reciprocals of the intercepts.
3. Clear fractions.

If an intercept is at infinity (that is, the plane is parallel to one of the axes), the reciprocal is 0. Several examples of planes that have been indexed are shown in Figure 15.10. If a plane cuts an axis on the negative side of the origin, the corresponding index is negative and is indicated with a minus sign over the index [i.e., $(\bar{h}kl)$].

In the hexagonal crystal system the lattice can be described by four hexagonal axes, a_1, a_2, a_3, c, and the indices are written ($hkil$). A symbol that is often used to enclose the Miller indices is the pair of braces, { }; it represents those planes that are equivalent by symmetry. Thus we represent the faces of a cube by {100} and this represents the (100), (010), (001) planes.

Indices of Direction

The direction in a crystal is often important and is also expressed as a set of indices where the integers are written between brackets, [hkl]. These are the smallest integers referred to the axes that have the ratio of the components of a vector in the

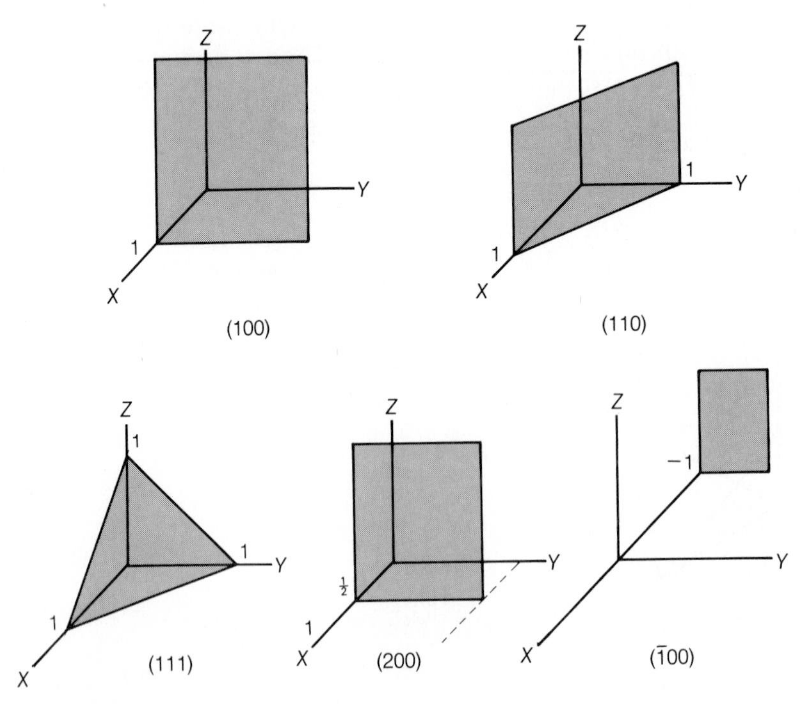

FIGURE 15.10 Miller indices of several planes in a cubic crystal.

desired direction. For example, in a cubic crystal the X axis is the [100] direction; the Y axis is the $[0\bar{1}0]$ direction. In cubic systems the direction $[hkl]$ is perpendicular to a plane (hkl) having the same indices.

The distance between planes is written as d_{hkl}. Specific formulas can be worked out relating d to a, b, c, and the Miller indices of the planes. For three simple cases we have

$$\text{Cubic cell: } \frac{1}{d^2} = \frac{h^2 + k^2 + l^2}{a^2} \qquad (a = b = c) \tag{15.3}$$

$$\text{Tetragonal cell: } \frac{1}{d^2} = \frac{h^2 + k^2}{a^2} + \frac{l^2}{c^2} \tag{15.4}$$

$$\text{Orthorhombic cell: } \frac{1}{d^2} = \frac{h^2}{a^2} + \frac{k^2}{b^2} + \frac{l^2}{c^2} \tag{15.5}$$

15.4 X-Ray Crystallography

In 1895 the German physicist Wilhelm Conrad Röntgen (1845–1923) discovered X rays. X rays are produced when high-energy electron beams (30–100 keV) strike a metal target. The X rays are high-energy photons ($\epsilon = h\nu$), the spectrum of which is generated by two different processes within the metal target. A continuous spectrum is formed when the high-speed electrons lose energy as they are slowed down rapidly by multiple collisions with the target. This process gives rise to *white radiation* or *Bremstrahlung,* so called because the radiation is continuous with a wide range of wavelengths. The second process superimposes on the continuous radiation spectrum discrete peaks that are characteristic of the target material. These peaks arise from electron transitions between the lowest-lying quantum levels. These levels have a special nomenclature. Corresponding to the energy levels $n = 1, 2, 3, \ldots,$ the groups of electrons are called the $K, L, M, \ldots$ shells. When one electron is knocked out of the K shell by a high-energy electron, an L-shell electron may replace it, thereby giving rise to a K_α spectral line with an energy equal to the energy difference between the L and the K states. Since two electrons are in the K shell, spin-orbit coupling occurs producing two closely spaced lines, the $K_{\alpha 1}$, and $K_{\alpha 2}$ lines. In a similar manner, if an electron from the M level replaces the one removed from the K level, the energy change results in a K_β line. The K_β line has greater energy (shorter wavelength) but is of lesser intensity because this transition is less favorable. Figure 15.11 shows how the intensity varies with the wavelength in the case of copper, a commonly used target material.

The fact that the X-ray spectra of elements are very similar to one another led the English physicist Henry Gwyn Jeffreys Moseley (1887–1915) to demonstrate* that the characteristic frequencies of the K_α lines increased steadily from the lighter to the heavier elements. This relationship defined the concept of **atomic number**.

It was proposed in the 1890s that crystals were composed of atoms arranged like closely packed spheres. After it was realized that X rays have wavelengths in

*H. G. J. Moseley, *Phil. Mag.*, 26, 1024 (1913); 27, 703 (1914).

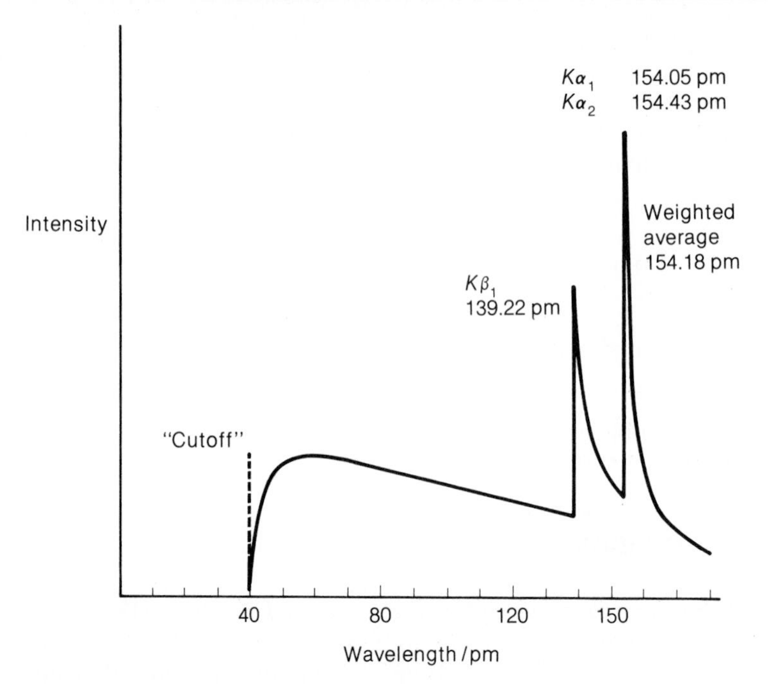

FIGURE 15.11 X-ray spectrum of copper at 35 kV showing the characteristic peaks. As the voltage of the exciting electrons increases, the cutoff and the peak intensity of the white radiation are shifted to shorter wavelengths.

the order of 100 pm, the German physicist Max Theodor Felix von Laue (1879–1960) suggested in 1912 that crystals could serve as gratings for the diffraction of X rays.* His colleagues W. Friedrich and P. Knipping† experimentally confirmed his theory using "white" X rays that had a wide range of wavelengths. At about the same time the English physicist Sir William Henry Bragg (1862–1942) and his son Sir (William) Lawrence Bragg (1890–1978) were experimenting with X rays having a narrow range of wavelengths. They were prevailed upon to test the theory that the minerals sylvite (KCl) and halite (NaCl) have face-centered cubic packing of spheres. The younger Bragg found that the observed X-ray pattern was that expected for the predicted structure and thereby revolutionized the physical sciences by pioneering X-ray methods for determining structures.

The Bragg Equation

The ability to scatter X rays depends on the number of electrons in an atom. Furthermore, since the atoms in a crystal are lined up in planes, each plane can diffract X rays. Since Friedrich used inhomogeneous ("white") X rays, his first diffraction pattern (Laue pattern) was very difficult to interpret.

---

*M. von Laue, *Ann. Physik, 41,* 989 (1912).

†W. Friedrich, P. Knipping, and M. Laue, *Sitzb. Kais. Akad. Wiss.*, Munchen, 303–22 (1912), *Ann. Physik, 41,* 971 (1912).

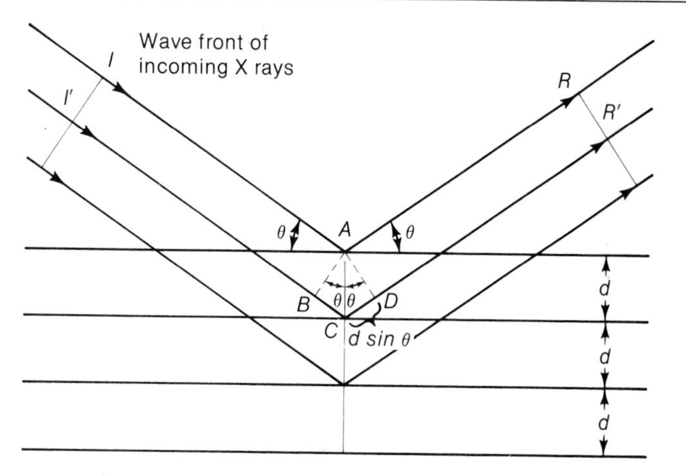

FIGURE 15.12 Derivation of the Bragg equation, $2d \sin \theta = n\lambda$. The diffraction of X rays may be considered as reflections from different planes.

The two Braggs simplified the situation by using X rays having a narrow wavelength range. They then found that the X rays were diffracted only at certain angles that depended on the wavelength and the interplanar spacings.

W. L. Bragg's explanation* for this phenomenon may be followed from Figure 15.12, where several parallel planes of a crystal are shown. X rays impinge on the crystal at an angle θ. The incident X rays of wavelength λ are reflected specularly (as if by a mirror), the angle of incidence being equal to the angle of reflection. A small part of the beam is reflected from surface atoms such as from point A. Some of the beam will penetrate to lower planes and will be reflected. Constructive interference (Section 11.1) of the beam at RR' occurs only if the difference in distances traveled by the two beams is an integral number of wavelengths, that is, only if the path IAR differs from that of $I'CR'$ by a whole number of wavelengths. Otherwise no diffraction will be observed.

The necessary condition for reinforcement is, therefore, that the distance $BC + CD$ be equal to an integral number of wavelengths ($n\lambda$). From Figure 15.12, $\sin \theta = CD/d$ or

$$CD = d \sin \theta \quad \text{and} \quad BC = d \sin \theta \tag{15.6}$$

This condition requires that

$$\boldsymbol{n\lambda = 2d \sin \theta} \tag{15.7}$$

which is the **Bragg equation.**

Homogeneous X rays of fixed λ are usually used and for a given set of lattice planes [e.g., (010) planes], d is fixed. Whether a diffraction maximum is found depends on θ, known as the *glancing angle* or *angle of incidence.* As θ increases, a series of maxima is obtained corresponding to n equal to 1, 2, 3, etc. Thus an X-ray spectrum is obtained in which successive maxima are called the first-order, second-

*W. L. Bragg, *Proc. Cambridge Phil. Soc.*, *17*, 43 (1913).

order, third-order, etc., reflections, depending on the value of n. As n increases, these reflections decrease in intensity until they generally become unobservable when $n = 5$ or greater. Second-order reflections are mathematically equivalent to a first-order reflection ($n = 1$) from planes with half the spacing. For example, if we are dealing with (100) planes, the second-order reflection is equivalent to a first-order reflection from the (200) planes (i.e., planes with half the spacing).

Note that $\sin\theta$ has a maximum value of 1. Thus for a first-order reflection $\lambda = 2d$. If $2d$ is smaller than λ, no reflection maximum can occur. On the other hand, if $\lambda \ll d$, the X rays are diffracted through extremely small angles.

EXAMPLE

Determine the angle of reflection when copper K_α radiation (0.154 nm) is incident on a cubic crystal with a lattice constant d of 0.400 nm.

SOLUTION

For the first-order reflection from (100) planes, we have, from Eq. 15.7,

$$0.154 \text{ nm} = 2(0.400 \text{ nm}) \sin\theta$$

$$\theta = \sin^{-1}\left(\frac{0.154 \text{ nm}}{0.800 \text{ nm}}\right) = 11°$$

■

From a knowledge of the experimentally observed diffraction angles it is possible to determine the interplanar spacings of a crystal. Then from a knowledge of all the spacings the unit cell size and shape can be determined. Since the intensities depend on the arrangement and nature of the basis within each unit cell, some complications do arise with respect to intensity. This problem will be discussed in Section 15.7.

15.5 The Reciprocal Lattice

Examination of Bragg's law written in the form

$$\sin\theta = \frac{n\lambda}{2}\frac{1}{d} \tag{15.8}$$

reveals that for structures with large interplanar spacing d the diffraction pattern will be compressed because of the inverse relation between θ and d. A direct relation based on $1/d$ is easier to work with and has practical significance, as we will see later. It is achieved by the formation of a *reciprocal lattice* in the following way. Construct from some lattice point, considered as the origin of the real or direct lattice, lines of length $1/d_{hkl}$ perpendicular to each respective lattice plane. In Figure 15.13 the planes of the direct lattice are shown as lines on a two-dimensional lattice. The reciprocal lattice points are designated by asterisks and are given the indices of the planes they represent. The set of points at the terminus of the normals to these planes constitutes a reciprocal lattice. The four points shown form one line of reciprocal lattice points having variable values of h and a constant value of k. Lines drawn from other direct lattice points give similar lines of points and thus form the complete reciprocal lattice. Thus *each plane in the direct lattice is represented by a point*

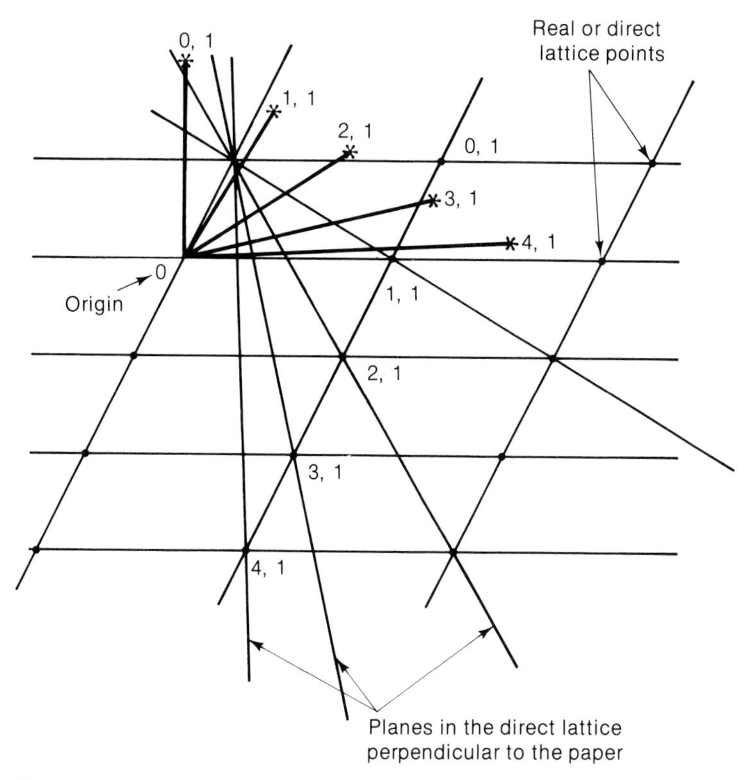

FIGURE 15.13 Formation of the reciprocal lattice. Planes cutting the direct lattice are represented by points (*) in reciprocal space. The drawing is based on a distance of 0.8 (arbitrary units) between (1, 1) planes.

in the reciprocal lattice. Since the reciprocal lattice points are derived from the actual planes of the direct lattice, their indices correspond to the Miller indices.

Such a reciprocal lattice can be formed for any lattice, and every crystal thus has both a direct crystal lattice and a reciprocal lattice. Just as a microscope image gives a view of a real structure, a *diffraction pattern gives a picture of the reciprocal lattice of the crystal.**

The relation between the two lattices depends on the angles between the primitive translation vectors of the direct lattice. The corresponding primitive translation vectors in the reciprocal lattice are $\mathbf{a}^*(= \mathbf{b} \times \mathbf{c})$, $\mathbf{b}^*(= \mathbf{c} \times \mathbf{a})$, $\mathbf{c}^*(= \mathbf{a} \times \mathbf{b})$ and the angles are α^*, β^* γ^*. (Gibbs invented the *reciprocal lattice vector* $\boldsymbol{\sigma}_{hkl}$, sometimes written as $\mathbf{S}$, the *scattering vector*), which is defined as

$$\boldsymbol{\sigma}_{\mathrm{hkl}} = h\mathbf{a}^* + k\mathbf{b}^* + l\mathbf{c}^* \tag{15.9}$$

This describes a vector from the origin in the reciprocal lattice to the point hkl. From the way in which we have constructed the lattice, it is clear that

$$\boldsymbol{\sigma}_{hkl} = \frac{1}{d_{hkl}} \tag{15.10}$$

*W. J. Moore, *Physical Chemistry*, Englewood Cliffs: Prentice-Hall, 1972, p. 854, discusses an interesting correlation between holograms and the reciprocal lattice.

Bragg's Law in Reciprocal Space

The reciprocal lattice allows easier visualization of Bragg's law when dealing with some of the experimental methods to be discussed in Section 15.6. Consider a crystal oriented in an X-ray beam of wavelength λ, the beam being parallel to the a^*b^* plane of the reciprocal lattice of the crystal. Our purpose is to establish the relationship of the diffracted beam to the reciprocal lattice points. In Figure 15.14 a line is drawn from A in the direction of the beam that passes through the crystal and the origin O of the reciprocal lattice. A circle of radius $1/\lambda$ is then constructed with its center on the line and its circumference on O.

If the reciprocal lattice point P lies on the circumference, then the line $\overline{BP}$ represents the surface of the crystal at the angle θ to the incident beam. The angle OPB is a right angle since it is inscribed within a semicircle. From trigonometry

$$\sin \theta = \frac{OP}{OB} = \frac{OP}{2(1/\lambda)} \tag{15.11}$$

Since the point P is a reciprocal lattice point, its distance from O is given by Eq. 15.10 as $1/d_{hkl}$. Therefore,

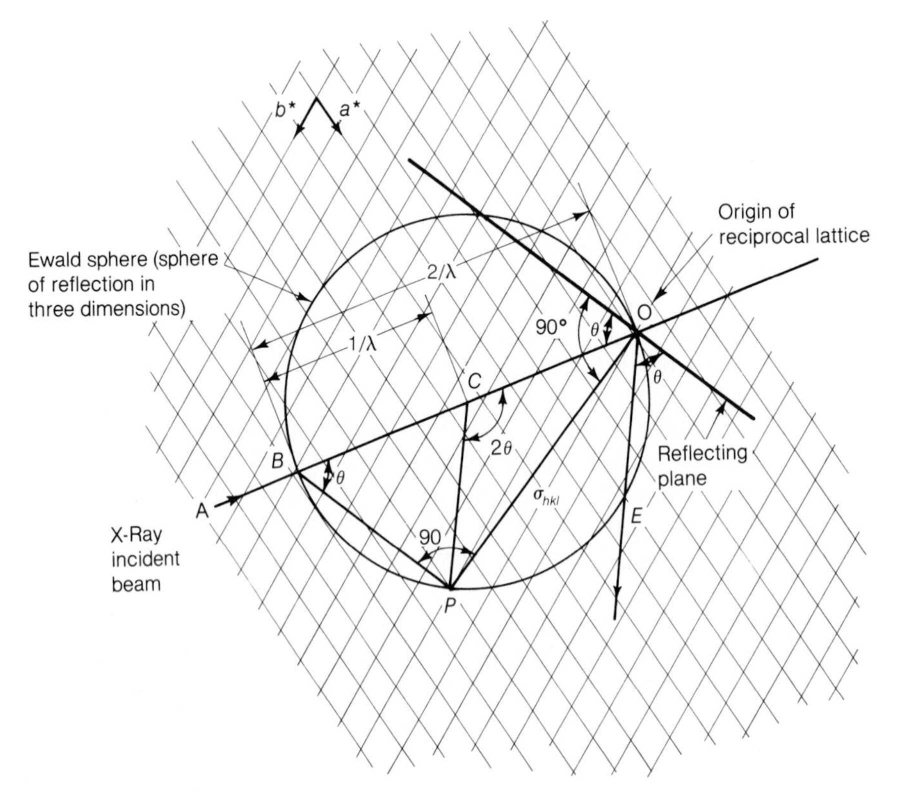

FIGURE 15.14 Construction showing diffraction and the sphere of reflection superimposed on a reciprocal lattice.

$$\sin \theta = \frac{\lambda}{2} \frac{1}{d_{hkl}} \tag{15.12}$$

or

$$2d_{hkl} \sin \theta = \lambda \tag{15.13}$$

This is Bragg's law with $n = 1$, and it implies that reflection occurs whenever a reciprocal lattice point coincides with the circle. When this happens, the reflecting plane of the crystal, which is parallel to BP, is perpendicular to OP and makes an angle θ with OB. The diffracted beam OE makes an angle 2θ with OB and is parallel to CP, which can also be considered as the diffracted X-ray beam.

Rotation around the diameter of the circle in Figure 15.14 generates a *sphere of reflection* or **Ewald sphere**. Any reciprocal lattice point falling on this sphere must also obey Bragg's law. Then $\mathbf{OP} = \boldsymbol{\sigma}_{hkl}$ is a translation vector of the reciprocal lattice where **OP** represents the direction of the diffracted beam.

For an arbitrary λ and position of the crystal, no reciprocal lattice point will fall on the sphere. However, rotating the crystal corresponds to rotating the lattice about its origin, and it brings various reciprocal lattice points into coincidence with the surface of the Ewald sphere. Another way of visualizing this focuses on the size of the sphere of reflection. The diameter of the sphere is $2/\lambda$ so that every reciprocal lattice point within a *limiting sphere* centered on O with radius $2/\lambda$ represents a potential reflection. If shorter wavelengths are used, the effect is to enlarge the sphere and increase the number of reflections. A corresponding reflection is observed as a spot on a photographic plate and can be related to the (hkl) indices of the reflecting plane.

15.6 Experimental Methods

Several X-ray techniques have been developed for structural studies, and each has its advantages and disadvantages. Photographic methods of recording data are described in order to present the principles in a more pictorial manner. Diffractometers are now available that detect the intensity of an X-ray beam by means of scintillation counters or proportional-counter tubes. Also, computer programs are now commonly used to automate the determinations of even quite complicated crystal structures from the measured intensity and angular relations.

The Laue Method

In the **Laue method** a single crystal is irradiated by a beam of continuous wavelength X rays. X rays are emitted at angles and wavelengths for which the Laue equations are satisfied. A flat film receives the diffracted beams and the diffraction pattern is composed of a series of spots that shows the symmetry of the crystal. This Laue method is used extensively for the rapid determination of crystal *orientation*, such as directions within a crystal, and of *symmetry*. In the former case the crystal is oriented in a goniometer and rotated until a desired direction is found, as indicated by the X-ray pattern. The Laue method is almost never used for crystal *structure* determinations because different orders may be reflected from a single plane on account of the continuous wavelengths employed. This hinders the determination of the reflected intensity, which is needed in crystal structure determinations.

The Powder Method

When single crystals are not available, or for a variety of purposes including the identification of crystalline materials, a simpler technique may be used. The crystalline material is ground to a powder which then presents all possible orientations to a collimated X-ray beam consisting primarily of K_α radiation. This technique was developed by Peter Joseph William Debye (1884–1966) and Paul Scherrer (b. 1890). The diffracted beam actually forms cones from each scattering plane and in the past has generally been recorded on a strip of X-ray film that has been rolled in a circle centered on the powdered specimen. The experimental apparatus is schematically drawn in Figure 15.15 and a typical X-ray powder pattern is shown in Figure 15.16. Such an arrangement is all that is required to differentiate between the three crystal possibilities in the cubic system. The diffracted rays form cones concentric with the original beam and have an angle of 2θ with the direction of the original beam. This is the same result that would result from mounting a single crystal and turning it through all possible orientations in the X-ray beam.

This technique is particularly important for the determination of lattice types and unit cell dimensions.

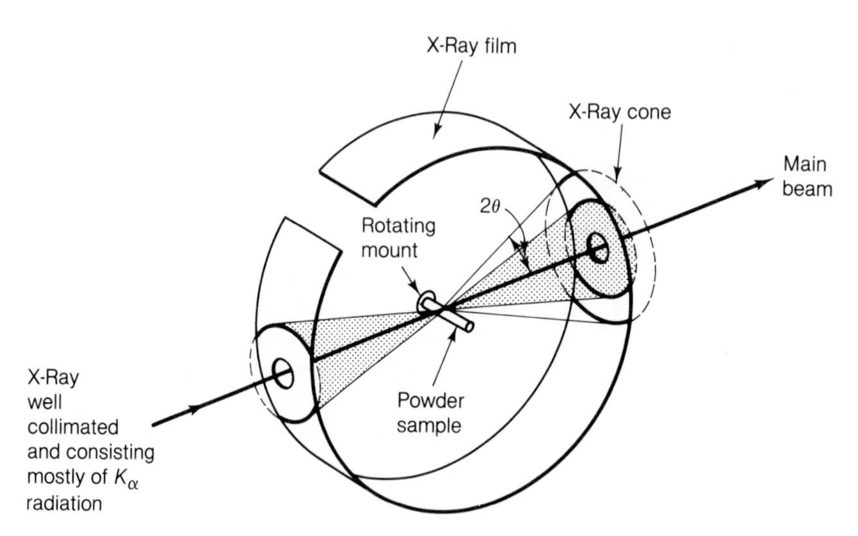

FIGURE 15.15 Schematic representation of the X-ray powder technique. X-ray film is held in the metal camera. Areas on film are segments of X-ray cones diffracted from powder sample.

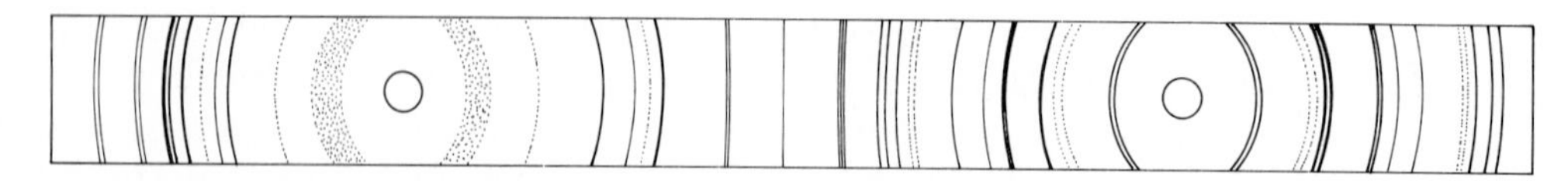

FIGURE 15.16 X-ray diffraction pattern of Fe_3N-BN taken with iron K_α radiation.

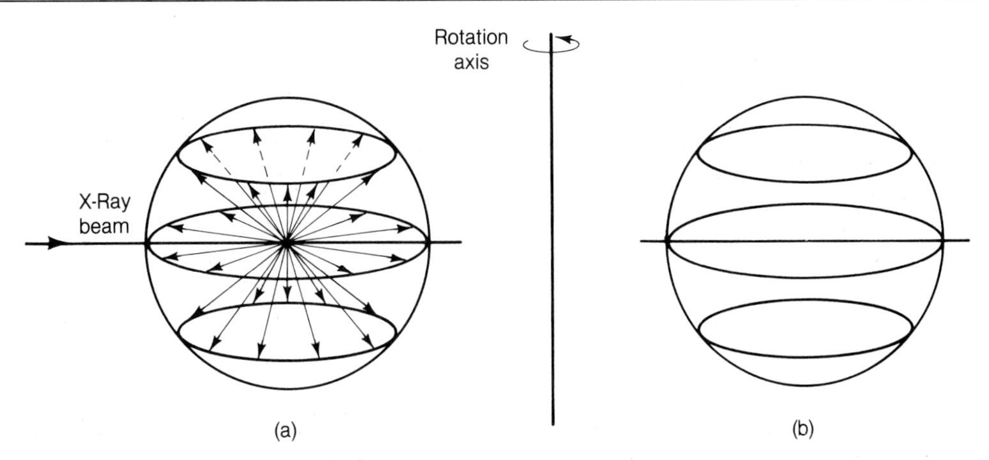

FIGURE 15.17 Relation between diffracted X rays in (a), the direct lattice, and in (b), the reciprocal lattice. The X-ray beam is perpendicular to a direct lattice axis. In (b) the diffracted X rays form circles with the sphere of reflection at three different levels.

The Rotating Crystal Methods

In the *rotating crystal methods* a single crystal is oriented so that a monochromatic X-ray beam is perpendicular to a direct-lattice axis. Rotation of the crystal causes different atomic planes to diffract X rays whenever the value of θ satisfies the Bragg equation. Since different levels of reciprocal lattice points are normal to the direct axis, the rotation about the axis causes each level of points to intersect the sphere of reflection in a circle, as shown in Figure 15.17a. The diffracted rays pass from the center of the sphere through these circles forming cones. The zero level is a flat cone with the other cone axes coincident and parallel to the rotating axis as shown in Figure 15.17b.

In order to record the diffracted X rays, X-ray film is mounted cylindrically around the sample and concentrically with the rotating sample holder. The X-ray beam passes through the crystal at right angles to the film and the crystal holder. X rays diffracted from all planes parallel to the rotation axis occur as reflected spots in a horizontal line on the film. Other planes cause diffraction both above and below this horizontal plane. These reflected spots occurring in lines are called *layer lines.*

Rotating crystal techniques include *oscillating-crystal* methods in which the crystal is oscillated through a limited angular range instead of making a full 360° rotation. This reduces the possibility of overlapping reflections. In general, these techniques are used to align crystals, to measure the cell edge, and to obtain crystal symmetry information.

Perhaps the most familiar device for making rotation and oscillation photographs is the *Weissenberg** camera. Another such device is the *precession* camera, designed in the 1940s by the American crystallographer Martin Julian Buerger† (b. 1903). For more detailed discussion of these methods, the reader is referred to the books listed in Suggested Reading. (p. 738).

*K.Weissenberg, *Z. Physik,* 23, 229 (1924).

†M. J. Buerger, *The Precession Method,* New York: Wiley, 1964.

Electron Diffraction

Electrons can also be diffracted by a crystal lattice. This was shown in 1927 by the American physicists Clinton Joseph Davisson (1881–1958) and Lester Halbert Germer (b. 1896). This was the first experimental proof of de Broglie's postulate on the wave nature of electrons (see Section 11.3).

Electron beams are far more efficiently scattered than are X-ray beams. Although X-ray diffraction patterns are analyzed on the basis that each X-ray photon has undergone just one deflection, this simplification is true for electron diffraction only when the crystalline samples are extremely thin. With solids, electron diffraction studies can only be made on surfaces, films and very thin crystals. The technique is also useful in determining the molecular structure and vibrations of gaseous molecules. The first electron diffraction pattern from a gaseous sample was obtained by R. Wierl* in 1931. In gas-phase experiments the sample gas is maintained between 10^{-4} to 10^{-6} mmHg pressure. The intensity of the atomic-scattering curve varies in a smooth fashion with the angle of scattering. The molecular scattering curve containing the desired information is superimposed on the previous curve and fluctuates fairly rapidly with θ when the energy of the beam is about 40 kV.

The de Broglie wavelength λ of an electron is related to its energy ϵ by the equation $\epsilon = h^2/2m\lambda^2$, where $m = 9.11 \times 10^{-31}$ kg is the mass of the electron. This expression comes from the kinetic energy of the particle, $\epsilon = p^2/2m$, with the substitution $\lambda = h/p$. A convenient relation is derived by setting $p = mu$ and $\frac{1}{2}mu^2 = \text{eV}$. Substitution gives

$$\lambda/\text{Å} = \frac{12.25}{(\epsilon/\text{eV})^{1/2}} \quad \text{or} \quad \lambda/\text{pm} = \frac{1225}{(\epsilon/\text{eV})^{1/2}} \tag{15.14}$$

For an electron that has been accelerated through a typical potential difference of 30–40 kV, the wavelength is approximately 6–7 pm (0.06–0.07 Å).

Neutron Diffraction

Other nuclear particles also have been diffracted. Neutrons are particularly useful because they interact with unpaired electron magnetic moments and with the nuclear magnetic moments. As a result neutron diffraction is complementary to the other two methods discussed.

A nearly monochromatic beam of neutrons is selected by a crystal monochromator from the range of wavelengths available (Figure 15.18). The wavelength of these neutrons also is governed by the de Broglie equation $\lambda = h/p$. The neutron energy is given by $\epsilon = h^2/(2M_n\lambda^2)$ where $M_n = 1.675 \times 10^{-24}$ kg is the mass of the neutron. This simplifies to

$$\lambda/\text{Å} = \frac{0.28}{(\epsilon/\text{eV})^{1/2}} \quad \text{or} \quad \lambda/\text{pm} = \frac{28}{(\epsilon/\text{eV})^{1/2}} \tag{15.15}$$

---

*R. Wierl, *Ann. der Physik, 8,* 521 (1931).

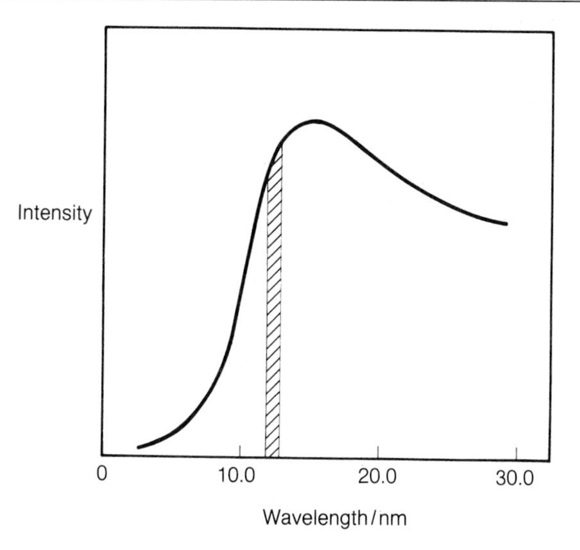

FIGURE 15.18 Intensity plotted against wavelength for a neutron beam. The beam from a reactor can have particular wavelengths selected (hatched area) by a crystal monochromator.

Neutron diffraction studies have been very useful in the location of protons, which scatter neutrons very efficiently. Hydrogen atoms, having only one electron, scatter X rays and electrons only weakly and, therefore, are not readily detected by them.

15.7 Interpretation of X-Ray Diffraction Patterns

The diffraction patterns of cubic systems are the simplest to treat. This is because reflections are not always observed from all sets of planes that can pass through the crystal. For example, if a set of planes is placed exactly between the planes of another set, interference of the rays reflected from the two sets may cause the absence of the line from the X-ray powder pattern. For example, in the simple cubic lattice, reflections from the (100), (110), and (111) planes all are observed. In the face-centered cubic (fcc) lattice, the (200) and (220) planes cause interference and consequently the (100) and (110) planes are not observed. Again in the body-centered cubic lattice, only (110) reflections are observed, for similar reasons.

Such considerations allow us to place restrictions on the values of h, k, and l for face-centered cubic fcc and body-centered cubic bcc systems. For simple cubic lattices there are no restrictions on the value of h, k, and l. In fcc lattices h, k, l are all odd or all even (including zero). For bcc lattices the sum $h + k + l$ must be even. These relations are demonstrated in Figure 15.19.

The distance between cubic planes is found from Eq. 15.3:

$$d_{hkl} = \frac{a}{\sqrt{h^2 + k^2 + l^2}} \tag{15.16}$$

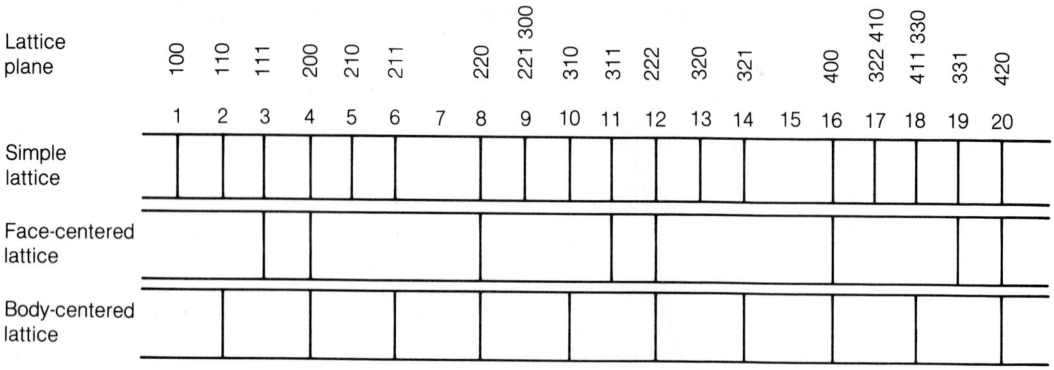

Lattice plane: 100, 110, 111, 200, 210, 211, 220, 221 300, 310, 311, 222, 320, 321, 400, 322 410, 411 330, 331, 420

1 2 3 4 5 6 7 8 9 10 11 12 13 14 15 16 17 18 19 20

Simple lattice

Face-centered lattice

Body-centered lattice

FIGURE 15.19 Powder-pattern lines expected for the three types of cubic lattices. Uniform spacing between actual lines does not occur since the spacings are modulated by the value of sin θ.

where a is the edge length of the unit cell. From the Bragg equation we find the angular relation

$$\sin^2 \theta = \frac{\lambda^2}{4a^2}(h^2 + k^2 + l^2) \tag{15.17}$$

Since θ and λ are known and λ and a are constant, the Miller indices of a line may be determined. Then the value of a may be found. In practice the (hkl) values are found by comparison to known X-ray patterns.

Such comparisons are not foolproof since, in compounds, the different atoms may cause certain reflections to have diminished intensity, sometimes almost to complete extinction. An example of this is found in a comparison of the X-ray patterns from sodium and potassium chloride. Both have interpenetrating fcc structures and are expected to have (111) planes. In NaCl the scattering factor for Na is much less than for Cl and the intensity of the (111) planes is slightly reduced by interference, but the number of electrons responsible for interference in potassium and chlorine is almost the same and the interference is almost complete. Potassium chloride thus appears to have a simple cubic lattice unless very careful measurements are made.

The number of particles per unit cell may also be calculated if the density and size of the cell are known. For example, the unit cell dimension for sodium chloride is found be be 0.564 nm (5.64 Å); its density is 2.163 g cm^{-3} at 25°C; its molar mass is 58.443 g mol^{-1}. Since the density of the crystal is the density of the unit cell, we may write

$$\text{density of cell} = D = \frac{\text{mass of cell}}{\text{volume of cell}} \tag{15.18}$$

If n is the number of chemical formula units in the cell having the mass M, Eq. 15.16 may be written as

$$D = \frac{nM}{V} = \frac{n \times \text{molar mass}}{L \times \text{volume of unit cell}} \tag{15.19}$$

Rearrangement and substitution of values for NaCl gives

$$n = \frac{2.163(\text{g/cm}^{-3})(5.64\text{Å})^3 10^{-24}(\text{cm}^3\text{Å}^{-3}) \times 6.022 \times 10^{23}(\text{mol}^{-1})}{58.443(\text{g mol}^{-1})}$$

$$= 3.999$$

The number of formula units in a unit cell must be an integer. Therefore, there must be four sodium and four chloride ions in a unit cell.

The fact that the (111) planes do occur in the X-ray pattern, as shown previously, is strong evidence that the structural units are ions, for if NaCl occurred as molecules, all the (111) planes would contain the same scattering centers and the interference would be complete.

15.8 Crystal Structure Determinations

When monochromatic radiation is used, the values of the interplanar spacings of a crystal can be determined from the observed diffraction angles. Thus it is possible to deduce the unit cell dimensions, as shown earlier. For simple crystals the symmetry and unit cell size may be enough information to arrive at the exact structure. More often, the problem remains of determining the arrangement of the atoms or ions within the unit cell. The information needed to establish this is contained in the intensities of the diffraction spots.

The Scattering Factor and Structure Factor

The crystal structure can alter the intensities of the various spots observed on X-ray film in the following way. Consider a lattice on which is placed a basis that consists of two different atoms, as shown in Figure 15.20. The Bragg equation gives the

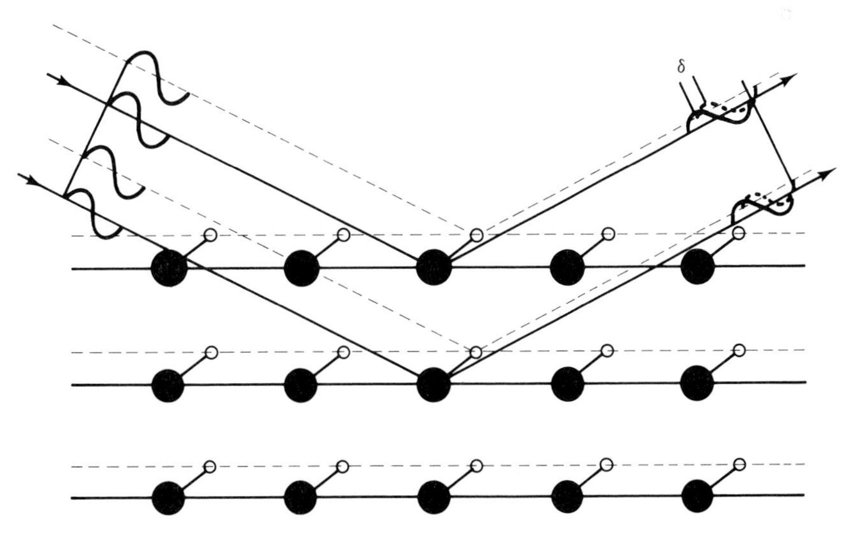

FIGURE 15.20 X rays scattered from a crystal with a basis of two different atoms in mutually displaced lattice arrays. The phase difference between the two waves is represented by δ.

angle of scattering in terms of the interplanar spacings. At some angle θ, the atoms of each type scatter X rays in phase with other atoms in the same array. However, because of the angular difference between the two arrays, the path length for scattered X rays is longer from one type of atom than it is from the other type. As a result, there is a phase difference between the waves generated from the arrays of the two types of atoms. Interference occurs, therefore, and reduces the intensity of the diffracted beam.

X-ray scattering is caused by the electrons in an atom and each different type of atom thus has a different ability to scatter X rays. Furthermore, because the electron distribution can be "smeared out" unsymmetrically under the influence of adjoining atoms, the ability of an atom to scatter X rays depends on the space group it is in.

In order to handle these problems, we introduce the scattering factor f_j that describes the scattering power of the jth atom. The value of f_j is directly proportional to the number of electrons in the atom; it decreases as θ increases because the electron distribution around the nucleus gives rise to a phase difference. Atomic scattering factors f_j have been tabulated for all the space groups.* This description represents the scattering factors in terms of electron density as given by the f_j in real space.

It is the amplitude of all the superimposed waves that is related to the intensity and, when the contribution from each atom in the unit cell is added, we obtain the **structure factor** $F(hkl)$:

$$F(hkl) = \sum_{j}^{N} f_j \cos 2\pi(hx_j + ky_j + lz_j) + i \sum_{j}^{N} f_j \sin 2\pi(hx_j + ky_j + lz_j) \qquad (15.20)$$

or

$$\mathbf{F(hkl)} = \sum_{j=1}^{N} f_j \exp[2\pi i(hx_j + ky_j + lz_j)] \qquad (15.21)$$

where N is the total number of atoms contained in the unit cell. Since the intensity is proportional to the amplitude squared,

$$I(hkl) \propto |F(hkl)|^2 \qquad (15.22)$$

The knowledge of the detailed crystal structure can lead to a prediction of the relative intensities of each diffraction spot. For instance, examination of Eq. 15.20 shows that unless the numbers h, k, and l are all even or all odd, the calculated intensity is zero. Similar considerations allow the calculation of intensities from known atom locations.

Equation 15.21 introduces i, the imaginary square root $\sqrt{-1}$, into the phase portion of the structure factor and gives a hint of the **phase problem,** namely that in general it is not possible to observe the phases. This problem is treated in the next subsection.

---

*_International Tables for X-Ray Crystallography_ (Vols. 1–5), Birmingham, England: Kynoch Press, 1952.

Fourier Synthesis

The goal in any crystal structure determination is to obtain a structure that leads to the prediction of the correct intensity of all the observed spots. Such a structure is almost certainly the correct structure. Starting with a presumed structure from which values of x_j, y_j, z_j can be assigned into Eq. 15.20 is at best a hit-or-miss method. Especially for large organic molecules, a method is needed to handle the large number of observed diffraction spots. The procedure used is based on the idea that a crystal has varying electron densities distributed three-dimensionally, resulting in the scattering of the X rays. If the electron density is $\rho(x, y, z)$, the number of electrons in the volume element dV is

$$\rho(x, y, z)\, dV \tag{15.23}$$

and the structure factor over all volume elements becomes

$$F(hkl) = \int_V \rho(x, y, z) \exp[2\pi i(hx + ky + lz)]\, dV \tag{15.24}$$

We now look for an electron-density function $\rho(x, y, z)$ to represent this electron-density distribution. Although this function can be quite complex for large molecules in a crystal, it is periodic and can be represented by a series of sums of sine and cosine functions, known as **Fourier series** after the French mathematician Jean Baptiste Joseph Fourier (1768–1830). In essence the Fourier series representation of a function uses adjustable constants to force a sufficiently large number of sine and cosine waves that may be written as an exponential term to reproduce the original function. The three-dimensional periodic electron density in a crystal is represented by

$$\rho(x, y, z) = \sum_{h'} \sum_{k'} \sum_{l'} A_{h'k'l'} \exp[2\pi i(h'x + k'y + l'z)] \tag{15.25}$$

where $A_{h'k'l'}$ are the Fourier coefficients and $h'k'l'$ are constants. Substitution of this expression into Eq. 15.24 gives

$$F(hkl) = \int_V h'k'l' A_{h'k'l'} \exp[2\pi i(h'x + k'y + l'z)] \exp[2\pi i(hx + ky + lz)]\, dV \tag{15.26}$$

The combined exponential terms are periodic and each integral is zero except for the terms where $h' = -h$, $k' = -k$, and $l = -l$. In this case

$$F(hkl) = \int_V A_{\bar{h}\bar{k}\bar{l}}\, dV = V_c A_{\bar{h}\bar{k}\bar{l}} \tag{15.27}$$

$$A_{\bar{h}\bar{k}\bar{l}} = \frac{1}{V_c} F(hkl) \tag{15.28}$$

where V_c represents the volume of the unit cell. Thus the Fourier coefficients $A_{\bar{h}\bar{k}\bar{l}}$ are the scattering amplitudes per unit volume. Substitution of Eq. 15.28 along with $-h = h'$, $-k = k'$, and $-l = l'$, into Eq. 15.25 gives the electron density:

$$\rho(x, y, z) = \frac{1}{V_c} \sum_h \sum_k \sum_l F(hkl) \exp[-2\pi i(hx + ky + lz)] \tag{15.29}$$

Comparison of this equation for electron density to that for the structure factor in Eq. 15.24 shows a marked similarity. They are *Fourier transforms** of each other. Equation 15.29 represents the electron density in direct space, whereas Eq. 15.23 represents the structure factors in terms of electron density in reciprocal space. In other words, when the phase is known, Eq. 15.29 allows us to recover the electron density of the crystal in real space from the observed diffraction amplitudes in reciprocal space.

There is an inherent problem in this method. Although $F(hkl)$ is the amplitude of the wave scattered by the *hkl* plane, the measured intensity of the beam is proportional to $|F(hkl)|^2$. Because of the way in which the original data are recorded, point by point, the phase relations are lost. Thus the transformation from Fourier space to real space is complicated by the determination of the phase. A number of techniques to circumvent this problem are in use today, and the interested reader is referred to more advanced texts.

15.9 Structure of Crystals

We have seen how X-ray diffraction techniques can give dimensional data on atomic, ionic, and molecular systems. Now we investigate how the elementary entities are held together in crystal systems. Two different approaches have been developed. Analogous to the valence-bond theory, the *bond model* of the solid state considers the crystal as an array of atoms, each atom possessing electrons used to form bonds with its neighbors. Such bonds extend over the entire crystal and may be of several types, including ionic and covalent bonds. The second method is known as the *band model* and is analogous to molecular orbital theory for molecules. In this approach the nuclei are fixed in the crystal lattice and then an electron "glue" is added. We will treat this method in Section 15.10. Both methods provide useful and complementary information.

The Bond Model

Several bond types may occur in crystalline solids depending on the nature of the entities making up the structure. Although the following classification is not always exact, it provides a useful framework from which to start.

---

*The function $f(x)$ has a Fourier transform defined by

$$F(r) = \int_{-\infty}^{+\infty} f(x) e^{-irx}\, dx$$

and the reverse transform is

$$f(x) = \int_{-\infty}^{+\infty} F(r) e^{irx}\, dr$$

Such transforms allow us to view a function in two different domains and are used primarily as an aid in analysis. In Fourier transform (FT) nmr spectroscopy, for instance, the wave form and the spectrum are Fourier transforms of each other.

1. *Van der Waals Bonds.* *Molecular crystals* is a name given to substances in the solid state that are held together by forces resulting from the interaction of inert atoms or essentially saturated molecules. Examples are solid nitrogen, carbon tetrachloride, and benzene.

2. *Ionic Bonds.* This type of bond occurs as a result of the interaction of charged ions. The ionic bond is both spherically symmetrical and undirected. There are no distinct ionic molecules but rather each ion is surrounded by as many oppositely charged ions as are spatially possible under the limitation of charge neutrality.

3. *Covalent Bonds.* Many crystals of nonmetals contain covalent bonds that result from the sharing of electrons between adjacent atoms. The crystal is constructed, therefore, to allow every atom to complete its electron octet. A wide variety of structures are possible. For example, carbon may exist in the *diamond lattice,* which as shown in Figure 15.21 may be represented by two interpenetrating face-centered cubic lattices. Carbon may also exist as graphite in which the planes have hexagonal symmetry.

Compounds may also crystallize in forms that allow the best use of space and bonding. For example, in the zinc blende (ZnS) structure each atom is surrounded tetrahedrally by unlike atoms. The *zinc blende structure* shown in Figure 15.21 is similar to that of diamond.

4. *Intermediate-Type Bonds.* Bonds of this type occur when a large ion, usually an anion, has its electron distribution distorted, or polarized, by an oppositely charged ion. Usually the effect is greatest for large anions and smallest for small cations since the positive charge tends to hold the electrons in place.

5. *Hydrogen Bonds.* These bonds play an important role in many crystal structures. Water is the most important example and is discussed in Section 16.2.

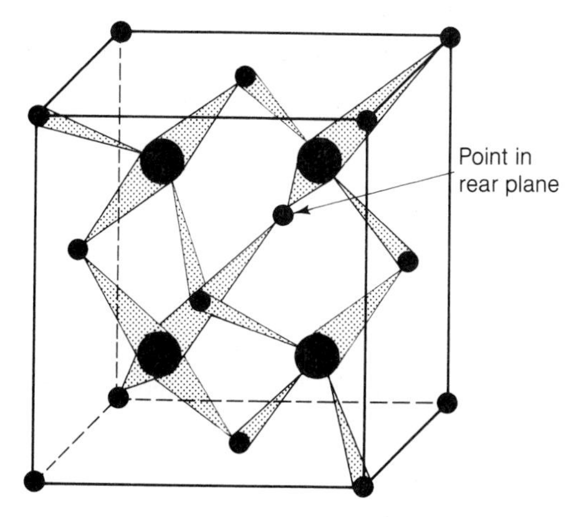

FIGURE 15.21 The zinc blende (ZnS) structure. Each sulfur atom is tetrahedrally surrounded by four zinc atoms. If all the atoms were the same, such as carbon, the diamond structure would result.

6. *Metallic Bonds.* In metals, covalent bonds are formed but are not highly directional and are spread over the entire crystal, a process that increases the stability of the crystal. This is covered in detail in Section 15.10.

Ionic, Covalent, and van der Waals Radii

Several generalizations regarding atomic size may be drawn from the precise data on the dimensions of crystals as provided by X-ray analysis. For example, if we consider oppositely charged ions in a crystal of salt, an equilibrium position will exist between the ions when the attractive and repulsive forces just cancel. This distance may be considered as the sum of the radii (**ionic crystal radii**) of two oppositely charged ions in contact in a stable crystal.

The **covalent radius** is an effective radius for an atom when it is covalently bonded in a compound. Its value is obtained from X-ray determinations by assigning half the homonuclear bond length to each of the atoms. Radii obtained in this manner have been found to be additive to within a few hundredths of a nanometer.

Ionic and covalent radii involve the distance between directly bonded atoms. The **van der Waals radius** is the distance of closest approach of two like atoms or molecules that are not bonded. Fairly accurate drawings of the shapes of molecules may be made using van der Waals radii to show how molecules fit together.

Binding Energy of Ionic Crystals

The stability and properties of ionic compounds are directly related to the interaction of the ions in the crystalline solid. Two points of reference are possible. If we compare the total energy of the solid with the energy of the same number of free neutral atoms at infinite distance from one another, the difference (free atom energy minus the crystal energy) is known as the **cohesive energy**. The crystal is stable if its total energy is lower in the crystal state than in the free state.

The second point of reference compares the energy of the cations and anions in a dilute gas phase, where the interactions are very small, to a crystalline state, where the interactions are quite large. This energy per mole is termed the **lattice energy** ΔE_c or, sometimes, the *crystal energy* of the compound. This energy represents the total potential energy of interaction of the ions at their specific locations in the given lattice structure. The lattice energy may be calculated from basic principles as well as measured experimentally.

The theoretical calculation of the lattice energy is based on two major contributions:

1. The *Electrostatic Interactions Between Ions.* The attractive potential of two ions with charges $z_i e$ and $z_j e$ has the form $-z_i z_j e^2/4\pi\epsilon_0 \epsilon r$. See Section 16.3 and the discussion leading to Eq. 16.12.

2. The *van der Waals Repulsive Energy.** This is modeled by the form $be^{-r/\rho}$ and operates to fix the equilibrium distance in a stable crystal. The values of b and ρ are

*For a quantum-mechanical discussion, see *Advances in Chemical Physics, 12: Intermolecular Forces,* J. O. Hirschfelder (Ed.), New York: Interscience, 1967.

determined from the requirement of equilibrium in the crystal. The expression for this repulsive term is often written as B/r^n. See Eq. 16.20.

The lattice energy ΔE_c of a single ion is a sum

$$(\Delta E_c)_i = \sum_j{}' E_{ij} \tag{15.30}$$

which includes all interactions involving the ith ion and the summation includes all ions except $j = i$. The lattice energy is found by summing Eq. 15.30 over all pairs of ions and includes the two types of contribution previously noted. The factor $\frac{1}{2}$ is introduced to ensure that each pair is counted only once. Thus

$$\Delta E_c = \frac{1}{2}\sum_{i,j}{}' \frac{z_i z_j e^2}{4\pi\epsilon_0 \epsilon r_{ij}} + \frac{1}{2} b \sum_{ij}{}' e^{-r_{ij}/\rho} \tag{15.31}$$

where r_{ij} is the interionic distance. This is known as the *Born-Mayer potential.*

The first sum may be thought of as the interaction of a positive charge, first with all negative charges forming a sphere closest to it and then with all other positive and negative charges in concentric spheres at increasing distances. Thus the interaction will consist of alternately plus and minus terms of diminishing magnitude. In effect, this relates all the interionic distances to the nearest-neighbor separation r_0 and the lattice constants d_{ij} of the crystal through the equation

$$r_{ij} = d_{ij} r_0 \tag{15.32}$$

In 1918 E. Madelung* calculated sums of the first type appearing in Eq. 15.31. The major electrostatic attraction for 1 mol of the ionic crystal is the *Madelung energy* E_M:

$$E_M = -\frac{L\alpha e^2}{4\pi\epsilon\epsilon_0 r_0} \tag{15.33}$$

where α is the **Madelung constant,** important in the theory of ionic crystals.† This constant is dimensionless and characteristic of the particular crystal structure. Substitution of Eq. 15.33 into Eq. 15.31 and minimization with respect to r_0 allows the result to be written as

$$\Delta E_c = -\frac{L\alpha e^2}{4\pi\epsilon_0 \epsilon r_0}\left(1 - \frac{\rho}{r_0}\right) \tag{15.34}$$

where ρ is a constant that can be evaluated from measurements of the compressibility. Consequently, a theoretical value for ΔE_c can be found.

The Born-Haber Cycle

It is appropriate to test the theoretical model just derived. Although the lattice energy ΔE_c for the reaction

$$M^+(g) + X^-(g) \longrightarrow MX(c) \tag{15.35}$$

---

*E. Madelung, *Physik Z., 19,* 524 (1918).

† See C. Kittel, *Introduction to Solid State Physics* (4th ed.), New York: Wiley, 1971, pp. 145ff for a description of the technique for finding lattice sums.

cannot be measured directly, it is possible to use measurable thermochemical data to arrive at a value for ΔE_c using the **Born-Haber cycle.** We start the cycle with metallic atom M and nonmetallic molecule X_2 in the gaseous state as shown in the diagram:

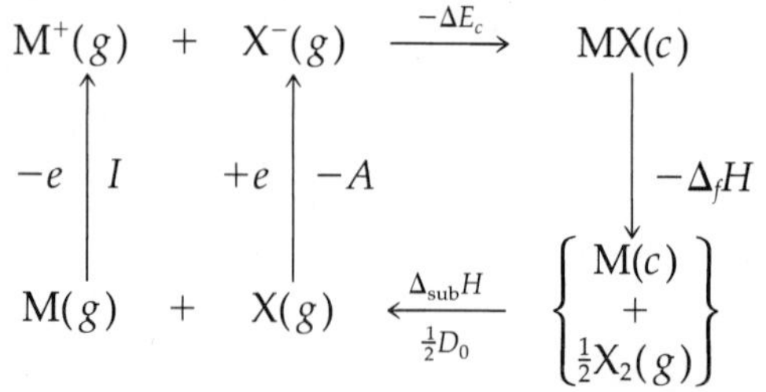

$$\begin{array}{ccccc}
M^+(g) & + & X^-(g) & \xrightarrow{-\Delta E_c} & MX(c) \\
-e \Big\uparrow I & & +e \Big\uparrow -A & & \Big\downarrow -\Delta_f H \\
M(g) & + & X(g) & \xleftarrow[\frac{1}{2}D_0]{\Delta_{\text{sub}}H} & \left\{ \begin{array}{c} M(c) \\ + \\ \frac{1}{2}X_2(g) \end{array} \right\}
\end{array}$$

The energy terms are $\Delta_f H$ = molar enthalpy of formation of $MX(c)$; $\Delta_{\text{sub}}H$ = molar enthalpy of sublimation of $M(c)$; D_0 = molar dissocation energy of $Cl_2(g)$ into atoms; A = electron affinity; and I = the ionization energy of Na. (All energies must be in the same units, especially as A and I are usually not in molar units.) In this cycle the system returns to its initial state and, consequently, the total energy absorbed must be zero. We may write, therefore,

$$\Delta U = 0 = -\Delta E_c - \Delta_f H + (\Delta_{\text{sub}}H + \tfrac{1}{2}D_0) + I - A \tag{15.36}$$

or

$$\Delta E_c = -\Delta_f H + \Delta_{\text{sub}}H + \tfrac{1}{2}D_0 + I - A \tag{15.37}$$

Of the quantities on the right side of this equation accurate values for electron affinities* have been most difficult to obtain. Values for I and D_0 are obtainable from spectroscopy. Practically speaking, sufficient data are available only for comparisons within the alkali-halide crystals. Good agreement between the two methods is obtained for some crystals, but where large deviations exist, large nonionic contributions to the lattice energy occur.

15.10 The Structure of Metals

Metals and other crystals that do not have highly directional bonds generally achieve their lowest energy when each entity is surrounded by the greatest possible number of neighbors. The structure of metals may, therefore, be discussed in terms of the way in which the atoms are packed together.

Closest Packing of Spheres

The ordering of atoms in a metal can most easily be seen by considering the atoms as spheres of identical size. *Closest packing* is the most efficient arrangement of spheres to fill an available space. A single layer of spheres occupies space most efficiently when each sphere is arranged to have six spheres around it. (See Problem 3.) Such an arrangement is shown in Figure 15.22a. The second layer of spheres

*See R. S. Berry and C. W. Reiman, *J. Chem. Phys.*, *38*, 1540 (1963).

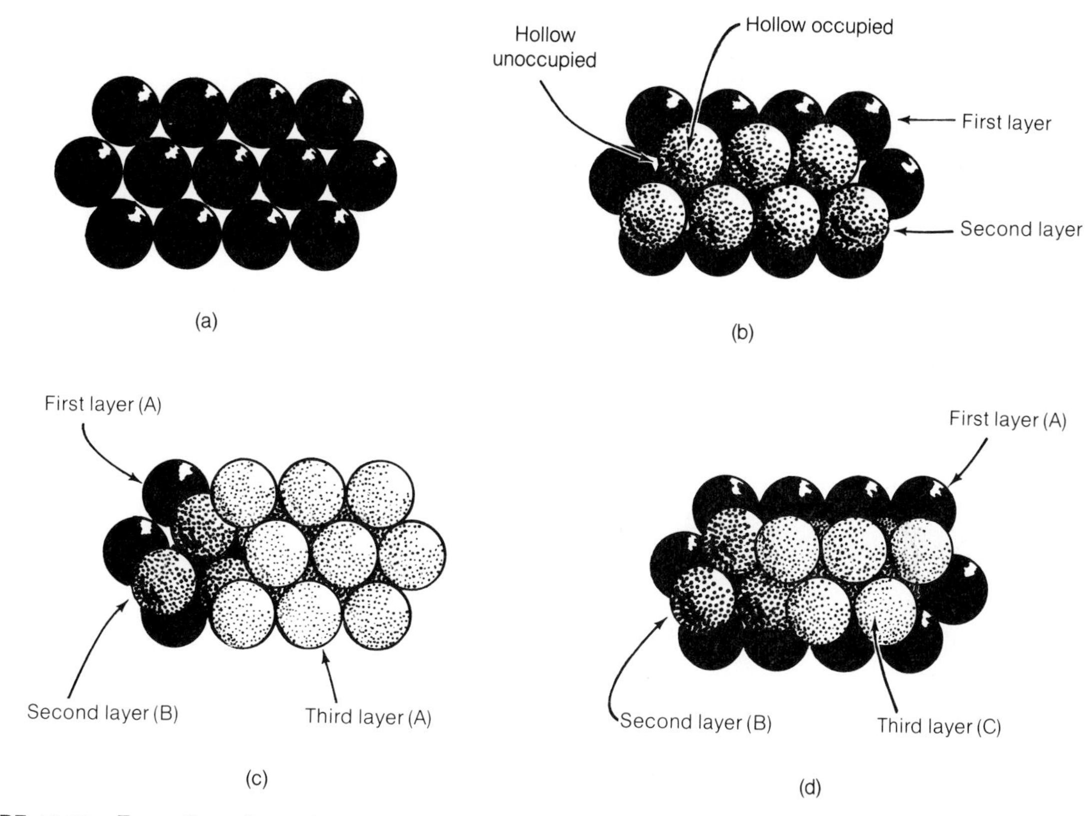

FIGURE 15.22 Formation of two closest-packed arrangements. (a) Base layer showing hexagonal symmetry. (b) Second layer superimposed on top of first showing some hollows of the first layer not occupied. (c) Third layer directly above first layer forming the hexagonal close-packed layers. (d) Third layer is set in alternate hollows forming the cubic close-packed layers.

cannot occupy all the hollows in the first layer as seen in Figure 15.22b. When the third layer is put in place, two different arrangements are possible. In Figure 15.22c the third layer is placed directly above the first layer. This results in **hexagonal close packing** and, if the order continues in the same fashion, the layer arrangement is designated *ABABAB* . . . , where letters represent the layers. The second arrangement places the third layer over the unfilled hollows of the first layer. This results in **cubic close packing** shown in Figure 15.22d. This is described by *ABCABC* . . . . Both methods of packing* result in the same fraction of space occupied and the same number of neighbors, 12.

These two types of closest packings give rise to two types of voids between layers. A **tetrahedral void** is formed when four spheres are arranged at the corners of a tetrahedron. This occurs when a triangular void in one plane has a sphere directly over it as shown in Figure 15.23a. If two triangular voids are joined at their bases to form a void surrounded by six spheres, an **octahedral void** results; the

*Both types of packing were suggested by W. Barlow, *Nature, 29*, 186, 205, 404 (1883), long before the advent of X-ray analysis.

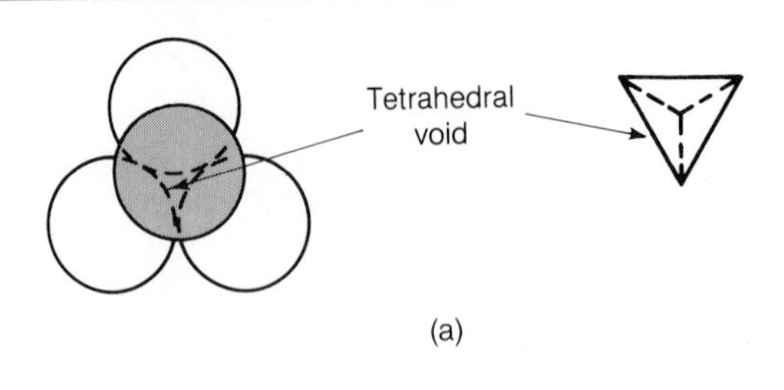

(a)

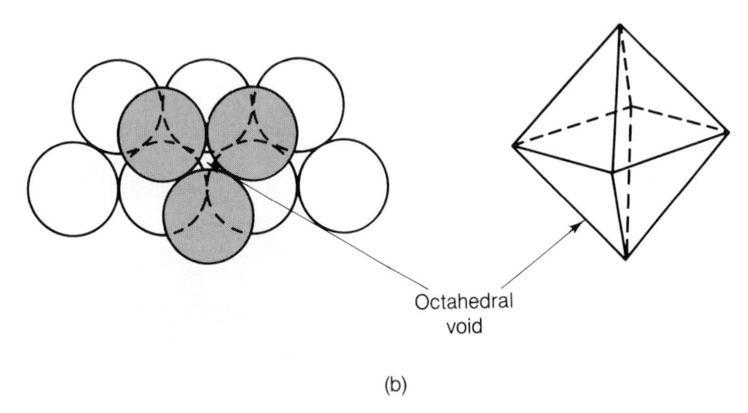

(b)

FIGURE 15.23 The formation of (a) tetrahedral and (b) octahedral voids.

spheres are arranged on the corners of an octahedron. The tetrahedral voids are comparatively small, accommodating a sphere with a radius of 0.23 that of the surrounding layer of spheres. Octahedral voids are large enough to accept spheres with a radius of 0.41 that of the surrounding spheres. The importance of the voids is easily seen. For instance, many oxides have either hexagonal or cubic closest packing of the oxygen atom with the smaller metal atoms in the voids. Furthermore, the properties of materials are greatly influenced by imperfections occurring in crystal lattices. An impurity atom (*interstitial atom*) that has been placed in a void to alter the orderly arrangement of atoms can cause changes in the electrical properties of the crystal.

Metallic Radii

Metallic radii are calculated from the unit cell dimensions of the metal using the assumption that the metal ions are in contact in the particular structure.

Cubic close-packed layers can be oriented to show that they have a face-centered structure. A face-centered cubic unit cell has metal-to-metal contact across the diagonal of the face as shown in Figure 15.24a. Examples of metals that crystallize in the fcc structure are copper, silver, and gold.

Metals often crystallize in the body-centered cubic arrangement shown in Figure 15.24b. This arrangement is not closest packing. Since there is a larger fraction of empty space in such systems, the metals with this structure tend to be more malleable and less brittle. Examples are sodium and the low-temperature form of iron, α-Fe.

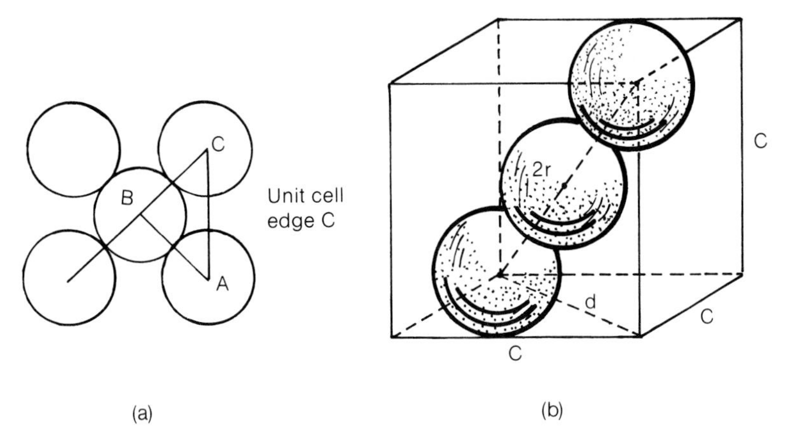

FIGURE 15.24 (a) A face-centered cubic unit cell showing the unit cell edge C. (b) A body-centered cubic system showing the unit cell length C.

15.11 Bonding in Metals, Semiconductors, and Insulators

The Free-Electron Gas Theory

Metals have high cohesive energies and high electrical and thermal conductivities. The *free-electron model* of metals, developed by the German physicist Paul Karl Ludwig Drude* (1863–1906) and the Dutch physicist Hendrik Antoon Lorentz† (1853–1928), can be used to explain these properties. In brief, the free-electron theory states that the valence electrons (*conduction electrons*) are free to move like a perfect gas about lattice points on which positive ions are fixed. In the presence of an electric field, the electrons are free to move along the potential gradient resulting in an electric current.

Drude and Lorentz were able to explain the electrical conductivity from a classical approach using the equations of motion of a particle in an electric field E. The velocity v_d of the electrons is the average velocity attained between collisions, and the relaxation time τ gives a measure of the time between collisions. At constant electric field the acceleration of an electron is $a = v_d/\tau = -eE/m_e$ and therefore

$$v_d = -\frac{eE\tau}{m_e} \tag{15.38}$$

Since the electric current density j is the charge on N electrons per unit volume, $-Ne$, multiplied by their velocity, v_d, we may write

$$j = -Nev_d \tag{15.39}$$

Combining these equations gives

$$j = \frac{Ne^2}{m_e}\tau E = \sigma E \tag{15.40}$$

---

* P. Drude, *Ann. Physik, 7*, 687 (1902).

† H. A. Lorentz, *Proc. Acad. Sci. Amst., 1*, 438, 585, 684 (1904).

where σ is the electrical conductivity. This expression is in accordance with Ohm's law. The drift velocity per unit electric field is known as the *mobility* μ,

$$\mu = \frac{v_d}{E} = -\frac{e\tau}{m_e} \tag{15.41}$$

and is related to the conductivity by

$$\sigma = Ne\mu \tag{15.42}$$

Although this theory explains some aspects of electrical conductivity in metals, it fails when applied to the calculation of heat capacities. Nor does the theory provide a good working model for the differences between metals, insulators, and semiconductors, in particular for the fact that the electrical conductivity of a metal decreases as the temperature is raised, whereas that of semiconductors and insulators increases. A more useful theory is now presented.

Band Theory

A relatively broad electron energy level may be treated in two ways: by the interaction of localized electrons and their energy levels or by an essentially free-electron approximation. The first approach is similar to the formation of two energy levels in the MO description of the hydrogen atom. For example, if five atoms are placed in a row with equal separations between the atoms, a Hückel MO-type treatment would yield five energy levels split from the originating energy level as shown in Figure 15.25. When each of the N atoms in a crystal contributes N orbitals, the

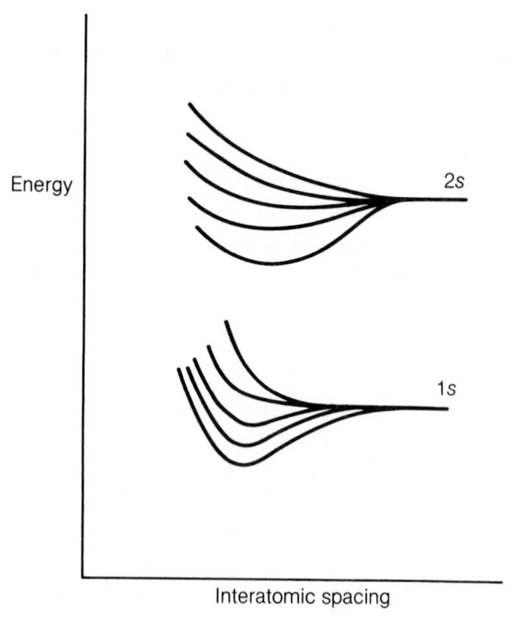

FIGURE 15.25 A potential-energy diagram for a row of five equally spaced atoms showing the formation of bands. Between levels a separation occurs but there is only a small separation of energy levels within a band.

energy may be written in terms of the Coulombic integral J and the exchange integral K:

$$\epsilon_n = J + 2K \cos \frac{2\pi m}{N} \tag{15.43}$$

where $m = 0, \pm 1, \pm 2, \ldots$. No matter how many atoms are present, the energy levels lie between the energy limits $J + 2K$ and $J - 2K$. As N becomes very large (in the order of 10^{22} cm^{-3}), the spacing between levels becomes so small that the energy appears like a continuum. Such a group of levels is called an **energy band**.

The second approach considers an assembly of electrons free to move in a crystal and then considers the manner in which they interact with the ionic lattice. This is essentially an application of the "particle-in-the-box" problem (Section 11.6) with periodic boundary conditions, the solutions of which can be written in the form

$$\psi = e^{ikx} \tag{15.44}$$

where $\boldsymbol{k}$, a *wave number*, is

$$\boldsymbol{k} = \frac{2\pi p}{h} = \frac{2\pi}{\lambda} \tag{15.45}$$

Here p and λ are the momentum and wavelength of the electron. The corresponding energy when the potential energy is assumed to be zero within the box is

$$E = \frac{\boldsymbol{k}^2 h^2}{8\pi^2 m} \tag{15.46}$$

and essentially one continuous band of energy levels is formed for different values of $\boldsymbol{k}$. When the periodicity of the ions in the lattice is considered, we find that the potential energy also varies periodically. The solution, ϕ_k, to this problem* is the product of the wave function of a free electron $\psi(x)$ and a function that has the periodicity of the lattice, $u(x)$:

$$\phi_k = e^{ikx} u(x) \tag{15.47}$$

These equations, developed in 1928, are known as the *Bloch functions,*† named after Felix Bloch (b. 1905). The effect of the periodicity of the potential is to introduce discontinuities into the shape of the energy curve because only certain electron energies are allowed and these must fall into a series of energy bands. The bands are separated by regions of forbidden energy or energy gaps. The first **Brillouin zone** is the filled energy band below the lowest lying energy gap.

The relation between energy bands and the corresponding *forbidden bands* or *energy gaps* is shown as an energy band diagram in Figure 15.26. The highest energy band that is completely filled with electrons is called the **valence band.** The electrons in this band are not free to move because they are involved in chemical bonding. The lowest lying band that is either empty or only partially filled with electrons is called the **conduction band** and in it the electrons are free to move

---

*R. de. L. Kronig and W. G. Penney, *Proc. Roy. Soc.* (London), *A 130*, 499 (1930).
†F. Bloch, *Z. Phys.*, *52*, 555 (1928).

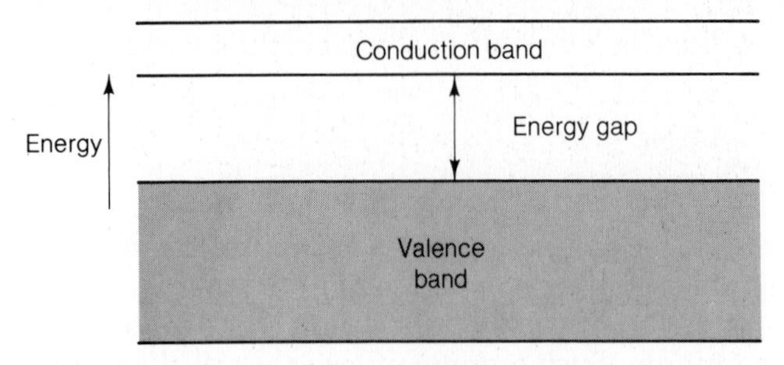

FIGURE 15.26 Energy band diagram for an insulator or intrinsic semiconductor. The conduction band contains no electrons and the valence band has all its energy states completely filled by electrons. The two bands are separated by a forbidden energy region.

throughout the solid with approximately zero activation energy. We can learn more about the nature of these bands by a study of heat capacities in Section 15.12.

15.12 Statistical Thermodynamics of Crystals: Theories of Heat Capacities

According to classical theory the heat capacity of a solid is independent of temperature. The molar heat capacity as found by Dulong and Petit in 1819 is

$$C_v \approx 3R \approx 25 \text{ J K}^{-1} \text{ mol}^{-1} \tag{15.48}$$

where R is the gas constant. Although this value is in good agreement with experimental values for most solids at room temperature, there are several exceptions: Boron, silicon, and carbon (diamond) all have values that are too low. The biggest problem however is that all substances have heat capacities that approach zero as the temperature approaches absolute zero.

The Einstein Model

In 1906 Einstein introduced* the idea that each atom in a solid is an independent harmonic oscillator and that each has the same fundamental frequency ν. Einstein then extended Planck's quantum theory to complete his model.

The average energy of a harmonic oscillator according to Planck's theory is given by Eq. 11.34, namely

$$\bar{\epsilon} = \frac{h\nu}{e^{h\nu/\mathrm{k}T} - 1} \tag{15.49}$$

For an Einstein solid consisting of N atoms, the average energy is

$$\bar{E} = 3N\frac{h\nu}{e^{h\nu/\mathrm{k}T} - 1} \tag{15.50}$$

where the factor of 3 comes from the fact that each atom has three degrees of freedom. In most solids the rotational and translational motions are eliminated by

*A. Einstein, *Ann. Physik, 22,* 180, 800 (1906).

the rigid structure of the lattice. This leaves $3N - 6$ vibrational degrees of freedom to consider, and because N is large, $3N - 6$ may be replaced by $3N$.

Since the heat capacity at constant volume is defined as $C_V = (\partial E/\partial T)_V$, **Einstein's heat capacity equation** is

$$C_V = 3N\mathbf{k}\left(\frac{h\nu}{\mathbf{k}T}\right)^2 \frac{e^{h\nu/\mathbf{k}T}}{(e^{h\nu/\mathbf{k}T} - 1)^2} \tag{15.51}$$

This equation qualitatively explains the breakdown of classical theory. In the limit of zero temperature, where the heat capacity must approach zero, Eq. 15.51 may be written as

$$C_V \approx 3N\mathbf{k}\left(\frac{h\nu}{\mathbf{k}T}\right)^2 e^{-h\nu/\mathbf{k}T} \qquad (T \to 0) \tag{15.52}$$

This approaches zero as $T \to 0$; however, it is known that C_v approaches zero as T^3, and Eq. 15.52 does not show this behavior. However, Eq. 15.51 does have a limiting value of $3R$ as $T \to \infty$, in accordance with experiment.

The Debye Model

When it was recognized that the heat capacity approached zero too rapidly according to the Einstein model, Peter Debye in 1912 introduced* his model of heat capacities. He considered a solid to be a three-dimensional isotropic continuum in which elastic waves could be excited with a continuous range of frequencies. This continuous range of frequencies was cut off at a particular frequency ν_D, which was determined by the properties of the solid. These vibrations are characterized by a distribution function $g(\nu)$ so defined that $g(\nu)\,d\nu$ gives the number of oscillators in the interval between ν and $\nu + d\nu$.

The logarithm of the partition function for this system may be written as

$$\ln Q = \sum_{i=1}^{i=\nu_D} g(\nu_i) \ln q(v_i) \tag{15.53}$$

where $q(v_i)$ is the partition function of the harmonic oscillator defined by Eq. 14.120. Since there is a continuum of frequencies in this case, the summation may be replaced by an integral. Using this latter expression, Debye found the vibrational energy of a solid to be

$$E = \int_0^{\nu_D} g(\nu) \frac{h\nu}{e^{h\nu/\mathbf{k}T} - 1} d\nu \tag{15.54}$$

Debye then applied the Rayleigh-Jeans relation (Section 11.1) for the distribution function

$$g(\nu)\,d\nu = c\nu^2\,d\nu \tag{15.55}$$

where c is a constant, and he was able to derive the equation

$$E = 9N\left(\frac{\mathbf{k}T}{h\nu_D}\right)^3 \mathbf{k}T \int_0^{x_m} \frac{x^3}{e^x - 1}\,dx \tag{15.56}$$

*P. Debye, *Ann. Physik, 39*, 789 (1912).

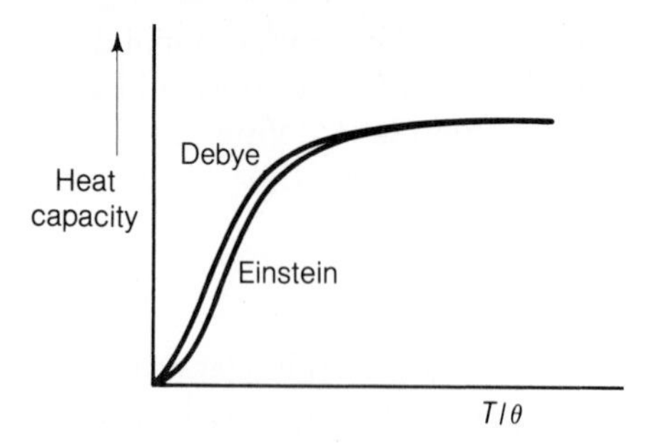

FIGURE 15.27 A comparison of the heat capacity as calculated by the Debye and Einstein theories. Experimental points fit the Debye curve fairly well.

where $x = h\nu/\mathbf{k}T$ and $x_m = h\nu_D/\mathbf{k}T$.

The heat capacity is then

$$C_V = \left(\frac{\partial E}{\partial T}\right)_V = 9N\mathbf{k}\left(\frac{T}{\Theta_D}\right)^3 \int_0^{x_m} \frac{x^4 e^x}{(e^x-1)^2}\,dx \tag{15.57}$$

where Θ_D, known as the **Debye temperature,** is defined by

$$\Theta_D = \frac{h\nu_D}{\mathbf{k}} \tag{15.58}$$

and is independent of temperature. The Debye temperature is characteristic of a particular substance. A few values of Θ_D are 1860 K for C, 150 K for Na, and 315 K for Cu.

The Debye model fits the experimental low-temperature heat capacities better than the Einstein model. At temperatures much less than the Debye temperature, Eq. 15.57 reduces to

$$C_V = \frac{12}{5}\pi^4 N\mathbf{k}\left(\frac{T}{\Theta_D}\right)^3 \tag{15.59}$$

This expression of the **Debye T^3 law** should hold at $T < \Theta_D/10$. For $T > \Theta_D/10$, $C_V = 464.4\ (T/\Theta_D)^3$. Tables* are available to facilitate calculations using Θ_D. A comparison plot of the heat capacity curves from the Debye and Einstein theories is shown in Figure 15.27.

Fermi-Dirac Statistics

In the electron theory of metals each atom gives one valence electron to the electron gas and these electrons are completely mobile. If there are N electrons, they should contribute $\frac{3}{2}N\mathbf{k}$ to the normal heat capacity of $\frac{3}{2}N\mathbf{k}$, but the observed electronic contribution at room temperature is generally less than 0.01 of this value. (We have just seen that the Debye theory fairly accurately describes the heat capacity.) How

*D. E. Gray (Ed.), *The American Institute of Physics Handbook,* New York: McGraw Hill, 1957.

can the electrons not contribute to the heat capacity? This question posed a difficulty for the early development of the theory.

The solution of the problem involved the Pauli exclusion principle and a new distribution function for electrons. If the quantum state for an electron includes a spin quantum number, then by the Pauli principle only one electron can occupy each quantum state. Thus the electrons in a metal at the absolute zero fill a succession of energy levels up to a particular level. The energy of the uppermost state is known as the **Fermi energy*** E_F, named after the Italian-American Enrico Fermi (1901–1954), and the level is known as the **Fermi level.** The Fermi energy divided by the Boltzmann constant is known as the **Fermi temperature** T_F. Even at 0 K some of the electrons in a metal occupy states of relatively high energy. The **Fermi-Dirac distribution function**

$$P(E) = \frac{1}{\exp[(E - E_F)/\mathbf{k}T] + 1} \tag{15.60}$$

gives the probability that a state with energy E will be occupied in an ideal electron gas. The value of E_F controls the tendency of electrons to move at an interface and is called the *electrochemical potential* of the electron. At the absolute zero the probability that all states below E_F are filled is one, and it is zero for all states above E_F. At any temperature above 0 K the probability that the Fermi level is occupied by an electron is $\frac{1}{2}$ when $E = E_F$. This is shown in Figure 15.28. At temperatures above the absolute zero, not every electron gains the energy $\mathbf{k}T$ but only those electrons that are within an energy range $\mathbf{k}T$ of the Fermi level. This explains the slight contribution of electrons to the heat capacity because only a fraction of the electrons, in the order of T/T_F, lie close enough to E_F. Here $T_F \equiv E_F/\mathbf{k} \approx 5 \times 10^4$ K.

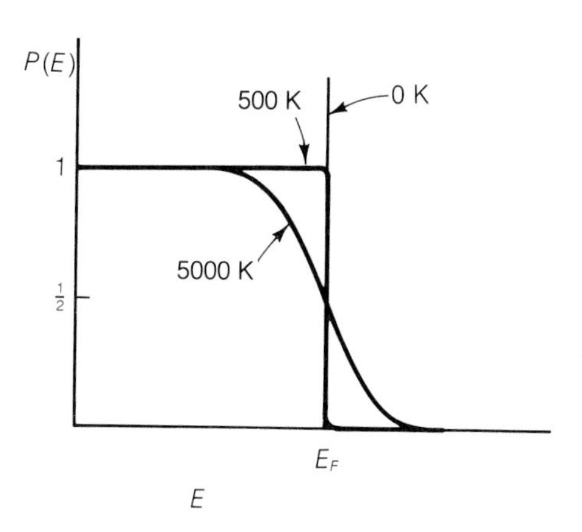

FIGURE 15.28 The Fermi-Dirac distribution function plotted for 500 K and 5000 K for the case

$$T_F = \frac{E_F}{\mathbf{k}} = 50\ 000 \text{ K}$$

Notice that even at 5000 K the probability of occupancy of higher energy states is small.

* E. Fermi, *Z. Physik*, *36*, 902–912 (1926). Also see H. A. Boorse and L. Motz, *The World of the Atom* (Vol. II), New York: Basic Books, 1966, pp. 1321–1331.

Experimentally the heat capacity may be written as a sum of the electronic and lattice terms

$$C = \gamma T + AT^3 \tag{15.61}$$

where γ and A are constants. Values of γ range from 0.00138 J mol^{-1} K^{-2} for sodium to 0.0092 J mol^{-1} K^{-2} for the γ form of manganese.

Electrons are not the only particles that follow Fermi-Dirac statistics. Electrons, protons, neutrons, and any species with an odd number of electrons, protons, and neutrons are called *fermions,* the half-integral spin being described by Fermi-Dirac statistics. The use of antisymmetric wave functions requires the use of these statistics.

The use of symmetric wave functions leads to **Bose-Einstein statistics** named in honor of the Indian statistical mechanician Raj Chandre Bose (b. 1901). Particles such as photons and species with an integral spin are called *bosons.*

15.13 Energy Bands and Defect Structures

The nature of the energy band determines several characteristics of solids, and the bands in turn can be influenced by impurities or by the defect structure of the material.

Metals, Semiconductors, and Insulators

Energy band diagrams may be used to illustrate the differences between metals, semiconductors, and insulators.

In metals the electrons completely fill the valence band. The higher energy conduction band is empty at absolute zero. This is shown in Figure 15.29a. There is no energy gap between the valence and conduction bands for metals; above absolute zero, electrons at the top of the highest occupied levels can gain thermal energy and move into the low lying empty levels of conduction bands. A substantial fraction of the electrons can be excited into singly occupied energy levels, even at relatively low temperatures. Such unpaired electrons contribute to the electrical conductivity of a metal, and the substance is called a *conductor.*

In *insulators* and *semiconductors,* shown in Figure 15.29b and c, the valence band is completely filled and an energy gap exists between it and the next higher energy band. If the energy gap is wide (large), there is little chance for the electron to be excited into an empty conduction band and the material is an insulator. If there is only a small energy gap, the material is an *intrinsic semiconductor,* the electron being excited rather easily into the conduction band. If the gap is wide but impurity atoms are added, it may be possible to establish levels within the gap that facilitate the movement of electrons into the conduction band. These latter systems are known as *impurity semiconductors,* or *extrinsic semiconductors.*

Doping of Semiconductors

Most technologically important semiconductors are of the extrinsic type in which the charge carrier production is determined by trace amounts of impurities or by lattice imperfections. For example, at room temperature gallium arsenide has a

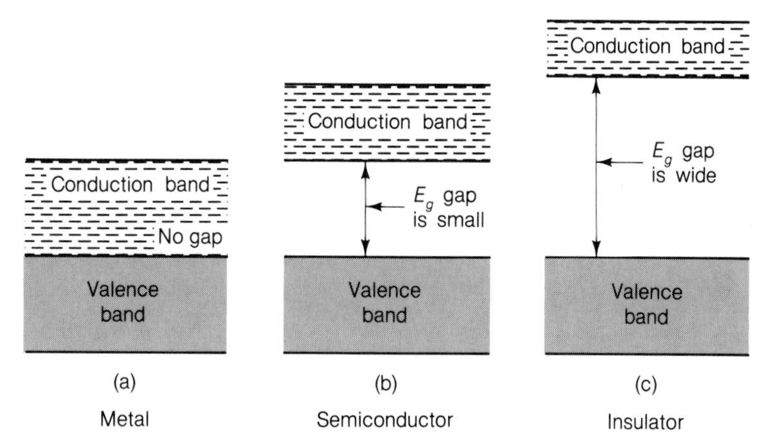

FIGURE 15.29 The relation of energy gaps and conduction bands to the valence band in the three types of solids.

band energy gap of about 1.4 eV corresponding to a T_F of 16 000 K. It would be classified as an insulator since no significant conductivity would exist until its melting point of 1511 K or above is reached. However, with the addition of an impurity (between about 0.1% to less than 1 ppm), the charge carrier type, its concentration, and the conductivity of the sample can be controlled.

When the crystal lattice is disturbed by the addition of impurities or even with irregularities in the lattice arrangement, isolated centers are generated. Some centers can contribute electrons to the conduction band of the material. Such defects are called **donors** or *donor centers*. If the primary charge carriers are electrons, the material is called an **n-type** semiconductor. If the centers remove electrons from the chemical bonds, they are called **acceptors** and produce electron vacancies. These vacancies behave as positive charge carriers in the valence band and are called **holes**. If holes are the predominant charge carrier, a **p-type** semiconductor results.

Imperfections normally exist in rather high concentrations in all real materials and may be classified as *point defects*, *dislocations* or *line defects*, and *plane defects* (grain boundaries).

15.14 Optical Properties of Solids

Absorption and emission of electromagnetic radiation by solids occur in the ultraviolet, visible, and near-infrared portions of the spectrum. Often this is caused by impurities or *activators* that have energy levels lying within the energy gap of the pure material. For example, pure Al_2O_3 is colorless but the addition of a small amount of Cr^{3+} ($Cr^{3+}/Al^{3+} \approx \frac{1}{2000}$) introduces a substitutional impurity replacing Al^{3+}. The result is ruby, which has a deep red color caused by the ligand field splitting of the *d* orbitals of chromium.

When ruby crystal is irradiated by ultraviolet light, it emits intense red light and we say that it is *luminescent*. See Section 13.6 for a discussion of the fluorescent and phosphorescent behavior of luminescent materials.

We first treat the absorption of radiation by solids. In metals, absorption is explained by excitation of electrons into higher energy bands. The energy radiated in the return to the ground state causes the metallic luster of most metals. In semiconductors, there is an *absorption edge* at which there is an abrupt and characteristic absorption of light. This corresponds to the excitation of electrons from the valence band across the energy gap to the conduction band. In insulators, absorption appears to occur at defects in the normally forbidden energy gap.

Color Centers: Nonstoichiometric Compounds

When alkali halides are heated in their alkali metal, atom vapor defects are introduced to make the crystal very slightly nonstoichiometric. As a result the crystals turn various colors: LiF, pink; NaCl, yellow; KCl, blue. The most common type of defect is the ***F* center.** It is formed in NaCl in the following way. A sodium metal atom ionizes on the surface of NaCl. The ion attracts a chloride ion from the surface nearby. This forms an anion vacancy on the surface that is attracted by a free electron. This combination is free to migrate into the bulk crystal. In essence the electron in an F center is surrounded by six Na^+ ions with which it is shared in hydrogenlike orbitals. The observed absorption bands correspond to an electron transition between a ground state and an excited state.

The energy levels formed are just below the conduction band and are excited states of an electron-hole pair, called an **exciton.** Exciton states are localized and do not contribute to the electrical conduction process.

Luminescence in Solids

Aside from fluorescence and phosphorescence emission, radiation from solids is observed in **thermoluminescence** and in lasers. In thermoluminescence the electrons in a solid are excited at low temperatures by the absorption of radiation, and electrons are trapped in the forbidden gap region. When the sample is heated, the electrons receive enough energy to escape the traps, and visible light is sometimes emitted. This property of thermoluminescence has been used to date the time at which pottery was fired.

Some of the characteristics of lasers have been touched on in Section 13.7.

Key Equations

Bragg equation:

$$n\lambda = 2d \sin \theta$$

where λ is the wavelength of the X ray, d is the interplanar spacing, and θ is the angle of incidence.

Wavelength from *electron diffraction:*

$$\lambda/\text{Å} = \frac{12.25}{(\epsilon/\text{eV})^{1/2}} \quad \text{or} \quad \lambda/\text{pm} = \frac{1225}{(\epsilon/\text{eV})^{1/2}}$$

Wavelength from *neutron diffraction:*

$$\lambda/\text{Å} = \frac{0.28}{(\epsilon/\text{eV})^{1/2}} \quad \text{or} \quad \lambda/\text{pm} = \frac{28}{(\epsilon/\text{eV})^{1/2}}$$

The *structure factor:*

$$F(hkl) = \sum_{j=1}^{N} f_j \exp[2\pi i(hx_j + ky_j + lz_j)]$$

Einstein's heat capacity equation:

$$C_v = 3N\mathbf{k}\left(\frac{h\nu}{\mathbf{k}T}\right)^2 \frac{e^{h\nu/\mathbf{k}T}}{(e^{h\nu/\mathbf{k}T} - 1)^2}$$

Debye's heat capacity equation

$$C_v = 9N\mathbf{k}\left(\frac{T}{\Theta_D}\right)^3 \int_0^{x_m} \frac{x^4 e^x}{(e^x - 1)^2}\, dx$$

where $\Theta_D = h\nu_D/\mathbf{k}$ and is known as the *Debye temperature.*

Debye's T^3 law:

$$C_v = \frac{12}{5}\pi^4 N\mathbf{k}\left(\frac{T}{\Theta_D}\right)^3$$

Problems

A Crystal Lattices, Unit Cells, Density

1. How many lattice points belong to a unit cell of
a. a face-centered lattice;
b. a body-centered lattice?

2. How many basis groups are there in
a. an end-centered lattice;
b. a primitive lattice?

3. **a.** Determine the efficiency of area utilization in packing circles onto the lattice points of a square lattice.
b. Compare that value with the efficiency of packing circles onto a triangular lattice.
c. Which packing uses area more efficiently and by how much?

4. Derive an equation to relate the density ρ of a right-angled unit cell to its edge lengths a, b, and c and the number of formula units z per unit cell.

5. Crystals of p,p'-dibromo-α,α'-difluorostilbene $[BrC_6H_4C(F){=}]_2$ are orthorhombic with edge lengths a = 28.32 Å; b = 7.36 Å; c = 6.08 Å. If there are four molecules in a unit cell, calculate the density of the crystal.

6. How many formula units exist in pure crystalline Si, which occurs in a face-centered cubic lattice, if its density is 2.328 99 g cm^{-3} and its cell length is a = 5.431 066 Å? The atomic mass of Si is 28.085 41 g mol^{-1}.

7. Sodium chloride crystallizes in a face-centered cubic lattice with four NaCl units per unit cell. If the edge length of the unit cell is 5.629 Å, what is the density of the crystal? Compare your answer to the value given in the CRC.

8. KCl is tetramolecular and crystallizes in a face-centered cubic lattice. If the edge length is 6.278 Å, what is the density of KCl? Compare your answer to the value in the CRC.

9. Calcium fluoride crystallizes in a face-centered cubic lattice where $a = b = c$, and it has a density of 3.18 g cm^{-3}. Calculate the unit cell length for CaF_2.

B Miller Indices and the Bragg Equation

10. Calculate the Miller indices of the parallel planes in a cubic lattice that intercept the unit cell length at $x = a$, $y = \frac{1}{2}a$, and $z = \frac{2}{3}a$.

11. Determine the distance (i.e., d value) of the closest plane parallel to the 100, 110, and 111 faces of the cubic lattice.

12. What are the Miller indices of the plane that cuts through the crystal axes at

a. $(2a, b, 3c)$;
b. $(2a, -3b, 2c)$;
c. $(a, b, -c)$?

13. Calculate the separation between planes in a cubic lattice with unit cell length of 389 pm when the indices are

a. 100,
b. 111,
c. $12\bar{1}$.

14. Copper sulfate single crystals are orthorhombic with unit cells of dimensions a = 488 pm, b = 666 pm, c = 832 pm. Calculate the diffraction angle from Cu K_α X rays (λ = 154 pm) for first-order reflections from the (100), (010), and (111) planes.

15. Single crystals of $FeSO_4$ are orthorhombic with unit cell dimensions a = 482 pm, b = 684 pm, c = 867 pm. Calculate the diffraction angle from Te K_α X rays (λ = 45.5 pm) from the (100), (010), and (111) planes.

16. Single crystals of $Hg(CN)_2$ are tetragonal with unit cell dimensions a = 967 pm and c = 892 pm. Calculate the first-order diffraction angles from the (100) and (111) planes when Cu K_α X rays (λ = 154 pm) are used.

17. A two-dimensional lattice is depicted with planes superimposed on it parallel to the third direction. Determine the Miller indices for each set of planes A, B, C, D.

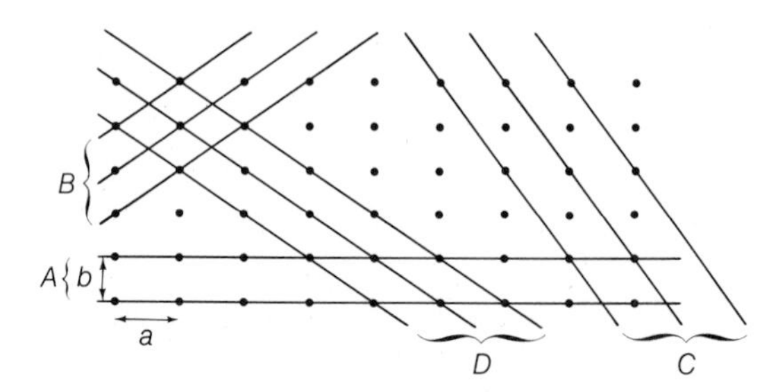

18. The layers of atoms in a crystal are separated by 325 pm. At what angle in a diffractometer will diffraction occur using

a. molybdenum K_α X rays (λ = 70.8 pm);
b. copper K_α X rays (λ = 154 pm)?

19. Calculate the wavelength of an electron that is accelerated through a potential difference of approximately 40 kV.

C Interpretation of X-Ray Data

20. A substance forms cubic crystals. A powder pattern shows reflections that have either all even or all odd indices. What type of unit cell does it have?

21. A powder pattern of a cubic material has lines that index as (110), (200), (220), (310), (222), (400). What is its type of unit cell?

***22.** The successive $\sin^2 \theta$ values obtained from a powder pattern for α-Fe are 1, 2, 3, 4, 5, 6, 7, 8, 9, etc.

a. If iron is in the cubic system, which type of unit cell is present?
b. If a copper X-ray tube is used (λ = 154.18 pm), calculate the side of the unit cell and the value of θ from (100) planes. The density of α-Fe is 7.90 g cm^{-3}.
c. What is the radius of the iron atom if the central atom in the cubic cell is asumed to be in contact with the corner atoms?

23. Potassium metal has a density of 0.856 g cm^{-3} and has a body-centered cubic lattice. Calculate the length of the unit cell a and the distance between (200), (110), and (222) planes. Potassium has an atomic mass of 39.102 g mol^{-1}.

24. Low angle lines in the Cu K_α powder pattern of KCl are found to be at $\theta = 14.18°$, 20.25°, and 25.10°. Find the crystal type from these data. (For Cu K_α, $\lambda = 154.18$ pm.) What other information is needed for a definitive determination?

25. The smallest observed diffraction angle of silver taken with Cu K_α radiation ($\lambda = 154.18$ pm) is 19.076°. This angle is associated with the (111) plane in the cubic close-packed structure of silver.

a. Determine the value of the unit cell length a.

b. If $\rho(\text{Ag}) = 10.500$ g cm^{-3} and $M = 107.87$ g mol^{-1}, calculate the number of atoms in the unit cell.

26. Sodium fluoride is known to form a cubic closed-packed structure. The smallest angle obtained with Cu K_α radiation ($\lambda = 1.5418$ Å) is 16.72° and is derived from the (111) planes. Find the value of a, the unit cell parameter.

D Bonding in Crystals and Metals

27. A tetrahedral void is shown in Figure 15.23. Determine the largest sphere of radius r that can fit into a tetrahedral void when the surrounding four spherical atoms of the lattice are in contact. Let the lattice atoms have radius R.

28. An octahedral void is surrounded by six spheres of radius R in contact. If one-sixth of each of the six coordinating spheres contribute to the volume of the octahedron surrounding the void, calculate the maximum radius of the void that can be accommodated.

29. Calculate the value of ΔE_c of RbBr from the following information: $\Delta_f H = -414$ kJ mol^{-1}; I (ionization energy of Rb) = 397 kJ mol^{-1}; $\Delta_{sub}H(\text{Rb}) = 84$ kJ mol^{-1}; D_0 (Br_2) = 192 kJ mol^{-1}; A (electron affinity, Br) = 318 kJ mol^{-1}.

E Supplementary Problems

30. Some of the d spacings for the mineral canfieldite (Ag_8SnS_6) are 3.23, 3.09, 3.04, 2.81, and 2.74 Å obtained with Cu K_α X rays ($\lambda = 1.5418$ Å).

a. Find the corresponding angles of diffraction.

b. This is a cubic system with $a = 21.54$ Å; determine the hkl values for the first 3 d spacings.

31. A copper selenide mineral (Cu_5Se_4) called athabascaite is orthorhombic with $a = 8.227$, $b = 11.982$, $c = 6.441$. Strong intensity lines using Cu K_α X rays ($\lambda = 154.18$ pm) are observed at 12.95°, 13.76°, and 14.79°. Determine the d spacings and assign hkl values to these lines.

32. Zinc blende is the face-centered cubic form of ZnS with Zn at 0, 0, 0; $\frac{1}{2}$, $\frac{1}{2}$, 0; 0, $\frac{1}{2}$, 0; $\frac{1}{2}$, $\frac{1}{2}$, 0 and with S at $\frac{1}{4}$, $\frac{1}{4}$, $\frac{1}{4}$; $\frac{1}{4}$, $\frac{3}{4}$, $\frac{3}{4}$; $\frac{3}{4}$, $\frac{1}{4}$, $\frac{3}{4}$; $\frac{3}{4}$, $\frac{3}{4}$, $\frac{1}{4}$.

a. Determine the structure factor from the (111) planes that give rise to the lowest angle reflection at $\theta = 14.30°$ using Cu K_α ($\lambda = 154.18$ pm).

b. Calculate the dimension a of the unit cell.

33. Calculate the Debye temperature of tungsten that is isotropic (an assumption of the Debye model). The cut-off frequency is given by

$$\nu_D = \left(\frac{9N}{4\pi V}\right)^{1/3} \left(\frac{1}{c_\ell^3} + \frac{2}{c_t^3}\right)^{-1/3}$$

where $c_\ell = 5.2496 \times 10^5$ cm s^{-1} and $c_t = 2.9092 \times 10^5$ cm s^{-1} are the longitudinal and transverse elastic wave velocities, respectively, in tungsten.

F Essay Questions

1. Explain why the initial X-ray investigation of the two face-centered cubic structures, NaCl and KCl, showed that NaCl was face centered whereas KCl was simple cubic.

2. If ΔH_c were required rather than ΔE_c, what modification of the Born-Haber cycle would be needed?

3. X-ray diffraction is often used to measure residual stress in metals. Suggest what change in the measured parameters allows this determination.

4. Gold diffuses faster in lead at 300°C than does sodium chloride in water at 15°C. Point defects based on vacancies can account for such high rates. For an ionic material, suggest ways in which such vacancies can occur without altering the stoichiometry of the crystal.

Suggested Reading

L. V. Azároff, *Introduction to Solids,* New York: McGraw-Hill, 1960.

M. J. Buerger, *Contemporary Crystallography,* New York: McGraw-Hill, 1970.

G. B. Carpenter, *Principles of Crystal Structure Determination,* New York: W. A. Benjamin, 1969.

M. I. Davis, *Electron Diffraction in Gases,* New York: Marcel Dekker, Inc., 1971

P. P. Ewald (Ed.), *Fifty Years of X-ray Diffraction,* Utrecht, The Netherlands: International Union of Crystallography, 1962, pp. 6–75.

C. Kittel, *Introduction to Solid State Physics* (4th ed.), New York: Wiley, 1971.

L. Pauling, *The Nature of Chemical Bond* (3rd ed.), Ithaca, N.Y.: Cornell University Press, 1960.

G. H. Stout and L. H. Jensen, *X-Ray Structure Determination,* Reading, Mass.: Addison-Wesley, 1968.

A. F. Wells, *Structural Inorganic Chemistry,* (3rd ed.), New York: Oxford University Press, 1961.

The Liquid State

Preview

Liquids are more difficult to treat than gases or solids, since there is neither complete disorder nor complete order. There are two alternative ways of dealing with the liquid state: We can start with the theory of nonideal gases or with the theory of solids.

If we start with the van der Waals equation for nonideal gases, the term a/V^2 is the internal pressure of the gas or liquid. The internal pressure varies considerably with the type of liquid and is strongly affected by the external pressure.

If liquids are considered as *disordered solids*, the important factor is that when a solid melts, there is a decrease in *long-range order.* Liquids exhibit little long-range order but have some *short-range order.* This is represented by a *radial distribution function,* which specifies the number of elementary entities found, on the average, at a given distance from a particular point. Information about the radial distribution function is provided by X-ray and neutron diffraction studies.

Liquid structure depends on the nature and strength of the *intermolecular forces.* In ordinary liquids the most important forces are the *dipole-dipole, dipole-*(*induced dipole*), and *dispersion forces.* In liquid water, for example, the dipole-dipole forces are by far the most important, and the dispersion forces are next in importance.

The theories of liquids can be classified as *lattice theories* and *distribution-function theories.* The lattice theories start with a description of the liquid structure and can be subdivided into the following:

1. *Free-volume,* or *cell,* theories, in which molecules are assumed to be present at lattice sites, all of which are occupied.
2. *Hole theories,* in which not all lattice sites are occupied.
3. *Monte-Carlo* and *molecular-dynamical theories,* which treat the average behavior of groups of molecules by numerical computation. The Monte-Carlo methods involve averaging over different *molecular configurations,* whereas the molecular-dynamical theories involve *time averaging.*

The *distribution-function theories,* instead of starting with a description of the liquid, arrive at the structure on the basis of the intermolecular forces.

16 The Liquid State

We have seen in Chapter 1 that gases at high temperatures and low densities behave ideally, which means that they are completely random on the molecular scale. In an ideal gas the interactions between the molecules can be neglected, and the molecules move independently of one another. It is therefore easy to develop a kinetic theory of ideal gases as done in Section 1.9. Moreover, since the energy of an ideal gas is simply the sum of the energies of the individual molecules, the thermodynamic and statistical treatment of the ideal gas presents no particular difficulty. In Sections 2.7 and 3.8 we have developed a number of thermodynamic relationships for ideal gases, and in Section 14.6 we have derived partition functions for ideal gases.

At lower temperatures and higher densities, gases show deviations from ideality, and the theoretical treatment then becomes somewhat more difficult. The intermolecular forces between the molecules now have to be taken into account. Nonideal gases are almost always treated by starting with the ideal-gas relationship and adding corrections to allow for the intermolecular forces. At sufficiently low temperatures and high densities, gases become liquids, and the intermolecular forces are then even more important. One approach to the theoretical treatment of liquids is to start with the ideal gas relationship and to make corrections. However, these corrections may be very large, and the treatment is correspondingly more difficult.

In contrast to the ideal gas, the ideal crystal is a completely orderly arrangement of elementary entities (i.e., atoms, molecules, or ions). Because of this, the theoretical treatment of the ideal crystal is not of great difficulty. The intermolecular forces are now large, but there is no translational energy, and rotational energy is usually unimportant. The elementary entities vibrate about equilibrium positions, and there is no difficulty about treating these motions. In Section 15.12, for example, we have developed a partition function for a perfect crystal. Solids that are not perfect crystals can be treated by starting with the completely ordered arrangement and by making suitable corrections. Liquids can also be dealt with from this point of view, but now the corrections are large and the treatment is usually extremely complicated.

The situation in a liquid is much harder to specify than that in a gas or a solid, and theoretical treatments are much more difficult to carry out. The intermolecular forces in a liquid are strong enough to lead to a condensed state but are not sufficiently strong to prevent translational motion of the molecular entities. Because of the thermal motions there is a certain amount of disorder, but the regularity of the arrangement is not completely destroyed as in a gas.

Crystals can conveniently be classified according to the kinds of forces that hold the atoms, ions, or molecules together. A similar classification is useful for liquids. Table 16.1 shows the main different types of solids and the corresponding types of liquids and indicates some of their more important properties. The intermolecular forces involved in holding liquids together are considered in Section 16.3.

We will start (Section 16.1) by considering liquids as modified dense gases and will then (Section 16.2) approach them from the direction of the solid state.

TABLE 16.1 Characteristics of Different Types of Solids and Liquids

Type of Solid	Properties	Examples	Type of Liquid	Properties	Examples	Constituent Elementary Entity (solid and liquid)	Binding Force
Ionic crystals	High melting points	NaCl ZnO	Ionic liquid	High boiling points	Molten salts	Ions	Electrostatic
Covalent crystals	Very high melting points	Diamond Si SiO_2	*	*	*	Atoms	Covalent bonds
Molecular crystals	Low melting points	Ar CH_4 CO_2	Molecular liquids	Low boiling points	H_2O, C_2H_5OH	Molecules	Van der Waals; dipole-dipole; hydrogen bonds
Metallic crystals	Moderate to high melting points	Li Al Fe	Metallic liquids	Moderate to high boiling points	Molten metals	Positive ions in electron gas	Electrostatic; resonance

*No corresponding liquids; covalent crystals such as diamond usually sublime and no liquid state exists.

16.1 Liquids Considered as Dense Gases

We recall from Chapter 1 that ideal gases obey the relationship $PV = nRT$ and that this equation can be modified in various ways to interpret nonideal behavior. The earliest and most famous modification was that of van der Waals, whose equation for 1 mol of gas is (Eq. 1.82)

$$\left(P + \frac{a}{V_m^2}\right)(V_m - b) = RT \tag{16.1}$$

The term b allows for the volume occupied by the molecules, and a/V_m^2 allows for the intermolecular forces. This equation corresponds to a P-V relationship such as shown in Figures 1.10 and 1.12. The left-hand regions of these curves, where the pressure increases rather strongly as the volume decreases, correspond to the liquid state, where the compressibility is much lower than that of the gas. The van der Waals equation therefore provides an interpretation of the liquid state and can be used to explain many of the interesting properties of liquids.

One of these, which we already noted in Section 1.12, is that liquids can exist in states of *negative* internal pressure. This arises because the S-shaped regions of the curves shown in Figure 1.12 can actually go below the volume axis corresponding to $P = 0$. Experimentally this has been known to be the case for many years. For example, if the space above the mercury in a simple barometer tube is completely filled with a pure liquid, and the tube brought to a vertical position, it is possible for the mercury to stand at a height greater than that corresponding to the barometric pressure. The liquid is therefore under tension, which means that its pressure is negative, and it can remain indefinitely in this condition. If the liquid is agitated, it breaks up into a vapor-liquid mixture, and the mercury level drops. It has been shown that mercury itself can withstand a negative pressure of about 100 atm. The condition for realizing negative pressures in liquids is that no nuclei (e.g., dust) should be present.

Another property of liquids and gases that can also be interpreted in terms of equations such as the van der Waals equation is the *law of the rectilinear diameter*. This we discussed in Section 1.11 (see Figure 1.11).

Internal Pressure

The term a/V^2 which appears in the van der Waals equation is the **internal pressure** P_i of the gas or liquid. We recall from Section 3.8 (Eq. 3.144) that

$$\left(\frac{\partial U}{\partial V}\right)_T = -P + T\left(\frac{\partial P}{\partial T}\right)_V \tag{16.2}$$

The quantity $(\partial U/\partial V)_T$ is the internal pressure, and $T(\partial P/\partial T)_V$ is sometimes known as the **thermal pressure;** thus

$$\text{external pressure } P = \text{thermal pressure } P_t - \text{internal pressure } P_i \tag{16.3}$$

For an ideal gas the right-hand side of Eq. 16.2 is equal to zero, and the internal pressure is therefore zero. For nonideal gases the internal pressure becomes appreciable, while for liquids it is usually much greater than the external pressure.

The internal pressure of a gas or liquid can be calculated from the P-V-T relationship. We showed (Eq. 3.143) that if the van der Waals equation applies,

$$\left(\frac{\partial U}{\partial V}\right)_T = \frac{a}{V_m^2} \tag{16.4}$$

and this equation is conveniently used to obtain a rough estimate of the internal pressure of a gas or a liquid.

EXAMPLE

Making use of the data given in Table 1.3, estimate the internal pressure of water vapor at 20°C and 1 atm pressure and of liquid water at 20°C; the density of water at 20°C is 1.00 g cm^{-3}.

SOLUTION

From Table 1.3, the van der Waals constant a is 0.5532 Pa m^6 mol^{-1}. The molar volume of water vapor at 20°C is

$$V = \frac{8.314 \times 293.15(\text{J mol}^{-1})}{1.013\ 25 \times 10^5(\text{Pa})} = 0.024\ 05\ \text{m}^3\ \text{mol}^{-1}$$

The internal pressure is therefore

$$P_i = \left(\frac{\partial U}{\partial V}\right)_T = \frac{0.5532(\text{Pa m}^6\ \text{mol}^{-2})}{(0.024\,05)^2\,(\text{m}^6\ \text{mol}^{-2})} = 960\ \text{Pa}$$

This is much less than the external pressure (1 atm = 1.013 25 × 10^5 Pa).

The molar mass of water is 18.008 g mol^{-1}, and since the density of liquid water is 1.00 g cm^{-3}, the molar volume is 18.008 cm^3 mol^{-1} = 1.8008 × 10^{-5} m^3 mol^{-1}. The internal pressure is therefore

$$P_i = \left(\frac{\partial U}{\partial V}\right)_T = \frac{0.5532(\text{Pa m}^6\ \text{mol}^{-2})}{(1.8008 \times 10^{-5})^2(\text{m}^6\ \text{mol}^{-2})} = 1.706 \times 10^9\ \text{Pa}$$

This pressure, equal to 16 840 atm, is now very much greater than the external pressure. ■

Water, because of its very strong hydrogen bonds, has an exceptionally high internal pressure. In general, dipolar liquids have larger internal pressures than nonpolar liquids because of the greater attractive forces between the molecules.

Since the internal pressure involves the intermolecular forces, it varies markedly with the external pressure, which affects the average distances between the molecules. At very high external pressures the repulsive forces become predominant, and the internal pressure can have a large negative value.

Internal pressure plays an important role in connection with solubilities, as first realized by Hildebrand.* If two liquids have similar internal pressures, they may obey Raoult's law (Section 5.4) fairly closely. Two liquids differing considerably in internal pressure usually show positive deviations from Raoult's law, which means that their mutual solubility is reduced.

Internal Energy

The internal energy of an ideal gas is entirely kinetic and depends on the temperature and not on the volume; for a monatomic gas, for example, the kinetic energy per mole is $\frac{3}{2}RT$. For a nonideal gas or a liquid, there is, in addition, potential energy due to the intermolecular forces, and this depends on the average separation between the molecules. If the volume changes the intermolecular forces are altered and therefore the potential energy depends on the volume as well as on the temperature.

The internal energy U of a liquid may be expressed as the sum of the kinetic and potential energies,

$$U = U_k + U_p \tag{16.5}$$

We can obtain an expression for U by integrating Eq. 16.2:

$$U = \int \left\{ T\left(\frac{\partial P}{\partial T}\right)_V - P \right\} dV + I(T) \tag{16.6}$$

where $I(T)$ is the constant of integration; by the rules of integration of partial derivatives the constant of integration is a function of temperature, but at a given temperature it is constant and corresponds to the constant of integration for ordinary derivatives. We can determine the value of $I(T)$ by noting that as the volume

*J. H. Hildebrand, *Solubility of Non-Electrolytes*, New York: Reinhold Pub. Co., 1936; J. H. Hildebrand and R. L. Scott, *Regular Solutions*, Englewood Cliffs, N.J.: Prentice-Hall, 1962.

becomes infinite, the fluid will behave as an ideal gas. We can denote the internal energy under these conditions by U_∞, which for a monatomic fluid is equal to $\frac{3}{2}RT$. Equation 16.6 may thus be written as

$$U = \int \left[T\left(\frac{dP}{dT}\right)_V - P \right] dV + U_\infty \tag{16.7}$$

This equation has been known as the **caloric equation of state**. It allows the internal energy to be calculated if P-V-T relationships for the gas are known.

The situation is particularly simple if the system obeys the van der Waals equation. In that case the expression within the integral is equal to a/V^2 (Eqs. 3.144 and 3.147). Equation 16.7 therefore becomes

$$U = \int \frac{a}{V^2}\, dV + U_\infty \tag{16.8}$$

$$= -\frac{a}{V} + U_\infty \tag{16.9}$$

EXAMPLE

Making use of the data in Table 1.3 for argon, estimate the internal energy of the following:

a. Gaseous argon at 273.15 K.
b. Gaseous argon at its normal boiling point, 87.3 K, at which the molar volume is 6.79 $dm^3\ mol^{-1}$.
c. Liquid argon at its boiling point, at which the molar volume is 0.0287 $dm^3\ mol^{-1}$.
d. Then estimate the enthalpy of vaporization of argon at its boiling point.

SOLUTION

a. The kinetic energy U_∞ at 273.15 K is

$$U_k = U_\infty = \tfrac{3}{2}RT = \tfrac{3}{2} \times 8.314 \times 273.15 \text{ J mol}^{-1}$$

$$= 3406.5 \text{ J mol}^{-1}$$

The molar volume at 273.15 K is 22.4 $dm^3\ mol^{-1}$ = 0.0224 $m^3\ mol^{-1}$, and the potential energy is therefore

$$U_p = -\frac{a}{V} = -\frac{0.2351 \text{ (Pa m}^6 \text{ mol}^{-2})}{0.0224 \text{ (m}^3 \text{ mol}^{-1})}$$

$$= -10.5 \text{ J mol}^{-1} \text{ (Pa m}^3 \equiv \text{kg m}^2\text{s}^{-2} \equiv \text{J)}$$

The total internal energy is thus

$$U = U_k + U_p = 3406.5 - 10.5 = 3396 \text{ J mol}^{-1}$$

b. For the gas at 87.3 K,

$$U_k = \tfrac{3}{2} \times 8.314 \times 87.3 = 1088.7 \text{ J mol}^{-1}$$

$$U_p = -\frac{0.2351}{0.006\,79} = -34.6 \text{ J mol}^{-1}$$

$$U = 1054 \text{ J mol}^{-1}$$

c. For the liquid at 87.3 K,

$$U_k = 1088.7 \text{ J mol}^{-1}$$

$$U_p = -\frac{0.2351}{2.87 \times 10^{-5}} = -8191.6 \text{ J mol}^{-1}$$

$$U = -7102 \text{ J mol}^{-1}$$

d. $\Delta_{vap}U = 1054 - (-7102) = 8156 \text{ J mol}^{-1}$

$$\Delta_{vap}H = \Delta_{vap}U + P(V_{vap} - V_{liq})$$

With $P = 1 \text{ atm} = 1.013\,25 \times 10^5$ Pa,

$$\Delta_{vap}H = 8156 + 1013\,25 \times 10^5\,(0.006\,79 - 0.000\,028\,7)$$

$$= 8156 + 685 = 8841 \text{ J mol}^{-1}$$

■

Different estimates of these quantities will, of course, be obtained if different equations of state are employed. The van der Waals equation gives a useful idea of magnitudes, but considerable improvement can be achieved by the use of some of the other equations.

If the liquid is not monatomic, the kinetic energy is greater than $\frac{3}{2}RT$ because of the contributions from vibrational and rotational energy. Usually, because of quantum effects, the vibrational contributions are unimportant. At ordinary temperatures there is usually a contribution of RT for a linear molecule and of $\frac{3}{2}RT$ for a nonlinear molecule.

EXAMPLE

Estimate the internal energy of liquid water at 20°C, assuming the translational and the rotational energies to be $\frac{3}{2}RT$ each and making use of the van der Waals constants given in Table 1.3; molar volume of liquid water at 20°C = 18.008 cm^3 mol^{-1}.

SOLUTION

$$U_k = \tfrac{3}{2}RT + \tfrac{3}{2}RT$$

$$= 3 \times 8.314 \times 293.15 = 7437 \text{ J mol}^{-1}$$

$$U_p = -\frac{a}{V} = -\frac{0.5532 \text{ (Pa m}^6\text{ mol}^{-2})}{1.8008 \times 10^{-5}\text{ (m}^3\text{ mol}^{-1})}$$

$$= -30\,720 \text{ J mol}^{-1}$$

The total internal energy is therefore

$$U = U_k + U_p = 7437 - 30\,720 = -23\,280 \text{ J mol}^{-1}$$

■

16.2 Liquids Considered as Disordered Solids

We must now see how the elementary entities in a liquid are arranged in space and how this arrangement affects the properties of the liquid. Properties such as potential energy, which depend on the arrangement of entities, are known as **configurational** properties.

The distinction between solids and liquids arises from the different configurations. At the absolute zero, according to classical physics, the elementary entities (atoms, molecules, or ions) in a perfect crystal are at rest in a regular space lattice, as we have seen in Chapter 15. As the temperature is raised, the entities vibrate about the rest positions, but these positions still remain in a regular geometrical arrangement; the *long-range order* is still maintained. As the temperature is raised still further, at a certain temperature the vibrational motions of the molecules have become so violent that the ordered lattice structure breaks down abruptly. This is the phenomenon of *melting*. At any temperature there is competition between the intermolecular attractions, which tend to produce the orderly arrangement that exists in the perfect crystal, and the kinetic energy, which tends to destroy this arrangement and produce the liquid (and eventually the gaseous) state.

Melting occurs very abruptly at a particular temperature; there is no continuous gradation of properties between crystal and liquid. It is possible for there to be a very limited amount of *short-range disorder* in a crystal. However, quite rigorous geometrical arrangements must be satisfied in a crystal structure, and if too much short-range disorder is introduced, there is a serious disruption of the long-range structure and melting will suddenly occur.* If we focus attention on a particular entity in a liquid, its immediate neighbors will not be grouped around it in a completely random fashion but will be in an arrangement that shows some resemblance to the structure in the solid. The next-nearest neighbors show a much smaller degree of order with respect to the entity in question, and at still greater distances there will be no correlation at all between the positions of the entities and that of the entity in question. Liquids therefore exhibit some short-range order but no long-range order. By contrast, solids exhibit both short-range and long-range order, whereas ideal gases exhibit no order at all. The fluidity of liquids arises as a result of the absence of long-range order.

The distance over which the short-range order exists in a liquid is usually a few atomic or molecular diameters. The distance decreases as the temperature is raised.

Radial Distribution Functions

Short-range order is conveniently represented by means of a **radial distribution function,** which specifies the number of elementary entities to be found, on the average, at a distance r from a particular entity.

If we consider first a perfect crystal at the absolute zero, the radial distribution function is determined precisely by the geometry of the lattice. Sodium, for example, crystallizes in a body-centered cubic lattice, as shown in Figure 16.1a, in which the dimensions are shown. It is to be seen that there are 8 atoms at a distance 367 pm from the central atom, 6 atoms at a distance 424 pm, 12 atoms at a distance 600 pm, and so on. These numbers N_r are represented as a function of distance in Figure 16.1b. This diagram, consisting of vertical lines, shows the radial distribution function in the solid at the absolute zero.

At higher temperatures the atoms are not at fixed positions, because of the vibrations. We must now concern ourselves with the average number $g(r)\,dr$ of

---

* This was clearly demonstrated, by the use of models, by J. D. Bernal, *Scientific American, 203,* 124 (1960); reprinted in *Readings in the Physical Sciences* (Vol. 2), San Francisco: W. H. Freeman, 1969.

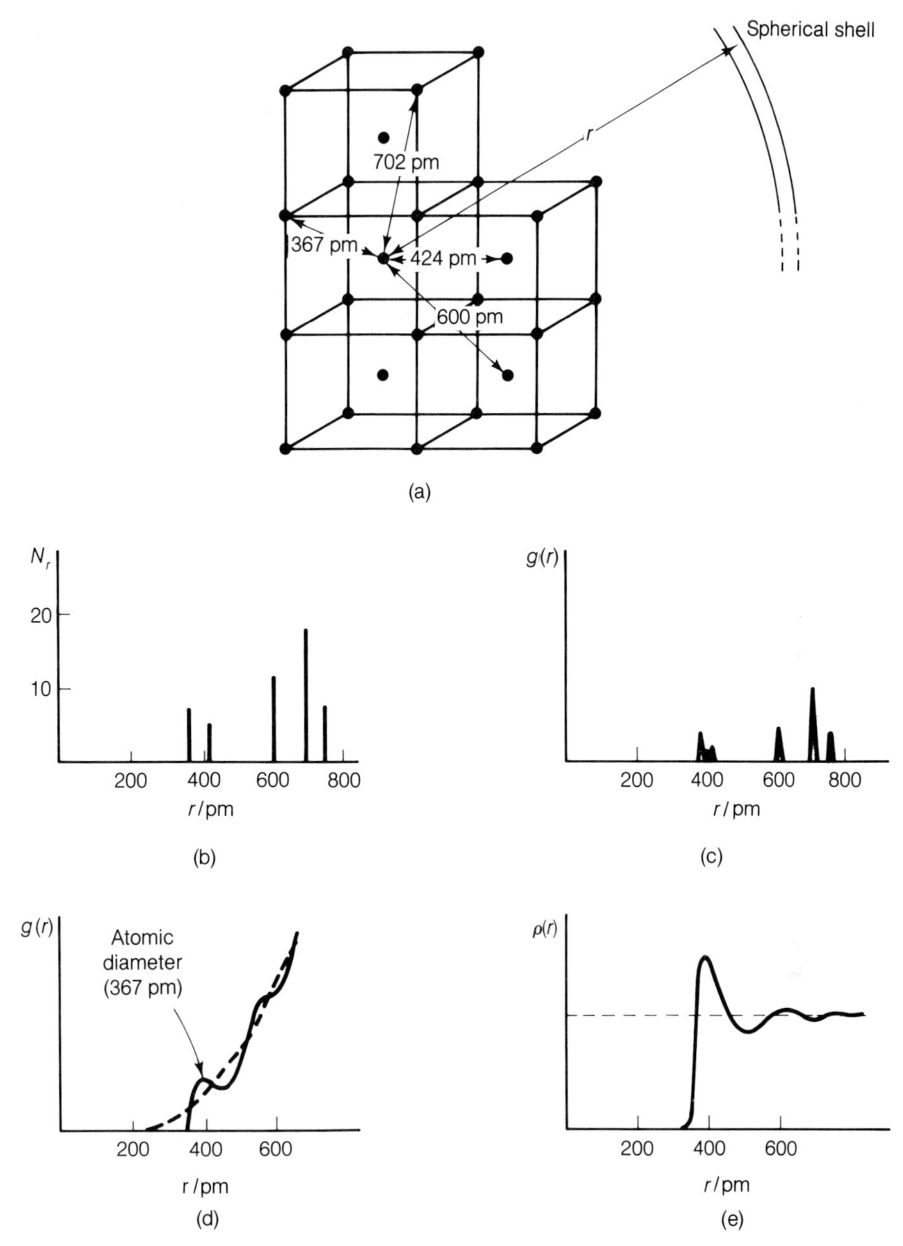

FIGURE 16.1 (a) A body-centered cubic lattice, showing the interatomic distances that apply to solid sodium. (b) The number N_r of sodium atoms at various distances from a particular atom in the solid at the absolute zero. (c) A radial distribution diagram for solid sodium at a higher temperature. (d) Radial distribution diagram for liquid sodium. (e) The number density $\rho(r)$ of sodium atoms plotted against distance r for liquid sodium.

atoms that lie within a spherical shell of thickness dr and at a distance r from the rest position of the central atom (Figure 16.1a). At the absolute zero, $g(r)$ is equal to N_r, but at higher temperatures the vertical lines are broadened into curves (Figure 16.1c) that increase in width as the temperature rises. The area under any of the peaks corresponds to the number of atoms on the corresponding spherical shell for the lattice at the absolute zero. We see from Figure 16.1c that the long-range order is still preserved in the solid.

The radial distribution function for the liquid is shown in Figure 16.1d. If there were complete disorder, the number of atoms whose centers lie between r and $r + dr$ from the central atom would be proportional to the volume of the spherical shell ($4\pi r^2\, dr$) and therefore to r^2. The dashed curve in Figure 16.1d, a parabola, shows such an r^2 dependence. The full curve gives the actual distribution in the liquid. At larger values of r the actual distribution curve is close to the curve for r^2, which means that there is no long-range order. At smaller r values, however, the curve deviates from the r^2 curve, showing that there is some short-range order. There is an initial maximum at about 400 pm, corresponding to a shell of nearest neighbors, and an inflexion at 500–600 pm corresponding to a second group of next-near neighbors. At greater distances the order is completely lost.

An alternative way of representing the situation is to plot the *number density* of the atoms for any distance r. This quantity, $\rho(r)$, is the number of atoms per unit volume and is the ratio of the number in a spherical shell to the volume of the shell:

$$\rho(r) = \frac{g(r)\, dr}{4\pi r^2\, dr} = \frac{g(r)}{4\pi r^2} \tag{16.10}$$

This function has the form shown in Figure 16.1e. At large distances, $\rho(r)$ is equal to the mean number density for the liquid as a whole, ρ_0. At short distances there are maxima and minima, reflecting the short-range order.

The previous considerations apply strictly only to spherical entities, in which the direction in space has no effect. For nonspherical molecules, such as benzene, there will be a tendency for them to become aligned in a particular orientation, and the distribution function will depend on angles as well as on r. The distribution functions obtained by X-ray and neutron diffraction, to be considered in the next subsection, are averaged over all angles.

X-Ray Diffraction

In Section 15.4 we have discussed the X-ray diffraction studies on solids. Shortly after von Laue's pioneering work on solids, several workers made similar studies on liquids. It was found that the diffraction pattern obtained from a liquid is not nearly so sharp as that from a crystal. Instead of the pattern of closely spaced spots forming a ring, found with a solid (Figure 15.16), a liquid only produces one or two diffuse rings surrounding an intense central spot. This lack of sharpness is due to the lack of long-range order, but the existence of the rings shows that there is some short-range order.

An X-ray diffraction experiment gives the intensity I_s of the diffracted rays as a function of the angle of diffraction. A typical example is shown in Figure 16.2a, for liquid mercury. The problem is then to derive the radial distribution function, and this is done by a procedure worked out by P. Zernicke and J. A. Prins,* involving the use of a Fourier inversion. Figure 16.2b shows radial distribution curves for liquid argon at two temperatures, obtained in this way by A. Eisenstein and N. S. Gingrich.† These curves show the number density $\rho(r)$ divided by ρ_0, the

*P. Zernicke and J. A. Prins, *Z. Physik*, *41*, 184 (1927).

†A. Eisenstein and N. S. Gingrich, *Phys. Rev.*, *58*, 307 (1940); *62*, 261 (1942).

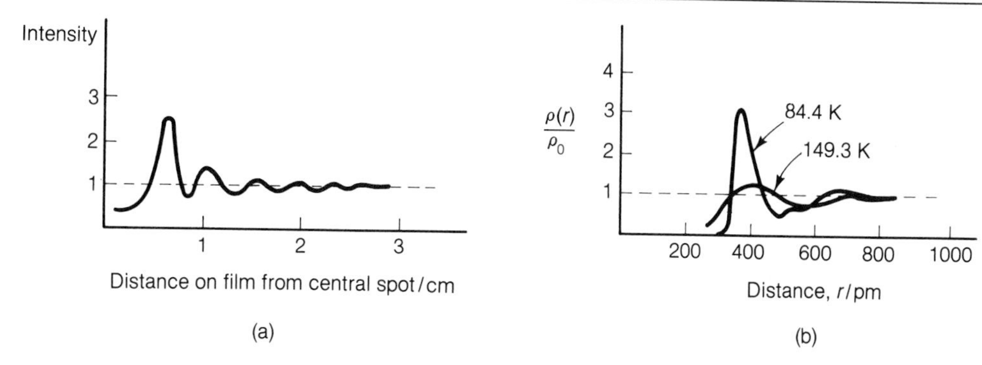

FIGURE 16.2 (a) A photometric tracing of an X-ray diffraction photograph, obtained with liquid mercury. (b) Radial distribution functions for liquid argon at 84.4 K and 149.3K.

value at large distances. At the lower temperature, 84.4 K, there is a well-defined maximum at about 340 pm, which is approximately the diameter of the argon atom. At twice this distance the order has essentially disappeared. At this temperature the number density at 340 pm is slightly more than 3 times the value for the liquid as a whole. At the higher temperature, 149.3 K, the short-range order is much less; the number density is now somewhat less than 1.5 times the value for the bulk of the liquid.

The first detailed X-ray diffraction study of liquid water was made in 1938 by J. Morgan and B. E. Warren*; and a more detailed study, covering a range of temperature, was made in 1966 by A. H. Narten, M. D. Danford, and H. A. Levy.† The radial distribution curves at 4°C and at 100°C are shown in Figure 16.3a. The function $\rho(r)$ is zero up to about 250 pm, which means that 250 pm is the effective molecular diameter. Above 700 pm at 4°C and a somewhat greater distance at 100°C, the function is essentially unity, which means that the local order does not extend beyond this distance. At both temperatures there is a well-defined peak at 290 nm. When $\rho(r)$ is integrated over the volume element $4\pi r^2\, dr$ in this shell, the number of nearest neighbors is calculated to be 4.4 at both temperatures; this value in fact applies from 4°C to 200°C. The $\rho(r)$ function also shows peaks at 350 pm, 450 pm, and 700 pm.

The structure of ordinary ice (ice-I) has been determined by X-ray diffraction and is shown in Figure 16.3b. Each H_2O molecule is surrounded tetrahedrally by four H_2O molecules, and the O—O distance is 276 pm. The strong peak at 290 pm for liquid water, with a coordination number of 4.4, therefore suggests a very similar tetrahedral arrangement in the liquid. The peak obtained at 350 pm with the liquid does not correspond to any distance in the solid. However, as indicated in Figure 16.3b, ice-I has interstitial sites at a distance of 350 pm from each oxygen atom. It therefore appears that when ice melts, some of the water molecules move from their tetrahedral sites into these interstitial sites, thus giving rise to the peak at 350 pm. There is a contraction in volume of about 9% when ice melts, and this is explained by a partial collapse of the tetrahedral structure, with a movement of some of the molecules into the interstitial sites.

*J. Morgan and B. E. Warren, *J. Chem. Phys.*, *6*, 666 (1938).
†A. H. Narten, M. D. Danford, and H. A. Levy, Oak Ridge National Laboratory Report ORNL-3997 (1966); *Discuss. Faraday Soc.*, *43*, 97 (1967).

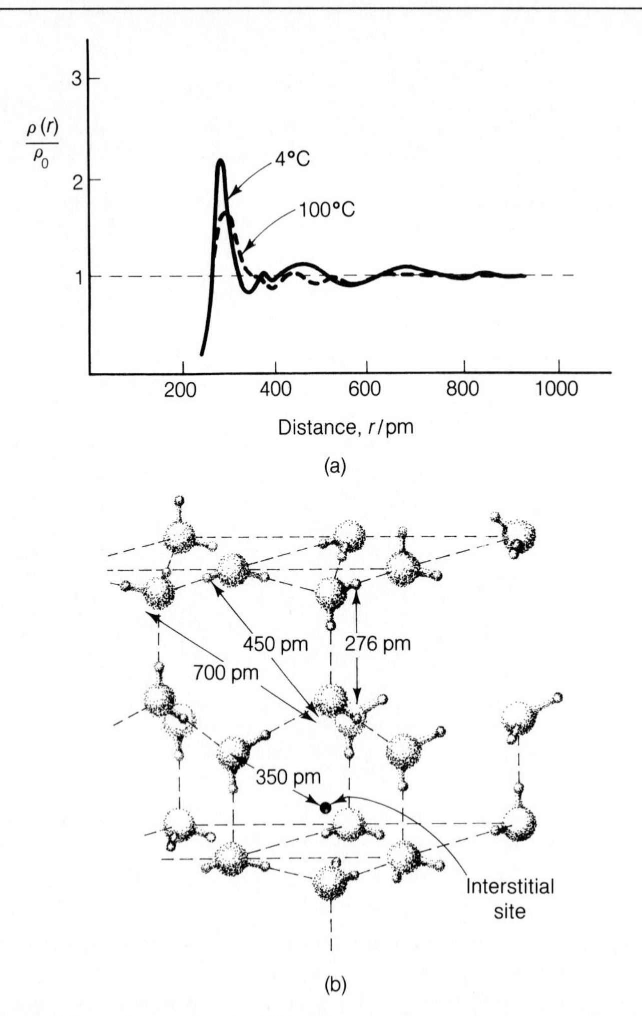

FIGURE 16.3 (a) Experimental radial distribution curves for water at 4°C and at 100°C. (b) The arrangement of water molecules in the crystal of ordinary ice (ice-I), showing some of the distances.

The peaks in the $\rho(r)$ function for water at ~450 pm and ~700 pm are consistent with the tetrahedral arrangement, as indicated in Fig. 16.3b.

Neutron Diffraction

Radial distribution functions for liquids may also be obtained by the study of the diffraction of neutrons. Like other particles, neutrons show wave properties and have a wavelength λ given by de Broglie's relationship $\lambda = h/mu$. The usual procedure is first to diffract a beam of neutrons by a single crystal of lead or copper so as to produce a monochromatic beam of wavelength about 0.11 nm. This beam then falls on the liquid, and scattered intensities are studied at various positions. Complications arise if the liquid contains an element with more than one isotope, since neutrons are scattered by the nuclei. An advantage of the neutron-scattering technique is that neutrons are scattered by light atoms such as H and He, for which X-ray scattering is negligible.

Glasses

Some materials occur in the *glassy* or *vitreous* state, and they present an interesting problem since they show properties of both solids and liquids. At sufficiently high temperatures they flow like liquids, but when cooled they do not solidify at a fixed temperature; instead their viscosity increases steadily until finally their physical properties are essentially those of solids. Glasses can usefully be described as *supercooled liquids.* Most liquids can be supercooled, but supercooled liquids are generally unstable and will solidify suddenly if they are further cooled or are agitated; glasses, however, can only be induced to crystallize with difficulty, if at all.

X-ray diffraction experiments on glasses have shown that they give the diffuse-ring pattern that is characteristic of liquids and that even at low temperatures they do not show the type of pattern found with solids. It is therefore concluded that, although having the outward appearance of solids, glasses are disordered on the atomic scale. This behavior can be understood from the standpoint of regarding a glass as a supercooled liquid. The viscosity of all liquids increases with decrease of temperature (see Section 18.1), so that if a highly viscous liquid is cooled sufficiently rapidly, atomic rearrangements are brought essentially to a standstill, and the disordered configuration of the liquid is retained. At sufficiently low temperatures the viscosity is so high that the material has the appearance of a solid in spite of having the atomic structure of a liquid.

16.3 Intermolecular Forces

When molecules are near enough to influence one another, forces of attraction and repulsion come into play. If there were no forces of attraction, all matter would be gaseous, since there would be nothing to bring the molecules together in the solid and liquid states. The behavior of matter in condensed phases is determined by the balance between the forces of attraction and repulsion.

The force F between two molecules is related to the potential energy E_p by the equation

$$F = -\frac{dE_p}{dr} \qquad (16.11)$$

where r is the distance between them. Potential energy is a relative quantity, and it is usual to take it to be zero when the molecules are separated by an infinite distance. It is usually more convenient to deal with energy rather than force, and Figure 16.4a shows the typical variation with distance r of the potential energy for two interacting molecules; Figure 16.4b shows the corresponding variation of the force.

There are a number of types of intermolecular forces, and these are listed in Table 16.2, which indicates how the force and the energy vary with distance and gives some typical values. Some of these forces will now be treated in a little more detail.

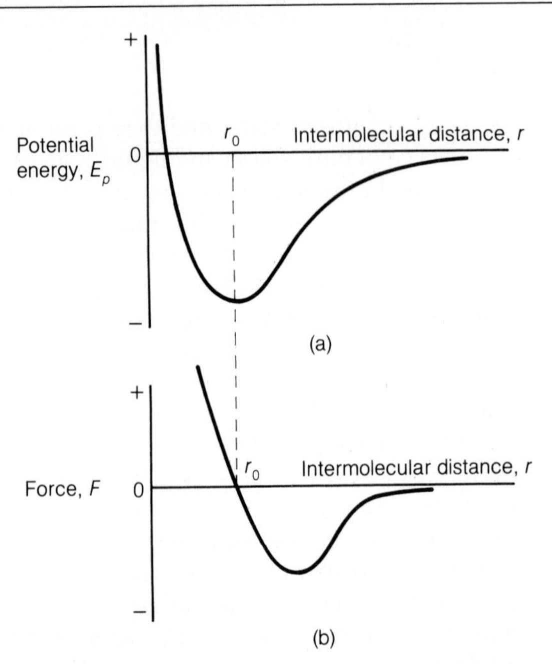

FIGURE 16.4 (a) The variation of potential energy with distance for two interacting molecules. (b) The corresponding variation of force.

Ion-Ion Forces

If two charges $z_A e$ and $z_B e$ are separated by a distance r, the force of attraction or repulsion obeys the inverse-square law, and the force and potential energy obey the equations

$$F = -\frac{z_A z_B e^2}{4\pi\epsilon_0 \epsilon r^2} \quad \text{and} \quad E_p = \frac{z_A z_B e^2}{4\pi\epsilon_0 \epsilon r} \tag{16.12}$$

TABLE 16.2 Types of Intermolecular Forces

Force	Dependence on Distance: Force	Dependence on Distance: Energy	Typical Potential-Energy Values*: E_p per Molecule (J)	Typical Potential-Energy Values*: E_p per Mole (kJ mol^{-1})
Ion-ion	r^{-2}	r^{-1}	-1.16×10^{-18}	-680
Ion-dipole	r^{-3}	r^{-2}	-1.20×10^{-19}	-72
Ion-(induced dipole)	r^{-5}	r^{-4}	-1.08×10^{-19}	-65
Dipole-dipole	r^{-7}	r^{-6}	-2.54×10^{-20}	-15.3
Dipole-(induced dipole)	r^{-7}	r^{-6}	-1.56×10^{-21}	-0.9
Dispersion	r^{-7}	r^{-6}	-7.34×10^{-20}	-44.2
Repulsion	r^{-13}	r^{-12}	—	—

*Calculated for the following values: unit (electronic) charge (1.602×10^{-19} C); dipole moment of 1 D (3.338×10^{-30} C m); polarizability of 1.5×10^{-30} m^3; intermolecular distance 2.00×10^{-10} m. The dispersion energy values are for water molecules (Table 16.3).

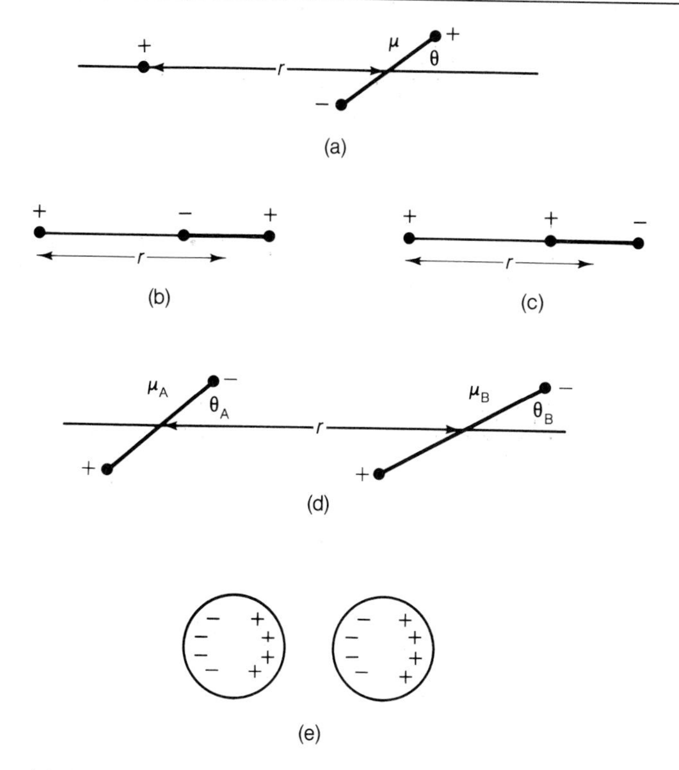

FIGURE 16.5 (a) An ion and a dipolar molecule. (b) The ion-dipole orientation giving maximum attraction. (c) The arrangement giving maximum repulsion. (d) Two permanent dipoles in the same plane. (e) Two nonpolar molecules inducing moments in one another.

Here ϵ_0 is the permittivity of a vacuum and ϵ is the relative permittivity or dielectric constant. For two ions in a vacuum, $\epsilon = 1$, but (as we have seen in the introduction to Chapter 7) if the ions are in solution, an appropriate value for ϵ must be used. The energies in this section are calculated for a vacuum.

Ion-Dipole Forces

The concept of dipole moment has been discussed in Section 12.3. Figure 16.5a shows a dipolar molecule with an ion at a distance r from its center of charge. The force between them depends on the angle θ; the arrangement in Figure 16.5b gives maximum attraction, and the arrangement in Figure 16.5c gives the maximum repulsion. In general it can be shown that, provided r is much larger than the length of the dipole, the force and potential energy are given by

$$F = \frac{z_A e \mu \cos \theta}{4\pi\epsilon_0 \epsilon r^3} \qquad E_p = -\frac{z_A e \mu \cos \theta}{4\pi\epsilon_0 \epsilon r^2} \tag{16.13}$$

where μ is the dipole moment.

EXAMPLE

A unit positive charge is situated at a distance of 200 pm from a molecule of dipole moment 3.338×10^{-30} C m (1.00 D). Calculate the maximum force and the corresponding potential energy.

SOLUTION
The maximum force is

$$F = \frac{e\mu}{4\pi\epsilon_0 r^3} = \frac{1.602 \times 10^{-19}(\text{C})\ 3.338 \times 10^{-30}(\text{C m})}{4\pi(8.854 \times 10^{-12})(\text{C}^2\ \text{N}^{-1}\ \text{m}^{-2})\ 2.00 \times 10^{-10}(\text{m})^3}$$

$$= 6.01 \times 10^{-10}\ \text{N}$$

The corresponding potential energy, relative to the separated species, is

$$E_p = -\frac{e\mu}{4\pi\epsilon_0 r^2} = -1.20 \times 10^{-19}\ \text{J} = -72.3\ \text{kJ mol}^{-1}$$

■

An atom, or a molecule that has no permanent dipole moment, can have a dipole moment induced in it by an electric field. If the field strength is E, the induced dipole moment is αE, where α is the electric polarizability of the molecule. The dipole is formed in the direction of the field, and there is therefore always attraction between an ion and an induced dipole. If an ion of charge $z_A e$ is at a distance r from a molecule of polarizability α, the potential energy is

$$E_p = -\frac{\alpha(z_A e)^2}{8\pi\epsilon_0 \epsilon r^4} \tag{16.14}$$

EXAMPLE
A unit charge is at a distance of 200 pm from a molecule having no dipole moment but a polarizability of $1.5 \times 10^{-30}\ \text{m}^3$. Calculate the potential energy.

SOLUTION
From Eq. 16.14,

$$E_p = -\frac{\alpha e^2}{8\pi\epsilon_0 r^4} = -\frac{1.5 \times 10^{-30}(\text{m}^3)[1.602 \times 10^{-19}(\text{C})]^2}{8\pi(8.854 \times 10^{-12})(\text{C}^2\ \text{N}^{-1}\ \text{m}^{-2})[2.00 \times 10^{-10}(\text{m})]^4}$$

$$= -1.08 \times 10^{-19}\ \text{J} = -65.0\ \text{kJ mol}^{-1}$$

Note that this is not much less than the energy calculated in the previous example for the ion-(permanent dipole) force.

■

Dipole-Dipole Forces

Figure 16.5d shows two permanent dipoles in the same plane. If the distance between them is much greater than the length of the dipoles, the potential energy is given by

$$E_p = -\frac{\mu_A \mu_B}{4\pi\epsilon_0 \epsilon r^3}(2 \cos \theta_A \cos \theta_B - \sin \theta_A \sin \theta_B) \tag{16.15}$$

When the dipoles are aligned in the same direction, the force of attraction is a maximum and the potential energy is

$$E_p = -\frac{\mu_A \mu_B}{2\pi\epsilon_0 \epsilon r^3} \tag{16.16}$$

In a gas or a liquid the potential energy will be an average over the various orientations of the molecules. Orientations giving rise to a lower potential energy are favored over those giving a higher potential energy, in accordance with the Boltzmann distribution. When this is taken into account, with an averaging over all orientations, the potential energy is found to be

$$E_p = -\frac{\mu_A^2 \mu_B^2}{24\pi^2 \epsilon_0^2 \epsilon^2 \mathbf{k} T r^6} \tag{16.17}$$

where **k** is the Boltzmann constant and T is the temperature. Note that the dependence is now on the inverse sixth power of the distance.

EXAMPLE

Calculate the potential energy at 25°C for two molecules of dipole moments 3.338×10^{-30} C m separated by 2.00×10^{-10} m.

SOLUTION

From Eq. 16.17,

$$E_p = -\frac{[3.338 \times 10^{-30}(\text{C m})]^4}{24\pi^2[8.854 \times 10^{-12}(\text{C}^2\ \text{N}^{-1}\ \text{m}^{-2})]^2 \quad 1.381 \times 10^{-23} \times 298.15\ (\text{J})[(2 \times 10^{-10}(\text{m})]^6}$$

$$= -2.54 \times 10^{-20}\ \text{J} = -15.3\ \text{kJ mol}^{-1}$$

■

A permanent dipole can induce a dipole in a neighboring molecule. If two molecules both have a dipole moment μ and a polarizability α, each will induce a dipole moment in the other, and the resulting potential energy is

$$E_p = -\frac{\alpha \mu^2}{2\pi\epsilon_0 \epsilon r^6} \tag{16.18}$$

If only one of the molecules has a permanent dipole moment, the potential energy is just half this value.

Dispersion Forces

The forces considered so far occur only if the molecules are charged or have permanent dipole moments. However, there must be attractive forces even between neutral molecules having zero dipole moments, since liquefaction always occurs at sufficiently low temperatures. The forces in such cases are known as **dispersion forces,** and the theory of them was worked out in 1930 on the basis of quantum mechanics by F. London.

The origin of the dispersion forces is as follows. Suppose that two completely nonpolar atoms or molecules are very close together, as shown in Figure 16.5e. On the average the electron clouds are arranged symmetrically, but at any given instant the electron distribution in one of the molecules may be unsymmetrical, as shown in the figure. The molecule is therefore momentarily a dipole, and it will induce a dipole in the neighboring molecule. Both molecules are hence dipoles at this instant, and the direction of the dipoles is such that they attract one another. Since the electron clouds are in rapid motion, there is a rapid fluctuation of dipoles, but at every instant the dipole in each molecule is inducing one in the other, and there is attraction. The effect is greatest if the molecules are of high polarizability.

London's quantum-mechanical treatment of the problem* led to the result that for two identical molecules of polarizability α the potential energy is

$$E_p = -\frac{3h\nu_0\alpha^2}{4r^6} \tag{16.19}$$

where ν_0 is the frequency of oscillation of the dipoles and h is the Planck constant.

Repulsive Forces

If two atoms or molecules are brought very close together, the electron clouds will interpenetrate and will no longer be able to shield the nuclei. There is then a repulsive force, and the energy rises as the intermolecular distance is reduced.

Theoretical treatments of this problem have led to the result that the potential energy due to this repulsive effect is best expressed by

$$E_p = \frac{B}{r^n} \tag{16.20}$$

where B and n are constants. For many molecules the best value for n is 12, and the value of B is usually determined by use of the experimental intermolecular distance.

Resultant Intermolecular Energies

Table 16.3 gives some idea of the magnitudes of the different kinds of attractive energies. The values are calculated from data given by F. London* and relate to an intermolecular separation of 500 pm. It is to be seen that the dipole-(induced dipole) forces are always quite small. For He and Xe, where there are no dipole forces, the only contribution is from the dispersion forces, which give a substantial attraction for Xe because of its very high polarizability. Carbon monoxide has only a very small dipole moment and the only significant force is the dispersion force. For the

TABLE 16.3 Attractive Energy Contributions for an Intermolecular Separation of 500 pm

			E_p/J mol^{-1}, due to		
Molecule	Dipole moment $\mu/10^{-30}$ C m	Polarizability $\alpha/10^{-30}$ m^3	Dipole-Dipole Forces	Dipole-(Induced Dipole) Forces	Dispersion Forces
He	0	0.20	0	0	−4.6
Xe	0	4.00	0	0	−850.0
CO	0.40	1.99	−0.012	−0.22	−260.0
HCl	3.43	2.63	−72.0	−2.0	−405.0
NH_3	5.01	2.21	−324.0	−39.0	−360.0
H_2O	6.14	1.48	−732.0	−39.0	−180.0

*F. London, *Z. Phys. Chem.* (B), *11*, 222 (1930); *Trans. Faraday Soc.*, *33*, 8 (1937).

molecules with larger dipole moments the dipole forces are more important, and for water the dipole-dipole force makes the largest contribution to the attraction.

For uncharged molecules the attractions all depend on r^{-6}, and if the repulsive contribution depends on r^{-12}, the net energy is

$$E_p = -\frac{A}{r^6} + \frac{B}{r^{12}} \tag{16.21}$$

where A and B are constants. This function is known as the **Lennard-Jones 6–12 function,** after the British theoretical physicist John Lennard-Jones (1894–1954),* who first suggested it. It corresponds to a curve of the form shown in Figure 16.4a, and it has been widely used in the study of interactions between uncharged molecules. If r_e is the equilibrium intermolecular separation and E_0 is the classical dissociation energy, $dE_p/dr = 0$ at $r = r_0$ and $E_p = -E_0$ at $r = r_e$; from these boundary conditions we find that

$$E_p = E_0\left\{\left(\frac{r_0}{r}\right)^{12} - 2\left(\frac{r_0}{r}\right)^{6}\right\} \tag{16.22}$$

This is often a more convenient form of the equation. Because of the theoretical difficulties of calculating the constants A and B, it is usual to make use of experimental values of E_0 and r_0.

16.4 Lattice Theories of Liquids

Theories of the liquid state fall into two main classes, which are conveniently referred to as *lattice* theories and *distribution function* theories. Lattice theories start with the assumption of a simplified liquid structure and, on that basis, develop expressions for the properties of liquid; the question of why the liquid assumes its particular structure is dealt with at a later stage. Such theories are closely related to theories of the solid state, which we have considered in Chapter 15.

The distribution-function theories begin with a consideration of the intermolecular forces between molecules and deal with how these forces lead to the particular structure that the liquid assumes. The equations obtained in theories of this class involve distribution functions that specify the probability that groups of molecules in the liquid are in particular configurations. These theories bear some resemblance to those dealing with gases.

The various theories of liquids represent alternative approaches to the problem, and the more they are developed, the closer do the different theories come together. Lattice theories are discussed in the present section; distribution-function theories are considered in Section 16.5. The lattice theories can be further subdivided as follows:

1. *Free-volume*, or *cell*, theories, in which the molecules are assumed to be present at lattice sites all of which are occupied. Whereas in the solid the atoms, ions, or molecules simply vibrate about their lattice positions, in the liquid they have more freedom and move within a "cell" created by the surrounding molecules.

---

*J. E. Lennard-Jones, *Proc. Roy. Soc., 106A*, 463 (1924).

2. *Hole* theories, which resemble the free-volume theories, but in which some of the lattice sites are considered to be unoccupied.

3. *Monte Carlo* and *molecular-dynamical* theories, which deal with groups of molecules and which treat, by direct numerical computation, the average behavior of the molecules within these cells.

Free-Volume Theories

A number of theoretical treatments have been based on the idea that each molecule in a liquid is confined by its neighbors to a small region of space. The local environment of each molecule is therefore much the same as it is in the solid, although there is no long-range order.

We shall here consider only the very simplest type of free-volume or cell theory. It will be assumed that all the "cages" in which the molecules are confined are identical and that the molecules move entirely independently of one another in their cages. The basic ideas in the treatment we will give are similar to those in the theories proposed in 1937 by Lennard-Jones and Devonshire*, and independently by Eyring and Hirschfelder.†

Suppose, for simplicity, that the molecules in the liquid are arranged in a simple cubic lattice. The molecules are assumed to be incompressible spheres of diameter d, and the distance between lattice positions is a (Figure 16.6a). If we consider motion along the X axis, we see that the distance in which the central molecule is free to move is $2a - 2d$. It can move this distance in three dimensions, so that the volume of the cube in which it can move (i.e. the free volume) is

$$v_f = (2a - 2d)^3 = 8(a - d)^3 \tag{16.23}$$

Alternatively, we might suggest that the molecule can move in a sphere of radius $a - d$, in which case the estimated free volume is

$$v_f = \frac{4\pi}{3}(a - d)^3 \tag{16.24}$$

which is somewhat smaller. The true volume probably lies somewhere in between. The value of d can be estimated from the known structure of the molecule, from the van der Waals b, or from viscosity measurements (see Table 18.1). The value of a can be estimated from the molar volume of the liquid; thus, for a simple cubic lattice, a^3 is simply the volume per molecule, or V_m/L, where V_m is the molar volume. For other lattices, a^3 is related to V_m/L by a numerical factor.

It is also possible to estimate free volumes of liquids on the basis of velocity of sound measurements. For most liquids the velocity of sound, u_l, is 5 to 10 times the velocity of sound, u_g, in the gas. The reason for this is illustrated in Figure 16.6a, which shows three molecules A, B, and C in a straight line. If a sound wave travels from left to right, molecule A collides with B and the signal is transmitted almost instantaneously to the opposite side of B. Thus, although molecule A travels only

*J. E. Lennard-Jones and A. F. Devonshire, *Proc. Roy. Soc., A163*, 53 (1937).
†H. Eyring and J. O. Hirschfelder, *J. Phys. Chem., 41*, 249 (1937).

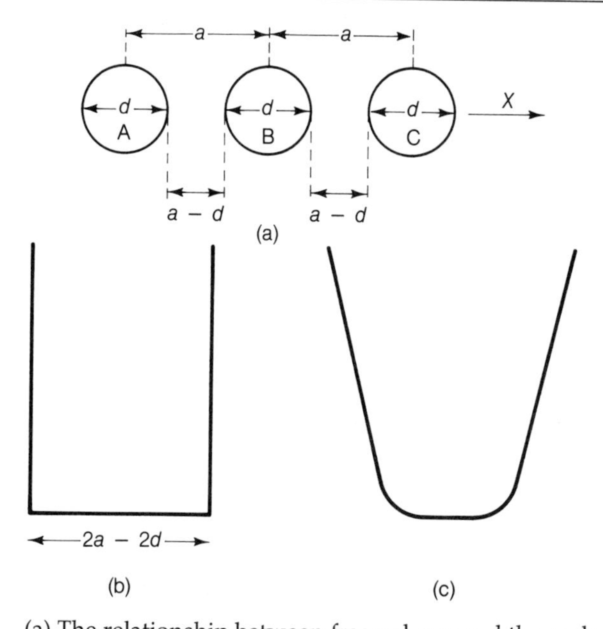

FIGURE 16.6 (a) The relationship between free volume and the molecular diameters and lattice distances. (b) The square-well potential-energy function assumed in the simplest application of free-volume theories of liquids; the molecule has free translational motion. (c) The type of potential-energy variation in the treatments of Lennard-Jones and Devonshire.

the distance $a - d$ in order to strike B, the sound wave has effectively traveled the distance a. The ratio of sound velocities is thus

$$\frac{u_l}{u_g} = \frac{a}{a - d} \tag{16.25}$$

and $a - d$, and hence the free volume, can therefore be estimated from the values of u_l, u_g, and a.

We will first consider the construction of a partition function for the liquid on the basis of the assumption that each molecule undergoes free translational motion in a free volume v_f (Figure 16.6b). The translational partition function per molecule (Eq. 14.103) is then

$$q_t = \frac{(2\pi m\mathbf{k}T)^{3/2}}{h^3} v_f \tag{16.26}$$

This must be multiplied by an internal partition function q_i that takes account of the rotational and vibrational motions of the molecules. In addition, we must multiply by the factor $e^{-\epsilon/\mathbf{k}T}$, where ϵ is the energy of the molecule inside its cage, with respect to the molecule in the gas phase. The complete partition function per molecule is therefore

$$q = \frac{(2\pi m\mathbf{k}T)^{3/2}}{h^3} v_f q_i e^{-\epsilon/\mathbf{k}T} \tag{16.27}$$

There are alternative procedures for expressing the partition function for an assembly of N molecules in the liquid phase. We have so far developed the argument on the assumption that the molecules are *localized*, being confined to cages at the various lattice sites. In reality, this is more like the behavior in a solid, but if we keep to this assumption, the partition function for the assembly is simply q raised to the Nth power (Eq. 14.62):

$$Q = q^N = \left[\frac{(2\pi m\mathbf{k}T)^{3/2}}{h^3} v_f q_i e^{-\epsilon/\mathbf{k}T}\right]^N \tag{16.28}$$

Alternatively, if we now relax this assumption and say that each molecule has access to the *entire* free volume of the liquid, we must replace v_f by $v_f N$. At the same time we must divide the partition function by $N!$, which is the number of ways of distributing the N molecules over all the N sites (Eq. 14.64). The partition function found on the basis of this assumption is

$$Q' = \frac{1}{N!}\left[\frac{(2\pi m\mathbf{k}T)^{3/2}}{h^3} N v_f q_i e^{-\epsilon/\mathbf{k}T}\right]^N \tag{16.29}$$

The ratio of these two partition functions is

$$\frac{Q'}{Q} = \frac{N^N}{N!} \tag{16.30}$$

and if we make use of Stirling's approximation (Eq. 14.25), this reduces to

$$\frac{Q'}{Q} = e^N \tag{16.31}$$

To help us consider which of the two formulations, Eqs. 16.28 and 16.29, is the more satisfactory, it is useful to calculate the entropy difference between them. In Table 14.3 we saw that the entropy is related to the partition function by Eq. 14.79:

$$S = \mathbf{k}T\frac{d \ln Q}{dT} + \mathbf{k} \ln Q \tag{16.32}$$

The difference in the entropies calculated from Eq. 16.28 and 16.29 is therefore

$$S' - S = \mathbf{k}T\frac{d \ln (Q'/Q)}{dT} + \mathbf{k} \ln \frac{Q'}{Q} \tag{16.33}$$

Since Q'/Q is e^N and is independent of T, the first term vanishes; the second is $N\mathbf{k}$:

$$S' - S = N\mathbf{k} \tag{16.34}$$

The molar entropy is obtained by dividing by the amount of substance n:

$$S'_m - S_m = \frac{N\mathbf{k}}{n} = L\mathbf{k} = R \tag{16.35}$$

Thus, the liquid model in which the molecules have complete access to the *entire* free volume of 1 mol of liquid gives an entropy of R greater than the model in which the molecules are confined to their *individual* free volumes. This additional entropy contribution associated with access to the entire free volume is known as the **communal entropy**. At one time it was thought that the partition function Q applies to the *solid* state and Q' to the *liquid* state and that on melting there is a sudden transition from one to the other with an increase of entropy due largely to the increase in communal entropy. Certain experimental results appeared to support this point of view. For example, for metals the constituent units are atoms and there are therefore no internal degrees of freedom to cause complications. Experimentally, entropies of fusion of metals lie in the range of 7.0 to 9.6 J K^{-1} mol^{-1} and, therefore, are similar to the value of R (8.314 J K^{-1} mol^{-1}). Moreover, substances for which there are internal degrees of freedom, which may be expected to play a more important role in the liquid state, generally have entropies of fusion in excess of R. It was therefore suggested that when melting occurs, there is an increase of entropy of R due to the communal entropy and an additional increase if the molecules have more rotational and vibrational freedom in the liquid than in the solid state.

However, in spite of this agreement, more detailed studies have shown that this view represents an oversimplification. In particular, it seems that the communal entropy does not appear sharply at the melting point but that the entire free volume becomes available to the molecules gradually, as the temperature is raised. The partition function Q' (Eq. 16.29) may therefore be a useful approximation for the liquid at higher temperatures, but it is not satisfactory just above the melting point.

The cell, or free-volume, theories of liquids have been developed and improved in a number of ways. In the outline we have given the molecules were assumed to be moving in a square potential-energy well (Figure 16.6b). Lennard-Jones and Devonshire* have improved the treatment considerably by taking into account the intermolecular forces. This means that the potential energy varies in the manner shown in Figure 16.6c, with two important consequences. One is that the free volume for each molecule is no longer a fixed quantity but varies with the molecular energy. The second is that the molecule is now not undergoing free translational motion, with a partition function given by Eq. 16.26, but has more of a vibrational character. This treatment has led to equations of state that are in much better agreement with experiment.

Other refinements that have been introduced allow for the occupancy of cells by more than one molecule; this may be significant at higher temperatures.

Once a partition function has been set up, all the thermodynamic properties can be calculated making use of the relationships listed in Table 14.3. It is also possible to calculate radial distribution functions. The thermodynamic values and radial distribution functions calculated from the partition function can then be compared with experiment, and the extent of agreement gives a useful indication of the reliability of the partition function.

Hole Theories

Another improvement to the cell or free-volume theories involves the assumption that some of the lattice sites may be unoccupied. In other words, the liquid contains *holes*, an idea first suggested by Henry Eyring.† Again, we will discuss the theory only in its simplest form.

---

*J. E. Lennard-Jones and A. F. Devonshire, *Proc. Roy. Soc.*, *A163*, 53 (1937); *A165*, 1 (1938); *A169*, 317 (1939); *A170*, 469 (1939).

†H. Eyring, *J. Chem. Phys.*, *4*, 283 (1936).

When most solids melt, the density decreases, and there is also a decrease in the number of nearest neighbors as determined by diffraction experiments. There is a still further decrease in the density and in the number of neighbors as the temperature of the liquid is raised. These facts suggest that when we employ a lattice model for a liquid, we ought to consider the possibility that some of the lattice sites are empty. The presence of holes, more or less randomly distributed in the lattice, gives an alternative interpretation of the entropy increase on melting. Moreover, there is evidence to suggest that the viscosity of most liquids depends mainly on the volume. This can be explained by the hypothesis that liquids contain holes and that diffusion and flow in liquids can be interpreted in terms of a movement of these holes.

Let us suppose, for simplicity, that on melting of n mol of a solid N_h holes are created and that the entropy of melting is due to the mixing of the N_h holes with N molecules. The entropy of melting is therefore given by use of Eq. 3.63:

$$\begin{aligned}\Delta S_m &= -N\boldsymbol{k} \ln \frac{N}{N + N_h} - N_h\boldsymbol{k} \ln \frac{N_h}{N + N_h} \\ &= N\boldsymbol{k}\left[\ln(1 + \gamma) + \gamma \ln \frac{1 + \gamma}{\gamma}\right] \qquad (16.36)\end{aligned}$$

where γ is the ratio of holes to molecules; that is,

$$\gamma = \frac{N_h}{N} \qquad (16.37)$$

The molar entropy of melting is obtained by dividing by the amount of substance n, and since $N\boldsymbol{k}/n = L\boldsymbol{k} = R$, we obtain

$$\Delta S_m = R\left[\ln(1 + \gamma) + \gamma \ln \frac{1 + \gamma}{\gamma}\right] \qquad (16.38)$$

The value of γ can therefore be evaluated by comparing this expression with experimental values of entropies of melting. A value of 0.5 for γ gives $\Delta S_m = 0.95R$, while $\gamma = 0.54$ gives $\Delta S_m = R$. Metals have entropies of melting of approximately R, so that it appears that melting introduces about half as many holes as there are molecules.

It is also possible to make an estimate of the sizes of the holes. The average volume occupied by a single molecule is taken to be V_s/L, where V_s is the volume of 1 mol of solid. If v_h is the average volume occupied by a hole, the ratio of the volume of a molecule to that of a hole, which we will write as ρ, is

$$\rho = \frac{V_s}{Lv_h} \qquad (16.39)$$

The number of holes is

$$N_h = \frac{V_l - V_s}{v_h} = \frac{V_l - V_s}{V_s/\rho L} \qquad (16.40)$$

and the ratio of holes to molecules is thus

$$\gamma = \frac{N_h}{L} = \rho\frac{V_l - V_s}{V_s} \qquad (16.41)$$

An estimate of ρ can therefore be obtained from γ and the values of V_l and V_s. For most metals ρ turns out to be about 20; the holes are therefore very much smaller than the molecules. For many other liquids, however, values of about 6 have been obtained for ρ.

The partition function for the liquid is obtained by combining the partition function for the molecules with the partition function for the holes. The former contribution is essentially the same as that for the f ee-volume theory of liquids. The holes are treated as if they were molecules occupying the volume V_h. The combined partition function is obtained by considering the number of ways in which the N molecules and N_h holes are distributed among the $N + N_h$ sites.

An important question is the amount of energy required to create a hole in a liquid, since this quantity must be incorporated into the partition function. This can be arrived at on the basis of the following simplified argument. Suppose that the N molecules in n mol of liquid are connected to each other by bonds and that on the average z bonds are attached to each molecule. If we count the total number of bonds by considering each molecule and multiplying by z, we count each bond twice; the total number of bonds is therefore $\frac{1}{2}zN$. If ϵ is the average strength of each bond, the energy required to vaporize N molecules is $\frac{1}{2}zN\epsilon$, and the energy required to vaporize one molecule is

$$\frac{\Delta V_{\text{vap}}}{L} = \frac{1}{2}z\epsilon \tag{16.42}$$

If alternatively we remove one molecule from the interior of the liquid and leave a hole the size of the molecule, we are breaking z bonds and the energy required is therefore $z\epsilon$. Half of this is required to vaporize the molecule, and the other half is thus the energy required to create a hole without vaporizing the molecule (i.e., by moving the molecules to other lattice positions). The energy required to create a hole of molecular size is thus equal to the energy required to vaporize a molecule.

We have seen that the holes in liquids are in some cases smaller than the molecules, and the energy required to create the holes is therefore smaller. Eyring and his co-workers have treated the viscosity of liquids (see also Section 18.1) on the hypothesis that the process involves the movement of molecules into adjacent holes. Activation energies for viscous flow are frequently one-third of heats of vaporization, and it is concluded that the holes are perhaps one-third the size of the molecules.

Monte Carlo and Molecular-Dynamical Theories

The molecules in a liquid are in ceaseless motion and they are constantly interacting with their neighbors. With the aid of modern computers it is possible to consider a group of a few hundred molecules that are approximately situated at lattice positions and to carry out calculations that take into account the intermolecular forces acting on each molecule. The properties of the liquid can then be calculated by an averaging procedure, and this can be done in two different ways. In the **Monte Carlo method** a large number of molecular configurations are selected in a particular way and the potential energy of each one is calculated; the various configurations are then weighted by the Boltzmann factor and the properties calculated on this basis. In the **molecular-dynamical method** the molecules are taken to have certain arbitrary initial positions and velocities, and the equations of motion are set up in terms of the intermolecular forces. The properties of the liquid are then obtained by averaging over time. The molecular dynamical method therefore involves *time averaging,* while the Monte Carlo method involves *averaging over different molecular configurations.*

Various computing devices have been used to make the Monte Carlo method more practicable. If molecular configurations are selected entirely at random, a very large majority will involve such interpenetration of orbitals that they are highly improbable. Computations made in this way are very time-consuming and inefficient. Much improvement has been brought about by the use of a procedure suggested by N. Metropolis and his collaborators.* In essence, the molecules move in a random fashion from one configuration to another, but if a molecule comes too close to its neighbors, it is required to bounce back to its original position. In this way the procedure avoids making computations for the very unlikely configurations for which the potential energies are excessively large. Another complication with the Monte Carlo method is that, since the number of molecules has to be limited, a substantial proportion are at the surface so that surface effects become important. This effect can be minimized by taking into account a group of surrounding molecules ("ghosts") that move in unison with the molecules under consideration.

In the molecular-dynamical method a group of molecules are assigned initial positions and velocities, and an assumption is made as to the dependence of the intermolecular force on the distance. The computer then performs the necessary calculations and determines the positions and velocities at various later times. Equilibrium is established fairly rapidly, after perhaps four collisions per molecule. The properties of the liquid can then be calculated from the positions, velocities, and potential energies at equilibrium. Corrections can again be made for surface effects.

Both the Monte Carlo and molecular-dynamical methods have given very satisfactory agreement with experiment. For example, it has proved possible to reproduce the P-V-T relationships for a number of liquids. The Monte Carlo method has been somewhat more useful for liquids at equilibrium, but the molecular-dynamical method has been valuable for interpreting the behavior of liquids in motion.

16.5 Distribution-Function Theories of Liquids

The lattice theories that we have outlined in Section 16.4 all start with a description of the structure of the liquid. The distribution-function theories, by contrast, place initial emphasis on the way in which the intermolecular forces determine the structure. The lattice theories bear more resemblance to the theories of solids, while the distribution-function theories are more closely related to the kinetic-molecular theories of gases.

The application of the distribution-function theories is very complicated, and the reader is referred to more advanced treatments for details. A relatively small

---

*N. Metropolis, A. W. Rosenbluth, M. N. Rosenbluth, and A. H. Teller, *J. Chem. Phys.*, *21*, 1087 (1953).

group of molecules is considered, and the molecules are dealt with in pairs; integrals representing the probability of each pair are formulated in terms of a function chosen for the intermolecular potential. The final distribution function is then obtained from these probability functions for all the pairs.

So far the distribution-function treatments have not been so successful as the lattice theories, but further work may well reverse this situation. The two methods are complementary to each other, and there is no fundamental reason for preferring one to the other.

Key Equations

Internal pressure of a fluid:

$$\text{internal pressure} \equiv \left(\frac{\partial U}{\partial V}\right)_T$$

If the van der Waals equation is obeyed,

$$\text{internal pressure} = \frac{a}{V_m^2}$$

Caloric equation of state:

$$U = \int \left\{ T\left(\frac{\partial P}{\partial T}\right)_V - P \right\} dV + U_\infty$$

If the van der Waals equation is obeyed,

$$U = -\frac{a}{V} + U_\infty$$

Intermolecular potential energies:

Ion-ion: $$E_p = -\frac{z_A z_B e^2}{4\pi\epsilon_0 \epsilon r}$$

Ion-dipole: $$E_p = -\frac{z_A e \mu \cos\theta}{4\pi\epsilon_0 \epsilon r^2}$$

Ion-(induced dipole): $$E_p = -\frac{\alpha (z_A e)^2}{8\pi\epsilon_0 \epsilon r^4}$$

(α = polarizability)

Dipole-dipole (average): $$E_p = -\frac{\mu_A^2 \mu_B^2}{24\pi^2 \epsilon_0^2 \epsilon^2 \mathbf{k} T r^6}$$

Dispersion (London's formula): $E_p = -\frac{3h\nu_0\alpha^2}{4r^6}$

Repulsive: $E_p = \frac{B}{r^{12}}$

The *Lennard-Jones 6–12 potential:* $E_p = -\frac{A}{r^6} + \frac{B}{r^{12}} = E_0\left[\left(\frac{r_0}{r}\right)^{12} - 2\left(\frac{r_0}{r}\right)^{6}\right]$

(r_0 = equilibrium separation)

Problems

A Thermodynamic Properties of Liquids

1. The density of liquid ethanol at 20°C is 0.790 g cm^{-3}, and the van der Waals constant a is 1.218 Pa m^6 mol^{-2}. Estimate the internal pressure and the potential-energy contribution to the internal energy.

2. In the example on page 742 we obtained the internal pressure of liquid water from the van der Waals constant a. A more reliable value is obtained by use of Eq. 16.2, from the *thermal pressure coefficient* $(\partial P/\partial T)_V$; this quantity is the ratio α/κ of the coefficient of expansion $\alpha\,[\equiv (1/V)(\partial V/\partial T)_P]$ to the compressibility $\kappa\,[\equiv -(1/V)(\partial V/\partial P)_T]$. For water at 1 atm pressure and 298 K the thermal pressure coefficient is 6.60×10^6 Pa K^{-1}. Calculate the internal pressure.

3. The density of liquid benzene at 0°C is 0.899 g cm^{-3}, and the van der Waals constant a is 1.824 m^6 mol^{-2} Pa. Estimate the internal pressure and the potential-energy contribution to the internal energy.

4. Make a better estimate of the internal energy of liquid benzene from its thermal pressure coefficient $(\partial P/\partial T)_V$, which at 298 K and 1 atm pressure is 1.24×10^6 Pa K^{-1}.

5. Calculate the internal pressures of the following liquids at 298 K and 1 atm pressure from their thermal pressure coefficients:

Hg:	4.49×10^6	Pa K^{-1}
n-Heptane:	8.53×10^5	Pa K^{-1}
n-Octane:	1.01×10^6	Pa K^{-1}
Diethyl ether:	8.06×10^5	Pa K^{-1}

6. The thermal pressure coefficient $(\partial P/\partial T)_V$, for CCl_4 vapor at 298 K and 10 Pa pressure, is 115 Pa K^{-1}. That for liquid CCl_4 at 298 K and 1 atm pressure is 1.24×10^6 Pa K^{-1}. Calculate the internal pressures of the vapor and the liquid under these conditions.

***7.** The following data apply to liquid acetic acid at 1 atm pressure and 293 K: density, d = 1.049 g cm^{-3}; coefficient of expansion, $\alpha = 1.06 \times 10^{-3}$ K^{-1}; compressibility, $\kappa = 9.08 \times 10^{-10}$ Pa^{-1}; van der Waals constant, a = 1.78 m^6 Pa mol^{-2}. Make two estimates of the internal pressure P_i, (a) using α and κ and (b) using a.

B Intermolecular Energies

8. A liquid having a molar volume of 50 cm^3 is converted into a vapor having a molar volume of 50 dm^3. By what factor does the average intermolecular energy change?

9. Calculate the maximum energy of attraction, in J and in kJ mol^{-1}, when a Ca^{2+} ion is separated from a molecule of dipole moment 6.18×10^{-30} C m (= 1.85 D; this is the dipole moment of water) by a distance of 500 pm in a vacuum.

10. Calculate the energy of attraction, in J and in kJ mol^{-1}, when a Ca^{2+} ion is separated in a vacuum from a Cl^- ion by a distance of 500 pm.

11. Calculate the energy of attraction, in J and in kJ mol^{-1}, when a Ca^{2+} ion is separated in a vacuum by a distance of 500 pm, from a molecule having no dipole moment but a polarizability of 2.0×10^{-30} m^3.

12. Calculate the average energy of attraction, in J and in kJ mol^{-1}, for two molecules of dipole moments 6.18×10^{-30} C m separated in a vacuum at 25°C by a distance of 500 pm.

***13.** The following values for A and B in the Lennard-Jones 6–12 function (Eq. 16.21) have been given for N_2:

$$A = 1.34 \times 10^{-5} \text{ J pm}^6$$

$$B = 3.42 \times 10^{10} \text{ J pm}^{12}$$

Calculate the equilibrium separation r_0 and the classical dissociation energy E_0, in J and in J mol^{-1}.

C Supplementary Problems

14. The following data apply to HBr: dipole moment, $\mu = 2.60 \times 10^{-30}$ C m; polarizability, $\alpha = 3.58 \times 10^{-30}$ m^3; oscillation frequency, $\nu_0 = 3.22 \times 10^{15}$ s^{-1}. Estimate the dipole-dipole, dipole-(induced dipole), and dispersion energies in J and in kJ mol^{-1} for two HBr molecules separated by 500 pm, at 25°C.

15. The following are the polarizabilities and oscillation frequencies for He, Ar, and Kr:

	Ne	Ar	Kr
Polarizability, $\alpha/10^{-30}$ m^3	0.39	1.63	2.46
Frequency, $\nu_0/10^{15}$ s^{-1}	5.21	3.39	2.94

Calculate the dispersion energies, for Ne, Ar, and Kr, corresponding to a separation of 500 pm. Values for He and Xe are given in Table 16.3; plot the five values against the boiling points of the noble gases:

	He	Ne	Ar	Kr	Xe
Boiling point, T_b/K	4.22	27.3	87.3	119.9	165.1

16. In Table 16.3 and Problem 15 the dispersion energies of noble gases were calculated for a constant interatomic distance of 500 pm. More realistic values are given here:

	He	Ne	Ar	Kr	Xe
Interatomic distance/pm	240	320	380	400	420

Recalculate the dispersion energies for these distances, and again plot the five values against the boiling points, which were given in Problem 15.

The experimental value for the enthalpy of vaporization of liquid argon is 6.7 kJ mol^{-1}. Make an estimate of the enthalpy of vaporization from your calculated value of E_p (at 380 pm), assuming the liquid to have a close-packed structure with each atom having 12 nearest neighbors.

17. Estimate the interaction energy between an argon atom and a water molecule at a separation of 600 pm, which is approximately the distance of closest approach. The necessary data are: H_2O: dipole moment, $\mu = 6.18 \times 10^{-30}$ C m; Ar: polarizability, $\alpha = 1.63 \times 10^{-30}$ m^3.

Argon in fact forms a solid hydrate, $Ar{\cdot}5H_2O$, but the binding energy between Ar and H_2O is about 40 kJ mol^{-1}, which is a good deal larger than the energy calculated from the dipole moment and polarizability. Suggest a reason for this discrepancy.

18. a. The Lennard-Jones potential

$$E = -\frac{A}{r^6} + \frac{B}{r^n}$$

can be formulated in a different way by expressing A and B in terms of the energy E_{min} at the minimum and the value r_0 of r at the minimum. Obtain the expression for E in this form.

b. If r^* is the value of r when $E = 0$, obtain the relationship between r^* and r_0.

c. The Lennard-Jones potential is often used with $n = 12$, and the equations are then simpler. Obtain E in terms of E_{min} and r_0 and in terms of E_{min} and r^*, for this special case of $n = 12$.

19. a. Derive the relationship

$$C_P - C_V = \frac{\alpha^2 VT}{\kappa}$$

where α is the coefficient of expansion and κ is the compressibility.

b. The value of $C_{V,m}$ for liquid CCl_4 at 298 K and at 1 atm pressure is 89.5 J K^{-1} mol^{-1}. Obtain the value of $C_{P,m}$ using the following data: $V_m = 97$ cm^3 mol^{-1}; $\alpha = 1.24 \times 10^{-3}$ K^{-1}; $\kappa = 10.6 \times 10^{-5}$ atm^{-1}.

c. Calculate $C_{P,m} - C_{V,m}$ for liquid acetic acid using the data given in Problem 7.

D Essay Questions

1. Explain qualitatively how intermolecular forces of attraction are related to the following properties of a liquid:

a. Vapor pressure.
b. Enthalpy of vaporization.
c. Normal boiling point.
d. Entropy of vaporization.

After you have studied Chapters 17 and 18, you can also deal with

e. Surface tension.

f. Viscosity.

g. Rates of diffusion.

2. Explain clearly the difference between dipole-dipole and London (dispersion) forces. With reference to a few examples, discuss the magnitudes of attractive energies arising from these forces.

Suggested Reading

J. A. Barker, *Lattice Theories of the Liquid State,* New York: Macmillan, 1963.

C. A. Croxton, *Liquid State Physics: A Statistical Mechanical Introduction,* Cambridge: University Press, 1974.

D. Eisenberg and W. Kauzmann, *The Structure and Properties of Water,* Oxford: Clarendon Press, 1969.

J. O. Hirschfelder, C. F. Curtiss, and R. B. Bird, *Molecular Theory of Gases and Liquids,* New York: Wiley, 1954.

J. A. Pryde, *The Liquid State,* London: Hutchinson Scientific Library, 1966.

J. S. Rowlinson, *Liquids and Liquid Mixtures,* London: Butterworth, 1959.

D. Tabor, *Gases, Liquids and Solids,* Harmondsworth, Middlesex: Penguin Books, 1969.

Chapter 17

Surface Chemistry and Colloids

Preview

A number of important effects arise from the fact that surfaces show properties that differ from those of bulk material. For example, substances tend to become *adsorbed* on surfaces, and the first part of this chapter deals with the basic laws of adsorption. *Adsorption* can be classified as *physical* or *van der Waals adsorption* and as *chemisorption,* which is much stronger since covalent bonding is involved.

The simplest type of chemisorption is that described by the *Langmuir adsorption isotherm.* This applies when the surface is homogeneous and when the forces between adsorbed molecules can be neglected. This type of adsoprtion is characterized by the fact that the surface can become *saturated* by the adsorbed molecules. Molecules on surfaces are often *dissociated.* Adsorption equilibria can be treated on the basis of a kinetic model and also by the methods of statistical mechanics.

Surface-catalyzed reactions involve the adsorption of the reacting molecules on the surface, and the Langmuir isotherms can be used to treat such reactions. When there are two reacting substances, there may be a *Langmuir-Hinshelwood mechanism,* in which reaction is between two adsorbed molecules. Alternatively, reaction may occur between an adsorbed molecule and a molecule in the gas phase, in which case we speak of a *Langmuir-Rideal* mechanism.

The structures of solid surfaces and adsorbed layers can be investigated by a number of experimental techniques. Various different atomic arrangements may exist at surfaces.

As far as liquid surfaces are concerned, an important property is the *surface tension,* which results from intermolecular attractions. One result of surface tension is the tendency of certain liquids to rise in a capillary tube, an effect known as *capillarity.*

Liquid films on surfaces have been studied particularly by the use of Irving Langmuir's *film balance,* which measures *surface pressure.* Liquid films may be *coherent,* in which case there is a critical area per molecule; such films behave like two-dimensional liquids. There are also *noncoherent* or *gaseous* films, which behave like two-dimensional gases.

A very important property of a surface between a solid and a liquid is that an *electric double layer* is formed. This was originally explained by Helmholtz as due to the attachment to the surface of a layer of ions from the solution. This *fixed-double-layer* theory was later superseded by a *diffuse-double-layer* theory, but it is now realized that it is necessary to postulate both a fixed double layer and a diffuse double layer. These theories lead to the concept of the *electrokinetic potential* or the ζ (Greek zeta) *potential*. This potential is the electric potential at the position of closest approach of the ions to the surface, with reference to the bulk solution.

Colloidal dispersions fall between true solutions and suspensions; the particles are either very large molecules or aggregates, but they are too small to be seen under the microscope. Their properties result from the fact that their ratio of surface area to volume is very large, so that the surface properties are important. There are a number of different types of collodial systems, including *sols*, *gels*, *emulsions*, and *micelles*. An important property of colloids is the *Tyndall effect*, which involves the scattering of light. The *ultramicroscope* is a useful instrument for studying colloidal systems, and in it *Brownian movement* can be observed.

The electrical properties of colloidal systems are important and arise from the attachment of ions to surfaces.

17 Surface Chemistry and Colloids

A plane that separates two phases is known as a *surface* or an *interface*. Surfaces show special properties that are different from those of the phases themselves. For example, the surface of a solid often shows a strong affinity for molecules that come into contact with it and which are said to be **adsorbed**. A simple way of visualizing adsorption is in terms of additional valence bonds at the surface, which are available for bonding; in reality, however, the situation is more complex.

In this chapter we will deal with the behavior of different kinds of surfaces and, in particular, with the nature of adsorption. We will first consider solid-gas interfaces, which are the easiest to understand and which provide a basis for understanding more complex surfaces. Attention will be given to the way in which molecules are attached to surfaces, to the thermodynamics and statistical mechanics of adsorption, and to the mechanisms of chemical reactions at surfaces. Colloidal systems are also considered in this chapter, since the properties of colloids are determined to a very considerable extent by surface effects.

Some of the earliest work on adsorption and surface catalysis was carried out by Michael Faraday*, who was particularly concerned with gas reactions on surfaces. He realized that such reactions occur in adsorbed films but thought that the main effect of the solid catalyst was to exert an attractive force on the gas molecules and to cause them to be present at much higher concentrations than in the main body of the gas. This idea is, however, shown to be false by the fact that in certain cases different surfaces give rise to different products of reaction; for example, ethanol decomposes mainly into ethylene and water on an alumina catalyst and mainly into acetaldehyde and hydrogen on copper. This and many other results clearly indicate that specific chemical forces must sometimes be involved when molecules become attached to surfaces.

---

*M. Faraday, *Phil. Trans.*, *124*, 55 (1834).

17.1 Adsorption

It is now recognized that two main types of adsorption may be clearly distinguished. In the first type the forces are of a physical nature and the adsorption is relatively weak. The forces correspond to those considered by J. H. van der Waals in connection with his equation of state for gases (Section 1.12) and are known as **van der Waals forces**. This type of adsorption is known as **physical adsorption**, *physisorption*, or **van der Waals adsorption**. The heat evolved when a mole of gas becomes physisorbed is usually small, less than 20 kJ. This type of adsorption only plays an unimportant role in catalysis, except for certain special types of reactions involving free atoms or radicals.

In the second type of adsorption, first considered in 1916 by the American chemist Irving Langmuir* (1881–1957), the adsorbed molecules are held to the surface by covalent forces of the same general type as those occurring between bound atoms in molecules. The heat evolved per mole for this type of adsorption, known as **chemisorption**, is usually comparable to that evolved in chemical bonding, namely 100 to 500 kJ.

An important consequence of this concept of chemisorption is that after a surface has become covered with a single layer of adsorbed molecules, it is *saturated*; additional adsorption can occur only on the layer already present, and this is generally weak adsorption. Langmuir thus emphasized that chemisorption involves the formation of a *unimolecular layer*. Many investigations on surfaces of known area have confirmed that adsorption ceases after a unimolecular layer is formed.

It was later suggested by the British† physical chemist Sir Hugh Taylor‡ (1890–1974) that chemisorption is frequently associated with an appreciable activation energy and may therefore be a relatively slow process. For this reason chemisorption is often referred to as *activated adsorption*. By contrast, van der Waals adsorption requires no activation energy and therefore occurs more rapidly than chemisorption.

Another important concept suggested by H. S. Taylor§ is that solid surfaces are never completely smooth and that adsorbed molecules will be attached more strongly to some surface sites than to others. This is particularly important in connection with catalysis, since chemical reaction may occur predominantly on certain sites, which Taylor referred to as *active centers*. Surface heterogeneity is discussed in further detail in Section 17.5.

17.2 Adsorption Isotherms

An equation that relates the amount of a substance attached to a surface to its concentration in the gas phase or in solution, at a fixed temperature, is known as an **adsorption isotherm**.

---

*I. Langmuir, *J. Amer. Chem. Soc., 38*, 2221 (1916).

†Although Taylor remained a British subject, his scientific work was done almost entirely in the United States, at Princeton University.

‡H. S. Taylor, *J. Amer. Chem. Soc., 53*, 578 (1931).

§H. S. Taylor, *Proc. Roy. Soc., A108*, 105 (1925).

The Langmuir Isotherm

The simplest isotherm was first obtained in 1916 by Irving Langmuir.* The basis of the derivation of the Langmuir isotherm is that all parts of the surface behave in exactly the same way as far as adsorption is concerned. Suppose that, after equilibrium is established, a fraction θ of the surface is covered by adsorbed molecules; a fraction $1 - \theta$ will not be covered. The rate of adsorption will then be proportional to the concentration [A] of the molecules in the gas phase or in solution and also to the fraction of the surface that is bare, because adsorption can only occur when molecules strike the bare surface. The rate of adsorption is thus

$$v_a = k_a[\mathrm{A}](1 - \theta) \tag{17.1}$$

where k_a is a constant relating to the adsorption process. The rate of desorption is proportional only to the number of molecules attached to the surface, which in turn is proportional to the fraction of surface covered:

$$v_d = k_d\theta \tag{17.2}$$

At equilibrium, the rates of adsorption and desorption are the same; thus

$$k_a[\mathrm{A}](1 - \theta) = k_d\theta \tag{17.3}$$

or

$$\frac{\theta}{1 - \theta} = \frac{k_a}{k_d}[\mathrm{A}] \tag{17.4}$$

The ratio k_a/k_d is an equilibrium constant and can be written as K; then

$$\frac{\theta}{1 - \theta} = K[\mathrm{A}] \tag{17.5}$$

or

$$\boldsymbol{\theta = \frac{K[\mathrm{A}]}{1 + K[\mathrm{A}]}} \tag{17.6}$$

A graph of θ against [A] is shown in Figure 17.1a. At sufficiently low concentrations we can neglect K[A] in comparison with unity, and then θ is proportional to [A]. We can write Eq. 17.6 as

$$1 - \theta = \frac{1}{1 + K[\mathrm{A}]} \tag{17.7}$$

so that at very high concentrations

$$1 - \theta = \frac{1}{K[\mathrm{A}]} \tag{17.8}$$

Note that the **Langmuir isotherm** (Eq. 17.6) is of exactly the same form as the Michaelis equation (Eq. 10.82). A distinctive feature of the isotherm is that the surface becomes *saturated* with molecules at high pressures.

---

*I. Langmuir, *J. Amer. Chem Soc.*, *38*, 2221 (1916); *40*, 1361 (1918).

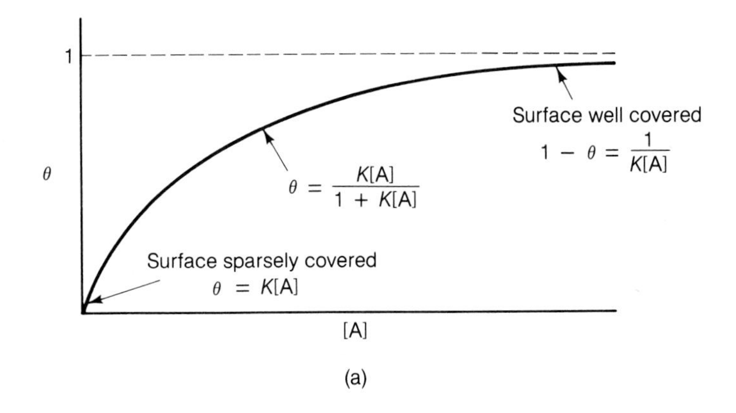

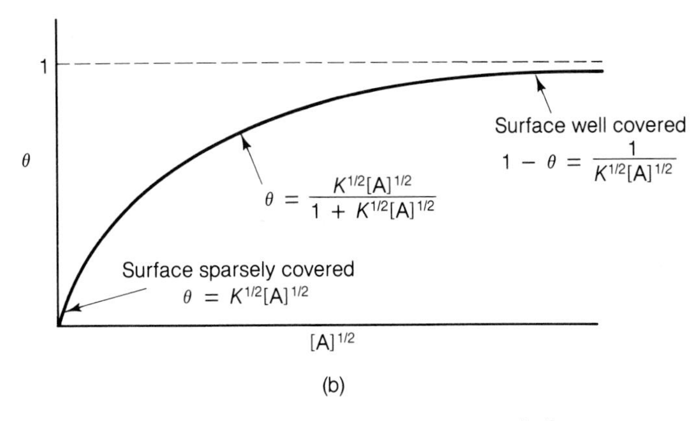

FIGURE 17.1 Plots of θ (fraction of surface covered) against [A] (concentration of a substance in the gas phase or in solution) for a system obeying the Langmuir isotherm. (a) Adsorption without dissociation (Eq. 17.6). (b) Adsorption with dissociation (Eq. 17.13).

Adsorption with Dissociation

The type of adsorption to which Eq (17.6) applies may be formulated as

$$\mathrm{A} + \text{—}\overset{|}{\mathrm{S}}\text{—} \rightleftharpoons \text{—}\overset{\overset{\mathrm{A}}{|}}{\mathrm{S}}\text{—}$$

where S represents a surface site and A the substance being adsorbed. In certain cases there is evidence that the process of adsorption is accompanied by the dissociation of the molecule when it becomes attached to the surface. For example, when hydrogen gas is adsorbed on the surface of many metals, the molecules are dissociated into atoms each of which occupies a surface site. This type of adsorption may be represented as

$$\mathrm{A_2} + \text{—}\overset{|}{\mathrm{S}}\text{—}\overset{|}{\mathrm{S}}\text{—} \rightleftharpoons \text{—}\overset{\overset{\mathrm{A}}{|}}{\mathrm{S}}\text{—}\overset{\overset{\mathrm{A}}{|}}{\mathrm{S}}\text{—}$$

The process of adsorption is now a reaction between the gas molecule and *two* surface sites, and the rate of adsorption is therefore

$$v_a = k_a[\mathrm{A}](1 - \theta)^2 \tag{17.9}$$

The desorption process involves reaction between *two* adsorbed atoms, and the rate is therefore proportional to the square of the fraction of surface covered,

$$v_d = k_d\theta^2 \tag{17.10}$$

At equilibrium the rates are equal, and therefore,

$$\frac{\theta}{1-\theta} = \left(\frac{k_a}{k_d}[\mathrm{A}]\right)^{1/2} \tag{17.11}$$

$$= K^{1/2}[\mathrm{A}]^{1/2} \tag{17.12}$$

where K is equal to k_a/k_d. This equation can be written as

$$\boldsymbol{\theta} = \frac{\boldsymbol{K^{1/2}[\mathrm{A}]^{1/2}}}{\boldsymbol{1 + K^{1/2}[\mathrm{A}]^{1/2}}} \tag{17.13}$$

A plot of θ against $[\mathrm{A}]^{1/2}$ is shown in Figure 17.1b. When the concentration is very small, $K^{1/2}[\mathrm{A}]^{1/2}$ is much smaller than unity, and θ is then proportional to $[\mathrm{A}]^{1/2}$. Equation 17.13 may be written as

$$1 - \theta = \frac{1}{1 + K^{1/2}[\mathrm{A}]^{1/2}} \tag{17.14}$$

so that at high concentrations, when $K^{1/2}[\mathrm{A}]^{1/2} \gg 1$,

$$1 - \theta = \frac{1}{K^{1/2}[\mathrm{A}]^{1/2}} \tag{17.15}$$

The fraction of the surface that is bare at high concentrations is therefore inversely proportional to the square root of the concentration.

Competitive Adsorption

The isotherm for two substances adsorbed on the same surface is of importance in connection with inhibition and with the kinetics of surface reactions involving two reactants. Suppose that the fraction of surface covered by molecules of type A is θ_A and that the fraction covered by B is θ_B. The fraction bare is $1 - \theta_\mathrm{A} - \theta_\mathrm{B}$. If both substances are adsorbed without dissociation, the rates of adsorption of A and B are

$$v_a^\mathrm{A} = k_a^\mathrm{A}[\mathrm{A}](1 - \theta_\mathrm{A} - \theta_\mathrm{B}) \tag{17.16}$$

and

$$v_a^\mathrm{B} = k_a^\mathrm{B}[\mathrm{B}](1 - \theta_\mathrm{A} - \theta_\mathrm{B}) \tag{17.17}$$

The rates of desorption are

$$v_d^\mathrm{A} = k_d^\mathrm{A}\theta_\mathrm{A} \tag{17.18}$$

$$v_d^\mathrm{B} = k_d^\mathrm{B}\theta_\mathrm{B} \tag{17.19}$$

Equating Eqs. 17.16 and 17.18 leads to

$$\frac{\theta_\mathrm{A}}{1 - \theta_\mathrm{A} - \theta_\mathrm{B}} = K_\mathrm{A}[\mathrm{A}] \tag{17.20}$$

where K_A is equal to k_a^A/k_d^A. From Eqs. 17.17 and 17.19 we obtain

$$\frac{\theta}{1-\theta_A-\theta_B} = K_B[B] \tag{17.21}$$

where K_B is k_a^B/k_d^B. Equations 17.20 and 17.21 are two simultaneous equations that can be solved to give, for the fractions covered by A and B, respectively,

$$\theta_A = \frac{K_A[A]}{1 + K_A[A] + K_B[B]} \tag{17.22}$$

$$\theta_B = \frac{K_B[B]}{1 + K_A[A] + K_B[B]} \tag{17.23}$$

Equation 17.22 reduces to Eq. 17.6 if $[B] = 0$ or if $K_B = 0$, which means that substance B is not adsorbed. It follows from Eqs. 17.22 and 17.23 that the fraction of the surface covered by one substance is reduced if the amount of the other substance is increased. This is because the molecules of A and B are competing with one another for a limited number of surface sites, and we speak of **competitive adsorption**. There is evidence that sometimes two substances are adsorbed on two different sets of surface sites, in which case there is no competition between them.

For gaseous systems, the concentration terms in all the equations derived in this section can, of course, be replaced by pressures; the equilibrium constants are then pressure equilibrium constants.

Other Isotherms

The various isotherms of the Langmuir type are based on the simplest of assumptions; all sites on the surface are assumed to be the same, and there are no interactions between adsorbed molecules. Systems that obey these equations are often referred to as showing **ideal adsorption;** the Langmuir equations have the same significance in connection with adsorption as has the ideal gas law $PV = nRT$ in connection with the behavior of gases. Systems frequently deviate significantly from the Langmuir equation. This may be because the surface is not uniform, and also there may be interactions between adsorbed molecules; a molecule attached to a surface may make it more difficult, or less difficult, for another molecule to become attached to a neighboring site, and this will lead to a deviation from the ideal adsorption equation. Nonideal systems can sometimes be fitted to an empirical adsorption isotherm due to the German physical chemist Herbert Max Finlay Freundlich* (1880–1941); according to this equation the amount of a substance adsorbed, x, is related to the concentration c by the equation

$$x = kc^n \tag{17.24}$$

where k and n are empirical constants.

This equation does not give saturation of the surface; the amount adsorbed keeps increasing as c increases. If Eq. 17.24 applies, a plot of $\log_{10} x$ against $\log_{10} c$ will give a straight line, of slope n.

*H. M. F. Freundlich, *Kappilarchemie,* Leipzig, 1909.

Another useful adsorption isotherm is that suggested by the Russian physical chemists A. Slygin and A. Frumkin.* It is

$$\theta = \frac{1}{f} \ln ac \tag{17.25}$$

where f and a are constants.

Both of these isotherms can be arrived at theoretically in terms of surface heterogeneity and also in terms of repulsive forces between adsorbed molecules.

17.3 Thermodynamics and Statistical Mechanics of Adsorption

A considerable number of measurements have been made of the enthalpy and entropy changes that occur on adsorption and desorption. As we have seen, van der Waals adsorption and chemisorption can be distinguished by the magnitudes of the enthalpy changes. Heat is always liberated on adsorption, and the enthalpy of adsorption is therefore always negative.

The reason that enthalpies of adsorption must be negative is that the adsorption process inevitably involves a decrease in entropy. This is because a molecule in the gas phase or in solution has more freedom of motion than one that is attached to a surface. In view of the thermodynamic relationship

$$\Delta G = \Delta H - T\,\Delta S \tag{17.26}$$

a process with a negative ΔS must have a negative ΔH if ΔG is to be negative (i.e., if the process is to occur to an appreciable extent).

By the methods of statistical mechanics (Chapter 14) it is possible to derive isotherms of the Langmuir form, and they express the constant K in terms of partition functions. This was first done by the British mathematical physicist Sir Ralph Howard Fowler (1889–1944)† We will consider only the case of the adsorption of a gas without dissociation, to which Eq. 17.6 applies. Suppose that the volume of the gas is V and that the area of the surface is S. The total number of molecules in the gas phase may be written as N_g, the number of adsorbed molecules as N_a, and the number of bare sites at equilibrium as N_s. The concentrations of these species follow:

Concentration in gas phase: $c_g = \dfrac{N_g}{V}$

Concentration of adsorbed molecules: $c_a = \dfrac{N_a}{S}$

Concentration of bare sites: $c_s = \dfrac{N_s}{S}$

---

*A. Slygin and A. Frumkin, *Acta Physiochim,* U.R.S.S., *3,* 791 (1935).

†R. H. Fowler, *Proc. Cambridge Phil. Soc., 31,* 260 (1935).

The equilibrium constant for the adsorption process is

$$K_c = \frac{c_a}{c_g c_s} = \frac{N_a/S}{(N_g/V)(N_s/S)} = \frac{N_a}{(N_g/V)N_s} \tag{17.27}$$

We saw in Section 14.8 that an equilibrium constant is equal to the ratio of the partition functions multiplied by an exponential term involving the difference between the zero-point levels of reactants and products:

$$K_c = \frac{q_a}{q_g q_s} e^{-\Delta E_0/RT} \tag{17.28}$$

The partition functions q_a and q_s to be used in this expression are those for unit surface area; q_g is for unit volume. Then

$$\frac{c_a}{c_s} = c_g \frac{q_a}{q_g q_s} e^{-\Delta E_0/RT} \tag{17.29}$$

If θ is the fraction of the surface that is covered,

$$\frac{c_a}{c_s} = \frac{\theta}{1 - \theta} \tag{17.30}$$

and, therefore,

$$\frac{\theta}{1 - \theta} = c_g \frac{q_a}{q_g q_s} e^{-\Delta E_0/RT} \tag{17.31}$$

The partition function q_g per unit volume may be written as

$$q_g = \frac{(2\pi m \mathbf{k}T)^{3/2}}{h^3} b_g \tag{17.32}$$

where b_g represents the rotational and vibrational factors in the partition function. The adsorption sites have very little freedom of motion and their partition function q_s may be taken as unity. The partition function q_a for the absorbed molecules only involves internal factors, which may be written as b_a. The adsorption isotherm thus becomes

$$\frac{\theta}{1 - \theta} = c_g \frac{h^3}{(2\pi m \mathbf{k}T)^{3/2}} \frac{b_a}{b_g} e^{-\Delta E_0/RT} \tag{17.33}$$

This equation has the same form as Eq. 17.5 but now the constant K is given in explicit form.

This equation is applicable to the situation where the adsorbed molecules are localized on the surface. This is usually the case with chemisorption, because of the strength of the binding of the adsorbed molecules to the surface. In some systems where there is weak binding and sparce surface coverage, there is evidence that the adsorbed molecules can move fairly freely on the surface; that is to say, they have two degrees of translational freedom. When this is the case, another equation applies (see Problem 23).

17.4 Chemical Reactions on Surfaces

An important concept in connection with surface reactions is the *molecularity*, which is the number of reactant molecules that come together during the course of reaction; we do not count the surface sites. The molecularity of a surface reaction is deduced from the kinetics on the basis of the experimental results and of theoretical considerations. Reactions involving a single reacting substance are usually, but not invariably, unimolecular. The mechanism of the surface-catalyzed ammonia decomposition is, for example, usually unimolecular. On the other hand, the kinetics of the decomposition of acetaldehyde on various surfaces can only be interpreted on the hypothesis that two acetaldehyde molecules, adsorbed on neighboring surface sites, undergo a bimolecular reaction. Reactions involving two reacting substances are usually bimolecular. When reactant molecules are dissociated on the surface, the reaction may involve interaction between an atom or radical and a molecule; for example, the exchange reaction between ammonia and deuterium on iron is a bimolecular interaction between a deuterium atom and an ammonia molecule.

Unimolecular Reactions

Surface reactions involving one molecule may be treated in terms of the Langmuir adsorption isotherm (Eq. 17.6). In the simplest case the rate of reaction is proportional to θ and is thus

$$v = k\theta = \frac{kK[\mathrm{A}]}{1 + K[\mathrm{A}]} \qquad (17.34)$$

This equation is of the same form as the Michaelis-Menten equation (Eq. 10.82), but we should note that it has been derived on the assumption of a rapid adsorption equilibrium followed by slow chemical reaction.

The dependence of v on [A], shown in Figure 17.2a, is exactly the same as that given by the Langmuir isotherm (Figure 17.1). At sufficiently high concentrations the rate is independent of the concentration, which means that the kinetics are zero order. At low concentrations, when $K[\mathrm{A}] \ll 1$, the kinetics are first order. A good example of this type of behavior is the decomposition of ammonia into nitrogen and hydrogen on a tungsten surface.*

Sometimes a substance other than the reactant is adsorbed on the surface, with the result that the effective surface area and, therefore, the rate are reduced. Suppose that a substance A is undergoing a unimolecular reaction on a surface and that a nonreacting substance I, known as an *inhibitor* or a *poison*, is also adsorbed. If the fraction of the surface covered by A is θ and that covered by I is θ_i, we have from Eq. 17.22 that

$$\theta = \frac{K[\mathrm{A}]}{1 + K[\mathrm{A}] + K_i[\mathrm{I}]} \qquad (17.35)$$

*C. N. Hinshelwood and R. E. Burk, *J. Chem. Soc.*, *127*, 1051, 1114 (1925).

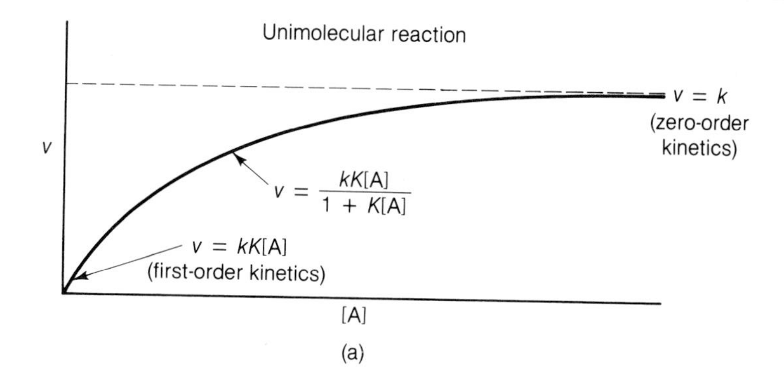

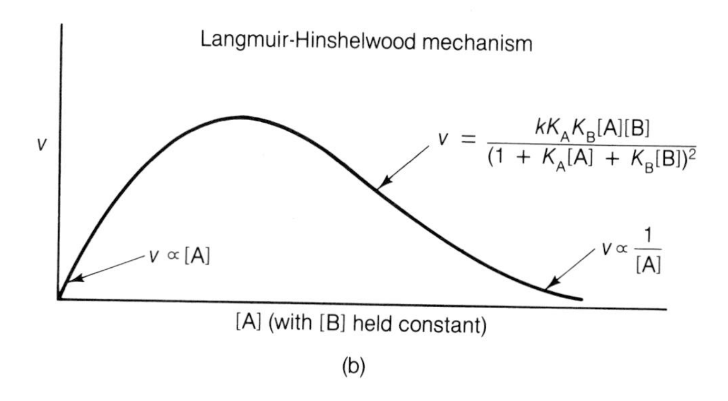

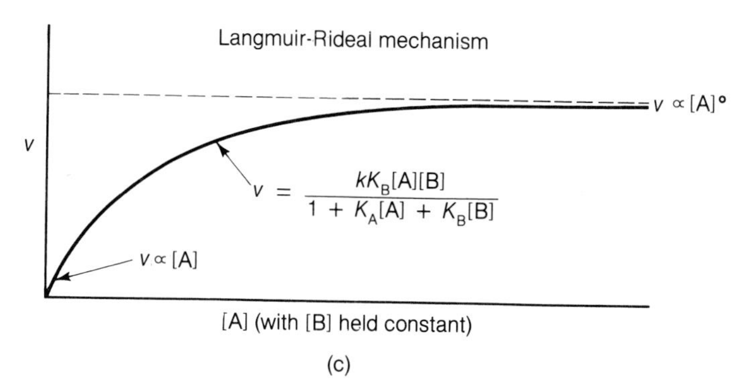

FIGURE 17.2 The variation of rate with concentration for various types of surface reactions. (a) Simple unimolecular processes. (b) A bimolecular reaction occurring by a Langmuir-Hinshelwood mechanism. (c) A bimolecular reaction occurring by a Langmuir-Rideal mechanism.

where K and K_i are the adsorption constants for A and I. The rate of reaction, equal to kK, is thus

$$v = \frac{kK[\mathrm{A}]}{1 + K[\mathrm{A}] + K_i[\mathrm{I}]} \tag{17.36}$$

In the absence of inhibitor this equation reduces to Eq. 17.34.

A case of special interest is when the surface is only sparsely covered by the reactant but is fairly fully covered by the inhibitor. In other words

$$K_i[\mathrm{I}] >> 1 + K[\mathrm{A}] \tag{17.37}$$

and the rate is then

$$v = \frac{kK[A]}{K_i[I]} \tag{17.38}$$

We can easily understand this relationship if we remember that when a surface is sparsely covered (as it now is by A), the coverage is proportional to [A] (Eq. 17.6); whereas if it is almost fully covered (as it now is by I), the fraction bare is inversely proportional to [I] (Eq. 17.8.) A good example of Eq. 17.38 is provided by the decomposition of ammonia on platinum,* the rate law for which is

$$v = \frac{k[NH_3]}{[H_2]} \tag{17.39}$$

Since hydrogen is a product of reaction, there is progressive inhibition as reaction proceeds. There is no appreciable inhibition by the other reaction product, nitrogen.

Bimolecular Reactions

When a solid surface catalyzes a bimolecular process, there are two possible mechanisms. One of these, first suggested by Langmuir† and further developed by Hinshelwood,‡ involves reaction between two molecules adsorbed on the surface. Such a mechanism may be formulated as follows.

$$A + B + \text{—S—S—} \rightleftharpoons \underset{\text{—S—S—}}{\overset{A\quad B}{|\quad\ |}} \longrightarrow \underset{\text{—S—S—}}{\overset{A\text{---}B^{\ddagger}}{\vdots\quad\ \vdots}} \longrightarrow \text{—S—S—} + Y + Z$$

activated complex

In the first step the two molecules A and B become adsorbed on neighboring sites on the surface. Reaction then takes place, by way of an activated complex, to give the reaction products.

When this **Langmuir-Hinshelwood mechanism** applies, the rate is proportional to the probability that A and B are adsorbed on neighboring sites, and this is proportional to the product of fractions of the surface, θ_A and θ_B, covered by A and B. These fractions were given in Eqs. 17.22 and 17.23, and the rate of reaction is therefore

$$v = k\theta\theta' \tag{17.40}$$

$$= \frac{kK_AK_B[A][B]}{(1 + K_A[A] + K_B[B])^2} \tag{17.41}$$

If [B] is held constant and [A] varied, the variation of rate is as shown in Figure 17.2b. The rate first rises, passes through a maximum, and then decreases toward zero. The physical explanation of the falling off of the rate at high concentrations is that one reactant displaces the other as its concentration is increased. The maximum rate corresponds to the presence of the maximum number of neighboring A-B pairs on the surface.

---

*C. N. Hinshelwood and R. E. Burk, *J. Chem. Soc.*, 1114 (1925).

†I. Langmuir, *Trans. Faraday Soc.*, *17*, 621 (1921).

‡C. N. Hinshelwood, *Kinetics of Chemical Change* (1st ed.) Oxford: Clarendon Press, 1926, p. 145.

At sufficiently low concentrations of A and B, Eq. 17.41 predicts second-order kinetics, and this has been observed in a number of systems. Rate maxima have also often been observed. A case of special interest is when one reactant (e.g., A) is weakly adsorbed and the other strongly adsorbed. This means that $K_B[B] \gg 1 + K_A[A]$, and it follows from Eq. 17.41 that

$$v = \frac{kK_A[A]}{K_B[B]} \tag{17.42}$$

The order is thus 1 for A and -1 for B. An example is the reaction between carbon monoxide and oxygen on quartz,* where the rate is proportional to the pressure of oxygen and inversely proportional to the pressure of carbon monoxide.

An alternative mechanism for a bimolecular surface process is for the reaction to occur between a molecule that is not adsorbed (e.g., A) and an adsorbed molecule (B). Such a mechanism was also considered by Langmuir† and was further developed by the British physical chemist Sir Eric K. Rideal‡ (1890–1974). This mechanism may be represented as

$$A + \underset{|}{\overset{B}{\text{—S—}}} \longrightarrow \underset{\text{activated complex}}{\overset{A\cdots B \; \ddagger}{\text{—S—}}} \longrightarrow \overset{|}{\text{—S—}} + \text{products}$$

The adsorption of A may occur; it is simply postulated, in this mechanism, that an adsorbed A does not react.

The rate for this **Langmuir-Rideal mechanism** is proportional to the concentration of A and to the fraction of the surface that is covered by B:

$$v = k[A]\theta_B \tag{17.43}$$

$$= \frac{kK_B[A][B]}{1 + K_A[A] + K_B[B]} \tag{17.44}$$

The variation of v with [A] is shown in Figure 17.2c; the same type of curve is found if v is plotted against [B]. There is now no maximum in the rate, and this provides a possible way of distinguishing between this mechanism and the Langmuir-Hinshelwood mechanism.

Not many ordinary chemical reactions occur by a Langmuir-Rideal mechanism. There is evidence, however, that radical combinations on surfaces sometimes occur in this way. The combination of hydrogen atoms, for example, is sometimes a first-order reaction, and it appears§ to occur by the mechanism

$$H + \overset{H}{\underset{}{\overset{|}{\text{—S—}}}} \longrightarrow \overset{|}{\text{—S—}} + H_2$$

*M. Bodenstein and F. Ohlmer, *Z. Physik, Chem., 53*, 166 (1905).
†I. Langmuir, *Trans. Faraday Soc., 17*, 621 (1921).
‡E. K. Rideal, *Proc. Cambridge Phil. Soc., 35*, 130 (1939); *Chem. Ind., 62*, 335, (1943).
§For a detailed discussion, see K. J. Laidler in P. H. Emmett, (Ed.), *Catalysis* (Vol. 1), New York: Reinhold, Pub. Co. (1954), pp. 178–180.

The fraction of the surface covered by hydrogen atoms is

$$\theta = \frac{K[H]}{1 + K[H]} \tag{17.45}$$

and the rate of combination is thus

$$v = k[H]\theta = \frac{kK[H]^2}{1 + K[H]} \tag{17.46}$$

At lower temperatures the surface may be fully covered, so that $K[H] >> 1$ and

$$v = k[H] \tag{17.47}$$

and the kinetics are first order. At higher temperatures the coverage decreases and if $1 >> K[H]$, the rate is

$$v = kK[H]^2 \tag{17.48}$$

An increase in order from 1 to 2 has in fact been observed as the temperature is raised.

In 1912, and later, Langmuir* made important investigations on the production of hydrogen atoms at hot tungsten surfaces, such as used in tungsten-filament incandescent lamps. The rate is proportional to the square root of the hydrogen pressure, and the mechanism is believed to be the reverse of that just given, namely

$$H_2 + \text{—}\underset{|}{\overset{|}{S}}\text{—} \longrightarrow \text{—}\overset{\overset{H}{|}}{S}\text{—} + H$$

17.5 Surface Heterogeneity

In our treatment of adsorption and of chemical reactions on surfaces we have so far been assuming that all the surface sites are of the same character. The Langmuir adsorption isotherm, for example, is based on this assumption of surface homogeneity. In reality, as H. S. Taylor first pointed out, surfaces are never absolutely smooth, and we have to take account of variations in surface activity.

Various lines of experimental evidence provide evidence for the *inherent heterogeneity* of surfaces. For example, if a metal is heated for a period of time, its capacity for adsorption and catalysis is usually decreased. This is due to sintering of the surface, resulting in a decrease in the number of atoms that constitute the most active centers.

There is also much kinetic evidence for the variability of surfaces. For example, the decomposition of ammonia on molybdenum is retarded by nitrogen, but as the surface becomes saturated by nitrogen, the rate of the decomposition does not fall to zero.† This suggests that the reaction can occur on certain surface sites on which the nitrogen cannot be adsorbed.

---

*I. Langmuir, *J. Am. Chem. Soc.*, *34*, 1310 (1912); *37*, 417 (1915).

†R. E. Burk, *Proc. Nat. Acad. Sci. U.S.*, *13*, 67 (1927).

In view of the fact that surfaces show such variability, it may at first seem surprising that fairly simple kinetic laws often apply to surface-catalyzed reactions. The reason is that because of the exponential term in the rate equation a reaction will occur predominantly on the most active surface sites. For example, suppose that a reaction can occur at ordinary temperatures on 10% of the surface sites with an activation energy of 50 kJ mol^{-1} and on the remaining 90% with activation energies ranging from 80 to 150 kJ mol^{-1}. In this situation the amount of reaction occurring on the 90% of less active sites will be negligible compared with that occurring on the 10% of most active sites. The variability of the 90% of the surface will therefore have little effect on the kinetic behavior.

Aside from an *inherent heterogeneity*, resulting in a true variation in surface sites, there can also be an *induced heterogeneity* resulting from the *interactions* between adsorbed molecules. Suppose, for example, that there is a significant repulsion between molecules that are adsorbed side by side on a surface. The result will be that as the substance is progressively adsorbed, the first molecules will not be close together, but subsequent molecules will necessarily be close to other molecules and will be adsorbed less strongly. The adsorption behavior is thus very similar to that occurring when the surface is inherently heterogeneous; the first molecules will be attached more strongly than the later ones.

The heat evolved on chemisorption usually falls as the surface is progressively covered. This can be due to inherent heterogeneity, to induced heterogeneity, or to a combination of the two.

Many surface catalysts are much more efficient if they are *promoted*. A **promoter** is a substance that in itself has little or no catalytic power but which when added to a catalyst greatly improves its performance. A pure iron surface, for example, is not a particularly good catalyst for the synthesis of ammonia from nitrogen and hydrogen; however, addition of small amounts of various materials, such as Al_2O_3, brings about a very considerable improvement in the catalytic performance. Promoters bring about their action in a variety of ways. Some of them act mainly by changing the structure of the surface, with an increase in the surface area. Promoters also act by producing surface regions, such as phase boundaries, which have a greater catalytic activity than other parts of the surface.

Most catalysts are easily poisoned by substances that are strongly adsorbed at the active centers on the surface. Since the proportion of active centers may be small, minute amounts of poison can cause a very considerable decrease in the effectiveness of a solid surface.

17.6 The Structure of Solid Surfaces and of Adsorbed Layers

A considerable amount of experimental and theoretical work has been directed toward understanding the structures of solid surfaces and of adsorbed layers. Only a very brief account can be given here; for further details the reader is directed to the many books and review articles on the subject, some of which are listed in Suggested Reading (p. 806).

Table 17.1 lists some of the experimental techniques that have been developed to investigate solid surfaces and gives an indication of what each technique in-

TABLE 17.1 Some Important Techniques for the Study of Surface Composition and Structure

Technique	Acronym	Comments
Angle-resolved photo-electron spectroscopy (see also UPS and XPS)	ARPES	Measurement of kinetic energies and angular distribution of photo-electrons, emitted when radiation strikes a surface.
Auger electron spectroscopy	AES	After the photoemission of an electron from a surface, there may be internal electronic reorganization, with the excess energy emitting a second electron (the Auger effect)
Field-ion spectroscopy	FIM	He^+ ions repelled from a surface (see text)
High-resolution electron-loss spectroscopy	HRELS	Inelastic backscattering of electrons; this technique provides information about absorbed molecules.
Ion-scattering spectroscopy	ISS	The scattering of ions from a surface
Low-energy electron diffraction	LEED	Backscattering of low-energy electrons from a layer of surface atoms (see text)
Ultraviolet photoelectron spectroscopy (see also ARPES)	UPS	Study of the emission of electrons when ultraviolet radiation strikes a surface
X-Ray photoelectron spectroscopy (see also ARPES)	XPS	Study of the emission of electrons when X radiation strikes a surface.

volves. Particularly important information has been revealed by **field-ion microscopy*** (FIM). A schematic diagram of the apparatus employed is shown in Figure 17.3. A stream of helium gas enters an evacuated chamber and impinges on a small crystal of the material under examination. An electric field of about 10^9 V cm^{-1} is applied to the crystal, and this is high enough to produce He^+ ions which are repelled radially from the surface. The ions are accelerated toward a fluorescent screen which displays a highly magnified image of the surface; the intensity at any point on the screen is proportional to the number of incident helium atoms. Analysis of the image permits an identification of the positions of the various atoms on the surface.

Another widely used technique is **low-energy electron diffraction**† (LEED), which involves the backscattering of low-energy electrons from a surface. The apparatus is shown schematically in Figure 17.4. In order for there to be effective diffraction, the wavelength of the radiation must be similiar to the interatomic distances. For a surface study it is also important that the electron beam should not penetrate much below the surface, which requires the electron beam to be of low

*E. W. Muler and T. T. Tsong, *Field Ion Microscopy*, New York: Elsevier, 1969.
†G. A. Somorjai and H. H. Farrell, "Low Energy Electron Diffraction," *Adv. Chem. Phys.*, *20*, 215 (1970).

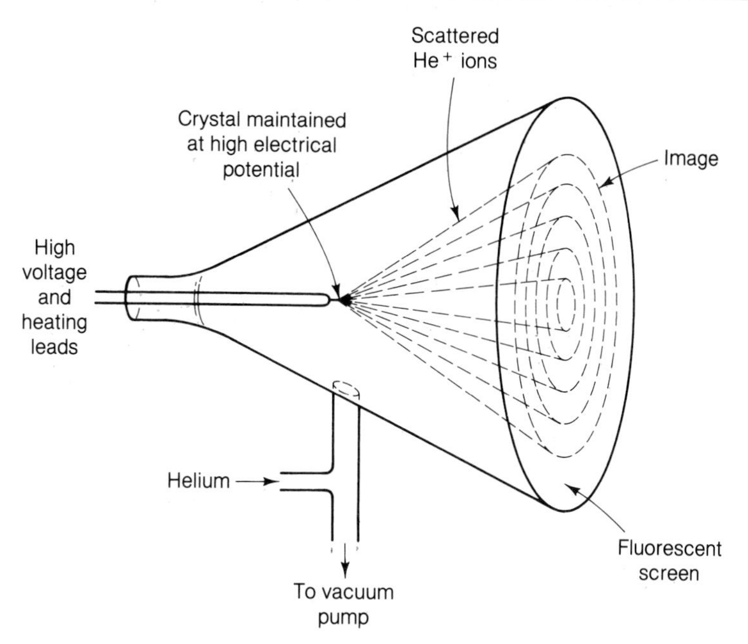

FIGURE 17.3 Schematic diagram showing the principle of the field-ion microscope for the investigation of surfaces.

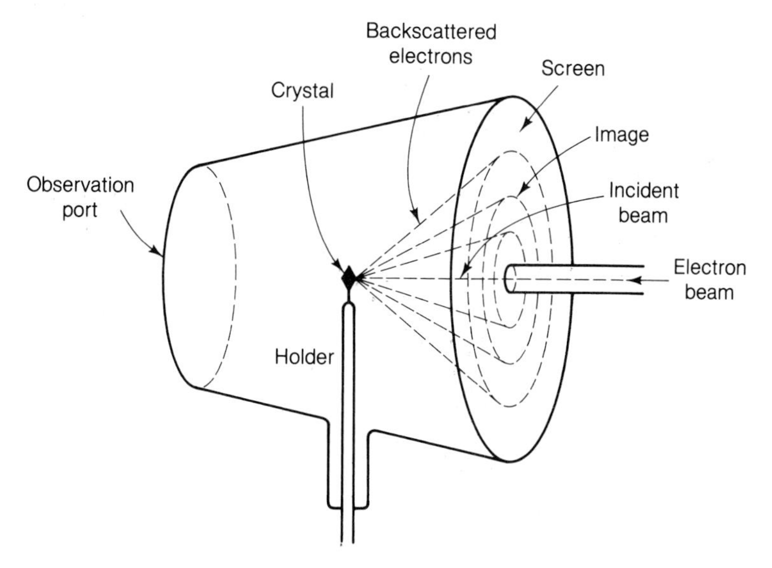

FIGURE 17.4 Schematic diagram showing the principle of the low-energy electron diffraction (LEED) technique for the study of surfaces.

energy. When these conditions are satisfied, the scattered beam reveals the properties of the surface atoms rather than those of the atoms in the bulk of the solid. The incident electrons have a narrow range of energies (i.e., are monochromatic), and the backscattered electrons are caused to impinge on a fluorescent screen. The structural information about the surface is provided by the elastically scattered electrons, which have not lost energy in the scattering process. If the atoms on the

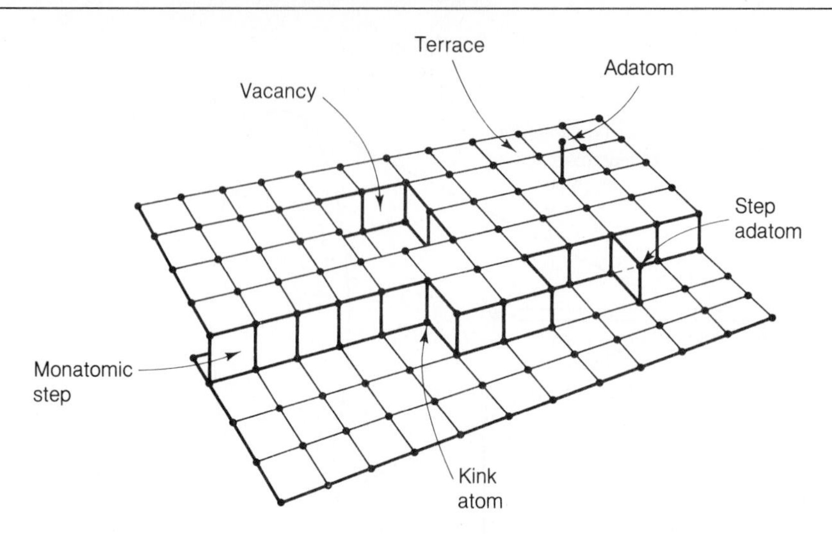

FIGURE 17.5 Schematic diagram of a solid surface, showing some of the features commonly observed.

surface are arranged in an ordered pattern, the diffraction spots on the fluorescent screen are small and of high intensity. Analysis of the distribution of the spots and of their intensity provides detailed information about the atomic configuration of the surface.

The various experimental studies of solid surfaces and adsorbed layers have revealed a number of very important results, which can be mentioned only very briefly. In the first place, surfaces are never smooth, but on the atomic scale they are highly heterogeneous. This is represented in Figure 17.5, which shows that there are various kinds of surface sites; there are atoms in terraces, atoms in steps, atoms at kinks, and adatoms, which project from the surface. These types of surface atoms differ in the number of near neighbors they have; adatoms have few, while atoms in terraces have more. Usually there are few adatoms and many more step, kink, and terrace atoms. The various types of atoms differ very markedly in their chemical behavior; there are large differences in heats of adsorption and in catalytic activity.

Under ordinary experimental conditions, surfaces are covered by a layer of adsorbed molecules, which are removed only by prolonged evacuation. Bare surface atoms are in a state of high Gibbs potential energy, and a lowering results from the formation of an adsorbed layer. Since adsorption is usually associated with a small activation energy, it occurs rapidly, so that a perfectly clean surface will at once become covered if it is exposed to the atmosphere.

There is a good deal of exchange between atoms and molecules that are adsorbed at the different surface sites.* This is because the activation energies for diffusion across a surface are considerably smaller than those for desorption into the gas phase. Under most conditions adsorbed species will soon establish themselves in a state of equilibrium.†

*C. Wagner and K. Hauffe, *Z. Elektrochem.*, *44*, 172 (1938); W. E. Garner, T. J. Gray, and F. S. Stone, *Discussions Faraday Soc.*, *8*, 246 (1950).

†E. Molinari and G. Parravano, *J. Am. Chem. Soc.*, *75*, 5233 (1953).

An important aspect of an adsorbed layer is that a surface dipole is usually produced, as a result of different electronegativities. Such dipoles are particularly important with ionic solids. When such dipoles exist, there is said to be an *electric double layer*, and we will later consider its significance in connection with electrokinetic effects (Sections 17.9 and 18.4).

17.7 Surface Tension and Capillarity

A molecule in the interior of a liquid is, on the average, attracted equally in all directions by its neighbors, and there is therefore no resultant force tending to move it in any direction. On the other hand, at the surface of a liquid that is in contact with vapor there is practically no force attracting the surface molecules away from the liquid, and there is therefore a net inward attraction on the surface molecules. This is represented in Figure 17.6a. If the surface area of a liquid is increased, more molecules are at the surface, and work must be done. A surface therefore has an excess Gibbs energy, relative to the interior of the liquid; the SI unit of this energy per unit surface is $\text{J m}^{-2} \equiv \text{kg s}^{-2} \equiv \text{N m}^{-1}$. This relationship shows that the excess surface energy per unit area is a force per unit length.

In 1805 the English physicist and physician Thomas Young (1773–1829) showed that surfaces behave as if a membrane were stretched over them. An analogy is provided by a soap bubble, which because of the tension of the membrane assumes a spherical form. In the same way a liquid, particularly if it is suspended in liquid with which it is immiscible so as to eliminate the effects of gravity, tends to become spherical. Very small drops of any liquid are almost exactly spherical; larger ones are flattened by their weights.

Figure 17.6b shows a thin film, such as a soap film, stretched on a wire frame having a movable side of length l. The film has two sides and the total length of the side of the film is $2l$. The force F required to stretch the film is proportional to $2l$,

$$F = \gamma(2l) \tag{17.49}$$

and the proportionality constant γ is known as the **surface tension**. Its SI unit is $\text{N m}^{-1} \equiv \text{J m}^{-2}$, and it is therefore the surface energy per unit area. Thus if the piston in Fig. 17.6b moves a distance dx, the area (on the two sides) increases by $2l$

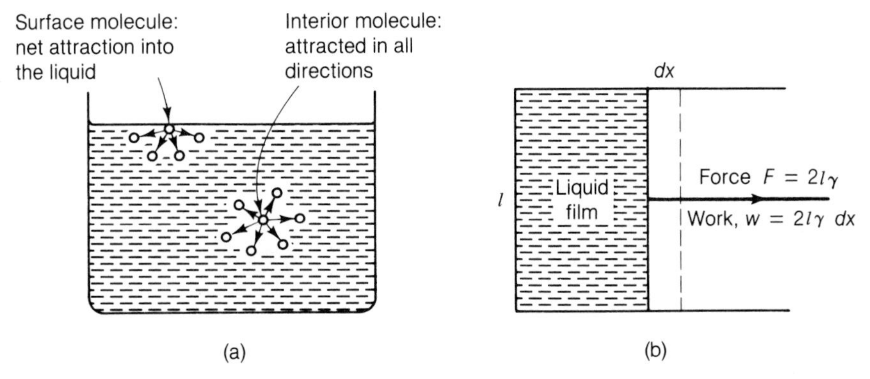

FIGURE 17.6 (a) Attractive forces acting on a surface molecule and on a molecule in the interior of a liquid. (b) A wire frame supporting a liquid film.

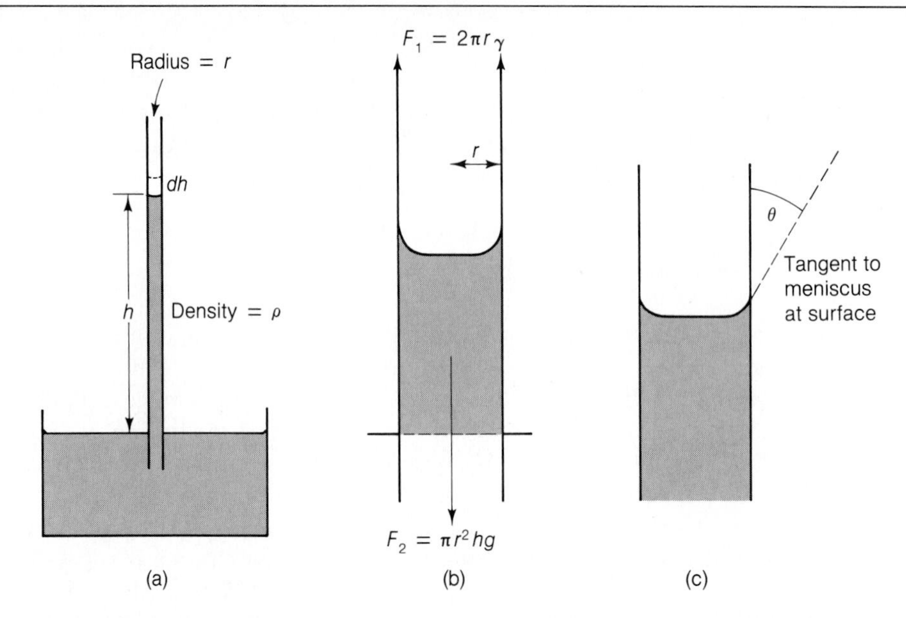

FIGURE 17.7 (a) The capillary-rise method for measuring surface tension. (b) The forces for the case in which the liquid completely wets the glass. (c) The liquid does not completely wet the surface.

dx, and the work done is $2l\gamma\, dx$. The ratio of the work done to the increase in surface area is therefore

$$\frac{\textbf{work done}}{\textbf{increase in surface area}} = \frac{2l\gamma\, dx}{2l\, dx} = \gamma \tag{17.50}$$

Note that this work done is the increase in Gibbs energy; the *increase in Gibbs energy is thus the increase in surface area multiplied by the surface tension* γ.

Several methods are used to measure surface tension. The simplest and most commonly employed is the *capillary-rise method*, illustrated in Figure 17.7a. A tube of small internal radius r is inserted vertically into a liquid, and if the liquid wets the glass, the level of liquid will rise in the tube. The simplest situation, found if there is complete wetting of the tube, is if the angle between the meniscus and the surface is zero, as shown in Figure 17.7b. If this is the case, the force F_1 is acting directly upward. This force is equal to the surface tension multiplied by the length of contact between the edge of the liquid and the tube; this length is equal to $2\pi r$ and therefore

$$F_1 = 2\pi r\gamma \tag{17.51}$$

At equilibrium this force is exactly balanced by the force F_2 due to the weight of the liquid in the capillary. The volume of this is $\pi r^2 h$, and if the density is ρ, the force is

$$F_2 = \pi r^2 h\rho g \tag{17.52}$$

where g is the acceleration of gravity. Therefore

$$2\pi r\gamma = \pi r^2 h\rho g \tag{17.53}$$

or

$$\gamma = \frac{rh\rho g}{2} \tag{17.54}$$

If the liquid does not completely wet the glass, there is an angle θ between the meniscus and the surface, as shown in Figure 17.7c, and when this is the case, Eq. 17.54 becomes

$$\boldsymbol{\gamma = \frac{rh\rho g}{2 \cos \theta}} \tag{17.55}$$

An interesting consequence of the existence of surface tension is that the vapor pressure of a spherical droplet of a liquid may be very considerably larger than that of the bulk liquid. Suppose that the ordinary vapor pressure of a liquid is P_0 and that when it is present in droplets of radius r, the vapor pressure is P. The Gibbs energy change when dn mol of liquid is transferred from a plane surface to a droplet is then

$$dG = dnRT \ln \frac{P}{P_0} \tag{17.56}$$

This change can also be calculated from the surface energy change that results from the increase in surface area. The volume of dn mol is $M\,dn/\rho$, where M is the molar mass and ρ is the density. The droplet has a surface area of $4\pi r^2$, and if the radius increases by dr, the increase in volume is $4\pi r^2\,dr$; thus

$$\frac{M\,dn}{\rho} = 4\pi r^2\,dr \tag{17.57}$$

and therefore

$$dr = \frac{M}{4\pi r^2 \rho}\,dn \tag{17.58}$$

The increase in surface area of the droplet is

$$dA = 4\pi(r + dr)^2 - 4\pi r^2 = 8\pi r\,dr \tag{17.59}$$

and the increase in surface Gibbs energy is this quantity multiplied by γ:

$$dG = 8\pi r\gamma\,dr \tag{17.60}$$

Insertion of the expression for dr in Eq. 17.58 leads to

$$dG = \frac{2\gamma M}{\rho r}\,dn \tag{17.61}$$

Equating Eqs. 17.56 and 17.61 gives

$$dnRT \ln \frac{P}{P_0} = \frac{2\gamma M}{\rho r}\,dn \tag{17.62}$$

or

$$\ln \frac{P}{P_0} = \frac{2\gamma M}{\rho rRT} \tag{17.63}$$

TABLE 17.2 The Vapor Pressure of Bulk Water and of Water Droplets at 25°C

Droplet Radius, r/m	Vapor Pressure, P/kPa	P/P_0
∞(bulk water)	3.167	1
10^{-6}	3.170	1.001
10^{-7}	3.202	1.011
10^{-8}	3.519	1.111
10^{-9}	9.12	2.88

Some values for water droplets, calculated from this relationship, are shown in Table 17.2.

The fact that the vapor pressure of a tiny droplet can be so much higher than that of the bulk liquid poses an interesting question regarding the process of condensation of a vapor. Suppose, for example, that air saturated with water is chilled. The vapor then becomes supersaturated and is in a metastable state. However, if the condensation first produces tiny droplets, these will have vapor pressures many times that of the bulk liquid, and they should immediately evaporate again. How, then, can condensation ever get started?

There are two possibilities. One is that, as a matter of chance, a large number of molecules may come together to form a droplet large enough that it does not at once reevaporate. This, however, is not very likely, and this explanation does not often apply. More often it is dust particles that act as nuclei for supersaturated vapors, or alternatively the condensation may occur at the surface of the vessel. It is well known that supersaturation is much more likely—that is, condensation is much more difficult—in the complete absence of dust particles.

17.8 Liquid Films on Surfaces

The spreading of oil on the surface of water has been of scientific interest for a very long time. In 1765 Benjamin Franklin made observations on a layer of oil on the surface of a pond in London and made an estimate of its thickness. In 1891 the German physicist A. Pockels* first studied the relationship between the area occupied by a film and the surface tension and found that the behavior was different above and below a certain critical area. In 1899 Lord Rayleigh† concluded that this critical area is that at which the molecules in the film are closely packed, in a *unimolecular layer*.

An important advance was made in 1917 when Irving Langmuir‡ devised a **film balance.** This apparatus, shown in Figure 17.8, measures the *surface pressure* π_s of the film, which is the *force exerted on it divided by the length of the edge along which the*

*A. Pockels, *Nature,* 43, 437 (1891) and later papers. Although Fräulein Pockels modestly described herself as "not a professional physicist," she made outstanding contributions to surface science.

†Lord Rayleigh, *Phil. Mag.,* *48,* 337 (1899).

‡I. Langmuir, *J. Am. Chem. Soc.,* *39,* 1848 (1917).

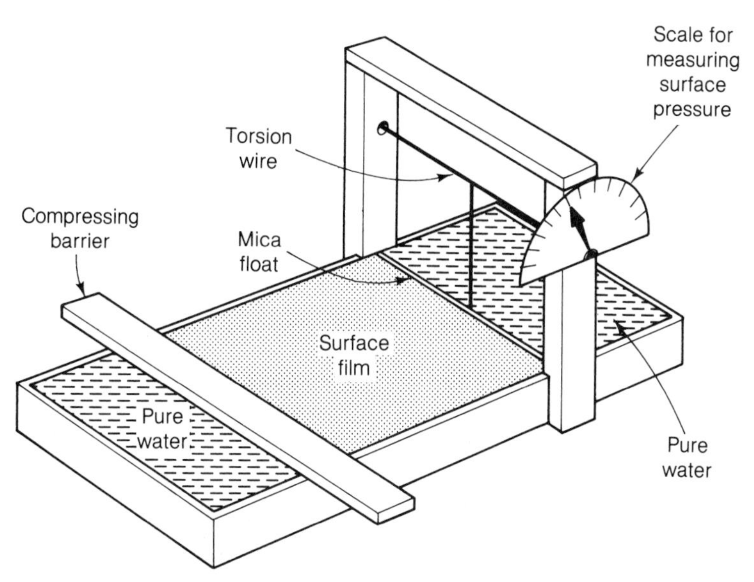

FIGURE 17.8 Diagrammatic representation of Langmuir's film balance.

force is exerted; its SI unit is N $m^{-1} \equiv$ kg s^{-2}, which is the same as that of the surface tension. In fact, we see from Figure 17.6b and Eq. 17.49 that for a soap film the surface pressure π_s (there equal to $F/2l$ since the film has two sides) is equal to the surface tension. For a film lying on the surface of water the surface pressure is equal to the *decrease* in surface tension brought about by the film.

The essential features of Langmuir's film balance are shown in Figure 17.8. A trough is filled to the brim with water, and a fixed barrier, which may be a strip of mica, floats on the surface; special devices are used to prevent leakage of the film past the float. The fixed barrier is suspended from a torsion wire by means of which the pressure is measured. A movable barrier rests on the sides of the trough and is in contact with the surface of the water. The area available to the film is varied by moving this barrier. Subsequent to Langmuir's pioneering work many studies of monolayers have been carried out with very similar apparatus.

Studies of the relationship between the surface area and the surface pressure have shown that two-dimensional monolayers can exist in states that are analogous to the solid, liquid, and gaseous states of three-dimensional matter. Surface films are found to be of two general kinds:

1. **Coherent Films.** For these there is a critical area per molecule; at larger areas the pressure is quite small, but when the critical area is reached, there is little further decrease in area with increase in pressure. An example of this type of behavior is shown in Figure 17.9a, which shows the type of pressure-area curve obtained with a long-chain fatty acid such as stearic acid. The critical area is about 0.2 nm^2. This behavior is analogous to that shown by a three-dimensional gas at a temperature below its critical temperature (compare Figure 1.12); when the gas is compressed sufficiently to form the liquid, the compressibility is then very much smaller than for the gas. The area of 0.2 nm^2 found with fatty acids is the area of cross section of the hydrocarbon chain, and when the critical area is reached, it is

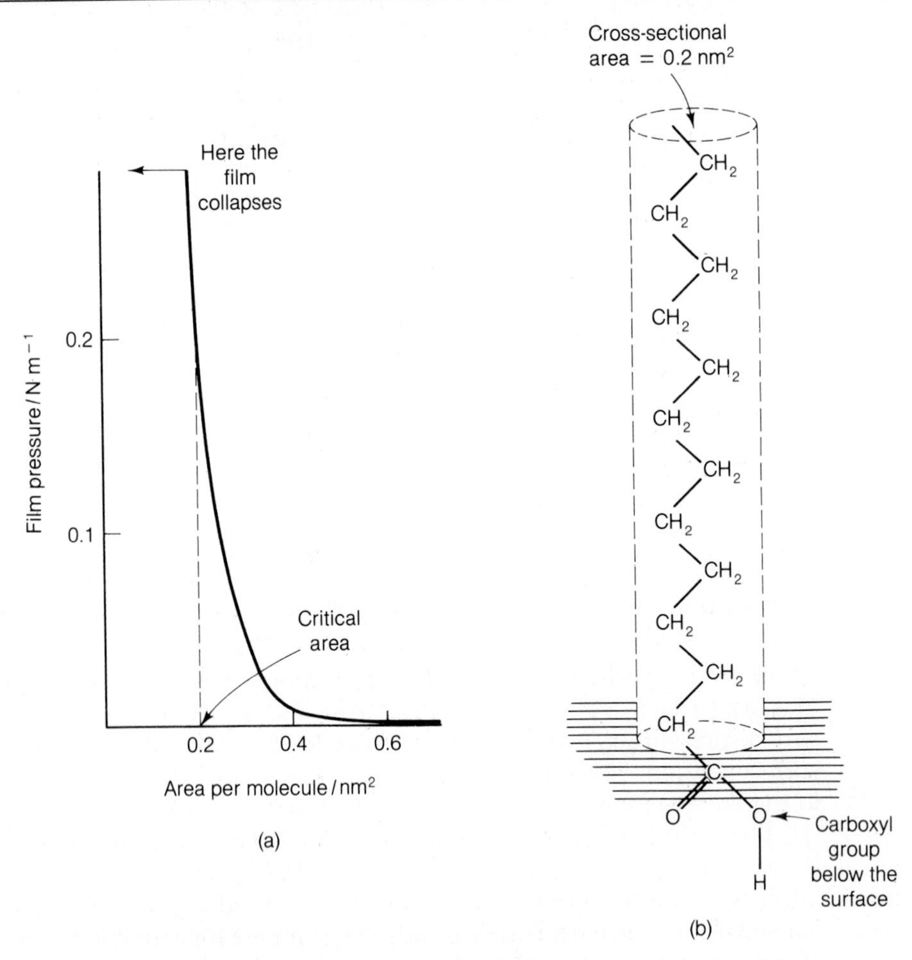

FIGURE 17.9 (a) Relationship between film pressure and molecular area, for a fatty acid such as stearic acid. (b) Diagrammatic representaiton of a fatty acid molecule at the surface of water.

concluded that the molecules are standing up with their polar —COOH groups in the water, as shown in Figure 17.9b.

When the surface pressure reaches a very high value, the area suddenly decreases. The reason for this is that when the molecules are pressed very tightly together, further compression causes them to pile on top of one another.

2. **Noncoherent Films.** For these there is no critical area. Such films obey, to a good approximation, the equation

$$\pi_s A = \mathbf{k}T \tag{17.64}$$

where π_s is the surface pressure and A is the area per molecule adsorbed. This equation is the two-dimensional analog of the ideal gas equation, and noncoherent films therefore behave like a two-dimensional gas; they are often called **gaseous films.**

There are not many examples of gaseous films, which tend to occur at higher temperatures; the molecular energies are then too high to permit the formation of the condensed phase.

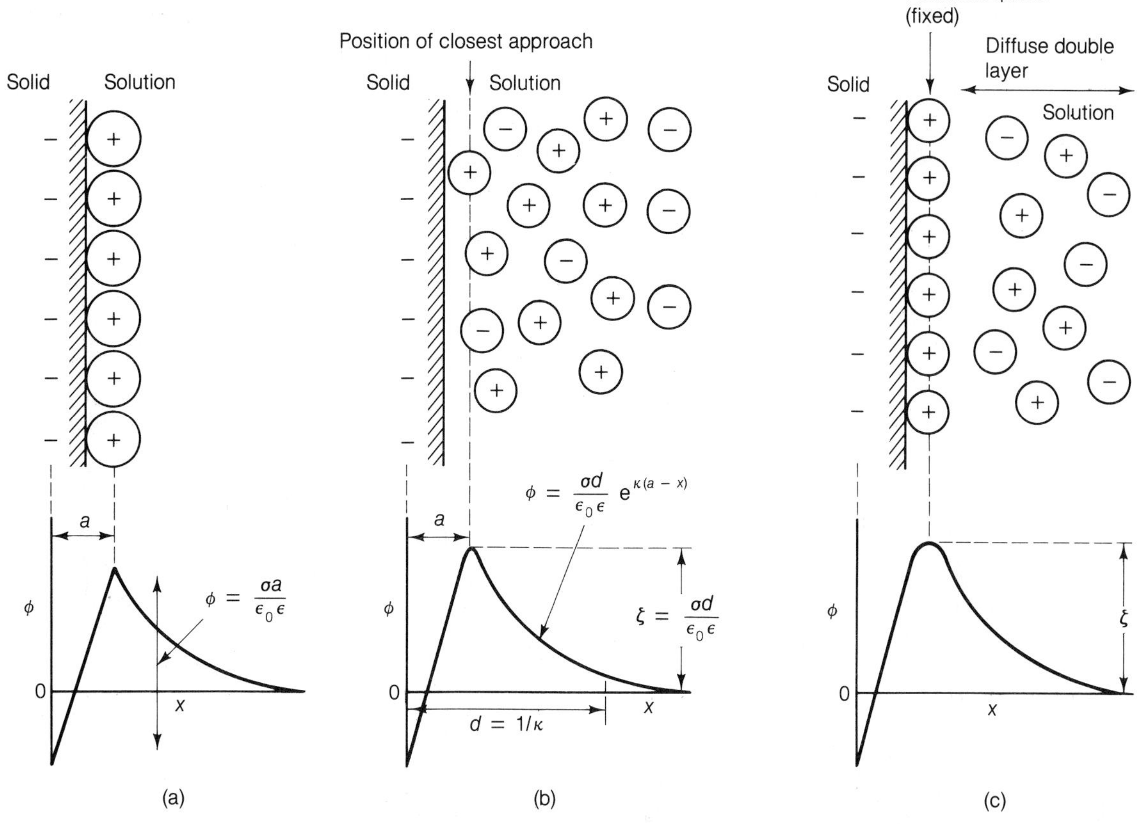

FIGURE 17.10 Three models for the structure of the electric double layer, showing the variations of electric potential ϕ with distance x from the negative charges on the surface. (a) Helmholtz model of the fixed double layer. (b) Diffuse double-layer model of Gouy and Chapman, showing Chapman's expressions for the variation of ϕ with x and for the ζ potential. (c) Stern's model, which is a combination of (a) and (b). The theory for this model is more complicated.

17.9 Solid-Liquid Interfaces

When there are interfaces between solids and liquids, certain properties are observed that indicate that potential differences are established between the solid and liquid phases. For example, if particles are suspended in a liquid and an electric potential is applied, the particles often move toward one of the electrodes. This process, known as *electrophoresis*, is treated in Section 18.4. Here we will consider some of the theories that have been put forward to explain the electric potentials.

The Electric Double Layer

The first and simplest theory was given in 1879 by Helmholtz.* According to his model, which is represented in Figure 17.10a, the surface of a solid can be regarded as bearing positive or negative charges. If these are negative, as in the diagram, a unimolecular layer of positive charges will be attracted to the surface from the

*H. L. F. von Helmholtz, *Wiss. Abhandl. Physik-tech. Reichsanstalt, 1*, 925 (1879).

solution. A **fixed double layer** is therefore formed, and this will correspond to an electric capacitor. If the distance between the layers of positive and negative charges is a and each layer has a charge density (i.e., a charge per unit area) of σ, it follows from electrostatic theory that the potential difference between the two layers is

$$\phi = \frac{\sigma a}{\epsilon_0 \epsilon} \tag{17.65}$$

where ϵ is the dielectric constant of the liquid medium.

This simple idea from Helmholtz, that a layer of ions from the solution becomes attached to the surface, was modified in 1910 by the French physicist Georges Gouy* (1854–1926) and in 1913 by the British chemist David Leonard Chapman† (1869–1959). These workers pointed out that the Helmholtz theory is unsatisfactory in neglecting the Boltzmann distribution of the ions. They suggested that on the solution side of the interface there is not a simple layer of ions but, instead, an ionic distribution that extends some distance from the surface; in other words, there is a **diffuse double layer** as shown in Figure 17.10b. Thermal agitation permits the free movement of the ions present in the solution, but the distribution of positive and negative ions is not uniform. In the example shown in the figure, the surface is negative, so that there are more positive ions in close proximity to the surface. The idea behind the Gouy-Chapman theory is very similar to that involved in the later Debye-Hückel theory of the ionic atmosphere surrounding an ion (Section 7.4). Indeed, 10 years before the formulation of that theory in 1923, Chapman had worked out an analogous treatment for the distribution of ions around a charged solid.

The expression for the potential ϕ to which Chapman's theory leads is shown in Figure 17.10b. As x becomes very large, ϕ approaches zero; when x is equal to a (the distance of closest approach of the ions to the surface), the potential is

$$\phi_a = \frac{\sigma}{\epsilon_0 \epsilon \kappa} = \frac{\sigma d}{\epsilon_0 \epsilon} \tag{17.66}$$

where $1/\kappa = d$ is the effective thickness of the double layer. This quantity is to be compared to the thickness of the ionic atmosphere in the Debye-Hückel theory (Section 7.4). The potential ϕ_a, which is the potential at $x = a$ with respect to the bulk solution, is known as the **electrokinetic potential** or **ζ potential** and is given the symbol ζ; thus

$$\zeta = \frac{\sigma d}{\epsilon_0 \epsilon} \tag{17.67}$$

The Gouy-Chapman theory did not prove entirely satisfactory, and in 1924 a considerable advance was made by the German-American physicist Otto Stern (b. 1888),‡ whose model is shown in Figure 17.10c. Essentially Stern combined the fixed double-layer model of Helmholtz with the diffuse-double-layer model of

---

*G. Gouy, *J. Phys.*, *9*, 457 (1910).
†D. L. Chapman, *Phil. Mag.*, *25*, 475 (1913).
‡O. Stern, *Z. Elektrochem.*, *30*, 508 (1924).

Gouy and Chapman. As shown in the figure, there is a fixed layer at the surface and also a diffuse layer. On the whole this treatment has proved to be very satisfactory, but for certain kinds of investigations it has been found necessary to develop more elaborate models.

17.10 Colloidal Systems

In a true solution, such as one of sugar or salt in water, the solute particles consist of individual molecules or ions. At the other extreme there are **suspensions,** in which the particles contain more than one molecule and are large enough to be seen by the eye or at least under a microscope. Between these extremes are to be found the **colloidal* dispersions,** in which the particles may contain more than one molecule but are not large enough to be seen in a microscope. It is impossible to draw a distinct line between colloidal dispersions, true solutions, and suspensions. The lower limit of microscopic visibility is about 2×10^{-7} m (0.2 μm or 200 nm), and this can be taken as the upper limit of the size of colloidal particles. The lower limit can be taken to be roughly 5×10^{-9} m (5 nm). This figure is comparable to the diameters of certain macromolecules, such as starch and proteins. Solutions of these substances therefore exhibit colloidal behavior, and although they may involve single molecules, they are conveniently classified as colloidal systems.

The essential properties of colloidal systems are due to the fact that the ratio of surface area to volume is very large. A true solution is a one-phase system, but a colloidal dispersion behaves as a two-phase system, since for each particle there is a definite surface of separation between it and the medium in which the particles are dispersed (the **dispersion medium**). At this surface certain characteristic properties, such as adsorption and electric potential, make themselves evident, since the total surface area in a colloidal dispersion can be very large. For example, the surface area of a cube of 1-cm edge is 6×10^{-4} m^2, but if the cube is subdivided into 10^{18} cubes of 10^{-8}-m edge, typical of colloidal systems, the total area is $10^{18} \times 6 \times 10^{-16} = 600$ m^2; the area has thus increased by a factor of 10^6. The fact that surface effects are important in colloid behavior is therefore not surprising.

The term **disperse phase** is used to refer to the particles that are present in the *dispersion medium.* Both the disperse phase and the dispersion medium may be solid, liquid, or gaseous. Since gases are always completely miscible, we cannot have a gas-in-gas colloidal dispersion, but all the other eight combinations are possible and are listed in Table 17.3, which gives the general name for each colloidal system and some well-known examples. In this section we will be mainly concerned with sols and emulsions.

---

*The word *colloid* (Greek *kolla,* glue; *-oeides,* like) was introduced in 1861 by the Scottish chemist Thomas Graham (1805–1869) to refer to substances that diffused slowly and would not pass through parchment.

TABLE 17.3 Types of Colloidal Systems

Dispersion Medium	Disperse Phase	Name of System	Examples
Gas	Liquid	Aerosol	Fog, mist, clouds
Gas	Solid	Aerosol	Smoke
Liquid	Gas	Foam	Whipped cream
Liquid	Liquid	Emulsion	Milk, mayonnaise
Liquid	Solid	Sol	Gold in water
Solid	Liquid	Gel	Jelly, cheese
Solid	Solid	Gel	Ruby glass, gold in glass
Solid	Gas	Solid foam	Pumice, floating soap

Lyophobic and Lyophilic Sols

Colloidal dispersions of solids in liquids (i.e., sols) can be roughly divided into two types:

1. **Lyophobic* Sols.** These can be called *hydrophobic sols* if the dispersion medium is water. The term *hydrophobic* was first used in 1905 by the French physicist Jean Baptiste Perrin (1870–1942) to denote a disperse phase, such as gold or arsenic sulfide, which has a low affinity for water; it is now applied more generally in surface chemistry to refer also to water-repellent surfaces.† Since there is low affinity for the solvent, lyophobic sols are relatively unstable. For example, when lyophobic systems are evaporated, solids are obtained that cannot easily be reconverted into sols, and the addition of electrolytes to lyophobic sols frequently causes coagulation and precipitation.

2. **Lyophilic** (liquid-loving) **Sols.** These are those in which there is a strong affinity between the disperse phase and the molecules comprising the dispersion medium. These sols are much more stable, and they behave much more like true solutions—which in the case of macromolecules they really are.

Colloidal dispersions are prepared in a variety of ways. Some substances are known as *intrinsic colloids* and readily form sols when they are brought into contact with a suitable dispersion medium. A colloidal solution of starch, for example, is easily prepared by introducing starch into boiling water. Intrinsic colloids are usually lyophilic, and they are usually either macromolecules (such as proteins) or long-chain molecules with polar end groups, which tend to aggregate and form

*From the Greek *lysis*, loosening, dissolving; *phobia*, fear of. Lyophobic therefore literally means *fear of dissolving* or *solvent fearing*.

†Note also the use of the word *hydrophobic* in the expression *hydrophobic bond* (Table 12.1). Hydrophobic bonding results from the fact that when nonpolar groups come together in aqueous solution there is a decrease in hydrogen bonding between water molecules, and a resulting increase in entropy.

particles of colloidal size, knows as **micelles**; soaps and other detergents are of this type. *Extrinsic colloids*, on the other hand, do not form colloidal dispersions readily, and special methods have to be used; lyophobic colloids are frequently of this kind. The methods used fall into two classes:

1. *Condensation Methods.* The materials are initially in true solution. Chemical reactions are used to produce the sol; care is taken—by controlling concentrations, for example—to prevent the growth of the particles and consequent precipitation. For example, sols of various oxides have been prepared by hydrolysis of salts.

2. *Dispersion Methods.* Material originally in massive form is disintegrated into particles of colloidal dimensions. In one procedure, known as **peptization** (Greek *pepticos*, promoting digestion), the disintegration is brought about by the action of a substance known as a *peptizing agent*. For example, cellulose is peptized by the addition of organic solvents, such as ethanol-ether mixtures, leading to the familiar "collodion" sol. In this example the solvent is itself the peptizing agent, but frequently something else must be added; thus, precipitates of certain metal hydroxides in water can be peptized by dilute alkali hydroxides. Physical dispersion methods are also useful for producing colloidal dispersions. For example, a *colloid mill*, through which the dispersion medium and the substance to be dispersed are passed, grinds the material into colloidal particles. Another technique is *electrical disintegration*, in which an arc is struck between metal electrodes under water, with the production of a metal sol; this is often known as *Bredig's method*, after the Polish-German physical chemist Georg Bredig* (1868–1944).

Light Scattering by Colloidal Particles

Although by definition colloidal particles are too small to be seen in the microscope, they can be detected by optical means. When light passes through a medium that contains no particles larger than about 10^{-9} m in diameter, the path of light cannot be detected and the medium is said to be *optically clear*. When colloidal particles are present, however, some of the light is scattered, and the incident beam passes through with weakened intensity. The first investigation of this phenomenon was made in 1871 by the British physicist John Tyndall† (1820–1893), and the scattering is known as the **Tyndall effect**; the path of light through the medium, made visible as a result of the scattering, is known as the *Tyndall beam*. A sunbeam is a well-known example of a Tyndall beam, the light being scattered by dust particles.

The type of apparatus used in a light-scattering experiment is shown schematically in Figure 17.11. The detector of the scattered light is mounted in such a way that the intensity of scattering can be measured at various scattering angles. Analysis of the scattering as a function of the angle provides valuable information about the sizes and shapes of colloidal particles; when these are single macromolecules, the technique is therefore useful in determining molar masses.

The theory of light scattering is very complicated and here we will merely outline the main ideas. The proportion of incident light that is scattered increases

---

*G. Bredig, *Z. Elektrochem.*, *4*, 514, 547 (1898).

†J. Tyndall, *Phil. Mag.*, *37*, 384 (1869).

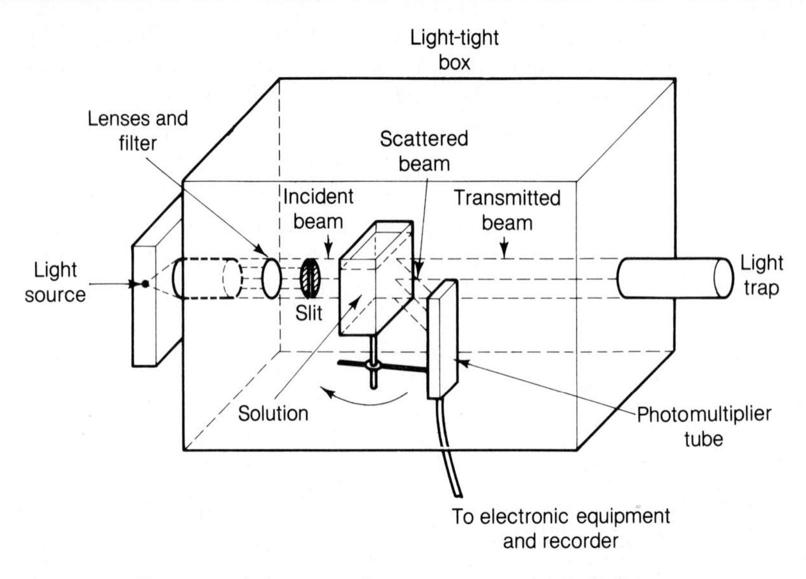

FIGURE 17.11 Schematic diagram of the type of apparatus used in a light-scattering experiment.

with an increase in the number and size of the particles. If the intensity of the incident radiation is I_0 and l is the length of the light path through the scattering medium, the intensity of the transmitted radiation is given by

$$I = I_0 e^{-\tau l} \tag{17.68}$$

where τ is known as the **turbidity**. In 1871 Lord Rayleigh* deduced that for spherical macromolecules of molar mass M, having dimensions much smaller than the wavelength λ of the radiation, the turbidity is given by

$$\tau = \frac{32\pi^3 n_0[(n - n_0)/c]^2 cM}{3\lambda^4 L} \tag{17.69}$$

where n and n_0 are the refractive indices of the solution and solvent, c is the concentration of the scattering particles, and L is the Avogadro constant. This has proved useful for measuring the molar masses of macromolecules that are believed to be spherical.

When the particle dimensions are not small compared to λ and the particles are not spherical, the theory is much more complicated. Important contributions to the theory have been made by Gustav Mie† in 1908, and in more recent years by Bruno H. Zimm‡ and Paul Doty.§

Less detailed studies can be carried out with an instrument known as an **ultramicroscope,** invented in 1903 by the Austro-German chemist Richard Adolf Zsigmondy (1865–1929) and his physicist colleague H. Siedentopf.‖ In this instru-

*J. W. Strutt (Lord Rayleigh), *Phil. Mag., 41,* 447 (1871); *47,* 375 (1988).
†G. Mie, *Ann. Physik* (4), *25,* 377 (1908).
‡B. H. Zimm, *J. Chem. Phys., 16,* 1093 (1948).
§A. M. Holzer, H. Benoit, and P. Doty, *J. Phys. Chem., 58,* 624 (1954).
‖H. Siedentopf and R. A. Zsigmondy, *Ann. Physik, 10,* 1 (1903).

ment a beam is passed through the colloidal system, and individual particles can be seen under the microscope as flashes of scattered light. Particles as small as 5–10 nm in diameter, much too small to be directly visible, can be detected in this way. This is a useful technique for counting particles in a sol; if the amount of material present as disperse phase is known, it is then possible to make an estimate of the size of the particles. Particles detected in an ultramicroscope undergo continuous and rapid motion in all directions, and this is known as **Brownian movement**, named after the Scottish botantist Robert Brown* (1773–1858) who first observed it in pollen grains seen under an ordinary microscope.

Electrical Properties of Colloidal Systems

When two electrodes are placed in a sol and an electric potential is applied across them, the particles in general move in one direction or another. This migration of colloidal particles in an electric field is called **electrophoresis,** and it is one example of a transport property (see Chapter 18). The motion occurs as a result of the ζ potential, the theory of which we met in Section 17.9; in Section 18.4 the theory of electrophoretic movement will be considered. Both lyophobic and lyophilic sols undergo electrophoresis.

The electric double layer and the ζ potential are important for colloids in other ways. The stability of a hydrophobic sol, for example, is very much dependent on the charges on the surface of the particles; repulsion between particles carrying the same charges prevents them from approaching one another and forming larger particles that will precipitate out. The charges at the surface of colloidal particles are influenced to a considerable extent by the adsorption of ions. For example, if a dilute solution of a silver salt is added to an excess of dilute sodium iodide solution, a silver iodide sol is formed in which the particles are negatively charged. This is because iodide ions become preferentially adsorbed on the surface of the silver iodide, as represented in Figure 17.12a; the Na^+ ions remain in solution but form a positively charged atmosphere because of the electrostatic attractions. If, on the other hand, dilute iodide solution is added to an excess of dilute silver nitrate solution, the particles in the sol are positively charged. Now the silver ions are preferentially adsorbed on the particles, the nitrate ions forming the atmosphere, as shown in Figure 17.12b.

The situation with lyophilic sols is a little different, in that the charges on the particles often result largely from the ionization of the material constituting the disperse phase.

Protein molecules in general bear positive or negative charges because of the ionizations of $—COOH$, $—NH_3^+$, $—OH$, and other groups. At low pH values the protein bears a net positive charge because groups like $—NH_3^+$ are present but there are no negatively charged groups; at high pH values negative groups like $—COO^-$ and $—O^-$ are present but there are no positively charged groups and the molecule has a net negative charge. At the same time the charge is influenced by the specific adsorption of ions. At some intermediate pH, the **isoionic point**, the protein does not migrate in an electric field.

*R. Brown, *Phil. Mag.*, *4*, 161 (1828); *6*, 161 (1829); *8*, 41 (1830).

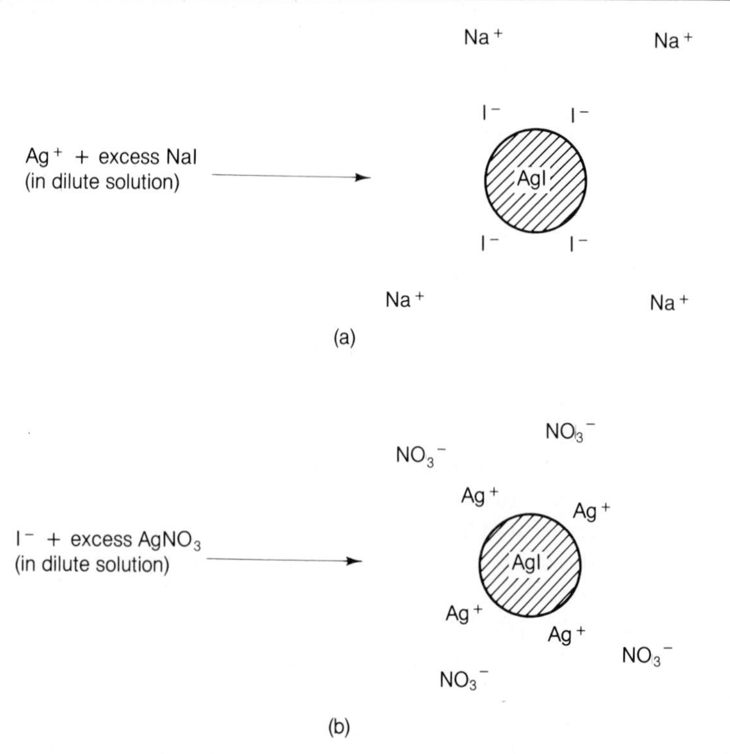

FIGURE 17.12 Silver iodide sols. (a) The particles are formed in the presence of excess iodide ions, which become adsorbed. (b) The particles are formed in the presence of excess silver ions, which are adsorbed.

Added electrolytes influence the stability of colloidal particles by affecting their charges. Again the behavior is different for lyophobic and lyophilic sols. Usually lyophobic particles are easily coagulated, with the formation of a visible precipitate, by the addition of electrolytes. The reason is that ions of opposite charge to that on the surface of the particles will easily become adsorbed, because of the electrostatic attraction, and will bring about a reduction in the charge of the particles, which will repel one another less strongly. This is illustrated in Figure 17.13a for a gold sol particle, precipitated by the addition of sodium chloride. Initially the particle has a negative charge, because of the adsorption of anions. When sodium chloride is added, the sodium ions reduce the surface charge and therefore the repulsion, and precipitation occurs.

The precipitating action of ions of opposite sign to those on the surface of the particle is greater the higher the valence of the ion. These statements regarding the influence of sign and valence on the precipitation of colloidal particles were made in 1882 by H. Schulze* and in 1900 by the English biologist Sir William Bate Hardy† (1864–1934) and are incorporated as the **Schulze-Hardy rule.**

Lyophilic sols differ from lyophobic sols in that much larger amounts of electrolytes are required to bring about precipitation. The reason is that lyophilic particles are to some extent protected by a layer of bound water molecules. This is illustrated

*H. Schulze, *J. Prakt. Chem.*, *25*, 431 (1882); *27*, 320 (1883).
†W. B. Hardy, *Proc. Roy. Soc.*, *66*, 110 (1900).

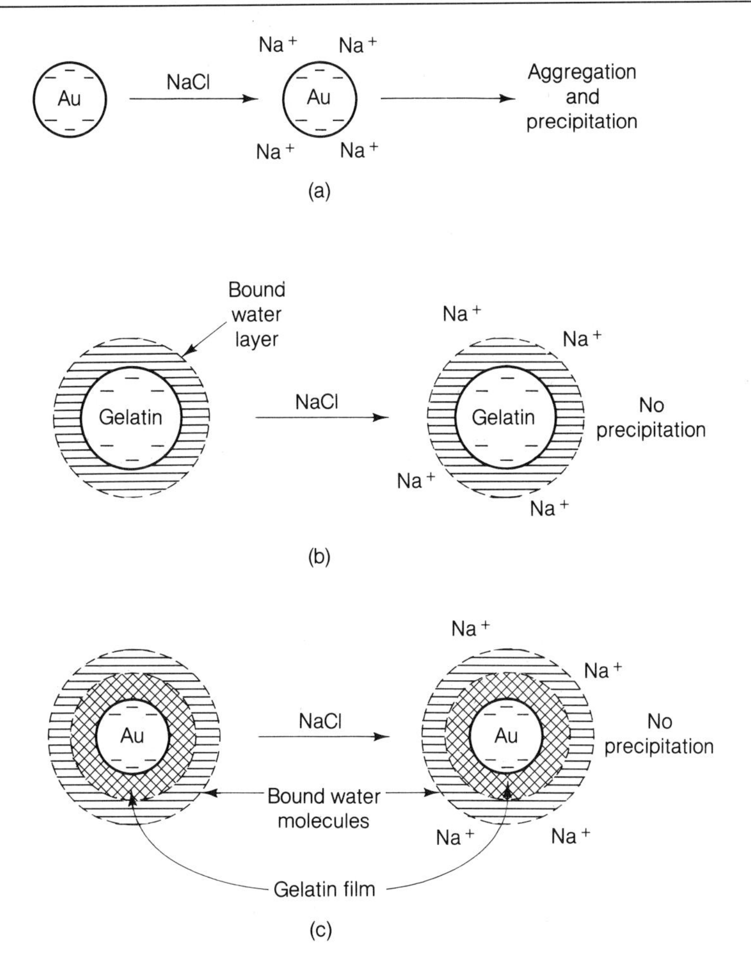

FIGURE 17.13 (a) Precipitation of a hydrophobic sol. The attachment of Na^+ ions to negatively charged gold particles allows them to aggregate. (b) The film of water surrounding a hydrophilic particle inhibits the attachment of ions. (c) Protective action of gelatin attached to a gold particle.

in Figure 17.13b for a protein such as gelatin, which can be regarded as surrounded by a film of water through which ions cannot easily penetrate. Higher concentrations of electrolytes do, however, exert a *salting-out effect,* as we have discussed in Section 7.11; this results from the tendency of the ions to bind water molecules which are thus less available to the substance being precipitated.

The high sensitivity of lyophobic particles to precipitation by electrolytes can sometimes be reduced by the addition of a lyophobic substance, which is said to be a **protective colloid** and to exert a *protective action.* As illustrated in Figure 17.13c, the protective substance becomes adsorbed on the surface of the lyophobic particles and protects them from the approach of ions of opposite sign.

Gels

Under certain circumstances it is possible to cause a lyophilic sol to coagulate and yield a semirigid jellylike mass that includes the whole of the liquid present in the sol. The product so obtained is known as a **gel,** and there are two types, *elastic gels* and *nonelastic gels.* A number of food products, such as jellies, jams, and cornstarch puddings, are elastic gels. A well-known example of an inelastic gel is that of silicic acid, commonly known as *silica gel.* The essential distinction between elastic and nonelastic gels is their behavior on dehydration. Partial dehydration of an elastic gel leads to the formation of an elastic solid from which the original sol can be regenerated by the addition of water. Dehydration of a nonelastic gel, on the other hand, leads to a glass or powder, which has little elasticity.

Another distinction between elastic and nonelastic gels relates to their ability to take up solvent. If an elastic gel such as gelatin is placed in water, it swells, water having been *imbibed* by the gel; the process is known as **imbibition.** Nonelastic gels, on the other hand, may take up solvent but they do not swell; the liquid enters the pores of the gel but, since the walls are rigid, the volume of the gel does not change.

Gel formation occurs in particular with molecules that can exist as extended chains. As the sol turns into a gel, the chains become interlocked so that the viscosity increases and eventually a semi-solid material is produced. The dispersion medium is held by capillary forces between the chains, some of the molecules exerting a specific solvating effect.

Emulsions

An emulsion consists of droplets of one liquid dispersed in another liquid. The droplets are usually from 0.1 to 1 μm in diameter and hence are larger than sol particles. Emulsions are generally unstable unless a third substance, known as an **emulsifying agent** or a *stabilizing agent,* is present. Soaps and detergents are effective emulsifying agents, particularly for oil-water emulsions. They consist of long-chain hydrocarbon molecules each having at one end a polar group such as a carboxylic acid or sulfonic acid group. These molecules are readily adsorbed at oil-water interfaces; the hydrocarbon chains become attached to the oil and the polar groups to the water, as shown schematically in Figure 17.14. The term **micelle** (New Latin *micella,* small crumb, diminutive of the Latin *mica,* crumb) was used by the Canadian-American chemist James William McBain* (1882–1953) and the American chemist William Draper Harkins† (1873–1951) to refer to particles that are stabilized by emulsifying agents in this manner.

The action of emulsifying agents is to reduce the interfacial tension between the two phases. The effect of surface tension, as we have seen in Section 17.7, is to cause the surface to become as small as possible. A high surface tension between the disperse phase and the dispersion medium will therefore tend to cause an emulsion to separate into two bulk phases (i.e., to *coagulate*). The adsorption of an emulsifying agent on an interface reduces this tension and therefore decreases the tendency of the emulsion to coagulate.

An emulsion with no stabilizing agent has properties similar to those of lyophobic sols; for example, they are easily coagulated by electrolytes. Stabilized emulsions, on the other hand, behave more like lyophilic sols and are only affected by electrolytes at high concentrations.

---

*J. W. McBain, *Advan. Colloid Chem.*, *1*, 124 (1942).

†W. D. Harkins, *J. Chem. Phys.*, *13*, 381 (1945); W. D. Harkins and R. S. Stearns, *ibid.*, *14*, 215 (1946).

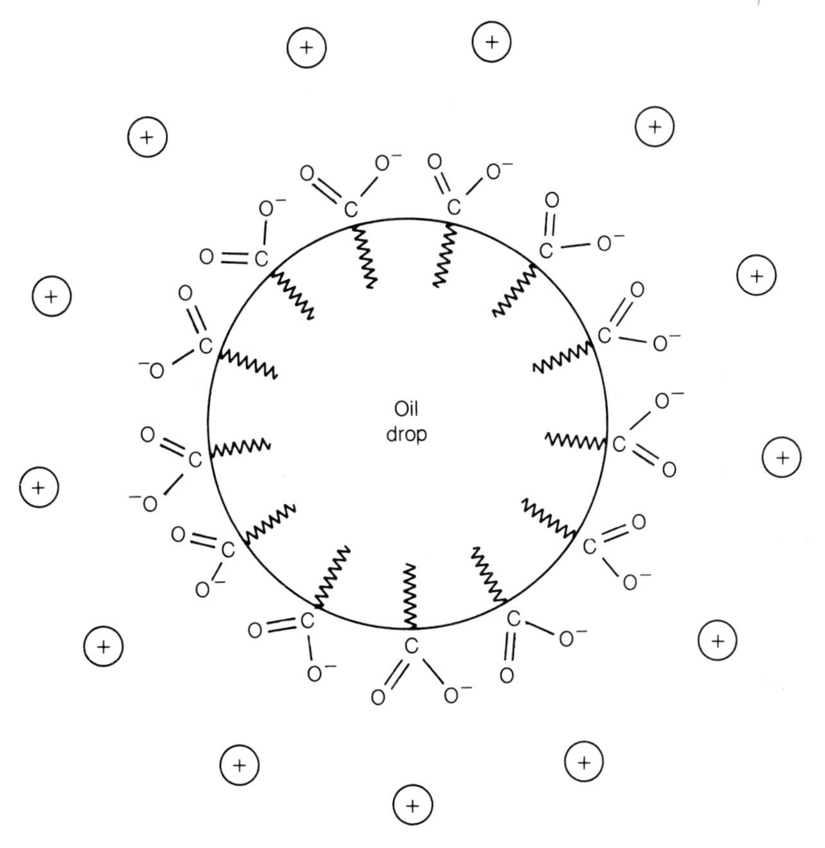

FIGURE 17.14 The structure of a micelle, in which a particle or drop is stabilized by molecules having an anionic group and a long hydrocarbon chain.

Key Equations

Langmuir isotherms for fraction θ of surface covered:

$$\text{Simple adsorption:} \quad \theta = \frac{K[A]}{1 + K[A]}$$

$$\text{Adsorption with dissociation:} \quad \theta = \frac{K^{1/2}[A]^{1/2}}{1 + K^{1/2}[A]^{1/2}}$$

$$\text{Competitive adsorption:} \quad \theta_A = \frac{K_A[A]}{1 + K_A[A] + K_B[B]}$$

Surface tension γ:

$$\text{Definition:} \quad \gamma \equiv \frac{\text{work done}}{\text{increase in surface area}}$$

$$\text{Capillary rise } h \text{ in tube of radius } r\text{:} \quad \gamma = \frac{rh\rho g}{2 \cos \theta}$$

Problems

A Adsorption Isotherms

1. A surface is half-covered by a gas when the pressure is 1 atm. The simple Langmuir isotherm (Eq. 17-6) applies:

a. What is K/atm^{-1}?

b. What pressures give 75%, 90%, 99%, 99.9% coverage?

c. What coverage is given by pressures of 0.1 atm, 0.5 atm, 1000 atm?

2. Show that, if the Langmuir isotherm is obeyed, a plot of P/V_a against P is linear; V_a is the volume of gas adsorbed, corrected to a standard temperature and pressure. Explain how, from such a plot, the volume V_a° corresponding to complete coverage and the isotherm constant K can be determined.

3. The following results were reported by Langmuir for the adsorption of nitrogen on mica at 20°C:

Pressure/atm	2.8	4.0	6.0	9.4	17.1	33.5
Amount of gas adsorbed/mm^3 at 20°C and 1 atm	12.0	15.1	19.0	23.9	28.2	33.0

a. Make a linear plot of these values in order to test the Langmuir isotherm, Eq. 17.6. If it applies, evaluate the constant K.

b. A maximum of about 10^{15} molecules usually covers 1 cm^2 of a surface. Make an estimate of the effective surface area in Langmuir's experiment.

B Kinetics of Surface Reactions

4. A first-order surface reaction is proceeding at a rate of 1.5×10^{-4} mol dm^{-3} s^{-1} and a rate constant 2.0×10^{-3} s^{-1}. What will be the rate and the rate constant if

a. The surface area is increased by a factor of 10?

b. The amount of gas is increased tenfold at constant pressure and temperature?

If these values of v and k apply to a reaction occurring on the surface of a spherical vessel of radius 10 cm,

c. What will be the rate and rate constant in a spherical vessel, of the same material, of radius 100 cm, at the same pressure and temperature?

d. Define a new rate constant k' that is independent of the gas volume V and the area S of the catalyst surface.

e. What would be its SI unit?

5. A zero-order reaction is proceeding at a rate of 2.5×10^{-3} mol dm^{-3} s^{-1} and a rate constant 2.5×10^{-3} mol dm^{-3} s^{-1}.

a. How will the changes a, b, and c in Problem 2 affect the rate and the rate constant in this case?

b. Again, define a rate constant which is independent of S and V.

c. What would be its SI unit?

6. The decomposition of ammonia on platinum,

$$2NH_3 = N_2 + 3H_2$$

is first order in NH_3 and the rate is inversely proportional to the hydrogen concentration (Eq. 17.39). Write the differential rate equation for the rate of formation of hydrogen, dx/dt, in terms of the initial concentration of ammonia, a_0, and the concentration x of hydrogen at time t.

***7.** On the basis of the mechanism given on page 781, derive an expression for the rate of formation of hydrogen atoms when hydrogen gas is in contact with hot tungsten. Under what conditions is the order of reaction one-half?

8. A unimolecular surface reaction is inhibited by a poison I and obeys Eq. 17.36. If E is the activation energy corresponding to the reaction of the adsorbed substrate molecule (i.e., corresponding to k) and ΔH_A and ΔH_I are the enthalpies of adsorption of A and I, what is the activation energy

  a. At very low concentrations of A and I?

  b. At a very high concentration of A and a very low concentration of I?

  c. At a very low concentration of A and a very high concentration of I?

***9.** Suppose that a reaction,

$$A \longrightarrow Y + Z$$

occurs initially as a homogeneous first-order reaction (rate constant k) but that the product Z is adsorbed on the surface and catalyzes the reaction according to a zero-order law (rate is $k_c z$). Obtain a differential equation for the rate of appearance of Z, and integrate it to give z as a function of time.

10. Suggest explanations for the following observations, in each case writing an appropriate rate equation based on a Langmuir isotherm:

  a. The decomposition of phosphine on tungsten is first order at low pressures and zero order at higher pressures, the activation energy being higher at the higher pressures.

  b. The decomposition of ammonia on molybdenum is retarded by the product nitrogen but the rate does not approach zero as the nitrogen pressure is increased.

  c. On certain surfaces (eq. Au) the hydrogen-oxygen reaction is first order in hydrogen and zero order in oxygen, with no decrease in rate as the oxygen pressure is greatly increased.

  d. The conversion of *para*-hydrogen into *ortho*-hydrogen is zero order on several transition metals.

C Surface Tension and Capillarity

11. The surface tension of water at 20°C is 7.27×10^{-2} N m^{-1} and its density is 0.998 g cm^{-3}. Assuming a contact angle θ of zero, calculate the rise of water at 20°C in a capillary tube of radius (a) 1 mm and (b) 10^{-3} cm. Take $g = 9.81$ m s^{-2}. (Capillaries in a tree have radii of about 10^{-3} cm, but sap can rise in a tree to much greater heights than obtained in this calculation. The reason is that the rise of sap depends to a considerable extent on osmotic flow; because of evaporation the leaves contain solutes of higher concentration than the trunk of the tree, and osmotic flow therefore occurs to the leaves.)

12. The density of liquid mercury at 273 K is 13.6 g cm^{-3} and the surface tension is 0.47 N m^{-1}. If the contact angle is 140°, calculate the capillary depression in a tube of 1-mm diameter.

13. The density of water at 20°C is 0.998 g cm^{-3} and the surface tension is 7.27×10^{-2} N m^{-1}. Calculate the ratio between the vapor pressure of a mist droplet having a mass of 10^{-12} g and the vapor pressure of water at a plane surface.

14. The two arms of a U-tube have radii of 0.05 cm and 0.10 cm. A liquid of density 0.80 g cm^{-3} is placed in the tube, and the height in the narrower arm is found to be 2.20 cm higher than that in the wider arm. Calculate the surface tension of the liquid, assuming $\theta = 0$.

15. A tube is placed in a certain liquid and the capillary rise is 1.5 cm. What would be the rise if the same tube were placed in another liquid that has half the surface tension and half the density of the first liquid? Assume that $\theta = 0$ in both cases.

16. When a certain capillary tube is placed in water, the capillary rise is 2.0 cm. Suppose that the tube is placed in the water in such a way that only 1.0 cm is above the surface; will the water flow over the edge? Explain your answer.

***17.** A layer of benzene, of density 0.8 g cm^{-3}, is floating on water of density 1.0 g cm^{-3}, and a vertical tube of internal diameter 0.1 mm is inserted at the interface. It is observed that there is a capillary rise of 4.0 cm and that the contact angle is 40°. Calculate the interfacial tension between water and benzene.

D Surface Films

18. A fatty acid was spread on the surface of water in a Langmuir film balance at 15°C, and the following results obtained:

Area/cm^2 μg^{-1}	5.7	28.2	507	1070	2200	11100
Surface pressure/10^{-3} N m^{-1}	30	0.3	0.2	0.1	0.05	0.01

Estimate the molecular weight of the acid and the area per molecule when the film was fully compressed.

19. N. K. Adam carried out surface film studies using a Langmuir film balance 14.0 cm in width having a floating barrier 13.8 cm long. In one investigation he introduced 52.0 μg of 1-hexadecanol ($C_{16}M_{33}OH$) on to the surface and measured the force on the float at varous lengths of the film, obtaining the following results:

Length	Force on Float
cm	10^{-5} N
20.9	4.14
20.3	8.56
20.1	26.2
19.6	69.0
19.1	108.0
18.6	234
18.3	323
18.1	394
17.8	531

Estimate the area per molecule when the film was fully compressed.

E Supplementary Problems

20. **a.** Show that for small coverages a system obeying the Langmuir isotherm will give a linear plot of ln (θ/P) against P, with a slope of -1.

b. What is the slope if ln (V_a/P) is plotted against V_a at small coverages? (V_a is the volume of gas adsorbed.)

21. A useful isotherm due to Brunauer, Emmett, and Teller, *J. Am. Chem. Soc.*, *60*, 309 (1938), is

$$\frac{P}{V_a} \cdot \frac{P^0}{P^0 - P} = \frac{1}{V_a^0 K} + \frac{P}{V_a^0}$$

where V_a is the volume of gas adsorbed and P^0 is the vapor pressure of the adsorbate (the substance adsorbed).

a. Suggest a method of testing the applicability of this equation, when V_a values are known at various pressures.

b. Show that this equation reduces to the Langmuir isotherm when $P^0 >> P$.

22. Derive the equation

$$\frac{\theta}{1-\theta} = c_g^{1/2} \frac{h^{3/2}}{(2\pi m \mathbf{k}T)^{3/4}} \frac{b_a^{1/2}}{b_g^{1/2}} e^{-\Delta E_0/2RT}$$

for the case of adsorption with dissociation; ΔE_0 is the energy of adsorption per mole.

23. Derive the equation

$$c_a = c_g \frac{h}{(2\pi m \mathbf{k}T)^{1/2}} \frac{b_a}{b_g} e^{-\Delta E_0/RT}$$

for the case of adsorption where the adsorbed molecules are completely mobile on the surface (i.e., have two degrees of translational freedom).

F Essay Questions

1. Describe some of the most important characteristics of a chemisorbed layer. In what ways does a physisorbed layer differ?

2. Derive the Langmuir adsorption isotherms for two substances competitively adsorbed on a surface. Show how these equations interpret the kinetics of bimolecular surface reactions, distinguishing between Langmuir-Hinshelwood and Langmuir-Rideal mechanisms.

3. Explain clearly the distinction between inherent and induced heterogeneity of surfaces.

4. Give a qualitative description of the electric double-layer theories of Helmholtz, Gouy and Chapman, and Stern.

5. Explain the difference between lyophilic and lyophobic sols, with reference to some of the properties in which they differ.

6. What information can be obtained from light-scattering experiments on colloidal particles in aqueous solution?

Suggested Reading

A. W. Adamson, *Physical Chemistry of Surfaces*, New York: Interscience, 1968.

J. Albery, *Electrode Kinetics*, Oxford: Clarendon Press, 1975.

B. E. Conway, *Theory and Principles of Electrode Processes*, New York: Ronald Press, 1965.

J. C. Dash, *Films on Solid Surfaces*, New York: Academic Press, 1975.

E. A. Flood (Ed.), *The Solid-Gas Interface*, New York; Marcel Dekker, 1966.

M. Kerker (Ed.), *Surface Chemistry and Colloids*, London: Butterworth, 1975.

K. J. Laidler, "The Kinetics of Electrode Processes," *J. Chem. Educ*, *47*, 600 (1970).

J. W. McBain, *Colloid Science*, Boston: D. C. Heath, 1950.

S. R. Morrison, *The Chemical Physics of Surfaces*, New York: Plenum, 1977.

G. A. Somorjai (Ed.), *The Structure and Chemistry of Solid Surfaces*, New York: Wiley, 1969.

G. A. Somorjai, *Principles of Surface Chemistry*, Englewood Cliffs, N.J.: Prentice-Hall, 1972.

G. A. Somorjai and L. L. Kesmodal, "The Structure of Solid Surfaces," in M. Kerker (Ed.), *Surface Chemistry and Colloids*, London: Butterworth, 1975.

C. Tanford, *The Physical Chemistry of Macromolecules*, New York: Wiley, 1961.

Chapter 18

Transport Properties

Preview

This chapter is concerned with properties that depend on the movement of molecules in fluids. One such property is *viscosity,* which is a measure of the frictional resistance in a fluid when there is an applied shearing force. The *coefficient of viscosity* is defined in terms of *Newton's law of viscous flow,* according to which the shearing force is proportional to the area and to the relative motion of two planes in the fluid. *Poiseuille's law* relates the rate with which liquid flows through a tube to various parameters, including the coefficient of viscosity.

The theory of the *viscosity of a gas* is very different from that of the viscosity of a liquid. In a gas the reason for the frictional force or drag between two moving parallel planes is that molecules are passing from one plane to another. The expression for the viscosity of a gas, derived on this basis in terms of the molecular diameters, shows that the viscosity increases with the square root of the absolute temperature. With a *liquid,* on the other hand, the viscosity decreases with temperature, in proportion to $\exp(E_{vis}/RT)$, where E_{vis} is the activation energy for the process in which layers of molecules move past one another. The viscosities of solutions are conveniently treated empirically.

Diffusion is a process in which solute molecules tend to flow from regions of higher concentration to regions of lower concentration. The laws of diffusion are summarized in two laws due to *Fick:* the first of these is concerned with the rate of flow across a concentration gradient, the second with the rate of change of concentration as a result of diffusion. These laws can be applied to some special situations, such as diffusion across a membrane and the "random walk." The *diffusion coefficient* can be related to the frictional force and, in the case of ionic diffusion, to the ionic mobility. In some cases the diffusion coefficient can be expressed in terms of the radius of the diffusing particle and the viscosity of the medium by the *Stokes-Einstein equation.*

Molecules in solution can be made to *sediment* by application of a large effective gravitational field, in an ultracentrifuge. On the basis of the theory of sedimentation, measurements in the ultracentrifuge allow molar masses to be determined.

Electrokinetic effects are those in which an electric potential brings about movement or in which movement in a fluid brings about an electric potential difference. For example, *electro-osmosis* is the movement of a fluid through a membrane when there is an electric potential difference across the membrane, and *electrophoresis* is the movement of colloidal particles in a solution. These effects depend on the *electrokinetic* or ζ potential, which we considered in Chapter 17.

18 Transport Properties

Properties that depend on rates of movement are known as **transport properties.** When aqueous systems are involved, we can also speak of *hydrodynamic* properties,* although this expression is often used for liquids other than water. Important transport properties are *viscosity, diffusion,* and *sedimentation.* A shearing force acting on a liquid produces a relative motion of different planes, and the extent of this motion is measured by a property known as the **fluidity,** the reciprocal of which is the **viscosity.** Diffusion is produced by a force resulting from a Gibbs energy difference between various positions in a solution in which there is a concentration gradient. Sedimentation occurs as a result of a gravitational field or of a centrifugal force which produces an effective gravitational field.

Transport can also be induced by the application of an electric field. We have already discussed, in Chapter 7, how simple ions move when a potential is applied; this motion contributes to the conductivity of a solution. Certain molecules, such as proteins, contain charged groups and the charges depend on the pH of the solution. Such molecules in solution may move in an electric field; this type of motion is referred to as **electrophoresis** (Greek *pherein,* to carry). A related type of transport occurs when an electric potential is applied across a charged membrane immersed in a solution; there is then movement of the solvent molecules, and we speak of *electroendosmosis* (Greek *endon,* within; *osmos,* impulse), or **electroosmosis.** These electrical transport properties are also considered in this chapter.

18.1 Viscosity

The *viscosity* of a fluid is a measure of the frictional resistance it offers to an applied shearing force. If a fluid is flowing past a surface, the layer adjacent to the surface is stagnant; successive layers have increasingly higher velocities. Figure 18.1 shows

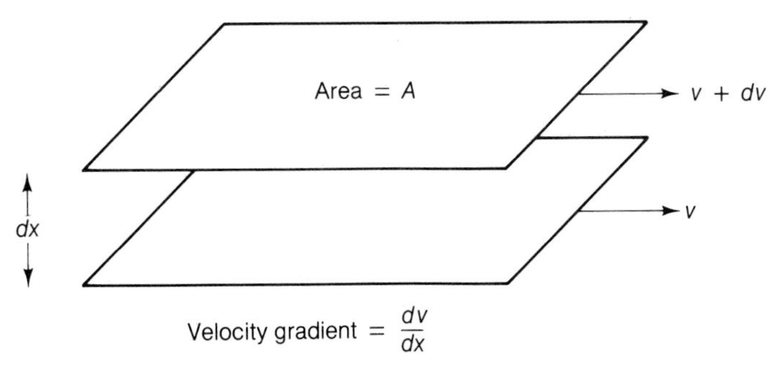

FIGURE 18.1 The definition of the coefficient of viscosity, η. Two parallel layers of fluid, of area A, are separated by a distance dx, and the difference between their velocities is dv.

*From the Greek *hydor,* water; *dynamis,* power. The term *fluid dynamics* is more appropriate than *hydrodynamics* when nonaqueous systems are involved.

two parallel planes in a fluid, separated by a distance dx and having velocities of flow differing by dv. According to **Newton's law of viscous flow,** the frictional force F, resisting the relative motion of two adjacent layers in the liquid, is proportional to the area A and to the velocity gradient dv/dx:

$$F = \eta A \frac{dv}{dx} \tag{18.1}$$

The proportionality constant η is known as the **coefficient of viscosity** or simply as the *viscosity*. Its reciprocal, the *fluidity*, is given the symbol ϕ.

The SI unit for the coefficient of viscosity η is kg m^{-1} s^{-1} or N s m^{-2}; a commonly used unit, the *poise* (g cm^{-1} s^{-1}), is one-tenth of the SI unit. The type of flow to which Eq. 18.1 applies is called **laminar,** *streamline,* or **Newtonian flow.** In flow of this kind there is superimposed on the random molecular velocities a net component of velocity in the direction of flow. Streamline flow is observed if the velocity of flow is not too large; with very rapid flow the motion becomes *turbulent* and Eq. 18.1 no longer applies.

Measurement of Viscosity

Viscosity is usually studied by allowing the fluid to flow through a tube of circular cross section and measuring the rate of flow. From this rate, and with the knowledge of the pressure acting and the dimensions of the tube, the coefficient of viscosity can be calculated on the basis of a theory developed in 1844 by the French physiologist Jean Leonard Poiseuille (1799–1869). Consider an incompressible fluid flowing through a tube of radius R and length l, with a pressure P_1 at one end and a pressure P_2 at the other (Figure 18.2a and b). The liquid at the walls of the tube is stagnant; the rate of flow increases to a maximum at the center of the tube. A cylinder of length l and radius r has an area of $2\pi rl$, and according to Eq. 18.1 the frictional force is

$$F = -\eta \frac{dv}{dr} 2\pi rl \tag{18.2}$$

where the velocity gradient dv/dr is a negative quantity. This force is exactly balanced by the force driving the fluid in this cylinder. This force is the pressure difference $P_1 - P_2$ multiplied by the area πr^2 of the cylinder; thus

$$-\eta \frac{dv}{dr} 2\pi rl = \pi r^2 (P_1 - P_2) \tag{18.3}$$

or

$$dv = -\frac{r}{2\eta l}(P_1 - P_2)\, dr \tag{18.4}$$

Integration of this gives

$$v = -\frac{(P_1 - P_2)}{4\eta l} r^2 + \text{constant} \tag{18.5}$$

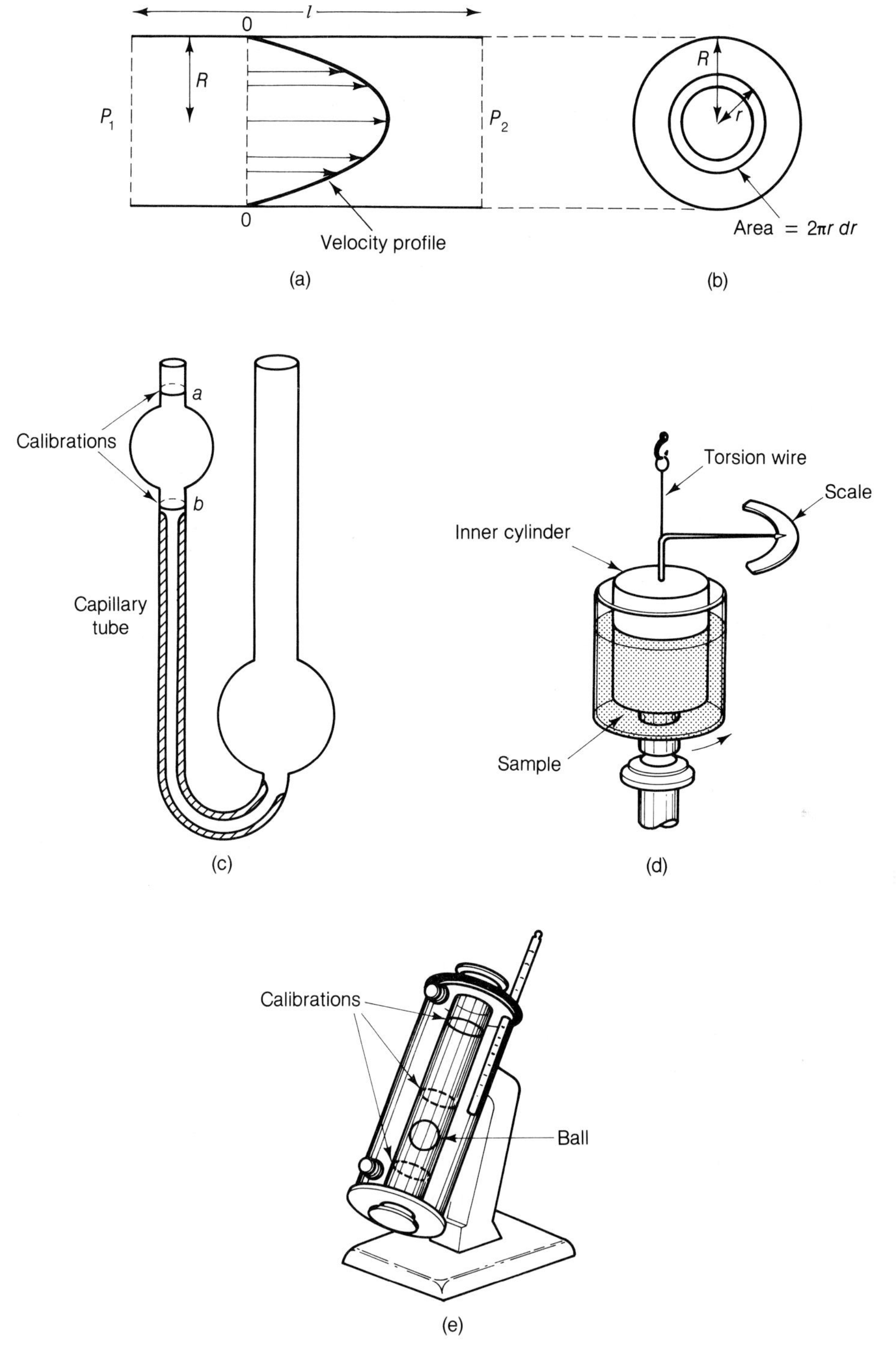

FIGURE 18.2 The measurement of viscosity. (a) Flow through a tube. (b) Cross section of tube. (c) An Ostwald viscometer for measuring the viscosity of a liquid. (d) Couette rotating-cylinder viscometer. (e) Falling-ball viscometer.

The velocity v is zero when $r = R$; the constant of integration is thus

$$\text{constant} = \frac{(P_1 - P_2)}{4\eta l} R^2 \tag{18.6}$$

and therefore

$$v = \frac{P_1 - P_2}{4\eta l}(R^2 - r^2) \tag{18.7}$$

The total volume of liquid flowing through the tube in unit time, dV/dt, is obtained by integrating over each element of cross-sectional area. Each element has an area of $2\pi r\, dr$ (Figure 18.2b) and therefore

$$\frac{dV}{dt} = \int_0^R 2\pi r v\, dr \tag{18.8}$$

$$= \frac{(P_1 - P_2)\pi}{2\eta l}\left\{R^2 \int_0^R r\, dr - \int_0^R r^3\, dr\right\} \tag{18.9}$$

$$= \frac{(P_1 - P_2)\pi R^4}{8\eta l} \tag{18.10}$$

This is the **Poiseuille equation,** and it enables η to be calculated from measurements of the rate of flow dV/dt in a tube of known dimensions, if the pressure difference $P_1 - P_2$ is known.

A commonly employed instrument for measuring the viscosity of a liquid is the *Ostwald viscosimeter* or *viscometer*, illustrated in Figure 18.2c. One measures the time that it takes for a quantity of liquid to pass through the tube, from one position to another, under the force of its own weight. Usually the instrument is calibrated by the use of a liquid of known viscosity.* Another piece of apparatus used for viscosity measurements is the *Couette rotating-cylinder viscometer* (Figure 18.2d). In this instrument a liquid is caused to rotate in an outer cylinder, and it causes a torque to be applied to the torsion wire attached to the inner cylinder. The apparatus is calibrated and the viscosity is calculated from the torque. Another device for measuring viscosity is the falling-ball viscometer (Figure 18.2e); the viscosity is calculated from the time required for the ball to fall from one position to another.

Viscosities of Gases

We have seen that viscosity arises because in a fluid there is a *frictional force*, or *drag*, between two parallel planes moving with different velocities. The theory of the viscosity of a gas is very different from the theory of the viscosity of a liquid. In the case of a gas, the origin of the frictional drag between the two parallel planes is that molecules are passing from one plane to the other. A helpful analogy is provided by two trains traveling in the same direction on parallel tracks but with speeds differing by Δv. Suppose that the passengers are eccentric enough to amuse themselves by jumping from one train to the other. A passenger of mass m jumping from

---

*The pressure difference $P_1 - P_2$ varies with the time but is proportional to the density; the densities of the calibrating liquid and of the sample must therefore be known.

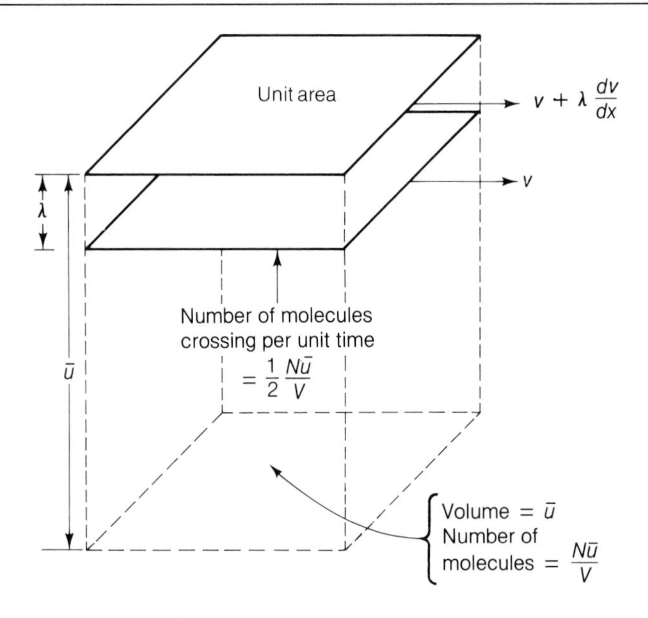

FIGURE 18.3 The kinetic theory of gaseous viscosity.

the faster train to the slower one transports momentum $m\,\Delta v$ to the slower train and therefore tends to increase its speed. Conversely, the passengers in the slower train who jump to the faster one remove $m\,\Delta v$ of momentum from it and tend to slow it down. As a result, the speeds of the trains tend to equalize, and the net effect is the same as if there were a frictional force acting between the trains.

The analogous situation in a flowing gas is represented in Figure 18.3. The lower plane is moving with velocity v. We consider the separation between the planes to be the mean free path λ of the gas (Section 1.9). The reason for this choice is that λ is the average distance that the gas molecules move between successive collisions, so that a molecule that has experienced a collision in one plane and is moving toward the other plane will, on the average, experience its next collision in that plane. In so doing it transports momentum from one plane to the other. If dv/dx is the velocity gradient, the difference in the velocities of the two planes, separated by λ, is $(dv/dx)\lambda$, and the transfer of momentum when the molecule moves from one plane to the other is

$$m\lambda \frac{dv}{dx}$$

Figure 18.3 shows two parallel planes of unit area, and we now consider how many molecules move from one plane to the other in unit time. The average molecular speed is $\overline{u}$, and in the figure we have constructed a rectangular prism of unit cross-sectional area and of length $\overline{u}$, indicated by dashed lines. The volume of that prism is $\overline{u}$, and if the gas has N molecules in a volume V, the number of molecules in the prism is $N\overline{u}/V$. Half of these molecules are moving in an upward direction and half in a downward direction, and on the average they travel the entire distance $\overline{u}$ in unit time. The number that cross unit area per unit time in one direction therefore is

$$\frac{1}{2} \cdot \frac{N\overline{u}}{V}$$

This is the number that jump from one plane to the other in unit time. Since each molecule transfers momentum of $m\lambda\,(dv/dx)$, the total momentum transported in unit time is

$$\frac{1}{2} \cdot \frac{N\overline{u}}{V} \cdot m\lambda \cdot \frac{dv}{dx}$$

Force is rate of change of momentum, and this expression is therefore the force F acting between the planes of unit area:

$$F = \frac{1}{2} \cdot \frac{N\overline{u}}{V} \cdot m\lambda \cdot \frac{dv}{dx} \tag{18.11}$$

Insertion of this expression into Eq. 18.1, with $A = 1$, gives

$$\frac{1}{2} \cdot \frac{N\overline{u}}{V} \cdot m\lambda \cdot \frac{dv}{dx} = \eta \frac{dv}{dx} \tag{18.12}$$

and therefore

$$\boldsymbol{\eta = \frac{Nm\overline{u}\lambda}{2V}} \tag{18.13}$$

Since the density ρ of the gas is Nm/V, this equation can be written as

$$\boldsymbol{\eta = \tfrac{1}{2}\rho\overline{u}\lambda} \tag{18.14}$$

In Equation 1.68 we saw that the mean free path λ of a gas is equal to $V/\sqrt{2}\pi d^2 N$, where d is the molecular diameter, and the insertion of this into Eq. 18.13 gives

$$\eta = \frac{m\overline{u}}{2\sqrt{2}\pi d^2} \tag{18.15}$$

Table 14.2 gave us the expression $(8\mathbf{k}T/\pi m)^{1/2}$ for $\overline{u}$, and therefore

$$\boldsymbol{\eta = \frac{(m\mathbf{k}T)^{1/2}}{\pi^{3/2}d^2}} \tag{18.16}$$

These equations for the coefficient of viscosity of a gas lead to some interesting predictions, all of which are confirmed by experiment. For example, we see from Eq. 18.16 that η depends on the molecular mass, the molecular diameter, and the temperature. The density or pressure of the gas does not affect the viscosity. This may seem surprising at first sight, but the explanation is not hard to find. At higher densities more molecules jump from one layer to the next, but λ is smaller and each jump involves the transport of less momentum. These two effects just counteract each other.

Second, from Eq. 18.16 we see that the viscosity increases with the square root of the temperature. We shall see in the next subsection that this prediction, supported by experiment, is in marked contrast to the behavior of a liquid, for which the viscosity decreases with increasing temperature. The explanation can be seen either from Eq. 18.14 or from Eq. 18.15. Raising the temperature does not change $\rho\lambda$, m, or d, but it increases $\overline{u}$, which is proportional to the square root of the absolute temperature. More molecules cross from one layer to the next as the temperature goes up, and this increases the drag and therefore the viscosity.

Finally, we can see from Eq. 18.16 that if we had two gases for which the molecules had identical masses but for which the molecular diameters d were different, the gas of higher d would have the lower viscosity. The reason for this is that an increase in d decreases the mean free path, so that each jump involves the transfer of less momentum and therefore produces less drag. Equation 18.16 would lead us to the conclusion that a hypothetical gas consisting of molecules of zero diameter would have infinite viscosity, which is obviously impossible. Molecules of zero size, of course, cannot collide with each other, and there is no change of momentum as a result of collisions; the equations we have derived therefore do not apply to this situation (see Problem 22).

Equation 18.15 provides us with a means of obtaining molecular diameters from viscosity measurements, and this method has often been used. Table 18.1 gives some diameters obtained in this way and compares them with values calculated from the van der Waals constant b (Section 1.12). From the diameters we can calculate mean free paths (Eq. 1.68), and some values are included in the table.

EXAMPLE

The viscosity of carbon dioxide at 25°C and 101.325 kPa is 13.8×10^{-6} kg m^{-1} s^{-1}. Estimate the molecular diameter.

SOLUTION

It is most convenient to use Eq. 18.16. The molecular mass is

$$m = \frac{44.01}{6.022 \times 10^{23}} = 7.308 \times 10^{-23}\ \text{g} = 7.308 \times 10^{-26}\ \text{kg}$$

TABLE 18.1 Viscosities of Gases, Molecular Diameters, and Mean Free Paths (at 25°C and 101.325 kPa)

Gas	Viscosity, η / 10^{-6} kg m^{-1} s^{-1}	Molecular Diameter/nm: From Viscosity	From van der Waals b	Mean Free Path/nm
He	18.6	0.225	0.248	180.6
H_2	8.42	0.281	0.276	115.8
N_2	16.7	0.386	0.314	61.4
O_2	18.09	0.383	0.290	62.7
CO_2	13.8	0.475	0.324	40.6

From Eq. 18.16,

$$d = \frac{(m\mathbf{k}T)^{1/4}}{\pi^{3/4}\eta^{1/2}}$$

$$= \frac{[7.308 \times 10^{-26}(\text{kg}) \times 1.381 \times 10^{-23}(\text{J K}^{-1}) \times 298.15(\text{K})]^{1/4}}{\pi^{3/4}[13.8 \times 10^{-6}(\text{kg m}^{-1}\ \text{s}^{-1})]^{1/2}}$$

$$= 4.75 \times 10^{-10}\ \text{m} = 0.475\ \text{nm}$$

(The units are $\text{kg}^{1/4}\ \text{J}^{1/4}\ \text{kg}^{-1/2}\ \text{m}^{1/2}\ \text{s}^{1/2}$; since $\text{J} = \text{kg m}^2\ \text{s}^{-2}$, this becomes $\text{kg}^{1/4}\ (\text{kg m}^2\ \text{s}^{-2})^{1/4}\ \text{kg}^{-1/2}\ \text{m}^{1/2}\ \text{s}^{1/2} = \text{m}$.) ■

Viscosities of Liquids

The viscous behavior of liquids is very different from that of gases. For example, whereas gas viscosities increase with rising temperature, the viscosities of liquids decrease. It was first shown empirically by the Spanish physical chemist J. de Guzman* that viscosity obeys a law of the Arrhenius type, a result that was later found by Arrhenius† himself. We have seen that fluidity is the reciprocal of viscosity, and the variation of fluidity ϕ with temperature can be expressed as

$$\phi = \frac{1}{\eta} = A_{\text{fl}}e^{-E_{\text{fl}}/RT} \qquad (18.17)$$

where A_{fl} is the pre-exponential factor for the flow process and E_{fl} is the activation energy. The viscosity itself can be expressed as

$$\eta = A_{\text{vis}}e^{E_{\text{vis}}/RT} \qquad (18.18)$$

where A_{vis} $(= 1/A_{\text{fl}})$ and $E_{\text{vis}}(= E_{\text{fl}})$ are the pre-exponential factor and activation energy, respectively. The latter is readily obtained from the slope of a plot of $\ln \eta$ against $1/T$.

The detailed theory of the viscosity of liquids is very complicated, and here we will present only the basic ideas and a simplified version of the treatment. Figure 18.4a shows two layers of molecules in a liquid; the upper layer moves to the right more rapidly than the lower. Additional molecules are present above and below these two layers, and in order for motion to occur the molecules will have to push neighboring molecules aside. As a result, there is an energy barrier to the flow process. Figure 18.4b shows a plot of potential energy against the distance moved by a given molecule. At positions A and Z the molecule is in a state of minimum potential energy, but because of the repulsions of neighboring molecules it has to surmount an energy barrier of ϵ_0. If d is the distance between the initial and final positions of the molecule, the molecule will have to travel a distance of $d/2$ to reach the top of the barrier, which will be symmetrical.

Suppose that a shearing force F per unit area is bringing about the relative displacement of the two planes. As far as a given molecule is concerned, this force acts on the effective area occupied by the molecule. For simplicity we will assume

*J. de Guzman, *Anales Soc. Españ. Fis. Quim.*, *11*, 353 (1913).

†S. Arrhenius, *Meddal. Vetenskapsaked. Nobelinst.*, 3 [20] (1916).

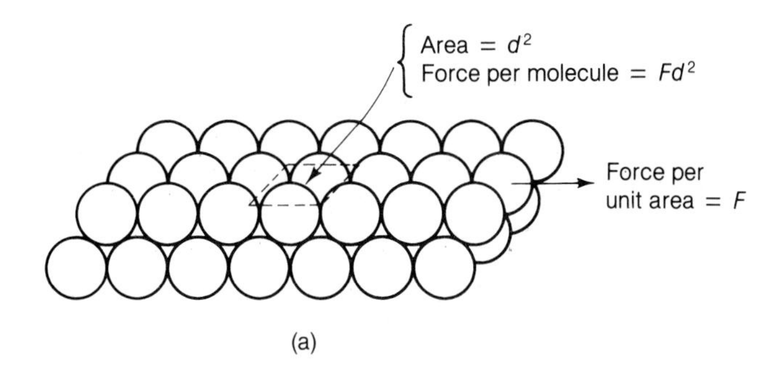

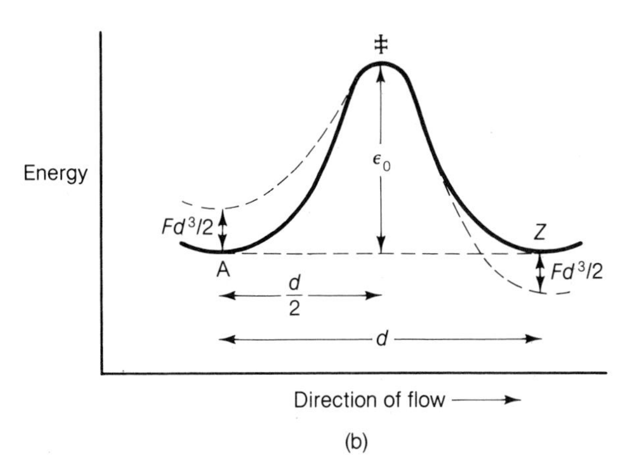

FIGURE 18.4 Liquid viscosity. (a) Two liquid layers moving with respect to each other under the influence of a force F per unit area. (b) Potential-energy diagram for viscous flow. The firm line shows the potential-energy profile when there is no shearing force and the dashed line shows the profile when there is a shearing force.

this area to be d^2. The force acting upon a single molecule is thus Fd^2, and in order for the molecule to reach the top of the potential-energy barrier this force will be exerted through a distance $d/2$. The effect of the force is thus to provide energy of $Fd^2 \times d/2 = Fd^3/2$. This means that, as shown by the dashed line in Figure 18.4b, the height of the barrier in the left-to-right direction is reduced by $Fd^3/2$, whereas that in the right-to-left direction is raised by $Fd^3/2$.

In the absence of the shearing force, the rate constant for the passage of a molecule over the potential-energy barrier, of height ϵ_0, can be written

$$k_0 = A'_{\text{vis}} e^{-\epsilon_0/\mathbf{k}T} \tag{18.19}$$

where A'_{vis} is the pre-exponential factor for the process. When the force is applied, the rate constant for the process from left to right is

$$k_1 = A'_{\text{vis}} \exp\left[-\frac{(\epsilon_0 - Fd^3/2)}{\mathbf{k}T}\right] = k_0 e^{Fd^3/2\mathbf{k}T} \tag{18.20}$$

while that for the right-to-left process is

$$k_{-1} = A'_{\text{vis}} \exp\left[-\frac{(\epsilon_0 + Fd^3/2)}{\mathbf{k}T}\right] = k_0 e^{-Fd^3/2\mathbf{k}T} \tag{18.21}$$

These rate constants represent the numbers of times a molecule crosses the barrier in the two directions. The rate of crossing is

$$k = k_1 - k_{-1} = k_0(e^{Fd^3/2\mathbf{k}T} - e^{-Fd^3/2\mathbf{k}T}) \tag{18.22}$$

Each time a crossing occurs, the molecule moves a distance d. Multiplication by d therefore gives the distance traveled in unit time, that is, the relative velocity Δv:

$$\Delta v = dk_0(e^{Fd^3/2\mathbf{k}T} - e^{-Fd^3/2\mathbf{k}T}) \tag{18.23}$$

In ordinary viscous flow F is sufficiently small that $Fd^3 << 2\mathbf{k}T$. We can therefore expand the exponentials and accept only the first terms:

$$\Delta v = dk_0\left[1 + \frac{Fd^3}{2\mathbf{k}T} - \left(1 - \frac{Fd^3}{2\mathbf{k}T}\right)\right] = \frac{k_0 F d^4}{\mathbf{k}T} \tag{18.24}$$

The velocity gradient dv/dx is $\Delta v/d$:

$$\frac{dv}{dx} = \frac{k_0 F d^3}{\mathbf{k}T} \tag{18.25}$$

Since F is the force per unit area, Eq. 18.1 becomes

$$F = \eta \frac{k_0 F d^3}{\mathbf{k}T} \tag{18.26}$$

from which it follows that

$$\eta = \frac{\mathbf{k}T}{k_0 d^3} \tag{18.27}$$

Introduction of the expression for k_0 (Eq. 18.19) gives

$$\eta = \frac{\mathbf{k}T}{A'_{\text{vis}} d^3} e^{\epsilon_0/\mathbf{k}T} \tag{18.28}$$

Since d^3 is the volume occupied by a single molecule in the liquid state, we can replace it by V_m/L, where V_m is the molar volume of the liquid and L is the Avogadro constant. Equation 18.28 can thus be written as

$$\eta = \frac{L\mathbf{k}T}{A'_{\text{vis}} V_m} e^{\epsilon_0/\mathbf{k}T} = \frac{RT}{A'_{\text{vis}} V_m} e^{E_{\text{vis}}/RT} \tag{18.29}$$

This is of the same form as the empirical equation (Eq. 18.18), and it predicts the same temperature dependence. The activation energy per mole, $E_{\text{vis}} = \epsilon_0 L$, proves to be about one-third to one-quarter of the molar enthalpy of vaporization of the liquid; some values are given in Table 18.2. In other words, for a molecule to push another aside in order to pass to a new equilibrium position, it has to expend one-third to one-quarter of the energy required to remove a molecule entirely from the liquid state.

Viscosities of Solutions

It is difficult to work out a satisfactory treatment of the viscosities of solutions based on fundamental principles. As a result, it is usually more satisfactory to proceed

TABLE 18.2 Activation Energies for Viscous Flow in Liquids, Compared with Enthalpies of Vaporization at the Boiling Point*

Liquid	E_{vis}/kJ mol^{-1}	ΔH_{vap}/kJ mol^{-1}	$\Delta_{vap}H/E_{vis}$
CCl_4	10.5	27.6	2.6
C_6H_6	10.6	27.9	2.6
CH_4	3.01	7.6	2.5
$CHCl_3$	7.4	27.8	3.8
CH_3COCH_3	6.9	26.8	3.9
H_2O	12.1	37.6	3.2

*Values taken from R. H. Ewell and H. Eyring, *J. Chem. Phys.*, 5, 726 (1937).

†The energy of activation for viscous flow in water varies considerably with temperature, as a result of the changes in hydrogen-bonded structure; E_{vis} is 21.1 kJ mol^{-1} at 0°C and 8.8 kJ mol^{-1} at 150°C.

empirically. In this way it has proved possible to obtain information about the sizes and shapes of solute molecules, particularly of macromolecules, from the viscosities of their solutions. When a solute is added to a liquid such as water, the viscosity is usually increased. Suppose that the viscosity of a pure liquid is η_0 and that the viscosity of a solution is η. The **specific viscosity,** defined as

$$\text{specific viscosity} = \frac{\eta - \eta_0}{\eta_0} \tag{18.30}$$

is the increase in viscosity $\eta - \eta_0$ relative to the viscosity of the pure solvent. Division of the specific viscosity by the mass concentration ρ of the solution gives what is known as the **reduced specific viscosity** [SI unit: m^3 kg^{-1}]:

$$\text{reduced specific viscosity} = \frac{1}{\rho} \cdot \frac{\eta - \eta_0}{\eta_0} \tag{18.31}$$

This quantity, however, has a contribution from the intermolecular interactions between the solute molecules. This contribution can be eliminated by extrapolating the reduced specific viscosity to infinite dilution; then we obtain what is known as the **intrinsic viscosity** [SI unit: m^3 kg^{-1}] given the symbol $[\eta]$:

$$[\boldsymbol{\eta}] = \lim_{\rho \to 0} \left[\frac{1}{\rho} \frac{\eta - \eta_0}{\eta_0} \right] \tag{18.32}$$

This quantity is the fractional change in the viscosity per unit concentration of solute molecules, at infinite dilution.

Various empirical relationships have been proposed to relate the intrinsic viscosity $[\eta]$ to the relative molecular mass M_r. The most successful of these is the equation

$$[\boldsymbol{\eta}] = kM_r^{\alpha} \tag{18.33}$$

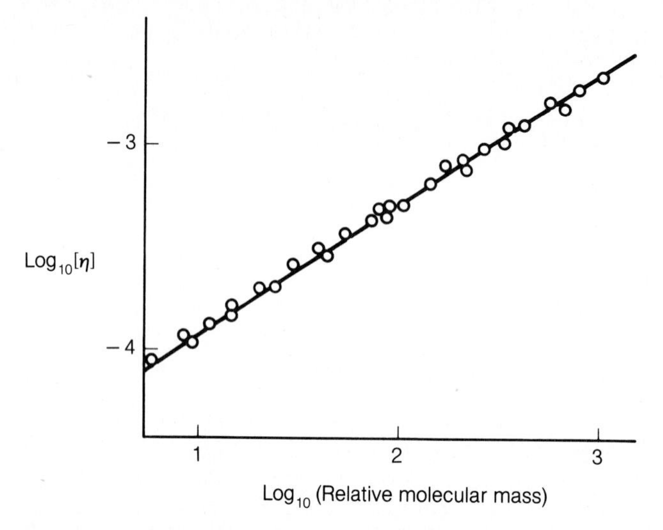

FIGURE 18.5 Logarithm of intrinsic viscosity of polybutylenes plotted against logarithm of relative molecular mass, for solutions in diisobutene at 20°C.

where K and α are constants. This equation was proposed independently by Herman F. Mark* in 1938 and by R. Houwink† in 1941 and is usually called the *Mark-Houwink equation.* If it is obeyed, a plot of log $[\eta]$ against log M_r will be a straight line. Figure 18.5 shows some results plotted in this way for various polyisobutenes in diisobutene as solvent. By the use of such calibration curves the relative molecular masses of unknown samples can be determined.

Attempts have been made to relate the value of α in Eq. 18.33 to the shape of the molecule. If the molecules are spherical, the intrinsic viscosity $[\eta]$ is independent of the size of the molecules, so that α is equal to zero. In agreement with this, all globular proteins, regardless of their size, have essentially the same $[\eta]$. If a molecule is elongated, its molecules are more effective in reducing the viscosity, and $[\eta]$ is larger; values of 1.3 or higher are frequently obtained for molecules that exist in solution as extended chains. Long-chain molecules that are coiled in solution give intermediate values of α, frequently in the range 0.6 to 0.75.

18.2 Diffusion

If solutions of different concentrations are brought into contact with each other, the solute molecules tend to flow from regions of higher concentration to regions of lower concentration, and there is ultimately an equalization of concentration. The driving force leading to diffusion is the Gibbs energy difference between regions of different concentration.

---

*H. F. Mark, *Der Jeste Körper,* Leipzig: Hirzel, 1938.

†R. Houwink, *J. Prakt. Chem.,* *157,* 15 (1941).

Fick's Laws

The German physiologist Adolf Eugen Fick* (1829–1901) formulated in 1855 two fundamental laws of diffusion. According to **Fick's first law,** the rate of diffusion dn/dt of a solute across an area A, known as the *diffusive flux* and given the symbol J, is

$$J = \frac{dn}{dt} = -DA\frac{\partial c}{\partial x} \tag{18.34}$$

where $\partial c/\partial x$ is the concentration gradient of the solute, and dn is the amount of solute crossing the area A in time dt.

The SI unit of dn/dt is mol s^{-1}, that of A is m^2, and that of $\partial c/\partial x$ is mol m^{-4}; that of the *diffusion coefficient* D is therefore m^2 s^{-1} and that of flux per unit area is mol m^{-2} s^{-1}. In practice, diffusion coefficients are commonly expressed as cm^2 s^{-1}, and in biological work the "fick" is sometimes used as a unit; 1 fick = 10^{-11} m^2 s^{-1}.

Fick also derived an equation for the rate of change of concentration as a result of diffusion. Figure 18.6 shows a system of cross-sectional area A, having a concentration c at position x and a concentration $c + dc$ at position $x + dx$. Because there is an increase in concentration as x increases, the net diffusion occurs from right to left in the diagram. The flux at x can be written as $J(x)$ and that at $x + dx$ as $J(x + dx)$, which is given by

$$J(x + dx) = J(x) + \frac{\partial J}{\partial x}dx \tag{18.35}$$

The net flux into the region between x and $x + dx$ is thus

$$J_{\text{net}} = J(x) - J(x + dx) = -\frac{\partial J}{\partial x}dx \tag{18.36}$$

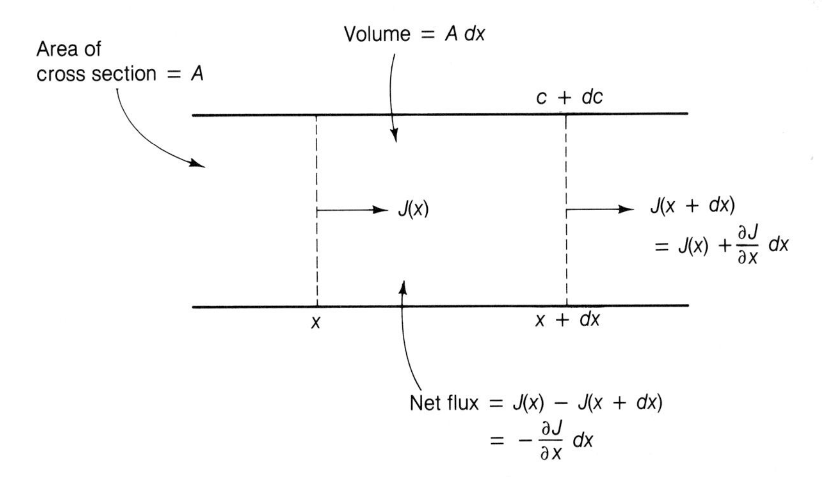

FIGURE 18.6 One-dimensional diffusion; a diagram illustrating Fick's second law.

* A. E. Fick, *Pogg. Ann.*, *94*, 59 (1855).

The net rate of increase in concentration in this element of volume is the net flux divided by the volume, which is $A\,dx$:

$$\frac{\partial c}{\partial t} = -\frac{1}{A} \cdot \frac{\partial J}{\partial x} \tag{18.37}$$

From Eq. 18.34, J is $-DA\,(\partial c/\partial x)$ and therefore

$$\frac{\partial c}{\partial t} = -\frac{1}{A} \cdot \frac{\partial}{\partial x}\left(-DA\frac{\partial c}{\partial x}\right) = \frac{\partial}{\partial x}\left(D\frac{\partial c}{\partial x}\right) \tag{18.38}$$

If D is independent of the distance x (as is always true to a good approximation), this equation reduces to

$$\frac{\partial c}{\partial t} = D\frac{\partial^2 c}{\partial x^2} \tag{18.39}$$

This is **Fick's second law of diffusion** for the special case of diffusion in one dimension (e.g., along the X axis). In liquids and solutions, which are said to be *isotropic*, D has the same value in all directions. Some solids are *anisotropic*, which means that D is not the same in all directions.

Solutions of Diffusion Equations

Equation 18.39 is a second-order, linear, and homogeneous differential equation. Its solution depends on the nature of the domain through which diffusion is taking place and on the initial conditions (i.e., the concentrations at various positions at some time that may be taken as $t = 0$). Some initial and boundary conditions present a difficult mathematical problem, and often explicit expressions cannot be obtained.

A particularly simple situation is when solutions at two different concentrations are separated by a porous diaphragm in such a way that the solutions can be stirred and maintained at uniform concentration. A convenient technique for measuring diffusion rates in water is to separate two solutions by a sintered glass membrane of thickness l, as shown schematically in Figure 18.7b. Such a membrane contains pores filled with water, and it is assumed that the diffusion through the membrane occurs at the same rate as in water (a correction can be made for the area of cross section occupied by the glass). When such a system is set up, it is found experimentally that a steady state is soon established; that is, the rate of diffusion does not change with time as long as the concentrations c_1 and c_2 on the two sides remain the same (this will be true to a good approximation if the solution volumes are large). In order for there to be a steady state, the concentration gradient must be uniform across the membrane; otherwise the rates of flow would vary across the membrane and there would be accumulation or depletion of material in certain regions of the membrane and therefore no steady state. In other words, the concentration within the membrane must fall linearly, as shown in Figure 18.7a. The concentration gradient is thus given by

$$-\frac{dc}{dx} = \frac{c_1 - c_2}{l} \tag{18.40}$$

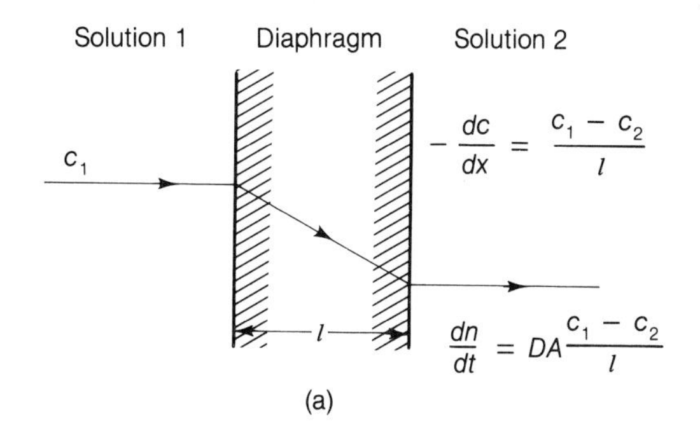

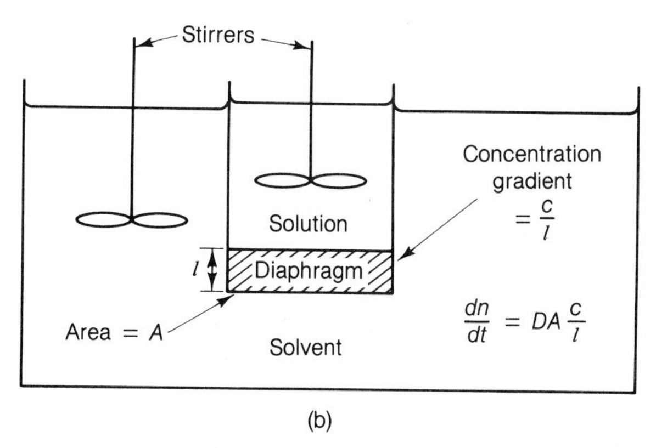

FIGURE 18.7 (a) Stirred solutions of concentrations c_1 and c_2 separated by a porous diaphragm of thickness l. In the steady state the concentration within the diaphragm changes linearly from c_1 to c_2. (b) Schematic diagram of simple apparatus for the measurement of diffusion constants.

and by Fick's first law the rate of flow through the membrane is

$$\frac{dn}{dt} = DA\frac{c_1 - c_2}{l} \tag{18.41}$$

The diffusion coefficient D is therefore calculated from the measurement of the rate of flow; the area A, thickness l, and concentration difference $c_1 - c_2$ are readily determined. Some diffusion coefficients are given in Table 18.3.

The case just considered is a very simple one in which the concentration has been forced to vary linearly from c_1 to c_2. Another case of interest is when there is an *instantaneous plane source* at a particular plane in a liquid. A simple way of arriving at the equations applicable to this case is to note that a general solution of Eq. 18.39 for Fick's second law is

$$c = \alpha t^{-1/2} e^{-x^2/4Dt} \tag{18.42}$$

where α is a constant. That this is a general solution may be verified by substitution

TABLE 18.3 Diffusion Coefficients at 20°C

Diffusing Substance	Relative Molecular Mass	Solvent	Diffusion Coefficient, D/m^2 s^{-1}
H_2	2	(gas)*	1.005×10^{-4}
O_2	32	(gas)*	1.36×10^{-5}
H_2O	18	H_2O†	25.4×10^{-10}
CH_3OH	32	H_2O	13.7×10^{-10}
H_2NCONH_2	60	H_2O	11.8×10^{-10}
Glycerol	92	H_2O	8.3×10^{-10}
Sucrose	342	H_2O	5.7×10^{-10}
Insulin	41 000	H_2O	8.2×10^{-11}
Horse hemoglobin	68 000	H_2O	6.3×10^{-11}
Urease	470 000	H_2O	3.5×10^{-11}
Tobacco mosaic virus	31 400 000	H_2O	5.3×10^{-12}
Phenol	94	Benzene	15.8×10^{-10}
Bromoform	353	Benzene	16.9×10^{-10}
I_2	254	Benzene	19.3×10^{-10}

*For self-diffusion at 101.325 kPa, calculated from viscosities (Table 18.1) using Eq. 18.57.

†Self-diffusion.

into Eq. 18.39.* When $t \to 0$, this function corresponds to $c = 0$ everywhere except at $x = 0$, where $c \to \infty$. In other words, this case corresponds to solute present at a plane at the origin $x = 0$; since it is present in zero volume, the concentration is infinite. The constant α is related to the "strength" of the source (i.e., to the amount n_0 of solute initially present at $x = 0$). Since the amount of solute remains the same at all times, n_0 is equal to the integral

$$A \int_{-\infty}^{\infty} c \, dx$$

*Thus, from Eq. 18.42,

$$\frac{\partial c}{\partial t} = \alpha\left(-\frac{1}{2}t^{-3/2}e^{-x^2/4Dt} + \frac{x^2}{4Dt^{5/2}}e^{-x^2/4Dt}\right)$$

$$\frac{\partial c}{\partial x} = -\frac{\alpha x}{2D}t^{-3/2}e^{-x^2/4Dt}$$

$$\frac{\partial^2 c}{\partial x^2} = -\frac{\alpha t^{-3/2}}{2D}e^{-x^2/4Dt} + \frac{\alpha x^2}{4D^2t^{5/2}}e^{-x^2/4Dt}$$

whence

$$\frac{\partial c}{\partial t} = D\frac{\partial^2 c}{\partial x^2}$$

at any time t. Introduction of the expression for c (Eq. 18.42) and integration leads to

$$n_0 = \alpha A \int_{\infty}^{\infty} t^{-1/2} e^{-x^2/4Dt}\, dx = 2\alpha(\pi D)^{1/2} A \tag{18.43}$$

The constant α is thus given by

$$\alpha = \frac{n_0}{2(\pi D)^{1/2} A} \tag{18.44}$$

and Eq. 18.42 becomes

$$c = \frac{n_0}{2(\pi D t)^{1/2} A} e^{-x^2/4Dt} \tag{18.45}$$

The concentration c in this expression is the amount of solute per unit distance; if the cross-sectional area is unity, c is the amount of solute per unit volume.

Figure 18.8 shows plots of c/n_0 against x, for three different values of Dt. This plot shows how the solute molecules spread out from the instantaneous plane source located at $x = 0$.

An interesting aspect of this problem is to focus attention on an individual solute molecule and to ask: what is the probability that it will diffuse a distance x in time t? We must, of course, allow a certain spread of distance, and we may call $P(x)\, dx$ the probability that the molecule has diffused a distance between x and $x + dx$. This probability is the number of solute molecules between x and $x + dx$ divided by the total number in the original source. Thus

$$P(x)\, dx = \frac{c(x)\, dx}{n_0} = \frac{1}{2(\pi D t)^{1/2}} e^{-x^2/4Dt}\, dx \tag{18.46}$$

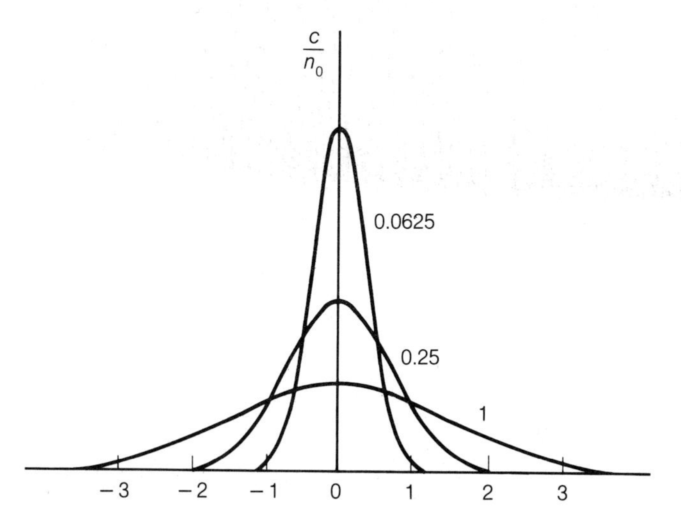

FIGURE 18.8 Plots of c/n_0 against distance from an instantaneous plane source at $x = 0$. At this plane there are initially n_0 molecules of solute, at infinite concentration (volume = 0). The numbers on the curves are values of DT.

We now ask: what is the mean square distance $\overline{x^2}$ traversed by a solute molecule in time t? (We do not ask what is the mean distance $\overline{x}$, since diffusion is equally probable in both directions, and $\overline{x} = 0$.) The mean square distance is given by

$$\overline{x^2} = \int_{-\infty}^{\infty} x^2 P(x)\, dx \tag{18.47}$$

Substitution of Eq. 18.46 into this, and evaluation of the integral, leads to

$$\overline{x^2} = 2Dt \tag{18.48}$$

An alternative derivation of Eq. 18.48 is also instructive. Figure 18.9a shows three parallel planes of unit area separated by distances x. If x is sufficiently small, the concentration gradient can be taken to be linear, and Figure 18.9b shows how the concentration varies from the coordinate $-x/2$ to $+x/2$. If the average concentration in the left-hand compartment is c, that in the right-hand compartment is

$$c + \frac{\partial c}{\partial x} x$$

and these are also the concentrations at the coordinates $-\frac{1}{2}x$ and $\frac{1}{2}x$, respectively.

Suppose now that x is the average distance that a molecule diffuses in time t. If we consider the flux J_1 from left to right across the cross section at $x = 0$, we see that half of the molecules that are in the left-hand volume (i.e., $\frac{1}{2}cx$ molecules) will cross in time t; thus

$$J_1 = \frac{1}{2}\frac{cx}{t} \tag{18.49}$$

The flux from right to left is one-half of the number of molecules in the right-hand volume, $\frac{1}{2}[c + (\partial c/\partial x)x]x$ divided by t:

$$J_{-1} = \frac{\frac{1}{2}\left(c + \frac{\partial c}{\partial x} x\right)x}{t} \tag{18.50}$$

The net flux from left to right, $J_1 - J_{-1}$, is therefore

$$J = \frac{\frac{1}{2}cx - \frac{1}{2}\left(c + \frac{\partial c}{\partial x} x\right)x}{t} \tag{18.51}$$

$$= -\frac{x^2}{2t}\frac{\partial c}{\partial x} \tag{18.52}$$

Comparison with Eq. 18.34, with $A = 1$, gives

$$D = \frac{x^2}{2t} \tag{18.53}$$

This is equivalent to Eq. 18.48, since it is the average value of x^2 that is given in this derivation.

Equation 18.48 is often referred to as the **random-walk equation** or, more colloquially, as the equation for the drunkard's walk, since it corresponds to com-

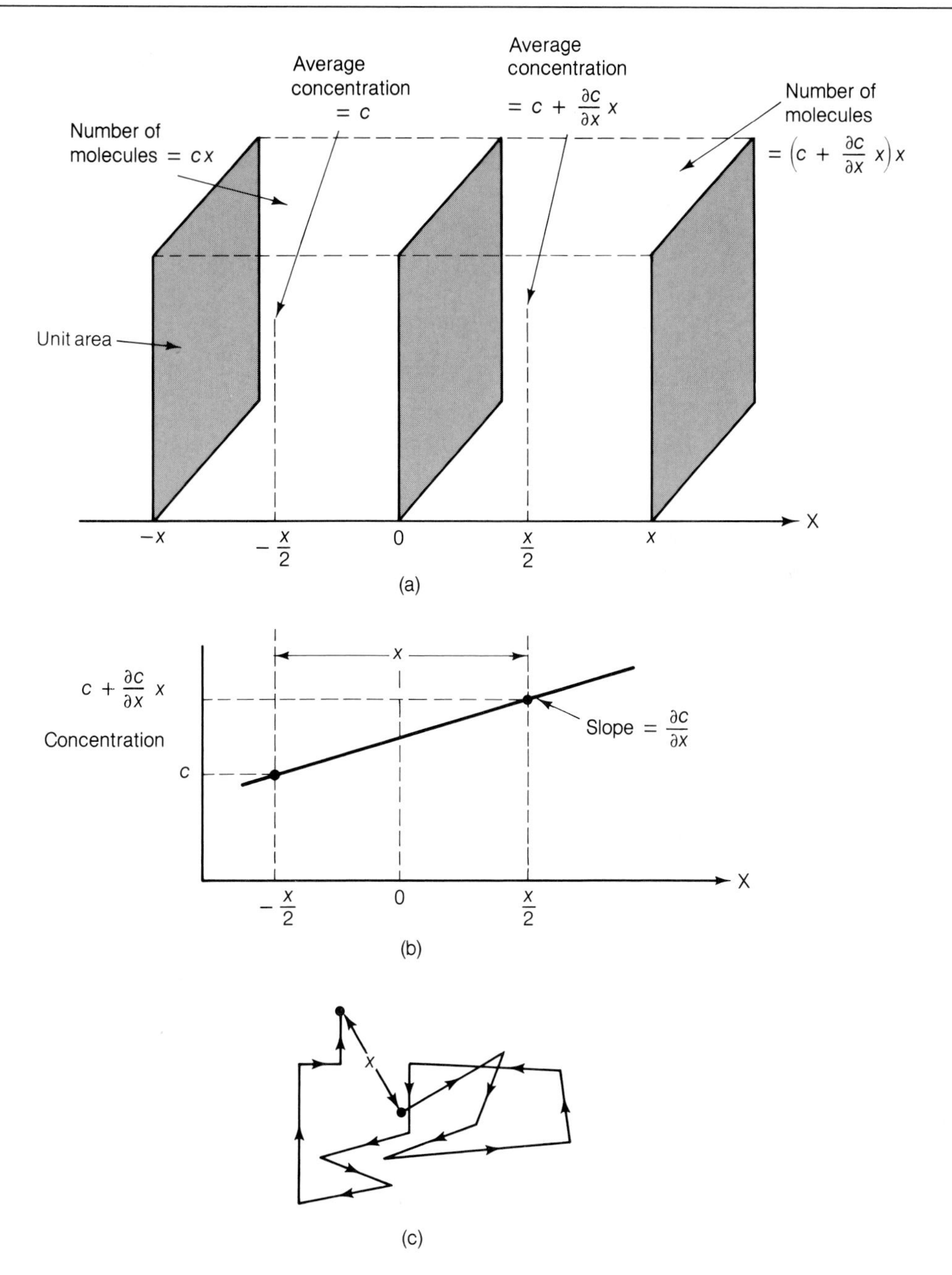

FIGURE 18.9 (a) Three parallel planes separated by distances x. (b) The corresponding concentration gradient. (c) Schematic representation of the random walk.

pletely random motion, with no sense of direction. Figure 18.9c shows the type of motion to which it corresponds. The equation provides an explanation for the Brownian motion of colloidal particles and is useful for providing estimates of mean diffusion distances.

EXAMPLE

The diffusion coefficient for carbon in α-Fe is 2.9×10^{-8} cm^2 s^{-1} at 500°C. How far would a carbon atom be expected to diffuse in 1 year (3.156×10^7 s)?

SOLUTION

From Eq. 18.48,

$$\overline{x^2} = 2 \times 2.9 \times 10^{-8} \times 3.156 \times 10^7 \text{ cm}^2$$

$$= 1.83 \text{ cm}^2$$

and therefore the expected distance is

$$(\overline{x^2})^{1/2} = 1.35 \text{ cm}$$

Such diffusion has important practical consequences, since diffusion of carbon into the area of a mechanical weld might lead to weakening of the weld and possible rupture. ■

Self-Diffusion of Gases

The problem of the diffusion of one gas into another is a somewhat difficult one, because of the different diameters and mean square velocities of the two kinds of molecules. Here we will be content to consider *self-diffusion*, in which all the molecules are identical in diameter and mass. Coefficients for self-diffusion can be measured by isotopically labeling some of the molecules and determining how the labeled gas diffuses into the unlabeled.

For a gas the mean free path λ represents a step that is random with respect to preceding and succeeding steps, and Eq. 18.48 will therefore apply to this step. The time it takes a molecule to travel λ is $\lambda/\overline{u}$, where $\overline{u}$ is the average speed. We therefore replace x^2 by λ^2 and t by $\lambda/\overline{u}$ in Eq. 18.48 and obtain

$$\lambda^2 = \frac{2D\lambda}{\overline{u}} \tag{18.54}$$

or

$$D = \tfrac{1}{2}\lambda\overline{u} \tag{18.55}$$

Introduction of the expressions for λ and $\overline{u}$ (Eq. 1.68 and Table 14.2) gives

$$D = \frac{V}{\pi d^2 N}\left(\frac{\mathbf{k}T}{\pi m}\right)^{1/2} = \frac{V_m}{\pi d^2 L}\left(\frac{\mathbf{k}T}{\pi m}\right)^{1/2} \tag{18.56}$$

Comparison of Eq. 18.55 with Eq. 18.14 for the coefficient of viscosity η gives the following relationship between D and η for a gas:

$$\boldsymbol{\eta = \rho D} \tag{18.57}$$

where ρ is the density (Nm/V).

EXAMPLE

If the molecular diameter for O_2 is 0.38 nm, estimate the self-diffusion coefficient and the viscosity of O_2 at 25°C and 101.325 kPa pressure.

SOLUTION

$PV = nRT = N\mathbf{k}T$ and therefore

$$\frac{V}{N} = \frac{\mathbf{k}T}{P} = \frac{1.381 \times 10^{-23}(\text{J K}^{-1}) \times 298.15(\text{K})}{1.013\ 25 \times 10^{5}(\text{Pa})} = 4.06 \times 10^{-26}\ \text{m}^3$$

The mass of the oxygen molecule is $32 \times 10^{-3}/6.022 \times 10^{23} = 5.31 \times 10^{-26}$ kg. Therefore from Eq. 18.56,

$$D = \frac{4.06 \times 10^{-26}(\text{m}^3)}{\pi \times [0.38 \times 10^{-9}(\text{m})]^2}\left[\frac{1.381 \times 10^{-23} \times 298.15(\text{J})}{\pi \times 5.31 \times 10^{-26}(\text{kg})}\right]^{1/2}$$

$$= 1.4 \times 10^{-5}\ \text{m}^2\ \text{s}^{-1}$$

From Eq. 18.57,

$$\eta = \rho D = \frac{NmD}{V} = \frac{5.31 \times 10^{-26}(\text{kg}) \times 1.4 \times 10^{-5}(\text{m}^2\ \text{s}^{-1})}{4.06 \times 10^{-26}(\text{m}^3)}$$

$$= 1.83 \times 10^{-5}\ \text{kg m}^{-1}\ \text{s}^{-1}$$ ■

Driving Force of Diffusion

Consider the diffusion across a distance dx over which there is a concentration change from c to $c + dc$ (see Figure 18.10). The force that drives the solute molecules to the more dilute region can be calculated from the difference between the molar Gibbs energy at concentration c and that at concentration $c + dc$. This molar Gibbs energy difference is

$$dG = G_{c+dc} - G_c = RT \ln\left(\frac{c + dc}{c}\right) = RT \ln\left(1 + \frac{dc}{c}\right) \tag{18.58}$$

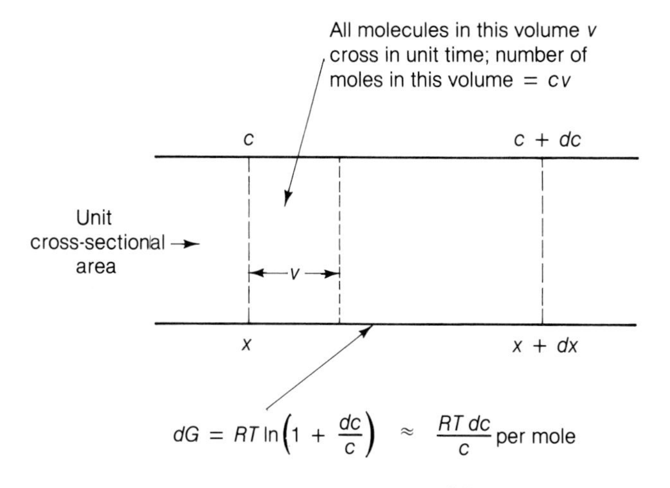

FIGURE 18.10 The driving force of diffusion.

Since dc/c is very small,

$$\ln\left(1 + \frac{dc}{c}\right) \approx \frac{dc}{c} \tag{18.59}$$

and therefore

$$dG = \frac{RT\,dc}{c} \tag{18.60}$$

This Gibbs energy difference is the work w done *on* the system in transferring a mole of solute from concentration c to $c + dc$. The work done *by* the system, $-w$, in transferring a *molecule* of solute from $c + dc$ to c is thus

$$-w = -\frac{RT}{L}\frac{dc}{c} \tag{18.61}$$

$$= -\frac{\mathbf{k}T\,dc}{c} \tag{18.62}$$

This work $-w$ is done over a distance dx, and if F_d is the driving force leading to diffusion

$$-w = F_d\,dx \tag{18.63}$$

The driving force per molecule is thus given by combining Eqs. 18.62 and 18.63:

$$F_d = -\frac{\mathbf{k}T}{c}\frac{dc}{dx} \tag{18.64}$$

When a driving force is applied to a molecule, its speed increases until the frictional force F_f acting on it is equal to the driving force. The molecule has then attained a *limiting speed.* Most theories of the frictional force on a molecule, such as that of Stokes (considered later), lead to the conclusion that the force is directly proportional to the speed; thus

$$F_f = fv \tag{18.65}$$

where f is known as the *frictional coefficient.* The limiting velocity is thus attained when

$$fv = -\frac{\mathbf{k}T}{c}\frac{dc}{dx} \tag{18.66}$$

or

$$cv = -\frac{\mathbf{k}T}{f}\frac{dc}{dx} \tag{18.67}$$

All molecules within a distance v of a given unit cross-sectional area will cross that area in unit time (Figure 18.10). The number of molecules in that volume is cv, which is therefore the flux J:

$$J = -\frac{\mathbf{k}T}{f}\frac{dc}{dx} \tag{18.68}$$

Comparison of this molecularly derived diffusion equation with Fick's first law, Eq. 18.34 with $A = 1$, shows that the diffusion coefficient is given by

$$D = \frac{\mathbf{k}T}{f} \tag{18.69}$$

This equation was first derived by Albert Einstein* (1879–1955) in 1905.

Diffusion and Ionic Mobility

We saw in Section 7.5 that the mobility of an ion in solution is defined as the speed with which the ion moves under a unit potential gradient. Since the diffusion coefficient of a species is related to the speed with which it moves under a unit concentration gradient, the mobilities of ions and, therefore, the molar ionic conductivities are proportional to their diffusion coefficients. We shall now obtain the proportionality factor.

A unit potential gradient is by definition one that exerts a unit force on a unit charge; it will therefore exert a force of Q^u N (kg m s^{-2}) on an ion having a charge Q^u C. The mobility† u_e of an ion having a charge Q is thus the speed with which it moves when a force Q acts upon it. Its speed when it is acted upon by a unit force is therefore u_e/Q. When an ion moves in an electric potential gradient, it soon attains a limiting speed at which the frictional force F_f is equal to the electric force, and this speed is

$$v = \frac{u_e}{Q} F_f \tag{18.70}$$

Comparison of this equation with Eq. 18.65 shows that the frictional coefficient f is given by

$$f = \frac{Q}{u_e} \tag{18.71}$$

and insertion of this into the Einstein equation (Eq. 18.69) gives

$$D = \frac{\mathbf{k}T}{Q} u_e \tag{18.72}$$

If the ion has a charge number of $|z_i|$, its charge Q is $F|z_i|/L$, where F is the Faraday constant and L is the Avogadro constant. By Eq. 7.64 its ionic conductivity $\lambda°$ is equal to Fu_e, so that $u_e = \lambda°/F$. Equation 18.72 therefore becomes

$$D = \frac{\mathbf{k}T}{F|z_i|/L} \cdot \frac{\lambda°}{F} = \frac{RT}{F^2|z_i|} \lambda° \tag{18.73}$$

The equation was first derived in 1888 by Nernst.‡ In view of its relationship to Einstein's equation (Eq. 18.69), Eq. 18.73 is often known as the **Nernst-Einstein**

---

* A. Einstein, *Ann. Physik*, *17*, 549 (1905); *19*, 371 (1906).

† In this section we use the symbol u_e to avoid confusion with u, the molecular velocity. Again we use the superscript u to indicate the *value* of the electric charge Q, which equals $Q^u C$.

‡ W. Nernst, *Z. Physik. Chem.*, *2*, 613 (1888).

equation. In practice, diffusion coefficients have to be measured for more than one type of ion; for electrolytes involving two ions of equal and opposite charges (e.g., NaCl, $ZnSO_4$), Nernst showed that the average diffusion coefficient is

$$D = \frac{2D_+D_-}{D_+ + D_-} \tag{18.74}$$

where D_+ and D_- are the individual diffusion coefficients.

EXAMPLE

Estimate the diffusion coefficient of NaCl in water at 25°C from the molar ionic conductivities given in Table 7.3.

SOLUTION

From Eq. 18.73, at 25°C,

$$D = \frac{8.314(\text{J K}^{-1}\ \text{mol}^{-1}) \times 298.15(\text{K})}{96\ 485^2(\text{C}^2\ \text{mol}^{-2})} \lambda°(\Omega^{-1}\ \text{cm}^2\ \text{mol}^{-1})$$

$$= 2.66 \times 10^{-7}\lambda°\ \text{J}\ \Omega^{-1}\ \text{C}^{-2}\ \text{cm}^2$$

$$= 2.66 \times 10^{-7}\lambda°\ \text{cm}^2\ \text{s}^{-1}$$

(since $\Omega = \text{V A}^{-1}$, $J = \text{V C}$, and $\text{C} = \text{A s}$).

Then, for Na^+,

$$D_+ = 2.66 \times 10^{-7} \times 50.1 = 1.33 \times 10^{-5}\ \text{cm}^2\ \text{s}^{-1}$$

For Cl^-,

$$D_- = 2.66 \times 10^{-7} \times 76.4 = 2.03 \times 10^{-5}\ \text{cm}^2\ \text{s}^{-1}$$

For NaCl, using Eq. 18.74,

$$D = \frac{2 \times 1.33 \times 10^{-5} \times 2.03 \times 10^{-5}}{(1.33 + 2.03) \times 10^{-5}}$$

$$= 1.61 \times 10^{-5}\ \text{cm}^2\ \text{s}^{-1}$$

■

Stokes's Law

The diffusion coefficient depends on the ease with which the solute molecules can move. In aqueous solution, the diffusion coefficient of a solute is a measure of how readily a solute molecule can push aside its neighboring water molecules and move into another position. An important aspect of the theory of diffusion is how the magnitudes of the frictional coefficients f and, hence, of the diffusion coefficients D, depend on the properties of the solute and solvent molecules.

Examination of the values given in Table 18.3 shows that diffusion coefficients tend to decrease as the molecular size increases. This is easy to understand, since a larger solute molecule has to push aside more solvent molecules during its progress and will therefore move more slowly than a smaller molecule. It is difficult to develop a precise theory of diffusion coefficients, but in 1851 the British physicist

Sir George Gabriel Stokes* (1819–1903) considered a simple situation in which the solute molecules are so much larger than the solvent molecules that the latter can be regarded as continuous (i.e., as not having molecular character). For such a system Stokes deduced that the frictional force F_f opposing the motion of a large particle of radius r moving at speed v through a solvent of viscosity η is given by

$$F_f = 6\pi r\eta v \tag{18.75}$$

The frictional coefficient is therefore

$$f = 6\pi r\eta \tag{18.76}$$

It then follows from Eq. 18.69 that when Stokes's law applies, the diffusion coefficient is given by

$$D = \frac{\mathrm{k}T}{6\pi r\eta} \tag{18.77}$$

This is often referred to as the **Stokes-Einstein equation.**

Measurement of D in a solvent of known viscosity therefore permits a value of the radius r to be calculated. Such a calculation is not very satisfactory for solute molecules, however, for several reasons. In the first place, Stokes's law is based on the assumption of very large spherical particles and a continuous solvent and involves some error even for approximately spherical macromolecules. Second, the solute molecules may not be spherical, and this introduces an additional error. Furthermore, solute molecules are commonly solvated, and in moving through the solution they transport some of their solvation layer. In spite of these drawbacks, Eq. 18.77 has proved useful in providing approximate values of molecular sizes.

EXAMPLE

The diffusion coefficient for glucose in water is 6.81×10^{-10} m^2 s^{-1} at 25°C. The viscosity of water at 25°C is 8.937×10^{-4} kg m^{-1} s^{-1}, and the density of glucose is 1.55 g cm^{-3}. Estimate the molar mass of glucose, assuming that Stokes's law applies and that the molecule is spherical.

SOLUTION

Since Stokes's law applies, D and η are related by Eq. 18.77 and the radius of the molecule is given by

$$r = \frac{\mathrm{k}T}{6\pi\eta D}$$

$$= \frac{1.38 \times 10^{-23}(\mathrm{J\ K^{-1}}) \times 298.15(\mathrm{K})}{6 \times 3.1426 \times 8.937 \times 10^{-4}(\mathrm{kg\ m^{-1}\ s^{-1}}) \times 6.81 \times 10^{-10}(\mathrm{m^2\ s^{-1}})}$$

$$= 3.59 \times 10^{-10}\ \mathrm{J\ kg^{-1}\ m^{-1}\ s^2}$$

$$= 3.59 \times 10^{-10}\ \mathrm{m} = 0.359\ \mathrm{nm}$$

*G. G. Stokes, *Trans. Cambridge Phil. Soc.*, Ser. 9, No. 8, (1851).

since 1 J = 1 kg m^2 s^{-2}. The estimated volume of the molecule is thus

$$\tfrac{4}{3}\pi(3.59 \times 10^{-10})^3 = 1.94 \times 10^{-28} \text{ m}^3$$

and its mass is

$$1.94 \times 10^{-28}(\text{m})^3 \times 1.55 \times 10^6(\text{g m}^{-3}) = 3.01 \times 10^{-22} \text{ g}$$

The molar mass M is this mass multiplied by the Avogadro constant L:

$$M = 3.01 \times 10^{-22} \times 6.022 \times 10^{23} = 181.3 \text{ g mol}^{-1}$$

This agrees very well with the true molar mass of 180.2 g mol^{-1}. ■

Diffusion Through Membranes

The speed with which molecules and ions can pass through membranes is a matter of great importance, particularly in biology. In work with membranes it is convenient to define a quantity known as the **permeability coefficient** P, which is defined as the flux through unit area when there is a unit concentration difference. We have previously considered the case of two stirred solutions, of concentrations c_1 and c_2, separated by a sintered glass diaphragm of thickness l. For such a system the rate of permeation is given by Eq. 18.41, and the flux through unit area is

$$J/A = D\frac{c_1 - c_2}{l} \qquad (18.78)$$

The permeability coefficient P is the flux per unit area per unit concentration difference and therefore

$$P \equiv \frac{J/A}{c_1 - c_2} = \frac{D}{l} \qquad (18.79)$$

Since the SI unit for D is m^2 s^{-1}, that for P is m s^{-1}.

It is not necessarily the case that the solute molecules will be just as soluble in the membrane as in the solvent. For example, a membrane might be composed of a certain amount of lipid material, and a solute molecule might have a certain proportion of nonpolar groups. Such a solute might well be more soluble in the membrane than in water. This effect can be represented by use of a partition or distribution coefficient K_p, which is simply an equilibrium constant; it is the ratio of the concentrations of the solute in the membrane and in the solvent. In this situation the concentrations at the surface of the membrane, as shown in Figure 18.11, are K_pc_1 and K_pc_2. The gradient is now given by

$$-\frac{dc}{dx} = \frac{K_pc_1 - K_pc_2}{l} = K_p\frac{c_1 - c_2}{l} \qquad (18.80)$$

and the rate of diffusion is

$$\frac{dn}{dt} = DAK_p\frac{c_1 - c_2}{l} \qquad (18.81)$$

Thus the rate, the apparent diffusion coefficient (DK_p), and the apparent permeability coefficient (DK_p/l) are all altered by the factor K_p. Since K_p may be greater or

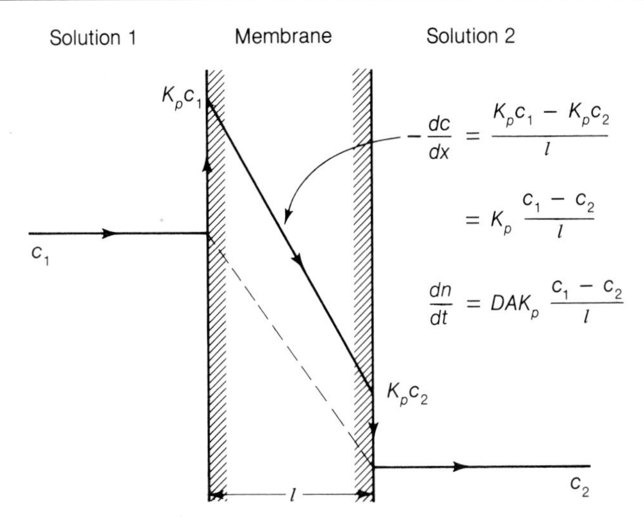

FIGURE 18.11 Stirred solutions of concentrations c_1 and c_2, separated by a membrane of thickness l. The solubility of the solute is different in the membrane and in the solvent (in the diagram the solubility is shown as higher in the membrane).

less than unity, there may be an enhancement or a diminution in the rate of permeation of the membrane. Sometimes this partitioning effect leads to rates that are several powers of 10 higher than would be obtained in its absence. Conversely, the rate of diffusion through the membrane will be abnormally small if the solute is much less soluble in the membrane than in the solution. For example, a carbohydrate such as sucrose is much less soluble in a fatty membrane than in water, where there is extensive hydrogen bonding; because of this effect it diffuses slowly through a fatty membrane.

Abnormally high rates of transport of solute molecules across membranes can arise from other causes than high solubility in the membrane. Sometimes solute molecules become attached to so-called *carrier* molecules, which remain in the membrane and transport the solute molecule from one side to another. If the solvent molecule is strongly attached to the carrier molecule, the concentration of the solute-carrier complex will be large, and the concentration gradient will be large for the complex. A high rate of permeation can result, and this effect is known as **facilitated transport.**

In facilitated transport the direction of flow is consistent with the concentration gradient. A completely different effect, frequently found in biological systems, is when the flow is contrary to the concentration gradient; the solute molecules move from the low-concentration side to the high-concentration side of the membrane. An example is the formation of the gastric juice in humans, where HCl flows from a solution of pH approximately 7 (i.e., $[H^+] \approx 10^{-7}M$) into a solution that is approximately 0.1 M in HCl.

This effect is known as **active transport.** It is brought about by a coupling of the diffusion process with an exergonic chemical reaction. Suppose, for example, that in a biological system at 37°C active transport were occurring against a 10/1 concentration gradient. The Gibbs energy difference per mole is

$$\Delta G = RT \ln 10 = (8.314 \times 310 \ln 10) \text{ J mol}^{-1} = 5.9 \text{ kJ mol}^{-1}$$

A reaction exergonic by this amount and coupled with the transport process would therefore be able to give transport against the 10/1 gradient. The details of such coupling processes in biological systems are the subject of much investigation.

18.3 Sedimentation

In Section 18.2 we have dealt with the movement of a solute as a result of a concentration gradient. Molecules in solution can also be made to move by subjecting them to other forces. For example, a solution may simply be allowed to stand. If the solute molecules are small, there will be no change in the distribution in space, since the thermal motion will counteract the tendency of the molecules to move in the gravitational field. Very large particles, however, will *sediment* on standing. In order for smaller particles to undergo *sedimentation,* it is necessary to increase the effective gravitational field by subjecting the solution to centrifugal motion. The velocity of sedimentation can lead to values of molar masses, as will now be explained.

Sedimentation Velocity

Suppose that a particle of mass m, having a specific volume (volume per unit mass) of V_1, is in a liquid of density ρ (mass per unit volume). The volume of liquid displaced by the particle is $V_1 m$ and the mass displaced is $V_1 m\rho$. The net force F_g acting on the particle as a result of the gravitational field is therefore

$$F_g = mg - V_1 m\rho g = (1 - V_1\rho)mg \tag{18.82}$$

where g is the acceleration of gravity. The particle will reach a limiting speed v when this force is equal to the frictional force F_f, which is equal to the frictional coefficient f multiplied by the speed:

$$F_f = fv = (1 - V_1\rho)mg \tag{18.83}$$

If Stokes's law applies to the particle, F_f is given by Eq. 18.75, and therefore

$$6\pi r\eta v = (1 - V_1\rho)mg \tag{18.84}$$

The limiting speed is

$$v = \frac{(1 - V_1\rho)mg}{6\pi r\eta} \tag{18.85}$$

Experiments on sedimentation under the earth's gravitational field were carried out in 1908 by the French physicist Jean Baptiste Perrin* (1870–1942), who observed under a microscope the movement of the particles of the pigment gamboge. In order to observe the sedimentation of smaller particles, such as proteins and other macromolecules, it is necessary to employ much higher fields. This is done by means of an **ultracentrifuge**, in which solutions are rotated at speeds up to 80 000 revolutions per minute, which produces fields up to $3 \times 10^5\, g$. The development of such ultracentrifugal techniques is largely due to the Swedish physical chemist

*J. B. Perrin, *Ann. Chim. Phys., 18,* 5 (1909); *Compt. Rend., 152,* 1380 (1911).

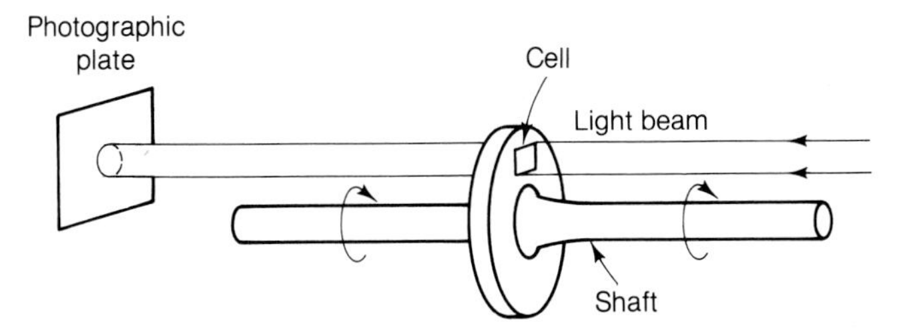

FIGURE 18.12 Schematic diagram of an ultracentrifuge. The shaft is rotated at very high speeds so that the effective gravitational field in the cell is very large and brings about sedimentation. The distribution of solute in the cell is determined from the blackening of the photographic plate at various positions. More satisfactorily, the distribution can be obtained from the refractive index, which varies approximately linearly with concentration. In the *schlieren* method, special methods of illumination are used that lead to darkening of the photographic plate corresponding to regions where the refractive index changes rapidly.

Theodor Svedberg* (1884–1971), whose work along these lines was started in 1923 and who devoted much study to the characterization of protein molecules and other macromolecules. A schematic diagram of an ultracentrifuge is shown in Figure 18.12. For a centrifugal field of force we replace g in Eq. 18.83 by $\omega^2 x$, where ω is the angular velocity and x is the distance from the center of rotation. Thus

$$v = \frac{(1 - V_1\rho)m\omega^2 x}{f} \tag{18.86}$$

The quantity

$$s = \frac{v}{\omega^2 x} \tag{18.87}$$

is known as the *sedimentation coefficient;* it is the sedimentation rate v when the centrifugal acceleration is unity. For a given molecular species in a given solvent at a given temeprature, s is a characteristic quantity. Its SI unit is the second but it is often expressed in *Svedberg units,* equal to 10^{-13} s.

If Stokes's law applies, f is $6\pi\eta r$ and therefore, from Eq. 18.86,

$$v = \frac{(1 - V_1\rho)m\omega^2 x}{6\pi\eta r} \tag{18.88}$$

The sedimentation coefficient is thus

$$s = \frac{(1 - V_1\rho)m}{6\pi\eta r} \tag{18.89}$$

Use of this equation, however, is unreliable, since Stokes's law is only valid for very large spherical particles. It is more satisfactory to express f by the use of Eq. 18.69, according to which it is $\mathbf{k}T/D$. Insertion of this expression into Eq. 18.86 then gives

$$v = \frac{D(1 - V_1\rho)m\omega^2 x}{\mathbf{k}T} \tag{18.90}$$

*T. Svedberg and C. Rinde, *J. Am. Chem. Soc.*, *45*, 943 (1923) and many subsequent publications.

or

$$s = \frac{D(1 - V_1\rho)m}{\mathbf{k}T} \tag{18.91}$$

$$= \frac{D(1 - V_1\rho)M}{RT} \tag{18.92}$$

The molar mass M can therefore be calculated from a measurement of the sedimentation coefficient and the diffusion coefficient:

$$M = \frac{RTs}{D(1 - V_1\rho)} \tag{18.93}$$

This equation, derived by Svedberg in 1929, has been the basis of many measurements of molar masses. For precise determination the values of s, D and V_1 should be extrapolated to infinite dilution.

EXAMPLE

A sample of human hemoglobin had a sedimentation constant of 4.48 Svedbergs in water at 20°C and a diffusion coefficient of 6.9×10^{-11} m^2 s^{-1}. The specific volume of human hemoglobin is 0.749 cm^3 g^{-1}, and the density of water at 20°C is 0.998 g cm^{-3}. Calculate the molar mass of human hemoglobin.

SOLUTION

The molar mass is obtained by inserting the following values into Eq. 18.93:

$R = 8.314$ J K^{-1} mol^{-1} $\qquad D = 6.9 \times 10^{-11}$ m^2 s^{-1}

$T = 293.15$ K $\qquad V_1 = 0.749$ cm^3 g^{-1}

$s = 4.48 \times 10^{-13}$ s $\qquad \rho = 0.998$ g cm^{-3}

Thus we have

$$m = \frac{8.314(\text{J K}^{-1}\text{ mol}^{-1}) \times 293.15(\text{K}) \times 4.48 \times 10^{-13}(\text{s})}{6.9 \times 10^{-11}(\text{m}^2\text{ s}^{-1}) \times (1 - 0.749 \times 0.998)}$$
$$= 62.6 \text{ J m}^{-2}\text{ s}^2\text{ mol}^{-1}$$

To convert this into the usual molar-mass units we note that 1 J = 1 kg m^2 s^{-2}; thus

$$M = 62.7 \text{ kg mol}^{-1} = 62\,700 \text{ g mol}^{-1}$$ ■

A particular advantage of the sedimentation-velocity technique is that a macromolecular solution containing more than one type of molecule is separated according to the molar masses of the components.

Sedimentation Equilibrium

An alternative method of using the ultracentrifuge to measure molar masses is to allow the distribution of particles to reach equilibrium. As sedimentation occurs in the ultracentrifuge, a concentration gradient is established, and this will cause the molecules to diffuse in the opposite direction. Eventually the system reaches a state of equilibrium at which the rate with which the solute is driven outward by the centrifugal force just equals the rate with which it diffuses inward under the influence of the concentration gradient.

The velocity v with which the particles travel as a result of the centrifugal field is given by Eq. 18.86. All particles within a distance v of a given cross-sectional area A will cross that area in unit time, and if the molecular concentration is c, there are vcA particles that cross; the sedimentation flux is thus vcA and, by Eq. 18.86, is

$$J(\text{sedimentation}) = vcA = \frac{(1 - V_1\rho)m\omega^2 xcA}{f} \tag{18.94}$$

The diffusive flux through area A is

$$J(\text{diffusive}) = -AD\frac{dc}{dx} \tag{18.95}$$

$$= -\frac{A\mathbf{k}T}{f}\frac{dc}{dx} \tag{18.96}$$

At equilibrium these two rates are equal, and we obtain

$$\frac{dc}{c} = -\frac{M(1 - V_1\rho)\omega^2 x\, dx}{RT} \tag{18.97}$$

where $M(= mL = mR/\mathbf{k})$ is the molar mass. Integration between two positions x_1 and x_2, at concentrations c_1 and c_2, leads to

$$M = \frac{2RT\ln(c_1/c_2)}{(1 - V_1\rho)\omega^2(x_2^2 - x_1^2)} \tag{18.98}$$

Thus, if measurements of the relative concentrations are made at two positions, after equilibrium has been established, this equation can be used to calculate the molar mass. The value so obtained is the mass-average molar mass (see Section 19.3).

The sedimentation-equilibrium method does not require an independent measurement of the diffusion coefficient, in contrast to the sedimentation-velocity method. However, the time required for complete equilibrium to be established is so long that the method is often inconvenient to use, especially if the relative molecular mass is greater than 5000.

In order to overcome this drawback, a modification to the sedimentation-equilibrium method was proposed in 1947 by the Canadian physicist William J. Archibald.* At the top meniscus of the cell and at the bottom of the cell, there can be no net flux, and the sedimentation-equilibrium equations apply at these sections at all times. Shortly after the ultracentrifuge is brought to its top speed, therefore, concentrations in these special sections can be determined and M calculated from Eq. 18.93. This modification of the sedimentation-equilibrium method greatly increases its applicability.

If a solution of a substance of low molar mass is ultracentrifuged, equilibrium is established within a fairly short period of time, and there will be a density

*W. J. Archibald, *J. Phys. Chem.*, *51*, 1204 (1947).

gradient in the solution. If a substance of high molar mass is added, it will float in this solution at the particular position at which its density is equal to the density of the solution. If the macromolecular substance is made up of fractions of different molar masses, it will separate into fractions that will remain at different planes in the cell. This technique, known as *density-gradient ultracentrifugation** has proved useful in establishing the different kinds of molecules in a macromolecular sample.

18.4 Electrokinetic Effects

The term *electrokinetic* is applied to those effects in which either an electric potential brings about movement or movement produces an electric potential. The occurrence of these electrokinetic effects is due to the existence of potential differences at solid-liquid interfaces or at the surfaces of macromolecules that are in solution.

In Section 17.9 we have considered various theories that apply to these potential differences. We have seen that a diffuse double layer exists at interfaces, as a result of ionic distributions. The potential difference between the bulk of the solution and the position of closest approach of the ions to the surface is known as the *electrokinetic* or *ζ potential*. The position of closest approach represents the boundary at which relative motion occurs, and the electrokinetic effects are therefore governed by the value of the ζ potential. In the Chapman-Gouy theory the ζ potential is given by Eq. 17.67, namely

$$\zeta = \frac{\sigma d}{\epsilon_0 \epsilon} \tag{18.99}$$

where σ is the charge density at the boundary at which relative motion occurs, $d(=1/\kappa)$ is the effective distance over which the ζ potential is operating, ϵ is the dielectric constant, and ϵ_0 is the permittivity of a vacuum. We shall now use this relationship to develop equations for some electrokinetic effects.

Electroosmosis

If a membrane separates two identical liquids or solutions and a potential difference is applied across the membrane, there results a flow of liquid through the pores of the membrane. This phenomenon is known as **electroendosmosis** or as **electroosmosis.** A simplified version of the theory is as follows.

When a liquid is forced by electroosmosis through the pores of a membrane, the rate of flow is determined by two opposing factors: the force of electroosmosis on the one hand and the frictional force between the moving liquid layer and the wall on the other. When the two forces are equal, there will be a uniform rate of flow. The situation in a single pore is represented in Figure 18.13a. On the solid side of the double layer the speed of flow is zero, whereas on the solution side it has attained the uniform speed v of the moving liquid; the velocity gradient, assumed to be uniform, is thus v/d. The force due to friction is the product of the velocity gradient and the coefficient of viscosity η of the liquid and is thus $\eta v/d$. The electric

*M. S. Meselson, F. W. Stahl, and J. Vinograd, *Proc. Nat. Acad. Sci.*, 43, 581 (1957); 44, 671 (1958).

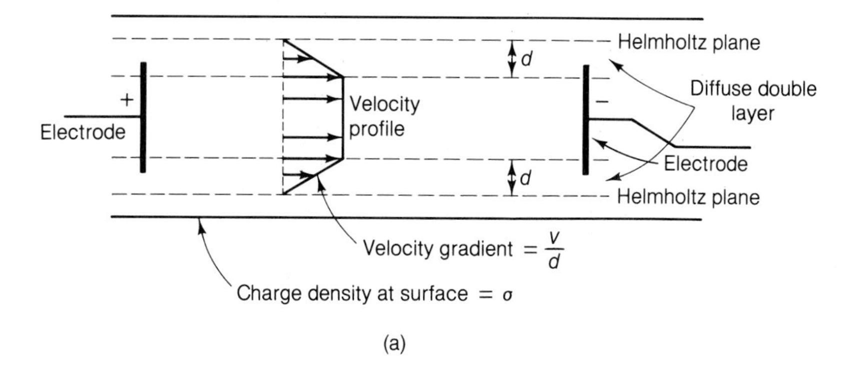

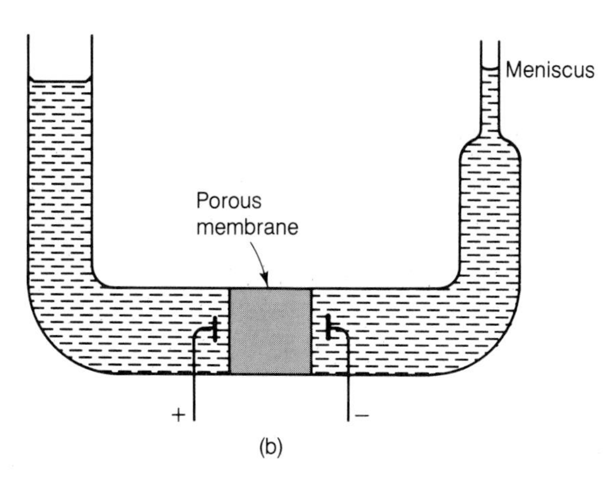

FIGURE 18.13 (a) Electroosmosis through a pore of a membrane. (b) Apparatus for studying electroosmosis; observations are made of the movement of the meniscus in the capillary tube.

force causing electroosmosis is equal to the product of the applied electric potential gradient V and the charge density σ at the boundary at which movement occurs. Thus, in the steady state,

$$\frac{\eta v}{d} = V\sigma \tag{18.100}$$

The ζ potential is related to the distance d by Eq. 18.99, and elimination of d between Eqs. 18.99 and 18.100 gives

$$v = \frac{\zeta \epsilon_0 \epsilon}{\eta} V \tag{18.101}$$

If the potential gradient V is unity, the uniform speed attained, v_0, is known as the *electroosmotic mobility* and is given by

$$v_0 = \frac{\zeta \epsilon_0 \epsilon}{\eta} \tag{18.102}$$

If A is the total area of cross section of all the pores in a membrane, the volume V_l of liquid transported electroosmotically per unit time is equal to Av, so that from Eq. 18.101

$$V_l = \frac{\zeta A \epsilon_0 \epsilon}{\eta} V \tag{18.103}$$

Electroosmosis may alternatively be studied using a single capillary tube instead of a membrane containing many pores; A is then equal to πr^2, where r is the radius of the tube, and Eq. 18.103 becomes

$$V_l = \frac{\pi \zeta r^2 \epsilon_0 \epsilon}{\eta} V \tag{18.104}$$

A simple type of apparatus for studying electroosmosis through a membrane is shown in Figure 18.13b.

Electrophoresis

In the derivation of Eq. 18.102 for the electroosmotic mobility under unit potential gradient, the moving liquid was regarded as a cylinder moving through a capillary tube. The positions of the liquid and wall can be reversed without affecting the argument, so that Eq. 18.102 also gives the velocity of movement of a solid cylindrical particle through a liquid under the influence of an applied field of unit potential gradient. This quantity is the **electrophoretic mobility;** thus

$$v_0 = \frac{\zeta \epsilon_0 \epsilon}{\eta} \tag{18.105}$$

for a cylindrical particle moving along its axis.

The treatment of particles of different shapes is more complicated, and various theories have been presented. In their theory of electrolytic conductance (Section 7.4) Debye and Hückel concluded in 1924 that for a spherical particle Eq. 18.105 should be replaced by

$$v_0 = \frac{2\zeta \epsilon_0 \epsilon}{3\eta} \tag{18.106}$$

This equation is only valid if the thickness of the double layer is large compared with the radius of the particle. The equation may thus be satisfactory for ions but less so for larger particles, and various improved equations have been suggested. In practice, however, electrophoresis experiments are usually carried out in an empirical manner, without reference to these equations.

The first important experiments on electrophoresis were carried out by the Swedish physical chemist Arne Wilhelm Kaurin Tiselius (1902–1971), who made electrophoresis a very powerful technique for studying mixtures of proteins.* He devised a special type of U-tube along which the protein molecules move under the influence of an electric potential; different proteins will move at different speeds. The tube consisted of portions fitted together at ground-glass joints, so that one of

* A Tiselius, *Nova Acta Reg. Soc. Sci. Upsaliensis*, 7 No. 4 (1930); *Trans. Faraday Soc.*, 33, 524 (1937).

a mixture of proteins could be isolated in one chamber. Optical methods are used to determine the quantity of each protein present in the mixture. The technique of electrophoresis supplements the ultracentrifuge, which separates according to molecular mass and shape; different macromolecules having the same molecular sizes and shapes behave identically in the ultracentrifuge but may have different electrical properties and, hence, can be separated by electrophoresis.

A difficulty with the use of the Tiselius tube is that a certain amount of local heating occurs, leading to convection currents which cause some mixing and disturb the separation. This problem is now frequently overcome by supporting the solution on a gel, such as a polyacrylamide gel. Another recent development in electrophoresis involves the addition of detergents, such as sodium dodecyl sulfate, to macromolecular solutions supported on gels. Proteins are then found to have electrophoretic mobilities that are proportional to their molecular masses, and as a result it is possible to separate proteins that are not easily separable in the absence of the detergent.

One special electrophoretic technique is *isoelectric focusing*. In this method a pH gradient is established so that the solution at the cathode is more basic and that at the anode is more acidic. Molecules will migrate toward the electrode of opposite charge, but as they do so, they will be subject to changing pHs that will tend to make them lose their charge. This process will continue until the species is concentrated at the position where the pH of the solution is the *isoelectric point*[*] of the molecule. If several species are present, each will concentrate at a position corresponding to its isoelectric point. With suitable pH gradients it is possible to separate species with very small differences in isoelectric points. Isoelectric focusing is also frequently carried out with the sample supported on a gel.

Reverse Electrokinetic Effects

In the two effects just mentioned, electrophoresis and electroosmosis, the application of an electric field brings about relative motion of two phases. If, on the other hand, movement is brought about, the displacement of charged layers with respect to each other brings about a potential difference between any two points in the direction of motion. For example, if a liquid is forced through the pores of a membrane or through a capillary tube, a potential difference is observed, its magnitude depending on the ζ potential. This phenomenon, the reverse of electroosmosis, is known as the **streaming potential.**

The reverse of electrophoresis occurs when small particles are allowed to fall through a liquid under the influence of gravity; a difference of potential is observed between two electrodes placed at different levels, and its magnitude again depends on the magnitude of the ζ potential. This phenomenon is known as the *sedimentation potential* or as the **Dorn effect,** after the German physicist Friedrick Ernst Dorn[†] (1848–1916), who discovered it in 1880.

---

[*] The *isoelectric point* is the pH at which a molecule or other particle has no net charge. It is not quite the same as the *isoionic point* (Section 17.10) which is the pH at which, as a result of *proton ionizations,* there is an equal number of positive and negative charges. The isoelectric point takes account of other charges, such as heavy metal ions attached to the particles.

[†] F. E. Dorn, *Ann. Physik, 10,* 70 (1880).

Key Equations

Viscosity:

Definition (when Newton's law applies): $F \equiv \eta A \dfrac{dv}{dt}$

Pouiseuille equation: $\dfrac{dV}{dt} = \dfrac{(P_1 - P_2)\pi R^4}{8\eta l}$

Kinetic theory expressions, for a gas:

$$\eta = \frac{Nm\bar{u}\lambda}{2V}$$

$$= \frac{1}{2}\rho\bar{u}\lambda$$

$$= \frac{(m\mathbf{k}T)^{1/2}}{\pi^{3/2}d^2}$$

Kinetic theory expression, for a liquid: $\eta = A_{\text{vis}}e^{E_{\text{vis}}/RT}$

Intrinsic viscosity of a solution: $[\eta] = \lim\limits_{\rho\to 0}\left[\dfrac{1}{\rho}\cdot\dfrac{\eta - \eta_0}{\eta_0}\right]$

Mark-Houwink equation: $[\eta] = kM_r^\alpha$

(M_r = relative molecular mass)

Diffusion:

Fick's first law: $J = \dfrac{dn}{dt} = -DA\dfrac{dc}{dx}$

Fick's second law: $\dfrac{\partial c}{\partial t} = D\dfrac{\partial^2 c}{\partial x^2}$

Random-walk equation: $\overline{x^2} = 2Dt$

Einstein equation: $D = \dfrac{\mathbf{k}T}{f}$ (f=frictional coefficient)

Nernst-Einstein equation: $D = \dfrac{RT}{F^2|z_i|}\lambda^\circ$

Stokes-Einstein equation: $D = \dfrac{\mathbf{k}T}{6\pi r\eta}$

Sedimentation:

Definition of sedimentation coefficient: $s \equiv \dfrac{v}{\omega^2 x}$

Svedberg's equation: $M = \dfrac{RTs}{D(1 - V_1\rho)}$

A Viscosity

1. In a normal adult at rest the average speed of flow of blood through the aorta is 0.33 m s^{-1}. The radius of the aorta is 9 mm and the viscosity of blood at body temperature, 37°C, is about 4.0×10^{-3} kg m^{-1} s^{-1}. Calculate the pressure drop along a 0.5-m length of the aorta, in millimetres of mercury.

2. A typical human capillary is about 1 mm long and has a radius of 2 μm. If the pressure drop along the capillary is 20 mmHg,

- **a.** Calculate the average linear speed of flow of blood of viscosity 4.0×10^{-3} kg m^{-1} s^{-1}.
- **b.** Calculate the volume of blood passing through each capillary per second.
- **c.** Estimate the number of capillaries in the body if they are supplied by the aorta described in Problem 1.

***3.** The viscosity of ethylene at 25.0°C and 101.325 kPa is 9.33×10^{-6} kg m^{-1} s^{-1}. Estimate

- **a.** the molecular diameter,
- **b.** the mean free path,
- **c.** the frequency of collisions experienced by a given molecule (Z_A), and
- **d.** the collision number (Z_{AA}).

***4.** For nonassociated liquids the fluidity ϕ (i.e., the reciprocal of the viscosity) obeys to a good approximation an equation of the Arrhenius form

$$\phi = Ae^{-E/RT}$$

where A and E are constants:

- **a.** For liquid CCl_4 the viscosity at 0°C is 1.33×10^{-3} kg m^{-1} s^{-1} and the activation energy E is 10.9 kJ mol^{-1}. Estimate the viscosity at 40.0°C.
- **b.** The Arrhenius law does not apply well to associated liquids such as water, but it can be used over a limited temperature range. At 20.0°C the viscosity of water is 1.002×10^{-3} kg m^{-1} s^{-1} and the activation energy for fluidity is 18.0 kJ mol^{-1}. Estimate the viscosity at 40.0°C.
- **c.** Over its entire liquid range the viscosity of water is represented to within 1% by the expression

$$\log_{10}\left(\frac{\eta_{20^\circ}}{\eta_{t^\circ}}\right) = \frac{1.370\,23(t^u - 20) + 8.36 \times 10^{-4}(t^u - 20)}{109 \times t^u}$$

where t^u is the value of the Celsius temperature. Use this expression to obtain a better estimate of the viscosity at 40.0°C, with $\eta_{20^\circ} = 1.002 \times 10^{-3}$ kg m^{-1} s^{-1}.

5. At 20.0°C the viscosity of pure toluene is 5.90×10^{-4} kg m^{-1} s^{-1}. Calculate the intrinsic viscosities of solutions, containing 0.1 g dm^{-3} of polymer in toluene, having the following viscosities:

- **a.** 5.95×10^{-4} kg $m^{-1}s^{-1}$
- **b.** 6.05×10^{-4} kg m^{-1} s^{-1}
- **c.** 6.27×10^{-4} kg m^{-1} s^{-1}

These solutions are sufficiently dilute that the reduced specific viscosity can be taken to be the intrinsic viscosity.

***6.** Suppose that solutions (a) and (c) in Problem 5 correspond to polymers of relative molecular masses 20 000 and 40 000, respectively. Assuming the Mark-Houwink equation to apply, make an estimate of the relative molecular mass of the polymer in solution (b).

B Diffusion

7. The molecular diameter of He is 0.225 nm. Estimate, at 0°C and 101.325 kPa,

- **a.** the viscosity of the gas,
- **b.** the self-diffusion coefficient,
- **c.** the mean speed of the molecules,
- **d.** the mean free path,
- **e.** the collision frequency Z_A, and
- **f.** the collision number Z_{AA}.

8. Calculate the mean square distance traveled by a molecule of H_2 at 20°C and 101.325 kPa in 10 s ($D = 1.005 \times 10^{-4}$ m^2 s^{-1}).

9. Solutions of (a) glucose ($D = 6.8 \times 10^{-10}$ m^2 s^{-1}) and (b) tobacco mosaic virus ($D = 5.3 \times 10^{-12}$ m^2 s^{-1}) were maintained at a constant temperature of 20°C and without agitation for 100 days. How far would a given molecule of each be expected to diffuse in that time?

10. Estimate the diffusion coefficient of cupric sulfate in water at 25°C from the molar conductivities given in Table 7.3 (p. 281).

11. Estimate the diffusion coefficient of sodium acetate in water at 25°C from the following mobility values:

Na^+: 5.19×10^{-4} cm^2 V^{-1} s^{-1}

CH_3COO^-: 4.24×10^{-4} cm^2 V^{-1} s^{-1}

12. The diffusion coefficient for horse hemoglobin in water is 6.3×10^{-11} m^2 s^{-1} at 20°C. The viscosity of water at 20°C is 1.002×10^{-3} kg m^{-1} s^{-1} and the specific volume of the protein is 0.75 cm^3 g^{-1}. Assume the hemoglobin molecule to be spherical and to obey Stokes's law, and estimate its radius and the relative molecular mass.

13. If the diffusion coefficient for insulin is 8.2×10^{-11} m^2 s^{-1} at 20°C, estimate the mean time required for an insulin molecule to diffuse through a distance equal to the diameter of a typical living cell ($\approx$10 μm).

***14.** A colloidal particle is spherical and has a diameter of 0.3 μm and a density of 1.18 g cm^{-3}. Estimate how long it will take for the particle to diffuse through a distance of 1 mm in water at 20.0°C ($\eta = 1.002 \times 10^{-3}$ kg^{-3} m^{-1} s^{-1}; density of water at 20.0°C = 0.998 g cm^{-3}). (See also Problem 18.)

C Sedimentation

15. Diphtheria toxin was found to have, at 20.0°C, a sedimentation coefficient of 4.60 Svedbergs and a diffusion coefficient of 5.96×10^{-7} cm^2 s^{-1}. The toxin has a specific volume of 0.736 cm^3 g^{-1}, and the density of water at 20°C is 0.998 g cm^{-3}. Estimate a value for the relative molecular mass of the toxin.

16. A protein has a sedimentation coefficient of 1.13×10^{-12} s^{-1} at 25.0°C and a diffusion coefficient of 4.2×10^{-11} m^2 s^{-1}. The density of the protein is 1.32 g cm^{-3} and that of water at 25.0°C is 0.997 g cm^{-3}. Calculate the molar mass of the protein.

***17.** A protein of molar mass 60 000 g mol^{-1} has a density of 1.31 g cm^{-3} and in water at 25.0°C ($\rho = 0.997$ g cm^{-3}; $\eta = 8.937 \times 10^{-4}$ kg m^{-1} s^{-1}) it has a sedimentation coefficient of 4.1×10^{-13} s^{-1}. Calculate the frictional coefficient f

a. from the sedimentation coefficient and
b. by the use of Stokes's law.

Suggest a reason why the two values are not quite the same.

***18.** How long will it take the particle from Problem 14 to *sediment* a distance of 1 mm in the earth's gravitational field ($g = 9.81$ m s^{-2})?

***19.** An aqueous colloidal solution contains spherical particles of uniform size and having a density of 1.33 g cm^{-3}. The diffusion coefficient at 25.0°C is 1.20×10^{-11} m^2 s^{-1}; make an estimate of the sedimentation coefficient ($\rho_{H_2O} = 0.997$ g cm^{-3}; $\eta_{H_2O} = 8.937 \times 10^{-4}$ kg m^{-1}s^{-1}).

D Supplementary Problems

20. a. The activation energy for the fluidity of n-octane is 12.6 kJ mol^{-1} and the viscosity at 0°C is 7.06×10^{-4} kg m^{-1} s^{-1}. Estimate the viscosity at 40.0°C.
b. A better temperature law for the viscosity of n-octane has been found to be

$$\eta = A(T/\text{K})^{-1.72} e^{543/(T/\text{K})}$$

Make another estimate of the viscosity at 40.0°C. What is the effective activation energy at 20°C?

21. In Problem 4c was given an empirical formula for the viscosity of water as a function of temperature. To what activation energies does this expression correspond at **a.** 20°C and at **b.** 100°C? Give a qualitative explanation for the difference between the two values.

22. Consider a hypothetical gas in which the molecules have mass but no size and do not interact with each other.

a. What would be the viscosity of such a gas?
b. Suppose instead that the molecules have zero size but attract one another. What can you then say about the viscosity?
c. If they repel one another, what would the viscosity be?
d. Give a clear explanation of your conclusions in all three cases.

E Essay Questions

1. Explain how the rate of diffusion through a membrane depends on

a. the size of the diffusing substance and
b. its solubility in the membrane.

2. Explain clearly the different mechanisms involved in the viscosity of gases and the viscosity of liquids.

Suggested Reading

R. S. Bradley, *The Phenomena of Fluid Motions*, Reading, Mass: Addison-Wesley, 1967.

S. Chapman and T. G. Cowling, *The Mathematical Theory of Non-Uniform Gases*, Cambridge: University Press, 1952.

J. Crank, *The Mathematics of Diffusion*, Oxford: Clarendon Press, 1970.

J. O. Hirschfelder, C. F. Curtiss, and R. B. Bird, *Molecular Theory of Gases and Liquids*, New York: Wiley, 1954.

J. H. Jeans, *Introduction to the Kinetic Theory of Gases*, Cambridge: University Press, 1959.

W. Jost, *Diffusion in Solids, Liquids, and Gases*, New York: Academic Press, 1960.

W. D. Stein, *The Movement of Molecules Across Cell Membranes*, New York: Academic Press, 1967.

P. G. Shewmon, *Diffusion in Solids*, New York: McGraw-Hill, 1963.

C. Tanford, *Physical Chemistry of Macromolecules*, New York: Wiley, 1961.

Chapter 19

Macromolecules

Preview

This chapter deals with some of the special properties of *macromolecules* or *polymers,* which are molecules having a molar mass of 10 000 g mol^{-1} or more.

There are various types of polymers and various mechanisms of polymerization. *Addition polymers* are usually formed by free-radical mechanisms, in which radicals add on to unsaturated molecules to form larger radicals. *Condensation polymers* are often formed by molecular mechanisms. Polymerization sometimes occurs by an *ionic mechanism,* which may be *cationic* or *anionic.* Catalysts, such as the *Ziegler-Natta catalysts,* have led to important developments in polymer technology.

Any process for preparing a polymer is bound to lead to a product in which there is a range of molar masses. Measurements of molar mass will therefore lead to some kind of an average value, and we will consider some of the different averages with reference to the methods of determination.

There are several experimental techniques that are used to determine the overall sizes and shapes of macromolecules. Most of these have been met in previous chapters, and in this chapter we summarize their usefulness for dealing with large molecules. Various experimental methods are used for studying finer details, such as bond distances and angles; the most widely used method is *X-ray diffraction.* These techniques have led to important information about the conformations of macromolecules. A number of theories have been proposed to explain protein conformations, and we will consider in particular the *random coils* that proteins sometimes form when they are in solution.

The three most important physical properties of solid polymers are *crystallinity, plasticity,* and *elasticity.* Many polymers show both elastic and plastic properties and are said to be *viscoelastic.* The contraction of a stretched polymer is frequently an *entropic* process; the enthalpy change is usually small, and the driving force is the tendency of the long molecules to shorten, a process that involves an entropy increase.

19 Macromolecules

The word **macromolecules** (Greek *macros,* long, large) was introduced by the German chemist Herman Staudinger (1881–1965) to refer to substances having relative molecular masses (formerly called "molecular weights") of more than 10 000. Macromolecular substances in solution show characteristic behavior with respect to properties such as viscosity, diffusion, the colligative properties, and sedimentation. In Chapter 17 we have considered the properties of colloids and in so doing have covered some important aspects of macromolecules. As we have seen, not all colloids are macromolecules; those that are not are called *association colloids,* while the term *molecular colloids* was introduced by Staudinger to refer to colloidal particles that consist of single macromolecules.

The word **polymer** (Greek *poly,* much, many; *meros,* part) is also used in this connection. It was coined in 1830 by the Swedish chemist Baron Jons Jacob Berzelius (1779–1848) to refer to a molecule of general formula M_n, made up by the repetition of n identical units M, the latter being known as the **monomer** (Greek *monos,* alone). However, over the years the meaning of the word *polymer* has become extended in various directions. It is no longer necessary for the polymer to consist of exactly an integral number of monomer units. Many macromolecular substances are formed by condensation reactions between monomer molecules, with the elimination of water molecules or other small molecules. The products of such reactions are now called *condensation polymers* even though, because of the spitting off of the small molecules, the formula of the macromolecule is not M_n. Furthermore, it is no longer necessary for the polymer to be formed from *identical* monomer molecules. The word **copolymer** is used to describe a molecule composed of two different units; if it is formed by addition processes (i.e., is an *addition polymer*), its formula will be M_nN_m, where M and N are the monomers and n and m are integers. The expression **homopolymer** (Greek *homos,* one and the same, jointly) can be used to describe polymers composed of only one type of monomer.

Nowadays the word *polymer* is even applied to substances that are made up of a considerable variety of units. For example, protein molecules are formed by condensation reactions involving over 20 different amino acids, and they are referred to as polymers. In some ways this extension of the meaning is unfortunate, since now there is essentially no difference between polymers and macromolecular substances, any of which can be called polymers in the broadest sense.

19.1 Mechanisms of Polymerization

Polymers can be classified as **addition polymers** and as **condensation polymers**. Addition polymers are usually prepared by reactions involving the participation of free radicals, whereas condensation polymers are usually formed from monomers by molecular reactions. In either case catalysts may be involved, and the processes may also be brought about by radiation of various kinds.

Olefinic substances, such as ethylene and styrene, usually form addition polymers. The initiation of free-radical polymerizations, leading to such polymers, can be brought about in various ways. For example, atoms or free radicals may be introduced to the monomer, and they will add on to the double bond:

$$R— + CH_2{=}CHR' \longrightarrow R—CH_2—CHR'—$$

This reaction has produced another radical, which in turn can add on to another monomer molecule:

$$R—CH_2CHR'— + CH_2{=}CHR' \longrightarrow R—CH_2CHR'—CH_2—CHR'—$$

This process can continue and form larger and larger radicals. Finally, two large radicals can combine together to form a polymer molecule. The radical R that has initiated the polymerization process can be introduced into the monomer system in a number of ways. It can be produced thermally or be formed by the action of a catalyst or be generated by photochemical or radiation-chemical processes as discussed in Sections 10.6 and 10.7.

An example of a condensation polymerization occurring by a *molecular* mechanism is the reaction between ethylene glycol and succinic acid:

$$\underset{\text{ethylene glycol}}{HO(CH_2)_2OH} + \underset{\text{succinic acid}}{HOOC(CH_2)_2COOH} \longrightarrow HO(CH_2)_2OCO(CH_2)_2COOH + H_2O$$

Since this product has two functional end groups, —OH and —COOH, it can react with two more monomer molecules, yielding a product that also has two functional end groups. The process can therefore continue indefinitely with the formation of a large copolymer. It can be brought about by the usual catalysts (e.g., acids and bases) for esterification reactions. Another example of a polycondensation reaction is the synthesis by the American chemist Wallace Hume Carothers (1896–1937), in 1934, of a form of *nylon,* which is a polyamide:

$$\underset{\text{adipic acid}}{COOH—(CH_2)_4—COOH} + \underset{\text{hexamethylene diamine}}{NH_2—(CH_2)_6—NH_2} \longrightarrow COOH—(CH_2)_4—CO—NH—(CH_2)_6—NH_2 + H_2O$$

The product has two functional end groups, —COOH and —NH_2, and the condensation reactions can continue and yield large molecules. The relative molecular masses of nylon molecules are typically about 15 000.

Polymerization processes in solution often occur by mechanisms in which the intermediates are *ions*. Such processes are catalyzed by acidic or basic substances, and their rates vary with the dielectric constant of the solvent in the manner expected of ionic processes. Polymerizations of this type can be classified as *cationic* or *anionic*, according to whether the processes are brought about by cationic or anionic species. An example is the polymerization of a substance such as isobutene. Such polymerizations are catalyzed by a variety of Lewis acids; examples are HCl, H_2SO_4, $AlCl_3$, I_2, and $AgClO_4$. All these substances, which we will write in general as MX, are electron acceptors, and they can add on to a monomer molecule as follows:

$$\mathrm{MX + CH_2{=}CHR \longrightarrow XM^-{-}CH_2{-}C^+HR}$$

The product of this reaction can add on to another olefin molecule:

$$\mathrm{XM^-{-}CH_2{-}C^+HR + CH_2{=}CHR \longrightarrow XM^-{-}CH_2{-}CHR{-}CH_2{-}C^+HR}$$

The process can continue until finally a large double ion splits off HX with the formation of a stable polymer molecule:

$$\mathrm{XM^-{-}(CH_2{-}CHR)_n{-}CH_2{-}C^+HR \longrightarrow M{-}(CH_2{-}CHR)_n{-}CH{=}CHR + HX}$$

Heterogeneous catalysis has proved to be very effective for bringing about polymerization. The pioneers in this field were the German chemist Karl Ziegler (1898–1973) and the Italian chemist Giulio Natta (1903–1979). They found that mixtures of $Al(C_2H_5)_3$ and $TiCl_4$ were very effective catalysts for the synthesis of polymers with specified stereoisomerism. They were successful, for example, in synthesizing a polymer that is identical with natural rubber. This material, which has elastic properties and is known as an *elastomer* (see Section 19.7), was studied by Michael Faraday in 1826 and is now known to be poly-*cis*-isoprene,

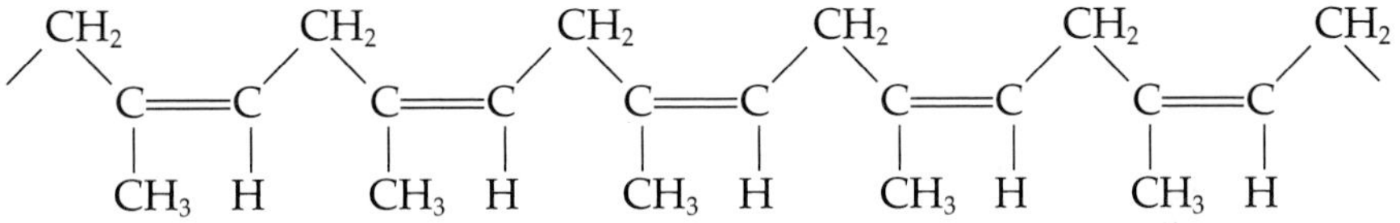

$$\begin{array}{ccccccccccc}
CH_2 & & CH_2 & & CH_2 & & CH_2 & & CH_2 & & CH_2 \\
/\ \backslash & & /\ \backslash & & /\ \backslash & & /\ \backslash & & /\ \backslash & & /\ \backslash \\
 & C{=}C & & C{=}C & & C{=}C & & C{=}C & & C{=}C & \\
 & |\quad\ | & & |\quad\ | & & |\quad\ | & & |\quad\ | & & |\quad\ | & \\
 & CH_3\ \ H & & CH_3\ \ H & & CH_3\ \ H & & CH_3\ \ H & & CH_3\ \ H &
\end{array}$$

Another naturally occurring polymer is gutta-percha, which is poly-*trans*-isoprene,

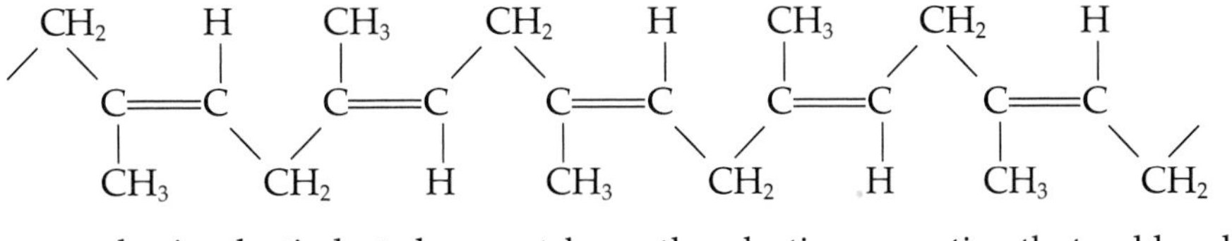

$$\begin{array}{cccccccccc}
CH_2 & H & CH_3 & CH_2 & H & CH_3 & CH_2 & H \\
C{=}C & & C{=}C & & C{=}C & & C{=}C & & C{=}C \\
CH_3 & CH_2 & H & CH_3 & CH_2 & H & CH_3 & CH_2
\end{array}$$

Gutta-percha is plastic but does not have the elastic properties that rubber has. Until 1955 all processes used to polymerize isoprene gave a mixture of poly-*cis*-isoprene and poly-*trans*-isoprene, mainly the latter, and this mixture is not a suitable artificial rubber. Use of the *Ziegler-Natta catalysts,* however, led in that year to the synthesis of the separate stereoisomeric forms and therefore to important developments in artificial rubber technology.

It is also possible, and sometimes very convenient, to bring about polymerizations in aqueous emulsions. A typical procedure that has been employed with monomers like methyl methacrylate

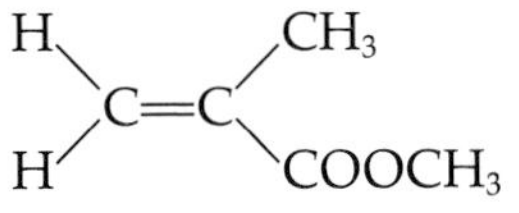

$$\begin{array}{ccc}
H & & CH_3 \\
 & C{=}C & \\
H & & COOCH_3
\end{array}$$

is first to form an aqueous emulsion with the monomer, by the use of an emulsifying agent such as cetyltrimethylammonium bromide. Addition of Fenton's reagent (a solution containing Fe^{2+} ions and hydrogen peroxide) will then bring about polymerization; this reagent generates free radicals, and the polymerization occurs by a free-radical mechanism. An advantage of emulsion polymerization is that the processes proceed much more rapidly than in bulk systems.

As we noted in Section 17.10, long-chain hydrocarbon residues having polar ends form *micelles*. When monomers such as methyl methacrylate are added to a micelle, some of it becomes *solubilized*, which means that monomer penetrates into the micelles; the remainder stays in the aqueous phase. The free radicals generated by the initiating medium (e.g., Fenton's reagent) are able to diffuse into the micelles, where they initiate polymerization. At the same time more monomer molecules enter the micelles, which increase in size and soon consist of polymer particles with the emulsifying agent adsorbed on the surface. The process goes on, and the micelle particles continue to grow, until almost all the monomer has been polymerized. This theory of emulsion polymerization was first put forward in 1947 by W. D. Harkins.*

19.2 Kinetics of Polymerization

Free-Radical Polymerization

The general reaction scheme for a free-radical polymerization can be represented as follows:

$$? \longrightarrow R_1 \qquad \text{initiation}$$

$$\left.\begin{array}{l} R_1 + M \longrightarrow R_2 \\ R_2 + M \longrightarrow R_3 \\ \text{-----------------} \\ R_{n-1} + M \longrightarrow R_n \end{array}\right\} \text{chain propagation}$$

$$R_n + R_m \longrightarrow M_{n+m} \qquad \text{termination}$$

The initiation reaction is a process in which a radical R_1 is produced; for the time being we will leave the nature of this reaction unspecified and will write the rate of formation of R_1 in this process as v_i. For simplicity we shall assume the rate constants of all the chain-propagating steps to be the same, k_p. Termination can involve any two radicals, including identical ones, and the rate constant for termination will be written as k_t.

The steady-state equation for R_1 is

$$v_i - k_p[R_1][M] - k_t[R_1]([R_1] + [R_2] + \cdots) = 0 \tag{19.1}$$

The final term is for the removal of R_1 by reaction with R_1, R_2, etc. The equation can be written as

*W. D. Harkins, *J. Amer. Chem. Soc.*, *69*, 1428 (1947); a quantitative formulation of the theory was put forward by W. V. Smith and R. H. Ewart, *J. Chem. Phys.*, *16*, 592 (1948).

$$v_i - k_p[R_1][M] - k_t[R_1] \sum_{n=1}^{\infty} [R_n] = 0 \tag{19.2}$$

Similarly for R_2,

$$k_p[R_1][M] - k_p[R_2][M] - k_t[R_2] \sum_{n=1}^{\infty} [R_n] = 0 \tag{19.3}$$

In general, for R_n,

$$k_p[R_{n-1}][M] - k_p[R_n][M] - k_t[R_n] \sum_{n=1}^{\infty} [R_n] = 0 \tag{19.4}$$

There is an infinite number of such equations, and the sum of all of them is

$$v_i - k_t\left(\sum_{n=1}^{\infty} [R_n]\right)^2 = 0 \tag{19.5}$$

This equation simply states that the rate of initiation is equal to the sum of the rates of all the termination processes, which must be true in the steady state.

The rate of disappearance of monomer is the sum of the rates of all the propagation reactions,

$$-\frac{d[M]}{dt} = k_p[M] \sum_{n=1}^{\infty} [R_n] \tag{19.6}$$

According to Eq. 19.5 the summation is given by

$$\sum_{n=1}^{\infty} [R_n] = \left(\frac{v_i}{k_t}\right)^{1/2} \tag{19.7}$$

and the rate of disappearance of monomer is therefore

$$-\frac{d[M]}{dt} = k_p\left(\frac{v_i}{k_t}\right)^{1/2}[M] \tag{19.8}$$

There are various special cases, according to the nature of the initiation process; we shall consider only a few of them. In a purely thermal initiation, in which the radicals are produced by heating the monomer, the initial radical formation may be second order in monomer:

$$v_i = k_i[M]^2 \tag{19.9}$$

The rate of polymerization is then

$$-\frac{d[M]}{dt} = k_p\left(\frac{k_i}{k_t}\right)^{1/2}[M]^2 \tag{19.10}$$

so that the process is second order. Alternatively, initiation might involve a second-order reaction between the monomer and a catalyst C,

$$v_i = k_i[M][C] \tag{19.11}$$

The polymerization rate is then

$$-\frac{d[M]}{dt} = k_p\left(\frac{k_i}{k_t}\right)^{1/2}[M]^{3/2}[C]^{1/2} \tag{19.12}$$

In a photochemical initiation the rate of initiation may be simply the intensity I of light absorbed, usually expressed in einsteins (mole quanta) per second (see Section 10.6),

$$v_i = I \tag{19.13}$$

The rate of polymerization is then

$$-\frac{d[\mathrm{M}]}{dt} = k_p\left(\frac{I}{k_i}\right)^{1/2}[\mathrm{M}] \tag{19.14}$$

Examples are known of all of these kinetic equations and of others we have not considered.

Condensation Polymerization

The kinetic equations for condensation polymerization are most easily formulated in terms of the concentrations of the functional groups (—OH, —COOH, etc.) that are undergoing reaction. Suppose that the initial concentration of these groups is c_0, and at time t the concentration is c. We shall assume that all the rate constants are the same, and therefore

$$-\frac{dc}{dt} = kc^2 \tag{19.15}$$

With the boundary condition $t = 0$, $c = c_0$, this integrates to

$$\frac{c_0 - c}{cc_0} = kt \tag{19.16}$$

(compare Eq. 9.35). The fraction f of functional groups that have disappeared at time t is

$$f = \frac{c_0 - c}{c_0} \tag{19.17}$$

Elimination of c between these equations gives

$$\frac{f}{1-f} = c_0kt \tag{19.18}$$

or

$$\frac{1}{1-f} = 1 + c_0kt \tag{19.19}$$

Plots of $f/(1 - f)$ or $1/(1 - f)$ should therefore be linear, and this has been verified for a number of reactions.

19.3 Relative Molecular Masses of Macromolecules

Most of the principles employed in the determination of relative molecular masses have been referred to in earlier chapters. Here we will discuss briefly some special problems associated with measuring the relative molecular masses of very large molecules.

Colligative Properties

The colligative properties (Section 5.8) are frequently used for molecular-mass determinations, but special difficulties arise with macromolecules. This is because the colligative properties depend on the molalities or molar concentrations of solutions, and these are necessarily small when the molecules are very large. This eliminates the vapor pressure, boiling point elevation, and freezing point depression methods, since the changes are much too small to be measured. For example, even with a 1 *M* solution in water the freezing point depression is only 1.86°C; with a macromolecular substance the depression may only be 10^{-6} of this, which is impossible to measure. A 1 *M* solution, however, produces an osmotic pressure of about 25×10^5 Pa (≈25 atm), and even one-millionth of this can be measured reasonably accurately.

A number of difficulties arise when the osmotic pressure method is applied to macromolecular material. One is that it may take an exceedingly long time for the system to reach osmotic equilibrium. This difficulty can be overcome by imposing a pressure on the solution side of the membrane and observing the flow of liquid as a function of the time; the osmotic pressure can be calculated from such results. Another difficulty is that an impurity of low relative molecular mass can greatly affect the results, since the molar concentration of such an impurity can be very large. Suppose, for example, that the substance under investigation has a relative molecular mass of 1 million, but that there is 1% of an impurity of relative molecular mass 100. If 1 g of the substance is present in 1 dm^3 of solution, its concentration is 10^{-6} *M*, but there is also present 10^{-3} g of the impurity at a concentration of 10^{-5} *M*. The relative molecular mass of the macromolecular substance under investigation would therefore be very seriously in error. A related difficulty also arises because of the Donnan effect (Section 7.12), but this can be minimized by carrying out the osmotic pressure measurements in the presence of a large excess of an electrolyte that diffuses freely through the membrane.

Another difficulty with osmotic pressure measurements is that there are serious deviations, even at very low concentrations of macromolecular substances, from the ideal equation

$$\pi = cRT \tag{19.20}$$

The American chemical physicists William George McMillan (b. 1919) and Joseph Edward Mayer (b. 1904) showed that the results can better be expressed by a power series in the mass ρ of solute per unit volume (e.g., g dm^{-3}):

$$\pi = RT\left(\frac{\rho}{M_r} + B\rho^2 + D\rho^3 + \cdots\right) \tag{19.21}$$

where M_r is the relative molecular mass and *B*, *D*, etc., are constants. The first term corresponds to Eq. 19.20, since $c = \rho/M_r$, and usually only the first two terms need be considered:

$$\frac{\pi}{\rho} = \frac{RT}{M_r} + RTB\rho \tag{19.22}$$

The ratio π/ρ can therefore be plotted against ρ, and the intercept at $\rho = 0$ gives RT/M_r, from which M_r can be calculated (see Problem 8).

In spite of these difficulties, relative molecular masses of up to 3 million have been determined by the osmotic pressure method.

Chemical Analysis

When a macromolecule contains one particular atom in very small proportions, a careful chemical analysis can lead to useful information about the relative molecular mass. This method is referred to as **elemental analysis**. A compound must contain in each molecule at least one atom of every element present, and the mass of the compound that contains 1 mol of a given element is thus the minimum possible value of its relative molecular mass. Therefore, if the percentage of the element is determined, the minimal relative molecular mass is given by

$$M_{min} = \frac{\text{atomic weight of element} \times 100}{\text{percentage of element in compound}} \tag{19.23}$$

The true relative molecular mass is nM_{min}, where n is the actual number of the atoms in the molecule.

The advantage of the method is that with careful analysis a very precise value of M_{min} can be obtained. The osmotic pressure method, for example, can give an approximate value of the relative molecular mass, and the value of n can then be obtained, leading to a more reliable value. The procedure is illustrated by the following example.

EXAMPLE

The following are the percentages of iron and sulfur (other than disulfide sulfur) found in a sample of hemoglobin:

Fe: 0.335%

S: 0.390%

An osmotic pressure determination leads to an approximate relative molecular mass of 70 000. Estimate a more reliable relative molecular mass.

SOLUTION

From the iron data (relative atomic mass of Fe = 55.85), using Eq. 19.23,

$$M_{min} = \frac{55.85 \times 100}{0.335} = 16\ 670$$

From the sulfur data (relative atomic mass of S = 32.06)

$$M_{min} = \frac{32.06 \times 100}{0.390} = 8220$$

The latter value is about half of the former; thus if we combine the two, we obtain a minimum of $\approx$ 16 500 for the relative molecular mass. This value would correspond to one atom of iron and two atoms of sulfur per molecule. The value is approximately one-quarter of 70 000, and a better relative molecular mass is therefore

$$M_{true} = n \times M_{min} = 4 \times 16\ 500 = 66\ 000$$

It follows that the molecule in fact contains four iron atoms and eight sulfur (other than disulfide) atoms. ■

This method is only useful if the molecule contains a particular element in a very small proportion. Many protein molecules do not satisfy this condition but sometimes contain a small proportion of a particular amino acid. A variation of the method of elemental analysis is to determine the percentage of an amino acid.

EXAMPLE
The wheat protein glutenin is found to contain 4.50% tyrosine. Calculate the minimal relative molecular mass. What is the true relative molecular mass if an approximate osmotic pressure determination leads to a value of ≈36 000?

SOLUTION
The relative molecular mass of tyrosine is 181.18, and the minimal relative molecular mass of the protein is therefore

$$M_{\text{min}} = \frac{181.18 \times 100}{4.50} = 4030$$

This is about one-ninth of the approximate relative molecular mass, and a more accurate value for glutenin is thus

$$M_{\text{true}} = n \times M_{\text{min}} = 9 \times 4030 = 36\ 300$$ ■

Other variants of the method of elemental analysis involve measuring the mass of the substance that combines with a known amount of a suitable reagent and measuring the proportion of groups in a polymer molecule. An example of the latter method is the estimation by the British organic chemist Walter Norman Haworth (1883–1950) of the relative molecular mass of cellulose, which consists of chains of glucose molecules. He completely methylated the hydroxyl groups and then, by gentle hydrolysis, separated the methylated glucose units. Most of the glucose units formed are 2, 3, 6-trimethyl glucose, but the units at the ends of the chains are 2, 3, 4, 6-tetramethyl glucose. The ratio of the two forms thus provides information about the lengths of the chains, and if there were no branching the relative molecular mass could be calculated.

In addition to these methods of determining relative molecular masses, there are a number of techniques that provide information about overall molecular sizes and shapes. These are considered in Section 19.4.

Types of Relative Molecular Masses

Macromolecular substances are always inhomogeneous in the molecular sense, the molecules having a range of sizes. The polymerization mechanisms that we have discussed lead to molecules of various sizes; in a free-radical mechanism, for example, polymer radicals of various lengths will be present in the system and will recombine in different ways.

The experimental methods for determining relative molecular masses will therefore measure some kind of an average value. The kind of average that is obtained is different for different methods. The colligative methods (e.g., the osmotic pressure method) essentially involve the counting of molecules, as do the analytical methods. They therefore provide the **number-average relative molecular mass**, which we will denote by the symbol M_n. If the number of molecules of the ith kind is N_i, the *number fraction*, or *mole fraction*, is given by

$$x_i = \frac{N_i}{\sum_i N_i} \tag{19.24}$$

where the summation is taken over all types of molecules present. The *number-average relative molecular mass* is then

$$\boldsymbol{M_n = \sum_i x_i M_i = \frac{\sum_i N_i M_i}{\sum_i N_i}} \tag{19.25}$$

EXAMPLE

Suppose that a protein sample consists of 30% of molecules of relative molecular mass 20 000, 40% of molecules of mass 30 000, and 30% of molecules of mass 60 000. Calculate the number-average relative molecular mass.

SOLUTION

The mole fractions are 0.3, 0.4, and 0.3, respectively. Therefore, from Eq. 19.25,

$$\begin{aligned} M_n &= (0.3 \times 20\ 000) + (0.4 \times 30\ 000) + (0.3 \times 60\ 000) \\ &= 36\ 000 \end{aligned}$$

■

Other methods of relative molecular-mass determination, such as diffusion, sedimentation, and light scattering, involve terms related to the masses or volumes of the molecules and give a relative molecular mass known as the mass-average relative molecular mass M_m. If m_i is the mass of the ith species, the mass fraction y_i is

$$y_i = \frac{m_i}{\sum_i m_i} \tag{19.26}$$

and the *mass-average relative molecular mass* is defined as

$$M_m = \sum_i y_i M_i = \frac{\sum_i m_i M_i}{\sum_i m_i} \tag{19.27}$$

Since $m_i = N_i M_i$, the **mass-average relative mass** can be expressed as

$$M_m = \frac{\sum_i N_i M_i^2}{\sum_i N_i M_i} \tag{19.28}$$

EXAMPLE

Using the data given in the previous example, calculate the mass-average relative molecular mass.

SOLUTION

We insert percentages into Eq. 19.28 instead of the number of molecules, N_i; the numbers are proportional to the percentages and the proportionality factor cancels out. Thus

$$M_m = \frac{(30)(20\ 000)^2 + (40)(30\ 000)^2 + (30)(60\ 000)^2}{(30)(20\ 000) + (40)(30\ 000) + (30)(60\ 000)} = 43\ 300$$ ■

Note that in this example M_m is greater than M_n. It can never be less than M_n. The two relative molecular masses would have been equal if the substance had consisted of molecular units all the same size; such a system is said to be *monodisperse.* If the substance consists of molecules of different sizes, as in the example given, it is said to be *polydisperse.* The M_m value is then greater than M_n, because in M_m the heavier units have been weighted more strongly than in M_n.

Other molecular relative masses are occasionally used. For example, there is the so-called *Z-average relative molecular mass* defined by

$$M_z = \frac{\sum_i N_i M_i^3}{\sum_i N_i M_i^2} \tag{19.29}$$

This is larger again than M_m, because the heavier components are weighted even more strongly.

19.4 Molecular Sizes and Shapes

In Section 19.3 we have discussed methods that are specifically designed to determine the relative molecular masses of macromolecules. There are also methods that are directed more toward determining the overall sizes and shapes of macromolecules. Most of these have already been mentioned in previous chapters, and here we will make only brief reference to them.

Flow Birefringence

One useful method is to measure *flow birefringence*. Certain solid substances in which the molecules are not spherical have different indices of refraction along different axes and are said to be *birefringent* or to exhibit *double refraction*. However, when a solution of such a substance is examined under a polarizing microscope, no birefringence is observed, since the molecules are oriented at random and the effects cancel out. If, on the other hand, the solution is caused to flow through a tube, ellipsoidal molecules are caused to orient in the direction of the flow, as shown in Figure 19.1. In practice, a flow birefringence experiment is usually carried out by enclosing the solution between two concentric cylindrical vessels, the outer one of which is rotated. Analysis of the results leads to an indication of the size and overall shape of the molecules, and hence an approximate relative molecular mass can be estimated.

Methods Based on Transport Properties

We have already discussed in previous chapters how overall molecular sizes and shapes can be estimated from studies of diffusion (Section 18.2), viscosity (Section 18.1), and sedimentation (Section 18.3). These methods are very suitable for macromolecules. The measurement of diffusion coefficients is convenient for relative molecular masses of up to 1 million but depends on assumptions as to the shape of the molecule. The sedimentation method is useful for even higher relative molecular masses, up to 50 million. The viscosity method involves a good deal of empiricism and is best used in conjunction with other methods.

Light Scattering

The light-scattering technique was discussed in Section 17.10. It is not suitable for very large molecules and is best for relative molecular masses up to 10 000.

Gel Filtration

Gel filtration is a technique that allows macromolecules of different sizes to be separated and provides approximate sizes of the molecules, by passing their solutions through gels having different pore sizes. A commercially available material, Sephadex, is made in a variety of pore sizes and is commonly employed. If a solution of macromolecular material is passed through a column containing such a gel, the various components appear at the outlet at different times. The molecular

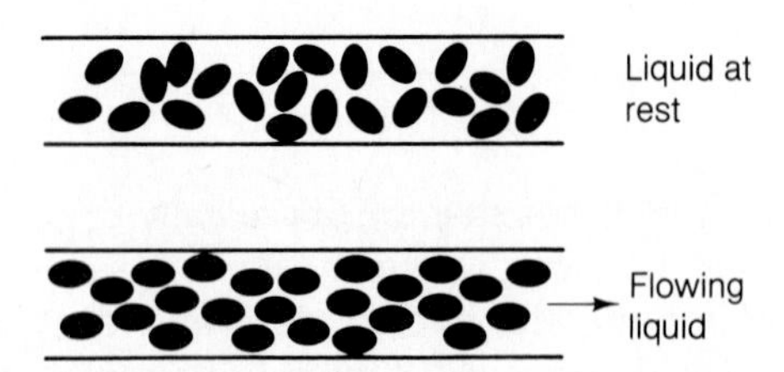

FIGURE 19.1 The orientation of ellipsoidal molecules in a flowing liquid.

TABLE 19.1 Methods for Determining Relative Molecular Masses and Molecular Sizes and Shapes

Method	Type of Relative Molecular Mass	Remarks
Chemical analysis	M_n	Requires an atom or group present in small proportions
Osmotic pressure	M_n	Applicable to relative molecular masses of up to 10^6
Vapor-pressure lowering	M_n	Applicable to relative molecular masses of up to 10^4
Diffusion	M_m	For relative molecular masses up to 10^6
Sedimentation	M_m	For relative molecular masses up to 5×10^7
Viscosity	M_v*	Best used in association with other methods
Light scattering	M_m	Applicable to relative molecular masses up to 10^4
Flow birefringence	M_m	Gives indication of the size and shape of molecules
Gel filtration	M_m	Gives only approximate relative molecular masses

*This is a special viscosity-average value; it is much closer to M_m than to M_n.

sizes estimated by this method are very approximate, and anomalies arise in the case of highly asymmetric molecules. The method does, however, have the advantage of being rapid, and it permits the sizes of different molecules in the same solution to be determined in one experiment.

Table 19.1 summarizes the various methods that can be used to determine relative molecular masses or to make estimates of molecular sizes and shapes. Figure 19.2 shows the sizes and shapes of some macromolecules, as determined by various methods.

19.5 Bond Distances and Angles

Thus far we have considered methods for determining relative molecular masses and overall sizes and shapes of macromolecules. There is also need for methods that provide information about much finer details of molecular structures, specifically about the lengths of individual chemical bonds, the angles between bonds, and the distances between nonbonded atoms. Studies of this kind allow a model of the macromolecule to be built, so that one has a very clear idea of its structure and can draw conclusions about its physical and chemical behavior.

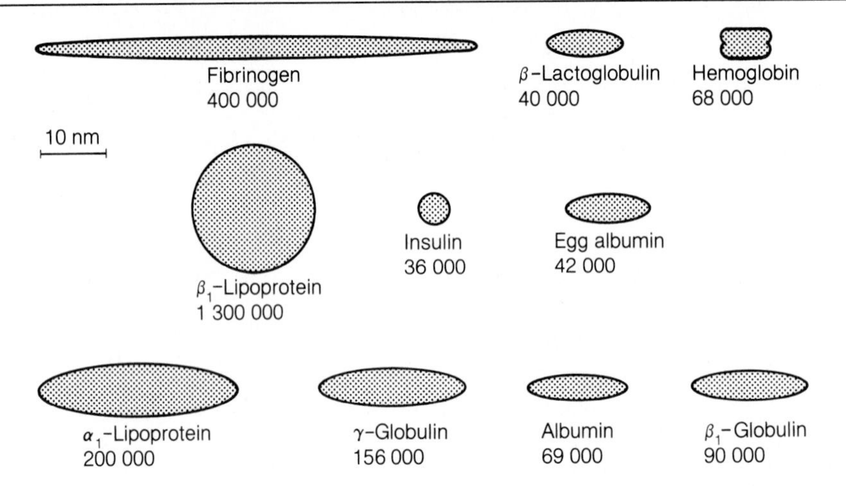

FIGURE 19.2 The sizes and shapes of various macromolecules. The approximate relative molecular masses are indicated.

The most powerful and widely used method is *X-ray diffraction*, which we have already discussed in some detail in Section 15.4. Here we will mention only a few points that particularly relate to macromolecules. The analysis of X-ray diffraction patterns is very difficult for large molecules because of the many distances involved. It is the electrons that scatter the X rays, and the image calculated from the diffraction pattern thus reveals the distribution of electrons within the molecule. The usual procedure with macromolecules is to use a computer to calculate the electron densities at a regular array of points and to make the image visible by drawing contour lines through points of equal electron density. These contour lines can be drawn on clear plastic sheets, and a three-dimensional image can be obtained by stacking the maps one above the other. With an instrument of high resolving power the atoms appear as peaks on the image map.

An important contribution to the analysis of X-ray patterns for macromolecules was made in 1953 by the British molecular physicist Max Ferdinand Perutz.* His method of *isomorphous replacement* involves the preparation and study of crystals into which heavy atoms, such as atoms of uranium, have been introduced without altering the crystal strucutre. This technique led rapidly to the detailed analysis of a number of macromolecules.

In spite of important technical developments, the establishment of a macromolecular structure by X-ray diffraction is still an extremely time-consuming process, often requiring several years before a conclusive result can be obtained. One difficulty is that the substance must be in pure crystalline form. The technique cannot be applied to macromolecules in aqueous solution, and there is always the question whether the structure of the molecule in solution is the same as that in the crystalline state. In spite of difficulties, the X-ray structures of a number of macromolecules have been determined, and the results have made a valuable con-

*D. W. Green, V. M. Ingram, and M. F. Perutz, *Proc. Roy. Soc.*, *A225*, 287 (1954); for a review, see C. C. F. Blake, *Adv. Protein Chem.*, *23*, 59 (1968).

tribution to the understanding of the behavior of such molecules, particularly in relation to biological systems.

A different X-ray approach to the problem of molecular structure was proposed in 1971 by the American physicists Dale E. Sayers and Farrel W. Lytie. Their method is based not on the diffraction of X rays by an array of atoms or ions but on the absorption of X rays by individual atoms and ions. This *X-ray absorption method* measures how the absorption is affected by the atoms in the immediate neighborhood of the atom that is absorbing the X rays and yields a knowledge of the atomic environment of each type of atom in the molecule. This technique promises to provide valuable information about the structures of macromolecules.

The diffraction of electron beams also provides information about macromolecular structure. An electron accelerated by 10 000 V has a velocity of 5.9×10^7 m s^{-1}, and its de Broglie wavelength (Section 11.3) is 12 pm. This is somewhat less than bond distances, and therefore it is possible to diffract an electron beam by a crystal used as a diffraction grating. Electron beams have the advantage over X rays that they can be brought into sharp focus by appropriate arrangements of electric and magnetic fields. This has led to the development of electron microscopes, which are capable of resolving images as small as 0.5 nm in diameter. This is not quite fine enough to allow interatomic distances to be measured, but valuable information about macromolecular structures has been obtained by the use of this technique.

19.6 Macromolecular Conformations

It is important to distinguish clearly between **configurations** and **conformations**. The term *configuration* refers to steric arrangements that cannot be converted into one another without the breaking of primary chemical bonds. Examples are the D and L configurations in molecules having asymmetric centers and the cis and trans isomers of olefinic compounds. The *conformations* of a given molecule, also referred to as *rotational isomers,* can be converted into one another without the breaking of primary bonds, for example by rotations about single bonds. It is permissible to break secondary bonds, such as hydrogen bonds; the interconversion of protein conformers, for example, involves the breaking of hydrogen bonds. Simple examples of two conformers are the "chair" and "boat" forms of cyclohexane.

The different conformations in which a molecule can exist arise from the fact that when, for example, there is rotation about a single bond, there is a variation in the potential energy of the molecule, as a result of the interactions between nonbonded atoms. Two different stable conformations correspond to minima in the potential energy, and these minima will be separated by a potential-energy barrier that must be overcome in order for interconversion to occur. At higher temperatures there will be rapid interconversion of conformers, and it may be impossible to separate them. At low temperatures the barriers are not easily overcome, and different conformers are more readily detected.

Various experimental techniques lead to information about macromolecular conformations. The most direct evidence is provided by X-ray diffraction, which, as we saw in Section 19.5, provides a detailed molecular structure. Other experimental methods also allow conclusions to be drawn about conformations. Spectroscopic techniques, including the study of absorption and fluorescence in the ultraviolet and absorption in the infrared, have been employed to some extent. For example, the absorption and fluorescence of proteins in the ultraviolet is largely due to tyrosine, tryptophan, and phenylalanine residues; from the results it is possible to deduce the extent to which these residues are buried inside the molecule. Infrared studies of the frequencies of vibrations of various groups can provide information about the environments of these groups. Nuclear magnetic resonance spectroscopy and electron spin resonance spectroscopy have also been used in this way.

Another technique for studying macromolecular conformations is **optical rotatory dispersion,*** which is concerned with the dependence of the optical rotation on the wavelength of the light employed. A plot of the specific rotation against the wavelength may show maxima and minima, with the specific rotation actually changing sign, as shown by the example in Figure 19.3. This behavior, discovered in 1896 by the French physicist A. Cotton, is known as the **Cotton effect.** The figure

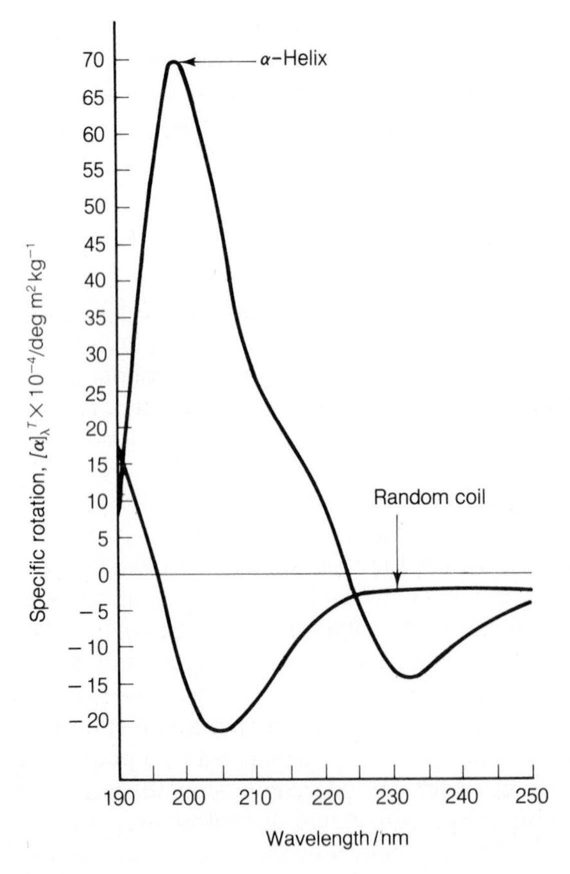

FIGURE 19.3 Optical rotatory dispersion curves for poly-L-lysine, in α-helical and random-coil conformations.

*For a review, see C. Djerassi, *Optical Rotatory Dispersion,* New York: McGraw-Hill, 1960.

shows the results for solutions of poly-L-lysine, in two conformations, a random coil and an α helix (see p. 866). The behavior of the two is very different and allows the conformation of an unknown peptide to be deduced from an experimental study of the optical rotatory dispersion. Unfortunately the theoretical interpretation of the results is complicated and sometimes ambiguous.

Protein Conformations

The conformations of protein molecules are of special importance and interest. A number of protein structures have been determined by the X-ray diffraction method, but before any such work had been done, some important suggestions had been made about protein conformations, many of which proved to be of great value in the development of the subject. One suggestion, made by the British physical chemist Sir Eric K. Rideal (1890–1974) and the American physical chemist Irving Langmuir (1881–1954), is that a protein molecule is "an oil drop with a polar coat."* We have already seen that nonpolar groups, such as alkyl groups, tend to stick together in an aqueous environment because of the formation of *hydrophobic bonds.* Some of the side groups on the amino acids that form the polypeptide chains are nonpolar groups, whereas polar groups such as —OH, —COO^-, and —NH_3^+ tend to form hydrogen bonds with water. The essence of Rideal and Langmuir's suggestion is that the polypeptide chains in proteins become folded in such a way that the nonpolar groups come into contact with each other as much as possible in the interior of the molecule, whereas polar groups are as far as possible on the exterior, where they can form hydrogen bonds with the surrounding water molecules.

The X-ray studies have shown that some proteins, known as *globular proteins,* have a more or less spherical form and that others, the *fibrous proteins,* are much more elongated. On the whole the globular proteins have a larger proportion of nonpolar groups than fibrous proteins. The reason that proteins having a larger proportion of nonpolar groups tend to take a globular form is almost certainly that numerous hydrophobic bonds are formed in the interior of the molecule.

The proteins of wool, hair, and muscle are fibrous and have a considerable amount of helical structure. In 1951 the American chemists Linus Pauling and Elias James Corey pointed out that certain helical structures formed by polypeptide chains allow a considerable amount of hydrogen bonding between the carboxyl group on one part of the chain and the amino group on another:

$$\rangle C{=}O \cdots\cdots H{-}N\langle$$

Pauling and Corey considered in detail the known bond lengths and angles in the flexible chains and concluded that two different helices provide the maximum numbers of hydrogen bonds (see Figure 19.4). X-ray work has shown that one of these helices, the so-called *α helix,* is quite common in protein structures, particularly in the fibrous proteins. In this arrangement, each N—H group is hydrogen

*This suggestion has been widely attributed to Rideal and Langmuir but does not seem to have been published by either of them. In reply to a letter from one of us enquiring about this, Sir Eric Rideal said that he thought he had mentioned it in lectures at Cambridge and that probably "some young gentleman" had copied it down and passed it on.

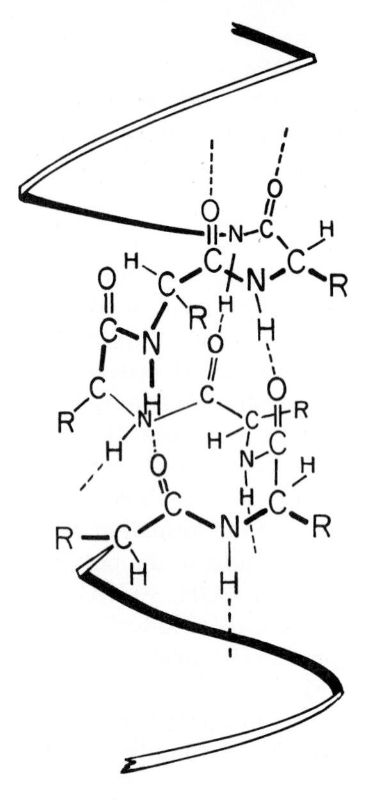

FIGURE 19.4 Schematic diagram of a right-handed α-helix, with amino acids in the L configuration.

bonded to the third C=O group beyond it along the helix; the result is that there are about 3.6 amino acid groups per turn of the helix. Such a helix may be regarded as a spiral staircase in which the amino acid residues form the steps; the height of each step is 0.15 nm. Just as a spiral staircase or a screw can be right-handed or left-handed, so can a protein helix. It appears that the right-handed helix occurs more commonly in protein structures. However, particularly with insulin, a left-handed helix sometimes occurs.

There are three important forces at work in producing protein conformations:

1. The tendency of nonpolar groups to form hydrophobic bonds and to remain in the interior of the molecule.
2. The tendency of polar groups to remain at the exterior of the molecule so that they can form hydrogen bonds with water molecules.
3. The tendency of C=O and N—H groups to form hydrogen bonds, with the formation of a helical structure.

The actual structure of a given protein is determined in a very subtle way by the sometimes conflicting demands of these different effects and thus depends to a considerable extent on the nature and positions of the various amino acid side groups.

A remarkable feature of the proteins is that, both in the crystalline state and in aqueous solution, one conformation is usually strongly favored over all others. In human hemoglobin, for example, almost all the molecules in a pure sample occur in the same conformation. This particular characteristic of the proteins results from the presence of a number of groups that can form hydrogen bonds and of other groups that form hydrophobic bonds with one another.

Random Coils

In contrast to the proteins, there are many polymers for which a considerable number of conformers are more or less equally stable. When such substances are in the solid state, or in solution, there will therefore be no unique conformer but a wide distribution of conformers. For such substances it is obviously important to understand the principles that relate to statistical distributions of conformers and to know which are the most probable forms.

The matter is quite complicated, and a number of factors have to be taken into account. One of these is the structure of the macromolecule itself; certain valence angles are strongly favored, and there may be restrictions to rotation about single bonds. The interactions between the solute and solvent molecules also have a considerable effect on the distribution of conformers.

We will begin by considering a hypothetical situation in which some of these effects are eliminated. We will suppose that there are no interactions between the solvent and solute molecules. We will also suppose that there are no preferred valence angles in the macromolecule, so that the angle between adjacent bonds can take any value with equal probability. The molecule will be regarded as made up of N segments each of length b, the segments being linked by universal joints that permit unrestricted bending. One conformation of such a molecule is shown schematically in Figure 19.5a, and our problem is to determine the mean end-to-end distance l, and related quantities.

The first question that may be asked is: What is the probability $P(l)\,dl$ that the distance between the two ends of the molecular chain lies between the values l and $l + dl$? A statistical treatment of the problem leads to the result that this probability is given by

$$P(l)\,dl = \frac{4}{\pi^{1/2}\lambda^3}\, l^2 e^{-l^2/\lambda^2}\, dl \tag{19.30}$$

where the parameter λ, which has the dimensions of length, is equal to

$$\lambda = \left(\frac{2N}{3}\right)^{1/2} b \tag{19.31}$$

The integral of $P(l)\,dl$ from 0 to ∞ is, of course, unity. Figure 19.5b shows a plot of $P(l)$ against l, for the case in which $N = 500$ and $b = 0.154$ nm; the latter is the normal C—C single-bond distance. We see that the probability $P(l)$ is zero at $l = 0$ and $l = \infty$ and passes through a maximum. It is easy to show, by differentiation of $P(l)$ and setting the derivative equal to zero, that its maximum value corresponds to

$$l = \lambda = \left(\frac{2N}{3}\right)^{1/2} b \tag{19.32}$$

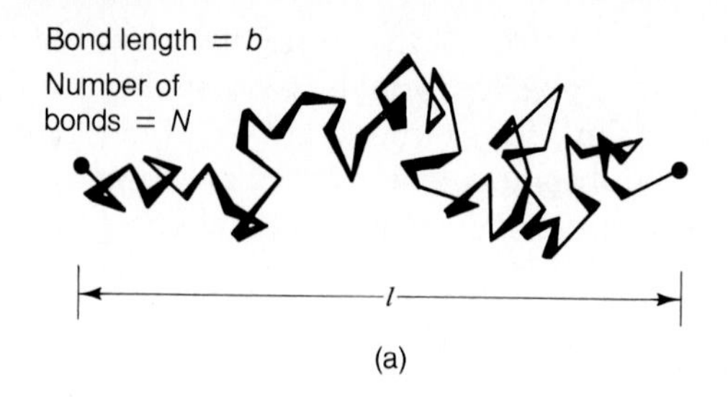

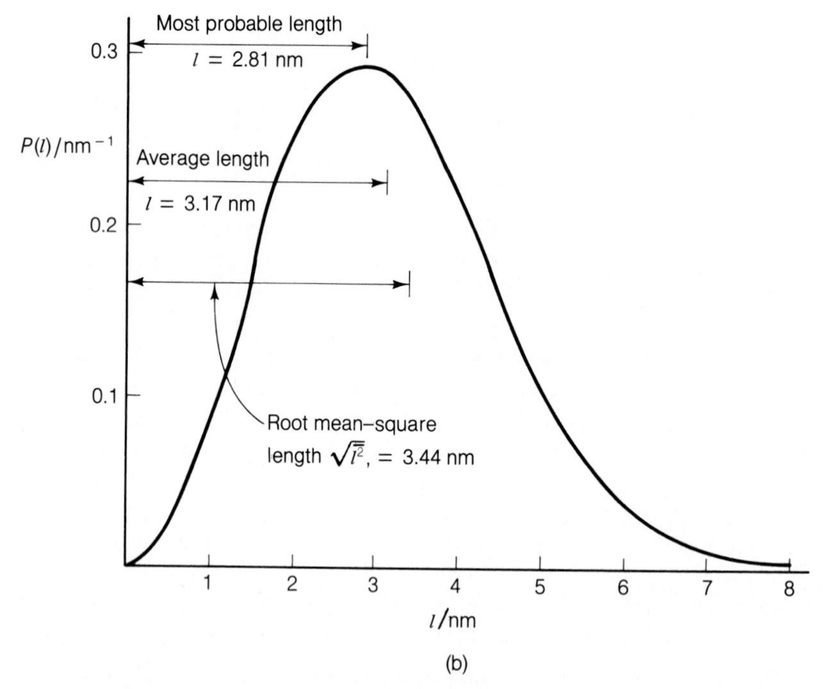

FIGURE 19.5 (a) A typical random coil, in three dimensions. (b) Plot of the probability against the length, for a random coil of 500 bonds of length 0.154 nm.

This is therefore the most probable value of l. In our example ($N = 500$, $b = 0.154$ nm), the most probable value of l is 2.81 nm.

We will now calculate the mean value of l and the mean square value of l. The latter is simpler and we will consider it first. It is given by

$$\overline{l^2} = \int_0^\infty l^2 P(l)dl \tag{19.33}$$

$$= \frac{4}{\pi^{1/2}\lambda^3} \int_0^\infty l^4 e^{-l^2/\lambda^2}\, dl \tag{19.34}$$

The integral is a standard one in statistical theory (see Table 14.1) and is equal to $3\pi^{1/2}\lambda^5/8$. The value of $\overline{l^2}$ is therefore

$$\overline{l^2} = \frac{3\lambda^2}{2} = Nb^2 \tag{19.35}$$

It is of interest to note the analogy between this equation and Eq. 18.48 which we derived for the mean square distance $\overline{x^2}$, traveled in time t by a particle of diffusion coefficient D:

$$\overline{x^2} = 2Dt \tag{19.36}$$

The mean square distance $\overline{x^2}$ is analogous to the mean square length $\overline{l^2}$ in the random-coil problem. The quantity $2D$ is the mean square distance that the particle travels in unit time and is analogous to the square of the distance b in the random-coil problem. Finally, the time t is analogous to the number N of segments. The two equations therefore closely parallel one another.

EXAMPLE

A polymer chain is composed of 500 C—C bonds, each having a length of 0.154 nm. On the assumption of complete flexibility, calculate (a) the length of the chain if it were completely extended and (b) the square root of the mean square length $\sqrt{\overline{l^2}}$.

SOLUTION

a. The length of the completely extended chain is

$$500 \times 0.154 = 77 \text{ nm}$$

b. For the random arrangement,

$$\sqrt{\overline{l^2}} = b\sqrt{N} = 0.154\sqrt{500} = 3.44 \text{ nm}$$

(see Figure 19.5b). ■

One might be tempted to think that an estimate of the volume of the molecule could be obtained by regarding $\sqrt{\overline{l^2}}$ as the diameter of a spherical molecule. In the previous example this would lead to an estimate of $\frac{4}{3}\pi(3.44/2)^3 = 21.3 \text{ nm}^3$. However, the true volume of the molecule is considerably smaller than this. In reality, the molecule does not remain in one conformation, but because of thermal motion it is constantly changing its shape. The volume just calculated is the volume of a sphere that, on the average, contains the molecule, which is a good deal smaller than the volume of this sphere. A more realistic estimate of the volume is obtained by taking $\sqrt{\overline{l^2}/6}$ as the *radius* of the sphere; in our example this leads to a volume of $\frac{4}{3}\pi(3.44^2/6)^{3/2} = 11.6 \text{ nm}^3$.

The expression for the mean length $\bar{l}$ is obtained in a similar way:

$$\bar{l} = \int_0^\infty lP(l)\, dl \tag{19.37}$$

$$= \frac{4}{\pi^{1/2}\lambda^3}\int_0^\infty l^3 e^{-l^2/\lambda^2}\, dl \tag{19.38}$$

The integral (see Table 14.1) is equal to $\lambda^4/2$, and it follows that

$$\bar{l} = \frac{2\lambda}{\pi^{1/2}} = \left(\frac{8N}{3\pi}\right)^{1/2} b \tag{19.39}$$

For $N = 500$ and $b = 0.154$ nm, this leads to $\bar{l} = 3.17$ nm. Figure 19.5b compares $\bar{l}$, $\sqrt{\overline{l^2}}$ and the most probable value of l for this particular polymer chain.

So far we have been assuming that the bonds in the molecule are connected by universal joints, with no restriction as to the angle between the bonds. The bond angles have been allowed to vary between 0° and 180°, which means that their average value is 90°. In reality the angle between carbon-carbon single bonds is the tetrahedral angle, 109°28′, and this has the effect of increasing the length of the polymer chain. An analysis of this problem leads to the result that the effect is the same as if the individual bond lengths were increased by a factor of $\sqrt{2}$. With this modification, all the relationships that we have considered (Eqs. 19.30–19.39) can be employed.

EXAMPLE

A polymer chain is composed of 500 C—C bonds of length 0.154 nm, with a bond angle of 109°28′. Calculate (a) the length of the chain if it were fully extended, (b) the mean length of the random coil, and (c) the root-mean-square length of the random coil.

SOLUTION

a. The maximum extension occurs when alternate carbon atoms are in a straight line, as shown in Figure 19.6a. A simple geometrical argument shows that the distance between alternate carbon atoms is $2\sqrt{2}b/\sqrt{3} = 1.633b = 0.251$ nm. The fully extended chain length is therefore

$$250 \times 0.251 = 62.8 \text{ nm}$$

b. The effective bond length is $\sqrt{2}b = 0.218$ nm. The average length of the random coil is therefore, from Eq. 19.39,

$$\bar{l} = \left(\frac{8 \times 500}{3\pi}\right)^{1/2} 0.218 = 4.49 \text{ nm}$$

c. From Eq. 19.35, the mean square length is

$$\overline{l^2} = 500 \times 0.218^2 = 23.8 \text{ nm}^2$$

and the root-mean-square length is

$$\sqrt{\overline{l^2}} = 4.87 \text{ nm}$$

■

Another adjustment that has to be made is for steric hindrance. Figure 19.6b shows a portion of a polyethylene chain, with some steric interaction between hydrogen atoms on alternate carbon atoms. The effect of such steric effects is to increase the length of the chain, to an extent that depends on the side groups. The degree of extension is a minimum with polyethylene where only hydrogen atoms are involved and larger where there are bulky side groups. This effect is difficult to treat theoretically, and usually an empirical factor is employed.

Finally, the nature of the solvent has an important effect on the conformation of a polymer molecule in solution, and again a precise theoretical treatment is difficult. If the solvent is such that the solvent molecules interact with groups on the polymer to an extent that is comparable with their interaction with one another, the treatment that we have outlined is fairly satisfactory. However, if the solvent molecules have more affinity to one another than they have to groups on the

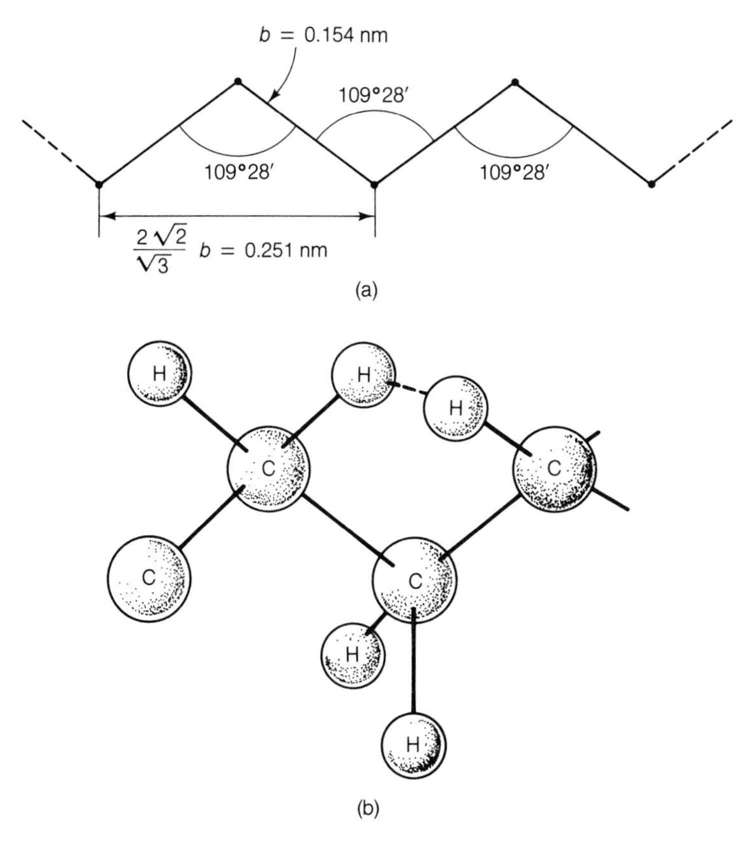

FIGURE 19.6 (a) The maximum extension of a hydrocarbon chain. (b) Steric hindrance between hydrogen atoms on a polymer chain.

polymer molecule, the latter will tend to roll up into a more or less spherical form, so as to minimize solvent-solute interactions. We have already seen that this occurs with protein molecules having a preponderance of hydrophobic side groups; these groups tend to turn inward so as to avoid the solvent molecules, and the result is a globular protein molecule.

19.7 Physical Properties of Solid Polymers

Solid polymers exhibit some rather interesting physical properties. They are neither completely crystalline nor completely amorphous but have some crystalline and some amorphous regions, and as a result they melt over a range of temperature. Many solid polymers can be molded and are said to be **plastic**. Some polymers show a surprising degree of **elasticity;** they can be extended to many times their original length without breaking, and when the stress is removed, they return to their original condition. We will now discuss these three characteristics, crystallinity, plasticity, and elasticity, in more detail.

Crystallinity

In a perfect crystal the atoms are arranged in a completely ordered pattern. The pure crystalline state represents an extreme that may be attained by some inorganic salts but which is less closely approached by polymeric material. The degree of crystallinity of a substance can be studied using an electron microscope or by X-ray diffraction. Single crystals of certain polymers can be grown from solution; for example, single crystals of polyethylene can be formed from solution in xylene. Examined under the electron microscope or by means of X-ray diffraction, such crystals are found to exhibit a regular pattern in certain regions but to be *amorphous* (Greek *a-*, without; *morphe,* form) in other regions, as illustrated schematically in Figure 19.7a.

There is a considerable variation in crystallinity between different polymers. Some are extremely amorphous, but even such materials show considerably more ordered structure than does a liquid. Fibrous proteins often have a high degree of order; this is true of natural materials such as hair and silk and of artificial polymers such as nylon and Teflon. Often these fibers have a well-defined periodicity along the axis but a more random arrangement in other directions. Some pioneering X-ray diffraction work on such fibers was done by the British physicist William Thomas Astbury (1898–1961).

The degree of crystallinity of a polymer depends to a considerable extent on the molecular structure and the degree of stretching. For example, polyisobutene,

$$\left(-\overset{\displaystyle CH_3}{\underset{\displaystyle CH_3}{\overset{|}{\underset{|}{C}}}}-CH_2- \right)_n$$

which is a type of rubber, has a very amphorous structure in the unstretched state, but when stretched it shows a very uniform X-ray diffraction pattern, indicating a considerable degree of crystallinity. In the unstretched state the polymer chains coil up in an irregular manner, as in Figure 19.7a, but the stretching causes a much greater degree of alignment, as shown in Figure 19.7b. Somewhat similar behavior is found with rubber.

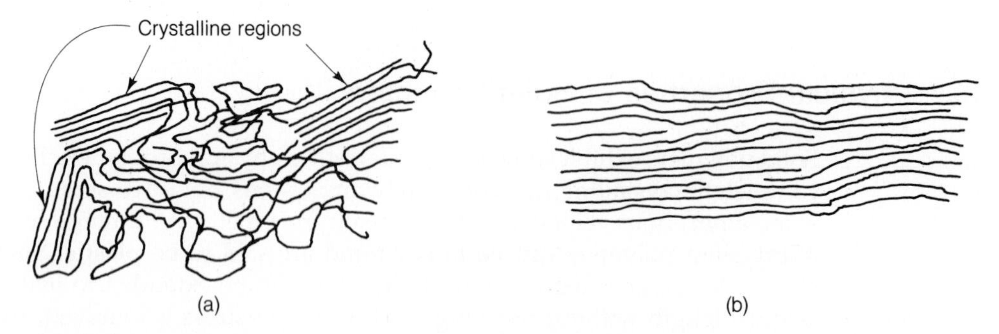

FIGURE 19.7 (a) Molecular chains in a polymer, showing crystalline and amorphous regions. (b) Molecular chains in a stretched polymer, showing a larger proportion of crystalline regions.

The degree of crystallinity is reduced when there are large side groups. Polystyrene, for example,

$$\left(\begin{array}{c} C_6H_5 \\ | \\ -C-CH_2- \\ | \\ H \end{array} \right)_n$$

has a glassy appearance and the X-ray analysis shows little crystallinity, even when the material is stretched. The large side groups prevent the molecules from becoming aligned in a regular pattern.

The degree of crystallinity has an important effect on the physical properties. A high degree of crystallinity leads to high tensile strength and low elasticity, whereas a more amorphous structure leads to greater elasticity and plasticity. Increasing the temperature reduces the crystallinity and therefore can convert a hard glassy polymer into one with much more plasticity.

Unlike substances of low relative molecular mass, high polymers do not melt at a definite temperature but gradually become more liquidlike. This is due to the fact that polymers always have some crystalline and some amorphous regions. The melting point is given by

$$T_m = \frac{\Delta H}{\Delta S} \tag{19.40}$$

where ΔH and ΔS are the enthalpy and entropy changes that occur on melting. Compared with the amorphous regions, the crystalline regions have lower enthalpies and entropies, and the ΔH and ΔS changes are hence larger when the crystalline regions melt. The effect on ΔH is greater than that on ΔS, and as a result the crystalline regions melt at a higher temperature than the amorphous regions. The polymer as a whole therefore melts over a range of temperature. These considerations explain why rubber that has been allowed to crystallize over a long period of time melts at a higher temperature than ordinary rubber.

Plasticity

Almost all polymeric substances show some degree of **plasticity** (Greek *plastikos,* moldable), by which is meant that they can be molded into a desired shape. Such polymers are often known as *plastics.* Natural plastics, such as shellac and bitumen, have been known since ancient times. The first artificial plastic, cellulose, was prepared in 1855 by the English chemist Alexander Parkes (1813–1890) who used it for moldings. However the modern synthetic plastics industry did not really begin until 1916 when the American chemist Leo Hendrik Baekeland (1863–1944) produced a phenol-formaldehyde plastic now known as *bakelite.*

Plastics fall into two main classes according to how they behave on heating. Some plastics can be repeatedly softened, hardened, and resoftened by appropriate heat treatment, care being taken to avoid decomposition. Such plastics are known as *thermoplastics*; examples are most natural waxes and resins and synthetic materials such as polyethylene, polystyrene, and nylon. Other plastics, known as *thermosetting plastics,* become permanently rigid when they are heated, since they undergo irreversible chemical change. Examples of thermoplastics are the phenolic and amino plastics.

The plasticity of polymeric material is increased if there are large substituent groups that force the chains apart, in this way making it easier for the substance to flow. The same effect can be brought about by the introduction of small amounts of solvent, insufficient to bring about solution but sufficient to reduce the forces between polymer chains. Substances that act in this way are known as *plasticizers*, and the effect is known as *plasticization*.

Elasticity

A substance is said to be **elastic** (Greek *elastikos*, driving, propulsive) if it spontaneously regains its former shape after being deformed, for example by compression or stretching. Gases are elastic, and so are many solid polymers, which are then called **elastomers.** Certain substances, like rubber, show a high degree of elasticity, being able to sustain a large deformation without rupture and to return rapidly to their original shape when the stress is removed. A material that is both elastic and plastic is said to exhibit **viscoelastic** properties.

Figure 19.8a shows the type of relationship that is usually obtained when a stress is applied to a material. **Stress** σ is defined as the force applied F divided by the cross-sectional area A to which it is applied,

$$\sigma = \frac{F}{A} \tag{19.41}$$

(see Figure 19.8b). The SI base unit of force is the newton, N ($=$ kg m s^{-2}), and the unit of stress is therefore N m^{-2} $=$ kg m^{-1} s^{-2}. The **strain** ϵ is the relative elongation of a bar of material; thus if a bar of length l_0 is elongated by Δl, the strain is

$$\epsilon = \frac{\Delta l}{l_0} \tag{19.42}$$

and is a dimensionless quantity. As shown in Figure 19.8a, at relatively small strains the stress is proportional to the strain, a relationship first formulated by the British physicist Thomas Young (1773–1829). At higher stresses the strain increases disproportionately; there is then structural breakdown followed by complete rupture.

The ratio of stress to strain in the region of proportionality is known as **Young's modulus** and given the symbol E:

$$E = \frac{\sigma}{\epsilon} \tag{19.43}$$

Table 19.2 shows some values of Young's modulus for a variety of materials; these do not necessarily exhibit elasticity; the stretching sometimes is irreversible. Many polymers (especially rubber) have extremely small values of Young's modulus, which means that small stresses bring about large extensions, and the stretching is usually reversible. Besides this high elasticity, rubber is capable of undergoing very large strains without rupture; a maximum elongation of 5 to 10 times the unstretched length is common. This behavior is shown schematically in Figure 19.8a.

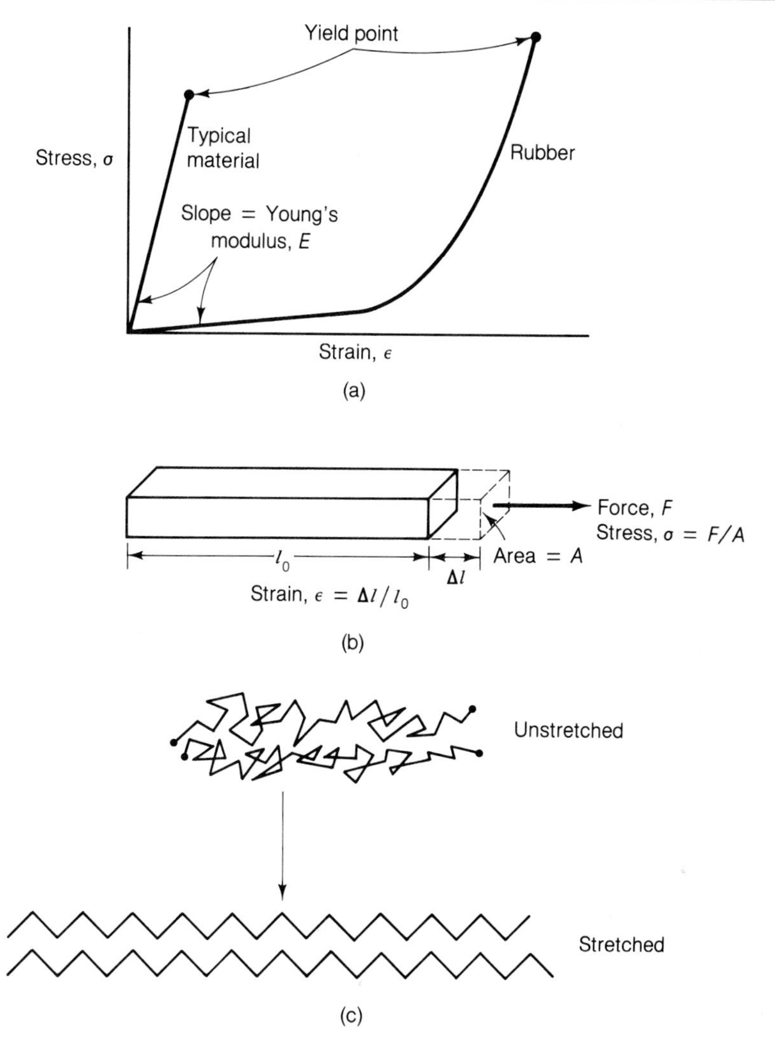

FIGURE 19.8 (a) Stress-strain curves for a typical material and for rubber. (b) The definitions of stress and strain. (c) Schematic diagram showing molecular conformations in unstretched and stretched rubber.

That some polymers are highly elastic is related to the fact that they consist of long molecular chains arranged in such a fashion that there is some internal mobility. As a result, extension can occur by uncoiling, without much change in interatomic distances, and therefore without much change in internal energy. The conformational changes that take place when an elastomer is stretched are shown schematically in Figure 19.8c. In the unstretched material, the polymer chains are twisted more or less at random. When tension is applied, the ends of each chain are forced apart, which means that the chains must become uncoiled and oriented along the direction of elongation. These molecular changes have been observed by X-ray diffraction and by electron microscopy. Polymers showing a high degree of crystallinity do not show high elasticity, since there is not sufficient internal mobility to permit the conformational changes to occur.

TABLE 19.2 Values of Young's Modulus

Material	Young's Modulus, E/N m^{-2}
Tungsten	4.2×10^{11}
Platinum	1.7×10^{11}
Copper	1.2×10^{11}
Cast iron	$1.0–1.3 \times 10^{11}$
Gold	8.0×10^{10}
Silver	7.9×10^{10}
Quartz	5.2×10^{10}
Glass	$5–8 \times 10^{10}$
Silk fiber	6.5×10^{9}
Thermoplastics	$1.5–3 \times 10^{9}$
Thermosetting plastics	$3.5–11 \times 10^{9}$
Soft vulcanized rubber	$1.5–5 \times 10^{6}$

Since the chains have become oriented along the direction of the stress, the system is more ordered than in its unstretched condition, when the chains had a much more random arrangement. The stretched state therefore has a lower entropy than the unstretched state. As a result, the process of contraction of a stretched elastomer is accompanied by an entropy increase. Spontaneity is determined by the Gibbs energy change, which at a given temperature T is

$$\Delta G = \Delta H - T\,\Delta S \tag{19.44}$$

With most elastomers ΔH is relatively small, so that to a good approximation

$$\Delta G \approx -T\,\Delta S \tag{19.45}$$

The entropy increase on contraction therefore leads to a Gibbs energy decrease, as required for a spontaneous process. The contraction of an elastomer therefore is an *entropic* process, being controlled almost entirely by the entropy increase. These considerations also apply to the contraction of a muscle fiber.

It is of interest to compare, from a thermodynamic point of view, the elasticity of a polymer with that of a gas. In Eq. 3.118 we saw that pressure is

$$P = -\left(\frac{\partial A}{\partial V}\right)_T \tag{19.46}$$

and since $A = U - TS$,

$$P = T\left(\frac{\partial S}{\partial V}\right)_T - \left(\frac{\partial U}{\partial V}\right)_T \tag{19.47}$$

For an ideal gas the second term is zero; for a gas obeying the van der Waals equation, $(\partial U/\partial V)_T = a/V_m^2$ (Eq. 3.143) and is usually fairly small. Thus for gases under most conditions

$$P \approx T\left(\frac{\partial S}{\partial V}\right)_T \tag{19.48}$$

The pressure of the gas is therefore determined by the change of entropy with volume and is proportional to the temperature. At a given temperature an increase in pressure requires $(\partial S/\partial V)_T$ to increase, and as a result the volume decreases.

In these equations for a gas the three-dimensional quantities P and V are involved. In the stretching of an elastomer we are concerned with a one-dimensional situation. The volume V is replaced by the length l and the pressure P by the force F that brings about the elongation. Equation 19.47 is thus replaced by

$$F = T\left(\frac{\partial S}{\partial l}\right)_T - \left(\frac{\partial U}{\partial l}\right)_T \tag{19.49}$$

Experimentally it is found that the force F is proportional to the temperature, so that the term involving the energy can be neglected as in the three-dimensional gas,

$$F \approx T\left(\frac{\partial S}{\partial l}\right)_T \tag{19.50}$$

Thus the entropy depends on the length, whereas the internal energy U and therefore the enthalpy H are practically independent of the extension.

If a constant force F is acting on an elastic polymer, an increase in temperature brings about a decrease in length. Since the force F that is required to maintain a given length is found experimentally to be proportional to T, it follows from Eq. 19.50 that $(\partial S/\partial l)_T$ is independent of temperature. The force required to keep the material at a constant extension increases as the temperature is raised, and if a constant force is applied, the material will contract.

Key Equations

Average molar masses:

$$\text{Number-average: } M_n \equiv \sum_i x_i M_i \equiv \frac{\sum_i N_i M_i}{\sum_i N_i} \qquad (x_i = \text{mole fraction})$$

$$\text{Mass-average: } M_m \equiv \sum_i y_i M_i \equiv \frac{\sum_i N_i M_i^2}{\sum_i N_i M_i} \qquad (y_i = \text{mass fraction})$$

$$Z\text{-average: } M_z \equiv \frac{\sum_i N_i M_i^3}{\sum_i N_i M_i^2}$$

Mechanical properties:

$$\text{Definition of stress: } \sigma \equiv \frac{\text{force, } F}{\text{area, } A}$$

$$\text{Definition of strain: } \epsilon \equiv \frac{\Delta l}{l_0}$$

$$\text{Young's modulus: } E \equiv \frac{\sigma}{\epsilon}$$

Problems

A Polymerization

1. The polymerization of styrene (M) catalyzed by benzoyl peroxide (C) obeys a kinetic equation of the form

$$-\frac{d[\mathrm{M}]}{dt} = k[\mathrm{M}]^{3/2}[\mathrm{C}]^{1/2}$$

Obtain an expression for the kinetic chain length, in terms of [M], [C], and the rate constants for initiation, propagation, and termination.

2. The polymerization of ethylene (M) photosensitized by acetone occurs by the mechanism

$$\mathrm{CH_3COCH_3} \xrightarrow{h\nu} \mathrm{CO} + 2\mathrm{CH_3}$$

$$\mathrm{CH_3} + \mathrm{C_2H_4} \xrightarrow{k_p} \mathrm{CH_3CH_2{-}CH_2{-}}$$

$$\mathrm{CH_3CH_2CH_2{-}} + \mathrm{C_2H_4} \xrightarrow{k_p} \mathrm{CH_3CH_2CH_2CH_2CH_2{-}}$$

$$\mathrm{R}_n + \mathrm{R}_m \xrightarrow{k_t} \mathrm{M}_{n+m}$$

Show that the rate equation is

$$-\frac{d[\mathrm{M}]}{dt} = k_p\left(\frac{2I}{k_t}\right)^{1/2}[\mathrm{M}]$$

where I is the intensity of light absorbed and k_p and k_t are the rate constants for the propagation and termination steps, respectively.

B Relative Molecular Masses

3. A polydisperse protein has 10% of molecules of molar mass 10 000 g mol^{-1}, 80% of mass 20 000 g mol^{-1}, and 10% of mass 40 000 g mol^{-1}. Calculate the number-average and mass-average molar masses.

4. The composition of a protein corresponds to 5 mol of molar mass 30 000 g mol^{-1} and 10 mol of molar mass 60 000 g mol^{-1}. Calculate the number-average and mass-average molar masses.

5. **a.** A polymer solution contains equal numbers of molecules of molar masses 20 000 g mol^{-1} and 30 000 g mol^{-1}. Calculate M_n and M_m.
b. A polymer solution contains equal masses of molecules of molar masses 20 000 g mol^{-1} and 30 000 g mol^{-1}. Calculate M_n and M_m.

6. The following osmotic pressures have been measured for a solution of a protein at 300 K and various concentrations:

ρ/g dm^{-3}	3.30	6.40	9.25	12.50	14.90
π/Pa	27.4	53.8	78.5	107.5	129.5

Calculate the molar mass of the protein.

7. The following results have been obtained with ox hemoglobin:

Fe(relative atomic mass 55.85) = 0.336%

S(relative atomic mass 32.06) = 0.48%

arginine content(molar mass 174.20 g mol^{-1}) = 4.24%.

What is the minimal molar mass consistent with these three values?

8. The following osmotic pressures have been measured for a solution of nitrocellulose in acetone at 20°C:

$\rho/g\ cm^{-3}$	4.52	9.37	16.0	23.7	29.8
π/Pa	246	572	1124	1940	2700

Calculate the molar mass of the sample.

C Supplementary Problems

9. A linear polyethylene molecule has a relative molecular mass of 50 000. Calculate

a. The root-mean-square end-to-end distance, assuming a random coil.

b. The mean end-to-end distance.

c. The length if the molecule were fully stretched.

(Assume the C—C bond length to be 0.154 nm and the bond angle to be the tetrahedral angle.)

10. A linear polystyrene molecule,

$$\left(\begin{array}{c} C_6H_5 \\ | \\ \text{—CH—CH}_2\text{—} \end{array} \right)_n$$

has a relative molecular mass of 75 000. Assume the molecule to be arranged as a random coil and calculate

a. The root-mean-square end-to-end distance.

b. The probability that the molecule has an end-to-end distance lying between 4.99 and 5.01 nm.

c. The most probable end-to-end distance.

11. A strip of rubber at 25°C supports a mass of 100 g. What force is the mass exerting on the rubber? If the temperature is raised to 50°C, what mass is required to maintain the rubber at the same extension, assuming the behavior to be completely entropic ($g = 9.81\ m\ s^{-2}$)?

12. Two solutions A and B are both 1% (by weight) solutions of polystyrene in toluene. The particles in solution A have a uniform molar mass of 10 000 g mol^{-1} and the intrinsic viscosity is $6.2 \times 10^{-4}\ kg\ m^{-1}\ s^{-1}$. The particles in solution B have a uniform molar mass of 100 000 g mol^{-1}, and the intrinsic viscosity is $9.5 \times 10^{-4}\ kg\ m^{-1}\ s^{-1}$. Estimate the intrinsic viscosity of a mixture of equal volumes of the two solutions.

D Essay Questions

1. What problems arise when colligative properties are used for molecular-mass determinations? How can these difficulties be circumvented?

2. Explain why isomorphous replacement is useful for determining crystal structure.

3. What techniques are available for the study of macromolecular conformations?

4. Define the terms *crystallinity*, *plasticity*, and *elasticity*.

Suggested Reading

F. W. Billmeyer, *Textbook of Polymer Science*, New York: Wiley, 1963.

D. C. Blackley, *Emulsion Polymerization*, London: Applied Science Pubs., 1975

J. D. Ferry, *Viscoelastic Properties of Polymers*, New York: Wiley, 1970.

P. J. Flory, *Principles of Polymer Chemistry*, Ithaca, N.Y.: Cornell University Press, 1953.

J. W. McBain, *Colloid Science*, Boston: D. C. Heath, 1950.

K. F. O'Driscoll, *The Nature and Chemistry of High Polymers*, New York: Reinhold Pub. Co., 1964.

J. K. Stile, *Introduction to Polymer Chemistry*, New York: Wiley, 1962.

C. Tanford, *The Physical Chemistry of Macromolecules*, New York: Wiley, 1961.

I. M. Ward, *Mechanical Properties of Solid Polymers*, London: Wiley, 1971.

A Units, Quantities, and Symbols: The SI/IUPAC Recommendations

In scientific work it is recommended that use should be made of a coherent system of units known as the *Système International d'Unités,* abbreviated as the SI units. A system is said to be coherent when the units for all derived physical quantities are obtained from certain base units by multiplication or division, without the use of any numerical factors. A number of articles have been written about the SI units, which represent an extension of the metric system.*

The International Union of Pure and Applied Chemistry (IUPAC) has strongly endorsed the use of SI units and has taken the matter much further by recommending terms, definitions, and symbols to be used in all branches of chemistry. An outline of their recommendation as they apply to physical chemistry is given in this appendix, which also makes more explicit some of the implications of the IUPAC recommendations. The latest official set of recommendations for physical chemistry is *Manual of Symbols and Terminology for Physicochemical Quantities and Units,* prepared for publication by D. H. Whiffen and published by Pergamon Press, Oxford, in 1979; it was also published in *Pure and Applied Chemistry,* Vol. 51, (1979), pp. 1–41. Other useful publications are listed in the footnote.† The student will find any of these publications extremely useful, as all of them contain the necessary information about both units and physical quantities, in a very compact form.

The SI base units and their symbols are as follows:

Physical Quantity	*Symbol for Quantity*	*SI Unit*	*Symbol for Unit*
length	l	metre	m
mass	m	kilogram	kg
thermodynamic temperature	T	kelvin	K
time	t	second	s
electric current	I	ampere	A
luminous intensity	I_v	candela	cd
amount of substance	n	mole	mol

*The official account of the SI units is *Le Système International d'Unités (SI),* published in 1973 by the Bureau International des Poids et Mesures and obtainable from OFFILIB, 48 rue Gay-Lussac, F 75005, Paris 5e. An authorized English translation of this, entitled *SI, The International System of Units,* was prepared jointly in 1973 by the U.S. National Bureau of Standards (Publication 330, available from the Government Printing Office, Washington, D.C. 20402) and by the U.K. National Physical Laboratory (available from Her Majesty's Stationery Office, P.O. Box 569, London SE1 OMH). Another useful publication is *The Metric Guide,* published in 1976 by the Council of Ministers of Education of Canada and available from OISE Publications Sales, 252 Bloor Street, West, Toronto, Ontario, M5S 1V6 (this publication begins with a very clear and elementary explanation of the metric system and later deals in some detail with SI units).

†*Quantities, Units and Symbols,* Royal Society, 6 Carlton House Terrace, London SW1Y 5AG, 1975; M. L. McGlashlan, *Physiochemical Quantities and Units* (2nd ed.), Royal Institute of Chemistry, London, 1971.

In addition there are two supplementary units for angles:

Physical Quantity	*Symbol for Quantity*	*SI Unit*	*Symbol for Unit*
plane angle	θ, ϕ, etc.	radian	rad
solid angle	Ω	steradian	sr

Note that in SI the symbol for a physical quantity is always printed in *italic* type, whereas a symbol for a unit is printed in roman type. In this book we use italics for two useful non-SI symbols, namely m ($\equiv$ mol kg^{-1}) and M ($\equiv$ mol dm^{-3}). Luminous intensity is seldom used in physical chemistry, but the other units are of importance and are considered in further detail later.

The following prefixes are used with SI units:

Fraction	*Prefix*	*Symbol*	*Multiple*	*Prefix*	*Symbol*
10^{-1}	deci	d	10	deca	da
10^{-2}	centi	c	10^{2}	hecto	h
10^{-3}	milli	m	10^{3}	kilo	k
10^{-6}	micro	μ	10^{6}	mega	M
10^{-9}	nano	n	10^{9}	giga	G
10^{-12}	pico	p	10^{12}	tera	T
10^{-15}	femto	f	10^{15}	peta	P
10^{-18}	atto	a	10^{18}	exa	E

Length

The metre* is defined as the length equal to 1 650 763.73 wavelengths in vacuum of the radiation of a spectral line in the orange-red region, corresponding to the transition between certain levels ($2p_{10}$ and $5d_5$) of the krypton-86 atom.

The following multiple units of length are also commonly used:

1 decimetre $= 10^{-1}$ metre (1 dm $= 10^{-1}$ m)

1 centimetre $= 10^{-2}$ metre (1 cm $= 10^{-2}$ m)

1 millimetre $= 10^{-3}$ metre (1 mm $= 10^{-3}$ m)

1 micrometre $= 10^{-6}$ metre (1 μm $= 10^{-6}$ m)

1 nanometre $= 10^{-9}$ metre (1 nm $= 10^{-9}$ m)

1 picometre $= 10^{-12}$ metre (1 pm $= 10^{-12}$ m)

1 kilometre $= 10^{3}$ metre (1 km $= 10^{3}$ m)

The micrometre was formerly called the micron, and the nanometre was formerly called the millimicron; the use of these older terms is discouraged. The angström (Å), a non-SI unit, is still commonly employed:

$$1\ \text{Å} = 10^{-10}\ \text{m} = 10^{-8}\ \text{cm} = 10^{-1}\ \text{nm} = 100\ \text{pm}$$

*Note the spelling of metre, even in U.S. scientific usage, metre being an international unit.

Although the angstrom is not recommended, it is still in very common use. Some workers are now expressing interatomic distances in picometres, such distances ranging from about 100 to 200 pm.

Volume

In SI the basic unit of volume is the cubic metre (m^3). The liter is now defined as equal to one cubic decimetre (dm^3). The cubic centimetre should be written as cm^3, not as cc.

Mass

The kilogram is the mass of a platinum-iridium block in the custody of the Bureau International des Poids et Mesures at Sèvres, France.

Although the kilogram is the base unit of mass, its multiples are expressed with reference to the gram (g); it would obviously be absurd to express a gram as a millikilogram! Thus 10^{-6} kg is written as 1 milligram (1 mg).

Time

The second is the duration of 9 192 770 periods of the radiation corresponding to the transition between the two hyperfine levels of the ground state of the cesium-133 atom.

The second should be used as far as possible, but it is sometimes convenient to use larger units:

1 minute (min)	= 60 s	1 day	= 86 400 s
1 hour	= 3600 s	1 year	= 3.156×10^7 s

Thermodynamic Temperature

The kelvin is strictly defined as the fraction 1/273.16 of the temperature interval between the absolute zero and the triple point of water. In practice one obtains the absolute temperature/K by adding 273.15 to the value of the temperature in degrees Celsius; for example,

$$25.0°C = 298.15 \text{ K}$$

The use of the symbols °K and deg is not recommended.

Electric Current

The ampere is that constant current which, if maintained in two straight parallel conductors of infinite length, of negligible cross section, and placed 1 metre apart in vacuum, would produce between these conductors a force equal to 2×10^{-7} newton per metre of length (1 newton = 1 kg m s^{-2}).

Amount of Substance

The mole is the amount of substance that contains as many elementary units as there are atoms in 12 g of carbon-12 ($^{12}_{6}C$). The elementary unit must be specified and may be an atom, a molecule, an ion, a radical, an electron, or any other elementary particle. It may also be any specified group or fraction of such entities such as a polymer or a fraction of a molecule. For example,

1 mol of HgCl has a mass of 236.04 g

1 mol of Hg_2Cl_2 has a mass of 472.08 g

1 mol of Hg has a mass of 200.59 g

1 mol of $\frac{1}{2}CuSO_4$ has a mass of 79.80 g

1 mol of $\frac{1}{2}Cu^{2+}$ has a mass of 31.77 g

1 mol of electrons has a mass of 5.486×10^{-4} g

The amount of substance should not be referred to as the "number of moles"; it is not a number but a quantity having units.

Concentration and Molality

The word *concentration* means the amount of substance divided by the volume of the solution. The SI unit is mol m^{-3}, but the usual unit is moles per liter (mol dm^{-3}), which is conveniently abbreviated as M. The use of equivalents and normalities (N) is not recommended. Using the word *molarity* to mean concentration is not recommended because of the possibility of confusion with *molality.*

The word *molality* means the amount of solute divided by the mass of the solvent. The usual units are mol kg^{-1}, for which the symbol m is frequently used; to avoid confusion with the symbol for metre, it is best to use italics.

Molar Quantities

The word *molar* used before the name of a quantity may mean "divided by amount of substance." For example, molar volume is the volume divided by the amount of substance. The word *molar* is also used to mean "divided by concentration," where "concentration" is "amount of substance divided by volume." A useful rule is that molar can be used whenever mol appears in the unit used for the quantity. The subscript m can be used to indicate a molar quantity; for example, molar volume can be written as $V_m (= V/n)$. If there is no danger of ambiguity, the subscript m can be omitted; this is always the case when a numerical value is given, since mol appears in the unit.

Derived Units

Table A-1 gives some SI derived units. The convention for writing the abbreviations should be noted. Prefixes are not separated from the unit (e.g., kg, not k g); otherwise units are separated (e.g., m s^{-1}, not ms^{-1}). The solidus can be used, as in m/s for m s^{-1}, but it should not appear more than once in the same expression unless parentheses are used. Thus the unit for entropy must not be written as J/K/mol. Acceptable forms for this unit are

$$\mathrm{J\ K^{-1}\ mol^{-1}} \qquad \mathrm{J(K\ mol)^{-1}} \qquad \mathrm{J/(K\ mol)}$$

and the correct pronunciation of the unit is "joule per kelvin mole," not "joule per kelvin per mole." The solidus is very useful for dividing quantities by units in tables

TABLE A-1 SI Derived Units

Physical Quantity	Name of Unit	Symbol	Definition	Alternative Form	Named after
Force	newton	N	kg m s^{-2}		Isaac Newton (1642–1727)
Pressure	pascal	Pa	kg m^{-1} s^{-2}	N m^{-2}	Blaise Pascal (1623–1662)
Energy	joule	J	kg m^2 s^{-2}	N m	James Prescott Joule (1818–1889)
Power	watt	W	kg m^2 s^{-3}	J s^{-1}	James Watt (1737–1819)
Electric charge	coulomb	C	A s		Charles Auguste de Coulomb (1736–1806)
Electric potential difference	volt	V	kg m^2 s^{-3} A^{-1}	J C^{-1}	Allesandro Volta (1745–1827)
Electric resistance	ohm	Ω	kg m^2 s^{-3} A^{-2}	V A^{-1}	Georg Simon Ohm (1787–1854)
Electric conductance	siemens	S	kg^{-1} m^{-2} s^3 A^2	Ω^{-1}	Werner von Siemens (1816–1892)
Electric capacitance	farad	F	kg^{-1} m^{-2} s^4 A^2	C V^{-1}	Michael Faraday (1791–1867)
Magnetic flux	weber	Wb	kg m^2 s^{-2} A^{-1}	V s	Wilhelm Weber (1804–1891)
Magnetic field strength	henry	H	kg m^2 s^{-2} A^{-2}	V A^{-1} s	Joseph Henry (1799–1878)
Magnetic flux density	tesla	T	kg s^{-2} A^{-1}	V s m^{-2}	Nikola Tesla (1856–1943)
Frequency	hertz	Hz	s^{-1}		Heinrich Hertz (1857–1894)
Activity (of radioactive source)	becquerel	Bq	s^{-1}		Antoine Henri Becquerel (1852–1908)
Absorbed dose (of radiation)	gray	Gy	m^2 s^{-2}	J kg^{-1}	Louis Harold Gray (1905–1965)

or figures, where the numerical value is needed. For example, a table heading for entropy changes could be ΔS/J K^{-1} mol^{-1}.

Some non-SI units are still in very common use, and some of these are listed in Table A-2, together with their SI equivalents. Since 1 atm is commonly chosen as a standard state for thermodynamic quantities, it is entirely acceptable to continue to use it. The other units in Table A-2, however, are being progressively abandoned.

PHYSICOCHEMICAL QUANTITIES AND SYMBOLS

The most important physicochemical quantities, most of the ones used in this book, are listed in Table A-3. This table is based on the IUPAC *Manual,* referred to earlier in this appendix, and the organization is much the same. The *Manual* gives some alternative symbols; here we give only those used in this book. We have included

TABLE A-2 Some non-SI Units

Physical Quantity	Name of Unit	Symbol	SI Equivalent
Pressure	(standard) atmosphere	atm	101 325 Pa
Pressure	millimetre of mercury	mmHg	101 325 Pa/760
Pressure	torr	torr	101 325 Pa/760
Pressure	bar	bar	10^5 Pa
Energy	thermochemical calorie	cal	4.184 J
Energy	electronvolt	eV	1.602×10^{-19} J
Energy	erg	erg	10^{-7} J
Length	angstrom	Å	10^{-10} m
Volume	liter	L	1 dm^3
Electric charge	electrostatic unit	esu	3.336×10^{-10} C
Force	dyne	dyn	10^{-5} N
Viscosity	poise	poise	0.1 kg m^{-1} s^{-1}
Diffusion coefficient	stokes	stokes	10^{-4} m^2 s^{-1}
Magnetic flux density	gauss	G	10^{-4} T
Dipole moment	debye	D	3.336×10^{-30} C m
Magnetic moment	Bohr magneton	μ_B	9.274×10^{-24} J T^{-1}

definitions of less well-known quantities when they can be expressed compactly; in other cases we have referred to the section of this book in which they are explained.

Each physical quantity has an SI unit, which is either an SI base unit or a combination of base units. For example, the SI unit of energy is the joule (J), which is kg m^2 s^{-2}. However it is quite appropriate, and may be more convenient, to use multiples or submultiples, such as kilojoules (kJ). Sometimes particular multiples are in common use, and when this is the case, they are indicated in the last column of Table A-3 (Customary Multiple). A blank in this column means that there is no strong preference for any multiple.

It is important to note that a *physical quantity is the product of a number and a unit.* For example, the Avogadro constant L is 6.022×10^{23} mol^{-1} and is the product of the number 6.022×10^{23} and the unit mol^{-1}. This being so, it should not be called the "Avogadro number"; it is not a number (i.e., a dimensionless quantity). If we need to refer to the number, another symbol (e.g., N) should be used. In this book we have used the superscript u to indicate the numerical value of a quantity; thus if an equilibrium constant K_c is 2.6 dm^3 mol^{-1}, $K_c^u = 2.6$.

The IUPAC *Manual* emphasizes that the use of a physical quantity or its symbol *should not imply any particular choice of unit.* For example, we may have a solution of concentration 1 mol dm^{-3}, which may alternatively be expressed as 10^3 mol m^{-3} or as 1 μmol mm^{-3} or as 10^{-3} mol cm^{-3}. The concentration is the same whichever unit we choose, and if we write an equation involving concentration, using the symbol

c, we must not assume that any particular unit is used. This being the case, it is unsatisfactory, for example, to define pH as

$$\mathrm{pH} = -\log_{10}[\mathrm{H}^+]$$

This is a very common procedure, but it presupposes that the reader knows that the unit of $[\mathrm{H}^+]$ is to be taken as mol dm^{-3} (which is not even the SI unit) and is then prepared to drop the unit in taking the logarithm. The only satisfactory definition of pH is

$$\mathrm{pH} = -\log_{10}([\mathrm{H}^+]/\mathrm{mol\ dm}^{-3})$$

Here we are dividing the concentration by the unit, which gives a number, and we are then taking the logarithm of the number.

In tables we are listing numbers, and in figures we are plotting numbers. Table headings and axes should therefore show the physical quantity divided by the unit. For example, in a plot of velocity against time, the axes could be labeled $v/\mathrm{m\ s}^{-1}$ and t/s.

TABLE A-3 Physicochemical Quantities, with their Symbols, Definitions, and Units

Vector quantities are shown in boldface type. An asterisk indicates that the quantity is dimensionless. Greek letters, with their names, are written inside the back cover.

A-3-1 Space, Time, and Related Quantities

Quantity	Symbol	Definition	SI Unit	Customary Multiple
Length	l		m	
Height	h		m	
Radius	r		m	
Diameter	d		m	
Path	s		m	
Wavelength	λ	Section 11.1	m	cm
Wave number	$\overline{\nu}$	$\overline{\nu} = 1/\lambda$	m^{-1}	cm^{-1}
Area	A		m^2	
Volume	V		m^3	m^3, dm^3 (L)
Time	t		s	s
Frequency	ν		$\mathrm{s}^{-1} \equiv \mathrm{Hz}$	s^{-1}, Hz, MHz, GHz
Relaxation time	τ		s	s
Velocity	v	$v = ds/dt$	$\mathrm{m\ s}^{-1}$	$\mathrm{m\ s}^{-1}$
Molecular velocity	u	$u = ds/dt$	$\mathrm{m\ s}^{-1}$	$\mathrm{m\ s}^{-1}$
Speed of light	c		$\mathrm{m\ s}^{-1}$	$\mathrm{m\ s}^{-1}$
Angular velocity	ω		$\mathrm{rad\ s}^{-1}$	$\mathrm{rad\ s}^{-1}$
Acceleration of free fall	g		$\mathrm{m\ s}^{-2}$	$\mathrm{m\ s}^{-2}$
Probability	P		*	*

TABLE A-3 Physicochemical Quantities, with their Symbols, Definitions, and Units (Continued)

A-3-2 Mechanical and Related Quantities

Quantity	Symbol	Definition	SI Unit	Customary Multiple
Mass	m		kg	kg, g
Reduced mass	μ	Section 11.6	kg	kg, g
Specific volume	v	$v \equiv V/m$	$m^3\ kg^{-1}$	$dm^3\ g^{-1}$
Density	ρ	$\rho \equiv m/V$	$kg\ m^{-3}$	$g\ dm^{-3}$, $g\ cm^{-3}$
Relative density	d	$d \equiv \rho/\rho_0$	*	*
Moment of inertia	I	$I \equiv \sum_i m_i r_i^2$	$kg\ m^2$	$kg\ m^2$
Momentum	p	$p \equiv mv$	$kg\ m\ s^{-1}$	$kg\ m\ s^{-1}$
Force	$\mathbf{F}$	$F \equiv dp/dt$	$N \equiv kg\ m\ s^{-2}$	N
Angular momentum	$\mathbf{L}$	$L \equiv r \times p$	$J\ s \equiv kg\ m^2\ s^{-1}$	$kg\ m^2\ s^{-1}$
Pressure	P	$P \equiv F/A$	$Pa \equiv kg\ m^{-1}\ s^{-2}$	Pa, kPa, atm
Isothermal compressibility	κ	$\kappa \equiv \frac{1}{V}\left(\frac{\partial V}{\partial P}\right)_T$	Pa^{-1}	Pa^{-1}
Coefficient of expansion	α	$\alpha \equiv \frac{1}{V}\left(\frac{\partial V}{\partial T}\right)_P$	K^{-1}	K^{-1}
Work	w	$w \equiv \int F\ ds$	$J \equiv kg\ m^2\ s^{-2}$	J, kJ
Energy	E	Section 1.2	$J \equiv kg\ m^2\ s^{-2}$	J, kJ
Potential energy	E_p	$\Delta E_p \equiv \int F\ ds$	$J \equiv kg\ m^2\ s^{-2}$	J, kJ
Kinetic energy	E_k	$E_k \equiv \frac{1}{2}mv^2$	$J \equiv kg\ m^2\ s^{-2}$	J, kJ
Hamiltonian	H	$H \equiv E_p + E_k$ (in terms of p)	$J \equiv kg\ m^2\ s^{-2}$	J
Power	P	$P \equiv dw/dt$	$W \equiv J\ s^{-1} \equiv kg\ m^2\ s^{-3}$	W
Viscosity	η	$F \equiv \eta A(dv/dx)$	$Pa\ s \equiv kg\ m^{-1}\ s^{-1}$	Pa s, $g\ cm^{-1}\ s^{-1}$ ($\equiv$ poise)
Intrinsic viscosity	$[\eta]$	Section 18.1	$m^3\ kg^{-1}$	$dm^3\ g^{-1}$
Fluidity	ϕ	$\phi \equiv 1/\eta$	$kg^{-1}\ m\ s$	
Kinematic viscosity	ν	$\nu \equiv \eta/\rho$	$m^2\ s^{-1}$	
Diffusion coefficient	D	$dn/dt \equiv -DA\ (dc/dx)$	$m^2\ s^{-1}$	$cm^2\ s^{-1}$
Surface tension	γ	dw/dA	$N\ m^{-1} \equiv kg\ s^{-2}$	$N\ m^{-1}$
Angle of contact	θ		rad	rad

3 Physicochemical Quantities, with their Symbols, Definitions, and Units (Continued)

ermodynamic and Related Quantities

	Symbol	Definition	SI Unit	Customary Multiple
namic or absolute re	T	Section 3.1	K	K
nperature	θ	$\theta \equiv T - 273.15$ K	K	K
s constant	R	$R \equiv \lim_{P\to 0} PV_m/T$	J K^{-1} mol^{-1}	J K^{-1} mol^{-1}
constant	$\mathbf{k}$	$\mathbf{k} \equiv R/L$	J K^{-1}	J K^{-1}
	q	Section 2.5	J	J, kJ
e on the system	w	$w \equiv \int F\, ds$	J	J, kJ
nergy	U	Section 2.5	J	J, kJ
rnal energy	U_m	$U_m \equiv U/n$	J mol^{-1}	kJ mol^{-1}
	H	$H \equiv U + PV$	J	J, kJ
nalpy	H_m	$H_m \equiv H/n \equiv U_m + PV_m$	J mol^{-1}	kJ mol^{-1}
molar enthalpy of	$\Delta_f H_m$	Section 2.6	J mol^{-1}	kJ mol^{-1}
acity at constant	C_V	$C_V \equiv (\partial U/\partial T)_V$	J K^{-1}	J K^{-1}
t capacity at volume	$C_{V,m}$	$C_{V,m} \equiv C_V/n$	J K^{-1} mol^{-1}	J K^{-1} mol^{-1}
eat capacity at volume	c_v	$c_v \equiv C_V/m$	J K^{-1} kg^{-1}	J K^{-1} g^{-1}
acity at constant	C_P	$C_P \equiv (\partial H/\partial T)_P$	J K^{-1}	J K^{-1}
at capacity at pressure	$C_{P,m}$	$C_{P,m} \equiv C_P/n$	J K^{-1} mol^{-1}	J K^{-1} mol^{-1}
neat capacity at pressure	c_p	$c_p \equiv C_P/m$	J K^{-1} kg^{-1}	J K^{-1} g^{-1}
neat capacities	γ	$\gamma \equiv C_p/C_V$	*	*
sion factor	Z	$Z \equiv PV_m/RT$	*	*
ompson coefficient	μ	$\mu \equiv (\partial T/\partial P)_H$	K Pa^{-1}	K Pa^{-1}
netric coefficient of e for products, for reactants)	ν_B	Section 2.6	*	*
netric sum	$\Sigma\nu$	$\Sigma\nu \equiv \sum_i \nu_i$	*	*
reaction	ξ	$\xi \equiv (n - n_0)/\nu$	mol	mol
	S	$dS \equiv dq_{rev}/T$	J K^{-1}	J K^{-1}
tropy	S_m	$S_m \equiv S/n$	J K^{-1} mol^{-1}	J K^{-1} mol^{-1}
nergy	G	$G \equiv H - TS$	J	J, kJ
bbs energy	G_m	$G_m \equiv G/n$	J mol^{-1}	kJ mol^{-1}

TABLE A-3 Physicochemical Quantities, with their Symbols, Definitions, and Uni

A-3-3 Molecular and Related Quantities

Quantity	Symbol	Definition	SI Unit	Customa Multiple
Relative atomic mass (formerly called "atomic weight")	A_r		*	*
Relative molecular mass (formerly called "molecular weight")	M_r		*	*
Molar mass	M		kg mol^{-1}	g mol^{-1}
Amount of substance	n	$n \equiv m/M$	mol	mol
Number of molecules	N		*	*
Avogadro constant	L	$L \equiv N/n$	mol^{-1}	mol^{-1}
Molar volume	V_m	$V_m \equiv V/n$	m^3 mol^{-1}	m^3 mol
Dipole moment	μ	Section 12.3	C m	
Symmetry number	σ	Section 14.6	*	*
Statistical weight	g	Section 14.4	*	*
Collision diameter	d, d_{AA}, d_{AB}	Section 1.9	m	nm, pr
Mean free path	λ	$\lambda = \sqrt{2}\pi d^2 L$	m	nm, pr
Mole fraction of B	x_B	$x_B \equiv n_B/\sum_i n_i$	*	*
Molality of B in solvent A	m_B	$m_B \equiv n_B/n_A M_A$	mol kg^{-1}	mol k
Concentration of B (formerly called "molarity")	c_B, [B]	$c_B \equiv n_B/V$	mol m^{-3}	mol d
Mass concentration of B	ρ_B	$\rho_B \equiv m_B/V$	kg m^{-3}	g dm
Partition function (assembly)	Q	Section 14.4	*	*
Partition function (molecule)	q	Section 14.4	*	*
Atomic number	Z	number of protons in nucleus	*	*
Neutron number	N	number of neutrons in nucleus	*	*
Mass number	A	$A \equiv Z + N$	*	*
Mass of electron	m_e		kg	kg
Bohr radius	a_0	$a_0 \equiv h^2\epsilon_0/\pi m_e e^2$	m	nm,
Magnetic moment	μ	Section 11.9	A m^2	A m
Bohr magneton	μ_B	$\mu_B \equiv eh/4\pi m_e$	A m^2	A m
Nuclear magneton	μ_N	$\mu_N \equiv eh/4\pi m_p$	A m^2	A m
Gyromagnetic ratio	γ	$\gamma \equiv \mu/L$	C kg^{-1} $\equiv$ A s kg^{-1}	C k
g factor	g	Section 13.10		*

TABLE A-3 Physicochemical Quantities, with their Symbols, Definitions, and Units (Continued)

A-3-4 Thermodynamic and Related Quantities (Continued)

Quantity	Symbol	Definition	SI Unit	Customary Multiple
Standard molar Gibbs energy change	ΔG_m°	Section 3.7	J mol^{-1}	kJ mol^{-1}
Helmholtz energy	A	$A \equiv U - TS$	J	J, kJ
Molar Helmholtz energy	A_m	$A_m \equiv A/n$	J mol^{-1}	kJ mol^{-1}
Standard molar Helmholtz energy change	ΔA_m°		J mol^{-1}	kJ mol^{-1}
Chemical potential of B	μ_B	$\mu_B \equiv (\partial G/\partial n_B)_{A,\ldots,T,P\ldots}$	J mol^{-1}	kJ mol^{-1}
Absolute activity of B	λ_B	$\lambda_B \equiv \exp(\mu_B/RT)$	*	*
Relative activity of B	a_B	$a_B \equiv \lambda_B/\lambda_B^*$	*	*
Fugacity	f		Pa	Pa
Osmotic pressure	π	Section 5.8	Pa	Pa
Activity coefficient, mole fraction basis	f	$f_B \equiv a_B/x_B$	*	*
Activity coefficient, molality basis	γ	$\gamma_B \equiv a_B/m_B$	*	*
Activity coefficient, concentration basis	y	$y_B \equiv a_B/c_B$	*	*
Equilibrium constant with respect to pressure (ideal gases)	K_P	$K_P \equiv (P_Y^y P_Z^z \ldots / P_A^a P_B^b \ldots)_{eq}$	Pa$^{\Sigma\nu}$ ($\Sigma\nu$=stoichiometric sum)	atm$^{\Sigma\nu}$
Equilibrium constant with respect to concentration (ideal systems)	K_c	$K_c \equiv ([Y]^y[Z]^z \ldots / [A]^a[B]^b \ldots)_{eq}$	(mol m^{-3})$^{\Sigma\nu}$	(mol dm^{-3})$^{\Sigma\nu}$
Dimensional-less equilibrium constant	K^u	Numerical value of K_p or K_c	*	*

A-3-5 Chemical Reactions

Quantity	Symbol	Definition	SI Unit	Customary Multiple
Rate of conversion	$\dot{\xi}$	$\dot{\xi} \equiv d\xi/dt$	mol s^{-1}	mol s^{-1}
Rate of reaction (at constant V)	v	$v \equiv (1/\nu_i)(dc_i/dt)$	mol m^{-3} s^{-1}	mol dm^{-3} s^{-1}
Rate of consumption of A (at constant V)	v_A	$v_A \equiv -dc_A/dt$	mol m^{-3} s^{-1}	mol dm^{-3} s^{-1}
Rate of formation of Z (at constant V)	v_Z	$v_Z \equiv dc_z/dt$	mol m^{-3} s^{-1}	mol dm^{-3} s^{-1}
Partial order	α, β	Section 9.3	*	*
Overall order	n	$n \equiv \alpha + \beta + \cdots$	*	*
Rate constant	k	$v \equiv k[A]^\alpha[B]^\beta \ldots$	(m^3 mol^{-1})$^{n-1}$ s^{-1}	(dm^3 mol^{-1})$^{n-1}$ s^{-1}

TABLE A-3 Physicochemical Quantities, with their Symbols, Definitions, and Units (Continued)

A-3-5 Chemical Reactions (Continued)

Quantity	Symbol	Definition	SI Unit	Customary Multiple
Chemical flux (chemiflux)	ϕ	Section 9.3	mol m^{-3} s^{-1}	mol dm^{-3} s^{-1}
Activation energy	E	$E \equiv R[\partial \ln k/d(1/T)]_P$	J mol^{-1}	kJ mol^{-1}
Pre-exponential factor	A	$k \equiv Ae^{-E/RT}$	(m^3 mol^{-1})$^{n-1}$ s^{-1}	(dm^3 mol^{-1})$^{n-1}$ s^{-1}
Gibbs energy of activation	$\Delta G^{\ddagger}$	$k \equiv \frac{kT}{h}e^{-\Delta G^{\ddagger}/RT}$	J mol^{-1}	kJ mol^{-1}
Enthalpy of activation	$\Delta H^{\ddagger}$	$k \equiv \frac{kT}{h}e^{\Delta S^{\ddagger}/R}e^{-\Delta H^{\ddagger}/RT}$	J mol^{-1}	kJ mol^{-1}
Entropy of activation	$\Delta S^{\ddagger}$		J K^{-1} mol^{-1}	J K^{-1} mol^{-1}
Collision frequency	Z_A	Section 1.9	s^{-1}	s^{-1}
Collision number	Z_{AA}, Z_{AB}	Section 1.9	m^{-3} s^{-1}	dm^{-3} s^{-1}
Half-life	$t_{1/2}$	Section 9.4	s	s
Relaxation time	τ	Section 9.5	s	s
Collision diameter	d_{AA}, d_{AB}	Section 1.9	m	nm, pm
Reaction probability	P_r		*	*
Reaction cross section	σ	$\sigma \equiv \pi d^2 P_r$	m^2	nm^2, pm^2

A-3-6 Electricity and Magnetism

Quantity	Symbol	Definition	SI Unit	Customary Multiple
Electric current	I		A	A, mA
Quantity of electricity	Q	$Q = \int I\,dt$	C ≡ A s	C
Elementary charge	e		C	C
Electric potential	V	$V \equiv w/Q$	V ≡ m^2 kg s^{-3} A^{-1}	V, mV, kV
Electric field strength	E	$E = F/Q$	V m^{-1}	V m^{-1}
Capacitance	C	$C = Q/\Delta V$	F ≡ m^{-2} kg^{-1} s^4 A^2	F
Permittivity of vacuum	ϵ_0	See p. 895	F m^{-1}	F m^{-1}
Permanent dipole moment	μ	Section 12.3	C m	C m
Electrical resistance	R	$R = \Delta V/I$	Ω ≡ V A^{-1}	Ω, mΩ
Electrical conductance	G	$G \equiv 1/R$	S ≡ Ω^{-1} ≡ A V^{-1}	S, Ω^{-1}
Magnetic flux density	B	Section 13.2	T ≡ kg s^{-2} A^{-1}	T
Magnetic field strength	H	Section 13.2	A m^{-1}	A m^{-1}
Magnetic flux	Φ	See p. 895	W ≡ V s	W
Permeability	μ	See p. 896	H m^{-1} ≡ kg m s^{-2} A^{-2}	H m^{-1}
Permeability of vacuum	μ_0	See p. 896	H m^{-1}	H m^{-1}

TABLE A-3 Physicochemical Quantities, with their Symbols, Definitions, and Units (Continued)

A-3-7 Electrochemistry

Quantity	Symbol	Definition	SI Unit	Customary Multiple
Faraday constant	F	$F \equiv Le$	$C\ mol^{-1}$	$C\ mol^{-1}$
Electromotive force	E	Section 8.1	V	V
Electrode potential	E	Section 8.2	V	V
Standard electrode potential	E^0	Section 8.2	V	V
Charge number of ion B	z_B	$z_B \equiv Q_B/e$	*	*
Charge number of cell reaction	z	Section 8.3	*	*
Electric mobility of B	u_B	Section 7.5	$m^2\ V^{-1}\ s^{-1}$	$cm^2\ V^{-1}\ s^{-1}$
Transport number of B	t_B	Section 7.6	*	*
Electrolytic conductivity	κ	Section 7.2	$S\ m^{-1} \equiv \Omega^{-1}\ m^{-1}$	$S\ cm^{-1} \equiv \Omega^{-1}\ cm^{-1}$
Molar conductivity of electrolyte	Λ	$\Lambda \equiv \kappa/c$	$S\ m^2\ mol^{-1}$	$S\ dm^2\ mol^{-1}$
Molar conductivity of ion B	λ_B	$\lambda_B \equiv Fu_B$	$S\ m^2\ mol^{-1}$	$S\ m^2\ mol^{-1} \equiv \Omega^{-1}\ m^2\ mol^{-1}$
Ionic strength	I	$I \equiv \frac{1}{2}\sum_i c_i z_i^2$	$mol\ m^{-3}$	$mol\ dm^{-3}$
Overpotential	η	$\eta \equiv E(I) - E(I = 0)$	V	V
Inner electric potential	ϕ	Section 17.9	V	V
Outer electric potential	ψ	Section 17.9	V	V
Electrokinetic potential	ξ	Section 17.9	V	V, mV
Thickness of diffusion layer	δ	Section 17.9	m	
Electro-osmotic or electrophoretic mobility	v	Section 18.4	$m^2\ V^{-1}\ s^{-1}$	$cm^2\ V^{-1}\ s^{-1}$

A-3-8 Electromagnetic Radiation

Quantity	Symbol	Definition	SI Unit	Customary Multiple
Planck constant	h	Section 11.1	J s	J s
Radiant excitance	M_e	Section 11.1	$W\ m^{-2}$	$W\ m^{-2}$
Intensity of light	I		cd	cd
Transmittance	T	$T \equiv I$ (transmitted)/ I(incident)	*	*
Absorbance	A	$A \equiv \log_{10}(1/T)$	*	*
(Linear decadic) absorption coefficient	a	$a \equiv A/l$	m^{-1}	cm^{-1}
Molar (decadic) absorption coefficient	ϵ	$\epsilon \equiv A/lc$	$m^2\ mol^{-1}$	$cm^2\ mol^{-1}$

TABLE A-3 Physicochemical Quantities, with their Symbols, Definitions, and Units (Continued)

A-3-8 Electromagnetic Radiation

Quantity	Symbol	Definition	SI Unit	Customary Multiple
Quantum yield	Φ	Section 10.6	*	*
Turbidity	τ	$I = I_0 e^{-\tau l}$	m^{-1}	cm^{-1}

Electrical Units

The units used in electricity and magnetism require some special discussion. The force of attraction or repulsion between charged bodies was first used by the French engineer Charles Coulomb (1736–1806) as the basis for the definition of the unit of electric charge. He defined the electrostatic unit (esu) of charge as

> **that point charge which, when placed 1 cm from a similar point charge in a vacuum, is repelled by a force of one dyne.**

The force is proportional to the product of the charges and inversely proportional to the distance of separation. Thus if two charges Q are separated by a distance r, Coulomb's definition can be expressed by the equation

$$F/\text{dyn} = \frac{(Q/\text{esu})^2}{(r/\text{cm})^2} \tag{A-1}$$

One dyne is the force that gives an acceleration of 1 cm s^{-2} to a mass of 1 gram, and is therefore equal to 10^{-5} newton (N); the newton is the force that produces an acceleration of 1 m s^{-2} in a mass of 1 kg. Thus

$$F/\text{N} = 10^{-5} F/\text{dyn} \tag{A-2}$$

The SI unit of charge is the coulomb (C) and the SI unit of distance is the metre. It can be shown that 1 coulomb represents the same physical situation as 1 esu unit of charge multiplied by 2.998×10^9, which is the speed of light divided by 1 dm s^{-1}. Thus 1 C = 2.998×10^9 esu and, therefore,

$$2.998 \times 10^9\ Q/\text{C} = Q/\text{esu} \tag{A-3}$$

Introduction of Eqs. A-2 and A-3 into Eq. A-1 gives

$$10^5\ F/\text{N} = \frac{(2.998 \times 10^9\ Q/\text{C})^2}{(100\ r/\text{m})^2} \tag{A-4}$$

and therefore

$$F/\text{N} = \frac{8.99 \times 10^9 (Q/\text{C})^2}{(r/\text{m})^2} \tag{A-5}$$

It has proved convenient to write this equation as

$$F = \frac{Q^2}{4\pi\epsilon_0 r^2} \tag{A-6}$$

where ϵ_0, known as the *permittivity of a vacuum*, is equal to $1/4\pi \times 8.99 \times 10^9$ C^2 N^{-1} m^{-2}(≡F m^{-1}):

$$\epsilon_0 = 8.854 \times 10^{-12}\ \mathrm{C^2\ N^{-1}\ m^{-2}} \quad (\equiv \mathrm{F\ m^{-1}})$$

The reason for the introduction of 4π is that certain equations, such as the Poisson equation, are simplified when this is done.

In a medium other than a vacuum, ϵ_0 in all the equations is replaced by $\epsilon_0 \epsilon$, where ϵ is known as the *relative permittivity* or more usually as the *dielectric constant*. Thus the force between two charges Q_1 and Q_2 is

$$F = \frac{Q_1 Q_2}{4\pi\epsilon_0 \epsilon r^2} \tag{A-7}$$

and the corresponding potential energy of interaction is

$$E_p = \frac{Q_1 Q_2}{4\pi\epsilon_0 \epsilon r} \tag{A-8}$$

There is an important fundamental difference between the electrostatic units and the SI units for electricity and magnetism. The former are *three-dimensional units*, by which is meant that in the electrostatic system all electrical and magnetic quantities can be expressed in terms of not more than *three* base units. In SI, however, which is based on the electromagnetic system, some quantities have to be expressed in terms of four base units, namely metre, kilogram, second, and ampere. The SI units of electricity and magnetism are therefore said to be *four dimensional*. An example is the ohm, which in terms of base SI units is kg m^2 s^{-3} A^{-2}. Because of this dimensional difference there is not in general a one-to-one correspondence between SI and electrostatic units.

Magnetic Units

As far as magnetic field is concerned, the property that is of most fundamental importance (e.g., from the standpoint of the effects of magnetic fields on magnetic moments arising from electron and nuclear spin) is the *magnetic flux density*, or *magnetic induction*, which is given the symbol B. The SI unit of this quantity is the tesla (T), which is kg s^{-2} A^{-1} or V s m^{-2}. The old unit is the gauss (G), which is 10^{-4} T.

The *magnetic flux* Φ is the magnetic flux density multiplied by the area through which the magnetic field is passing, and its SI unit is therefore

$$\mathrm{T\ m^2 \equiv kg\ m^2\ s^{-1}\ A^{-1} \equiv V\ s}$$

This unit is given a special name, the weber (Wb). The old unit of magnetic flux is the maxwell (≡ gauss × cm^2), equal to 10^{-8} Wb.

Another magnetic quantity that is frequently encountered is the *magnetic field strength* H. Its SI unit is A m^{-1}. The magnetic field strength is related to the magnetic flux density by the equation

$$H = \frac{B}{\mu} \tag{A-9}$$

where μ is the *permeability* of the medium through which the field is passing. The SI unit of permeability is

$$\frac{\text{kg s}^{-2}\ \text{A}^{-1}}{\text{A m}^{-1}} \equiv \text{kg m s}^{-2}\ \text{A}^{-2}$$

A special unit, the henry, is the unit of *inductance* and it is defined as kg $\text{m}^2\ \text{s}^{-2}\ \text{A}^{-2}$; it is given the symbol H. The unit of permeability can therefore be written as H m^{-1}. If the field is passing through a vacuum, the permeability is given the symbol μ_0. The *permeability of a vacuum* has the value of $4\pi \times 10^{-7}\ \text{H m}^{-1}$ exactly.

B Physical Constants

Quantity	Symbol	Value
Avogadro constant	L	6.022×10^{23} mol^{-1}
Elementary charge	e	1.602×10^{-19} C
Gas constant	R	8.314 J K^{-1} mol^{-1}
Boltzmann constant	$\mathbf{k} \equiv R/L$	1.381×10^{-23} J K^{-1}
Planck constant	h	6.626×10^{-34} J s
Speed of light in vacuum	c	2.998×10^{8} m s^{-1}
Zero of the Celsius scale	T_0	273.15 K exactly
Faraday constant	$F \equiv Le$	96 485 C mol^{-1}
Permittivity of vacuum	$\epsilon_0 \equiv 1/\mu_0 c^2$	8.854×10^{-12} F m^{-1}
Permeability of vacuum	μ_0	$4\pi \times 10^{-7}$ H m^{-1} exactly
Rest mass of electron	m_e	9.110×10^{-31} kg
Rest mass of proton	m_p	1.673×10^{-27} kg
Rest mass of neutron	m_n	1.675×10^{-27} kg
Rydberg constant	R	1.097×10^{7} m^{-1}
Hartree energy	E_h	4.360×10^{-18} J
Bohr radius	a_0	5.292×10^{-11} m
Bohr magneton	μ_B	9.274×10^{-24} J T^{-1}
Nuclear magneton	μ_N	5.051×10^{-27} J T^{-1}
Electron magnetic moment	μ_0	9.285×10^{-24} J T^{-1}
Landé g factor for free electron	g_e	2.0023
Proton gyromagnetic ratio	γ_P	2.675×10^{8} s^{-1} T^{-1}
Normal atmospheric pressure	P_0	$1.013\ 25 \times 10^{5}$ Pa exactly
Standard molar volume of ideal gas	V_0	2.2414×10^{-2} m^{3} mol^{-1}
Standard acceleration of free fall	g_n	9.806 65 m s^{-2} exactly
Ratio of circumference to diameter of circle	π	3.1416
Base of natural logarithms	e	2.7183

C Some Mathematical Relationships

In this appendix we list, without proof, some mathematical relationships that are used in this book.*

Differentials

$$\frac{d}{dx}(yz) = y\frac{dz}{dx} + z\frac{dy}{dx} \qquad \frac{d}{dx}\left(\frac{y}{z}\right) = \frac{z\,(dy/dx) - y\,(dz/dx)}{z^2}$$

$$\frac{df(y)}{dx} = \frac{df(y)}{dy} \cdot \frac{dy}{dx}$$

$$\frac{de^x}{dx} = e^x \qquad \frac{de^y}{dx} = e^y\frac{dy}{dx}$$

$$\frac{d\ln x}{dx} = \frac{1}{x} \qquad \frac{d\ln y}{dx} = \frac{1}{y} \cdot \frac{dy}{dx}$$

Integrals

If an integration is performed, and no additional information is provided, the result involves a constant of integration I. For example,

$$\int x^2\,dx = \frac{x^3}{3} + I \tag{C-1}$$

The constant of integration can be evaluated if the value of the integral is known for a particular value of the variable x. An integral of the kind shown in Eq. C-1 is known as an *indefinite integral.*

Alternatively, the integral may be a *definite integral,* which means that the integration is performed between certain limiting values of the variable. For example, the definite integral

$$\int_b^a x^2\,dx$$

means that x can have an upper limit of a and a lower limit of b. The solution is

$$\int_b^a x^2\,dx = \left.\frac{x^3}{3}\right|_b^a = \frac{a^3 - b^3}{3} \tag{C - 2}$$

*There are many sources of additional mathematical information; for example, CRC *Handbook of Chemistry and Physics,* (62nd ed.) CRC Press, Boca Raton, Fla., 1981.

Some useful integrals are listed here:

$$\int x^n\, dx = \frac{x^{n+1}}{n+1} + I \quad \text{if } n \neq -1 \qquad \int \frac{dx}{x} = \ln x + I$$

$$\int e^{ax}\, dx = \frac{1}{a} e^{ax} + I \qquad \int \sin x\, dx = -\cos x + I$$

$$\int \cos x\, dx = \sin x + I \qquad \int x\, dy = xy - \int y\, dx \quad \text{(integration by parts)}$$

$$\int \ln ax\, dx = x \ln ax - x + I$$

Definite integrals useful for statistical problems are to be found in the appendix to Chapter 14 (p. 685).

Integration of Differential Equations

In all the differential equations in this book, the variables can be separated and the integrations can therefore be performed independently. Suppose, for example, that we have the differential equation

$$a\frac{dx}{x} + by\, dy = 0 \qquad \text{(C-3)}$$

where x and y are variables and a and b are constants. The variables are already separated and integration proceeds as follows:

$$a\int \frac{dx}{x} + b\int y\, dy = 0 \qquad \text{(C-4)}$$

$$a \ln x + \frac{by^2}{2} = I \qquad \text{(C-5)}$$

The constant of integration can be obtained from a known value of y at some known value of x. This is often a *boundary condition* (i.e., the value of one variable is known when the other is zero or infinity or has some other "boundary" value).

For examples of the integration of differential equations, see Eqs. 2.84–2.86 and 9.32–9.35.

Partial Derivatives

If a function z depends on two or more variables, the *partial derivative* relates to the dependence of z on one variable, with all other variables held constant. For example, if $z(x, y)$ is a function of x and y, the partial derivative

$$\left(\frac{\partial z}{\partial x}\right)_y$$

expresses the dependence of z on x when y is held constant.

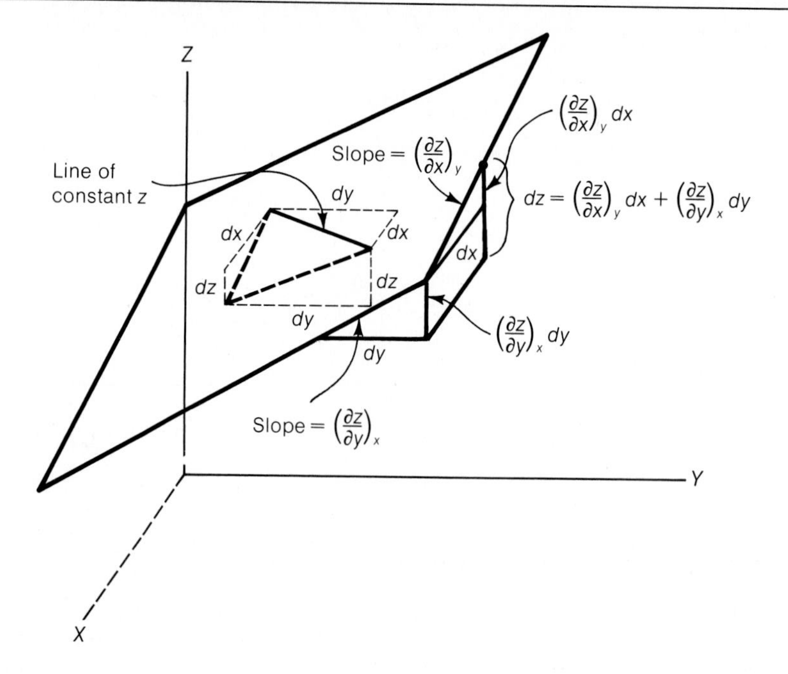

FIGURE C-1

In such a case, the differential dz can be expressed as

$$dz = \left(\frac{\partial z}{\partial x}\right)_y dx + \left(\frac{\partial z}{\partial y}\right)_x dy \tag{C-6}$$

Figure C-1 gives an interpretation of this relationship. Examples of its use are to be found in Eqs. 2.113–2.114, 3.112–115, and 14.16.

According to *Euler's reciprocity theorem,* in the case of exact differentials (e.g., state functions) the order of differentiation does not matter. Thus if z is a function of the two variables x and y,

$$\left[\frac{\partial}{\partial x}\left(\frac{\partial z}{\partial y}\right)_x\right]_y \equiv \left[\frac{\partial}{\partial y}\left(\frac{\partial z}{\partial x}\right)_y\right]_x \tag{C-7}$$

These double differentials may be written more compactly as

$$\frac{\partial^2 z}{\partial x\, \partial y} \equiv \frac{\partial^2 z}{\partial y\, \partial x} \tag{C-8}$$

Examples of the use of this theorem are to be found in Eqs. 3.121–3.125. Another important relationship is that

$$\left(\frac{\partial z}{\partial x}\right)_y = \frac{1}{(\partial x/\partial z)_y} \tag{C-9}$$

If z is a function of x and y, Eq. C-6 applies, and we can hold z constant and divide by dx:

$$0 = \left(\frac{\partial z}{\partial x}\right)_y + \left(\frac{\partial z}{\partial y}\right)_x \left(\frac{\partial y}{\partial x}\right)_z \tag{C-10}$$

Then applying Eq. C-9,

$$\left(\frac{\partial x}{\partial y}\right)_z \left(\frac{\partial y}{\partial z}\right)_x \left(\frac{\partial z}{\partial x}\right)_y = -1 \tag{C-11}$$

or

$$\left(\frac{\partial z}{\partial x}\right)_y = -\left(\frac{\partial z}{\partial y}\right)_x \left(\frac{\partial y}{\partial x}\right)_z \tag{C-12}$$

This is *Euler's chain relationship.* An interpretation of it is also shown in Figure C-1.

Exact and Inexact Differentials

Some functions such as U and H depend only on the state of the system. The integral of the differential of such a state function is the difference between the values at the two limits. The value of the integration is path independent and the differential is said to be *exact.*

If the integral of a differential depends on the path chosen in going from A to B, the differential is said to be *inexact.* The symbol đ is used to indicate an inexact differential.

Proof that a differential is exact is based on Euler's *criterion for exactness.* The total differential dz of z is related to the differentials dx and dy by the equation

$$dz = M(x, y)\, dx + N(x, y)\, dy \tag{C-13}$$

where M and N are functions of the variables x and y. If z has a definite value at each point in the X-Y plane, it also must be a function of x and y. This allows us to write

$$dz = \left(\frac{\partial z}{\partial x}\right)_y dx + \left(\frac{\partial z}{\partial y}\right)_x dy \tag{C-14}$$

which is the same as Eq. C-6. Comparing Eqs. C-13 and C-14 shows that

$$M(x, y) = \left(\frac{\partial z}{\partial x}\right)_y; \qquad N(x, y) = \left(\frac{\partial z}{\partial y}\right)_x \tag{C-15}$$

Since mixed second derivatives are equal from Eq. C-7, then

$$\left(\frac{\partial M}{\partial y}\right)_x = \left(\frac{\partial N}{\partial x}\right)_y \tag{C-16}$$

This equation must be satisfied if dz is an exact differential. Thus, $dz = y\, dx$ is not an exact differential, but the sum of two inexact differentials might be. For example, $dz = y\, dx + x\, dy$ is an exact differential.

Answers to Problems

Chapter 1

1. 298 kJ; 476 kJ
2. $t = 885°C$
3. 4.13×10^{-23} J
4. 92.6 W
5. 1 atm = 101.325 0144 kPa
6. 0.47 dm^3
7. 0.5 dm^3
8. **(a)** $\frac{n}{V} = 0.0409$ mol dm^{-3}; 2.463×10^{22} molecules dm^{-3} **(b)** 4.03×10^{-11} mol dm^{-3}; 2.43×10^{13} molecules dm^{-3}
9. **(a)** 0.225 dm^3 **(b)** 107.6 g mol^{-1}
10. 31.7 g mol^{-1}
11. 28.4 g mol^{-1}
12. 96.82 kPa
13. 1020 m^3
14. 1240 g (Ar), $P_{total} = 20$ atm
15. 148 g mol^{-1}
16. 2400 J
17. **(a)** $n = 0.1226$ mol **(b)** Number of molecules $= 7.385 \times 10^{22}$ **(c)** $\sqrt{\overline{u^2}} = 515.2$ m s^{-1} **(d)** $\bar{\epsilon}_k = 6.174 \times 10^{-21}$ J **(e)** E_k(total) = 455.8 J
18. 1.15
19. $\lambda = 6.54 \times 10^{-8}$ m
    $Z_A = 7.26 \times 10^9$ s^{-1}
    $Z_{AA} = 8.9 \times 10^{34}$ m^{-3} s^{-1}
20. $\lambda = \dfrac{RT}{\sqrt{2}\pi d^2 LP}$
21. **(a)** 1.044×10^{-4}m **(b)** 1.37×10^{-7} m **(c)** 1.39×10^{-10} m
22. 3.60×10^{18} m
23. Curves similar to Figure 1.10
24. Ideal gas: 6.91 atm
    Van der Waals gas: 6.41 atm
25. $T_r = 1.15$, $P_r = 0.75$, $Z = 0.815$, $V = 0.44$ dm^3
26. $a = 2\, V_c\, RT_c$, $b = \dfrac{V_c}{2}$
27. $P \rightarrow$zero, $Z = 1$
28. $a = 0.5563$ Pa m^6 mol^{-2}; $b = 0.0638 \times 10^{-3}$ m^3 mol^{-1}
29. Ideal gas: 8.91 atm
    Van der Waals gas: 8.56 atm
    Dieterici: 8.45 atm
    Beattie Bridgeman: 8.50 atm
30. $b = 0.162 \times 10^{-3}$ m^3 mol^{-1}
31. Second virial coefficient: $-\dfrac{a}{V_m - b}$
    Third virial coefficient: $\dfrac{a}{2RT(V_m - b)}$
32. 49.7°C
33. **(a)** $\ln \dfrac{P}{P_0} = -\dfrac{M}{RT}(9.807z - 5 \times 10^{-6}z^2)$
    (b) $P = 2.73 \times 10^{-5}$ atm

34. $L = 7.44 \times 10^{23}$ mol^{-1}
35. **(a)** 1 mmHg = 1.000 000 14 torr **(b)** 3.24×10^7 particles/m^3

Chapter 2

1. 1.404 kJ
2. 10.07 kJ
3. 6.025 kJ mol^{-1}; 0.165 J mol^{-1}
4. 37.57 kJ mol^{-1}; 3.06 kJ mol^{-1}
5. −5.65 kJ mol^{-1}
6. **(a)** −972.7 kJ mol^{-1} **(b)** −975.2 kJ mol^{-1}
7. **(a)** −3274.9 kJ mol^{-1} **(b)** −3278.6 kJ mol^{-1}
8. **(a)** 72.7°C **(b)** 4.34 kg
9. 3.79 kJ; zero
10. 0.00608 J
11. 75.4 J K^{-1} mol^{-1}
12. 314 kJ; 314 s
13. 64.9 kJ mol^{-1}
14. **(a)** −727.7 kJ mol^{-1} **(b)** −237.5 kJ mol^{-1} **(c)** −202.2 kJ mol^{-1}
15. −1559.9 kJ mol^{-1}
17. 53.1 kJ mol^{-1}
18. −44.1 kJ mol^{-1}
19. **(a)** −203.4 kJ mol^{-1} **(b)** −291.2 kJ mol^{-1}
20. −492.7 kJ mol^{-1}
21. −2.05 kJ mol^{-1}
22. −908.8 kJ mol^{-1}
23. 4.88 kJ mol^{-1}
24. −395.0 kJ mol^{-1}
25. −105.9 kJ mol^{-1}
26. **(a)** 4 atm **(b)** 4.54 kJ **(c)** 21.1 J K^{-1}
27. **(a)** Zero **(b)** 4.22 kJ **(c)** 4.22 kJ **(d)** 5.47 atm = 554.2 kPa **(e)** 6.21 kJ **(f)** 5.89 kJ
28. **(a)** 15.3 dm^3 **(b)** −1.66 kJ **(c)** 5.88 kJ **(d)** 5.88 kJ **(e)** 4.22 kJ
29. **(a)** Zero **(b)** 8 atm = 810.6 kPa **(c)** 3.15 kJ **(d)** 3.15 kJ **(e)** Zero
30. **(a)** 3633 J **(b)** Zero **(c)** 3633 J
31. **(a)** 7.31 kJ **(b)** Zero **(c)** 7.31 kJ
32. **(a)** 0.552 atm; 182.9 K **(b)** −434 J; −610 J
33. **(a)** 2.898 dm^3, 16.81 dm^3 **(b)** 204.8 K **(c)** −398 J K^{-1}; −521 J K^{-1}
34. **(a)** $C_{P,m}$/J K^{-1} mol^{-1} = 29.84 + 8.2 × 10^{-3} (T/K) **(b)** Zero
37. 173.5 K
38. −1.5 kJ mol^{-1}; −2.1 kJ mol^{-1}
39. **(a)** 0.62 atm; 372.2 K **(b)** 900 J mol^{-1}; −100 J mol^{-1}
40. **(a)** 9.76 kJ mol^{-1} **(b)** 10.16 kJ mol^{-1}
41. **(a)** 1.15 kJ mol^{-1} **(b)** 1.21 kJ mol^{-1}
46. 4760 K
47. 221 K; −3030 J; −11 200 J

Chapter 3

1. **(a)** 80% **(b)** 30 kJ **(c)** 150 J K^{-1} **(d)** 150 J K^{-1} **(e)** Zero **(f)** Zero **(g)** −150 kJ **(h)** 160 kJ
2. 7010 J
3. **(b)** 40 kJ; −30 kJ; 400 K
4. 87.3 J K^{-1} mol^{-1}; 88.0 J K^{-1} mol^{-1}; 108.8 J K^{-1} mol^{-1}; 109.7 J K^{-1} mol^{-1}
5. **(a)** −242.4 J K^{-1} mol^{-1} **(b)** −456.2 J K^{-1} mol^{-1}
6. **(a)** 3.52 J K^{-1} **(b)** 2.11 J K^{-1}
7. 21.9 J K^{-1}
8. 26.0 J K^{-1}
9. 4.61 J K^{-1} mol^{-1}
10. −344.7 J K^{-1} mol^{-1}
11. **(a)** 19.1 J K^{-1} mol^{-1}; −19.1 J K^{-1} mol^{-1} **(b)** 19.1 J K^{-1} mol^{-1}; zero; net ΔS = 19.1 J K^{-1} mol^{-1}
12. **(a)** + **(b)** + **(c)** − **(d)** +
13. $n \int_{T_1}^{T_2} \frac{C_{P,m}}{T} dT = nC_{P,m} \ln \frac{T_2}{T_1}$ if $C_{P,m}$ is constant.
14. 42.4 J K^{-1}
15. **(a)** −6.28 J K^{-1} **(b)** 7.10 J K^{-1} **(c)** 0.82 J K^{-1}
16. 1.66 J K^{-1}; zero
17. 36.8 J K^{-1} mol^{-1}
18. Irreversible; 0.4
19. −237.6 kJ mol^{-1}
20. 3060 J mol^{-1}; ΔU = 37.5 kJ mol^{-1}; ΔG = 0; ΔS = 108.8 J K^{-1} mol^{-1}
21. Zero; −213 J mol^{-1}
22. Zero; zero; 19.14 J K^{-1}; −5.7 kJ mol^{-1}; −5.7 kJ mol^{-1}
23. **(a)** −34.14 kJ **(b)** 16.92 kJ; 85.00 kJ; 500.6 K
24. 214.5 J K^{-1}; 44.0 kJ mol^{-1}; −19.9 kJ mol^{-1}
25. **(a)** ΔU, ΔH **(b)** ΔS **(c)** ΔH **(d)** ΔH **(e)** None **(f)** ΔG **(g)** ΔU **(h)** None
26. 1.50 kJ mol^{-1}
27. −3.89 kJ mol^{-1}
28. 500 J mol^{-1}; zero; zero; −1.73 kJ mol^{-1}; 5.76 J K^{-1} mol^{-1}; q_{rev} = −1.73 kJ mol^{-1}; w_{rev} = 1.73 kJ mol^{-1}
29. Zero; 30 kJ mol^{-1}; 33.1 kJ mol^{-1}; 114.6 J K^{-1} mol^{-1}; −9.66 kJ mol^{-1}
30. −40.6 kJ mol^{-1}; −103 J K^{-1} mol^{-1}; −2.15 kJ mol^{-1}
31. −1.12 kJ mol^{-1}; −1.50 kJ mol^{-1}; 2.22 J K^{-1} mol^{-1}
32. 48.2%; 28.7%
33. 37.1 W
34. **(a)** 59.6% **(b)** 11.9% **(c)** 6.6%

35. 2.03×10^5 kJ
36. 22.6
37. 78 kJ; 412 kJ

Chapter 4

1. 2.7 mol dm^{-3}
2. 42 mol
3. 26 mol
4. 2.45 atm
5. 3.77×10^{-5} mol dm^{-3}; 2.78×10^{-3} atm
7. 0.5; 0.0321 mol dm^{-3}; 0.79 atm; 0.79; none
8. 0.165 mol dm^{-3}; 4.04 atm; 11.5
9. 7.94×10^{-9} atm; 2.59×10^{-10} mol dm^{-3}; 3.97×10^{-9}
10. 1.5; 4
11. 0.063 mol
12. **(a)** 14.42 kJ mol^{-1} **(b)** 0.01 mol dm^{-3}; 0.045 mol dm^{-3}
13. 60.4 kJ mol^{-1}
14. **(a)** 0.321 **(b)** 408 **(c)** -12.6 kJ mol^{-1}
15. **(a)** 2.04×10^9 atm^{-2} **(b)** 9.2×10^{40} atm^{-2} **(c)** 5.0×10^{17} atm^{-1}; 9.5×10^{-13}
16. **(a)** 28.5 kJ mol^{-1}; 16.1 kJ mol^{-1} **(b)** 1.994 mol; 1.994 mol; 6.30×10^{-3} mol; 6.30×10^{-3} mol
17. 2.27×10^{-3} dm^3 mol^{-1}; 1.66×10^5 mol dm^{-3}; 377
18. **(a)** Zero **(b)** Positive **(c)** 0.0490 mol dm^{-3}; 7.92 kJ mol^{-1} **(d)** >1 atm **(e)** Negative
19. **(a)** Yes **(b)** Yes **(c)** Positive
20. **(a)** -101.01 kJ mol^{-1}; -136.94 kJ mol^{-1}; -120.5 J K^{-1} mol^{-1}; 1 atm **(b)** 4.98×10^{17} atm^{-1} **(c)** 1.22×10^{19} dm^3 mol^{-1} **(d)** 108.94 kJ mol^{-1} **(e)** -93.9 J K^{-1} mol^{-1} **(f)** 7.51×10^{12} atm^{-1}
21. **(a)** -457.1 kJ mol^{-1}; -483.7 kJ mol^{-1}; -89.0 J K^{-1} mol^{-1} **(b)** 1.24×10^{80} atm^{-1} **(c)** -281.3 kJ mol^{-1}; 2.92×10^6 atm^{-1}
22. **(a)** 35.1 J K^{-1} mol^{-1} **(b)** 1.66×10^5 mol dm^{-3} **(c)** -30.6 kJ mol^{-1}; 2.27×10^5 mol dm^{-3}
23. **(a)** -6.84 kJ mol^{-1} **(b)** 0.060 atm; 0.940 atm
24. **(a)** 56.8 kJ mol^{-1} **(b)** -56.8 kJ mol^{-1}
25. 55.9 kJ mol^{-1}
26. **(a)** 2.26 atm^{-1}; -4.6 kJ mol^{-1}; 92.8 kJ mol^{-1}; 145 J K^{-1} mol^{-1} **(b)** 0.041 mol dm^{-3}; 17.9 kJ mol^{-1} **(c)** 0.065 mol dm^{-3}; 0.070 mol dm^{-3}; 0.065 mol dm^{-3}; 0.035 mol dm^{-3}
27. -173.12 kJ mol^{-1}; -29.21 kJ mol^{-1}; -482.7 J K^{-1} mol^{-1}; 1.31×10^5 atm^{-3}
28. -102.8 kJ mol^{-1}; -74.1 kJ mol^{-1}; -98.3 J K^{-1} mol^{-1}
29. **(a)** -7.3 kJ mol^{-1} **(b)** -18.7 kJ mol^{-1}; from left to right
30. **(a)** 41.13 kJ mol^{-1}; 28.62 kJ mol^{-1}; 41.96 J K^{-1} mol^{-1} **(b)** 9.64×10^{-6} **(c)** $\Delta H°(T)/\text{J} = 42\,200 - 12.55(T/\text{K}) + 1.17 \times 10^{-3}(T/\text{K})^2 + \dfrac{76.6 \times 10^4}{T/\text{K}}$ **(d)** $\ln K_p = 14.03 - \dfrac{5076}{T/\text{K}} - 1.51 \ln(T/\text{K}) - 1.41 \times 10^{-4}(T/\text{K}) - \dfrac{4.61}{(T/\text{K})^2}$ **(e)** 0.20
31. 1.37×10^{-12} atm; 7.81×10^{-12} atm; 5.23×10^{-11} atm; 609 kJ mol^{-1}; 316.8 kJ mol^{-1}; 210 J K^{-1} mol^{-1}
32. **(a)** 0.0638; 0.202; 0.415 **(b)** 1.36×10^{-4} mol dm^{-3}; 16.0×10^{-4} mol dm^{-3}; 92.6×10^{-4} mol dm^{-3} **(c)** 1.216 kPa; 17.0 kPa; 98.0 kPa **(d)** 149 kJ mol^{-1}; 138 kJ mol^{-1} **(e)** 18.9 kJ mol^{-1}; 102 J K^{-1} mol^{-1}
37. 3.98 kJ mol^{-1}, 0.22; 2.20 kJ mol^{-1}, 0.43; 0.42 kJ mol^{-1}, 0.85; -1.36 kJ mol^{-1}, 1.67; -3.14 kJ mol^{-1}, 3.24; -4.93 kJ mol^{-1}, 6.27; 44.4°C

Chapter 5

1. $\Delta_{vap}H_m = 42.67$ kJ mol^{-1} versus 40.57 kJ mol^{-1} (CRC Handbook)
2. 984 Pa
3. Second order; discontinuous
4. 46.8 kJ mol^{-1}
5. $\Delta_{vap}H_m = 55.06$ kJ mol^{-1}; difference is result of nonideal behavior.
6. $\Delta_{vap}H = 30.105$ kJ mol^{-1}; this is a 5.7% error.
7. 328.49 K
8. 373.11 K
9. $\ln \dfrac{P_2}{P_1} = \dfrac{\Delta_{vap}H_m}{M} \ln\left[\left(\dfrac{T_2}{T_1}\right)\left(\dfrac{RT_1 + M}{RT_2 + M}\right)\right]$
10. $P_T = 6.01$ kPa; 0.357
11. $x_2 = \dfrac{m_2M_1}{1 + m_2M_1} = \dfrac{(m_2/\text{mol kg}^{-1})(M_1/\text{g mol}^{-1})}{1000 + (m_2/\text{mol kg}^{-1})(M_1/\text{g mol}^{-1})}$
12. $x_2 = \dfrac{c_2M_1}{\rho + c_2(M_1 - M_2)} = \dfrac{(c_2/\text{mol dm}^{-3})(M_1/\text{g mol}^{-1})}{(1000\,\rho/\text{g cm}^{-3}) + (c_2/\text{mol dm}^{-3})[(M_1 - M_2)/\text{g mol}^{-1}]}$
13. $c_2 = \dfrac{m_2\rho}{(1 + m_2M_2)}$; $c_2/\text{mol dm}^{-3} = \dfrac{1000(m_2/\text{mol kg}^{-1})(\rho/\text{g cm}^{-3})}{1000 + (m_2/\text{mol kg}^{-1})(M_2/\text{g mol}^{-1})}$
14. $V^*_{H_2O} = 18.068 - 0.015\,744\ m^2 + 0.001\,7016\ m^3$
15. $V_2 = \dfrac{M_2}{\rho} - (M_1x_1 + M_2x_2)\dfrac{x_1}{\rho^2}\dfrac{d\rho}{dx_2}$
16. $V_2 = 0.037\,57$ dm^3 mol^{-1}

17. P_{total} = 19.41 kPa; $x_{prop\ dib}$ = 0.527; $x_{ethyl\ dib}$ = 0.473
18. **(a)** $x(N_2) = 1.06 \times 10^{-5}$; $x(O_2) = 5.15 \times 10^{-6}$ **(b)** $c(N_2) = 5.88 \times 10^{-4}$ *M*; $c(O_2) = 2.86 \times 10^{-4}$ *M*
19. 182.18 g mol^{-1}
20. 0.160 kg mol^{-1} = 160 g mol^{-1}
21. Ratio $= \dfrac{M_2}{M_1} = 1.80$
22. 97.3 mol % pure
23. $a_1 = 0.69$, $a_2 = 0.34$; $f_1 = 1.03$, $f_2 = 1.03$
24. $x_1 = 0.966$; $a_1 = 0.941$; $f_1 = 0.974$
25. 0 to −1.73 kJ mol^{-1}
26. $f_{H_2O} = 0.980$
27. −0.003 66 K
28. π = 249.4 kPa
29. $a_1 = 0.9690$; $f_1 = 0.995$
30. 39.4 g mol^{-1}
31. 94% dissociated
32. 310 g mol^{-1}

Chapter 6

1. Number of degrees of freedom equals 2.
2. **(a)** 2 degrees of freedom **(b)** 3 degrees of freedom **(c)** 2 degrees of freedom
3. 2 components
4. 2 components, 1 if only $CaCO_3$ is present.
5.

T/K	*P*/kPa	Phases Present
200	100	A, B, gas
300	300	A, B, liquid
400	400	B, liquid, gas

6. **(a)** 60°C **(b)** Vapor is 88% B. **(c)** 53% B in still **(d)** Mass of A is 21% or 23.4 g; that of B is 79% or 88.1 g.
7. 1.2
8. 112.4 compared to 112.6
9. $\dfrac{w_A}{w_B} = \dfrac{n_A M_A}{n_B M_B} = \dfrac{P_A^* M_A}{P_B^* M_B}$
10. 5.51 kg
11. 1.93 to 1 or 65.9 wt% chlorobenzene
12. $y_{IAA} = 0.172$; $y_{IBA} = 0.828$
13. As stated in problem
14. $\alpha = 2.126 \times 10^{-4}\ K^{-1}$
15. 38.1 atm
16. 42% solid and 58% liquid; composition of liquid is $0.31x_{Si}$.
17. Curve is similar in appearance to Figure 6.16.
18. **(a)** Two salts are first in equilibrium with liquid of composition *b*; then solid salt A disappears. **(b)** Two solid salts are in equilibrium with solution of composition *b*. **(c)** Always a one-phase system.
19. **(a)** Peritectic point **(b)** Eutectic point **(c)** Melting point **(d)** Incongruent melting **(e)** Phase transition
20. Stable compound: $Fe_2O_3 \cdot Y_2O_3$; unstable compound: $3Y_2O_3 \cdot 5Fe_2O_3$
21. Peritectic-type diagram
22. Simple compound (Al_2Se_3) phase diagram
23. Peritectic and simple compound formation indicated
24. Decreasing step diagram indicated for pressures
25. A: 95.5% toluene, 4% acetic acid, and 0.5% H_2O
    B: 1% toluene, 37% acetic acid, and 62% H_2O
    The ratio is 4B:1A.
26. 4 composition triangles indicated
27. 2 stable triple points, a metastable triple point *W*, *L*, *V*
28. 2 compounds and one peritectic reaction
29. A peritectic reaction coupled with a simple eutectic
30. **(a)** K_2CO_3 in equilibrium with water-rich A*Ea* or alcohol-rich A*c*B saturated solutions. Region A*ac*: K_2CO_3 in equilibrium with conjugate liquids *a* and *c*. Region *abc*: two conjugate liquids

Chapter 7

1. 33.8 mA
2. 2.38 mA
3. 1.51×10^{-3} mol dm^{-3}
4. 9.12 μM
5. 4.07×10^{-6} mol^2 dm^{-6}
6. 1.87×10^{-5} mol dm^{-3}
7. **(a)** 30.5 nm **(b)** 0.673 nm
8. 129.9 Ω^{-1} cm^2 mol^{-1}
9. 0.0403
10. **(a)** 0.317; 0.683 **(b)** $\lambda^{\circ}_{Li^+} = 36.4\ \Omega^{-1}\ cm^2\ mol^{-1}$; 78.6 Ω^{-1} cm^2 mol^{-1}; 3.77×10^{-4} cm^2 V^{-1} s^{-1}; 8.15 cm^2 V^{-1} s^{-1}
11. 0.442; 0.558
12. 3.63×10^{-3} cm^2 V^{-1} s^{-1}; 7.90×10^{-4} cm^2 V^{-1} s^{-1}
13. 5.19×10^{-2} cm s^{-1}; 7.92×10^{-2} cm s^{-1}
14. 0.070 cm s^{-1}; 0.138 cm s^{-1}
15. **(a)** 2.30×10^{-19} J **(b)** 2.30×10^{-25} J **(c)** 2.30×10^{-19} J
16. −407.1 kJ mol^{-1}; −877.8 kJ mol^{-1}; −394.1 kJ mol^{-1}
17. −1051.4 kJ mol^{-1}; −372.3 kJ mol^{-1}; −1828.7 kJ mol^{-1}; −4500.6 kJ mol^{-1}; −355.7 kJ mol^{-1}; −341.9 kJ mol^{-1}

18. 0.1 M; 0.3 M; 0.4 M; 0.3 M; 1.0 M
19. 0.026
20. 55.9 kJ mol^{-1}; 1.46×10^{-5} M
22. 123.8 kJ mol^{-1}
23. 0.0147 g dm^{-3}; 2.71×10^{11} mol^3 dm^{-9}
24. Palmitate side: $[Na^+] = 0.18\ M$; $[Cl^-] = 0.08\ M$; other side: $[Na^+] = [Cl^-] = 0.12\ M$
25. 0.025 M; 0.033 M; 0.0067 M; 0.0167 M
26. **(a)** 1.14×10^{-3} mol dm^{-3} **(b)** 8.08×10^{-5} mol dm^{-3}
27. 0.090 mol dm^{-3}
28. 30 Ω^{-1} cm^2 mol^{-1}; 4.0×10^{-5} mol dm^{-3}
29. 0.358 nm; 4.96 kJ mol^{-1}
30. 31 J K^{-1} mol^{-1}
31. **(a)** 14.3 kJ mol^{-1} **(b)** 12.5 kJ mol^{-1}

Chapter 8

1. **(a)** $H_2 \rightarrow 2H^+ + 2e^-$; $Cl_2 + 2e^- \rightarrow 2Cl^-$
   $H_2 + Cl_2 \rightarrow 2H^+ + 2Cl^-$;
   $E = E^\circ - \frac{RT}{2F} \ln (a^2_{H+} a^2_{Cl-})^u$
   (b) $2Hg + 2Cl^- \rightarrow Hg_2Cl_2 + 2e^-$;
   $2H^+ + 2e^- \rightarrow H^+$
   $2Hg + 2H^+ + 2Cl^- \rightarrow Hg_2Cl_2 + H_2$;
   $E = E^\circ + \frac{RT}{2F} \ln (a_{H+}\, a_{Cl-})^u$
   (c) $Ag + Cl^- \rightarrow AgCl + e^-$;
   $2e^- + Hg_2Cl_2 \rightarrow 2Hg + 2Cl^-$
   $2Ag + Hg_2Cl_2 \rightarrow 2AgCl + Hg$; $E = E^\circ$ (no concentration dependence)
   (d) $\frac{1}{2}H_2 \rightarrow H^+ + e^-$; $AgI(s) + e^- \rightarrow Au + I^-$
   $AuI(s) + \frac{1}{2}H_2 \rightarrow Au + H^+ + I^-$;
   $E = E^\circ - \frac{RT}{F} \ln (a_{H+}\, a_{I-})^u$
   (e) $Ag + Cl^-(a_1) \rightarrow AgCl + e^-$;
   $AgCl + e^- \rightarrow Ag + Cl^-(a_2)$
   $Cl^-(a_1) \rightarrow Cl^-(a_2)$; $E = \frac{RT}{F} \ln \frac{a_1}{a_2}$
2. **(a)** AH_2 is oxidized by B. **(b)** 0.44 V **(c)** None
3. 0.905 V
4. 9.15×10^7 dm^6 mol^{-2}
5. 5.2×10^{-11}
6. −24.1 kJ mol^{-1}
7. −237.4 kJ mol^{-1}
8. 1.66×10^6; half will form Cu^{2+} and half Cu.
9. 0.0178 V
10. −0.157 V
11. **(a)** −0.036 V **(b)** −0.162 V
12. 0.462 mol dm^{-3}
13. 10.4 mV
14. −11.8 mV
15. **(a)** 2.26×10^{-14} C **(b)** 1.5×10^{-6}
16. **(a)** −0.152 V **(b)** −0.147 V
17. $Cu \rightarrow Cu^{2+} + 2e^-$;
    $AgCl(s) + e^- \rightarrow Ag(s) + Cl^-$
    $2AgCl(s) + Cu \rightarrow 2Ag(s) + Cu^{2+} + 2Cl^-$
    $y_\pm = 0.50$
18. **(a)** $2Tl + 2Cl^-(0.02\ m) \rightarrow 2TlCl(s) + 2e^-$
    $Cd^{2+}(0.01\ m) + 2e^- \rightarrow Cd$
    $2Tl + Cd^{2+}(0.01\ m) + 2Cl^-(0.02\ m) \rightarrow 2TlCl(s) + Cd$ **(b)** −0.054 V; 0.105 V
19. 0.216 V; −41.7 kJ mol^{-1}; −40.4 kJ mol^{-1}; 4.21 J K^{-1} mol^{-1}
20. −196.5 kJ mol^{-1}; −199.4 kJ mol^{-1}; −9.65 J K^{-1} mol^{-1}
21. **(a)** −603.8 kJ mol^{-1} **(b)** −780.0 kJ mol^{-1}
22. **(a)** 0.58 V **(b)** −119 kJ mol^{-1}
23. 5.01×10^{-13} mol^2 kg^{-2}; 7.08×10^{-7} mol kg^{-1}
24. 2.49
25. 8.45×10^{-17} mol^2 kg^{-2}; 9.19×10^{-9} mol kg^{-1}
26. **(a)** 2750 kJ mol^{-1} **(b)** 2740 kJ mol^{-1}
27. Zero
28. −0.118 V
29. $E = \frac{2t_- RT}{F} \ln \frac{m_2}{m_1}$; 0.839; 0.161
30. 8.2×10^{-8} mol^2 kg^{-2}

Chapter 9

1. **(a)** 3 **(b)** Both rates are 3.6×10^{-3} mol dm^{-3} s^{-1} **(c)** None **(d)** Decreased by a factor of 8; none.
2. $\alpha = 1$, $\beta = 1$; 6.21×10^{-4} dm^3 mol^{-1} s^{-1}
3. $\alpha = 2$, $\beta = 1$; 1.64×10^{-3} dm^6 mol^{-2} s^{-1}
4. **(a)** 5.18 hr; 0.669
5. 6.64 times
6. **(a)** 77 ps **(b)** 7.7 ns
7. 0.539 μg; 0.177 μg
8. 4.32×10^{-4} mol dm^{-3} s^{-1}
9. 2
10. 5.61×10^{-7} s^{-1}; **(a)** 61.6% **(b)** 37.9% **(c)** 0.78%
11. 87.0 days; 9.22×10^{-8} s^{-1}; **(a)** 2653 **(b)** 233
12. 290 s
13. 51.2 kJ mol^{-1}
14. 143.0 kJ mol^{-1}
15. **(a)** 6.68×10^3 **(b)** 6.62
16. 49.5 kJ mol^{-1}
17. 120.7 kJ mol^{-1}
18. −1.5
19. 409
20. 138.8 kJ mol^{-1}; −76.7 J K^{-1} mol^{-1}; 190.4 kJ mol^{-1}; 1.02×10^{10} dm^3 mol^{-1} s^{-1}

21. 84.1 kJ mol^{-1}; 81.6 kJ mol^{-1}; 100.7 kJ mol^{-1}; $7.48 \times 10^9\ s^{-1}$; $-64.1\ J\ K^{-1}\ mol^{-1}$
22. 108.8 kJ mol^{-1}; 90.0 kJ mol^{-1}; 87.5 kJ mol^{-1}; $4.94 \times 10^9\ s^{-1}$; $-68.03\ J\ K^{-1}\ mol^{-1}$
23. 348.6 kJ mol^{-1}; 730 J K^{-1} mol^{-1}
24. **(a)** $3.04 \times 10^{31}\ m^{-3}\ s^{-1}$ **(b)** $8.58 \times 10^{-7}\ dm^3\ mol^{-1}\ s^{-1}$; $-62.1\ J\ K^{-1}\ mol^{-1}$
25. 2
26. 1.92 $dm^3\ mol^{-1}\ s^{-1}$; $z_A z_B = -2$
27. $\log_{10} k = \log_{10} k_0 - B(z_A^2 + z_B^2)\sqrt{\frac{I}{\text{mol dm}^{-3}}}$; no
28. $-8.54\ cm^3\ mol^{-1}$
29. $-18.7\ cm^3\ mol^{-1}$
30. $-14.3\ cm^3\ mol^{-1}$
33. $k = \frac{1}{t(2b_0 - a_0)} \ln \frac{a_0(b_0 - x)}{b_0(a_0 - 2x)}$
34. $k = \frac{2a_0x - x^2}{8a_0^2(a_0 - x)^2t}$; $\frac{3}{8a_0^2k}$
37. **(a)** $k_1 = \frac{x_e}{(2a_0 - x_e)t} \ln \frac{a_0x_e + x(a_0 - x_e)}{a_0(x_e - x)}$
(b) $1.51 \times 10^{-4}\ s^{-1}$; $3.73 \times 10^{-4}\ dm^3\ mol^{-1}\ s^{-1}$
(c) $7.46 \times 10^{-3}\ dm^6\ mol^{-2}\ s^{-1}$

Chapter 10

1. $v = k_1[A][B]$
2. $v = k_2\left(\frac{k_1}{k_{-1}}\right)^{1/2}[A]^{1/2}[B]$
3. $v = \frac{k_1k_2[A][B]}{k_{-1} + k_2[B]}$; **(a)** $v = \frac{k_1k_2}{k_{-1}}[A][B]$
(b) $v = k_1[A]$
4. $2A \rightarrow X$ (very slow)
$X + 2B \rightarrow 2Y + 2Z$ (very fast)
5. Two simultaneous reactions
6. Two consecutive reactions
7. $v = \frac{k_1k_2[N_2O_5][NO]}{k_{-1}[NO_2] + k_2[NO]}$
8. $v = \frac{k_1k_2[NO]^2[O_2]}{k_{-1} + k_2[O_2]}$; $v = \frac{k_1k_2}{k_{-1}}[NO]^2[O_2]$
9. $v = k_2\left(\frac{k_1}{k_{-1}}\right)^{1/2}[Cl_2]^{1/2}[CH_4]$
10. $v_{O_3} = \frac{2k_1k_2[O_3]^2}{k_{-1}[O_2] + k_2} = 2k_1[O_3]^2$ in absence of O_2
11. 306 nm
12. 2 molecules; $HI + h\nu \rightarrow HI^*$, $HI^* + HI \rightarrow H_2 + I_2$
13. 47 s
14. 0.76 mol
15. $v = k_2(2/k_4)^{1/2} I_a^{1/2}[CHCl_3]$
16. H^+: 18.8 eV; 7.4 eV for O^-
17. 4.64×10^{-4} mol dm^{-3}
18. $v = k_2\left(\frac{k_1}{k_{-1}}\right)^{1/2}[I_2]^{1/2}[CH_3CHO]$
19. $\alpha = \rho/\rho'$
20. 2.48×10^6 s
21. $v = \frac{k_1k_2[Co(NH_3)_5Cl^{2+}][OH^-]}{k_{-1} + k_2}$
22. 2.0 mmol dm^{-3}
23. 16 μmol dm^{-3}; 0.22 μmol dm^{-3}
24. **(a)** 8.2 kJ mol^{-1}; 5.7 kJ mol^{-1}; 42.9 kJ mol^{-1}; $-124\ J\ K^{-1}\ mol^{-1}$ **(b)** -19.6 kJ mol^{-1}; 14.2 kJ mol^{-1}; 113 J K^{-1} mol^{-1}
25. 5.8 mmol dm^{-3}; 4.0×10^{-5} mol $dm^3\ s^{-1}$; 12.5 s^{-1}
26. 90.5 kJ mol^{-1}; 87.9 kJ mol^{-1}; 108.7 kJ mol^{-1}; $-66.5\ J\ K^{-1}\ mol^{-1}$
27. $v = \frac{k_1k_2k_3[E]_0[A][B]}{k_{-1}k_3 + k_1k_3[A] + k_2k_3[B] + k_1k_2[A][B]}$
28. $v = \frac{k_2[E]_0[S]}{K_m[1 + ([I]/K_i)] + [S]}$;
$\varepsilon = \frac{K_m[I]/K_i}{K_m[1 + ([I]/K_i)] + [S]}$
29. $v = \frac{[k_2k_3/(k_2 + k_3)][E]_0[S]}{[(k_{-1} + k_2)/k_1]\cdot[k_3/(k_2 + k_3)] + [S]}$
$k_c = \frac{k_1k_2}{k_{-1} + k_2}$ $\quad K_m = \frac{k_{-1} + k_2}{k_1}\cdot\frac{k_3}{k_2 + k_3}$
32. $v = \frac{k_1k_2[A]^2}{k_{-1}[A] + k_2}$
33. $v = \frac{k_1k_2k_3[E]_0[A][B]}{k_{-1}(k_{-2} + k_3) + k_1(k_{-2} + k_3)[A] + k_2(k_{-2} + k_3)[B] + k_1k_2[A][B]}$
34. $f = \frac{c_0kt}{1 + c_0kt}$

Chapter 11

1. **(a)** $9.22 \times 10^{14}\ s^{-1}$ **(b)** $3.07 \times 10^4\ cm^{-1}$
(c) 6.085×10^{-19} J; 3.80 eV; 366 kJ mol^{-1}
(d) 2.039×10^{-27} kg m s^{-1}
2. **(a)** 1.526 m **(b)** 1.302×10^{-25} J; 8.127×10^{-7} eV; 0.078 J mol^{-1} **(c)** 4.34×10^{-34} kg m s^{-1}
3. $1.73 \times 10^{12}\ s^{-1}$
4. 0.3 s^{-1}
5. **(a)** 0.88 N m^{-1} **(b)** 0.021 m s^{-1}
6. kT
7. 1.39×10^{20}; 1.204×10^{-27} kg m s^{-1}
8. 1.82 eV; 0.23 eV
9. **(a)** 12.1 pm **(b)** 29.3 pm **(c)** 6.65 fm
(d) 2.58 pm
10. 10^{-10} m s^{-1}
11. **(a)** 1.875×10^6 m s^{-1}; 338 pm **(b)** 1.875×10^7 m s^{-1}; 38.8 pm **(c)** 5.93×10^8 m s^{-1}; 1.23 pm

13. **(a)** $\frac{1}{\sqrt{2}}(\psi_1 + \psi_2)$ **(b)** $\frac{1}{\sqrt{2}}(\psi_1 - \psi_2)$
(c) $\frac{1}{\sqrt{3}}(\psi_1 + \psi_2 + \psi_3)$ **(d)** $\frac{1}{\sqrt{3}}\psi_1 - \frac{1}{\sqrt{6}}\psi_2 + \frac{1}{\sqrt{2}}\psi_3$

14. Yes; a^2

18. **(a)** 0 **(d)** k **(f)** ik

19. **(a)** 1.13×10^4 eV **(b)** 1.13×10^{12} eV; the electron would be emitted as a β particle.

20. **(a)** $\sqrt{ab/(b-a)}$ **(b)** $[ab/(b-a)] \ln (b/a)$

21. 5; 11

22. They are not.

23. 2.34×10^{-20} J; 7.03×10^{-20} J

24. $-(h^2/8\pi^2 I)\, \partial^2/\partial\phi^2$

25. 2.18×10^{-18} J = 13.6 eV

26. 2.19×10^6 m s^{-1}; 3.32×10^{-10} m (332 pm); $\lambda = 2\pi a_0 n$; n

27. 4.05×10^{-6} m; 4.90×10^{-20} J

28. -10.9×10^{-18} J; $-$ 0.25 au

29. 1.27

30. 1.87

31. **(a)** 6.15 **(b)** 4.9 **(c)** 2.2

32. $E = \frac{n_x^2 h^2}{8ma^2} + \frac{n_y^2 h^2}{8mb^2} + \frac{n_z^2 h^2}{8mc^2}$

33. 9.1046×10^{-31} kg; 9.1070×10^{-31} kg; **(a)** Wavelengths slightly shorter for D **(b)** 656.30 nm

34. $(4r^2/a_0^3)\, e^{-2r/a_0}\, dr$

36. **(a)** 1.53×10^{-17} J = 95.3 eV **(b)** 1.53×10^{-9} J = 9.53×10^9 eV

Chapter 12

1. 336 pm; -340.5 kJ mol^{-1}

2. 55%

3. **(a)** 0.47 **(b)** 6%

5. 17.7%; 11.8%; 5.5%; 1.8%

6. 1, 3, 2, 3, 2.5, 3, 1.5, 1, 2; BN, BO, OF and OF^+ are paramagnetic, the rest diamagnetic.

7. **(a)** C_2, CN **(b)** O_2, F_2, NO

9. D_{3h}, C_{2v}, $D_{\infty h}$

10. $CHCl_3$; E, C_3, $3\sigma_v$; C_{3v}
CH_2Cl_2: E, C_2, 2σ; C_{2v}
Naphthalene: E, C_2, $2\bar{C}_2$, 3σ, i; D_{2h}
NO_2: E, C_2, $2\sigma_v$; C_{2v}
Cyclopropane: E, C_3, $3C_2$, $3\sigma_v$, σ_h; D_{3h}
CO_3^{2-}: E, C_3, $3C_2$, $3\sigma_v$, σ_h; D_{3h}
C_2H_2: E, C_∞, ∞C_2, $\infty\sigma_v$, σ_h, i; $D_{\infty h}$

11. Yes, but the two forms rapidly interconvert, and optical activity is not observed.

12. **(a)** σ_g^+; **(b)** a_1, b_2, a_1 **(c)** σ_g^+, σ_u^+ **(d)** a_1' **(e)** a_1; **(f)** σ^+

14. $\frac{1}{\sqrt{3}}$; $\frac{\sqrt{2}}{\sqrt{3}}$, $\frac{\sqrt{2}}{\sqrt{3}}$

15. $(1/\sqrt{2})(s + p_z)$; $(1\sqrt{2})(s - p_z)$

16. $S_3 = C_3\sigma_h$; $S_5 = C_5\sigma_h$; C_1, C_2, C_{2v}, C_{3v}, $C_{\infty v}$

Chapter 13

1. 0.152; 0.705; 3.04 dm^3 mol^{-1} cm^{-1}

2. 12.0 dm^3 mol^{-1} cm^{-1}

3. 2.91×10^{-4} mol dm^{-3}

4. NADH: 3.46×10^{-5} mol dm^{-3}; NAD^+: 1.04×10^{-5} mol dm^{-3}

5. 0.17%

6. $^2S_{1/2}$; $^2P_{3/2}$ and $^2P_{1/2}$

7. $^1S_0$, $^1P_1$, $^1D_2$, $^3D_3$, $^3D_2$, $^3D_1$, $^3P_2$, $^3P_1$, $^3P_0$, $^3S_1$

8. **(a)** $^2P_{3/2}$, $^2P_{1/2}$ **(b)** $^2D_{5/2}$ $^2D_{3/2}$, $^2D_{1/2}$

9. $^1P : J = 1$
$^3P : J = 2, 1, 0$
$^4P : J = \frac{5}{2}, \frac{3}{2}, \frac{1}{2}$
$^1D : J = 2$
$^2D : J = \frac{5}{2}, \frac{3}{2}$
$^3D : J = 3, 2, 1$
$^4D : J = \frac{7}{2}, \frac{5}{2}, \frac{3}{2}, \frac{1}{2}$

10. $\frac{2}{3}$; 1.24 cm^{-1}

11. 0.93 cm^{-1}

12. 164 pm

13. 92 pm; 22.1 cm^{-1}; 15.4 cm^{-1}

14. 113 pm

15. **(a)** HCl, H_2O, CH_3Cl, CH_2Cl_2, H_2O_2, NH_3 **(b)** All except H_2 **(c)** All except CH_4 and SF_6 **(d)** All

16. 511.6 kg s^{-2}

17. 199 pm

18. 457.1 kJ mol^{-1}; 21.7 mol^{-1}, 17.7 kJ mol^{-1}; 435.4 kJ mol^{-1}, 439.4 kJ mol^{-1}

19. A—B—B

20. 2400 cm^{-1}

21. 1973.7 cm^{-1}

22. 314.3 kg s^{-1}

24. 4.7×10^5 K

25. **(a)** 2.7×10^{-10} s **(b)** 2.7×10^{-11} s **(c)** 2.7×10^{-12} s **(d)** 4.05×10^{-10} s

26. 0.85 eV = 82.0 kJ mol^{-1}

27. 0.76 eV = 73.7 kJ mol^{-1}

28. 1.324 T

29. 2.0026

30. 4.329×10^{-27} J T^{-1}; 13.28×10^{-27} J T^{-1}; 4.170×10^{-27} J T^{-1}; 3.455×10^{-27} J T^{-1}

31. 1.41 T; 14.4 T

32. 456 Hz

33. 1.99 MHz

34. 4.39 T

35. 3.98×10^5 Hz; 1.33×10^{-5} cm^{-1}

36. *cis*-$C_2H_2Cl_2$: The point group is C_{2v}:
(a) is a_1; active in both infrared and Raman.
(b) is a_2; inactive in infrared but active in Raman.
(c) is b_1; active in infrared and Raman.
(d) is a_1; active in both infrared and Raman.
trans-$C_2H_2Cl_2$: The point group is C_{2h}:
(e) is a_g; inactive in infrared, active in Raman.
(f) is a_u; active in infrared, inactive in Raman.
(g) is b_u; active in infrared, inactive in Raman.
(h) is a_g; inactive in infrared, active in Raman.
Benzene: The point group is D_{6h}:
(i) is a_{1g}; inactive in infrared; active in Raman.
(j) is a_{2u}; active in infrared, inactive in Raman.
(k) is b_{1u}; inactive in both infrared and Raman.
38. a_2'' and e' (not a_1')
39. 8.96×10^{13} s^{-1}

Chapter 14

1. (a) $\sqrt{\frac{3\pi}{8}} = 1.085$ (b) $2/\sqrt{\pi} = 1.128$
2. (a) 10 900 K (b) 173 000 K
3. (a) 8.77×10^{-2}; no effect of m or T (b) 0.89; no effect of m
4. They are equal.
5. (a) 0.922 (b) 9.29×10^{10}
6. $C_p = \left\{\frac{\partial}{\partial T}\left[\mathbf{k}T^2\left(\frac{\partial \ln Q}{\partial T}\right)_V\right]\right\}_P + N\mathbf{k}$
7. (a) $U - U_0 = N\mathbf{k}T^2\left(\frac{\partial \ln q}{\partial T}\right)_V$; $S = N\mathbf{k}T\left(\frac{\partial \ln q}{\partial T}\right)_V + N\mathbf{k} \ln q$; $A - U_0 = -N\mathbf{k}T \ln q$; $H - U_0 = N\mathbf{k}T^2\left(\frac{\partial \ln q}{\partial T}\right)_V + N\mathbf{k}T$; $G - U_0 = -N\mathbf{k}T \ln \frac{q}{e}$ (b) $U - U_0 = N\mathbf{k}T^2 \left(\frac{\partial \ln q}{\partial T}\right)_V$; $S = N\mathbf{k}T\left(\frac{\partial \ln q}{\partial T}\right)_V + N\mathbf{k} \ln \frac{qe}{N}$; $A - U_0 = -N\mathbf{k}T \ln \frac{qe}{N}$; $H - U_0 = N\mathbf{k}T^2\left(\frac{\partial \ln q}{\partial T}\right)_V + N\mathbf{k}T$; $G - U_0 = -N\mathbf{k}T \ln \frac{q}{N}$
8. (a) and (b) $P = N\mathbf{k}T\left(\frac{\partial \ln q}{\partial V}\right)_T = nRT\left(\frac{\partial \ln q}{\partial V}\right)_T$
9. 463.7 J K^{-1} mol^{-1}
10. 2.48 kJ mol^{-1}
11. 8.314 J K^{-1} mol^{-1}
12. $U_m - U_{0,m} = \frac{3}{2}RT$
13. (a) 1.45×10^{32}; 1.223×10^{25} (b) 7.47×10^{31}; 1.183×10^{25} (c) 6.75×10^{32}; 1.315×10^{25}
14. 51.9; 2.38×10^{24}
15. (a) 1.09 (b) 42.7
17. 154.7 J K^{-1} mol^{-1}
18. (a) 1.26 (b) 6.90
19. 2, 12, 6, 6, 3, 6, 12, 1, 2
21. 181.0 J K^{-1} mol^{-1}
22. 150.30 J K^{-1} mol^{-1}; 47.2 J K^{-1} mol^{-1}; zero
23. Zero
24. (a) S_m = constant + $\frac{3}{2}R \ln M_r$; $\frac{3R}{2M_r}$; (b) No dependence (c) S_m = constant + $\frac{5}{2}R \ln T$; $\frac{5R}{2T}$
25. −228.6 kJ mol^{-1}
26. 0.90
27. (a) 5.93×10^5 atm^{-2} (b) 3.34×10^{-7} atm^{-2}
28. (a) 2 (b) 4 (c) 4 (d) 3 (e) 2
29. $(2m/\pi\mathbf{k}T)^{1/2}e^{-mu_x^2/2\mathbf{k}T}\,du_x$; zero
30. $(\pi\epsilon_x\mathbf{k}T)^{-1/2}e^{-\epsilon_x/\mathbf{k}T}\,d\epsilon_x$; $\frac{1}{2}\mathbf{k}T$
31. $(m/\mathbf{k}T)e^{-mu^2/2\mathbf{k}T}u\,du$; $e^{-\epsilon^*/\mathbf{k}T}$
32. (a) $\frac{1}{2}L\epsilon$ (b) $R \ln 2$ (c) RT (d) $RT(1. - \ln 2)$
33. $S_m = 2R + R \ln\left(\frac{2\pi m\mathbf{k}TA}{Nh^2}\right)$; 162 J K^{-1} mol^{-1}
34. $K_c = 1.70 \times 10^{-3}$ mol dm^{-3}; $K_p = 0.178$ atm
35. $K_c = 0.0202$ mol dm^{-3}; $K_p = 1.66$ atm
36. $K_c = 2.14 \times 10^{-7}$ mol dm^{-3}; $K_p = 2.11 \times 10^{-5}$ atm

Chapter 15

1. (a) 4 lattice points (b) 2 lattice points
2. (a) 2 basis groups (b) 1 basis group
3. (a) 78.5% (b) 90.7% (c) triangular, 1.16 times
4. $\rho = \frac{Mz}{abcL}$
5. 1.96 g cm^{-3}
6. 8
7. 2.176 g cm^{-3}
8. 2.001 g cm^{-3}
9. 546 pm
10. (243)
11. $d_{100} = a$; $d_{110} = a/\sqrt{2}$; $d_{111} = a/\sqrt{3}$
12. (362), $(3\bar{2}3)$, $(11\bar{1})$
13. $d_{100} = 389$ pm; $d_{111} = 225$ pm; $d_{121} = 159$ pm
14. $\theta_{100} = 9.08°$; $\theta_{010} = 6.64°$; $\theta_{111} = 12.49°$
15. $\theta_{100} = 2.70°$; $\theta_{010} = 1.91°$; $\theta_{111} = 3.63°$
16. $\theta_{100} = 4.57°$; $\theta_{111} = 8.15°$
17. A, (0, 1, 0); B, (−1, 1, 0); C, (2, 1, 0); D, (1, 1, 0)
18. (a) $\theta = 6.25°$ (b) $\theta = 13.70°$
19. $\lambda = 6.1$ pm
20. *fcc*

21. *bcc*
22. **(a)** *bcc* **(b)** $a = 286$ pm; $\theta = 32.6°$ **(c)** 123.8 pm
23. (200), 266.7 pm; (110), 377.1 pm; (222), 154.0 pm
24. Cubic system
25. $N(\text{Ag}) = 4.0$; *fcc*
26. $a = 4.64$ Å
27. $r = 0.225R$
28. $r = 0.414R$
29. $\Delta E_c = 668$ kJ mol^{-1}
30. **(a)** $d = 3.23$ Å, $\theta = 13.81°$; $d = 3.09$ Å, $\theta = 14.45°$; $d = 3.04$ Å, $\theta = 14.69°$; $d = 2.84$ Å, $\theta = 15.75°$; $d = 2.74$ Å, $\theta = 16.33°$ **(b)** $d = 3.23$ Å, $hkl = (622)$; $d = 3.09$ Å, $hkl = (444)$; $d = 3.04$ Å, $hkl = (543)$
31. $d_{12.95°} = 3.44$ Å, $hkl = (201)$ $d_{13.76°} = 3.24$ Å, $hkl = (002)$ $d_{14.79°} = 3.02$ Å, $hkl = (040)$
32. **(a)** $4f_{\text{Zn}} - 4if_{\text{S}}$ **(b)** $a = 540.5$ pm
33. $\Theta_D = 383$ K

Chapter 16

1. 3.58×10^8 Pa = 3530 atm; -20.9 kJ mol^{-1}
2. 1.97×10^9 Pa = 19 400 atm
3. 9.95×10^8 Pa = 9820 atm; -42.6 kJ mol^{-1}
4. 3.69×10^8 Pa = 3650 atm
5. **(a)** 1.34×10^9 Pa = 13 200 atm **(b)** 2.54×10^8 Pa = 2500 atm **(c)** 3.01×10^8 Pa = 2970 atm **(d)** 2.40×10^8 Pa = 2370 atm
6. 3.43×10^4 Pa = 0.339 atm; 3.69×10^8 Pa = 3646 atm
7. **(a)** 3.42×10^8 Pa = 3380 atm **(b)** 5.44×10^8 Pa = 5370 atm
8. 10^{-6}
9. -7.12×10^{-20} J = -42.9 kJ mol^{-1}
10. -9.23×10^{-19} J = -555.6 kJ mol^{-1}
11. -1.48×10^{-20} J = -8.91 kJ mol^{-1}
12. -1.22×10^{-21} J = -740 J mol^{-1}
13. -1.31×10^{-21} J = -790 J mol^{-1}
14. -3.83×10^{-23} J = -23 J mol^{-1}; -2.79×10^{-23} J = -16.8 J mol^{-1}; -1.31×10^{-21} J = -790 J mol^{-1}
15. -2.52×10^{-23} J = -15.2 J mol^{-1}; -2.86×10^{-22} J = -172 J mol^{-1}; -5.66×10^{-22} J = -341 J mol^{-1}
16. 376 J mol^{-1}; 221 J mol^{-1}; 892 J mol^{-1}; 1.30 kJ mol^{-1}; 2.42 kJ mol^{-1}; 5.3 kJ mol^{-1}.
17. -1.20×10^{-23} J = -7.22 J mol^{-1}
18. **(a)** $\dfrac{E}{E_{\text{min}}} = -\dfrac{n}{6-n}\left(\dfrac{r_0}{r}\right)^6 + \dfrac{6}{6-n}\left(\dfrac{r_0}{r}\right)^n$ **(b)** $\left(\dfrac{r^*}{r_0}\right)^{n-6} = \dfrac{6}{n}$ **(c)** $\dfrac{E}{E_{\text{min}}} = 2\left(\dfrac{r_0}{r}\right)^6 - \left(\dfrac{r_0}{r}\right)^{12} = 4\left[\left(\dfrac{r^*}{r}\right)^6 - \left(\dfrac{r^*}{r}\right)^{12}\right]$
19. **(b)** 132.0 J K^{-1} mol^{-1} **(c)** 20.7 J K^{-1} mol^{-1}

Chapter 17

1. **(a)** 1 **(b)** 3 atm; 9 atm; 99 atm; 999 atm **(c)** 0.91; 0.33; 0.999.
3. **(a)** 3.72 dm^3 mol^{-1} **(b)** 10^{-1} m^2
4. **(a)** 1.5×10^{-3} mol dm^{-3} s^{-1}; 2.0×10^{-2} s^{-1} **(b)** 1.5×10^{-5} mol dm^{-3} s^{-1}; 2.0×10^{-4} s^{-1} **(c)** 1.5×10^{-5} mol dm^{-3} s^{-1}; 2.0×10^{-4} s^{-1}; $k' = kV/S$; m s^{-1}
5. **(a)** 2.5×10^{-2} mol dm^{-3} s^{-1}; 2.5×10^{-2} mol dm^{-3} s^{-1} **(b)** 2.5×10^{-4} mol dm^{-3} s^{-1}; 2.5×10^{-4} mol dm^{-3} s^{-1} **(c)** 2.5×10^{-4} mol dm^{-3} s^{-1}; 2.5×10^{-4} mol dm^{-3} s^{-1}; $k' = kV/S$; mol m^{-2} s^{-1}
6. $\dfrac{dx}{dt} = \dfrac{ka_0}{x} - \dfrac{2k}{3}$
7. $v = \dfrac{k[\text{H}]}{1 + K^{1/2}[\text{H}_2]^{1/2}}$; when $\theta \to 1$.
8. **(a)** $v = kK[A]\,\alpha\, e^{-(E+\Delta H_A)RT}$ **(b)** $v = k\,\alpha\, e^{-E/RT}$ **(c)** $v = \dfrac{kK}{K_i}\dfrac{[A]}{[I]}\,\alpha\, e^{-(E+\Delta H_A-\Delta H_I)/RT}$
9. $\dfrac{dz}{dt} = k(a_0 - z) + k_s z$ $\quad z = \dfrac{ka_0}{k_s - k}[1 - e^{-(k_s-k)t}]$
10. **(a)** $v = kK[\text{A}]$; $v = k$ **(b)** Eq. 17.36 **(c)** $v = k[\text{H}_2]$ **(d)** $v = \dfrac{k}{K}$
11. **(a)** 1.49 cm **(b)** 1.49 m
12. -10.4 mm
13. 1.0017
14. 0.086 N m^{-1}
15. 1.5 cm
16. No
17. 2.58×10^{-3} N m^{-1}
18. 216 g mol^{-1}; 0.204 nm^2
19. 0.19 nm^2
20. Slope = $-\dfrac{1}{V_a^0}$

Chapter 18

1. 0.49 mmHg
2. **(a)** 0.33 mm s^{-1} **(b)** 4.19×10^{-15} m^3 s^{-1} **(c)** 2×10^{10}

3. **(a)** 0.516 nm **(b)** 34.3 nm **(c)** 1.38×10^{10} s^{-1} **(d)** 1.7×10^{35} m^{-3} s^{-1}
4. **(a)** 7.20×10^{-4} kg m^{-1} s^{-1} **(b)** 6.25×10^{-4} kg m^{-1} s^{-1} **(c)** 6.53×10^{-4} kg m^{-1} s^{-1}
5. **(a)** 0.084 dm^3 g^{-1} **(b)** 0.254 dm^3 g^{-1} **(c)** 0.627 dm^3 g^{-1}
6. 29 200
7. **(a)** 1.78×10^{-5} kg m^{-1} s^{-1} **(b)** 9.97×10^{-5} m^2 s^{-1} **(c)** 1202 m s^{-1} **(d)** 1.655×10^{-7} m **(e)** 7.263×10^9 s^{-1} **(f)** 9.75×10^{34} m^{-3} s^{-1}
8. 4.5 cm
9. 0.96 cm
10. 8.9×10^{-6} cm^2 s^{-1}
11. 1.20×10^{-5} cm^2 s^{-1}
12. 132 000
13. 0.61 s
14. 3.5×10^5 s
15. 70 860
16. 272 500 g mol^{-1}
17. **(a)** 5.80×10^{-11} kg s^{-1} **(b)** 4.43×10^{-11} kg s^{-1}
18. 1.12×10^5 s
19. 3.43×10^{-11} s
20. **(a)** 3.48×10^{-4} kg m^{-1} s^{-1} **(b)** 4.33×10^{-4} kg m^{-1} s^{-1}; 8.7 kJ mol^{-1}
21. **(a)** 17.5 kJ mol^{-1} **(b)** 12.1 kJ mol^{-1}
22. **(a)** Zero **(b)** and **(c)** A viscosity that decreases with increasing T.

Chapter 19

1. $\frac{k_p}{(k_i k_t)^{1/2}} \cdot \frac{[M]^{1/2}}{[C]^{1/2}}$
3. 21 000 g mol^{-1}; 23 333 g mol^{-1}
4. 50 000 g mol^{-1}; 54 000 g mol^{-1}
5. **(a)** 25 000 g mol^{-1}; 26 000 g mol^{-1} **(b)** 24 000 g mol^{-1}; 25 000 g mol^{-1}
6. 305 000 g mol^{-1}
7. 33 300 g mol^{-1}
8. 50 800 g mol^{-1}
9. **(a)** 9.21 nm **(b)** 8.48 nm **(c)** 223.8 nm
10. **(a)** 5.85 nm **(b)** 3.5×10^{-3} **(c)** 4.78 nm
11. 0.981 N; 108:4 g
12. 8.5×10^{-4} kg m^{-1} s^{-1}

Index

A page number in **boldface** print indicates that a physical quantity is defined or an equation or concept is defined and explained. (A) indicates that a quantity is defined, and its units given, in Appendix A (pp. 881–895). A table of values is indicated by (T).

THE GREEK ALPHABET

Α, α . . . *Alpha*	Η, η . . . *Eta*	Ν, ν . . . *Nu*	Τ, τ . . . *Tau*
Β, β . . . *Beta*	Θ, ϑ, θ . . *Theta*	Ξ, ξ . . . *Xi*	Υ, υ . . . *Upsilon*
Γ, γ . . . *Gamma*	Ι, ι . . . *Iota*	Ο, ο . . . *Omicron*	Φ, φ, ϕ . . *Phi*
Δ, δ . . . *Delta*	Κ, κ . . . *Kappa*	Π, π . . . *Pi*	Χ, χ . . . *Chi*
Ε, ϵ . . . *Epsilon*	Λ, λ . . . *Lambda*	Ρ, ρ . . . *Rho*	Ψ, ψ . . . *Psi*
Ζ, ζ . . . *Zeta*	Μ, μ . . . *Mu*	Σ, σ . . . *Sigma*	Ω, ω . . . *Omega*

SI PREFIXES

Fraction	Prefix	Symbol	Multiple	Prefix	Symbol
10^{-1}	deci	d	10	deka	da
10^{-2}	centi	c	10^{2}	hecto	h
10^{-3}	milli	m	10^{3}	kilo	k
10^{-6}	micro	μ	10^{6}	mega	M
10^{-9}	nano	n	10^{9}	giga	G
10^{-12}	pico	p	10^{12}	tera	T
10^{-15}	femto	f	10^{15}	peta	P
10^{-18}	atto	a	10^{18}	exa	E

SI BASE UNITS

Physical quantity	Symbol for quantity	SI unit	Symbol for unit
Length	l	metre	m
Mass	m	kilogram	kg
Time	t	second	s
Thermodynamic temperature	T	kelvin	K
Electric current	I	ampere	A
Luminous intensity	I_v	candela	cd
Amount of substance	n	mole	mol

PHYSICAL CONSTANTS*

Speed of light in vacuum	c	2.998×10^{8} m s^{-1}
Elementary charge	e	1.602×10^{-19} C
Avogadro constant	L	6.022×10^{23} mol^{-1}
Atomic mass unit	u	1.661×10^{-27} kg
Electron rest mass	m_e	9.110×10^{-31} kg
Proton rest mass	m_p	1.673×10^{-27} kg
Faraday constant	F	9.6485×10^{4} C mol^{-1}
Planck constant	h	6.626×10^{-34} J s
Rydberg constant	R_∞	1.097×10^{7} m^{-1}
Gas constant	R	8.314 J K^{-1} mol^{-1}
		1.987 cal K^{-1} mol^{-1} dm^{3}
		0.082 06 dm^{3} atm K^{-1} mol^{-1}
Boltzmann constant	$\mathbf{k}$	1.381×10^{-23} J K^{-1}
Permittivity of vacuum	ϵ_0	8.854×10^{-12} C^{2} N^{-1} m^{2}
	$1/4\pi\epsilon_0$	0.8988×10^{10} N m^{2} C^{-2}

*More values are given in Appendix B.